GUIDE TO THE
FLOWERS OF WESTERN CHINA

GUIDE TO THE FLOWERS OF WESTERN CHINA

CHRISTOPHER GREY-WILSON
& PHILLIP CRIBB

EDITED BY VICTORIA MATTHEWS

Kew Publishing
Royal Botanic Gardens, Kew

Kew
ROYAL BOTANIC GARDENS

© The Board of Trustees of the Royal Botanic Gardens, Kew 2011
Illustrations and photographs © the artists and photographers as stated in the captions

The authors have asserted their rights to be identified as the authors of this work in accordance with the Copyright, Designs and Patents Act 1988

All rights reserved. No part of this publication may be reproduced, stored in a retrieval system, or transmitted, in any form, or by any means, electronic, mechanical, photocopying, recording or otherwise, without written permission of the publisher unless in accordance with the provisions of the Copyright Designs and Patents Act 1988.

Great care has been taken to maintain the accuracy of the information contained in this work. However, neither the publisher nor the editors can be held responsible for any consequences arising from use of the information contained herein.

First published in 2011 by
Royal Botanic Gardens, Kew,
Richmond, Surrey, TW9 3AB, UK
www.kew.org

ISBN 978-1-84246-169-3

British Library Cataloguing in Publication Data
A catalogue record for this book is available from the British Library

Editor: Victoria Matthews

Production editor: Sharon Whitehead
Typesetting and page layout: Christine Beard
Publishing, Design & Photography, Royal Botanic Gardens, Kew

Cover design: Lyn Davies

Front cover: *Rhododendron* forma *prattii* forms impressive stands near the Hailuogou glacier in western Sechuan. Photograph by Harry Jans.

Title page: *Primula thearosa*, Galung La, Bomi, south east Tibet. Photograph by Harry Jans.

Printed and bound by
Firmengruppe APPL, aprinta druck
Wemding, Germany

For information or to purchase all Kew titles please visit
www.kewbooks.com or email publishing@kew.org

MIX
Paper from responsible sources
FSC™ C004592

Kew's mission is to inspire and deliver science-based plant conservation worldwide, enhancing the quality of life.

Kew receives half of its running costs from Government through the Department for Environment, Food and Rural Affairs (Defra). All other funding needed to support Kew's vital work comes from members, foundations, donors and commercial activities including book sales.

Contents

Foreword vi

Introduction
    Scope of the book 1
    Area covered 2
    Selection of species for inclusion 7
    Nomenclature 8
    Description of species 8
    Photography 10
    Species in cultivation 11
    China 15
        When to go 15
        Topography 15
        Climate 20
        Floristic regions 26
        Chinese and other place names 45
    Further reading 46
    Maps 47

Species descriptions
    Gymnosperms 51
    Angiosperms 67

Acknowledgements 614

Selected bibliography 614

Glossary 617

Index 621

Foreword

by Roy Lancaster

It is now 18 years since I last set foot in western China and 31 years since that never-to-be-forgotten autumn day when I realised a lifelong ambition in visiting Mt Omei (Emei Shan) in Sichuan province, long acknowledged by botanists and plant explorers as one of the richest locations for native flora in the whole of that vast country. I spent just three days on this timeless mountain where every step I took on the leaf-strewn path to the summit revealed ever more plants, some familiar from cultivation, most not. It was enough to convince me that Sichuan was a plantsman's paradise, and that would have been that had I not returned the following spring as a member of the Sino-British Expedition to the Cangshan (Dali Mountains) of western Yunnan. This time it was for three weeks, time enough both to see a huge number of plants of all kinds and more importantly to get to know them better, helped by our Chinese colleagues whose local knowledge and expertise made identifications possible and further information more readily accessed.

Sadly, such valuable assistance was not always available. I remember on subsequent visits to both Yunnan and Sichuan struggling with sundry volumes of national and regional floras in which line drawings provided my only hope of identifications in the field while the accompanying descriptions in Chinese remained, to my eyes certainly, tantalising yet unintelligible. If only the present book had been available to me then. This is not to say that identifications in the field are everything. I well remember the excitement and increasing sense of anticipation I experienced in returning home from China and the first opportunity to sit quietly with my reference books and my notebook open, ready to check the many descriptions of nameless plants encountered. In many cases, I was successful in identifying these plants but even now there remain those still nameless after many years.

There is no doubt that good photographs of plants seen in the wild are both helpful in identification and treasured reminders of a special day and place, and the present volume is a testimony to that. Both authors are veterans of botanical travel, especially in western China, as evidenced by their many high-quality photographs from wide-ranging locations. They are also long experienced in providing accurate and critical information on a wide range of plants in a format that readers, whether amateur or professional, will find helpful and easy to follow. Obviously, given the immense variety of plants found wild in western China, the plants in this book are no more than a selection, albeit an impressive and well-balanced one.

The photographs form the largest collection of its kind ever published and say much for the authors' patience and dedication as well as their influence in attracting the contributions of so many other authorities on and observers of the Chinese flora. In addition to its core intention of providing a means of helping travellers in western China to identify the plants they see, and by extension providing a wonderful adventure to arm-chair travellers, this book is offered as a celebration of one of the temperate world's great floras, a flora that has fascinated explorers (and gardeners) for 150 years and will continue to do so for as long as the spirit of adventure and the demands of a curious mind remain. Just as important, this book reminds us of a beautiful yet fragile world whose continued existence lies in our hands.

For those who wonder why plant lovers are so passionate about their subject, the following pages provide the answer.

March 2011

Introduction

SCOPE OF THE BOOK

China has a very rich flora, indeed one that is unrivalled in temperate latitudes of the world. About the same size as the USA, China is estimated to contain 12 per cent (almost one eighth) of the world's plant biodiversity, about 30,000 species of higher plants distributed in c. 353 families and 3,184 genera. By comparison, North America (USA and Canada) has about 17,000 species of higher plants. The species diversity within China reflects the wide range of habitats and altitudes within China: no other country in the world boasts such diversity of habitat from tropical and subtropical regions to temperate and boreal forests, as well as extensive alpine areas well above the treeline in the west. In addition, the effect of climate, particularly the influence of the summer monsoon, means that some parts of China receive a very high summer rainfall, whereas others are dry and desert-like. In winter, some areas have deep snow cover whereas others are dry and intensely cold. Within these varied and diverse regions, a wide range of plants has evolved, including many endemic species. It is estimated that China has c. 220 endemic genera, and at species level, 56 per cent of species (c. 16,800) are endemic.

The astonishing richness of the Chinese flora, particularly the temperate element, was soon realised by plant hunters in the late nineteenth and early part of the twentieth centuries. Collectors from Russia, Britain, France, North America and elsewhere scoured the country for interesting new plants for their scientific collections and also, importantly, sought out hardy plants suitable for introduction into cultivation. The rich and colourful flora gave them ample scope and today our gardens, parks and botanic institutes are full of plants of Chinese origin. Trees and shrubs, perennials, bulbs, ferns and numerous alpine plants have greatly enriched our gardens, indeed it is fair to say that our gardens would be very much poorer without Chinese plants.

Chinese botany took a great step forward in the twentieth century with botanists working in many parts of the country. Today, each province in China has its own botanical institute and numerous fine Chinese botanists are working on the local flora. This has culminated in the Flora of China Project, which aims to document all the plants found within China, a huge and complicated undertaking involving Chinese as well as foreign botanists. This work is based on the comprehensive *Flora Reipublicae Popularis Sinicae*, a massive 80-volume work (published in 126 parts from 1959 to 2004) that involved hundreds of Chinese botanists; however, that work, which is of great importance internationally, was written in Chinese and is unavailable to the vast majority of foreign scientists. The Flora of China Project, now nearing completion, brings this valuable work to the attention of English-speaking scientists, at the same time allowing updates and revisions to be carried out. This complex undertaking has been made possible by the establishment of a number of editorial centres in China, the USA, the UK and France, with the prime managerial centre at the Missouri Botanical Garden. The published volumes are accompanied by volumes of line illustrations. All of the volumes completed to date can be viewed on the Flora of China website (http://hua.huh.harvard.edu/china/).

Today, most parts of China are open to foreign travellers and many people venture there to enjoy the amazing variety of scenery, the mix of ethnic races, the charming and hospitable welcome and the rich flora and fauna. Many foreign scientists visit the country annually, working and collaborating with their Chinese colleagues, and this is especially so with botanists. In China, all collecting (plants, animals, minerals and so on) is prohibited without the permission of the Chinese authorities: this can at times be very difficult to obtain, even for *bone fide* scientists. To collect without permission is to break the law. This should in no way inhibit the casual visitor who wishes to see and study the wild flowers of a particular area; the camera and pen can record the details of the plants, leaving them undisturbed and available for others to enjoy. The majority of garden-worthy Chinese plants are already available in nurseries in the West and Japan.

Anyone viewing the Chinese flora for the first time will be awed by the sheer size and number of species. To give examples of familiar genera within China: *Clematis* has 147 species (93 endemic); *Gentiana* 248 species (the majority endemic); *Pedicularis* 352 species (271 endemic); *Primula* c. 300 species (the majority endemic); *Rhododendron* 571 species (409 endemic); *Salix* 275 species (189 endemic); *Saxifraga* 216 species (139 endemic) and there are many more examples of large Chinese genera.

AREA COVERED

Nowhere in China is the floral richness better seen than in the west of the country. With its diverse scenery, lush subtropical and temperate forests, huge rivers and massive divides crowned in places by high snowy peaks, it is scarcely surprising that western China (particularly Yunnan and Sichuan) has been the centre of plant exploration for many years. This book can include only a small proportion of the great wealth and diversity of plant life in the area, but it strives to be a celebration of its wonderful richness. While it can be used as a field guide to many of the more common and colourful elements of the flora, it also includes many local endemics and some great rarities. It is not likely that this book will cover all of the plants that might be seen on a botanical trek in western China but it does, we hope, include many of those that the casual traveller is likely to come across. The book covers western China in the broad sense from Qinghai and Gansu in the north, southwards through Sichuan, west Hubei and Shaanxi to Yunnan, west Guangxi and Guizhou, as well as the extreme eastern and south-eastern fringes of Tibet (Xizang).

It was originally envisaged that east and south-east Tibet would be covered in as great a detail as the rest of the area. Indeed the distribution of plants included in this book extends into these areas where relevant. However, with the publication in 2005 of Toshio Yoshida's magnificent photographic treatise *Himalayan Plants Illustrated*, a slight change in tactics was adopted. Toshio Yoshida's work contains many plants found in southern and south-eastern Tibet and, in the main, we have tried not to overlap significantly with his work; the two volumes thus complement each other in many ways and between them cover a very large number of plants. At the time of writing, *Himalayan Plants Illustrated* is only available in Japanese, although much information can be gleaned by from the maps and photographic captions. An English version of this important work would be extremely valuable.

OPPOSITE: Mixed broadleaved deciduous forest at Shennongjia in west Hubei (PC)

An alpine meadow dominated by *Trollius ranunculoides*, north of Jigzhi, south-eastern Qinghai (CGW)

Wet, seasonally flooded meadows at 3,500 m at the north end of the Zhongdian Plateau, north-western Yunnan (CGW)

INTRODUCTION | 5

LEFT: Lush, low-altitude subtropical forest at Xishuangbanna in south-western Yunnan close to the Laos and Myanmar frontiers (CGW)
RIGHT: Shilin (Stone Forest) in central Yunnan with its weather-sculptured rocks has become an important tourist attraction (CGW)

Semi-desert, eroded landscape near Guide at 2,360 m in Qinghai (HJ)

Ancient Buddhist monasteries dominate towns and villages throughout Tibetan regions. This monastery at Zhongdian in north-western Yunnan was restored after being sacked during the Cultural Revolution. Tibetan monks and doctors as well as villagers have traditionally used a very large number of native plants for medicinal purposes (HJ)

High passes throughout the Buddhist regions of Tibet and western China are marked by cairns of stones and prayer-flags; here the Ya Jia La in west Sichuan (TK)

SELECTION OF SPECIES FOR INCLUSION

We were faced with the impossible task of trying to represent the extremely rich and varied flora of western China in this single volume. The tree and shrub flora alone would comfortably fill the pages before considering the far richer and colourful herb flora. The selection of species for inclusion has been restricted to some extent by the photographs available to us. (A small number of important species has been included, that are not accompanied by a photograph.) Although we have endeavoured to give as wide and even coverage as possible to illustrate the immense richness of the area, certain genera demand a more in-depth treatment because they play such an important role in the flora and are such a conspicuous feature of it. As a result, many species of *Arisaema, Cypripedium, Gentiana, Impatiens, Iris, Lilium, Primula* and *Rhododendron* are included. There are many small genera with just one or two colourful species that are also included, as well as important, localised, rare or endangered species. Our prime aim has been not only to celebrate the wealth of the western Chinese flora but also, we hope, to enable those travelling there to be able to identify many of the species that they are likely to encounter in the wild.

This book also contains a great deal of information of use to the armchair traveller, horticulturist and gardener. As has been already pointed out, a large number of western Chinese species are cultivated in gardens, including botanic gardens and parks, and this volume also serves as a reference work to many of these plants, containing as it does, the largest collection of photographs of Chinese plants ever published.

NOMENCLATURE

We have endeavoured to give the most up-to-date scientific names for the plants included in this work. We have tried as far as possible to follow those used in the *Flora of China* or in the most recent monographs and revisions of various genera, and these publications are listed in the bibliography towards the end of the book.

Important synonyms follow the accepted name for a particular species, subspecies, etc. and these are also listed in the index; they are of particular relevance to those who know a plant under a synonym and wish to find it in the text. Space has not allowed us to list all synonyms of some species and we would refer those who wish to find a fuller synonymy to the *Flora of China* website.

DESCRIPTION OF SPECIES

The descriptions used in the book are necessarily brief, but we have endeavoured to include enough factual information to aid identification alongside the photographs, each complimenting the other. Certain key information, common to groups of species or individuals, is presented in each account in order to condense the information and further help identification of those species included in the text. Be aware, though, that many species are not included and for these more detailed reference works will need to be consulted.

The key information follows a conventional dichotomous botanical key, wherever possible using simple terms.

Unless they are particularly significant, alien and widely distributed weed species are not included in the text, neither are most cultivated plants (especially crop plants) grown in western China, of which there is a large and diverse range. Grasses (including the all-important bamboos), sedges, rushes and ferns are also omitted as they would have added considerably to what is already a very large text.

Measurements are metric and presented in metres (m), centimetres (cm) and millimetres (mm). All indicate the normal range, whether for plant height, leaf size or flower diameter, but sometimes include more unusual measurements, for example, leaves (3–)5–7(–9) cm long means that the leaves are normally 5–7 cm long, but may occasionally be as little as 3 cm or as much as 9 cm in length. In the field, aberrations may well occur that do not necessarily fit the description. This applies particularly to flower colour, which can vary immensely in some species, whereas it is very uniform in others. Where plants appear to vary considerably in their general details and do not 'fit' the description comfortably, then hybridisation might be suspected; however, it has to be accepted that some species are inherently very variable.

A plant's status is given at the beginning of each description or, if all the species in a particular genus are the same, in the generic description. This information describes, for example, whether the plant is an annual, biennial or perennial herb, a shrub or a tree and, if either of the latter two, whether it is evergreen or deciduous.

In addition to the descriptions, other information is given under each species:

Distribution. The distribution of a species, subspecies or variety is given immediately after the description. Where a particular plant occurs outside the area covered by the book, whether in other parts of China or elsewhere, this information is included in square brackets (e.g. cChina, Himalaya, eRussia). An asterisk (*) immediately following the accepted name of a species indicates that it is endemic to China (including Tibet (Xizang)).

Habitat. Condensed habitat information follows the distribution and indicates the main habitats in which that plant is most likely to be found (e.g. coniferous forests, shrubberies, alpine meadows, etc.). Rock or soil type may be given when particularly relevant. More detailed habitat information can be found on the *Flora of China* website for those families already completed.

Altitude. This information follows the habitat information and indicates (in metres) the altitudinal range at which the plant is most likely to be found. Such information is gleaned from the various regional Chinese floras, from monographs and revisions or the *Flora of China* or, where no such information is readily available, from herbarium specimens. Altitudinal information should be taken as a general guide, as available data are often very incomplete.

Flowering time. The time of the year at which a particular plant is most likely to be seen in flower is given after the altitude data. Flowering time may vary from year to year to some extent because it is dependent on various factors such as temperature, available moisture and other inter-related factors. Aberrant flowering is common in many species, especially in alpine plants, which can sometimes be 'held back' by unusually deep winter snow, sometimes failing to flower until well into the summer. Flowering time is indicated as numbered months (I = January, II = February, III = March, IV = April and so on).

Other information. Medicinal botany has been extremely important in China for many centuries and this practice continues to this day. Most towns and villages have at least one pharmacy in which traditional medicine plays a very significant part. Medicinal plants can be found at markets and street vendors throughout China. The stalls often also include the bones, mummified bodies, skin or otherwise preserved parts of a large range of animals, sometimes including endangered bear and tiger. The *Flora of China*, which can easily be accessed at its website (http://hua.huh.harvard.edu/china/), indicates the medicinal or other economic properties of the plants described.

New information. The authors are aware that information on habitat, altitude and flowering time can only be based on existing records and is sometimes very speculative. They are keen to learn of any relevant new records or observations regarding the plants in this book, particularly records of species outside their stated distributions.

Identifying plants. For most naturalists, it is easiest to try to identify a particular plant in the field rather than from a specimen or photograph. This can at times be very difficult in China, where plants may be unfamiliar and information rather scanty. Sketches of relevant characters can be useful to aid identification, and digital cameras make it possible to record many details at very little expense. As a last resort, but only where permission is granted, small pieces of plant can be removed from the field for later analysis. No pieces should be removed from any reserve or national park, and certainly not from any protected species, unless you are undertaking *bone fide* scientific work in conjunction with Chinese scientists. It is worth reiterating that it is illegal to collect plants or plant material (this includes seeds) anywhere in China without specific permission from the relevant Chinese authorities. This in no way inhibits one's enjoyment at seeing China's amazing, rich and colourful flora in the wild, indeed one's enjoyment is all the greater for knowing that the flowers remain for others to appreciate.

The following characteristics may help those trying to identify species in the wild. A ×10 hand lens is recommended to help observation of the fine details of plants, in particular hairs and glands.

- Height and nature of the plant (whether a tree, a shrub, herbaceous, cushion-forming, etc.).
- Shape and arrangement of the leaves (whether stalked or stalkless, toothed, lobed or entire).
- The presence of leaf surface hairs or scales, especially on the under-surface of the leaf.
- Flower shape and colour (the numbers of the various flower parts such as sepals, petals, stamens and styles, whether the sepals or petals are fused and to what degree).
- Shape, nature and colour of the fruit (whether dry or fleshy, whether they split on ripening, and if so how and into how many divisions).
- Habitat type (whether the plant is growing in forests or meadows, or on alpine screes etc., or if it appears to be tolerant of a wide range of habitat types).

- Locality (where the plant is found can be a useful clue to its identity, ruling out species not found in a particular region; however, careful and accurate observation can sometimes result in new localities being found for a particular plant).
- Altitude (as with locality, this can at times be a useful guide to accurate identification).

It is often a combination of all these factors that eventually leads to an identification, although some species can be readily identified by simply using the information in the text and the accompanying photograph.

PHOTOGRAPHY

As far as possible all of the photographs included in this book were taken in the wild within the area covered by the book. This was not always feasible, however, and hence some of the photographs were taken outside the area covered (either in other parts of China, southern Tibet, the Himalaya or even Europe). Where cultivated plants have been photographed, we have tried to ensure that they are of known wild origin, although again, this has not always been possible. Each photograph is accompanied by a short caption giving the scientific name of the plant: localities are given when the photograph was taken in the wild, where the plant photographed was in cultivation, the caption included the abbreviation 'cult.'.

The majority of photographs have been taken by the authors; however, some very important photographs have been solicited from other sources. We are extremely grateful to those who provided photographs that have helped to fill serious gaps in the coverage. All photographic captions end with the photographer's initials in brackets as follows:

AC = Anne Chambers
AD = Alan Dunkley
AF = Aljos Farjon
AGS = Alpine Garden Society
CGW = Christopher Grey-Wilson
CS = Christopher Sanders
DC = Duncan Coombs
DD = David DuPuy
DJ = Doug Joyce
DL = David Lang
DR = Dieter Rückebrott
DTE = Derek Turner Ettlinger
DZ = Dieter Zschummel
EL = Erica Larkcom
GD = Gary Dunlop
HB = Hilary Birks
HH = Håkan Hallander
HJ = Harry Jans
HP = Holger Perner
IB = Ian Butterfield
JA = Joe Atkin

JB = John Birks
JF = Jeanette Fryer
JG = Jim Gardiner
JL = Jane Leeds
JM = John Mitchell
JR = John Richards
KC = Kenneth Cox
LYB = Luo Yi-bo in Plant Bank of China
MF = Maurice Foster
MFL = Mark Flanagan
MG = Mike Grant
MN = Margaret North
MO = Mikinori Ogisu
MW = Mark Woods
MWA = Martin Walsh
NR = Norman Robson
PC = Phillip Cribb
PCH = Peter Corkhill
PCX = Peter Cox
PHD = Paul Harcourt Davis

RBGE = Royal Botanic Garden Edinburgh
RBGK = Royal Botanic Gardens, Kew
RD = Ray Drew
RE = Richard Evernden
RL = Roy Lancaster
RLS = Rod Leeds
RM = Ray Morgan
RMc = Ron McBeath
RR = Robert Rolfe
RS = Rosie Steele
SC = Simon Crutchley
SG = Stephan Gale
TC = Tom Cope
TK = Tony Kirkham
TS = Tony Schilling
TY = Toshio Yoshida
VM = Victoria Matthews
YM = Yong Ming Yuan

SPECIES IN CULTIVATION

The Chinese have been prolific growers of plants for many centuries, using both native and imported species and their derivatives. Parks and gardens, especially those associated with temples and monasteries, are full of plants grown for their decorative qualities, such as camellias and peonies. Certain plants, such as the *Ginkgo*, some conifers and many selections of bamboo, play a significant role in Chinese culture. Bamboo is grown extensively in central and southern China where there are several large bamboo gardens or parks.

From a western perspective, the Chinese flora has had a very marked effect on our gardens and there have been countless Chinese plant introductions over the centuries (particularly the nineteenth and twentieth), primarily at the hands of intrepid explorers from North America, Britain, France, Russia, Sweden and many other countries. We do not have space here to detail the many exploits of these plant hunters, but these are well covered in a number of books, most notably in the introductory pages to Roy Lancaster's *Travels in China – Plantsman's Paradise* (1989, 2nd edition 2008). Suffice it to say that many of the plants growing in our gardens are of Chinese origin and the most significant genera are listed below (p. 13). Furthermore, selections from many of these plants have been made, and many hybrids created.

Mixed deciduous woodland with *Larix potaninii* and *Betula* in west Sichuan, with the Siguniangshan in the background (TK)

Almost impenetrable *Rhododendron* shrubberies, often covering large areas, are a marked feature of many mountains in swChina and eTibet: *R. aganniphum* on the upper slopes of Daxueshan, north-western Yunnan (HJ)

Rich mountain meadows, with *Trollius*, *Pedicularis*, *Polygonum* and a medley of other species; Hongyuan Plateau, north-westernSichuan (HJ)

CONIFERS: *Abies, Ginkgo, Juniperus, Larix, Metasequoia, Picea, Pinus*.
DECIDUOUS AND EVERGREEN TREES AND SHRUBS: *Acer, Berberis, Betula, Buddleja, Camellia, Clematis, Clerodendrum, Cornus, Corylopsis, Cotoneaster, Daphne, Davidia, Deutzia, Euonymus, Forsythia, Hydrangea, Hypericum, Ilex, Indigofera, Jasminum, Lonicera, Magnolia, Mahonia, Malus, Osmanthus, Paeonia, Philadelphus, Pieris, Populus, Pyracantha, Rhododendron, Rosa, Rubus, Salix, Schisandra, Sorbus, Spiraea, Syringa, Tilia, Trachelospermum, Vaccinium, Viburnum, Wisteria*.
PERENNIALS: *Aconitum, Adenophora, Aquilegia, Astilbe, Cimicifuga, Clematis* (herbaceous), *Dracocephalum, Epimedium, Euphorbia, Geranium, Impatiens, Incarvillea, Lysimachia, Meconopsis, Paeonia, Paris, Primula, Rehmannia, Rheum, Rodgersia, Salvia, Stellera, Thalictrum, Trollius*.
ALPINES: *Adonis, Androsace, Anemone, Corydalis, Draba, Gentiana, Omphalogramma, Petrocosmea, Primula, Saxifraga, Scutellaria*.
BULBS AND THEIR RELATIVES: *Allium, Arisaema, Cardiocrinum, Fritillaria, Hedychium, Iris, Lilium, Nomocharis, Polygonatum, Roscoea*.
ORCHIDS: *Calanthe, Cymbidium, Cypripedium, Coelogyne, Dendrobium, Paphiopedilum, Phaius, Pleione*.

Users of this book will find that it includes a very large number of plants that are familiar in gardens. Indeed, this work is intended to give information on how these plants grow in the wild that may be of use to those cultivating them.

On heavily grazed meadows unpalatable plants such as *Euphorbia jolkinii* and *Iris bulleyana* proliferate at the expense of more appetising meadow herbs; Zhongdian plateau, north-western Yunnan (CGW)

Corydalis melanochlora growing on rugged scree at 4,400 m on Baimashan in north-western Yunnan (HJ)

CHINA

When to go

Travel in western China at almost any time of the year and you are certain to see interesting plants in flower or fruit. The north of the region is cold and rather inhospitable during the winter months, but the far south-west is subtropical. Touring groups or individuals travelling in China to 'see the sights' tend to go almost at any time, but those wishing to travel to see flowers and other wildlife are more restricted. Those going during the monsoon period (particularly late June through to September) can find the travelling tricky. The weather can be very good at times but prolonged and heavy rain can restrict movement around the country and hamper the enjoyment of the flora. Having said that, many of the most interesting plants come into flower during the monsoon period, although the pre-monsoon period can be especially good for certain plant groups, rhododendrons in particular.

For those wishing to combine sightseeing with some natural history, then perhaps the best times to visit western China are April–June and September–October, when the weather can be excellent, walking easy and the views clear for photography. The serious naturalist will need to don waterproofs and head into the mountains during the monsoon in order to see many of the plants in flower: at higher elevations numerous alpine plants are at their best during July and August. Even during these months, the peak of the monsoon period, mornings are often dry and sunny, and many days sometimes weeks can go by without significant rain. Generally, the further north and west one travels, the more likely it is to be dry; whereas the south (particularly the south-west) can be extremely wet and humid during the same period.

Topography

After Canada and Russia, China is the third largest country in the world, with a total land area of 9.6 million sq. km. It measures over 5,200 km from east to west and over 5,500 km north to south. The country is bordered by 12 neighbouring countries, and has a coastline in the east and south-east that extends for more than 14,500 km.

The country's topography is varied and complex, ranging from the world's highest mountains to lowland basins, and from extensive plateaus and hills to broad, flat and fertile plains.

The region covered by this book lies in the west of the country, and includes east and south-east Qinghai, south Gansu, and west and central Shaanxi in the north, east and south-east Tibet (Xizang), Sichuan and west Hubei in the middle, and Yunnan, Guizhou and west Guangxi in the south. Most of this region is mountainous, and includes Gonggashan (Minya Konka), which at 7,556 m is the highest peak within the Chinese frontier. The largest lowland area is occupied by the Sichuan Basin: other lowland areas can be found in south-west and south-east Yunnan and in Guangxi.

Rivers

China is drained by more than 1,500 rivers: most of the large rivers rise in the west. Of the country's total land area, 64 per cent is drained by rivers that flow into the sea; the remaining rivers flow into inland lakes or disappear into deserts or salt marshes.

The large river systems that dominate south-west China are the Yangtze River (known as the Jinshajiang in western China, and the Changjiang in central and eastern China), the Pearl River (Zhujiang) and the Mekong River (Langcangjiang), which all flow east or south-east and empty into the Pacific Ocean, and the Salween River (Nujiang), which flows south into the Indian Ocean.

All of these arise on the Tibetan Plateau (in Tibet (Xizang Autonomous Region) and Qinghai province). China's largest river, the Jinshajiang–Changjiang (which is 6,300 km long with a catchment area of 1.809 million sq. km) heads south-east from Qinghai into eastern Tibet and north-west Yunnan, before hitting the mass of Yulongxueshan (Jade Dragon Snow Mountain) in north-west Yunnan. This mountain deflects the Jinshajiang at Shigu to the east, so that (as the Changjiang) it eventually flows out into the South China Sea. The upper reaches are sparsely populated, but the middle and lower parts have a warm and humid climate, plentiful rainfall and fertile soil. These make the area an important agricultural region and, as a result, it is densely populated.

The Jinshajiang flows in parallel with two other great rivers, the Lancangjiang and the Nujiang, all three heading more or less south-east from the south-eastern corner of Tibet through west and north-west Yunnan. Unlike the Jinshajiang, the courses of the Lancangjiang and Nujiang continue southwards through deep trenches. Together, they form one of the most extraordinary topographical areas in western China, where very deep, often treacherous valleys are interspersed with very steep, high ranges or divides. These divides, which are subject to the monsoon, are botanically extremely rich and were the focus of many early plant collectors and naturalists, such as Père Delavay, George Forrest, Heinrich Handel-Mazzetti, Frank Kingdon Ward and Joseph Rock.

The spectacular waterfalls at the World Heritage Site of Jiuzhaigou in north-western Sichuan, tumble over impressive deposits of tufa from a series of lakes (CGW)

The 4,200 km-long Lancangjiang crosses Yunnan and forms both the border between Myanmar and Laos and most of the border between Laos and Thailand, before flowing across Cambodia and southern Vietnam into the rich Lancangjiang delta, where it empties into the South China Sea. In Yunnan, the river is full of rapids and runs in a deep gorge at about 1,200 m elevation, skirting east of Meilixueshan in north-west Yunnan.

The Nujiang, 2,800 km long and mainland South East Asia's longest undammed river, runs from the Tibetan Plateau through Yunnan and the Shan and Kayah States of Myanmar, along the Thai–Myanmar border and eventually flows into the Indian Ocean.

Mountains
Several mountainous areas lie within the region covered by this book. These include the eastern edge of the Qinghai–Tibet Plateau, the highest and largest plateau on earth and popularly called the 'roof of the world'. The plateau covers 2.2 million square kilometres and averages 4,000 m above sea level. Mt Everest (Zhumulamgma in Chinese, Qomolangma in Tibetan), the highest peak in the world at 8,848 m, and other mighty Himalayan peaks lie on its southern margin, but are outside the area covered by this book.

To the east of the Qinghai–Tibet Plateau lies a complex of high mountain ranges intersected with plateaus and basins with elevations from 2,000 to 4,000 m. This area includes the Minshan, Nushan and Gaoligongshan, the Yunnan–Guizhou plateaus and the Sichuan Basin.

Small areas of cultivated land along the Lancangjiang, dwarfed by the mountains: Bangda to Chamdo, south-eastern Tibet (HJ)

The Minshan, a high, little-explored snowy range, dominates the skyline on the north Sichuan–south Gansu border (CGW)

Glacial lakes and marshy areas with *Rhododendron nivale* on Stone Mountain (Gologshan) near Jigzhi, south-eastern Qinghai (CGW)

East of these mainly mountainous regions, the land falls away to the fertile and heavily cultivated plains of eastern China. To the north and north-west, aridity increases towards the Mongolian steppe and the deserts of Central Asia. To the south, tropical and monsoon forests predominate in the few areas where the land has not been converted to agriculture.

The mountains of the south-west are characterised by extremely complex topography, ranging from less than 1,000 m on some valley floors to the 7,556 m of Gonggashan in west Sichuan. The mountain ridges are generally orientated in a north–south direction, at right angles to the main Himalayan range, and are effectively the eastern downturned edge of that range. The region includes the Hengduanshan, Gaoligongshan and Nushan of western Yunnan, several ranges at the south-east edge of the Tibetan Plateau, the Shalulishan, Daxueshan (including Gonggashan), Cholashan and Qionglaishan systems of Sichuan, and the Minshan on the Gansu–Sichuan border. These constitute the South-west China Biodiversity Hotspot, which covers over 262,400 sq. km of temperate to alpine mountains, lying between the eastern-most edge of the Tibetan Plateau and the Central Chinese Plain. (The Ailaoshan and Wuliangshan of central Yunnan form part of the Indo-Burma Biodiversity Hotspot.)

In the upper Gangheba, Yulongxueshan, north-western Yunnan, moraines and screes give way to cliffs and bare exposed rocks (CGW)

Namcha Barwa, seen between Bomi and Tongmai in south-eastern Tibet (HJ)

Climate

Western China is dissected by large fast-flowing rivers in deep gorges, producing a complex topography that results in a wide range of climatic conditions. Parts of Yunnan are frost-free throughout the year and the northern boundary of the region has short, frost-free periods, but the high mountain peaks of Sichuan, Yunnan and Tibet have permanent glaciers. Annual average rainfall in the region exceeds 1,000 mm on south-western slopes at higher elevations in Yunnan. By contrast, parts of the north-west of the region, in the rain-shadow of the Tibetan Plateau, rarely receive more than 400 mm per year.

Although the climate of China exhibits great variety, it is essentially continental and influenced by both topography and the monsoons. Over most of the country, the four seasons are distinct.

From September until April, cold, dry winds from Siberia and Mongolia in the north gradually weaken as they reach the southern part of the country, resulting in cool autumns and cold, dry winters with great differences in temperature.

The warm, moist monsoons, lasting from late May until September, bring abundant rainfall off the oceans as well as high temperatures, with little difference in temperature between the south and north. China's complex and varied climate results in a great variety of temperature and rainfall zones. Temperatures vary considerably across the region, influenced by latitude and the monsoon. The region lies mostly to the south of the Huaihe–Qinling Mountains and the south-east Qinghai–Tibet Plateau, and thus winter temperatures lie below zero. Further south, the mean temperature does not fall below zero.

In monsoon-dominated regions of swChina, racks are used for drying hay; Napahai, Zhongdian Plateau, north-western Yunnan (CGW)

Even in summer-monsoon areas, the inner and deeper valleys can be relatively dry and this rain-shadow effect produces a very different flora to that of the higher, more exposed, wetter surrounding ridges and mountains: a view of the deep dry valley of the Jinshajiang above Benzilan in north-western Yunnan (HJ)

In summer, temperatures throughout the region are mostly above 20°C, despite the high Qinghai–Tibet Plateau and various mountain chains. By contrast, average temperatures of over 32°C can be found in summer along the Jinshajiang–Changjiang, for instance in the cities of Chongqing and Wuhan.

Five climatic zones can be identified in south-west China, although these merge into one another at their extremes:

- Cold-temperate Zone. Dominated by extremely cold winters and warm summers with a low average rainfall and most of the precipitation falling in the summer. Includes eastern Qinghai and adjacent parts of Tibet. The high mountain regions of south Gansu, west Sichuan and north-west Yunnan above 5,000 m can receive heavy winter snow and have a decidedly more alpine climate, with cool summers during which rainfall is relatively high, when compared with the plateau areas to the west.
- Mid-temperate Zone. Sandwiched between the cold temperate zone to the west and north and the warmer lower altitude temperate zones to the south and east are the principal mountain areas of Hubei, west Sichuan and west and north-west Yunnan, at altitudes of about 2,800 to 3,500 m. Although there are rain-shadow areas within mountainous western China, most places in this zone have cold, relatively dry long winters and warm summers when most of the precipitation falls.

Rice paddies by a village close to Szemao in south-western Yunnan (CGW)

A mosaic of terraces along the Jinshajiang (Yangtse) south of Shigu, north-western Yunnan (CGW)

Rice paddies along the margin of Erhai, Dali, north-western Yunnan (CGW)

- Warm-temperate Zone. Regions with warm wet summers and cold, relatively dry short winters. This encompasses most of the mid-altitude mountainous zones (1,500–2,800 m) in Hubei, Sichuan, Yunnan and extreme south-east Tibet.
- Subtropical Zone. Regions with warm humid summers, during which most of the precipitation falls during the monsoon, and the cool, relatively dry winters. Much of the lowlands and hilly regions of Guangxi, Guizhou, Hubei, Sichuan and Yunnan are subtropical. These regions include the main rice-growing areas of western China.
- Tropical Zone. In the south of the region (roughly south of the Tropic of Cancer), high summer monsoon precipitation and humid hot weather is offset by relatively dry, yet warm winters. Only relatively small areas in the extreme south of Yunnan and Guangxi are truly tropical.

TYPICAL TEMPERATURES AND PRECIPITATION LEVELS

CITY PROVINCE	Xining Qinghai	Lanzhou Gansu	Yinchang Hubei	Chengdu Sichuan	Kunming Yunnan	Guiyang Guizhou
TEMPERATURES						
January Low	-16 to -14°C (4–7°F)	-13 to -11°C (9–12°F)	0–2°C (32–36°F)	2–4°C (36–39°F)	1–3°C (34–37°F)	1–3°C (34–37°F)
January High	0–2°C (32–36°F)	0–2°C (32–36°F)	7–9°C (45–48°F)	9–11°C (48–52°F)	14–16°C (57–61°F)	8–10°C (46–50°F)
July Low	11–13°C (52–55°F)	16–18°C (61–64°F)	24–26°C (75–79°F)	18–20°C (64–68°F)	16–18°C (61–64°F)	20–22°C (68–72°F)
July High	24–26°C (75–79°F)	28–30°C (82–86°F)	32–34°C (90–93°F)	29–31°C (84–88°F)	23–25°C (73–77°F)	28–30°C (82–86°F)
PRECIPITATION						
January	<5 mm (<0.2 in)	<5 mm (<0.2 in)	18–23 mm (0.7–0.9 in)	5–10 mm (0.2–0.4 in)	10–15 mm (0.4–0.6 in)	20–25 mm (0.8–1 in)
July	80–85 mm (3.1–3.4 in)	65–70 mm (2.6–2.8 in)	221–226 mm (8.6–8.8 in)	235–240 mm (9.2–9.4 in)	210–215 mm (8.2–8.4 in)	175–180 mm (6.8–7 in)

Rainfall

In the summer, a south-east monsoon, originating over the Pacific Ocean, sweeps into south-east China, generally arriving in May. At the same time, the equatorial Indian monsoon arrives from the south-west. Although southern Yunnan may receive heavy rainfall in late May and early June, the main rainy season in much of western China occurs in July and August and generally begins to wane in the second half of September. In a particularly heavy monsoon year, the rain may not disperse until well into October. The south and south-east parts of the country, therefore, receive the heaviest rainfall and this declines considerably northwards and westwards. Very heavy rain can occur, however, where the heavy monsoon clouds are forced against the high mountains of west Yunnan, particularly those close to the Myanmar border. Likewise, the high mountains of west and south-west Sichuan can receive considerable summer rainfall.

Rainfall declines after the monsoon ends and the late autumn and winter months are noticeably dry, although heavy snow can fall on the higher mountains.

The mighty Jinshajiang (Yangtse) near Hutaio Xia in north-western Yunnan supports a subtropical agriculture: much of the land is terraced except where the mountains are too steep (CGW)

Floristic regions

Climatic, topographic and edaphic conditions produce a wide variety of vegetation types across the South-west China Biodiversity Hotspot, including broadleaved and coniferous forests, bamboo groves, scrub communities, savannahs, meadows, prairies, freshwater wetlands, alpine scrub and scree and rock communities. The major regions within the area covered in this book are dealt with in order here, listed approximately from west to east and from south to north.

South-east Tibet

This region covers the eastern part of the Tibet–Qinghai Plateau and the higher areas that extend along mountain crests in the river gorge area of south-east Tibet and west Sichuan, between 3,500 and 6,000 m. It has a range of alpine vegetation types including meadow, steppe, cold desert and high elevation cushion-plant communities. It is too cold to support forests but the rainfall (more than 500 mm per year) produces meadows in places that either are free of snow and exposed to wind during winter or are prone to flooding. Dense shrubberies occur in places that are sheltered and well-drained. To the south-east, forests are found in the river gorges that lie between steep mountain ridges.

LEFT: Hay meadows, unspoilt by the use of herbicides, are rich in flowering plants in early summer as here on the Zhongdian Plateau, north-western Yunnan (CGW)

RIGHT: *Fritillaria delavayi* is a specialised and local plant of high coarse screes; photographed on Daxueshan, north-western Yunnan (JM)

Cliff-crevices and ledges provide a habitat for many specialised and non-specialised plants, especially alpines at higher altitudes: *Paraquilegia microphylla*, one of the most colourful alpine cliff-dwellers, is common on cliffs on Baimashan, north-western Yunnan between 3,700 and 4,200 m (HJ)

Extensive screes, like these on the Daxueshan in north-western Yunnan, harbour an interesting variety of alpine plants, especially on the more stabilised areas (CGW)

Much of eastern Tibet Plateau and its margins receive too little rainfall to support forests and are dominated by vast areas of yak-grazed grassland. At 3,500 m south of Aba in north-western Sichuan, *Stellera chamaejasme* is a conspicuous plant on grassy slopes (CGW)

Rugged mountains dominate wChina, especially those close to the edge of the Tibetan Plateau; here part of Daxueshan in north-western Yunnan at c. 5,000 m (CGW)

Heavily grazed meadows on many Tibetan mountains give a bleak effect, although plants like *Meconopsis integrifolia* and many other meadow plants survive in such habitats: on the Kongbo Pa La in south-eastern Tibet, the *Meconopsis* is a dwarf ecotype far smaller than the plants found at lower altitudes (AC)

A drier climate around Muli in south-western Sichuan has resulted in a landscape of fragmented *Rhododendron* moorland and coniferous woodland (PC)

The alpine zone in Tibet includes those areas where the average temperature during July, the warmest month of the year, is no more than 10°C. Within this region, moisture determines whether meadow, steppe or alpine desert vegetation are present. As rainfall decreases towards the north-west, the vegetation changes from dense scrub to alpine meadow to steppe, and finally to semi-desert and true desert. North-facing slopes, which hold the snow in winter, support evergreen sclerophyllous shrubs such as small-leaved *Rhododendron* species. South-facing slopes are often bare and those with vegetation tend to support deciduous genera such as *Caragana*, *Potentilla* and *Salix*. The widespread *Kobresia* (sedge) meadows of the south-east Tibetan Plateau are very distinctive and harbour a wide range of other species, especially herbaceous perennials.

These moist alpine regions have some of the richest alpine landscapes on earth and form a colourful tapestry during the spring and summer. Families include numerous members of the Compositae, Fumariaceae, Gentianaceae, Liliaceae, Papaveraceae, Primulaceae, Ranunculaceae, Rosaceae, Scrophulariaceae and Umbelliferae. These regions are a centre of diversity for several alpine genera, including *Corydalis*, *Meconopsis*, *Pedicularis*, *Primula* and *Saxifraga*.

Aspect can greatly influence vegetation, especially at altitude, as here on the Daxueshan in north-western Yunnan: the right moister, cooler, north-facing slopes have a dense cover of dwarf rhododendrons, while on the other sunnier and drier side of the valley, the slopes are comparatively bare (HJ)

Rich mountain meadows, with *Trollius, Pedicularis, Polygonum* and a medley of other species; Hongyuan Plateau, north-western Sichuan (HJ)

Gaoligongshan

The Gaoligongshan form the divide between the Nujiang to the east and the Irrawaddy to the west. The 300-km-long Gaoligong Nature Reserve, located above 2,000 m along the crest of the narrow Gaoligongshan, includes a complete range of vegetation types from subtropical evergreen broadleaved forests to subalpine coniferous forests, large tracts of which are relatively undisturbed. Below 2,000 m, warm-temperate, evergreen, broadleaved forests containing species of Fagaceae and Lauraceae are found. Above 3,000 m, these are replaced by cold-temperate, deciduous forests and coniferous forests.

Nujiang–Lancangjiang gorges

The Yunnan Plateau is separated from Myanmar and India by several parallel river gorges, themselves separated by a series of steep, high mountain ridges or divides. The rivers flow at subtropical elevations (600–2,000 m), whereas the ridge crests attain a maximum elevation of about 4,000 m in the south, increasing to more than 6,000 m in the north. Much of this region remains very inaccessible and the forest cover that remains today is taller than that in most other parts of China. The steep slopes and higher elevations continue to support some of the country's richest biological diversity, including amongst many others, several species of conifer that are endemic to these gorges. With steep topography and isolated ridges and valleys, this area also has distinctive ecological communities: both the steepness and the relative remoteness result in a higher proportion of relict natural forests than is found in most other parts of China.

The Nujiang and Lancangjiang flow in deep valleys through a landscape with a wide range of habitats from semi-arid river valleys to alpine ridges. They pass through an elevation and latitude gradient that ranges from almost tropical (25°N and 600 m elevation) to cold-temperate (30° N and more than 6,000 m elevation). The mountains and river valleys in the southern part of this area are the richest and most distinct: the nearly intact forests include a mixture of Indo-Malayan as well as Palearactic species. Xishuangbanna in south-west Yunnan is the best known and most readily accessible region for the casual visitor.

The gorge area is notably rich in temperate conifers, with about 20 being found there, some being endemic. One of the most notable is *Cupressus duclouxiana*, which grows throughout the dry river gorge valleys of this ecoregion, especially on basic, limestone-derived soils. In the Nushan and Gaoligongshan in the western Hengduanshan, the rare *Taiwania flousiana* is locally dominant at 2,200–2,400 m in some forests, flourishing in humid areas where the rainfall is between 1,000 and 1,700 mm a year. Other conifers include *Tsuga dumosa* and *Pinus griffithii*. Broadleaved species include *Cyclobalanopsis*, *Magnolia* and *Schima,* as well as various members of the Lauraceae.

Species of *Abies*, *Juniperus*, *Larix*, *Picea* and *Pinus* are abundant at higher elevations. Between 2,700 and 4,000 m *Abies delavayi,* a highly decorative fir, often forms pure stands, especially in the Cangshan, Nushan and southern Gaoligongshan, all relatively close to the Yunnan–Myanmar border.

The abrupt contrast in zones on Daxueshan in north-western Yunnan, between the upper forest (primarily *Abies delavayi* and rhododendrons) and the exposed limestone ridges and peaks is most striking (HJ)

Several unusual forest types are found in the gorge area: sclerophyllous broadleaved evergreen forests on sunny slopes between 2,800 and 4,200 m are often dominated by subalpine oak species such as *Quercus aquifolioides*. *Pinus densata* often dominates secondary forest growing in sunny aspects below 3,500 m, especially where *Picea* has been removed. On some steep hillsides, secondary forests of *Populus bonatti* and *Betula utilis* can be found while, in places, damp subalpine valleys may support *Juniperus wallichiana,* and higher still, *Cupressus duclouxiana*, especially on basic soils. There is sometimes an understorey of scrub oak, such as *Quercus guyavifolia* and *Quercus monimotricha*.

Yunnan Plateau
The Yunnan Plateau enjoys a mild climate, often said to be spring-like throughout the year. This region is an expansive upland area of low hills and broad intermontane basins, lying between the limestone hills of south-east China and the Tibetan Plateau. At an elevation of about 2,000 m, the summers are very wet but not particularly hot, and the winters are cool but not too cold. Much of the Plateau has been either converted to agriculture or stripped of its primary forest. Much primary forest has been replaced by sparse stands of *Pinus yunnanensis*, which may grow together with *Michelia yunnanensis*, *Camellia pitardii* and various species of *Rhododendron*. Mountains, especially those in the western part, support luxuriant evergreen broadleaved forests, and small pockets of old forests remain intact in some mountain ranges. In the famous Xishan (Western Hills) near Kunming, there are remnant stands of subtropical evergreen broadleaved forest. Such forest can also be seen in several other places, such as Jizushan near Dali and in small temple-protected forests throughout much of the region.

Above 1,800 m, temperate crops such as wheat, beans and potatoes are extensively cultivated, especially in the valley bottoms as here at Wengshui, north-western Yunnan (CGW)

Most of the Yunnan Plateau receives about 1,200 mm of rain per year, 80 per cent during the summer monsoon. Winter can bring cold weather and light snow can fall in the mountains, although snowfall is very rare in the lower basin areas.

The characteristic vegetation over much of the Plateau, seasonally humid, evergreen broadleaved forest, is adapted to wet summers and an extended cool, dry season that lasts through the winter and early spring (November to April). On the crests of some of the higher hills, temperate cloud-forest communities occur, whereas on the low hills and in the basins, seasonally dry, subtropical forests are characteristic, with Fagaceae (*Castanopsis*, *Cyclobalanopsis* and *Lithocarpus* spp.) and Lauraceae (*Lindera* and *Persea* spp.) dominating.

The western part of the Plateau includes outlier ranges of the Hengduanshan that exceed 2,500 m. The higher ridges, with their cool, humid climate, are often cloud-shrouded during the summer months and may receive some snowfall during the winter. These ridges harbour temperate cloud forest, which is different from the subtropical forests of the Plateau. Here, trees can be festooned with epiphytic orchids and ferns along with a plethora of mosses and lichens. Small bamboos such as *Sinarundinaria* spp., ferns such as *Dryopteris*, and many shade-tolerant plants, including many species of *Arisaema* and *Impatiens*, form a rich understorey.

Wheat bundles drying on racks at Lijiang, north-western Yunnan, a common practice in swChina where the summer monsoon can cause drying difficulties (CGW)

LEFT: At Deqin in north-western Yunnan, the confined valley allows very little room for cultivation (CGW)
RIGHT: *Abies–Rhododendron* (here *R. rex* subsp. *fictolacteum*) forest on the eastern slopes of Cangshan near Dali, north-western Yunnan (CGW)

Hengduanshan

The Hengduanshan is a complex of mountain ridges and river valleys lying between the western edge of the Yunnan Plateau and Tibet. It is one of China's few north–south ranges, cutting more or less at right angles to the Himalaya; the name *Hengduan* means 'transect'. The northern part lies to the west of the Sichuan Basin and includes the Minshan and Qionglaishan, whereas the southern part borders the river gorge country. Deep, narrow river valleys are separated by parallel ranges. Below 2,000 m, these can be surprisingly dry, supporting a drought-resistant, small-leaved scrub. In the south, the subalpine zone is colonised at lower altitudes (2,800–4,000 m) by broad-leaved species of genera such as *Acer*, *Betula*, *Rhododendron* and *Sorbus*, whereas at higher latitudes, one or two species of conifer may form a dominant forest.

In north-western Yunnan, *Picea* forests, dominated by *Picea likangensis* and *Picea brachytyla*, are found at 3,100–3,500 m, whereas *Abies georgei* occurs at higher elevations, 3,500–4,000 m. *Larix potaninii*, often in association with *Abies* and *Picea*, can be found throughout the region, sometimes as pure stands. These forests often colonise moraines and unstable slopes, often densely so.

In sparse conifer forests, evergreen and deciduous broadleaved trees may form a closed canopy beneath the conifers. Here *Acer*, *Rhododendron* and bamboo (especially *Fargesia* spp.) may be dominant. Ferns and the lichen *Usnea longissima*, which forms long traceries from branches, can infest moist forests that are cloudy through much of the summer, the lichen gathering a great deal of moisture from the abundant clouds and mist. Sunny slopes often support evergreen *Quercus*, such as *Quercus aquifolioides*, and this sometimes extends into the subalpine zone to 3,500 m, where it mixes with *Abies* and *Larix*.

In general, forests dominated by pale bluish green *Picea* are found below the *Abies* species (which appear almost black in the landscape) at 3,100–3,600 m, especially in those places with persistent summer cloud and reliable winter snow. The highest subalpine forests (3,500–4,100 m) are dominated by one of four species of *Abies* (especially *Abies delavayi* and *Abies georgei*), often as almost pure stands.

Nature Reserves in the southern Hengduanshan include the Yulongxueshan (Jade Dragon Snow Mountain) Nature Protection Area, a scenic area that is easily accessible from Lijiang, and Habaxueshan (Haba Snow Mountains) Nature Reserve. Napahai Nature Reserve, near Zhongdian, supports a large wetland area that provides important wintering grounds for black-necked cranes (*Grus nigricollis*) and other waterfowl. Lugu Hu (Lugu Lake), in the same general vicinity, is a very important reserve for overwintering wildfowl. This reserve encompasses a deep natural lake set in a landscape of pine-covered highlands at about 2,400 m.

LEFT: *Abies–Rhododendron* (here *R. oreotrephes* and *R. wardii*) forests on the slopes of Daxueshan, north-western Yunnan (CGW)
RIGHT: The rugged mountains of west China have numerous deep valleys and gorges that are often densely wooded and difficult to negotiate (CGW)

Rich, mixed broadleaved mid-altitude forest on the Lancangjiang–Jinshajiang divide, west Yunnan (CGW)

An exposed ridge top on Baimashan in north-western Yunnan, where low cushion-forming plants like *Chionocharis hookeri* fill an important niche (HJ)

Arenaria polytrichoides forms neat cushions on exposed mountain moorland on Baimashan, north-western Yunnan (HJ)

Abies and *Rhododendron* form the upper limit of forest on the Daxueshan, north-western Yunnan at c. 4,000 m (CGW)

Guizhou Plateau

The Guizhou Plateau (1,000–1,400 m) is a scenic region of limestone karst, riddled with caves and sinkholes, and intermontane basins that lies to the east of the higher Yunnan Plateau, south of the Changjiang. Its steep hills and isolated pinnacles are deeply dissected by river valleys. In the karst areas, many of the rivers and streams are partly subterranean. The porous rock and mild weather (more than 200 days are overcast during the year) account for the distinctive ecology of the region. Despite the high rainfall, water quickly drains through the porous rocks leaving plants easily stressed during dry periods. This attractive area, dominated by its typical limestone pinnacles, is easily reached via Guilin in Guangxi, and has become a major tourist attraction; however, the characteristic formations are found throughout Guangxi, south Guizhou and south-east Yunnan, stretching southwards into Vietnam.

The vegetation of the Guizhou Plateau consists of subtropical calcicolous evergreen broadleaved forests, originally dominated by Fagaceae (*Castanopsis*, *Cyclobalanopsis* and *Quercus*), Lauraceae (*Eugenia* and *Phoebe*) and Theaceae (*Camellia* and *Schima*). Lower down, seasonal tropical forests abound, dominated in the main by species of *Erythrina*, *Eugenia*, *Ficus*, *Helicia* and *Sterculia*. Secondary forests shelter many plants including species of *Albizia*, *Cornus*, *Liquidambar* and *Rhus*. Conifers are commonly found on the summit areas of the limestone peaks and ridges.

Coniferous forests are found in the east and north of the Guizhou Plateau, dominated especially by *Pinus massoniana* in the north and east, with *Abies* and *Picea* at higher elevations, and *Pinus yunnanensis* in the south of the region. *Cunninghamia lanceolata* is found in the north. On Fanjingshan and other high mountains, the summits are home to Himalayan-type forests comprising *Abies fargesii*, *Tsuga chinensis*, *Acer flabellatum*, *Enkianthus chinensis*, *Rhododendron argyrophyllum* subsp. *hypoglaucum* and *Prunus serrulata* amongst many others. Fanjingshan supports fine stands of *Davidia involucrata*, with trees up to 25 m tall.

The original forest of the Guizhou Plateau has now largely been replaced by secondary scrub or sparse woodland with an association of ericaceous species (*Lyonia*, *Rhododendron* and *Vaccinium* in particular) as well as *Eurya*, *Myrica*, *Myrsine* and *Quercus*.

Sichuan Basin

The Sichuan Basin (also known as the Red Basin) is a fertile, subtropical region of plains interspersed by low hills encircled by mountains. The Tibetan marches lie to the west and north, the Yunnan Plateau to the south, and several hundred kilometres of mid-elevation hills cut by the eastward-flowing Changjiang to the east. At one time, the area supported extensive subtropical evergreen broadleaved forests, but much of the land has been extensively cultivated for thousands of years, today leaving just a few remnant patches of original forest in more remote and inaccessible places.

Much of central and east Sichuan is low lying and subtropical: often called the 'ricebowl of China', this large province produces one third of the rice grown in China (CGW)

In the main, the Sichuan Basin is a fertile area of alluvial plains with large areas of red sandstones and purple shales, all producing fertile agricultural soils capable of supporting several crops each year. The limestone hills that dissect the area rise to about 700 m in places. As the rivers, which originate on the Tibetan Plateau, flow into the basin, they are checked and deposit vast amounts of fertile alluvium in the west of the Basin especially.

The climate of the Sichuan Basin is subtropical with cool, cloudy winters and very warm, hazy summers. Temperatures in the Basin are mild with very warm summers (26–29°C) and mild winters (5–8°C). Cold air, which streams down from Central Asia in the north during the winter months, is moderated by the influence of the northern mountain barrier. In summer, the skies are often overcast and a thick haze obscures much of the landscape as the heat builds up each day. In the winter, fog forms on many days. Chengdu averages more than 300 days of foggy weather during the year: it is said "dogs bark when the sun appears".

A few temple forests and holy mountains, such as Leshan and Emeishan, support remnant patches of climax forest vegetation, and some floodplains include semi-wild wetland areas, which are important feeding and resting grounds for various migratory bird species. Apart from these, the most extensive natural forests are to be found on the lower slopes of the mountains fringing the Basin: these include Emeishan which, because it is a holy mountain, has long been saved from deforestation.

In the dry inner valleys of south-western Sichuan, cultivation is confined primarily to the valley floors that are irrigated from nearby streams and rivers; photograph taken to the east of Derong (CGW)

Subtropical Fagaceae (*Castanopsis* and *Quercus*), Lauraceae (*Cinnamomum*, *Lindera*, *Litsea* and *Machilus*) and Theaceae (*Schima*) probably constitute the original vegetation of the Sichuan Basin. Today, uncultivated areas support a scrub of *Myrica nana*, *Rhododendron* spp. and *Vaccinium bracteatum*. Less disturbed sites may have thin stands of *Pinus massoniana* or *Cryptomeria japonica*, but these have often been planted. Limestone areas support scrub and forest vegetation that is florally distinctive, consisting of *Platycarya* spp., *Zanthoxylum planispinum* and *Rosa* spp.

Emeishan, greatly revered by Chinese Buddhists, is one of the best places to see the remnants of the lush Sichuan Basin subtropical forest with its associated animals. Amongst the rich assortment of trees and shrubs can be found *Davidia involucrata*, *Gingko biloba*, *Nothaphoebe omeiensis* and *Rosa omeiensis*. The rare conifer *Cathaya argyrophylla* is found in south-east Sichuan.

Qionglaishan–Minshan

The Qionglaishan–Minshan separate Amdo on the eastern edge of the Tibetan Plateau from the low-elevation Sichuan Basin, and are some of the highest and steepest mountains on earth. Gonggashan, the highest summit at 7,556 m, has glaciers on its steep eastern face that run well below the treeline, often at a very sharp angle, while adjacent river valleys may be as low as 1,000 or 2,000 m. Despite the fact that these mountains are located at subtropical latitudes, significant winter snow accumulates above 3,000 m, and substantial

At c. 4,200 m on Daxueshan in north-western Yunnan, the *Abies–Rhododendron* forests dwarf rapidly and finally give way to alpine meadows (CGW)

coniferous forests are found in the middle and upper elevations. *Pinus* species (primarily *Pinus tabuliformis* in the north and *Pinus armandii* in the south) are found in the drier, lower elevation valleys. Above 3,000 m, other conifers dominate, particularly species of *Abies*, *Picea* and *Tsuga* as well as *Larix potaninii*. At higher elevations, particularly on calcareous soils, species of *Juniperus* are found, often growing in surprisingly rocky places.

The zones of forest here are similar to those in the Hengduanshan and on the Yunnan Plateau to the south, except that they are found here at a lower elevation due to the latitude. Here, the effects of the summer monsoon are much less marked, whereas winter rainfall is greater.

In these coniferous forests, deciduous broadleaved trees, including species of *Acer*, *Betula* and *Sorbus*, and shrubs such as species of *Daphne*, *Mussaenda* and *Viburnum*, can form a lower canopy through which the conifers protrude. Thickets of small to medium-sized bamboos also grow here and provide a critical food resource for giant pandas and red pandas.

Mountain ridges above 4,000 m support alpine vegetation that is ecologically and floristically similar to that in adjacent south-east Tibet.

The dry valleys and gorges of the Minjiang and its tributaries, between 1,500 and 2,500 m, have vegetation consisting of a sparse cover of dwarf, thorny, aromatic shrubs, including *Acacia* spp., *Caryopteris* spp., *Rosa* spp., *Vitex* spp. and *Pistacia weinmannifolia*, often scattered in otherwise rather bare or rocky places. One of the American prickly pears, *Opuntia monocantha*, is naturalised and very abundant in some of these semi-arid areas.

Autumn colours strike the slopes, between Pasho and Rawu, 3,950 m, in south-east Tibet (HJ)

In relatively recent times, some of the panda reserves such as Wolong and the Huanglong–Wanglang–Juizhaigou complex, which are extremely important in the conservation of the giant panda, have been developed for tourism. The number of tourists from China and many other parts of the world has increased enormously in recent years.

Human population density is low in this area, primarily because the terrain is too steep, moist and cool for all but limited agriculture. In many parts, the landscape is often very wild and the vegetation unspoilt.

Dabashan and Qinlingshan

The biologically rich Dabashan and Qinlingshan form an important divide between the country's two greatest rivers, the Jinshajiang–Changjiang and the Huanghe (Yellow River) and run east-west, separating the Sichuan Basin from the extensive plains and loess plateaus of northern China. They also provide a biogeographic barrier between mostly evergreen, subtropical forests to the south and temperate, mostly deciduous, forests to the north. In addition, the Dabashan supports dense evergreen forests containing many endemic species, including threatened trees such as the deciduous *Metasequoia glyptostroboides*.

The Dabashan lies on the north-eastern rim of the Sichuan Basin and extends eastwards toward the Changjiang Plain. This mountain range has a significant effect on the regional climate, keeping it milder than that of areas at the same altitude further north. As a result, the plains to the south of the Dabashan have a warm temperate to subtropical, moisture-loving flora. The foothills boast a mixed evergreen and deciduous mix of Fagaceae (*Quercus acutissima* and *Quercus variabilis*). Higher elevations support warm-temperate forests of *Pinus massoniana* whereas, higher still, *Pinus armandii* makes an appearance. The Shennongjia Mountain Nature Reserve, rising to 3,100 m with an area of some 704 sq. km, has the most intact habitat in this region and is renowned for its dense forest cover and the age-long gathering of medicinal herbs. Here extensive primary old-growth forest is conserved, unfortunately some of the last in this part of China. Its flora, numbering more than 2,600 species, includes *Cercidiphyllum japonicum*, *Davidia involucrata*, *Emmenopterys henryi*, *Eucommia ulmoides* and *Tetracentron sinense*.

The Qinlingshan, which rises to 3,700 m, forms a barrier between the Sichuan Basin and the steppes and plains of north-central China. It is subject to strong, cold winter winds from the north, and thus its forests are characterised by temperate plants. The annual rainfall is about 700–950 mm depending on location.

The deciduous forests of the foothills are replaced by coniferous forests at lower elevations than those in the more sheltered Dabashan. On the whole, however, these forests are better preserved in the Qinlingshan than in the lowlands to the north. The foothill forests are dominated by northern temperate deciduous trees such as *Quercus acutissima*, *Quercus variabilis*, *Juglans regia* and species of *Acer*, *Celtis* and *Fraxinus*. Evergreen species include both broadleaved trees, such as *Castanopsis sclerophylla* and *Cyclobalanopsis glauca*, and various conifers.

Pinus armandii grows in association with *Betula*, *Carpinus* and *Quercus* at mid-altitudes. Higher, at about 2,600–3,000 m, these give way to a subalpine association primarily dominated by *Abies fargesii*, *Larix chinensis* and *Betula*, with *Rhododendron fastigiatum* abundant on the forest floor, especially in more open areas.

Taibaishan (3,767 m), a sacred mountain, is probably the most accessible location. Its forests have been protected over the centuries.

Chinese and other place names

It is sometimes very difficult to trace Chinese place names, especially those used in older literature. In 1958, the Government of the People's Republic of China (PRC) replaced the Wade-Giles Mandarin romanisation system by the Pinyin system, which was adopted by the International Organisation for Standardisation in 1982. Some important and classic localities in western China, particularly those known to the earlier European and American plant-hunters and naturalists, have changed and we have endeavoured to present the important changes in the list below. The list also includes English or Anglicised names (including translations) and provides the modern (Pinyin) Chinese equivalent.

Atendse = Deqin
Atuntsi = Deqin
Bangdsera = Benzilan
Beimashan = Baimashan
Bema-schan = Baimashan
Betahai = Bitahai
Big Snow Mountain = Daxueshan
Burma = Myanmar
Camellia Temple = Yufengsi
Canton = Guangzhou
Chienchuan = Jian Chuan
Chomolungma = Zhumulangma
Chung-ch-ing = Chongqing
Chungking = Chongqing
Chungtien = Zhongdian
Dadjienlou = Kangding
Dadu River = Dadujiang
Dali Range = Cangshan
Dalifu = Erhai
Dardo = Kangding
Dechen = Deqin
Deqen = Deqin
Diqin = Deqin
Donggrergo = Huanglong
Dschöngdu = Chengdu
Dschungdien = Zhongdian
Erhlang Shan = Erlangshan
Everest = Zhumulangma
Fan-ching Shan = Fanjingshan
Fang Hsien = Fangxian
Four Maidens Mountain = Siguniangshan
Four Sisters Mountain = Siguniangshan
Ganghoba = Gangheba
Gaocheng = Litang
Genyen Shan = Genyanshan
Guenhenshan = Genyanshan
Guenyenshan = Genyanshan

Gwanghsi = Guangxi
Ha-la-ma = Hongyuan
Heng Tuan = Hengduanshan
Hianglong = Huanglongsi
Hoking = Heqing
Hsiagwan = Xiaguan
Hsia-Kuan = Xiaguan
Hsingyi = Anlong
Hsinhwa = Xinhua
Hsinlung = Xinlong
Hupeh = Hubei
Hurama = Hongyuan
Ichang = Yichang
Inner Mongolia = Nei Mongol
Jade Dragon Snow Mountain = Yulongxueshan
Jone = Zhuoni
Kagurpo = Kawagebo
Kansu = Gansu
Kao-li-kung Shan = Gaoligongshan
Karkapo = Kawagebo
Kawa Karpo = Kawagebo
Kwangsi = Guangxi
Kweichow = Guizhou
Lanchou = Lanzhou
Lanchow = Lanzhou
Leishan = Leshan
Li-chiang = Lijiang
Lidjiang = Lijiang
Likiang = Lijiang
Litiping = Lidiping
Little Snow Mountain = Xiaoxueshan
Little Zhongdian = Xiaozhongdian
Liu-pa = Liuba
Lower Yangtze River = Changjiang
Lu-ting = Luding
Mekong River = Lancangjiang
Min Kiang = Minjiang

Min River = Minjiang
Minya Konka = Gonggashan
Moupin = Baoxing
Mt Emei = Emeishan
Mt Everest = Zhumulangma
Mt Genyuen = Genyanshan
Mt Omei = Emeishan
Omei Shan = Emeishan
Pearl River = Zhujiang
Pei-ma-shan = Baimashan
Qomolangma = Zhumulangma
Salween River = Nujiang
Schigu = Shigu
Setschwan = Sichuan
Shangri-La = Zhongdian (the name Shangri-La was proposed in 2001 to promote tourism)
Shan-hsi = Shaanxi (and also Shanxi)
Shansi = Shaanxi (and also Shanxi)
Shensi = Shaanxi
Shi-ch'ang = Xichang
Shih-kou = Shigu
Shikou = Shigu
Shiku = Shigu
Shweli = Lungchuanchiang
Siakwan = Xiaguan
Sinkiang = Xinjiang
Stone Forest = Shilin
Sung-p'an = Songpan

Szechuen = Sichuan
Szechwan = Sichuan
Tachienlu = Kangding
Ta-chien-lu = Kangding
Tali = Dali
Tali Range = Cangshan
Talifu = Erhai
Tatsienlu = Kangding
Teng-yueh = Tengchong
Tien Bao = Tianbao
Tiger Lake = Laohuhai
Tiger Leaping Gorge = Hutiao Xia
Tsangpo River = Yarlung Zangbojiang
Tulung = Dulong
Upper Yangtze River = Jinshajiang
Weishi = Weixi
Wengsui = Wengshui
Western Hills = Xishan
Woolong = Wolong
Xishwanbanna = Xishuangbanna
Ya-chou-fu = Ya'an
Yangpi = Yangbi
Yangtze River = Jinshajiang (in upper part), Changjiang (in lower part)
Yellow River = Huanghe
Yulong Shan = Yulongxueshan
Yunnanfu = Kunming
Zitsa Degu = Juizhaigou

FURTHER READING

A more detailed account of the diverse ecosystems in China can be found on the web (http://www.nationalgeographic.com). This website provides a useful listing for the habitats encountered in western China.

Davis, S. D., Heywood, V. H. & Hamilton, A. C. (1995). Centres of Plant Diversity: a Guide and Strategy for Their Conservation, Volume 2: Asia, Australasia and the Pacific. World Wide Fund for Nature and IUCN — The World Conservation Union.

Li, W. (1993). Forests of the Himalayan–Hengduan Mountains of China and Strategies for Their Sustainable Development. International Centre for Integrated Mountain Development (ICIMOD), Kathmandu, Nepal.

Mackinnon, J., Sha, M., Cheung, C., Carey, G., Xiang, Z. & Melville, D. (1996). A Biodiversity Review of China. World Wide Fund for Nature, Hong Kong.

Wu, Z. and the Editorial Board of the Kunming Institute of Ecology (1995). Vegetation Ecological Landscapes of Yunnan. Forestry Press of China, Beijing, People's Republic of China.

Zhao, J., Zheng G., Wang H. & Xu J. (eds) (1990). The Natural History of China. McGraw-Hill, USA and HarperCollins, UK.

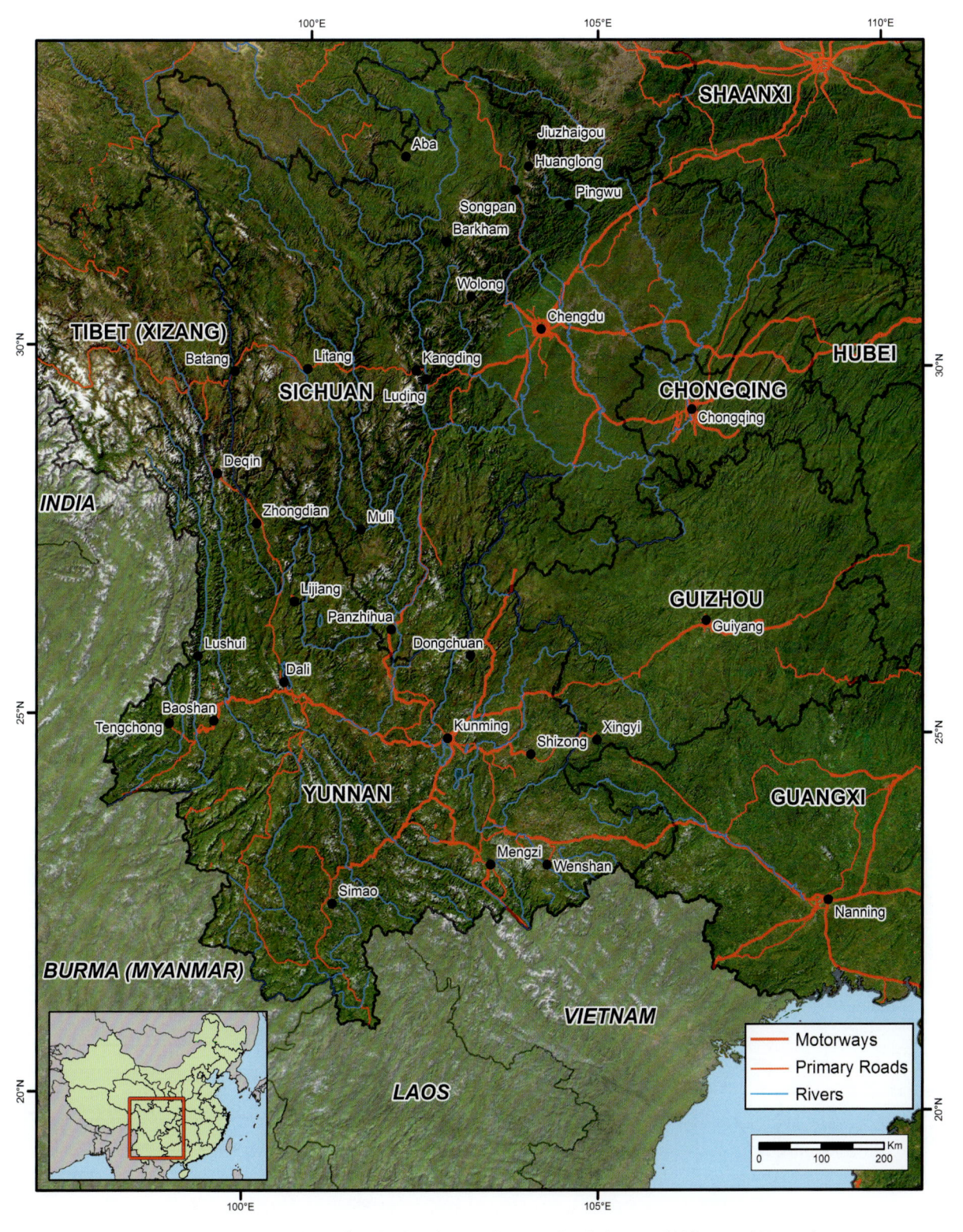

Map 1. Major roads and cities in western China. (See back endpapers for more detailed maps of Sichuan and Yunnan.)

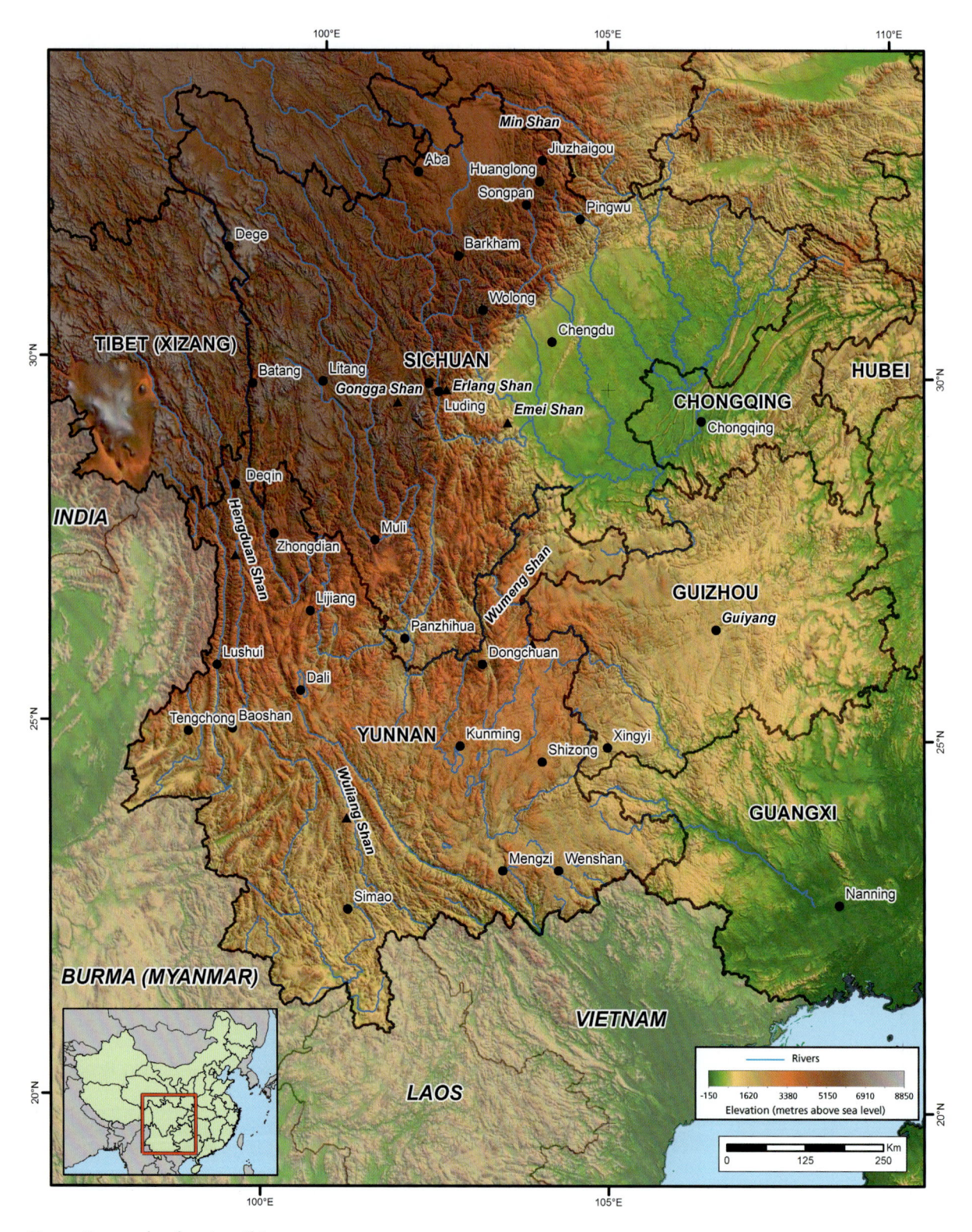

Map 2. Topography of western China.

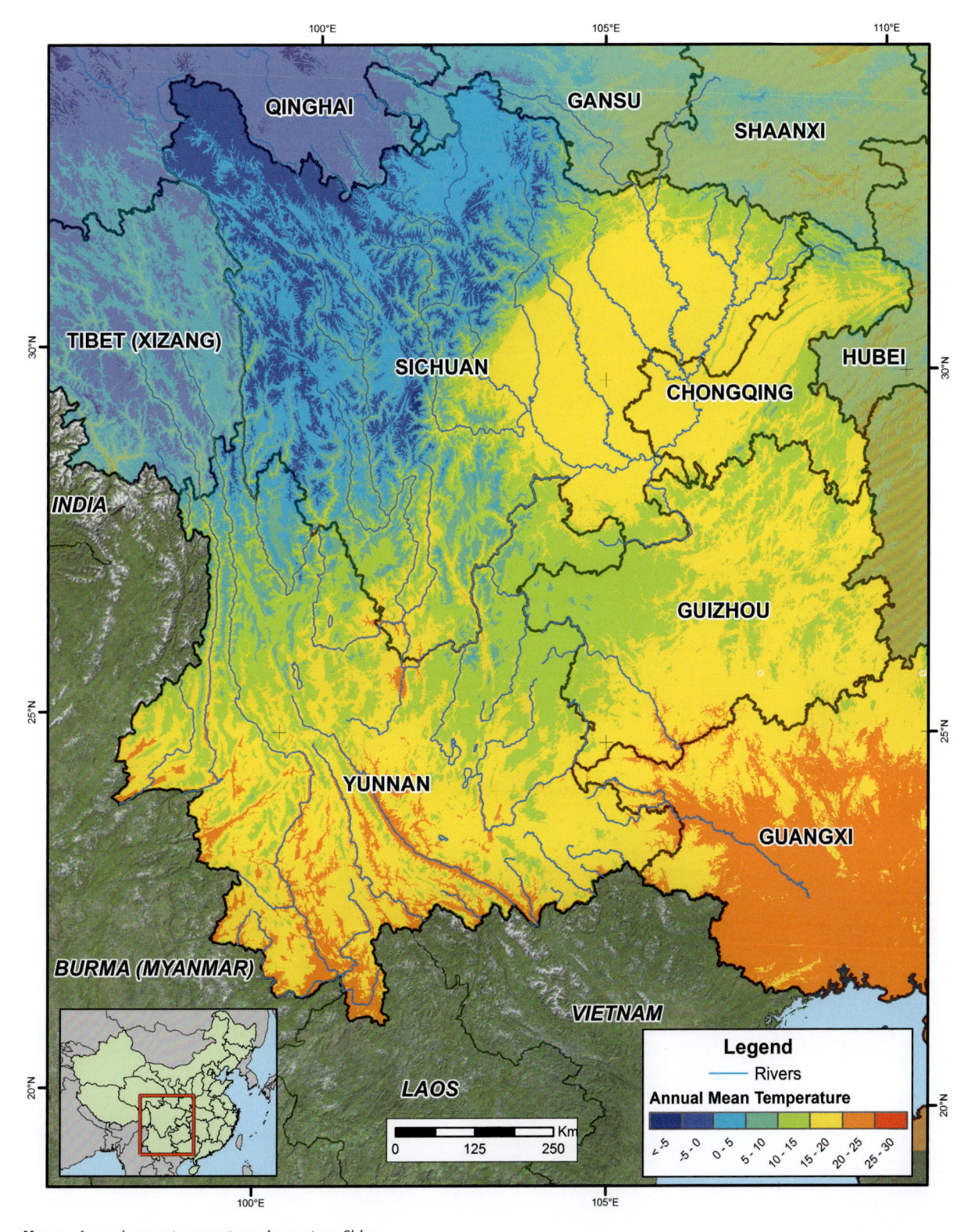

Map 3. Annual mean temperatures in western China.

Gymnosperms

GINKGOACEAE

GINKGO

A genus with a single species endemic to China, possibly extinct in the wild but widely planted around temples and as a street tree. Large, dioecious, **deciduous tree** with finely shredding grey bark with cross fissures, and erect, suberect or spreading main branches. **Leaves** borne on long and short shoots, with venation in 2 opposite rows. Male and female organs carried on short shoots; male catkins cylindrical, pendent. **Fruit** olive-like, yellowish green, rancid or foul-smelling when ripe, containing a single large **seed**.

Ginkgo biloba L.* **Tree** to 40 m conical when young but with a broad, spreading crown at maturity, trunk up to 4 m in diam.; **shoots** green, turning pale grey. **Leaves** fan-shaped, 5–8 cm long and broad, notched in middle of apex, bright green, butter-yellow before falling. **Male catkins** short-stalked, ivory-coloured, 12–22 mm long. **Fruits** yellow-green, 3 cm long, long-stalked. Old trees are common around temples throughout w & swChina [eChina]; 300–1,100 m.

CEPHALOTAXACEAE

CEPHALOTAXUS

A genus of c. 10 species (6 in China) in e and seAsia. Dioecious, **evergreen shrubs** or small **trees**, 2–15 m tall, with *Taxus*-like leaves in 2 ranks. **Male catkins** small globose cones. **Fruits** plum- or olive-like, dull green, borne at base of previous year's shoots.

Cephalotaxus fortunei Hook. **Shrub** or small **tree** 10–15 m tall with red-brown, shredding bark. **Leaves** linear, sharply and stiffly pointed, 40–90 × 3–5 mm, mid-green to yellow-green with 2 matt white bands beneath. **Fruits** egg-shaped, c. 25 mm long, on a 3–12 mm stalk, aril turning purple when ripe. sGansu, wHubei, Shaanxi, w, nw & swSichuan, nwYunnan; mixed coniferous and broadleaved forests, shrubberies, roadsides, 200–3,700 m. The typical variety (**var.** *fortunei*) is found in most of wChina [s & cChina, nMyanmar]. **Var.** *alpina* H. L. Li*, with stalkless **male cones**, **leaves** 1.5–3 mm and an aril with prominent longitudinal ridges, is found in sGansu, nShaanxi, n & wSichuan, nYunnan; 1,800–3,700 m.

Cephalotaxus sinensis (Rehd. & Wils.) H. L. Li Similar to the previous species, but **leaves** more densely arranged, generally smaller, 18–50 × 2–3.5 mm, sharply pointed, mid-green above, yellow-green beneath. **Fruits** 1–5 together, obovoid, 16–25 mm long, on 3–8 mm stalks; aril red to red-purple. sGansu, wGuangxi , neGuizhou, wHubei, Sichuan, w & cShaanxi, Yunnan [c & eChina, Taiwan]; montane coniferous forests and shrubberies, in valley bottoms near streams, 600–3,200 m.

CUPRESSACEAE

CALOCEDRUS

Just 3 species (1 in China) found in South East Asia, Myanmar, the USA and Mexico. **Evergreen trees**, ± pyramidal, branchlets arranged in a plane, side shoots 4-angled. **Leaves** scale-like, in false whorls of 2 pairs with the lateral pair slightly larger than the other. **Cones** with few scales; **seeds** with only 1 wing.

Calocedrus macrolepis Kurz **Tree** to 35 m with a broadly conical crown at maturity; trunk to 1.5 m in diam., bark smooth, grey-brown to reddish brown, smooth when young and whiter, fissured and exfoliating with age. **Leaves** scale-like, rather thin-textured, 3–4 mm long. **Male cones** yellow, 4–8 mm long. Mature **female cones** reddish brown, 6–20 mm long, with blunt scales. wGuangxi , sGuizhou, se & cYunnan [sChina, neMyanmar, nIndia, Laos, neThailand, nVietnam]; forests, field margins, 300–2,000 m.

CRYPTOMERIA

A genus of a single species from Japan and seChina. Tall, **evergreen trees** with red-brown to orange-brown, fibrous, soft bark. **Leaves** spirally arranged, awl-shaped, small. **Male cones** axillary, produced in early spring; **female cones** terminal on shoots 1 cm long, globular, scales with 4–7 tooth-like projections at apex.

Cryptomeria japonica (Thunb. ex L. f.) D. Don. **Tree** to 30 m with a conical crown when young, spreading with age; bark peeling in long strips at maturity, but in small square flakes when young. **Leaves** awl-shaped, tapering to a point, 5–15 mm long, mid-green to dark green. **Male cones** 10 mm long; **female cones** globular, 15–20 mm in diam. Sichuan, Yunnan [seChina, Japan]; widely planted as a timber crop and as an ornamental, 800–2,500 m.

CUNNINGHAMIA

A genus with just 1 species in China, Taiwan and Vietnam. Small, **evergreen tree** with a columnar habit and sparse crown; bark thick and fibrous. **Leaves** spirally arranged. **Male cones** broadly obovoid, in clusters of 8–20; **female cones** 1–3, on short shoots, terminal, consisting of 30 or more scales.

Cunninghamia lanceolata (Lamb.) Hook. **Tree** to 50 m but usually much smaller, with trunk up to 3 m in diam., lower branches often persistent and dead, but **shoots** green for several years. **Leaves** recurved, lanceolate, tapering to an pointed, sharp apex, 10–70 × 1.5–5 mm, decurrent on shoots, dark green above but silvery-white beneath. **Male cones** clustered at shoot-tips; **female cones** egg-shaped-conical, rounded at apex, 20–45 mm in diam., scales ovate, long-pointed. Gansu, Guizhou, Guangxi, Hubei, Sichuan, Shaanxi, Yunnan [sChina, Taiwan, Vietnam]; mixed broadleaved forests or in small, pure stands, roadsides, also widely planted, 200–2,800 m.

1. *Gingko biloba*; Shennonjia, wHubei (PC)
2. *Gingko biloba*; Jie-tai Temple, near Beijing (CGW)
3. *Gingko biloba* in fruit; cult. (PC)
4. *Cephalotaxus fortunei*; Jiuzhaigou, nwSichuan (PC)
5. *Cunninghamia lanceolata*, Shennongjia, wHubei (PC)
6. *Cunninghamia lanceolata*; cult. (PC)
7. *Cryptomeria japonica*; cult. (PC)
8. *Calocedrus macrolepis*; cYunnan (MF)

1. *Cupressus duclouxiana*; Lancangjiang valley, nwYunnan (AF)
2. *Cupressus duclouxiana*; north of Zhongdian, nwYunnan (PC)
3. *Juniperus chinensis* var. *sargentii*; cult.(PC)
4. *Juniperus pingii* var. *wilsonii*; Baimashan, nwYunnan (PC)
5. *Juniperus squamata*; Muli, swSichuan (PC)
6. *Juniperus squamata*; Baimashan, nwYunnan (PC)

CUPRESSUS

A genus with 17 species (9 in China) found throughout north temperate regions of both the Old and New Worlds. Small to large, resinous, **evergreen trees** with ascending branchlets that are usually decussately arranged. **Leaves** scale-like, whorled, 4-ranked but in 2 almost equal pairs. **Male cones** egg-shaped to oblong; **female cones** terminal, woody, with peltate scales, ripening at end of second season.

Cupressus chengiana S. Y. Hu* **Tree** to 30 m with a columnar habit, trunk up to 1 m in diam. at maturity with the bark peeling vertically into strips between fine fissures. Foliage grey-green in flattened open sprays, **leaves** 1–1.5 mm long, pointed, incurved, glandular on reverse. **Cones** brown or reddish brown, globose, 12–20 mm in diam., with 8–12 scales. sGansu, nw & wSichuan; mountain slopes and valleys, 800–2,900 m.

Cupressus duclouxiana Hickel* **Tree** to 25 m with a columnar to narrowly conical habit and finely fissured bark, trunk eventually up to 80 cm in diam. **Shoots** pale yellowish green with the bluish green or grey-green foliage arranged in a series of short flattened sprays. **Leaves** closely appressed, 1–2 mm long, with incurved, pointed tips, obscurely glandular. **Cones** globose, 15–32 mm in diam., dark brown or purplish brown. swSichuan, seTibet, Yunnan; forests on mountain slopes and ridges, 1,400–3,300 m.

JUNIPERUS

A genus of c. 54 species (23 in China) found in north temperate and montane, subtropical regions of the Old and New Worlds. Monoecious or dioecious, prostrate or erect **shrubs**, or small to large, resinous, **evergreen trees** with **leaves** of 2 types: juvenile awl-shaped or needle-like leaves in whorls of 3, and scale-like adult leaves in 2s or 3s. **Cones** berry-like, consisting of closely appressed scales.

1. Both juvenile and mature leaves present ....
   .... *J. chinensis*, *J. tibetica*
1. Only juvenile leaves present ....
   .... *J. pingii*, *J. recurva*, *J. squamata*

Juniperus chinensis L. **Tree**, usually dioecious, to 25 m with an egg-shaped-conical crown and brown bark peeling in long, twisting strips; branchlets straight or slightly curved, 4-angled or terete. Juvenile **leaves** very prickly, 2–5 mm long, found in 2s or 3s at base of mature shoots and on separate juvenile shoots, 2–5 mm long; scale-leaves in decussate pairs, closely appressed, blunt, 1–3 mm long. **Cones** globose to egg-shaped, 6–8 mm long, brown when ripe, with a glaucous bloom. Throughout China, except far south [Myanmar, Japan, Korea, eTaiwan, eRussia]; mountain slopes, 1,400–2,300 m.

Juniperus tibetica Kom.* (syn. *J. distans* Florin; *J. potaninii* Kom.; *J. zaidamensis* Kom.; *Sabina tibetica* (Kom.) W. C. Cheng & L. K. Fu) Monoecious **tree** to 30 m with densely or loosely arranged, terete or slightly 4-angled branchlets. **Leaves** both scale-like and needle-like; needle-leaves (on seedlings and young plants) in whorls of 3, 4–8 mm long; scale-leaves decussate or in whorls of 3, ovate-rhombic, blunt, 1–3 mm long, with a conspicuous gland beneath. **Cones** erect, egg-shaped or almost globose, 9–16 mm long, brown, black, or purplish black when ripe. sGansu, sQinghai, Sichuan, e & seTibet [swTibet]; forests on mountain slopes or in valleys, 2,700–4,800 m.

Juniperus pingii W. C. Cheng ex Ferré* Monoecious **shrub** or small **tree** to 30 m with grey-brown fissured bark; branchlets pendulous, 6-angled. **Leaves** in whorls of 3, ovate-lanceolate, 3–5 mm long, up to 1.5 mm broad, with recurved tips, yellow-green to mid-green on outer surface, glaucous on inner. **Cones** egg-shaped, 7–9 mm long, black and lustrous when ripe. swSichuan, nwYunnan; forests and thickets on mountain slopes, 2,600–3,800 m. **Var. *wilsonii* (Rehd.) Silba*** is shrubby with branchlets that are neither pendulous nor 6-angled. sGansu, nwHubei, sQinghai, sShaanxi, Sichuan, eTibet, Yunnan; 2,600–4,900 m.

Juniperus recurva Buch.-Ham. ex D. Don. Readily distinguished from *J. squamata* by the pendulous branch-tips. Monoecious **tree** or **shrub** with light greyish brown or brown bark; **leaves** in whorls of 3, all needle-like, straight to slightly incurved, 3–10 mm long, greyish white to somewhat glaucous beneath; **cones** slightly glaucous at first but maturing purple-black, egg-shaped, 6–12 mm long. seTibet, nwYunnan [Himalaya, nMyanmar]; forests and thickets, 1,800–3,900 m. The typical variety (**var. *recurva***) has cones 7–12 mm long and egg-shaped seeds; **var. *coxii* (A. B. Jacks.) Melville** has **cones** 6–8 mm long and conical-egg-shaped seeds. This species is similar to *J. pingii* but its **leaves** lack longitudinal grooves and are not keeled beneath.

Juniperus squamata Buch.-Ham. ex D. Don. Very variable prostrate to suberect monoecious **shrub** or small **tree** to 5 m, with flaking, smooth, reddish brown bark and erect or spreading shoots. **Leaves** all awl-like, in 3s, spreading forwards, 4–8 mm long, pointed to long-pointed, bluntly ridged, green to bluish green with a silvery-white inner face. **Cones** egg-shaped, 10 mm in diam., reddish brown turning purplish black. Guizhou, wHubei, Sichuan, Shaanxi, Tibet, Yunnan [cChina, Himalaya, Taiwan]; mountain slopes, ravines, woodland margins, 1,600–4,500 m.

METASEQUOIA

A genus with a single species endemic to cChina. Large, deciduous **trees** with a conical habit and with deciduous branchlets; buds borne in opposite pairs. **Leaves** soft-textured, linear. **Cones** small, with scales in opposite pairs.

Metasequoia glyptostroboides Hu & W. C. Cheng* **Tree** to 40 m, crown at maturity narrowly conical with ascending branches, bark fibrous, stringy, orange-brown to reddish brown, with base of trunk often becoming fluted. **Leaves** fresh green, soft-textured, linear, pointed, 15–25 mm long. **Cones** brown, egg-shaped, c. 2 cm long, pointed, on a stalk 2–4 cm long. swHubei, neSichuan [Hunan]; ravines and streamsides, woodland margins, very rare in the wild but widely planted. Easily confused with *Taxodium distichum*, which is widely planted as a street tree or on the edge of rice paddies throughout s & cChina; *Taxodium* is most readily distinguished by the alternate, rather than opposite, twigs and buds.

TAIWANIA

A genus with 1 species in China, Taiwan and Vietnam. Large, **evergreen tree** with a columnar habit and bark peeling in strips. **Shoots** green for several years because of decurrent leaf-bases, **leaves** on young trees radial, awl-shaped, while fruiting shoots have smaller scale-like leaves. **Male cones** egg-shaped or egg-shaped-ellipsoid, 6–35 in racemes, produced in spring; **female cones** terminal, solitary or aggregated, egg-shaped, each scale 2-seeded, ripening during the first autumn.

Taiwania cryptomerioides Hayata Large **tree** to 40 m with grey-brown bark and a **trunk** up to 2 m in diam. **Leaves** on young trees sickle-shaped, incurved, tapering to a point, to 20 mm long, those on coning shoots smaller and scale-like, incurved, to 4 mm long. **Cones** 13–16 mm long, each with 15–21 scales. nwYunnan [Taiwan, nVietnam]; forests on well-drained soils in moist, warm conditions, 1,100–2,500 m. Very rare but forming pure stands where it is found; occasionally planted as a timber tree or around temples.

PINACEAE

ABIES

A genus of 49 species (15 in China) in the northern hemisphere, mainly boreal and temperate. Large, **evergreen trees**, generally with a conical habit, with irregularly whorled branches. **Leaves** linear, spirally arranged on shoots but can be twisted to lie apparently in 1 plane, pectinate, leaving an oval or rounded, depressed scar when they fall. **Male cones** borne on underside of weaker shoots of previous year; **female cones** arising from lateral buds, erect, ripening in a year, the scales deciduous but leaving the central axis.

1. Branchlets thin; cones brown to yellow-brown, female cone-scales thickest at or below middle . . . **A. ernestii**
1. Branchlets thick; cones purple, purple-brown or black, female cone-scales thickest at apex 2

2. Leaf-margin back-rolled, partly concealing the silver stomatal bands beneath . . . . **A. delavayi**
2. Leaf-margin flat or almost so, the silver stomatal bands beneath fully revealed . . **A. fabri, A. fargesii, A. forrestii**

Abies ernestii Rehd.* (syn. *A. recurvata* Mast. var. *ernestii* (Rehd.) Kuan) **Tree** to 60 m with a conical habit and longitudinally flaking, dark grey bark, trunk up to 2 m in diam.; **shoots** red-brown to purple, glabrous or with red-brown hairs. **Leaves** linear, 10–70 × 2–2.5 mm, apex pointed or long-pointed. **Cones** oblong-egg-shaped, 4–14 × 3–3.5 cm. swGansu, wHubei, wSichuan, eTibet, nwYunnan; mountains slopes, usually in mixed forests of *Quercus* and *Pinus*, 2,500–3,800 m.

Abies delavayi Franch. **Tree** 10–25 m tall, candelabrum-like at maturity, with smooth, grey bark that fissures towards the base, **trunk** up to 1 m in diam.; **shoots** maroon to reddish brown, with globular, orange-brown or red-brown winter buds. **Leaves** lax, radially arranged, linear to S-shaped, 15–30 × 1.5–2.5 mm, notched at apex, with strongly back-rolled margins, fresh to mid-green, with 2, partly hidden, silver bands of stomata beneath. **Cones** narrowly cylindrical, 6–10 × 3–4 cm, short-stalked, black when mature, glaucous; bracts protruding. seTibet, w & nwYunnan [neIndia, nMyanmar]; mountain slopes, often forming extensive stands, 3,000–4,300 m.

Abies fabri (Mast.) Craib* **Tree** to 15 m with a conical habit and fissured, grey bark, **trunk** up to 1 m in diam.; buds conical, purple. **Shoots** pale brown to grey-brown, glabrous. **Leaves** linear, 15–30 × 2–2.5 mm, apex rounded and notched. **Cones** egg-shaped, 5–11 × 3–4.5 cm, slightly glaucous; bracts protruding. w & nwSichuan; mountain slopes and in river basins, often forming extensive stands, 1,500–4,000 m.

Abies fargesii Franch.* **Tree** to 40 m with a columnar habit and finely flaky, dark grey or dark grey-brown bark, **trunk** to 1.5 m in diam.; buds egg-shaped or almost globose, resinous, purple. **Shoots** red-brown to purple, glabrous or with red-brown hairs. **Leaves** linear, 10–25 × 1.5–4 mm, apex pointed to long-pointed. **Cones** oblong-egg-shaped, 4–10 × 2.5–4 cm, rich dark purple, slightly glaucous; bracts protruding. sGansu, wHubei, Sichuan, sShaanxi]; coniferous woodland, mountain slopes, ravines, 1,500–3,900 m.

Abies forrestii Coltm.-Rog.* **Tree** to 20 m with a conical habit and smooth, dark grey bark and a **trunk** up to 80 cm in diam., buds globose or egg-shaped, covered with white resin; **shoots** reddish brown, glabrous or hairy. **Leaves** linear, 20–35 × 2.5 mm, apex rounded and notched, margin not back-rolled. **Cones** stalkless, egg-shaped or shortly cylindrical, 7–12 × 3.5–6 cm, dark brown-purple to blackish brown; bracts protruding. swSichuan, seTibet, nwYunnan; mountain slopes, often making pure stands, 2,500–4,200 m.

1. *Metasequoia glytostroboides*; cult. UK (CGW)
2. *Metasequoia glytostroboides* trunk; cult. Kunming, cYunnan (CGW)
3. *Metasequoia glytostroboides*; cult. (AF)
4. *Taiwania cryptomerioides*; Kunming Botanic Garden, cYunnan (PC)
5. *Abies fargesii*; Shennonjia, wHubei (PC)
6. *Abies delavayi*; Yulongxueshan, Lijiang, nwYunnan (CGW)
7. *Abies delavayi*; Ganghoba, Yulongxueshan, nwYunnan (CGW)
8. *Abies delavayi*; Muli, swSichuan (PC)
9. *Abies forrestii*; near Deqin, nwYunnan (PC)
10. *Abies forrestii*; near Deqin, nwYunnan (AF)

1. *Cathaya argyrophylla*; cult. Kunming, Yunnan (PC)
2. *Cedrus deodara*; cult. (PC)
3. *Keteleeria evelyniana*; Hengduanshan, nwYunnan (PC)
4. *Larix potaninii*; Gangheba, Yulongxueshan, nwYunnan (CGW)
5. *Larix potaninii*; Gangheba, Yulongxueshan, nwYunnan (CGW)
6. *Larix potaninii*; wSichuan (TK)
7. *Larix potaninii*; Gangheba, Yulongxueshan, nwYunnan (CGW)

CATHAYA

A genus with a single species endemic to China. Medium-sized, **evergreen tree** with long and short shoots. **Leaves** evergreen, borne singly, juvenile leaves ciliate, hairy on upper side. **Male cones** 1–3, from axillary buds on branchlets; **female cones** axillary, stalkless, erect at first but then pendent with age, carried on long shoots, rather than on short shoots as in *Larix* and *Cedrus*.

Cathaya argyrophylla Chun & Kuang* **Tree** to 20 m, trunk to 40 cm in diam. with dark grey, irregularly flaking bark; leading **shoots** erect. **Leaves** linear, 40–60 × 2.5–3 mm, dark glossy green above, silvery beneath. **Cones** 3–5 × 1.5–3 cm, persisting for several years, green turning dark brown with age. neGuangxi, Guizhou, seSichuan [Hunan]; open slopes and ridges in mountains, 900–1,900 m.

CEDRUS

A genus of 4 closely related species (2 in China, 1 introduced) in nwAfrica, w & cAsia, and with 1 extending into Tibet. Large, **evergreen trees** with spreading to suberect branches; **shoots** ascending, level or descending. **Leaves** linear, needle-like, carried singly on long shoots but in false whorls on short shoots, borne on short pegs. **Male catkins** and young **female cones** produced in autumn; cones erect, with deciduous scales.

Cedrus deodara (Roxb.) D. Don. **Tree** to 35 m often with a massive **trunk**, branches spreading in older trees and with shoot-tips descending, giving the tree a weeping appearance; bark vertically fissured, dark brown or blackish. **Leaves** grey-green, quadrangular, pointed, 3–5 cm long. **Male cones** 6–7 cm long, shedding pollen in autumn; **female cones** 8–12 × 5–6 cm, brown when mature but glaucous when young. Widely planted in mountains of wChina [s & swTibet, wHimalaya].

KETELEERIA

A genus with 5 species (3 in China, endemic) in s & cChina, Taiwan, Laos and Vietnam. Large, **evergreen trees** with broad crowns and longitudinally fissured bark. Bud-scales persisting as a rosette around the base of the shoots. **Leaves** short, linear, spirally arranged on shoots. **Male cones** carried in clusters arising from a single bud; **female cones** erect, borne on short leafy shoots and ripening after one season; scales rounded or rhombic at tip.

Keteleeria davidiana (Bertr.) Beissn. **Tree** 10–50 m with a conical crown when young but widely spreading with age; **trunk** to 2.5 m in diam. with grey, furrowed, bark that is shed in scales to reveal red-brown areas beneath. **Shoots** chestnut-brown, densely hairy, with globular, reddish buds. **Leaves** linear, blunt, 15–50 × 3–4.5 mm, glossy green above, paler beneath. **Cones** egg-shaped-cylindrical, 8–21 × 4–6 cm, scales obovate or obovoid to almost heart-shaped, blunt, entire, slightly bent outward. seGansu, nGuangxi, Guizhou, wHubei, seSichuan, sShaanxi, n & nwYunnan [swHunan, Taiwan]; hills, mountain slopes and hot, dry valleys, 200–1,500 m.

Keteleeria evelyniana Mast. Slighter **tree** than preceding species, **trunk** not exceeding 1 m in diam.; cones cylindrical, 8–21 × 4.5–6 cm in diam., scales widely spreading, rhombic-ovate, longer than wide, almost acute, irregularly toothed, reflexed. wGuizhou, wSichuan, Yunnan [Laos, Vietnam]; hills, mountain slopes and hot, dry valleys, 700–2,900 m.

LARIX

A genus with 11 species (6 in China) in the northern hemisphere, mostly in the mountains. Large, resinous, **deciduous trees**. **Leaves** linear, soft-textured, spirally arranged, borne on both long and short shoots, those on short shoots borne in clusters. **Cones** erect, ripening at end of first season, bracts shorter or longer than seed-scales.

Larix griffithii Hook. f. var. *speciosa* (W. C. Cheng & Y. W. Law) Farjon* (syn. *L. speciosa* W. C. Cheng & Y. W. Law) **Tree** to 20 m with a narrowly conical habit and a **trunk** to 80 cm in diam., bark red-brown turning light grey, with scaly fissures and ridges. **Shoots** orange-brown, finely downy when young, buds egg-shaped, red-brown, resinous. **Leaves** linear, almost acute to blunt, 25–55 × 1–1.8 mm, green with 2 silvery stripes beneath. **Cones** cylindrical, 7–9 × 2–3 cm, purple, turning purple-brown with age; scales reflexed, protruding in part, 7.5–10 mm long. seTibet, nwYunnan; mountain slopes, 2,600–4,100 m.

Larix potaninii Batalin (syn. *L. thibetica* Franch.) **Tree** to 50 m, sometimes much less, with a columnar habit and purple-grey bark with long vertical fissures and ridges. **Shoots** orange-brown, ageing to grey-brown after 2 or 3 years, buds red-brown, resinous. **Leaves** 12–35 × 1–1.5 mm, green with 2 silvery stripes beneath. **Cones** egg-shaped to oblong-egg-shaped, 3–7.5 × 1.5–3.5 cm, violet-purple turning brown with age; scales broadly oblong, the protruding part narrowly conical, not recurved. sGansu, sShaanxi, w & swSichuan, e & seTibet, nwYunnan [swTibet, Myanmar]; mountain slopes and river valleys, 2,500–4,300 m.

PICEA

A genus of 34 species (14 in China) found in the mountains of the north temperate zone. Medium-sized to large, resinous, **evergreen trees** with needle-like **leaves**, spirally arranged on long shoots, entire, never notched, borne on a peg that remains when the leaf falls, 4-sided, square, rhombic or diamond-shaped in cross-section. **Cones** arising from terminal clusters of buds, erect at first but becoming pendent, ripening in the first season, releasing **seeds** by scales opening.

1. Mature cones purple-brown . . . . *P. likiangensis*
1. Mature cones yellow-brown to dull brown 2
2. Leaves bluish green or glaucous . . . . *P. asperata*
2. Leaves glossy green . . . . . . *P. brachytyla, P. wilsonii*

Picea likiangensis (Franch.) E. Pritz **Tree** to 50 m with a conical habit and pale grey bark, becoming fissured, **trunk** up to 2.5 m in diam. **Shoots** pale yellow or pale yellow-brown, bearing purple-brown buds. **Leaves** pointed, 6–15 × 1–1.5 mm, glossy green. Young **cones** bright red-purple, mature ones cylindrical, purple-brown, 4–12 cm long, 1.7–3.5 cm in diam.; scales with wavy margins. s & seQinghai, w, nw & swSichuan, seTibet, nwYunnan, [Bhutan]; mountain slopes, ravines and river valleys, stabilised moraines, 2,500–4,100 m.

Picea asperata Mast.* **Tree** to 45 m with a columnar or conical habit and purplish grey bark with moderately flaky scales, **trunk** to 1 m in diam. **Shoots** stout, prominently ridged with pegs, yellow-buff or buff, becoming ash-grey, bearing pale brown, slightly resinous buds. **Leaves** very stiff, stout, blunt, 12–25 × 1–1.5 mm. **Cones** cylindrical, pale brown to dull brown, 5–16 cm long; scales rounded. s & eGansu, eQinghai, Sichuan, swShaanxi; mountain slopes, river basins, 2,400–3,600 m.

Picea brachystyla (Franch.) E. Pritz **Tree** to 30 m with a domed habit and grey to purplish grey bark with reddish cracks, **trunk** up to 1 m in diam.; **shoots** white, yellow-buff or pale yellow, turning orange-brown, bearing pale brown, slightly resinous buds. **Leaves** very stiff, stout, blunt, 1–2.5 × 1–1.5 mm, with 2 silvery or bluish lines underneath. **Cones** cylindrical, green or purple, turning pale brown to dull brown with age, 6–12 cm long, 3–4 cm in diam.; scales rounded. sGansu, nwHubei, Sichuan, sShaanxi, seTibet, nwYunnan [neIndia, nMyanmar]; mountain slopes and valleys, 1,500–3,800 m.

Picea wilsonii Mast.* **Tree** to 50 m with a columnar or conical habit and pinkish brown or grey-brown bark with large papery scales, **trunk** up to 1.3 m in diam.; **shoots** yellow-green or yellow-grey, becoming shiny ash-grey, buds brown, slightly resinous. **Leaves** very stiff, stout, blunt, 0.8–1.3 × 1.2–1.7 mm. **Cones** oblong, green maturing to yellow-brown, pale brown or dull brown, 5–8 cm long, 2.5–4 cm in diam. Gansu, Hubei, Qinghai, n & wSichuan, Shaanxi [Hebei, Shanxi, Nei Mongol]; mountain slopes, river valleys, 1,400–2,800 m.

PINUS

A genus with c. 108 species (39 in China) widely distributed in the northern hemisphere, south to Sumatra in Asia and cAmerica in the New World. Small to large, resinous, **evergreen trees** with branches in whorls. **Leaves** of 2 types: most obvious leaves needle-like, in bundles of 2–5 on short shoots, other leaves small brown scales arranged singly and spirally on leading shoots but bearing short shoots in their axils. **Male cones** borne at base of current year's growth; pollen shed in early summer. **Female cones** woody, small to large, erect or pendent, usually terminal or almost so on current year's growth; scales with a terminal or dorsal, short, often prickle-like projection.

1. Scale-projection terminal on scales of mature cone (subgenus *Strobus*) 2
1. Scale-projection dorsal on scales of mature cone (subgenus *Pinus*) 3
2. Bark peeling in jigsaw-like flakes; needle leaves in 3s; cones not exceeding 7 cm long . . . . *P. bungeana*
2. Bark fissured, often flaky; needle leaves in 5s; cones 10 cm long or more . . . . *P. armandii, P. wallichiana*
3. Needle leaves in 3s, 10–30 cm long; cones 5–11 cm long . . . . *P. yunnanensis*
3. Needle leaves in 2s, occasionally 3s, 7–15 cm long; cones 2.5–9 cm long . . . . *P. densata, P. tabuliformis*

Pinus bungeana Zucc. ex Endl.* **Tree** to 20 m; bark with a flaky camouflage pattern, smooth, grey-green, exfoliating to reveal creamy to pale yellow areas that turn greenish, olive-brown, reddish or purple with age. **Shoots** olive-green at first. **Leaves** 6–8 cm long, dark yellowish green with grey-green lines of stomata on inner faces. **Male cones** 1 cm long; **female cones** egg-shaped with spiny short projections, 4–7 × 3.5 cm, green turning brown at maturity. sGansu, Hubei, nSichuan, sShaanxi [c & nChina]; hills and mountain slopes, 500–1,800 m. Widely planted in temperate China as an ornamental.

Pinus armandii Franch. **Tree** to 40 m with a conical crown when young but a conical to rounded, open crown of whorled branches when mature; bark greyish green, smooth in young trees but darker and fissured into small scales with age. **Branches** ascending, the **shoots** olive-green at first. **Leaves** 110–200 × 1 mm, glossy grey-green to glaucous green. **Male cones** 1 cm long; **female cones** pendent, ellipsoid, often curved, 10–30 × 3.5 cm, green turning pale brown at maturity. The **seeds** are edible. sGansu, nw & cGuizhou, wHubei, sShaanxi, Sichuan, seTibet, Yunnan [cChina, nMyanmar, Taiwan]; mountain slopes, river basins, often rather dry habitats, 1,000–3,300 m. Widely planted in swChina.

Pinus wallichiana A. B. Jacks. **Tree** to 35 m with a conical crown when young but more rounded and open when mature; bark greyish green, smooth in young trees but darker and fissured into small scales with age. **Branches** ascending, the **shoots** olive-green, becoming grey-green. **Leaves** in bundles of 5, arching-floppy, 110–200 × 1 mm, grey-green to glaucous green. **Male cones** 1 cm long. **Female cones** pendent, ellipsoid, often curved, 10–30 × 3.5 cm, green turning pale brown at maturity. seTibet, nwYunnan [swTibet, Himalaya]; mountain slopes, temperate rainforest, generally in rather dry habitats, 1,600–3,300 m.

1. *Picea likiangensis*; Gangheba, Yulongxueshan, nwYunnan (CGW)
2. *Picea likiangensis*; Tianchi, nwYunnan (PC); (inset) *Picea likiangensis*; Gangheba, Yulongxueshan, nwYunnan (PC)
3. *Picea brachystyla*; north of Songpan, nwSichuan (CGW)
4. *Picea wilsonii*; Wolong, wSichuan (CGW)
5. *Pinus bungeana*; Jietaisi, near Beijing (CGW)
6. *Pinus bungeana*; Forbidden City, Beijing (PC)
7. *Pinus armandii*; by Yufengsi, Lijiang, nwYunnan (CGW)
8. *Pinus wallichiana*; Paro, Bhutan (CGW); (inset) *Pinus wallichiana* (PC)

1. *Pinus yunnanensis*; north of Zhongdian, nwYunnan (CGW)
2. *Pinus yunnanensis*; Yulongxueshan, nwYunnan (CGW)
3. *Pinus tabuliformis*; Muli, sSichuan (PC)
4. *Tsuga dumosa*; Heshui, Yulongxueshan, nwYunnan (CGW)
5. *Tsuga chinensis*; Zhongdian, nwYunnan (PC)
6. *Podocarpus forrestii*; Dali, nwYunnan (PC)
7. *Podocarpus macrophyllus*; Leshan, wSichuan (PC)
8. *Podocarpus macrophyllus*; Longdu, Guizhou (PC)
9. *Podocarpus macrophyllus*; Maolan, seGuizhou (PC)

Pinus yunnanensis Franch.* **Tree** to 30 m with a rounded crown when mature; bark greyish brown, fissured into small plates. **Branches** spreading, **shoots** stout, pinkish brown, becoming brown. Buds red-brown with white-fringed scales, not resinous, 1 cm long. **Leaves** in bundles of 3, slender, 100–300 × 1–1.2 mm, bright to greyish green. **Male cones** 1 cm; **female cones** pendent, egg-shaped-conical, sometimes clustered, 5–11 × 3–5 cm, dark brown when mature, scales gently rounded with a slightly impressed or protruding short projection. eGuangxi, eGuizhou, s & swSichuan, seTibet, Yunnan; mountain slopes, river basins, dry and sunny places, often forming extensive stands, 400–3,100 m.

Pinus densata Mast.* (syn. *P. tabuliformis* Carrière var. *densata* (Mast.) Rehd.) Very similar to *P. tabuliformis* but **leaves** in bundles of 2 or 3, slender, 90–150 × 1–1.5 mm, mid-green. **Male cones** 1–1.8 cm long, yellowish brown; **female cones** solitary or paired, obliquely broadly egg-shaped–conical, sometimes clustered, 4–6 × 4–7 cm, scales hard, stiff, woody, flat, with a backward-pointing prickle. sQinghai, w, nw & seSichuan, seTibet, Yunnan; open, montane forests, often forming pure stands, 2,600–3,500 m.

Pinus tabuliformis Carrière **Tree** to 30 m with a conical crown when young, flat-topped and spreading at maturity; bark reddish brown, scaly, fissured and greyish at base. Branches spreading, the **shoots** yellow-brown to brown; buds chestnut-brown, slightly resinous. **Leaves** in bundles of 2, slender, 60–150 × 1–1.5 mm, mid-green. **Male cones** 5–9 mm long; **female cones** broadly egg-shaped–conical, sometimes clustered, 4–9 × 4–9 cm, scales hard, stiff, woody, flat with a small central spine. Gansu, Hubei, Qinghai, Sichuan, Shaanxi [n & cChina, Korea]; hills and mountains, 100–2,600 m. **Var. henryi (Mast.) Kuan.*** (syn. *P. henryi* Mast.) is distinguished by the smaller **cones** not more than 5 × 4.5 cm, and scales with a rounded or depressed short projection. wHubei, neSichuan, Shaanxi [Hunan]; mountain slopes, 1,100–2,000 m.

TSUGA

A genus of 9 species (4 in China) in North America and in Asia from the Himalaya to China, Taiwan and Japan. Large **evergreen trees** with irregularly whorled **branches**. **Shoots** with a small peg-like projection at base of each leaf; **leaves** linear, flat, sometimes of irregular length, usually arranged in a pectinate manner, often with 2 silver lines on lower surface. **Male cones** solitary, growing from lateral buds; **female cones** small, pendent, terminal on second year branches.

Tsuga chinensis (Franch.) E. Pritz (incl. *T. yunnanensis* Mast.) Smaller tree than *T. dumosa*, to 20 m, **shoots** pale brown, ridged on decurrent leaf-bases, with a red-brown mark on crest of ridge just behind leaf-base, shiny, with short black hairs in the grooves. **Leaves** linear, tapered at tip, notched or rounded, 5–27 × 2–3 mm, fresh green above, with 2 silvery bands beneath. **Cones** egg-shaped, yellow-brown, 1.5–4 × 1.2–2.5 cm; scales thick, densely arranged. Gansu, Guizhou, Hubei, Sichuan, seTibet, Yunnan [neTibet, c & eChina, Taiwan]; mountains in mixed forests, but also valleys and river basins, 1,000–3,500 m.

Tsuga dumosa (D. Don.) Eichler **Tree** to 40 m with a conical crown at maturity; **trunk** to 2.7 m in diam., with pinkish brown bark that eventually turns grey-brown, scaly and ridged in old trees. **Shoots** pale brown, pinkish brown at maturity, finely hairy. **Leaves** linear, tapered at tip, pointed to blunt, 10–30 × 1.5–3 mm, bluish green above, with 2 silvery bands beneath. **Cones** broadly egg-shaped, shiny brown, 1.5–3 × 1–2 cm; scales thin-textured, rounded or obovoid, to 1 cm broad. seSichuan, seTibet, nwYunnan [Himalaya]; mountain slopes and valleys, 2,300–3,500 m.

Tsuga forrestii Downie* Very similar to *T. chinensis* but generally a smaller **tree** to 15 m with reddish brown bark and distinctive reddish brown, hairy branchlets that become shiny and pinkish brown. Guizhou, swSichuan, nwYunnan; habitats and altitudes similar to those of *T. chinensis*.

TAXACEAE

PODOCARPUS

A genus of 106 species (5 in China) widely distributed in the tropics and subtropics mainly of the southern hemisphere and reaching the mountains in the temperate zone. Dioecious, **evergreen shrubs** or **trees**. **Leaves** solitary, borne spirally or in a pectinate arrangement, needle-like to lanceolate, glossy dark green above, paler beneath. **Fruits** 1-seeded, borne on an axillary shoot, with an enlarged, basal, olive- or plum-like receptacle bearing the seed at the apex.

Podocarpus forrestii Craib & W. W. Sm.* In contrast to *P. macrophyllus*, a **shrub** not more than 4 m tall with narrowly winged leaf-stalks and elliptic to linear-elliptic **leaves** not more than 9 × 1 cm, with 30–50 stomatal lines beneath. nwYunnan (Dali). Possibly no more than a form of *P. macrophyllus* (see below).

Podocarpus macrophyllus (Thunb.) Sweet Small **tree**, 2–20 m tall. **Leaves** spirally arranged on the shoots, crowded, erect or spreading, linear-oblanceolate to oblanceolate or almost elliptic, slightly curved, 6–10 × 0.9–1 cm, light red when young, green above with 2 broad glaucous lines beneath at maturity. **Fruit** with a 13 mm long, dark violet receptacle, **seeds** egg-shaped, 9–10 mm long. Guangxi, Guizhou, Hubei, Sichuan, Yunnan [s, c & eChina, Japan, Myanmar]; forests, thickets, roadsides, to 1,800 m. Widely cultivated as a pot plant or in gardens.

TAXUS

A genus with 10 species (3 in China) in the northern hemisphere, with 1 species venturing south of the equator into Malaysia. Long-lived, **evergreen shrubs** or **trees**. **Leaves** alternate, linear, pectinately arranged, dark green above, paler beneath. **Male cones** catkin-like, scattered on second year branchlets. **Fruits** with a solitary, green seed enclosed in a bright red or orange, sticky aril. Poisonous.

Taxus wallichiana Zucc. (incl. *T. yunnanensis* W. C. Cheng & L. K. Fu) **Tree** to 30 m with a broadly conical to domed crown, **trunk** up to 1.3 m in diam., bark grey-brown to purple-brown, smooth, scaly and fluted. **Shoots** green for several years, 3–9 cm long, becoming brown. **Leaves** 2-ranked, narrowly sickle-shaped, long-pointed, 1.5–3.5 × 2–4 mm, margins slightly incurved. **Male cones** 8 mm long, clustered along undersides of previous season's shoots. **Fruits** at end of previous year's shoots; **seed** 7 mm long, aril bright red. sGansu, Guizhou, wHubei, Sichuan, seTibet, Yunnan [s, c & eChina, Bhutan, nIndia, nLaos, nMyanmar, Taiwan, nVietnam]; broadleaved and coniferous forests, ravines, rocky places, 1,000–3,500 m. **Var. *chinensis* (Pig.) Rehd.*** with linear, thin-textured **leaves** 1.5–2.2 cm long, is found in sGansu, Guizhou, wHubei, Sichuan, eYunnan [s, c & eChina].

EPHEDRACEAE

EPHEDRA

A genus of c. 40 species (14 in China) from e & nAfrica, Asia, Europe and the Americas. Small, dioecious or rarely monoecious **herbs**, **shrubs** or **subshrubs**. **Stems** erect, procumbent or rarely climbing, with opposite or whorled herbaceous branches. **Leaves** very small, opposite or in whorls of 3, basally fused and reduced to sheaths. **Cones** terminal and axillary: **male cones** solitary or clustered, stalkless or shortly stalked; **female cones** opposite or in whorls, stalkless or not, egg-shaped to almost globose, 8–11 mm long, red and fleshy at maturity; containing 1–2 seeds.

Ephedra likiangensis Florin* Small **shrub** or **subshrub** to 1.5 m with erect or sometimes procumbent **stems**. Herbaceous branchlets ascending, green or brownish green with **leaves** opposite or rarely in whorls of 3, fused for half to three-quarters of their length. **Male cones** clustered at nodes, stalkless or short-stalked, with 5–8 protruding stamens; **female cones** solitary or opposite at the nodes, stalkless or not, egg-shaped to almost globose, 8–11 mm long, red and fleshy at maturity. wGuizhou, sw & wSichuan, seTibet, nwYunnan [neTibet]; sparsely vegetated rocky terrain and grassland in mountains, 2,300–4,200 m.

Ephedra monosperma Gmelin ex C. A. Mey. In contrast to *E. likiangensis*, a small **shrublet** not more than 20 cm tall, often with creeping runners, with much-branched **stems**, lower nodes congested; **male cones** with 6–8 stalkless stamens; **female cones** 6–9 mm long. Gansu, Qinghai, Sichuan, Tibet [n & nwChina, cAsia, wHimalaya, Mongolia, Russia]; rocky and dry habitats, 1,400–4,800 m. V–VI.

CYCADACEAE

CYCAS

A genus of c. 60 species (16 in China, half of them endemic) from eAfrica, s & eAsia, nAustralia and Pacific islands. Small to large, palm-like **trees** with branched or simple **trunks** and pinnate, sometimes multipinnate, **leaves**; leaf-stalks bearing paired spines; leaflets paired, leathery, linear, pointed, often glossy dark green. **Male sporophylls** cone-like, egg-shaped-cylindrical to spindle-shaped; **female sporophylls** many, leaf-like, often densely hairy or woolly, pinnately lobed, bearing marginal **seeds**.

> 1. Pairs of spines on leaf-stalk 5–18; seeds orange to red
> .... *C. panzhihuaensis*, *C. revoluta*
> 1. Pairs of spines on leaf-stalk 25–50; seeds pale yellow
> .... *C. szechuanensis*

Cycas panzhihuaensis L. Zhou & S. Y. Yang* **Trunk** unbranched, to 2 m. **Leaves** 70–130 cm long; leaflets in 70–120 pairs, 12–20 cm long; leaf-stalk spines in 5–13 pairs. **Male cones** brown, egg-shaped-cylindrical to spindle-shaped, 25–45 cm long; **female fronds** densely woolly; **seeds** 2–4, orange to red, obovoid to ellipsoid, 4–5 cm long. sSichuan, nYunnan (Jinshajiang); grassy places and among shrubs on limestone, sandy places in river valleys, 1,100–2,000 m.

Cycas revoluta Thunb. **Trunk** often branched, to 3 m. **Leaves** 70–140 cm long; leaflets in 60–150 pairs, 10–20 cm long; leaf-stalk spines in 6–18 pairs. **Male cones** pale yellow, egg-shaped–cylindrical, 30–60 cm long; **female fronds** more than 30, densely brown-woolly; seeds bright orange-red, globose, 2–4 cm in diam. Widely planted in sw & wChina [Fujian, Japan, at low altitudes]; sea level to 500 m.

Cycas szechuanensis W. C. Cheng & L. K. Fu* (*C. guizhouensis* K. M. Lan & R. F. Zhou; *C. multiovula* D. Y. Wang) **Trunk** unbranched, to 3 m. **Leaves** 100–250 cm long; leaflets in 60–120 pairs; leaf-stalk spines 15–35 cm long. **Male cones** spindle-shaped to cylindrical, 25 cm long; **female fronds** more than 30, densely yellow-brown-woolly; **seeds** pale yellow, globose, 2.5–3 cm in diam. nwGuangxi , swGuizhou (Maling), seYunnan (Nanpanjiang); amongst shrubs, open forests in hot, dry river valleys, 400–1,500 m.

1. *Ephedra likiangensis*; Yulongxueshan, nwYunnan (PC)
2. *Ephedra likiangensis*; Gangheba, Yulongxueshan, nwYunnan (PC)
3. *Ephedra monosperma*; Muli, swSichuan (PC)
4. *Cycas panzhihuaensis*; Dukou, sSichuan (PC)
5. *Cycas panzhihuaensis*; Dukou, sSichuan (PC)
6. *Cycas revoluta*; cult. (JG)
7. *Cycas revoluta*; cult. (JG)

Angiosperms

MAGNOLIACEAE

MAGNOLIA

A genus of c. 120 species from Asia (including Japan), the Himalaya and North America (28 species in China of which the majority are endemic). **Evergreen** or **deciduous shrubs** or **trees** with simple, alternate **leaves**; stipules free or joined to leaf-stalk. **Flowers** showy, terminal, solitary, hermaphrodite, white, creamy white, pink or purplish; tepals 9–21, free, arranged in 3s to 5s; stamens numerous, spirally arranged; carpels free, few to many, spirally arranged. **Fruit** a cone-like collection of follicles, often unevenly developed, each splitting dorsally.

1. Flowers erect 2
1. Flowers semi-nodding or pendent, almost globose, produced with the leaves (section *Oyama*) .... **M. globosa, M. sinensis, M. wilsonii**
2. Evergreen tree; flowers opening almost flat amongst the leaves, lasting a day (sect. *Gwillimia*) ..... **M. delavayi**
2. Deciduous trees; flowers tulip-like, produced generally before the leaves emerge, long-lasting (section *Yulana*) 3
3. Tepals ± equal in size ... **M. campbellii, M. sprengeri**
3. Outer 3 tepals smaller than the others .... **M. dawsoniana, M. sargentiana**

Magnolia delavayi Franch.* Flat-topped, evergreen **tree** 12–18 m tall with large leathery, ovate to elliptic, pointed **leaves**, 14–26 × 7–15.5 cm, coarsely reticulate, leaf-stalks yellow-hairy at first, 4–6.5 cm long. **Flowers** slightly fragrant, creamy white, 15–20 cm across; tepals 7–9.5 cm long. **Fruit**-cones egg-shaped, 7–15 cm long, on a stalk up to 4 cm long. swGuizhou, wSichuan, n, nw, w & cYunnan; broadleaved forests, 1,500–2,800 m. IV–VI, occasionally later. Often planted.

Magnolia campbellii Hook. f. & Thomson var. *mollicomata* (W. W. Sm.) Kingdon Ward* Large deciduous **tree** to 30 m with ± downy young shoots and thinly papery, elliptic, ovate or obovate **leaves**, 10–23 × 4.5–10 cm, apex pointed, base sometimes slightly heart-shaped. **Flowers** large, opening to a wide bowl shape, 15–25 cm across, mauve pink, slightly fragrant, borne on a downy stalk; tepals up to 10 cm long. **Fruit**-cones erect, 8.5–17 cm long, on a stalk up to 5 cm long. Sichuan, seTibet, nw & wYunnan; mixed forests, 2,500–3,500 m. III–V. The typical variety (var. *campbellii*), from the Himalaya, has white to pale rose-pink flowers on glabrous stalks, and egg-shaped flower-buds.

Magnolia sprengeri Pamp.* Smaller **tree** than *M. campbellii* (to 20 m) with glabrous shoots. **Flowers** appearing with young leaves, each up to 20 cm across, tepals white flushed with rose-purple at base. **Fruit**-cones 6–10 cm long, on a short, 10–20 mm stalk. Guizhou, Hubei, Sichuan, Shaanxi [cChina]; evergreen broadleaved and mixed forests, 1,200–2,000 m. VI–VII.

Magnolia dawsoniana Rehd. & Wils.* Similar to next previous species, but a smaller **tree** or bush to 13 m, with deep, lustrous green, rather leathery **leaves**, downy only along midrib beneath; **flowers** smaller, tepals pale rose-pink suffused with purple outside, not more than 10 cm long. wSichuan (Kangding area); broadleaved forests, 1,800–2,300 m. IV–V.

Magnolia sargentiana Rehd. & Wils.* **Tree** to 20 m with thinly leathery, rather pale green, obovate **leaves**, 10–17.5 × 5–10 cm, apex pointed or rounded, grey-downy beneath. **Flowers** 15–20 cm across, tepals purplish red or purplish pink outside, white to pale rose-pink inside, 11–12.5 cm long; stamens purplish, 14–17 mm long. **Fruit-cones** cylindrical, 10–20 cm long. w & swSichuan, nYunnan; broadleaved forests, 1,600–2,500 m. IV–V.

Magnolia globosa Hook. f. & Thomson* **Deciduous tree** 3–10 m tall with membranous, ovate to elliptic **leaves**, 12.5–21 × 6–11.5 cm, apex pointed or rounded, slightly heart-shaped at base, glaucous and brownish-hairy beneath, densely so at first, with 12–15 pairs of lateral veins, leaf-stalks 2–5 cm long, densely hairy. **Flowers** fragrant, cup-shaped or globose, 6–8 cm across; tepals 9–12, creamy white, 4–7.5 cm long; stamens 12–18 mm long, crimson. **Fruit-cones** red, cylindrical, 2.5–9 cm long. swSichuan, seTibet, nwYunnan; mixed forests and thickets, 1,900–3,300 m. V–VI.

Magnolia sinensis (Rehd. & Wils.) Stapf* Similar to *M. globosa*, but a **shrub** or small **tree** with 6–8 pairs of lateral veins per leaf and larger, saucer-shaped **flowers**, 10–13 cm across, with creamy white stamens tipped rose-purple; tepals usually 9. **Fruit-cones pink**, 5–6 cm long. wSichuan; similar habitats, 2,280–2,745 m. V–VII.

Magnolia wilsonii (Finet & Gagnep.) Rehd.* Closely allied to *M. globosa*, but with the young buds pale brown-felted, **leaves** distinctly pointed at apex, reticulated, with 10–12 pairs of lateral veins, **flowers** saucer-shaped, pure white with bright red to violet stamens. **Fruit-cones** purplish pink, cylindrical-egg-shaped, 5–8 cm long. Guizhou, c & wSichuan, n, nw & wYunnan; similar habitats, 1,900–3,000 m. V–VII.

MICHELIA

A genus of 40 species (34 in China, many endemic) in south and south east Asia, east to Borneo and Japan. **Evergreen shrubs** or **trees** with simple, stipulate **leaves**. **Flowers** hermaphrodite, on short stalks at leaf-axils; tepals 6–21, free; anthers many, with connective extended into a short to long appendage. Carpels few to many. Included in *Magnolia* by some botanists but readily distinguished by the lateral rather than terminal flowers.

1. *Magnolia delavayi*; nr Lijiang, Yunnan (HJ)
2. *Magnolia delavayi*; nr Kunming, cYunnan (CGW)
3. *Magnolia campbellii* var. *mollicomata*; wSichuan (HP)
4. *Magnolia campbellii* var. *mollicomata*; cult. (RBGK)
5. *Magnolia sprengeri*; cult. (JG)
6. *Magnolia sargentiana*; cult. (JG)
7. *Magnolia dawsoniana*; cult. (JG)
8. *Magnolia dawsoniana*; cult. (JG)
9. *Magnolia globosa*; cult. (MF)
10. *Magnolia sinensis*; cult. (JG)
11. *Magnolia sinensis*; cult. (CGW)
12. *Magnolia wilsonii*; cult. (MF)

1. *Michelia yunnanensis*; Shizong, eYunnan (PC)
2. *Michelia yunnanensis*; Shizong, eYunnan (PC)
3. *Michelia doltsopa*; Bhutan (MF)
4. *Michelia doltsopa*; cult. (JG)
5. *Michelia doltsopa*; Bhutan (MF)
6. *Manglietia insignis*; Guilin, neGuangxi (PC)
7. *Liriodendron chinense*; cult, RBG Kew (TK)
8. *Liriodendron chinense*; Sichuan (CGW)
9. *Illicium majus*; Tien'e, nwGuangxi (PC)
10. *Illicium jiadifengii*; Fanjingshan, neGuizhou (PC)

1. Shrub or small tree not exceeding 5 m; stamens 5–10 mm long . . . . **M. yunnanensis**
1. Trees 15–30 m tall; stamens 8–15 mm long . . . . **M. doltsopa, M. wilsonii**

Michelia yunnanensis Franch. ex Finet & Gagnep.* **Shrub** or small **tree** 2–5 m tall. **Leaves** oblong to narrowly obovate or elliptic, 2–10 × 1–4 cm, reticulately veined, base rounded, apex pointed. **Flowers** often rather crowed on the branchlets, fragrant, 2.2–3.5 cm long; tepals often 6–9, white. Guizhou, Sichuan, Yunnan (common on hills around Kunming); forests, thickets, 1,100–2,300 m. IV–V.

Michelia doltsopa Buch.-Ham. ex DC. **Evergreen tree** to 30 m with oblong to narrowly ovate-elliptic **leaves**, base rounded, apex long-pointed, firm-textured, 9–22 × 4–8 cm, glossy green above, paler beneath. **Flowers** solitary, very fragrant, 4–5 cm long, very short-stalked; tepals 12–16, white to cream, obovate; stamens 8–15 mm long. seTibet, Yunnan [eHimalaya]; broadleaved forests, 1,600–2,400 m. IV–V.

Michelia wilsonii Finet & Gagnep* Similar to previous species, but not more than 20 m tall with obovate to narrowly ovate-elliptic, long-pointed **leaves**, 11–14 × 4–6 cm. **Flowers** smaller, fragrant, 2–3 cm long, yellowish white; stamens 10–12 mm long. Guizhou, wHubei, Sichuan [cChina]; broadleaved forests, temple gardens, 1,000–1,500 m. IV–V.

MANGLIETIA

A genus of c. 25 species (18 in China) in subtropical and tropical Asia. **Evergreen**, small to large **trees**. **Leaves** simple; stipules free or joined to leaf-stalk. **Flowers** terminal, solitary, hermaphrodite, usually showy, white, creamy white, pink or purplish; tepals 9–13, free, arranged in 3s; stamens many, spirally arranged; carpels free, few to many, spirally arranged, each containing 4–12 ovules. **Fruit** a cone-like collection of follicles, each splitting dorsally.

Manglietia insignis (Wall.) Blume **Tree** to 25 m with leathery, narrowly ovate to elliptic, long-pointed or mucronate **leaves**, 14.5–26.5 × 4.5–8 cm. **Flowers** fragrant, 11–14 cm across; tepals spathulate, 5–6.5 cm long, white or purple; stamens 1.3–1.6 cm long. **Fruit-cones** egg-shaped-oblong, 6.5–11 cm long. Guangxi, Guizhou, Hubei, Sichuan, s & seTibet, Yunnan [Hunan, Himalaya]; mixed forests, usually growing on yellow loamy soils, 600–2,000 m. V–VI. Often cultivated for the flowers.

LIRIODENDRON

A genus of just 2 species, one in North America, the other in China. Large **deciduous trees** with obscurely 4-lobed **leaves**, truncate or inverted V-shaped at apex. **Flowers** terminal, tulip-like, with 9–17 tepals; stamens numerous, pale creamy yellow; ovary stalkless with many free carpels.

Liriodendron chinense (Hemsl.) Sarg.* **Tree** to 40 m, **leaves** 4-lobed with a prominently 2-lobed apex, 5.5–12 × 3–9 cm, leaf-stalks slender, 3–8 cm long. **Flowers** erect, cup-shaped, 2.5–4.5 cm tall, pale green or yellowish, longitudinally marked with yellow or orange outside; stamens 2–2.2 cm long. **Fruit** erect, egg-shaped, 4.5–9 cm long. Guizhou, Hubei, Sichuan, Shaanxi, Yunnan [c & sChina]; mixed forests, on sandy and rocky slopes, 500–1,700 m, often very local. VI–VII. Planted as an ornamental.

ILLICIACEAE

ILLICIUM

A genus of c. 40 species (8 in China) distributed in eastern Asia, east USA, Mexico and the West Indies. **Evergreen trees** and **shrubs** closely related to Magnoliaceae, but differing primarily in the **fruit** that consists of a single whorl of carpels arranged in a star shape, very different from the cone-like structure found in Magnoliaceae.

Illicium majus Hook. f. & Thomson **Tree** to 20 m with **leaves** in clusters of 3–6, oblong-lanceolate to oblanceolate, 10–20 × 2.5–7 cm, long-pointed, stalks 1–2.5 cm long. **Flowers** axillary or ± terminal, on stalks 2–3 cm long (to 4.5 cm in fruit); tepals white to pale yellow, 15–21, elliptic to elliptic-oblong, 6–15 mm long; stamens 12–21; carpels 11–14. **Fruit** consisting of 10–14 follicles, black-brown when ripe, tasting of mace. Guangxi, Guizhou, Yunnan [sChina, seAsia, Myanmar, Vietnam]; mixed forests, thickets, riverbanks, rocky slopes, 300–2,500 m. IV–VI.

Illicium jiadifengii B. N. Chang* **Tree** to 20 m. **Leaves** in clusters of 3–5, narrowly elliptic, 7–16 × 2–4.5 cm, tailed to long-pointed, stalks 1.5–3 cm long. **Flowers** axillary or ± terminal, on stalks 1.5–3 cm long; tepals white to pale yellow, 33–55, tongue-like, 14–17 mm long; stamens 28–32; carpels 12–14. **Fruit** with 12–14 follicles. Guizhou, Guangxi, Hubei, e & sSichuan [c & eChina]; mixed forests, 1,000–2,000 m. III–VI.

SCHISANDRACEAE

SCHISANDRA

A genus of c. 30 species (19 in China) distributed in east and south east Asia with one species in south-east North America. Dioecious, or rarely monoecious, **climbers** the young twigs often spur-like. **Leaves** thin-textured with translucent glands, base decurrent onto leaf-stalk. **Flowers** axillary, solitary, paired, or in clusters of up to 8; tepals 5–12, outer and inner tepals smaller than middle ones. **Male flowers** with 5–60 stamens; **female flowers** with 12–120 spirally arranged carpels. **Fruit** elongated, pendulous, like a string of fleshy beads.

1. Flowers orange-yellow or white .... **S. grandiflora, S. sphenanthera**
1. Flowers red .... **S. rubriflora**

Schisandra grandiflora (Wall.) Hook. f. & Thomson Similar to next species, but twigs purple to purplish brown, maturing to grey, **leaves** narrowly elliptic or elliptic, to obovate-elliptic or ovate, 8–14 cm long, margin finely toothed to almost entire; tepals 7–10, in 3 whorls, white, broadly elliptic to obovate, almost equal, conspicuously glandular. seTibet, swYunnan [swTibet, c & eHimalaya, nIndia, Myanmar, Thailand]; forested slopes and shrubberies, 1,800–3,100 m. IV–VI.

Schisandra sphenanthera Rehd. & Wils.* **Climber** to 5 m; twigs reddish brown, with dense lenticels. **Leaves** obovate to obovate-oblong, sometimes orbicular, rarely elliptic, 5–11 cm long, shortly pointed to long-pointed, dark green above, grey-green with white dots beneath, margin finely toothed, leaf-stalks red, 1–3 cm long, narrowly winged. **Flowers** borne basally on twigs; tepals 5–9, orange-yellow, almost equal, elliptic to oblong-obovate, middle whorl 6–12 mm long, glandular outside and ciliate. Gansu, Guizhou, Hubei, Sichuan, Shaanxi, neYunnan [c & eChina]; shrubberies in wet places and on slopes, open woodland, 600–3,000 m. IV–VII.

Schisandra rubriflora (Franch.) Rehd. & Wils.* **Twiner** to 4 m, sometimes more, with purplish brown twigs that mature black. **Leaves** obovate, elliptic-obovate or oblanceolate, rarely elliptic or ovate, 6–15 cm long, apex long-pointed, margin toothed, midrib and lateral veins pale red beneath. Tepals 5–8, red, outer elliptic to obovate, almost equal, ciliate, middle ones 10–17 mm long. **Fruit** to 20 cm long, green at first but maturing reddish purple. sGansu, Hubei, Sichuan, seTibet, sw & wYunnan; forests, often in ravines or on rocky slopes, dense shrubberies, 1,000–1,300 m. V–VI.

CHLORANTHACEAE

CHLORANTHUS

A genus of c. 17 species (13 in China, 9 endemic) found in temperate and tropical Asia. **Subshrubs** and perennial **herbs** with opposite or whorled **leaves**, bases of leaf-stalks fused, forming a transverse ridge. **Inflorescence** a spike or panicle with very small hermaphrodite **flowers** without a perianth and with 1 or 3 stamens. **Fruit** a drupe, often pear-shaped.

1. Leaves stalkless .... **C. sessilifolius**
1. Leaves stalked .... **C. henryi, C. multistachys**

Chloranthus sessilifolius K. F. Wu* Perennial rhizomatous **herb** to 70 cm; **stems** erect with paired, scale-like **leaves** below; leaves opposite, uppermost in a whorl of 4, rhombic to elliptic or obovate, 12–20 × 7–12 cm, sometimes suffused reddish purple beneath or just on veins. **Flowers** in 1–4 slender, terminal, drooping greenish spikes; **flowers** whitish, with 3 stamens. **Drupes** brown when ripe, 2–2.5 cm. Guangxi, Guizhou, Sichuan [Fujian, Guangdong, Jiangxi]; wet habitats in forests and shrubberies, 600–1,200 m. III–IV.

Chloranthus henryi Hemsl.* **Perennial herb** to 70 cm, from a stout rhizome, with 1–several stems, 4–5-noded. **Leaves** opposite, 4 uppermost whorled at stem apex, obovate to elliptic, narrowing to apex, 12–20 × 7–12 cm, dark green above, pale green to purple beneath, scurfily hairy on veins beneath; leaf-stalks 5–12 mm long. **Flower-spikes** terminal, with 2–4 pendent branches; **flowers** white; stamens 3, 2–2.5 mm long. Guizhou, Sichuan [c & sChina]; damp places in evergreen forests and thickets, 600–2,000 m. IV–VI. **Var. *hupehensis* (Pamp.) K. F. Wu*** has obovate to almost orbicular, glabrous **leaves** and lateral as well as terminal **inflorescences** on short stalks 2.5–5 cm long. Gansu, Hubei, Shaanxi.

Chloranthus multistachys Pei* Differs from previous species in having many axillary and terminal **spikes**, and **flowers** with 1–2 stamens with small connectives. Gansu, Guizhou, Hubei, Sichuan, Shaanxi [c & sChina]; damp places in evergreen forests, thickets and on slopes, 400–1,700 m. V–VII.

LAURACEAE

An economically important family. *Cinnanmomum camphora* and some other species yield camphor and essential oil, which are used for perfume or pharmcologically. The fruits of several genera contain oils and fats that are used in industry. The timber of some genera is very valuable. The bark of *Cinnamomum cassia* and the root of *Lindera aggregata* are famous in Chinese traditional medicine.

LINDERA

A genus of c. 80 species (30 in China), mostly in east and south Asia. **Evergreen** or **deciduous**, dioecious **shrubs** or **trees**. **Leaves** alternate, aromatic, simple or 3-lobed. **Inflorescences** a dense cluster of **flowers** surrounded by an involucre of 4 bracts that encloses the unisexual flowers in bud; sepals 6, petaloid; petals absent; anthers 2-celled. **Fruits** fleshy or becoming dry and splitting at maturity, 1-stoned.

Lindera obtusiloba Blume (syn. *L. cercidifolia* Hemsl.) **Deciduous shrub** or tree, 6–10 m tall, with brown young **shoots**, marked with pale lenticels. **Leaves** broadly ovate, 6–12.5 × 3.7–10 cm, 3-veined, pointed to broadly 3-lobed at apex, heart-shaped at base, glossy yellow-green above, pale and downy on veins beneath, midrib downy; leaf-stalks 1.2–2.5 cm long. **Flowers** yellow, 7–8 mm long, covered in silky hairs, numerous, in dense clusters produced before leaves. **Fruits** globose, glossy black, 8 mm long. Gansu, Hubei, Sichuan, nwYunnan [c & nChina, Korea, Japan]; mixed and broadleaved forests, shrubberies, 200–3,300 m. III–IV.

1. *Schisandra grandiflora*; Dali, nwYunnan, (RS)
2. *Schisandra sphenanthera*; cult. (CS)
3. *Schisandra rubriflora*; Kangding, wSichuan (PC)
4. *Schisandra rubriflora*; cult. (CGW)
5. *Schisandra rubriflora* in fruit (CGW)
6. *Chloranthus sessilifolius*; Maolan, seGuizhou (PC)
7. *Lindera obtusiloba*; Shennongjia, wHubei (PC)

***Lindera megaphylla* Hemsl.** In contrast to the previous species, an **evergreen shrub** or **tree** 4–20 m tall, with dark purplish young shoots; terminal buds woolly, **leaves** aromatic, oblong to oblanceolate, 12–25 × 2.5–7 cm, pointed, pinnately veined. **Flowers** numerous, yellow, in short-stalked, axillary umbels 2.2–2.5 cm across. **Fruits** egg-shaped, black, 20 mm long. Guizhou, wHubei, Sichuan, Yunnan [c & sChina, Taiwan]; laurel forest, in deep ravines, sea level to 1,500 m. III–V.

CINNAMOMUM

A genus of c. 250 species (49 in China), distributed in tropical and subtropical east Asia, Australia and the Pacific islands. **Evergreen**, fragrant **trees** or **shrubs** with alternate, almost opposite or opposite **leaves**, sometimes aggregated at top of branchlets, leathery, often 3-nerved. **Inflorescence** paniculate, axillary, terminal or almost so, composed of 3-many-flowered cymes, **flowers** small to medium-sized, yellow or white, hermaphrodite, cup-shaped or campanulate with 6 almost equal lobes, perianth-tube short; fertile stamens 9, in 3 whorls; staminodes 3; ovary always as long as slender style, stigma capitate or disc-like, sometimes 3-lobed. **Fruit** fleshy, subtended by a conical, cup-shaped or campanulate perianth-cup.

***Cinnamomum camphora* (L.) Presl** (syn. *Laurus camphora* L.; *Persea camfora* Spreng.; *Camphora officinarum* Nees) Broad, roundly crowned, strongly camphor-scented, **evergreen tree**, to 30 m with yellow-brown fissured bark and terete branchlets. **Leaves** alternate, somewhat leathery, ovate-elliptic, 6–12 × 2.5–5.5 cm, pointed, shiny green or yellow-green above, yellow- to grey-green beneath. **Inflorescence** 3.5–7 cm long, **flowers** greenish white or yellowish, c. 3 mm long. **Fruit** egg-shaped or almost globose, 6–8 mm across, purple-black when ripe. Guangxi, Guizhou, seSichuan, Yunnan [Vietnam, Korea, Japan]; hills and valleys, to 2,000 m. IV–V.

***Cinnamomum glanduliferum* (Wall.) Meissn.** (syn. *Laurus glandulifera* Wall.; *Camphora glandulifera* Nees; *Machilus mekongensis* Diels) **Evergreen**, camphor-scented **tree**, 5–15 m tall with grey-brown, longitudinally fissured bark, peeling, red-brown inside. **Branchlets** angled. **Leaves** alternate, leathery, elliptic to ovate-elliptic or lanceolate, 6–15 × 4–6.5 cm, pointed to somewhat long-pointed, glossy dark green above, glaucous beneath; leaf-stalks 1.5–3 cm long. **Inflorescences** shorter than leaves, 4–10 cm long. **Flowers** small, to 3 mm long, yellowish. **Fruit** globose, to 10 mm across, black when ripe. sGuizhou, s & swSichuan, seTibet, c & nYunnan [seAsia, c & eHimalaya, India, Myanmar]; evergreen broadleaved hill forests, 1,500–2,500(–3,000) m. III–V.

LITSEA

A genus of c. 200 species (73 in China, mostly in the south and south-west) distributed in tropical and subtropical Asia, Australia and the Americas. **Evergreen** or **deciduous**, dioecious **trees** or **shrubs** with alternate, rarely opposite or whorled, pinnately nerved **leaves**. **Inflorescence** an umbellate cyme or panicle, solitary or clustered at leaf-axils; involucral bracts 4–6, decussate, persistent at flowering. **Flowers** unisexual; perianth-tube long or short, segments in 2 whorls of 3, equal or unequal; **male flowers** with 9 or 12 fertile stamens in 3–4 whorls, anthers dehiscing inwards (towards centre of flower), 4-celled; **female flowers** with 9–12 staminodes, a superior ovary, and conspicuous style. **Fruit** seated on a shallowly disc-like or deeply cup-shaped perianth-tube.

1. Leaf-stalks 2–3 cm long . . . . **L. populifolia**
1. Leaf-stalks less than 2 cm long 2
2. Branchlets covered with hairs . . . . **L. veitchiana**
2. Branchlets glabrous . . . . **L. cubeba, L. rubescens**

Litsea populifolia* (Hemsl.) Gamble (syn. *Lindera populifolia* Hemsl.; *Lindera obovata* Franch.; *Litsea longipetiolata* Lecomte) Small **deciduous tree** to 3–5 m. **Leaves** alternate, usually clustered at tip of branchlets; leaf-blade orbicular to broadly obovate, 6–8 × 5–7 cm, rounded at apex, glaucous beneath. **Umbels** crowded at tip of branchlets, 9–11-flowered, flowering with the leaves; perianth-segments 6, ovate or broadly ovate, yellow, **male flowers** with 9 fertile stamens. **Fruit** globose, 5–6 mm across. Sichuan, seTibet, neYunnan; sunny mountain slopes, riverbanks in valleys, shrubberies, dry secondary forests, 750–2,000 m. IV–V.

Litsea veitchiana* Gamble **Deciduous shrub** or small **tree** to 4 m with yellow-white-silky branchlets when young. **Leaves** alternate, obovate or obovate-oblong, 4–15 × 2.5–5.5 cm, pointed or blunt, densely hairy on both surfaces when young; leaf-stalks 1–1.2 cm, densely hairy when young. **Umbels** solitary at tip of branchlets of previous year, flowering before or with the leaves, 10–13-flowered, **male flowers** with 6, elliptic or rounded, 3-nerved and glandular perianth-segments and 9 fertile stamens. **Fruit** globose, c. 5 mm across, black when ripe. Guizhou, Hubei, Sichuan, nwYunnan; roadsides, shrubberies on mountains, 400–3,800 m. III–V.

***Litsea cubeba* (Lour.) Pers.** **Deciduous shrub** or small **tree** to 8–10 m with alternate, lanceolate, oblong or elliptic **leaves**, 4–11 × 1.1–2.4 cm, long-pointed or pointed, glabrous, glaucous or silky-downy beneath; leaf-stalks 6–20 mm, glabrous. **Umbels** solitary or clustered, 4–6-flowered, flowering before or with the leaves, **male flowers** with 6, broadly ovate perianth-segments and 9 fertile stamens. **Fruit** almost globose, c. 5 mm, black when ripe. Guizhou, Hubei, Sichuan, seTibet, Yunnan [s & seAsia, c & sChina, Taiwan]; sunny slopes, shrubberies, sparse forests, roadsides, stream- and riverbanks, 500–3,200 m. III–V.

Litsea rubescens* Lecomte **Deciduous shrub** or small **tree**, 4–10 m tall, with red branchlets. **Leaves** alternate, elliptic, lanceolate-elliptic or rounded-elliptic, 4–6 × 1.7–3.5 cm, pointed to blunt, glabrous; leaf-stalks 12–16 mm long, red when young. **Umbels** axillary, male ones 10–18-flowered, **male flowers** with 6, yellow, broadly elliptic perianth-segments and 9 fertile stamens. **Fruit** globose, c. 8 mm across. Guizhou, Hubei, Sichuan, sShaanxi, seTibet, Yunnan [Hunan]; margins and glades in evergreen broadleaved forests, thickets, 700–3,800 m. IV–VI.

SAURURACEAE

HOUTTUYNIA

A genus with 1 species restricted to east and South East Asia characterised by the spike-like inflorescence subtended by a whorl of petal-like bracts.

Houttuynia cordata **Thunb.** Patch-forming, rhizomatous, **herbaceous perennial** to 60 cm, foul-smelling when crushed, **stems** green or purplish red. **Leaves** heart-shaped, 3–10 × 2–6 cm, rather fleshy, densely glandular but usually glabrous, green or mottled, sometimes flushed red or purple. **Inflorescence** a dense, erect spike 15–30 mm long subtended by 4–(6–8) conspicuous, white, oblong or obovate involucral bracts to 15 mm long; **flowers** white at maturity, very small, with 3(–4) stamens and 3 pistils. **Fruit** a small dehiscent capsule. Gansu, Guangxi, Guizhou, Hubei, Sichuan, Shaanxi, seTibet, Yunnan [swTibet, widespread in China, the Himalaya, India, Myanmar, Japan, Korea, Thailand]; damp and wet habitats in forests, wet meadows, ravines, streamsides, ditches and field margins, roadsides, to 2,500 m. IV–IX.

ARISTOLOCHIACEAE

SARUMA

A single species endemic to China with characteristic 3-petalled **flowers**; stamens 12 in 2 series; **fruit** consisting of 6 carpels fused close to the base.

Saruma henryi **Oliv.*** Tufted, rhizomatous, **herbaceous perennial** with erect to ascending **stems** bearing alternate, heart-shaped, rather dull green, stalked **leaves**, blades to 15 × 13 cm. **Flowers** solitary on slender downy stalks in upper leaf-axils; calyx with 3 elliptic-ovate lobes, 8–9 mm long; petals butter-yellow or greenish yellow, heart- to kidney-shaped, 10–11 mm long, short-clawed. Gansu, Guizhou, Hubei, n & eSichuan, Shaanxi; forests, shrubby ravines, streamsides, 600–1,000 m. IV–VII.

ASARUM

A northern hemisphere genus with c. 90 species (39 in China, the majority endemic), concentrated in South East Asia. Rhizomatous, evergreen, **herbaceous perennials**, usually with aromatic roots and rhizomes. **Leaves** solitary or paired, often arrow-, kidney- or heart-shaped, sometimes patterned, long-stalked. **Flowers** solitary or paired, terminal, often hidden at base of plant below the foliage at ground level; sepals forming a tube or cup fused to the ovary or not, often ribbed or chequered, with a constricted mouth at the top and 3 spreading to deflexed lobes; stamens 12. **Fruit** a fleshy, sometimes rather spongy capsule, splitting irregularly when ripe.

1. *Houttuynia cordata*; Cangshan, nw Yunnan (CGW)
2. *Houttuynia cordata*; Wolong, w Sichuan (PC)
3. *Saruma henryi*; cult. (CGW)

1. Calyx with a distinct cup-shaped or campanulate tube beyond attachment to ovary, glabrous outside 2
1. Calyx without a distinct, or with a very short, tube beyond attachment to ovary, sepals free or somewhat fused near the base, downy outside 3
2. Leaves glossy green or purplish red beneath .... ***A. delavayi***
2. Leaves not as above ..... ***A. maximum, A. splendens***
3. Sepals reflexed, pointed to blunt .... ***A. caulescens, A. himalaicum, A. pulchellum***
3. Sepals erect to spreading 4
4. Sepals forming a very short tube beyond attachment to ovary .... ***A. debile, A. caudigerellum***
4. Sepals free beyond attachment to ovary .... ***A. caudigerum***

1. *Asarum splendens*; cult. (DJ)
2. *Asarum maximum*; cult. (DJ)
3. *Asarum delavayi*; cult. (RR)
4. *Asarum delavayi*; cult. (RR)
5. *Aristolochia kwangsiensis*; Guilin, neGuangxi (PC)
6. *Aristolochia moupinensis*; Wolong, wSichuan (PC)
7. *Nelumbo nucifera*; Xishuanbanna, swYunnan (CGW)

Asarum delavayi Franch.* Rather robust, patch-forming **perennial** with horizontal rhizomes. **Leaves** solitary, blades ovate with heart-shaped base, to arrow-shaped, 7–15 × 6–11 cm, bright shiny green above and slightly hairy, sometimes with white blotches, apex long-pointed. **Flowers** campanulate, 4–6 cm across, on stalks 10–35 mm long, purplish green, tube ± cylindrical, chequered inside, strongly constricted at mouth, lobes broad-ovate, with an ovate rough papillose zone at base. swSichuan, nwYunnan; forests, shrubberies, damp rocky and shady places on mountain slopes, 800–1,600 m. IV–VI.

Asarum maximum Hemsl.* Patch-forming **perennial** with horizontal rhizomes and solitary, ovate to arrow-shaped **leaves** 6–13 × 7–15 cm, dull green and hairy above, sometimes with white patches. **Flowers** on stalks up to 5 cm long; calyx dark purple, 4–6 cm across, campanulate, tube c. 25 × 15–20 mm, swollen in the middle with a whitish girdle-like ring, slightly constricted at mouth, lobes spreading widely apart, ovate, with transversely corrugated ridges at base surrounding a cushion-like white zone. Hubei, Sichuan; humus-filled pockets in forests, 600–800 m. IV–V.

Asarum splendens (F. Maek.) C. Y. Cheng & C. S. Yang* Patch-forming **perennials** with horizontal rhizomes and solitary **leaves**, blades heart- to arrow-shaped, 6–10 × 5–9 cm, deep green above, often with white blotches. **Flowers** solitary on a stalk 8–10 mm long; **calyx** purplish green, campanulate, 2.5–3 cm across, slightly constricted at throat, tube c. 14 × 20 mm, the ovate lobes somewhat undulate, often mottled, with a semicircular corrugated whitish area at base. Guizhou, Hubei, Sichuan, neYunnan; shrubberies, damp grassy places, damp slopes, 800–1,300 m. IV–V.

Asarum caulescens Maxim.* Patch-forming **perennial**, sparsely white-hairy (staying white when dried). **Leaves** paired, almost heart-shaped, long-pointed, 4–9 cm long, dark green, stalks 6–12 cm long, sparsely hairy. **Calyx** pink to purple, cup-shaped, tube 3–6 mm long, long-hairy outside, lobes reflexed, triangular, blunt or pointed, 3–6 mm long; stamens 12, stamens and styles slightly protruding from calyx-tube. Gansu, Guizhou, Hubei, Sichuan, Shaanxi; mixed woods on mountain slopes, 700–1,700 m. IV–V.

Asarum himalaicum Hook. f. & Thomson Patch-forming **perennial**, sparsely white-hairy (staying white when dried). **Leaves** solitary on vegetative shoots, heart-shaped, somewhat long-pointed, 4–8 cm long, dark green, stalks 10–25 cm long, becoming hairless. **Calyx** purplish, somewhat campanulate, tube 6–8 mm long, 8–14 mm across, long-hairy outside, lobes reflexed, triangular, blunt or pointed, 3–5 mm long; stamens 12, stamens and styles not protruding from calyx-tube. Gansu, Guizhou, Hubei, Sichuan, seTibet [swTibet, Himalaya]; damp places in mixed woods, along streams, 1,300–3,100 m. IV–VI.

Asarum pulchellum Hemsl.* Tufted plant densely covered with long white hairs (turning black when dried). **Leaves** paired, broadly ovate to ovate-heart-shaped, 5–8 cm long, dark green, stalks 10–20 cm long, hairy. **Calyx** purple with a white upper surface, urn-shaped, with a tube 8–12 mm long, 8–14 mm across, long-hairy outside, lobes reflexed, triangular, 4–8 mm long, blunt or pointed; stamens 9–12. Guizhou, Hubei, Sichuan, neYunnan [cChina]; mixed woods on mountain slopes, 700–1,700 m. IV–V.

Asarum debile Franch.* Patch-forming or tufted plant. **Leaves** paired on flowering stems, solitary on sterile stems, heart-shaped, 2.5–4 cm long, dark green, downy on veins, leaf-stalks 5–12 cm long, glabrous. **Flowers** small, less than 10 mm long; calyx purple, campanulate, tube 3–5 mm long, 5–7 mm across, lobes erect, triangular, 3–4 mm long; stamens 6–9. Hubei, Sichuan, Shaanxi, Yunnan [cChina]; mixed woods on mountain slopes, along streams, in rock-crevices, 1,300–2,300 m. V–VI.

Asarum caudigerellum C. Y. Cheng & C. S. Yang* **Leaves** paired on flowering stems, solitary on sterile stems, heart-shaped, 3–7 cm long, dark green, downy on veins beneath, leaf-stalks 4–10 cm long, glabrous. **Flowers** more than 15 mm long; **calyx** brownish purple, campanulate, tube 6–12 mm long, 9–15 mm across, lobes spreading, triangular, 7–10 mm long; stamens 12. Guizhou, Hubei, Sichuan, neYunnan; thickets and mixed woods on mountain slopes, along streams, 1,600–2,100 m. IV–V.

Asarum caudigerum Hance Patch-forming **perennial** with vertical rhizomes and paired, heart-shaped to triangular-ovate dark green **leaves** with white blotches, 4–10 × 3.5–10 cm. **Flowers** on a stalk 10–20 mm long; calyx green or purplish green, campanulate, 2–5 cm across, lobes erect to spreading, ovate but narrowed at top into a slender tail up to 25 mm long. Guangxi, Guizhou, Hubei, Sichuan, Yunnan [Vietnam, Taiwan]; forests, mountain slopes, streamsides, 300–1,700 m. III–V.

ARISTOLOCHIA

A genus of c. 400 species (45 in China), temperate, subtropical and tropical, cosmopolitan. **Shrubs**, **herbs** or **climbers**, usually with tuberous roots. **Leaves** alternate, entire or lobed. **Flowers** axillary, solitary or in bunches, zygomorphic consisting of fused calyx-lobes, tube often dilated near base but apically cylindrical or funnel-shaped, 1–2-lobed; stamens 6 in 1 series; ovary inferior, 6-celled. **Fruit** a dry, 6-valved capsule.

1. Twining herb; perianth-tube straight or slightly curved .... ***A. delavayi***
1. Climbing shrubs; flowers horseshoe-shaped or with a knee-like curve in middle in side view 2
2. Leaves with 2–3 pairs of veins from the base (palmately veined) .... ***A. kunmingensis, A. kwangsiensis, A. moupinensis***
2. Leaves with 6–8 pairs of veins from the base (palmately veined) .... ***A. forrestiana***

ARISTOLOCHIACEAE: Aristolochia

Aristolochia delavayi Franch.* Pungent plant with terete stems that are pruinose, grooved and densely knobby. **Leaves** ± stalkless, ovate, 2–8 cm long, densely dotted. **Flowers** solitary, axillary; **calyx** yellow with a deep purple throat, tube c. 15 mm long, limb strap-like, 2–3.5 cm long. Sichuan, nwYunnan; thickets on limestone mountain slopes, 1,600–1,900 m. V–VIII.

Aristolochia kunmingensis C. Y. Cheng & J. S. Ma* Similar to previous species, but with ovate to narrow-ovate **leaves**, 6–14 × 4–8 cm and awl-shaped branchlets 5–8 mm long. Guizhou, Yunnan; thickets, forests on mountain slopes, 2,000–2,200 m. IV–V.

Aristolochia kwangsiensis W. Y. Chun & F. C. How ex C. F. Liang* Similar to previous 2 species, but generally with larger **leaves** with a deeper, 3–5 cm, basal sinus, **flowers** dark purple with yellow in throat, tube horseshoe-shaped and densely hairy, limb 3.5–4.5 cm across, warty. **Fruit**-capsule 8–10 cm long. Guangxi, Guizhou, Sichuan, Yunnan [eChina]; similar habitats, 600–1,600 m. III–V. (See photo 5 on p. 76.)

Aristolochia moupinensis Franch.* **Climbing shrub** to 4 m, with grey-yellow, densely hairy **stems**. **Leaves** heart-shaped, 6–16 × 5–12 cm, hairy below like the stems, more sparsely so above, stalks 3–8 cm long. **Flowers** lateral, paired or solitary on a short hairy stalk; **calyx** yellowish with purple veins, 4–5 cm, with an abruptly bent, downy, pale tube and a 3–3.5-cm disc-like limb, which is shallowly 3-lobed. **Fruit**-capsule cylindrical, 6–8 cm long, dehiscing from base. Guizhou, Sichuan, Yunnan [cChina]; forests, shrubberies, ravines, streamsides, 2,000–3,200 m. V–VI. (See photo 6 on p. 76.)

Aristolochia forrestiana J. S. Ma* **Climbing shrub** with grooved, glabrous stems. **Leaves** ovate, often narrowly so, 9–13 × 3–5 cm, with a shallow, heart-shaped base, leathery, glabrous, stalk 2–3 cm long. **Flowers** solitary in leaf-axils; **calyx** purple with an abruptly curved, downy tube and a 6 cm, shallowly 3-lobed limb. **Fruit** unknown. w & nwYunnan; forests, shrubberies, 2,900–3,600 m. IV–VI.

NYMPHACEAE

NUPHAR

A genus of c. 10 species (2 in China), widespread in north temperate regions. **Perennial herbs** with creeping, branched rhizomes. **Leaves** dimorphic, either floating with long stalks and thick leathery blades or submerged with short stalks and thin-textured blades. **Flowers** held above the water, regular; sepals 4–7, yellow or orange, petaloid, oblong to obovate, persistent; petals numerous, yellow, small and stamen-like; carpels lacking marginal appendages. **Fruit** egg-shaped to urn-shaped, irregularly dehiscent.

Nuphar pumila (Timm) DC. Creeping **perennial herb** with broad-ovate to ovate (seldom elliptic) floating **leaves**, 6–17 cm long, with a heart-shaped base, leaf-stalks 20–50 cm long, downy. **Flowers** 10–45 mm across, on downy 40–50 cm stalks; sepals yellow, oblong to elliptic, 10–25 mm long; petals narrowly wedge-shaped to broadly linear, notched, 5–7 mm long; stigmatic disc deeply lobed, 4–7.5 mm across. Guizhou, Hubei, Sichuan, Shaanxi, Yunnan [c & sChina, nEurope, Japan, Korea, Russia]; lakes, ponds and broad ditches, 600–2,800 m. V–IX.

EURYALE

A single species in eastern Asia including China. **Annual** or **short-lived perennial, rhizomatous herb** with both submerged and floating, centrally peltate **leaves** with entire margin. **Flowers** floating or often partially or entirely submerged, cleistogamous or with a spreading perianth; sepals 4, greenish, not petaloid, persistent; petals numerous, in c. 5 series, showy, gradually transitional with the stamens; carpels 7–16, completely united, without a style. **Fruit** irregularly dehiscent.

Euryale ferox Salisb. **Floating leaves** with prickles on stalks and along veins, broadly elliptic to orbicular, to 1.3 m across, green above, purple beneath; **submerged leaves** arrow-shaped or elliptic, not prickly, 4–10 cm long. **Flowers** 3–5 cm across, with a stout, densely prickly stalk; sepals triangular-ovate, 10–15 mm long, prickly outside; petals oblong-lanceolate, 10 mm long, outer purple-violet, inner white. Gansu, Guangxi, Guizhou, Hubei, Sichuan, Shaanxi, Yunnan, [China except Nei Mongol, Qinghai, Tibet; India, Japan, Korea, eRussia]; ponds, lakes, wide ditches, 800–3,400 m. VI–VIII.

NYMPHAEA

A genus containing c. 50 species (5 in China) widespread in tropical and subtropical freshwater habitats. **Perennial, rhizomatous herbs** with mostly floating **leaves**, base heart- to arrow-shaped, margin entire to toothed. **Flowers** floating or above the water; sepals 4, greenish, not petal-like; petals 8-many, large, showy, often grading into stamens; true stamens shorter than sepals and petals; carpels partially or completely united with a stalkless stigma. **Fruit** irregularly dehiscent. In China, several native species, as well as *Nymphaea mexicana* Zucc. and *N. alba* L. var. *rubra* Lönnroth, are cultivated.

Nymphaea tetragona Georgi Tufted **perennial** with ovate-heart-shaped to ovate-elliptic, scarcely peltate **leaves**, 5–12 × 3.5–9 cm, base deeply heart-shaped, margin entire. **Flowers** floating, 3–6 cm across; sepals broad-lanceolate to narrow-ovate, 2–3.5 cm long, persistent; petals 8–15, white, broad-lanceolate, oblong, or obovate, 2–2.5 cm long. **Fruit** globose, 2–2.5 cm across. Guizhou, Hubei, Sichuan, Tibet, Yunnan [c & sChina, Asia, Europe, North America]; ponds and lakes, to 4,000 m. VI–VIII.

NELUMBOACEAE

NELUMBO

A genus with 2 species (1 in China) distributed in east and South East Asia, north Australia, and Central and North America. **Rhizomatous**, **perennial**, **aquatic herbs** with swollen tubers. **Leaves** arising directly from rhizomes, floating or aerial. **Flowers** solitary, long-stalked, hermaphrodite, regular, borne well above water surface; tepals numerous, petal-like. **Fruit** top-like, bearing numerous nut-like seeds loosely embedded in cavities on the flattened top.

Nelumbo nucifera Gaertn. (syn. *N. komarovii* Grossh.; *N. nucifera* var. *macrorhizomata* Nakai; *Nelumbium speciosum* Willd.; *Nymphaea nelumbo* L.) **Patch-forming**, **rhizomatous**, **perennial** 1–2 m tall with orbicular **leaves**, 25–90 cm across, glabrous, glaucous and water-repellent, margin entire. **Flowers** waterlily-like, 10–23 cm across, on stalks longer than leaf-stalks; tepals soon falling, pink or white, oblong-elliptic to obovate, 5–10 cm long. **Fruit** 5–10 cm across, green at first, becoming brown. China except Qinghai, Tibet [widespread in tropical, subtropical Asia, Australia]; commonly cultivated in lakes and ponds. VI–VIII. (See photo 7 on p. 76.)

BERBERIDACEAE

BERBERIS

A genus of c. 500 species (180 in China) with a cosmopolitan distribution, predominantly in temperate regions and mountains in the tropics and subtropics. **Evergreen** or **deciduous** shrubs, stems purple or red when young (yellow when cut), usually bearing spines. **Leaves** simple, generally arranged in clusters or whorls, stalkless or short-stalked, margins entire or spiny. **Inflorescence** 1-flowered, clustered, umbellate, racemose or paniculate. **Flowers** campanulate, yellow to orange, fragrant; sepals in 2 whorls, inner longer than outer; petals in a single whorl; stamens 6, attached to base of petals. Ovary with 1–10 ovules. **Fruit** an ellipsoid to globose berry, red to purple or black when ripe.

1. Leaves obovate, margins entire or toothed 2
1. Leaves ovate to lanceolate, elliptic or linear-elliptic . . 6
2. Leaf-margins entire or occasionally very slightly toothed
 . . . *B. forrestii*, *B. mouillacana*, *B. vernae*, *B.wilsoniae*
2. Leaf-margins toothed to spine-edged 3
3. Flowers not more than 5 mm across . . . . . *B. potaninii*
3. Flowers 5–9 mm across 4
4. Spines on stems solitary; flowers in racemes or clusters
 . . . . *B. dawoensis*, *B. dictyoneura*,
 B. tenuipedicellata, *Berberis* sp.
4. Spines on stems 3-branched, rarely simple; flowers solitary or in racemes 5
5. Flowers usually solitary . . . .
 . . . . . . *B. dictyophylla*, *B. muliensis*, *B. stiebritziana*
5. Flowers in racemes . . . . *B. henryana*
6. Deciduous shrubs with ovate leaves . . . . *B. jamesiana*
6. Evergreen shrubs with lanceolate to linear-elliptic or oblong-elliptic leaves . . . . *B. davidii*, *B. julianae*,
 B. lijiangensis, *B. taliensis*

Berberis forrestii Ahrendt* **Shrub** 2–4.5 m tall; stem-spines 3-branched, 1–2 cm long. **Leaves** 1.3–4.5 × 0.6–2 cm, greyish above, grey-bloomy beneath, margin entire or shallowly toothed. **Inflorescence** racemose, 2.5–7 cm long; **flowers** 3–4 mm long, canary-yellow. **Berries** ellipsoid, c. 10 mm long, bright red. sSichuan, nwYunnan (Lijiang); forests, scrub, open pasture, amongst rocks on mountain slopes, 3,600–4,000 m. V–VI.

Berberis mouillacana C. K. Schneid.* (syn. *B. boschanii* C. K. Schneid.) **Deciduous shrub**, 1.5–3 m tall with dark grey, grooved branches and sometimes reddish, non-warty shoots; spines simple, sometimes 3-branched or absent, pale yellow, 3–18 mm long. **Leaves** obovate or oblong-obovate, 1–6 × 0.5–3.5 cm, shiny deep green above, paler beneath, margin entire, occasionally with 1–8 tiny spine-tipped teeth on each side. **Inflorescence** usually a raceme with a few clustered **flowers** at base, occasionally almost umbellate; sepals in 1–2 whorls; petals broadly elliptic, c. 4.5 × 3 mm, notched. **Berry** egg-shaped-ellipsoid, 9–10 × 5–6 mm, coral-red or scarlet when ripe. Gansu, Qinghai, Sichuan, nwYunnan; forests, forest margins, shrubberies, roadsides on slopes, 2,000–3,500 m. IV–V.

1. *Berberis forrestii*; Muli, swSichuan (PC)
2. *Berberis mouillacana*; Baishui, nwYunnan (HJ)

1. *Berberis vernae*; Juizhaigou, nwSichuan (PC)
2. *Berberis wilsoniae*; Juizhaigou, nwSichuan (CGW)
3. *Berberis wilsoniae*; Juizhaigou, nwSichuan (PC)
4. *Berberis potaninii*; Juizhaigou, nwSichuan (CGW)
5. *Berberis dawoensis*; Napahai, nwYunnan (PC)
6. *Berberis tenuipedicillata*; Kangding, wSichuan (PC)
7. *Berberis dictyophylla*; cult. nwYunnan (CGW)
8. *Berberis stiebritziana*; cult. nwYunnan (CGW)
9. *Berberis henryana*; Huanglongsi, nwSichuan (PC)
10. *Berberis jamesiana*; Meilixueshan, nwYunnan (TY)

Berberis vernae C. K. Schneid.* **Shrub** 1–2.6 m tall, with solitary yellow spines 0.8–4 cm long. **Leaves** obovate, rounded, 1.5–3 × 0.4–0.8 cm, margins entire; leaf-stalks 5–25 mm long. **Inflorescence** ascending, racemose, 2.5–4 cm long, bearing yellow **flowers** 2–3 mm long. **Berries** globose, 5–6 mm, pale whitish pink. Gansu, n & neSichuan; forests, scrub, open pasture on mountain slopes, 2,600–3,100 m. VI–VII.

Berberis wilsoniae Hemsl.* (syn. *B. parvifolia* Sprague; *B. stapfiana* C. K. Schneid.; *B. subcaulialata* (C. K. Schneid.) C. K. Schneid.; *B. wilsoniae* var. *parvifolia* (Sprague) Ahrendt) **Shrub** 0.3–1 m tall, with arching brownish grey, slightly black-warty branches; **stems** grooved, dark red, bearing 3-branched, slender spines 10–20 mm long. **Leaves** semi-evergreen, deep green above, grey and with a slight bloom beneath, obovate-spathulate, 1–2.5 × 0.2-0.6 cm, margin entire. **Inflorescence** a cluster of 4–7 **flowers**, each flower 5–6 mm long, lemon-yellow; sepals in 2 whorls. **Berries** globose, 4–6 mm, pinkish red, soft when ripe, with a slight bloom. Gansu, Hubei, seQinghai, n, nw, w & swSichuan, Shaanxi, e & seTibet, nw & wYunnan; cliffs, amongst rocks on mountain slopes, roadsides, forest margins, streamsides, 1,100–4,000 m. VI–IX (fruit ripe I–II).

Berberis potaninii Maxim.* (syn. *B. liechtensteinii* C. K. Schneid.) **Shrub** 1–1.5 m tall. **Stems** smooth, dark red with a white bloom; spines 3–6 mm long. **Leaves** evergreen, rigid, obovate, pointed, mucronate, 1–3.5 × 0.2–1.5 cm, margin 3–9-toothed. **Inflorescence** a 4–12-flowered raceme, **flowers** 4.5–5 mm long. **Berries** egg-shaped, 7–8 mm long, pale red when ripe. Gansu, nw & wSichuan; dry scrub by streams and rivers, 2,000–3,000 m. V–VI.

Berberis dawoensis K. Meyer* Differs from *B. tenuipedicellata* in having **stems** with red young shoots and spines 6–15 mm long, and **leaves** pruinose when young, oblong-obovate, blunt, 1–3 × 0.7–1 cm, margins 3–7-spined. **Inflorescence** 5–10-flowered, almost clustered to umbellate-racemose, **flowers** 5–8.5 mm long. **Berries** oblong-egg-shaped, red, with a slightly bluish white bloom when ripe. seTibet, nwYunnan (Deqin area); shrubberies, open woodland, 3,700–4,300 m. VI–VII.

Berberis dictyoneura C. K. Schneid.* **Shrub** 1–2 m tall with sparsely warty **stems** that are green when young; spines solitary, 1–2 cm long. **Leaves** deciduous, rigid, obovate, blunt, 1–2.5 × 0.4–1 cm, grey-green, margins 7–15-spined. **Inflorescence** a 7–14-flowered raceme, **flowers** 5–6 cm long. **Berries** egg-shaped, 7–8 mm long, pink when ripe. nwSichuan; shrubberies, scrub on mountain slopes, roadsides, 2,800–3,200 m. V–VI.

Berberis tenuipedicellata T. S. Ying* Similar to *B. dictyoneura*, but with evergreen, obovate **leaves**, 0.8–2 × 0.5–1.2 cm, margins minutely toothed, spinulose, and with almost globose **fruit** pitted at apex. wSichuan; thickets, roadsides, 2,450–3,100 m. V–VI.

Berberis sp.* Similar in general characteristics to *B. dictyoneura*, but **leaves** with very fine bristle-like marginal spines and **flowers** in erect to ascending, cylindrical racemes, each with c. 40 flowers. nwSichuan (Juizhaigou); similar habitats, 2,400–2,700 m. VII.

Berberis dictyophylla Franch.* **Deciduous shrub** 1–3 m tall with terete to somewhat grooved, dark red, pruinose **stems**; spines 10–30 mm long. **Leaves** rigid, obovate, blunt, 1–2 × 0.4–1 cm, finely reticulate, entire, glaucous beneath. **Flowers** usually solitary in each leaf-cluster, 8–9 mm long, golden yellow tipped with red, on a very short stalk. **Berries** egg-shaped with persistent style, 7–8 mm long, dull red with a white bloom when ripe; style absent. Sichuan, seTibet, Yunnan; pastures, scrub, open *Picea* forest on mountain slopes, 2,800–4,400 m. V–VI.

Berberis muliensis Ahrendt* Like *B. dictyophylla*, but with a larger **fruit** lacking a persistent style. swSichuan, seTibet; open alpine meadows, 3,300–3,600 m. VI–VII.

Berberis stiebritziana C. K. Schneid.* Similar to *B. dictyophylla*, but spines 10–25 mm long, **leaves** lacking reticulate venation and **berries** 1–1.2 cm long. Sichuan, seTibet, nwYunnan; shrubberies on mountain slopes, 3,000–4,000 m. VI–VII.

Berberis henryana C. K. Schneid.* **Deciduous shrub**, 2–3 m tall with greyish yellow or dark brown branches and reddish, simple or 3-branched spines 10–30 mm long, sometimes absent. **Leaves** stalked, dark green above, grey-green beneath, 1.5–3 × 0.8–1.8 cm, margin with 10–20 very fine, spine-tipped teeth on each side. **Inflorescence** a 10–20-flowered raceme, 2–6 cm; sepals in 2 whorls; petals oblong-obovate, 5–6 × 4–5 mm, with an incised apex. **Berries** ellipsoid, c. 9 × 6 mm, red when ripe, without a bloom. Gansu, Guizhou, Hubei, Sichuan, Shaanxi [cChina]; thickets, forests, forest margins, weedy places; 1,000–2,500 m. V–VI.

Berberis jamesiana Forrest & W. W. Sm.* **Shrub** 2–4 m tall with purple young shoots that mature to lustrous red; spines 3-branched, 3–5 cm long. **Leaves** thick, semi-leathery and rigid, 3–10 × 1–5 cm, closely reticulate, white beneath, margins entire or finely toothed; petals 5–25 mm long. **Inflorescence** racemose, 2.5–10 cm long, with 15–40 flowers, each 3–4 mm long, deep yellow. **Berries** globose, 7–9 mm, creamy white turning pale red. Sichuan, seTibet, nwYunnan; forests, scrub, open pasture, amongst rocks on mountain slopes, 3,600–4,600 m. IV–VI.

Berberis davidii Ahrendt* (syn. *B. densa* C. K. Schneid.; *B. wallichiana* DC. forma *parvifolia* Franch.) **Shrub** to 1 m with brownish grey, grooved, black, warty branches; spines slender, 3-branched, to 20 mm long. **Leaves** elliptic or elliptic-lanceolate, 1–5 × 0.6–1.5 cm, dark green above, yellowish green beneath, with a back-rolled margin with 3–9 spinulose,

1. *Berberis lijiangensis*; Gangheba, Yulongxueshan, nwYunnan (CGW)
2. *Berberis davidii*; Heishui, Yulongxueshan, nwYunnan (CGW)
3. *Mahonia dolichostylis*; Shizong to Xingyi, swGuizhou (PC)
4. *Mahonia lomariifolia*; cult. (JG)
5. *Mahonia lomariifolia*; cult. (CGW)
6. *Mahonia lomariifolia*; wYunnan (MF)
7. *Nandina domestica*; cult. (CGW)

minute teeth on each side. **Flowers** in clusters of 6–8; sepals in 3 whorls; petals obovate-elliptic, c. 5.5 × 3.2 mm, entire. **Berries** ellipsoid, 8–9 mm long, red when ripe, with a whitish bloom. Yunnan; grassy slopes, grassland, riversides, forest glades, shrubberies; 2,000–3,500 m. V–VI.

Berberis julianae C. K. Schneid.* **Shrub** 1–2.5 m tall; spines 3-branched, 10–25 mm long. **Leaves** elliptic to lanceolate, pointed to blunt, 5–8 × 1–2.3 cm, margins toothed; leaf-stalks 5–6 mm long. **Inflorescence** a cluster of 6–10 **flowers**, each 4–5 mm long, yellow. **Berries** ellipsoid, 10–12 mm long, purple when ripe. Hubei, Sichuan; open *Abies fabri* forest, scrub, 2,600–3,000 m. V–VII.

Berberis lijiangensis C. Y. Wu ex S. Y. Bao* **Evergreen shrub**, 1–2 m tall with brownish grey branches and straw-coloured shoots; spines 3-branched, 15–30 mm long. **Leaves** almost stalkless, oblong-elliptic or narrowly elliptic, 3–5 × 1.4–1.8 cm, deep green with a waxy bloom above, olive-green beneath, margin sometimes slightly back-rolled, with 3–4 spine-tipped teeth on each side. **Flowers** in clusters of 3–6; sepals in 2 whorls; petals primrose-yellow, oblong-obovate, c. 6 × 4 mm. **Berries** oblong, c. 12 × 5–7 mm, black when ripe, with a waxy bloom. w & nwYunnan; forests, forest margins, thickets, shrubberies, 2,700–3,400 m. V–VI.

Berberis taliensis C. K. Schneid.* Small **shrub** 0.3–1.6 m tall, with warty, grooved **stems**; spines 7–12 mm long. **Leaves** stalkless, linear-elliptic, pointed, with back-rolled margin, 2.3–4 × 0.4–0.8 cm, white beneath. **Flowers** in clusters of 2–5, each flower 5.5–6 mm long, bright yellow. **Berries** ellipsoid, c. 11 mm long, purplish black. nwYunnan; woodland, boulder-strewn shrubby hillsides, 2,500–3,900 m. VI–VII.

MAHONIA

A genus of c. 120 species (43 in China) in temperate regions of the Old and New Worlds. **Evergreen shrubs** with scaly, rarely smooth, **stems** that lack spines. **Leaves** tufted at apex of erect branches, stiff, leathery, pinnate, with few to many opposite leaflets with toothed to sharply spiny margins. **Inflorescences** dense, cylindrical, many-flowered spikes, several in a cluster at top of stem. **Flowers** usually yellow, fragrant. **Fruit** a berry, egg-shaped to ellipsoid, blue-black when ripe.

1. Bracts 2–4 times as long as flower-stalks .... *M. longibracteata*
1. Bracts shorter or as long as flower-stalks 2
2. Leaves with a short stalk not more than 2.5 cm long .. .. *M. bealei, M. dolichostylis*
2. Leaves with a long stalk more than 3 cm long .... *M. lomariifolia, M. veitchiorum*

Mahonia bealei (Fortune) Carr.* **Shrub** 1.5–2.3 m tall. **Leaves** almost stalkless, 25–50 cm long; leaflets 11–17, linear-lanceolate to narrow-ovate, 6–10 × 2.7–4.5 cm, with 3–6-spined margins. **Inflorescences** 5–12 cm long; **flowers** bright yellow, 6–8 mm long, very fragrant. Hubei, n & eSichuan [cChina]; woods and shrubberies on hillsides, 900–2,000 m. I–II.

Mahonia dolichostylis Takeda* (syn. *M. duclouxiana* Gagnep.) **Shrub** 1.5–8 m tall. **Leaves** almost stalkless, 25–40 cm long; leaflets 11–17, obliquely oblong-ovate, pointed, 4.5–12 × 2.7–4.5 cm, with 8–12-spined, toothed margins. **Inflorescences** densely clustered, 10–18 cm long; **flowers** bright yellow, 7.5–9 mm long, fragrant. swGuizhou, Yunnan; woods and secondary shrubberies on dry hillsides, 1,600–2,600 m. II–III.

Mahonia lomariifolia Takeda Imposing **shrub** 2.5–4 m tall with long-stalked **leaves** 25–60 cm long; leaflets 8–40, linear-lanceolate to oblong-ovate, long-pointed, 3.2–10 × 1.2–2.6 cm, with 3–7-spined margins. **Inflorescences** 9–20 cm long; **flowers** bright yellow, 8–9 mm long, fragrant. **Fruit** ellipsoid, 16–18 mm long, blue-black when ripe. s & swSichuan, Yunnan [neMyanmar, Taiwan]; mixed woods on slopes, 2,600–3,900 m. IX–XII.

Mahonia veitchiorum Hemsl. & Wils.* **Shrub** 0.6–2 m tall with long-stalked **leaves** 20–40 cm long; leaflets 9–13, obliquely elliptic, pointed, 4–11 × 1.7–4.5 cm, with 10–18-spined, toothed margins. **Inflorescences** 6–12 cm long; **flowers** bright yellow, 5.5–7 mm long, fragrant. sSichuan (Emeishan), Yunnan; woods and open shrubberies on hillsides, 600–2,300 m. VI–VII.

Mahonia longibracteata Takeda* **Shrub** 0.6–3.3 m tall. **Leaves** 20–25 cm long, dark green above, yellow-green beneath; leaflets obliquely ovate, long-pointed, 4–7.5 × 2–3.2 cm, with 3–5-spined toothed margins. **Inflorescences** 8–10 cm long; **flowers** yellow, 4.5–6 mm long, fragrant. wYunnan; woods and bamboo thickets on slopes, 2,600–3,500 m. V–VIII.

NANDINA

A genus with a single species found from the east Himalaya to Japan and Indonesia. Sometimes placed in a separate family, Nandinaceae, on account of the numerous spirally arranged sepals and anthers that open by longitudinal slits.

Nandina domestica Thunb. A rather elegant, bamboo-like, **evergreen shrub** to 2.5 m. **Leaves** clustered towards stem-tips, 2–3-pinnate, leaflets almost leathery, linear-lanceolate, 3–10 cm long, often red-tinged in spring and purplish in autumn. **Flowers** in erect panicles at stem-tips, fragrant, 8–12 mm across, white with large yellow anthers. **Berries** globose, 7–8 mm, red or purple-red. Widely planted as an ornamental [c & eChina], 600–2,600 m. V–VII.

PODOPHYLLACEAE

Often treated within the Berberidaceae, as herbaceous representatives of that family.

PODOPHYLLUM

A genus of c. 14 species (12 in China) of woodland plants primarily found in the Sino-Himalayan region. **Herbaceous perennials** with scaly rhizomes and unbranched **stems**. **Leaves** 1–3, alternate, folded umbrella-like upon emerging, opening widely, peltate or palmately lobed. **Flowers** solitary or clustered, nodal or on stalk of uppermost leaf; sepals 3–6, greenish or the inner coloured; petals often 6, overlapping; stamens 6–12. **Fruit** a fleshy berry, rupturing irregularly, pulpy.

1. Mature leaves with a single deep sinus reaching to top of leaf-stalk .... ***P. hexandrum***
1. Mature leaves peltate, with the blade completely encircling the top of the leaf-stalk 2
2. Leaf-margin with uneven-sized mucronate teeth; petals reflexing with age .... ***P. aurantiocaule***
2. Leaf-margin with even teeth; petals erect to spreading, never reflexed 3
3. Flowers small, bowl-shaped, petals not more than 25 mm long .... ***P. difforme***, ***P. mairei***
3. Flowers larger, rounded or funnel-shaped, petals more than 25 mm long .... ***P. delavayi***, ***P. delavayi*** var. ***longipetalum***, ***P. versipelle***

Podophyllum hexandrum **Royle** (syn. *P. emodi* Wall.) A very variable tufted plant to 60 cm, often less, **stems** with a waxy bloom, often flushed orange or reddish at first. **Leaves** palmate (peltate in juvenile plants), to 25 cm across, with 3–5 coarsely toothed lobes, deep plain green or variously patterned with purplish black or purplish crimson. **Flowers** solitary, held above or beneath leaves, generally appearing with the partly developed leaves; sepals 6, soon falling as a cap; petals 6 oval, white to pink or rich flamingo-rose, 15–30 mm long. **Fruit** pendent, plum-like, red with a waxy bloom, to 60 mm long. sGansu, wHubei, Sichuan, seTibet, Yunnan [swTibet, Himalaya]; coniferous and broadleaved forests, shrubberies, rocky slopes, grassy meadows, 1,800–4,500 m. IV–VI.

Podophyllum aurantiocaule **Hand.-Mazz**. **subsp.** ***furfuraceum*** **(S. Y. Bao) J. M. H. Shaw** (syn. *Dysosma furfuracea* S. Y. Bao) Plant to 90 cm with glabrous, grooved **stems**. **Stem-leaves** 1–3, peltate with a sunken centre, to 30 cm across, 5–9-lobed, downy at first and then often mottled with purple, otherwise deep green. **Flowers** 2–9, rarely solitary; sepals pale green, soon falling; petals 6, white to cream or pink, 15–20 mm long, spreading widely apart to reflexed. **Fruit** pear-shaped or globose, 20–30 mm across, reddish orange when ripe. seTibet, nwYunnan [swTibet, Bhutan, nMyanmar]; damp places in coniferous and broadleaved or mixed forests, bamboo thickets, shrubberies, grassy slopes, 1,200–2,700 m. III–IV.

Podophyllum difforme **Hemsl. ex Wils.*** (syn. *Dysosma difformis* (Hemsl. ex Wils.) T. H. Wang) Plant to 30 cm, with glabrous **stems** bearing 1–3 arrow-shaped to triangular or almost 2-lobed **leaves**, to 18 × 10 cm, margin often shallowly lobed, lower leaf larger than the upper, deep velvety green, often with paler or darker variegation, margin sparsely toothed. **Flowers** pendulous, 1–3, below the leaves; petals 3–6, salmon-pink to reddish purple, obovate, 15–25 mm long, occasionally larger. **Fruit** rounded to egg-shaped, 10–20 mm, somewhat warty. Guangxi, Guizhou, wHubei, Sichuan [Hunan]; hill and valley forests, 1,200–1,800 m, local. IV–V.

Podophyllum mairei **Gagnep.*** (syn. *Dysosma aurantiocaulis* sensu T. S. Ying; *D. mairei* (Gagnep.) Hiroe) Similar to previous species but with glabrous **leaves** and leaf-stalks and a finely toothed leaf-margin, the 5–7 **flowers** always borne on leaf-stalks. nwYunnan, possibly adjacent parts of Sichuan; similar habitats and flowering time.

Podophyllum delavayi **Franch.*** (syn. *P. veitchii* Hemsl. & Wils.; *Dysosma delavayi* (Franch.) Hu) Plant to 30 cm, with grooved, sometimes hairy, **stems**. **Stem-leaves** 1–2, peltate, often with dark green or purple-brown mottling when young, eventually to 15 cm across, divided to about halfway into 5–8 triangular lobes, with a few irregular teeth along the margin, generally glabrous. **Flowers** 1–3, borne in the fork of the leaf-stalks, pendulous; sepals 6, soon falling; petals 6(–9), deep pink to deep purple or purple-red, rarely white, lanceolate to spathulate, 35–70 mm long. **Fruit** olive-shaped, to 35 mm long, dark red when ripe. Guizhou, Sichuan, sShaanxi, Yunnan; dense forests, wooded ravines, 1,200–2,200 m. IV–V. A variant with extra-long petals, 80–100 mm long, endemic to Emeishan in sSichuan, is distinguished as **var.** ***longipetalum*** **J. L. Wu & P. Zhuang ex J. M. H. Shaw**.

Podophyllum versipelle **Hance** A more robust plant than the previous species, 20–80 cm tall, with **stems** bearing 1–3 dark green leaves. **Leaves** to 40 cm across, sparsely hairy to downy beneath, margins of the 6–8 lobes finely toothed. **Flowers** up to 10, sometimes more, borne on stalk of uppermost leaf (held just beneath the leaf); petals dark red-purple, oblong, 25–35 mm long. Guangxi, Guizhou, Hubei, Sichuan, Yunnan [sChina, Vietnam]; humus-rich forests and other shady habitats, 500–2,400 m. IV–V. Most plants in w & swChina are referable to **subsp.** ***boreale*** **J. M. H. Shaw*** which has more deeply lobed **leaves** with up to 5 lobes; **flowers** in clusters of up to 19 (not 4–9) and flower-stalks usually glabrous.

1. *Podophyllum hexandrum*; Songpan to Jiuzhaigou, nwSichuan (PC)
2. *Podophyllum hexandrum*; Napahai, nwYunnan (PC)
3. *Podophyllum hexandrum*; Bitahai, Zhongdian, nwYunnan (CGW)
4. *Podophyllum hexandrum*; Gangheba, Yulongxueshan, nwYunnan (CGW)
5. *Podophyllum hexandrum*; cult. (CGW)
6. *Podophyllum hexandrum* in fruit; cult. (CGW)
7. *Podophyllum hexandrum*; Napahai, nwYunnan (CGW)
8. *Podophyllum delavayi*; cult. (CGW)
9. *Podophyllum delavayi*; cult. (CGW)
10. *Podophyllum versipelle*; cult. (CGW)

1. *Diphylleia sinensis*; Tianbao, Zhongdian, nwYunnan (PC)
2. *Diphylleia sinensis*; Shennongjia, wHubei (PC)
3. *Epimedium ecalcaratum*; cult. (RR)
4. *Epimedium brevicornu*; cult. (RR)
5. *Epimedium sagittatum*; Fanjingshan, neGuizhou (PC)
6. *Epimedium pubescens*; cult. (RR)
7. *Epimedium fargesii*; cult. (RR)
8. *Epimedium stellulatum*; cult. (RR)

DIPHYLLEIA

A genus of 3 species (1 in China) from south-east USA, west China and Japan. Related to *Podophyllum*, but with distinctive cymose clusters of flowers held above the leaves.

Diphylleia sinensis H. L. Li* **Herbaceous perennial** to 80 cm, with erect **stems** bearing 2 well-separated leaves as well as separate basal pedate leaves; **stem-leaves** palmately lobed, slightly peltate, lobes coarsely toothed, deep green. **Flowers** usually 15 or more in a terminal cymose cluster, each flower with 6 transitory sepals and 6 flat, white petals. **Fruit** a rounded, dark blue berry. sGansu, Hubei, Sichuan, Shaanxi, Yunnan; open woodland, shrubberies, rocky places, 2,600–3,600 m. V–VI.

EPIMEDIUM

A genus with c. 54 species scattered from south Europe to Japan, with the majority in China. Tufted, rhizomatous, **perennial herbs** with basal or stem-leaves, mostly 2-ternate. **Inflorescence** simple or cymose, often panicle-like; sepals 4 + 4, the outer unequal, the inner spreading or reflexed, petaloid; petals 4, produced outwards with a pouch or spur; stamens 4. **Fruit** a 2-parted capsule, the larger half containing the seeds.

1. Petals spurless or slightly pouched at base . . . . *E. ecalcaratum*, *E. platypetalum*
1. Petals with a short to long spur or at least deeply pouched 2
2. Flowers small, not more than 20 mm across, spur not exceeding 4 mm long, shorter than sepals and rather inconspicuous 3
2. Flowers large, 20–60 mm across, spurs more than 10 mm long, longer than sepals 4
3. Stem-leaves with 9, occasionally 5, leaflets . . . . *E. brevicornu*, *E. sagittatum*
3. Stem-leaves with 3, rarely 5, leaflets . . . . *E. pubescens*, *E. pubescens* subsp. *cavaleriei*, *E. fargesii*, *E. stellulatum*
4. Stamens exposed, not enclosed by the petal-limbs . . . . . . *E. leptorrhizum*, *E. brachyrrhizum*, *E. acuminatum*, *E. franchetii*, *E. membranaceum*
4. Stamens hidden, enclosed by the petal-limbs . . . . *E. davidii*, *E. fangii*, *E. flavum*

Epimedium ecalcaratum G. Y. Zhong* Short-**rhizomatous** plant to 60 cm, often less, flowering **stem** bearing 1–3 **leaves** with 3, 5 or 7 leaflets; leaflets ovate with a heart-shaped base and short (1 mm) marginal spines. **Inflorescences** simple or branched, with up to 30 small, campanulate, yellow **flowers** with red inner sepals; petals larger than sepals, 8–9 mm long, flat, slightly pouched at base, rarely with a short spur. wSichuan (Baoxing region); forests and forest margins in damp, shady places, 1,400–2,200 m. IV–V.

Epimedium platypetalum K. Meyer* Similar to previous species, but a smaller plant with rounded to broadly ovate **leaves** and simple **inflorescence**; **flowers** primrose-yellow, usually with white inner sepals. wSichuan, similar habitats. IV–V.

Epimedium brevicornu Maxim.* Plant **clump-forming**, to 60 cm, often only half that height, with glandular **stems** bearing an opposite pair of 2-ternate **leaves**; leaflets usually 9, occasionally less, broad-ovate, with spiny-toothed margin, often rather yellow-green and papery at maturity. **Inflorescences** branched, many-flowered; **flowers** 14–15 mm across, white or yellowish with orange petals; sepals spreading, lanceolate, 8–9 mm long, much longer than petals. Gansu, nwSichuan, Shaanxi; wooded, bushy and rocky places, ravines, 800–2,100 m. IV–V.

Epimedium sagittatum (Sieb. & Zucc.) Maxim.* Similar to preceding species, but leaflets narrower, lanceolate to narrow-ovate, dark lustrous green, and flowers smaller, only 8–9 mm across, inner sepals only 4 mm long, slightly longer than petals. Guizhou, Hubei [c & eChina]; forests, ravines, 900–1,700 m. V–VII.

Epimedium pubescens Maxim.* **Tufted**, short-rhizomatous plant to 60 cm, with glandular **stems** generally bearing a pair of opposite **leaves**; leaflets 3, occasionally solitary, ovate to lanceolate with a heart-shaped base and a very spiny-toothed margin, downy beneath. **Inflorescences** lax and branched, glandular, many-flowered; **flowers** star-like, white, with spreading narrow-lanceolate, inner sepals 5–7 mm long, much exceeding the small and inconspicuous, pouched, brownish petals; stamens 3–4 mm long. Guizhou, Sichuan, Shaanxi [Anhui]; forests and forested ravines, 750–1,400 m. IV–V. Plants from Guizhou belong to **subsp.** *cavaleriei* (Stearn) Stearn, which differs from subsp. *pubescens* in having a non-glandular **inflorescence** and **flowers** with almost black (rather than purplish) outer sepals.

Epimedium fargesii Franch.* Like *E. pubescens* but inflorescences sometimes unbranched and flowers larger, inner sepals 15–18 mm long, spreading to reflexed, white or pale purple, petals larger, to 9 mm long, dark purple with a whitish tip to the spur; stamens 7–10 mm long. n & neSichuan; similar habitat, 900–1,600 m. IV–V.

Epimedium stellulatum Stearn* Like *E. pubescens*, but inner sepals always white, 11–12 mm long, and petals minute, brownish with an orange base. wHubei (Wudangshan), n & neSichuan, [Jiangxi]; rocky forests, ravines, ruins, 800–1,200 m. IV–V.

Epimedium leptorrhizum Stearn* Plant long-rhizomatous, **creeping**, to 30 cm, with reddish-hairy **stems** bearing a single leaf. **Leaves** with 3, occasionally 1, ovate, long-pointed leaflets that are markedly heart-shaped at base and with a very spiny margin, leathery at maturity. **Inflorescence** simple, with 4–8

PODOPHYLLACEAE: Epimedium

flowers that are white flushed with rose-pink; inner sepals narrow-lanceolate, 11–20 mm long, equalling the downcurved, horn-like spurs. Guizhou, seSichuan; forests and forest margins, banks, damp mossy places, 800–1,500 m. IV–V.

Epimedium brachyrrhizum **Stearn*** Like *E. leptorrhizum*, but plants more **clump-forming** with a far more compact rhizome, leaflets more spiny-margined (i.e. with more marginal spines) and with lanceolate rather than elliptic inner sepals; **flowers** white, pink at the base of the petals, spurs downcurved, 22–26 mm long. neGuizhou (Fanjingshan); forests, wooded slopes, 600–1,200 m. IV–V.

Epimedium acuminatum **Franch.*** Similar to *E. leptorrhizum*, but a taller plant to 50 cm with flowering **stems** bearing 2(–3) leaves and lax, many-flowered, branched **inflorescences**; **flowers** white, rose-purple to violet or occasionally yellow, the horn-shaped, downcurved spurs sometimes very darkly coloured, 15–25 mm long, greatly exceeding the 8–12 mm-long sepals. Guizhou, c, s & seSichuan (including Emeishan), Yunnan; forests and forest margins, banks, damp, often mossy places, 400–1,400 m. IV–V.

Epimedium franchetii **Stearn*** Similar to *E. acuminatum*, but leaflets often narrower and more densely spiny on margin; **inflorescence** more compact, unbranched, with pale sulphur-yellow flowers. Guizhou, Hubei; forests, ravines, banks, 950–1,400 m. IV–V.

Epimedium membranaceum **K. Meyer*** Similar in general appearance to *E. acuminatum* and *E. franchetii*, but leaflets broader, ovate, with a less extended tip, downy beneath; **flowers** in a branched **inflorescence**, whitish or pale yellow, sometimes with a pink flush, the inner sepals 6–7 mm long. e & sSichuan, nYunnan; similar habitats, altitude and flowering time.

Epimedium davidii **Franch.*** Plant **short-rhizomatous**, to 50 cm, with glandular flowering **stems** usually bearing a pair of leaves. **Leaves** with 3 or 5 leaflets, with tufts of reddish hairs at nodes; leaflets ovate with a heart-shaped base and a spiny-toothed margin. **Inflorescence** branched, with up to 24 yellow **flowers** with reddish inner sepals; petals with rounded blades forming a cup around stamens, spurs 10–15 mm long, downcurved. wSichuan (centred on Baoxing); forests and forest margins, damp shady places, ravines, 850–1,500 m. IV.

Epimedium fangii **Stearn*** Similar in general appearance to the preceding species, but a more spreading plant with glabrous and near-horizontal spurs. sSichuan (Emeishan); forests, ravines, 850–1,100 m. IV–V.

Epimedium flavum **Stearn*** Similar to *E. davidii*, but plant rarely exceeding 30 cm tall, with blunt-tipped leaflets, and paler, sulphur-yellow **flowers** with spreading spurs that are half enveloped by the larger, 10–11 mm long, horizontal inner sepals. wSichuan (Erlangshan); habitats similar to those of *E. davidii*.

LARDIZABALACEAE

HOLBOELLIA

A genus of 20 species (9 in China) from the Himalaya and China. Vigorous, **evergreen** and **deciduous climbers** with alternate, compound **leaves**. Flowers in clustered, generally short racemes, with several **female flowers** below more numerous **male flowers**, all with 6 large sepals and 6 minute petals, male flowers with 6 stamens, female with 6 staminodes and 3 carpels. **Fruit** sausage-like, fleshy, non-splitting with seeds embedded in pulp.

Holboellia latifolia **Wall.** Rampant **evergreen**, twining **climber** to 6 m with alternate, compound, long-stalked, palmate **leaves**; leaflets 3–9, dark shiny green, elliptic to lanceolate, 4–13 × 1–7 cm, with entire or slightly toothed margin. **Inflorescences** clustered, few-flowered racemes; **male flowers** greenish white, 11–15 mm long, outer 3 oblong, blunt, inner 3 more lanceolate and pointed; **female flowers** purple, 16–22 mm long, inner 3 sepals shorter and narrower than outer 3, with minute staminodes. **Fruit** oblong, rather irregular, 5–7 cm long, reddish purple when ripe. Guizhou, Sichuan, seTibet, Yunnan [Himalaya]; forests and forest margins, shrubberies, damp shady places, ravines, 600–3,000 m. IV–V.

Holboellia coriacea **Diels*** Readily distinguished from the previous species by the thick, leathery **leaves** with 3 leaflets, terminal leaflet larger than lateral, 3-veined from the base; **male flowers** white striped purple, 9–10 mm long; **female flowers** reddish purple, 12–14 mm long. Guizhou, Hubei, Sichuan, Shaanxi [cChina]; similar habitats, 500–2,000 m. IV–V.

AKEBIA

A genus of 5 species (4 in China) from China, Korea and Japan. **Deciduous** or **semi-evergreen**, **twining climbers** with alternate, palmately compound **leaves**. Flowers in lateral racemes with 1–2 large **female flowers** at the base and more numerous **male flowers** above, all usually with 3, occasionally 6, sepals but no petals; male flowers with 6 stamens, female with 3–9 carpels. **Fruit** a cluster of fleshy follicles splitting along 1 side, **seeds** embedded in pulp.

Akebia quinata **(Houtt.) Decne** Vigorous, **deciduous climber** to 6 m, with slender grey-brown twining **stems**. **Leaves** dark green with a paler glaucous reverse; leaflets normally 5, occasionally 7, obovate to almost elliptic, 2–5 × 1.5–2.5 cm, terminal leaflet rather larger than the others. **Racemes** up to 12 cm long, faintly fragrant; **male flowers** up to 11, 6–8 mm long, sepals purple, rarely greenish white; **female flowers** 10–20 mm long, longer-stalked, sepals the same colour as male flowers or darker, usually with 3–6 carpels. **Fruit** sausage-shaped, to 8 cm long, purplish when ripe. Sichuan [eHubei, e & neChina, Korea, Japan]; forests and forest margins, shrubberies, mountain slopes, ravines, streamsides, 300–1,500 m. IV–V.

1. *Epimedium brachyrrhizum*; cult. (RR)
2. *Epimedium acuminatum*; Anlong, swGuizhou (PC)
3. *Epimedium acuminatum*; Anlong, swGuizhou (PC)
4. *Epimedium flavum*; cult. (RR)
5. *Epimedium acuminatum*; cult. (RR)
6. *Epimedium davidii*; cult. (RR)
7. *Holboellia latifolia*; cult. (CGW)
8. *Akebia quinata*; cult. (CGW)

1. *Decaisnea insignis*; cult. (CGW)
2. *Decaisnea insignis*; wSichuan (TK)
3. *Clematis potaninii* subsp. *potaninii*; Daxueshan, nwYunnan (CGW)
4. *Clematis potaninii* subsp. *fargesii*; near Longriba, nwSichuan (CGW)
5. *Clematis armandii*; cult. (CGW)
6. *Clematis armandii*; cult. (CGW)
7. *Clematis finetiana*; between Shizong and Xingyi, eYunnan (PC)
8. *Clematis uncinata*; Jiuzhaigou, nwSichuan (CGW)

Akebia trifoliata **(Thunb.) Kordz.** Very similar to the previous species, but readily distinguished by the **leaves** with 3(–5) leaflets, leaflets larger and more ovate, 3–8 × 1.5–6 cm, pale green beneath, margin usually wavy and with rounded teeth; **male flowers** up to 30, pale to mid-purple; **female flowers** dark purple to purplish brown. seGansu, Guizhou, Hubei, sShaanxi, Yunnan [c & eChina, Japan]; similar habitats, especially in semi-deciduous forests, 200–2,100 m. IV–V.

DECAISNEA

A Sino-Himalayan genus of just 1 species with odd-pinnate **leaves** and terminal panicles of 6-sepalled **flowers**; petals absent; **male flowers** with 6 stamens fused into a tube below, **female flowers** with 3 straight carpels and a ring of staminodes. **Fruit** large, sausage-like.

Decaisnea insignis **(Griffith) Hook. f. & Thomson** (syn. *D. fargesii* Franch.) Erect, monoecious, **deciduous shrub** with simple or sparingly branched **stems**. **Leaves** with 13–25 ovate to almost oblong leaflets, each 6–14 × 3–7 cm, pointed to long-pointed, membranous, green above, glaucous beneath. Panicles with long raceme-like branches to 30 cm; **flowers** yellowish green, 17–24 mm long, with elliptic, pointed sepals. **Fruit** pendent, cylindrical, 5–10 cm long, bluish black when ripe, edible. sGansu, eGuangxi, Guizhou, Hubei, Sichuan, sShaanxi, seTibet, Yunnan [c & eChina, Himalaya, nMyanmar]; mixed forests, shrubberies, ravines, generally in rather wet places, 900–3,600 m. IV–VI.

RANUNCULACEAE

CLEMATIS

A cosmopolitan genus with c. 300 species (147 in China, 93 endemic) distributed worldwide. **Woody climbers**, **shrubs**, **subshrubs** and **herbaceous perennials**, usually with paired **leaves**; the climbing species have twining leaf-stalks; generally with hermaphrodite **flowers**, but sometimes **dioecious**. **Leaves** simple to ternate, pinnate or 2-pinnate; leaflets entire to toothed or lobed. **Flowers** solitary or in terminal or lateral cymes or panicles, variously shaped from flat to cupped, campanulate, tubular or urn-shaped; sepals 4 or more, petaloid; petals usually absent, occasionally present and transitional with the numerous stamens, with staminodes often present in such flowers. **Fruit** a collection of achenes, each often with a long feathery, persistent tail (style).

Synopsis of groups

Group One. Flowers in lateral or terminal clusters (cymes or panicles), flat to slightly cupped, upright to ascending or horizontal; stamens glabrous, p. 91.

Group Two. Flowers solitary in axils of leaves or winter buds, flat to slightly cupped, upright to ascending or horizontal; stamens glabrous, p. 92.

Group Three. Flowers basically pale to deep yellow, solitary or in well-developed inflorescences, deeply cupped to campanulate or lantern-shaped, generally nodding; stamens hairy, at least in part, p. 95.

Group Four. Flowers basically white, pale violet-purple to rose-purple or purplish brown, or red, solitary or in well-developed inflorescences; stamens hairy, at least in part, p. 96.

> **Group One**
> 1. Flowers white, sometimes creamy white at first, relatively large, 20–50 mm across; leaflets leathery, untoothed, 3–5-veined .. ***C. potaninii*, *C. potaninii* subsp. *fargesii*, *C. armandii*, *C. finetiana*, *C. uncinata***
> 1. Flowers cream, creamy white, greenish or greenish yellow, 7–25 mm across; leaflets not leathery, toothed, net-veined .... ***C. brevicaudata*, *C. grata*, *C. gouriana***

Clematis potaninii **Maxim.*** Vigorous **deciduous climber** to 7 m with strongly ribbed, downy young **stems**. **Leaves** to 30 cm, 1–2-pinnate with ovate, coarsely toothed and lobed, mid-green, thin leaflets. **Flowers** white or cream, 45–70 mm across, in 1–3-flowered cymes; sepals normally (5–)6(–7) oval sepals. **Achenes** glabrous with a feathery tail to 30 mm long. w & swSichuan, e & seTibet, nwYunnan; woodland margins, shrubberies, ravines, streamsides, 1,400–3,660 m. VI–VII. **Subsp. *fargesii* (Franch.) Grey-Wilson*** is less vigorous with fewer leaf divisions, **flowers** 25–40 mm across, in cymes with 3–7 flowers. sGansu, n & nwSichuan, s & swShaanxi.

Clematis armandii **Franch**. In contrast to *C. potaninii*, a very vigorous **evergreen climber** to 10 m with deep glossy green, leathery, ternate **leaves**; leaflets oblong-lanceolate, to 16 cm long, with 3(–5) main veins from the base, margin entire. **Flowers** white, sometimes flushed pink, fragrant, in lateral clusters of 7 or more, 30–50 mm across, appearing before new leaves emerge; sepals usually 4, elliptic. **Achenes** hairy with a feathery tail to 30 mm long. Guizhou, Hubei, Sichuan, sShaanxi, e & seTibet, Yunnan [nMyanmar, nVietnam]; woodland and woodland margins, shrubberies, streamsides, in subtropical and warm-temperate habitats, 500–1,800 m. III–V.

Clematis finetiana **Lévl. & Vaniot*** Similar to *C. armandii* but leaflets thinner and more lanceolate or narrow-ovate, not more than 10 cm long. **Flowers** very fragrant, 20–40 mm across, white with a greenish reverse, in clusters of 3–7. Guizhou, Hubei, c & eSichuan, eYunnan; habitats similar to those of *C. armandii*, primarily subtropical, 100–1,200 m. IV–VI.

Clematis uncinata **Champ. ex Benth.** Superficially very like *C. armandii*, but **leaves** pinnate with 5 leaflets, to 10 × 3.8 cm, glaucous beneath, the leaf- and leaflet-stalks with a distinct articulation; **flowers** 20–30 mm across, creamy white, fragrant. sGansu, Hubei, Sichuan, Yunnan [c & eChina, Taiwan, nVietnam, Japan]; shrubberies, hedgerows, rocky places and ravines, streamsides, 700–2,500 m. V–VII.

Clematis brevicaudata **DC.** Vigorous **deciduous climber** to 15 m, often forming a dense entanglement, with ribbed **stems** and 2-ternate **leaves**; leaflets ovate to lanceolate, coarsely toothed and sometimes 3-lobed, glabrous or minutely downy. **Flowers** cream flushed green or greenish yellow, 12–20 mm across, fragrant, numerous, in 'foamy' lateral or terminal sprays; sepals 4, downy on both sides. **Achenes** minutely hairy with a feathery tail to 20 mm long. Gansu, Hubei, Qinghai, Sichuan, Shaanxi, nwYunnan [nChina, Korea, Mongolia, e & seRussia]; open forests, shrubberies, hedgerows, rocky places, in rather dry temperate regions, 460–2,800 m. VII–X.

Clematis grata **Wall.** Similar to *C. brevicaudata*, but young growth densely downy and **leaves** 1–2-pinnate; **flowers** 15–25 mm across, greenish white to cream, with 4(–6), oval sepals that are glabrous inside. **Achenes** hairy with a pale yellowish feathery tail to 35 mm long. Guizhou, s & seSichuan, seTibet, Yunnan [swTibet, sChina, Himalaya]; forest margins, shrubberies, hedgerows, river boulders, 600–2,500 m. VII–IX.

Clematis gouriana **Roxb.** Differs from *C. brevicaudata* and *C. grata* in having mainly pinnate (rarely 2-ternate) **leaves** with 5 or 7 leaflets, and in the numerous small creamy white **flowers**, 7–18 mm across; sepals downy on both sides. **Achenes** finely hairy with a feathery tail to 60 mm long. Guizhou, wHubei, Sichuan, seTibet, c & sYunnan [sChina, Himalaya, Philippines]; subtropical and warm broadleaved forests and forest margins, shrubberies, streamsides, 400–2,600 m. IX–X.

Group Two
1. Slender subshrub, not climbing, rarely more than 1.5 m tall .... *C. chrysocoma*
1. Rampant climbers, 3–12 m tall .... *C. montana*, *C. montana* var. *grandiflora*, *C. montana* var. *rubens*, *C. montana* var. *sterilis*, *C. montana* var. *trichogyna*, *C. montana* var. *wilsonii*, *C. venusta*, *C. gracilifolia*, *C. gracilifolia* var. *dissectifolia*, *C. spooneri*

Clematis chrysocoma **Franch.*** **Subshrub** to 1.5 m, rarely taller, often only 20–50 cm, with young **stems** and **leaves** densely covered in golden or tawny hairs. **Leaves** ternate, ovate leaflets coarsely toothed to 3-lobed. **Flowers** usually borne in axils of current year's shoots on stalks up to 20 cm long, flat or slightly cupped, pale pink, rose-pink, rose-purple or occasionally white; sepals 4 (rarely up to 8), oval. **Achenes** with a curved feathery tail to 25 mm long, golden brown-hairy overall. wGuizhou, swSichuan, n, nw & wYunnan; open forests, shrubberies, rocky and banks or slopes, 1,500–3,200 m. V–VII.

Clematis montana **Buch.-Ham.** A very variable, vigorous **climber** to 12 m, often entangling small trees and bushes, with ribbed **stems** and ternate **leaves**; leaflets deep green above, paler and shiny beneath, oval to lanceolate, slightly lobed to coarsely toothed, tip pointed. **Flowers** solitary or in few-flowered clusters from lateral buds of previous year, fragrant or not, 40–60 mm across, white, sometimes flushed pink beneath; sepals 4, usually oval to elliptic; anthers 1.5–3 mm long. **Achenes** glabrous with a feathery tail 25–40 mm long. sGansu, Guizhou, Hubei, sQinghai, Sichuan, Shaanxi, e & seTibet, Yunnan [swTibet, sChina, Himalaya, Taiwan]; mixed or deciduous woodland, shrubberies, streamsides, rocky places, 1,200–4,000 m. IV–VI. Many variants have been recognised but the most significant in the region are: **var.** *grandiflora* **Hook.*** (syn. *C. anemoniflora* D. Don) a less vigorous plant to 3–4 m bearing large **flowers**, 7–12 cm across, sepals often with a pointed apex; anthers 3–4 mm long, often pinkish. sGansu, Guizhou, w Hubei, Sichuan, sShaanxi, seTibet [swTibet], 1,100–3,500 m; **var.** *rubens* **Wils.*** is similar to the typical plant but with deep purple stems, **leaves** with a purple flush when young and reddish pink **flowers**; probably confined to wHubei, c & eSichuan; **var.** *sterilis* **Hand.-Mazz.*** is a slighter plant to 4–5 m with small entire leaflets and denser clusters of small white **flowers** mostly just 20–35 mm across. swSichuan, seTibet, nwYunnan, 2400–3000 m; **var.** *trichogyna* **M. C. Chang*** has white sepals with a fine 2–3 mm-long point at the apex and appressed-hairy **fruit**. nwYunnan, 2,550–3,200 m; **var.** *wilsonii* **Sprague*** is very similar to typical *C. montana*, but is a late-flowering variant with white **flowers**, flushed green beneath, 40–80 mm across, sepals rounded or somewhat notched at apex. wHubei, w & eSichuan, nYunnan, 2,400–3,600 m. VI–VII.

Clematis venusta **M. C. Chang*** Very close to *C. montana*, but leaflets lanceolate and usually entire; **flowers** white, 50–80 mm across, often greenish in bud. nwYunnan (Lijiang-Zhongdian region); similar habitats, 2,300–2,700 m. IV–V.

1. *Clematis brevicaudata*; Jietaisi, near Beijing (CGW)
2. *Clematis grata*; Benzilan, nwYunnan (CGW)
3. *Clematis chrysocoma*; Lijiang, nwYunnan (CGW)
4. *Clematis chrysocoma*; Gangheba, Yulongxueshan, nwYunnan (CGW)
5. *Clematis chrysocoma* form with extra sepals; Heishui, Yulongxueshan, nwYunnan (PC)
6. *Clematis montana*; Napahai, Zhongdian, nwYunnan (CGW)
7. *Clematis montana*; Yulongxueshan, nwYunnan (CGW)
8. *Clematis montana* var. *grandiflora*; Yulongxueshan, nwYunnan (CGW)
9. *Clematis montana* form; Wolong, wSichuan (CGW)
10. *Clematis montana* var. *wilsonii*; cult. (CGW)
11. *Clematis montana* form with extra sepals; Kangding, wSichuan (PC)
12. *Clematis montana*; cult. (CGW)
13. *Clematis venusta*; near Sandauwan, Yulongxueshan, nwYunnan (CGW)

Clematis macropetala **Ledeb.** (syn. *Atragene macropetala* (Ledeb.) Ledeb.; *Clematis alpina* (L.) Mill. var. *macropetala* (Ledeb.) Maxim.) A very distinctive species climbing to 4 m, often less, with 2-ternate **leaves**; leaflets ovate to lanceolate, 2–5 × 1–4.5 cm, margin toothed. **Flowers** solitary, lantern-shaped, 3–6 cm across; sepals 4, blue, violet-blue or purplish blue, elliptic, 3–4.8 cm long, pointed; staminodes present, outer petal-like, similarly coloured to sepals, but narrower; inner progressively smaller and transitional to the stamens, softly hairy. **Achenes** sparsely hairy with a feathery tail to 45 mm long. Gansu, eQinghai, Shaanxi [n & neChina, eRussia]; open woodland and shrubberies, occasionally on rocks, 900–2,000 m. VI–VIII.

RANUNCULUS

A cosmopolitan genus with c. 550 species (125 in China, 66 endemic), concentrated particularly in northern temperate regions. **Annual** or **perennial herbs** usually with basal and **stem-leaves**, these simple or variously lobed or dissected. **Flowers** solitary or in cymose **inflorescences**; sepals and petals often 5, sometimes fewer or up to 10; sepals usually smaller than petals, usually green; petals showy, generally shiny, often yellow or white, with a nectar-pit inside near base; stamens numerous. **Fruit** a rounded or elongated head of achenes.

Ranunculus similis **Hemsl.*** (syn. *R. involucratus* Maxim.; *Oxygraphis involucrata* (Maxim.) Riedl) **Perennial** herb, not more than 8 cm tall. **Basal leaves** 2–4, rounded to kidney-shaped, 3-lobed or with 3, 5 or 7 rounded teeth, to 10 × 16 mm, leaf-stalk to 3.8 cm long; **stem-leaves** 2–3, crowded below flowers, fan-shaped, stalkless. **Flowers** solitary, 12–14 mm across, 5-parted; sepals elliptic, about half the length of the 5 obovate petals. **Achenes** in a head to 8 mm across, on a glabrous receptacle. Qinghai, nwSichuan, eTibet [nTibet, Xinjiang].

OXYGRAPHIS

A genus of 4 species (all in China) in Central Asia, the Himalaya and China, closely related to *Ranunculus*. Small **tufted perennials** with all **leaves** basal and generally with solitary, scapose **flowers**; sepals 5–8, persistent and enlarging in fruit, occasionally falling, normally green; petals 5–19, yellow, often whitening with age, elliptic to oblong, with a nectary-pit close to the base; stamens numerous. **Fruit** a cluster of achenes on a cone-like receptacle.

1. Sepals persistent and rather leathery; scapes glabrous . . . . ***O. glacialis*, *O. tenuifolia***
1. Sepals deciduous, papery; scapes minutely downy . . . . ***O. delavayi***

Oxygraphis glacialis **Fisch. ex DC.** (syn. *Ranunculus kamchaticus* DC.) Small glabrous plant not more than 5 cm tall in flower, with ovate to elliptic or obovate, stalked **leaves**, blades to 30 × 28 mm, base wedge-shaped to somewhat heart-shaped. **Flowers** 15–26 mm across, borne on scapes 2–5 cm long; sepals 5, oblong to rounded; petals 12–19, 7–12 mm long. s & swGansu, Qinghai, Sichuan, sShaanxi, Tibet, nwYunnan [nwChina, Kazakhstan, Himalaya, Mongolia, eRussia]; short grassy and stony alpine meadows and slopes, moraines, streamsides, open low shrubberies, 2,700–5,000 m. IV–IX.

Oxygraphis tenuifolia **W. E. Evans*** Readily distinguished by the linear to linear-lanceolate **leaves** not more than 4 mm wide; **flowers** only 10–12 mm across, with 9–11 petals. sw & wSichuan, nwYunnan (Zhongdian); similar habitats, 3,400–4,300 m. VI–IX.

Oxygraphis delavayi **Franch.*** Small plant similar to *O. glacialis*, but **leaves** indistinctly 3–9-lobed, to 38 mm long; **flowers** solitary or occasionally 2–3 on a common stem, 11–22 mm across; sepals 5, papery, oblong to narrow-elliptic; petals 5–10, elliptic to oblong, yellow. swSichuan, nwYunnan, neighbouring part of Tibet; gravelly and grassy alpine slopes, 3,500–5,000 m. IV–VIII.

ADONIS

A genus of 30 species (10 in China, 3 endemic) in the temperate northern Old World. Small, herbaceous **perennial** and **annual herbs**, often with scale-like **basal leaves**, the **stem-leaves** finely divided; bracts absent. **Flowers** solitary or in branched cymes, *Ranunculus*-like; sepals 5–8; petals 5–24; stamens numerous. **Fruit** a collection of achenes, often with raised veins.

1. Flowers yellow . . . . ***A. sutchuenensis***
1. Flowers white to pale blue to pale purple . . . . ***A. davidii*, *A. coerulea***

Adonis sutchuenensis **Franch.*** **Tufted perennial** to 40 cm, with or without branched **stems** which have sheath-like **leaves** at base; middle and upper stem-leaves with finely divided pinnate segments, ultimate divisions sharply toothed. **Flowers** erect, deeply cupped, 3.5–4.5 cm across, with 6 pale green sepals and 8–12 yellow, oblong petals. n & neSichuan, sShaanxi; grassy and shrubby places, open forests, 1,100–3,300 m. IV–VI.

Adonis davidii **Franch.*** (syn. *A. brevistyla* Franch.; *A. delavayi* Franch.) **Tufted perennial** forming hummocks to 50 cm, often less, with branched **stems** clothed at base with membranous scales. **Leaves** bright green, ferny, with pinnately divided finely toothed segments. **Flowers** cup- to saucer-shaped, 2–4 cm across; sepals 5–7, whitish or purplish; petals 7–14, oblong, white, generally suffused with purple or lilac outside. sGansu, sGuizhou, wHubei, n, w & swSichuan, sShaanxi, seTibet, nwYunnan; open forests and forest margins, grassy places, streamsides, ravines, 1,900–3,500 m. IV–VII.

Adonis coerulea **Maxim.*** Similar to the previous species, but the plant smaller, to only 15 cm, with spreading **stems**; leaf-blades oblong rather than triangular or pentagonal in outline. **Flowers** 10–18 mm across; sepals 5–7; petals 8–9, purple to pale blue or whitish, narrow-obovate. Gansu, c & seQinghai, n & wSichuan, neTibet; alpine scrub and grassy habitats, 2,300–5,000 m. IV–VII.

CALLIANTHEMUM

A genus of 12 species (5 in China) distributed in Europe and Asia. Small glabrous **perennial herbs**. **Leaves** mainly basal, often rather fleshy, pinnately lobed. **Flowers** solitary with 5 sepals, 5–13 narrow petals and numerous stamens. **Fruit** a collection of achenes.

Callianthemum farreri **W. W. Sm.*** Small **herbaceous perennial** not more than 8 cm tall, with spreading stems and leaves. **Leaves** all basal, only partly developed at flowering time, somewhat fleshy, pinnately lobed, with 2–3 pairs of fan-shaped lateral leaflets. **Flowers** 25–30 mm across; sepals whitish or pale green; petals 8–9, narrow-obovate, white or pale bluish purple, sometimes spotted. sGansu, nwSichuan; grassy slopes, open forests, scrub, 3,500–4,000 m. V–VI.

Callianthemum pimpinelloides **(D. Don) Hook. f. & Thomson** (syn. *C. cashmirianum* Cambess.; *C. tibeticum* Witasek; *Ranunculus pimpinelloides* D. Don) Generally smaller than the preceding species, with **flowers** only 11–14 mm across, sepals 3–6 mm long, petals 5–7, white, pink or pale purple, oblong to almost linear. Qinghai, Sichuan, Tibet, w & nwYunnan [Himalaya, cAsia]; mountain grassland and moors, 3,200–5,600 m. IV–VI.

ANEMONE

A genus with c. 150 species (53 in China, 22 endemic) scattered across the temperate northern hemisphere and in temperate South America. **Tufted**, **rhizomatous** and **tuberous perennials** with undivided to variously divided **basal leaves**; upper **stem-leaves** forming an involucre below the flowers, often bract-like. **Flowers** solitary or in cymes or umbels; sepals often 5–6, to 18, petaloid; petals absent; stamens numerous. **Fruit** a collection of glabrous or variously hairy achenes.

SYNOPSIS OF GROUPS

Group One. Involucral bracts stalked, sometimes shortly so; achenes woolly, glabrous or minutely hairy, p. 100.

Group Two. Involucral bracts stalkless; achenes glabrous or downy; blades of bracts similar in size to basal leaves, achenes glabrous, p. 103.

Group Three. Involucral bracts stalkless; achenes glabrous or downy; blades of bracts smaller than basal leaves, achenes downy, p. 103.

> **Group One**
> 1. Achenes glabrous or minutely hairy ..... *A. rivularis*, *A. rivularis* var. *flore-minore*, *A. davidii*, *A. prattii*
> 1. Achenes woolly 2
> 2. Flowers solitary or paired .... *A. rupicola*
> 2. Flowers numerous ....
> .... *A. hupehensis*, *A. tomentosa*, *A. vitifolia*

Anemone rivularis **Buch.-Ham. ex DC.** (including *A. esquirolii* H. Lévl. & Vaniot; *A. leveillei* Ulbr.) A very variable erect, tufted **perennial** to 60 cm, sometimes taller. **Basal leaves** 3–5, with leaf-stalk up to 15 cm, blade pentagonal to somewhat kidney-shaped in outline, 3–10 × 5–15 cm, with 3 segments, that are further lobed and toothed. Scapes 1–3 per leaf-rosette, each 2–3-branched and many-flowered; sepals 5–10, white, blue, purplish or mauve, elliptic, 10–16 mm long. **Achenes** glabrous. c & sGansu, wGuangxi, Guizhou, swHubei, e & seQinghai, Sichuan, Shaanxi, se & eTibet, Yunnan [swTibet, Himalaya, India, Sri Lanka, Sumatra]; forest margins, shrubberies, grassy places, stream and lake margins, 800–4,900 m. V–VIII. Variants with very small, 5–6-sepalled **flowers** with sepals 6–10 mm long, found mainly in the north of the range, are distinguished as **var.** *flore-minore* **Maxim.***

Anemone davidii **Franch.*** In contrast to *A. rivularis*, a rhizomatous, **patch-forming** plant to 40 cm, with several scale-like **leaves** at stem-base, then 2–5-ternate leaves with lobed segments, almost glabrous or minutely hairy. **Flowers** only 1–3, each subtended by a leaf-like bract; sepals usually 5, white, elliptic to obovate, 15–20 mm long. Guizhou, wHubei, Sichuan, seTibet, nwYunnan; shady places in mixed forests and shrubberies, bamboo thickets, rocky places, streamsides, 1,000–3,500 m. V–VI.

Anemone prattii **Huth ex Ulbr.*** Similar to *A. davidii*, but a smaller plant to 30 cm (generally no more than 20 cm), usually with solitary white flowers, sometimes with a blue or lilac flush on the reverse; sepals 8–10 mm long. w & swSichuan, n & nwYunnan; shady places in forests, humus pockets in rocks, 1,700–2,400 m. IV–V.

Anemone rupicola **Camb.** Rhizomatous, **patch-forming** plant to 30 cm, occasionally more, with stalked, deep green, 3-parted **leaves**, 2–6 × 2–10 cm, segments sharply toothed, sparsely and minutely hairy; involucral bracts similar to basal leaves but very short-stalked. **Flowers** 32–45 mm across; sepals 5, occasionally as many as 9, white, cream, sometimes with a lilac or purple reverse, broadly elliptic to obovate, 15–20 mm long. w & swSichuan, seTibet, nwYunnan [swTibet, Himalaya]; rocky and stony places, rock-ledges and cliffs, open rocky woodland, 2,400–4,200 m. VI–VIII.

1. *Adonis coerulea*; near Henan, eQinghai (RS)
2. *Adonis coerulea*; above Jiuzhaigou, nwSichuan (CGW)
3. *Callianthemum farreri*; above Jiuzhaigou, nwSichuan (CGW)
4. *Anemone rivularis* on the edge of cultivated land; Cangshan, nwYunnan (PC)
5. *Anemone rivularis*; Napahai, Zhongdian, nwYunnan (CGW)
6. *Anemone rivularis*; Yulongxueshan, nwYunnan (CGW)
7. *Anemone davidii*; Emeishan, sSichuan (PC)
8. *Anemone prattii*; cult., from nwYunnan (CGW)
9. *Anemone rupicola*; upper Gangheba, Yulongxueshan, nwYunnan (CGW)
10. *Anemone hupehensis*; Chongjiang, nwYunnan (PC)

1. *Anemone tomentosa*; Jiuzhaigou, nwSichuan (CGW)
2. *Anemone tomentosa*; Laohu hai (Tiger Lake), Jiuzhaigou, nwSichuan (HJ)
3. *Anemone vitifolia*; Dali, nwYunnan (CGW)
4. *Anemone demissa* var. *major*; Gangheba, Yulongxueshan, nwYunnan (CGW)
5. *Anemone demissa* var. *major*; Gangheba, Yulongxueshan, nwYunnan (CGW)
6. *Anemone demissa* form; Huanglongsi, nwSichuan (CGW)
7. *Anemone demissa* var. *major*; Napahai, Zhongdian, nwYunnan (CGW)
8. *Anemone narcissiflora* subsp. *protracta*; Shennongjia, wHubei (PC)
9. *Anemone* aff. *cathayensis*; Balangshan, Wolong, wSichuan (CGW)

Anemone hupehensis **(Lemoine) Lemoine*** (syn. *A. japonica* (Thunb.) Sieb. & Zucc. var. *hupehensis* Lemoine) A rather vigorous stoloniferous, patch-forming **herbaceous perennial**, 30–120 cm tall. **Basal leaves** with stalks up to 35 cm long and with a ternate blade, leaflets toothed (largest c.10 cm long), covered in sparse bristly hairs; involucral bracts similar but short-stalked. **Inflorescence** a 2–3-branched, many-flowered cyme; sepals 5, white, pink, pinkish purple or reddish purple, obovate, 20–30 mm long. nGuangxi, Guizhou, wHubei, Sichuan, sShaanxi, Yunnan [c & sChina]; shrubberies, grassy slopes and banks, streamsides, 400–2,600 m. VII–X. Double-flowered forms sometimes occur in the wild. (See photo 10 on p. 101.)

Anemone tomentosa **(Maxim.) C. P'ei*** (syn. *A. japonica* (Thunb.) Sieb. & Zucc. var. *tomentosa* Maxim.) Similar in general appearance to the previous species, but an even more robust plant to 1.5 m; **leaves** densely white-downy beneath; sepals white or pink, not more than 20 mm long. wHubei, e & seQinghai, w & nwSichuan, Shaanxi; open shrubberies, grassy places, lake and stream margins, 700–3,400 m. VI–IX.

Anemone vitifolia **Buch.-Ham. ex DC.** Readily distinguished from the previous 2 species by 3–5-lobed, not ternate, leaf-blades, 10–20 × 10–20 cm; **flowers** rarely more than 9, sepals white, sometimes with a lilac or purplish reverse, ovate to obovate, 15–20 mm long. swSichuan, seTibet, Yunnan [swTibet, Himalaya]; open woods, shrubberies, grassy places, streamsides, 1,200–2,700 m. VII–X.

Group Two

Anemone demissa **Hook. f. & Thomson** Very variable **perennial** to 30 cm in flower, sometimes as short as 10 cm. Basal **leaves** in a lax, spreading basal rosette, white-hairy leaf-stalks up to 30 cm long, leaf-blades pentagonal in outline or kidney-shaped, 5–8 × 4–6 cm, basically 3-parted, with lobed and toothed, woolly segments; involucral bracts similar, short-stalked. **Inflorescences** 2–8-flowered umbels, ascending, several per leaf-rosette. **Flowers** 18–40 mm across; sepals 5–8, generally blue, purple or red, sometimes white, elliptic to obovate, 8–18 mm long. swGansu, s & seQinghai, nw, w & swSichuan, e & seTibet [swTibet, Himalaya]; open forests and forest margins, shrubberies, grassy places, streamsides, often gregarious, 3,000–4,600 m. V–VII. Varieties that have been recognised include: **var. *major* W. T. Wang*** with larger **leaves** to 12 cm across and scapes 20–50 cm tall. w & swSichuan, seTibet, nwYunnan [swTibet]; and **var. *villosissima* Brühl**, with particularly densely long-woolly leaf-stalks and scapes. sGansu, nw, w & swSichuan, seTibet, nwYunnan [swTibet, Himalaya].

Anemone narcissiflora **L. subsp. *protracta* (Ulbr.) Ziman & Fedor.** Very similar to the previous species, but **inflorescence** erect, generally just 1 per leaf-rosette, occasionally paired, leaf-blades wider than long, 3–7 × 4–12 cm; sepals white or cream, rarely with a pink flush, obovate, 12–18 mm long. wHubei, w & nwYunnan [widespread in northern hemisphere outside China]; *Picea* forests, shrubberies, bamboo thickets, alpine meadows and other grassy places, 1,800–4,000 m. V–VII.

Anemone smithiana **Lauener & Panigrahi** Readily separated from previous 2 species by the larger **flowers**, sepals usually 5, bright pink or purple-pink, 15–20 mm long; leaf-blades wider than long. seTibet [swTibet, c & eHimalaya]; grassy banks, streamsides, open shrubberies, 3,800–4,300 m. VI–VII.

Anemone aff. *cathayensis* **Kit. ex Ziman & Kad.*** **Tufted** plant to 40 cm, rather similar to the previous 2 species, but **leaves** rounded in outline, similar in size to, or smaller than, the deeply lobed involucral bracts. **Flowers** 4–6, nestling in centre of involucral ruff; sepals white with pinkish lilac in centre and on reverse. wSichuan (Balangshan); alpine meadows, streamsides, 3,400–3,800 m. VII. *A. cathayensis* is recorded from the provinces of Hebei, Henan, Shanxi, as well as being native to Korea.

Group Three

1. Leaf-bases of rosette-leaves truncated or slightly heart-shaped . . . . **A. yulongshanica**, **A. yulongshanica var. truncata**
1. Leaf-bases of rosette-leaves heart- to wedge-shaped or attenuate 2
2. Leaves with closely overlapping segments and lobes . . . . **A. imbricata**
2. Leaves without closely overlapping segments and lobes 3
3. Leaves 2-ternately divided; achenes glabrous, borne on an ellipsoid receptacle . . . . **A. rupestris**, **A. rupestris subsp. gelida**
3. Leaves 3-lobed to entire; achenes usually downy, borne on a globose receptacle 4
4. Blades of basal leaves wider than long, rounded or heart-shaped at base 5
4. Blades of basal leaves generally longer than wide, attenuate or wedge-shaped at base . . . . **A. trullifolia**, **A. coelestina**
5. Stalks of basal leaves 1–2 mm wide; bracts usually divided; sepals downy . . . . **A. obtusiloba**, **A. patula**, **A. polycarpa**
5. Stalks of basal leaves more than 2 mm wide; bracts often undivided, sometimes 3-lobed; sepals glabrous or minutely downy . . . . **A. rockii**, **A. rockii var. pilocarpa**, **A. subpinnata**, **A. geum**

Anemone yulongshanica **W. T. Wang*** Closely allied to *A. trullifolia*, differing primarily in the leaf-base characteristic noted above, **leaves** with 3 main divisions or 3-lobed, with the central one itself 3-lobed; bracts 3-lobed. **Flowers** solitary, 16–30 mm across, with 6 white oval sepals, and silky-hairy carpels. swSichuan, nwYunnan; open coniferous forests, rocky and grassy places, 2,600–3,900 m. V–VII. **Var. *truncata* (Comber) W. T. Wang*** differs in the short, densely hairy leaf-stalks, unlobed or only slightly lobed middle leaf-segments and in the 5–6-sepalled white, blue or primrose-yellow **flowers**. Confined to swSichuan, nwYunnan (Lijiang area).

Anemone imbricata **Maxim.*** (syn. *Anemone obtusiloba* D. Don subsp. *imbricata* (Maxim.) Brühl) Small **rosette-forming perennial** not more than 12 cm tall in flower, with up to 7, 3-parted, softly hairy basal **leaves**, central segment short-stalked, the lateral unevenly 3-parted. **Scapes** up to 5, 1–2-flowered, with involucre of small, 3-parted, leaf-like bracts; sepals 5–9, purple to blackish purple, obovate, 8–13 mm long. swGansu, Qinghai, nw & wSichuan, eTibet; open shrubberies, grassy places, 3,200–5,300 m. V–VII.

Anemone rupestris **Wall. ex Hook. f. & Thomson** Very variable, small, **rosette-forming**, somewhat hairy **perennial** to 20 cm. Basal **leaves** almost glabrous, rounded in outline, with 3 main segments that are further divided into 3 lobes, central segment long-stalked; bracts undivided. **Flowers** solitary, small, 10–14 mm across; sepals 6–8, white or cream, occasionally reddish, outer broader than inner; filaments somewhat dilated in the middle. w & swSichuan, seTibet, nwYunnan [swTibet, c & eHimalaya]; shrubberies, especially of *Rhododendron*, alpine meadows, rocky places, 2,700–4,800 m. V–VIII. **Subsp. *gelida* (Maxim.) Lauener** is smaller in most parts, with 3-lobed bracts. Similar distribution but apparently at higher altitudinal range, 4,800–5,000 m.

Anemone obtusiloba **D. Don** Very variable **rosette-forming perennial**, often densely hairy, 5–20 cm tall. Basal **leaves** generally 3-segmented, with heart-shaped base, segments further lobed and toothed; bracts 3-parted to 3-toothed or sometimes undivided, central segment with short stalk. Scapes up to 5, bearing 1–3 **flowers**, each 16–24 mm across; sepals 5–6, white, blue, mauve, purplish or yellow, elliptic, downy beneath; filaments linear. Gansu, seQinghai, sw & wSichuan, Shaanxi, e & seTibet, nwYunnan [swTibet, Himalaya, nMyanmar]; alpine meadows, open shrubberies, forest margins, occasionally in coniferous forests, banks, 2,900–4,000 m. V–VIII.

Anemone patula **C. C. Chang ex W. T. Wang*** Very similar to the preceding species, but with all basal leaf-segments stalkless; **flowers** solitary with filaments that are clearly dilated in middle region. seQinghai, nw & wSichuan; similar habitats, 3,500–4,000 m. V–VII.

Anemone polycarpa **W. E. Evans*** Also very like *A. obtusiloba*, but base of basal **leaves** rounded rather than heart-shaped, central segment with stalk 10–15 mm long; **flowers** with staminodes present (absent in the preceding 2 species); achenes with hooked styles (straight or curved in related species). swSichuan, seTibet, nwYunnan; grassy and rocky places, open shrubberies, 3,600–4,800 m. V–VIII.

Anemone rockii **Ulbr.** (syn. *A. obtusiloba* D. Don subsp. *rockii* (Ulbr.) Lauener). **Rosette-forming perennial**, rather sparsely downy and becoming hairless, to 30 cm. Basal **leaves** with 3, broad-ovate, toothed or somewhat lobed segments, central segment with a stalk 1–2 mm long. **Flowers** solitary on erect to ascending scapes, 14–24 mm across; sepals 7–8, oblong, white, occasionally blue, outer larger than inner; filaments linear. **Achenes** glabrous. swGansu, seQinghai, w & nwSichuan, seTibet, nwYunnan [swTibet, Himalaya]; grassy and rocky slopes, open shrubberies, 2,100–4,000 m. V–VIII. **Var. *pilocarpa* W. T. Wang*** from swSichuan, nwYunnan at 2,100–2,300 m, has **flowers** with 8–9 sepals and densely hairy achenes.

Anemone subpinnata **W. T. Wang*** Differs from both *A. rockii* and *A. obtusiloba* in the pinnately lobed basal **leaves** and in the normally undivided, occasionally 3-lobed bracts. **Flowers** can be white, blue, purplish or violet; filaments linear. swSichuan (Muli); alpine meadows, 3,200–3,750 m. V–VII.

Anemone geum **H. Lévl.** (including subsp. *ovalifolia* (Brühl) Chaud.) Similar to *A. rockii*, but with more hairy **leaves**, central segment with a stalk 5–10 mm long; **flowers** on spreading to ascending scapes, 10–22 mm across, sepals often 5 (but sometimes up to 8), white, yellow, mauve or blue, ovate, even; filaments slightly dilated in the middle. sGansu, seQinghai, nw, w & swSichuan, eTibet, nwYunnan [c & eHimalaya]; alpine meadows, open scrub, moorland, 2,000–4,800 m. V–IX.

Anemone trullifolia **Hook. f. & Thomson** (syn. *A. obtusiloba* D. Don subsp. *trullifolia* (Hook. f. & Thomson) Brühl) A very variable **rosette-forming perennial**, often densely hairy, to 25 cm. Basal **leaves** wedge-shaped or rhombic, with an attenuate base, sometimes obovate, 3-lobed to unlobed, often with a few teeth towards top; **bracts** similar to leaves, unstalked. **Flowers** 1–3 on spreading to ascending scapes, 12–24 mm; sepals 5–7, white, blue, yellow or reddish, downy beneath, sometimes with a dark spot at base; filaments dilated towards base. **Achenes** hairy. Gansu, seQinghai, Sichuan, e & seTibet, w & nwYunnan [swTibet, eHimalaya]; open forests, forest margins, meadows, stony slopes, 2,500–4,800 m. V–IX.

Anemone coelestina **Franch.*** (syn. *A. obtusiloba* D. Don subsp. *coelestina* (Franch.) Brühl) Very like *A. trullifolia*, but basal **leaves** generally unlobed, oblong to oblong-linear and scapes 1-flowered; sepals violet-blue or reddish blue, rarely white; filaments linear. sGansu, nw, w & swSichuan, seTibet, nwYunnan [swTibet]; alpine meadows, open shrubberies, amongst rhododendrons, 2,500–4,800 m. V–IX.

1. *Anemone imbricata*; Zhaorigenshan, seQinghai (RS)
2. *Anemone rupestris*; Napahai, Zhongdian, nwYunnan (CGW)
3. *Anemone rupestris*; Napahai, Zhongdian, nwYunnan (CGW)
4. *Anemone obtusiloba*; eBhutan (CGW)
5. *Anemone obtusiloba*; Yulongxueshan, nwYunnan (CGW)
6. *Anemone rockii*; north of Songpan, nwSichuan (CGW)
7. *Anemone rockii*; seQinghai (RS)
8. *Anemone geum*; near Songpan, nwSichuan (CGW)
9. *Anemone trullifolia*; cult., from nwYunnan (CGW)
10. *Anemone coelestina*; between Songpan and Huanglong, nwSichuan (CGW)

1. *Anemoclema glaucifolium*; Habashan, nwYunnan (AD)
2. *Hepatica henryi*; Shennongjia, wHubei (PC)
3. *Pulsatilla chinensis*; cult., from neChina (CGW)
4. *Thalictrum delavayi*; Gangheba, Yulongxueshan, nwYunnan (CGW)
5. *Thalictrum delavayi* var. *acuminatum*; Tianbao, nwYunnan (PC)
6. *Thalictrum alpinum*; unknown locality (AGS)

ANEMOCLEMA

A genus with a single species endemic to south-west China, closely related to *Anemone*, but with pinnately lobed **leaves**.

Anemoclema glaucifolium (Franch.) W. T. Wang* (syn. *Anemone glaucifolia* Franch.) Tufted **herbaceous**, rhizomatous **perennial** to 60 cm, occasionally more, with lax rosettes of glaucous, slightly hairy, oblanceolate, pinnately lobed **leaves**, to 25 cm long, with 6–9 pairs of lateral lobes, these further lobed or toothed. **Inflorescence** scapose, usually with 1–3 flowers; **flowers** cupped, 4–9 cm across; bracts 3 in a whorl, lanceolate, pointed, toothed or not; sepals 5, mauve to lilac-pink, ovate, silky outside; stamens numerous. **Achenes** hairy. swSichuan, nwYunnan; open forests, rocky places, grassy slopes, 1,700–3,000 m. VI–VII.

HEPATICA

A genus with 7 species (2 in China, 1 endemic) in the temperate northern hemisphere. Similar to *Anemone*, but differing primarily in the calyx-like **bracts** immediately beneath, and close to, the flowers.

Hepatica henryi (Oliv.) Steward* Small tufted, rhizomatous, hairy plant to 12 cm, with rounded to kidney-shaped, shallowly to deeply 3-lobed **leaves**, lobes pointed. **Bracts** ovate, 5–11 mm long, entire to 3-toothed. **Flowers** 18–25 mm across; sepals 6, white, oblong to elliptic, 8–12 mm long. wHubei, Sichuan, Shaanxi [nHunan]; forests, grassy slopes, 1,300–2,500 m. IV–V.

PULSATILLA

A genus of 11 species (2 in China) in the temperate northern hemisphere. Similar to *Anemone*, but **leaves** palmately or pinnately divided, and **achenes** adorned with a long feathery appendage (persistent style).

Pulsatilla chinensis (Bunge) Regel (syn. *Anemone chinensis* Bunge) Tufted **perennial** to 35 cm tall in flower, taller in fruit. **Leaves** in basal rosettes, to 14 cm long, with 3 leaflets; leaflets ovate, 2–3-lobed, densely hairy beneath like the stalks, glabrous above. **Flowers** on hairy scapes above the leaves, pendent or semi-pendent, with a ruff of 3-lobed bracts beneath; sepals erect to semi-spreading, violet to violet-blue, 28–46 mm long; anthers yellow. **Achenes** with a style to 6.5 cm long. sGansu, Hubei, se & eQinghai, wSichuan, Shaanxi [n & neChina, Korea, Mongolia]; forest margins, grassy slopes, 1,200–3,200 m. VI–VII.

Pulsatilla millefolium (Hemsl. & Wils.) Ulbr.* (syn. *Anemone millefolium* Hemsl. & Wils.) Markedly different from preceding species, a **tufted** plant to 20 cm in flower, with 3-pinnately divided **leaves** (like those of carrots), slightly hairy beneath. **Flowers** small, with a ruff of leaf-like bracts immediately below; sepals pale greenish yellow, yellowish or reddish purple, 10–20 mm long; anthers yellow. **Achenes** with a feathery style 30–35 mm long. n, s & wSichuan, neYunnan; open forests, shrubberies, rocky slopes, roadsides, 2,200–3,300 m. V–VII.

THALICTRUM

A genus of c. 150 species (about half in China, 49 endemic) found worldwide. **Perennial herbs** with erect to arching **stems** and ternately or pinnately compound **leaves**, lower stalked, upper usually stalkless. **Flowers** few to many in spike-like or paniculate, cymose or umbellate **inflorescences**; sepals 4–10, sometimes large, colourful and petal-like, but often small, greenish or whitish and soon falling; true petals absent; stamens up to 70, prominent, sometimes with brightly coloured filaments. **Fruit** a collection of small achenes, with or without a persistent style.

1. Sepals large and showy, longer than stamens . . . . *T. delavayi*, *T. delavayi* var. *acuminatum*, *T. delavayi* var. *decorum*, *T. grandiflorum*
1. Sepals small, often rather inconspicuous or quickly falling, shorter than or equalling stamens 2
2. Inflorescence spike-like . . . . *T. alpinum*
2. Inflorescence a diffuse panicle . . . . *T. yunnanense*, *T. cultratum*, *T. macrorhynchum*

Thalictrum delavayi Franch.* Lanky, glabrous plant to 2 m, but as short as 60 cm, with branched **stems**, **basal leaves** generally withered by flowering time. **Leaves** 3–4-pinnate, stalked; leaflets ovate to rounded to obovate or elliptic, to 30 × 25 mm, with a rounded to heart-shaped or somewhat wedge-shaped base, generally 3-lobed in upper part. **Inflorescence** a large, rather diffuse panicle, usually many-flowered; sepals 4–5, purple or bluish purple, oblong to elliptic, 6–11 × 2.2–5 mm; stamens 5–7 mm long. **Achenes** 15–22, with a short style. wGuizhou, w & swSichuan, seTibet, Yunnan; open forests, forest margins, shrubberies, damp rocky slopes, stream- and ditchsides, 1,800–3,400 m. VI–IX. **Var.** *acuminatum* Franch.* has narrowly elliptic to lanceolate sepals 9–14 × 4–5 mm. swSichuan, nwYunnan, c. 1,800 m. **Var.** *decorum* Franch.* has broader ovate sepals, mostly 5–8 mm wide. nwYunnan, c. 3,000 m.

Thalictrum grandiflorum Maxim.* Similar to the previous species, but a much smaller plant not more than 40 cm with ternately compound **leaves**, leaflets often larger, to 23 × 14 mm; sepals pink, narrow-ovate, 13–20 × 5–10 mm. sGansu, n & nwSichuan; grassy and rocky slopes, shrubberies, c. 1,000 m. VII–X.

Thalictrum alpinum L. Rather delicate, glabrous plant to 40 cm, but sometimes only 10 cm. **Leaves** all basal, pinnate with 2-ternate divisions, leaflets ovate to rounded or obovate, to 5 × 5 mm, occasionally larger, 3- or 5-lobed at apex. **Flowers** in a raceme-like, sometimes branched, **inflorescence**; sepals green flushed purple, oval, 2–3 mm long, quickly falling; stamens 7–12, longer than sepals, with purplish filaments and yellow anthers. Gansu, Qinghai, Sichuan, Shaanxi, Tibet, Yunnan [Europe, Asia including Himalaya]; rocky and grassy slopes, meadows, damp places, moraines, 2,400–5,300 m. VI–VIII.

Thalictrum yunnanense W. T. Wang* **Tufted**, downy plant to 60 cm, often half that height, with smooth, erect, branched, leafy **stems**. **Leaves** 2–3-ternate, leaflets rounded to obovate, to 30 × 40 mm, prominently veined beneath, base rounded to wedge-shaped, apex 3-lobed. **Flowers** in terminal and lateral clusters; sepals greenish white, ovate to elliptic, 3–4 mm long; stamens numerous, filaments whitish, dilated at top and as broad as the anthers; stigmas purple. **Achenes** 10–50. Yunnan; open forests, forest margins, ravines, c. 2,000 m. VI–VIII.

Thalictrum cultratum Wall. Similar to *T. yunnanense*, but a bigger plant to 1.2 m, basal and lower **stem-leaves** generally withered by flowering time, leaflets not normally more than 10 × 10 mm; **flowers** in lax panicles, nodding; filaments purplish, linear, not dilated, narrower than anthers; achenes in clusters of 4–9. Gansu, Sichuan, s, se & eTibet, nwYunnan [eHimalaya]; rocky places, meadows, open scrub, but normally in wet habitats, 1,700–3,800 m. VI–VII.

Thalictrum macrorhynchum Franch.* Similar to *T. yunnanense*, but plant glabrous, leaflets with a rounded to heart-shaped base; sepals white; filaments dilated at top and broader than anthers; **achenes** 10–20. Gansu, Hubei, n, ne & eSichuan, Shaanxi; habitats similar to those of *T. yunnanense* as well as shrubberies, 900–2,900 m. VI–VII.

CALTHA

A genus of 15 species (4 in China) scattered in the colder regions of the northern and southern hemispheres. Tufted **perennial herbs** primarily with heart- or kidney-shaped **leaves**, these all basal, or basal and on the **stem**. **Flowers** solitary, sometimes scapose, or several in a lax to rather dense, often branched **inflorescence**; sepals large and petal-like but true petals absent; stamens numerous, with linear filaments. **Fruit** a collection of up to 40 follicles.

1. Flowers clustered, generally 4 or more per inflorescence, borne on leafy stems .... *C. palustris*, *C. palustris* var. *barthei*, *C. palustris* var. *himalaica*, *C. palustris* var. *umbrosa*
1. Flowers mostly solitary or occasionally paired, often scapose .... *C. scaposa*, *C. sinogracilis*, *C. sinogracilis* forma *rubriflora*

Caltha palustris L. An extremely variable and complex species in the typical form (var. *palustris*), forming clumps 40–70 cm tall, with hollow **stems** 6 mm in diameter at flowering. **Leaves** almost orbicular to kidney-shaped or somewhat triangular, to 7 × 9 cm, with a deep heart-shaped base, dull or glossy green, **basal leaves** long-stalked with a toothed margin; **stem-leaves** similar, diminishing in size upwards, short-stalked to stalkless. **Flowers** yellow, 3–7 per cluster, in a simple or branched inflorescence, 22–46 mm across; sepals 5, occasionally 6–7, obovate, 10–22 mm long. **Fruit** with 7–25 follicles. s & swGansu, Guizhou, Sichuan, e & seTibet, n, nw & cYunnan [swTibet, Europe, Asia]; marshy places, damp grassy meadows, stream and lake margins, open forests, seepage zones, sometimes growing in shallow water, 600–4,000 m. V–VIII. **Var.** *barthei* Hance* (syn. *C. fistulosa* Schipcz.) is a more robust plant, generally 80–120 cm tall, with thick, hollow **stems** to 12 mm in diameter at flowering and leaves of ± equal size, the **flowers** yellow or red. swGansu, s, w & nwSichuan, eTibet, nwYunnan; 1,000–3,800 m. **Var.** *himalaica* Tamura*, by contrast, has the dimensions of var. *palustris* but solid **stems** and **leaves** generally rounded with a deep heart-shaped base. seTibet [swTibet]; 2,800–3,100 m. **Var.** *umbrosa* Diels* is like var. *palustris* in dimensions, although often on the small side, the uppermost **stem-leaves** palmately divided with rather pointed divisions, and the **inflorescence** less dense with each branch often ending in a solitary flower. swSichuan, nw & cYunnan; 2,500–3,000 m.

Caltha scaposa Hook. f. & Thomson Variable low, **tufted** plant to 15 cm tall in flower, twice the height in fruit; **stems** simple or somewhat branched. **Leaves** mostly basal, glossy green, ovate- to heart-shaped, margin entire to somewhat scalloped or rarely toothed, apex rounded; **stem-leaves**, when present, smaller and short-stalked or stalkless. **Flowers** yellow, generally solitary, sometimes paired, saucer-shaped, 20–36 mm across; sepals 5–8, ovate to obovate, occasionally elliptic, 9–17 mm long. **Fruit** with 6–11 (occasionally more) short-stalked follicles. s & swGansu, s & seQinghai, Sichuan, e & seTibet, nwYunnan [eHimalaya, nMyanmar]; wet mountain and valley meadows, lake and stream margins, open shrubberies, 2,800–4,100 m. VI–IX.

Caltha sinogracilis W. T. Wang* Readily distinguished from the previous species by its smaller stature, the **leaves** more kidney-shaped, not more than 17 × 35 mm, margin toothed. **Flowers** always solitary and scapose, petals yellow, elliptic, 9–14 mm long; follicles 5–10, stalkless. seTibet, nwYunnan; similar habitats, 3,200–4,100 m. VI–IX. Plants with red, reddish purple or magenta **flowers** are named forma *rubriflora* (B. L. Burtt & Lauener) W. T. Wang* (syn. *C. rubriflora* B. L. Burtt & Lauener).

1. *Thalictrum yunnanense*; Baishui, Yulongxueshan, nwYunnan (PC)
2. *Thalictrum macrorhynchum*; near Aba, nwSichuan (CGW)
3. *Caltha palustris*; Balangshan, Wolong, wSichuan (CGW)
4. *Caltha palustris* var. *umbrosa*; Yulongxueshan, nwYunnan (CGW)
5. *Caltha palustris* var. *barthei*; Jiuzhaigou, nwSichuan (CGW)
6. *Caltha palustris* var. *barthei*; Jiuzhaigou, nwSichuan (CGW)
7. *Caltha palustris* var. *barthei*; north of Songpan, nwSichuan (HJ)
8. *Caltha palustris* var. *barthei*; north of Songpan, nwSichuan (HJ)
9. *Caltha scaposa*; Baimashan, nwYunnan (CGW)
10. *Caltha scaposa*; Huanglong, nwSichuan (CGW)
11. *Caltha sinogracilis* forma *rubriflora*; Galung La, near Bomi, seTibet (HJ)

TROLLIUS

A genus of c. 30 species (16 in China, 8 endemic) scattered in north temperate regions. **Perennial herbs** with palmately lobed leaves, mainly in basal tufts. **Flowers** buttercup-like, usually yellow, solitary or several in a cyme; sepals 5–many, petaloid, with a nectary-groove close to the base; petals present, generally narrow, usually shorter than sepals; stamens numerous. **Fruit** a cluster of 5–many follicles.

1. Flowers usually 2–3; petals 7–8 mm long . . . .
   . . . . **T. buddae, T. yunnanensis**
1. Flowers solitary; petals 2–6 mm long 2
2. Sepals with a purple or purple-brown reverse . . . .
   . . **T. pumilus, T. farreri, T. farreri var. major, T. vaginatus**
2. Sepals with a yellow reverse . . . .
   . . . . **T. ranunculoides, T. micranthus**

Trollius buddae **Schipcz.*** (syn. *T. stenopetalus* Stapf) Tufted, erect **perennials**, usually to 70 cm with 1–3 basal, long-stalked **leaves**, blade pentagonal in outline and with 3 main lobes, these further lobed and toothed; stem-leaves usually 2–3, uppermost smaller than lower. **Flowers** 26–35 mm across, plain bright yellow; sepals 5, obovate to elliptic, 12–16 mm long; petals linear-elliptic, 7–8 mm long, about as long as stamens. **Fruit** with 20–30 follicles, with curved styles. sGansu, w & nSichuan, sShaanxi; damp grassy and marshy meadows, open shrubberies, 1,800–2,400 m. VI–VII.

Trollius yunnanensis **(Franch.) Ulbr.*** Similar to the previous species, but **flowers** larger, generally 40–55 mm across; sepals 5–7, broadly obovate with a rounded, truncated or slightly notched apex, 17–25 mm long; petals shorter than stamens; follicles 7–25, with an erect, straight style. sGansu, Sichuan, w & nwYunnan; habitats similar to those of *T. buddae*, 1,900–3,900 m. VI–IX.

Trollius pumilus **D. Don** Variable, small **tufted** plant, generally 5–15 cm tall at flowering time, taller in fruit. **Basal leaves** 1–3 with a stalk up to 5 cm, blade pentagonal in outline, with a heart-shaped base, basically 3-parted, segments slightly lobed, toothed; **stem-leaves** 2–3, similar to basal, generally near base of stem. **Flowers** solitary, buttercup-like, 15–22 mm across, yellow or whitish inside, with purplish reverse; sepals 5, ovate to obovate, 6.5–10 mm long, apex rounded; petals linear-spathulate, small, not more than 3 mm long, shorter than stamens. **Fruit** with 6–16 follicles. s & swGansu, e & seQinghai, Sichuan, Tibet [Himalaya, nMyanmar]; damp grassy places and marshes, open shrubberies, 2,300–4,800 m. Late VI–VIII.

Trollius farreri **Stapf*** Similar to the previous species, but **leaves** more deeply cut, with neat triangular marginal teeth, all basal or sub-basal (**stem-leaves** in middle of stem in **var. major W. T. Wang**). **Flowers** often larger, 18–34 mm across; sepals dark purple on reverse; petals 4–5 mm long; follicles 6 or more. sGansu, e & seQinghai, w & nwSichuan, sShaanxi, e & seTibet, nwYunnan; habitats similar to those of the typical variety, 3,500–4,700 m. VI–VII.

Trollius vaginatus **Hand.-Mazz.*** Similar to *T. farreri*, but **stem-leaves** generally long-stalked, the central segment 3-lobed almost to base; **flowers** 25–36 mm across, generally purplish brown on reverse, petals almost as long as stamens; follicles only 4–6. swSichuan, nwYunnan; similar habitats, 3,000–4,200 m. VI–VII.

Trollius ranunculoides **Hemsl.*** Small **tufted perennial** 6–18 cm tall, taller in fruit. **Basal leaves** long-stalked, blade rounded or pentagonal in outline, basically 3-lobed, the lobes further lobed and toothed; **stem-leaves** 2–3, smaller than basal. **Flowers** buttercup-like, solitary, 22–36 mm across; sepals 5, occasionally 6–8, obovate with a rounded to truncated apex, 10–17 mm long; petals linear-spathulate, 4–6 mm long, shorter than stamens. **Fruit** with 7–9 follicles, each with a persistent, erect style. sGansu, s & seQinghai, w & nwSichuan, eTibet, nwYunnan; grassy slopes and meadows, 2,900–4,100 m. V–VII.

Trollius micranthus **Hand.-Mazz.*** Differs from the previous species in having only sub-basal **stem-leaves**, and smaller **flowers**, only 12–15 mm across with sepals narrow-obovate to oblong, 5–8 mm long. seTibet, nwYunnan (Zhongdian-Deqin); habitats similar to those of *T. ranunculoides*, 3,900–4,200 m. VI–VIII.

PARAQUILEGIA

A genus from eastern and Central Asia with 5 species (3 in China). Tufted **perennial cliff-dwellers**, with short stout rhizomes covered with old leaf-base remains. **Leaves** long-stalked, ternately compound, often grey-green or bluish green. **Flowers** cup- or saucer-shaped, solitary, with a pair of bracts towards the top of flower-stalks; sepals 5, petal-like; petals 5, sometimes more, reduced to a ring of yellow or orange, small, pouched nectaries surrounding the boss of many stamens. **Fruit** a head of 5–8 follicles.

---

1. *Trollius buddae*; Balangshan, Wolong, wSichuan (CGW)
2. *Trollius yunnanensis*, with *Primula poissonii*; Napahai, Zhongdian, nwYunnan (CGW)
3. *Trollius pumilus*; north of Songpan, nwSichuan (CGW)
4. *Trollius pumilus*; north of Jigzhi, seQinghai (CGW)
5. *Trollius farreri* var. *major*; near Aba, nwSichuan (CGW)
6. *Trollius farreri*; Danglexiaoshan, seQinghai (RS)
7. *Trollius ranunculoides*; Wutoudi, Yulongxueshan, nwYunnan (CGW)

Paraquilegia microphylla **(Royle) Drumm. & Hutch.** Very variable **perennial** to 30 cm, often less, forming dense tufts. **Leaves** 2-ternate, glabrous; leaflets green or glaucous, fan-shaped to rhombic, 3-parted or -lobed, often further divided or lobed, not more than 10 mm long. **Flowers** solitary, deeply cupped, semi-pendulous to ascending; sepals 5, prominent, blue, purplish, lilac or white, oval to obovate, 14–25 mm long; petals much smaller, often reduced to yellow or orange nectaries. Follicles 5–8; **seeds** brown, smooth, with a narrow wing. swGansu, n & wSichuan, e & seTibet, nwYunnan [cAsia, Himalaya]; rock-crevices and -ledges, occasionally on screes, 2,700–4,300 m. V–VII.

Paraquilegia anemonoides **(Willd.) Ulbr.** Very similar to the previous species, but **flowers** tending to be rather smaller, sepals bluish or purplish. **Seeds** grey-brown, densely wrinkled. n & cGansu, n & eQinghai [Himalaya, cAsia]. VI–VII. Often considered to be conspecific with *P. microphylla*.

DICHOCARPUM

A genus of 15 species (11 in China, all endemic) in the Himalaya and eastern Asia. **Rhizomatous perennial**, with thin, few-leaved **stems**. **Leaves** with 3 primary divisions, usually with up to 13 leaflets. **Flowers** in small cymes, usually white, with 5 petaloid sepals and 5 much smaller yellow or golden petals; stamens 5–25. **Fruit** a pair of follicles united at the base.

Dichocarpum fargesii **(Franch.) W. T. Wang & P. K. Hsiao*** Glabrous, rather pale plant to 35 cm, generally less, with few **basal leaves**, these long-stalked with 5–9 ovate leaflets, 5–9 × 5–9 mm, bluntly lobed or toothed; **stem-leaves** 2, opposite, like the basal but short-stalked. **Flowers** white, 2–4 on delicate stalks, 8–10 mm across; sepals elliptic, 4–5 mm long; petals funnel-shaped, yellow, not more than 2.5 mm long; stamens usually 10. Follicles linear, 12–15 mm long, widely diverging. seGansu, Guizhou, wHubei, eSichuan, sShaanxi [cChina]; shady damp habitats, in forests and amongst rocks, 1,300–1,600 m. V–VI.

CIMICIFUGA

A genus of c. 18 species (8 in China, 3 endemic) distributed in the temperate northern hemisphere. **Perennial herbs** with ternately or pinnately compound **leaves**. **Inflorescence** a slender, often dense, raceme, generally 8–45 cm long, branched or not, bearing small **flowers**; sepals 4–5, petaloid, white; petals often with 2 empty anther-cells; stamens numerous, with yellow anthers. **Fruit** consisting of 1–8 beaked follicles.

1. Inflorescence well branched . . **C. foetida**, **C. brachycarpa**
1. Inflorescence generally unbranched or with 1 or 2 short branches towards base . . . **C. simplex**, **C. yunnanensis**

Cimicifuga foetida **L.** Very variable **rhizomatous perennial** to 2 m with erect, branched **stems** bearing compound-pinnate **leaves**, the lower long-stalked, upper short-stalked or stalkless; leaflets ovate, the terminal rhombic, 2.5–10 × 1–7 cm, margin lobed and toothed. **Inflorescence** racemose, to 45 cm long with up to 20 branches; **flowers** only c. 4 mm; sepals white or greenish white, 3–4 mm long; petals elliptic, c. 3 mm long, entire or 2-lobed; carpels grey-downy. Follicles 2–5, oblong, 8–14 mm long, downy. Gansu, Hubei, Qinghai, Sichuan, Shaanxi, Tibet, Yunnan [cChina, Himalaya, Mongolia, Myanmar, eRussia]; forests, forest margins, ravines, grassy slopes, 1,700–3,600 m. VII–IX.

Cimicifuga brachycarpa **P. K. Hsiao*** (syn. *Actaea brachycarpa* (P. K. Hsiao) Compton) Similar to *C. foetida* but **inflorescence** with no more than 6 branches, carpels and follicles glabrous or almost so; follicles usually 2–3. Hubei, Sichuan, neYunnan (Zhenxiong Xian) [Henan, Shanxi]; habitats similar to those of *C. foetida*, c. 2,000 m. VII–IX.

Cimicifuga simplex **(DC.) Wormsk. ex Turcz.** Robust **rhizomatous perennial** to 1.5 m often less, with compound-pinnate **leaves**, the lower long-stalked; leaflets lanceolate to ovate, the terminal often rhombic, 3–8.5 × 1.5–5.5 cm, often 3-lobed, toothed, with a few white hairs on the veins above. **Inflorescence** a simple raceme to 35 cm long, occasionally with several short branches at base, rachis and flower-stalks (mostly 5–10 mm long) white- or grey-downy; sepals elliptic, c. 4 mm long; petals smaller, 2-lobed. Follicles 2–8, 7–9 mm long, glabrous or appressed-hairy. Gansu, Sichuan, Shaanxi [n, e & cChina, Japan, Korea, Mongolia, eRussia]; forest margins, shrubberies, grassy slopes, open scrub, to 3,200 m. VIII–IX.

Cimicifuga yunnanensis **P. K. Hsiao*** Smaller plant than *C. simplex*, 40–90 cm tall, stems densely downy above, leaflets not exceeding 3.7 × 3 cm and somewhat downy above; leaf-stalks 1–4 mm long. nwYunnan; forest margins, grassy slopes, 2,900–4,100 m. VII–VIII.

ACTAEA

A genus of c. 8 species (2 in China) distributed in north temperate regions. Very similar to *Cimicifuga*, but **inflorescences** short, not more than 5 cm long, and **fruit** a berry.

Actaea asiatica **H. Hara** Clump-forming herbaceous, rhizomatous **perennial** to 80 cm, with long-stalked, compound-pinnate **leaves** to c. 30 cm wide, upper smaller than lower; leaflets ovate (terminal one often almost rhombic), 4–10 × 2–8 cm, almost glabrous above, margin toothed. **Inflorescence** racemose, 2–5 cm long in flower but elongating in fruit; sepals obovate, 2.5–3 mm long, white; petals shorter, yellowish. **Berries** purplish black, 5–6 mm, on thickened stalks. Gansu, Hubei, Qinghai, Sichuan, Shaanxi, e & seTibet, Yunnan [n & cChina, Japan, Korea, eRussia]; forests, shrubberies, to 3,100 m. V–VI.

1. *Paraquilegia microphylla*; Baimashan, nwYunnan (CGW)
2. *Paraquilegia microphylla*, Baimashan, nwYunnan (HJ)
3. *Paraquilegia microphylla*; Baimashan, nwYunnan (CGW)
4. *Paraquilegia microphylla*; Gangheba, Yulongxueshan, nwYunnan (CGW)
5. *Paraquilegia microphylla*; Balangshan, Wolong, wSichuan (CGW)
6. *Dichocarpum fargesii*; Shennongjia, wHubei (PC)
7. *Actaea asiatica*; Wolong, wSichuan (PC)

1. *Aconitum hemsleyanum*; Gangheba, Yulongxueshan, nwYunnan (PC)
2. *Aconitum gymnandrum*; Kangyang, swGansu (HJ)
3. *Aconitum gymnandrum*; Kangyang, swGansu (HJ)
4. *Aconitum sinomontanum*; east of Aba, nwSichuan (CGW)
5. *Aconitum brachypodum*; Zhongdian, nwYunnan (PC)
6. *Aconitum carmichaelii*; cult., from Kunming, cYunnan (PC)

ACONITUM

A genus with 400 species (211 in China, 166 endemic) distributed in the temperate northern hemisphere. Primarily **herbaceous perennials**, some climbing with twining **stems**, occasionally **annuals**, with palmately divided, occasionally undivided **leaves**. **Flowers** solitary or in racemes, sometimes branched, zygomorphic; sepals 5, petal-like, uppermost generally larger, helmet- or hood-like; petals 2, clawed, limb usually spurred; stamens numerous. **Fruit** a small collection of usually 3–5 follicles. All parts of the plants are poisonous.

1. Plants with twining stems . . . . *A. hemsleyanum, A. vilmorinianum, A. tsaii*
1. Plants non-twining, usually with erect stems 2
2. Annuals; stamens protruding through sepal gaps . . . . *A. gymnandrum*
2. Perennials; clump-forming; stamens contained within sepals 3
3. Upper sepal tall and rather narrow, cylindrical . . . . *A. sinomontanum, A. scaposum*
3. Upper sepal shallow, rounded and boat-shaped . . . . *A. brachypodum, A. carmichaelii*

Aconitum hemsleyanum E. Pritz. A very variable **climber** to 4 m, with branched, slender, purplish, twining **stems**. **Leaves** long-stalked, ± pentagonal in outline, with 3 segments separated to middle of leaf, these further lobed and toothed. **Flowers** up to 12 in a terminal raceme; sepals dark blue, shiny, glabrous to downy, lower sepal 12–13 mm long, lateral rather larger and broader, upper 20–24 mm long, forming a helmet-like hood. **Fruit** usually with 5 follicles, 12–15 mm long. nwGuizhou, Hubei, Sichuan, sShaanxi, e & seTibet, w & nwYunnan [cChina, Myanmar] (in this area 11 varieties are recognised); forests, forest margins, shrubberies, rocky and grassy slopes, 1,700–3,500 m. VIII–X.

Aconitum vilmorinianum Kom.* Similar to *A. hemsleyanum*, but **stems** normally sparsely hairy, **leaves** parted almost to the base and racemes with no more than 6 smaller purple-blue **flowers**, sepals downy outside; upper sepal 15–16 mm high, with a short beak; follicles 16–18 mm long. wGuizhou, swSichuan, n, nw & cYunnan; habitats similar to those of *A. hemsleyanum*, 2,100–3,000 m. VIII–X.

Aconitum tsaii W. T. Wang* Similar to *A. hemsleyanum* and *A. vilmorinianum*, but **leaves** 3-parted almost to the base, the central segment shallowly pinnately lobed, lower bracts like the leaves, grading into the obovate upper bracts; **flowers** up to 10 per inflorescence, with greenish or pale purplish sepals, glabrous outside, upper sepal 18–23 mm high with an upcurved beak. nwYunnan (Bijiang Xian); shrubberies, grassy and rocky places, 3,200–4,100 m. VIII–IX.

Aconitum gymnandrum Maxim.* **Annuals** to 80 cm, generally 15–50 cm, branched and variably hairy. Basal **leaves** withered by flowering time, stem-leaves evenly spaced, stalked, 3-parted, the segments further divided into narrow-ovate or lanceolate lobes. **Inflorescences** with up to 16 flowers on downy stalks; bracts leaf-like, the upper 3-lobed or linear; sepals purple-blue, occasionally white, slightly downy outside, rather gappy, lower sepal 11–13 mm long, lateral larger and wider, 15–18 mm long, upper boat-shaped, 17–19 mm long, clawed. **Fruit** with 6–13 downy follicles. sGansu, Qinghai, w & nwSichuan, eTibet, Yunnan; grassy and stony slopes, streamsides, open shrubberies, sometimes cultivated, 500–3,800 m. VI–VIII.

Aconitum sinomontanum Nakai* Rather robust **perennial** to 1.5 m, **stems** simple or branched, downy above. Lower **leaves** long-stalked, middle and upper short-stalked or stalkless; blade rounded or kidney-shaped in outline, 3-parted, to 15 × 28 cm, segments shallowly lobed and with triangular teeth. **Inflorescence** many-flowered, to 50 cm long; **flowers** pale blue to purple-blue; lower and lateral sepals 10–12 mm long, the lateral far broader, upper forming a tall cylindrical hood, 16–24 mm high, 4–8 mm in diameter. **Fruit** with 3 glabrous follicles, 11–17 mm long. sGansu, neGuangxi, Guizhou, wHubei, e & seQinghai, Sichuan, Shaanxi [cChina]; open forests, forest margins, shrubberies, scrub, damp grassy places, 1,000–3,700 m. VI–IX.

Aconitum scaposum Franch. Similar to the previous species, but rarely more than 70 cm, with few (2–4) **stem-leaves**, sometimes none; sepals purple-blue, the upper 13–18 mm high, beaked; follicles 3, unequal, slightly hairy. sGansu, Guizhou, Hubei, Sichuan, sShaanxi, n & neYunnan [cChina, c & eHimalaya, Myanmar]; habitats similar to those of *A. sinomontanum*, 1,200–3,900 m. VIII–IX.

Aconitum brachypodum Diels* Erect **perennial** to 80 cm with simple or branched, slightly hairy **stems**. **Leaves** rather crowded, progressively smaller up the stem, blade 3-parted almost to the base, to 5.5 × 8 cm, segments divided into linear lobes. **Inflorescence** a few–many-flowered narrow raceme; sepals pale to mid-blue or purple-blue, often with deeper veins, lower 13–16 mm long, lateral somewhat larger and broader, upper forming a semi-moon-shaped hood, 20–30 mm long with a short beak, downy. **Fruit** with 3 or 5 downy follicles. Sichuan, Yunnan; open scrub, grassy places, 2,800–4,300 m. IX–X.

Aconitum carmichaelii Debeaux A rather variable plant to 1.5 m, readily distinguished from the previous species by the shallowly 3-parted **leaves** with broad, incisely lobed and toothed leaf-segments and by the deep purple-blue somewhat larger **flowers** in rather dense racemes. seGansu, nGuangxi, Guizhou, wHubei, Sichuan, Shaanxi, Yunnan [c, s & seChina, nVietnam]; forest margins, open shrubberies, grassy slopes, streamsides, to 2,200 m. IX–X.

DELPHINIUM

A genus with c. 350 species (173 in China, 150 endemic), primarily in the northern hemisphere but with a few in equatorial Africa. Mainly herbaceous **perennials**, occasionally **annuals** or **biennials**, sometimes with twining **stems**, often viscid. **Leaves** palmately, rarely pinnately lobed, sometimes confined to base of plant. **Flowers** solitary, or more often in racemes or branched corymbs; sepals 5, petal-like, uppermost spurred; petals 2, stalkless, clawed, with an expanded limb, smaller than sepals; stamens numerous. **Fruit** a collection of 3–10, many-seeded follicles.

1. Leaves pinnately lobed . . . . *D. anthriscifolium*
1. Leaves palmately lobed 2
2. Plants at least 25 cm tall in flower . . . *D. grandiflorum*, ... *D. caeruleum, D. thibeticum, D. kamaonense*
2. Small plants not more than 25 cm tall in flower, often smaller . . . . *D. chrysotrichum, D. monanthum, D. beesianum*

Delphinium anthriscifolium Hance Erect, thin-stemmed **annual** to 80 cm with alternate, pinnately lobed **leaves** with the main lobes further divided; basal leaves withered by flowering time. **Inflorescence** a 2–10-flowered, finely downy raceme; **flowers** violet or purple; sepals elliptic, 8–16 mm long, uppermost with a slender upcurved spur to 22 mm long, generally far less; petals broadened towards apex, glabrous. Gansu, Guangxi, Guizhou, Hubei, Sichuan, sShaanxi, ne, e & seYunnan [c & seChina, nVietnam]; open forests, forest margins, shrubberies, grassy places, streamsides, to 1,700 m. III–VI.

Delphinium grandiflorum L. Tufted **perennial** to 65 cm, with branched **stems**. **Leaves** with a heart-shaped base, divided almost to the middle into 3–5 primary segments, these further divided into linear-lanceolate lobes. **Flowers** in 3–10-flowered racemes; sepals blue to purple-blue, oval-elliptic, 10–19 mm long, uppermost with a straight to somewhat curved spur, 15–22 mm long; petals entire, not divided, often whitish in centre. Follicles 3. Gansu, eQinghai, nwSichuan, Shaanxi, nw & cYunnan [n & cChina, Mongolia, eRussia]; open forests, scrub, grassy places, field boundaries, roadsides, to 3,500 m. V–XI.

Delphinium caeruleum Jacq. Similar to *D. grandiflorum* but **leaves** grey-green, **flowers** often very bright blue or purple-blue, with narrower, obovate sepals, petal-limbs uniformly coloured, like sepals, spur to 28 mm long. Follicles 4–5. Gansu, Qinghai, wSichuan, nwYunnan, seTibet [swTibet, c & eHimalaya]; low scrub, alpine meadows, grassy and gravelly slopes, 2,100–5,000 m. VII–X. A very variable species: high altitude forms can be quite dwarf.

Delphinium thibeticum Finet & Gagnep.* Readily distinguished *D. caeruleum* by the slender spike-like racemes of rather narrow purple-blue **flowers** that often point in one direction, spur tapered to a fine, recurved point, 25–35 mm long; **fruit** usually with 3 follicles. w, nw & swSichuan, seTibet, nwYunnan; open forests, shrubberies, rocky and grassy slopes and valleys, 2,800–3,800 m. VII–IX.

Delphinium kamaonense Huth **Perennial** to 45 cm with palmate, grey-green **leaves** divided almost to the base into lanceolate to linear, pointed lobes. **Inflorescence** racemose, often branched below, with up to 14 flowers per raceme, bracts linear, lowermost sometimes divided, the stalks up to 6.5 cm long with bracteoles borne in middle or towards top; sepals pale to deep blue, occasionally white, oval to almost oblong, uppermost with a slightly curved, ascending spur 12–25 mm long; petals undivided, glabrous, with a white or yellowish mark towards base. Follicles 3, downy. swGansu, sQinghai, sw, w, nw & nSichuan, e & seTibet [swTibet, w & cHimalaya]; meadows, grassy slopes, open shrubberies, 2,500–4,200 m. VI–VIII.

Delphinium chrysotrichum Finet & Gagnep.* Small **tufted**, downy plant to 20 cm tall in flower, usually half that height, **leaves** rounded and rather *Geranium*-like, divided to over halfway into generally 5, blunt-lobed segments. **Flowers** cupped, up to 5 in a compact, downy corymb; sepals purple or blue-purple, often yellowish outside, ovate, uppermost with a narrow conical, slightly curved spur 13–24 mm long; staminodes prominent, deep brownish black. Follicles usually 3. w & swSichuan, nwYunnan, neighbouring parts of Tibet; stony and grassy slopes, rock-crevices, 3,000–5,000 m. VIII–IX.

Delphinium monanthum Hand.-Mazz.* (syn. *D. candelabrum* Ostenf. var. *monanthum* (Hand.-Mazz.) W. T. Wang) Similar to *D. chrysotrichum*, but **leaves** divided almost to base into narrower divisions. **Flowers** often solitary or paired, sepals spreading widely apart, purple-blue with deeper veins; spur narrower, 18–30 mm long, slightly recurved. swGansu, e & seQinghai, w & nwSichuan, neTibet, nwYunnan; habitats similar to those of *D. chrysotrichum*, 4,100–5,000 m. VII–X.

Delphinium beesianum W. W. Sm.* Similar to *D. monanthum*, but sepals, petals and staminodes all the same colour, blue or purple-blue; **flowers** up to 5 in a corymb, spur 24–28 mm long; staminodes entire to cleft, with a yellowish beard. w & swSichuan, e & seTibet, nwYunnan; habitats similar to those of *D. monanthum*, as well as alpine meadows, 3,500–4,800 m. IX–XI.

1. *Delphinium anthriscifolium*; Shennongjia, wHubei (PC)
2. *Delphinium grandiflorum*; Gangheba, Yulongxueshan, nwYunnan (EL)
3. *Delphinium caeruleum*; Nepal (CGW)
4. *Delphinium kamaonense*; Genyanshan, south side, swSichuan (AD)
5. *Delphinium chrysotrichum*; Daxueshan, nwYunnan (HJ)
6. *Delphinium monanthum*; Baimashan, nwYunnan (RM)
7. *Delphinium beesianum*; Baimashan, nwYunnan (JM)

1. *Aquilegia ecalcarata*; Jiuzhaigou, nwSichuan (PC)
2. *Aquilegia ecalcarata*; Jiuzhaigou, nwSichuan (CGW)
3. *Aquilegia rockii*; Gangheba, Yulongxueshan, nwYunnan (CGW)
4. *Souliea vaginata*; Gonggashan, wSichuan (PC)
5. *Souliea vaginata*; Tianchi, nwYunnan (RS)
6. *Beesia calthifolia*; Muli, swSichuan (PC)
7. *Helleborus thibetanus* in fruit; Wanglang, nwSichuan (PC)
8. *Helleborus thibetanus*; cult., from nwSichuan (CGW)

AQUILEGIA

A genus of 70 species (13 in China, 4 endemic) found in the temperate northern hemisphere. **Perennial herbs**. **Leaves** mostly basal, ternately divided with numerous lobed leaflets; stem-leaves similar but smaller and short-stalked. **Flowers** nodding, in branched clusters with leaf-like bracts; sepals 5, petal-like, spreading; petals erect, with a broad blunt blade, narrowed at base into an upward projecting spur; stamens numerous. **Fruit** a collection of follicles.

1. Spurs absent or very short, not more than 2 mm long .. .. *A. ecalcarata*
1. Spurs present and prominent, at least 15 mm long ... . *A. rockii*, *A. oxysepala*

Aquilegia ecalcarata **Maxim.*** (syn. *Semiaquilegia ecalcarata* (Maxim.) Sprague & Hutch.) Rather delicate, sparsely hairy, short-lived **perennial**, with erect, somewhat branched **stems**. **Leaves** mostly in basal part of plant, 2-ternate, leaflets not more than 4 cm long, lateral unevenly 2-lobed, terminal 3-lobed. **Flowers** rather small and nodding, purple; sepals narrow-ovate, to 14 × 6 mm, spreading; petals more oblong, not spreading and slightly shorter than sepals; spur genially absent. Follicles 4–5. Gansu, Guizhou, Hubei, Qinghai, n & wSichuan, Shaanxi, eTibet; shrubberies, open forests, grassy places, roadsides, 1,800–3,500 m. V–VII.

Aquilegia rockii **Munz*** **Perennial herb** to 80 cm, with erect **stems** that are somewhat branched and densely glandular-downy above. **Leaves** mostly in lower part of plant, 2-ternate; leaflets ovate to obovate, lateral 2-lobed, terminal leaflets usually 3-lobed. **Flowers** in very lax cymose **inflorescence**, pendulous, purple to blue overall; sepals narrow-ovate, 2–3 cm long; petals oblong, with a hairy, straight to slightly incurved spur 16–20 mm long. Follicles usually 5, densely glandular-downy. swSichuan, seTibet, n & nwYunnan; mixed forests, shrubberies, rocky places, road- and streamsides, 2,500–3,500 m. VI–VIII.

Aquilegia oxysepala **Trautv. & C. A. Mey.*** Similar to the previous species, but **flowers** bicoloured, sepals purple, petals yellowish white with spur sharply hooked at tip. Gansu, Hubei, e & seQinghai, Sichuan, sShaanxi, nYunnan; habitats similar to those of *A. rockii*, 400–2,700 m. VI–VIII.

SOULIEA

A genus with just 1 species found from the eastern Himalaya to western China. Rhizomatous **perennial** with basal stem-scales, and 2–3-ternate **leaves**. **Flowers** in racemes with 5 petaloid sepals, 5 toothed petals and numerous stamens. **Fruit** consisting of 1–3 narrow-elliptic follicles.

Souliea vaginata **(Maxim.) Franch.** (syn. *Isopyrum vaginatum* Maxim.; *Actaea vaginata* (Maxim.) Compton) Erect plant to 70 cm tall in fruit, much shorter at flowering stage, glabrous or almost so, with 2–4 membranous scales at stem base. **Leaves** 2, on the stem, long-stalked, triangular in outline, to 24 cm long but only partly developed at flowering, divided into ovate-triangular lobes. **Inflorescence** a 4–6-flowered raceme; **flowers** 12–14 mm across; sepals white, oval, 8–11 mm long, 3-veined; petals half the length, white or pink, many-veined. Follicles erect, solitary or paired, 3.5–7 cm long. sGansu, e & seQinghai, Sichuan, sShaanxi, seTibet, nwYunnan [c & eHimalaya, Myanmar]; forests, forest margins, grassy places, road- and tracksides, 2,800–4,000 m. V–VI.

BEESIA

A genus with 2 species (both in China, 1 endemic) in central and southwest China and Myanmar. **Perennial** rhizomatous **herbs** with up to 4 long-stalked basal **leaves**, blades heart-shaped to almost triangular. **Flowers** flat, 5-parted, in small clusters on a scape with a basal membranous sheath; petals absent; stamens numerous. **Fruit** a solitary long, narrow follicle.

Beesia calthifolia **(Maxim. ex Oliv.) Ulbr.** Tufted **rhizomatous herb** to 60 cm tall in fruit. **Leaves** basal, long-stalked, blade ovate to kidney-shaped, often deeply heart-shaped at base, to 10 × 16 cm, with finely toothed margin, usually glabrous. Scapes erect, bearing a cymose **inflorescence**, with a membranous basal sheath; bracts small, lanceolate; sepals white or pink, 5, petal-like, elliptic or lanceolate, 3–6 mm long; petals absent; stamens numerous. **Fruit** 11–17 mm long. sGansu, nGuangxi, Guizhou, wHubei, Sichuan, nwYunnan [nMyanmar]; damp or wet places in forests and shrubberies, particularly in the deeper valleys, 1,400–3,500 m. V–VII.

HELLEBORUS

(sometimes put in a separate family **Helleboraceae**)

A genus of c. 20 species (just 1 in China) in Europe and west Asia. **Evergreen** or **herbaceous perennials** with or without leafy **stems**. **Leaves** pedately compound. **Flowers** in stalked cymes accompanied by leaf-like bracts; sepals usually 5, petal-like, green or coloured; petals reduced to a ring of yellowish or orange, tubular nectaries; stamens numerous. **Fruit** a head of many-seeded follicles.

Helleborus thibetanus **Franch.*** Tufted, summer-dormant **herbaceous perennials**. Basal **leaves** bluish green, with 3 primary divisions, outer 2 further divided, segments elliptic to oblanceolate, margin toothed. **Flowers** usually 2–3, 32–65 mm across, on **stems** to 50 cm tall; sepals pale pink to pinkish red, elliptic, spreading; nectaries yellowish green, 5–6 mm long. **Fruit** with 2–3 follicles, occasionally more. sGansu, nwHubei, nwSichuan, sShaanxi; forests, forest margins, shrubberies, in dappled shade, always local, 1,100–3,700 m. III–V.

CIRCAESTERACEAE

CIRCAEASTER

A Sino-Himalayan genus with a single species; a small **herb** of wet and shady habitats.

Circaeaster agrestis **Maxim.** Small **annual** not more than 10 cm tall, with leaves in a whorl at top of **stem**. **Leaves** stalked, pale green, flushed with pink beneath, blade ± diamond-shaped to spathulate or obovate, 4–23 × 1–11 mm, finely toothed in upper part. **Flowers** in fascicles, tiny; sepals narrow-ovate, c. 0.5 mm long; petals absent; stamens 2–6. **Fruit** an oblong to spindle-shaped achene, to almost 4 mm long, covered in hooked spines. sGansu, e & sQinghai, Sichuan, sShaanxi, eTibet, nwYunnan [swTibet, Xinjiang, Himalaya]; forests, wet rocks and rock-ledges, wet grassland in shade, 2,100–5,000 m. IV–VII.

PAEONIACEAE

PAEONIA

A genus of c. 40 species (15 in China, 10 endemic) from western USA, Europe, west, central and north-east Asia and Japan. **Herbaceous perennials** and pithy-stemmed **shrubs** with alternate, mostly ternately divided leaves with lobed or deeply dissected segments. **Flowers** usually large and terminal, sometimes supported by a few lateral buds; sepals usually 5, green; petals 5–13 (more in cultivated forms), brightly coloured; stamens numerous, borne on a distinctive disc which in the shrubby tree peonies is extended to envelop the follicles; follicles usually 3–5, occasionally solitary, containing large black or bluish black **seeds** often accompanied by smaller, sterile red ones. Peonies, both herbaceous and shrubby, are widely cultivated in China, particularly in parks and monastery or temple gardens.

1. Shrubs or subshrubs with at least some woody stems; flowers with a well-developed disc that partly or wholly envelops the follicles 2
1. Herbaceous perennials; flowers with a poorly developed annular disc 4
2. Flowers solitary, terminal, large, 12–22 cm across, basically white or pale pink .... *P. rockii*, *P. decomposita*, *P. ostii*, *P. suffruticosa*
2. Flowers usually 2–3 per shoot, relatively small, 7–12 cm across, basically red, yellow, orange, greenish yellow or purplish red 3
3. Follicles 2–5; filaments and stigmas often not yellow .. .. *P. delavayi*, *P. delavayi* subsp. *lutea*, *P. delavayi* forma *trollioides*
3. Follicles generally solitary; filaments and stigma always yellow .... *P. ludlowii*
4. Leaf-divisions glabrous .... *P. mairei*
4. Leaf-divisions with bristles along veins above .... *P. lactiflora*, *P. veitchii*

Paeonia rockii **(S. G. Haw & Lauener) T. Hong & J. J. Li*** **Deciduous shrub** to 2 m, sometimes suckering, with erect to ascending, sparingly branched, woody **stems**. **Leaves** 3-ternate to 2–3-pinnate, with 19–33 lanceolate to ovate, often lobed leaflets. **Flowers** solitary, large, to 20 cm across, erect, bowl-shaped; petals 8–12, white, sometimes with a pink flush, with a conspicuous deep purple or maroon basal blotch; stamens yellow with white filaments. Follicles often 5, densely downy, completely enveloped by the pale yellow disc. sGansu, wHubei, c & sShaanxi; broadleaved forests, forest margins, shrubberies, rocky gullies, 1,100–2,800 m. IV–V.

Paeonia decomposita **Hand.-Mazz.*** (syn. *P. szechuanica* W. P. Fang) Readily distinguished from *P. rockii* by the more finely divided **leaves** with 33–63 leaflets and by glabrous follicles that are only covered by the whitish disc in the lower two-thirds. **Shrub** to 1.5 m with rose-pink flowers not more than 15 cm across, petals often somewhat toothed at margin. nwSichuan; broadleaved and *Cupressus* forests, shrubberies, 2,000–3,100 m. IV–V.

Paeonia ostii **Hong Tao et al.*** Similar to *P. rockii* and *P. decomposita*, but **leaves** with only 9–15 grey-green, ovate-lanceolate leaflets that are mostly unlobed, and large **flowers**, 13–15 cm across, that are pure white and unmarked, rarely with a hint of purple at the petal base; stamens with purplish filaments, ovary partly surrounded by a purplish sheath. Gansu, Shaanxi [Henan, Hunan], 1,100–1,400 m. V–VI.

Note. *P. suffruticosa* Haworth is widely cultivated in the region, particularly in parks and temple gardens; numerous cultivars exist in a wide range of colours and include plants with single, semi-double and fully double **flowers**.

Paeonia delavayi **Franch.*** Very variable **shrub** generally 0.8–2 m tall, with greyish brown, erect to ascending, mostly unbranched **stems**. **Leaves** 2-ternate, with the primary leaflets further segmented, deep or bright green, sometimes with a purplish or reddish flush. **Flowers** 1–3 per shoot, nodding (held below the leaves), deeply cupped, 6–9 cm across; petals usually 7–11, red to purple-red, shiny; filaments pinkish red to dark purple. Follicles 2–5, occasionally more, glabrous, to 4 × 1.5 cm. w & swSichuan, seTibet, n, nw & cYunnan; dryish *Quercus* and coniferous woods, shrubberies, stony banks, roadsides, 2,000–3,600 m. V–VI. **Subsp. *lutea* (Delavay ex Franch.) B. A. Shen*** (syn. *P. lutea* Delavay ex Franch.) is very similar, but with yellow **flowers**, petals sometimes with a reddish or brownish basal blotch; filaments yellow or greenish yellow. Similar habitats and distribution: when subsp. *lutea* grows in close proximity to subsp. *delavayi*, hybrid swarms occur with a range of flower colour including greenish yellow, orange, orange-brown or various shades of red, occasionally bicoloured. **Forma *trollioides* (Stapf ex Stern) S. G. Haw*** (syn. *P. potaninii* Kom.; *P. trollioides* Stapf ex Stern; *P. potaninii* var. *trollioides* (Stapf ex Stern) Stern) is, in contrast,

1. Peony nursery near Xining, Qinghai (RS)
2. *Paeonia rockii*; cult. (CGW)
3. *Paeonia rockii* fruit; cult. (CGW)
4. *Paeonia ostii*; cult. (CGW)
5. *Paeonia suffruticosa*; cult. (CGW)
6. *Paeonia suffruticosa*, double-flowered cultivar; cult. (CGW)
7. *Paeonia delavayi* subsp. *delavayi*; Gangheba, Yulongxueshan, nwYunnan (CGW)
8. *Paeonia delavayi* subsp. *lutea*; cult., from nwYunnan (CGW)
9. *Paeonia delavayi* subsp. *lutea*; Cangshan, nwYunnan (PC)
10. *Paeonia delavayi* forma *trollioides*; near Geza, nwYunnan (CGW)

1. *Paeonia* hybrids (*P. delavayi* subsp. *delavayi* × subsp. *lutea*); Napahai, Zhongdian, nwYunnan (CGW)
2. *Paeonia* hybrid (*P. delavayi* subsp. *delavayi* × subsp. *lutea*); Napahai, Zhongdian, nwYunnan (CGW)
3. *Paeonia mairei*; cult. (CGW)
4. *Paeonia lactiflora*; cult. (CGW)
5. *Paeonia veitchii*; north of Songpan, nwSichuan (PC)
6. *Paeonia veitchii*; Wonglang, nwSichuan (PC)
7. *Paeonia veitchii*; Wonglang, nwSichuan (CGW)
8. *Paeonia veitchii*; Bama, Makehe, Qinghai (HJ)
9. *Paeonia veitchii*; Wolong, wSichuan (PC)

a low suckering, patch-forming **subshrub** rarely more than 70 cm, with smaller, more finely divided **leaves**, and red, orange or yellow **flowers** that are often held just above the foliage. w & swSichuan, nwYunnan.

Paeonia ludlowii **(Stern & Tayl.) D. Y. Hong*** Essentially similar to *P. delavayi*, but a more robust **shrub** to 3 m, with leaf-segments generally 3-lobed. **Flowers** larger, to 12 cm across; petals pure yellow, spreading apart at maturity; filaments always yellow. Follicles generally solitary, eventually to 7 × 3.3 cm. seTibet; open forests, shrubberies, 2,900–3,500 m. V–VI.

Paeonia mairei **H. Lévl.*** Clump-forming, **herbaceous perennial** to 1 m, with glabrous **stems**. **Leaves** 2-ternate with up to 19 leaflets some of which may be further segmented. **Flowers** solitary, 7–14 cm, erect, shallowly cup-shaped; petals 7–9, pink to rose-pink or red; filaments purple-red. Follicles 2–3, yellow-hairy, occasionally glabrous. seGansu, nwGuizhou, swHubei, c & sSichuan, sShaanxi, neYunnan; broadleaved forests, 1,500–2,700 m. IV–V.

Paeonia lactiflora **Pallas*** **Herbaceous** clump-forming **perennials** to 70 cm, with erect, unbranched, glabrous **stems**. **Leaves** 2-ternate, with up to 15 lanceolate to ovate, unlobed leaflets. **Flowers** several per shoot, erect, bowl-shaped, 8–13 cm across; petals 9–13, white or pink; filaments yellow. Follicles 3–5, green or purplish, usually glabrous. sGansu, Shaanxi [widely cultivated in China]; grassy places, open woods, shrubberies, 400–2,300 m. V–VI.

Paeonia veitchii **Lynch*** (syn. *P. anomala* L. subsp. *veitchii* (Lynch) D. Y. Hong & K. Y. Pan) Differs in the more finely divided **leaves** which have more than 20 linear-lanceolate leaflets that are further lobed. **Flowers** usually 2–4, often with several additional undeveloped buds in upper leaf-axils, ascending to semi-nodding, saucer-shaped to almost flat, 7–14 cm across; petals 6–9 pink to rose-red or red, sometimes white. Follicles 2–5, yellowish brown, hairy. s & cGansu, e & seQinghai, n, nw & wSichuan, sShaanxi, eTibet, nYunnan; shrubberies, forest margins, open grassy places, field boundaries, 1,200–3,900 m. IV–VI.

PAPAVERACEAE

MECONOPSIS

A genus of c. 55 species (43 in China, 23 endemic) in the Himalaya and western China, and with 1 species in western Europe. **Monocarpic** or **perennial herbs**, stemless or with a distinct **stem** and with simple, lobed or unlobed **leaves**, variously adorned with hairs, bristles or spines. **Flowers** solitary, scapose (sometimes several scapes fused in lower part), or borne in distinct racemes or umbels, sometimes paniculate, often nodding in bud; sepals 2, soon falling; petals often 4 but up to 9, occasionally more; stamens numerous; ovary generally with a distinct style but this sometimes rudimentary or absent. **Fruit** a capsule, dehiscing by 4 or more apical pores or slits, variously adorned with hairs, bristles or spines, rarely glabrous.

SYNOPSIS OF GROUPS

Group One. Flowers yellow, p. 123.

Group Two. Flowers never yellow; plants armed with stiff, sharp bristle-hairs, p. 124.

Group Three. As Group Two, but plants with flexuous or soft hairs, p. 127.

Group One

Meconopsis integrifolia **(Maxim.) Franch.*** **Monocarpic** plant to 1 m, sometimes as little as 20 cm in flower, leaves and stems covered with golden yellow or reddish brown hairs. **Stem** simple, with a single whorl of leaf-like bracts at top. **Leaves** mostly basal, oblanceolate to obovate or elliptic, to 37 × 5 cm, prominently 3-veined from the base, margin entire. **Flowers** 3–10, erect to almost so, to 20 cm in diameter; petals 6–8, deep to mid-yellow; ovary with a broad (10–14 mm), 4–7-rayed stigma, without a style. s & swGansu, seQinghai, w, nw & swSichuan (not south of Muli), e & seTibet [cTibet]; alpine meadows, moorland, mountain tundra, open shrubberies, 2,800–4,400 m. V–VII.

Heavy grazing of meadows on many Tibetan mountains gives a bleak effect, although plants like *Meconopsis pseudointegrifolia* and many other meadow plants survive in such habitats. On the Kongbo Pa La in se Tibet, the *Meconopsis* is a dwarf ecotype far smaller than the plants found at lower altitudes (AC).

Subsp. *lijiangensis* Grey-Wilson* has generally paler yellow **flowers** and a distinct, yet short, style 2–4 mm long, stigma only 5–7 mm in diameter. swSichuan (south of Muli), seTibet, nwYunnan, apparently not growing with *M. pseudointegrifolia*.

Meconopsis pseudointegrifolia Prain Similar to previous species, but plants to 50 cm with pinnately veined **leaves**; bracts held close to or amongst the rosette-leaves. **Flowers** appearing to be scapose, sulphur-yellow, nodding or semi-nodding, saucer- or shallowly bowl-shaped; style well developed, 3–6 mm long, protruding beyond the boss of stamens, stigma with 7–10 rays, style and stigmatic rays short-bristly. **Capsule** 12–20 mm long, with stout spreading spines. seTibet, nwYunnan [nMyanmar]; alpine meadows and moorland amongst low shrubs, 3,200–4,500 m. V–VII. **Subsp. *robusta* Grey-Wilson** is a generally larger plant to 1.2 m, with a well-developed **stem** bearing scattered **leaves** as well as a whorl of leaf-like bracts beneath the principal flowers; **flowers** to 25 cm across, style 7–11 mm long, lacking bristles. swSichuan, seTibet, nwYunnan [swTibet, nMyanmar]; similar altitudes but generally growing in marshy places in shrubberies and on forest margins. **Subsp. *daliensis* Grey-Wilson*** is a dwarf variant not more than 25 cm tall, very silky-hairy overall, with erect **flowers**; style 2–5 mm long, glabrous. seTibet, nwYunnan (Cangshan).

Group Two

1. Flowers all or mostly on basal scapes ... ***M. horridula***
1. Flowers in elongated racemes 2
2. Leaves glaucous, with dark pimples ....
   .... ***M. rudis, M. balangensis***
2. Leaves green or grey-green, usually without dark pimples ...... ***M. prattii, M. prainiana, M. racemosa, M. speciosa***

Meconopsis horridula Hook. f. & Thomson Small **monocarpic**, very sharp-bristly plant not more than 25 cm tall in flower. **Leaves** all basal, in a lax spreading rosette, elliptic to elliptic-oblanceolate, with or without dark pimples, margin undulate to somewhat toothed. **Flowers** solitary, scapose, 40–82 mm across, pendent to semi-pendent; scapes 7–25 cm long, bristly, sometimes fused together in lower part; petals 6–10, blue, purplish, occasionally with a pink or lilac flush; anthers yellow; style 4–6 mm long. **Capsule** to 20 × 11 mm, with dense spreading to erect spines. wSichuan, Tibet [Himalaya]; alpine meadows and tundra, moorland, moraines, screes, 3,970–4,800 m. VI–IX.

Meconopsis rudis (Prain) Prain* **Monocarpic**, sparsely to moderately bristly plant to 30 cm tall in flower. **Leaves** all in a basal rosette, bluish green with sparse dark pimples, elliptic to obovate, margin undulate or somewhat lobed. **Flowers** 38–84 mm across, up to 17 per raceme; petals 5–8, blue, purple, violet-blue or pinkish; style 1–3 mm long. **Capsule** 8–15 mm long, with stout erect to ascending spines. w & swSichuan, nwYunnan; rocky meadows, screes, moraines, rock-ledges and -crevices, 3,350–4,800 m. V–IX. Note: Similar plants from wSichuan with dilated inner filaments represent an undescribed species ***M. balangensis.***

Meconopsis prattii (Prain) Prain* Stout **monocarpic**, bristly plant to 1 m, more often 50–80 cm; bristles all of similar length. **Leaves** narrow-elliptic to lanceolate, erect to ascending, basal and on lower half of stem. **Flowers** 14–32, horizontal to semi-nodding, 38–70 mm across; petals 5–8, pale blue, bluish purple, or purple to wine-purple; style short, stigma pale green or creamish, included within the boss of stamens. **Capsule** 17–20 mm long, with spreading bristles. swSichuan, seTibet, nwYunnan; rocky slopes, screes, open woodland, shrubberies, always below treeline, 3,380–3,910 m. V–VII. Note: similar plants from seTibet with 4-petalled, pale blue or primrose-yellow flowers are referable to ***M. prainiana* Kingdon Ward**.

Meconopsis racemosa Maxim. Very similar to *M. prattii*, but leaf- and stem-bristles mixed long and short, the lower **stem-leaves** generally subtending a flower; petals various shades of blue, sometimes flushed with pink, purple or lilac; stigma whitish, protruding beyond the boss of stamens. **Capsule** 9–12 mm long. c & sGansu, Qinghai, w, nw & swSichuan, Tibet, nwYunnan [Himalaya, nMyanmar]; alpine tundra, mountain meadows, rocky slopes, stabilised screes, above treeline, 3,500–4,700 m. V–VIII.

Meconopsis speciosa Prain* Similar to preceding species but immediately distinguished by the pinnately lobed **leaves**, lobes generally rather blunt; **flowers** fragrant, pale to mid-azure-blue, sometimes purplish, with 4–8 petals; stamens with very dark filaments. seTibet, nwYunnan; rocky places, screes, rocky alpine meadows, 3,650–5,180 m. VII–VIII.

---

1. *Meconopsis integrifolia* subsp. *integrifolia*; above Longriba, nwSichuan (CGW)
2. *Meconopsis integrifolia* subsp. *integrifolia*; Huanglong, nwSichuan (CGW)
3. *Meconopsis integrifolia* subsp. *integrifolia*; near Dawu, Qinghai (RS)
4. *Meconopsis integrifolia* subsp. *lijiangensis*; Daxueshan, nwYunnan (CGW)
5. *Meconopsis pseudointegrifolia*; Tianchi, nwYunnan (PC)
6. *Meconopsis pseudointegrifolia*; nr Lancangjiang-Jinshajiang divide, Lidiping, nwYunnan (CGW)
7. *Meconopsis pseudointegrifolia*; Tianchi, nw Yunnan (CGW)
8. *Meconopsis horridula*; nNepal (CGW)
9. *Meconopsis rudis*; upper Gangheba, Yulongxueshan, nwYunnan (CGW)
10. *Meconopsis balangensis*; Balangshan, Wolong, wSichuan (CGW)
11. *Meconopsis* aff. *racemosa*; Napahai, Zhongdian, nwYunnan (HJ)
12. *Meconopsis racemosa*; between Longriba and Aba, nwSichuan (CGW)
13. *Meconopsis racemosa*; Huanglong, nwSichuan (CGW)
14. *Meconopsis speciosa*; Yading Nature Reserve, swSichuan (AD)

1. *Meconopsis punicea*; north of Jigzhi, seQinghai (CGW)
2. *Meconopsis punicea*; north of Songpan, nwSichuan (CGW)
3. *Meconopsis ×cookei*; seQinghai (JM)
4. *Meconopsis quintuplinervia*; north of Jigzhi, seQinghai (CGW)
5. *Meconopsis delavayi*; Gangheba, Yulongxueshan, nwYunnan (HJ)
6. *Meconopsis impedita*; Serkhyem La, SE Tibet (HJ)
7. *Meconopsis* aff. *pulchella*; Balangshan, wSichuan (CGW)
8. *Meconopsis henrici* var. *psilonomma*; Huanglong, nwSichuan (CGW)
9. *Meconopsis sinomaculata*; above Jiuzhaigou, nwSichuan (CGW)
10. *Meconopsis wilsonii* subsp. *australis*; Pianma, nwYunnan (JA)
11. *Meconopsis betonicifolia*; Jiangchuang, nwYunnan (JA)
12. *Meconopsis baileyi*; near Serkhyem La, SE Tibet (HJ)

PAPAVERACEAE: Meconopsis

Group Three

1. Flowers borne on basal scapes 2
1. Flowers in a bracteate or ebracteate inflorescence .. 5
2. Flowers red, pendent .... **M. punicea, M. × cookei**
2. Flowers blue, violet-blue, purple, lilac or pink, ascending to pendent 3
3. Filaments linear 4
3. Filaments clearly dilated towards base .... **M. henrici,** .... **M. henrici** var. **psilonomma, M. sinomaculata**
4. Leaves with longitudinal veins .... **M. quintuplinervia**
4. Leaves without longitudinal veins (pinnately veined) .. .. **M. delavayi, M. impedita, M. pulchella**
5. Plant more than 45 cm tall (to 2 m) in flower 6
5. Plant not more than 40 cm tall in flower, often less ... . **M. forrestii, M. lancifolia**
6. Plant with an evergreen, overwintering leaf-rosette; inflorescence paniculate .... **M. wilsonii,** **M. wilsonii** subsp. **australis**, **M. wilsonii** subsp. **wilsonii**
6. Plant with resting buds but no overwintering leaf-rosette; inflorescence racemose or flowers solitary .... **M. betonicifolia, M. baileyi, M.lancifolia, M. forrestii**

Meconopsis punicea Maxim.* **Monocarpic** or **perennial** tufted plant to 80 cm. All **leaves** basal, elliptic to oblanceolate, densely hairy, with an entire margin and 3–5 parallel veins. **Flowers** solitary, scapose, on long slender stalks; petals 4, rarely 5–6, intense scarlet-red, rarely pinkish red or white, to 12 cm long. s & swGansu, seQinghai, w & nwSichuan, eTibet; damp alpine meadows and moorland, open woodland, shrubberies, 2,895–4,420 m. V–VIII. Hybrid plants with *M. quintuplinervia* (**M. × cookei** Tayl.), with fleshy pink to pinkish purple **flowers**, can sometimes be found when the parents grow in close proximity.

Meconopsis quintuplinervia Regel* Clump-forming **perennial** to 40 cm. **Leaves** all basal, in lax, crowded rosettes, elliptic to oblanceolate, with 3–5 longitudinal veins and an entire margin, hairy all over. **Flowers** pendent, campanulate, 32–38 mm long, borne on slender, erect scapes high above the foliage; petals 4, rarely 6, pale to deep lavender-blue to rich purplish blue. s & swGansu, seQinghai, n & nwSichuan, wShaanxi neTibet; alpine meadows and moorland, open shrubberies, banks, margins of cultivation, 2,285–4,268 m. V–VIII.

Meconopsis delavayi (Franch.) Franch.* Small tufted **perennial** not more than 20 cm tall in flower. **Leaves** deep green, glaucous beneath, ovate to oblanceolate, to 14 cm long, glabrous or with a few stiff hairs. **Flowers** 44–60 mm across, scapose, 1 per leaf-rosette, pendent or semi-pendent, usually borne on long, slender stalks well above the foliage; petals 4, occasionally more, purple-blue to violet-purple, rarely rose-purple. nwYunnan (centred on Yulongxueshan); alpine meadows, moraines, open shrubberies, 3,050–4,270 m. V–VII.

Meconopsis impedita Prain Similar to *M. delavayi* but **monocarpic** with a single leaf-rosette giving rise to 8 or more flowers. **Leaves** elliptic to oblanceolate, entire to incisely lobed at margin, sparsely to densely stiff-hairy. **Flowers** borne on basal scapes up to 25 cm tall; petals generally 6–10, reddish violet to dark violet or blackish violet, spreading widely. **Capsule** to 37 mm long, bristly. swSichuan, seTibet, nwYunnan [nMyanmar]; rocky alpine meadows, screes, granitic cliff-ledges and -crevices, 3,050–4,575 m. VI–VII. A related species, *M. pulchella* Yoshida, Sun & Boufford, with half-nodding, 4-petalled, flowers and entire leaves has been recently described from W Sichuan (Mianning Xian).

Meconopsis henrici Bur. & Franch.* **Monocarpic** plant to 40 cm tall in flower, with basal rosettes of oblanceolate, entire or slightly toothed **leaves** that are sparsely to densely covered with stiff hairs. **Flowers** up to 11, 78–100 mm across, borne on basal scapes or very rarely in a bractless **inflorescences**; petals 5–9, violet-blue to deep purple, not more than 50 mm long, spreading widely. wSichuan, (?)eTibet; alpine meadows, open shrubberies, 3,660–4,575 m. V–VII. Var. **psilonomma** (Farrer) G. Taylor* bears rather larger, solitary **flowers** on thickish basal scapes; sGansu, seQinghai, nwSichuan; 3,385–3,965 m.

Meconopsis sinomaculata Grey-Wilson* Similar in growth and stature to *M. henrici*, but **flowers** larger, deeply cupped and with a rather floppy appearance; petals 4–6, 50–70 mm long, azure, often flushed purple or lilac, with a prominent maroon-black basal blotch. seQinghai, nwSichuan; sloping stony meadows, low montane shrubberies, 3,350–3,900 m. VI–VII.

Meconopsis wilsonii Grey-Wilson Stout **monocarpic** plant to 2 m tall in flower, with a large leaf-rosette in the first 2–3 years; **rosette-leaves** to 50 cm long, pinnately lobed, beset with a covering of stiff hairs, **stem-leaves** similar but increasingly smaller and less divided up the plant. **Inflorescence** a many-flowered, narrow panicle, with out-facing **flowers** on ± erect stalks, deeply cupped, 40–55 mm across; petals 4, wine-purple to maroon-wine, or occasionally white flushed wine-red. **Capsule** oblong, 14–20 mm long, bristly. wSichuan, seTibet, nwYunnan [neMyanmar]; forest margins and clearings, shrubberies, below cliffs, grassy slopes, below treeline, 3,050–4,000 m. VI–VIII. **Subsp. australis** Grey-Wilson differs in the basal and lower **leaves** having only 4–5 pairs of primary lobes and flower-stalks spreading-hairy. seTibet, nwYunnan [neMyanmar]. **Subsp. wilsonii*** has basal and lower **leaves** with 6–8 pairs of primary lobes and the hairs on flower-stalks ± appressed, never spreading. wSichuan (scarce).

Meconopsis betonicifolia Franch. Tufted **perennial** to 1.7 m tall, with rigid, leafy **stems**. **Leaves** stiff-hairy, paddle-shaped with a rounded to heart-shaped base, a toothed margin and narrow leaf-stalks. **Flowers** 8.5–11 cm across, semi-nodding, in few-flowered racemes; petals 4–6, sky-blue to rose-lavender, spreading widely. **Capsule** narrow-oblong in outline, 4–7-valved, glabrous. nwYunnan; forest margins, shady glades, thickets, watercourse margins, 3,050–3,965 m. VI–VII. Plants with a whorl of upper leaf-like bracts and bristly fruits are referable to **M. baileyi** Prain. seTibet [NE India]. (See photo 12 on p. 126.)

Meconopsis lancifolia **(Franch.) Franch. ex Prain** Small **monocarpic** plant. **Leaves** mid-green, mostly in a lax basal rosette, oblanceolate to elliptic or linear-lanceolate, margin entire or slightly lobed, glabrous to somewhat bristly. **Inflorescence** an erect, ebractrate raceme, sometimes also with a few basal scapes; **flowers** up to 12, 50–70 mm across, semi-nodding to out-facing; petals 4–8, violet-blue to purple or pinkish purple, rarely blue. **Capsule** narrow-oblong in outline, 16–22 mm long, bristly, 5–6-valved, with distinct style. swGansu, w, nw & swSichuan, e & seTibet, nwYunnan [nMyanmar]; stony alpine meadows, moorland, grassy ledges, 3,355–4,875 m. V–VIII.

Meconopsis forrestii **Prain*** Not unlike a slender version of *M. lancifolia*, but **flowers** slightly smaller, not more than 7 and only borne on upper part of **stem**, petals 4, pale blue, sometimes with a purplish flush. **Capsule** narrow-cylindrical, 40–65 mm long, 2–4-valved, style absent. swSichuan, nwYunnan; open shrubberies, alpine meadows, forest margins, amongst rocks, 3,050–4,420 m. VI–VII.

EOMECON

A distinctive genus with a single species confined to China, with relatively large basal **leaves** and poppy-like **flowers** in almost umbellate, scapose **inflorescences**.

Eomecon chionantha **Hance*** Rhizomatous, glabrous **perennial** to 50 cm, with large, rather fleshy, basal, grey-green **leaves**, heart- to kidney-shaped, 10–20 cm across, scalloped on margin, on long stalks well above the ground. **Inflorescences** with scale-like bracts, leafless, **flowers** 3–4 cm across; sepals 2, soon falling; petals 4, white, oval; stamens numerous, with yellow anthers. Guangxi, Guizhou [sc & eChina]; woodland, in deep leafy soils, shady banks, 1,400–1,800 m. V–VII.

DICRANOSTIGMA

A genus of 8 species (3 in China) from the central & eastern Himalaya, closely related to *Glaucium*, also with clasping **stem-leaves**, but differing in the shorter **fruit** capsules that lack a horny, arrow-like tip.

Dicranostigma franchetiana **(Prain) Fedde*** **Annuals** to 1 m, with erect, branched, smooth **stems**. **Leaves** mostly basal, mealy, oblong in outline, to 20 cm long, pinnately lobed, lower lobes smaller than upper; stem-leaves shorter, stalkless, fewer-lobed or unlobed. **Flowers** solitary, semi-nodding, cup-shaped, 3–4 cm across; sepals 2, mealy, soon falling; petals 4, orange-yellow, oval; stamens numerous, yellow, with pale filaments. **Capsule** cylindrical-oblong, 8–14 mm long. w & swSichuan, n & nwYunnan; rocky and sandy places, riverbanks, 2,200–4,100 m. VI–VIII.

Dicranostigma leptopodum **(Maxim.) Fedde*** Very like a small version of preceding species, with spreading rosettes of mainly basal **leaves** and pale to mid-yellow **flowers**, 2–2.8 cm across. c & sGansu, (?)Hubei, n & nw Sichuan, Shaanxi; habitats similar to those of *D. franchetiana*, as well as banks, roadsides and stabilised screes. VI–IX.

MACLEAYA

A very distinctive genus with 2 species restricted to China and Japan. Stout **herbaceous perennial** with erect, leafy **stems**; **leaves** rounded to heart-shaped, alternate, stalked. **Flowers** in large terminal panicles, numerous but small, each comprising 4 spreading sepals, no petals and 8–30 stamens. **Fruit** a small elliptic, 2-parted capsule.

Macleaya microcarpa **(Maxim.) Fedde*** Stout plant to 2.5 m, with large, sculptured **leaves** to 30 cm across, grey-green above, pale whitish grey beneath, margin with rounded and toothed lobes. **Inflorescence**s with numerous **flowers** making feathery panicles, buff or creamish overall, sometimes with a pinkish flush, each flower with 8–12 stamens. **Fruit** 1-seeded. Gansu, Hubei, Sichuan, Shaanxi; open woodland, woodland margins, open shrubberies, road-, lake- and riversides, 400–1,600 m. VI–VIII.

Macleaya cordata **(Willd.) R. Br.** Very similar to the previous species, but flowers and **leaves** often with a bronzy flush when young, **flowers** with 25–30 stamens, **fruit** with 4–6 **seeds**. Hubei, Shaanxi [c & eChina, Japan]; similar habitats. VI–VIII.

FUMARIACEAE

CORYDALIS

A complex genus with almost 500 species scattered worldwide (except for Arctic regions) with more than half the species in China. Ranging from sappy **annual** and **perennial herbs** to **subshrubs** with both **basal** and **stem-leaves**, rarely the latter absent, ternately to pinnately divided; rootstock a taproot, or tuberous, rhizomatous (often scaly) or fibrous. **Flowers** in racemes, corymbs or pseudopanicles, zygomorphic, bracteate or not; sepals 2, often small and membranous, fringed or not; petals 4, inner 2 smaller and partly concealed by outer, the uppermost petal extended backwards into a pouch or spur, the lower petal often clawed. **Fruit** a 2-valved capsule, more rarely contorted or jointed.

SYNOPSIS OF GROUPS

Group One. Flowers basically yellow, greenish yellow or white; spurs upcurved, at least at tip. (p. 131)
Group Two. Flowers basically yellow, greenish yellow or white; spurs horizontal to downcurved. (p. 131)
Group Three. Flowers basically blue, pink, lilac or purple; spurs straight to upcurved, or upward-pointing. (p. 135)
Group Four. Flowers basically blue, pink, lilac or purple; spurs downcurved, at least at tip, horizontal to downward-pointing. (p. 139)

1. *Meconopsis lancifolia* form; Balangshan, Wolong, wSichuan (HJ)
2. *Meconopsis lancifolia*; Baimashan, nwYunnan (CGW)
3. *Meconopsis forrestii*; Tianchi, nwYunnan (PC)
4. *Eomecon chionantha*; Fanjingshan, neGuizhou (PC)
5. *Dicranostigma franchetiana*; Bitahai, Zhongdian, nwYunnan (PC)
6. *Dicranostigma leptopodum*; south of Songpan, Minjiang valley, nwSichuan (CGW)
7. *Macleaya microcarpa*; north of Songpan, nwSichuan (TK)

1. *Corydalis ophiocarpa*; Kangding, wSichuan (PC)
2. *Corydalis cheilanthifolia*; cult. (CGW)
3. *Corydalis cytisiflora*; Chuanzhusi, Gongganlin, wSichuan (HJ)
4. *Corydalis cytisiflora*; Huanglong, nwSichuan (CGW)
5. *Corydalis linarioides*; Matang, Zhegushan, nwSichuan (HJ)

Group One

1. Large leafy biennial with small flowers, corolla 10–14 mm long; fruit capsules contorted ..... **C. ophiocarpa**
1. Plants not as above and flowers larger; fruit capsules not contorted 2
2. Plants evergreen, scapose .... **C. cheilanthifolia, C. moupinensis**
2. Plants dying down to a resting bud in autumn, stems slender, leafy, not scapose 2
3. Leaflets elliptic to obovate, not back-rolled at margin; corolla plain yellow .... **C. cytisiflora, C. yunnanensis, C. prattii**
3. Leaflets narrow and stiff, slightly back-rolled at margin; corolla veined green .... **C. linarioides, C. densispica**

Corydalis ophiocarpa Hook. f. & Thomson Large leafy **biennials** with erect to spreading **stems** to 1.2 m. **Leaves** ferny, 2-pinnately divided, leaflets small, obovate to elliptic, lobed or not, semi-translucent, blue-grey with bronze or purple flush; all leaves stalked. **Inflorescence** congested and spike-like, to 30 cm long; **flowers** ± pendent, corolla pale yellow, cream or whitish, often purple-tipped, 10–14 mm long. Gansu, Guizhou, Hubei, Qinghai, Sichuan, Shaanxi, seTibet, Yunnan [c & eChina, Himalaya, Japan, Taiwan]; mixed and coniferous forests, meadows, roadsides, ravines, 1,100–3,300(–4,000) m. V–VII.

Corydalis cheilanthifolia Hemsl.* Tufted, green **perennials** forming low, rounded hummocks to 25 cm. **Leaves** narrow-triangular in outline, pinnately divided, with finely dissected fern-like leaflets, deep green, sometimes flushed with pink or bronze; bracts narrow-lanceolate. **Inflorescence** racemose, many-flowered, held just above the foliage; corolla golden yellow, 12–16 mm long, with a short blunt, slightly upturned spur about half the length of the flower. Gansu, Guizhou, wHubei, e & neSichuan, eYunnan; rocky and stony places, generally along ravines and riverbanks, 800–1,700 m. IV–VI.

Corydalis moupinensis Franch.* Similar to previous species, but a blue-green plant with broader, blunter leaflets and larger flowers, these 18–22 mm long. wSichuan, nwYunnan; habitats similar to those of *C. cheilanthifolia*, 1,000–2,500 m. V–VII.

Corydalis cytisiflora (Fedde) Lidén ex C. Y. Wu, H. Chuang & Z. Y. Su* (syn. *C. curviflora* Maxim. ex Hemsl. var. *cytisiflora* Fedde) Small 'bulbous' **perennial** closely related to the blue-flowered *C. pachycentra* and of similar dimensions but often taller, to 50 cm. **Basal leaves** stalked, glaucous green, often 2-ternately divided, sometimes with just 5 divisions, leaflets elliptic to elliptic-lanceolate to obovate, to 30 × 4 mm; **stem-leaves** 1–2, stalkless with 3–5, occasionally more, linear-elliptic leaflets; bracts entire, lanceolate to elliptic. **Inflorescence** a fairly congested raceme with up to 20 flowers; **flowers** creamy yellow to primrose-yellow, darker at apex, 17–22 mm long (including spur), with a narrowly cone-shaped, upcurved spur half the length of flower. s & swGansu, seQinghai, w & nwSichuan; sparse grassy and stony habitats, open shrubberies, banks, 2,865–4,000 m. V–VIII.

Corydalis yunnanensis Franch.* (syn. *C. delphinoides* Fedde; *C. pterygopetala* Hand.-Mazz.) Tufted **perennial** to 60 cm, with branched **stems** and *Thalictrum*-like, bright green, 2-pinnately dissected **leaves**; bracts small, linear or linear-lanceolate. **Inflorescence** a many-flowered, lax, airy raceme generally supported by lateral branches at base; **flowers** bright yellow, 13–17 mm long (including spur), with a slender, slightly tapered, upcurved spur that is almost two-thirds the length of the flower; sepals white, scale-like, toothed. swSichuan, w & nwYunnan; forest glades and margins, open shrubberies, streamsides, 2,100–3,400 m. VI–IX.

Corydalis prattii Franch.* Rather similar to *C. cytisiflora*, but rarely more than 35 cm, basal **leaves** often with 7 elliptic leaflets, stem-leaves with narrower, linear-lanceolate leaflets to 65 × 6 mm; **flowers** in rather lax racemes; corolla golden yellow, 12–13 mm long (including spur). wSichuan; open shrubberies, grassy slopes, 2,745–4,000 m. VI–VII.

Corydalis linarioides Maxim.* (syn. *C. schlechteriana* Maxim.) Variable tufted **perennial** to 50 cm, sometimes more, with erect, slender, sparsely leafy stems. **Leaves** mostly stalked, pinnately lobed, with rather stiff, linear to linear-elliptic leaflets that are slightly back-rolled at margin, sometimes with minute hairs on veins beneath; stem-leaves 2–3, alternate, with 3 pairs of leaflets; bracts simple, linear-lanceolate, about as long as flower-stalks, lowermost pinnately lobed. **Inflorescence** a lax raceme of up to 15 flowers; **flowers** yellow with greenish veins, 16–22 mm long (including spur), spur somewhat blunt-ended; sepals whitish, scale-like, margin irregularly slashed. sGansu, seQinghai, Sichuan, Shaanxi, eTibet [Ningxia]; alpine meadows, low alpine scrub, grassy banks, 2,100–4,000 m. VI–VII.

Corydalis densispica C. Y. Wu* Very similar to the previous species, but a more slender plant, **stems** bearing several ternately divided, stalked **leaves**; **bracts** mostly lobed or incised; **flowers** up to 50, small, 13–16 mm long (including spur), lower petal with a corrugated margin. swSichuan, seTibet, nwYunnan; shrubberies, alpine scrub, 3,200–4,200 m. VI–VII.

Group Two

1. Flowers basically white .... **C. trachycarpa**
1. Flowers basically yellow or greenish yellow 2
2. Corolla deep yellow- or chocolate-tipped ... **C. hamata, C. pseudohamata, C. conspersa, C. scaberula, C. corymbosa**
2. Corolla with a plain, greenish yellow or deep yellow tip 3
3. Tufted cliff-dwelling plants with branching stems and 2-pinnate leaves ... **C. wilsonii, C. saxicola, C. tomentella**
3. Plants not as above (meadow and woodland dwellers), with pinnate to ternate or almost finger-like leaves .. 4

4. Coarse, tufted, leafy plants; inflorescence often branched 5
4. Plants with leaves mainly at base; inflorescence simple 6
5. Spur at least half the length of the corolla .... **C. rheinbabeniana, C. madida**
5. Spur very short, less than a quarter the length of the corolla .... **C. adunca, C. stricta**
6. Leaves greenish or grey-green, ternate to pinnate with long, slender leaflets .. **C. pseudocristata, C. eugeniae, C. atuntsuensis**
6. Leaves very glaucous, 1–2-pinnate, leaflets with broad overlapping lobes; spurs green-tipped and veined .... **C. pachypoda, C. dasyptera**

Corydalis trachycarpa Maxim.* Low tufted **perennial** to 20 cm tall in flower with a lax rosette of glaucous, 2-pinnately dissected, stalked **leaves**, leaflets rather thin, elliptic to elliptic-obovate; stem-leaves becoming increasingly smaller and with fewer leaflets, short-stalked to almost stalkless; bracts lobed. **Inflorescence** a rather dense many-flowered raceme; **flowers** fragrant, white or cream with a purple brown apex, 19–27 mm long (including spur), spur slender, tapered and downcurved, about two-thirds the length of the flower. sGansu, s & seQinghai, w & nwSichuan, eTibet; rocky alpine slopes and screes, established moraines, alpine meadows, 3,000–4,800 m. VI–IX.

Corydalis hamata Franch.* (syn. *C. fluminicola* W. W. Sm.) Low **perennial** with yellow, cord-like roots and **stems** to 30 cm, with a decumbent base. **Leaves** 2-pinnately divided, very glaucous, with small elliptic to obovate leaflets, basal leaves in rosettes, stalked, stem-leaves increasingly smaller, short-stalked to almost stalkless; bracts longer than wide. **Inflorescence** a many-flowered, dense, erect raceme; **flowers** fragrant, bright citron-yellow with dark blue apex, 13–16 mm long (including spur); margin of outer petals entire; spur whitish with a downcurved tip. w & swSichuan, eTibet, nwYunnan; damp rocky places, often by streams or seepages, sometimes growing in shallow running water, 3,960–5,335 m. VI–VIII.

Corydalis pseudohamata Fedde* Very similar in general appearance to *C. hamata* (and sometimes included in that species), but with a more upright habit, to 40 cm, and with larger, more finely cut, thinner **leaves**; **flowers** similarly coloured but generally more pendent and slightly less brightly coloured; margin of outer petals toothed; bracts toothed to almost entire at apex. sGansu, seQinghai, nwSichuan; marshy places, stream and pool margins, 3,600–4,800 m. V–VII.

Corydalis conspersa Maxim. Very similar to *C. hamata*, but distinguished primarily by the broad-spathulate, reddish- or purple-flushed **bracts** with a finely irregular (appearing nibbled) margin; **flowers** creamy white with a yellow or pale yellow apex with a bold, deep blue mark; bracts wider than long. swGansu, seQinghai, w & nwSichuan, eTibet [c & eTibet, Nepal]; similar habitats, 3,800–5,000 m. VI–VII.

Corydalis scaberula Maxim.* Readily distinguished from *C. hamata*, *C. pseudohamata* and *C. conspersa* by the very dwarf habit (plants not more than 8 cm tall in flower), with grey, 2-pinnate, softly scurfy **leaves** spreading close to the ground, leaflets ovate to elliptic, entire; stem-leaves 2, opposite; **flowers** 15–20 mm long (including spur), in congested **racemes** amongst the leaves, lemon-yellow tipped chocolate-brown, spur about half the length of the flower; **bracts** deeply dissected, frond-like; sepals kidney-shaped, with tiny teeth. s, se & cQinghai, nwSichuan, neTibet; screes, moraines and other rocky habitats, alpine meadows, 3,900–5,600 m. VI–IX.

Corydalis corymbosa C. Y. Wu & Z. Y. Su* Readily distinguished from other *Corydalis* species in China by the dwarf tufted habit, the plant only 5–15 cm in flower, with congested, rather fleshy, pinnately lobed, glaucous **leaves**, with numerous small, closely overlapping, elliptic, crisped leaflets; lower bracts lobed, upper entire. **Flowers** in small clusters amongst the leaves; corolla ascending, bright yellow tipped brownish red; sepals ovate, toothed. seTibet, nwYunnan (Baimashan in particular); screes and moraines, 4,250–5,100 m. VI–VIII.

Corydalis wilsonii N. E. Br.* Low, tap-rooted **perennials** forming basal rosettes or tufts of rather fleshy, very glaucous-blue, glabrous, 2-pinnately lobed **leaves** with numerous elliptic leaflets. **Flowers** in long dense **racemes** tinged with green; corolla bright yellow, 18–24 mm long (including spur), spur rather short with a blunt downcurved tip; sepals brownish, oval, 4–6 mm long, entire. **Capsules** glabrous. nwHubei (Fangxian); limestone cliff-crevices and -ledges, 3,000–3,050 m. IV–VII.

Corydalis saxicola G. S. Bunting* (syn. *C. thalictrifolia* Franch., not Regel) A coarser plant than *C. wilsonii*, with *Thalictrum*-like foliage, with broader, fresh green leaflets to 35 × 30 mm; **flowers** larger, 25–31 mm long (including spur), golden yellow, in few-flowered, 1-sided racemes; bracts elliptic to ovate, entire. Guangxi, Guizhou, wHubei, Sichuan, Shaanxi, Yunnan [Zhejiang]; limestone cliffs and rocks, 1,600–3,900 m. IV–VII.

Corydalis tomentella Franch.* Closely related to *C. wilsonii* and rather like a more delicate version with smaller **flowers** (16–19 mm long), **leaves** characteristically minutely white-downy giving a distinctly furry appearance. **Inflorescence** upto 10-flowered. **Capsules** finely hairy. wHubei, e, s & seSichuan, Shaanxi; dry rocky habitats, cliffs, 700–1,000 m. IV–VI.

Corydalis rheinbabeniana Fedde* Lax, tufted **perennial** to 50 cm tall in flower, with greyish **stems**. **Leaves** pinnately lobed with relatively few, long, linear to linear-elliptic, unlobed leaflets to 35 × 6 mm, lower leaves stalked, upper ± stalkless;

1. *Corydalis trachycarpa*; Huanglong, nwSichuan (CGW)
2. *Corydalis hamata*; Baimashan, nwYunnan (CGW)
3. *Corydalis pseudohamata*; Huanglong, nwSichuan (CGW)
4. *Corydalis conspersa*; Huanglong, nwSichuan (CGW)
5. *Corydalis scaberula*; Danglexiaoshan, seQinghai (RS)
6. *Corydalis corymbosa*; Daxueshan, nwYunnan (CGW)
7. *Corydalis saxicola*; cult. (CGW)

1. *Corydalis rheinbabeniana*; north of Aba, nwSichuan (CGW)
2. *Corydalis madida*; Wolong, wSichuan (JM)
3. *Corydalis adunca*; nwSichuan (JM)
4. *Corydalis stricta*; Dahebahe, Qinghai (RS)
5. *Corydalis stricta*; Dahebahe, Qinghai (RS)
6. *Corydalis pseudocristata*; Balangshan, Wolong, wSichuan (CGW)
7. *Corydalis eugeniae*, with *Fritillaria cirrhosa*; Gangheba, Yulongxueshan, nwYunnan (CGW)
8. *Corydalis atuntsuensis*; Gonggashan pass, wSichuan (PC)
9. *Corydalis pachypoda*; upper Gangheba, Yulongxueshan, nwYunnan (CGW)
10. *Corydalis dasyptera*; Huanglong, nwSichuan (JM)
11. *Corydalis dasyptera*; Huanglong, nwSichuan (CGW)

bracts with a few lobes; stem-leaves 1–5. **Inflorescence** a many-flowered, dense raceme, generally with a few short branches at base; **flowers** bright yellow with greenish marking at apex, 18–23 mm long (including spur), spur downcurved towards tip, slightly more than half the length of flower; lower bracts dissected or toothed. swGansu, seQinghai, nwSichuan; dry rocky sunny habitats, roadside banks, open shrubberies, 3,100–4,270 m. VI–VIII.

Corydalis madida Lidén & Z. Y. Su* Tufted, rather diffuse **perennial** with reddish, leafy **stems** and markedly glaucous, 2-ternately divided **leaves**, leaflets thin, elliptic to obovate, blunt, lobed or not; upper leaves less divided, grading into the bracts. **Inflorescence** a congested raceme, elongating eventually, sometimes with 1 or 2 short lateral branches; **flowers** directed downwards, pale creamy yellow, darker at apex, purple-flushed on ageing, 26–29 mm long (including spur), with a slender cylindrical spur curved to downcurved towards tip; sepals tiny, toothed at apex. wSichuan (Wolong); damp shady rocks in gorges and ravines, in seepage zone, 2,100–2,200 m. VI–IX.

Corydalis adunca Maxim.* (syn. *C. albicaulis* Franch.) Large, rather diffuse **perennial** to 80 cm in flower, with basal rosettes of finely divided, pinnately dissected, markedly glaucous **leaves**, leaflets very small, elliptic to obovate; stem-leaves similar but decreasing in size up the stem, short-stalked. **Inflorescence** a lax many-flowered, terminal or lateral raceme, often branched; **flowers** canary-yellow, 13–15 mm long (including spur), spur almost straight, only 2–3 mm long, somewhat thickened towards rounded tip. Gansu, s & seQinghai, Sichuan, Shaanxi, eTibet, nwYunnan [nChina]; dry rocky and sunny places, banks, river gravels, roadsides, 1,000–3,900 m. VI–VIII.

Corydalis stricta Steph. ex Fisch. (syn. *C. astragalina* Hook. f. & Thomson) Similar to the previous species, but **leaves** grey-green, thicker and more coarsely dissected; **flowers** rather larger, brighter yellow, often brownish beneath, 16–18 mm long, in a far denser, short raceme. swGansu, s & seQinghai, nwSichuan, neTibet [nwChina, Himalaya, Mongolia, sRussia]; habitats similar to those of *C. adunca*, 3,450–4,400 m. VI–VIII.

Corydalis pseudocristata Fedde* Tuberous-rooted **perennial** with a simple, erect **stem** to 20 cm. Basal **leaves** few, long-stalked, 2-ternate or pinnately divided with lanceolate to elliptic leaflets to 34 × 5 mm; stem-leaves usually a pair towards the top, similar to basal, with 5–9 leaflets, bluish green, sometimes dark at apex, almost stalkless. **Inflorescence** a rather congested 1-sided raceme with up to 20 flowers; **flowers** cream, yellow at apex, 24–35 mm long (including spur), spur pronouncedly incurved, blunt, more than half the length of the flower; bracts elliptic to ovate, or lowermost pinnately lobed. wSichuan; alpine meadows, low open alpine scrub, 2,895–4,573 m. VI–VII.

Corydalis eugeniae Fedde* Readily distinguished from *C. pseudocristata* by the smaller, brighter yellow, green-veined **flowers**, 14–16 mm long (including spur), spur narrower, almost horizontal, turned down at tip; **leaves** pinnate, with 5 or 7 linear-elliptic leaflets to 30 × 2 mm. w, nw & swSichuan, nwYunnan; damp alpine meadows, rocky grassland, forest margins, 3,400–4,500 m. VI–VIII.

Corydalis atuntsuensis W. W. Sm.* Closely related to the next species, but with deep green **leaves** with very slender stiff leaflets, rarely more than 1 mm wide, and with slightly back-rolled margins, stem-leaves often 2, almost opposite; **flowers** golden yellow, 14–17 mm long. sQinghai, w & swSichuan, eTibet, nwYunnan; similar habitats, 3,900–5,000 m. VI–IX.

Corydalis pachypoda Franch.* (syn. *C. tibetica* Hook. f. & Thomson var. *pachypoda* Franch.) Tufted **perennial** to 20 cm with primarily basal **leaves** in a lax rosette, these very *Ruta*-like, 2-pinnately lobed, with numerous small, thick, somewhat overlapping, ovate to elliptic leaflets. **Inflorescence** a lax, ± scapose raceme with up to 16 flowers; **flowers** yellow with greenish markings and veins, 18–24 mm long (including spur), spur almost straight, slightly downcurved at tip; bracts elliptic, entire or the basal 3–5-lobed. nwYunnan; damp rocky habitats, stabilised screes and moraines, low scrub, 3,000–4,600 m. V–VII.

Corydalis dasyptera Maxim.* Very similar to the previous species, but a dwarfer, more compact plant not more than 18 cm tall in flower, with narrower, less dissected **leaves** and dense, more rounded clusters of brighter yellow **flowers** marked purple-brown, flowers set close to the ground; corolla 19–22 mm long; **bracts** toothed to entire. s & swGansu, seQinghai, nwSichuan, eTibet; alpine rocky meadows and screes, 2,700–4,200 m. VI–VII.

Group Three

1. Plants at least 20 cm tall in flower, usually 30–50 cm, occasionally to 1 m 2
1. Small plants rarely more than 20 cm tall in flower . . . 5
2. Plants with large 2-ternate to pinnately compound leaves with large, toothed leaflets; inflorescence arching . . . . **C. temulifolia**
2. Plants with relatively small leaves with leaflets rarely more than 15 mm wide, deeply and often incisely lobed 3
3. Plants patch-forming with scaly stolons or rhizomes at soil level; bracts mostly toothed or lobed . . . . **C. flexuosa, C. elata, C. calycosa**
3. Plants diffuse, spreading and branched, generally tap-rooted; bracts entire to lobed 4
4. Annuals, biennials or short-lived perennials, initially forming a leafy rosette . . . **C. smithiana, C. linstowiana**
4. Tufted perennial . . **C. flaccida, C. kokiana, C. taliensis, C. quantmeyeriana, C. ellipticarpa**
5. Spur upcurved . . . . **C. pachycentra, C. curviflora, C. shensiana, C. hemidicentra**
5. Spur straight . . . . **C. oxypetala, C. trifoliata**

***Corydalis temulifolia* Franch.** Large leafy **perennial** to 80 cm with pinnately or ternately compound, plain green **leaves**, leaflets large, to 50 × 30 mm, occasionally larger, elliptic to elliptic-rhomboid, shallowly toothed and lobed; all leaves stalked. **Inflorescence** an arching rather lax, many-flowered, 1-sided raceme; **flowers** relatively large, 34–38 mm long (including spur), corolla sky- to mid-blue, sometimes with a pinkish flush, spur narrow-conical, less than half the length of the flower; **bracts** usually toothed. seGansu, Guangxi, Guizhou, wHubei, Sichuan, sShaanxi, seYunnan [Vietnam]; damp forests, forest margins, streamsides, ravines, roadsides, 1,300–2,700 m. III–VI.

Corydalis flexuosa* Franch. (syn. *C. balsamiflora* Prain) Very variable, stoloniferous or **rhizomatous** species with scaly rhizomes and erect leafy, often purplish **stems** to 40 cm; bulbils sometimes present in leaf-axils. **Leaves** pinnately or 2-ternately divided, rather thin, deep green to bluish green, often with purple markings. **Racemes** dense at first but becoming laxer, often supported by 1 or several, lateral bracts; lower bracts divided; **corolla** pale to deep blue, sometimes with a purplish flush, 21–25 mm long (including spur), lower lip narrow, elliptic, spur straight to upcurved, tapered to tip. **Capsules** linear, pendent. w & swSichuan; damp forests, forest margins, shrubberies, river margins, tracksides, 1,900–4,300 m. V–VII.

Corydalis elata* Franch. Very similar to the previous species, but a taller plant to 60 cm with broader, often rather bright green, leaflets; bulbils absent; **flowers** azure-blue or lilac-blue to indigo, sometimes with a hint of purple, with white markings at apex. **Capsules** narrowly obovoid-elliptic rather than linear. w & sSichuan (especially Kanding-Luding); habitats similar to those of *C. flexuosa*, 2,900–3,800 m. V–IX.

Corydalis calycosa* H. Chuang (syn. *C. flexuosa* Franch. subsp. *kuanhsienensis* C. Y. Wu; *C. gemmipara* H. Chuang var. *ecristata* H. Chuang) Similar to the next 2 species but seldom more than 30 cm, with bulbils in leaf-axils; **flowers** densely clustered, deep violet-blue with whitish throat, lower lip broad, almost rounded, spur straight or slightly curved at tip. wSichuan (Balangshan); alpine meadows, streamsides, 3,400–4,500 m. VI–VIII.

Corydalis smithiana* Fedde Diffuse **biennial** or short-lived **perennial**, 10–30 cm tall in flower, forming a small leaf-rosette in first year; **stems** red-striped, those of flowering plants much-branched and leafy. **Leaves** bluish green, often flushed with purple, 2-pinnately lobed with elliptic, pointed leaflets; lower bracts lobed, upper entire. **Flowers** reddish purple or deep purple, occasionally pale whitish pink or bluish, 15–22 mm long (including spur), in rather lax, 4–7-flowered racemes, with a strong musky fragrance; sepals scale-like, fringed, soon falling; spur narrowed and slightly downcurved towards tip. swSichuan, nwYunnan; banks, stony places and gravels, often by tracks and roads, 3,100–3,900 m. IV–VII.

Corydalis linstowiana* Fedde In contrast to *C. smithiana*, **annual** or **biennial**, to 20 cm, with broader, greener **leaves** with more blunt leaflets; all **bracts** divided; **corolla** purplish blue to lavender, often rather pale, 15–20 mm long, with relatively large, fringed, white sepals. wSichuan; stony and rocky habitats, 2,900–3,600 m. V–VII.

***Corydalis flaccida* Hook. f. & Thomson** Coarse tap-rooted **perennial** with spreading **stems** to 1 m, often less, stems greenish, sometimes purple-flushed. **Leaves** large, rather *Thalictrum*-like, 2-pinnately lobed, with numerous rounded to elliptic leaflets. **Bracts** ovate-elliptic, green, lower leaf-like, upper entire. **Flowers** in lax racemes or pseudopanicles, purple to pinkish purple, indigo or whitish, sometimes variegated with blue, 18–26 mm long (including downcurved spur which is a third of the length of the flower); sepals heart-shaped, with irregular (appearing nibbled) margin. swSichuan, seTibet, w & nwYunnan [swTibet, c & eHimalaya, Myanmar]; damp habitats in forests and shrubberies, amongst damp rocks, streamsides, grassy places, generally in partial shade, 2,900–4,000 m. V–IX.

Corydalis kokiana* Hand.-Mazz. Tufted **perennial** to 30 cm, with a rather spreading, much-branched habit, **stems** often red-flushed. **Leaves** grey-green, 2-pinnately divided, with lanceolate to elliptic, pointed leaflets. **Flowers** in lax terminal and lateral racemes of up to 26, occasionally more, soft blue to purplish or lilac, 15–21 mm long (including spur), spur strongly recurved towards tip and nearly half length of flower; sepals scale-like, soon falling; bracts slightly pinnately lobed to entire. w & swSichuan, eTibet, nwYunnan; dry rocky places, banks, forest margins, 2,900–4,500 m. V–VII.

Corydalis taliensis* Franch. (including *C. bulleyana* Diels) Similar to *C. kokiana* but forming a generally far larger, more cascading plant with attractive olive-green **leaves** with lighter markings and more blunt leaflets; **flowers** in racemes to 10 cm long, lilac to purple or pink, indigo or reddish, 20–25 mm long (including spur), in rather congested racemes, sweetly scented; sepals white; spur straight or downturned at tip, one-third to half the length of the flower; bracts ovate, scale-like, entire or the lowermost slightly divided. Yunnan; rocky places, forest margins, old walls, rooftops, 1,500–1,800 m. III–IX.

Corydalis quantmeyeriana* Fedde Similar in general appearance to *C. taliensis*, but **leaves** with more rounded, less deeply toothed leaflets, often flushed with purple, leaf-stalk sheaths usually blackish purple; **inflorescence** with divided bracts, **flowers** rose-pink, generally larger, 27–32 mm in length (including spur), in many-flowered racemes; spur rather narrow, downturned at tip, more than half the length of the flower. seQinghai, nwSichuan; damp places in forests and shrubberies, streamsides, ravines, 2,400–4,000 m. VI–VIII. Plants from sGansu and nSichuan with cream, often purple-flushed, dumpier, smaller flowers are distinguished as ***C. ellipticarpa* Z. Y. Su.**

1. *Corydalis temulifolia*; Shennongjia, wHubei (PC)
2. *Corydalis flexuosa*; Wolong, wSichuan (CGW)
3. *Corydalis flexuosa*; Wolong, wSichuan (CGW)
4. *Corydalis flexuosa* form; cult., from Wolong, wSichuan (CGW)
5. *Corydalis elata*; Emeishan, sSichuan (CGW)
6. *Corydalis calycosa*; Balangshan, Wolong, wSichuan (CGW)
7. *Corydalis smithiana*; Napahai, Zhongdian, nwYunnan (CGW)
8. *Corydalis kokiana*; near Bitahai, Zhongdian, nwYunnan (CGW)
9. *Corydalis kokiana*; near Bitahai, Zhongdian, nwYunnan (CGW)
10. *Corydalis quantmeyeriana*; Huanglong, nwSichuan (CGW)

Corydalis pachycentra Franch.* Very variable plant ranging in height from 4–20 cm tall in flower, sometimes spreading by means of slender underground **stems**. **Leaves** rather bluish green, basal long-stalked, 2-ternately lobed, with oval, blunt leaflets; stem-leaves 1–4, ± stalkless, palmately lobed, often with just 3 elliptic leaflets; bracts oval to elliptic, entire. **Flowers** pale to deep azure-blue to indigo, occasionally whitish, in a rather congested raceme of up to 12; corolla 12–17 mm long (including spur), lip oval with a whitish, grooved centre; spur horn-shaped, upcurved to a blunt tip. w & swSichuan, eTibet, nwYunnan; alpine meadows, amongst alpine scrub, forest margins, shrubberies, 3,500–4,500 m. V–IX.

Corydalis curviflora Maxim. ex Hemsl.* Very similar to *C. pachycentra*, but **stem-leaves** with more (often 5–9) linear to narrow-elliptic leaflets and lobed bracts, the lower at least as long as, or exceeding, the flowers. **Flowers** in 10–15-flowered racemes; corolla pale to deep ultramarine-blue or lavender, 12–16 mm long (including narrowly conical spur); sepals slightly cut irregularly towards base, soon falling. sGansu, e & seQinghai, n, nw & cSichuan [Ningxia]; habitats similar to those of *C. pachycentra*, 3,650–5,335 m. VI–VII. As with *C. pachycentra*, high altitude forms can be very dwarf. A variable species within which 2 subspecies are recognised.

Corydalis shensiana Lidén ex C. Y. Wu, H. Chuang & Z. Y. Su* Closely related to *C. curviflora*, but bracts entire or almost so, much shorter than **flowers** which have a narrower and shorter spur; **stem-leaves** palmately lobed and bracts lanceolate to linear. swGansu, Shaanxi [wHenan]; habitats similar to those of *C. curviflora*, 1,300–3,300 m. V–VII.

Corydalis hemidicentra Hand.-Mazz.* Readily distinguished from *C. shensiana* in this artificial grouping by the relatively large and fleshy, *Hepatica*-like **leaves** with 3 leaflets, leaflets rounded, often reddish brown or greyish green and marbled, to 29 × 27 mm; bracts obovate, small. **Flowers** fragrant, up to 12 in an almost umbellate raceme well above the leaves, sky-blue to purplish blue or indigo, 20–28 mm long (including spur), spur upcurved, blunt-ended; sepals rounded, entire to slightly toothed. **Capsules** pendent (erect to ascending in the other species). seTibet, nwYunnan; screes, generally limestone, 3,500–5,300 m. VI–VIII.

Corydalis oxypetala Franch.* Small plant rarely exceeding 25 cm tall in flower, with slender **stems** and deep green, rather thick, blunt-lobed basal **leaves**; stem-leaves similar to basal leaves but smaller and fewer-lobed, usually with 5 or 7 leaflets, stalkless; bracts shorter than flowers, oval or elliptic, entire or the lowermost slightly lobed. **Flowers** up to 15 in a raceme, sky- to azure-blue with darker markings in the mouth, or occasionally rose-purple, 17–24 mm long (including slender, straight, conical spur); sepals very small, scale-like, fringed. nwYunnan (mainly Cangshan); rocky and grassy places, forest margins, shrubberies, 2,440–3,050 m. V–IX.

Corydalis trifoliata Franch. (syn. *C. quadriflora* Hand.-Mazz.) Readily distinguished from the previous species by the **stem-leaves** with 3 leaflets and blue **flowers** with a cylindrical spur. seTibet, nwYunnan [swTibet, c & eHimalaya, nMyanmar]; similar habitats, 3,000–4,300 m. VI–IX.

> **Group Four**
> 1. Flower-stalks recurved in fruit, or fruit pendent . . . .
>   *C. benecincta*, *C. benecincta* subsp. *trilobipetala*
> 1. Flower-stalks erect in fruit 2
> 2. Flowers small, not more than 13 mm long . . . .
>   . . . . *C. pseudoadoxa*
> 2. Flowers larger, 17 mm long or more .. *C. melanochlora*,
>   *C. calcicola*

Corydalis benecincta W. W. Sm.* Low tufted **perennial** not more than 10 cm with relatively large, fleshy, grey, mottled, **leaves** with 3 leaflets, sometimes with a purplish flush; leaflets obovate to rounded, to 50 × 32 mm. **Flowers** up to 15 in a rather congested raceme amongst the leaves, purple to pink or purplish red, with dark bluish purple flushing towards apex, 18–23 mm long (including spur), spur broad, strongly downcurved towards tip; stalks recurved in **fruit**. swSichuan, w & nwYunnan; screes, rocky banks, cliff-ledges, generally on limestone, 3,400–6,000 m. VI–VII. Variants from the same area with a waxy bloom to the **leaves**, fewer and more congested, pale mauve or purplish blue **flowers**, the outer petals with a 3-lobed limb are distinguished as **subsp. *trilobipetala* (Hand.-Mazz.) Lidén*** (syn. *C. trilobipetala* Hand.-Mazz.).

Corydalis pseudoadoxa C. Y. Yu & H. Chuang* Tufted **perennial** to 15 cm tall in flower, **basal leaves** stalked, 2-ternately compound, green flushed with brown, leaflets elliptic to obovate, entire to somewhat lobed, blunt. **Inflorescence** a 8–12-flowered raceme; corolla bright ultramarine-blue, tipped white, 10–13 mm long; **bracts** mostly entire, lowermost sometimes lobed. seTibet, nwYunnan; meadows and other grassy places, low *Rhododendron* scrub, 3,700–4,300 m. V–VII.

---

1. *Corydalis pachycentra*; Daxueshan, nwYunnan (CGW)
2. *Corydalis pachycentra*; Baimashan, nwYunnan (CGW)
3. *Corydalis pachycentra* (high altitude form); Daxueshan, nwYunnan (CGW)
4. *Corydalis curviflora*; south of Aba, nwSichuan (CGW)
5. *Corydalis curviflora*; south of Jiuzhaigou, nwSichuan (CGW)
6. *Corydalis curviflora* (dwarf form); Songpan, nwSichuan (CGW)
7. *Corydalis shensiana*; Maya to Jianni, swGansu (CGW)
8. *Corydalis hemidicentra*; Baimashan, nwYunnan (CGW)
9. *Corydalis oxypetala*; Cangshan, nwYunnan (CGW)
10. *Corydalis benecincta*; Daxueshan, nwYunnan (CGW)
11. *Corydalis benecincta*; Habashan, nwYunnan (CGW)

Corydalis melanochlora* Maxim. (syn. *C. adrienii* Prain) Low 'scaly-bulbous' **perennial** not more than 12 cm, with a basal tuft of thick, glaucous, 2-pinnately divided **leaves**, leaflets elliptic to elliptic-obovate, closely overlapping; bracts overlapping, lobed. **Inflorescence** a compact 4–8-flowered raceme held amongst or just above leaves; **flowers** fragrant, blue, rose-purple, purplish blue or whitish, often bicoloured, and with darker markings at apex, 18–24 mm long (including spur), spur downcurved towards tip, about half the length of the flower; bracts mostly lobed like fingers. Gansu, seQinghai, w & swSichuan, seTibet, nwYunnan; alpine screes and other rocky habitats, 3,660–4,575(–5,500) m. VI–IX.

Corydalis calcicola* W. W. Sm. Closely related to the white-flowered *C. trachycarpa*, but leaflets glaucous or green with a short sharp point at apex, and **flowers** pink or purplish with deep purple or purple-maroon tip, smaller, 16–18 mm long (including spur), spur almost twice the length of the rest of the flower, blunt. w & swSichuan, nw Yunnan; limestone or schistose rocks and screes, rocky scrub, 2,900–4,800 m. VI–IX.

ICHTYOSELMIS

A genus of 1 species found in China and north Myanmar.

***Ichtyoselmis macrantha* (Oliv.) Lidén** (syn. *Dicentra macrantha* Oliv.) Lax-tufted **perennial** to 60 cm with 2-ternate **leaves**, leaflets ovate to elliptic, 3–8 × 2–6 cm, with pointed, toothed lobes, often golden along the margins, glaucous beneath. **Flowers** 3–14, in lax cymose clusters beneath the leaves, pendent on slender stalks, long-heart-shaped; petals cream or yellow to greenish white, 35–50 mm long, narrowed in middle. **Capsule** narrow-elliptic, 3–4 cm long, dehiscent. Guizhou, Hubei (Jianshi), s & swSichuan, nwYunnan [nMyanmar]; damp habitats in forests, 1,500–2,700 m. IV–VII.

DACTYLICAPNOS

A genus of 11 species (10 in China, 1 endemic) in west China and the Himalaya. Very close to *Dicentra* but **herbs** with alternate **leaves**, **climbing** by tendrils.

***Dactylicapnos scandens* (D. Don) Hutch.** (syn. *Dicentra scandens* D. Don; *D. thalictrifolia* (Wall.) Hook. f.) Rather fleshy **climber** to 3 m, with branched, slender, rather pale, glabrous **stems**. **Leaves** alternate, 2–3-ternate, ending in a branched tendril, leaflets ovate to elliptic, to 3.5 × 1.8 cm. **Flowers** pendent, in lax clusters of up to 14; sepals ovate-triangular, 3–4 mm long, entire; petals yellow, sometimes pinkish towards tip, outer 2 coherent in lower half, pouched at base, 18–21 mm long, inner 2 invisible except at tip. **Fruit** narrowly spindle-shaped, 14–20 mm long, fleshy, greenish white flushed purple, often mauve when ripe, non-dehiscent. wGuangxi, sYunnan [sTibet (Chumbi), Himalaya, Myanmar, Sri Lanka, nVietnam]; forests, shrubberies, ravines, ditchsides, 1,600–3,000 m. VII–IX.

***Dactylicapnos lichiangensis* (Fedde) Hand.-Mazz.** (syn. *Dicentra lichiangensis* Fedde) Differs from *D. scandens* in the angular **stems** and bracts and sepals with fringed margins; leaflets with a short, downturned terminal mucro; sepals 4–6 mm long; petals pale yellow, 12–16 mm long, with large basal auricle. **Fruit** green, linear-oblong, 3–6 cm long with thin, flat valves, dehiscent. swSichuan (Muli), seTibet, c & nwYunnan [India (Assam)]; forest margins, dryish shrubberies, grassy places, 1,700–3,000 m. VI–X.

DROSERACEAE

DROSERA

A genus of c. 100 species (6 in China) distributed in temperate and subtropical regions, especially Australasia. **Insectivorous** plants with sticky gland-covered **leaves**, and 5–9-parted **flowers**. **Fruit** a 2–6-valved capsule.

***Drosera peltata* Sm. ex Willd.** (including *D. lunata* Buch.-Ham. ex DC.) Tufted **perennial herb** with slender annual aerial **stems**, erect or clambering to 30 cm, often branched in upper part. **Leaves** yellowish green, the basal, when present, in a dense rosette, stem-leaves alternate, with rounded to lunar-shaped blade, 2–5 mm long, glandular-hairy, especially on margin. **Inflorescence** terminal, with up to 22 flowers; sepals 5–7, united near base, 2–4 mm long; petals white, rarely pink, oblong, 4–6 mm long; styles 3. Gansu, Guangxi, Guizhou, Hubei, swSichuan, seTibet, Yunnan [swTibet, c & sChina, Himalaya, seAsia, Australia]; meadows, streamsides, open *Pinus* forests, scrub, to 3,700 m. VI–IX.

HAMAMELIDACEAE

HAMAMELIS

A genus of c. 6 species (1 in China) in temperate east Asia and North America. **Deciduous shrubs** or **small trees** with alternate, *Corylus*-like, simple, toothed **leaves**. **Flowers** produced in winter or early spring before leaves appear, with 4 thin, linear, ribbon-like, yellow petals; anthers 2-celled.

Hamamelis mollis* Oliv. **Deciduous shrub** or **small tree** with stout, zigzag, spreading branches, downy when young. **Leaves** broadly obovate, abruptly long-pointed, 7.5–12.5 × 5.5–10 cm, broadly and shallowly toothed, leaf-stalks 8 mm long, downy. **Inflorescence** many-flowered, stalkless clusters of very fragrant **flowers**, borne on previous year's twigs; sepals rich red-brown, hairy; petals yellow, linear, ribbon-like, 15–17 mm long. Guangxi, Hubei, eSichuan [c & eChina]; shrubberies, open forests, 300–800 m. II–IV.

1. *Corydalis melanochlora* growing on rugged scree at 4,400 m on Baimashan, nwYunnan (HJ)
2. *Corydalis melanochlora*; Baimashan, nwYunnan (CGW)
3. *Corydalis melanochlora*; Baimashan, nwYunnan (CGW)
4. *Corydalis calcicola*; Baimashan, nwYunnan (AD)
5. *Corydalis calcicola*; Baimashan, nwYunnan (HJ)
6. *Corydalis calcicola*; Yading Nature Reserve, swSichuan (AD)
7. *Ichtyoselmis macrantha*; Emeishan, sSichuan (PC)
8. *Dactylicapnos scandens*; eNepal (CGW)
9. *Dactylicapnos scandens*; cNepal (CGW)
10. *Dactylicapnos lichiangensis*; cult., from nwYunnan (CGW)
11. *Drosera peltata*; Cangshan, nwYunnan (CGW)
12. *Hamamelis mollis*; cult. (CGW)

1. *Liquidambar formosana*; cult. (PC)
2. *Corylopsis yunnanensis*; Cangshan, nwYunnan (PC)
3. *Corylopsis yunnanensis*; Cangshan, nwYunnan (PC)
4. *Corylopsis sinensis*; cult. (JG)
5. *Corylopsis sinensis*; wSichuan (TK)
6. *Corylopsis sinensis*; Erlangshan, wSichuan (MF)
7. *Corylopsis veitchiana*; cult. (JG)
8. *Corylopsis willmottiae*; cult. (MF)

LOROPETALUM

A genus of 3 species (all in China, 2 endemic), also found in India and Japan. **Evergreen**, stellate-hairy **shrubs** or **small trees** with alternate, simple **leaves**, margin entire or finely toothed. **Flowers** in clusters of 6–8, with the leaves, with 4–5, thin, linear, ribbon-like, white petals; ovary inferior, 4-celled. **Fruit** a woody capsule with 2 tips. Closely related to *Hamamelis*, which has 2-celled anthers and 4-parted flowers.

Loropetalum chinense (R. Br.) Oliv. (syn. *Hamamelis chinense* R. Br.) Twiggy, **evergreen shrub**, 1.5–2 m, often broader than tall; twigs crooked, wiry, often with brown stellate hairs. **Leaves** ovate or elliptic, pointed, oblique at base, 2.5–6.4 × 1.6–3.2 cm, margin finely toothed; petals 3–4 mm long, hairy. **Flowers** like those of *Hamamelis* but white or rarely pink; sepals and petals stellately white-downy, petals 4, linear, 1.8–2 cm × 1.5–2 mm. **Fruit** woody, egg-shaped, nut-like. Guangxi, Guizhou, Hubei, Sichuan, Yunnan [c & sChina, eHimalaya, Japan]; broadleaved forests, shrubberies, 1,000–1,200 m. II–IV.

LIQUIDAMBAR

A genus of 4 species (2 in China) in east Asia, North America and the Middle East. Large, dioecious, **deciduous trees** with opposite, palmately lobed, *Acer*-like leaves. **Inflorescences** pendent, globular, many-flowered; **flowers** greenish or yellowish; male in racemes or heads, consisting of stamens only; female flowers solitary, with sepals and 2–3 carpels only, and with needle-like staminodes that persist in the fruit; styles coiled backwards. **Fruit** a woody capsule.

Liquidambar formosana Hance (syn. *L. acerifolia* Maxim.) **Deciduous tree**, 10–40 m tall. **Leaves** palmately 3(–5)-lobed, 7–12.5 × 8–15 cm, lobes pointed to long-pointed, finely toothed, turning golden yellow to reddish bronze in autumn; petals 8–12 cm long, glabrous or hairy. **Fruit** globular, 2.5–3.7 cm across, in clusters of 24 or more, each capsule beaked and subtended by an awl-shaped scale. Guizhou, Hubei, Sichuan [s, c & eChina, Korea, Taiwan, nVietnam] widely planted in region and elsewhere; 300–1,100 m. V–VI.

CORYLOPSIS

A genus with nearly 30 species (20 in China, 19 endemic) distributed from India to China, Korea and Japan. **Deciduous** or **semi-evergreen shrubs** and **small trees** with slender branches that are usually adorned with stellate hairs. **Leaves** alternate, stalked, membranous, with a toothed margin. **Flowers** usually fragrant, in compact bracteate racemes, pendent or partly so, appearing before leaves, with 2–3 leaves at base of inflorescence-stalks; sepals 5, sometimes persistent; petals 5, yellow, usually rounded; stamens 5; style solitary, with capitate stigma. **Fruit** a 4-valved capsule.

1. Style shorter than petals, 1–2.5 mm long . . . .
   . . . . **C. yunnanensis, C. omeiensis**
1. Style equalling or exceeding petals, 5–7 mm long . . . .
   . . . . **C. sinensis, C. veitchiana, C. willmottiae**

Corylopsis yunnanensis Diels* **Shrub** to 3 m with oblong, glabrous **buds**. **Leaves** with short 10–12 mm long stalks, blades obovate to rounded, 5–8 × 3–6 cm, hairy along the veins beneath, with heart-shaped base and 8 pairs of lateral veins. **Racemes** 1.5–2.5 cm long, with 2 basal leaves on the short stalk; petals spathulate, 6–7 mm long, pale yellow. w & nwYunnan; forests, forest glades and margins, c. 1,500 m. IV–V.

Corylopsis omeiensis Yang* Readily distinguished from previous species by the very small **flowers** with petals c. 2 mm long; in addition, **leaves** are greyish white beneath and have 6–7 pairs of lateral veins. swGuizhou (Pan Xian), sSichuan (Emeishan); montane forests, c. 1,500 m. III–V.

Corylopsis sinensis Hemsl.* **Deciduous shrub** to 3 m with spreading **branches** and glabrous or downy **buds**. **Leaves** with 5–10 mm long, downy stalks, blade ovate to obovate or almost oblong, 3–9 × 2–6 cm, grey-brown, downy beneath, base asymmetrically heart-shaped, with 7–9 pairs of lateral veins. **Racemes** 3–4 mm long; petals greenish yellow, spathulate, 5–6 mm long; stamens shorter than petals, with yellow anthers, style slightly protruding; floral cup and ovary glabrous. Guangxi, Guizhou, Hubei, Sichuan [c & eChina]; montane forests, 1,000–1,500 m. III–V.

Corylopsis veitchiana Bean* Very similar to previous species, but bracts and petals always glabrous, the **leaves** with hairs confined to main veins beneath; **flowers** primrose-yellow with slightly broader petals with reddish brown anthers protruding slightly beyond them. Hubei, eSichuan [Anhui]; forests, c. 1,200 m. IV–VI.

Corylopsis willmottiae Rehd. & Wils.* Readily distinguished from previous 2 species by the stellate-downy floral cup and ovaries; **shrub** to 5 m, **leaves** with 10–15 mm long, glabrous stalks, blade glabrous or soon becoming so; **flowers** soft green-yellow, the petals and styles 3–4 mm long. wSichuan; montane forests, c. 1,200 m. III–V.

CERCIDIPHYLLACEAE

CERCIDIPHYLLUM

A genus with 1 species found in China and Japan. Large, **deciduous**, dioecious **trees** with pendulous or semi-pendulous branches bearing both long and short shoots. **Leaves** mostly opposite, simple, of 2 types, one on long shoots, the other on short shoots, stalked, turning red or yellow in the autumn; stipules falling early. **Inflorescences** appearing before leaves on short shoots, in bundles; **flowers** lacking sepals and petals; male flowers 4, stalkless; female flowers 2–6, short-stalked. **Fruit** a follicle with flattened, winged seeds.

Cercidiphyllum japonicum Sieb. & Zucc. Spreading **tree**, 10–20 m tall. **Leaves** of long shoots ovate, elliptic or obovate, 3.2–4.5 cm long, stalks 1.5–2 cm long; leaves of short shoots ovate to kidney-shaped, 3.7–9 cm long, stalk 1.4–4.7 cm long. **Male flowers** with stamens 9 mm long; **female flowers** with carpels 1–1.5 mm long. **Fruit** black to brown, 1–1.8 cm long. sGansu, neGuizhou, Hubei, Sichuan, sShaanxi, neYunnan [c & eChina, Japan]; forests and forest margins, often by streams, 600–2,700 m. IV–V.

PLATANACEAE

PLATANUS

A genus of 6 species (3 in China) distributed from south-east Europe to Iran, with 1 in South East Asia. **Deciduous trees** with characteristically irregularly flaking bark. **Leaves** alternate, usually palmately lobed, occasionally simple. **Flowers** in dense pendulous heads, wind-pollinated; sepals and petals 3–4. **Fruit** in dense, hairy heads, wind-dispersed.

Platanus orientalis L. Large monoecious, **deciduous tree** to 30 m with yellow-brown downy branchlets; bark peeling in large irregular flakes to leave a pattern. **Leaves** alternate, maple-like, 9–18 × 8–16 cm, downy at first, 5–7-lobed, long-stalked. Male and female **flowers** borne on separate branches, in dense globose inflorescences; male flowers 3–8-parted with tiny sepals and petals; female flowers with 3–8 carpels. **Fruit** in dense green (later brown), pendulous, globose clusters 2–2.5 cm in diameter. Widely planted tree in China, near habitation, along streets [seEurope, wAsia], at low to mid-altitudes. III–V.

TETRACENTRACEAE

TETRACENTRON

A genus with a single species in the Himalaya, China and Vietnam. Large **evergreen trees** with long and short shoots. **Leaves** alternate, simple, on one-year-old branches or on short older branchlets, stalked; stipules joined to leaf-stalks. **Inflorescences** many-flowered catkins; **flowers** hermaphrodite, in whorls of 4, 4-parted; ovary superior. **Fruit** a follicle with flattened, winged seeds.

Tetracentron sinense Oliv. **Tree** up to 40 m with grey-brown bark. **Leaves** broadly ovate, long-pointed, 7–16 × 4–12 cm, minutely toothed on margins. **Inflorescences** 80–125-flowered; **flowers** green-yellow, 1–2 mm across, male with protruding stamens. **Fruit** brown, 2.5–5 mm across, with 4–6, spindle-shaped **seeds**. sGansu, Guizhou, wHubei, Sichuan, sShaanxi, seTibet, Yunnan [swTibet, cChina, Himalaya, nMyanmar, nVietnam]; forests, ravines, streamsides, 1,100–3,500 m. IV–VII.

BETULACEAE

BETULA

A genus of c. 60 species (32 in China) in temperate regions of the northern hemisphere. **Deciduous shrubs** or **trees** with leathery or papery, conspicuously parallel-veined, simple **leaves**, gland-spotted on underside, margin toothed. **Flowers** unisexual: male catkins pendent, elongated-cylindrical, with overlapping bracts, each subtending 3 flowers, flowers with 2 stamens; female catkins ellipsoid or cylindrical, 1–several in a spike-like arrangement, with leathery 3-lobed bracts, each subtending 3 flowers. **Fruit** a small nutlet, with or without membranous wings.

1. Leaves leathery; nutlets without wings; shrubs or small trees not exceeding 5 m .... **B. calcicola, B. potaninii**
1. Leaves papery; nutlets with or without wings; trees at least 8 m tall 2
2. Nutlets without wings .... **B. delavayi, B. delavayi** var. **polyneura**
2. Nutlets with wings 3
3. Branchlets without resinous glands; female inflorescence 10–25 mm wide .... **B. austrosinensis**
3. Branchlets with sparse to dense resinous glands; female inflorescences not more than 10 mm wide ... 4
4. Bark reddish brown to orange or orange-red; leaves with 8–15 pairs of lateral veins .... **B. utilis, B. utilis** var. **prattii, B. albosinensis, B. albosinensis** var. **septentrionalis, B. luminifera**
4. Bark greyish white, silvery-white or greyish black; leaves with 3–8 pairs of lateral veins .... **B. platyphylla, B. insignis**

Betula calcicola (W. W. Sm.) P. C. Li* (syn. *B. delavayi* Franch. var. *calcicola* W. W. Sm.) **Deciduous shrub** to 4 m, generally only 1–2 m, with grey-black bark and brown, densely yellow-hairy branchlets. **Leaves** ovate, elliptic-ovate or ovate-lanceolate, pointed or long-pointed, 2–3.2 × 1.5–2.5 cm, corrugated, doubly pointedly toothed, densely white- or yellow-hairy beneath. **Fruit**: catkins oblong-cylindrical, 1.5–2 cm long. swSichuan, w & nwYunnan; rocky shrubberies, woodland margins, thickets on damp cliffs, moraines, 1,900–3,100 m. VI–VII.

Betula potaninii* Batalin (syn. *B. wilsonii* Bean) **Shrub** or **tree** to 5 m with grey-brown bark and brown, densely white- or yellow-hairy branchlets. **Leaves** ovate-lanceolate or oblong-lanceolate, pointed or long-pointed, 2–2.5 × 1–2.5 cm, doubly pointedly toothed, many-veined, densely yellow-brown or white silky-hairy along veins beneath. **Fruit**: catkins oblong-cylindrical, 1–2 cm long. seGansu, n, nw & swSichuan, Shaanxi; rocky shrubberies and thickets on damp cliffs, 1,900–3,100 m. V–VI.

Betula delavayi* Franch. **Tree** to 8 m with dark grey bark and brown, densely yellow-hairy branchlets. **Leaves** ovate, elliptic or oblong, long-pointed, 2–7 × 1–4 cm, doubly toothed, white-downy along veins beneath, with 10–14 pairs of lateral veins. **Fruit**: catkins oblong, pendent, 1–2.5 cm long. Gansu, wHubei, Qinghai, w, nw & swSichuan, eTibet, nwYunnan; broadleaved forests, shrubberies, 2,400–4,000 m. V–VI. **Var. *polyneura* Hu ex P. C. Li***, from nwYunnan (Lijiang), has 19–21 pairs of lateral veins per leaf.

Betula austrosinensis* Chun ex P. C. Li **Tree** to 25 m with brown to grey-brown, fissured bark and yellow-brown, somewhat hairy branchlets. **Leaves** ovate to elliptic or almost lanceolate, to 14 × 7 cm, hairy along veins and in axils beneath, resin-dotted beneath, margin finely doubly toothed, with 12–14 pairs of lateral veins. **Fruit**: catkins erect, oblong-cylindrical, to 5.5 cm long, downy. Guangxi, Guizhou, Hubei, Sichuan, Yunnan [cChina]; broadleaved forests, ravines, 700–1,900 m. V–VI.

1. *Platanus orientalis*; cult. (PC)
2. *Betula calcicola*; Gangheba, Yulongxueshan, nwYunnan (CGW)
3. *Betula calcicola*; Gangheba, Yulongxueshan, nwYunnan (CGW)
4. *Betula delavayi*; cult., from Yulongxueshan, nwYunnan (CGW)
5. *Betula austrosinensis*; cult. (TK)

Betula utilis D. Don (syn. *B. bhojpattra* Lindl.) **Tree** to 35 m with dark reddish brown or coppery bark, occasionally with a hint of pink, peeling in thin flakes, and brown resinous-glandular, downy branchlets. **Leaves** ovate, ovate-elliptic or ovate-oblong, long-pointed, 4–9 × 2.5–5.6 cm, doubly toothed, downy and resinous beneath; veins in 9–12 pairs. **Fruit**: catkins 1–3, pendent, 3–5 cm long. Gansu, Qinghai, w, nw & swSichuan, Shaanxi, e & seTibet, nwYunnan [Himalaya]; temperate broadleaved forests, riverbanks, 2,500–3,800 m. VI–VII. A valuable timber tree. **Var. *prattii* Burkill*** has striking bright orange **bark**, occasionally orange-brown or with a grey flush, and dark green, lustrous **leaves** that are downy beneath. Gansu, n, w & swSichuan, nwYunnan.

Betula albosinensis Burkill* (syn. *B. utilis* D. Don var. *sinenesis* (Franch.) Winkl.) Similar to *B. utilis* but **bark** orange-red or orange-brown, with a whitish bloom, peeling off in thin strips; branchlets glabrous. **Leaves** downy but not bearded in axils of lateral veins beneath, becoming hairless; veins in 9–14 pairs. sGansu, wHubei, Qinghai, Sichuan, sShaanxi [n & cChina]; temperate broadleaved forests, 1,000–3,400 m. V–VI. A valuable timber tree. **Var. *septentrionalis* C. K. Schneid.*** has smooth orange-grey **bark** with a marked white bloom, and more oblong **leaves**. Gansu, Sichuan, Shaanxi.

Betula luminifera Winkl.* Differs from the previous 2 species in the bright reddish brown **branches**, twigs covered with pale hairs. **Leaves** ovate, with a rounded to heart-shaped base, 5–10 × 2.5–6 cm, dark dull green above, paler and brighter beneath due to a mass of tiny luminous glands, simply toothed, each tooth terminating in an abrupt slender point; lateral veins in 10–14 pairs. **Female catkins** 3.2–8 cm long, solitary, occasionally paired. Gansu, Guangxi, Guizhou, Hubei, Sichuan, Shaanxi, Yunnan; woodland, river margins, 900–2,900 m. V–VI.

Betula platyphylla Sukaczev (syn. *B. alba* L. var. *tauschii* Regel; *B. japonica* Sieb. ex Winkl.) **Tree** to 30 m with greyish white or silvery-white **bark** (the white comes off on the hands!), peeling in sheets, and brown, sparsely resinous, glandular branchlets. **Leaves** ovate-triangular or broadly ovate, pointed or long-pointed, 3–9 × 2–7.5 cm, doubly toothed, glabrous and resinous beneath, bluish green; lateral veins in 5–8 pairs. **Fruit**: catkins pendent, 2–5 cm long, solitary. Gansu, Qinghai, w, nw & swSichuan, Shaanxi, seTibet, nwYunnan [Japan, Korea, Mongolia, eRussia]; temperate broadleaved forests, 700–4,200 m. VI–VII. An important timber tree. The western Chinese material is attributable to **var. *szechuanica* (Schneid.) Redh.*** (syn. *B. szechuanica* (Schneid.) Jansson); however, botanists do not agree on the status of this plant.

Betula insignis Franch.* **Tree** to 25 m with greyish black, fragrant **bark** and fragrant, brown branchlets. **Leaves** ovate-lanceolate to elliptic, long-pointed, 8–13 × 3–6 cm, doubly mucronately toothed, resinous beneath. **Fruit**: catkins erect or recurved, oblong, 2.5–4 cm long. Guizhou, wHubei, Sichuan; temperate broadleaved forests, 2,500–3,800 m. V–VI.

CARPINUS

A genus of c. 45 species (33 in China) in temperate regions of the northern hemisphere. **Deciduous shrubs** or **trees** with zigzag twigs. **Leaves** conspicuously parallel-veined, simple, toothed or doubly toothed. **Flowers** unisexual: male catkins pendent, borne on old wood, with flowers bearing numerous stamens; female inflorescences erect, terminal on young shoots. **Fruit-clusters** becoming pendent; **fruit** a ribbed nut, enclosed in an oblique, toothed bract.

1. Largest leaves more than 10 cm long, with at least 24 pairs of lateral veins; fruit-clusters densely overlapping .... **C. fangiana**
1. Largest leaves generally less than 10 cm long, with no more than 18 pairs of lateral veins; fruit-clusters loosely overlapping 2
2. Leaf-stalks densely yellow-hairy ...... **C. monbeigiana**
2. Leaf-stalks glabrous, sparsely silky-hairy or becoming hairless .... **C. viminea, C. pubescens, C. henryana**

Carpinus fangiana Hu* **Tree** to 15 m with sparsely hairy branchlets. **Leaves** elliptic-ovate, 10–27 × 2.5–8.5 cm, pointed, with 24–34 pairs of lateral veins, margin doubly toothed; stalks 10–15 mm long, glabrous. **Fruit-clusters** 13–50 cm long, bracts 15–25 mm long, light brown. nGuangxi, Guizhou, Sichuan, eYunnan; forests and thickets on rocky mountain slopes, 900–2,700 m. IV–VI.

Carpinus monbeigiana Hand.-Mazz.* **Tree** to 16 m with slightly hairy, dark purplish shoots. **Leaves** oblong-lanceolate or elliptic-lanceolate, 5–10 × 2.5–4 cm, pointed, doubly bristly toothed, softly hairy on veins on underside, with 14–18 pairs of veins; leaf-stalks 5–10 mm long. **Fruit-clusters** 5–8 cm long, bracts 16–20 mm long. seTibet, c & nwYunnan; subtropical broadleaved forests and thickets on slopes, 1,700–2,800 m. V–VI.

---

1. *Betula utilis* var. *prattii* bark; Heishui, Yulongxueshan, nwYunnan (CGW)
2. *Betula utilis* var. *prattii* bark; cult., ex Sichuan (CGW)
3. *Betula albosinensis*; Datonghe, eQinghai (RS)
4. *Betula albosinensis* bark; Jiuzhaigou, nwSichuan (PC)
5. *Betula albosinensis* var. *septentrionalis* bark; cult., from Sichuan (CGW)
6. *Betula luminifera*; cult. (CGW)
7. *Betula luminifera*; cult. (CGW)
8. *Betula platyphylla* var. *szechuanica*, with *B. delavayi*; Gangheba, Yulongxueshan, nwYunnan (CGW)
9. *Betula platyphylla* var. *szechuanica* bark; cult., from nwYunnan (CGW)
10. *Betula insignis*; Shennongjia, wHubei (PC)
11. *Carpinus fangiana*; cult. (RL)
12. *Carpinus viminea*; cult. (RL)

***Carpinus viminea* Lindl.** (syn. *C. fargesii* Franch.) **Shrub** or small **tree** to 20 m. **Leaves** ovate, 6–11 × 3–5 cm, doubly toothed, pointed to long-pointed, with 12–15 pairs of lateral veins; leaf-stalks 15–30 mm long, glabrous. **Fruit-clusters** 5–15 cm long, **bracts** 15–25 mm long. Guizhou, wHubei, eSichuan, seTibet, Yunnan [swTibet, c & sChina, Himalaya, Thailand, Vietnam]; woods and thickets on slopes, 400–2,600 m. IV–VI. (See photo 12 on p. 146.)

***Carpinus pubescens* Burkill** (syn. *C. seemaniana* Diels) **Shrub** or small **tree** to 17 m with very hairy branchlets. **Leaves** oblong, elliptic-ovate, 3–10 × 1.2–3.5 cm, pointed, doubly toothed, with 12–14 pairs of lateral veins; leaf-stalks 4–15 mm long, sparsely hairy and becoming hairless. **Fruit-clusters** 5–7 cm long, **bracts** 10–25 mm long. Guizhou, Hubei, sSichuan, Shaanxi, eYunnan [nVietnamn]; woods in valleys and thickets on limestone mountain summits and slopes, 450–2,600 m. V–VI.

Carpinus henryana* (Winkl.) Winkl. Differs from previous species in being a **tree** to 20 m with narrowly lanceolate to ovate-lanceolate, doubly toothed **leaves**, 5–8 × 2–3 cm, gland-dotted, silky-hairy on veins beneath; lateral veins in 14–16 pairs. eGansu, wGuizhou, wHubei, eSichuan, sShaanxi, ne & eYunnan; temperate mixed forests, 1,600–2,900 m. V–VI.

CORYLACEAE

CORYLUS

A genus of c. 20 species (7 in China) in temperate regions of the northern hemisphere. **Deciduous**, monoecious **shrubs** or **trees** with alternate, simple, unevenly toothed **leaves**. **Flowers** unisexual, appearing before the leaves; **male catkins** pendent, in clusters of 2–5, flowers with 4–8 stamens; **female inflorescences** bud-like, erect, 2-flowered, each with a bract and 2 bractlets. **Fruit** stalkless, a nut, enclosed in a leafy covering of enlarged bracts (husk).

1. Fruit with spiny husks .. *C. ferox, C. ferox* var. *tibetica*
1. Fruit with glabrous or hairy husks ....*C. heterophylla,*
    ... *C. heterophylla* var. *sutchuanensis, C. yunnanensis*

***Corylus ferox* Wall.** **Shrub** or **tree** to 20 m with glabrous, dark brown-hairy **shoots** and softly hairy **buds**. **Leaves** broadly obovate or ovate, 5–15 × 3.2–9 cm, long-pointed, irregularly toothed, hairy and glandular beneath; leaf-stalks 1–3 cm long. **Male catkins** 5–7.5 cm long. **Nuts** in clusters of 3–6, covered by spiny bracts, resembling a chestnut. sGansu, Guizhou, wHubei, Sichuan, Shaanxi, seTibet, Yunnan [swTibet, Himalaya]; thickets and mixed woodland margins, on mountain slopes, 1,500–3,800 m. **Var.** *thibetica* **(Batalin) Franch.** has glabrous **buds** and obovate or elliptic **leaves**. Sichuan, seTibet, Yunnan [swTibet, Ningxia, Myanmar].

Corylus heterophylla* Trautv.** **Shrub** or **tree** to 7 m with sparsely softly hairy branchlets. **Leaves** oblong to obovate, 4–13 × 2.5–10.5 cm, long-pointed, irregularly toothed, hairy on veins beneath; leaf-stalks 0.5–2 cm long. **Male catkins** 4–9.5 cm long. **Nuts** 2–6, covered by 1.5–2.5 cm long, entire bracts, downy and glandular near base, slightly longer than the nuts. Gansu, Guizhou, Hubei, neSichuan, Shaanxi [cChina, Japan, (?)Korea, eRussia]; thickets and temperate broadleaved woodland, 550–2,770 m. II–IV. **Var.** *sutchuanensis* **Franch. differs in having elliptic-obovate to almost orbicular **leaves** rounded and mucronate at apex and toothed bract-lobes. sGansu, Guizhou, Hubei, neSichuan, Shaanxi, Yunnan [c & eChina].

Corylus yunnanensis* (Franch.) A. Camus Like *C. heterophylla* but with yellow-hairy shoots and leaf-stalks, lobulate **leaves**, and **bracts** as long as the **nut**. wGuizhou, Hubei, w & swSichuan, Yunnan.

FAGACEAE

CASTANEA

A genus of c. 12 species (3 in China) in temperate regions of the Old and New Worlds. **Deciduous trees** with furrowed **bark**. **Leaves** spirally arranged, parallel-veined. **Flowers** unisexual; male in catkins, flowers in clusters of 1–3, each with 10–12 stamens; female flowers 3, borne on lower part of inflorescence, surrounded by numerous united spiny bracts. **Fru**it a nut 1–3 in a 2–4-valved husk. Both *C. mollissima* and *C. henryi* are cultivated for their edible nuts.

***Castanea mollissima* Blume** **Deciduous tree** to 20 m with velvety young shoots and elliptic-oblong to oblong-lanceolate **leaves**, 10–17 cm long, pointed to long-pointed, coarsely toothed, lacking glands on underside. Male **inflorescence** 10–13 cm long, solitary in leaf-axils. **Nuts** 2–3, enclosed in a husk 2–3 cm across, densely covered with soft spine-like bracts. Gansu, Guizhou, Hubei, Qinghai, Sichuan, Shaanxi, seTibet, Yunnan [c & eChina, Korea]; wooded slopes, near habitation, often cultivated, to 2,800 m. IV–VI.

Castanea seguinii* Dode Differs from the previous species in having **leaves** covered in scale-like glands and downy along the veins. Distribution similar to that of *C. mollissima* but not in Korea; mixed woods and thickets, 400–2,000 m. V–VII.

Castanea henryi* (Skan) Rehd. & Wils. **Deciduous tree** to 20 m, sometimes more, with rather dark, glabrous branchlets. **Leaves** dark green above, paler beneath, oblong-lanceolate, tapered to a slender point, 10–20 × 2.5–6 cm, with a bristle-toothed margin, glabrous except for a few hairs on veins beneath. **Male catkins** solitary in leaf-axils, 10 cm long or more. **Nuts** solitary, enclosed in a prickly husk c. 2.5 cm wide; usually 2 or 3 fruits together. Hubei, Sichuan; wooded slopes, ridges, close to habitation, 100–1,800 m. V–VI.

CASTANOPSIS

A genus of c. 120 species (58 in China) in tropical and subtropical Asian forests and woods. **Evergreen trees** with **bark** splitting or peeling longitudinally. **Leaves** alternate, in 2 opposite rows or spirally arranged, entire, simple, toothed or lobed, stellate-hairy when young. Male and female **flowers** on same tree but in separate inflorescences; male flowers in fascicles of 3–7 in erect catkins, each flower with 9–12 stamens; female flowers solitary or in clusters of 3–5, inconspicuous. **Fruit** a nut, 1–3, in bristly or spiny husks enclosing the nuts, maturing in the second year.

Castanopsis fargesii Franch. **Tree** 10–30 m tall with rusty scale-like hairy branchlets. **Leaves** leathery, oblong, ovate, or lanceolate, 7–15 × 2–5 cm, pointed or long-pointed, with thick and mealy hairs beneath, margins entire or with a few teeth; leaf-stalk 1–2 cm long. **Female inflorescence** solitary, to 30 cm long. **Nuts** 1 per husk, conical or almost globose, 1–1.5 cm long, husk 2.5–5 cm across, red-brown-hairy; bract spiny. Guizhou, Hubei, Sichuan, Yunnan [c & sChina, Taiwan]; broadleaved forests, 200–2,100 m. IV–V.

Castanopsis delavayi Franch.* **Tree** 10–20 m tall with glabrous branchlets. **Leaves** leathery, ovate, obovate, or elliptic, 5–13 × 3–7 cm, pointed or rounded, grey or silvery-white beneath, margin toothed or just wavy; leaf-stalks 1–2 cm long. **Male spikes** erect, yellow, to 12 cm long; **female inflorescence** solitary, 10–15 cm long. **Nuts** 1 per husk, egg-shaped, 1.3–1.4 cm across, husk 1.5–2 cm long, yellow-brown-hairy; bract spiny. swGuizhou, swSichuan, Yunnan [Guangdong]; mixed and broadleaved, evergreen forests, 1,500–2,800 m. IV–V.

FAGUS

A genus of c. 10 species (2 in China) in temperate regions of the Old and New Worlds. **Deciduous trees** with smooth, grey **bark**. **Leaves** alternate, parallel-veined. **Flowers** unisexual but borne on same tree; male flowers clustered, with 8–16 stamens, numerous in slender-stalked, globose heads; female flowers 2, surrounded by numerous united bracts. **Fruit** egg-shaped-triangular nuts, 1–2 in a woody, prickly or bract-like, 4-valved husk.

Fagus engleriana Seeman* **Deciduous tree** 8–25 m tall, often multi-trunked. **Leaves** ovate to elliptic, 5–10 × 2.5–5 cm, long-pointed, rather glaucous when young, margins wavy. Husks 1.5–1.8 cm long, covered with downy, thread-like bracts, **nuts** as long as husks. sGuizhou, nwHubei, eSichuan, Shaanxi, Yunnan [cChina]; broadleaved and mixed forests, 1,500–2,500 m. IV–V.

Fagus longipetiolata Seeman A larger **tree** than the previous species with larger, longer-stalked **leaves**, 9–15 cm long; husks larger, 2–2.5 cm long, curly-bristled, on long, downy stalks. Guizhou, Hubei, Sichuan, Shaanxi, Yunnan [c & sChina, Vietnam]; similar habitats, 300–2,400 m. IV–V.

1. *Corylus ferox* var. *thibetica*; cult. (MFL)
2. *Corylus heterophylla*; Shennongjia, wHubei (PC)
3. *Castanea mollissima*; Jiangkou, neGuizhou (PC)
4. *Castanea henryi*; Shennongjia, wHubei (PC)
5. *Castanea henryi*; Shennongjia, wHubei (PC)

LITHOCARPUS

A genus of c. 300 species (123 in China) in South and South East Asia, 1 species in North America. **Evergreen shrubs** or **trees** with smooth **bark**. **Leaves** spirally arranged, entire, simple, toothed or lobed, stellate-hairy when young. Male and female flowers on same plant but in separate inflorescences; **male inflorescence** erect, simple or branched catkins, flowers with 10–12 stamens; **female flowers** solitary or in clusters of 3, inconspicuous. **Fruit** a nut (acorn) enclosed in a small to large, corky, woody, horny or crustose cup.

Lithocarpus dealbatus **(Hook. f. & Thomson ex Miq.) Rehd.** **Tree** to 20 m with pale brown, hairy branchlets. **Leaves** leathery, ovate, ovate-elliptic or lanceolate, 7–14 ×2–5 cm, pointed or long-pointed, grey with waxy scales beneath, margins entire or undulate; stalk 1–2 cm long. **Male inflorescences** clustered, up to 15 cm long. **Female inflorescences** in terminal clusters, up to 20 cm long. **Nuts** globose, flattened above, 1–2 × 0.8–1.6 cm, cups 1–1.8 cm across with triangular, overlapping bracts. Guizhou, swSichuan, seTibet, Yunnan [eHimalaya, nLaos, nVietnam]; mixed forests, often with *Picea* and *Pinus yunnanensis*; 1,000–2,800 m. V–VI.

Lithocarpus variolosus **(Franch.) Chun** **Tree** to 20 m with blackish branchlets. **Leaves** leathery, broadly ovate, ovate-elliptic or lanceolate, long-pointed and sickle-shaped, 6–15 × 3–5 cm, with wax scales beneath; stalk 1–1.2 cm long. **Male inflorescences** solitary or branched; **female inflorescences** in terminal clusters, 3–6 cm long. **Nuts** globose, flattened above, 1–2 × 1.2–2.6 cm, cups 1.5–2.5 cm across with bracts in continuous or broken rings. swSichuan, nwYunnan [Vietnam]; mixed forests, often with *Picea*, *Abies* and subalpine *Quercus*, 2,500–3,000 m. V–VI.

QUERCUS

A genus of c. 450 species (102 in China) in northern temperate and subtropical forests and woods of both the Old and New Worlds. **Evergreen** or **deciduous shrubs** or **trees** with **bark** splitting or peeling longitudinally. **Leaves** spirally arranged, entire, simple, toothed or lobed, stellate-hairy when young. Male and female flowers on same tree but in separate inflorescences; **male flowers** with 4–7 stamens, in pendent catkins; **female flowers** few-many, inconspicuous. **Fruit** a nut (acorn) in a small to large, scaly, bristly or concentrically ringed cup.

1. Evergreen trees or shrubs; nuts with awl-shaped scales ..... ***Q. baronii, Q. monimotricha, Q. semecarpifolia, Q. guyavifolia, Q. aquifolioides, Q. rehderiana, Q. spinosa***
1. Deciduous trees or shrubs; nuts with long scales ... 2
2. Fruit-cups without concentric rings .... ***Q. dentata, Q. yunnanensis, Q. variabilis***
2. Fruit-cups with concentric rings .... ***Q. lamellosa, Q. schottkyana*** (both treated in the genus *Cyclobalanopsis* in *Flora of China*)

1. *Quercus monimotricha*; Lijiang, nwYunnan (CGW)
2. *Quercus semecarpifolia*; Kangding, wSichuan (PC)
3. *Quercus guyavifolia*; Heishui, nwYunnan (PC)
4. *Quercus aquifolioides*; nwYunnan (RS)
5. *Quercus dentata*; Jiuzhaigou, nwSIchuan (CGW)

Quercus baronii Skan* (syn. *Q. dielsiana* Seeman) **Evergreen shrub** or **tree** 2–15 m, with grey stellate hairs on shoots. **Leaves** ovate-lanceolate to oblong, 1.8–6.7 × 0.9–2.8 cm, pointed, with yellow-grey down beneath, margins spiny; leafstalks short, grey-yellow-hairy. **Nuts** 1–several, short-stalked, egg-shaped, 1–1.8 cm long, cup 0.8–1 cm across, enclosing two-thirds of nut, with downy awl-shaped bracts. Gansu, Hubei, nwSichuan, Shaanxi [cChina]; mixed forests on steep dry slopes above rivers, often on limestone, 500–2,200 m. VI–VII.

Quercus monimotricha (Hand.-Mazz.) Hand.-Mazz. (syn. *Q. spinosa* David ex Franch. var. *monimotricha* Hand.-Mazz.) **Evergreen shrub** 0.5–2 m, branchlets nearly whorled, with brown, tufted hair-covering. **Leaves** elliptic to obovate, 2–3.5 × 1.2–3 cm, blunt to mucronate, with conspicuously tufted, matted hairs when young, base rounded to shallowly heart-shaped, margin with long needle-like teeth, sometimes entire. **Nuts** egg-shaped, 1–1.3 × 0.8–1 cm, cup 3–4 × c. 10 mm, covering base of nut, with a 5–10 mm stalk. wSichuan, nwYunnan [Myanmar]; dry mountain slopes, rocky places, 2,000–3,500 m. VI–VII.

Quercus semecarpifolia R. Br. **Evergreen shrub** or **tree** to 30 m with soft red-brown stellate down on shoots. **Leaves** leathery, elliptic or oblong, 5–12 × 3–7 cm, rounded at apex, dark green above, pale fawn stellate-downy beneath; margin of leaves on young plants toothed, but on older ones entire. **Male catkins** 3–7.5 cm long. **Nuts** 1–2 on short downy stalk, globose or egg-shaped, 2–3 cm long, cup thin, shallow, disc-like, with triangular, erect, ciliate scales. wSichuan, seTibet [Himalaya]; dry slopes in mountains, often with *Juniperus* and rhododendrons, 2,400–4,000 m. V–VII.

Quercus guyavifolia H. Lévl.* (syn. *Q. ilex* L. var. *rufescens* Franch.; *Q. pannosa* Hand.-Mazz.) **Tree** or **shrub** to 15 m, very like *Q. aquifolioides*, but leaf-blades persistently hairy beneath (not becoming glabrous) and cup with an expanded, wavy rim; nuts 15–18 mm in diameter as opposed to 10–15 mm. Guizhou, sSichuan, Yunnan; montane forests, 2,500–4,000 m. V–VII.

Quercus aquifolioides Rehd. & Wils. Like *Q. guyavifolia*, but leaves at first with red or orange-brown down beneath, becoming glabrous at maturity, to 7 × 3.5 cm, nuts rather smaller, 10–15 mm in diameter. Guizhou, w & swSichuan, seTibet, Yunnan [swTibet, eHimalaya, nMyanmar]; forests up to subalpine levels, 2,000–4,500 m. V–VI.

The following closely allied species, treated by some botanists as conspecific with *Q. semecarpifolia*, have cups that are more deeply cupped: *Q. rehderiana* Hand.-Mazz., from Guizhou, Sichuan, Tibet, Yunnan [Thailand], and *Q. spinosa* Franch. (syn. *Q. gilliana* Rehd. & E. H. Wilson), from Gansu, Guizhou, Hubei, Sichuan, Shaanxi, Tibet, Yunnan [Myanmar], have **leaves** with only stellate hairs beneath, with **male catkins** up to 16 cm long in the former, but less than 3 cm long in the latter.

Quercus dentata Thunb. (syn. *Q. obovata* Bunge) **Deciduous tree** to 25 m, branchlets yellow-grey stellate-hairy. **Leaves** obovate, 10–30 × 6–30 cm, dark green above, densely greyish brown-downy beneath, base rounded, margin somewhat wavy or with a few coarse teeth on each side; leafstalks 2–5 mm long, brown-downy. **Female inflorescences** axillary on apical part of young shoots, 1–3 cm long. **Nuts** egg-shaped, to 23 × 15 mm, glabrous, cup deep, 12–20 × 20–50 mm including the reddish brown scales, enclosing up to two-thirds of the nut. Gansu, Guizhou, Hubei, Sichuan, Shaanxi, Yunnan [c & nChina, Japan, Korea]; mixed broadleaved forests, to 2,700 m. IV–V.

Quercus yunnanensis Franch.* Similar to *Q. dentata* but with shorter triangular cup-scales. swSichuan, nwYunnan; similar habitats and flowering time.

Quercus variabilis Blume (syn. *Q. bungeana* Forbes; *Q. chinensis* Bunge, not Abel) **Tree** 5–30 m tall with glabrous, greyish brown young shoots. **Leaves** ovate-lanceolate to oblong-elliptic, 7–17 × 2.5–6 cm, pointed, with bristle-like marginal teeth, with greyish stellate felt beneath; leaf-stalks 0.7–3 cm long. **Male catkins** up to 10 cm long. **Nuts** almost globose, 1–1.5 cm in diameter, cup with long, blunt, spreading, awl-shaped, hairy scales. Gansu, Guizhou, Hubei, Sichuan, Shaanxi, seTibet, Yunnan [c & nChina, Korea, Japan, Taiwan]; mixed forests, 500–2,500 m. V–VII.

Quercus lamellosa Sm. (syn. *Cyclobalanopsis lamellosa* (Sm.) Oerst.) Large **deciduous tree** to 40 m. **Leaves** narrowly elliptic to oblong, 15–25 × 6.2–10 cm, pointed, sharply and conspicuously toothed on margin. **Nuts** stalkless, 1–4 on a short, stout spike, flattish, 2.5–3 cm in diameter, cup of 10 concentric rings, almost enclosing nut. seTibet wYunnan [Himalaya]; deciduous forests, 2,000–3,600 m. VI–VII.

Quercus schottkyana Rehd. & Wils. (syn. *Cyclobalanopsis glaucoides* Schottky) A smaller tree than *Q. lamellosa* with glaucous, elliptic-lanceolate, pointed leaves with shallowly toothed margin, cup with concentric rings, 1 cm in diameter. wSichuan, seTibet, c & sYunnan [Myanmar]; steep valleys in mixed thickets and forests, 1,500–2,800 m. V–VII.

CARYOPHYLLACEAE

DIANTHUS

A genus of c. 300 species (16 in China), mostly Eurasian, but some in Africa, the majority confined to the mountains. **Annuals**, **biennials** and **perennials**, often tufted, but also cushion-forming. **Leaves** opposite and flowers in cymes, or solitary, fragrant; calyx tubular, with 2 or more pairs of bracts clasping the base; petals with a long claw and expanded, entire to toothed or fringed limb. **Fruit** a many-seeded capsule.

1. Petals with toothed margin . . . . **D. chinensis**
1. Petals with margin deeply cut into narrow lobes . . . . . .
   . . . . **D. superbus, D. longicalyx**

Dianthus chinensis L. (syn. *D. amurenese* Jacq.; *D. versicolor* Fisch. ex Link) A very variable **perennial herb** to 50 cm, generally less. **Leaves** linear-lanceolate, pointed, green or grey-green, to 50 × 4 mm, with prominent midvein. **Flowers** solitary or several in lax clusters, 25–35 mm across, scented; calyx cylindrical, 15–25 mm long, with 2 pairs of oval bracts in lower half; petals pink, red, purplish or white, obovate to almost triangular, generally with a dark ring in centre of flower and with some spotting and bearding. Gansu, Qinghai, Shaanxi [n & neChina, Korea, Mongolia, eRussia]; dry slopes, grassy and sandy places, forest margins, shrubberies, rocky places and ravines, banks, c. 3,400 m. V–VI.

Dianthus superbus L. Laxly tufted **perennial herb** to 60 cm with slender **stems**, branched towards base. **Leaves** bluish green or green, linear-lanceolate, to 100 × 5 mm. **Flowers** solitary or 2, mostly terminal, 30–50 mm across, scented; calyx cylindrical, 25–30 mm long, often reddish purple, with 2–3 pairs of obovate bracts at base, covering lower quarter of calyx; petals pink or white, unspotted, deeply cut for about half the length. Gansu, Guangxi, Guizhou, Hubei, Qinghai, Sichuan, Shaanxi [temperate Asia, Europe]; meadows and grassy banks, forest margins, shrubberies, streamsides, 400–3,700 m. VI–IX.

Dianthus longicalyx Miq.* Similar to previous species, but often taller; **leaves** up to 10 cm long, margin often minutely toothed; **flowers** 2-several in a lax cyme; calyx green, 30–40 mm long, with 3–4 pairs of bracts covering lower fifth of calyx-tube. Gansu, Guangxi, Guizhou, Hubei, Sichuan, Shaanxi [n & cChina]; similar habitats, 800–2,100 m. VI–VIII.

ARENARIA

A genus of c. 300 species (102 in China of which three-quarters are endemic) in the temperate northern hemisphere. **Annual** and **perennial herbs**, some forming dense cushions. **Leaves** generally opposite, flattish and entire. **Flowers** in dense to lax cymes or solitary; sepals 4–5, free; petals 4–5, free, entire to deeply notched, sometimes toothed; stamens 8 or 10, rarely 2–5; styles 2 or 3. **Fruit** a small 3- or 5-valved capsule.

1. Plants forming dense green, often moss-like, cushions or hummock 2
1. Plants not as above, generally loosely tufted 4
2. Sepals equalling or longer than petals . . . **A. kansuensis**
2. Sepals shorter than petals 3
3. Petals 7 mm long or more; sepals 5 mm long or more . .
   . . . . **A. oreophila, A. smithiana, A. qinghaiensis**
3. Petals not more than 5 mm long; sepals not more than 3 mm long . . . . **A. polytrichoides, A. densissima**
4. Petals notched, or at least truncated at apex . . . .
   . . . . **A. roseiflora, A. napuligera**
4. Petals entire 5
5. Flowers semi-nodding . . . . **A. eudonta**
5. Flowers ascending to erect 6
6. Leaves aggregated at stem-apex . . . . **A. longistyla**
6. Leaves not aggregated at stem-apex . . **A. glanduligera,**
          A. glanduligera var. cernua, A. rhodantha,
               A. melandryoides, A. roborowski

Arenaria kansuensis Maxim.* **Cushion-forming** plant to 30 cm across, making deep green, low mounds or hemispheres. **Leaves** tightly congested into small rosettes, linear-lanceolate with a hard pointed tip, triangular in cross-section, to 20 × 1 mm, with a few tiny marginal teeth near base. **Flowers** solitary on very short (2–4 mm) stalks; sepals lanceolate, 5–6 mm long, glandular-hairy, margin membranous; petals 5, white or cream, obovate, 4–5 mm long; anthers brownish. sGansu, Qinghai, w, nw & swSichuan, e & seTibet, nwYunnan; grassy and stony alpine slopes, gravels, old moraines, 3,500–5,300 m. VI–VIII.

Arenaria oreophila Hook. f.* (syn. *A. lichiangensis* W. W. Sm.) Dense **cushion-forming** hemispherical plant, with densely glandular-hairy **stems** and flower-stalks. **Leaves** linear to linear-elliptic, to 20 × 1.5 mm, with raised midvein and whitish margin, tip hard, pointed; stem-leaves shorter and broader, ovate-lanceolate, not more than 5 mm long, ciliate. **Flowers** solitary on stalks 5–8 mm long; sepals elliptic, 5 mm long, glandular-hairy; petals 5, white, narrow-obovate, 7–8 mm long; anthers yellow. seQinghai, w & swSichuan, nwYunnan [Sikkim]; stony and grassy alpine meadows, screes, 3,500–5,000 m. VI–VII.

Arenaria smithiana Mattf.* Similar to previous species but generally larger, the **cushions** to 18 cm high, leaf-margins inrolled and flower-stalks at least 10 mm long; **flowers** larger, sepals 6–10 mm long and petals 12–15 mm long, white or straw-coloured. seTibet, nwYunnan; similar habitats, 4,000–4,500 m. VI–VIII.

Arenaria qinghaiensis Y. W. Tsui & L. H. Zhou* Similar to *A. oreophila*, forming dense green **cushions** to 40 cm across; **leaves** linear, to 15 × 1 mm, with slightly reflexed, pointed tip; **flowers** on short glabrous stalks not more than 1 mm long; sepals linear-lanceolate, 7–8 mm long, petals slightly longer, blunt. Qinghai; alpine meadows, moors, stony places, c. 4,200 m. VI–VII.

---

1. *Dianthus chinensis*; Badaling, Beijing (CGW)
2. *Dianthus superbus*; near Hongyuan, nwSichuan (CGW)
3. *Arenaria kansuensis*; Huanglong, nwSichuan (CGW)
4. *Arenaria kansuensis*; Huanglong, nwSichuan (CGW)
5. *Arenaria oreophila*; Baimashan, nwYunnan (CGW)
6. *Arenaria smithiana*; Baimashan, nwYunnan (CGW)
7. *Arenaria qinghaiensis*; Shenshan, Jigzhi, seQinghai (HJ)

1. *Arenaria polytrichoides*; Daxueshan, nwYunnan (HJ)
2. *Arenaria polytrichoides*; near Batang, wSichuan (HJ)
3. *Arenaria polytrichoides*; Baimashan, nwYunnan (CGW)
4. *Arenaria longistyla*; Gangheba, Yulongxueshan, nwYunnan (CGW)
5. *Arenaria glanduligera*; eNepal (CGW)
6. *Arenaria rhodantha*; Huanglong, nwSichuan (CGW)
7. *Arenaria melandryoides*; nBhutan (RS)
8. *Arenaria roseiflora*; Baimashan, nwYunnan (CGW)
9. *Thylacospermum caespitosum*; Daxueshan, nwYunnan (CGW)

Arenaria polytrichoides Edgew.* Rather robust **cushion-plant** forming low domes to 50 cm across. **Leaves** densely overlapping, terminating in small, closely packed rosettes, linear-lanceolate, to 10 × 1 mm, with veins raised beneath, margin slightly inrolled and tip with abrupt point. **Flowers** solitary, tiny, stalkless; sepals ovate to elliptic, 3 mm long; petals 5, white, obovate to ovate, scarcely longer than sepals; anthers yellow. seQinghai, w & swSichuan, seTibet, nwYunnan; stony and grassy alpine meadows, screes, moraines, 3,500–5,300 m. VI–VII.

Arenaria densissima Wall. ex Edgew. Readily distinguished from previous species by **flowers** with stalks 2–4 mm long, petals c. 5 mm long, (twice as long as sepals), and violet anthers. seQinghai, w & nwSichuan, seTibet [swTibet, Himalaya]; similar habitats, 3,600–5,300 m. VI–VIII.

Arenaria eudonta W. W. Sm.* **Tufted perennial** with erect to spreading, sparsely hairy **stems** to 20 cm, sometimes more, often with some sterile shoots in some leaf-axils. **Leaves** opposite, short-stalked, linear to linear-lanceolate, to 25 × 2 mm, lowermost leaves often ovate to lanceolate. **Flowers** solitary or in cymes of up to 5, on glandular-hairy stalks; sepals lanceolate, 8–11 mm long; petals white or pink, oblanceolate to obovate, at least twice as long as sepals, generally 2–3-toothed at apex; anthers red; styles 2. nwYunnan; stony and grassy alpine meadows, 3,000–4,200 m. VI–VIII.

Arenaria longistyla Franch.* Small **tufted herb** to 8 cm; **stems** hairy, usually in 2 lines. **Leaves** linear to linear-lanceolate, to 20 × 2 mm, apex mucronate, base of each pair united into a sheath. **Flowers** axillary, on glandular-hairy stalks 6–50 mm long; sepals 5, lanceolate, 5–10 mm long, margin white, membranous; petals 5, white, oval to rounded, 5–10 mm long, blunt. w & swSichuan, seTibet, nwYunnan; alpine meadows, screes and stabilised moraines, streamside rocks, stony meadows, forest margins, 2,800–5,000 m. VI–VII.

Arenaria glanduligera Edgew. Small plant to 8 cm, forming rounded, greenish tufts, rarely more than 15 cm across. **Leaves** paired, ovate to lanceolate or almost rounded, flat, to 6 × 3 mm, glandular-hairy, ciliate. **Flowers** solitary or paired, on slender, white-hairy stalks, 5–20 mm long; sepals 5, lanceolate, 4–5 mm long, spreading; petals pink to violet-pink, darker at base, ovate to obovate, 5–7 mm long, spreading; stamens with violet or purplish filaments and pale violet to greenish anthers. seTibet [swTibet, Himalaya]; alpine meadows, screes, moraines, 4,500–5,500 m. VI–IX. **Var. cernua** F. N. Williams (syn. *A. forrestii* Diels) is a laxer plant to 15 cm tall, with **stems** hairy along 1 side only and **leaves** glabrous except for margin near the base; petals pink or white; anthers usually yellow. sGansu, seQinghai, w & swSichuan, seTibet, nwYunnan [Nepal].

Arenaria rhodantha Pax & K. Hoffm.* Similar to previous species, but flower-stalks glandular-hairy and petals violet, 8–9 mm long; anthers violet. w & nwSichuan, e & seTibet; similar habitats, 4,000–5,000 m. VI–VII.

Arenaria melandryoides Edgew. Readily distinguished from previous 2 species by generally having deep green, often violet-flushed **leaves**, 2 styles and a 2-valved capsule; **flowers** on recurved stalks, petals pink or white. seTibet, nwYunnan [swTibet, eHimalaya]; alpine meadows, 3,700–4,800 m. VI–VII.

Arenaria roborowski Maxim.* Distinguished from the previous 3 species by linear **leaves**, and small white **flowers** on glabrous stalks, petals only 4–5 mm long; sepals sharply pointed but with a broadened base, slightly longer than petals. sQinghai, w & nwSichuan, eTibet; similar habitats, 4,200–5,100 m. VII–VIII.

Arenaria roseiflora Sprague* **Tufted herb** to 15 cm, with spreading violet **stems**, brown-glandular-hairy above. **Leaves** paired, lanceolate to elliptic, to 15 × 5 mm, with rounded base and apex, ciliate. **Flowers** solitary, on recurved, hairy stalks; sepals lanceolate, 5–10 mm long, covered with violet glandular hairs; petals 5, pink or white, obovate, 10–18 mm long, apex shallowly notched; anthers violet. nwYunnan; rocky and grassy alpine meadows, screes, moraines, 300–4,500 m. VI–VII.

Arenaria napuligera Franch.* Readily distinguished from previous species by the **annual** habit, plants not more than 15 cm tall, **flowers** solitary (var. *monocephala* W. W. Sm.) or several in a cyme, on erect stalks; petals 12–16 mm long, usually notched or cleft; styles 2–3. w & swSichuan, seTibet, nwYunnan; stony and grassy alpine meadows, 3,000–5,100 m. VI–VII.

THYLACOSPERMUM

A genus with a single species distributed in the Tianshan, Himalaya and western China, which looks very similar to the cushion-forming species of *Arenaria*, but has sepals fused in lower half into a tube, petals entire, styles 2–3 and **capsule** leathery, opening with 4–5 teeth.

Thylacospermum caespitosum (Cambess.) Schischk. (syn. *Arenaria caespitosa* (Cambess.) Kozhevn.; *Bryomorpha rupifraga* Kar. & Kir.; *Thylacospermum rupifragum* (Kar. & Kir.) Schrenk) **Cushion-** or **mat-forming perennial** making moss-like, glabrous hummocks up to 1 m across, sometimes more. **Leaves** ovate to lanceolate, glossy green, 2–4 mm long, forming tiny crammed rosettes, with a pointed tip, ciliate. **Flowers** solitary, stalkless, 3–4 mm across; calyx green, tubular with 4–5 lanceolate lobes, 3-veined; petals white, ovate. Gansu, Qinghai, Sichuan, Tibet, nwYunnan [Himalaya, Tianshan]; rocky and stony mountain meadows and slopes, moraines, stabilised screes, rock-crevices, 4,300–6,000 m. VI–VII.

SILENE

A genus with c. 500 species (110 in China) in the northern hemisphere. Ranging from weedy, lowland **annuals** and coarse tufted **perennials** to **tuft-**, **mat-** and **cushion-forming** alpines. **Leaves** opposite, linear to elliptic or lanceolate. **Flowers** solitary, or more often in branched cymes; calyx tubular to campanulate; petals clawed, with an expanded notched or lobed limb; stamens usually 10; stigmas 3 or 5. **Fruit** a valved capsule.

1. Flowers solitary, rarely 2–3, occasionally more 2
1. Flowers numerous in branched cymes . . . . **S. esquamata**, **S. dawoensis**
2. Plants cushion- or mat-forming . . . . **S. davidii**
2. Plants neither cushion- nor mat-forming, generally tufted . . . . **S. nigrescens**, **S. nigrescens** subsp. *latifolia*, **S. himalayensis**, **S. gonosperma**

Silene davidii **(Franch.) Oxel. & Lidén*** (syn. *Lychnis davidii* Franch.) Bright green **perennial** forming an intricately branched many-crowned cushion. **Basal leaves** linear-oblanceolate, to 25 × 2–3 mm, glabrous, shortly ciliate, tip pointed; **stem-leaves** absent or in 1 or 2 pairs. **Flowers** solitary, ascending to erect, 15–20 mm across, held amongst the leaves; calyx dark violet, tubular-campanulate, 13–18 mm long; petals lilac, deep pink or pale red, with narrowly wedge-shaped claw and deeply notched rounded limb. seQinghai, w, nw & swSichuan, eTibet, nwYunnan; alpine meadows and moors, 4,100–4,700 m. VII–VIII.

Silene nigrescens **(Edgew.) Maj.** **Tufted perennial** herb to 15 cm, with several ascending, glandular-hairy **stems**. **Leaves** mostly at base of plant, deep green or grey-green, linear to narrow-oblanceolate, 30–60 × 2–4 mm, sparsely hairy, ciliate at base; stem-leaves in 1–2 pairs, smaller and narrower than basal leaves. **Flowers** solitary or 2–3, nodding, on glandular-hairy stalks; calyx inflated-globose, 18–20 mm long, membranous, with 10 prominent, dark violet veins; petals blackish violet or reddish violet, protruding 3–5 mm beyond calyx, notched. swSichuan, seTibet, nwYunnan [swTibet, Himalaya, Myanmar]; alpine meadows and screes, rocky and gravelly places, 3,000–4,500 m. VII–IX. Plants with broader leaves (5–10 mm wide) and petals that often protrude further are referable to **subsp.** *latifolia* **Bocq.*** swSichuan, nwYunnan.

Silene himalayensis **(Rohrb.) Maj.** (syn. *Lychnis himalayensis* (Rohrb.) Edgew.; *Melandrium himalayense* (Rohrb.) Y. Z. Zhao; *Silene gonosperma* (Rupr.) Bocq. subsp. *himalayensis* (Rohrb.) Bocq.) Very similar to previous species, but often taller, to 50 cm, usually with 3–6 pairs of **stem-leaves** and 3–7 **flowers** per stem, sometimes solitary; calyx smaller, 10–12 mm, with the violet veins characteristically joined to form a network; petals only slightly protruding beyond calyx. wHubei, Sichuan, Shaanxi, Tibet, w & nwYunnan [Hebei, Himalaya]; alpine meadows, open alpine scrub, 2,000–5,000 m. VI–VII.

Silene gonosperma **(Rupr.) Bocq.** (syn. *Melandrium pumilum* (Benth.) Walp.) Similar to *S. nigrescens*, but generally more densely downy, with linear-oblanceolate **basal leaves**, 30–60 × 4–8 mm, and 1–3 pairs of **stem-leaves**; calyx smaller, campanulate-globose, 13–15 mm long, with brown-violet to violet veins, not joined to form a network; petals violet, not usually protruding. Gansu, Qinghai, Sichuan, neTibet, nwYunnan [n & nwChina, cAsia]; alpine meadows, stony places, 1,600–4,400 m. VI–VII.

The above 3 species represent a very complex taxonomic grouping (section *Physolychnis*) whose relationship is not fully understood.

Silene esquamata **W. W. Sm.*** Erect **perennial herb** to 60 cm with slender, branched **stems** that are sticky above and hairy below. **Basal leaves** spathulate, soon withering; **stem-leaves** elliptic to oblanceolate, 2–7 × 0.5–1.5 cm, rough beneath, glabrous above, tip pointed to long-pointed. **Inflorescence** a lax, regularly branched thyrse; **flowers** erect, 18–24 mm across, on slender, often purplish stalks; calyx narrow-tubular (club-shaped in fruit), 12–17 mm long, with green or violet veins, almost glabrous; petals pale red to pink, the claw about the same length as calyx, limb obovate, deeply notched; stamens and styles protruding. w & swSichuan, nwYunnan; mountain grassland, stony pastures, open scrub, 1,800–3,300 m. VI–VIII.

Silene dawoensis **Limpr.*** Similar to previous species, but often taller, **stems** stouter and arising from a woody base; **stem-leaves** linear to linear-lanceolate, not more than 5 mm wide; **flowers** larger, 22–30 mm across, calyx 25–35 mm long. w & swSichuan, nwYunnan; similar habitats, sometimes also on cliffs and in rocky places, 1,400–3,100 m. VI–VIII.

AMARANTHACEAE

ALTERNANTHERA

A genus of c. 80 species (5 in China, only 1 native) primarily located in tropical and subtropical America. **Annual** and **perennial herbs** with creeping to ascending **stems** and opposite, entire-margined **leaves**. **Flowers** in stalked or sessile heads: tepals 5; stamens 3–5.

Alternanthera sessilis **(L.) R. Br. ex DC.** (syn. *A. nodiflora* R. Br.; *Gomphrena sessilis* L.) **Perennial herb** to 45 cm, generally less, with spreading or creeping **stems**, green or purplish, striped. **Leaves** linear-lanceolate to oblong or obovate, 1–8 × 0.2–2 cm, glabrous or somewhat hairy, margin entire to toothed. **Bracts** and bracteoles white, long-pointed. **Flowers** dense, borne on a white-hairy axis; tepals white, ovate, 2–3 mm long; stamens 3. Guangxi, Guizhou, Hubei, Sichuan, Yunnan [c & sChina, seAsia, Himalaya, India, Taiwan]; damp and swampy places, roadsides, near habitation, fallow fields, sea level to 1,200 m. V–VII.

1. *Silene davidii*; Gaoersi Pass, near Kangding, wSichuan (HJ)
2. *Silene nigrescens*; eNepal (CGW)
3. *Silene nigrescens*; Yading Nature Reserve, swSichuan (AD)
4. *Silene himalayensis*; Baimashan, nwYunnan (CGW)
5. *Silene gonosperma*; Meilixueshan, east side, nwYunnan (AD)
6. *Silene esquamata*; Cangshan, nwYunnan (CGW)
7. *Alternanthera sessilis*; Deqin, nwYunnan (PC)

1. *Phytolacca acinos*; Kunming, cYunnan (PC)
2. *Bistorta amplexicaulis*; cult. (CGW)
3. *Bistorta amplexicaulis*; cult. (CGW)
4. *Bistorta paleacea*; Wutoudi, Yulongxueshan, nwYunnan (CGW)
5. *Bistorta macrophylla*; Cangshan, nwYunnan (CGW)
6. *Bistorta vivipara*; Jiuzhaigou, nwSichuan (CGW)
7. *Bistorta emodi*; Cangshan, nwYunnan (CGW)

PHYTOLACCACEAE

PHYTOLACCA

A genus of 25 species (1 endemic to China), mostly cosmopolitan. **Herbs** or occasionally **trees** with rather thick stems and often fleshy or papery, alternate leaves. **Inflorescence** racemose to paniculate, terminal or opposite **leaves**; tepals 5, persistent; stamens 6–33, inserted at base of tepals. **Fruit** a berry with 5–16 free or fused carpels.

Phytolacca acinos Roxb. (syn. *P. esculenta* Van Houtte) Glabrous **perennial** to 1.5 m, with erect green or purplish **stems**. **Leaves** stalked, blade elliptic to lanceolate, 10–30 × 4.5–15 cm. **Inflorescence** an erect raceme on a stalk 1–4 cm long; **flowers** hermaphrodite, 7–9 mm across; tepals 5, white or greenish yellow; stamens 8–10 with pink anthers; carpels 8; **berries** 7–8 mm, purplish black when mature. Guangxi, Guizhou, Hubei, Sichuan, Shaanxi, seTibet, Yunnan [s, c & eChina, c & eHimalaya, Japan, Korea, Myanmar, Vietnam]; open forests, hillslopes, roadsides, in and around cultivation, often as a weed, 500–3,400 m. V–VIII.

POLYGONACEAE

BISTORTA

(included in *Polygonum* by some botanists)

A genus with 85 species (46 in China) widely distributed in the north temperate zone. **Perennial herbs** with entire **leaves** and simple or branched **inflorescences**; ocrea present at each node, green or membranous. **Flowers** in spikes or heads, often dense, sometimes with accompanying bulbils; flowers small, generally with 5 perianth-segments and often 8 stamens, sometimes fewer. **Fruit** a 3-angled nutlet.

1. Perennial herbs with dense cylindrical spikes of flowers 2
1. Subshrubs with creeping and rooting branches; flowers in lax, slender spikes .... *B. emodi*
2. Stem-leaves (at least the upper), stalkless and clasping stem .... *B. amplexicaulis*, *B. amplexicaulis* var. *pendula*
2. Stem-leaves not as above, generally short-stalked or almost stalkless 3
3. Inflorescence with flowers not replaced by bulbils .... *B. macrophylla*, *B. coriacea*, *B. paleacea*
3. Inflorescence with lower flowers replaced by bulbils .. .. *B. vivipara*

Bistorta amplexicaulis (D. Don) Greene (syn. *Polygonum amplexicaule* D. Don) Clump-forming, **herbaceous perennial** to 1 m tall in flower, usually less, with robust, erect, branched **stems**. **Leaves** ovate to oblong-ovate, 4–10 × 2–5 cm, with heart-shaped base, long-pointed apex and slightly back-rolled margin, basal leaves long-stalked; upper stem-leaves stalkless, half-clasping stem. **Inflorescence** a dense to lax terminal or lateral, erect to nodding (**var. pendula** H. Hara) spike; **flowers** red, perianth-segments elliptic, 3–5 mm long; styles 3. Gansu, Hubei, Sichuan, Shaanxi, Tibet, Yunnan [Himalaya]; damp places in grassy habitats, open shrubberies, mountain slopes, forest margins, often in partial shade, 1,000–3,300 m. VI–IX.

Bistorta macrophylla (D. Don) Soják (syn. *Polygonum macrophyllum* D. Don) Rhizomatous, **patch-forming perennial** with simple, erect **stems** to 30 cm. **Leaves** green with a grey-green reverse, basal stalked, with oblong to lanceolate blade, 3–11 × 0.2–3 cm, base slightly heart-shaped to wedge-shaped, margin back-rolled and with thickened vein-tips; stem-leaves few, stalkless or almost so, linear to lanceolate. **Inflorescence** a dense, erect spike to 2.5 cm long and half as wide; **flowers** pink or whitish, perianth-segments elliptic, 2.5–3 mm long; anthers purplish black, protruding; styles 3. Gansu, Guizhou, Hubei, seQinghai, Sichuan, Shaanxi, Tibet, Yunnan [Himalaya]; alpine meadows and other grassy open places, 2,000–5,000 m. VI–IX.

Bistorta coriacea Samuelsson & Hand.-Mazz.* Similar to previous species, but leaf-blades more leathery, ovate to elliptic-ovate, 4–14 × 2–4.5 cm; **stem-leaves** smaller, ovate-lanceolate, short-stalked; **spikes** 4–5 cm long, purplish red, perianth-segments 4–5 mm long, anthers blue-black. Guizhou, Sichuan, e & seTibet, Yunnan; forest margins, shrubberies, grassy slopes, 2,800–5,000 m. VII–IX.

Bistorta paleacea (Wall. ex Hook. f.) Yonekura & H. Ohashi (syn. *Polygonum paleaceum* Wall. ex Hook. f.) Similar to previous 2 species but a taller plant to 60 cm with thinner **leaves** that are grey-green beneath, oblong to lanceolate, 6–18 × 2–3 cm, glabrous to somewhat hairy, vein-tips thickened along the somewhat back-rolled margin. **Inflorescence** spike-like, 4–6 cm long; **flowers** pinkish or white. Guangxi, Guizhou, Sichuan, Yunnan [neIndia, nThailand]; meadows and grassy slopes, forest margins, 1,500–4,000 m. VI–VIII.

Bistorta vivipara (L.) S. F. Gray (syn. *Polygonum viviparum* L.) A slighter plant than the previous 2 species, but often reaching 50 cm, with slender **stems**, readily distinguished by the slender **spikes** of white or pale pink **flowers** in which the lower flowers are replaced by bulbils. Gansu, Guizhou, Hubei, Qinghai, Sichuan, Shaanxi, Tibet, Yunnan [northern hemisphere]; meadows, slopes, forest margins, shrubberies, streamsides, alpine steppe, 1,200–5,100 m. V–VIII.

Bistorta emodi (Meisn.) Petrov (syn. *Polygonum emodi* Meisn.) Spreading **subshrub** not more than 25 cm tall, with wiry branches rooting at nodes. **Leaves** in tufts, grey-green, lanceolate, 3–15 × 0.3–3.5 cm, with a short stalk and back-rolled margin. **Inflorescence** a terminal spike 2–6 cm long; **flowers** dull reddish purple with 5 perianth-segments. Sichuan, Tibet, Yunnan [Himalaya]; rock-crevices and -ledges, thickets, forests, 1,300–3,000 m. VI–VIII.

ACONOGONON

A genus of 40 species (21 in China) from the temperate Old World. Like *Bistorta*, but **flowers** in paniculate, corymbose, or clustered heads. Sometimes included in the genus *Persicaria*.

1. Plants 90 cm or more tall in flower .... **A. polystachyum, A. molle**
1. Plants not more than 90 cm tall in flower .... **A. campanulatum**

Aconogonon polystachyum **(Wall. ex Meisn.) M. Kral.** (syn. *Polygonum polystachyum* Wall. ex Meisn.) Stout, **tufted perennial** to 1 m, with rather angular, much-branched stems, usually downy. **Leaves** grey-green, short-stalked, lanceolate to oblong-lanceolate, to 17 × 7 cm, long-pointed, downy beneath, slightly so above, margin entire. **Inflorescence** paniculate, terminal, with spreading branches; **flowers** white or pinkish, 5-parted; largest perianth-segments 3 mm long. **Fruit** brown, 3-angled, hidden within the persistent perianth. Sichuan, seTibet, Yunnan [swTibet, Afghanistan, Himalaya, Myanmar]; forests, ravines, thickets, river- and streamsides, 2,200–4,500 m. VII–IX.

Aconogonon molle **(D. Don) Hara** (syn. *Polygonum molle* D. Don) Similar to previous species, but often taller, to 2 m, **leaves** elliptic to elliptic-lanceolate, to 20 × 6 cm, pale beneath; **flowers** always white, perianth-segments not more than 2 mm long. **Fruit** slightly longer than perianth, shiny, black. Guangxi, Guizhou, Tibet, Yunnan [Himalaya, nMyanmar, Thailand]; similar habitats, 1,200–3,500 m. VIII–IX.

Aconogonon campanulatum **(Hook. f.) Hara** (syn. *Polygonum campanulatum* Hook. f.) Patch-forming **perennial herb** to 90 cm, with branched, downy **stems**. **Leaves** stalked, ovate, to 15 × 5 cm, downy, ciliate, apex long-pointed to tailed. **Inflorescence** a lax panicle with small ovate bracts, each subtending 2–3 **flowers**; perianth campanulate, pink or white, 5-parted, with purple anthers not projecting. Guizhou, wHubei, Sichuan, Tibet, Yunnan [c & eHimalaya]; slopes, rocky places, open forests, forest margins, streamsides, 1,400–4,100 m. VII–VIII.

KOENIGIA

A genus of 4–5 species (4 in China) in the temperate northern hemisphere and the Himalaya. Small **annual** or **perennial herbs**, with alternate to almost opposite **leaves**. **Flowers** solitary or clustered, terminal and lateral, with 3 unwinged perianth-segments and 3 stamens; styles 2. **Fruit** a triangular nut.

Koenigia forrestii **(Diels) Mesicek & Soják** (syn. *Polygonum forrestii* Diels) Small **patch-forming**, creeping **perennial** with hairy **stems**. **Leaves** stalked, blade rounded to kidney-shaped, to 4 cm across, base heart-shaped, margin wavy, densely ciliate. **Inflorescence** terminal, corymbose; **flowers** white or cream, 4–5-parted, perianth-segments obovate, 4–5 mm long; stamens 6–8, with purple anthers. Guizhou, Sichuan, seTibet, Yunnan [swTibet, Himalaya]; grassy and rocky alpine meadows, roadsides, banks, 3,500–4,800 m. VII–IX.

Koenigia islandica **L**. Very small tufted, erect to spreading, reddish, glabrous **annual**. **Leaves** usually alternate, elliptic to obovate or rounded, to 5 × 4 mm, margin entire. **Flowers** in small lateral clusters, greenish, 3-parted, segments c. 1 mm long; styles 2–3. Gansu, Qinghai, Sichuan, Tibet, nwYunnan [nwChina, cAsia, Himalaya, Arctic northern hemisphere, Russia]; damp grassy and stony alpine meadows, pool margins, 2,000–4,900 m. VII–IX.

PERSICARIA

A genus of c. 80 species (38 in China) in the north temperate zone. Closely related to *Polygonum* but distinguished by having **inflorescence** terminal, often paniculate or almost capitate, rather than small lateral flower-clusters. *Bistorta* differs in having dense, short to long, cylindrical flower-spikes.

Persicaria capitata **(Buch.-Ham. ex D. Don) H. Gross** (syn. *Cephalophilon capitatum* (Buch.-Ham. ex D. Don) Tzvelev; *Polygonum capitatum* Buch.-Ham. ex D. Don) **Patch-forming**, creeping **perennial** with prostrate **stems** rooting at nodes. **Leaves** short-stalked, ovate to elliptic, 1.5–3 × 1–2.5 cm, green or blue-green, often with a V-shaped dark mark above, margin entire. **Inflorescence** terminal, capitate, to 10 mm in diameter, **flowers** pinkish, 2–3 mm long. Guangxi, Guizhou, Hubei, Sichuan, Tibet, Yunnan [Himalaya, India, Sri Lanka, Thailand, Vietnam]; shady places, rocky slopes, damp ledges, field and rice paddy margins, 600–3,500 m. VI–IX.

Persicaria runcinata **(Buch.-Ham. ex D. Don) H. Gross** (syn. *Cephalophilon runcinatum* (Buch.-Ham. ex D. Don) Tzvelev; *Polygonum runcinatum* Buch.-Ham. ex D. Don) A more upright plant than the previous species, generally 30–60 cm tall in flower, with distinctive **leaves** that have 1–3 pairs of lobes below a large terminal lobe, 4–8 × 2–4 cm; **flower-clusters** 5–15 mm, pink or whitish. Gansu, Guangxi, Guizhou, Hubei, Sichuan, Shaanxi, Tibet, Yunnan [seAsia, Himalaya]; similar habitats, 800–3,900 m. VI–IX.

Persicaria chinensis **(L.) H. Gross** (syn. *Polygonum chinense* L.) Very variable, stout, **rhizomatous perennial** to 1 m, with erect, much-branched **stems**, rather woody at base. **Leaves** short-stalked, stalks often with auriculate base, blade ovate to lanceolate, 4–16 × 1.5–8 cm, base truncated or somewhat heart-shaped, apex somewhat long-pointed. **Inflorescence**

1. *Aconogonon polystachyum*; eNepal (CGW)
2. *Aconogonon polystachyum*; cult. (CGW)
3. *Aconogonon molle*; Cangshan, nwYunnan (PC)
4. *Aconogonon campanulatum*; eNepal (CGW)
5. *Koenigia forrestii*; Baimashan, nwYunnan (CGW)
6. *Koenigia forrestii*; cult. (CGW)
7. *Persicaria capitata*; Cangshan, nwYunnan (CGW)

terminal or lateral, in 3–5 mm clusters making up a panicle-like structure; **flowers** white or pale pink, perianth-segments ovate, 3–4 mm long, becoming enlarged and blue-black in **fruit**; stamens not protruding. s & swGansu, Guangxi, Guizhou, Hubei, Sichuan, sShaanxi, Tibet, Yunnan [s, c & eChina, seAsia, Himalaya, India, Japan]; wet grassy places, ditches, riversides, forests, shrubberies, mountain slopes, to 3,000 m. VII–XI.

RHEUM

A genus of c. 60 species (38 in China, half endemic) scattered in temperate and subtropical Asia. Stout-rooted **perennials** with hollow **stems** and often large simple, deeply toothed or palmately lobed rather fleshy **leaves**, mainly in a basal rosette or on lower part of stem; stems with large membranous ocreae. **Inflorescence** spike-like or paniculate, with or without large enveloping bracts; **flowers** with a persistent perianth of 6 tepals, green, greenish yellow or reddish; stamens usually 9; styles 3. **Fruit** a triangular, winged nutlet.

1. Inflorescence concealed by large, overlapping leaf-like bracts . . . . **R. nobile, R. alexandrae**
1. Inflorescence exposed, without bracts 2
2. Plant stemless; infloresecence spike-like amongst the leaves . . . . **R. przewalskyi**
2. Plants with a stem and a few stem-leaves; inflorescence a narrow to broad panicle 3
3. Leaf-blade palmately lobed . . . . **R. tanguticum, R. palmatum**
3. Leaf-blade unlobed 4
4. Plants not more than 30 cm in flower; leaves not more than 6 cm long . . . . **R. delavayi, R. pumilum**
4. Plants 40–90 cm in flower; leaves at least 8 cm long . . . . **R. acuminatum, R. forrestii, R. likiangensis**

Rheum nobile **Hook. f. & Thomson** Large **monocarpic herb** to 2 m in flower, sometimes as little as 50 cm, with a stout, erect, lined **stem**, flowering after some years as a basal leaf-rosette. **Basal leaves** heart-shaped to rounded, to 35 × 35 cm, rather leathery, deep green or sometimes flushed with bronze, reddish or purplish, with a stout stalk; **stem-leaves** crowded, similar to basal leaves but decreasing in size upwards and transitional with the bracts. **Inflorescence** paniculate, cone-shaped, completely enveloped in large, downturned, overlapping, creamy or pale yellow bracts; **flowers** yellowish green, in clusters of up to 9 in the panicle, tepals c. 2 mm long. seTibet, nwYunnan [swTibet, eHimalaya]; exposed rocky slopes, stony alpine meadows, cliff-ledges, 3,700–4,800 m. VI–VII.

Rheum alexandrae **Batalin*** Readily distinguished from previous species by the rhizomatous, **patch-forming** habit, each clump generally producing several **inflorescences** to 80 cm; **basal leaves** up to 6, bright green, blade ovate to ovate-elliptic, 9–14 × 6.5–9 cm; **stem-leaves** merging with the elliptic to obovate, cream or greenish bracts. w & swSichuan, e & seTibet, nwYunnan; marshy and swampy ground, especially in open thickets or around lakes, 3,000–4,600 m. VI–VII.

Rheum przewalskyi **Losinsk.*** Stout short **herb** with all leaves basal, yellow-green, often purplish beneath. **Leaves** ovate to ovate-rhombic, 10–20 × 9–17 cm, rather leathery, rough, base almost heart-shaped, margin entire; leaf-stalks reddish purple, to 10 cm long. **Inflorescence** spike-like with 2–4 branches; **flowers** yellowish white. **Fruit** egg-shaped, 8–10 mm long with wings 3 mm wide. Gansu, Qinghai, nwSichuan; stony slopes, open shrubberies, open forests, 1,500–5,000 m. VI–VII.

Rheum tanguticum **(Maxim. ex Regel) Maxim. ex Balf.*** (syn. *R. palmatum* L. subsp. *tanguticum* Maxim. ex Regel) Stout **herb** to 2 m in flower, **stem** hollow, glabrous or downy. **Leaves** large, basal palmately 5-lobed, to 60 cm long, lobes pointed; stem-leaves few, much dissected into narrow-lanceolate, pointed lobes. **Inflorescence** a large, much-branched panicle; **flowers** small, purple-red, occasionally pale red, tepals c. 1.5 mm long. **Fruit** oblong, 8–10 mm long, wings c. 2.5 mm wide. Gansu, Qinghai, Shaanxi, eTibet; shrubberies, valley meadows, field margins, 1,600–3,000 m. VI–VII.

Rheum palmatum **L.*** (syn. *R. potaninii* Losinsk.) Similar to previous species but all **leaves** palmate with broad-triangular, pinnately lobed divisions, not deeply dissected, downy beneath; stem-leaves similar but decreasing in size upwards. **Inflorescence** a large, dense, rather narrow panicle, downy. Gansu, Hubei, Qinghai, Sichuan, Shaanxi, e & seTibet, Yunnan; rocky slopes and valley sides and bottoms, 1,500–4,400 m. VI–VII.

Rheum delavayi **Franch.** Short **herb** to 30 cm in flower, with an erect, purplish, solid **stem** to 5 mm in diameter. **Basal leaves** not usually more than 4, blade dark green, oblong-elliptic to ovate, 3–6 × 2.5–5 cm, base somewhat heart-shaped, margin entire; **stem-leaves** only 1–2. **Flowers** in a narrow panicle, purplish, 4–5 mm; tepals c. 2.5 mm long, with a deep purplish red margin. **Fruit** 8–9 mm long. w & swSichuan, n & nwYunnan [Bhutan, Nepal]; rocky slopes, open shrubberies, 3,000–4,800 m. VI–VIII.

1. *Rheum nobile*; Daxueshan, nwYunnan (HJ)
2. *Rheum alexandrae*; Zheduo Pass, wSichuan (HJ)
3. *Rheum alexandrae*; Kangding, wSichuan (PC)
4. *Rheum przewalskyi*; Huashixia, cQinghai (RS)
5. *Rheum tanguticum*; near Dawu, cQinghai (RS)
6. *Rheum palmatum*; wSichuan (TK)
7. *Rheum delavayi*; Daxueshan, nwYunnan (CGW)

Rheum pumilum **Maxim.*** Similar to the previous species, but leaf-blades always ovate-elliptic, not exceeding 5 × 3 cm, downy beneath, **fruit** smaller, 5–6 mm long. Gansu, Qinghai, Sichuan, eTibet; similar habitats, 2,800–4,500 m. VI–VII.

Rheum acuminatum **Hook. f. & Thomson** Tufted **herb** to 80 cm, with erect, glabrous **stems**, often flushed purple-red. **Leaves** with purple-red stalks, blade heart-shaped, 13–20 × 12–19 mm, long-pointed, dark green above, usually purple-red beneath, margin entire; **stem-leaves** 1–3, similar but smaller. **Inflorescence** a lax panicle, often 2-branched near middle; **flowers** purple-red, tepals to 2.5 mm long. **Fruit** nearly rhomboid or broad-ellipsoid, 7–8 mm long, with narrow purple-red wings. sGansu, Sichuan, Tibet, Yunnan [Himalaya, nMyanmar]; open forests, forest margins, stony slopes, open shrubberies, 2,800–4,000 m. VI–VII.

Rheum forrestii **Diels*** Erect **herb** to 90 cm in flower with a hollow **stem** to 14 mm in diameter, yellowish-downy in upper part. **Basal leaves** 3–5, blades mid-green, ovate with heart-shaped base, 12–20 × 10–18 cm, rough-hairy beneath; **stem-leaves** 1–2. **Inflorescence** narrow, branched from middle; **flowers** yellowish green; tepals lanceolate, c. 2 mm long, inner 3 with dark green median stripe. seTibet, nwYunnan; rocky slopes, 3,000–3,500 m. VI–VII.

Rheum likiangensis **Sam.*** A generally smaller plant than the previous species with **leaves** not exceeding 16 × 14 cm and with 5–7 (not 5–9) basal veins; tepals greenish white, broad-elliptic. swSichuan, seTibet, nwYunnan; forest margins, open forests, meadows, often amongst shrubs, 2,500–4,000 m. VI–VII.

OXYRIA

A genus of 2 species (both in China, 1 endemic) in Asia, Europe and North America. **Tufted perennials** with kidney- to heart-shaped, alternate leaves; ocreae present. **Flowers** numerous, in panicles or whorled racemes, greenish or whitish, perianth 4-parted with 2 outer and 2 inner tepals, the latter enlarging and persisting in fruit; stamens 6. **Fruit** a winged achene with membranous wings.

Oxyria sinensis **Hemsl.*** **Patch-forming perennial** with stout, woody rhizomes, to 50 cm with rough-hairy **stems**. **Leaves** kidney-shaped to rounded-heart-shaped, to 4 × 5 cm, rough-hairy on veins beneath, otherwise glabrous, margin wavy. **Inflorescence** paniculate, much-branched, with brown, papery bracts; **flowers** in clusters of 5–8, unisexual. Guizhou, Sichuan, e & seTibet, Yunnan; dry stony banks, grassy places, roadsides, 1,600–3,800 m. IV–X.

Oxyria digyna **(L.) Hill** Readily distinguished from the previous species by the slighter habit, not more than 30 cm tall, by smaller and thinner non-wavy **leaves** (to 3 × 4 cm) and by hermaphrodite **flowers**. Qinghai, Sichuan, Shaanxi, Tibet, Yunnan [widespread temperate northern hemisphere]; alpine meadows, rocky places, flushes, wet rocks, 1,300–4,900 m. VI–X.

RUMEX

A genus of c. 200 species (27 in China, 1 endemic) found in the north and south temperate zones. **Annual** and **perennial herbs** and occasional **subshrubs**, often with spear- or arrow-shaped **leaves**. **Fruit** papery with 1 or several wings derived from inner tepals, enlarged and wing-like.

Rumex hastatum **D. Don** **Shrub** to 90 cm with slender, much-branched purple-brown branches, green when young. **Leaves** spear-shaped, rather leathery, to 30 × 2 mm, with linear to triangular lobes. **Inflorescence** a terminal, rather lax panicle bearing flowers on slender stalks; **male flowers** with uniform tepals, but **female flowers** with outer tepals reflexed as **fruit** develops and inner enlarging. Fruit-valves pinkish, rounded to kidney-shaped. s & swSichuan, seTibet, Yunnan [swTibet, Himalaya]; dry stony and rocky slopes, rocky gullies, 600–3,200 m. IV–V.

THEACEAE

CAMELLIA

An Asian genus with c. 120 species (97 in China, 76 endemic) in north-east India, the Himalaya, South East Asia, North Korea and south Japan. **Evergreen shrubs** or small to medium-sized **trees**. **Leaves** leathery, simple, margin entire or toothed. **Flowers** solitary, hermaphrodite, terminal or axillary, often showy, with or without a stalk; bracts grading into sepals; petals 5–12, joined at base; stamens numerous in 2–6 whorls; ovary superior, 1–5-celled. **Fruit** thin- to thick-coated, dehiscent; **seeds** globose to polyhedral.

1. Flowers yellow . . . . ***C. petelotii***
1. Flowers white, pink or purple 2
2. Bracts (bud-scales) and sepals differentiated, bracts persistent; flowers white; fruit 3–5-celled . . . . ***C. sinensis, C. taliensis***
2. Bracts (bud-scales) and sepals not differentiated; flowers red, pink or white; fruit usually 3- or 5-celled 3
3. Fruit 5-celled . . . . ***C. yunnanensis***
3. Fruit 3-celled 4
4. Bracts velvety . . . . ***C. reticulata***
4. Bracts not velvety 5
5. Flowers mainly pink to red . . . . ***C. pitardii, C. pitardii*** var. ***yunnanica, C. saluenensis, C. saluenensis*** form, ***C. omeiensis***
5. Flowers white . . . . ***C. cuspidata, C. forrestii***

1. *Rheum pumilum*; Yading Nature Reserve, swSichuan (AD)
2. *Rheum acuminatum*; Yulongxueshan, nwYunnan (DC)
3. *Rheum likiangensis*; Muli, swSichuan (PC)
4. *Oxyria sinensis*; north of Dali, nwYunnan (PC)
5. *Oxyria sinensis*; cult. (CGW)
6. *Rumex hastatum*; near Xiaguan, nwYunnan (PC)
7. *Camellia petelotii*; cult. (PC)

1. *Camellia yunnanensis*; wYunnan (MF)
2. *Camellia reticulata*; above Shigu, nwYunnan (PC)
3. *Camellia pitardii*; Shizong, eYunnan (PC)
4. *Camellia saluenensis*; sSichuan (MF)
5. *Camellia saluenensis* form; cult., from nwYunnan (JM)
6. *Camellia cuspidata*; Fanjingshan, neGuizhou (PC)
7. *Schima wallichii*; eNepal (CGW)

Camellia petelotii (Merr.) Sealy (syn. *Thea petelotii* Merr.; *Camellia chrysantha* (Hu) Tuyama; *C. nitidissima* C. W. Chi; *Theopsis chrysantha* Hu) **Shrub** or small **tree** to 5 m, often less, with narrow-oblong to lanceolate, leathery **leaves**, 8–11 × 3–4.5 cm, base wedge-shaped, apex long-pointed, margin minutely toothed. **Flowers** deeply cupped, 2.5–4 cm across, borne laterally on stalks to 15 mm long, bracts usually 5, yellowish green; petals 8–10, pale primrose-yellow to golden, rather waxy; stamens numerous, with yellow anthers. **Fruit** large, 4–5 cm. Guangxi [nVietnam]; forests, river valleys, streamsides, 200–900 m. I–II. (See photo 7 on p. 165.)

Camellia sinensis (L.) O. Kuntze **Evergreen shrub** or small **tree** 1–4 m tall with densely softly hairy twigs. **Leaves** lanceolate, 4–11 × 2–3.5 cm, blunt to long-pointed, margin shallowly toothed. **Flowers** 1–3, stalked, 2.5–3.8 cm across, white, sweetly scented. **Fruit** egg-shaped, 4–6 cm across. Yunnan [nIndia]; thickets, forests, 2,200–3,000 m, but widely planted elsewhere in China as a crop (tea). XII–III.

Camellia taliensis (W. W. Sm.) Melchior Very like the previous species, but **leaves** 7.5–15 cm long, elliptic or obovate, and **flowers** larger white, 5–6.3 cm across. nwYunnan (east of Erhai) [nMyanmar, nThailand]; mountain and valley forests, 1,300–2,400 m. X–XI.

Camellia yunnanensis (Pitard ex Diels) Cohen-Stuart.* **Evergreen shrub** or small **tree** 1–4 m tall with densely softly hairy twigs. **Leaves** elliptic to broadly lanceolate, 4–6.5 × 2–3 cm, blunt to long-pointed, entire, densely woolly. **Flowers** stalked, 3–5 cm across, white, sweetly scented; sepal-margin membranous; petals 7–9, obovate, 1.5–2.5 cm long, reflexed above middle. **Fruit** egg-shaped, 4–6 cm across. Guizhou, Sichuan, Yunnan; woods and thickets on hillsides on plateaus, 2,000–2,600 m. XII–III.

Camellia reticulata Lindl.* Large **evergreen shrub** or **tree** to 11 m with glabrous twigs. **Leaves** leathery, obovate to elliptic, 5–11.5 × 2.5–5.5 cm, blunt to somewhat long-pointed, margin finely toothed. **Flowers** 6–11 cm across, terminal or lateral; petals 5–8, rose-pink, rarely white, obovate, 5–6 cm long; ovary downy. **Fruit** globular, 4–9 cm across. w & nwYunnan; woods and scrub on hillsides, 1,900–2,900 m. II–V. Many of the temple-grown camellias (e.g. in Kunming and Lijiang), are selected forms of this species.

Camellia pitardii Cohen-Stuart* **Evergreen shrub** or **tree** to 7 m with smooth twigs. **Leaves** oblong to oblanceolate-elliptic, 4.5–12.5 × 1.8–3.5 cm, pointed to tailed, margin sharply toothed. **Flowers** 5–6 cm across; petals 5–6, rose-red to white, obovate, 4–5 cm long; ovary downy. **Fruit** globular, 3 cm across. Guangxi, Guizhou, sSichuan, Yunnan [Hunan]; thickets and woods on plateaus and hillside, 1,500–2,800 m. XII–IV. **Var. yunnanica** Sealy* has downy twigs and brown-hairy sepals. wYunnan.

Camellia saluenensis Stapf ex Bean* Well-branched, **evergreen shrub** 1–5 m tall with downy twigs. **Leaves** leathery, narrowly oblong to elliptic, 5–5.5 × 1–2.5 cm, pointed, finely toothed, each tooth with black gland at tip. **Flowers** 4–5 cm across, on short stalks; petals 6–7, pale blush-pink, obovate, notched. **Fruit** globular, 2–2.5 cm across. wGuangxi, s & swSichuan, c & wYunnan; woods and scrub on hillsides, 1,900–2,500 m. III–V.

Camellia saluenensis form* Small dense **evergreen shrub** to 0.75 m, twigs slightly woolly at first. **Leaves** elliptic to elliptic-oblanceolate, 1.8–3.5 cm long, pointed or almost so, margin minutely toothed. **Flowers** stalkless, 3–4 cm across; sepals greenish, glabrous; petals pink at first, eventually fading to almost white, obovate, 2.5–3.5 cm long, notched. nwYunnan (nw of Kunming); degraded *Pinus* woodland, 2,100–2,200 m. II–IV.

Camellia omeiensis Chang* Readily distinguished from *C. saluenensis* by glabrous twigs and larger elliptic **leaves**, 9–12 × 4–5.5 cm, abruptly pointed. **Flowers** terminal, stalkless, 8–9 cm across; petals 7–9, red, broadly orbicular to obovate, 4–5.5 cm long. sSichuan (Emeishan); similar habitat and altitude. III–V.

Camellia cuspidata Veitch* **Evergreen shrub** 1–2 m tall with glabrous twigs. **Leaves** narrowly to broadly elliptic, 5–9 × 1.4–3 cm, long-pointed to tailed, dark or purplish green beneath, margin finely toothed. **Flowers** stalked, 2–3.5 cm across, sweetly scented; sepals green; petals 6–7, white; stamens yellow. Guizhou, Hubei, Sichuan, Shaanxi, Yunnan [s & seChina]; coniferous and broadleaved forests, 1,500–2,500 m. III–V.

Camellia forrestii (Diels) Cohen-Stuart **Evergreen shrub** 1–5 m tall with densely woolly twigs. **Leaves** ovate to oblong-ovate, 3–5 × 1–2.5 cm, long-pointed, densely woolly, margin entire. **Flowers** stalked, small, 1–2 cm across, sweetly scented; sepals downy; petals 5–6, white, obovate, 0.5–1 cm long; ovary downy. **Fruit** egg-shaped, 1–2 cm across. cYunnan [Vietnam]; woods and thickets, 1,500–3,000 m. I–II.

SCHIMA

A genus of c. 20 species (13 in China) ranging from the Himalaya to South East Asia. **Evergreen trees** with alternate, entire **leaves**. **Flowers** in terminal clusters, 5-parted, with a small 5-toothed calyx and large showy petals; buds globular; stamens numerous. **Fruit** a 5-valved capsule, with flat, winged **seeds**.

Schima wallichii (DC.) Korth. **Evergreen tree** to 25 m with rugged, dark grey **bark** and leathery, deep green, elliptic-oblong stalked **leaves**, 10–18 × 3.5–7 cm, usually entire, with reddish veins beneath. **Flowers** in showy terminal clusters, fragrant, 3–4.5 cm across; calyx small, green, 5-toothed; petals 5, broad-ovate, white; stamens numerous, yellow. **Fruit** a globular capsule. (?)Guangxi, sSichuan, seTibet, Yunnan [c & eHimalaya]; forests, forest margins, borders of cultivation, 200–2,000 m. V–VI.

ACTINIDIACEAE

ACTINIDIA

A genus of c. 30 species (52 in China) in eastern Asia, particularly China. Mainly dioecious, twining, **deciduous climbers**, with **flowers** in lateral clusters on current year's shoots, these often pendent or semi-pendent; sepals and petals often 5, but sometimes 4 or 6–7; stamens 10 or more; anthers dehiscing by pores. **Fruit** a succulent berry.

1. Some leaves splashed with white in part or all over ... . **A. kolomikta, A. pilosula**
1. Leaves plain green or bluish green 2
2. Flowers white to cream, relatively large, at least 30 cm across, flattish to saucer-shaped .... **A. chinensis**
2. Flowers white, pink or red, smaller, not more than 15 mm across, cupped ..... **A. coriacea, A. henryi, A. callosa, A. rubricaulis**

Actinidia kolomikta (Maxim. & Rupr.) Maxim. Vigorous **slender climber** to 7 m. **Leaves** ovate to ovate-oblong, to 15 × 10 cm, rounded or heart-shaped at base, somewhat bristly at first, plain green, but some leaves white at tip or in upper half, occasionally white overall, this turning pink with age. **Flowers** with lily-of-the-valley (*Convallaria*) fragrance, 1–3 in leaf-axils, cupped, 11–13 mm across; petals white or pinkish, oval. **Fruit** ovoid, 25–30 mm long, brownish when ripe. sGansu, Hubei, Sichuan, Shaanxi, Yunnan [neChina, Japan, Korea]; forests, forest margins, shrubberies, river valleys, 1,800–2,900 m. VI–VII.

Actinidia pilosula (Finet & Gagnep.) Stapf ex Hand.-Mazz.* Similar to previous species, but a more downy plant overall, especially on leaf undersurface, **leaves** smaller, generally not more than 12 × 6 cm, elliptic to elliptic-lanceolate, base wedge-shaped to truncated, some leaves half or completely white above; **flowers** in clusters of up to 7. Hubei, Sichuan, Yunnan [cChina]; similar habitats and altitudes. VI–VII.

Actinidia chinensis Planch.* Vigorous, dioecious **climber** with reddish-bristly, twining young branches. **Leaves** heart-shaped to rounded, stalked, to 20 × 15 cm, stiffly hairy on margin and beneath. **Flowers** on short lateral branches from year-old wood, accompanied by rather smaller leaves; white at first but soon turning cream and then buff-yellow, 30–37 mm across, saucer-shaped; calyx with rounded, woolly lobes; petals obovate; stamens numerous in male flowers. **Fruit** walnut-sized, green, covered in stiff reddish brown hairs. sGansu, wHubei, Sichuan, Yunnan [cChina]; *Quercus* and other deciduous forests, shrubberies, to 2,400 m. VI–VII. Widely cultivated for the edible fruit, which are often considerably larger than the wild form; hermaphrodite clones are known.

Actinidia coriacea Dunn* Vigorous unisexual **climber** to 7 m with glabrous young shoots, dark brown later and covered with white lenticels. **Leaves** stalked, leathery, lanceolate to oblanceolate, 7.5–14 × 2–4.5 cm, dark green above, paler beneath, margin finely toothed. **Flowers** 1–3 in leaf-axils, lurid red to reddish purple, 12–13 mm across; sepals greenish; petals 5, occasionally more, concave, oval. **Fruit** globose to egg-shaped, brown with white dots, 17–20 mm long. nwGuangxi, Guizhou, wHubei, Sichuan, Yunnan [wHunan]; forests, thickets, ravines, 800–2,200 m. VI–VII.

Actinidia henryi (Maxim.) Dunn* (syn. *A. callosa* Lindl. var. *henryi* Maxim.) Like previous species, but **leaves** narrow-ovate with a heart-shaped base, to 16 × 6.8 cm, paler or slightly glaucous beneath, and **flowers** pearly white, 9–11 mm across, in rounded clusters. Hubei, Sichuan, Shaanxi, Yunnan; similar habitats, 1,200–2,600 m. VI–VII. The closely related *A. callosa* Lindl. (native to the Himalaya and with which it has been much confused) has coarser **leaves** and white to pale yellow **flowers**.

Actinidia rubricaulis Dunn* Similar to the previous 2 species, but **leaves** elliptic-ovate, rather shiny pale green beneath and **flowers** more open, white, 4-petalled. nwGuangxi, Guizhou, Hubei, Sichuan, Yunnan [wHunan]; similar habitats, 300–2,900 m. VI–VII.

STACHYURACEAE

STACHYURUS

A genus of 8 species (7 in China, 4 endemic) confined to eastern Asia. **Deciduous** or **evergreen trees** and **shrubs**, sometimes climbing. **Leaves** simple, alternate. **Inflorescence** forming in autumn and flowering in spring before leaves appear, spike-like or racemose, erect to pendent; **flowers** small, regular, 4-parted, with 8 stamens in 2 series. **Fruit** a small many-seeded berry.

Stachyurus chinensis Franch.* **Shrub** to 2.5 m with arching, dark green to reddish brown shoots. **Leaves** stalked, ovate to oblong-ovate, glabrous, 7–17 cm long, long-pointed, margin toothed. **Flowers** up to 30 in pendent racemes like strings of beads, each flower very pale yellow and c. 7–8 mm across, with style just protruding. Guangxi, Hubei, Sichuan, Yunnan; forest margins and glades, shrubberies, ravines, 400–3,000 m. III–IV. Sometimes included within the Japanese *S. praecox* Sieb. & Zucc.

1. *Actinidia kolomikta*; Wolong, wSichuan (PC)
2. *Actinidia kolomikta*; Emeishan, sSichuan (PC)
3. *Actinidia pilosula*; Jiuzhaigou, nwSichuan (CGW)
4. *Actinidia pilosula*; Wolong, wSichuan (PC)
5. *Actinidia chinensis*; Shennongjia, wHubei (PC)
6. *Actinidia chinensis*; Shennongjia, wHubei (PC)
7. *Actinidia coriacea*; Yutang, neGuizhou (PC)
8. *Actinidia henryi*; above Jiuzhaigou, nwSichuan (CGW)
9. *Actinidia rubricaulis*; Shennongjia, wHubei (PC)
10. *Stachyurus chinensis*; Wangxiantai, wGuizhou (PC)
11. *Stachyurus chinensis*; cult. (CGW)

SYMPLOCACEAE

SYMPLOCOS

A genus of c. 250 species (42 in China) in tropical and subtropical regions of America, Australia and Asia. **Evergreen** or rarely **deciduous shrubs** or **trees**. **Leaves** spirally arranged or in 2 opposite rows, simple, margin entire, toothed, or glandular-toothed; stipules absent. **Inflorescences** spikes, racemes, panicles, or clusters, rarely flowers solitary. **Flowers** regular, usually hermaphrodite, supported by 1 bract and 2 bracteoles; calyx-lobes usually 5, persistent, valvate or overlapping; corolla white or yellow, divided nearly to base, lobes usually 5, overlapping. Stamens many, rarely 4 or 5, joined to base of corolla-tube; anthers almost globose, 2-celled. Ovary inferior to half inferior, usually 3-celled, usually with an apical 5-glandular, annular, cylindrical, or 5-lobed disc. Style 1, thread-like; stigma small, capitate or 2–5-lobed. **Fruit** a drupe.

Symplocos sumuntia Buch.-Ham. ex D. Don (syn. *S. botryantha* Franch.; *S. caerulea* H. Lévl.; *S. caudata* Wall.; *S. cavaleriei* H. Lévl.; *S. decora* Hance; *S. dolichostylosa* Y. F. Wu; *S. fuboensis* M. Y. Fang; *S. leucophylla* Brand; *S. macrostroma* Hayata; *S. ovatibracteata* Y. F. Wu; *S. prunifolia* Sieb. & Zucc.; *S. punctata* Brand; *S. rachitricha* Y. F. Wu; *S. sasakii* Hayata; *S. somai* Hayata; *S. sozanensis* Hayata; *S. subconnata* Hand.-Mazz.; *S. swinhoeana* Hance; *S. urceolaris* Hance) **Shrub** or **small tree** to 6 m. **Leaves** somewhat leathery, elliptic, narrowly ovate or ovate, 2–10 × 0.7–4.5 cm, long-pointed, glossy dark green above, paler beneath, margin toothed. **Inflorescence** axillary, racemose, up to 6 cm long. **Flowers** opening widely, white; calyx-lobes ciliate; corolla-lobes 4–8 mm long; stamens up to 40, longer than corolla-lobes. **Fruit** black. Guizhou, Hubei, Sichuan, Yunnan [throughout c & sChina, Taiwan, Japan, Korea, seAsia, Himalaya]; in montane and hill forests and thickets, up to 2,200 m. II–XI.

CLUSIACEAE (GUTTIFERAE)

HYPERICUM
(formerly in Hypericaceae)

A cosmopolitan genus with c. 460 species (64 in China, 33 endemic). **Perennial herbs**, **shrubs** or rarely **trees**, with opposite, gland-dotted, entire **leaves**. **Inflorescences** cymose; **flowers** regular, golden yellow to lemon-yellow, lacking nectar; sepals 5, free or partly united, often gland-dotted; petals 5, free, contorted in bud, showy and thin-textured; stamens many in 5 bundles opposite the petals, filaments long, slender; ovary 1–5-celled, with 3–5 styles projecting beyond stamens. **Fruit** a capsule containing numerous small **seeds**.

1. Perennial herbs .... *H. daliense, H. perforatum*
1. Shrubs 2
2. Mat-forming plants .... *H. reptans*
2. Shrubs up to 3 m 3
3. Stamens in bundles of 50–70 ... *H. patulum, H. henryi, H. hookerianum*
3. Stamens in bundles of 45–50 3
4. Leaves stalkless or almost so .... *H. subsessile*
4. Leaves distinctly (often very shortly so) stalked .... *H. lancasteri, H. latisepalum, H. bellum, H. forrestii, H. pseudohenryi*

Hypericum daliense N. Robson* Plants to 40 cm, erect, from creeping, rooting base; **stems** single or few, unbranched, 4-angled when young, becoming terete, eglandular. **Leaves** stalkless, elliptic-oblong to lanceolate, 2.2–4.5 cm long, apex rounded, gland-dots on leaf-blade dense and pale, except for those inside margin which are black, margin entire. **Inflorescence** 5–14-flowered, from 1 or 2 nodes, nearly flat-topped, congested. **Flowers** 1–1.5 cm in diameter, star-shaped; petals bright yellow, oblong-oblanceolate, 6–8 mm long; stamens 25–35. nwYunnan (Dali and Lijiang ranges); open places amongst scrub, 2,400–3,100 m. VII–VIII.

Hypericum perforatum L. (syn. *H. perforatum* var. *confertiflorum* Debeaux) Differs from the previous species in usually being taller, to 70 cm, much-branched in upper part, internodes markedly 2(–4)-lined, lines with black glands; **leaves** smaller, 1–2.5 cm long, oblong to oblong-elliptic; petals golden yellow, to 1.5 cm long, capsule-valves with central linear and lateral vesicular glands. Gansu, Guizhou, Hubei, Sichuan, Shaanxi, Yunnan [c & nChina, n & wAsia, nwAfrica, Europe, Atlantic islands]; a weed of cultivation and waste places, also in sparse woods, meadows, grassland and steppe, usually in dry or well-drained habitats, sea level to 2,800 m. VI–IX.

Hypericum reptans Hook. f. & Thomson ex Dyer Prostrate or spreading **subshrub**, to 30 cm, forming clumps or mats to 1 m across, **stems** 4-lined, branching pinnately, rooting at nodes. **Leaves** elliptic or elliptic-oblong to oblanceolate or more rarely obovate, 7–16 mm long, paler or glaucous beneath, apex blunt. **Inflorescences** 1-flowered; **flowers** 20–30 mm across, ± deeply cupped; sepals reflexed in bud, but spreading in fruit; petals deep golden yellow, sometimes tinged red, broadly obovate, 11–18 mm long. seTibet, nwYunnan [c & eHimalaya, nMyanmar]; grassy slopes, rocky places, streamsides, forest margins, 2,500–3,500 m. VII–VIII.

Hypericum patulum Thunb.* **Evergreen** or **semi-evergreen shrub** to 1 m with 4-lined or 4-angled **stems** when young, these soon becoming 2-lined or even terete. **Leaves** lanceolate to oblong or ovate, 1.5–6 cm long, blunt-apiculate, rather glaucous beneath, stalks 0.5–2 mm long. **Flowers** up to 15 in rather flat-topped clusters from uppermost 1–2 nodes, each flower deeply cupped, 2.5–4.5 cm across; sepals erect, almost equal to unequal, often reddish, broad-ovate to elliptic or almost rounded, 5–10 mm long; petals golden yellow, not red-tinged, oblong-obovate to broadly obovate, 12–18 mm long. nGuizhou, Sichuan; probably introduced elsewhere in w, c & sChina; open forests, shrubberies, cliffs, roadsides, 300–2,400 m. V–IX.

Hypericum henryi H. Lévl. & Vaniot Similar to previous species, but generally taller, to 3 m, with rather smaller **leaves** that are markedly glaucous beneath; **inflorescences** up to 7-flowered; stamens in bundles of usually 40–60. c & swGuizhou, c & sSichuan, n & nwYunnan [Myanmar, Sumatra, Thailand, Vietnam]; similar habitats, 1,300–3,000 m. IV–VIII.

Hypericum hookerianum Wight & Arnott (syn. *H. patulum* Thunb. var. *hookerianum* (Wight & Arnott) Kuntze) Round-topped **shrub** to 1.75 m with ovate **leaves**, 2.5–7.8 cm long, paler or ± glaucous beneath, leaf-stalks 1–4 mm long. **Inflorescences** 1–5-flowered; **flowers** 3–6 cm in diameter, ± deeply cupped; petals waxy, deep golden to pale yellow. seTibet [swTibet, Himalaya, Bangladesh, nThailand, nVietnam]; shrubberies on slopes, forest margins, 1,900–3,400 m. IV–VII.

Hypericum subsessile N. Robson* **Shrub** to 1–1.5 m with narrow-elliptic **leaves**, 3.5–6.5 cm long, paler to glaucous beneath. **Inflorescences** 1–8-flowered; **flowers** not waxy, 3.5–4.5 cm in diameter, shallowly cupped to star-shaped, bright yellow, sometimes tinged red; stamens more than a third longer than petals. wSichuan (Hanyuan), nwYunnan (Dali); shrubberies and open areas in mixed forests on mountain slopes, 2,400–3,000 m. VII–X.

Hypericum lancasteri N. Robson* **Shrub** to 1.5 m with suberect to spreading branches and purplish red young shoots; **stems** 2–4-lined. **Leaves** oblong-lanceolate to lanceolate or rarely ovate, 3–6 cm long, paler to very

1. *Hypericum reptans*; cult. (NR)
2. *Hypericum henryi*; cult. Lijiang, nwYunnan (CGW)
3. *Hypericum hookerianum*; cult. (NR)
4. *Hypericum hookerianum*; cult. (NR)
5. *Hypericum lancasteri*; cult. (NR)
6. *Hypericum lancasteri*; Xiaguan, nwYunnan (RL)

glaucous beneath, net-veined, dotted with glands and short streaks, leaf-stalks 1–1.5 mm long. **Inflorescence**s lax, 1–11-flowered, borne from apical nodes; **flowers** 3–5.5(–6.5) cm across, shallowly cupped to almost star-like; sepals wide-spreading to recurved, narrow-oblong or narrowly oblong-lanceolate, 8–11 mm long; petals golden yellow, oblong-obovate, 1.7–2.8 cm long; stamen-bundles each with 45–50 **stamens**. swGuizhou, sSichuan, nw & cYunnan (especially around Dali and Kunming); dry banks, grassy slopes, 1,700–2,600 m. V–VII.

Hypericum latisepalum (N. Robson) N. Robson (syn. *H. bellum* H. L. Li subsp. *latisepalum* N. Robson) **Shrub** to 1.5 m with 4-angled **stems** when young, very soon terete. **Leaves** lanceolate to broadly ovate, oblong or triangular-ovate, to 6.7 cm long or more, paler or somewhat glaucous beneath, leaf-stalks 1.5–3 mm long. **Inflorescences** 1–14-flowered, borne from apical nodes, nearly flat-topped; **flowers** cupped, 4–6 cm across; sepals erect, almost equal, ovate to broadly elliptic, 0.8–1.3 cm long, with gland-lines and an entire margin; petals golden yellow, broadly obovate, 2.3–3.7 cm long, 3 times as long as sepals, eglandular, margin entire. seTibet, w & nwYunnan [swTibet (Zayü), neIndia, nMyanmar]. 2,500–3,700 m. VI–X.

Hypericum bellum H. L. Li Similar to *H. latisepalum* but a dwarfer **semi-evergreen shrub** to 70 cm; **leaves** only 1.1–1.6 as long as broad, sepals narrowly elliptic or oblong to obovate, apex usually rounded; **flowers** 2.5–3.5 cm in diameter; stamens a third to half as long as petals; capsule narrowly egg-shaped. wSichuan, seTibet, nwYunnan [eHimalaya, neIndia]; open forests, forest margins, shrubberies, grassy slopes, (1,400–)1,900–3,500 m. VI–VIII.

Hypericum forrestii (Chitt.) N. Robson (syn. *H. patulum* Thunb. var. *forrestii* Chitt.; *H. patulum* forma *forrestii* (Chitt.) Rehd.) Differs from the preceding 2 species in being an almost **deciduous shrub** to 2 m, generally less, with terete **stems** and triangular-lanceolate or lanceolate-oblong to ± broadly triangular-ovate **leaves**; sepals rounded, with distinctly transparent margin; ovary and capsule pyramidal-egg-shaped. wSichuan (Kangding, Tianquan), n, nw & swYunnan [nMyanmar]; stony hillsides, margins of *Pinus* forests, open shrubberies, sometimes beside streams, 1,500–3,000(–4,000) m. V–IX.

Hypericum pseudohenryi N. Robson* Like *H. forrestii*, but **stems** 4-lined and **leaves** ovate or ovate-oblong to lanceolate or lanceolate-oblong, 2–6.6 cm long; **flowers** 3–5.5 cm in diameter, with pointed sepals, petals golden yellow, sometimes becoming reflexed. w & swSichuan, nw & neYunnan; similar habitats, 1,400–3,800 m. VI–IX. Plants with characters intermediate between this species and *H. forrestii* are known where they overlap in the wild.

TILIACEAE

TILIA

A genus with c. 35 species (half in China) distributed in eastern North America, Europe and temperate Asia. **Deciduous shrubs** or more commonly **trees** with simple, stalked **leaves** with toothed margins. **Inflorescence** a pendent long-stalked cyme, shorter to longer than **leaves**, accompanied by a long, membranous, oblong to oblanceolate bract, often borne in middle of stalk; **flowers** 2-several, 5-parted, often fragrant, yellowish white to dull white. **Fruit** nut-like, ribbed or smooth, containing up to 3 **seeds**.

1. Leaf-margin toothed, with bristles .... *T. henryana*
1. Leaf-margin toothed but not with bristles 2
2. Inflorescence few-flowered (rarely more than 5) ......
   .... *T. chinensis*, *T. intonsa*
2. Inflorescence 7-flowered or more ....*T. paucicostata*,
   T. paucicostata var. *yunnanensis*, *T. oliveri*

Tilia henryana Szysz.* **Tree** 10–25 m tall, with branchlets that are stellate-downy at first. **Leaves** broadly oblong-ovate to heart-shaped, 3.5–15 × 4–14 cm, long-pointed, brownish-downy beneath, margin toothed, lateral leaf-veins ending in bristles 2–6 mm long, leaf-stalks 2.5–4 cm long. **Inflorescence** longer than leaves, bearing numerous whitish **flowers**; bracts 8–14 cm long. Hubei, Shaanxi [c & eChina]; mixed forests, often near water, 1,900–3,000 m. VI–VIII.

Tilia chinensis Maxim.* **Shrub** or **tree** 3–20 m tall with glabrous branchlets and ovate **leaves**, 3–24 × 3–17 cm, long-pointed, base heart-shaped to very obliquely truncate, green with red veins and leaf-stalks, margins finely toothed; leaf-stalks 4.5–8 cm long. **Inflorescences** few-flowered, shorter than leaves, yellowish white; bracts oblanceolate, 5–11 cm long. **Fruit** egg-shaped, 9–10 mm long, strongly ribbed. Gansu, Hubei, Sichuan, Shaanxi, seTibet, Yunnan [Henan]; mixed forests, woods in gorges, streamsides, 1,400–3,800 m. VI–VIII.

---

1. *Hypericum latisepalum*; cult. (NR)
2. *Hypericum bellum*; nwYunnan (DL)
3. *Hypericum forrestii*; Cangshan, nwYunnan (CGW)
4. *Hypericum pseudohenryi*; Kangding, wSichuan (RL)
5. *Tilia henryana*; cult.(RL)
6. *Tilia henryana*; cult., from Hubei (CGW)
7. *Tilia chinensis*; near Benzilan, nwYunnan (DC)

Tilia intonsa **Wils. ex Rehd.*** (syn. *T. tonsura* hort.) Similar to previous species, but young branchlets densely clothed with short yellow-brown hairs, becoming hairless in second season, **leaves** greyish downy beneath with stellate hairs. **Fruit** 5-angled rather than strongly ribbed. w & nwSichuan; similar habitats and altitudes. VI–VII.

Tilia paucicostata **Maxim.*** **Tree** 8–15 m tall with glabrous young shoots and obliquely heart-shaped **leaves**, 4.5–6.5 × 3.5–6 cm, long-pointed, deep green above, bluish green beneath, with tufts of hairs in vein-axils, margin toothed; leaf-stalks 2–5 cm long. **Inflorescences** with 7–15 flowers, bracts oblong, 5.5–7 cm long, glabrous. Gansu, Hubei, Sichuan, n & nwYunnan [cChina]; mixed forests, riverbanks, 2,500–3,500 m. VI–VIII. Plants from Gansu, Sichuan and Yunnan that differ in having larger **leaves** (up to 8.5 × 7 cm), and bracts 5–9.5 cm long are distinguished as var. *yunnanensis* **Diels***.

Tilia oliveri **Szysz.*** (syn. *T. pendula* Engl. ex C. K. Schneid.) **Tree** 6–15 m tall with glabrous branchlets. **Leaves** obliquely broadly ovate to heart-shaped, 5–9.5 × 4–9 cm, long-pointed, obliquely truncate at base, green but paler, downy white beneath, margin finely toothed; leaf-stalks 2–7.5 cm long. **Inflorescences** with up to 20 cream **flowers**; bracts stalkless, oblanceolate, 4.5–10.5 cm long, blunt. Gansu, Hubei, eSichuan, Shaanxi [cChina]; mixed forests, streamsides, 1,200–2,000 m. VI–VII.

1. *Tilia intonsa*; Jiuzhaigou, nwSichuan (PC)
2. *Tilia oliveri*; Shennongjia, wHubei (PC)
3. *Ulmus pumila*; Mongolia (MF)
4. *Ulmus pumila* bark; Mongolia (MF)
5. *Zelkova serrata*; cult. (PC)
6. *Abelmoschus moschatus*; near Guilin, neGuizhou (CGW)

ULMACEAE

ULMUS

Ulmus bergmanniana C. K. Schneid.* **Tree** to 26 m with greyish white **bark** and purplish brown twigs. **Leaves** elliptic, narrowly elliptic, obovate-oblong or ovate, 6–18 × 3–8.5 cm, downy on veins beneath. **Fruit** glabrous, pale brown or tan, obovate, 1.2–1.8 cm long. Gansu, Hubei, Sichuan, Shaanxi, seTibet, Yunnan [cChina]; broadleaved forests, 1,500–2,000 m. II–IV.

Ulmus pumila L. **Tree** to 25 m with dark grey **bark** and yellowish grey twigs. **Leaves** elliptic-ovate to elliptic-lanceolate, 2–8 × 1.2–3.5 cm, becoming hairless beneath. **Fruit** glabrous, pale tan, orbicular, 1–2 cm long. Gansu, eQinghai, Sichuan, Shaanxi, eTibet [n & cChina, Korea, Mongolia, eRussia, cAsia]; valleys and plains, hillslopes, 1,000–2,000 m. III–V. Widely cultivated in China.

ZELKOVA

A genus of 5 species (3 in China) from south-east Europe, south-west and South East Asia. **Deciduous trees** with exfoliating **bark**. **Leaves** simple, in 2 opposite rows, stalked, toothed or scalloped; stipules linear-lanceolate, soon falling. **Inflorescences** axillary; **flowers** unisexual and hermaphrodite, appearing with the leaves, campanulate, with 4–6 tepals; male flowers clustered, axillary, with 4–6 stamens borne opposite the tepals; female and hermaphrodite flowers solitary with a stalkless ovary. **Fruit** a drupe.

Zelkova serrata (Thunb.) Makino **Tree** to 30 m with pale to brownish grey, exfoliating **bark** and brown twigs. **Leaves** elliptic-ovate to elliptic, 3–10 × 1.5–5 cm, glabrous or downy along veins beneath. **Fruit** pea-green, orbicular, 2.5–3.5 cm across, ridged. Gansu, Guizhou, Hubei, Sichuan, Shaanxi [n & cChina, Taiwan, Japan, Korea, eRussia]; valley streamsides, 500–2,000 m. IV.

CELTIS

A genus of c. 60 species (11 in China) from the tropics and subtropics. **Deciduous** or **evergreen trees**. **Leaves** simple, alternate, in several ranks, stalked, entire to toothed, 3-veined from base; stipules membranous or papery. **Inflorescences** simple or branched, many-flowered; **flowers** small, unisexual or hermaphrodite, appearing before the leaves in spring, campanulate, with 4–5 tepals; male flowers borne directly on the branches (cauliflorous) or clustered in leaf-axils, with 4–5 stamens borne opposite tepals; female and hermaphrodite flowers borne apically in polygamous inflorescences, ovaries 1-celled. **Fruit** a drupe, smooth or reticulately patterned.

1. Fruit 4–7 mm in diameter . . . . . **C. bungeana, C. biondii**
1. Fruit 10–13 mm in diameter . . . . **C. cerasifera**

Celtis bungeana Blume **Tree** to 10 m with grey **bark** and brown twigs. **Leaves** ovate, narrowly ovate or oblong-ovate, 3–7 × 2–4 cm, glabrous. **Fruit** globose, 4–5 mm in diameter, blackish blue when ripe. seGansu, Guizhou, Hubei, Qinghai, Sichuan, Shaanxi, e & seTibet, Yunnan [n & cChina, Korea]; broadleaved forests, scrub, roadsides, mountain slopes, 100–2,300 m. IV–V.

Celtis biondii Pamp. **Tree** to 18 m with grey **bark** and yellow-brown twigs. **Leaves** ovate-elliptic, ovate-elliptic or ovate, 2.5–8 × 2–4 cm, with inconspicuous appressed hairs. **Fruit** globose, yellow to reddish orange, 5–7 mm in diameter. seGansu, Guangxi, Guizhou, Hubei, Sichuan, sShaanxi, Yunnan [c & sChina, Taiwan, Japan]; broadleaved forests and scrub on limestone, sea level to 2,000 m. IV–V.

Celtis cerasifera C. K. Schneid.* **Tree** to 35 m with grey-brown **bark** and green twigs. **Leaves** ovate-elliptic or ovate, 5–15 × 2.5–7.5 cm, glabrous or with hairs along veins beneath. **Fruit** globose, 10–13 mm in diameter, blackish blue when ripe. Guangxi, Guizhou, Sichuan, sShaanxi, seTibet, Yunnan [cChina], broadleaved forests, scrub in valleys and on slopes, 800–2,400 m. IV.

MALVACEAE

ABELMOSCHUS

A genus of c. 15 species (6 in China, 1 endemic) found mainly in the Old World tropics and subtropics. **Annual** and **perennial herbs**, generally rather bristly with palmately divided **leaves**. **Flowers** solitary or in racemes with an epicalyx of 4–16 segments and a spathe-like calyx; petals 5. **Fruit** a capsule. *A. esculentus* (L.) Moench., grown primarily for the large edible fruits, is widely cultivated in Asia (including China), and is occasionally naturalised along field boundaries and roadways and in waste places.

Abelmoschus manihot (L.) Medik Bristly **annual** or **perennial** to 2 m, generally less, with erect **stems** that are often black-dotted. **Leaves** deep green, long-stalked, alternate, deeply palmately lobed, generally with 3 or 5 toothed lobes. **Flowers** solitary, lateral on short stiff stalks, 11–14 cm across; calyx 5-lobed, the 4–8-segmented epicalyx persistent in fruit; corolla pale to bright yellow, maroon in centre. **Fruit-capsule** to 5 cm long, 5-angled, bristly, on a stalk to 8 cm long. Guangxi, Guizhou, Sichuan, Yunnan [s & seChina, seAsia, Himalaya]; shrubberies, waste land, fields and field boundaries, to 2,400 m. VII–XI.

Abelmoschus moschatus Medik. Very similar to the previous species, but with 6–10 linear epicalyx segments and a larger **capsule**, 5–8 cm long, borne on a stalk 7–12 cm long. Distribution similar to that of *A. manihot*, but widespread in subtropical and tropical Asia. Pink- and red-flowered forms are cultivated.

BOMBACACEAE

BOMBAX

A genus with c. 50 tropical species (3 in s & swChina), mainly in America, also in Asia, Africa and Australia. **Deciduous trees**, usually with spiny young **trunks**. **Leaves** palmate. **Flowers** solitary or in bundles, axillary or terminal, usually red, sometimes orange and white, large, appearing before leaves; **calyx** usually cup-shaped, leathery, hairy on upper surface, falling with petals and stamens; petals 5, obovate or obovate-lanceolate; stamens numerous, joined into a tube, anthers kidney-shaped, dorsifixed; ovary 5-celled. **Fruit** a capsule dehiscent into 5 leathery valves. **Seeds** black, small, enclosed in silky wool.

Bombax ceiba L. **Tree** to 25 m with greyish **bark**; **trunk** of young trees usually very spiny. **Leaflets** 5–7, oblong to oblong lanceolate, 10–16 × 3.5–5.5 cm, glabrous. **Flowers** solitary, terminal, 10 cm across, red or orange-red; calyx cup-shaped, densely yellow-hairy outside; petals obovate-oblong, 8–10 × 3.4 cm. Stamens in several series, outer in 5 bundles, each with more than 10 stamens, inner series shorter. **Capsule** oblong, 10–15 cm long, densely grey-white-hairy. Guizhou, Sichuan, Yunnan [sChina, seAsia, Australia, Cambodia, India, Laos, Myanmar, Nepal, Sri Lanka, Thailand, Vietnam]; dry river valleys, savannas, sea-level to 1,400 m. III–IV.

CANNABACEAE

CANNABIS

A genus of 1–2 species originally from Central Asia but the true origin obscured by long use and development by man (1 species naturalised and cultivated in China, possibly native in Xinjiang). The plants can be either dioecious or monoecious.

Cannabis sativa L. Strong-smelling **annual** 1–3 m tall, with furrowed **stems**, white-hairy branches and alternate, stalked, finger-like leaves; **leaflets** 7–11 normally, lanceolate to linear-lanceolate, 7–15 × 0.5–1.5 cm, toothed, with scattered brown resinous glands. **Inflorescences** paniculate, male to 25 cm long, female smaller and denser; **male flowers** yellowish green, nodding, sepals 2.5–4 mm long, petals absent; **female flowers** green, stalkless, with accompanying leaf-like bracts. **Achenes** egg-shaped but flattened, 2–5 mm long. Cultivated, naturalised or casual throughout China, mainly below 1,500 m. V–VII. Probably native to cAsia, but widespread or naturalised in Asia.

URTICACEAE

PILEA

A genus with c. 400 species (80 in China, 31 endemic), primarily in tropical and subtropical regions. Monoecious or dioecious **herbs**, **shrubs** or **subshrubs** mostly with opposite, occasionally alternate, **leaves**. **Inflorescences** solitary or paired, densely to laxly cymose, with tiny flowers; **male flowers** with a 4–5-parted perianth, **female flowers** usually 3-parted. **Fruit** an achene.

Pilea peperomioides Diels* Dioecious, tufted, **evergreen**, rhizomatous **perennial** with rough-woody **stems** to 40 cm, terminating in a lax rosette of thick, fleshy, bright green, long-stalked **leaves**; leaf-blades ovate to almost rounded, peltate, 2.5–9 × 2–8 cm, margin entire. **Flowers** tiny, in branched cymes at upper nodes, male and female generally on separate plants, male up to 20 cm, purplish, female smaller and often greenish; perianth-lobes 4; stamens 4. **Fruit** a tiny purplish achene. swSichuan, nwYunnan; shady damp or wet rocks, cliffs by waterfalls, ravines, 1,500–3,000 m. IV–VII. Despite being rare and endangered in the wild, it is commonly grown as a pot plant in China and elsewhere.

DEBREGEASIA

A genus with 6 species (all found in China) distributed in North Africa, tropical South East Asia and eastern Australia. Monoecious or dioecious **shrubs** or small **trees** with alternate, stalked **leaves**, with deciduous membranous stipules located between leaf-stalks, leaf-blades 3-veined. **Flowers** greenish or yellowish green, in dense clusters, unisexual; male flowers with 3–5 lobes and 3–5 stamens and a rudimentary ovary; female flowers with a 3–4-toothed perianth-tube, often fleshy and enlarging in fruit, and a short style. **Fruit** a small drupe-like achene, enclosed in the fleshy, persistent perianth.

Debregeasia orientalis C. J. Chen Usually dioecious, **evergreen shrub** to 4 m with slender reddish branchlets, appressed-downy or almost glabrous. **Leaves** oblong to linear-lanceolate, 5–18 × 1–3 cm, dark green and rather rough above, paler, greyish- or whitish-downy beneath, margin finely toothed. **Inflorescences** on previous year's shoots, often before the new leaves; **male flowers** with (3–)4 lobes; **female flowers** 4-toothed at tube apex. **Fruits** orange when ripe, 1–2 mm long. sGansu, Guangxi, Guizhou, Hubei, Sichuan, sShaanxi, seTibet, Yunnan [eChina, Himalaya, Japan, Taiwan]; shady damp places in forests and by streams, 300–2,800 m. II–IV.

Debregeasia longifolia (Burm. f.) Weddell (syn. *D. velutina* Gaudich.) Similar to the previous species, but branchlets with spreading hairs, and with at least some **inflorescences** borne on current season's shoots. seGansu, Guangxi, Guizhou, wHubei, sShaanxi, seTibet, Yunnan [swTibet, Guangdong, seAsia, Himalaya, India, Sri Lanka]; similar habitats, 500–3,200 m. VIII–XII.

BALANOPHORACEAE

BALANOPHORA

A genus of c. 20 species (12 in China, 1 endemic) distributed in the Old World tropics and subtropics. Monoecious or dioecious **parasitic plants** with much-branched rhizomes, and with paired, fleshy, often scale-like **leaves** without chlorophyll. **Inflorescences** spadix-like, varying from cylindrical to globose, held close to ground level; **male flowers** stalked, with a 3- or 6-lobed perianth and numerous stamens bearing white pollen; **female flowers** densely crowded, consisting of a pistil with an elongated style.

Balanophora laxiflora **Hemsl.** (syn. *B. formosana* Hayata; *B. spicata* Hayata) Plant dioecious, generally red or red-brown, sometimes flushed purplish throughout, with up to 14 alternate, elliptic-oblong **scale-leaves**. Scapes to 10 cm tall; **male inflorescence** cylindrical, to 18 cm long; **female inflorescence** egg-shaped, gradually elongating, not more than 6 cm long. Guangxi, Guizhou, Hubei, Sichuan, seTibet, Yunnan [s & seChina, Laos, Thailand, Taiwan, Vietnam]; dense forests, 600–1,700 m. IX–XI.

LORANTHACEAE

TAXILLUS

A genus of 18 species (half endemic to China) in south and South East Asia. **Parasitic shrubs** growing on a variety of hosts. **Stems** rather angular, bearing opposite or alternate **leaves** that are downy with stellate or whorled hairs. **Flowers** 2–5 in lateral racemes, with a small bract below each flower, 4(–5)-parted; calyx egg-shaped to elliptic in outline, base rounded with entire to minutely toothed rim; corolla tubular in bud, split along 1 side when open and with a slightly curved tube, lobes all reflexing away from the split side; stamens joined to base of corolla-lobes. **Fruit** an egg-shaped to ellipsoid berry.

Taxillus delavayi **(Tiegh.) Danser** Laxly branched **parasitic shrub** to 1 m with angular, glabrous **stems**. **Leaves** alternate or almost opposite, leathery, short-stalked, blade ovate to lanceolate, to 5 × 2 cm, with 3–4 pairs of lateral veins. **Flowers** usually in clusters of 2–4, short-stalked, glabrous; calyx ellipsoid, c. 2.5 mm, with a ± entire rim; corolla red, with a slightly curved tube and short lanceolate lobes, 20–30 mm

1. *Cannabis sativa*; Dali, nwYunnan (PC)
2. *Pilea peperomioides*; cult., from Dali, nwYunnan (CGW)
3. *Pilea peperomioides*, male inflorescences; cult., from Yunnan (CGW)
4. *Debregeasia orientalis*; Maolan, seGuizhou (PC)
5. *Balanophora laxiflora*; Wengxiantai, swGuizhou (PC)

long overall. **Berries** ellipsoid, 8–10 mm, orange or yellow at maturity. nwGuangxi, wGuizhou, Sichuan, e & seTibet, Yunnan [Myanmar, nVietnam]; forests, forest margins, ravines, shrubberies, growing on a various hosts, 1,500–3,500 m. II–VII.

Taxillus thibetensis **(Lecompte) Danser*** Readily distinguished from previous species by yellowish brown-downy (white in **var. albus Jiarong Wu**) young **stems** and opposite **leaves**, blade often more oblong, to 10 × 5.5 cm, downy beneath and with 5–8 pairs of lateral veins; corolla slightly swollen at base, downy outside. Guizhou, swSichuan, seTibet, Yunnan; similar habitats, on a variety of hosts including *Prunus*, *Pyrus*, *Salix* and *Quercus*, 1,700–3,000 m. V–IX.

1. *Taxillus delavayi*; above Shigu, nwYunnan (CGW)
2. *Arceuthobium pini* growing on *Pinus yunnanensis*; Baishui, Yulongxueshan, nwYunnan (CGW)
3. *Viola kunawarensis*; Baimashan, nwYunnan (HJ)
4. *Viola philippica*; Gangheba, nwYunnan (RS)
5. *Viola collina*; Beijing, neChina (CGW)
6. *Viola biflora*; north of Songpan, nwSichuan (CGW)

SCURRULA

A genus of 50 species (10 in China, 2 endemic) from south and South East Asia. Very similar to *Taxillus* but calyx turban- or pear-shaped, narrowed at base, and **fruit** with a narrowed attenuate base.

Scurrula parasitica **L.** Laxly branched **parasitic shrub** to 1 m, with densely downy young branches. **Leaves** opposite or almost so, stalked, blade ovate to oblong, to 8 × 4 cm, rather thin and with 5–6 pairs of lateral veins. **Flowers** in clusters of 3–7, sometimes several clusters together; calyx turban-shaped, 2–3 mm, rim untoothed; corolla red to greenish yellow (**var. graciliflora (Roxb. ex J. H. Schultes) H. S. Kiu**), with a somewhat curved tube that is slightly inflated towards top and with short reflexed lanceolate lobes, 14–25 mm long. **Berries** 8–10 mm, reddish yellow at maturity. Guangxi, Guizhou, Sichuan, seTibet, Yunnan [s & eChina, seAsia, eHimalaya, Myanmar, Thailand, Vietnam]; forests and forest margins, hillslopes, on a variety of broadleaved trees and shrubs, rarely on conifers, to 2,100 m. X–I.

Scurrula pulverulenta Wall. Similar to the previous species but branchlets grey, white-downy when young; **leaves** ovate-oblong, to 16 × 9 cm. **Flowers** in clusters of 8–16; calyx pear-shaped; corolla yellow or creamish, 30–40 mm long. Yunnan [Himalaya, Myanmar, Thailand]; similar habitats, parasitic on mainly subtropical trees such as *Mallotus*, *Litsea* and *Vitex*, and also on citrus, to 1,800 m. VIII–III.

Scurrula buddleioides (Desr.) G. Don. A larger **parasitic shrub** than the previous 2 species, to 2 m, with greyish yellow-downy young branchlets and more leathery oblong **leaves** to 10–8 cm, with 4–5 pairs of lateral veins; **flowers** in clustered bundles; calyx pear-shaped, 2–3 mm, rim ciliate; corolla red, downy outside, with slightly curved tube and reflexed lobes, not more than 20 mm long. Sichuan, seTibet, Yunnan [India]; similar habitats, parasitic on a wide range of broadleaved families including Caprifoliaceae, Fagaceae, Rosaceae and Tiliaceae, 1,100–2,200 m. I–XII.

VISCACEAE

ARCEUTHOBIUM

A genus of 45 species (5 in China, 3 endemic), distributed from south Europe to North Africa, Asia and North America. Intricately branched, **semi-parasitic** (on conifers), dioecious **subshrubs** with paired or whorled **branches**. **Leaves** reduced to scales borne in 4 ranks. **Flowers** tiny, yellow or greenish, male with 3–4 lobes and stalkless anthers, female with a short tube and a 2-lobed, persistent perianth fused to ovary. **Fruit** a small egg-shaped berry, explosively dehiscent at maturity.

Arceuthobium pini Hawksw. & Wiens* Greenish or yellowish green plant to 20 cm across, with opposite or sometimes whorled **branches**. **Scale-leaves** 0.5–1.5 mm long. **Male flowers** terminal, solitary or paired, yellow, terminal, 2–2.5 mm across, usually with 3 perianth-lobes and with a cup-shaped bract beneath; **female flowers** solitary, terminal or lateral, greenish, 1 mm long with a red style. **Berries** 3–4 mm long, greenish yellow in lower part. swSichuan, seTibet, nwYunnan [swTibet]; coniferous and mixed forests, especially parasitic on *Pinus densata* and *P. yunnanensis*, 2,600–3,800 m. IV–VII.

Arceuthobium sichuanense (H. S. Kiu) Hawksw. & Wiens Smaller plant than the previous species, not more than 6 cm across with short **branches** not exceeding 10 mm long, with smaller solitary **male flowers** 1.5–2 mm across. Qinghai, Sichuan, seTibet [Bhutan]; coniferous forests, parasitic on *Picea* species such as *P. likiangensis*, 3,400–4,100 m. VI–VII.

VIOLACEAE

VIOLA

A genus with c. 550 species (96 in China, 35 endemic), primarily found in the temperate northern hemisphere and South America, and with a few scattered on African mountains and in South East Asia. **Annuals** or tufted, sometimes stoloniferous **perennials**, sometimes **subshrubby**. **Leaves** alternate or all basal, variously shaped, sometimes lobed, stipulate. **Flowers** solitary, lateral, zygomorphic, 5-parted; sepals small and usually narrow; petals prominent, 2 upper, 2 lateral, and 1 lower that forms a lip, spurred or pouched at base. **Fruit** a 3-parted capsule.

1. Flowers pink, violet or white .... *V. kunawarensis*, *V. philippica*, *V. collina*
1. Flowers yellow ...... *V. biflora*, *V. biflora* var. *hirsuta*, *V. rockiana*, *V. cameleo*

Viola kunawarensis Royle Dwarf tufted **perennial** to 6 cm with a short, thick rhizome. **Leaves** all in a basal rosette, ovate to elliptic or oblong, to 2 × 0.5 cm, stalked, margin entire or slightly scalloped, base wedge-shaped, glabrous; stipules whitish, fringed. **Flowers** purple-blue, with deeper veining on lip, 8–12 mm across, the 4 upper petals obovate, larger than lip; spur very short and pouched. Gansu, Qinghai, nwSichuan, Tibet, nwYunnan [Afghanistan, cAsia, Himalaya, Mongolia, Russia]; alpine and subalpine meadows, thickets, 2,900–4,800 m. V–VII.

Viola philippica Cav. Larger plant than preceding species, to 15 cm tall in flower, with tufts of oblong to ovate or ovate-triangular **leaves**, to 4 × 1 cm, glabrous or minutely hairy, particularly on veins beneath, margin scalloped. **Flowers** larger, 18–24 mm across, purple-violet or purple, occasionally white, with deeper veining on petals, pale in throat; spur tubular, 3–8 mm long. Gansu, Guangxi, Guizhou, Hubei, Sichuan, Shaanxi, Yunnan [much of China, seAsia, Korea, Japan, Philippines]; grassy places, fields, forest margins, road- and pathsides, sea level to 1,700 m. IV–V.

Viola collina Besser Readily distinguished from the previous 2 species by finely hairy **leaves** and leaf-stalks, rarely becoming hairless, leaves more heart-shaped, and by the bearded base to the 2 lateral petals; plant to 10 cm tall in flower, taller in fruit; **flowers** pinkish or purplish with pale throat and purple veining on lip; spur whitish, short, c. 3.5 mm long. Guizhou, Hubei, Sichuan, Shaanxi, nwYunnan [c & neChina, widespread Europe to Japan]; shady damp habitats, forests and forest margins, grassy places, roadsides, thickets, sea level to 2,800 m. III–V.

Viola biflora L. Small tufted **perennial herb** to 10 cm. **Basal leaves** few, long-stalked, kidney-shaped, 10–30 mm across, hairy or not, margin scalloped; stipules oval; **stem-leaves** few, short-stalked, otherwise similar to basal. **Flowers** solitary or paired, 12–15 mm across, bright yellow with brown or purple streaks on lip, upper petals elliptic, lateral and lower oval to obovate, equal to the upper; spur very short, 1.5–2.5 mm long. Gansu, Hubei, Sichuan, Tibet, Yunnan [n & cChina, Europe, w & cAsia, Himalaya, Japan, nMyanmar, North America]; meadows, rocky slopes, forests, shrubberies, 2,400–4,500 m. V–VIII. Hairy plants are generally assigned to **var. *hirsuta* W. Beck.**

***Viola rockiana* W. Beck.** (syn. *V. biflora* L. var. *rockiana* (W. Beck.) Y. S. Chen) Similar to previous species, but **leaves** heart-shaped to rounded, often with semi-translucent veins and a more toothed margin; petals broader, both lateral and lower with brown streaks towards base, lower shorter than the others; spur shortly pouched, not more than 1.5 mm long. Gansu, Qinghai, w & swSichuan, eTibet, nwYunnan [(?)nMyanmar]; similar habitats and altitudes. V–VII.

Viola cameleo* H. Boisieu Slighter plant than preceding 2 species, with thin bluish green, narrow-ovate to triangular-ovate **leaves**, to 3 × 3.5 cm, base shallowly heart-shaped, margin coarsely scalloped; basal leaves 1–2, stem-leaves few, somewhat smaller. **Flowers** solitary in upper leaf-axils, golden yellow with purplish marks on lip, petals oblong to oblong-elliptic, lip more obovate; spur tubular, 3–4 mm long. wHubei, w & swSichuan, nwYunnan; rocky and stony banks, margins of thickets, 1,800–3,800 m. V–VII.

TAMARICACEAE

MYRICARIA

A genus of **shrubs** and **subshrubs** with 13 species (10 in China, 4 endemic) in Europe and west, central and east Asia. **Leaves** small, simple, alternate, stalkless. **Flowers** hermaphrodite, in terminal or lateral racemes or panicles; sepals 5, margin membranous; petals 5; stamens 10, 5 long and 5 short. **Fruit** a 3-parted capsule.

***Myricaria squamosa* Desv.** (syn. *M. germanica* (L.) Desv. var. *squamosa* (Desv.) Maxim.) Erect **shrub** to 5 m, more usually 2–3 m, with purplish brown or red-brown old **stems**, yellowish green when young. **Leaves** lanceolate to ovate or oblong, to 5 × 2 mm. **Racemes** lateral, borne on old branches before leaves appear, solitary or clustered, dense at first, with numerous overlapping scales at base; sepals ovate-lanceolate, to 4 × 1 mm; petals rose-pink to purple-red, obovate to elliptic, 4–5 mm long. Gansu, Qinghai, Sichuan, Tibet [cAsia, Himalaya]; damp mountain slopes, margins of rivers, streams and lakes, 2,400–4,600 m. V–VII.

Myricaria laxiflora* (Franch.) P. Y. Zhang & Y. J. Zhang Similar to the previous species but **shrub** not more than 1.5 m tall and with denser **leaves** not more than 4 × 1 mm and **racemes** generally laxer and terminal, 6–12 cm long; petals pink or purple, obovate, 5–6 mm long. Hubei, Sichuan; similar habitats, roadsides. VI–VIII.

BEGONIACEAE

BEGONIA

A genus of c. 1,400 species (c. 160 in China, primarily in south-east Yunnan and south-west Guangxi) widely distributed in tropical and subtropical regions, especially in Central and South America. Succulent **perennial**, usually monoecious **herbs**, with **tubers** or **rhizomes** and erect, often very short, **stems**, rarely climbing. **Leaves** simple, rarely palmately lobed, alternate or all basal, usually obliquely asymmetric at base, margin often irregularly toothed and divided, occasionally entire, venation usually palmate; stipules membranous, deciduous. **Flowers** unisexual, 2–several, rarely numerous, in dichotomous cymes or sometimes a panicle; tepals petaloid; **male flowers** with 2–4 valvate or 4 decussate tepals and numerous stamens; **female flowers** with 2–5 free tepals and an inferior ovary. **Fruit-capsule** usually dry and unequally 3-winged, rarely wingless, 3–4-horned; **seeds** minute, numerous.

1. Leaves scattered along distinct stems 2
1. Leaves all basal 3
2. Leaves unlobed . . . . ***B. grandis***
2. Leaves shallowly to deeply lobed . . . . ***B. polytricha*, *B. palmata***
3. Leaf-blade peltate . . . . ***B. cavaleriei***
3. Leaf-blade not peltate 4
4. Leaves half to deeply palmately lobed or shallowly and regularly lobed . . . . ***B. taliensis*, *B. pedatifida*, *B. wilsonii***
4. Leaf-blade entire or slightly lobed towards apex 5
5. Leaves green, sometimes purplish beneath . . . . ***B. asperifolia*, *B. henryi*, *B. leprosa***
5. Leaves brownish green . . . . ***B. labordei*, *B. limprichtii*, *B. morsei***

***Begonia grandis* Dry.** (syn. *B. evansiana* Andrews) Plant to 60 cm, red-tinged on **stem** and leaves. **Leaves** obliquely ovate-heart-shaped, 5–19 cm long, apiculate, purple-red beneath, margin irregularly toothed; leaf-stalks glabrous. **Inflorescence** elongated, branched, many-flowered. **Flowers** pink; **male flowers** with 4 tepals, the larger ones elliptic-ovate, the other 2 much smaller; **female flowers** similar. **Capsules** unequally 3-winged. Guizhou, Hubei, s, e & wSichuan, sShaanxi, Yunnan [c & sChina, Japan]; damp shady places on ledges, cliffs, streamsides, 300–3,400 m. VII–X. A very variable species with several recognised varieties throughout the range.

Begonia polytricha* C. Y. Wu **Herb** 20–30 cm tall with elongated **rhizomes** and densely red-hairy **stems**. **Leaves** basal and on the stem, asymmetrically ovate, 6–7 × 4–5 cm, long-hirsute beneath, sparingly so above, base oblique-heart-shaped, margin toothed, shallowly lobed; leaf-stalks 4–12 cm long, red-hairy. **Inflorescence** densely red-hairy, borne on a stalk 1–3 cm long; **male flowers** with 4 white tepals, outer 2 oblong-ovate, 9–13 × 7–10 mm, red-hirsute beneath, inner 2 smaller, oblong; **female flowers** with 5, orbicular to narrowly oblong tepals, largest 10 × 9 mm, hairy outside, ovary densely red-hairy. seYunnan; broadleaved forests, in shady and damp places, 1,800–2,200 m. VI–X.

Begonia palmata **D. Don** (syn. *B. laciniata* Roxb. ex Wall.) Readily distinguished from the previous species by larger, palmately 3–7-lobed **leaves** 12–20 cm long, lobes narrowly to broadly triangular, long-pointed, dark green above, paler beneath, leaf-stalks 5–10 mm long, softly brown-hairy. **Flowers** white, pink or rose, usually 4–several, on densely brown-downy branches. seTibet, s & wYunnan [Himalaya, neIndia, nMyanmar, Vietnam]; rocks and cliffs in damp shady places, streamsides, dense evergreen broadleaved forests, shrubberies, 1,300–3,200 m. VIII–IX.

Begonia cavaleriei **H. Lévl. Perennial** herb with an elongated, stoloniferous **rhizome**. **Leaves** obliquely ovate, 8–15 cm long with long-pointed apex, brownish green above, paler beneath, shortly hairy; leaf-stalks up to 25 cm long. **Inflorescence** up to 20 cm tall, glabrous, bearing several flowers; **male flowers** with 4, rounded tepals 17–20 mm long; **female flowers** with 3 tepals. **Capsules** pendent, oblong, with 3 unequal moon-shaped wings. Guangxi, sGuizhou, seYunnan [Vietnam]; evergreen forests in damp shady and places, streamsides, often on rocks, 700–1,000 m. V–VII. (See photo 1 on p. 182.)

Begonia taliensis **Gagnep.*** (syn. *B. muliensis* T. T. Yu) **Leaves** 1–2, basal, hand-like with triangular lobes, up to 25 cm long and broad, deep green, occasionally purplish above and sparsely hirsute, pale green beneath; leaf-stalk 21–41 cm long, angular. **Inflorescence** 19–30 cm tall, several-flowered, **flowers** pink; **male flowers** with 4 tepals, 7–12 mm long;

1. *Viola rockiana*; Daxueshan, nwYunnan (CGW)
2. *Viola cameleo*; Baimashan, nwYunnan (CGW)
3. *Viola cameleo*; Kangding, wSichuan (PC)
4. *Myricaria squamosa*; Wolong, wSichuan (CGW)
5. *Begonia grandis*; cult. (CGW)
6. *Begonia polytricha*; Laojunshan, seYunnan (PC)

BEGONIACEAE: Begonia

female flowers with 3 tepals, 5–10 mm long. **Capsules** egg-shaped-oblong, 12–15 mm long, with 3 unequal wings, the largest wing obliquely triangular, to 2.4 cm long. sSichuan, w & nwYunnan; shrubberies, mixed forests, c. 2,400 m. VIII.

Begonia pedatifida H. Lévl.* Similar to previous species, but **leaves** smaller, not more than 17 cm long, with bristly down, margin toothed, leaf-stalks 12–20 cm long, with soft brown hairs. **Inflorescence** 7–15 cm tall with larger white or pink **flowers** on hairy stalks, tepals of **male flowers** 18–25 mm long, female tepals 18–20 mm long. Guizhou, Hubei, Sichuan [Hunan]; rocks and cliffs in damp shady places by streams in evergreen broadleaved forests, 350–1,700 m. I–VII.

Begonia wilsonii Gagnep.* **Leaves** 1–2, obliquely ovate-heart-shaped to rhombic, 12–25 cm long and wide, with long-tailed apex and 3–7(–9) shallow, triangular lobes, deep green above, paler or occasionally dark purple beneath, margin triangular-toothed; leaf-stalks 11–25 cm long, almost glabrous. **Inflorescence** 4–12 cm tall, glabrous, **flowers** pink, 5–10 on 2–4 branches; **male flowers** with 4 ovate tepals, 8–14 mm long; **female flowers** with 3 tepals. **Capsules** spindle-shaped, 10–12 mm long, glabrous, wingless. sSichuan (Emeishan, Hongxi); damp rocks by streams and damp shady rocks in montane forests, 300–1,500 m. VII–VIII.

Begonia asperifolia Irmsch.* Plant 15–30 cm tall, with 1–2, obliquely heart-shaped **leaves** 15–34 cm long, that are shallowly triangularly lobed in upper half, deep green and sparsely hairy above, pale green or purplish beneath, palmately 6–8 veined, margin unequally doubly toothed, each tooth with a short awn. **Inflorescence** 12–15 cm tall, bearing several pink **flowers**; **male flowers** with 4 tepals; **female flowers** with 5 unequal tepals. **Capsules** pendent,

1. *Begonia cavaleriei*; Longhua, swGuangxi (PC)
2. *Begonia henryi*; Leshan, sSichuan (HJ)
3. *Begonia leprosa*; Longhua, swGuangxi (PC)
4. *Begonia morsei*; Pogang, swGuangxi (PC)
5. *Thladiantha cordifolia*; Xishan, Kunming, cYunnan (CGW)
6. *Thladiantha longisepala*; swGuangxi (PC)

oblong to obovoid-oblong, 1–2 cm long, with 3 unequal wings. seTibet, w & nwYunnan; damp rocks and shady banks by streams in mixed broadleaved forests, rarely epiphytic, 1,800–3,600 m. VIII–X.

Begonia henryi Hemsl.* (syn. *B. delavayi* Gagnep.; *B. mairei* H. Lévl.) **Herbaceous**, tuberous-rooted **perennial** to 15 cm. **Leaves** basal, solitary (occasionally 2), triangular-ovate or broad-ovate, rarely ± orbicular, 3.5–6 × 4–7.5 cm, sparsely hairy, with an oblique heart-shaped base, margin scalloped; leaf-stalks 6–13 cm, brown-hairy. **Inflorescence** 7.5–12 cm tall, hairy or almost glabrous, few-flowered; **male flowers** with 2, pink, rounded to broad-ovate tepals, 8–12 × 10–13 mm, with numerous stamens; **female flowers** with 2 rounded tepals 6–8 × 7–8 mm and a glabrous ovary. **Capsules** pendent, oblong, to 11 × 5 mm, unequally 3-winged. nGuangxi, seGuizhou, Hubei, Sichuan, Yunnan; rocks and rock-crevices, beneath banks and outcrops in moist, shady places, 800–2,600 m. IX–X.

Begonia leprosa Hance* **Leaves** minutely wrinkled, broadly ovate, blunt, 4.5–9 cm long, glabrous, reddish brown when young; leaf-stalks usually longer than leaves, red, hairy. **Inflorescence** much longer than leaves, red-stalked, branching, many-flowered; flowers pink. **Capsules** with 3 rounded lobes, unwinged. Guangxi, s & seYunnan [Guangdong]; shady damp banks in evergreen forests, to 1,800 m. VI–X.

Begonia labordei H. Lévl.* Plant with obliquely ovate-heart-shaped **leaves**, occasionally slightly lobed near apex, 5–12 cm long, long-pointed, awned, brownish green and sparsely softly hairy above, pale beneath, palmately 7-veined, margin shallowly and remotely toothed; leaf-stalks to 35 cm long, densely and softly brown-hairy. **Inflorescence** 8–27 cm tall, **flowers** several, white or pink; **male flowers** with 4 tepals, outer 2 broadly ovate, 18–20 mm long, with long soft hairs outside, inner 2 narrow and half the size; **female flowers** with 4–5, strongly unequal, tepals, the largest almost orbicular to broad-ovate, 15–16 mm long. **Capsules** pendent, obovoid-oblong, 10–13 mm long, with unequal wings. Guizhou, sSichuan, Yunnan; damp rocks by streams, cliff-ledges, 850–3,600 m. VI–VII.

Begonia limprichtii Irmsch.* Readily distinguished from the previous species by smaller **leaves**, not more than 8 cm long, and by the white, rarely pink **flowers**. sSichuan (Emeishan), n & nwYunnan; damp, shady places in shrubberies and evergreen forests, 500–1,650 m. VI–VII.

Begonia morsei Irmsch. Like the previous species this has small ovate **leaves**, not more than 7 cm long, often patterned with pale stripes, with a heart-shaped base and rusty-brown-hairy stalks. **Inflorescence** slightly overtopping leaves, 8–11 cm tall, bearing just 2 pink **flowers**. swGuangxi, sYunnan [nVietnam]; damp shady banks and rocks in evergreen forests, caves, to 1,450 m. VI–XI.

CUCURBITACEAE

THLADIANTHA

A genus of c. 22 species (all found in China) distributed from north India to Vietnam, Japan and Korea. Dioecious **climbers** or spreading **perennial herbs** usually with tubers, herbaceous **stems**, simple or 2-fid tendrils and mostly simple, (rarely 3–7-lobed), heart-shaped **leaves**, with a toothed margin. **Flowers** 5-parted, with a campanulate, yellow corolla; **male flowers** in cymes or racemes, rarely solitary; **female flowers** solitary or 2–3 together on a short stalk; ovary smooth or warty, with 3 2-lobed stigmas. **Fruit** fleshy, berry-like, smooth or warty, many-ribbed or unribbed.

1. Tendrils simple . ***T. cordifolia*, *T. sessilifolia*, *T. villosula***
1. Tendrils 2-fid . . . . ***T. longisepala*, *T. capitata***

Thladiantha cordifolia (Blume) Cogn. **Stems** much-branched, robust, angular, downy, to 4 m, sometimes more. **Leaves** ovate-heart-shaped, 8–15 × 6–11 cm, long-pointed, margin irregularly toothed, densely pale yellow-downy. **Flowers** butter-yellow; male flowers 3-several in a dense, short raceme, with fan-shaped bracts, corolla 15–17 mm long; **female flowers** solitary. **Fruit** oblong, 3–5 cm long, with rough, grooved rind. Guangxi, Yunnan [Guangdong, India, Laos, Vietnam]; streamsides, forests, shrubberies, 800–2,600 m. V–XI.

Thladiantha sessilifolia Hand.-Mazz.* Differs from the previous species in its slender, **branches** that become hairless, and in its papery, lanceolate or ovate-lanceolate **leaves**, 8–16 × 2–4.5 cm, densely downy beneath, margin slightly toothed; leaf-stalks less than 15 mm long; **male flowers** 1–5 in a short-stalked raceme, lacking bracts. **Fruit** ovate-oblong. swSichuan, Yunnan; similar habitats and altitudes. V–VIII.

Thladiantha villosula Cogn.* Readily distinguished from previous 2 species by the much-branched, densely glandular-softly hairy **stems**, and in having membranous **leaves**, ovate-heart-shaped to almost circular, somewhat long-pointed, 6–12 cm long, margin toothed. **Male flowers** 2–7 in a raceme, lacking bracts; inflorescence-stalk usually bearing a densely downy, leafy, toothed appendage. **Fruit** oblong, red-brown, reticulate. Gansu, Guizhou, wHubei, Sichuan, sShaanxi [sHenan]; ditchsides, forests, thickets, 2,000–2,800 m. V–VIII.

Thladiantha longisepala C. Y. Wu, A. M. Lu & Z. Y. Zhang* Plant to 3 m with **stem** and **branches** angular-grooved, sparsely bristly, becoming hairless. **Leaves** membranous, ovate-heart-shaped or broadly ovate-heart-shaped, 8–14 cm long, somewhat long-pointed, densely bristly above, minutely downy beneath, margin toothed. **Flowers** yellow; **male inflorescence** much-branched, 5–13 cm long, calyx more than 10 mm long, corolla-segments ovate, long-pointed, 20–22 mm long; **female flowers** 2–5, on a short stalk. **Fruit** oblong, yellow-brown-downy and rough with bristles. Guangxi, s & sw Sichuan, nwYunnan; shrubberies on mountain slopes, stream- and riversides, 2,400–3,500 m. VI X.

Thladiantha capitata Cogn.* Similar to the previous species, but **leaves** broadly ovate or ovate-triangular, with sparsely toothed margin. **Male flowers** smaller, in congested umbels of up to 15, calyx less than 10 mm long, corolla-segments 10–12 mm long; **female flowers** solitary or 2–3 together at top of inflorescence-stalk, corolla 15–28 mm long. **Fruit** oblong, 4 cm long. wSichuan; margins of forests and shrubberies, 1,000–2,700 m. V–XI.

HERPETOSPERMUM

A genus with 1 species in south-west China, India and Nepal. Dioecious **annual climber**, lacking tubers, stems with simple tendrils. **Male flowers** in a raceme or rarely solitary, calyx-tube funnel-shaped with awl-shaped lobes, corolla broadly campanulate, 5-parted, with entire lobes; stamens 3, not projecting from calyx-tube, anthers joined; **female flowers** solitary, similar to male, but with 3-celled, oblong ovary and 3 dilated stigmas. **Fruit** broadly oblong, 3-valved from apex nearly to base.

Herpetospermum pedunculosum (Seringe) Baill. Plant to 2.5 m, **stems** and **branches** thread-like, minutely downy at first. **Leaves** membranous, ovate-heart-shaped, tailed to long-pointed, 6–12 × 4–9 cm, margin toothed. **Male inflorescence** 12–40 cm long, with 5–10 yellow flowers, corolla-lobes elliptic, 20–22 × 12–15 mm; **female flowers** similar, ovary with dense yellow, stiff hairs. **Fruit** broadly oblong, 3-valved to base, fibrous, 7–8 cm long. seTibet, w & nwYunnan [swTibet, India, Nepal]; thickets and forest margins on mountain slopes, 2,300–2,500 m. VI–X.

TRICHOSANTHES

A genus of c. 80 species (33 in China) distributed in Asia and north Australia. **Climbing**, dioecious, **annual** or **perennial herbs**. **Leaves** simple, unlobed, palmate or rarely with 3–5 leaflets, margin usually finely toothed. **Tendrils** usually 2–5-branched, rarely simple. **Flowers** usually white; **male flowers** in racemes, rarely solitary or in axillary pairs, calyx-tube cylindrical, sepals 5, entire, toothed or cut into narrow lobes, petals 5, usually long-fringed, stamens 3, inserted on calyx-tube; **female flowers** solitary, very rarely in racemes, calyx and corolla as in the male flowers, ovary inferior. **Fruit** fleshy, globose, egg-shaped or spindle-shaped, many-seeded, indehiscent, usually glabrous and smooth, with seeds packed in pulp.

Trichosanthes rosthornii Harms Vigorous **climber** to 4 m with grooved, sparsely minutely downy **stems**. **Leaves** broad-ovate to almost orbicular, deeply 3–7-lobed, 8–12 × 7–11 cm; lobes linear-lanceolate to oblanceolate, long-pointed, short-hairy beneath, glabrous above, margin toothed. **Male racemes** solitary or in axillary pairs; bracts rhombic-obovate, 6–14 mm long, margin irregularly filamentous-fringed; calyx-tube narrowly cylindrical, 2.5–3(–3.5) cm long; **female flowers** solitary, calyx-tube cylindrical, 20–25 mm long, ovary minutely downy. **Fruit** globose or oblong, 8–16 cm long, glabrous, orange-yellow when ripe. Guangxi, Guizhou, wHubei, s, e & wSichuan, Yunnan [sAnhui, nGuangdong, neJiangxi]; dense forests along valleys, shrubberies, grassy mountain slopes, 400–1,850 m. Flowering VI–VIII, fruiting VIII–X.

SCHIZOPEPON

A genus with 8 species (all found in China), distributed from the Himalaya to east Asia. **Climbing** or **scrambling**, monoecious or dioecious **herbs** with slender **stems** and **branches**. **Tendrils** 2-fid. **Leaves** ovate-heart-shaped or broadly ovate-heart-shaped, rarely spear-shaped, usually 5–7-lobed, margin irregularly toothed. **Flowers** small, unisexual or rarely hermaphrodite; **male flowers** usually in a raceme; **female flowers** solitary or few in a raceme; calyx-tube cup-shaped or campanulate, with 5 lobes; corolla-lobes 5, white, ovate; stamens 3, free or joined; ovary 2–3-celled with a short 3- or 5-fid style. **Fruit** small, ovate or conical, pointed or long-pointed, smooth or dotted, 3-valved or indehiscent, 1–3-seeded.

Schizopepon dioicus Cogn. ex Oliv.* Scrambling **herb** to 2.5 m. **Leaves** membranous, broadly ovate-heart-shaped, 5–9 × 3–7 cm, long-pointed, usually 2–3-angled on each side, glabrous, margin toothed. **Male flowers** in a raceme, **female flowers** solitary or few at top of a short stalk; calyx-lobes awl-shaped or narrowly lanceolate, 1–1.2 mm long; corolla rotate, white, lobes lanceolate or oblong-lanceolate, 2–3.5 mm long; stamens 3; ovary glabrous, 3-celled. **Fruit** broadly egg-shaped, pointed, 11–12 cm long, warty, pale brown when mature, 3-valved, usually with 2 seeds. wHubei, eSichuan, sShaanxi [nwHunan]; grassy places, banks, shrubberies, 1,000–2,400 m. VII–X.

Schizopepon longipes Gagn.* Similar to *S. dioicus*, but leaves ovate, irregularly 3–5-lobed, long-pointed, 4–6 cm long, minutely downy, middle lobe larger and triangular, margin toothed; **male flowers** with lanceolate calyx-lobes, 1.8–2 mm long, corolla-lobes 3–5 mm long. wSichuan, nwYunnan; similar habitats, 2,000–3,000 m. VII–X.

---

1. *Trichosanthes rosthornii*; Yichang to Shennongjia, wHubei (PC)
2. *Populus rotundifolia*; Zhongdian, nwYunnan (EL)
3. *Populus yunnanensis*; cult., from nwYunnan (CGW)

SALICACEAE

POPULUS

A genus of c. 100 species (71 in China, 47 endemic) in the northern hemisphere. Usually large, fast-growing, **deciduous trees** with spirally arranged, simple **leaves**, those on lateral shoots often smaller and differently shaped to those on the main shoots. **Flowers** in pendent catkins in early spring, appearing before the leaves, male and female usually on separate trees, **male flowers** generally denser, with red or purple anthers, **female flowers** catkins elongating in fruit, ovaries egg-shaped, surmounted by 2–4 stigmas. **Seeds** surrounded by a tuft of white cottony hairs, blown long distances in the wind.

1. Bracts with long marginal hairs; leaves small, blade seldom more than 8 cm long .... *P. rotundifolia*, *P. adenopoda*, *P. davidiana*
1. Bracts glabrous; leaves larger, mostly 8–30 cm long .. 2
2. Leaf-blade rarely more than 7 cm wide, twice as long as wide .... *P. yunnanensis*
2. Leaf-blade usually more than 8 cm wide (to 15 cm), often as wide as long or almost so 3
3. Buds and leaf-stalks glabrous .... *P. wilsonii*
3. Buds and leaf-stalks downy .... *P. lasiocarpa*, *P. szechuanica*, *P. ciliata*

Populus rotundifolia **Griff.** **Tree** to 20 m, but often thicket-forming due to coppicing, with smooth grey **bark** and brown-downy young **branches**; buds downy, often sticky. **Leaves** fluttery, triangular-ovate to almost orbicular, to 8.5 × 8 cm, green above, grey-green beneath, downy at first, with a wavy-toothed margin; leaf-stalks slender. **Female catkins** 4–7 cm long, elongating to 10 cm in fruit. Gansu, Guizhou, Sichuan, Shaanxi, se & eTibet, Yunnan [Bhutan]; valley sides, mountain slopes, open woodland, shrubberies, 1,200–2,800 m. III–IV.

Populus adenopoda **Maxim.*** Closely related to the previous species, but distinct in having glabrous buds and shiny dark green **leaves** with a shallowly toothed margin and long drawn out, tail-like tips, leaf-base with a pair of conspicuous glands; female catkins with characteristic palmately lobed bracts. Guangxi, Guizhou, Hubei, Sichuan, Shaanxi, Yunnan [c & eChina]; similar habitats, 300–2,500 m. III–IV. Widely used as a timber tree.

Populus davidiana **Dode** Suckering **tree** to 25 m, with greyish or greyish white, smooth **bark**, rough towards base on older trees; young branchlets reddish brown; **buds** glabrous. **Leaves** triangular-ovate to almost orbicular, to 6 × 5.5 cm, with finely toothed margin and pointed to slightly long-pointed apex. **Female catkins** 4–7 cm long, to 12 cm in fruit. Gansu, Guangxi, Guizhou, Hubei, Qinghai, Sichuan, Shaanxi, Tibet, Yunnan [n & neChina, Korea, Mongolia, eRussia]; mountain slopes, woodland, 100–3,800 m. III–IV.

Populus yunnanensis **Dode*** Rather upright **tree** to 20 m, often less, with furrowed grey **bark**, purplish brown older **branches** and reddish or yellowish brown branchlets; **buds** glabrous. **Leaves** ovate to ovate-elliptic, to 16 × 7.5 cm, green and shiny above, greyish or whitish beneath, apex long-pointed, margin finely toothed, ciliate when young; leaf-stalks, like the midrib beneath, reddish. **Female catkins** 10–15 cm long, ovary with 3–4 stigmas. Guizhou, Sichuan, Yunnan; mountain slopes, forests, widely planted along roads, 1,300–3,700 m. III–IV.

1. *Populus wilsonii*; cult., from Sichuan (CGW)
2. *Populus lasiocarpa*, female; cult. (CGW)
3. *Populus lasiocarpa*; wSichuan (TK)
4. *Salix lindleyana*, female; Yulongxueshan, nwYunnan (DC)
5. *Salix serpyllum*, male; eNepal (CGW)
6. *Salix variegata*; Yulongxueshan, nwYunnan (PC)
7. *Salix fargesii*; cult. (CGW)
8. *Salix fargesii*, female; cult. (CGW)

Populus wilsonii C. K. Schneid.* Bold, handsome **tree** to 25 m with slightly furrowed reddish brown **bark**, branchlets purplish, or brownish at first, but eventually grey-brown; **buds** large, reddish brown or purplish brown. **Leaves** ovate to elliptic-ovate or almost rounded, to 20 × 15 cm, bluish green above but paler grey-green beneath, margin bluntly gland-toothed, apex rounded. **Female catkins** up to 15 cm long in fruit. Gansu, Hubei, Sichuan, Shaanxi, e & seTibet, Yunnan; broadleaved forests, 1,300–3,400 m. IV–V.

Populus lasiocarpa Oliv.* Bold, handsome **tree** to 20 m with dark grey, furrowed bark and yellowish brown to purplish brown, angled, downy branchlets; buds large, downy. **Leaves** ovate, to 30 × 15 cm, long-pointed, green above, paler beneath, base heart-shaped, usually with a pair of glands, margin gland-toothed; leaf-stalks red, up to 8 cm long. Female catkins up to 24 cm long in fruit. Guizhou, Hubei, Sichuan, Shaanxi, Yunnan; mountain slopes, forests, riversides, 1,300–3,500 m. IV–V.

Populus szechuanica C. K. Schneid.* Taller **tree** than the previous species, to 40 m, distinguished by greyish or whitish **bark**, which is fissured towards base of trunk; **buds** purplish, glabrous or downy. **Leaves** often rather reddish, ovate to ovate-lanceolate, to 18 × 15 cm, with a wedge-shaped, rounded or slightly heart-shaped base, margin toothed, finely ciliate at first. **Female catkins** not more than 20 cm long in fruit. Gansu, Sichuan, Shaanxi, e & seTibet, Yunnan; mountain forests, 1,100–4,600 m. IV–V.

Populus ciliata Wall. ex Royle Similar to *P. lasiocarpa*, but with brown, glabrous young branchlets and glabrous **buds**; **leaves** not more than 15 × 12 cm, dull green above, grey-green beneath, glabrous but with densely ciliate margin. seTibet, w & nwYunnan [swTibet, Himalaya, Myanmar]; similar habitats, sometimes in open situations, 2,200–3,400 m. IV–V.

SALIX

A genus of c. 520 species (275 in China, 189 endemic) in temperate and Arctic regions of the northern hemisphere, a few being found south of the equator. **Dioecious trees** or **shrubs** with leaf-buds enclosed in a single scale. **Leaves** alternate, toothed or entire, usually stalked; stipules present. **Flowers** few to many, in erect to ascending, rounded to elongated or linear catkins, appearing with or before the leaves, bracts entire; **male flowers** with 1–5 (occasionally more) stamens; **female flowers** with 1-celled ovary and a single 2-lobed stigma. **Fruit** a 2-valved capsule containing numerous tiny **seeds**, each with a plume of white hairs at base.

1. Dwarf hummock-forming or prostrate shrubs with leaves up to 2 cm long ... **S. lindleyana, S. brachista, S. souliei, S. serpyllum**
1. Upright shrubs or small trees with leaves mostly 4 cm long or more (except see *S. variegata*) 2
2. Leaves small, not more than 2 cm long at maturity .... **S. variegata**
2. Leaves at least 4 cm long at maturity 3
3. Branchlets stout, shiny, mahogany .... **S. fargesii**
3. Branchlets not as above, often pruinose or hairy .... **S. magnifica, S. moupinensis, S. wilsonii, S. balfouriana**

Salix lindleyana Wall. ex Andersson Prostrate **shrub** forming a dense carpet with russet-coloured branchlets and yellowish green **buds**. **Leaves** oblong to lanceolate or obovate, 1.2–1.6 × 0.4–0.6 cm, occasionally larger, bright shiny green above, paler beneath and hairy at first, margin entire, often slightly back-rolled. **Catkins** small, egg-shaped, few-flowered, with purplish red bracts, appearing with the leaves; filaments hairy at base. Tibet, nwYunnan [c & eHimalaya]; damp rock-crevices, moraines, alpine meadows, above 4,000 m. VI–VII.

Salix brachista C. K. Schneid.* In contrast to the previous species, makes a mound with creeping lateral branchlets, **leaves** more ovate or obovate-elliptic, with densely hairy stalks and undersurface. w & swSichuan, e & seTibet, w & nwYunnan; thickets, mountain slopes, rocky places, 2,600–3,900 m. VI–VII.

Salix souliei Seemen* Related to *S. brachista*, but stamens with glabrous filaments; **leaves** bright green and rather wrinkled above, glaucous and silky-white beneath when young, stalks 4–7 mm long, hairy. seQinghai, w, nw & swSichuan, e & seTibet, nwYunnan; rocky montane habitats, moraines, 4,200–4,800 m. VI–VII.

Salix serpyllum Andersson (syn. *S. hylematica* C. K. Schneid.) Readily distinguished from the previous species by bright shiny green, elliptic to oblanceolate **leaves**, to 1 × 0.4 cm, glaucous beneath and minutely toothed towards apex; **catkins** numerous, c. 1 cm long, red. seTibet [swTibet, Himalaya]; rocky alpine meadows, banks, moraines, streamsides, 3,200–4,500 m. V–VI.

Salix variegata Franch.* **Shrub** to 1 m tall with pinkish purple branchlets, hairy at first. **Leaves** short-stalked, oblong-oblanceolate to oblong, to 1.5 × 0.4 cm, appressed-silvery-hairy beneath, less hairy above, margin entire to slightly toothed. **Catkins** appearing with young leaves; **male catkins** 1.5–2.5 cm long with a pair of stamens, filaments joined, bracts half the length of filaments; **female catkins** the same length but stouter, stigma 2-cleft. seGansu, Guizhou, wHubei, Sichuan, sShaanxi, eTibet, n & nwYunnan; gullies, streamsides, wet shrubberies, 1,200–3,200 m. IV–V.

Salix fargesii Burkill* Distinctive wide-spreading **shrub** to 3 m, occasionally more, with shiny mahogany, stout branchlets that have a few filamentous hairs at base. **Leaves** elliptic to lanceolate-ovate, 7–11 × 4–6 cm, pale greenish below with whitish hairs along veins, dull deep green above, finely glandular-toothed along margin; lateral veins 16–20 on each side. **Catkins** 6–8 cm long, with downy bracts. **Fruit-catkins** 9–12 cm long. Gansu, wHubei, Sichuan, Shaanxi; mountain shrubberies, wet places, streamsides, 1,400–1,600 m. IV–V.

Salix magnifica **Hemsl.*** **Shrub** or small **tree** to 5 m, with glabrous branchlets that have a bloom when young. **Leaves** rather magnolia-like, with a reddish stalk 2–4 cm long, blade ovate to elliptic or oblong, 15–20 × 6–11 cm, rather leathery, glaucous beneath, margin entire to finely toothed; lateral veins 12–15 on each side. **Catkins** 6–10 cm long, glabrous, on a branchlet 4–7 cm long with 2–5 leaflets. **Fruit-catkins** 16–23 cm long. n, w & nwSichuan; wet woodland, river- and streamsides, valleys, mountainsides, 2,100–3,000 m. V–VI.

Salix moupinensis **Franch.*** Very similar to the previous species, but young branchlets silky-hairy at first, becoming reddish brown or yellowish; **leaves** rather smaller with stalks not more than 1.7 cm long, usually with a gland at top; **catkins** with a hairy, not glabrous, rachis. w & swSichuan, n & nwYunnan; mountain shrubberies, streamsides, 1,500–3,000 m. IV–V.

Salix wilsonii **Seemen ex Diels*** Readily distinguished from the previous 2 species by dull brown branchlets, while the smaller, elliptic to oblong or lanceolate **leaves**, 4–5 × 2–3 cm, have a pointed apex and are pale green beneath; **catkins** 2–6 cm long, male longer than female. Hubei [c & eChina]; flat scrubby areas and riverbanks, to 1,800 m. III–IV.

Salix balfouriana **C. K. Schneid.*** Large **shrub** or small **tree** to 5 m, with reddish black branchlets, woolly at first. **Leaves** short-stalked, elliptic to elliptic-oblong or obovate, to 8 × 4 cm, sometimes larger, pale beneath and downy at first, dark green and glabrous above at maturity except sometimes for the veins, margin glandular-toothed. **Catkins** 2–3 cm long, appearing with the very young leaves, usually with 1–2 leaves at base, these sometimes absent; **male catkins** with 2 stamens per flower, filaments with long soft hairs; **female catkins** with a reddish purple ovary and a 2-lobed style. w & swSichuan, nwYunnan; mountain slopes, thickets, 2,800–4,000 m. IV–V.

CRUCIFERAE (BRASSICACEAE)

CARDAMINE

A genus of c. 200 species (48 in China, 24 endemic) distributed worldwide. **Annual** or **perennial herbs** with simple or variously lobed **leaves**. **Inflorescence** a raceme; **flowers** white, pink or purple. **Fruit** linear, compressed, many-seeded, the 2 valves springing open suddenly when ripe.

1. Stem-leaves undivided . . . . *C. circaeoides*
1. Stem-leaves pinnate, rarely with 3 leaflets 2
2. Stem-leaves large, leaflets at least 30 mm long, coarsely toothed . . . . . *C. macrophylla, C. tangutorum*
2. Stem-leaves small, leaflets rarely more than 22 mm long, entire or somewhat lobed but not toothed 3
3. Rhizomes with bulbils; stem thread-like and fragile at point of attachment to rhizome . . . . . . *C. franchetiana, C. pulchella, C. loxostemonoides*
3. Rhizomes without bulbils; stem relatively stout at point of attachment to rhizome 4
4. Stem-leaves not clasping stem at their bases . . . . *C. delavayi, C. purpurascens, C. microzyga*
4. Stem-leaves clasping stem at their bases . . *C. griffithii*

Cardamine circaeoides **Hook. f. & Thomson** Tufted **rhizomatous perennial** to 40 cm, basal **leaves** usually simple, occasionally with small lateral leaflets, blade heart-shaped to almost orbicular, to 5.5 × 4.5 cm, margin almost entire to scalloped or slightly lobed; stem-leaves usually simple, similar to lower, but smaller. **Inflorescence** racemose, elongating as fruit develop; sepals oblong, 2–3.5 mm long; petals white, spathulate-obovate, 5–8 mm long. **Fruit** linear, 13–30 mm long, constricted between seeds. Gansu, Guangxi, Sichuan, Yunnan [Hunan, Himalaya, Taiwan]; damp places in forests and along streams, ravines, rocky places, 400–3,300 m. III–VII.

Cardamine macrophylla **Willd.** Very variable and widespread, **clump-forming perennial** to 90 cm, often less, with ascending to erect leafy **stems**. **Leaves**, including basal ones, pinnate with a lanceolate to oblong or ovate terminal leaflet, and 2–6 pairs of similar but slightly smaller lateral leaflets; all leaflets with toothed to almost entire margin, glabrous or downy. **Inflorescence** a dense to lax, many-flowered raceme, sometimes branched; sepals oblong, erect, less than half length of petals; petals purple, pink or lilac, obovate to spathulate, 10–17 mm long. **Fruit** linear, 25–60 mm long, glabrous or slightly downy. Gansu, Guizhou, Hubei, Qinghai, Sichuan, Shaanxi, Tibet, Yunnan [cChina, eHimalaya, Mongolia, eRussia]; marshy places in forests and meadows, stream and pool margins, wet shrubberies, rocky places, ravines, 1,200–4,200 m. IV–VIII.

Cardamine tangutorum **O. E. Schulz*** Very similar to the previous species, but typically a smaller plant rarely more than 35 cm tall, with usually no more than 4 pairs of **stem-leaves**; base of leaflets not running down rachis; **inflorescence** with up to 15 purple flowers. Gansu, Qinghai, w & nwSichuan, Shaanxi, eTibet [n & cChina]; similar habitats, 1,300–4,400 m. V–VII.

Cardamine franchetiana **Diels*** Slender **herb** to 20 cm, hairy or glabrous, with simple **stems** and 1–2 basal pinnate **leaves**; stem-leaves usually 2–3, like the basal but generally smaller, with a oblong, slightly lobed to lobed terminal leaflets and 3–6 pairs of smaller, elliptic to oblong leaflets. **Inflorescence** a 3–8-flowered raceme; sepals erect, green with a membranous margin; petals white, rarely pale lavender, obovate, 6–9 mm long, apex rounded. **Fruit** linear, 15–25 mm long, glabrous. s & seQinghai, Sichuan, e & seTibet, n & nwYunnan; stony and rocky meadows, screes, rock-crevices, damp pastures; 2,300–4,800 m. VI–VIII.

1. *Salix magnifica*, female; Wolong, wSichuan (PC)
2. *Salix magnifica*, female; Wolong, wSichuan (PC)
3. *Salix moupinensis*; Kangding, wSichuan (PC)
4. *Salix balfouriana*; Gonggashan Pass, wSichuan (PC)
5. *Cardamine circaeoides*; Huanglong, nwSichuan (PC)
6. *Cardamine macrophylla*; Emeishan, sSichuan (CGW)
7. *Cardamine macrophylla*; Balangshan, Wolong, wSichuan (CGW)
8. *Cardamine tangutorum*; Huanglong, nwSichuan (CGW)]
9. *Cardamine franchetiana*; upper Gangheba, Yulongxueshan, nwYunnan (CGW)

1. *Cardamine pulchella*; Tianchi, nwYunnan (RS)
2. *Cardamine loxostemonoides*; Daxueshan, nwYunnan (CGW)
3. *Cardamine loxostemonoides*; Baimashan, nwYunnan (PC)
4. *Cardamine purpurascens*; Daxueshan, nwYunnan (CGW)
5. *Cardamine microzyga*; south of Aba, nwSichuan (CGW)
6. *Cardamine griffithii*; Cangshan, nwYunnan (PC)
7. *Megacarpaea delavayi*; near Litang, wSichuan (HJ)
8. *Megacarpaea delavayi*; Daxueshan, nwYunnan (CGW)
9. *Erysimum forrestii*; upper Gangheba, Yulongxueshan, nwYunnan (CGW)

Cardamine pulchella (Hook. f. & Thomson) Al-Shehbaz & Yang (syn. *Loxostemon pulchellus* Hook. f. & Th.) Similar to the previous species, but **stem-leaves** with only 1–2 pairs of entire oblong to narrow-elliptic leaflets, leaf-axils bearing tiny bulbils; racemes usually with 2–5 **flowers**, petals deep purple or mauve. **Fruit** not more than 17 mm long. Qinghai, Sichuan, e & seTibet, n & nwYunnan [Himalaya]; similar habitats, 3,400–4,600 m. V–VIII.

Cardamine loxostemonoides O. E. Schulz (syn. *Loxostemon loxostemoides* (O. E. Schulz) Y. Z. Lan & T. Y. Cheo) Distinguished from the previous 2 species by **leaves** with a rounded, often 3-lobed terminal leaflet, lateral leaflets in 2–5 pairs, also often lobed; **flowers** larger, petals purple with darker veins, 8–12 mm long. **Fruit** linear, 25–35 mm long. seTibet, nwYunnan [swTibet, Himalaya]; damp grassy places and screes, steam margins, stony slopes, 2,900–5,500 m. VI–VII.

Cardamine delavayi Franch. **Perennial herb** to 40 cm, glabrous to somewhat hairy, especially at nodes, with erect, generally unbranched **stems**. **Basal leaves** simple, kidney- to heart-shaped, sometimes somewhat 5-lobed; **stem-leaves** with 3 leaflets, stalked, leaflets oblong to oval, terminal slightly larger than lateral. **Racemes** with up to 15 **flowers**; sepals erect, greenish with membranous margin; petals white, obovate, with rounded apex, 5–7 mm long. **Fruit** linear, 20–30 mm long, glabrous. w & swSichuan, nwYunnan [Bhutan]; damp forest clearings, stony streamsides, 2,100–4,000 m. V–VII.

Cardamine purpurascens (O. E. Schulz) Al-Shehbaz et al.* (syn. *Loxostemon purpurascens* (O. E. Schulz) Fang ex Y. Z. Lan & T. Y. Cheo) Readily distinguished from the previous species by pinnate **basal leaves** and **stem-leaves** with 4–7 pairs of often slightly lobed lateral leaflets, terminal leaflet narrow-elliptic to almost linear; petals purple, lavender to almost magenta, 7–11 mm long. w & swSichuan, nwYunnan; similar habitats as well as marshes and open shrubberies, 3,100–4,700 m. V–VII.

Cardamine microzyga O. E. Schulz* Readily distinguished from the previous 2 species by **stem-leaves** that have 6–9 pairs of small slightly toothed to lobed leaflets, terminal leaflet obovate; **flowers** purple or white; petals obovate, 6.5–10 mm long. w & nwSichuan, eTibet; streamsides, damp grassy places and meadows, 2,600–4,600 m. VI–IX.

Cardamine griffithii Hook. f. & Thomson **Perennial herb** with erect leafy **stems** to 80 cm. **Leaves** stalkless, lower and middle ones with a rounded to obovate terminal leaflet and 2–4 pairs of somewhat smaller, slightly lobed, ciliate leaflets, basal pair of leaflets clasping stem. **Inflorescence** an elongating raceme, with up to 20 **flowers**; sepals greenish, erect; petals purple or lavender, spathulate to obovate, 6–9 mm long, with rounded apex. **Fruit** linear, 20–40 mm long, smooth. Sichuan, e & seTibet, w & nwYunnan [Himalaya]; marshy and damp grassy places, streamsides, open forests, shady rocks, 2,400–4,500 m. V–VIII.

MEGACARPAEA

A genus of c. 9 species (1 in western China) found in central Asia and the Himalaya. **Perennial herbs** with a stout caudex armed with the remains of previous year's leaf-stalks; hairs simple. **Flowers** in racemes or panicles, petals entire or variously toothed; stamens 6 or up to 24. **Fruit** an indehiscent, 2-lobed silicula, with narrowly to broadly winged, leathery, 1-seeded valves.

Megacarpaea delavayi Franch. Rather stout, erect, very variable, *Cardamine*-like **perennial** to 80 cm with pinnately lobed **leaves**, basal with a flattened stalk and up to 13 pairs of entire to variously toothed lobes; **stem-leaves** similar but stalkless and with fewer lobes. **Inflorescence** a simple or branched raceme; petals purplish, pink or lavender, oblong to obovate, 3–4 m long, entire to slightly or deeply 3- or 5-lobed. **Fruit** 10–12 mm long with 2 oval, narrowly winged lobes. sGansu, Qinghai, Sichuan, e & seTibet, w & nwYunnan [nMyanmar]; damp grassy places, swampy meadows, damp slopes, open shrubberies, 3,300–4,800 m. V–VIII.

ERYSIMUM

A genus with c. 150 species (17 in China, 5 endemic) mainly in the temperate northern hemisphere. **Annual**, **biennial** and **perennial** herbs, occasionally subshrubby, with branched hairs. **Leaves** in rosettes or alternate, simple or lobed. **Flowers** in racemes, mostly yellow or orange, occasionally purple or violet. **Fruit** linear in outline, often 4-angled, many-seeded.

1. Flowers yellow or orange-yellow . . . .
. . . . *E. forrestii, E. wardii, E. handel-mazzettii*
1. Flowers pink or purple . . . . . *E. roseum, E. funiculosum*

Erysimum forrestii (W. W. Sm.) Pol.* (syn. *Parrya forrestii* W. W. Sm.) Small **perennial herb** with a basal leaf-rosette or tuft, to 20 cm, generally with several **stems**. **Leaves** grey-green, elliptic to oblong or ovate, to 60 × 15 mm, narrowed at base, margin with a few coarse teeth, adorned with branched hairs. **Flowers** bright yellow, in a condensed rounded raceme, elongating somewhat in fruit, fragrant, each flower with a bract; sepals erect, 7–10 mm long, outer 2 with pouched base; petals 17–25 mm long, long-clawed. **Fruit** linear, flattened, 40–90 mm long, downy. w & nwYunnan; limestone rocks, screes and moraines, rocky meadows, river gravels, 3,600–4,900 m. V–VII.

Erysimum wardii* Pol. Similar to the preceding species, but a taller plant, 20–90 cm; **leaves** linear-lanceolate, not usually more than 7 mm wide, and only lowermost flowers in **inflorescence** with a bract; petals yellow or orange-yellow; **fruit** up to 11 cm long but not more than 2 mm wide. wSichuan, e & seTibet, w & nwYunnan; similar habitats, 3,000–4,600 m. V–VII.

Erysimum handelmazzettii* Pol. (syn. *Cheiranthus acaulis* Hand.-Mazz.) Similar to *E. forrestii*, but a smaller, more compact plant not more than 10 cm with several leaf-rosettes; **leaves** oblong to lanceolate, entire to slightly toothed; **flowers** in almost umbellate clusters amongst the foliage; **fruit** linear, slightly 4-angled, 2.5–5 cm long. w & swSichuan, nwYunnan; screes and gravelly places, stabilised moraines, rock-crevices, 4,100–4,800 m. V–VII.

Erysimum roseum* (Maxim.) Pol. (syn. *Cheiranthus roseus* Maxim.) Tufted, few-branched **perennial** to 35 cm with leafy **stems**; basal **leaves** in a rosette; blade grey-green, oblong to obovate, to 70 × 10 mm, pointed to almost blunt, margin slightly toothed to entire, lower stalked, upper stalkless. **Inflorescence** corymbose at first but elongating in fruit, only lowermost **flowers** with bracts; sepals narrow-oblong, persistent in young fruit; petals pink or purple, obovate, 16–25 mm long. **Fruit** linear, 15–40 mm long, 4-angled, somewhat hairy. Gansu, Qinghai, Sichuan, eTibet, nwYunnan; stony alpine meadows, screes, cliffs, 3,200–4,900 m. V–VII.

***Erysimum funiculosum* Hook. f. & Thomson** Readily distinguished from the previous species in being a small plant not more than 8 cm tall, with narrower **leaves** mainly in basal rosettes; **inflorescence** not elongating in fruit, petals pink, narrow-spathulate, 7–9 mm long. Gansu, Qinghai, Tibet [Sikkim]; similar habitats as well as stony beaches, 3,400–5,500 m. VI–VII.

BRAYA

A genus with 6 species (3 in China) distributed in the Arctic and the temperate northern hemisphere. **Tufted** or **cushion-forming** plants, with most **leaves** in basal rosettes, generally entire. **Inflorescence** racemose, elongating in fruit; **flowers** pink, purple or white, rarely yellow; petals blunt. **Fruit** linear to lanceolate or oblong in outline, each valve with a distinct midvein.

***Braya forrestii* W. W. Sm.** Dense **cushion-forming perennial** to 6 cm, adorned with simple hairs. **Leaves** in rosettes, stalked, blade linear to linear-oblanceolate, 1–3 × 0.5–2 mm, entire. **Inflorescence** a scapose raceme, bractless; sepals oblong, 2–5 mm long, purple near top; petals pink, purple or white, obovate, 4–6 mm long, apex rounded. **Fruit** egg-shaped, 4–7 mm long, glabrous or minutely hairy. w & swSichuan, e & seTibet, nwYunnan [Bhutan]; alpine meadows, rocky pastures, 3,700–5,000 m. V–VII.

PEGAEOPHYTON

A genus of 6 species (4 in China, 1 endemic) from the Himalaya and north India to China. **Perennials** with basal tufts or rosettes of **leaves** borne at the end of a simple or branched caudex, stemless, glabrous or with simple hairs. **Flowers** solitary on long stalks in leaf-axils; petals longer than sepals; stamens 6 with nectar-glands at base. **Fruit** a silicula with membranous or papery dehiscent valves.

***Pegaeophyton scapiflorum* (Hook. f. & Thomson) C. Marq. & Airy Shaw** Perennial **herb** to 15 cm with rosettes of somewhat fleshy, elliptic to oblong, oblanceolate or linear-lanceolate **leaves**, to 8 × 1.2 cm, with entire to toothed margin, glabrous or slightly downy. **Flowers** on slender stalks to 15 cm long; sepals ovate, 2.5–6 mm, glabrous or slightly hairy; petals white, occasionally pink or blue, with greenish or bluish base, obovate to spathulate, 5–7 mm long, not notched. **Fruit** oblong to ovate or rounded, flattish, 5–20 mm long, valves papery and glabrous. Gansu, Qinghai, Sichuan, Tibet, w & nwYunnan [nwChina, Himalaya, nMyanmar]; damp places, alpine moorland, wet moraines, river- and streamsides, rocky places, snow hollows, glacier margins, swampy habitats, 3,500–5,400 m. V–IX. **Subsp. robustum (O. E. Schulz) Al-Shehbaz et al.** (syn. *P. sinense* (Hemsl.) Hayek & Hand.-Mazz.) is a stouter plant overall, with a thick, generally unbranched caudex and larger **flowers** with petals 8–12 mm long. w & swSichuan, seTibet, w & nwYunnan [swTibet, Bhutan]; similar habitats.

Pegaeophyton angustiseptatum* Al-Shehbaz et al. Similar to the previous species, but caudex unbranched, **leaves** smaller oblanceolate to spathulate, not more than 25 × 10 mm and **flowers** smaller, petals not more than 5 mm long; **fruit** oblong to rounded, 4–7 mm long with thinly keeled valves, borne on a strongly curved stalk that pushes fruit into the ground. nwYunnan; damp and swampy habitats, 3,200–4,800 m. VI–VII.

DONTOSTEMON

A genus of 11 species (all in China, 1 endemic) in Central Asia, the Himalaya, India, Japan and eastern Russia. **Annual** and **perennial herbs** with simple hairs, sometimes mixed with glandular ones. **Leaves** sometimes in basal rosette, generally pinnately lobed; stem-leaves similar to basal. **Flowers** in bractless racemes that elongate as fruit develop; sepals not pouched at base; petals with a heart-shaped or oblanceolate blade, and a claw equal to, or longer than, sepals; stamens with median filament pairs united. **Fruit** a siliqua, linear in outline, somewhat constricted between seeds, valves with prominent midveins.

Dontostemon tibeticus* (Maxim.) Al-Shehbaz Small **biennial herb** to 20 cm, usually less, branched from base. **Leaves** with a short stalk, lanceolate to oblong, pinnately lobed (with up to 11 pairs of lobes), to 30 × 10 mm, slightly fleshy. **Flowers** in

1. *Erysimum wardii*; Baimashan, nwYunnan (PC)
2. *Erysimum handelmazzettii*; Habashan, nwYunnan (MN)
3. *Erysimum roseum*; Danglexiaoshan, seQinghai (RS)
4. *Erysimum funiculosum*; Animaqingshan, cQinghai (HJ)
5. *Braya forrestii*; Baimashan, nwYunnan (CGW)
6. *Braya forrestii*; Baimashan, nwYunnan (HJ)
7. *Pegaeophyton scapiflorum*; Cangshan, nwYunnan (CGW)
8. *Pegaeophyton scapiflorum*; Baimashan, nwYunnan (CGW)
9. *Dontostemon tibeticus*; Animaqingshan, cQinghai (HJ)

1. *Taphrospermum verticillatum*; Baimashan, nwYunnan (HJ)
2. *Solms-laubachia minor*; Daxueshan, nwYunnan (CGW)
3. *Solms-laubachia minor*; Daxueshan, nwYunnan (CGW)
4. *Solms-laubachia xerophyta*; Baimashan, nwYunnan (JM)
5. *Solms-laubachia xerophyta*; Yading Nature Reserve, swSichuan (MN)
6. *Solms-laubachia retropilosa*; Baimashan, nwYunnan (HJ)
7. *Solms-laubachia retropilosa*; Baimashan, nwYunnan (HJ)
8. *Solms-laubachia pulcherrima*; Wutoudi, Yulongxueshan, nwYunnan (CGW)
9. *Solms-laubachia pulcherrima*; upper Gangheba, Yulongxueshan, nwYunnan (CGW)

condensed **racemes** at first, later elongating; petals white with pink or purple claw, 8–11 mm overall, blade rounded, apex notched. **Fruit** straight, 10–16 mm long, constricted between seeds, warty. Gansu, Qinghai, neTibet; screes, rocky slopes, stony disturbed meadows, gravelly and sandy places, 3,200–5,200 m. VI–VII.

Dontostemon pinnatifidus **(Willd.) Al-Shehbaz & H. Ohba**
Differs from the previous species in the pinnately lobed to toothed **basal leaves**, and middle and upper **stem-leaves** toothed (linear and untoothed in **subsp.** *linearifolius* **(Maxim.) Al-Shehbaz & H. Ohba**), leaves ciliate, and petals plain white, unnotched. **Fruit** 15–40 mm long, glandular. Gansu, Qinghai, n, nw & swSichuan, Tibet, nwYunnan [India, Nepal, Mongolia, Russia]; similar habitats, 2,500–4,500 m. VI–VIII.

TAPHROSPERMUM

A genus with 7 species (6 in China, 3 endemic) distributed in Central Asia, the Himalaya and China. **Perennial** or **biennial herbs** with prostrate to ascending **stems**. **Leaves** not in rosettes, lowermost whorled or alternate, simple. **Flowers** white, in racemes, with bracts present; petals rounded to notched at apex. **Fruit** cylindrical to oblong or heart-shaped.

Taphrospermum verticillatum **(Jeffrey & W. W. Sm.) Al-Shehbaz*** (syn. *Braya verticillata* (Jeffrey & W. W. Sm.) W. W. Sm.) Small **biennial** or short-lived **perennial** rarely more than 15 cm tall, with 1–several **stems**. Lowermost **leaves** in a whorl, the others opposite or alternate, oblong, to 15 × 9 mm, rounded at apex, glabrous. **Racemes** many-flowered, congested at first but elongating in fruit; sepals oblong, 2.5–3.5 mm long, falling, with a few apical hairs; petals white, occasionally pale lavender, obovate, 7–9 mm long, clawed, apex notched. **Fruit** ovate to oblong, 7–13 mm long, smooth, glabrous. seTibet, w & nwYunnan; stony moorland, screes, stabilised moraines, cliff-ledges, 3,800–5,200 m. VI–VII.

Taphrospermum fontanum **(Maxim.) Al-Shehbaz & G. Yang*** (syn. *Dilophia fontanum* Maxim.) Similar in general proportions to the previous species, but all **leaves** alternate, rarely more than 10 × 4 mm; **flowers** smaller, white or pale lavender, petals 2–6 mm long, apex slightly notched. **Fruit** reverse heart-shaped, 3–5 mm long. Gansu, Qinghai, Sichuan, Tibet [Xinjiang]; similar habitats, as well as disturbed alpine meadows, damp shingles, alpine permafrost swamps, 3,600–5,300 m. VI–VIII.

SOLMS-LAUBACHIA

A genus of 9 species (all in China) found in the Himalaya and west China. **Perennial**, rosette-forming plants with a thick caudex bearing the previous year's leaf-remains. **Flowers** solitary, borne on scapes, blue, pink, purple or white, with 2 nectar-glands at base of ovary. **Fruit** linear to oblong or lanceolate in outline with papery, net-veined valves.

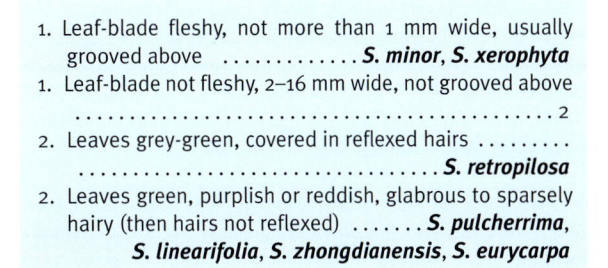

1. Leaf-blade fleshy, not more than 1 mm wide, usually grooved above .... **S. minor, S. xerophyta**
1. Leaf-blade not fleshy, 2–16 mm wide, not grooved above 2
2. Leaves grey-green, covered in reflexed hairs .... **S. retropilosa**
2. Leaves green, purplish or reddish, glabrous to sparsely hairy (then hairs not reflexed) .... **S. pulcherrima, S. linearifolia, S. zhongdianensis, S. eurycarpa**

Solms-laubachia minor **Hand.-Mazz.*** Low **tufted herb** not more than 3 cm with a thick, closely branched caudex. **Leaves** deep green, reddish or purplish, erect to ascending, linear, to 10 × 1 mm, ciliate, apex pointed. **Flowers** often numerous, borne singly on erect, long, slender scapes directly from leaf-rosettes; sepals 4–6 mm long; petals mauve-pink, often paler lilac or purple in centre, 10–12 mm long, with a spreading obovate limb. **Fruit** lanceolate, 15–20 mm long, downy. w & swSichuan, nwYunnan; rocky alpine meadows, stabilised screes, crevices, cliffs, 2,500–4,600 m. V–VII.

Solms-laubachia xerophyta **(W. W. Sm.) Comber*** Very similar to the previous species, but **leaves** longer (20–45 mm) and with a thickened stalk, and fruiting stalks 15–35 mm long (not 2–7 mm); sepals 6–8 mm long; petals pink or pale blue, 17–22 mm long. w & swSichuan, nwYunnan; similar habitats, 3,700–5,200 m. VI–VIII.

Solms-laubachia retropilosa **Botsch.*** **Tufted herb** to 10 cm with a tough, branching caudex covered in old leaf-remains. **Leaves** linear-lanceolate to linear-oblanceolate, to 45 × 6 mm, narrowed below into a thin stalk, covered in grey reflexed hairs. **Flowers** often numerous, borne singly on erect, long, slender scapes directly from leaf-rosettes; sepals 5–8 mm long; petals purple or pinkish purple, 12–18 mm long, limb obovate, equalling the claw. **Fruit** lanceolate-oblong, 20–60 mm long, glabrous or downy. w & swSichuan, seTibet, nwYunnan; screes, rocky meadows, low open shrubberies, 4,200–5,100 m. V–VII.

Solms-laubachia pulcherrima **Muschl.*** (syn. *Parrya ciliaris* Bur. & Franch.) Tough, **tufted herb** to 10 cm with a broad caudex covered in old leaf-remains. **Leaves** rather thick, lanceolate to oblanceolate, sometimes linear-lanceolate, to 55 × 7 mm, narrowed below into a thickened stalk, ciliate, otherwise surface glabrous or with a few crispy hairs. **Flowers** often numerous, borne singly on erect, long, slender scapes directly from leaf-rosettes relatively large; sepals 5–10 mm long, downy; petals turquoise-blue to pink, 17–25 mm long, with a broad-obovate limb. **Fruit** lanceolate, 25–35 mm long, glabrous or hairy along margin. w & swSichuan, seTibet, nwYunnan; rocky places, stabilised screes, moraines, cliffs, gravels, generally on limestone, 3,300–5,200 m. V–VII.

Solms-laubachia linearifolia (W. W. Sm.) O. E. Schulz.* (syn. *Parrya linearifolia* W. W. Sm.) Similar to the previous species, but **leaves** rarely more than 3.5 mm wide, linear to linear-oblanceolate, sparsely to densely hairy and ciliate. **Fruit** larger, 4–7 cm long, usually hairy along margin. Similar distribution and habitats, 3,400–4,700 m. V–VII.

Solms-laubachia zhongdianensis J. P. Yue, Al-Shehbaz & H. Sun* Very close to the previous species, but a lax **cushion-forming** plant, **leaves** needle-like, incurved, crowded into clusters. **Flowers** smaller, white flushed with pinkish mauve. nwYunnan; screes, sandy areas, 3,200–4,500 m. VI–VII.

Solms-laubachia eurycarpa (Maxim.) Botsch.* (syn. *Parrya eurycarpa* Maxim.) Readily distinguished from previous 2 species by broader **leaves**, mostly 15–50 × 10–16 mm, that have a thickened, 'corky', often purplish stalk; **flowers** pale lavender to pink; **fruit** lanceolate to linear-lanceolate, 5–8.5 cm long. Gansu, Qinghai, w, nw & swSichuan, e & seTibet, nwYunnan; similar habitats, 3,800–4,900 m. VI–VII.

LEPIDOSTEMON

A Sino-Himalayan genus of 5 species (3 in China, 2 endemic). **Annual** and **perennial herbs** with rosettes of **leaves** and few-many-flowered **racemes**. **Flowers** with winged filaments. **Fruit** a siliqua with papery valves and with prominent marginal veins.

Lepidostemon rosularis (K. C. Kuan & Z. X. An) Al-Shehbaz* (syn. *Christolea rosularis* K. C. Kuan & Z. X. An) Small **herb** to 4 cm, at first with rosettes of small entire **leaves** (dry by flowering time), these succeeded by dense rosettes of larger oblanceolate to ovate somewhat toothed leaves to 10 × 6 mm, ciliate. **Flowers** in dense cluster-like racemes; sepals oblong, 2–3.5 mm long; petals white, obovate, 3–5 mm, with a short claw, apex rounded. **Fruit** linear-oblong, 10–20 mm long, flattened. seTibet; screes, stony places, dry gullies, 4,200–5,100 m. VI–VII.

HEMILOPHIA

A genus of 4 species restricted to south-west China. Low **rhizomatous herbs** with simple, mainly basal **leaves** in lax rosettes. **Flowers** few in short racemes; petals yellow, white, pink or purple, or bicoloured, apex notched. **Fruit** oblong in outline, rounded in cross-section, glabrous.

Hemilophia rockii O. E. Schulz* Small **rhizomatous herb** no more than 6 cm tall with oblanceolate to ovate, finely hairy **stem-leaves**, to 8 × 2.5 mm, with a short stalk 1–2 mm long; **basal leaves** slightly larger, lanceolate. **Flowers** in dense semi-rounded, bracteate, racemose clusters that elongate in fruit; sepals greenish with thin membranous margin; petals cream with a greenish yellow centre and green veins, obovate, apex deeply notched (almost 2-lobed), 8–10 mm long overall, persistent in fruit. swSichuan, eYunnan; limestone screes and gravels, 3,900–4,900 m. VI–VII.

Hemilophia sessilifolia Al-Shehbaz* Similar to preceding species, but **stem-leaves** stalkless, **racemes** not elongating in fruit; sepals membranous, soon falling; petals not persisting in fruiting stage. w & nwYunnan; similar habitats and flowering time, 4,300–4,600 m.

DRABA

A genus with 350 species worldwide (48 in China). Mainly **perennial herbs**, many **cushion-forming** or **tufted evergreen alpines**, but including some **annual** and **biennial** species, leafy or scapose; **stems** and **leaves** covered with simple, branched or stellate hairs. **Inflorescence** a bracteate or ebracteate raceme, congested but often elongating in fruit; nectar-glands present at base of stamens. **Fruit** an ellipsoid, egg-shaped or lanceolate silicula, sometimes twisted.

1. Flowers borne on leafless scapes 2
1. Flowers borne at ends of leafy stems 3
2. Petals yellow .. ***D. oreades, D. involucrata, D. jucunda***
2. Petals white .... ***D. glomerata***
3. Petals white .... ***D. ladyginii***
3. Petals yellow .... ***D. yunnanensis, D. surculosa, D. polyphylla***

Draba oreades Schrenk Very variable, **cushion-forming perennial** with numerous small leaf-rosettes crammed together. **Leaves** lanceolate to obovate or almost rounded, 5–20 × 2–8 mm, downy, stalkless or with a short stalk, margin usually untoothed. **Flowers** in almost umbellate racemes that scarcely elongate in fruit, with up to 25 flowers; sepals erect, 1.5–2.5 mm long, soon falling; petals bright yellow, obovate to elliptic, 2.5–5 mm long, with rounded or slightly notched apex. **Fruit** ovate to rounded in outline, 4–9 mm long, inflated, with pointed apex and style not more than 1 mm long, stalk glabrous at least on 1 side. Gansu, Qinghai, Sichuan, Shaanxi, Tibet, Yunnan [nChina, cAsia, Mongolia, Himalaya]; rock-crevices, screes, moraines, stony and rocky meadows, cliffs, gravels, 2,300–5,500 m. VI–VIII.

Draba involucrata (W. W. Sm.) W. W. Sm.* Similar to *D. oreades*, but **leaves** not more than 5 × 4 mm, with predominantly stellate hairs, and **fruit** always flattened, not more than 5 mm long, sometimes twisted, with a blunt apex, stalks hairy all round. w, nw & swSichuan, e & seTibet, w & nwYunnan; similar habitats as well as damp mossy rocks and streamsides, 3,300–5,100 m. VI–VIII.

Draba jucunda W. W. Sm.* Readily distinguished from previous 2 species by larger **flowers**, petals 5–8 mm long; **fruit** oblong to elliptic or almost rounded in outline, 6–12 mm long, slightly twisted or not, with persistent style 1.5–3.5 mm long. seTibet, nwYunnan; screes, gravelly and rocky places, 3,400–4,600 m. VI–VIII.

Draba glomerata Royle Dense **cushion-forming perennial** not more than 10 cm tall. **Leaves** in tight basal rosettes, oblong to lanceolate, 2–8 × 1–2 mm, covered in dense stellate hairs, margin entire. **Racemes** almost umbellate, 2–8-flowered, on slender downy **stems** above the cushion; sepals oblong, erect, 1–2 mm long; petals spathulate, 1.8–3 mm long, apex with slight notch. **Fruit** ovate, 3–5 mm long, flat, not twisted, generally glabrous. Gansu, Qinghai, Sichuan, Tibet [nwChina, Himalaya]; stony and gravelly habitats, streamsides, rocky meadows, 2,900–5,500 m. VI–VIII.

Draba ladyginii Pohle* **Tufted perennial** to 45 cm with simple, rarely slightly branched, slender **stems**. **Basal leaves** in rosettes, stalked, elliptic to lanceolate, 10–25 × 2–7 mm, pointed, downy, margin entire to finely toothed; **stem-leaves** ovate to elliptic, similar in size to basal but stalkless. **Inflorescence** many-flowered, racemose, elongating in fruit; sepals oblong, 1–1.5 mm, erect; petals white, spathulate, 2–3 mm long, apex notched. **Fruit** linear-elliptic in outline, 7–12 mm long, glabrous. Gansu, Hubei, Qinghai, Sichuan, Shaanxi, eTibet, nwYunnan [n, nw & cChina]; woodland margins, sandy places, roadsides, open shrubberies, alpine grassland, 2,100–4,700 m. VI–VIII.

1. *Solms-laubachia linearifolia*; Kongbo Pa La, seTibet (AC)
2. *Solms-laubachia zhongdianensis*; Zhongdian, nwYunnan (DZ)
3. *Solms-laubachia eurycarpa*; Baimashan, nwYunnan (CGW)
4. *Lepidostemon rosularis*; Cha La, seTibet (AC)
5. *Hemilophia sessilifolia*; Daxueshan, nwYunnan (CGW)
6. *Draba oreades*; nwSichuan (JM)
7. *Draba involucrata*; Baimashan, nwYunnan (HJ)
8. *Draba ladyginii*; Shennongjia, wHubei (PC)

CRUCIFERAE: Draba

Draba yunnanensis* Franch. **Tufted** or **cushion-forming perennial** to 30 cm tall in flower, often far less. **Basal leaves** in a rosette, oblong or linear-oblong to oblanceolate, 10–35 × 2–6 mm, stalkless, entire, stellate-downy; **stem-leaves** similar with a narrowed, non-clasping base. **Racemes** with up to 60 flowers, much-elongating in fruit, lowermost flowers with leaf-like bracts; sepals oblong, 2–3 mm long; petals bright to deep primrose-yellow, obovate, 5–8 mm long, apex notched. **Fruit** ovate to elliptic, flat and often slightly twisted, glabrous, with a style 0.4–1.2 mm long. w & swSichuan, seTibet, nwYunnan; rock-crevices, cliffs, rocky slopes, open rocky coniferous or mixed forests, 2,300–5,500 m. V–VII.

Draba surculosa* Franch. Closely related to the previous species, but an often taller plant to 45 cm, with somewhat larger, toothed to entire-margined **leaves**; **stem-leaves** similar but with a clasping base. Similar distribution and habitats, as well as grassy and rocky meadows, shrubberies, forest margins and ravines, 2,600–4,600 m. VI–VIII.

***Draba polyphylla* O. E. Schulz** Like *D. surculosa*, but plant non-stoloniferous and with **stems** simple at the base, **stem-leaves** not more than 3.5 cm long; **fruit** with up to 12 seeds (not 16–24). seTibet, nwYunnan [swTibet, c & eHimalaya]; open forests, streamsides, rocky slopes and screes, damp meadows, 900–5,000 m. V–VII.

BAIMASHANIA

A genus of 2 species restricted to south-west China. Tight **cushion-forming** plants with tiny, simple **leaves** in rosettes. **Flowers** solitary or 2–3 per stem; petals pink, blunt. **Fruit** linear in outline.

Baimashania pulvinata Al-Shehbaz* Small tight, **cushion-forming** plant to 15 cm across with leaves in basal rosettes, bearing simple and forked hairs. **Leaves** ovate to oblong, to 4 × 2 mm, hairy, with ciliate base and entire margin. **Flowers** solitary on very short stalks (to 5 mm long in fruit); sepals oblong, 1.5–2.5 mm long; petals pink, spathulate, 3–4 mm long, apex slightly notched. **Fruit** linear, 4–8 mm long, glabrous. nwYunnan (Baimashan); limestone rocks and crevices, rocky meadows, 4,200–4,600 m. VI–VII.

CAPPARACEAE

CAPPARIS

A genus of c. 350 species (c. 25 in China) primarily found in tropical and subtropical regions with a few in cooler climates. **Shrubs**, **trees** or **climbers**, often with paired stipules at the nodes. **Leaves** alternate, simple, entire. **Flowers** solitary or in umbels or panicles; sepals 4 (2 + 2); petals 4, posterior pair joined along the inner margin; ovary borne on a gynophore. **Fruit** berry-like.

Capparis cantoniensis Lour. (syn. *C. pumila* Champ. ex Benth.) Clambering **shrub** to 5 m, often less, with pale brown to greyish branchlets that are angled at first, armed with pairs of hooked or straight spine-like stipules at each node, 3–5 mm long. **Leaves** leathery, oblong to ovate-lanceolate, 5–10 × 1.5–4 cm, often flushed red-orange when young; leaf-stalks 4–6 mm long. **Flowers** fragrant, in small lateral panicles, up to 11 on a stalk to 3 cm long; sepals 4–5 mm long, inner 2 somewhat smaller than outer 2 and with a membranous margin; petals white, obovate to oblong, 4–6 mm long. **Fruit** rounded to elliptic in outline, 10–15 mm. Gansu, Guangxi, Guizhou, s & seYunnan [e & seChina, seAsia, Himalaya, India]; wet, generally semi-shaded, habitats, hillsides, shrubberies, open forest, sea level to 1,100 m. II–X.

---

1. *Draba yunnanensis*; Baimashan, nwYunnan (HJ)
2. *Draba surculosa*, with *Leontopodium caespitosum*; Cangshan, nwYunnan (CGW)
3. *Draba polyphylla*; Tianchi, nwYunnan (CGW)
4. *Baimashania pulvinata*; Baimashan, nwYunnan (HJ)
5. *Clethra delavayi*; cult. (RL)
6. *Clethra fargesii*; cult. (CGW)

CLETHRACEAE

CLETHRA

A genus of c. 83 species (7 in China) in America, Asia and Madeira. **Deciduous** or **evergreen shrubs** with alternate **leaves**. **Flowers** in ascending to horizontal, slender racemes or panicles at shoot-tips; calyx with 5 short lobes; corolla often pendent, urn-shaped to campanulate, with 5 petals only fused together at very base. **Fruit** a small, many-seeded capsule enclosed in persistent calyx.

> 1. Deciduous, with papery leaves with numerous marginal teeth (22–60 on each side) .... **C. delavayi**, **C. fargesii**
> 1. Evergreen, with thick, sometimes almost leathery leaves with relatively few marginal teeth (usually to 20 on each side) .... **C. bodinieri**, **C. fabri**

Clethra delavayi Franch. (incl. *C. cavaleriei* H. Lévl.; *C. esquirolii* H. Lévl.; *C. purpurea* W. P. Fang & L. C. Hu) Very variable **tree** or **shrub** to 8 m, usually far less, with ovate to elliptic or oblanceolate **leaves**, 4.2–18 × 1.4–8 cm, glabrous to downy beneath, with 7–24 pairs of lateral veins. **Racemes** up to 21 cm long, rachis and flower-stalks stellate-downy, the latter 3–15 mm long; calyx campanulate, white to reddish pink, occasionally purplish, 5–12 mm long, margin entire, non-thickened; stamens somewhat protruding or not. Guangxi, Guizhou, Hubei, Sichuan, seTibet, Yunnan [eHimalaya, Vietnam]; mixed and coniferous forests, shrubberies, montane slopes, 300–4,000 m. VI–IX.

Clethra fargesii Franch.* Differs from previous species in bunched **racemes** (3–8 normally per shoot), tawny brown or rusty flower-stalks and inflorescence-rachis, and smaller flowers, the corolla 3–5 mm long with margin fringed and thickened towards top, stamens normally protruding. Can be a small **bush** or a **tree** to 15 m. Guizhou, Hubei, Sichuan; broadleaved forests and open woodland, shrubberies, 700–2,100 m. VII–VIII.

Clethra bodinieri H. Lévl.* **Shrub** to 5 m with elliptic to oblanceolate **leaves**, 4–13 × 1.2–3.2 cm, often glaucous beneath and rough-hairy on veins, with 7–10 pairs of lateral veins. **Racemes** 1 per shoot, to 14 cm long, stalks 2–8 mm long; corolla white to pinkish red, 4–5 mm long, lobes slightly thickened and often fringed towards top; stamens slightly protruding or not. Guangxi, Guizhou, Yunnan [Hainan, Hunan]; forests on hillslopes and in valleys, shrubberies, 200–1,700 m. VI–VIII.

Clethra fabri Hance Similar to the previous species (but can be a small **tree** to 7 m) with rather broader **leaves** with up to 13 pairs of lateral veins, **racemes** 2–7 per shoot, each up to 20 cm long, the stalks and inflorescence-rachis with stellate hairs; corolla white or cream, 2.8–4.5 mm long. Guangxi, Guizhou, Yunnan [Hainan, Vietnam]; similar habitats and grassy slopes, 300–2,000 m. VII–VIII.

ERICACEAE

RHODODENDRON

A genus with c. 1,000 species (571 in China, 409 endemic) in the northern hemisphere, South East Asia and northern Australia. **Evergreen** and **deciduous shrubs** and **trees**, rarely entirely glabrous, usually adorned with various types of hairs or scales. **Leaves** small to very large, alternate or in false whorls near shoot-tips, entire, sometimes with a back-rolled margin, usually stalked. **Inflorescences** mostly terminal, occasionally lateral (axillary), lax to dense, umbel-like racemes, rarely flowers solitary; **calyx** 5-lobed (sometimes more), or reduced to a cup-shaped structure or rim; **corolla** often vividly coloured, tubular to campanulate or trumpet-shaped to almost rotate or saucer-shaped, 5–8-lobed, with or without nectar-pouches at base of tube; stamens often 10, occasionally fewer, or up to 27, with glabrous or hairy filaments; ovary normally 5-celled, but up to 12-celled in some species, glabrous to hairy or scaly. **Fruit** a capsule, woody and opening by pores near top, or thinner and opening by long slits for most of its length.

Rhododendron is the single largest genus of flowering **trees** or **shrubs** in western China. As a result, it is a very important constituent of mountain forests and shrubberies, sometimes making almost pure stands, while it can also be a very important element on mountains, moorland and on slopes well above the treeline. The classification of *Rhododendron* is complex but its understanding is key to accurate species identification. Although the whole classification is not required here, all the important elements are included below to enable a better understanding of the genus in western China. The species are accordingly listed under their appropriate subgeneric, sectional and subsectional groupings and are reached by the page number presented in the subgeneric classification (which follows the key to subgenera and sections). The classification closely follows that of James Cullen in *Hardy Rhododendron Species: a Guide to Identification* (2005).

KEY TO SUBGENERA AND SECTIONS

1. Plants with scales at least on leaf undersurface, often also on young shoots, upper leaf surface, flower-stalks, calyx and even corolla; leaves flat or curved in bud: **(subgenus *Rhododendron*)** 2
1. Plants without scales on any part; leaves back-rolled in bud 3
2. Corolla campanulate to tubular, normally with projecting stamens and style; bud-scales of inflorescence not fringed with branched hairs ... **section *Rhododendron***
2. Corolla parallel-sided or with a narrow flaring tube and spreading limb; stamens and style included within corolla-tube; bud-scales of inflorescence fringed with branched hairs .... **section *Pogonanthum***
3. Plants evergreen, not dimorphic; new shoots produced from buds below the flower-buds .... **subgenus *Hymenanthes***
3. Plants dimorphic, with spring leaves and evergreen summer leaves; new shoots arising from the same bud as the flowers, the latter always terminal .... **subgenus *Tsutsusi***

Subgeneric classification covering the *Rhododendron* species in this book

SUBGENUS RHODODENDRON

SECTION *Rhododendron* **Shrubs** or small **trees** with scales at least on some parts, these entire (occasionally wavy or vesicular). **Flowers** usually terminal, sometimes lateral, small to large, calyx varying in size from minute to large, corolla usually 5(–10)-lobed; stamens 10–27. (p. 204)

Subsection *Edgeworthia* Straggling, often epiphytic **shrubs**. **Leaves** densely hairy, the hairs obscuring the scales but the latter visible on flower-stalks and calyces; stamens 10; styles deflexed at base. 2 species; 1 in China. (p. 204)

Subsection *Maddenia* Straggling **shrubs**, generally rock-dwelling or epiphytic. **Leaves** often pale, scaly beneath. **Flowers** in clusters of up to 12, occasionally solitary, usually fragrant; corolla broadly funnel-shaped to almost campanulate, usually large, with 5 or more lobes; stamens 10–25; ovary shouldered at top with style arising from a small depression. 26 species; 15 in China. (p. 204)

Subsection *Moupinensia* **Shrubs**, usually non-epiphytic; like subsection *Maddenia*, but generally smaller and with smaller **flowers**; stamens 10; ovary tapering into style. 3 species, all in China. (p. 204)

---

1. *Rhododendron wardii* and *R. oreotrephes*; near Weixi, nwYunnan (CGW)
2. *Rhodoendron wardii*; Napahai, Zhongdian, nwYunnan (RS)
3. *Rhododendron phaeochrysum*; Tianchi Lake, nwYunnan (HJ)
4. *Rhododendron oreotrephes* and *R. wardii*; Daxueshan, nwYunnan (CGW)
5. *Rhododendron fastigiatum*; Daxueshan, nwYunnan (CGW)
6. *Rhododendron adenogynum*; Wutoudi, Yulongxueshan, nwYunnan (CGW)
7. *Rhododendron saluenense*; swSichuan (AD)

Subsection *Triflora* Evergreen or **semi-evergreen**, often rather upright to straggling **shrubs**. **Leaves** with entire scales beneath. **Flowers** in usually terminal, sometimes lateral clusters of 1–15; calyx very small; corolla usually funnel-shaped; stamens 10, long, protruding; style set in a small depression at top of ovary. 20 species; 16 in China. (p. 204)

Subsection *Scabrifolia* Closely related to subsection *Triflora*, but small **shrubs** with lateral (axillary) clusters or racemes of **flowers**. 6 species, all in China. (p. 208)

Subsection *Heliolepida* **Shrubs** or small **trees** with scaly, often purplish or reddish young shoots. **Leaves** evergreen, with dense, broadly rimmed scales beneath, very aromatic when crushed. **Flowers** in terminal clusters; stamens 10; style variable in length, arising from a small depression in top of ovary. 4 species, all in China. (p. 208)

Subsection *Lapponica* Small to medium **shrubs** mainly growing on moorland, mostly evergreen. **Leaves** scaly, small, often aromatic when crushed; scales often with wavy rims. **Flowers** small, usually terminal, generally few or solitary; calyx conspicuous, 5-lobed, scaly, lobes ciliate; corolla broadly funnel-shaped; stamens 5–10; style arising from a depression in top of ovary. 28 species; 26 in China. (p. 208)

Subsection *Saluenensia* Small **shrubs** not more than 1.5 m tall, with densely scaly young shoots, often also with bristles. **Leaves** small, densely scaly beneath and with overlapping scalloped-margined scales. **Flowers** in small terminal clusters of 2–5, occasionally solitary; calyx relatively large, 5-lobed, coloured and hairy; corolla broadly funnel-shaped, scaly outside; stamens 10; style usually red. 2 species, both in China. (p. 212)

Subsection *Cinnabarina* Medium to tall, usually **evergreen shrubs**. **Leaves** usually glaucous, with dense, narrowly rimmed, non-touching scales beneath. **Flowers** in terminal or lateral clusters, generally rather few and pendulous; calyx small, disc-like; corolla trumpet-shaped, lobes semi-spreading or not, tube with 5 nectar drops at base within; stamens 10; ovary scaly with style arising from small depression in top. 3 species; 1 species in seTibet and c & eHimalaya. (p. 212)

Subsection *Virgata* Small **evergreen shrubs** with scales on young shoots and both sides of **leaves**. **Flowers** solitary in axils of upper leaves; calyx 5-lobed; corolla funnel-shaped; stamens 10; ovary scaly. 1 species in wChina and c & eHimalaya. (p. 212)

Subsection *Micrantha* Medium-sized **evergreen shrub** with scaly and downy young shoots and evergreen **leaves** that are scaly beneath. **Flowers** small, in dense terminal clusters; corolla funnel-shaped, deeply lobed; stamens 10; ovary scaly. 1 species in China and Korea. (p. 215)

Subsection *Boothia* Evergreen **shrubs**, sometimes epiphytic. Lower surface of leaves with minute whitish hairs (papillae) and tiny bladder-like (vesicular) scales sunk in pits. **Flowers** terminal, solitary or clustered; corolla salverform to campanulate, white or yellow; stamens 10; ovary scaly. 6 species; all in China. (p. 215)

Subsection *Glauca* Small **evergreen shrubs**, often with smooth coppery, scaly bark. **Leaves** grey- or white-papillose beneath, with tiny golden scales mixed with a few larger dark brown scales; calyx 5-lobed; corolla usually campanulate; stamens 10, often unequal; ovary scaly. 6 species; 4 in China. (p. 215)

Subsection *Campylogyna* Prostrate to cushion-forming **evergreen shrubs** with scaly young shoots. **Leaves** papillose, often silvery-white beneath and with scattered bladder-like (vesicular) scales. **Flowers** solitary or few in a terminal cluster; calyx 5-lobed; corolla campanulate with a whitish bloom; stamens 10. 1 species in swChina, neIndia and neMyanmar. (p. 215)

Subsection *Lepidota* Small **evergreen** or **deciduous shrubs**. **Leaves** scaly beneath, scales with broad translucent rims. **Flowers** terminal, solitary or up to 5 in a cluster; calyx 5-lobed; corolla broad-campanulate with a short tube, scaly outside; stamens 10; ovary scaly. 3 species, 1 species in wChina and Himalaya. (p. 215)

Subsection *Trichoclada* **Deciduous** or rarely **evergreen shrubs** with generally bristly and scaly young shoots. **Leaves** usually distinctly bristly, with golden, brownish or purplish bladder-like (vesicular) scales beneath. **Flowers** small, yellow, campanulate; stamens usually 10; ovary scaly. 5 species all in China. (p. 215)

SECTION *Pogonanthum* Evergreen **shrubs**, often dense, prostrate to erect or hummock-forming, with small, strongly and sweetly aromatic leaves that are densely scaly beneath, as are the young stems. **Flowers** small, in dense terminal clusters, often fragrant; calyx conspicuously 5-lobed, densely scaly; corolla small, *Daphne*-like, with narrow tube and wide-spreading lobes, scaly outside; stamens 5–10, hidden within corolla-tube; ovary scaly. 12 species; 9 in China. (p. 216)

SUBGENUS HYMENANTHES

Subsection *Fortunea* **Shrubs** or small **trees** with rough bark; young shoots white- or grey-downy at first. **Leaves** glabrous at maturity or with some fluffy or stellate hairs beneath, especially along midrib. **Flowers** in lax clusters, frequently scented; calyx tiny or well developed; corolla narrow- to broad-campanulate, usually 5–8-lobed, white to pink or purple, usually without nectar-pouches; stamens 10–16; ovary with or without short-stalked glands. 18 species, 16 in China. (p. 216)

Subsection *Grandia* Big **shrubs** to large **trees** with stout branches and glabrous or hairy young shoots. **Leaves** large, leathery, often with a dense felted covering of rosulate or dendroid hairs beneath. **Flowers** in dense rounded clusters of 10–30; calyx minute, not more than 2 mm long; corolla 6–10-lobed, usually oblique-campanulate or tubular-campanulate, usually without nectar-pouches; stamens 12–18; ovary usually hairy and glandular. 11 species; 9 in China. (p. 219)

Subsection *Falconera* Very similar to subsection *Grandia* but **leaves** and **stems** with a woolly covering of distinctive cup-shaped hairs, rather than a matted hair-covering, although this sometimes also present. 10 species, 8 in China. (p. 219)

Subsection *Campylocarpa* **Shrubs** or small **trees**, sometimes rather straggly with age, young shoots glandular or glabrous. **Leaves** oval to orbicular, glabrous at maturity, sometimes with tiny red glands beneath. **Flowers** in lax clusters of up to 15; calyx variable; corolla 5-lobed, campanulate to saucer-shaped, without nectar-pouches; stamens 10; ovary and style glandular or style glabrous. 4 species, all in China. (p. 220)

Subsection *Selensia* **Shrubs** or small **trees**, young shoots with short-stalked glands or bristles. **Leaves** glabrous beneath or with a thin covering of dendroid hairs. **Flowers** in clusters of up to 10; calyx variable, sometimes tiny; corolla campanulate to almost funnel-shaped, 5-lobed; stamens 10; ovary with short-stalked glands. 7 species, all in China. (p. 220)

Subsection *Argyrophylla* Rather dense shrubs to small trees with hairy to scurfy young shoots. **Leaves** usually rather narrow, glabrous above, with a thin grey, whitish or fawn felt beneath, composed of rosulate or woolly hairs. **Flowers** in lax clusters of up to 30, sometimes more; calyx tiny; corolla broadly funnel-shaped to almost campanulate, usually without nectar-pouches; stamens 10–15; ovary glabrous or variably hairy or glandular. 13 species, 12 in China, 1 in Taiwan. (p. 220)

Subsection *Arborea* Large **shrubs** to sizeable **trees** with rough bark and densely hairy young shoots. Leaves with a spongy felt of dendroid hairs beneath. **Flowers** in dense clusters of up to 25; corolla usually campanulate, 5-lobed, with nectar-pouches at base of tube; stamens 10; ovary densely hairy. 3 species; 2 in China. (p. 223)

Subsection *Taliensia* **Shrubs**, small or large, to small **trees** with glabrous to densely hairy young shoots. **Leaves** linear to elliptic, often glabrous above but woolly or felted, often densely so, beneath; hairs varying from branched to stellate, or in bundles. **Flowers** in compact clusters of 4–20; calyx minute; corolla 5–7-lobed, campanulate to ± funnel-shaped, without nectar-pouches at base, white to pink or purple; stamens usually 10; ovary hairy to glabrous. 40 species, nearly all in China. (p. 223)

Subsection *Fulva* **Shrubs** or small **trees** with hairy young shoots. **Leaves** with a dense felted covering beneath, composed of dendroid hairs, sometimes accompanied by an upper layer of long hairs. **Flowers** in clusters of up to 30; calyx minute; corolla campanulate, 5-lobed, without nectar-pouches; stamens 10; ovary glabrous. 2 species, both in China. (p. 224)

Subsection *Barbata* **Shrubs** or small **trees** with smooth peeling bark and bristly young shoots. **Leaves** glabrous or sparsely bristly, sometimes with a thin covering of dendroid hairs beneath. **Flowers** in dense clusters of 10–25; calyx minute to larger and cup-shaped; corolla campanulate, with nectar-pouches at base of tube; stamens 10; ovary hairy and glandular or glabrous. 4 species, confined to Himalaya and sTibet. (p. 227)

Subsection *Neriiflora* **Shrubs**, sometimes dwarf and creeping, to small **trees**, with sparsely to densely felted young shoots. **Leaves** glabrous or with a covering of rosulate or dendroid hairs beneath, often incomplete. **Flowers** up to 20 in a lax to dense cluster; calyx minute to larger and cup-shaped; corolla tubular-campanulate, 5-lobed, with nectar-pouches at base of tube; stamens 10; ovary hairy, sometimes also with glands, or glabrous. 26 species, nearly all in China. (p. 227)

Subsection *Thomsonia* Evergreen shrubs and small trees with smooth, peeling bark and glabrous or sparsely glandular young shoots. **Leaves** elliptic to oval, ± glabrous at maturity or with a sparse hair-covering, or red-dotted beneath. **Flowers** in lax clusters of up to 5; calyx usually relatively large, cup-shaped; corolla funnel-campanulate, with 5 nectar-pouches at base of tube; stamens 10; ovary glabrous, hairy or glandular. 13 species; 10 in China. (p. 227)

SUBGENUS TSUTSUSI

SECTION *Tsutsusi* A small section that contains the familiar pot or Indian azaleas characterised by their much-bred dwarf to medium habit with **evergreen**, **semi-evergreen** or occasionally deciduous, variously hairy leaves; hairs flattened. **Leaves** dimorphic; spring leaves larger and deciduous, summer leaves smaller and semi-evergreen. **Flowers** in small clusters; calyx 5-lobed; corolla usually tubular-campanulate to funnel-shaped, 5-lobed; stamens 10; ovary hairy. 24 species, primarily in Japan, Korea and Taiwan; 3 in China. (p. 227)

SUBGENUS *Rhododendron* SECTION *Rhododendron*

Plants with scales at least on leaf undersurface, often also on young shoots, upper leaf-surface, leaf-stalks, calyx and even corolla; **leaves** flat or curved in bud; bud-scales of **inflorescence** not fringed with branched hairs. **Calyx** campanulate to tubular, normally with projecting stamens and style.

Subsection *Edgeworthia*

Rhododendron edgeworthii Hook. f.   Low to rather leggy **evergreen shrub** to 3.3 m, with thick-woolly branchlets. **Leaves** bright to deep green, oblong-ovate to lanceolate or elliptic, 6–15 × 2.5–5 cm, rough and deep-veined above, white to rusty brown beneath with a thick felt obscuring the scales. **Flowers** 1–5 in lax clusters, powerfully fragrant; calyx large and deeply 5-lobed; corolla widely funnel-shaped, 3–7.5 cm long, white, occasionally flushed pink, sometimes with yellow basal blotch within; ovary hairy. seTibet, nw & cYunnan [eHimalaya]; epiphytic or rock-dwelling, often on cliffs, 1,800–4,000 m. IV–V.

Subsection *Maddenia*

1. Stamens 15–25 per flower . . . . **R. maddenii**
1. Stamens 10–13 per flower 2
2. Calyx large, 15–26 mm long . . **R. nuttallii, R. megacalyx**
2. Calyx small, not more than 12 mm long . . . .
    . . . . **R. pachypodum, R. dalhousiae, R. lindleyi**

Rhododendron maddenii Hook. f. subsp. *crassum* (Franch.) Cullen   Sprawling to rather straggly **shrub** to 4 m with flaking bark. **Leaves** deep shiny green above, elliptic, 9–18 × 5.5–8 cm, glabrous above, with dense, often overlapping scales beneath. **Flowers** in clusters of up to 11; **calyx** to 16 mm long, glabrous; corolla 4.5–12.5 cm long and wide, white with pink lines or pink, rose, purple or orange, sweetly fragrant, scaly outside. seTibet, Yunnan [neIndia, Myanmar, Thailand]; dense forest, open rocky ridges, cliffs, rarely epiphytic, 1,500–3,700 m. IV–V.

Rhododendron nuttallii Booth ex Nutt.   Large **shrub** or small **tree** to 10 m, often rather leggy with age. **Leaves** deep green, puckered and rather shiny above, large, elliptic, 13–26 × 5.5–12 cm, with dense uneven scales beneath and a network of veins. **Flowers** in clusters of 2–7; **calyx** large, to 26 mm long, 5-lobed, usually ± glabrous; corolla large, to 12.5 cm long and across, powerfully fragrant, white, sometimes pink-tinged, with yellow blotch within, lobes with wavy, reflexed margins. seTibet, nwYunnan [eHimalaya]; rainforest margins, rocky places, cliffs and also epiphytic, 1,100–3,700 m. IV–V.

Rhododendron megacalyx Balf. f. & Kingdon-Ward   Similar to *R. nuttallii*, but **leaves** with sunken veins beneath and scales of 2 sizes, the smaller sunken in pits; **flowers** somewhat smaller, white from purple-flushed buds. e & seTibet, nwYunnan [nMyanmar]; similar habitats, 1,800–4,000 m. IV–VI.

Rhododendron pachypodum Balf. f. & W. W. Sm. (including *R. ciliicalyx* Franch.)   Rather dense to straggly **shrub** 1–7.6 m. **Leaves** matt mid-green, obovate to elliptic-oblanceolate, 3–10 × 1.3–3 cm, with dense to rather sparse scales beneath. **Flowers** in lax clusters of up to 4, fragrant; calyx small, not more than 10 mm long, sometimes with 1 lobe longer that the others; corolla large, 5–10 cm long and wide, white to rose-pink, sometimes with a yellow or brownish blotch within, minutely scaly outside, lobes broad, blunt and widely reflexing. wYunnan [Guangdong, nMyanmar]; forest edges, rocky scrub, cliffs, ridges, 1,800–3,700 m. IV–V.

Rhododendron dalhousiae Hook. f.   Straggly, sometimes sprawling **shrub** to 3 m with bristly young shoots and peeling cinnamon bark. **Leaves** somewhat shiny, deep green, only borne at shoot-tips, obovate to oblanceolate, 6–20 × 2.5–7.5 cm, glabrous above, grey or brownish green, with tiny scattered reddish scales. **Flowers** in clusters of 2–6; calyx glabrous; corolla cream (greenish yellow in bud), 8–10 cm long, moderately scented. seTibet [c & eHimalaya]; usually rock-dwelling or epiphytic, in rainforest, 1,800–2,700 m. IV–V.

Rhododendron lindleyi T. Moore   Very similar to *R. dalhousiae*, but with rather smaller, more elliptic **leaves** and sweetly scented white **flowers** to 11.5 cm long, usually with a yellow or orange base. seTibet [c & eHimalaya]; similar habitats, 1,800–3,400 m. IV–V.

Subsection *Moupinensia*

Rhododendron moupinense Franch.*   Compact **evergreen shrub** to 1.5 m, becoming more leggy and open with age. **Leaves** leathery, dark shiny green, oval to elliptic, 2–4.5 × 1–2.3 cm, paler and with scattered hairs beneath; leaf-stalks bristly. **Flowers** solitary or paired; calyx to 4 mm, obvious; corolla broadly funnel-shaped, to 4.5 cm long, white, often pink-tinged, to deep rose-pink, sometimes spotted within; stamens 10. Guizhou, cSichuan, neYunnan; epiphytic in broadleaved forests, and rock-dwelling, 2,000–3,300 m. II–IV.

Subsection *Triflora*

1. Flowers yellow . . . . **R. lutescens, R. ambiguum, R. triflorum**
1. Flowers blue, pink, purple or white 2
2. Leaves with hairs/bristles on midrib beneath, at least in lower part . . . . **R. augustinii** subsp. **augustinii, R. augustinii** subsp. **chasmanthum, R. augustinii** subsp. **hardyi, R. trichanthum**
2. Leaves with glabrous midribs 3
3. Leaves pale green to glaucous . . . . **R. oreotrephes, R. rigidum**
3. Leaves deep to mid-green . . . . **R. concinnum, R. yunnanense, R. tatsienense, R. davidsonianum**

1. *Rhododendron edgeworthii*; Cangshan, nwYunnan (RL)
2. *Rhododendron nuttallii*; seTibet (RL)
3. *Rhododendron megacalyx*; new Dulong road, nwYunnan (RL)
4. *Rhododendron pachypodum*; Cangshan, nwYunnan (PC)
5. *Rhododendron dalhousiae*; Bhutan (TS)
6. *Rhododendron lindleyi*; Bhutan (TS)
7. *Rhododendron moupinense*; cult. (RL)

Rhododendron lutescens Franch.* Lax, rather upright **shrub** to 6 m, often less. **Leaves** lanceolate to ovate-lanceolate, with sharply pointed apex, somewhat scaly above, more densely so beneath with yellowish to brownish scales. **Flowers** 1–3, usually in lateral inflorescences; corolla broadly funnel-shaped, 13–26 mm long, up to 4.5 cm across, pale to deep primrose-yellow, with green spots within. Guizhou, Hubei, w & sSichuan, neYunnan; lower to upper tree belts and in shrubberies near treeline, 550–3,200 m. III–V.

Rhododendron ambiguum Hemsl.* Dense rounded **shrub** usually, 0.5–5.7 m tall. **Leaves** small, lanceolate to ovate-lanceolate or elliptic, 2.2–3.8 × 1.2–3.2 cm, dark green and scaly above, paler and glaucous beneath, with dark brown uneven scales. **Flowers** in terminal clusters of 2–7; corolla broadly funnel-shaped, 2–2.6 cm long, greenish to pale yellow, scaly outside. s & wSichuan; open forests, thickets, rocky places, 2,600–4,500 m. V–VII.

Rhododendron triflorum Hook. f. Similar to *R. ambiguum*, but a more upright **shrub** with larger, somewhat aromatic **leaves** to 7 cm long but not more than 2 cm wide, not scaly above. **Flowers** in clusters of up to 4, cream to pale yellow or greenish yellow, often flushed with pink or bronze. seTibet [c & eHimalaya]; similar habitats, 2,100–4,000 m. V–VI.

Rhododendron augustinii Hemsl.* Rather erect **shrub** to 7 m, young leaves usually downy. **Leaves** mid-green to yellowish green, oblong to lanceolate, with sparse scattering of golden or brownish scales beneath and hairs on lower part of midrib. **Flowers** in clusters of 2–6; calyx usually with a fringe of long hairs; corolla broadly funnel-shaped, 2–4.3 cm long, pale to deep blue, lavender-blue or violet-mauve, rarely white or pink, blotched or spotted with orange, green or purple. wHubei, Sichuan, e & seTibet, nwYunnan; rocky places, riversides, forest margins, 1,200–3,660 m. IV–VI. Three subspecies are recognised here: **subsp. *augustinii****,

---

1. *Rhododendron lutescens*; Wolong, wSichuan (CGW)
2. *Rhododendron ambiguum*; Emeishan, sSichuan (CGW)
3. *Rhododendron triflorum*; Tang Chu, Bhutan (CGW)
4. *Rhododendron augustinii* subsp. *chasmanthum*; Wolong, wSichuan (CGW)
5. *Rhododendron augustinii* subsp. *chasmanthum*; Wolong, wSichuan (CGW)
6. *Rhododendron augustinii* subsp. *hardyi*; cult. (CGW)
7. *Rhododendron trichanthum*; cult. (CGW)
8. *Rhododendron oreotrephes* and *Piptanthus concolor*; Daxueshan, nwYunnan (CGW)
9. *Rhododendron oreotrephes*; Lancangjiang–Jinshajiang divide, nwYunnan (CGW)
10. *Rhododendron concinnum*; Kangding, nwSichuan (PC)
11. *Rhododendron concinnum*; Jiuzhaigou, nwSichuan (CGW)

wHubei, n, e & sSichuan; **subsp. *chasmanthum*** (Diels) Cullen* which has glabrous leaf-stalks and tighter clusters of **flowers**, w & swSichuan, seTibet, nwYunnan; **subsp. *hardyi*** (Davidian) Cullen*, a **near-deciduous** plant with white or greenish white **flowers** spotted with yellowish green, eTibet, nwYunnan, from higher altitudes.

Rhododendron trichanthum Rehd.* Upright **bush** with larger **leaves** than preceding species, to 11 × 3.7 cm, bristly and scaly on both surfaces, as well as on stalks. **Corolla** bluish mauve to pale purple or reddish purple, of a similar size. wSichuan; open woodland, forest margins, often thicket-forming, 1,600–3,150 m. IV–VI.

Rhododendron oreotrephes W. W. Sm. Upright **shrub**, sometimes broader and more rounded, 1–7.6 m tall. **Leaves** generally glaucous, pale green when young, elliptic to oblong or ovate, 1.8–8.9 × 1.2–4.3 cm, lower surface paler with scattered (non-touching) greyish, purplish or reddish brown scales. **Flowers** in terminal or lateral clusters of 1–3, occasionally more; calyx tiny and rim-like, slightly scaly; corolla broadly funnel-shaped to almost campanulate, 1.8–4 cm long, pink to rose-pink or bright purple, occasionally whitish, spotted or not. Sichuan, seTibet, n & nwYunnan [nMyanmar]; open forests (especially coniferous), forest margins, shrubberies, rocky slopes, ridge-tops, 2,700–4,300 m. IV–VI. (See also photos on pp. 36, 201 and 260.)

Rhododendron rigidum Franch.* Similar to *R. oreotrephes*, but a more compact bush not exceeding 3 m, with narrower **leaves**; **flowers** narrower, funnel-shaped, white to pink, rose-pink or lavender, generally marked with olive-brown, reddish brown or gold within. s & swSichuan, n & nwYunnan; similar habitats, 1,200–3,400 m. IV–V.

Rhododendron concinnum Hemsl.* Rather dense **shrub** to 4.5 m, often less, very similar in appearance to *R. ambiguum* apart from flower colour. **Leaves** deep to mid-green above and somewhat shiny, oblong-lanceolate to elliptic or ovate, 2.5–8.5 × 1.2–3.5 cm, scaly above, more densely so beneath, with scattered grey or brown, relatively large scales. **Flowers** in terminal or lateral clusters of 2–5; corolla broadly funnel-shaped, 2–3 cm long, deep purple, reddish purple or ruby-red to lavender-pink, with or without greenish or crimson spots, scaly outside. sGansu, Hubei, w, nw & cSichuan, Shaanxi [Henan]; forest margins, open areas in woodland, rocky slopes, cliffs, 1,500–4,400 m. IV–VI.

Rhododendron yunnanense Franch. Variable, erect, often rather sparse **evergreen** to almost **deciduous shrub** to 3.6 m, often less. **Leaves** thinnish, deep to mid-green, elliptic to oblanceolate, 2.5–10.4 × 0.8–2.8 cm, sparely scaly above, with scattered brownish scales and variable amounts of bristles beneath. **Flowers** fragrant, in clusters of 2–6, occasionally solitary; calyx minute, often scaly and ciliate; corolla broadly funnel-shaped, 1.8–3.4 cm long, white, pinkish, rose-pink, rose-purple to lavender-blue, usually with a large rose-pink, crimson or greenish blotch within. nwGuizhou, swSichuan, seTibet, nw & neYunnan [n & neMyanmar]; open woodland, shrubberies, rocky slopes, ridges, stabilised screes, 900–4,300 m. IV–VI.

Rhododendron tatsienense Franch.* Similar to *R. yunnanense*, but smaller in all parts, with reddish **stems** and stiffer more leathery **leaves**; corolla not more than 2.3 cm long, generally with red spots within. w & swSichuan, nwYunnan; open rocky places, shrubberies, 2,100–3,700 m. IV–V.

Rhododendron davidsonianum Rehd.* Closely related to *R. yunnanense*, but a more leafy, fully **evergreen shrub** to 5 m. **Leaves** deep green and rather leathery, lanceolate to oblong, 2.3–7.8 × 0.8–2.6 cm, with a characteristic V-shaped cross-section, with dense, often touching, brown scales beneath. **Flowers** in clusters, usually of up to 6; corolla broadly funnel-shaped, 1.9–3.3 cm long, pale to deep pink, lavender or mauve, rarely white, with or without purple blotch or spots within. sw & cSichuan; open coniferous forests, forest margins, cliffs, streamsides, 1,800–3,500 m. IV–VI.

Subsection *Scabrifolia*

Rhododendron racemosum Franch.* Very variable compact to lax and leggy **shrub**, to 4 m, often less than 2 m, usually with reddish branches. **Leaves** small, deep green to bluish green, obovate to elliptic, ± glabrous above, but glaucous and dotted with scales beneath. **Flowers** 1–4 in upper leaf-axils; corolla through all shades of pink to purplish pink or white, funnel-shaped, 8–23 mm long. wGuizhou, swSichuan, n & nwYunnan; dry moorland, low shrubberies, rocky places, forest margins, locally abundant, 800–4,300 m. III–VI.

Subsection *Heliolepida*

Rhododendron rubiginosum Franch. Spreading to erect **shrub**, often becoming leggy with age, to 9 m, with scaly young **branches**. **Leaves** deep shiny green, oblong to elliptic-lanceolate, 6–11.5 × 2.5–4.5 cm, reddish brown to brownish green beneath with dense flaky scales that also envelope the stalks, faintly aromatic when crushed. **Flowers** 4–10 in open-topped clusters; calyx tiny, scaly; corolla broadly funnel-shaped, 3–4 cm long, white flushed pink or purple, to pink, lavender or mauve-pink, with purple, crimson or brownish blotching and spotting within; stamens 10; style bent downwards. seSichuan, seTibet, Yunnan [nMyanmar]; shrubberies, forest glades, bamboo thickets, 2,300–4,300 m. IV–VI.

Rhododendron heliolepis Franch. Far less vigorous than the previous species, generally 1.5–3 m, with more strongly aromatic **leaves** that are covered beneath in glistening golden or brownish scales; **flowers** rather smaller, 2–3.5 cm long, white, pink or purplish, spotted or blotched with purple, red, green or brown. sw & wSichuan, seTibet, w & nwYunnan [neMyanmar]; coniferous forests, shrubberies, 2,400–3,800 m. VI–IX.

Subsection *Lapponica*

1. Calyx very small, not more than 3 mm long ....
   .... *R. intricatum*, *R. complexum*, *R. tapetiforme*, *R. thymifolium*, *R. orthocladum*, *R. telmateium*, *R. hippophaeoides*, *R. capitatum*, *R. nitidulum*
1. Calyx larger, 3–12 mm long 2
2. Leaves relatively large, usually more than 5 mm wide ..
   .... *R. cuneatum*, *R. russatum*, *R. rupicola*, *R. rupicola* var. *chryseum*, *R. rupicola* var. *muliense*, *R. flavidum*, *R.* × *wongii*
2. Leaves small, usually 5 mm wide or less ....
   .... *R. fastigiatum*, *R. impeditum*, *R. nivale*, *R. nivale* subsp. *boreale*

Rhododendron intricatum Franch.* Compact, bushy **shrub** to 1.5 m. **Leaves** pale grey-green, elliptic to oblong, 5–14 × 3–7 mm, with golden scales above, pale straw-coloured or buff beneath, with dense pale scales. **Flowers** 2–8 in dense clusters; calyx 0.5–2 mm long; corolla tubular, 8–14 mm long, with spreading lobes, pale lavender to dark blue or mauve; stamens usually 10, very short but longer than style. c & swSichuan, n & nwYunnan; forest margins, damp meadows, rocky slopes, 2,800–4,900 m. V–VII.

Rhododendron complexum Balf. f. & W. W. Sm.* Compact to upright **shrub** to 0.6 m. **Leaves** grey-green, elliptic to ovate, 3–11 × 2–6 mm, covered beneath in dense, uniform rusty scales. **Flowers** in clusters of 3–5; calyx minute; corolla narrow, funnel-shaped, 9–13 mm long, pale lilac to rose-purple; stamens 5–6; style very short. nwYunnan; moorland, stony slopes, 3,400–4,600 m. IV–V.

---

1. *Rhododendron yunnanense*; Yulonxueshan, nwYunnan (HJ)
2. *Rhododendron yunnanense*; lower Gangheba, Yulongxueshan, nwYunnan (CGW)
3. *Rhododendron yunnanense*; pass from Jian Chuan to Lijiang, nwYunnan (CGW)
4. *Rhododendron yunnanense*; pass from Jian Chuan to Lijiang, nwYunnan (CGW)
5. *Rhododendron tatsienense*; Zhongdian to Habashan, nwYunnan (CGW)
6. *Rhododendron racemosum*; Zhongdian, nwYunnan (CGW)
7. *Rhododendron racemosum*; nwYunnan (RS)
8. *Rhododendron rubiginosum*; Daxueshan, nwYunnan (CGW)
9. *Rhododendron rubiginosum*; Daxueshan, nwYunnan (CGW)
10. *Rhododendron heliolepis*; Gangheba, Yulongxueshan, nwYunnan (CGW)
11. *Rhododendron complexum*; Zhongdian, nwYunnan (CGW)

Rhododendron tapetiforme **Balf. f. & Kingdon-Ward** Low rounded or mat-forming **shrub** to 90 cm. **Leaves** dark green, elliptic to rounded, 4–15 × 3–8 mm, blunt, with transparent scales above, paler with dense rusty brown scales beneath. **Flowers** 1–4; calyx to 2 mm long; corolla broadly funnel-shaped, 9–16 mm long, purple to purple-blue, rarely rose-pink or violet; style generally longer than the 5–10 stamens. e & seTibet, nwYunnan [n & neMyanmar]; moorland, often forming extensive colonies, 3,500–4,600 m. V–VII.

Rhododendron thymifolium **Maxim.*** Erect **shrub** to 1.2 m. **Leaves** deep to mid-green or grey-green, narrow-obovate, 5–12 × 2–5 mm, with dense pale fawn scales beneath. **Flowers** usually solitary, occasionally paired; calyx very small, 0.5–1.2 mm long, often rim-like; corolla broadly funnel-shaped, 7–11 mm long, purple to lavender; stamens 10, shorter or longer than style. sGansu, Qinghai, nSichuan, eTibet; moorland, rocky slopes, 2,600–4,600 m. V–VII.

Rhododendron orthocladum **Balf. f. & Forrest*** Similar to *R. thymifolium*, but plants denser and more compact, **leaves** with a mixture of pale yellowish or fawn scales and darker ones beneath; flowers 1–5 per cluster, style not longer than stamens. swSichuan, nwYunnan; low shrubberies, rocky places, forest margins, 3,440–4,300 m. IV–VI.

Rhododendron telmateium **Balf. f. & W. W. Sm.*** Similar to *R. tapetiforme*, but often a more erect **shrub** with matt grey-green, lanceolate to oval **leaves** (to 12 × 5 mm) that have a distinct point at tip, covered on both surfaces by pale golden scales, beneath mixed with dark reddish brown ones; style shorter to longer than stamens. **Corolla** purple, to rose-pink or lavender. swSichuan, n & nwYunnan; moorland, marshy places, open woodland, 2,900–5,000 m. IV–VII. Low, compact variants with neat, oval to elliptic **leaves** (noted especially from Yulongxueshan, nwYunnan) are referable to *R. telmateium* Diacritum Group.

---

1. *Rhododendron tapetiforme*; Daxueshan, nwYunnan (CGW)
2. *Rhododendron telmateium*; Tianchi, nwYunnan (PC)
3. *Rhododendron* aff. *telmateium*; Yulongxueshan, nwYunnan (CGW)
4. *Rhododendron hippophaeoides*; Zhongdian, nwYunnan (PC)
5. *Rhododendron hippophaeoides*; Zhongdian, nwYunnan (PC)
6. *Rhododendron capitatum*; Songpan to Jiuzhaigou, nwSichuan (CGW)
7. *Rhododendron nitidulum*; Emeishan, sSichuan (CGW)
8. *Rhododendron cuneatum*; Gangheba, Yulongxueshan, nwYunnan (CGW)
9. *Rhododendron russatum*; Lancangjiang–Nujiang divide, nwYunnan (CGW)
10. *Rhododendron russatum*; Baimashan, nwYunnan (HJ)
11. *Rhododendron rupicola* var. *rupicola*; Baimashan, nwYunnan (CGW)

Rhododendron hippophaeoides **Balf. f. & W. W. Sm.*** Variable dense, spreading to upright **shrub**. **Leaves** mildly aromatic, grey-green, elliptic to oblong, 8–38 × 5–10 mm, densely dotted beneath with yellowish or buff transparent scales. **Flowers** in clusters of 4–8; calyx 0.5–1.8 mm long; corolla broadly funnel-shaped, 10–13 mm long, purple-blue to lavender-blue, more rarely rose-pink or white; stamens 10. swSichuan, Yunnan except e & se; moorland, stony slopes, marshy ground, 2,400–4,800 m. IV–VII.

Rhododendron capitatum **Maxim.*** Erect **shrub** to 1.5 m with elliptic, pale, shiny green **leaves** to 25 × 9 mm, pale brown beneath, speckled with pale- and dark-centred scales. **Flowers** in clusters of 3–5; calyx variable in size, with ciliate, oblong lobes; corolla broadly funnel-shaped, 10–15 mm long, lavender to deep purple; stamens 10, shorter than or equalling style. Gansu, seQinghai, Sichuan, Shaanxi, eTibet; forests, shrubberies, meadows, 3,000–4,300 m. VI–VII.

Rhododendron nitidulum **Rehd. & Wils.*** Very similar in general appearance to the previous species, but an often more spreading **shrub** with ovate to elliptic **leaves** not more than 11 × 7 mm, with dense pale scales beneath; **flowers** solitary or paired; corolla funnel-shaped, violet-purple to rose-lilac, hairy within; style longer than stamens. nw & cSichuan; similar habitats, 3,000–5,000 m. IV–V.

Rhododendron cuneatum **W. W. Sm.*** Upright to spreading **shrub** to 2 m. **Leaves** sometimes aromatic, elliptic, 10–70 × 5–26 mm, deep green above, paler and with dense brown, fawn or cream scales beneath. **Flowers** in clusters of 3–6; calyx 4–12 mm long; corolla ± funnel-shaped, 1.6–3.4 cm long, rose-purple to rose-lavender; stamens 10, generally shorter than style. swSichuan, n & wYunnan; upper forest zone and above treeline, 2,700–4,300 m. V–VII.

Rhododendron russatum **Balf. f. & Forrest*** Variable dense to rather upright, straggly **shrub**, to 1.8 m, but often less than half that height. **Leaves** mildly aromatic, dark shiny green, elliptic to oblong, 16–40 × 6–17 mm, with dense mixed pale and dark reddish brown scales beneath. **Flowers** 4–7; calyx 3–6 mm long, fringed with long hairs; corolla broadly funnel-shaped, reddish purple to indigo-blue, pink or lavender; style hairy, longer than the 10 stamens. swSichuan, n & nwYunnan; alpine moorland, amongst other low shrubs, forest margins, 3,400–4,300 m. IV–VI. Very difficult to distinguish from *R. rupicola* out of flower.

Rhododendron rupicola **W. W. Sm.** Dense to more open straggly **shrub**, rarely more than 50 cm. **Leaves** deep shiny green above, elliptic, 6–21 × 3–12 mm, with mixed pale and dark scales beneath. **Flowers** 2–6; calyx 4–5 mm long, usually with a band of pale scales; corolla broadly funnel-shaped, 10–17 mm long, very deep purple or reddish purple, to crimson; stamens 5–10, slightly shorter than glabrous style. swSichuan, seTibet, n, nw & cYunnan [nMyanmar]; alpine moorland and meadows, rocky slopes, 3,000–4,600 m. V–VII. Yellow-flowered plants are referable to var. *chryseum* (Balf. f. & Kingdon-Ward) M. N. Philipson & Philipson* (seTibet, nwYunnan) and var. *muliense* (Balf. f. & Forrest) M. N. Philipson & Philipson* (swSichuan, with broader **leaves** and hair-fringed calyces).

Rhododendron flavidum Franch.* Similar to *R. rupicola* var. *chryseum* in the yellow **flowers**, but plants more erect with shiny, dark green **leaves** sparsely dotted beneath with red scales; flowers solitary or in clusters of 2–3 with 8–10 stamens that are shorter than style. w & nwSichuan; shrubberies, forest margins, grassy slopes, 3,000–4,000 m. V–VI. Hybrids between this species and *R. ambiguum*, (**R.** × *wongii* Hemsl. & E. H. Wilson), with pale creamy yellow **flowers**, are occasionally found in wSichuan.

Rhododendron fastigiatum Franch.* Small **shrub** to 0.6 m, often less, spreading or mound-forming. **Leaves** greyish blue or greenish blue, oblong to ovate, 5–16 × 3–6 mm, paler and grey or fawn beneath with pale scales. **Flowers** 1–5; calyx 2.5–5.5 mm long; corolla broadly funnel-shaped, 9–18 mm long, pale to deep purple or purplish blue, rarely pink; stamens normally 10, shorter than style. n, nw & cYunnan; upper forest zone, stony pastures, screes and other rocky places, 3,200–4,900 m. V–VII. (See also photo 5 on p. 201.)

Rhododendron impeditum Balf. f. & W. W. Sm.* Very similar to *R. fastigiatum*, but **leaves** dark green above, with dense rust-coloured or dark brown scales beneath; corolla similar in shape and size, violet to purple or rose-lavender. swSichuan, nYunnan; similar habitats, 2,700–4,900 m. V–VII.

Rhododendron nivale Hook. f. Very variable prostrate to mound-forming or erect **shrub**, not more than 1 m. **Leaves** very small, green or grey-green, elliptic to ovate, 5–9 × 2–5 mm, dotted beneath with pale and a few dark scales. **Flowers** 2–3 or solitary; calyx 2–4 mm, usually with band of dark scales along margin; corolla broadly funnel-shaped, 9–13 mm long, pink to purple or magenta or lilac; stamens usually 10, usually shorter than style. seQinghai, seTibet [swTibet, Himalaya]; moorland, rocky slopes and screes, often forming extensive colonies, 3,000–4,500 m. V–VII. **Subsp.** *boreale* Philipson & Philipson* is generally rather taller and laxer with rudimentary or absent calyx-lobes; Sichuan, e & seTibet, nw & cYunnan. (See also photo on p. 18.)

Subsection *Saluenensia*

Rhododendron saluenense Franch. Erect **shrub** to 1.5 m, with bristly young growth. **Leaves** deep shiny green, oblong to elliptic, apex blunt, 8–36 × 2–7 mm, scaly and often bristly above, densely scaly beneath and with scattered bristles on midrib. **Flowers** 1–3, on bristly stalks; corolla 17–31 mm long, purple to reddish purple, occasionally pink; ovary scaly. swSichuan, seTibet, nwYunnan [neMyanmar]; forest margins, moorland, open slopes, 3,400–5,200 m.

V–VII. **Subsp.** *chameunum* (Balf. f. & Forrest) Cullen is a prostrate or mound-forming plant not more than 0.6 m tall, **leaves** without scales above, often taking on reddish or coppery tones in autumn. Similar distribution, more common. (See also photo 7 on p. 201.)

Rhododendron calostrotum Balf. f. & Kingdon-Ward Similar to *R. saluenense*, but branchlets, leaf-stalks and leaf-midribs and ovaries glabrous or becoming so; **leaves** sometimes glaucous above; **flowers** 1–5, corolla pink, purple, magenta or almost crimson. seTibet, nw & wYunnan [swTibet, nMyanmar]; moorland, meadows, rocky slopes, cliffs, 3,050–4,550 m. V–VII.

Subsection *Cinnabarina*

Rhododendron cinnabarinum Hook. f. Rather upright **shrub** to 7 m, often less, with usually purplish, scaly **branches**. **Leaves** mid-green to deep glaucous green, usually shiny, elliptic to lanceolate, 3–11.5 × 1.8–5.5 cm, without scales, paler and glaucous beneath and densely scaly. **Flowers** 2–9, semi-pendent; calyx tiny, 1–2 mm long; corolla tubular-campanulate, 2.6–5 cm long, red to orange or yellow or bicoloured. seTibet [c & eHimalaya]; open forests, shrubberies, hillslopes, often gregarious, 2,100–4,000 m. IV–VI. **Subsp.** *xanthocodon* (Hutch.) Cullen (syn. *R. concatenans* Hutch.) has markedly glaucous **leaves** broader than those of the typical plant (**subsp.** *cinnabarinum*) that are scaly above, and yellow, orange, apricot or sometimes purple **flowers**; seTibet [eHimalaya].

Subsection *Virgata*

Rhododendron virgatum Hook. f. Laxly branched, often rather leggy, **evergreen shrub** to 2.5 m. **Leaves** deep green, lanceolate to oval, recurved, to 8 × 2 cm, pale and with uneven brown scales beneath. **Flowers** solitary, usually in upper leaf-axils; calyx to 3 mm long, with rounded lobes; corolla funnel-

---

1. *Rhododendron rupicola* var. *chryseum*; Baimashan, nwYunnan (CGW)
2. *Rhododendron rupicola* var. *chryseum*; Baimashan, nwYunnan (PC)
3. *Rhododendron flavidum*; Kangding, wSichuan (PC)
4. *Rhododendron* × *wongii*; Kangding, wSichuan (PC)
5. *Rhododendron fastigiatum*; Daxueshan, nwYunnan (CGW)
6. *Rhododendron nivale* subsp. *boreale*; Gonggashan pass, Kangding, wSichuan (PC)
7. *Rhododendron nivale* subsp. *boreale*; Balangshan pass, Wolong, wSichuan (CGW)
8. *Rhododendron saluenense* subsp. *chameunum*; Genyanshan, swSichuan (AD)
9. *Rhododendron saluenense* subsp. *chameunum*; Lancangjiang–Jinshajiang divide, Lidiping, nwYunnan (CGW)
10. *Rhododendron calostrotum*; Zibenshan, seTibet (PCX)
11. *Rhododendron cinnabarinum* subsp. *xanthocodon*; Poskong La, seTibet (KC)
12. *Rhododendron virgatum*; Paro valley, Bhutan (CGW)

shaped, 2.5–4 cm long, pink to mauve or white. seTibet [c & eHimalaya]; dryish slopes, rocky places, forest margins, 2,000–4,000 m. III–V. The Yunnan material is referable to **subsp. oleifolium (Franch.) Cullen***, which tends to have pale **flowers** less than 2.5 cm long, occurs in Yunnan and Tibet.

Subsection *Micrantha*

Rhododendron micranthum **Turcz.** Rather leggy **evergreen bush** to 1.8 m. **Leaves** mid-green, somewhat shiny, oblanceolate, 3–4.5 × 0.6–1.3 cm, brown-scaly beneath. **Flowers** more than 20 in dense terminal and lateral clusters; calyx tiny, 1–2 mm long, fringed with hairs, scaly; corolla campanulate, not more than 8 mm long, white, scaly outside; stamens 10, protruding. sGansu, wHubei, w & nwSichuan, neYunnan (rare) [c & eChina, Korea]; shrubberies, thickets, 1,600–3,200 m. VI–VII.

Subsection *Boothia*

Rhododendron leptocarpum **Nutt.** Rather lax, generally spreading **shrub**. **Leaves** bright green, elliptic, often narrowly so, 3–8 × 1.5–3.6 cm, greyish green with scattered scales beneath. **Flowers** up to 8 in lax clusters; calyx 2–4 mm long with reflexed lobes; corolla ± salverform, 8–14 mm long, pale yellow or white, stamens protruding. seTibet, wYunnan [eHimalaya]; epiphytic and rock-dwelling, 2,400–4,300 m. V–VII.

Rhododendron sulfureum **Franch.** Differs from *R. leptocarpum* in the hairy young growth, shiny dark green **leaves** which are usually somewhat glaucous beneath, and in the more campanulate, greenish yellow to deep yellow **corollas** up to 20 mm long; calyx 3–6 mm long. seTibet, nw & wYunnan [nMyanmar]; epiphytic and rock-dwelling, especially on cliffs, 2,100–4,000 m. IV–V.

Subsection *Glauca*

Rhododendron brachyanthum **Franch.** Rather dense, sometimes leggy **shrub** to 2 m, often far less, with peeling bark. **Leaves** dark green, elliptic to oblanceolate, 2–6.5 × 1–2.6 cm, glaucous beneath and with mixed brown and smaller yellowish or whitish scales. **Flowers** nodding, in lax clusters of up to 10; calyx with 5 deep lobes, 3–5 mm long; corolla broadly campanulate, pale yellow or greenish yellow; stamens

---

1. *Rhododendron micranthum*; Jiuzhaigou, nwSichuan (CGW)
2. *Rhododendron sulfureum*; Dapingdi, Cangshan, nwYunnan (PCX)
3. *Rhododendron brachyanthum*; Cangshan, nwYunnan (PC)
4. *Rhododendron brachyanthum*; Zhongdian, nwYunnan (CGW)
5. *Rhododendron lepidotum*; nwYunnan (CGW)
6. *Rhododendron mekongense*; Lancangjiang–Jinshajiang divide, Lidiping, nwYunnan (CGW)
7. *Rhododendron trichocladum*; Huadianba, Cangshan, nwYunnan (PCX)
8. *Rhododendron lepidostylum*; cult. (CGW)

10. seTibet, nwYunnan [neMyanmar]; dry shrubberies, forest margins and clearings, 2,700–4,400 m. V–VII.

Subsection *Campylogyna*

Rhododendron campylogynum **Franch.** Small mounded to prostrate **evergreen shrub**, not more than 1.3 m, with scaly young **stems**. **Leaves** mid-green to dark shiny green, narrowly elliptic to obovate, 0.7–3.7 × 0.3–1.2 cm, generally without scales above, beneath pale green or glaucous, with sparse scattered scales. **Flowers** 1–3; calyx 1–6 mm long; corolla campanulate, 10–23 mm long, red, purple, purplish black, claret, pink or cream, held just clear of the foliage; stamens 8–10; ovary scaly. seTibet, c & wYunnan [swTibet, neIndia, neMyanmar]; rocky places, shrubby hillslopes, cliffs, 2,400–4,900 m. V–VII.

Subsection *Lepidota*

Rhododendron lepidotum **Wall. ex G. Don** Stoloniferous, dense to open and rather leggy **evergreen shrub** to 2 m, often far less. **Leaves** deep green, elliptic to obovate, to 3.8 × 1.2 cm, greyish green beneath, densely scaly on both surfaces. **Flowers** 1–3; calyx small, to 4 mm, scaly; corolla ± salverform with short tube, 12–24 mm across, pink, red, yellow or occasionally white, often with darker spotting; stamens 10, protruding; style deflexed. seTibet, nwYunnan [Himalaya]; rocky slopes, shrubberies, forest glades, streamsides, gullies, 2,400–4,900 m. IV–VII.

Subsection *Trichoclada*

Rhododendron mekongense **Franch.** Upright, often rather leggy, **deciduous shrub** to 2.5 m, often less, with bristly branches. **Leaves** rather bright green, often bronze-tinged, oval to elliptic, to 5.5 × 2 cm, bristly beneath and with uneven scales. **Flowers** in mostly terminal clusters of up to 5; calyx scaly, sometimes bristly, to 7 mm long; corolla yellow to greenish yellow, 14–23 mm long, occasionally red-tinged. seTibet, c & nwYunnan [swTibet, c & eHimalaya]; moorland, shrubberies, forest margins, 2,700–4,500 m. IV–VI. **Var. rubrolineatum (Balf. f. & Forrest) Cullen** has scarcely bristly **leaves** and cream or yellow, red-flushed **flowers**. Distribution similar to that of var. *mekongense* but not cYunnan.

Rhododendron trichocladum **Franch.** Similar to the previous species, but a more bristly plant, especially on **stems**, leaf-margins and flower-stalks, whereas calyx-lobes are fringed with long hairs. Many intermediates can be found in the wild. w & cYunnan [nMyanmar], at similar altitudes and in similar habitats. V–VI.

Rhododendron lepidostylum **Balf. f. & Forrest*** Readily distinguished from the previous two species by markedly glaucous, more bristly foliage and rather larger **flowers**, 20–33 mm long, spotted or not. swYunnan; rocky places, particularly cliffs and ledges, 3,000–3,700 m. V–VI.

SUBGENUS *Rhododendron* SECTION *Pogonanthum*

Plants with scales at least on leaf-undersurface, often also on the young shoots, upper leaf-surface, leaf-stalks, calyx and even corolla; **leaves** flat or curved in bud; bud-scales of inflorescence fringed with branched hairs. **Corolla** parallel-sided or with a narrow flaring tube and spreading limb; stamens and style included within corolla-tube.

> 1. Bud-scales persistent . . . . **R. cephalanthum**
> 1. Bud-scales quickly deciduous 2
> 2. Stamens 5 . . . . **R. primuliflorum** Bur. & Franch., **R. trichostomum** Franch.
> 2. Stamens usually 6–8 . . . . **R. anthopogon**

Rhododendron cephalanthum Franch.   Prostrate to upright **shrub**. **Leaves** deep shiny green, oval to elliptic, 10–47 × 5–23 mm, with recurved margins, with dense reddish brown to fawn scales beneath. **Flowers** 5–10; calyx 3–7 mm long, with hairy margin; corolla 6–14 mm long, white to pink or rose; stamens normally 5. seTibet, n, nw & cYunnan [swTibet, eHimalaya]; alpine meadows, moorland, rocky places, cliffs, 2,700–4,900 m. IV–VII.

Rhododendron primuliflorum Bur. & Franch.*   Low spreading to more upright, straggly **shrub** to 1.8 m, often far less. **Leaves** dark shiny green above, lanceolate to oval, 8–35 × 5–14 mm, with dense fawn or brown scales beneath. **Flowers** in dense clusters of 5–10; calyx 2.5–5 mm long; corolla 10–16 mm long, white to pink, usually with a yellowish base. ?Gansu, n, w & swSichuan, seTibet, n & nwYunnan [swTibet]; alpine meadows, rocky places, cliffs, forest margins, 3,400–4,600 m. IV–VI. (See also photo 9 on p. 222.)

Rhododendron trichostomum Franch.*   Similar to *R. primuliflorum*, but **leaves** deep matt green, linear to linear-lanceolate with recurved margins, 8–34 × 2–8 mm. **Flowers** in clusters of 8–20; calyx tiny, 0.5–2 mm long; corolla 8–16 mm long, white to rose-pink. sw & cSichuan, n & nwYunnan; rocky slopes, low scrub, forest margins, 3,000–4,000 m. V–VII.

Rhododendron anthopogon D. Don   Rather dense, highly aromatic **shrub** to 1.5 m. **Leaves** shiny dark green or grey-green, elliptic to ovate or almost rounded, 10–38 × 8–25 mm, with dense reddish brown scales beneath. **Flowers** in clusters of 5–9; calyx 2–5 mm long, hairy and scaly; corolla 8–16 mm long, white, cream, pale yellow, pink or rose, described as looking like 'damp tissue paper'. seTibet [swTibet, Himalaya]; rocky places and moorland, often forming extensive colonies, 2,700–4,900 m. IV–VII, sometimes later.

SUBGENUS *Hymenanthes*

Evergreen trees and **shrubs** without scales, glabrous or with a covering of hairs on the young parts, especially leaves and stems; new **shoots** produced from buds below the flower-buds; **leaves** back-rolled in bud.

Subsection *Fortunea*

> 1. Style glabrous; shoots slender . . . .
>   . . . . **R. oreodoxa**, **R. oreodoxa** var. *fargesii*,
>   **R. oreodoxa** var. *shensiense*, **R. orbiculare**
> 1. Style glandular or hairy, at least in part; shoots stout . . 2
> 2. Calyx without glands . . . .
>   . . . . **R. calophytum**, **R. sutchuenense**
> 2. Calyx glandular . . **R. decorum**, **R. vernicosum**, **R. fortunei**

Rhododendron oreodoxa Franch.*   Upright to rounded **bush** or small **tree** to 6 m with slender **shoots**. **Leaves** relatively small, dark green, slightly shiny, elliptic to oblanceolate-elliptic, 6–8.5 × 2.2–4 cm, glaucous beneath. **Flowers** in clusters of 6–12; calyx minute, glabrous to glandular; corolla 7–8-lobed, broadly campanulate, 3–4 cm long, white with lilac flush, to pale pink, often with purple spots within, unscented; ovary glabrous or glandular. sGansu, nw & wHubei, Sichuan, eTibet; woodland and woodland margins, 2,100–3,000 m. IV–V. **Var. *fargesii* (Franch.) D. F. Chamb.***, from sGansu, wHubei, n & eSichuan, has rather broader **leaves**, **corolla** lilac-pink to deep pink, 6–7-lobed, deeply speckled, and ovaries with stalked glands. **Var. *shensiense* D. F. Chamb.***, from sGansu, wHubei, nwSichuan [sShaanxi], has distinctive 5-lobed **corollas** and ovaries with stalked glands.

Rhododendron orbiculare Decne*   A distinctive species forming a rather compact to open **shrub** to 4.5 m, with green glabrous **shoots**. **Leaves** leathery, matt-green above, glaucous beneath, rounded to ovate, to 12 × 10 cm, base somewhat heart-shaped, apex rounded. **Flowers** usually in lax clusters of 7–10, on glandular stalks; calyx minute, glabrous; corolla rose-pink, campanulate, 3.2–3.5 cm, glabrous, with 6–8 notched lobes. neGuangxi, w & swSichuan; forests, rocky slopes, 1,400–4,000 m. V–VI.

Rhododendron calophytum Franch.*   Broad **shrub** or small **tree**, 4–9 m tall with branches and new growth white-downy. **Leaves** leathery, shiny deep green, oblong-lanceolate to oblanceolate, 14–30 × 4–8 cm, generally becoming hairless beneath at maturity. **Flowers** in open-topped clusters of 5–30; calyx glabrous; corolla 5–7-lobed, broadly campanulate, swollen on 1 side, 5–6 cm long, white to pink, pale mauve or purple, with a deep red blotch and spots within; stamens 15–20, very short. e, se & cSichuan, nwYunnan; forests and forest margins, 1,800–4,000 m. IV–VI.

1. *Rhododendron cephalanthum*; nwYunnan (MN)
2. *Rhododendron primuliflorum*; Gangheba, Yulongxueshan, nwYunnan (CGW)
3. *Rhododendron trichostomum*; Geza, nwYunnan (CGW)
4. *Rhododendron anthopogon*; cult. (RL)
5. *Rhododendron oreodoxa*; Kangding, wSichuan (PC)
6. *Rhododendron oreodoxa* var. *fargesii*; Shennongjia, wHubei (PC)
7. *Rhododendron orbiculare*; near Kangding, wSichuan (HJ)
8. *Rhododendron calophytum*; Yulongxueshan, nwYunnan (CGW)

Rhododendron sutchuenense* Franch. Similar to *R. calophytum*, but **leaves** with recurved margins, midrib usually thinly hairy beneath and **flowers** slightly larger, spotted within but never blotched; stamens 13–15; stigma characteristically reddish. wHubei, eSichuan; forests, often amongst evergreen *Quercus* or in bamboo thickets, 1,500–2,400 m. III–V.

***Rhododendron decorum* Franch.** Lax to rather straggly **shrub** or small **tree** to 9 m. **Leaves** dark matt green, elliptic to oblanceolate, 7–30 × 2.2–11 cm, glabrous. **Flowers** very fragrant, in open-topped clusters of 7–14 ; calyx 1–3 mm long, glandular; corolla 6–8-lobed, broadly funnel-shaped, 6.5–10 cm long, white to pale pink, rose-pink or lavender, with a yellow, greenish or crimson basal tinge; stamens 12–20. Sichuan, Yunnan [nMyanmar]; forest margins, *Pinus* woodland, shrubberies, rocky places, 1,800–4,000 m. V–VII.

Rhododendron vernicosum* Franch. Similar to *R. decorum*, but **leaves** not more than 12 × 5 cm, often oblong or oval, with minute hairs beneath; **flowers** unscented, smaller, not more than 5 cm long, pale pink to bright rose-pink or lavender-pink, with crimson spots within; stamens usually 14, glabrous; ovary and style with characteristic dense, dark red glands. sGansu, Sichuan, seTibet, Yunnan; forest margins and clearings, shrubberies, streamsides, 2,700–4,300 m. IV–VI. Hybridises readily in the wild with *R. decorum* and *R. wardii*, amongst others.

Rhododendron fortunei* Lindl. Similar to the preceding two species, but **leaves** dark green, oblanceolate to obovate, to 17 × 8 cm, pale beneath, stalks characteristically reddish, purplish or bluish; corolla 4–5 cm long with 7 somewhat recurved, wavy-margined lobes, white with a lilac flush, lilac-pink to pink; stamens 14–16. Guangxi, Guizhou, Sichuan [c & eChina]; woodland, ravines, shrubberies, 600–2,100 m. IV–V.

---

1. *Rhododendron decorum*; Zhongdian, nwYunnan (CGW)
2. *Rhododendron decorum*; Yulongxueshan, nwYunnan (CGW)
3. *Rhododendron decorum*; Yulongxueshan, nwYunnan (CGW)
4. *Rhododendron decorum*; Yulongxueshan, nwYunnan (CGW)
5. *Rhododendron vernicosum*; Napahai, Zhongdian, nwYunnan (CGW)
6. *Rhododendron vernicosum*; Napahai, Zhongdian, nwYunnan (PC)
7. *Rhododendron fortunei*; cult. (JG)
8. *Rhododendron sinogrande*; cult. (RL)
9. *Rhododendron sinogrande*; cult. (CS)
10. *Rhododendron praestans*; Lancangjiang–Nujiang divide, nwYunnan (PCX)
11. *Rhododendron grande*; cult. (RL)
12. *Rhododendron rex* subsp. *fictolacteum*; Xingzu, Lancangjiang–Jinshajiang divide, nwYunnan (CGW)
13. *Rhododendron rex* subsp. *fictolacteum*; Xingzu, Lancangjiang–Jinshajiang divide, nwYunnan (CGW)

Subsection *Grandia*

1. Ovary and calyx hairy .... ***R. sinogrande*, *R. praestans***
1. Ovary and calyx glandular or glabrous ..... ***R. grande***

***Rhododendron sinogrande* Balf. f. & W. W. Sm.** Stout **tree** to 14 m with wide-spreading lower **branches**. **Leaves** very large, shiny dark green and rather rough above, oblanceolate to oval, 20–70 × 8–30 cm, with a silvery to fawn hair-covering beneath. **Flowers** in dense clusters of 12–30; corolla 8–10-lobed, 6–7.5 cm long, creamy white to pale yellow, with large crimson blotch within. seTibet, n, nw & wYunnan [neIndia, nMyanmar]; usually in mixed forests, 2,100–3,400 m. IV–VI.

***Rhododendron praestans* Balf. f. & W. W. Sm.** Large **bush** or small **tree** to 9 m. **Leaves** oblanceolate to obovate, 18–38 × 7–14 cm, deep shiny green and rather rough above, with a shiny greyish white to bronzy brown hair-covering beneath stalks winged. **Flowers** in heads of 10–25; corolla 8-lobed, 3.5–4.5 cm long, white to pink with pink to magenta-rose flush, usually with spots or a blotch within. seTibet, nwYunnan [nMyanmar]; coniferous and mixed woodland, 2,700–4,300 m. III–V.

***Rhododendron grande* Wight** Large laxly branched **tree** 6–15 m tall, often single-trunked. **Leaves** mid-green, slightly shiny, large, elliptic to oblanceolate, 15–46 × 8–13 cm, rather rough above, with a loose white, silvery or fawn down beneath. **Flowers** in large clusters of 15–25; corolla oblique-campanulate, 5–7 cm long, 8-lobed, cream to pale yellow or pink with purple spotting or blotching within; stamens 15–16. seTibet [c & eHimalaya]; evergreen and mixed forests, often in gullies, 2,100–3,200 m. III–IV.

Subsection *Falconera*

Rhododendron rex* H. Lévl. Large **shrub** or small **tree** to 12 m, though often less, with greyish- or whitish-downy young **branches**. **Leaves** usually dark green and shiny, obovate, 12–37 × 5.5–13.5 cm, glabrous at maturity above, but with dense fawn wool beneath. **Flowers** in large rounded clusters of up to 30; calyx tiny, downy; corolla campanulate with an oblique mouth, 5–7 cm long, white flushed pink, generally with a crimson blotch and spotting within, the 8 lobes frilled; stamens 14–16. swSichuan, neYunnan; coniferous or mixed forests, forest margins, 3,000–4,300 m. IV–VI. **Subsp. *fictolacteum* (Balf. f.) D. F. Chamb.** is far more widespread: it has narrower oblanceolate **leaves** that have a reddish brown to buff hair-covering beneath and rather smaller **flowers** ranging from white to pale lilac, mauve-pink or rose-pink. seTibet, nw & wYunnan [nMyanmar]. (See also photo on p. 35.)

Rhododendron coriaceum* Franch. Similar to the preceding species, but **leaves** pale, with a grey or pale fawn hair-covering beneath; **corolla** 5–7-lobed, not more than 4 cm long; stamens 10–14. seTibet, nwYunnan; coniferous forests, shrubberies, 3,000–4150 m. IV–V.

Rhododendron galactinum* Balf. f. **Shrub** or small **tree** to 9 m with rough, grey-brown **bark** and very distinctive downy **buds** with short scales. **Leaves** ovate to oblanceolate, 12–21 × 5–8 cm, glossy mid-green above with yellow-green midrib, with a thin buff to pale reddish brown hair-covering beneath. **Flowers** in clusters of 9–15; calyx very small, 1 mm, downy; corolla campanulate, 28–32 mm long, white to pale rose-pink, blotched and flecked within; stamens 14; ovary and style usually glabrous. c & wSichuan; forests, ravines, 2,300–3,300 m. V–VI.

Subsection *Campylocarpa*

1. Corolla saucer-shaped, white, often flushed pink, or pink to rose-pink .... *R. souliei*
1. Corolla broadly campanulate, yellow, cream, rarely white .... *R. wardii*, *R. wardii* var. *puralbum*, *R. campylocarpum*

Rhododendron souliei* Franch. **Shrub** or leggy **tree** to 5 m. **Leaves** glaucous when young, later deep green, ovate to ovate-elliptic or almost rounded, 3.5–8.2 × 2.2–5 cm, glabrous above and beneath. **Flowers** 3–8; calyx 3–8 mm long, with rounded ciliate lobes; corolla 2.5–3.5 cm long; style glandular. sw & cSichuan; open forests of *Quercus* or conifers, shrubberies, 2,700–4,300 m. V–VI.

Rhododendron wardii* W. W. Sm. Compact to open, rounded **shrub** to small **tree**, to 7.6 m. **Leaves** glaucous at first, then deep green, rounded to ovate or oblong-elliptic, 3–12 × 2–6 cm, pale green or glaucous and minutely dotted beneath, glabrous overall. **Flowers** 5–14; calyx 4–12 mm long; corolla pale to deep yellow, 2–4 cm long, with or without crimson blotch within, buds sometimes flushed apricot or orange-red; style glandular. swSichuan, seTibet, nwYunnan; open forests, thickets, shrubberies, open slopes, 2,700–4,300 m. IV–VI. **Var.** ***puralbum* (Balf. f. & W. W. Sm.) D. F. Chamb.*** has somewhat larger **leaves** and more deeply cupped white, pink-flushed **flowers**. swSichuan, seTibet, nwYunnan; similar habitats. (See also photo on p. 36 and photos 2 and 4 on p. 206.)

Rhododendron campylocarpum* Hook. f. Similar to *R. wardii*, differing mainly in the tiny calyx and style, which is eglandular or at least not glandular in upper half, but also in possessing a curved, rather than straight, **fruit-capsule**. eTibet, wYunnan [swTibet, c & eHimalaya]; *Rhododendron* or coniferous forests, forest margins, rocky places, cliffs, 3,400–4,300 m. IV–VI. **Subsp.** ***caloxanthum* (Balf. f. & Forrest) D. F. Chamb.** with rounded **leaves** and cream, rather than yellow, **flowers** is found in eTibet, nwYunnan [Myanmar].

Subsection *Selensia*

Rhododendron selense* Franch. Variable rounded to rather erect and straggly **shrub** to 6 m, often less. **Leaves** dark green or grey-green, obovate to elliptic, 2.6–9 × 1.5–4 cm, glabrous above and beneath or with a thin hair-covering beneath, leaf-stalks often bristly glandular. **Flowers** in lax terminal clusters of 3–8; calyx 1–8 mm long, glandular; corolla campanulate, 2.5–4 cm long, white to deep pink, sometimes spotted within; stamens 10; ovary densely glandular. Plants with **leaves** glaucous beneath and tiny calyces (1–2 mm long) are referable to **subsp.** ***selense**** swSichuan, seTibet, nwYunnan; coniferous forests, shrubberies, bamboo thickets, sometimes above treeline, 2,700–4,400 m. IV–V. Plants without glaucous leaf-undersurfaces and with larger calyces (2–3 mm long) are differentiated as **subsp.** ***dasycladum* (Balf. f. & W. W. Sm.) D. F. Chamb.*** swSichuan, seTibet, nwYunnan; similar habitats and altitudes. IV–VI. Plants similar to subsp. *selense* but with broader, pink **corollas**, and calyces 3–6 mm long are referable to **subsp.** ***jucundum* (Balf. f. & W. W. Sm.) D. F. Chamb.*** nwYunnan (Cangshan); *Rhododendron*–conifer forests, 3,350–3,660 m. IV–V.

Subsection *Argyrophylla*

1. Leaf-tip almost blunt to almost pointed, not long-pointed .... *R. argyrophyllum*, *R. argyrophyllum* subsp. *hypoglaucum*, *R. argyrophyllum* subsp. *nankingense*, *R. argyrophyllum* subsp. *omeiense*, *R. pingianum*
1. Leaf-tip markedly long-pointed or sharply pointed .... *R. longipes*

Rhododendron argyrophyllum* Franch. Very variable erect, rather open, rounded **shrub**, or small **tree** to 6 m, occasionally more. **Leaves** matt green to deep shiny green, elliptic to oblanceolate, 6–16 × 1.5–6 cm, glabrous above, with a thin silvery, white or fawn down beneath, sometimes markedly glaucous. **Flowers** in clusters of 4–12; calyx 1–2 mm, downy; corolla 2–4 cm long, white to deep rose-pink, spotted within or not. **Subsp.** ***argyrophyllum**** has **leaves** usually 6–9 cm long and the ovary lacking glands. e & neGuizhou, neYunnan, sShaanxi; **subsp.** ***hypoglaucum* (Hemsl.) D. F. Chamb.*** is like subsp. *argyrophyllum* but with glandular flower-stalks and ovary. wHubei, eSichuan; **subsp.** ***nankingense* (Cowan) D. F. Chamb.*** has large **leaves**, 11–16 cm long, rough above, and **flowers** rather large and deep pink to purple. Guizhou; **subsp.** ***omeiense* (Rehd. & E.H. Wilson) D. F. Chamb.*** has **leaves** that are smaller and narrower, fawn-downy beneath. c & sSichuan, especially Emeishan. All subspecies grow in forests, shrubberies, ravines, rocky slopes, 1,500–3,600 m. IV–VI.

Rhododendron pingianum* W. P. Fang Very similar to *R. argyrophyllum*, differing only in the relatively broader **leaves** and deeper-coloured **flowers**, as well as having reddish brown-felted ovaries. sSichuan; forests and forest margins, 2,000–2,700 m. V–VI.

1. *Rhododendron galactinum*; Balangshan, Wolong, wSichuan (CGW)
2. *Rhododendron souliei*; cult. (CGW)
3. *Rhododendron wardii*; Napahai, Zhongdian, nwYunnan (HJ)
4. *Rhododendron wardii*; Lancangjiang–Jinshajiang divide, nwYunnan (CGW)
5. *Rhododendron wardii*; Lancangjiang–Jinshajiang divide, nwYunnan (CGW)
6. *Rhododendron campylocarpum*; west of Ugyen Choling, Bhutan (CGW)
7. *Rhododendron selense* subsp. *dasycladum*; Lancangjiang–Jinshajiang divide, nwYunnan (CGW)
8. *Rhododendron pingianum*; Emeishan, sSichuan (PC)
9. *Rhododendron argyrophyllum* subsp. *omiense*; Emeishan, sSichuan (PC)

1. *Rhododendron arboreum* subsp. *delavayi*; Dali to Lijiang, nwYunnan (JM)
2. *Rhododendron lacteum*; Xingyi to Kunming, cYunnan (PC)
3. *Rhododendron roxieanum*; Napahai, Zhongdian, nwYunnan (JM)
4. *Rhododendron roxieanum*; Napahai, Zhongdian, nwYunnan (PC)
5. *Rhododendron phaeochrysum*; Lancangjiang–Jinshajiang divide, nwYunnan (CGW)
6. *Rhododendron phaeochrysum*; nwYunnan (HJ)
7. *Rhododendron phaeochrysum*; Lancangjiang–Jinshajiang divide, nwYunnan (CGW)
8. *Rhododendron aganniphum*; Daxueshan, nwYunnan (HJ)
9. *Rhododendron aganniphum* with *R. primuliflorum*; Genyanshan, swSichuan (MN)
10. *Rhododendron aganniphum*; Tianchi, nwYunnan (PC)
11. *Rhododendron przewalskii*; seQinghai (JM)
12. *Rhododendron przewalskii*; nwSichuan (TK)

Rhododendron longipes **Rehd. & E.H. Wilson*** Erect **shrub** to 4 m, often less, with branchlets white-downy at first. **Leaves** mid-green, somewhat shiny, lanceolate to oblanceolate, 5–13 × 1.5–3.5 cm, with a thin pale brown felt beneath. **Flowers** in lax clusters of 8–15; calyx 1–2 mm; corolla 2.5–4 cm long, pale pink to rose-pink or purple, generally unspotted. Guizhou, se & swSichuan, neYunnan; broadleaved forests, occasionally on cliffs, 1,300–2,500 m. IV–V.

Subsection *Arborea*

Rhododendron arboreum **Sm. subsp.** ***delavayi*** **(Franch.) D. F. Chamb.** (syn. *R. delavayi* Franch.) Large **bush** or small **tree** to 9 m, with a single or several **trunks** and rough **bark**; branchlets densely hairy. **Leaves** thick, hard-leathery, deep shiny olive-green, lanceolate to elliptic-oblanceolate, 5–16 × 1.8–3 cm, with thin spongy whitish or fawn hair-covering beneath, tip pointed. **Flowers** up to 20 in a dense terminal head; calyx tiny, 1–3 mm, usually hairy; corolla tubular-campanulate, 3–5 cm long, deep crimson to carmine; stamens 10. nwGuangxi, wGuizhou, swSichuan, seTibet, Yunnan [eHimalaya, Myanmar, nThailand, nVietnam]; mixed and coniferous forests, remnant forests, forest margins, 1,500–3,200 m. III–V.

Subsection *Taliensia*

1. Flowers yellow .... **R. lacteum**
1. Flowers white, pink, rose-pink, purple or reddish ... 2
2. Calyx very small to tiny, not more than 3 mm long ... 3
2. Calyx larger, 5–15 mm long 4
3. Leaves with downturned or inrolled margins ....
   .... **R. roxieanum, R. proteoides**
3. Leaves with flat margins .... **R. phaeochrysum,**
    R. phaeochrysum var. levistratum, R. aganniphum,
    R. aganniphum var. flavorufum,
    R. przewalskii, R. taliense
4. Calyx glabrous, not glandular .... **R. beesianum**
4. Calyx sparsely to densely glandular 5
5. Corolla neither spotted nor blotched within ....
   ...... **R. adenogynum, R. balfourianum, R. bureavii,**
   R. bureavioides
5. Corolla usually spotted or blotched within ....
   .... **R. faberi, R. prattii**

Rhododendron lacteum **Franch.*** **Shrub** or small **tree** to 7 m, though often less. **Leaves** leathery, dull green above, elliptic to obovate, 8–17 × 4.5–10 cm, becoming hairless above at maturity, grey-brown felted beneath. **Flowers** in dense rounded clusters of up to 30; calyx 5–10 mm long, usually glabrous; corolla broadly campanulate, 4–5 cm long, clear yellow, sometimes with a reddish brown blotch within. ne & wcYunnan; forests and forest margins, shrubberies, often around treeline, 3,000–4,000 m. V–VI.

Rhododendron roxieanum **Forrest*** Very variable compact to straggly **shrub** or small **tree**, 0.3–4 m tall. **Leaves** dark shiny green above, linear to elliptic, 5–12 × 0.6–4 cm, glabrous above, with a thick fawn or reddish brown, sometimes spongy, felt beneath. **Flowers** in rounded clusters of up to 15; calyx small, 0.5–2 mm long, hairy and glandular; corolla funnel-shaped to almost campanulate, 2–4 cm long, white to cream, often with a pink flush and generally spotted within. swSichuan, seTibet, nwYunnan; upper forest zone and above treeline, 3,400–4,300 m. V–VII.

Rhododendron proteoides **Balf. f. & W. W. Sm.*** Small spreading, rather dense **shrub** usually not more than 30 cm. **Leaves** rather dull green above, elliptic, 2–4 × 0.7–1.5 cm, with strongly inrolled margins, becoming hairless above at maturity, with a dense woolly reddish brown felt beneath. **Flowers** in dense rounded clusters of up to 10; calyx tiny, 0.5 mm long; corolla campanulate, 2.5–3.5 cm long, white to cream flushed rose-pink, finely spotted within. swSichuan, seTibet, nwYunnan; rocky slopes, cliffs, 3,700–4,600 m. V–VII. Intergrades with *R. roxieanum* in some areas.

Rhododendron phaeochrysum **Balf. f. & W. W. Sm.*** Variable, generally compact, rather upright **shrub** or small **tree**, to 4.5 m. **Leaves** dark green and somewhat shiny above, elliptic to ovate-oblong, 4–14 × 1–6.5 cm, ± glabrous above, with a felt of fawn or reddish brown hairs beneath. **Flowers** in rounded or flat-topped clusters of up to 15; calyx small, only 1–2 mm long, usually glabrous; corolla broadly funnel-shaped, 30–50 mm long, white to pink, spotted within, sometimes densely so. sGansu, c & nSichuan, seTibet, n, nw & wYunnan [swTibet]; close to or above treeline, sometimes on bog margins, often extensive, 3,000–4,700 m. V–VII. **Var. *levistratum* (Balf. f. & Forrest) D. F. Chamb.*** is readily distinguished by the fawn to yellow-brown or reddish brown felt on undersurface of **leaves** and by smaller **flowers**, the corolla 20–35 mm long. Gansu, c & nSichuan, n & nwYunnan.

Rhododendron aganniphum **Balf. f. & Kingdon-Ward*** Similar to *R. phaeochrysum* with which it hybridises, differing in **leaves** which often have a white bloom above and a pale white to greyish white hair-covering beneath. n, w & swSichuan, seTibet, nwYunnan; shrubberies and rocky places above treeline, 2,700–4,700 m. VI–VII. **Var. *flavorufum* (Balf. f. & Forrest) D. F. Chamb.*** has a deep reddish brown hair-covering on underside of **leaves**, which splits into an irregular netted pattern. Similar distribution, altitude and flowering time. (See also photo on p. 12.)

Rhododendron przewalskii **Maxim.*** Very similar to *R. phaeochrysum*, but **leaves** generally rather smaller (less than 10 × 4.5 cm) with a tell-tale yellowish midrib above, and glabrous, or thinly downy or becoming hairless, beneath; **flowers** smaller, 15–35 mm long, usually white. seGansu, seQinghai, n & nwSichuan; at or just above treeline, often in extensive colonies, 3,000–4,300 m. V–VII.

Rhododendron taliense Franch.*   Compact to rather leggy **shrub** to 4 m, often less. **Leaves** matt or somewhat shiny dark green above, lanceolate to oblong-ovate, 5–11 × 2–4 cm, glabrous above, with a thick reddish brown felt beneath. **Flowers** in rounded to flat-topped clusters of 10–20; calyx small 0.5–2 mm long, glabrous; corolla ± campanulate, 3–3.5 mm long, white or cream, often with a pink flush, spotted within. wYunnan; forests and forest margins, shrubberies, at or above treeline, 3,000–4,000 m. V–VI.

Rhododendron beesianum Diels   Large **shrub** or small **tree**, 2–9 m tall with sticky **buds**. **Leaves** elliptic to oblanceolate, 9–29 × 2.6–8.2 cm, ± glabrous above, with a thin grey, fawn or brownish hair-covering beneath. **Flowers** in rounded to flat-topped clusters of up to 20; calyx 5–10 mm long, glabrous; corolla broadly campanulate, 3.5–5.3 cm long, white to rose-pink, rose-purple or reddish, spotted or not within. swSichuan, seTibet, nwYunnan [nMyanmar]; conifer belt, or in rocky gullies where it is locally common, 3,000–4,400 m. V–VII.

Rhododendron adenogynum Diels*   Compact to spreading **shrub**, 1–4 m tall, often rather dense. **Leaves** leathery, dark green above at maturity, narrow-elliptic to elliptic, 6–11 × 2–4 cm, almost glabrous above, with a dense woolly, rather spongy yellowish hair-covering beneath, turning olive-brown at maturity. **Flowers** in a lax clusters of up to 12; calyx large, to 15 mm long, glandular, with uneven lobes; corolla funnel-shaped, 3–4.5 cm long, varying from white, to pink, rose-pink or rose-purple. swSichuan, seTibet, w & nwYunnan; generally forming extensive colonies above treeline on rocky exposed slopes and in gullies, 3,000–4,300 m. V–VII. (See also photo 6 on p. 201.)

Rhododendron balfourianum Diels*   Compact, generally rounded **shrub** to 4.5 m, often less, similar to *R. adenogynum*, but **leaves** with a shiny dense, spongy hair-covering, silvery-white at first but becoming greyish brown or pinkish cinnamon at maturity. **Flowers** with a smaller calyx, 6–10 mm long; corolla generally rose-pink, more unusually white, pale pink or purplish pink. swSichuan, nw & wYunnan; upper forest zone and above treeline, 3,000–4,600 m. V–VII.

Rhododendron bureavii Franch.*   Rounded **shrub** or occasionally small **tree**, 2–7.6 m tall. **Leaves** dark green and shiny above at maturity, elliptic, 4.5–19 × 2–7 cm, usually glabrous above, with a dense salmon-pink felt beneath which turns rusty red at maturity. **Flowers** in lax clusters of up to 20; calyx relatively large, to 10 mm long, densely glandular; corolla tubular-campanulate, 2.5–4 cm long, white flushed with pink to rose-pink; ovary and style glandular-downy. nc & nwYunnan; local in open forests and shrubberies, 3,000–3,900 m. V–VI.

Rhododendron bureavioides Balf. f.*   Rather similar to *R. bureavii*, but with distinctive short, bent leaf-stalks, not more than 10 mm long, corollas slightly larger, 4–4·5 cm long, and glandular, not hairy, ovaries and styles. nw & wcSichuan; mixed and deciduous woodland, forested ridges and slopes, 3,000–3,500 m. V–VI.

Rhododendron faberi Hemsl.*   Rather dense **shrub** to 6 m. **Leaves** deep shiny green above, oblong-elliptic to lanceolate, 6–11 × 2.8–4.8 cm, glabrous above, with a dense reddish brown to reddish felt beneath which is partly shed to reveal a whitish sublayer. **Flowers** in rounded or flat-topped clusters of not more than 12; calyx 7–12 mm long, sparsely glandular; corolla ± campanulate, 3–4 cm long, white to pale pink, usually spotted or blotched within, occasionally plain. sSichuan; rocky ridges and peaks with trees and other shrubs, 3,000–3,400 m. V–VI.

Rhododendron prattii Franch.*   Very similar to *R. faberi*, but with larger, broader, more oblong **leaves**, 8–19 × 5–8 cm, often with a greenish yellow midrib, and corolla up to 5.6 cm long, white spotted with red; calyx 5–10 mm long, glandular and hairy. s, c & wSichuan; forests, rocky slopes, 2,700–4,300 m. IV–VI.

Subsection *Fulva*

Rhododendron fulvum Balf. f. & W. W. Sm.   Rounded **shrub** or small **tree** to 9 m, young **branches** with a white to rusty coloured down. **Leaves** shiny dark green to matt mid-green above (**subsp. *fulvoides* (Balf. f. & Forrest) D. F. Chamb.**), elliptic to oblanceolate, to 27 × 9 cm, with granular fawn to cinnamon down beneath. **Flowers** in rounded clusters of up to 20; calyx tiny, not more than 2 mm, glabrous; corolla campanulate, to 5 cm long, white to pale pink, generally blotched within; ovary and style usually glabrous. seTibet, nw, w & cYunnan [nMyanmar]; coniferous and mixed forests, 2,500–4,500 m. IV–V.

---

1. *Rhododendron taliense*; Cangshan, nwYunnan (PCX)
2. *Rhododendron beesianum*; east of Weixi, Lancangjiang–Jinshajiang divide, nwYunnan (CGW)
3. *Rhododendron adenogynum*; Wutoudi, Yulongxueshan, nwYunnan (CGW)
4. *Rhododendron adenogynum*; Wutoudi, Yulongxueshan, nwYunnan (CGW)
5. *Rhododendron balfourianum*; Napahai, Zhongdian, nwYunnan (PCX)
6. *Rhododendron bureavii*; cult. (CGW)
7. *Rhododendron bureavioides*; Kangding, wSichuan (PC)
8. *Rhododendron faberi*; Emeishan, sSichuan (PC)
9. *Rhododendron prattii*; Balangshan, Wolong, wSichuan (PC)
10. *Rhododendron fulvum*; Yulongxueshan, nwYunnan (PCX)

1. *Rhododendron uvariifolium*; Napahai, Zhongdian, nwYunnan (PCX)
2. *Rhododendron barbatum*; Gurkha Himal, Nepal (CGW)
3. *Rhododendron forrestii*; Namwa La, seTibet (PCX)
4. *Rhododendron neriiflorum*; Cangshan, nwYunnan (PC)
5. *Rhododendron thomsonii*; Ugyen Choling, Bhutan (CGW)
6. *Rhododendron stewartianum*; Namwa La, seTibet (PCX)
7. *Rhododendron microphyton*; Cangshan, nwYunnan (PC)
8. *Rhododendron simsii*; Cangshan, nwYunnan (PC)

ERICACEAE: Lyonia

Rhododendron uvariifolium Diels Very similar to preceding species, but **leaves** with a more matted white to ash-grey down beneath, and leaf-tip often more pointed. swSichuan, nwYunnan [neIndia]; coniferous and mixed forests, 2,100–4,300 m. IV–VI.

Subsection *Barbata*

Rhododendron barbatum Wall. Large bush or upright **tree** to 9 m, branching from base, with bristly branchlets and sticky **buds**; **bark** characteristically peeling, purple to plum-coloured. **Leaves** elliptic to elliptic-lanceolate, 10–20 × 2.5–7 cm, usually deep green, glabrous overall at maturity, with bristly stalks. **Flowers** 10–20 in dense rounded heads; calyx relatively large, 10–15 mm long, glabrous; corolla tubular-campanulate, 4–5 cm long, scarlet to scarlet-crimson, unmarked; ovary densely glandular; style glabrous. seTibet [c & eHimalaya]; mixed forests and forest margins, 2,400–3,700 m. IV–V.

Subsection *Neriiflora*

Rhododendron forrestii (Balf. f.) Diels Low mound-forming to creeping **shrub** not more than 15 cm. **Leaves** deep shiny green, rounded to obovate, to 5 × 3 cm, often less, glabrous above and beneath or somewhat glandular. **Flowers** usually solitary, occasionally paired; calyx minute; corolla campanulate with a rather narrow tube, 3–3.5 cm long, scarlet to crimson; stamens 10. seTibet, nwYunnan [neIndia, nMyanmar]; rocky slopes, banks, streamsides, pastures, 3,000–4,600 m. IV–VI.

Rhododendron neriiflorum Franch. Distinctive species forming a rather compact **shrub** to 6 m, but often only 1–3 m tall. **Leaves** deep to mid-green and somewhat shiny, elliptic to oblong, to 9 × 3.5 cm, generally glabrous and rather glaucous beneath. **Flowers** in lax to rather dense terminal clusters of 5–12; calyx variable, 2–15 mm long, without glands; corolla scarlet to crimson or rose-red, 3.5–4.5 cm long. seTibet, w & nwYunnan [nMyanmar]; forests, shrubberies, 2,100–3,400 m. IV–VI. **Subsp. phaedropum (Balf. f. & Forrest) Tagg** (syn. *R. neriiflorum* var. *appropinquans* (Tagg & Forrest) W. K. Hu) has narrower **leaves**, a somewhat glandular calyx and **flowers** in shades of rose-pink, pale yellow or orange. seTibet, nwYunnan [swTibet, eHimalaya] at similar altitudes.

Subsection *Thomsonia*

Rhododendron thomsonii Hook. f. Rather upright **shrub** or small **tree** to 7 m with multicoloured peeling **bark**. **Leaves** leathery, mid-green to deep matt or shiny green but generally markedly glaucous when young, rounded to broadly elliptic, 3–11.5 × 2–7.5 cm, glabrous, glaucous beneath. **Flowers** up to 12; calyx cup-shaped, to 20 mm long, greenish, reddish or purplish; corolla rather fleshy and campanulate, 3.5–6 cm long, deep red to dark wine, generally with a marked bloom outside. seTibet [swTibet, eHimalaya]; coniferous forests and shrubberies, occasionally in more open marshy places, 2,400–4,300 m. IV–VI.

Rhododendron stewartianum Diels Similar to preceding species, but a **bush** not more than 2.5 m, bud-scales on new growth bright red and **leaves** rather narrower and more oblong, usually with some bristles along midrib beneath. **Flowers** somewhat smaller, corolla very variable in colour from white to pink, purple, rose-pink, red or yellow, often flushed giving a bicolored appearance, generally spotted within. seTibet, nwYunnan [nMyanmar]; shrubberies, bamboo thickets, rocky slopes, forest margins, 3,000–4,300 m. III–V.

SUBGENUS *Tsutsusi* SECTION *Tsutsusi*

A small section that contains the familiar pot or Indian azaleas characterised by their much-branched dwarf to medium habit with **evergreen**, **semi-evergreen** or occasionally **deciduous**, variously hairy **leaves**, scaleless; leaves back-rolled in bud, dimorphic, spring leaves mainly deciduous, summer leaves mainly evergreen, usually smaller; new shoots arising from the same bud as the **flowers**, the latter always terminal.

Rhododendron microphyton Franch. Rather dense **evergreen shrub** to 2 m, often only 0.3–1 m, with lanceolate to elliptic **leaves** to 4 × 1.5 cm, covered in appressed brown hairs, especially dense beneath. **Flowers** in clusters of 3–6; calyx to 5 mm long, brownish; corolla funnel-shaped, 13–19 mm across, pink to rose-purple, occasionally whitish, spotted within; stamens 5. swSichuan, Yunnan [Myanmar, ?nThailand]; dry scrubby slopes, ravines, gullies, 1,800–2,500 m. III–V.

Rhododendron simsii Planch. Readily distinguished from previous species by the more straggly habit, to 3 m, by dimorphic **leaves**, spring leaves 3–7 cm long, summer ones only 1–2 cm long, all with pale dense short hairs beneath; **flowers** larger, red, spotted, usually with 10 stamens. Guizhou, Hubei, Sichuan, seTibet, Yunnan [c & eChina, sJapan, neMyanmar, Taiwan, Thailand, Vietnam]; cliffs, dry rocky places, riverbanks, open shrubberies, 300–2,400 m. III–VI, sometimes later.

LYONIA

A genus of 35 species (5 in China) from Asia and North America. **Deciduous** or **evergreen trees** and **shrubs**, often patch-forming; **buds** usually with 2 large overlapping scales. **Leaves** spirally arranged, entire in Asian species. **Flowers** in lateral racemes, generally 5-parted but up to 8-parted in some instances; sepals usually valvate in bud; corolla tubular or urn-shaped, with short lobes; stamens with flattened filaments, often spurred at the anther-tip. **Fruit** a small many-seeded capsule.

***Lyonia ovalifolia* (Wall.) Drude** Very variable **evergreen** or **deciduous shrub** or small **tree**, not more than 5 m, with glabrous or downy branchlets. **Leaves** with stalks 4–9 mm long, blade ovate to elliptic or occasionally almost rounded, 3–20 × 2–12 cm, usually papery, sparsely to moderately hairy on both surfaces, base rounded to almost heart-shaped or wedge-shaped, apex long-pointed. **Racemes** 5–20 cm long, many-flowered, **flowers** nodding, lower bract leaf-like; calyx-lobes 2–6 mm long; corolla tubular, 8–11 mm long; stamens with filaments with a pair of spurs at top. **Capsule** and ovary glabrous. sGansu, Guangxi, Guizhou, Hubei, Sichuan, Shaanxi, seTibet, Yunnan [swTibet, Himalaya, Thailand, Vietnam, Malaysia]; forests and forest margins, shrubberies, tracksides and more open places, ravines, 200–3,400 m. V–VII. **Var. *hebecarpa* (Franch. ex Forbes & Hemsl.) Chun** (syn. *Pieris ovalifolia* (Wall.) D. Don var. *hebecarpa* Franch. ex Forbes & Hemsl.) is distinguished by having densely hairy ovaries and capsules, and **leaves** generally not more than 12 × 6 cm. Similar distribution but not in Himalaya or Tibet; similar habitats and altitudes. **Var. *lanceolata* (Wall.) Hand.-Mazz.** differs in having **leaves** lanceolate to elliptic-lanceolate, usually to 13 × 5 cm and calyx-lobes lanceolate rather than triangular. Guangxi, Guizhou, Hubei, Sichuan, seTibet [neIndia, Myanmar]; similar habitats, 700–2,400 m.

***Lyonia macrocalyx* (J. Anth.) Airy Shaw** Similar to the previous species, but with leathery **leaves**, smaller **racemes** (not more than 7 cm long), **flowers** with a larger calyx, the lobes 6–11 mm long, corolla urn-shaped and filaments unspurred. seTibet, nwYunnan [nMyanmar]; forests, particularly *Picea*, cliffs within forests and shrubberies, 1,800–3,500 m. VII–VIII.

***Lyonia villosa* (Wall. ex C. B. Clarke) Hand.-Mazz.** (syn. *L. ovalifolia* (Wall.) Drude var. *pubescens* Franch.; *Pieris villosa* Wall.) Readily distinguished from previous two species by possessing smaller **leaves**, 1.5–7 × 1–4 cm that have a blunt, mucronate tip; **racemes** short, mostly 1–5 cm long, brownish-downy, corolla 5–8 mm long. Guizhou, Sichuan, seTibet, Yunnan [Himalaya]; mixed and *Rhododendron* forests, shrubberies, 1,000–3,900 m. V–VII.

PIERIS

A genus of 7 species (3 in China) distributed in east Asia, eastern North America and the Caribbean. **Evergreen shrubs** and **trees** with leathery **leaves**. **Flowers** 5-parted; calyx valvate; corolla urn-shaped or tubular-urn-shaped, with 5 short lobes; anthers spurred on back at junction with connective. **Fruit** a 5-valved capsule.

***Pieris formosa* D. Don** (including *P. bodinieri* H. Lévl.; *P. forrestii* Harrow) **Evergreen shrub** or small **tree** to 5 m, occasionally more, with glabrous or downy twigs and spirally arranged or pseudowhorled **leaves**, these lanceolate to elliptic, occasionally oblanceolate, 3–14 × 1–2 cm, finely veined, minutely toothed, often bright red when young. **Inflorescence** racemose or paniculate, to 10 cm long, occasionally longer, many-flowered; **corolla** white, tubular-urn-shaped, 5–8 mm long. Gansu, Guangxi, Guizhou, Hubei, Sichuan, Shaanxi, seTibet, Yunnan [swTibet, c & sChina, c & eHimalaya, Myanmar, Vietnam]; open slopes, forest margins, shrubberies, *Rhododendron* thickets, 900–3,500 m. V–VI.

ENKIANTHUS

A genus of 12 species (7 in China, 2 endemic) from the eastern Himalaya to Japan. **Deciduous shrubs** and small **trees** with **leaves** clustered towards shoot-tips, stalked, toothed or not. **Flowers** in terminal umbels or short racemes, more rarely paired or solitary; calyx with 5 short lobes; corolla broadly campanulate with 5 short lobes; stamens shorter than corolla. **Fruit** a small capsule.

***Enkianthus deflexus* (Griff.) C. K. Schneid.** **Shrub** or **tree** to 5 m with elliptic to oblong-elliptic **leaves**, 3–7 × 1.5–3 cm, papery, rough-hairy on veins, becoming hairless, margin finely toothed. **Inflorescence** with many flowers in a corymbose cluster, with densely hairy stalks that are often also glandular; calyx 3–4 mm, lobes ciliate; **corolla** white, brick-red or pale yellow, 8–15 mm long, with erect, triangular lobes. sGansu, Guizhou, Hubei, Sichuan, seTibet, Yunnan [swTibet, cHimalaya, Myanmar]; forests of *Pinus* or *Quercus* or mixed, shrubberies, ravines, 1,000–3,300 m. V–VII.

Enkianthus chinensis* Franch. Similar to the previous species, but **leaves** glabrous and flower-stalks downy or glabrous; calyx often non-ciliate; corolla yellowish, orange-striped and red-flushed, smaller, 7–10 mm, with slightly recurved lobes. Guangxi, Guizhou, Hubei, Sichuan, Yunnan [c & eChina]; similar habitats, 900–2,100 m. V–VII.

GAULTHERIA

A genus of 135 species (32 in China, 15 endemic) widely distributed in the northern hemisphere, South East Asia and Australia. Creeping to erect **evergreen shrubs** with leathery, spirally arranged, usually toothed **leaves**. **Flowers** (4–)5-parted, solitary, racemose or in panicles; corolla urn-shaped to campanulate or tubular; stamens not protruding. **Fruit** a berry.

1. Prostrate, mat-forming shrubs with leaves no more than 16 mm long . . . . ***G. trichophylla*, *G. nummarioides*, *G. sinensis***
1. Spreading to erect shrubs or small trees with leaves at least 30 mm long . . . . ***G. hookeri*, *G. fragrantissima*, *G. wardii***

***Gaultheria trichophylla* Royle** Much-branched, prostrate **shrub** forming intricate mats. **Leaves** dense shiny, deep green, elliptic to oblong, leathery, 5–13 × 2–5 mm, ciliate, minutely toothed at margin. **Flowers** solitary in leaf-axils; calyx glabrous; corolla white, campanulate, 5–6 mm long, deeply 5-

ERICACEAE: Gaultheria

lobed. **Berries** blue when ripe, 6–10 mm, fleshy, glabrous. c & wSichuan, seTibet, nwYunnan [swTibet, Himalaya]; mountain moors, alpine meadows, rocky places, stabilised moraines, 3,000–4,700 m. V–VI.

Gaultheria nummularioides **D. Don** Differs from the previous species in the relatively broader, ovate to elliptic **leaves**, mostly 5–10 × 3–9 mm, bristly beneath, corolla white, pink or crimson, c. 5 mm long, and black or blue-purple **berries**, 6–7 mm at maturity. wcSichuan, seTibet, n, nw & seYunnan [swTibet, Himalaya, Java, Sumatra]; coniferous forests, often in rocky places, occasionally epiphytic, 1,000–3,400 m. VII–X.

Gaultheria sinensis **J. Anth.** Much-branched, rather twiggy procumbent **shrub** to 15 cm tall. **Leaves** rather dense, deep green, leathery, elliptic, 6–16 × 2–6 mm, apex pointed, margin slightly downturned, minutely toothed. **Flowers** solitary in axils of upper leaves; calyx-lobes ovate-triangular, 2 mm, ciliate; corolla white, urn-shaped, 4–5 mm long, with 5 very small, erect to recurved lobes. **Berries** dark blue or white at maturity, 5–8 mm in diameter nw & wSichuan, seTibet, nw & cYunnan [c & eHimalaya, nMyanmar]; coniferous forests, *Rhododendron* thickets, alpine moors, rocky meadows, large boulder and rock ledges, 3,000–4,300 m. V–VII.

1. *Lyonia ovalifolia*; Jian Chuan to Lijiang, nwYunnan (PC)
2. *Pieris formosa*; Fanjingshan, neGuizhou (PC)
3. *Pieris formosa*; Jian Chuan to Lijiang, nwYunnan (CGW)
4. *Pieris formosa*; Gurkha Himal, Nepal (CGW)
5. *Enkianthus deflexus*; Lancangjiang–Jinshajiang divide, nwYunnan (CGW)
6. *Gaultheria trichophylla*; south of Kambachen, Nepal (CGW)
7. *Gaultheria sinensis*; Cangshan, nwYunnan (PC)

1. *Gaultheria hookeri*; above Paro, Bhutan (CGW)
2. *Gaultheria fragrantissima*; Cangshan, nwYunnan (PC)
3. *Gaultheria fragrantissima*; Cangshan, nwYunnan (CGW)
4. *Cassiope pectinata*; Baimashan, nwYunnan (CGW)
5. *Cassiope pectinata*; Tianchi, nwYunnan (HJ)
6. *Cassiope selaginoides*; Huanglong, nwSichuan (CGW)
7. *Cassiope selaginoides*; Huanglong, nwSichuan (HJ)
8. *Diplarche multiflora*; Galung La, near Bomi, seTibet (HJ)

Gaultheria hookeri **C. B. Clarke** Spreading to erect **shrub** to 1 m with peeling **bark** and reddish brown-hairy **shoots**. **Leaves** leathery, short-stalked, blade elliptic to lanceolate, 3–11 × 1–4 cm, dotted and hairy beneath, glabrous above. **Flowers** in terminal and lateral racemes to 5 cm long; calyx glabrous with triangular lobes; corolla white or pink, rounded urn-shaped, 4–5 mm long, with very short lobes. **Berries** fleshy, blue-black when ripe, 5–8 mm. swGuizhou, s & swSichuan, seTibet, n, nw & seYunnan [eHimalaya]; forests and forest margins, shrubberies, rocky places, ravines, mountain summits, 1,000–3,000 m. V–VI.

Gaultheria fragrantissima **Wall.** (syn. *G. forrestii* Diels) Larger than previous species, usually 1–3 m, glabrous except for racemes, **leaves** elliptic to ovate or obovate, 5–17 × 2–6.5 cm; **racemes** 3–9 cm long, **corolla** white, never pink, rather more tubular. **Berries** hairy, bluish purple when ripe. seTibet, Yunnan [swTibet, eHimalaya, nVietnam, Sri Lanka]; similar habitats, 1,000–3,200 m. I–V.

Gaultheria wardii **C. Marq. & Airy Shaw** Readily distinguished from previous two species by densely hairy **leaves**, 3–7 × 1–2.3 cm, and by shorter **racemes** or terminal **panicles** 1–3.5 cm long; **corolla** white, small, 3–5 mm long. **Berries** blue-black when ripe, 6–8 mm, silky with hairs. seTibet, nwYunnan [neIndia, nMyanmar]; forests and forest margins, shrubberies, 1,000–2,700 m. VI–VIII.

CASSIOPE

A genus of c. 17 species (11 in China, 6 endemic), circumboreal in distribution. Dwarf **evergreen shrubs** with slender **stems**, usually branched towards base, bearing closely overlapping, scale-like, veinless **leaves** in 4 ranks. **Flowers** solitary, nodding on slender stalks, usually 5-parted; calyx with appressed lobes; corolla campanulate, white, rarely pink, with 5 short, triangular lobes; stamens not protruding, anthers 2-awned. **Fruit** a small capsule.

1. Leaves furrowed on upper surface (inside) . . . .
   . . . . **C. pectinata, C. nana, C. selaginoides**
1. Leaves not furrowed on upper surface (inside) 2
2. Leaves spreading . . . . **C. palpebrata**
2. Leaves erect, pressed closely to stem . . . .
   . . . . **C. myosuroides, C. membranifolia**

Cassiope pectinata **Stapf** Patch-forming **shrub** to 30 cm, generally lower, with erect flowering **shoots**, finely hairy and glandular. **Leaves** deep grey-green, oblong, narrow boat-shaped in side view, 5–7 mm long, base 2-lobed, apex apiculate. **Flowers** on slender, downy stalks up to 20 mm long; calyx greenish, pinkish or purple-red, lobes with membranous margin; corolla white, 6–7 mm long, glabrous. swSichuan, seTibet, nwYunnan [nMyanmar]; *Rhododendron* moorland and scrub, alpine meadows, rocky places, 3,600–4,100 m. VI–VIII.

Cassiope nana **T. Z. Hsu*** Very similar to the previous species, but not more than 10 cm, **leaves** smaller, 2–4 mm long and more elliptic, furrow above glabrous; sepals with a minutely hairy margin. nwYunnan; woodland margins, alpine meadows, moorland, 2,000–3,800 m. VI–VII.

Cassiope selaginoides **Hook. f. & Thomson** Similar to *C. pectinata*, but with more slender, deep green **stems**, **leaves** laxer, less densely overlapping, triangular-ovate, only 2–3 mm long, margin fringed with hairs, eventually becoming hairless; calyx crimson or reddish brown; corolla white, reddish within. nw, w & swSichuan, seTibet, neYunnan [Himalaya, nMyanmar]; *Rhododendron* shrubberies, alpine meadows, rocky places, cliffs, 3,000–4,200 m. V–VII.

Cassiope palpebrata **W. W. Sm.** Dwarf **shrub** not more than 6 cm tall with erect to spreading **stems**. **Leaves** spreading widely apart, elliptic to oblong, 1.5–3.5 mm long, deep green, glabrous. **Flowers** on downy stalks to 25 mm long; calyx purple; corolla white, 5–8 mm long. nwYunnan [neMyanmar]; rocky and grassy habitats, moorland, 3,000–4,300 m. V–VII.

Cassiope myosuroides **W. W. Sm.** Dwarf **shrub** not more than 7 cm with ascending to erect **stems**. **Leaves** elliptic, 2–2.5 mm long, minutely bristly inside, with a membranous margin and pointed apex. **Flowers** on slender, downy stalks 7–12 mm long; calyx dark purple, sepals with a membranous, minutely hairy margin; corolla white, often red-flushed, 5–8 mm long. nwYunnan [n & eMyanmar]; alpine meadows and slopes, rocky places, 4,000–4,500 m. V–IX.

Cassiope membranifolia **R. C. Fang*** Very similar to the previous species, but **leaves** smaller, not more than 1.8 mm long, ± glabrous on inside, apex with a long membranous sharp point, often longer than blade. nwYunnan; alpine meadows, mossy rocks, c. 3,600 m. VII–VIII.

DIPLARCHE

A genus with just two species distributed in the eastern Himalaya and south-west China. Dwarf, dense, **evergreen shrubs** with crowded, sessile, narrow **leaves** that are often gland-tipped. **Flowers** in racemes accompanied by leaf-like bracts; corolla with a short tube and 5 spreading lobes; stamens 10 (5 basal and 5 attached to corolla-tube). **Fruit** a small 5-parted capsule.

Diplarche multiflora **Hook. f. & Thomson** Small tufted plant to 15 cm, somewhat glandular-downy. **Leaves** narrowly ovate-lanceolate, 6–7 × 1–1.5 mm, margin bristly, tip pointed, shiny, gland-tipped. **Racemes** with up to 20 flowers; calyx with 5 narrow-oblong lobes; **corolla** rose-pink, 6–7 mm long, lobes rounded. seTibet, nwYunnan [Myanmar]; alpine meadows, shrubby and rocky places, cliffs, 3,500–4,100 m. VI–IX.

MONOTROPACEAE
(sometimes included in ERICACEAE)

MONOTROPA

A genus of just 2 species (both in China) widespread in the northern hemisphere and northern South America. Small **saprophytic perennial herbs** without chlorophyll, with fleshy white or reddish **stems**, usually bearing scale-like leaves. **Flowers** solitary or several in semi-nodding racemes (erect in fruit), bracteate; sepals 4–5, generally pressed to the petals; petals 3–6, oblong, forming a tube, often pouched at base; stamens 8–12. **Fruit** a small, many-seeded capsule with 5 cells.

Monotropa hypopitys L. Very variable, small yellow-brown, fleshy **herb** to 20 cm, more in fruit. **Racemes** 2–11-flowered, emerging from soil in a nodding position with some of the bracts below soil surface, the others ovate to oblong, stalkless, 7–15 mm long; **flowers** tubular-campanulate, 10–15 mm long; sepals 3–5; petals 4–6, finely downy or glabrous outside, white or yellowish, occasionally brownish yellow; style thin, with yellow stigma. Gansu, wHubei (Shennongjia), Qinghai, nSichuan, Shaanxi [widespread in n & eChina and other parts of northern hemisphere]; deep leaf-litter in deciduous and mixed forests, to 2,500 m. VII–VIII.

Monotropa uniflora L. Readily distinguished from previous species by overall white **inflorescence** bearing a solitary **flower** with a thick style, which emerges from the soil in a nodding position; petals 3–8, 12–22 mm long. Gansu, Hubei, Qinghai, Sichuan, Shaanxi, e & seTibet, Yunnan [Himalaya, Taiwan, India, Japan, Americas]; similar habitats, to 1,500 m. VIII–X.

MONOTROPASTRUM

A genus with just 2 species (both in China) in south, east and South East Asia and the Malay Archipelago. It is closely related to *Monotropa*, differing primarily in the **fruit** which is a berry, and the ovary with a single compartment.

Monotropastrum humile (D. Don) H. Hara (syn. *Monotropa humilis* D. Don; *M. uniflora* L. var. *pentapetala* Mak.; *Cheilotheca humilis* (D. Don) H. Keng) Plant to 20 cm, pale amethyst in colour. **Inflorescence** scapose, 1-flowered, **flowers** nodding, tubular-campanulate; sepals 1–5, oblong, 8–20 mm long, usually glabrous; petals 3–5, usually oblong, 10–20 mm long, slightly longer than sepals, downy outside. **Berries** erect to nodding, rounded, 9–20 mm in diameter, white. Hubei, seTibet, Yunnan [s & seChina, seAsia, Malay Archipelago]; damp deciduous and mixed forests, 100–2,500 m. IV–VIII.

PYROLACEAE
(sometimes included in ERICACEAE)

PYROLA

A genus of 35 species (26 in China, 15 endemic) scattered in north temperate regions, extending south to north Sumatra. Erect, glabrous, **evergreen**, **patch-forming herbs** with unbranched ascending **stems**, scaly at base. **Leaves** long-stalked, in basal rosettes, margin entire or toothed. **Flowers** nodding, many, in slender racemes borne on scapes that are scaly at base; calyx 5-lobed, persistent; petals 5, free; stamens glabrous, anthers opening by pores; style protruding, usually curved and dilated towards tip. **Fruit** a capsule, nodding, valves connected by fibres at the valve-margins.

1. Flowers 15–25 mm in diameter 2
1. Flowers small, not more than 10 mm in diameter .... *P. szechuanica, P. sororia*
2. Petals white, yellowish or yellowish green .... *P. rotundifolia, P. calliantha*
2. Petals greenish yellow .... *P. forrestiana*

Pyrola rotundifolia L. Plant 15–30 cm tall, with ribbed **stems** bearing 1 or 2 sheathing, scale-like, brownish bracts. **Leaves** 4–7, leathery, orbicular to ovate, rounded, to 6 cm long, margin slightly scalloped or nearly entire; stalks twice as long as blades. **Raceme** 8–15-flowered, 6–16 cm long. **Flowers** 15–20 mm in diameter, out-facing or slightly nodding, opening widely, scented of lily-of-the-valley (*Convallaria*); sepals ovate-lanceolate to lanceolate, rounded, 3.5–5.5 mm long, reflexed at tip; petals incurved, pure white or occasionally pinkish, orbicular-ovate, 6.5–10 mm long, rather thick. Gansu, w, nw & swSichuan, seTibet, w & nwYunnan [swTibet, c & nChina, Eurasia]; coniferous forests, often in deep shade, thickets, grassy and rocky slopes, 1,400–3,200 m. VI–VII.

Pyrola calliantha Andres* (syn. *P. rotundifolia* L. var. *chinensis* Andres) Similar to the previous species, but **leaves** elliptic to ovate, blade purplish and often glaucous beneath, **flowers** with strap-like sepals that are 5–7.5 mm long, petals obovate. Gansu, Guizhou, Hubei, Qinghai, Sichuan, seTibet, Yunnan [swTibet, c & eChina]; mixed and coniferous forests, 700–4,100 m. VI–VIII.

Pyrola forrestiana Andres* Plant 20–27 cm tall, with basal rosettes of 3–7, thick-leathery, rather rough **leaves**, blade triangular-ovate, obovate, or almost orbicular, 2.5–4.5 cm long, blunt, margin conspicuously scalloped. **Racemes** 9–12-flowered, 6–9 cm long. **Flowers** nodding, saucer-shaped, 10–15 mm in diameter, on stalks 5–10 mm long; sepals triangular or triangular-ovate, 1.5–2.5 mm long, apex blunt; petals greenish yellow, ovate or almost orbicular, 7–9 mm long. Hubei, Sichuan, eTibet, Yunnan [Hunan]; shady places in damp mixed forests, 1,500–3,800 m. VII–VIII.

Pyrola szechuanica **Andres*** Plant 15–25 cm tall, rosettes with 3–7, thinly leathery broadly ovate to broadly oblong **leaves**, 4–6 cm long, clustered at base of **stem**, apex blunt, margin with remote teeth; leaf-stalks 6–7.5 cm long. **Raceme** 6–12-flowered, 4–6 cm long. **Flowers** ± nodding, opening widely, saucer-shaped, 10 mm in diameter, on stalks up to 10 mm long; sepals strap-like, 4–6 mm long, apex with sudden small point; petals yellow or yellowish white, broadly ovate, rounded, 6–10 mm long. wSichuan; evergreen and mixed forests, 1,400–2,700 m. VI–VII.

Pyrola sororia **Andres*** Similar to the previous species, but **leaves** 6–8, ovate to almost circular, **flowers** smaller, generally 7–9 mm in diameter, on stalks 3–4 mm long; petals white or yellowish green. seTibet, w & nwYunnan; similar habitats, 2,700–3,900 m. VII–VIII.

1. *Monotropa hypopitys*; eNepal (CGW)
2. *Monotropa uniflora*; cTaiwan (PC)
3. *Monotropastrum humile*; Laojunshan, seYunnan (PC)
4. *Pyrola rotundifolia*; Galung La, near Bomi, seTibet (HJ)
5. *Pyrola rotundifolia*; wSichuan (HJ)
6. *Pyrola calliantha*; Jiuzhaigou, nwSichuan (CGW)
7. *Pyrola sororia*; Cangshan, nwYunnan (DC)

1. *Diapensia purpurea*; Baimashan, nwYunnan (CGW)
2. *Diapensia purpurea*; Baimashan, nwYunnan (HJ)
3. *Diapensia purpurea* (white-flowered form); Baimashan, nwYunnan (HJ)
4. *Diapensia himalaica*; Galung La, near Bomi, seTibet (HJ)
5. *Diapensia himalaica* (yellow-flowered form); Doshong La, seTibet (AC)
6. *Berneuxia thibetica*; cult., from Sichuan (RR)
7. *Diospyros kaki*; cult. (CGW)
8. *Diospyros lotus*; cult. (CGW)
9. *Diospyros lotus*; Shennongjia, wHubei (PC)

MONESES

A single species widely distributed in Europe, Asia and North America. A small, glabrous, stoloniferous, **evergreen**, **perennial herb** with stalked, leathery **leaves**. **Flowers** solitary, nodding, terminal on a long scape; sepals 5, deeply lobed, persistent; petals 5 spreading, white or pink, orbicular; stamens 10; ovary globose or almost so, with a straight style.

Moneses uniflora (L.) A. Gray (syn. *Pyrola uniflora* L.; *Moneses grandiflora* Salisb.; *M. rhombifolia* (Hayata) Andres) **Patch-forming** plant to 10 cm, with **leaves** in 3 or 4 whorls, each with 2 or 3 leaves, these alternating with 5 scales; blade shiny, dark green, broadly rhombic, rounded to blunt, 10–16 mm long; stalks to 15 mm long. **Flowers** nodding, regular, 13–22 mm in diameter, opening fully; petals white, orbicular to broadly ovate, stalkless. Gansu, n & wSichuan, nwYunnan [nChina, widespread in cool northern hemisphere]; damp mossy coniferous forests, more rarely in wet *Betula* or *Pinus* woods, 1,000–2,500 m. VI–VII.

DIAPENSIACEAE

DIAPENSIA

A genus of 4 species (3 in China) in the Himalaya and west China, and circumboreal. Low or prostrate **shrublets**, often patch- or mat-forming, with crowded leathery, entire **leaves**. **Flowers** solitary, almost stalkless to short-stalked; calyx 5-lobed, with several tiny bracts at base; corolla campanulate to cupped, with 5 rounded lobes fused near middle; stamens 5; style solitary, capitate. **Fruit** a small 3-parted capsule, surrounded by persistent calyx.

Diapensia purpurea Diels **Hummock-** or **mat-forming** plant up to 80 cm across but no more than 10 cm tall, with dull, deep green, minutely hairy, oblong to elliptic-spathulate **leaves**, 3–5 × 1.5–3 mm, without stomata on upper surface, apex rounded and slightly downcurved. **Flowers** almost stalkless; sepals spathulate to oblong; corolla pinkish red to pinkish purple, 14–18 mm across, with a broad short tube 2–6 mm long. w & swSichuan, seTibet, nwYunnan [nMyanmar]; bare rocky places, stony banks, low heath, stabilised moraines, cliff-ledges, 2,600–4,500 m. V–VII. Plants in nwYunnan (especially Cangshan) with yellowish or white **flowers** have been distinguished in the past as *D. bulleyana* Forrest ex Diels, but are barely distinguishable on other criteria.

Diapensia himalaica Hook. f. & Thomson Very similar to the previous species, but **leaves** generally narrower, 1–2.5 mm wide, deep shiny green, not minutely hairy, with stomata on upper surface; calyx purplish red, **corolla** reddish pink, rose-purple, rarely white or yellow, with a slightly longer (4–8 mm) tube. seTibet, nwYunnan [c & eHimalaya]; similar habitats as well as damp rock faces, 3,200–5,000 m. V–VII.

Diapensia wardii W. E. Evans* Distinguished from previous two species by the larger stature to 15 cm tall, larger ovate to elliptic **leaves**, 6–10 × 4–7 mm, and **flowers** borne on a short yet distinct scape 10–20 mm long; corolla purplish red. seTibet; rock-faces beneath shrubby rhododendrons, 3,200–3,400 m. VI–VII.

BERNEUXIA

A genus with 1 species endemic to China. Flowers in capitate **racemes**; leaf-stalks with a pair of bracteoles in the middle; **flowers** 5-parted, with persistent sepals; corolla soon falling with stamens inserted at base; **fruit** a small 3-parted capsule.

Berneuxia thibetica Decne* Tufted, **rhizomatous herb** to 25 cm. **Leaves** basal, leathery, obovate to elliptic-obovate, 3–10 × 1.7–4 cm, with entire margin, deep glossy green above, pale and greyish beneath. **Inflorescence** with up to 12 flowers borne on a stalk that elongates as fruit develop; calyx reddish or greenish, sepals 4–5 mm long; **corolla** white or flushed with rose-pink, campanulate, 9–10 mm long. nwGuizhou, Sichuan, seTibet, nwYunnan; wet broadleaved and coniferous (mainly *Abies*) forests, damp shrubberies, 1,700–3,500 m. IV–VI.

EBENACEAE

DIOSPYROS

A genus with c. 485 species (60 in China), primarily pantropical but with a few in temperate regions. **Dioecious**, **evergreen** or **deciduous shrubs** or **trees** without terminal buds. **Leaves** alternate. **Flowers** lateral, male in small clusters, female usually solitary; calyx 3–5-lobed; corolla urn-shaped, 3–5-lobed. **Fruit** a small to substantial berry, often leathery but fleshy within.

Diospyros kaki Thunb. Sizeable **deciduous tree** to 27 m, with hairy to glabrous branchlets and small black winter **buds**. **Leaves** stalked, blade ovate to elliptic or lanceolate, 5–18 × 2.5–9 cm, glossy deep green. **Flowers** lateral; male in clusters of 3–5, with a hairy calyx equalling the 4-lobed white, yellowish or red, 6–10 mm corolla, and with 16–24 stamens; female flowers solitary with a large 3–5 cm calyx and smaller 4-lobed, yellowish or white corolla with recurved lobes, and 8 or 16 staminodes. **Fruit** globose, often somewhat flattened above, 2–9 cm, orange or yellow when ripe. Gansu, Guangxi, Guizhou, Hubei, Sichuan, Yunnan [widespread in China, Taiwan]; open woodland, slopes, widely cultivated both in China and elsewhere, and naturalised, to 2,200 m. V–VI. Many cultivated forms exist.

Diospyros lotus L. Similar to the previous species, but **leaves** often rather smaller, **male flowers** with urn-shaped, red or pale yellow corollas, c. 4 mm, **female flowers** with corolla green or reddish, a similar shape but larger, 4–5-lobed. **Fruit** smaller, 1–2 cm, yellow at first but turning bluish black, with a glaucous bloom. Gansu, Guizhou, Hubei, Sichuan, Shaanxi, seTibet, Yunnan [widespread in China, c & swAsia, sEurope]; forests, shrubberies, valley slopes, 500–2,500 m. V–VI.

STYRACACEAE

STYRAX

A genus of 130 species (31 in China) distributed in the Mediterranean region, the Americas and east Asia. **Deciduous** or **evergreen trees** and **shrubs**, generally with alternate **leaves**. **Flowers** in lateral or terminal racemes or panicles, sometimes solitary, generally pendulous; calyx cup-shaped, usually 5-toothed; corolla campanulate with a short tube and 5–7 spreading lobes, usually white; stamens (8–)10(–13). **Fruit** 3-valved, but indehiscent.

1. Margin of corolla-lobes narrow, inrolled or valvate . . . . **S. formosanus**
1. Margin of corolla-lobes not inrolled, overlapping . . . . **S. grandiflorus, S. limprichtii, S. wilsonii**

Styrax formosanus Matsum. **Deciduous shrub** to 3 m with brownish, downy branchlets and alternate elliptic to obovate **leaves**, mostly 2–5 × 1.5–2.5 cm, with toothed margin and 3–5 pairs of lateral veins; leaf-stalk 3–4 mm long. **Flowers** in racemes of 3–5, each with a short stalk 8–12 mm long; calyx obscurely 5-toothed; corolla 17–24 mm across, with a short tube 2–3 mm long; stamens shorter than corolla. Guangxi [e & sChina, Taiwan]; thickets, shrubberies, ravines, 500–1,300 m. IV–VII.

Styrax grandiflorus Griff. (syn. *S. cavaleriei* H. Lévl; *S. duclouxii* Perkins) **Deciduous shrub** or small **tree** to 7 m with grey bark. **Leaves** alternate, oblong to almost elliptic, 3–7 × 2–4 cm, with 5–7 pairs of lateral veins and finely toothed margin, glabrous except on veins beneath; leaf-stalk 3–7 mm long. **Flowers** in lax clusters of up to 9, each pendent on a downy stalk 25–50 mm long; calyx obscurely 5-lobed, downy; corolla 25–35 mm across with a short tube 3–5 mm long; stamens usually 10; style equalling corolla-lobes or somewhat shorter. Guangxi, Guizhou, seTibet, Yunnan [swTibet, Himalaya, Japan, Myanmar, Taiwan]; forests, thickets, shrubberies, ravines, 1,000–2,100 m. IV–VI.

Styrax limprichtii Lingelsh. & Borza* **Deciduous shrub** to 2.5 m with grey-stellate-hairy branchlets and alternate **leaves**, blade papery, elliptic to obovate, 35–70 × 20–45 mm, margin finely toothed towards pointed to somewhat long-pointed apex, densely stellate-downy beneath; leaf-stalk very short. **Flowers** 13–15 mm across, in 3–6-flowered terminal racemes to 4 cm long; calyx furry with yellow-brown stellate hairs; corolla white, campanulate, with a tube 4 mm long and elliptic lobes 9–11 mm long. swSichuan, nwYunnan; forests and forest margins, 1,700–2,400 m. III–IV.

Styrax wilsonii Rehd.* Differs from the previous species in the smaller leaf-blades, usually 10–25 × 7–20 mm, with a 2–4-lobed or coarsely toothed margin, grey-downy beneath; racemes shorter, corolla-lobes with yellowish stellate down. wSichuan (Baoxing area mainly); open forests, shrubberies, 1,300–1,700 m. IV–VII.

PTEROSTYRAX

A genus with 4 species (2 in China) in east Asia. Readily distinguished from *Styrax* by large drooping panicles of small **flowers** with an inferior ovary and protruding stamens.

Pterostyrax psilophyllus Diels ex Perkins* (syn. *P. cavaleriei* Guill.) Small **tree** to 15 m with stellate-hairy branchlets and leaves; **leaves** elliptic to oblong-obovate, 5–15 × 5–9 cm, pointed to long-pointed, sometimes 3-lobed, margin toothed. **Inflorescences** terminal and lateral, to 15 cm long, **flowers** 12–14 mm long; calyx campanulate; petals white, oblong, pointed. **Fruit** spindle-shaped, 5–10-ribbed, yellow-grey-downy. wGuangxi, Guizhou, swHubei, eSichuan, neYunnan; damp forests, ravines, 600–2,500 m. IV–V. Formerly included in *P. hispida* Sieb. & Zucc., which is now regarded as being confined to Japan.

1. *Styrax formosanus*; Tian'e, nwGuangxi (PC)
2. *Styrax grandiflorus*; north-west of Tengchong, wYunnan (PC)
3. *Pterostyrax psilophyllus*; cult. (TK)

PRIMULACEAE

PRIMULA

A genus with c. 500 species (c. 300 in China, many endemic), scattered right across the temperate northern hemisphere, but with outliers in Ethiopia, Java and Sumatra, as well as in South America. **Tufted**, occasionally **mat-forming** plants, with or without meal (farina). **Leaves** in basal rosettes. **Flowers** normally heterostylous with pin-eyed (long-styled) and thrum-eyed (short-styled) individuals, solitary or in umbels, superposed whorls, heads or spikes; calyx tubular to campanulate, 5-toothed; corolla salverform with a long tube (not restricted at throat), a flat to cupped, 5-lobed limb, and sometimes with an annulus; stamens inserted in throat or on middle of tube. **Fruit** a 5-parted capsule.

The genus is divided into subgenera, sections and subsections: the outline of those set out below follows John Richards's *Primula* monograph (Richards, 2002). *Primula* is an important horticultural genus with many hybrids: many species are grown in western gardens for their ornamental value.

SUBGENUS AURICULASTRUM

SECTION *Amethystina* **Deciduous** with involute vernation and with rather fleshy glabrous **leaves**, lacking meal; leaf-stalks poorly differentiated. **Inflorescence** a simple, few-flowered umbel of usually pendent **flowers**; calyx campanulate, short; corolla narrowly campanulate, rarely annulate; capsule equalling calyx. 8 species, all but 1 in China. (p. 239)

SUBGENUS AUGANTHUS

SECTION *Monocarpicae* **Annuals** or **biennials**, often winter- flowering, or **monocarpic perennials** with thin, hairy and often glandular **leaves**, usually forming a lax basal rosette, leaf-blades broad, lobed and toothed; vernation revolute. Scapes, bracts and calyx often mealy. **Inflorescence** with up to 7 whorls of **flowers**, one above the other, sometimes in spikes; calyx narrow, ribbed, and joined to fruit; corolla salverform, annulus absent. **Capsule** 5-valved. 13 species, all but 1 in China. (p. 240)

SECTION *Obconicolisteri* Differs from the previous section in being **perennial**, often stoloniferous and lacking meal, with **flowers** in a simple umbel, calyx cup-shaped, scarcely lobed, corolla with or without annulus, **capsule** without valves, eventually crumbling. 16 species, 12 in China. (p. 240)

SECTION *Malvacea* **Herbaceous**, softly hairy **perennials** without meal, readily distinguished from the other sections in the subgenus by the large open calyx that forms a distinctive disc at the fruiting stage; corolla with an annulus. 5 species, all found in swChina. (p. 240)

SECTION *Cortusoides* Usually **deciduous** species without meal, with soft-hairy, often rather pleated, rounded to oval, often heart-shaped **leaves**, with pinnate veins (subsection *Cortusoides*) or palmate veins (subsection *Geranioides*). Scapes usually exceeding leaves, bearing an umbel or rarely several whorls of **flowers** one above the other; calyx narrow with erect lobes; corolla salverform with a narrow tube, often with a yellow eye. 23 species, 12 in China. (p.240)

SECTION *Bullatae* **Evergreen perennials**, often rather woody at base. **Leaves** leathery, wrinkled, margin finely toothed, back-rolled, sticky-glandular and with a sweet, sickly fragrance. **Flowers** in umbels or sometimes solitary, on sticky stalks; corolla salverform, sometimes rather asymmetrical, annulus absent. 7 species, all in China. (p. 243)

SECTION *Dryadifolia* Differs from the other sections listed above in being long-lived, **mat-forming**, **evergreen perennials**. **Leaves** small, leathery glabrous, with back-rolled margin, with or without meal, persisting once dead. **Inflorescence** 1–3-flowered, calyx relatively large and loose, corolla usually with annulus. 4 species, all in China. (p. 243)

SUBGENUS ALEURITIA

SECTION *Petiolares* **Perennials** with persistent bud-scales and rosettes of rather narrow, often oval **leaves**, margin finely toothed, back-rolled (**evergreen** in subsection *Petiolares*, **deciduous** in subsections *Edgeworthii* and *Sonchifolia*), often glandular, but glabrous. **Flowers** in large posy-like clusters, with or without a short scape; corolla salverform, usually with annulus (in subsections *Petiolares* and *Sonchifolia*). **Capsule** without valves, crumbling at maturity. 31 species, 22 in China. (p. 243)

SECTION *Davidii* Differs primarily from section *Petiolares* in bearing multicellular hairs on the **leaves** and in the presence of a well-developed, usually hairy scape; corolla usually somewhat funnel-shaped. 17 species, all in China. (p. 244)

SECTION *Crystallophlomis* (syn. section *Nivales*) **Deciduous** plants with overwintering resting buds sitting at or below ground level. **Leaves** with back-rolled margins and sheath-like stalks, usually mealy but glabrous. **Flowers** in scapose inflorescences, umbellate or with several whorls one above the other; corolla-lobes usually entire. **Capsule** often exceeding calyx and with apical teeth. 46 species, 34 in China. (p. 244)

Subsection *Crystallophlomis* Plants with generally small resting buds below ground surface, and corollas flat-faced, often with annulus; **capsule** exceeding calyx. 28 species, 19 in China. (p. 244)

1. A primula meadow on the Yunnan–Sichuan border north of Daxueshan, with *P. secundiflora* and *P. sikkimensis* intermingled (CGW)
2. Primulas can form extensive colonies in favoured places, such as here, *P. sikkimensis* near Litang in wSichuan (AD)
3. *Primula secundiflora* and *P. sikkimensis* in a meadow at Zhongdian, nwYunnan (CGW)
4. Meadows in nw Sichuan, alive with a mass of *Primula nutans* (TK)
5. *Primula amethystina*; Tianchi, nwYunnan (CGW)
6. *Primula amethystina*; Tianchi, nwYunnan (CGW)

Subsection _Calliantha_ Plants with large, egg-shaped resting **buds** sitting on the ground and mostly funnel-shaped **corollas**, not annulate; **capsule** not exceeding calyx. 5 species, 3 in China. (p. 247)

Subsection _Agleniana_ Very similar to subsection _Calliantha_ but **corolla** usually yellow and lacking an annulus; the **capsules** dehisce by a membrane (not valvate as in other subsections). 3 species, all in China centred on seTibet. (p. 247)

Subsection _Maximowiczii_ Plants with large resting **buds** at ground level, lacking meal; **corolla** with characteristic narrow reflexed lobes; **capsule** exceeding calyx. 7 species, all in China. (p. 247)

SECTION _Proliferae_ Primarily robust **perennials** with **deciduous** or **evergreen** rosettes of scarcely stalked **leaves**, glabrous or glandular but not hairy, margin back-rolled; resting **buds** small, inconspicuous. **Flowers** often in several tiers or whorls (candelabrum-like), often mealy in upper part; corolla salverform or campanulate. **Capsules** rounded. 19 species, 14 in China. (p. 248)

SECTION _Sikkimensis_ Deciduous **perennials** with broad rosettes of non-hairy, non-mealy **leaves** often with a heart-shaped base. **Scapes** exceeding leaves, bearing a single whorl of pendent **flowers**; corolla campanulate, with mealy face, especially around the eye, and with entire lobes; annulus absent. 9 species, 5 in China. (p. 251)

SECTION _Armerina_ Rather delicate, small to medium-sized species with lax rosettes or tufts of spathulate, glabrous **leaves**, generally lacking meal except on scapes and sometimes calyces. **Scapes** slender, usually exceeding leaves, bracts small, pouched at base; **flowers** usually relatively few in a simple umbel; corolla variable in colour, with a white or yellow eye, tube usually much exceeding calyx, as does the 5-valved capsule. 17 species, about half in China. (p. 251)

SECTION _Yunnanensis_ Dwarf, generally short-lived **perennials**, with glabrous **leaves** that are densely mealy, at least beneath. Bracts relatively large and leaf-like. **Flowers** in a small 1-sided cluster; corolla lilac, rose-pink or purple, usually without an annulus, lobes often deeply 2-fid. 15 species, 13 in China. (p. 252)

SECTION _Pulchella_ Leaves narrow-elliptic to narrow-oblanceolate, shiny deep green above but thickly mealy beneath, the margins downturned. **Flowers** few on long stalks, corolla often dark blue or purple, sometimes pink, usually without an annulus. 15 species, 11 in China. (p. 252)

SECTION _Minutissimae_ Dwarf mat- or patch-forming species overwintering as small resting buds at ground level, hairy only in corolla-mouth, with very small, usually toothed leaves that are often mealy. **Bracts** not pouched at base. **Flowers** few, with or without a short scape; corolla in shades of blue, violet, rose-pink or lavender, mouth (in the 2 species included here) filled with a tuft of hairs, lobes 2-fid usually with annulus. 21 species, 19 in China, predominantly in s & seTibet. (p. 255)

SECTION _Denticulata_ Plants often **clump-forming**. **Leaves** deciduous with large persistent bud-scales at base outside, and sparsely downy, mealy beneath when young. **Flowers** in spherical heads on long scapes. All flowers fertile, open; calyx mealy; corolla often lilac or lavender. 13 species, 9 in China. (p. 255)

SECTION _Capitatae_ Similar to section _Denticulata_, but rosettes lacking basal bud-scales and the central **flowers** in the disc-like head sterile and failing to open. 2 species, both in China. (p. 255)

SECTION _Muscarioides_ Plants **deciduous**, overwintering by small **buds** at or below the ground. **Leaves** in small lax rosettes, usually without meal but with multicellular hairs. **Flowers** fragrant, pendent or semi-pendent, borne in spikes or heads on scapes that usually exceed leaves, often mealy in upper part; corolla with a narrow tube (longer than the unequally lobed, campanulate calyx), with spreading, often notched lobes. 19 species, 16 in China. (p. 255)

SECTION _Soldanelloides_ Mat-**forming** or **tufted** species, overwintering as tiny dormant buds. Close to section _Muscarioides_, but **scapes** usually rather short and few-flowered; **corolla** usually campanulate with a relatively short tube and erect to suberect lobes. **Capsule** subglobose, included in calyx. 20 species, 9 in China. (p. 256)

SUBGENUS AURICULASTRUM

SECTION _Amethystina_

Primula amethystina **Franch.*** Small **tufted** plant not more than 15 cm with obovate, **deciduous leaves** without meal, with horny, somewhat toothed margin. **Scapes** exceeding leaves, bearing up to 20 **flowers** in an umbel; corolla campanulate, 12–15 mm long, deep purple-blue to violet-blue or deep blue, very rarely white, lobes generally fringed or with short teeth. w & swSichuan, seTibet, nw & wYunnan; wet meadows, dwarf _Rhododendron_ scrub, edges of thickets, sometimes abundant, 3,500–5,000 m. V–VII. The most widespread and robust variety is **var. _brevifolia_ (Forrest) W. W. Sm. & Fletcher***; **var. _amethystina_** is confined to the Cangshan in nwYunnan and **var. _argutidens_ (Franch.) W. W. Sm. & Fletcher***, a smaller plant with deeply toothed **leaves**, is restricted to swSichuan.

Primula silaensis Petitm. Similar to the preceding species, but a more delicate plant with broadly oval to rounded **leaves** with tiny marginal teeth and smaller **flowers**, 8–10 mm long, that are pale violet-blue, deeper in centre, and with a less crimped corolla. seTibet, w & nwYunnan [neIndia, nMyanmar]; wet stony slopes, mossy meadows, 4,000–4,800 m. V–VI.

SUBGENUS AUGANTHUS

SECTION *Monocarpicae*

Primula malacoides Franch. Very variable hairy **annual**, generally with mealy scapes and **flowers**, to 40 cm tall in flower. **Leaves** pale to mid-green, rounded to oval or heart-shaped, with toothed and lobed margin and long stalks. **Flowers** in up to 6 whorls (each with c. 20 flowers) one above the other, lowermost whorl shorter than leaves; corolla salverform, the flat limb to 20 mm across, rose-pink, lavender, pink, purple or white, with notched lobes, scented. wGuizhou, Yunnan [eMyanmar]; primarily a weed of arable fields and field boundaries, ditches, 1,500–2,100 m. III–VI.

Primula forbesii Franch. Similar to the preceding species, but a smaller more compact plant with less deeply lobed, short-stalked **leaves**; lowermost whorl of flowers clear of the leaves. sSichuan, Yunnan [ne & eMyanmar]; marshy and damp places, margins of irrigation, rice paddies, 2,000–3,300 m. II–VI.

Primula duclouxii Petitm.* Closely related to *P. malacoides*, but readily distinguished by the small stature, plant not more than 6 cm tall, **flowers** small with corolla only 10 mm across, and also by the strongly deflexed **fruit**. cYunnan (Xishan, by Kunming); shady limestone cliff-crevices, c. 2,250 m. I–III.

SECTION *Obconicolisteri*

Primula obconica Hance Tufted **evergreen**, hairy, somewhat sticky **perennial** to 30 cm, often less, with heart-shaped, rather dull green, slightly toothed **leaves**. **Inflorescence** with up to 12 flowers in a simple umbel; corolla salverform, 18–25 mm across, pink, rose-pink, lavender or white. Guizhou, Hubei, wSichuan, seTibet, Yunnan [nThailand]; shrubberies, streamsides, beneath trees, generally in leafy soil, 1,800–3,200 m. IV–VI.

SECTION *Malvacea*

Primula malvacea Franch.* Variable **tufted** hairy species to 40 cm in flower, with soft, somewhat fleshy and wrinkled, kidney-shaped to ovate, long-stalked **leaves** to 30 cm long, generally far less, deeply heart-shaped at base, margin toothed, lobed or not. **Inflorescence** erect, sticky-hairy, with several whorls or pseudowhorls of flowers; calyx campanulate, 8–15 mm long, the ovate lobes entire to toothed; **corolla** pale to dark rose-pink, pink or purple, occasionally white (**var.** *alba* **Forrest**),

tube equalling or longer than calyx, limb 15–25 mm across with notched lobes. swSichuan, nwYunnan; open woods, dry calcareous or slaty rocks, 2,000–3,700 m. V–VII.

SECTION *Cortusoides*

1. Leaves pinnately veined; calyx 8–12 mm long . . . . ***P. polyneura***
1. Leaves palmately veined; calyx not more than 7 mm long 2
2. Tall plants with flowers held well above leaves . . . . ***P. heucherifolia, P. septemloba***
2. Dwarf plants with flowers held at leaf height 3
3. Leaf-stalks more than 10 mm long; leaf-blades thin, shallowly lobed . . . . ***P. alsophila***
3. Leaf-stalks less than 10 mm long; leaf-blades relatively thick, deeply lobed . . . . ***P. palmata, P. latisecta***

Primula polyneura Franch.* Tufted **perennial** to 50 cm in flower, generally far less, with rather thick **leaves**, with 7–11 shallow, rather angular, toothed lobes; leaves often with a dense whitish or reddish down beneath. **Flowers** usually in a simple umbel of up to 12; corolla to 25 mm across, pale pink to purple. seGansu, nw, w & sSichuan, eTibet, nwYunnan; damp shady woodland (generally coniferous), glades, banks, streamsides, 2,300–4,300 m. IV–VI.

Primula heucherifolia Franch.* **Rhizomatous**, shortly spreading white- or tawny-hairy plant. **Leaves** soft green, blade rounded-heart-shaped with up to 11 rounded, blunt-toothed lobes. **Flowers** few (rarely more than 8) in a 1-sided umbel on a scape to 30 cm; corolla 15–25 mm across, deep purple to lilac-mauve, lobes slightly notched. w & wcSichuan, nwYunnan; shady places in damp leafy soils, particularly amongst rocks and tree-roots, 2,000–3,200 m. IV–VI.

Primula septemloba Franch.* Taller plant than previous species, to 50 cm in flower; **inflorescences** with several whorls of flowers one above the other; corolla funnel-shaped, rosy purple, 14–17 mm across. swSichuan, nwYunnan; damp shady forests, gullies, bamboo thickets, 3,000–4,300 m. V–VII.

Primula alsophila Balf. f. & Farrer* **Creeping**, **patch-forming** plant not more than 8 cm tall in flower, with runners. **Leaves** rather bright green, thin, parasol-like, usually with 5–7 shallow lobes, flower-stalks slender. **Flowers** 2–3 on a thin scape above leaves; corolla pink to purple with a pale eye, 13–17 mm across. sGansu, (?)seQinghai, nSichuan, eTibet; dense, damp shady forests, often in mossy places, 3,300–4,000 m. V–VII.

Primula palmata Hand.-Mazz.* **Patch-forming** plant rarely more than 12 cm tall, with rounded **leaves** divided for three-quarters into narrow, sharply pointed lobes. **Flowers** 1–3 on a scape, borne at leaf height; corolla 13–17 mm, rose-pink to pinkish purple. w & nwSichuan; mixed and deciduous woodland, leafy or mossy soils, 2,300–2,700 m. V–VI.

1. *Primula malacoides*; cult. (JR)
2. *Primula duclouxii*; Xishan, Kunming, cYunnan (PC)
3. *Primula obconica*; above Shigou, nwYunnan (PC)
4. *Primula malvacea*; cult. (CGW)
5. *Primula polyneura*; Baimashan, nwYunnan (PC)
6. *Primula heucherifolia*; Sandauwan, Yulongxueshan, nwYunnan (PC)
7. *Primula septemloba*; road to Geza, nwYunnan (CGW)
8. *Primula alsophila*; Jiuzhaigou, nwSichuan (CGW)
9. *Primula palmata*; Jiuzhaigou, nwSichuan (PC)

1. *Primula bracteata*; near Geza, nwYunnan (HJ)
2. *Primula bracteata*; cult. (CGW)
3. *Primula forrestii*; Gangheba, Yulongxueshan, nwYunnan (CGW)
4. *Primula forrestii*; Gangheba, Yulongxueshan, nwYunnan (CGW)
5. *Primula forrestii* (pale form); Gangheba, Yulongxueshan, nwYunnan (CGW)
6. *Primula dryadifolia*; above Wutoudi, Yulongxueshan, nwYunnan (JM)
7. *Primula dryadifolia*; Baimashan, nwYunnan (HJ)
8. *Primula dryadifolia* subsp. *congestifolia*; Galung La, near Bomi, seTibet (HJ)
9. *Primula gracilipes*; below Rupina La, Nepal (CGW)
10. *Primula sonchifolia*; Tianchi, nwYunnan (CGW)
11. *Primula sonchifolia*; Napahai, nwYunnan (CGW)
12. *Primula moupinensis*; cult. (JR)

Primula latisecta **W. W. Sm.*** Similar to the previous species, but leaves divided for only two-thirds into broader, less pointed lobes, the leaf-stalks with dense pale hairs. seTibet; dense forests, in moist leafy soils, 3,500–4,000 m. V–VI.

SECTION *Bullatae*

1. Flowers solitary or several, clustered on a very short scape amongst the leaves . . . . . . *P. bracteata, P. rockii*
1. Flowers 5 or more, clustered on a distinct scape at or above leaf height . . . *P. forrestii, P. bullata, P. redolens*

Primula bracteata **Franch.*** Low cushion-forming plant without meal but sticky with glands, **leaves** elliptic, pointed, with back-rolled margins. **Flowers** solitary on short stalks amongst the leaves; corolla 15–20 mm across, pink, rose-pink or white with a darker centre surrounding a yellow eye, lobes deeply notched. swSichuan, nwYunnan; limestone and sandstone cliff-crevices and -ledges, 2,600–3,700 m. IV–VI.

Primula rockii **W. W. Sm.*** Similar to the previous species, but with **leaves** more spathulate in shape and yellow or orange **flowers**, sometimes up to 5 on a very short scape. swSichuan (Kulu and Muli regions); similar habitats, 3,000–4,425 m. V–VI.

Primula forrestii **Balf. f.*** Rather robust **subshrubby** plant forming long woody **stems** over time, to 25 cm tall in flower. **Leaves** deep olive-green, hairy (especially when young), often yellow-mealy when young, especially beneath, oval with heart-shaped base and finely toothed margin. **Flowers** up to 25 in a simple, somewhat 1-sided umbel, sweetly scented, on long-hairy, sticky scapes; corolla golden yellow with an orange eye, more rarely cream, lobes deeply notched. nwYunnan (particularly Yulongxueshan); dry places amongst limestone rocks and ledges, 2,899–3,200 m. IV–early VI. Smaller plants referable to *P. bullata* **Franch.*** are found north of Dali (Heechanmen) and have **leaves** that are glabrous or short-hairy above, the blade with an attenuated base.

Primula redolens **Balf. f. & Kingdon-Ward*** (syn. *P. forrestii* Balf. f. var. *redolens* (Balf. f. & Kingdon-Ward) A. J. Richards) Similar to *P. forrestii*, but with pink or lilac **flowers** with a yellow eye. nwYunnan (near Weixi, Lancangjiang Gorge); 3,000–3,200 m. IV–V.

SECTION *Dryadifolia*

Primula dryadifolia **Franch.** Low **evergreen**, **patch-** or **mat-forming**, variable plant with prostrate, somewhat woody **stems**. **Leaves** leathery, deep shiny green above, margin inrolled, with white or yellow meal beneath. **Flowers** up to 5 on scapes to 10 cm, semi-nodding; corolla 18–25 m across, deep pink to rose-crimson with a darker, occasionally yellowish, eye. w & swSichuan, extreme seTibet, n & nwYunnan [nBhutan]; stony meadows, screes, moraines, cliff-ledges, amongst boulders, sometimes carpeting the ground, 3,000–5,500 m. V–VII. **Subsp. congestifolia (Forrest) W. W. Sm. & Forrest*** differs from the typical plant in its more carpeting habit, smaller **leaves** and solitary, up-facing **flowers**. seTibet, nwYunnan.

Primula jonardunii **W. W. Sm.*** Differs from the previous species in being altogether dwarfer with solitary **flowers** held low amongst the **leaves** on very short **stems**; corolla hairy within. seTibet; wet stony meadows, fringes of low shrubberies, 3,900–4,700 m. V–VII.

SUBGENUS ALEURITIA

SECTION *Petiolares*

1. Plants evergreen . . . . *P. gracilipes*
1. Plants deciduous, with resting buds 2
2. Corolla relatively large, limb 22 mm or more across, with star-shaped orange eye . . . . *P. sonchifolia, P. moupinensis, P. hoffmanniana, P. tanneri, P. tanneri* var. *porrecta*
2. Corolla relatively small, limb not more than 12 mm across . . . . *P. hookeri*

Primula gracilipes **Craib** Low **tufted** plants with rather dense, deep green rosettes; **leaves** obovate, mealy when young and with reddish stalks, margin with irregular sharp teeth. Scape absent; **flowers** up to 30, crowded; corolla with flat limb 22–30 mm across, pink with a yellow or greenish eye surrounded by a white zone. seTibet (c & eHimalaya); wet forests, mossy banks, 2,500–4,200 m. IV–VI.

Primula sonchifolia **Franch.** Plants with large egg-shaped resting **buds**, often several together. **Leaves** ovate-lanceolate, to 7 cm long, in a spreading rosette, mealy at first, greatly enlarging after flowering. **Flowers** up to 30, sometimes borne on a short scape amongst the leaves; corolla with a slightly cupped limb 22–30 mm across, pale to deep blue, lilac, pinkish or white, with a star-shaped orange eye that is often bordered in white, lobes with or without tiny fringing teeth. w & swSichuan, seTibet, w & nwYunnan [nMyanmar]; forests, often on sloping ground or banks, damp meadows, grassy slopes in damp shady places, wet flushes, streamsides, gullies, 3,300–4,600 m. III–VI.

Primula moupinensis **Franch.** Differs from the previous species by the small resting buds and often stoloniferous habit. **Flowers** flesh-pink, rarely pale lilac-blue, lobes clearly notched at apex. sw, c & wSichuan, seTibet, nwYunnan [nMyanmar]; similar habitats but rarely inside forests, 1,300–4,700 m. IV–VI. Plants from nwYunnan with greener **leaves** and more strictly rotate **flowers** with more deeply notched lobes, are sometimes separated as *P. hoffmanniana* **W. W. Sm.**

Primula tanneri King subsp. *tsariensis* (W. W. Sm.) A. J. Richards* Differs from the previous two species in lacking meal overall. **Leaves** at flowering time with a distinctly winged stalk. **Corolla** rich purple, lobes wavy, notched. seTibet; forests and grassy places, banks, amongst tree roots, 3,200–4,900 m. VI–VII. The plant pictured is probably referable to **var. *porrecta* W. W. Sm.**, which has especially large and deep-coloured **flowers**. **Subsp. *tanneri*** is confined to Himalaya.

Primula hookeri Watt A very distinctive species with **leaves** no more than 4 cm long at flowering time, oblong, with broad stalk and deeply toothed margin. **Flowers** in basal clusters which elongate on a thick whitish **stem** as fruit develop, individual **flowers** at first on very short stalks 1–2 mm long, which again elongate and thicken after flowering; **corolla** deeply cupped, 10–12 mm across, white, cream or violet, with white eye. w & swSichuan, seTibet, nwYunnan [Himalaya, nMyanmar]; mossy forest floors, stony places, snow hollows, snow-sodden areas, 3,300–5,000 m. III–V.

SECTION *Davidii*

Primula ovalifolia Franch.* Plant to 18 cm tall in flower, with primrose-like, rough, deep green, rather leathery **leaves**, which are oval, finely toothed or untoothed, glabrous or only slightly hairy above. **Flowers** up to 9 in a lax umbel; corolla 18–25 mm across, violet-blue with a prominent white eye, lobes deeply notched. wHubei, s, w & cSichuan, n & eYunnan; broadleaved forests, ravines, 1,200–2,500 m. III–V.

Primula taliensis Forrest* Differs from the previous species in the smoother, more pronouncedly toothed, obovate **leaves** that are covered in rusty hairs. **Flowers** up to 12, corolla pink or blue, each lobe 3-toothed. nwYunnan (especially Cangshan and mountains along the Myanmar frontier); woodland, stony pastures, 3,000–3,700 m. IV–V.

SECTION *Crystallophlomis* (syn. section *Nivales*)

Subsection *Crystallophlomis*

1. Flowers yellow . . . . *P. orbicularis, P. soongii*
1. Flowers purple, blue, lilac, purple-violet or white . . . 2
2. Dwarf plants not more than 8 cm tall . . . . *P. brevicula, P. minor*
2. Larger plants rarely less than 10 cm tall 3
3. Corolla-tube not darker than limb . . . . . *P. chionantha, P. chionantha* subsp. *sinopurpurea*, *P. chionantha* subsp. *sinoplantaginea*
3. Corolla-tube darker than limb 4
4. Plants lacking meal . . . . *P. russeola, P. optata*
4. Plants mealy . . . . *P. limbata*

Primula orbicularis Hemsl.* Rather stout plant to 30 cm in flower with rosettes of rather pale bright green, erect to ascending, parallel-margined **leaves** with strongly back-rolled margin; white-mealy, especially beneath and on **inflorescence**. **Flowers** bright yellow, usually 15 or more, in a compact, rounded head, corolla-lobes narrow-elliptic; calyx 7–10 mm long. sGansu, seQinghai, w & nwSichuan; wet meadows, moorland, open shrubberies, seepage zones, locally common, 3,400–3,800 m. VI–VII.

Primula soongii F. H. Chen & C. M. Hu* Similar to the previous species, but taller with more spreading, rounded to oval **leaves**. **Flowers** pale yellow with oval corolla-lobes; calyx relatively long, 10–14 mm. (?)seQinghai, nw & wSichuan; open shrubberies, damp meadows, 3,200–4,600 m. V–VII.

Primula brevicula Balf. f. & Forrest* Like a dwarf *P. chionantha* (see p. 247). Plants rarely exceeding 8 cm tall in **flower**. **Leaves** oval to elliptic, borne in a lax rosette, narrowed abruptly into a short purplish, unwinged stalk, margin finely toothed. **Flowers** slaty blue (sometimes with a pink flush) with a white eye, up to 6 on a scape just clear of the leaves; corolla-tube somewhat longer than calyx. (?)swSichuan, seTibet, nw & wYunnan; thin alpine grassland, moorland, stabilised screes and moraines, 3,900–4,700 m. V–VII.

Primula minor Balf. f. & Kingdon-Ward* Very similar to the previous species, but **leaves** narrow-elliptic, margin usually toothed, with dense creamy meal beneath; **corolla** with slender tube twice the length of calyx. nwYunnan, neighbouring parts of Tibet; similar habitats, 4,000–5,000 m. VI–VII.

Primula russeola Balf. f. & Forrest* Tufted, often clump-forming plant with spreading rosettes of bright green, elliptic, rather smooth **leaves** (mostly more than 5 cm long) without meal, with flat, finely toothed margin, apex pointed. **Flowers** up to 7 in a 1-sided cluster on a scape somewhat longer than leaves; corolla lilac with a dark tube and a broad dark (occasionally white) eye. swSichuan, seTibet, nwYunnan; rock-ledges, damp banks, snowmelt hollows, 3,800–4,400 m. VI–VII.

Primula optata Farrer* Similar to *P. russeola*, but **leaves** smaller, less than 5 cm long, rather dull green, apex rounded. **Flowers** usually more than 10 per cluster; corolla pale lilac or rose-lilac with a small dark eye. sGansu, seQinghai, nwSichuan; rock debris and crevices, screes, generally limestone, 3,900–4,600 m. VI–VII. Sometimes grows in association with *P. limbata* with which it can be easily confused.

Primula limbata Balf. f. & Forrest* Tufted plant to 25 cm in flower, with spreading rosettes of rather thick, grey-green, elliptic to oblanceolate **leaves** with a characteristic beaded, mealy margin. **Flowers** up to 14, in a rather dense cluster on mealy scapes well clear of the leaves, sometimes with an extra whorl of flowers close above; calyx mealy; corolla rose-lilac, limb 18–20 mm, with a small dark eye. sGansu, seQinghai, w & nwSichuan; damp grassy slopes on limestone, 3,700–4,300 m. VI–VII.

1. *Primula tanneri* subsp. *tsariensis*; Galung La, near Bomi, seTibet (HJ)
2. *Primula ovalifolia*; cult. (JR)
3. *Primula orbicularis*; Huanglong, nwSichuan (CGW)
4. *Primula soongii*; Huanglong, nwSichuan (CGW)
5. *Primula brevicula*; Daxueshan, nw Yunnan (CGW)
6. *Primula minor*; Daxueshan, nwYunnan (CGW)
7. *Primula russeola*; Daxueshan, nwYunnan (CGW)
8. *Primula optata*; Jiuzhaigou, nwSichuan (CGW)
9. *Primula limbata*; Huanglong, nwSichuan (CGW)

Primula chionantha Balf. f. & Forrest* A rather robust, **clump-forming perennial** to 50 cm in flower, often only half that height. **Leaves** oblanceolate, ascending to erect, to 10–25 × 2–6 cm, margin entire to finely toothed, generally back-rolled, with variable amounts of yellowish meal. **Inflorescence** with 1–3 close-set whorls, each with up to 15 spreading **flowers**, equalling or overtopping leaves; calyx narrow-campanulate, 7–10 mm long, mealy; corolla white or cream, fragrant, limb 25–30 mm across, with oval lobes. nwYunnan (confined to Zhongdian Plateau); wet meadows, open shrubberies, woodland margins, seepage areas, 3,100–3,450 m. V–VI. **Subsp. *sinopurpurea* (Balf. f.) A. J. Richards*** (syn. *P. sinopurpurea* Balf. f.) is very similar, but plants have white or yellowish meal and purple **flowers** with a dark or whitish eye; calyx often flushed with lilac or purple, or blackish. sGansu, sw & wSichuan, seTibet, w & nwYunnan; 3,100–4,500 m, in similar habitats. V–VII. **Subsp. *sinoplantaginea* (Balf. f.) A. J. Richards*** (syn. *P. sinoplantaginea* Balf. f.) is an altogether dwarfer plant, seldom exceeding 15 cm, with narrow-elliptic **leaves** not more than 10 mm wide, with yellowish meal; **flowers** purple, with a dark eye, generally in a simple whorl. Similar distribution to subsp. *sinopurpurea*, but a high altitude plant (4,500–5,300 m) in alpine meadows, rocky places, moorland. VI–VII.

Subsection *Calliantha*

Primula calliantha Franch. Plant with a spherical resting **bud** from which the bright green, obovate to elliptic **leaves** arise, blade with or without greenish meal, margin finely toothed. **Flowers** up to 10 in a simple umbel on scape to 30 cm tall; calyx blackish, mealy, 11–15 mm long; corolla dark blue or purple-violet, with or without prominent yellowish, mealy eye, lobes notched, the upper 2 curving back. seTibet, nwYunnan [n & neMyanmar]; leafy soils and grassy glades in *Abies* forests and beneath rhododendrons, 3,500–4,000 m. V–VI.

1. *Primula chionantha*; Tianchi, nwYunnan (PC)
2. *Primula chionantha* subsp. *sinopurpurea*; Daxueshan, nwYunnan (CGW)
3. *Primula chionantha* subsp. *sinopurpurea*; road to Tianchi, nwYunnan (CGW)
4. *Primula chionantha* subsp. *sinoplataginea*; Baimashan, nwYunnan (CGW)
5. *Primula calliantha*; Deqin, Baimashan, nwYunnan (HJ)
6. *Primula bryophila*; Daxueshan, nwYunnan (CGW)
7. *Primula boreiocaliantha*; Baimashan, nwYunnan (PC)
8. *Primula boreiocaliantha*; Daxueshan, nwYunnan (CGW)
9. *Primula thearosa*; Galung La, Bomi, seTibet (HJ)
10. *Primula thearosa*; Galung La, Bomi, seTibet (HJ)
11. *Primula tangutica*; north of Songpan, nwSichuan (CGW)
12. *Primula tangutica*; north of Songpan, nwSichuan (PC)
13. *Primula woodwardii*; north of Jigzhi, seQinghai (CGW)
14. *Primula euprepes*; seTibet (HJ)

Primula bryophila Balf. f. & Farrer* Differs from the previous species in being a smaller plant with broad-oval **leaves** with a more deeply toothed margin and white meal beneath; **corolla** large (34–40 mm across), campanulate, rose-pink with a whitish or yellowish eye. seTibet, nwYunnan [neMyanmar]; deep leafy soils beneath rhododendrons, 3,800–4,200 m. V–VII.

Primula boreiocalliantha Balf.f. & Forrest* (syn. *P. hongshanensis* Z. D. Fang, H. Sun & Rankin) Closely related to the previous 2 species but **leaves** elliptic-lanceolate to narrow-oblanceolate, green above, markedly white-mealy beneath and with a finely toothed margin; **inflorescence** with flowers in 1–2 superposed whorls (each with up to 5 flowers); **corolla** 26–32 mm across, amethyst-pink with a wide white-mealy eye, lobes rather square-ended, deeply notched. nwYunnan (Zhongdian, Daxueshan, Baimashan); *Rhododendron* woodland, shrubberies, beneath *Abies*, 4,000–4,200 m. VII–VIII.

Subsection *Agleniana*

Primula agleniana Balf. f. & Forrest Plant to 40 cm in flower with a reddish 'bulbous' base, **leaves** lanceolate, to 30 cm long at maturity, pointed-tipped and with toothed margin, with greenish yellow meal beneath. **Inflorescence** an umbel of up to 8 flowers on a long slender scape; **corolla** soft yellow or sometimes orange-yellow, relatively large, 30–40 mm across. seTibet, nwYunnan [eMyanmar]; shrubberies, banks and streamsides on dark humusy soils, 3,200–4,000 m. VI–VII.

Primula thearosa Kingdon-Ward. Similar to the previous species but with soft to bright pink flowers. seTibet [nMyanmar].

Subsection *Maximowiczii*

1. Flowers red, blackish, violet-red, violet-purple or violet .... *P. tangutica*, *P. woodwardii*, *P. euprepes*
1. Flowers yellow .... *P. szechuanica*

Primula tangutica Duthie* Small to medium plant to 40 cm, with rosettes of deep, rather bright green, oblong-elliptic, practically stalkless **leaves**. **Flowers** in simple clusters or with 2–3 distant whorls, scented; corolla blackish red to purplish red, with very narrow, linear lobes, annulus slight. sGansu, seQinghai, n, nw & wSichuan; grassy and stony alpine meadows, open shrubberies, marshy places, 3,700–4,300 m. VI–VIII.

Primula woodwardii Balf. f.* Shorter plant than the previous species, not more than 15 cm, with 3–5 flowers in a simple 1-sided **umbel. Flowers** with purple tube and spreading, violet, somewhat reflexed corolla-lobes, mouth of tube with a pronounced 10-lobed annulus. sGansu, seQinghai, (?)nwSichuan; alpine meadows, banks, 3,000–4,000 m. VI–VII.

Primula euprepes W. W. Sm.* (syn. *P. advena* W. W. Sm. var. *euprepes* (W. W. Sm.) Chen & C. M. Hu) Similar to the others in this group but leaf-stalks broadly winged and **flowers** chocolate-coloured, mealy, with broad-oval, slightly reflexed lobes. seTibet (Kongbo to PemaKo); rhododendron scrub, rocky forests and meadows, 4,000–4,600 m. VI–VII.

Primula szechuanica **Pax*** Apart from flower colour, very similar to preceding 2 species; **leaves** toothed and stalked and fragrant bright yellow **flowers** with rather broader lobes (c. 3 mm), these sometimes strongly reflexing. w, nw & swSichuan, eTibet, nwYunnan; mountain meadows, marshy places, low shrubberies, 2,800–4,000 m. VI–VII.

SECTION *Proliferae*

1. Corolla salverform; flowers not nodding 2
1. Corolla funnel-shaped or campanulate; flowers nodding 5
2. Plants not mealy . . . . *P. poissonii, P. wilsonii*
2. Plants mealy in part, generally on scapes and calyces . . 3
3. Flowers yellow or orange (purple with a yellow eye in *P. bulleyana* subsp. *beesiana*) 4
3. Flowers carmine to reddish purple or orange-scarlet, with a dark eye . . . . . *P. pulverulenta, P. cockburniana*
4. Flowers yellow; plants evergreen . . . . *P. prolifera*
4. Flowers red or purple; plants deciduous . . . . *P. bulleyana, P. bulleyana* subsp. *beesiana, P. chungensis*
5. Corolla yellow and white; plants without meal . . . . *P. serratifolia*
5. Corolla dark crimson or violet-crimson; plants with some meal, especially on inflorescence . . . . . *P. secundiflora*

Primula poissonii **Franch.*** Relatively robust plant to 50 cm in flower, with rosettes of deep bluish green oblanceolate **leaves** with finely toothed margin, almost stalkless. **Flowers** in 2–6 distant, spreading whorls, each with up to 9 flowers; corolla 24–30 mm across, bright reddish purple, occasionally flesh-coloured or white, with a golden eye. swSichuan, nwYunnan; marshy places, meadows, open shrubberies, streamsides, often abundant, 3,200–4,000 m. V–VII.

Primula wilsonii **Dunn*** Similar to the preceding species, but **leaves** aromatic when crushed and flowers on **scapes** up to 90 cm tall, **corolla** more cupped, reddish to blackish, 14–16 mm across. seSichuan, nw & wYunnan; similar habitats, 3,500–4,300 m. V–VII.

Primula prolifera **Wall.** Robust **perennial** to 1 m, often less, with spreading rosettes of deep green, often shiny, oblanceolate to diamond-shaped, toothed **leaves**. **Flowers** numerous on stout scapes with up to 7 whorls, each whorl with up to 12 flowers; corolla pale to golden yellow, limb 17–20 mm across. seTibet, wYunnan (near Myanmar border) [eHimalaya, nMyanmar, Java, Sumatra]; marshes, wet pastures, streamsides, open shrubberies, 2,000–3,500 m. V–VIII.

Primula bulleyana **Forrest*** Robust plant to 80 cm in flower, with oblanceolate, finely toothed **leaves** with red midribs. Flowers in stout **inflorescences** with up to 5 whorls; **corolla** yellow to orange-yellow or golden, limb 18–24 mm across. s & swSichuan, nwYunnan; wet pastures, marshes, open shrubberies, streamsides, 2,500–3,300 m. V–VII. **Subsp. beesiana (Forrest) A. J. Richards*** (syn. *P. beesiana* Forrest) has purple or rose-carmine **flowers** with a yellow eye. Sometimes grows with subsp. *bulleyana* and other species (especially *P. poissonii*), when hybrid swarms arise in a medley of bright flower colours.

Primula chungensis **Balf. f. & Kingdon-Ward*** Similar to *P. bulleyana* subsp. *bulleyana*, differing in the white leaf-midribs, short (not finely pointed) calyx-lobes and somewhat smaller **flowers** borne on stalks to 20 mm long. w & swSichuan, nwYunnan; similar habitats, 2,900–3,200 m. VI–VII.

Primula pulverulenta **Duthie*** Very robust **tufted** plant to 1 m in flower, with clusters of leaf-rosettes, **leaves** oblanceolate to oval, margin irregularly toothed. Flowers in stout **inflorescences** with up to 8 whorls (each with up to 10 flowers), borne on mealy stalks like the scape; **corolla** carmine-red to reddish purple, limb 22–30 mm across. wSichuan (especially near Kangding); marshy places, streamsides, open shrubberies, forest margins, 2,000–3,200 m. V–VII.

Primula cockburniana **Hemsl.*** An altogether slighter plant than the previous species, with the general proportions of *P. chungensis* but **flowers** orange-scarlet with contrasting silvery-mealy corolla. w & swSichuan (Kangding and Kulu regions); wet alpine meadows, marshy places, 2,900–3,200 m. V–VII.

Primula serratifolia **Franch.** **Tufted deciduous**, non-mealy plant to 30 cm in flower, with elliptic to oblanceolate, matt green, rather rough **leaves** with finely toothed margin. Flowers usually 3–7 in simple **umbel**, sometimes with a second whorl above, borne on a scape above the foliage; **corolla** deep yellow, lobes with a broad cream or pale yellow margin, slightly notched. nwYunnan and neighbouring part of Tibet [nMyanmar]; shrubberies, particularly of rhododendrons, marshy places, streamsides, 3,000–4,000 m. V–VI.

1. *Primula szechuanica*; Munigou, Kangding, wSichuan (PC)
2. *Primula szechuanica*; nwYunnan (JM)
3. *Primula poissonii*; Heishui, Yulongxueshan, nwYunnan (PC)
4. *Primula poissonii*; Xiaozhongdian, nw Yunnan (CGW)
5. *Primula poissonii*; Cangshan, nwYunnan (CGW)
6. *Primula prolifera*; cult. (CGW)
7. *Primula prolifera*; Ugyen Choling, Bhutan (CGW)
8. *Primula bulleyana* subsp. *beesiana*; Yulongxueshan, nwYunnan (CGW)
9. *Primula bulleyana* subsp. *beesiana*; north of Lijiang, nwYunnan (HJ)
10. *Primula bulleyana* subsp. *bulleyana*; Sandauwan, Yulongxueshan, nwYunnan (CGW)
11. *Primula bulleyana* ×*P. poissonii*; Heishui, Yulongxueshan, nwYunnan (PC)
12. *Primula chungensis*; cult. (CGW)
13. *Primula pulverulenta*; cult. (CGW)
14. *Primula cockburniana*; Kangding, wSichuan (PC)

Primula secundiflora Franch.* Relatively robust **evergreen** plant to 90 cm in flower, with deep bright green rosettes of oblanceolate, finely toothed **leaves**. **Flowers** in a drooping whorl of up to 20, sometimes with a second smaller whorl above, upper scape, flower-stalks and calyces white-mealy; **corolla** brilliant dark crimson, campanulate, with rounded or squarish overlapping lobes. swSichuan, seTibet, nwYunnan; marshes, meadows, open woodland, shrubberies, stream and lake margins, often abundant, 3,200–4,300 m. IV–VII. (See also photos 1 and 3 on p. 238.)

SECTION *Sikkimensis*

1. Base of leaf-blade attenuate into a winged stalk . . . . .
   . . ***P. sikkimensis*, *P. sikkimensis* var. *pseudosikkimensis***
1. Base of leaf-blade heart-shaped, only upper part of leaf-stalk winged 2
2. Corolla-lobes entire to slightly toothed . . . ***P. florindae***
2. Corolla-lobes clearly notched . . . . ***P. alpicola*, *P. alpicola* var. *alba*, *P. alpicola* var. *violacea*, *P. ioessa***

Primula sikkimensis Hook. f. Plant very variable, often robust, **clump-forming**. **Leaves** bright to deep shiny green, elliptic to oblanceolate, up to 40 cm long, margin finely toothed. **Inflorescence** with 20 or more pendent, fragrant flowers, usually in a simple umbel on erect scape to 90 cm tall; **corolla** campanulate, sulphur-yellow to primrose-yellow or occasionally ivory, 22–30 mm long. sGansu, seQinghai, w & swSichuan, seTibet, w & nwYunnan [swTibet, Himalaya, nMyanmar]; wet meadows, open woodland, shrubberies, seepage zones, streamsides, 2,900–4,500 m. V–VII. **Var.** ***pseudosikkimensis* (Forrest) A. J. Richards*** (syn. *P. pseudosikkimensis* Forrest) is a high altitude variant, not more than 20 cm tall, with disproportionately large **flowers**. nwYunnan (apparently confined to Yulongxueshan at 4,500–5,200 m); alpine meadows and moorland, moraines. VI–VII. (See also photos 1, 2 and 3 on p. 238.)

---

1. *Primula secundiflora* with *P. sikkimensis*; north of Daxueshan, sSichuan (CGW)
2. *Primula secundiflora*; Zhongdian, nwYunnan (CGW)
3. *Primula sikkimensis*; Napahai, Zhongdian, nwYunnan (CGW)
4. *Primula sikkimensis*; Zhongdian, nwYunnan (HJ)
5. *Primula sikkimensis* var. *pseudosikkimensis*; Wutoudi, Yulongxueshan, nwYunnan (CGW)
6. *Primula florindae*; cult. (CGW)
7. *Primula alpicola*; cult. (JR)
8. *Primula alpicola* var. *violacea*; cult. (JR)
9. *Primula ioessa* var. *hopeana*; Serkhyem La, seTibet (HJ)
10. *Primula fasciculata*; Bitahai, Zhongdian, nwYunnan (CGW)
11. *Primula gemmifera*; Balangshan, Wolong, wSichuan (CGW)

Primula florindae Kingdon-Ward* Large cabbagey, **clump-forming** plant to 120 cm in flower, usually shorter. **Leaves** oval with heart-shaped base, to 20 × 15 cm; stalk often reddish. **Inflorescence** stout, held well above leaves, with large umbel of up to 80 pendent, fragrant **flowers**; calyx campanulate, 7–10 mm long, with cream meal; corolla campanulate, sulphur-yellow, to 20 mm long. seTibet (Yarlung Zangbojiang (Tsangpo) basin); forest bogs, streamsides, seepage zones, 3,800–4,200 m. V–VII.

Primula alpicola (W. W. Sm.) Stapf Similar to *P. sikkimensis* but leaf-blades rough, dark olive-green, oval with rounded base and **corolla** broadly campanulate (more saucer-shaped), very variable in colour from white (**var.** ***alba* W. W. Sm.**), cream and yellow (**var.** ***alpicola***), to rose-pink, purple, wine-red and violet (**var.** ***violacea* (Stapf) W. W. Sm. & Fletcher**), white-mealy within. Similar distribution (as well as [nBhutan]) and flowering time to *P. florindae* and often growing with it.

Primula ioessa W. W. Sm. Differs from *P. alpicola* in deep green, smooth **leaves** with a wide whitish midrib, scape to 45 cm tall in **flower**; corolla lilac-blue or white with a mealy eye. seTibet (especially Tsari) [sTibet, Bhutan, eNepal]; wet meadows, often with other *Primula* species, 3,600–4,300 m. V–VII. **Var.** ***hopeana* (Balf. f. & Cooper) A. J. Richards** (syn. *P. hopeana* Balf. f. & Cooper) has neatly scalloped-toothed **leaves** and ivory **flowers**. seTibet [cHimalaya]. It is sometimes treated as a distinct species.

SECTION *Armerina*

1. Flowers often borne singly or 2–3 together amidst foliage . . . . ***P. fasciculata***
1. Flowers 4 or more borne on a scape well above foliage 2
2. Corolla rotate, not zygomorphic . . . . ***P. gemmifera*, *P. gemmifera* var. *monantha***
2. Corolla weakly zygomorphic 3
3. Scapes and calyces mealy; bracts not pouched at base . . . . ***P. zambalensis*, *P. conspersa***
3. Scapes and calyces not mealy; bracts pouched at base . . . . ***P. involucrata*, *P. involucrata* subsp. *yargongensis*, *P. nutans***

Primula fasciculata Balf. f. & Kingdon-Ward* Plant forming low tufts or mats not more than 4 cm tall. **Leaves** spathulate, bright green, with inturned, entire margin. **Flowers** rose-pink with whitish or yellow eye, the waxy corolla with a rather crinkled limb 10–12 mm wide, lobes deeply notched. sGansu, w, nw & swSichuan, e & seTibet, nwYunnan; marshy meadows, tracks, seepage zones, 3,500–4,000 m. V–VII.

Primula gemmifera Batalin* Small tufted plant with short-jointed rhizomes that break up into 1-bud portions during the winter. **Leaves** forming lax rosettes, oblanceolate to elliptic, smooth, with entire or slightly toothed, back-rolled margin, without meal. **Flowers** up to 10 on an ascending mealy scape; calyx campanulate, 5–10 mm long, purplish and mealy; corolla pink to purple, ringed, 15–25 mm across, lobes deeply notched. sGansu, seQinghai, w, nw & swSichuan, eTibet, n & nwYunnan; rocks, screes, peaty grassland, streamsides, 3,000–4,200 m. V–VII. Very small plants with solitary flowers without an annulus are assigned to var. *monantha* W. W. Sm. & Fletcher* restricted to nwYunnan (Lancangjiang–Jinshajiang divide).

Primula zambalensis Petitm.* Tufted, often **clump-forming** plant to 20 cm in flower, with deep green, ascending, oval, toothed **leaves**, with long slender stalks, lacking meal. **Flowers** 3–8, on slender wiry scape well above foliage; corolla lilac-blue to sky-blue, without a ring, 15–22 mm across, with shallowly notched lobes. nwYunnan, (?)w & swSichuan; moraines, earthy banks, screes, open alpine turf, 4,000–4,800 m. VI–VII.

Primula conspersa Balf. f. & Purdom* Readily distinguished from the previous species, and the related *P. gemmifera*, by flowers in 2–3 whorls in the **inflorescence**; **corolla** lilac, limb 10–15 mm across. sGansu, nwSichuan; damp alpine meadows, open woodland, 2,400–3,000 m. VI–VII.

Primula involucrata Wall. subsp. *yargongensis* (Petitm.) W. W. Sm. & Fletcher Slender, tufted plant to 30 cm, often less, with tufts of elliptic to spathulate, long-stalked **leaves** with entire or minutely toothed margin. **Flowers** up to 10 in a lax 1-sided umbel borne on a slender scape well above foliage; corolla pink to purple, 18–22 mm across, with golden or whitish annulus, weakly zygomorphic, lobes with a broad, shallow notch, upper 2 often reflexed. sGansu, seQinghai, w, nw & swSichuan, seTibet, nwYunnan [swTibet, Himalaya, nMyanmar]; marshy meadows, seepage zones, streamsides, 2,740–5,000 m. V–VII. **Subsp. *involucrata*** is widespread in Himalaya and in sTibet.

Primula nutans Georgi A more delicate plant than the preceding species, to 15 cm, with few-leaved rosettes of spathulate **leaves**, back-rolled at margin; scapes with 1–4 almost erect **flowers**; corolla pink or mauve-pink to whitish with small pale yellow ring, 12–16 mm across, calyx often black-dotted. Gansu, wHubei, Qinghai, nwSichuan [widespread in Arctic regions and c Asia]; alpine meadows, marshy turf, damp moraines and gravels, 2,000–4,500 m. V–VII. (See also photo 4 on p. 238.)

SECTION *Yunnanensis*

1. Leaf-margin entire or almost so . . . . . . **P. yunnanensis**
1. Leaf-margin toothed . . . . **P. florida, P. rupicola**

Primula yunnanensis Franch. Small, rather delicate, short-lived plant, not more than 10 cm and often with a solitary leaf-rosette. **Leaves** spathulate to oblanceolate, with slender stalk, not more than 3.5 cm long overall, with some yellow meal, especially beneath. **Flowers** 1–5 in a lax cluster above leaves; corolla relatively large, 20–25 mm in diameter, pink to lilac with whitish eye, lobes deeply notched, upper 2 often somewhat deflexed. Sichuan, nwYunnan, adjacent parts of Tibet [nMyanmar]; rocky places, especially banks and cliffs, often by streams and rivers, locally common, 2,800–5,000 m. V–VI.

Primula florida Balf. f. & Forrest* Small, short-lived **perennial**, often with a solitary leaf-rosette, usually with yellow or white meal, sometimes non-mealy. **Leaves** to 10 cm long, oblong to oval, deeply and incisely lobed, abruptly contracted below into a slender stalk. **Flowers** up to 10 in a lax umbel, borne on erect scape to 25 cm (more in fruit); corolla pink to lilac-purple, or bluish pink, broadly campanulate, 14–20 mm in diameter, lobes deeply notched. w & swSichuan, nwYunnan; stony and rocky places on limestone or sandstone, cliff-ledges, sometimes on forest margins, 3,200–4,500 m. VI–VII.

Primula rupicola Balf. f. & Forrest* Similar to preceding species, but plants forming tufts with many rosettes, leaf-margin deeply toothed, tapered below into a winged stalk. Corolla bright rose-pink with a yellowish, cream or greenish eye, flat-faced. nwYunnan; rocky places, cliff ledges, riverbanks, 3,000–4,000 m. VI–VII.

SECTION *Pulchella*

1. Flowers blue, bluish lavender or pink . . . . **P. pulchella, P. pulchella subsp. prattii, P. stenocalyx**
1. Flowers yellow . . . . **P. flava**

Primula pulchella Franch.* **Perennial** to 30 cm in flower, often less, with a large yellowish-mealy resting bud at ground level. **Leaves** forming a spreading rosette, narrow-elliptic to oblanceolate, to 20 × 2.5 cm, deep bluish green above with paler veins, yellow with meal beneath. **Flowers** up to 15 in an umbel on a mealy scape well above the foliage; corolla deep blue with yellow eye, occasionally lavender, flat-faced, 15–20 mm in diameter, lobes shallowly notched. swSichuan, nwYunnan, adjacent parts of Tibet; dryish stony pastures,

---

1. *Primula zambalensis*; Baimashan, nwYunnan (HJ)
2. *Primula conspersa*; Huanglong, nwSichuan (CGW)
3. *Primula involucrata* subsp. *yargongensis*; pass above Geza, nwYunnan (CGW)
4. *Primula nutans*; Qinghaihu, Qinghai (RS)
5. *Primula nutans*; Qinghaihu, Qinghai (RS)
6. *Primula yunnanensis*; Baishiu, Yulongxueshan, nwYunnan (CGW)
7. *Primula florida*; Daxueshan, nwYunnan (CGW)
8. *Primula rupicola*; Yulongxueshan, nwYunnan (PC)
9. *Primula pulchella*; Yulongxueshan, nwYunnan (PC)
10. *Primula pulchella*; Gangheba, Yulongxueshan, nwYunnan (CGW)

1. *Primula stenocalyx*; Wolong, wSichuan (PC)
2. *Primula flava*; Qinghai (RS)
3. *Primula bella*; Cangshan, nwYunnan (CGW)
4. *Primula nanobella*; Baimashan, nwYunnan (CGW)
5. *Primula walshii*; seTibet (AC)
6. *Primula denticulata* subsp. *sinodenticulata*; Cangshan, nwYunnan (PC)
7. *Primula denticulata* subsp. *sinodenticulata*; Cangshan, nwYunnan (PC)
8. *Primula capitata*; below Ghunsa, Nepal (CGW)

alpine meadows, sometimes in large drifts, 2,800–4,500 m. V–VII. **Subsp. *prattii* (Hemsl.) Halda*** is a slighter plant with yellow flowers, from wSichuan (particularly Kangding).

Primula stenocalyx **Maxim.*** Small tufted plant with lax rosettes of oblanceolate to spathulate, glandular **leaves**, to 5 × 1.5 cm, margin entire to finely toothed, with or without white meal. **Flowers** usually clear of the foliage on scape to 15 cm, up to 16 in an umbel; calyx narrow-cylindrical, greenish black with white mealy stripes between the teeth; corolla bluish lavender to pink, limb 15–20 mm in diameter, with whitish eye, lobes deeply notched. c & sGansu, w & nwSichuan, neighbouring parts of Qinghai & Tibet; limestone rocks, ledges, dry grassy banks, 2,700–4,300 m. IV–VI.

Primula flava **Maxim.*** Small tufted plant with spathulate **leaves**, blade sometimes almost rounded, margin shallowly toothed, with dust-like cream meal. **Flowers** solitary or up to 7 in an umbel held just clear of foliage; corolla primrose-yellow with deeper eye, tube slender, at least twice the length of calyx, limb 14–18 mm, with broad, almost rounded, deeply notched lobes. w & swGansu, seQinghai, eTibet; limestone cliffs, dry calcareous silts, generally in deep shade, 2,700–4,500 m. VI–VII.

SECTION *Minutissimae*

> 1. Throat of corolla filled with a wad of white or violet hairs
>   .... ***P. bella, P. nanobella***
> 1. Throat of corolla without a wad of white hairs ....
>   .... ***P. walshii***

Primula bella **Franch**. Very variable small tufted **perennial**, with a single or several rosettes, with or without yellow or whitish meal. **Leaves** oval to elliptic or rounded, to 20 × 5 mm, margin deeply incised/toothed. **Flowers** solitary or 2–3 borne on slender scape to 40 mm long; corolla relatively large, 18–25 mm in diameter, violet, purple or rose-pink, salverform, with a characteristic pompom of white hairs filling the mouth, lobes deeply notched. swSichuan, w & nwYunnan (especially Cangshan), neighbouring parts of Tibet [n & neMyanmar]; damp places, shady rocks, snow hollows, wet rocks and cliffs, often gregarious, 4,000–5,300 m. V–VIII. **Subsp. *cyclostegia* (Hand.-Mazz.) W. W. Sm. & Forrest*** is more robust and confined to the Zhongdian Plateau in nwYunnan; **subsp. *bonatiana* (Petitm.) W. W. Sm.*** is a non-mealy variant from the Yunnan–seTibet border region.

Primula nanobella **Balf. f. & Forrest*** Similar to the preceding species, but a high alpine plant, often **mat-forming**, lacking meal; **flowers** generally solitary on a short stalk, corolla smaller, 8–15 mm in diameter, amethyst, pink to purple, with a dark central ring surrounding a pink or violet pompom of hairs. seTibet, nwYunnan; wet passes and mountain ridges, alpine tundra, mossy rocks and banks, 4,200–5,500 m. VI–VIII.

Primula walshii **Craib** Plant lacking meal, not more than 2 cm tall in flower, with a solitary rosette of rather bright green **leaves**, these oblong and semi-leathery, rough, to 15 × 5 mm, margin entire, apex pointed to blunt. **Flowers** 1–4 on very short scapes that elongate to 3 cm in fruit; corolla pink to pale bluish purple, relatively small, limb 5–8 mm wide, lobes deeply notched. w & nwSichuan, e & seTibet [c & eHimalaya]; grassy and stony alpine slopes, rock-pockets, 3,800–5,400 m. VI–VII.

SECTION *Denticulata*

Primula denticulata **Sm**. **subsp. *sinodenticulata* (Balf. f. & Forrest) W. W. Sm. & Forrest** **Tufted**, very variable, clump-forming **perennial**, overwintering as large leathery resting buds at ground level. **Leaves** in large spreading rosette, elliptic to oblanceolate, eventually up to 35 × 5 cm, tough, deep to mid-green, sparsely hairy and with finely toothed margin, thinly mealy beneath when young. **Inflorescence** a 'drumstick' dense head of numerous **flowers** on a thick scape to 40 cm, sometimes more; corolla lilac to lavender or purple, occasionally white, pink or reddish, 15–20 mm in diameter, lobes notched. Guizhou, w, s & swSichuan, seTibet, Yunnan [swTibet, Myanmar]; damp open habitats, meadows, open woodland, forest clearings, streamsides, pool margins, locally abundant, 1,500–4,500 m. IV–VII. Some plants from sTibet are referable to **subsp. *denticulata*** which is also found in Himalaya, Afghanistan: the **corollas** are only 10–15 mm in diameter.

SECTION *Capitatae*

Primula capitata **Hook. f.** Very variable, rather robust, **tufted perennial** to 45 cm (taller in fruit), with spreading rosettes of non-mealy (in China at least) elliptic to oblanceolate **leaves**, to 13 × 2 cm, with a finely toothed, rather wavy margin. **Inflorescence** a tight globose or disc-like, many-flowered head borne on a stiff, rather thick scape well above the leaves; corolla funnel-shaped or salverform, 7–10 mm in diameter, purple to bluish purple, with notched lobes. seTibet, w & nwYunnan [swTibet, c & eHimalaya]; damp slopes, streamsides, open woodland or above treeline, 3,300–4,500 m. V–VII. **Subsp. *sphaerocephala* (Balf. f. & Forrest) W. W. Sm. & Forrest*** in seTibet, Yunnan has globose **flower-heads** and **leaves** that lack meal; **subsp. *crispata* (Balf. f. & Forrest) W. W. Sm. & Forrest** from seTibet [eHimalaya], also lacks mealy **leaves** but the **flower-head** is disc-like; **subsp. *craibeana* (Balf. f. & Forrest) W. W. Sm. & Forrest**, from sTibet [SK], has narrow **leaves** that are mealy beneath, and globose **flower-heads**.

SECTION *Muscarioides*

> 1. Flowers generally more than 30; calyx bright red .....
>   .... ***P. vialii***
> 1. Flowers less than 30; calyx green or purple 2
> 2. Corolla-limb 15–25 mm across, mealy within ....
>   .... ***P. flaccida, P. spicata***

2. Corolla-limb not exceeding 15 mm across, not mealy within 3
3. Scape hairy, without meal .... **P. deflexa**
3. Scape glabrous, often mealy in part 4
4. Corolla-limb 8–10 mm across .. **P. cernua, P. pinnatifida**
4. Corolla-limb not more than 7 mm across 5
5. Leaves almost glabrous, shiny above . **P. muscarioides**
5. Leaves hairy, matt above ..... **P. apoclita, P. violacea**

Primula vialii Delavay ex Franch.* (syn. *P. littoniana* Forrest) Rather robust plant to 60 cm in flower, with tufts of erect, elliptic **leaves** with clearly recurved, finely toothed margin. **Flowers** numerous (sometimes more than 120) in a dense, cylindrical **spike**, red in bud; corolla violet-blue, with rather narrow, pointed, unnotched lobes. swSichuan, nwYunnan; open shrubberies, damp or stony meadows, apparently now rare, 2,850–4,000 m. VI–VII.

Primula flaccida N. P. Balakr.* Rather robust plant to 50 cm in flower, with elliptic, ± erect **leaves** with finely toothed margin, moderately to sparsely hairy. **Flowers** up to 20 in a dense cone-like cluster, with dark green to purplish calyx; corolla lavender-blue, with unnotched to slightly notched lobes, tube more than 4 mm wide. wGuizhou, swSichuan, n, nw & eYunnan; rocky meadows, open *Pinus* woodland, open shrubberies, 2,700–3,600 m. IV–VII.

Primula spicata Franch.* Less robust than the previous species with ± stalked, toothed **leaves**, **flowers** few, in a lax, 1-sided **spike**; corolla blue, funnel-shaped, tube not more than 2 mm wide. nwYunnan; stony meadows and banks, cliff-ledges, 3,000–3,700 m. IV–VI.

Primula deflexa Duthie* Elegant plant to 30 cm in flower, often less, with elliptic to oval, ascending **leaves** with finely toothed margin, hairy on veins. **Flowers** in a short spike borne on long scapes well clear of foliage; corolla 12–15 mm long, purplish blue to mid-blue, occasionally white with bluish eye, tube less than 2 mm wide. swSichuan, nwYunnan; open shrubberies, forest margins, damp grassy sites, stream- and pathsides, 3,300–4,300 m. V–VII.

Primula cernua Franch.* **Tufted** plant to 40 cm in flower, often less, with ascending, oblanceolate, finely toothed to entire-margined **leaves** that are almost stalkess. **Flowers** in an oblong spike, borne on mealy scapes well above foliage, spreading to semi-pendent; corolla violet to violet-blue, with a relatively large cupped limb. swSichuan, nwYunnan; open, relatively dry habitats, generally in *Pinus* woodland, 2,500–4,000 m. V–VII.

Primula pinnatifida Franch.* **Tufted** plant not more than 15 cm tall in flower, with distinctive lobed to pinnately lobed, hairy **leaves**, in a spreading rosette. **Flowers** in a rather dense cluster on mealy scapes above foliage; corolla pale to deep blue or amethyst-pink, with flared, bowl-shaped limb. swSichuan, nwYunnan; rocky alpine pastures, snow hollows, moraines, upper margins of coniferous woodland, 3,500–5,000 m. V–VII.

Primula muscarioides Hemsl.* Rather elegant, lanky plant to 40 cm, with rosettes of oblong to elliptic, blunt-toothed, stalked **leaves**. **Flowers** in a short **spike**, scapes far exceeding leaves, white-mealy near top, as is the purplish calyx; corolla purple-blue, with narrow, spreading lobes. (?)seQinghai, sw & nwSichuan, seTibet, nwYunnan; open shrubberies, often swampy, damp pastures, streamsides, 3,200–4,000 m. V–VII.

Primula apoclita Balf. f. & Forrest* **Tufted** plant not more than 16 cm tall in flower, with rosettes of elliptic to oblong, bluntly toothed **leaves**. **Flowers** in a rather dense head, on scapes as long as or exceeding leaves, both upper scape and deep purple calyces with yellowish meal; corolla purplish blue, limb ± equalling tube. swSichuan, n & nwYunnan; stony alpine pastures, low montane shrubberies, 4,000–4,700 m. V–VII.

Primula violacea W. W. Sm. & Kingdon-Ward* Similar to the previous species, but **leaves** with rough bristly hairs and **flowers** larger, 10–12 mm long, deep violet. w & swSichuan; rocky places, amongst low bushes, 4,000–4,300 m. V–VII.

SECTION *Soldanelloides*

Primula cawdoriana Kingdon-Ward* Compact plant often with root buds, forming clumps of small neat rosettes; **leaves** elliptic-oval, to 5 cm long, crisped and with a neatly shallowly lobed margin. **Flowers** up to 8 in a tight cluster at the top of a long, slender, slightly mealy scape; corolla pointing downwards, blue to pale mauve, very slender and tubular, to 30 mm long, lobes deeply dissected into 2–3 slender, sometimes toothed segments. seTibet (Tsari to PemaKo); rocky places, mossy banks, cliffs and ledges, 4,000–4,700 m. VI–VII.

OMPHALOGRAMMA

A genus of 13 species (9 in China) from the eastern Himalaya and north Myanmar. Closely related to *Primula*, but **leaf-rosettes** with a single, ebracteate scape and **flowers** always solitary, not heterostylous; calyx and corolla 6–(5–7)-lobed, corolla usually funnel-shaped, slightly zygomorphic. **Fruit** a capsule dehiscing with short terminal valves.

1. Corolla salverform with a narrow cylindrical tube .... **O. vinciflora, O. aff. viola-grandis**
1. Corolla funnel-shaped, narrowing towards base 2
2. Leaf-base tapered into winged stalk .... **O. souliei, O. elwesiana**
2. Leaf-base rounded to heart-shaped, with a distinct stalk 3
3. Leaves glabrous above .... **O. forrestii**
3. Leaves hairy above (sometimes sparsely so) and below ..... **O. delavayi, O. elegans, O. tibeticum, O. minus**

1. *Primula vialii*; cult. (CGW)
2. *Primula flaccida*; cult. (JR)
3. *Primula spicata*; Cangshan, nwYunnan (PC)
4. *Primula deflexa* with *Adonis davidii*; Tianchi, nwYunnan (HJ)
5. *Primula cernua*; cult. (JR)
6. *Primula pinnatifida*; Yulongxueshan, nwYunnan (CGW)
7. *Primula muscarioides*; south of Aba, nwSichuan (CGW)
8. *Primula apoclita*; Baimashan, nwYunnan (CGW)
9. *Primula cawdoriana*; Serkhyem La, seTibet (HJ)

1. *Omphalogramma vinciflora*; Lancangjiang–Jinshajiang divide, Lidiping, nwYunnan (CGW)
2. *Omphalogramma* aff. *viola-grandis*; Balangshan, Wolong, wSichuan (PC)
3. *Omphalogramma elwesiana*; cult. (RBGE)
4. *Omphalogramma forrestii*; Tianchi, nwYunnan (HJ)
5. *Omphalogramma delavayi*; Cangshan, nwYunnan (CGW)
6. *Omphalogramma tibeticum*; Galung La, near Bomi, seTibet (DZ)
7. *Omphalogramma minus*; seTibet (DZ)
8. *Bryocarpum himalaicum*; Chelai La, Bhutan (PC)
9. *Cortusa matthioli* subsp. *pekinensis*; Shennongjia, wHubei (PC)

Omphalogramma vinciflora (Franch.) Franch.* (syn. *Primula vinciflora* Franch.) Plants to 35 cm in flower. **Rosette-leaves** bright green, oblong to oblanceolate, to 14 × 3 cm, hairy, attenuated at base into a winged stalk, margin usually entire, occasionally slightly and minutely scalloped. Scapes erect and rather slender, hairy. **Corolla** deep indigo-blue, with 3–4.5 cm limb and slender cylindrical tube 2–3 cm long; lobes 6–8, rounded to obovate, notched, downy outside. sGansu, n, nw & swSichuan, e & seTibet, nwYunnan; damp meadows, open scrub, 2,200–4,600 m. V–VI.

Omphalogramma aff. *viola-grandis* (Farrer & Purdom) Balf. f.* Similar to the previous species, but plants shorter and stouter, **leaves** grey-green, densely downy and with a toothed margin. Scapes stouter and densely downy. **Corolla** rather larger with broader, obovate lobes that are clearly notched at apex. w & nwSichuan (particularly Balangshan); damp alpine meadows, 3,200–4,000 m. VI–VII.

Omphalogramma souliei Franch.* (syn. *Primula franchetii* Pax) Tufted **herbaceous perennial** with a scaly stock to 30 cm, taller in fruit. **Leaves** elliptic to oblong, developing with or after the flowers, to 8 × 5 cm, glabrous or slightly hairy, margin entire. Scapes erect, glandular-hairy. **Corolla** bluish purple to reddish purple, limb 4–6 cm wide, tube 3–4 cm long, funnel-shaped; lobes 6, oblong to obovate, deeply notched and with entire to toothed margin; style protruding. swSichuan, seTibet, nwYunnan; forest margins (normally of *Pinus*), *Rhododendron* shrubberies, 3,300–4,500 m. VI–VII.

Omphalogramma elwesiana (King ex Watt) Franch. Similar to the preceding species, but corolla-tube not exceeding 25 mm with style not protruding; filaments glabrous. seTibet [swTibet, Himalaya]; damp meadows, 3,800–4,000 m. VI–VII.

Omphalogramma forrestii Balf. f.* Tufted **perennial** with leaves and flowers developing consecutively, to 30 cm tall in flower. **Leaves** oblong to oblong-elliptic, to 10 × 5 cm at flowering time, with a somewhat wavy to minutely toothed margin. Scapes erect, hairy, densely so towards top. **Corolla** dark purple, with limb 4.5–5.5 cm wide and a funnel-shaped tube 3.5–5 cm long; lobes 6, elliptic to ovate, notched at apex and with a slightly hairy margin; style glabrous. swSichuan, nwYunnan; meadows, *Rhododendron* shrubberies, 3,500–4,000 m. VI–VII.

Omphalogramma delavayi (Franch.) Franch.* Tufted **perennial** to 15 cm in flower, twice the height in fruit. **Leaves** oblong to ovate or rounded, only partly developed at flowering time, to 10 × 7 cm eventually, margin often slightly toothed. **Flowers** often numerous, with erect hairy scapes; corolla rose-purple, with limb 3–3.5 cm wide and tube 4–5 cm long, downy on outside; lobes usually 6, with sharply toothed margin; style hairy in lower part. wYunnan; grassy places, low alpine shrubberies, 3,300–4,000 m. VI–VII.

Omphalogramma elegans Forrest Similar to *O. delavayi*, but **leaves** usually developing after the flowers; **corolla** deep violet, larger, limb 4–6 cm across with notched, not toothed, lobes; style glabrous. seTibet, nwYunnan [nMyanmar]; shrubberies, peaty places along woodland margins, 3,200–4,700 m. VI–VII.

Omphalogramma tibeticum Fletcher* Closely related to *O. elegans*, but **leaves** often partly developed at flowering time and flowers deep violet with whitish or yellowish throat; **corolla** glandular-downy outside; style glandular-downy. seTibet (Bomi Xian); grassy glades and banks, open woodland, shrubberies, c. 4,000 m. VI–VII.

Omphalogramma minus Hand.-Mazz.* Similar to the preceding 3 species, but **flowers** small, corolla-tube not exceeding 9 mm wide and limb 20–30 mm across. Flowers deep indigo-blue to purplish pink. swSichuan, e & seTibet, nwYunnan; low shrubberies on rocky slopes, 3,500–4,000 m. VI–VII.

BRYOCARPUM

A genus with a single species primarily endemic to the eastern Himalaya, closely related to *Omphalogramma*, but with a **fruit-capsule** that is circumscissile (opening by a lid).

Bryocarpum himalaicum Hook. f. & Thomson Tufted rhizomatous **perennial** to 20 cm in flower, with a rosette of long-stalked, ovate to almost oblong **leaves**, 3–9 × 2–4 cm, glandular and finely hairy. **Flowers** solitary, scapose, up to 3 per leaf-rosette; calyx 7–12 mm, divided to base, with fine black glands; corolla yellow, 7-lobed, pendent, 15–25 mm long, lobes linear, fused near middle. seTibet (Medog Xian) [c & eHimalaya]; mixed or *Pinus* forests, in damp leaf litter, 3,000–4,000 m. V.

CORTUSA

A genus of c. 10 species (1 in China) from central Europe to Central and northern Asia. Very similar to *Primula*, but stamens fused into a membranous ring towards base, and anthers apiculate.

Cortusa matthioli L. subsp. *pekinensis* (V. A. Richter) Kitag Tufted **perennial herb** to 30 cm in flower, with rosettes of long-stalked, rounded to kidney-shaped **leaves**, 3.5–8 × 4–8 cm, with 7–11 sharply toothed lobes, downy or becoming hairless. **Inflorescence** a 5–10-flowered umbel, **flowers** pendent or almost so; corolla funnel-shaped, purple-violet, 8–12 mm long with protruding style. Gansu, Hubei, Shaanxi [n & nwChina, Korea, Russia]; forests and forest margins, shrubberies, streamsides, damp banks, 1,500–3,000 m. V–VII. **Subsp. *matthioli*** is found in nChina, Asia, Europe and has ± blunt leaf-lobes.

ANDROSACE

A genus of c. 100 species (73 in China of which many are endemic) distributed in temperate regions of the northern hemisphere. **Tufted** or **cushion-forming** plants, (sometimes rosette-forming monocarpic herbs), with lax to dense rosettes of **leaves** that can be tiny to large, entire to variously toothed or lobed. **Flowers** solitary or in umbels on a distinct scape, flowers 5-parted, homostylous; calyx usually campanulate, shallowly to deeply lobed; corolla with short tube, equalling or shorter than calyx, usually constricted at mouth and with 5 spreading lobes. **Fruit** a small, 5-parted capsule.

Synopsis of groups

Group One. Flowers solitary or paired. (p. 260)

Group Two. Flowers in distinct umbels; plants usually with a single leaf-rosette. (p. 260)

Group Three. Flowers in distinct umbels; plants with many rosettes, forming mats, tufts or cushions; umbels many-flowered, borne on erect scapes 8–25 cm long. (p. 263)

Group Four. As Group 3 but umbels few-flowered (normally 2–6), borne on short scapes not exceeding 7 cm long. (p. 263)

Group One

1. Leaves glabrous except for a ciliate margin .. *A. delavayi*
1. Leaves hairy ....
   .... *A. tapete, A. selago, A. tangulashanensis*

Androsace delavayi **Franch**. Plants forming low **mounds** or small **cushions** of tightly packed globular grey-green or yellowish green rosettes; obovate, 4–5 mm long, ciliate. **Flowers** borne on very short (2–4 mm) stalks just clear of rosettes, almond-scented; corolla pink or white, 7–8 mm in diameter, with yellowish or greenish eye. swSichuan, seTibet, nwYunnan [swTibet, Himalaya, nMyanmar]; rock-crevices and other rocky places, consolidated moraines, low alpine scrub, 3,000–4,800 m. VI–VII.

Androsace tapete **Maxim**. **Tap-rooted** plants forming neat grey-green or silvery-grey **cushions** of tightly packed, globular leaf-rosettes; **leaves** elliptic, 2–3.5 mm long, silky with hairs in upper part, especially beneath. **Flowers** always solitary and stalkless; corolla white, 3–5 mm in diameter, with yellow or greenish eye which reddens after pollination. s & swGansu, Qinghai, w, nw & swSichuan, se & eTibet, nwYunnan [swTibet, sXinjiang, nHimalaya]; dry shaly and rocky slopes, sparse alpine meadows, sometimes in open alpine scrub, 3,000–5,500 m. V–VII.

Androsace selago **Hook. f. & Thomson ex Klatt.** Very similar to the previous species and forming dense **cushions**, but **flowers** solitary or paired on short stalks 2–7 mm long, with a bract at base of calyx; corolla 4–6 mm in diameter w & nwSichuan, se & eTibet [swTibet, eHimalaya]; rocky places, cliff-ledges, 3,600–5,400 m. VI–VII.

Androsace tangulashanensis **Y. C. Yang & R. F. Huang*** Very similar to *A. selago*, but plants tending to form **mats** rather than cushions; **leaves** 2.5–3 mm long, lanceolate, glabrous; **corolla** 6–7 mm in diameter. s & swQinghai [nTibet]; grassy tundra, bleak stony slopes, riverbanks, 4,000–5,000 m. VII–VIII.

Group Two

Androsace bulleyana **Forrest*** (syn. *A. aizoon* Duby var. *coccinea* Franch.) Plants usually **biennial**, forming a spreading, deep green **rosette** up to 15 cm across, often smaller, the stiff spathulate **leaves** with pointed apex and papery margin adorned with short cilia. Scapes up to 9 per plant, to 20 cm long (generally half that height), bearing a semi-hemispherical **umbel** of up to 25 flowers; **corolla** crimson to bright scarlet, 8–15 mm in diameter, with small yellow eye. swSichuan, w & nwYunnan; dry rocky slopes, open coniferous forests, roadside banks, 1,800–3,200 m. V–VI.

Androsace integra **(Maxim.) Hand.-Mazz.*** (syn. *A. aizoon* Duby var. *integra* Maxim.) Similar to *A. bulleyana*, but plants with blunter **leaves** and up to 18 scapes, varying in length on the same plant; **corolla** flesh-pink to bright pink, orange-pink

OPPOSITE: *Androsace rigida* with *Rhododendron oreotrephes*; valley below Daxueshan, nwYunnan (CGW)

1. *Androsace delavayi*; Baimashan, nwYunnan (CGW)
2. *Androsace delavayi*; Wutoudi, Yulongxueshan, nwYunnan (CGW)
3. *Androsace tapete*; Daxueshan, nwYunnan (HJ)
4. *Androsace tapete*; Daxueshan, nwYunnan (HJ)
5. *Androsace selago*; pass to Huanglong, nwSichuan (PC)
6. *Androsace selago*; Huanglong, nwSichuan (CGW)
7. *Androsace tangulashanensis*; Qinghai (HJ)
8. *Androsace tangulashanensis*; Qinghai (HJ)
9. *Androsace bulleyana*; Zhongdian, nwYunnan (CGW)
10. *Androsace integra*; Songpan, nwSichuan (PC)

1. *Androsace henryi*; Wolong, wSichuan (CGW)
2. *Androsace limprichtii*; swSichuan (HJ)
3. *Androsace spinulifera*; Napahai, Zhongdian, nwYunnan (HJ)
4. *Androsace spinulifera*; Xiaozhongdian, nwYunnan (CGW)
5. *Androsace alchemilloides*; Gangheba, Yulongxueshan, nwYunnan (CGW)
6. *Androsace wardii*; Baimashan, nwYunnan (HJ)
7. *Androsace rigida*; Daxueshan, nwYunnan (CGW)
8. *Androsace rigida*; above Xiaozhongdian, nwYunnan (CGW)
9. *Androsace minor*; Cangshan, nwYunnan (CGW)
10. *Androsace minor*; near Kanding, wSichuan (HJ)
11. *Androsace mariae*; Hezuozhen, Qinghai (HJ)
12. *Androsace mariae*; north of Markham, seTibet (DZ)

or reddish pink, only 6–10 mm in diameter. seQinghai, w & nwSichuan, neTibet, nwYunnan; stony and earthy banks, open shrubberies, 2,500–3,000 m. VI–VII.

Group Three
1. Leaves kidney-shaped or rounded, toothed .... **A. henryi**
1. Leaves not as above, untoothed 2
2. Plants stoloniferous .... **A. limprichtii**
2. Plants non-stoloniferous, clump-forming .. **A. spinulifera**

Androsace henryi Oliv. Rather robust plant to 25 cm, forming a cluster of open **rosettes** of bluish green, long-stalked, hairy **leaves** with neatly toothed blades up to 6 cm across. **Umbels** held well above foliage, hemispherical, each with up to 25 flowers; **corolla** white, 8–10 mm in diameter, with notched lobes. sGansu, wHubei, mountainous wSichuan, seTibet, nwYunnan [eHimalaya, Myanmar]; dense damp forests and glades in shade or semi-shade, 1,550–3,200 m. V–VI.

Androsace limprichtii Pax & K. Hoffm.* Stoloniferous, mat-forming plant. **Rosettes** with trimorphic **leaves**: outer small and spathulate, withered by flowering time, middle **leaves** stalkless, oval with rounded tip, silky-hairy, inner 3 times the size, up to 30 mm long, oar-shaped with slender stalk. **Umbels** on scapes up to 15 cm long, with up to 17 flowers; **corolla** white or pale pink, 8–9 mm in diameter, with rounded lobes. w & swSichuan, nwYunnan, neighbouring parts of Tibet; open forests and shrubberies, rocky slopes, cliffs, 3,800–4,500 m. V–VI.

Androsace spinulifera (Franch.) R. Knuth* Clump-forming **perennial** to 30 cm in flower, overwintering as tight, globular, *Sempervivum*-like rosettes, 20–25 mm across. **Winter leaves** ovate to lanceolate, hairy only along margin, tip spine-like; **summer leaves**, in contrast, are oblanceolate, erect and up to 18 cm long, narrowed towards base and minutely bristly all over. Scapes erect, bearing many-flowered hemispherical **umbels**; **corolla** bright rose-pink to red or rose-purple, 8–10 mm in diameter, with yellow eye which reddens after pollination. w & swSichuan, nwYunnan; open coniferous and deciduous forests, glades in shrubberies, grassy or gravelly slopes, 2,900–4,500 m. V–VI.

Group Four
1. Leaves lobed .... **A. alchemilloides**
1. Leaves not lobed 2
2. Flowers pink or red, rarely white 3
2. Flowers white or yellow, occasionally pink in bud ... 4
3. Plant tufted or cushion-forming, non-stoloniferous .... **A. wardii**
3. Plants stoloniferous .... **A. rigida, A. minor, A. mariae, A. mairei**

4. Leaves thickened at margin and appearing 2-furrowed .... **A. bisulca, A. bisulca** var. **bisulca, A. bisulca** var. **aurata, A. bisulca** var. **brahmaputrae**
4. Leaves neither thickened at margin nor appearing 2-furrowed .... **A. zambalensis, A. yargongensis, A. brachystegia, A. stenophylla**

Androsace alchemilloides Franch.* Tufted **perennial** rarely more than 6 cm with spreading **branches** covered in old leaf-remnants. **Leaves** in terminal rosettes, kidney-shaped, but deeply divided into 3 stalked segments that are further divided into narrow untoothed lobes. **Umbels** few-flowered, on scapes to 40 mm long; **corolla** white or pale pink, 7–9 mm in diameter, lobes rounded to slightly notched. swSichuan, seTibet, nwYunnan; rock-clefts, particularly in limestone cliffs, in partial shade, 3,000–4,500 m. V–VI.

Androsace wardii W. W. Sm.* Tufted or laxly cushion-forming plant to 7 cm in flower. **Rosettes** lax and rather open, **leaves** dimorphic, outer obovate, to 10 mm long, white-hairy beneath, inner up to 3 times larger, oblanceolate, narrowed into a stalk at base and hairy all over. **Umbels** with up to 8 flowers; **corolla** cherry-pink to deep red, 6–8 mm in diameter, lobes rounded to slightly notched. swSichuan, seTibet, nwYunnan; rocky places, amongst dwarf rhododendrons, 3,400–4,600 m. VI–VII.

Androsace rigida Hand.-Mazz.* Very variable mat- to lax cushion-forming plant with open leaf-rosettes. **Leaves** somewhat fleshy, ciliate, variable, overwintering ones linear-spathulate, to 20 mm long, summer (inner) leaves broadly elliptic to spathulate, to 20 mm long, tip with a short point. **Umbels** with up to 7 flowers on scapes to 45 mm long; **corolla** pink, 7–10 mm in diameter, lobes slightly notched. w & swSichuan, w & nwYunnan; coniferous forests and forest margins, *Rhododendron* shrubberies, mossy banks in semi-shade, scrubby alpine meadows, 2,500–3,800 m. V–VII.

Androsace minor (Hand.-Mazz.) C. M. Hu & Y. C. Yang* Smaller plant than previous species, **leaves** up to 6 mm long with rigid hairs but tip without a sharp point. Scapes to 25 mm, bearing a tight **umbel** of up to 10 flowers; **corolla** pink or white, 5–6 mm in diameter. w & swSichuan, nwYunnan; open coniferous forests and forest margins, meadows, 3,600–4,700 m. V–VI.

Androsace mariae Kanitz* (syn. *A. tibetica* Knuth) Differs from the previous 2 species in having usually glabrous inner **leaves**, except for the ciliate, pale, papery margin, and tip pointed but not with a sudden point. Scapes up to 10 cm long, but very variable, with not more than 5 flowers; **corolla** pink to rose-red, purple or blue-purple, rarely white, with yellowish eye that becomes crimson after pollination. eQinghai, w & nwSichuan, Shaanxi, seTibet [Nei Mongol]; rocky places, banks, open shrubberies, forest margins, 2,000–5,000 m. VI–VII.

Androsace mairei H. Lévl.* Differs from *A. rigida* in the outer narrow-spathulate **leaves** being no more than 5 mm, the inner oblanceolate to elliptic, to 16 mm long, sparsely hairy but with a white-ciliate margin, and without a short point at tip. **Umbels** usually 5–6-flowered, on scapes 20–30 mm long; **corolla** pink, 5–6 mm in diameter. wHubei, neSichuan, nw & neYunnan; rocky places in open forests, 3,000–3,500 m. VI–VII.

Androsace bisulca Bur. & Franch.* Cushion-forming plant with dense, small leaf-rosettes; **leaves** elliptic to lanceolate, 4–12 mm long, outer smaller than inner but all ciliate, with a tuft of hairs at tip and a prominent midrib beneath. **Umbels** 2–8-flowered, on scapes up to 30 mm long; **corolla** white with yellow eye (**var. bisulca**) or egg-yolk-yellow (**var. aurata** (Petitm.) Y. C. Yang & R. F. Huang*, restricted to swSichuan), or pink (**var. brahmaputrae** Hand.-Mazz.*, restricted to seTibet), 8–10 mm in diameter, with rounded lobes. w & swSichuan, e & seTibet; dry grassy places, slopes along forest margins, sandy banks by rivers, 3,000–4,500 m. VI–VII.

Androsace zambalensis (Petitm.) Hand.-Mazz. Plants forming tight mats of small globular, rather yellowish green **rosettes**. **Leaves** oblong-elliptic, curving inwards, 7–12 mm, keeled beneath, covered with short hairs on both surfaces and with bristly-ciliate margin. **Umbels** 3–5-flowered, on short scapes not more than 18 mm long; **corolla** white, rarely pink, 6–8 mm in diameter, with yellow eye that turns red after pollination. w & swSichuan, seTibet, nwYunnan [swTibet, Himalaya]; stony slopes, consolidated moraines, snow hollows, shaly slopes, 4,000–5,200 m. VI–VII.

Androsace yargongensis Petitm.* Dense mat-forming plant with closely packed globular **rosettes**; **leaves** rather leathery, elliptic, 3–9 mm long, inner often longer and narrower than outer, often reddish towards blunt tip, hairy beneath, margin with soft hairs at right angles to plane of leaf. **Umbels** 3–6-flowered, on stalks to 30 mm long; **corolla** white, sometimes pink in bud, 7–10 mm in diameter, lobes usually slightly notched. swGansu, w, nw & swSichuan, nwYunnan, neighbouring parts of Tibet; rocky places, moraines, stony slopes, 3,600–5,000 m. VI–VII.

Androsace brachystegia Hand.-Mazz.* Similar to the previous species, but bracts are shorter (not longer) than the accompanying stalks, and **inflorescences** are only 1–3-flowered, on scapes up to 40 mm long; **flower** white or pale pink. s & seGansu, seQinghai, nwSichuan; open alpine scrub and grassy slopes, 4,000–4,500 m. VI–VII.

Androsace stenophylla (Petitm.) Hand.-Mazz.* Taller plant (5–20 cm) than previous species, with narrow-lanceolate to spathulate **leaves** to 12 mm long, minutely hairy or glabrous. **Umbels** on long, slender, hairy stalks, usually 6–12-flowered; **corolla** white or pale pink, limb 6–8 mm in diameter. w & nwSichuan, e & seTibet; open woodland, grassy slopes, cliffs, 2,900–4,200 m. VI–VII.

UNNAMED SPECIES

Androsace aff. *bisulca* Bur. & Franch.* Small laxly tufted plant not more than 5 cm tall, with **rosettes** of small linear-elliptic, deep green, slightly hairy **leaves** with a pointed apex. **Flowers** in **umbels** of 3–7, corolla white with tiny yellow eye that changes to red, limb 7–8 mm in diameter, lobes elliptic-oval. nwSichuan (Huanglongsi); exposed limestone rocks and ledges, c. 3,990 m. VI–VII. Almost certainly an undescribed species.

Androsace aff. *mariae* Kanitz* Lax cushion- or mat-forming plant up to 20 cm across, composed of numerous small **leaf-rosettes**; **leaves** elliptic-spathulate, grey-hairy, 7–10 mm long, apex mucronate. **Flowers** in rounded, rather dense umbels of up to 12 flowers, on short, 10–30 mm stalks; corolla white with tiny yellow eye that soon turns red, limb 8–10 mm in diameter, with oblong to rounded, entire lobes. nwSichuan (Huanglongsi); open grassy slopes, roadside banks, c. 3,400 m. VI–VII.

POMATOSACE

A genus of a single species endemic to China with **flowers** very similar to *Androsace*, but **leaves** pinnately lobed and **capsule** with a cap rather than valves.

Pomatosace filicula Maxim.* **Annual** or **biennial herb** to 15 cm tall at most, forming flat rosettes of pinnately lobed **leaves**, lobes toothed or not. **Inflorescence** a 6–12-flowered umbel with linear bracts; calyx cup-shaped with 5 triangular lobes; **corolla** white, tube very short and limb 2–3 mm in diameter, 5-lobed. e & seQinghai, nwSichuan, neighbouring parts of Tibet; grassy slopes, banks, open shrubberies, stony places, 2,800–4,500 m. V–VII.

---

1. *Androsace mairei*; Gangheba, Yulongxueshan, nwYunnan (CGW)
2. *Androsace mairei*; Gangheba, Yulongxueshan, nwYunnan (CGW)
3. *Androsace bisulca* var. *aurata*; north of Daxueshan, nwSichuan (CGW)
4. *Androsace bisulca* var. *brahmaputrae*; west of Gyamda, seTibet (DZ)
5. *Androsace zambalensis*; Dungda La, seTibet (HJ)
6. *Androsace zambalensis*; Dungda La, seTibet (HJ)
7. *Androsace yargongensis*; Baimashan, nwYunnan (CGW)
8. *Androsace brachystegia*; Huanglong, nwSichuan (CGW)
9. *Androsace brachystegia*; Zheduo, swSichuan (JR)
10. *Androsace stenophylla*; Huanglong, nwSichuan (CGW)
11. *Androsace* aff. *bisulca*; pass above Huanglong, nwSichuan (CGW)
12. *Androsace* aff. *mariae*; Huanglong, nwSichuan (CGW)
13. *Pomatosace filicula*; Songpan, nwSichuan (PC)

PRIMULACEAE: Lysimachia

LYSIMACHIA

A genus of 180 species (two-thirds found in China, many endemic) scattered in the temperate and subtropical northern hemisphere, extending south of the equator in places. Procumbent to erect **herbs**; **leaves** alternate, opposite or whorled. **Flowers** solitary, or in racemes or panicles, 5–9-parted; corolla white, yellow or rarely pink, campanulate to rather flat, petals fused only at base. **Fruit** a valved capsule.

1. Flowers yellow .... *L. drymarifolia*
1. Flowers white or pink ....
   .... *L. prolifera, L. taliensis, L. delavayi*

Lysimachia drymarifolia Franch.* **Tufted perennial** to 20 cm, often far less, with procumbent **stems** covered in rust-coloured hairs. **Leaves** paired, stalked, blade ovate to heart- or kidney-shaped, 10–25 × 8–22 mm, with black glandular streaks. **Flowers** solitary in leaf-axils, on stalks to 40 mm long; corolla butter-yellow, with short (2 mm) tube and limb 10–16 mm across, the ovate to oblong lobes with a few black streaks. swSichuan, n, nw & cYunnan; damp habitats along streams and in forests, 1,400–3,500 m. IV–V.

Lysimachia prolifera Klatt. **Perennial** patch-forming **herb** to 25 cm, **stems** often prostrate at base, finely glandular-hairy. **Leaves** grey-green, paired, with a narrowly winged stalk, blade ovate to oblanceolate, 7–20 × 6–12 mm, glabrous, with sparse dark dot-like glands or streaks. **Flowers** in lateral clusters of 3–5; calyx-lobes lanceolate, covered in dark purple or black glands or streaks; corolla white or pink, 12–16 mm across, with a short, 1.5 mm tube and spreading obovate lobes. w & swSichuan, seTibet, nwYunnan [swTibet, Himalaya]; wet habitats, open woodland, seepage zones, mountain slopes, gravels, 2,700–3,200 m. V–VI.

Lysimachia taliensis Bonati* Readily distinguished from previous species by being an erect **perennial** to 80 cm; in having stalkless, lanceolate to linear **leaves** to 8 cm long in whorls of 3–4; and **flowers** in terminal racemes to 7 cm long; calyx-lobes linear-lanceolate. w, nw & cYunnan; grassy slopes, open forests and forest margins, shrubberies, 2,600–3,800 m. IV–V.

Lysimachia delavayi Franch.* Easily distinguished by the square **stems** to 50 cm, and mainly alternate (rarely opposite or whorled) lanceolate to linear **leaves** to 80 mm long; **flowers** in terminal, cone-shaped racemes that gradually elongate, calyx-lobes ciliate, with glandular stripes outside; corolla pink. nYunnan; *Pinus* and *Quercus* woodland, 2,100–2,900 m. V–VI.

1. *Lysimachia drymarifolia*; Lancangjiang–Jinshajiang divide, Lidiping, nwYunnan (CGW)
2. *Lysimachia prolifera*; Paro, Bhutan (CGW)
3. *Ceratostigma willmottianum*; wSichuan (TK)
4. *Ceratostigma willmottianum*; cult. (PC)
5. *Prunus serrula* bark; cult. (CGW)
6. *Prunus wilsonii*; Emeishan, sSichuan (PC)

MYRSINACEAE

MYRSINE

A genus of c. 300 species (11 in China), distributed from the Azores and Africa to Asia. **Shrubs** or small **trees** with entire **leaves**, and hermaphrodite or unisexual, 4–5(–6)-parted **flowers**.

Myrsine semiserrata **Wall.** (syn. *Celastrus cavaleriei* H. Lévl.; *C. seguinii* H. Lévl.; *Myrsine semiserrata* var. *brachypoda* Z. Y. Zhu) **Shrub** or **tree** 3–7 m tall with slightly angular, glabrous branchlets. **Leaves** slightly asymmetric, glossy, elliptic to lanceolate, 5–9(–14) × 2–2.5(–4) cm, pointed to long-pointed, glabrous, margin remotely and minutely toothed to middle; stalks 6–8 mm long, decurrent at base. **Inflorescences** axillary, umbellate or in bundles, stalkless, with ciliate, dotted bracts. **Flowers** white to yellowish, 2 mm across, 4-parted; petals prominently dotted to middle, ciliate. **Fruit** red, becoming purple-black, globose, 5–7 mm in diameter, densely dotted. Guizhou, Hubei, Sichuan, Tibet, Yunnan [sChina, c & eHimalaya, Myanmar]; broadleaved forests, limestone hillsides, mountain slopes, road- and streamsides, 500–2,700 m. II–IV.

PLUMBAGINACEAE

CERATOSTIGMA

A genus of 8 species (5 in China) from east Africa and Asia. Small **shrubs** or **perennial herbs** with alternate **leaves** that have incurved hairs along margin. **Flowers** in short spikelets making terminal or lateral clusters surrounded by leaf-like bracts; calyx tubular, splitting between the ribs at fruiting stage; corolla salverform with a slender tube extending well beyond calyx, 5-lobed; stamens 5. **Fruit** a capsule enclosed within the calyx.

Ceratostigma willmottianum **Stapf*** Spreading, rhizomatous, **herbaceous perennial** to 1 m, occasionally more, with rather fragile branchlets. **Leaves** ovate to obovate or diamond-shaped, 2–5 × 1.2–2 cm, those below flowers generally narrower; leaf-stalks with clasping bases, leaving a ring-like scar when they fall. **Inflorescence** 3–7-flowered, terminal; calyx 10–15 mm long, sparsely bristly along ribs; **corolla** blue with reddish purple tube, limb 20–24 mm across, lobes slightly notched at apex. Gansu, wGuizhou, s, w & swSichuan, seTibet, Yunnan; forest margins, shrubberies, sunny banks, roadsides, 700–3,500 m. VII–X.

Ceratostigma minus **Stapf ex Prain*** In contrast to the previous species, a small **deciduous shrub** to 1.5 m with **leaves** not exceeding 3 × 1.6 cm; **flowers** smaller but more numerous, calyx 6–9 mm long, corolla-limb 14–18 mm across. sGansu, w & swSichuan, e & seTibet, w & nwYunnan; dry slopes along valleys, open shrubberies, 1,000–4,800 m. VI–X.

ROSACEAE

PRUNUS

A genus of c. 330 species (92 in China) distributed in temperate Asia, Europe and North America. **Deciduous shrubs** or **trees**, the branchlets with or without spines. **Leaves** convolute, alternate or in bundles, simple, margin toothed (teeth often gland-tipped); leaf-stalks usually with 2 apical nectaries. **Inflorescences** axillary, 1–2-flowered, in bundles or corymbose. **Flowers** opening before or with the leaves, regular, stalked; hypanthium campanulate or tubular; sepals 5, reflexed or erect; petals 5, spreading or erect; stamens 15–50; carpel solitary, glabrous or hairy. **Fruit** a 1-seeded drupe.

The genus has been divided into 6 genera (*Amygdalus*, *Armeniaca*, *Cerasus*, *Laurocerasus*, *Padus* and *Prunus*) in the recent *Flora of China* treatment, but the traditional treatment is preferred here.

Prunus serrula **Franch.** (syn. *Cerasus serrula* (Franch.) T. T. Yu & C. L. Li) Small to medium-sized **tree**, 2–12 m tall with lustrous, greyish to purplish brown **bark**, peeling in horizontal papery strips. **Leaves** lanceolate to lanceolate-ovate, 3.5–7 × 1–2 cm, long-pointed, margin pointedly toothed or doubly toothed; stalks 5–8 mm long, hairy. **Inflorescence** 1–2-flowered; **flowers** 10–12 mm across; sepals ovate-triangular, 3 mm long; petals white, obovate-elliptic, 4–5 mm long; stamens 38–44. **Fruit** egg-shaped, 1 cm long, purplish black when ripe. Guizhou, Qinghai, Sichuan, e & seTibet, Yunnan [c & eHimalaya, nMyanmar]; forests and scrub on mountain slopes, ravines, mountain meadows, 1,200–4,000 m. IV–V.

Prunus wilsonii **(C. K. Schneid.) Diels ex Koehne*** (syn. *P. sericea* (Batalin) Koehne; *Padus wilsonii* C. K. Schneid.) **Deciduous tree** to 30 m with purplish brown or blackish brown **branches**. **Leaves** elliptic to oblong or obovate, 6–14 × 3–6 cm, dark green above, sometimes purple-flushed, pale and downy beneath, margin usually toothed; stalks 7–8 mm long with a pair of nectaries at top. **Flowers** many, in narrow lateral racemes to 14 cm long with 2–5 leaves at base, each flower 6–8 mm across; petals white, oblong; stamens c. 20. **Fruit** rounded to egg-shaped, 8–10 mm, reddish brown ripening to blackish purple. Gansu, Guangxi, Guizhou, Hubei, Sichuan, Shaanxi, e & seTibet, Yunnan [c & eChina]; valley slopes, open woodland, forest margins, glades, 900–2,500 m. IV–V.

MADDENIA

A Himalayan genus of c. 7 species (6 in China, 5 endemic). **Deciduous trees** and **shrubs** with unarmed **branches** and alternate simple **leaves** with gland-tipped teeth. **Inflorescence** many-flowered racemes terminating short branchlets. **Flowers** hermaphrodite or female; sepals and petals ± identical, 10–12 small segments; stamens 20–40 in 2 whorls; style solitary and slender. **Fruit** a small, 1–2-seeded drupe.

Maddenia himalaica Hook. f. & Thomson **Shrub** or small **tree** to 8 m with reddish brown softly hairy **shoots**, becoming purplish brown and hairless with age. **Leaves** dark green and glabrous above, paler and usually softly hairy beneath, blade elliptic to oblong, 5–15 × 1.8–5 cm. **Racemes** rather dense, to 6 cm long; **flowers** greenish or reddish brown, segments soon falling; stamens 20–30, equalling style. Drupe purple when ripe, 8–9 mm. e & seTibet [c & eHimalaya]; forests, shrubberies, 2,800–4,200 m. V–VI.

Maddenia hypoleuca Koehne* Readily distinguished from other species by **leaves** that are whitish glaucous, glabrous beneath; **tree** to 7 m with purplish branchlets and leaves usually not more than 9 × 4 cm; drupe purplish black when ripe. Gansu, wHubei, Shaanxi; open forests, shrubberies, 1,000–1,800 m. IV–V.

Maddenia wilsonii Koehne* Similar to *M. himalaica*, but young branchlets hairy, although not softly hairy, eventually becoming hairless and shiny brown or purple-brown, winter **buds** purplish brown (not purplish red) and style noticeably longer than the 30–40 stamens; drupe black when ripe. sGansu, Guizhou, wHubei, Sichuan, Shaanxi; open slopes, shrubberies, river margins in open sunny places, 1,500–3,600 m. IV–VI.

PRINSEPIA

A genus of c. 5 species (4 in China) in the Himalaya, from Pakistan to north Myanmar and west China. Erect, spreading or climbing **shrubs**, branches with axillary spines and pith. **Leaves** alternate or in bundles, simple, stipulate with a short stalk. **Inflorescences** 1-flowered or racemose, solitary or in bundles on short shoots in leaf-axils of the previous year's branches. **Flowers** hermaphrodite, regular, stalkless or almost so; hypanthium mouth with an annular disc; sepals 5, unequal, persistent; petals 5, almost orbicular; stamens 10 or more in 2 or 3 whorls attached to hypanthium rim; ovary superior, glabrous, 1-seeded. **Fruit** a drupe with a fleshy mesocarp.

Prinsepia utilis Royle Spreading **shrub**, 1–5 m tall with arching, robust, grey-green, angled **branches**; spines to 3.5 cm long. **Leaves** oblong to ovate-lanceolate, 3.5–9 × 1.5–3 cm, pointed to long-pointed, margin minutely toothed. **Racemes** 3–6 cm long, many-flowered, stalk and bracts brown-hairy. **Flowers** 9–10 mm across; sepals broadly ovate, brown-hairy; petals white, broadly obovate, 4–5 mm long. Drupe oblong-obovoid-egg-shaped, 8 mm long, dark purplish black when ripe. Guizhou, Sichuan, Tibet, Yunnan [Himalaya, nIndia, nMyanmar]; mountain slopes, valleys, thickets and shrubberies, roadsides, 1,000–2,600 m. IV–VI.

RUBUS

A cosmopolitan and complex genus with over 700 species (200 in China). Small to large, **deciduous** or **semi-evergreen shrubs** or **subshrubs**, more rarely dwarf **creeping herbs**. **Stems** erect to arching or prostrate, leafy, usually with prickles and/or bristles. **Leaves** alternate, stalked, simple, pinnate or palmate, toothed, glabrous or hairy; stipules persistent, ± joined to stalks. **Inflorescence** 1–many-flowered, racemose, paniculate or corymbose. **Flowers** small to medium, white, pink, red or purple; sepals 5, persistent; petals 5, inserted on rim of hypanthium; stamens many; carpels many. **Fruit** fleshy, consisting of a conical or rounded collection of drupelets easily detached from the torus.

The genus contains many edible 'berries' including the familiar blackberry and raspberry.

1. Plants shrubby or suffruticose, creeping, stems densely bristly; leaves simple; stipules free .... ***R. tricolor***
1. Large bushy shrubs with prickly branches; leaves pinnate, stipules joined to leaf-stalk 2
2. Stems and branchlets white .. ***R. biflorus, R. thibetanus***
2. Stems and branchlets not white .... ***R. rosifolius***

Rubus tricolor Focke* **Shrub** 1–4 m tall, often **patch-forming**, with climbing or creeping **stems**, branchlets brown or reddish brown, with bristles and glandular hairs. **Leaves** lustrous deep green, ovate to oblong, often somewhat 3-lobed, 6–12 × 3–8 cm, long-pointed, greyish yellow-woolly. **Inflorescence** terminal, few-flowered, or flowers solitary in leaf-axils, **flowers** 2–3 cm across; sepals lanceolate, 8–12 mm long, yellow-brown-woolly; petals white, obovate, 7–9 mm long. **Fruit** almost globose, 1.5–1.7 cm, bright red when ripe. Sichuan, Yunnan; forests and thickets on slopes, 1,800–3,600 m. V–VII.

Rubus biflorus Buch.-Ham. ex Sm. **Climbing** or **clambering shrub**, 1–3 m tall with arching whitish branches adorned with sparse curved prickles. **Leaves** odd-pinnate, with 3 or 5 leaflets, 2.5–5 × 1.5–4 cm, coarsely toothed or doubly toothed, densely grey or yellowish grey beneath, terminal leaflet broad-ovate to almost orbicular, lateral ovate or elliptic. **Inflorescence** terminal, corymbose, 4–6 cm across, 4–8-flowered, or 2–several-flowered in leaf-axils. **Flowers** 15–20 mm across; petals white, almost orbicular, 7–8 mm long. **Fruit** partly enclosed in calyx, globose, 10–15 mm, yellow when ripe. Gansu, Guizhou, Sichuan, Shaanxi, Tibet, Yunnan [c & eHimalaya, nMyanmar]; valleys, thickets, forests, roadsides, by rivers and streams, 1,500–3,500 m. V–VII.

1. *Maddenia himalaica*; Bhutan (CGW)
2. *Prinsepia utilis*; seTibet (CS)
3. *Prinsepia utilis* in fruit; Heishui, Yulongxueshan, nwYN (PC)
4. *Rubus biflorus*; Wolong, wSichuan (CGW)
5. *Rubus biflorus* (with *Tsuga dumosa*); Heishui, Yulongxueshan, nwYunnan (CGW)
6. *Rubus thibetanus*; cult. (CGW)
7. *Rubus rosifolius*; Wolong, wSichuan (PC)
8. *Rubus tricolor*; cult. (RL)
9. *Rubus tricolor*; Cangshan, nwYunnan (PC)

1. *Sorbus megalocarpa*; cult. (RL)
2. *Sorbus megalocarpa*; cult. (CS)
3. *Sorbus pallescens*; near Liuba, wSichuan (RL)
5. *Sorbus discolor*; Deqin, nwYunnan (PC)
4. *Sorbus discolor*; cult. (CGW)
6. *Sorbus sargentiana*; cult. (RL)
7. *Sorbus rehderiana*; Deqin, nwYunnan (PC)
8. *Sorbus forrestii*; cult. (PC)
9. *Sorbus forrestii*; Baimashan, nwYunnan (PC)
10. *Sorbus forrestii*; cult. (CGW)

Rubus thibetanus Franch.* In contrast to the previous species, an erect **shrub**, 2–3 m tall, with reddish brown branchlets with a white bloom, densely softly hairy, with sparse prickles. **Leaves** with 7, 9 or 11 obliquely ovate to ovate-orbicular, long-pointed leaflets. **Flowers** 10–12 mm across; petals pink to purple, orbicular-ovate, 3–4.5 mm long; stamens numerous, purple. **Fruit** almost globose, 8–10 mm, purple-black or dark red when ripe. Gansu, Sichuan, Shaanxi, e & seTibet; grassland, landslips, thickets, open shrubberies, forests, roadsides, often in dry places, 900–2,100 m. V–VII.

Rubus rosifolius Sm. **Erect** or **climbing shrub**, 2–3 m tall with glaucous, grey-brown or dark reddish brown, softly hairy branches adorned with straight or curved prickles. **Leaves** odd-pinnate, with 5 or 7 ovate to ovate-elliptic, long-pointed leaflets, 4–7 × 1.5–5 cm, with long soft hairs and yellow glands on both surfaces. **Inflorescence** terminal or in leaf-axils, 1–2-flowered, flowers 2–3 cm across; petals white, oblong to obovate, 8–15 mm long. **Fruit** egg-shaped-globose, 10–15 mm across, red when ripe. Guizhou, Hubei, Sichuan, Shaanxi, Yunnan [India, Japan, seAsia]; grassland, landslips, shrubberies, forests, roadsides, to 2,800 m. V–VIII. Double-flowered forms are cultivated. (See photo 7 on p. 269.)

SORBUS

A complex genus of c. 100 species (67 in China) in northern temperate Asia, Europe and North America. **Shrubs** or small to large **deciduous trees**, with rather large, egg-shaped to conical **buds**. **Leaves** alternate, entire or pinnate, margins toothed; stipules present. **Inflorescence** compound or rarely simple, corymbose or paniculate, many-flowered. **Flowers** white, creamy white to pink (many are apomictic); hypanthium urn-shaped or campanulate; sepals 5, ovate to triangular, persistent, glabrous or hairy; petals 5, free; stamens 15–40; ovary semi-inferior or inferior, 2–5-celled. **Fruit** a small pome, white, yellow, pink, red or brown when ripe, several-seeded. Many species produce fiery autumn leaf colours.

1. Leaves simple .... **S. megalocarpa, S. pallescens**
1. Leaves odd-pinnate 2
2. Trees 5–15 m tall .... **S. discolor, S. pseudohupehensis, S. sargentiana, S. rehderiana, S. forrestii**
2. Shrubs or small trees up to 4 m tall ..... **S. koehneana, S. pseudovilmorinii, S. gonggashanica, S. prattii**

Sorbus megalocarpa Rehd.* **Shrub** or small **tree**, 5–8 m tall with dark purple branchlets, turning blackish brown when mature. **Leaves** 10–16 cm long, elliptic-obovate, 10–18 × 5–9 cm, long-pointed, margin scalloped to toothed, greyish green, paler and cottony-hairy beneath. **Inflorescence** terminal, compound, corymbose, many-flowered, 9–13 cm across. **Flowers** 5–8 mm; petals white, broadly ovate, blunt, 3 mm long; stamens 20, as long as petals. **Fruit** egg-shaped to globose, 2–3 cm across, dark red when ripe. Guangxi, Guizhou, Hubei, Sichuan, n & nwYunnan [Hunan]; valleys, by rivers and streams and in forests, often on rocky terrain, 1,200–2,700 m. V–VI.

Sorbus pallescens Rehd.* Small **tree** to 7 m with red- to purple-brown branchlets, sparsely hairy when young. **Leaves** elliptic, elliptic-obovate or ovate, 5–10 × 2–5 cm, pointed, margins doubly toothed, woolly on both surfaces when young, only along veins when mature. **Inflorescence** terminal, compound, corymbose, many-flowered, 4–4.5 cm across. **Flowers** 9–12 mm; petals white, triangular-ovate to ovate, blunt, 4–4.5 mm long; stamens 20, almost as long as petals. **Fruit** almost globose, 6–8 mm across, white flushed with red when ripe. w & swSichuan, seTibet, nwYunnan; mixed forests and thickets, streamsides, 2,000–3,300 m. V–VI.

Sorbus discolor (Maxim.) E. Goetze* (syn. *S. hupehensis* C. K. Schneid.) Small to medium-sized **tree** 5–10 m tall with purple-brown branchlets, turning greyish brown when mature, slightly white-hairy when young. **Leaves** dark green, 10–15 cm long; leaflets 9–17, oblong to oblong-lanceolate, 3–5 × 1–1.8 cm, margin toothed in upper half. **Inflorescence** many-flowered, 6–10 cm across. **Flowers** 5–7 mm across; petals white, ovate, blunt, 3–4 mm long; stamens 20; styles 3–5. **Fruit** globose, 5–8 mm across, white, sometimes tinged with pink, when ripe. Gansu, Hubei, Qinghai, Sichuan, Shaanxi, nwYunnan [c & neChina]; dense forests, thickets, gullies, ravines, mountain slopes, to 3,800 m. V–VI. Similar plants from nwYunnan (Lijiang), but with bluish grey leaflets and flowers with 3–4 styles are probably referable to ***S. pseudohupehensis*** MacAllister*

Sorbus sargentiana Koehne* Differs from previous species in having robust, stout grey-brown twigs with very sticky buds; 9–15 large lanceolate leaflets, to 13 × 4.5 cm, and large, persistent stipules; **fruit** large, bright orange, in very dense clusters. wSichuan; forests, ravines, shrubberies, open slopes, 2,000–3,200 m. V–VI.

Sorbus rehderiana Koehne Small **tree** (3–)4–8 m tall with reddish brown to greyish brown branchlets, hairy when young. **Leaves** shiny green, odd-pinnate, 10–15 cm long; leaflets 15–19, oblong to oblong-lanceolate, 2.5–5 × 1–1.5 cm, pointed, hairy and with red-brown hairs on midvein beneath. **Inflorescence** compound, corymbose, many-flowered, 3–5 cm across. **Flowers** 5–7 mm across; petals white, broadly ovate, blunt, 3–5 mm long; stamens 20. **Fruit** globose, 6–8 mm across, crimson to pinkish white when ripe. Qinghai, w & swSichuan, seTibet, nwYunnan [nMyanmar]; forests, thickets, ravines, gullies, mountain slopes, 2,800–4,300 m. V–VI.

Sorbus forrestii MacAllister & Gillham* Distinguished from previous species by **buds** that are white-hairy at tip, as are the bud-scales, and by the even-sized elliptic-oblong, scarcely toothed leaflets to 20 mm long. nwYunnan (Baimashan); open coniferous and *Rhododendron* forests, mountain slopes, 3,200–3,800 m. VI–VII.

Sorbus koehneana C. K. Schneid.* **Shrub** or small **tree**, 1.5–4 m tall with dark to black-grey, glabrous branchlets. **Leaves** odd-pinnate, 10–16 cm long; leaflets 17–25, oblong to oblong-lanceolate, 1.5–3 × 0.5–1 cm, pointed, glabrous or minutely downy beneath. **Inflorescence** compound, corymbose, many-flowered, 4–6 cm across. **Flowers** 8–9 mm across; petals white, broadly ovate, blunt, 4–6 mm long; stamens 20. **Fruit** globose, 6–8 mm across, white when ripe. Gansu, Hubei, Qinghai, Sichuan, Shaanxi, wYunnan [cChina]; forests, thickets, gullies and ravines, in valleys and on mountain slopes, 2,300–4,000 m. V–VI. Plants from swSichuan, nwYunnan and neighbouring parts of Tibet, that have leaflets toothed only at apex, with impressed veins beneath, and **fruit** with reddish brown hairs near base, are referable to *S. pseudovilmorinii* MacAllister: the fruit can be crimson at first but mature to white with crimson flecks.

Sorbus gonggashanica MacAllister* Readily distinguished from the previous species by the **leaves**, which have 17–23 pairs of ovate leaflets, by the larger **flowers**, 11–12 mm across, and by the inverted pear-shaped **fruit** 11–12 across, white with a pink flush around calyx. wSichuan (Liuba, Gonggashan); shrubberies, open woodland, rocky slopes, 2,300–3,300 m. V–VI.

Sorbus prattii Koehne Similar to the previous 2 species, but with dark to brown-grey branchlets that are sparsely hairy when young, **leaves** with 19–27, oblong to oblong-ovate leaflets, stipules lanceolate to ovate, those of upper leaves green and leaflet-like; petals broadly ovate, 4–5 mm long, blunt; **fruit** globose, 7–8 mm, white when ripe. wSichuan, seTibet, nwYunnan [swTibet, Bhutan, nIndia]; coniferous and mixed forests on mountains, 2,000–4,500 m. V–VII.

CRATAEGUS

A genus with c. 100 species (18 in China) scattered in north temperate regions of the Old and New Worlds. **Deciduous shrubs** or small **trees** with armed, thorny **branches**. **Leaves** simple, stipulate, stalked, usually lobed, margin toothed. **Inflorescence** solitary, or more usually corymbose, few–many-flowered. **Flowers** white, rarely flushed with pink, sweetly scented; sepals and petals 5; stamens 5–25; ovary inferior or semi-inferior. **Fruit** a pome with persistent sepals at apex, cells 1-seeded.

Crataegus chungtienensis W. W. Sm.* **Shrub** or small **tree** to 6 m with grey-brown branchlets and largish purplish brown **buds**. **Leaves** broadly ovate, 4–7 × 3.5–5 cm, densely hairy on veins beneath. **Inflorescence** 3–4 cm across, corymbose, many-flowered, flowers 9–10 mm across; petals white, broadly obovate, 6 mm long. **Fruit** ellipsoid, 5–6 mm across, red. nwYunnan; shrubberies and woodland on steep rocky slopes, streamsides, 2,500–3,500 m. V–VI.

ERIOBOTRYA

A genus of c. 30 species (14 in China) in east Asia. **Evergreen shrubs** or **trees** with simple **leaves** with entire or toothed margins, stipulate, stalked. **Inflorescence** terminal, paniculate, many-flowered. **Flowers** regular; hypanthium cup-shaped or obconic; sepals 5, persistent; petals 5, free; stamens 20; ovary inferior. **Fruit** a pome, fleshy or dry, 2–5-celled, each cell containing 1–2 seeds.

Eriobotrya japonica (Thunb.) Lindl.* Small **tree** to 10 m with yellow-brown, densely rusty-hairy branchlets. **Leaves** deep grey-green, lanceolate, oblanceolate or elliptic-oblong, 12–30 × 3–9 cm, pointed or long-pointed, toothed towards apex, densely greyish rusty-hairy beneath. **Inflorescence** 10–19 cm long, many-flowered, stalk rusty-woolly. **Flowers** 1.2–2 cm across, fragrant; sepals triangular-ovate, 2–3 mm long, densely rusty-hairy; petals white, oblong to ovate, 5–9 mm long, blunt or notched. **Fruit** edible, plum-shaped, 1–2.5 cm, yellow or orange-yellow, containing large, black **seeds**. Hubei, widely planted elsewhere, to 2,000 m. V–VI.

OSTEOMELES

A genus with 5 species (3 in China) in east Asia. Small, **evergreen** or **deciduous subshrubs** or **shrubs** with odd-pinnate, stipulate **leaves** with a narrowly winged rachis; leaflets opposite, entire, stalkless or short-stalked. **Inflorescence** corymbose, many-flowered. **Flowers** white; sepals and petals 5; stamens 20; ovary inferior, 5-celled, each cell with 1 ovule. **Fruit** a small pome with persistent sepals.

Osteomeles schwerinae C. K. Schneid.* Small, **deciduous** or **evergreen shrub**, 1–4 m tall with reddish brown branchlets covered with dense whitish grey hairs when young. **Leaves** with 7–15 pairs of leaflets, these elliptic to oblong-elliptic, 5–10 × 2–4 mm, pointed to mucronate. **Inflorescence** 2–3 cm across, 3–5-flowered. **Flowers** 9–10 mm across; petals white, ovate-lanceolate, 5–7 mm long. **Fruit** egg-shaped to almost globose, 6–8 mm across, bluish black when ripe. Gansu, Guizhou, Shaanxi, Tibet, Yunnan; shrubberies, mixed woodland, fields, roadsides, rocky slopes, 1,000–3,000 m. IV–VI.

PYRACANTHA

A genus of c. 10 species (7 in China) in south Europe, west Asia and the Himalaya. **Evergreen shrubs** or small **trees**; branches armed with thorns. **Leaves** simple, alternate or in bundles, short-stalked, margin minutely scalloped, toothed or entire; stipules minute. **Inflorescence** compoundly corymbose, many-flowered. **Flowers** white, rarely pink-flushed, sweetly scented; hypanthium short; sepals and petals 5; stamens 5–25; ovary semi-inferior, 5-celled. **Fruit** a berry-like pome with persistent incurved sepals at apex, red, orange or yellow when ripe; cells 2-seeded.

1. *Sorbus koehneana*; Muli, swSichuan (PC)
2. *Sorbus koehneana*; cult. (CGW)
3. *Sorbus gonggashanica*; cult. (CGW)
4. *Crataegus chungtienensis*; Zhongdian, nwYunnan (PC)
5. *Crataegus chungtienensis*; Napahai, Zhongdian, nwYunnan (PC)
6. *Eriobotrya japonica*; cult., Kunming, cYunnan (CGW)
7. *Osteomeles schwerinae*; north of Dali, nwYunnan (PC)
8. *Pyracantha angustifolia*; Lijiang, nwYunnan (CGW)
9. *Pyracantha angustifolia*; near Dali, nwYunnan (CGW)

1. *Pyracantha crenulata*; cult. (CGW)
2. *Pyracantha crenulata*; Dali, nwYunnan (CGW)
3. *Pyracantha crenulata*; Dali, nwYunnan (CGW)
4. *Pyracantha fortuneana*; Shigu, nwYunnan (DC)
5. *Photinia davidiana*; south of Tuguancun, nwYunnan (PC)
6. *Photinia davidiana*; south of Tuguancun, nwYunnan (PC)
7. *Cotoneaster qungbixiensis*; cult., from nwYunnan (JF)
8. *Cotoneaster qungbixiensis*; cult., from nwYunnan (JF)
9. *Cotoneaster induratus*; cult., from Yulongxueshan, nwYunnan (CGW)

1. Leaves densely white-downy beneath .. *P. angustifolia*
1. Leaves glabrous or somewhat hairy beneath ....
   .... *P. crenulata, P. fortuneana*

Pyracantha angustifolia (Franch.) C. K. Schneid.* **Shrub** to 5 m with short, thorn-tipped side **branches**; young branchlets rusty-hairy. **Leaves** obovate to obovate-oblong, 15–60 × 5–20 mm, blunt or notched, toothed. **Inflorescence** lax, many-flowered, 3–4 cm across. **Flowers** 8–10 mm in diameter; petals white, almost orbicular, 4 mm long; stamens 20; ovary densely hairy. **Fruit** almost globose, 5 mm across, red or orange-red when ripe. Guizhou, Hubei, Sichuan, Shaanxi, e & seTibet, Yunnan [c & eChina]; thickets, streamsides, roadside banks, 500–2,800 m. IV–VI. (See photos 8 and 9 on p. 273.)

Pyracantha crenulata (D. Don) M. Roem. (syn. *P. rodgersiana* hort. ex L. H. Bailey) **Shrub** or small **tree** up to 5 m with short, thorn-tipped side **branches**; branchlets rusty-hairy when young. **Leaves** oblong to oblanceolate, 20–70 × 8–18 mm, blunt, pointed or apiculate, minutely scalloped, glabrous to slightly hairy beneath. **Inflorescence** rather lax, many-flowered, 3–5 cm across. **Flowers** 6–9 mm in diameter; hypanthium glabrous; petals white, orbicular, 4–5 mm long; stamens 20; ovary densely white-hairy. **Fruit** almost globose, 3–8 mm across, yellow-orange or orange-red when ripe. Gansu, Guizhou, Hubei, Sichuan, Shaanxi, seTibet, Yunnan [swTibet, c & eChina, Himalaya]; thickets, shrubberies, streamsides, roadside banks, hedges, 700–2,500 m. IV–V.

Pyracantha fortuneana (Maxim.) H. L. Li* Similar to the previous species, but **leaves** glabrous with toothed margin, obovate to oblong-obovate; **flowers** 9–10 mm across; **fruit** c. 5 mm, red to dark red when ripe. Guangxi, Guizhou, Hubei, Sichuan, Shaanxi, seTibet, Yunnan; shrubberies, stream-, road- and pathsides, rocky places, 500–2,800 m. III–V.

PHOTINIA

A genus of 66 species (48 in China) in east and South East Asia, also in Mexico. **Evergreen shrubs** or **trees** with simple, leathery, stalked, stipulate **leaves**, margin toothed or wavy. **Inflorescences** corymbose or almost umbellate, few-many-flowered, terminal or axillary. **Flowers** regular; hypanthium cup-shaped or campanulate, half-joined to ovary; sepals 5, erect, short; petals 5, spreading; stamens 20; ovary semi-inferior, hairy, 4–5-celled. **Fruit** a pome, fleshy, with persistent sepals.

The *Flora of China* account considers *Stranvaesia* to be distinct from *Photinia* but, on morphological grounds, the generic distinction cannot be upheld.

Photinia davidiana (Decne) Cardot (syn. *Stranvaesia davidiana* Decne) **Evergreen shrub** or small **tree**, 1–10 m tall with grey-brown **bracts** when mature, densely woolly when young. **Leaves** oblong to oblanceolate, 5–12 × 2–4.5 cm, grey-brown-hairy along midvein above and beneath, red when young. **Inflorescences** terminal, compoundly corymbose, rachis and flower-stalks softly hairy. **Flowers** 5–10 mm across; sepals 5, triangular-ovate, 2–3 mm long; petals white, almost orbicular, 4–5 mm long; anthers red-purple; style 5, mostly joined. **Fruit** almost globose, 7–8 mm across, red when ripe. Gansu, Guizhou, Hubei, Sichuan, Shaanxi, Yunnan [cChina, nVietnam]; valleys and mountain slopes in damp forests, woodland, shrubberies, 900–3,000 m. V–VI.

COTONEASTER

A genus of c. 100 species (c. 65 in China) distributed in Asia, Europe and north-west Africa, also in Mexico. Small to large, **evergreen, semi-evergreen** or **deciduous shrubs** or **trees** with terete, unarmed **branches**. **Leaves** alternate, simple, short-stalked; stipules small, awl-shaped, soon falling. **Inflorescences** terminal or axillary, cymes or corymbs, or **flowers** sometimes solitary; hypanthium bell- to top-shaped, joined to ovary; sepals 5, short; petals 5, spreading or almost erect, overlapping, white, pink or purple; stamens 10–20, inserted on mouth of hypanthium; ovary inferior or semi-inferior. **Fruit** a small, 1-seeded, berry-like pome.

1. Petals erect to somewhat incurved, pink and red .....
   .... *C. qungbixiensis, C. induratus, C. rehderi*
1. Petals spreading, white or cream 2
2. Leaves relatively large, most at least 20 mm long ... 3
2. Leaves small, rarely more than 18 mm long, mostly under 10 mm 4
3. Evergreen shrubs .... *C. lacteus, C. brickellii*
3. Deciduous shrubs .... *C. bullatus, C. frigidus*
4. Ascending to upright shrubs ....
   ...... *C. buxifolius, C. lidjiangensis, C. hodjingensis*
4. Low spreading to prostrate shrubs ....
   .... *C. cochleatus, C. horizontalis*

Cotoneaster qungbixiensis J. Fryer & B. Hylmö* Spreading, rather lax **deciduous** or **semi-evergreen shrub** to 3 m, with arching, reddish brown **branches**. **Leaves** dark shiny green above, grey-downy beneath, ovate to narrow-elliptic, 2.8–4.2 × 1.3–2.1 cm, pointed, with 2–4 pairs of veins. **Flowers** 7–8 mm long, in clusters of 3–9, occasionally more, on short, 4–6-leaved lateral shoots; petals pink with dark red base; stamens 20 with dark red filaments and white anthers. **Fruit** globose to obovoid, 8–10 mm long, orange-red and shiny when ripe. nwYunnan; woodland margins, shrubberies, rocky slopes, 2,400–3,300 m. VI–VII.

Cotoneaster induratus J. Fryer & B. Hylmö* Similar to the previous species, but branchlets more erect and purplish brown; **leaves** smaller, elliptic to ovate, 1.9–2.7 × 1–1.4 cm. **Flowers** 5–7 mm long, not more than 5 per cluster; petals reddish maroon and pink, margin off-white. **Fruit** orange-red when ripe. nwYunnan; similar habitats, 2,800–3,600 m. VI–VII.

Cotoneaster rehderi Pojark.* **Shrub** to 5 m, often less, with erect to spreading maroon branchlets, rough-hairy at first. **Leaves** large, almost leathery, elliptic to obovate, 7–21 × 4.5–9 cm, pointed to long-pointed, mid-green and shiny, markedly puckered above, pale beneath, with 8–11 pairs of veins. **Flowers** 4–6 mm long, in cymose clusters of up to 30 on 4-leaved shoots to 12 cm long; petals erect to incurved, red to maroon with pink margin; stamens 20 with pink filaments and white anthers. **Fruit** globose, sometimes somewhat flattened above, 8–11 mm in diameter, bright scarlet red and shiny when ripe. w & sw Sichuan, nwYunnan; woodland margins, shrubberies, rocky places, ravines, 1,900–3,400 m. V–VI.

Cotoneaster lacteus W. W. Sm.* **Shrub** or small **tree** to 8 m with erect to spreading, purple-black, rather slender branchlets, yellow-downy at first. **Leaves** leathery, obovate to elliptic, 4.2–12 × 2–6 cm, dull dark green and rather rough above, yellowish-downy beneath, with pointed apex and 7–9 pairs of impressed veins. **Flowers** 6–8 mm across, in large branched, cymose clusters of up to 150 on usually 4-leaved shoots to 10 cm long; petals white, occasionally pink-flushed, spreading, sometimes with hair-tufts; stamens 20, with white filaments and purple anthers. **Fruit** globose, 6–7 mm in diameter, crimson when ripe, in drooping clusters. w & sw Sichuan, nwYunnan; shrubberies, woodland margins, ravines, 2,000–3,000 m. VII.

Cotoneaster brickellii J. Fryer & B. Hylmö* Readily distinguished from previous species by smaller **leaves**, not more than 3.5 × 1.9 cm with 5–7 pairs of veins, white-downy beneath; **flowers** in similarly large clusters, cream in bud, 6 mm across when fully open, glabrous. **Fruit** globose, 5–6 mm in diameter, orange-red when ripe, sparsely hairy. nwYunnan; similar habitats, 1,700–2,500 m. VII.

Cotoneaster bullatus Bois* **Deciduous shrub** to 3 m with greyish black branchlets. **Leaves** oblong-ovate, elliptic or oblong-lanceolate, 3.5–7 × 2–4 cm, long-pointed, deep shiny green and conspicuously puckered above, glabrous or downy beneath, with 5–8 pairs of veins. **Inflorescences** corymbose, 2.5–5 cm across, 5–15-flowered, rachis and flower-stalks hairy, flowers 7–8 mm across; sepals triangular, 1–1.5 mm long; petals pink, erect, obovate, blunt, 4–4.5 mm long; stamens 20–22, shorter than petals. **Fruit** globose to obovoid, 6–8 mm long, red when ripe. Hubei, Sichuan, seTibet, Yunnan; mountain slopes and in valleys in open forests, shrubberies, riverbanks, 900–3,200 m. V–VI.

Cotoneaster frigidus Wall. ex Lindl. (syn. *C. himalaiensis* hort. ex Zabel) Readily distinguished from the previous species by narrow-elliptic to ovate-elliptic **leaves**, 3.5–10 × 1.5–3 cm, with impressed veins above but not puckered, and by the larger, 20–40-flowered corymbs with 6–7 mm white **flowers**; stamens 18–20. **Fruit** ellipsoid, 4–5 mm, bright red when ripe. seTibet [swTibet, Himalaya]; broadleaved forests, shrubberies, river valleys, 2,800–3,300 m. III–V.

Cotoneaster buxifolius Wall. ex Lindl. **Evergreen** or **semi-evergreen shrub** to 2 m with greyish brown or brownish branchlets. **Leaves** elliptic or elliptic-obovate, 5–10 × 4–9 mm, pointed or mucronate, densely grey-hairy above, appressed-hairy beneath, becoming hairless with age. **Inflorescences** 3–5-flowered, rachis hairy. **Flowers** 7–9 mm across; sepals ovate-triangular, 1.5–2 mm long; petals white, spreading, almost orbicular or broadly ovate, 3–4 mm long; stamens 20, almost as long as petals. **Fruit** almost globose, 5–6 mm long, red when ripe. Guizhou, Sichuan, seTibet, Yunnan [swTibet, Himalaya, nMyanmar]; rocky mountain slopes, cliffs, thickets, roadsides, 1,000–3,900 m. IV–VI.

Cotoneaster lidjiangensis G. Klotz* Rather dense **evergreen shrub** to 1.5 m with wide-spreading, maroon branchlets, densely rough-hairy at first. **Leaves** leathery, elliptic to obovate-elliptic, 10–18 × 5–8 mm, blunt, dull dark green above, sometimes slightly shiny, white-downy beneath, with 3–5 pairs of veins. **Flowers** 8–9 mm across, in small clusters of up to 8, on short 4–5-leaved shoots; petals half- to fully spreading, white, sometimes pinkish in bud; stamens 17–20, with mauve anthers. **Fruit** globose to obovate, 5–6 mm in diameter, red to crimson when ripe. nwYunnan; woodland margins, shrubberies, rocky places, 1,700–3,400 m. VI.

Cotoneaster hodjingensis G. Klotz* Small, rather dense **evergreen shrub** to 1.5 m, branchlets eventually spreading widely, maroon, densely downy at first. **Leaves** lanceolate to almost elliptic, 5–11 × 2–5 mm, pointed to almost blunt, dark shiny green above, grey-white-downy beneath, with 2–4 pairs of impressed veins. **Flowers** 7–8 mm across, solitary or 2–3 borne on short 4-leaved shoots; petals white; stamens 20, with white filaments and violet anthers. **Fruit** globose or almost so, 6–7 mm in diameter, crimson-red when ripe. nwYunnan; shrubberies, rocky places, ravines, woodland, streamsides, 1,700–3,000 m. VI.

Cotoneaster cochleatus (Franch.) G. Klotz* Low carpeting **evergreen** to 0.5 m with prostrate, rooting, red to purple-black branches. **Leaves** leathery, shiny deep green, obovate to

1. *Cotoneaster rehderi*; cult. (JF)
2. *Cotoneaster rehderi*; cult., from wSichuan (CGW)
3. *Cotoneaster lacteus*; cult. (JF)
4. *Cotoneaster lacteus*; cult. (CGW)
5. *Cotoneaster brickellii*; cult., from nwYunnan (JF)
6. *Cotoneaster brickellii*; cult., from nwYunnan (JF)
7. *Cotoneaster bullatus*; cult. (CGW)
8. *Cotoneaster frigidus*; cult. (CS)
9. *Cotoneaster lidjiangensis*; cult., from nwYunnan (JF)
10. *Cotoneaster lidjiangensis*; cult., from nwYunnan (JF)
11. *Cotoneaster hodjingensis*; cult., from nwYunnan (JF)
12. *Cotoneaster cochleatus*; Bomi, Chumdo, seTibet (HJ)
13. *Cotoneaster cochleatus*; Yulongxueshan, nwYunnan (CGW)

1. *Cotoneaster horizontalis*; cult. (CGW)
2. *Cotoneaster horizontalis*; cult. (CGW)
3. *Malus rockii*; Napahai, Zhongdian, nwYunnan (CGW)
4. *Malus rockii*; Napahai, Zhongdian, nwYunnan (CGW)
5. *Malus rockii* in fruit; cult., from Zhongdian, nwYunnan (CGW)
6. *Malus hupehensis*; cult. (CS)
7. *Malus yunnanensis*; cult., from Zhongdian, nwYunnan (CGW)
8. *Malus yunnanensis* in fruit; cult., from Zhongdian, nwYunnan (CGW)
9. *Pyrus pashia*; cult. (JG)
10. *Pyrus pashia* in fruit; cult. (JG)

almost orbicular, 5–14 × 3–9 mm, apex blunt to rounded, sometimes notched, spirally arranged on sterile shoots. **Flowers** mostly solitary, 8–10 mm across, pink in bud, borne on short 3–4-leaved shoots; petals white, spreading; stamens usually 20. **Fruit** almost globose, 7–9 mm in diameter, crimson when ripe. Sichuan, seTibet, w & nwYunnan [swTibet]; rocky mountain slopes, rocky thickets, cliffs, roadsides, 2,300–4,000 m. V–early VII.

Cotoneaster horizontalis Decne Readily distinguished from the other species in this account by the flattened branched fans, with the blackish branchlets arranged in a strict herringbone pattern, procumbent or fanning out against rocks, to 50 cm tall. **Leaves** deciduous, elliptic to rounded, deep shiny green (reddening in autumn), 6–14 × 4–9 mm. **Flowers** 5–7 mm, with white, pink or reddish, erect petals. **Fruit** 5–7 mm, bright red when ripe. Gansu, Guizhou, Hubei, Sichuan, Shaanxi, seTibet, Yunnan [swTibet, c & eChina, Nepal, Taiwan]; rocky places, cliffs, ravines, thickets, 1,500–3,500 m. V–VI.

MALUS

A north temperate genus of c. 55 species (26 in China). **Shrubs** to medium-sized, usually **deciduous trees**, less commonly **semi-evergreen**, with alternate, stalked, stipulate, simple **leaves**, often with a toothed, sometimes lobed, margins. **Inflorescences** corymbose-racemose, often somewhat umbellate, few–many-flowered. **Flowers** stalked, white to pinkish white or rose-pink; sepals and petals 5; stamens 15–50, with yellow anthers; ovary inferior, 3–5-celled. **Fruit** a fleshy pome, each cell 1–2-seeded. Those listed below were previously included in *Pyrus*.

1. Leaf-blades convolute in bud, never lobed; fruit without stone cells .... **M. rockii, M. hupehensis**
1. Leaf-blades folded in bud, often shallowly lobed; fruit usually with a few stone cells ....
 .... **M. yunnanensis, M. prattii**

Malus rockii Rehd. Small to medium-sized **tree** to 8–10 m with a domed or spreading habit, branchlets dark brown, glabrous. **Leaves** ovate-elliptic to oblong-ovate, 6–12 × 3.5–7 cm, long-pointed, margins irregularly toothed, green but turning yellow and orange in autumn. **Inflorescences** umbel-like, 4–6 cm across, 4–8-flowered. **Flowers** 25–30 mm across; sepals sparsely hairy or glabrous; petals white, almost orbicular, rounded at apex, 12–15 mm long. **Fruit** egg-shaped to almost globose, 10–15 mm across, red when ripe. swSichuan, seTibet, nwYunnan [Bhutan]; mixed forests in valleys, woodland margins, open shrubberies, grassy slopes, 2,400–3,800 m. V–VI.

Malus hupehensis (Pamp.) Rehd.* Differs from the previous species in the minutely downy young branchlets, the ovate-elliptic **leaves**, 5–10 × 2.5–4 cm, base wedge-shaped, and in the larger **flowers** (35–40 mm across), white, pink in **bud**. **Fruit** small, 9–10 mm, red and cherry-like. Gansu, Guizhou, Hubei, Shaanxi [c & eChina]; wooded slopes, open valley shrubberies, field boundaries, to 2,900 m. IV–V.

Malus yunnanensis (Franch.) C. K. Schneid. Small to medium-sized **tree** up to 10 m, with a domed or spreading habit and dark purplish brown branchlets that are hairy when young. **Leaves** ovate to almost heart-shaped, 6–12 × 4–7 cm, pointed, often shallowly lobed, margin doubly toothed. **Inflorescence** a corymb-like raceme, 5–9 cm across, 8–12-flowered. **Flowers** 12–15 mm across; sepals densely hairy; petals white, almost orbicular, rounded at apex, 7–8 mm long. **Fruit** globose, 10–15 mm across, red with whitish dots when ripe, on downy stalks. Guizhou, Hubei, Sichuan, Shaanxi, eTibet, Yunnan [swTibet, nMyanmar]; mixed forests on steep slopes or by water in valleys, woodland margins, 1,600–3,800 m. V–VI.

Malus prattii (Hemsl.) C. K. Schneid.* Similar to the previous species, but leaf-margin unlobed, minutely toothed, blade glabrous or sparsely minutely downy beneath, and yellow or red **fruit** on glabrous stalks. w & swSichuan, nwYunnan; mixed open forests, shrubberies, grassy slopes, 1,400–3,500 m. V–VI.

PYRUS

A genus of 25 species (14 in China, 8 endemic) from Europe, North Africa and Asia. Very similar to *Malus*, but differing primarily in the styles being free (not united towards base), and in the usually pear-shaped **fruit** that contain numerous small 'gritty' stone-cells.

Pyrus pashia Buch.-Ham. ex D. Don **Deciduous tree** to 12 m, with purplish brown branchlets that are often armed. **Leaves** stalked, ovate to elliptic, 4–7 × 2–5 cm, downy when young, margin toothed. **Flowers** 20–35 mm across, each on a stalk 20–30 mm long, in umbel-like clusters of up to 13; sepals triangular, downy; petals white, obovate, 8–10 mm long; stamens 25–30; styles 3–5. **Fruit** egg-shaped to almost rounded, 10–15 mm, brown with pale dots. Guizhou, Sichuan, seTibet, Yunnan [swTibet, Himalaya, Myanmar, Thailand, Vietnam]; shrubberies, scrubby slopes, 600–3,000 m. III–IV.

ROSA

A genus of c. 200 species (95 in China, 65 endemic) in north temperate and subtropical regions of the Old and New Worlds. Small to large, **evergreen** or **deciduous**, **shrubs** or rambling **climbers** with erect to arching **stems** that are usually prickly and sometimes also bristly. **Leaves** alternate, odd-pinnate, with stipules at base of leaf-stalk and generally joined to it. **Flowers** showy, white, pink, purple, red or yellow, flat or saucer-shaped, often fragrant, in panicles or corymbs, sometimes compound, or solitary or 2–3; sepals and petals 4–5, borne on a disc at apex of hypanthium; stamens many, in whorls; ovary of many, free carpels, enclosed within the fleshy hypanthium; styles terminal or lateral, sometimes united into a short column. **Fruit** a hip, top- to urn- or flask-shaped, orange, red or black when ripe.

SYNOPSIS OF GROUPS

Group One. Shrubs; stipules joined to leaf-stalk; flowers solitary; sepals and petals usually 4, rarely 5, the latter white or cream. (below)

Group Two. Climbing shrubs; stipules free, often falling; flowers numerous in corymbs or compound panicles; sepals and petals usually 5 (except in double-flowered forms). (below)

Group Three. Vigorous evergreen or deciduous, rambling, generally prickly climbers, occasionally shrubby; stipules joined to leaf-stalk; flowers numerous in corymbs or panicles, often compound; sepals 5, deciduous; petals 5 (many petalled in some derived forms); styles free or united into a column, protruding, longer than stamens. (below)

Group Four. Evergreen or deciduous prickly shrubs; stipules joined to leaf-stalks; flowers solitary or in corymbs or panicles; sepals and petals usually 5; styles free, not protruding, shorter than stamens. (p. 284)

Group One

Rosa sericea **Lindl.** Erect, prickly **shrub** to 2 m with purplish red or purplish brown **stems**; prickles absent or paired, with broad base. **Leaves** with 7–11 leaflets each ovate to elliptic, 8–30 × 4–10 mm, glabrous or downy beneath, margin toothed. **Flowers** axillary, 25–50 mm across; sepals persistent; petals 4, creamy white to white, broadly obovate to ovate, apex notched. Styles free, shorter than stamens, hairy. **Hips** pear-shaped to globose, 8–15 mm long, purple-brown to red, rarely orange, glabrous, with slender, non-thickened stalk. Gansu, Guizhou, Hubei, Qinghai, Sichuan, Shaanxi, Tibet, Yunnan [Ningxia, Himalaya]; margins of broadleaved and coniferous woods, shrubberies and scrub, road- and tracksides, 700–4,000 m. V–VII. Considered by some botanists to be conspecific with the following species.

Rosa omeiensis **Rolfe** Very similar to the previous species, although it can reach 4 m. Prickles absent to dense, with broad base, sometimes flared into a wide wing on **stems** (**var. pteracantha Franch.**). Leaflets 9–17. **Flowers** white, rarely pinkish or yellow. Hips similar to those of *R. sericea*, 6–20 mm long, yellow, pale to deep red, with thick, fleshy stalk. Gansu, Guizhou, Hubei, Qinghai, Sichuan, Shaanxi, Tibet, Yunnan [Ningxia, Himalaya]; similar habitats and altitude. V–VII.

Rosa sikangensis **T. T. Yu & T. C. Ku*** Closely related to previous 2 species, but a small **shrub** 1–1.5 m tall, with prickly and bristly **stems**. **Leaflets** 7–9, oblong to obovate, apex rounded to truncate, margin doubly toothed, downy on both surfaces, glandular beneath. **Flowers** solitary, axillary, 2.5–5 cm across, white with densely glandular sepals; hips almost globose, 8–10 mm across, red when ripe, glandular-hairy. Sichuan, Tibet, Yunnan; thickets and scrub, roadsides, 2,900–4,200 m. IV–VI.

Group Two

1. Flowers solitary; hips bristly .... **R. laevigata**
1. Flowers in corymbs or umbels; hips smooth ....
   .... **R. banksiae, R. cymosa**

Rosa laevigata **Mich.** (syn. *R. triphylla* Roxb.) **Evergreen climbing shrub** to 5 m with purple-brown, robust **branches** armed, when young, with dense glandular bristles. **Leaves** with 3(–4) elliptic to ovate or obovate leaflets, 2–6 × 1.2–3.5 cm, rather leathery, margin toothed, glabrous above. **Flowers** 5–10 cm across; sepals persistent, leaf-like, ovate-lanceolate, almost as long as petals, glabrous outside; petals white, 5 (but semi-double and double forms exist), obovate, apex notched. **Hips** pear-shaped to obovoid, 8–15 mm across, purple-brown when ripe, densely bristly. Guangxi, Guizhou, Sichuan, sShaanxi, Yunnan [c, s & seChina, Taiwan, nVietnam]; thickets and shrubberies, open slopes, roadsides, hedgerows, 200–1,600 m. IV–VI.

Rosa banksiae **R. Br.*** Vigorous **climbing evergreen shrub** to 6 m with red-brown branchlets and scattered, short, curved prickles, or unarmed. **Leaves** leathery, glabrous and shiny green above, hairy on veins beneath; leaflets 3–5, elliptic-ovate to oblong-lanceolate, 2–5 × 0.8–1.8 cm, pointed. **Inflorescence** a simple umbel or corymb, 4–15-flowered. **Flowers** 15–25 mm across; petals 5, white to pale yellow, obovate, apex rounded. **Hips** globose to egg-shaped, 5–7 mm across, orange to blackish brown when ripe. Gansu, Guizhou, Hubei, Sichuan, Yunnan [cChina]; thickets and shrubberies in valleys, especially by streams and rivers, also on roadside banks, 500–2,200 m. IV–VI. Widely cultivated, especially the double white-flowered **var. *banksiae*** and the double yellow-flowered 'Lutea'.

Rosa cymosa **Tratt.** (syn. *R. bodinieri* H. Lévl. & Vaniot; *R. indica* L.) Very similar to the previous species, but flowers in compound **inflorescences** (corymbs or panicles) and sepals normally pinnately lobed; **leaves** with 3, 5 or 7 leathery leaflets to 6 × 2.5 cm; **flowers** slightly larger, 20–25 mm across, petals white or pale yellow, apex slightly notched; **hips** 4–7 mm across, reddish brown to purple or black when ripe. Guangxi, Guizhou, eHubei, Sichuan, sShaanxi, Yunnan [s & eChina, Laos, Taiwan, Vietnam]; hillslopes, shrubberies, ravines, stream- and roadsides, to 1,800 m. V–VI.

Group Three

1. Styles free, slightly protruding, shorter than, or as long as stamens 2
1. Styles united to form a column, protruding, shorter than to as long as the stamens 3

1. *Rosa sericea*; Gangheba, Yulongxueshan, nwYunnan (CGW)
2. *Rosa sericea*; Napahai, Zhongdian, nwYunnan (PC)
3. *Rosa sericea*; Gangheba, Yulongxueshan, nwYunnan (CGW)
4. *Rosa omiensis*; Daxueshan, nwYunnan (CGW)
5. *Rosa omiensis* var. *pteracantha*; Minjiang valley, south of Songpan, nwSichuan (CGW)
6. *Rosa laevigata*; Tian'e, nwGuangxi (PC)
7. *Rosa cymosa*; Fushui, swGuangxi (PC)
8. *Rosa cymosa*; Fushui, swGuangxi (PC)

2. Flowers white, large, 8–10 cm across .. **R. odorata var. gigantea, R. odorata var. odorata, R. odorata var. erubescens, R. odorata var. pseudoindica**
2. Flowers pale yellow, white to pink, rose-pink or red, not more than 5 cm across ....
.... **R. chinensis, R. chinensis var. spontanea**
3. Stipules with a toothed or pectinate margin ....
.... **R. multiflora, R. lichiangensis**
3. Stipules entire, generally glandular-downy 4
4. Leaflets glabrous .... **R. longicuspis, R. filipes, R. soulieana, R. soulieana var. microphylla**
4. Leaflets downy, at least beneath .. **R. brunonii, R. rubus**

Rosa odorata (Andrews) Sweet var. *gigantea* (Collett ex Crépin) Rehd. & E. H. Wilson (syn. *R. gigantea* Collett ex Crépin) Erect, **evergreen** to **semi-evergreen**, **climbing** or **scrambling shrub**, with long, robust, tapering branches; prickles few to many, curved, flat and stout. **Leaves** leathery, with 3–5 elliptic, ovate to oblong-ovate leaflets, 2–7 × 1.5–3 cm, glabrous, margins appressed-toothed; stipules joined to leaf-stalks. **Flowers** large, solitary or 2–3 together, very fragrant, 4–10 cm across; petals 5, white, obovate, apex notched; styles free, protruding, nearly as long as stamens. **Hips** globose, flattened above, 10–20 mm long, red when ripe, glabrous. Yunnan [nMyanmar, nThailand, nVietnam]; mixed forests, thickets and scrub on mountain slopes, roadsides, 1,400–2,700 m. III–V. Widely planted in c & wChina as varieties with white tinged with pink, yellow or orange, semi-double or double flowers (**var. *odorata*, var. *erubescens* (Focke) T. T. Yu & T. C. Ku** and **var. *pseudoindica* (Lindl.) Rehd.**).

Rosa chinensis Jacq.* Very different from the previous species, being a **climber** or erect **bush** with purplish brown branchlets with few to many, curved, flat, stout prickles; leaflets somewhat larger, glabrous; **flowers** 4–5, clustered, smaller, 4–5 cm across, slightly fragrant, white, pale to rose-pink; **hips** egg- to pear-shaped, 10–20 mm across, red when ripe, glabrous. Guizhou, Hubei, Sichuan; thickets and scrub, roadsides, 1,300–2,800 m, but widely cultivated as semi-double or double forms throughout China. IV–VI. **Var. *spontanea* (Rehd. & E. H. Wilson) T. T. Yu & T. C. Ku*** has 1–3 single **flowers** that open pink and darken to red, and orange hips. wHubei, nwSichuan.

1. *Rosa chinensis*; cult. (RL)
2. *Rosa multiflora*; Lijiang, nwYunnan (CGW)
3. *Rosa multiflora*; Lijiang, nwYunnan (CGW)
4. *Rosa longicuspis*; south of Lijiang, nwYunnan (CGW)
5. *Rosa longicuspis*; Lijiang, nwYunnan (CGW)
6. *Rosa filipes*; near Longriba, nwSichuan (CGW)
7. *Rosa filipes*; upper Minjiang valley, nwSichuan (CGW)
8. *Rosa soulieana*; Minjiang valley, south of Songpan, nwSichuan (CGW)
9. *Rosa soulieana* var. *microphylla*; Minjiang valley, south of Songpan, nwSichuan (CGW)

Rosa multiflora Thunb.* Variable, vigorous **climbing shrub** to 6 m, **stems** armed with curved prickles in pairs below leaves. **Leaves** with 5–9 ovate to oblong leaflets, 1–5 × 0.8–2.8 cm, downy beneath. **Inflorescence** corymbose with numerous **flowers**, these 15–40 cm across, white, pale pink to rose-pink, fragrant; flower-stalks glabrous to glandular-hairy; sepals deciduous; petals 5, obovate, apex notched. **Hips** almost globose, 6–8 mm, brown-red and shiny when ripe, glabrous. Gansu, Guizhou, Yunnan; thickets and shrubberies, by streams and rivers, on open slopes, roadsides, hedgerows, 300–2,000 m. Widely planted in c & sChina, Japan, Korea. V–VII. Various forms are cultivated including those with semi-double and double flowers.

Rosa lichiangensis T. T. Yu & T. C. Ku* Readily distinguished from previous species by its less vigorous nature, being a **climbing shrub** to 2 m tall with scattered, slightly curved prickles. **Leaves** with only 3–5 obovate to elliptic leaflets, 1–2.3 × 0.5–1.3 cm. **Flowers** 25–30 mm across, pink, fragrant, 2–4 in a cluster. **Hips** almost globose, 6–8 mm, red when ripe, glabrous. nwYunnan; habitats similar to those of *R. multiflora*, 2,300–2,700 m. VI–VII.

Rosa longicuspis H. Lévl. & Vaniot Vigorous **climbing** or **scrambling**, **evergreen shrub** 1.5–6 m tall, sometimes more, with purplish brown young **branches** and sparse, scattered, curved prickles. **Leaves** dark green and lustrous, glabrous; leaflets (5–)7–9, leathery, ovate, elliptic or oblong-ovate, 3–7 × 1–3.5 cm, margin with pointed teeth. **Inflorescence** a many-flowered corymb. **Flowers** 30–40 mm across, fragrant; sepals deciduous, downy; petals 5, white or creamy white, broadly obovate, irregularly nibbled at apex. **Hips** pear-shaped, 10–12 mm across, dark red and shiny when ripe, glabrous. Guizhou, Sichuan, Yunnan [neIndia]; mixed evergreen forests, thickets, shrubberies, open dry slopes, roadsides, 400–2,700 m. V–VI.

Rosa filipes Rehd. & E. H. Wilson* Related to *R. longicuspis*, but **leaves** not so lustrous, leaflets thinner and more papery, lanceolate or oblong, 4–7 × 1.5–3 cm, dotted with glands beneath; **flowers** 30–40 mm across, white, fragrant; **hips** almost globose, 7–9 mm, dark red and shiny when ripe. Gansu, Sichuan, Shaanxi, eTibet, Yunnan; thickets, shrubberies, roadsides, 1,300–2,300 m. V–VII.

Rosa soulieana Crépin* Related to *R. filipes*, but leaflets smaller, grey-green, not more than 3 × 2 cm, lacking glands beneath; **flowers** generally numerous (solitary in **var. *microphylla* T. T. Yu & T. C. Ku***), 30–50 mm across, white with yellowish centre; **hips** c. 10 mm across, orange-red at first, turning blackish purple at maturity. nw, w & swSichuan, e & seTibet, nwYunnan [sAnhui]; shrubberies, rocky slopes, hedgerows, road- and streamsides, 2,500–3,700m. VI–VII.

***Rosa brunonii* Lindl.** Readily distinguished from previous 3 species by red-brown branchlets, and rather smaller leaflets, 3–5 × 1–1.5 cm, that are sparingly downy beneath; **flowers** 30–50 mm across, white, fragrant; sepals usually with lateral lobes. **Hips** egg-shaped, 9–10 mm long, purplish brown or dark red and shiny when ripe. Sichuan, se & eTibet, Yunnan [swTibet, Himalaya, nMyanmar]; forests, thickets and shrubberies, by streams and rivers, valleys on open slopes, roadsides, 1,900–2,800 m. V–VII.

Rosa rubus* H. Lévl. & Vaniot Similar to *R. brunonii*, but leaflets usually 5 (sometimes 3), elliptic to obovate, 3–8 × 2–4.5 cm, and sepals entire; **flowers** up to 25 per panicle, each 2.5–3 cm across, fragrant. **Hips** rounded, 8–10 mm, bright red to orange- or purple-brown when ripe. Gansu, Guangxi, Guizhou, Hubei, Sichuan, Shaanxi, Yunnan [e & seChina]; rocky and grassy slopes, cliffs, riverbanks, roadsides, 500–1,300 m. IV–VI.

Group Four

1. Hypanthium globose, flattened above (top-shaped), densely bristly .... **R. roxburghii**
1. Hypanthium globose to urn- or flask-shaped, glabrous or sparsely bristly 2
2. Inflorescence many-flowered .. **R. setipoda, R. davidii, R. multibracteata**
2. Inflorescences with 2–3(–5) flowers in a cluster, or flowers solitary 3
3. Leaflets not more than 7 per leaf ...... **R. forrestiana**
3. Leaflets 7–15 per leaf 4
4. Flowers small, 15–30(–35) mm across ...... **R. farreri, R. prattii, R. sertata**
4. Flowers larger, 30–50 mm across 5
5. Bracts absent .... **R. graciliflora**
5. Bracts present .... **R. willmottiae, R. macrophylla, R. moyesii, R. sweginzowii**

***Rosa roxburghii* Tratt.** Open **shrub** 1–2.5 m tall with purple-brown branchlets turning to greyish brown with age, armed with straight prickles. **Leaves** mid-green, slightly shiny, with 9–15 elliptic to oblong leaflets, 1–2 × 0.6–1.2 cm, glabrous, margin with pointed teeth. **Flowers** solitary or several together at branchlet-tips, 40–60 cm across, pink to rose-purple; sepals broad-ovate, densely bristly; petals 5, obovate. **Hips** 15–20 mm across, densely bristly, greenish red when ripe. sGansu, Guangxi, Guizhou, Hubei, sShaanxi, Sichuan, eTibet, Yunnan [c & eChina, Japan]; montane forests, thickets and shrubberies, by streams, field margins, 500–1,500 m. IV–VI.

Rosa setipoda* Hemsl. & E. H. Wilson **Deciduous shrub** up to 3 m with glabrous branchlets; prickles sparse or absent, scattered, straight or slightly curved. **Leaves** greyish green, with 5–9 elliptic or ovate leaflets, blunt to pointed, 2.5–5.2 × 1.2–3 cm, downy and glandular beneath, margins doubly toothed. **Inflorescence** a lax, many-flowered corymb. **Flowers** 35–50 mm across on glandular stalks; sepals deciduous, ovate, long-tailed, pinnately lobed; petals 5, pink to rose-purple, broadly obovate, apex notched, slightly downy outside; styles free, much shorter than stamens. **Hips** flask-shaped, 10–20 mm across, dark coral-red when ripe. Hubei, Sichuan; shrubberies and thickets on mountain slopes, 1,800–2,600 m. V–VII.

Rosa davidii* Crépin Distinguished from the previous species by the more spreading habit, smaller leaflets, not more than 10 mm broad, glabrous or downy beneath, and smaller pink **flowers**, 20–30 mm across with leafy sepals; **hips** 18–25 mm across, orange-red to deep red when ripe. Gansu, Sichuan, Shaanxi, e & seTibet, Yunnan [Ningxia]; margins of broadleaved and coniferous woods, 1,500–3,000 m. VI–VII.

Rosa multibracteata* Hemsl. & E. H. Wilson Intricately branched **shrub** 1–2.5 m tall; prickles in pairs below **leaves**, numerous, long and straight. **Leaves** deeply lustrous green, with 7–9 rather small, ovate to almost orbicular leaflets, 8–15 × 5–10 mm, margin toothed, glabrous or downy beneath. **Flowers** axillary, solitary or several together, c. 25–30 mm across, pale to rose-pink, on glandular-hairy stalks; petals 5, obovate, apex notched. **Hips** oblong to almost globose, 8–10 mm long, orange-red and shiny when ripe, glabrous. Gansu, seQinghai, nwSichuan, Shaanxi; thickets and shrubberies, by streams and rivers, open rocky slopes, roadsides, 1,300–3,800 m. VI–VII.

Rosa forrestiana* Boul. Small **shrub** 1–2 m tall, with curving **stems**; prickles in pairs below leaves, yellowish, straight. **Leaves** with 5–7 almost orbicular, ovate or obovate leaflets, 6–18 × 4–15 mm, sparsely downy or glandular beneath, margins doubly toothed. **Inflorescence** 1–5-flowered. **Flowers** 20–35 mm across, on glandular-hairy stalks; petals 5, rose-red, broadly obovate, apex notched; styles free, slightly protruding, nearly equalling stamens. **Hips** egg-shaped, 9–13 mm across, red when ripe, with a ruff of broad bracts, also with persistent sepals. Sichuan, nwYunnan; thickets and shrubberies on slopes, roadsides, 2,400–3,000 m. V–VII.

Rosa farreri* Cox Small, sparsely prickly, rather dainty **shrub** 1–2 m tall. **Leaves** with 7–9 ovate to elliptic leaflets, 5–18 × 3–10 mm, sparsely downy on midvein beneath. **Flowers** solitary, axillary, 15–20 mm across; petals 5, pink to white, obovate to oblong, apex notched. **Hips** ellipsoid to egg-shaped-oblong, 8–12 mm long, red when ripe, glabrous. Gansu, n & nwSichuan; broadleaved woods and scrub, often on margins, 1,500–2,800 m. VI–VII.

Rosa prattii* Hemsl. **Shrub** 1–2.5 m tall, with purplish brown branchlets and scattered, straight prickles that taper to a broad base. **Leaves** with 7–15 elliptic or oblong leaflets, 6–20 × 4–10 mm, glabrous, but downy along midvein beneath, margin minutely toothed. **Inflorescence** an umbel-like, 2–7-flowered corymb. **Flowers** small, c. 20 mm across, with glandular stalks;

1. *Rosa brunonii*; near Lidiping, Lancangjiang–Jinshajiang divide, Lidiping, nwYunnan (CGW)
2. *Rosa rubus*; Tian'e, nwGuangxi (PC)
3. *Rosa roxburghii*; Tian'e, nwGuangxi (PC)
4. *Rosa roxburghii* in fruit; near Dali, nwYunnan (TK)
5. *Rosa setipoda*; cult. (CGW)
6. *Rosa multibracteata*; Minjiang valley, south of Songpan, nwSichuan (CGW)
7. *Rosa farreri*; north of Songpan, nwSichuan (CGW)

1. *Rosa sertata*; north of Songpan, nwSichuan (CGW)
2. *Rosa graciliflora*; Jiuzhaigou, nwSichuan (PC)
3. *Rosa willmottiae*; near Songpan, nwSichuan (JM)
4. *Rosa willmottiae*; near Songpan, nwSichuan (PC)
5. *Rosa macrophylla*; cNepal (CGW)
6. *Rosa macrophylla* in fruit; cult. (CGW)
7. *Rosa moyesii*; Napahai, Zhongdian, nwYunnan (CGW)
8. *Rosa moyesii* in fruit; Napahai, Zhongdian, nwYunnan (EL)
9. *Rosa sweginzowii*; Balangshan, Wolong, wSichuan (CGW)
10. *Fragaria nilgherrensis*; Sandauwan, Yulongxueshan, nwYunnan (PC)
11. *Fragaria pentaphylla*; Danyun, nwSichuan, (PC)

sepals deciduous; petals 5, pink, broadly obovate, apex notched; styles free, much shorter than stamens. **Hips** egg-shaped to ellipsoid, 5–8 mm across, dark red when ripe. Gansu, Sichuan, nwYunnan; broadleaved and coniferous forests, montane scrub thickets, 1,900–3,000 m. VI–VIII.

Rosa sertata Rolfe* **Shrub** to 2 m with slender, glabrous branches, prickles arranged in pairs below leaves, straight and to 8 mm long. **Leaves** with 7–15 elliptic to ovate-elliptic leaflets, 1–2.5 × 0.7–1.5 cm, glabrous, or slightly hairy on midrib beneath, margin simple-toothed. **Flowers** several or solitary, pale pink to rose-pink, 20–35 mm across; sepals leafy, glabrous outside, downy within; petals broad-obovate, notched. **Hips** egg-shaped, 12–20 mm long, red when ripe, usually glabrous. Gansu, Hubei, Sichuan, Shaanxi, Yunnan [cChina]; open woodland, shrubberies, mountain slopes, riverbanks, tracks, roadsides, 1,400–2,200 m. VI–VII.

Rosa graciliflora Rehd. & E. H. Wilson* **Deciduous shrub** to 4 m with slender **branches** armed with sparse straight prickles. **Leaves** greyish green, with (7–)9–11 ovate to elliptic leaflets, 8–20 × 7–12 mm, margin simple to doubly toothed. **Flowers** solitary, axillary, 25–35 mm across; sepals leafy, downy inside; petals 5, pink to deep red, obovate, apex notched; styles free, shorter than stamens, downy. **Hip** pear- to flask-shaped, 20–30 mm long, red when ripe. Sichuan, e & seTibet, nwYunnan; coniferous forests and forest margins, shrubberies, 3,300–4,500 m. V–VII.

Rosa willmottiae Hemsl.* Erect, prickly **shrub** 1–3 m tall, **stems** with prickles in pairs borne below leaves, numerous, straight and with flared base. **Leaves** with 7–9 ovate to elliptic leaflets, 6–17 × 4–12 mm, margins toothed, glabrous or downy beneath. **Flowers** axillary, mostly solitary, 28–32 mm across, on glandular-hairy stalks; sepals deciduous; petals 5, pale pink to rose-pink, obovate, apex notched. **Hips** oblong to almost globose, c. 10 mm across, orange-red and shiny when ripe, glabrous. Gansu, Qinghai, w & nwSichuan; thickets and shrubberies, by streams and rivers, on open slopes, tracks, roadsides, 1,300–3,800 m. VI–VII.

Rosa macrophylla Lindl. Differs from the previous species by the more vigorous habit, to 4.5 m with purplish brown branchlets, and by the 9–11 oblong to ovate-elliptic, larger (25–65 mm long) leaflets, densely downy beneath. **Flowers** solitary or in fascicles of 2–3, semi-nodding, deeply cupped, 35–50 mm across; sepals leafy; petals 5, deep pink, obtriangular, apex notched. **Hips** oblong to flask-shaped, 15–35 mm long, purplish red when ripe. seTibet, w & nwYunnan [c & eHimalaya]; margins of forests, thickets, shrubberies, 2,400–3,700 m. V–VII.

Rosa moyesii Hemsl. & E. H. Wilson* Small to large **shrub** 1–4 m tall, with erect and arching **stems**; prickles in pairs below leaves, straight or slightly curved, tapering to broad base. **Leaves** with 7–13 ovate, elliptic or oblong-ovate leaflets, 10–50 × 8–25 mm, densely hairy all over, or along veins beneath, margin toothed. **Flowers** solitary or 2–3 together, 30–50 mm across, rather flat to saucer-shaped; sepals 5, ovate, leafy, erect and persistent in fruit; petals 5, deep red to rose-purple, broadly obovate, apex notched; styles free, shorter than stamens, downy. **Hips** almost globose to flask-shaped, 25–50 mm long, purple- to orange-red when ripe. Sichuan, Shaanxi, nwYunnan; scrub and thickets on mountain slopes, 2,700–3,800 m. VI–VII.

Rosa sweginzowii Koehne* Distinguished from the previous species by **leaves** that are glandular or downy beneath and with doubly toothed margins, the clustered pink **flowers**, and the relatively broader and smaller **hips**, oblong to obovoid-oblong in outline, 15–25 mm long, purplish red when ripe, somewhat bristly. Gansu, Hubei, Qinghai, Sichuan, Shaanxi, e & seTibet, Yunnan [Ningxia]; margins of *Pinus* and *Picea* forests, thickets, shrubberies, tracks, roadsides, 2,300–4,000 m. VI–VII.

FRAGARIA

A genus of c. 20 species (8 in China), mostly in the northern hemisphere but also in alpine regions of South America. **Perennial**, **stoloniferous**, **dioecious herbs**, with stolons rooting at nodes. **Leaves** alternate, stalked, with 3 or 5 leaflets, stipulate, stipules membranous, joined to leaf-stalk. **Inflorescence** erect, several-flowered, cymose or corymbiform. **Flowers** regular, hermaphrodite; hypanthium concave or hemispherical; sepals 5, valvate, with epicalyx of 5 smaller segments alternating with the sepals; petals 5, white, rarely yellow; stamens numerous; carpels numerous, free, borne on convex receptacle. **Fruit** a fleshy, juicy, swollen receptacle bearing achenes on the surface.

Fragaria nilgherrensis Schltr. ex J. Gay. Robust **herb** 5–25 cm tall with densely tufted, yellow-brown, silky-hairy **stems**. **Leaves** with 3 leaflets, leaflets obovate to elliptic, 1–4.5 × 0.8–3 cm, apex rounded, margin toothed, waxy, silky-hairy below. **Inflorescence** cymose, 2–5-flowered. **Flowers** 10–20 mm across; petals white, orbicular to obovate; anthers yellow. **Fruit** globose, white, sometimes tinged with red. Guizhou, Hubei, Sichuan, Shaanxi, Yunnan [c & eChina, Himalaya, Taiwan, nVietnam]; glades in woods and forests, meadows, open areas by streams, 700–3,000 m. IV–VII.

Fragaria orientalis Losinsk. Distinguished from the previous species by having normally hermaphrodite **flowers**, and by the purple **fruit** with spreading or reflexed (not appressed) sepals. Gansu, Qinghai, Shaanxi [n & neChina, Korea, Mongolia, eRussia]; forests, meadows, mountain slopes, 600–4,000 m. V–VII.

Fragaria pentaphylla Losinsk.* Readily distinguished from the previous 2 species by **leaves** which generally have an additional, smaller pair of leaflets on upper half of stalk but below the 3 terminal leaflets; **flowers** hermaphrodite; **fruit** egg-shaped, with reflexed sepals, ripening red or occasionally white (**forma alba** Staudt & Dickoré). Gansu, Sichuan, Shaanxi; forests, forest margins, open shrubberies, stony places, 1,000–2,700 m. IV–VII.

SANGUISORBA

A genus of c. 20 species (7 in China) in Europe and Asia. **Perennial herbs** with pinnately lobed **leaves**. **Flowers** small, 4-parted, lacking petals, male, female or hermaphrodite, in dense globose or elongated heads.

Sanguisorba filiformis **(Hook. f.) Hand.-Mazz.** Rather slender **tufted perennial** to 35 cm with thread-like **stems**. **Leaves** glabrous, basal with 3–5 pairs of ovate to rounded, toothed leaflets, 4–15 mm long, stem-leaves similar but uppermost with fewer leaflets. **Inflorescence** capitate, rounded, 3–7 mm across, with several female **flowers** surrounded by the male, uppermost opening first; sepals 4, white, elliptic to obovate, male flowers with 7–8 protruding stamens, female with a thread-like style to 1.5 mm long and a dilated stigma. Sichuan, Tibet, Yunnan [eHimalaya]; wet meadows and marshes, seepage areas, streamsides, often gregarious, 1,200–4,500 m. VI–IX.

POTENTILLA

A genus of 500 species (88 in China), mostly in the northern hemisphere but also in alpine regions of the southern hemisphere. **Annual**, **biennial** or **perennial herbs**, or **perennial** suffruticose or large **shrubs**, with erect, ascending to horizontal, leafy **stems**. **Leaves** pinnate to palmately compound, stipulate, stipules fused to the stalks. **Inflorescence** 1-flowered, cymose or cymose-paniculate. **Flowers** regular, hermaphrodite; hypanthium concave or hemispherical; sepals 5, valvate, epicalyx present; petals 5, white, yellow or purple; stamens c. 30, in 3 whorls; carpels numerous, free. **Fruit** a collection of achenes borne on a swollen receptacle.

Synopsis of groups

Group One. Perennial herbs or cushion-forming perennials with a woody base; leaves pinnate, palmate or with 3 leaflets; leaflets often toothed on margin; ovary glabrous. (below)

Group Two. Plants shrubby; woody stems above ground; leaves pinnate, rarely with 3 leaflets; ovary densely hairy. (p. 291)

Group One

1. Flowers basically white, occasionally reddish . . . .
   . . . . *P. coriandrifolia, P. salesoviana*
1. Flowers yellow 2
2. Leaves pinnate with more than 5 (up to 51) leaflets . . 3
2. Leaves with 3 or 5 leaflets 4
3. Plants markedly stoloniferous . . . .
   . . . . *P. anserine, P. leuconota*
3. Plants not stoloniferous . . *P. stenophylla, P. polyphylla,*
   P. discolor, P. bifurca
4. Dense cushion-forming plants . . . . *P. biflora*
4. Lax tufted to spreading plants, often with a woody base
   . . . . *P. eriocarpa, P. cuneata, P. simulatrix,*
   P. saundersiana

Potentilla coriandrifolia **D. Don** Low, tufted **perennial herb** 4–13 cm tall, with erect or ascending **stems**, hairy or not. **Leaves** 11–16 cm long, pinnate, leaflets opposite, 5–17, lanceolate, pointed, sparsely appressed-hairy above, densely so beneath, margin dissected. **Inflorescence** terminal, 1–3-flowered. **Flowers** 8–15 mm across; sepals triangular-ovate, pointed; petals white with purplish red base, 4–7 mm long, obovate, apex notched. w & swSichuan, Tibet, nwYunnan [c & eHimalaya, Bhutan, Myanmar]; alpine meadows and amongst rocks, 3,300–4,200 m. V–VII.

Potentilla salesoviana **(Steph.) Asch. & Graeb.** (syn. *Comarum salesovianum* Steph.) Robust **perennial herb** to 1 m, more usually 40–70 cm, with red-brown-hairy **stems**. **Leaves** deep green, pinnate with 7–11, oblong-lanceolate to ovate-lanceolate leaflets, to 3.5 × 1.2 cm, margin sharply toothed, often white-mealy beneath; upper leaves with 1 or 3 leaflets. **Inflorescence** laxly cymose. **Flowers** 2–3.2 cm across; sepals purplish, triangular, equalling petals; petals white, occasionally reddish, obovate, apex rounded. Gansu, Qinghai, Tibet [nChina, Himalaya, cAsia, Mongolia, Russia]; meadows, rocky slopes, riversides, 3,600–4,000 m. VI–VIII.

Potentilla anserina **L.** Patch-forming, markedly **stoloniferous perennial** with prostrate **stems** rooting at nodes. **Leaves** pinnate, 2–20 cm long; leaflets 11–23, elliptic, obovate-elliptic or oblong-elliptic, 10–25 × 5–10 mm, pointed, silvery-hairy beneath, margin toothed. **Inflorescence** 1-flowered. **Flowers** 20–28 mm in diameter; sepals triangular-ovate; petals yellow, obovate, 8–12 mm long, apex rounded. Achenes smooth, 1.5 mm long, glabrous. Gansu, Qinghai, Sichuan, Shaanxi, Tibet, Yunnan [c & nChina, cAsia, Europe, North America, Nepal, Russia]; meadows, grassland and on slopes, amongst rocks, streambeds and ditches, roadsides in damp spots, to 4,100 m. IV–XI. A cosmopolitan weed.

Potentilla leuconota **D. Don*** Distinguished from previous species by the pseudoumbellate **inflorescence**, and smaller **flowers** 15–25 mm in diameter. Gansu, Qinghai, Sichuan, Shaanxi, Tibet, Yunnan [c & nChina]; similar habitats, to 4,100 m. V–VIII.

Potentilla stenophylla **(Franch.) Diels** **Tufted perennial** to 25 cm tall in flower, often only 3–12 cm, with pinnate **leaves** 2–23 cm long; leaflets 11–51, crowded, oblong, 3–15 × 2–8 mm, decreasing in size towards base of leaf, densely silvery-hairy below and on the sharply toothed margins; stem-leaves like basal but stalkless. **Inflorescence** cymose. **Flowers** 15–25 mm

---

1. *Sanguisorba filiformis*; Sanba, nwYunnan (PC)
2. *Potentilla coriandrifolia*; eNepal (CGW)
3. *Potentilla salesoviana*; near Tongren, seQinghai (RS)
4. *Potentilla anserina*; sFrance (CGW)
5. *Potentilla biflora*; Baimashan, nwYunnan (JM)
6. *Potentilla biflora*; Baimashan, nwYunnan (CGW)

in diameter; petals yellow, oblong, apex rounded. w & swSichuan, seTibet, nwYunnan [swTibet, Sikkim, Myanmar]; alpine meadows, grassland and on slopes, forest margins, 3,200–5,800 m. VII–IX.

Potentilla polyphylla **Wall. ex Lehm.** Readily distinguished from the previous species by being a larger plant, to 40 cm tall in flower; **leaves** with 11–25 ovate to obovate or linear-lanceolate leaflets, these 10–50 × 8–16 mm; **flowers** 10–20 mm in diameter. Sichuan, Yunnan [Himalaya, India, Java]; meadows, open forests, forest margins, 2,500–4,500 m. VII–IX.

Potentilla discolor **Bunge** Like *P. polyphylla*, but a more densely hairy plant; **leaves** with just 5–9, lanceolate, pointed or blunt leaflets, 10–50 × 5–8 mm, sparsely white-hairy above, densely white- or grey-hairy beneath; leaf-stalks densely softly hairy. **Inflorescence** cymose, laxly few–many-flowered. **Flowers** 10–20 mm in diameter; petals obovate, 5–9 mm long, apex notched or rounded. Gansu, Sichuan, Shaanxi, Tibet, Yunnan [n & cChina, Japan, Korea]; open forests, shrubberies, alpine meadows, ravines, to 3,000 m. V–IX.

Potentilla bifurca **L.** Tufted plant to 20 cm tall in flower with ascending to prostrate **stems**. Readily distinguished from previous 3 species by having grey-green **leaves** with 7–17 leaflets that are entire except for 4–6 teeth at apex, sometimes entire overall; **flowers** small, 7–15 mm across, solitary or in small clusters. Gansu, Qinghai, Sichuan, Shaanxi, Tibet, nwYunnan [Europe, n & cAsia]; open forests, mountains slopes, loess areas, dry meadows, field edges, road-, track- and riversides, 400–4,000 m. V–X.

Potentilla biflora **Willd. ex Schltr.** Dense cushion-forming plant to 1 m across, usually less; **stems** erect, 4–12 cm long, hairy. **Leaves** with 5 leaflets, 2–6 cm long; leaflets linear, pointed or long-pointed, 8–17 × 1–3 mm, hairy above, densely softly white-hairy beneath, margin entire, recurved. **Inflorescence** terminal, 1–3-flowered. **Flowers** 12–18 mm in diameter; petals yellow, oblong-obovate, 5–10 mm long, apex notched. Gansu, Sichuan, Shaanxi, Tibet [Xinjiang, Mongolia, Nepal, eRussia, cAsia, North America]; alpine meadows, grassy places amongst rocks, gravel-beds, stabilised moraines, 2,300–4,800 m. VI–X.

Potentilla eriocarpa **Wall. ex Lehm.** Patch-forming **perennial** with stout, elongated, woody rhizomes bearing erect **stems** to 12 cm, white-hairy. **Leaves** 3–7 cm long, with 3 leaflets; leaflets obovate-elliptic or obovate-wedge-shaped, pointed, sparsely to densely softly white-hairy on both surfaces, 5–7-toothed at apex. **Inflorescence** terminal, 1–3-flowered, laxly racemose. **Flowers** 20–25 mm in diameter; petals golden yellow, usually with an orange spot at base, broadly obovate, apex notched. Sichuan, Shaanxi, Tibet, w & nwYunnan [Himalaya]; rocky alpine meadows and slopes in grassland, amongst rocks, rock-ledges, ravines, scrub, open forests, 2,700–5,000 m. VII–X.

Potentilla cuneata **Wall. ex Lehm.** Readily distinguished from *P. eriocarpa* by more slender, stoloniferous, tufted habit with smaller **leaves**, 2–3 cm long, leaflets less hairy, often becoming hairless, apex truncated or 3-toothed; petals plain yellow. w & swSichuan, seTibet, nwYunnan [swTibet, Himalaya]; alpine meadows, rocky places, forest margins, shrubberies, 2,700–3,600 m. VI–X.

Potentilla simulatrix **Th. Wolf*** Low-growing, **stoloniferous perennial** with thread-like stolons. Basal **leaves** tufted, to 10 cm long, with 3 leaflets, leaflets wedge-shaped to obovate, toothed, hairy, terminal leaflet larger than lateral; stolon leaves with short stalks. **Flowers** bright yellow, solitary, 7–10 mm across, on slender stalks. Gansu, Qinghai, Sichuan, Shaanxi [nChina]; damp habitats, open forests, meadows, streamsides, 300–2,200 m. IV–X.

Potentilla saundersiana **Royle** Mat-forming or tufted **perennial** to 20 cm with white-hairy **stems**. **Leaves** palmately lobed with 3 or 5 leaflets, these oblong-obovate, to 20 × 10 mm, often smaller, grey-green above, white-woolly beneath, margin coarsely toothed. **Flowers** in lax, few-flowered cymes or solitary, 10–14 mm across; sepals triangular-ovate, apex often toothed; petals yellow, obovate, longer than sepals, apex notched. Gansu, Qinghai, Sichuan, Shaanxi, Tibet, Yunnan [c & nChina, Himalaya]; alpine meadows, thickets, stony and gravelly places, 2,600–5,200 m. VI–VIII. Prostrate plants with solitary or paired **flowers** are referable to **var. *caespitosa* (Lehm.) Th. Wolf**.

---

1. *Potentilla eriocarpa*; cNepal (CGW)
2. *Potentilla cuneata*; Balangshan, wSichuan (CGW)
3. *Potentilla cuneata*; Galung La, near Bomi, seTibet (HJ)
4. *Potentilla saundersiana*; Huanglong, nwSichuan (CGW)
5. *Potentilla fruticosa*; Gangheba, Yulongxueshan, nwYunnan (CGW)
6. *Potentilla fruticosa*; Jiuzhaigou, nwSichuan (PC)
7. *Potentilla fruticosa* var. *arbuscula*; Wutoudi, Yulongxueshan, nwYunnan (CGW)
8. *Potentilla glabra*; Huanglong, nwSichuan (CGW)
9. *Potentilla glabra*; north of Songpan, nwSichuan (CGW)
10. *Potentilla parvifolia* with *P. glabra* and cream hybrid; seQinghai (RS)

Group Two

Potentilla fruticosa **L.** Very variable, erect or prostrate, intricately branched **shrub** to 1.5 m with reddish brown or grey peeling **bark**. **Leaves** pinnate with 3 or 5, oblong, obovate-oblong or ovate-lanceolate leaflets, 3–20 × 3–10 mm, sparsely to densely hairy below, margin entire. **Flowers** 10–30 mm across, solitary or few in a terminal cluster, primrose-yellow to golden yellow; sepals ovate, pointed; petals broadly obovate, apex rounded. Gansu, Sichuan, Shaanxi, Tibet, Yunnan [n & cChina, temperate Asia, Europe, North America]; alpine meadows, grassy slopes, amongst rocks, cliffs, scrub, shrubberies, open forests, 400–5,000 m. V–IX. Plants forming low, spreading mounds with very slender **stems**, leaflets not exceeding 5 × 4 mm, and solitary **flowers** are distinguished as **var. *pumila* Hook. f.** (seTibet, (?)nwYunnan [swTibet, Himalaya]); variants making erect **shrubs** with very silvery-hairy, grey leaflets with flat margins are distinguished as **var. *albicans* Rehd. & E. H. Wilson*** (Sichuan, e & seTibet, Yunnan); variants closely resembling the typical plant (**var. *fruticosa***) but with leaflets white-woolly beneath, raised veins on upper surface and a strongly back-rolled, rather than flat margin, are named **var. *arbuscula* (D. Don) Maxim.** (syn. *P. arbuscula* D. Don) (w & swSichuan, seTibet, nwYunnan [swTibet, c & eHimalaya]).

Potentilla glabra **Lodd.** (syn. *P. davurica* Nestl.; *P. fruticosa* L. var. *davurica* (Nestl.) Seringe) Very variable, erect or prostrate, much-branched **shrub** to 2 m, similar to *P. fruticosa* but **leaves** with 5 or 7 leaflets, generally sparsely hairy on both surfaces, with flat or slightly back-rolled margin; **flowers** white. Gansu, Hubei, Qinghai, Sichuan, Shaanxi, Yunnan [n & cChina, Korea, Mongolia, Russia]; similar habitats, 1,200–4,200 m. V–X. Hybridises with *P. fruticosa* in the wild.

Potentilla parvifolia **Fisch. ex Lehm.** Variable, prostrate or erect, much-branched **shrub**, 0.3–1.5 m, similar to *P. fruticosa*, but **leaves** with 5 or 7, lanceolate to linear-lanceolate leaflets (lower 2 pairs usually palmately arranged), not more than 10 × 5 mm; **flowers** 12–22 mm across; petals yellow, broadly obovate, apex notched or rounded. Gansu, Qinghai, Sichuan, Tibet, Yunnan [nChina, Mongolia, Russia]; similar habitats, 900–5,000 m. IV–IX.

SPENCERIA

A genus with 1 species restricted to Bhutan and west China which looks superficially like a *Potentilla*, but **achenes** are borne on an urn-shaped receptacle and are hidden within the calyx as the **fruit** develops (1 achene per receptacle).

Spenceria ramalana **Trimen** Small tufted **perennial herbs** 18–32 cm tall, with reddish brown, sparsely leafy, white-hairy, erect **stems**. **Leaves** 4.5–13 cm long, odd-pinnate; leaflets 13–21, papery, 1–2.5 × 0.5–1 cm, 2–3-lobed at apex, margin entire. **Inflorescence** erect, 5–20 cm long, 12–15-flowered.

Flowers 10–20 mm across, each with a 7–8-lobed involucre beneath; sepals 5, 3–6 mm long, glandular near tip; petals yellow, 5, free, 4–10 mm long, oblanceolate to obovate, apex rounded; stamens 30–40. w & swSichuan, seTibet, nwYunnan [Bhutan]; alpine meadows, limestone slopes, banks, 3,000–5,000 m. V–VII. (See photos 1 and 2 on p. 293.)

SIBBALDIA

A genus with c. 20 species (13 in China, 4 endemic) mainly in cold regions of the northern hemisphere. Tufted **perennial herbs**, often rather woody at base, with palmately divided **leaves** with 3 or 5, stalkless leaflets, generally silky-hairy, especially beneath; stipules present. **Flowers** solitary or numerous in dense cymes, not more than 7 mm across, unisexual or hermaphrodite; sepals (4–)5, with smaller epicalyx-scales outside; petals (4–)5, variously coloured, smaller to longer than sepals; stamens usually 5, occasionally 4 or 10. **Fruit** a head of tiny achenes.

Sibbaldia cuneata **Hornem. ex Kuntze** Tufted **perennial** to 15 cm at the most. **Leaves** with 3 leaflets; leaflets grey-green, elliptic to obovate, 8–25 × 6–18 mm, with a truncated, 3- or 5-toothed apex, sparsely hairy overall; stem-leaves 1 or 2, like the basal but smaller. **Flowers** 5–7 mm across, in compact cymes; sepals ovate, pointed; petals yellowish, obovate, ± equalling sepals. Qinghai, Sichuan, Tibet, Yunnan [nwChina, Himalaya, seRussia, Taiwan]; alpine meadows, rock-crevices, 3,400–4,500 m. V–X.

Sibbaldia purpurea **Royle** Readily distinguished from previous species by **leaves** with 5 leaflets with silky white hairs, and by purplish red **flowers**. Sichuan, Shaanxi, Tibet, Yunnan [Himalaya]; similar habitats as well as forest margins, 3,600–4,700 m. VI–VIII.

TAIHANGIA

A genus with a single species endemic to China. Closely related to *Potentilla*, but **leaves** generally simple and unlobed; **flowers** either male or hermaphrodite, on the same or separate plants. Previously recorded only from sHebei and nHenan.

Taihangia aff. *rupestris* **T. T. Yu & C. L. Li*** Small **rhizomatous perennial herb** to 15 cm. **Leaves** in a lax basal rosette, stalked, the stalks 3–5 cm long, downy; blade ovate to heart-shaped or elliptic, 2.5–7 × 2–6.5 cm, with scalloped margin, covered in silky hairs. **Flowers** solitary, held just above foliage, on a stalk with a whorl of small linear-lanceolate, or somewhat lobed, leafy bracts in the middle; sepals 5, ovate, with small, alternating, linear epicalyx segments; petals 5, white, rounded to reverse heart-shaped, clearly notched at top. neGuizhou (Fanjingshan) [cChina]; damp shady rocks, c. 2,388 m.

GEUM

A genus of c. 70 species (3 in China), mostly in the northern hemisphere. **Perennial**, **rhizomatous**, sometimes **stoloniferous**, dioecious **herbs**. **Leaves** alternate, stalked, stipulate, basal pinnate or pseudopinnate, stem-leaves bract-like or with 3 leaflets, occasionally pinnate; stipules membranous, joined to leaf-stalk. **Inflorescences** 1-flowered, or erect, several-flowered cymes. **Flowers** regular; sepals 5, valvate, epicalyx present, the segments alternating with the sepals; petals 5, white, red or yellow; stamens numerous. **Fruit** a collection of achenes, each hooked at tip.

Geum aleppicum **Jacq.** Robust **herb**, 30–100 cm tall with densely tufted, yellow-brown, silky-hairy **stems**. **Leaves** 5–15 cm long, stem-leaves pinnate, rigidly hairy on both surfaces; leaflets rhombic-ovate, apex blunt or pointed, margin coarsely toothed. **Inflorescence** terminal, lax, few-flowered. **Flowers** 10–17 mm across; petals yellow, orbicular to obovate. **Fruit** obovoid. Gansu, Guizhou, Hubei, Sichuan, Shaanxi, Tibet, Yunnan [c & nChina, temperate Asia, Europe]; glades in woods and forests, alpine meadows, open areas by streams, 200–3,500 m. V–X.

Geum japonicum **Thunb. var. *chinense* F. Bolle*** Hairier plant, readily distinguished from the previous species by simple, entire to 3-lobed stem-leaves and slightly larger flowers. Gansu, Guangxi, Guizhou, Hubei, Sichuan, Shaanxi, Yunnan [c, e & neChina]; similar habitats, 200–2,300 m. V–X. **Var.** *japonicum* occurs in Korea and Japan.

SIBIRAEA

A genus of 4 species (3 in China) in temperate Asia and Europe. Small, dioecious, deciduous **shrubs** with stout, erect, terete branches bearing egg-shaped purplish brown **buds**. **Leaves** alternate, without stipules, almost stalkless, entire, not toothed. **Inflorescence** paniculate, terminal, erect to almost so, stalked, cylindrical, catkin-like, many-flowered. **Flowers** small, short-stalked; sepals 5, erect, persistent in fruit; petals 5, white, longer than sepals; stamens 20–25, vestigial in female flowers; carpels 5, joined at base. **Fruit** a collection of follicles.

Sibiraea tomentosa **Diels*** Small **shrub** to 1 m with erect branches, hairy when young. **Leaves** clustered on short branches, narrowly oblong-obovate to oblanceolate, 5–7 × 2–2.5 cm, pointed, densely silvery-hairy on both surfaces, somewhat succulent. **Inflorescence** paniculate, 5–8 cm long, with downy bracts. **Flowers** 5–8 mm across; sepals greenish, reddish or purplish, triangular, 1.5 mm long, hairy; petals yellowish white, spathulate, blunt. nwYunnan; stream-beds and -banks, damp rocky places, stabilised moraines, outwash gravels, 3,200–4,000 m. V–VI.

Sibiraea angustata* (Rehd.) Hand.-Mazz. Taller plant than previous species, to 2.5 m with narrowly lanceolate to oblanceolate **leaves** that become hairless when mature, and with white, rather than yellowish, **flowers**. Gansu, Qinghai, w, nw & swSichuan, e & seTibet, nwYunnan; habitats and flowering time similar to those of *S. tomentosa*.

NEILLIA

A genus of c. 17 species (15 in China) in temperate Asia. Small, **deciduous shrubs** with spreading, terete or angular branchlets and egg-shaped **buds**. **Leaves** often 2-ranked, simple, shallowly 3- or 5-lobed, margin doubly toothed; stipules deciduous, conspicuous. **Inflorescences** terminal or axillary, racemose or paniculate, many-flowered. **Flowers** hermaphrodite; sepals 5, erect, persistent; petals 5, white or pink, ± equalling sepals; stamens 10–30; carpels 1–5. **Fruit** a collection of follicles enclosed by persistent hypanthium.

Neillia gracilis* Franch. Small **shrub** to 50 cm with recurved, angled **branches**. **Leaves** ovate to triangular-ovate, 3- or 5-lobed, 2.5–3 × 2–3 cm, downy on both sides; stipules ovate, pointed, 4–6 mm long, ciliate. **Inflorescences** axillary, 1–1.8 cm long, 3–7-flowered. **Flowers** c. 6 mm across; hypanthium campanulate, hairy; sepals triangular-ovate, long-pointed, 2–3 mm long; petals white or pinkish, orbicular, 4 mm long; stamens 15–20; ovary densely long-hairy. w & swSichuan, nYunnan; alpine meadows, damp slopes, 2,800–3,000 m. V–VII.

Neillia thibetica* Bur. & Franch. Larger **shrub** than previous species, 1–3 m tall, with larger, ovate to ovate-elliptic, irregularly 3- or 5-lobed **leaves**, 5–10 × 3–6 cm. **Inflorescences** larger, 5–15 cm long, 15–25-flowered. **Flowers** c. 4 mm across; hypanthium cylindrical, downy; sepals triangular, tailed, 2–3 mm long, densely downy; petals pinkish white, obovate, 3 mm long; ovary hairy only at apex. w & swSichuan, nwYunnan; mixed forests, often by streams, banks, pathsides, 1,500–3,000 m. V–VI.

1. *Spenceria ramalana*; swSichuan (AD)
2. *Spenceria ramalana*; Zhongdian, nwYunnan (PC)
3. *Sibbaldia cuneata*; Muli, swSichuan (PC)
4. *Sibbaldia cuneata*; Baimashan, nwYunnan (PC)
5. *Sibbaldia purpurea*; Baimashan, nwYunnan (PC)
6. *Taihangia* aff. *rupestris*; Fanjingshan, ne Guizhou (PC)
7. *Sibiraea angustata*; Gangheba, Yulongxueshan, nwYunnan (HJ)
8. *Neillia thibetica*; Wolong, wSichuan (PC)

KERRIA

A genus with 1 species confined to temperate China and Japan. **Deciduous shrub** with slender, erect **branches** bearing alternate, entire **leaves**; stipules linear-tapering, downy, deciduous. **Flowers** large, flat, golden yellow, petals with minutely toothed or entire margin. **Fruit** a collection of small brownish black, obovoid to hemispherical, rough-surfaced achenes.

Kerria japonica (L.) DC. Deciduous **shrub**, often forming a clump or thicket, 1–3 m with erect, leafy, green branchlets that eventually turn brown. **Leaves** green, ovate, 4–11 × 3–6 cm, long-pointed, hairy when young, margin doubly toothed. **Flowers** 3–5 cm across; petals golden yellow, elliptic, 1.5–2 cm long. **Achenes** obovoid, 4–4.5 mm long. Gansu, Hubei, Sichuan, Shaanxi, Yunnan [c & eChina, Japan]; thickets on steep wooded slopes, to 3,000 m. IV–VI. A double-flowered form is frequently cultivated in China and elsewhere.

SORBARIA

A genus of c. 9 species (3 in China) in temperate Asia. Large **deciduous shrubs** or **subshrubs** with yellow-green young branches, reddish to yellow-brown when mature. **Leaves** alternate, stipulate, pinnate; leaflets opposite, stalkless or almost so. **Inflorescence** large, paniculate, bearing numerous small **flowers**; hypanthium shallowly cup-shaped; sepals 5, reflexed, short, persistent in fruit; petals 5, overlapping, white, orbicular to ovate; carpels 5, opposite sepals, joined at base, glabrous; stamens 20–50. **Fruit** a collection of several-seeded, glabrous follicles.

Sorbaria arborea C. K. Schneid.* **Shrub** to 6 m with spreading, red-brown, stellate-hairy **branches**. **Leaves** 2-pinnate; leaflets pleated, opposite, 13–17, oblong-lanceolate to lanceolate, 4–9 × 1–3 cm, long-pointed, ± glabrous; stipules ovate, 8–10 mm long. **Panicles** lax, many-flowered, 20–30 cm long, *Astilbe*-like. **Flowers** 6–7 mm across; sepals oblong-ovate; petals white, almost orbicular, apex blunt, 3–4 mm long; Gansu, Guizhou, Sichuan, Shaanxi, e & seTibet, Yunnan [cChina]; dense mixed forests, forest margins, stream- and roadsides, 1,600–3,500 m. VI–IX.

Sorbaria kirilowii (Regel & Tiling) Maxim.* A generally smaller, more upright **shrub** than *S. arborea* with up to 21 leaflets per **leaf**; **panicles** erect, denser; fruit-stalks erect (not recurved). Gansu, Qinghai, Shaanxi [nChina]; open mixed forest, forest margins, shrubberies, 200–1,300 m. VI–IX.

ARUNCUS

A genus of c. 4 species (2 in China) in northern temperate Asia, Europe and Alaska. Monoecious, **herbaceous perennials**, sometimes with a woody base, growing from robust rhizome; **stems** erect, angled. **Leaves** lacking stipules, 1–3-pinnate; leaflets with sharp double teeth. **Inflorescence** paniculate, erect, large, many-flowered. **Flowers** stalkless or almost so, unisexual (female flowers with short sterile stamens and 3–4 carpels, male with 15–20 stamens and no carpels); hypanthium cup-shaped with ring-like disc on rim; sepals 5, persistent in fruit; petals 5, white, obovate. **Fruit** a group of follicles, pendent, each 2-seeded.

Aruncus sylvester Kostel. ex Maxim. Large **herb** up to 3 m with dark purple, glabrous **stems** and 2–3-pinnate **leaves**; leaflets ovate-lanceolate to narrowly elliptic, long-pointed, 5–13 × 2–8 cm, ± glabrous, margin irregularly toothed. **Panicle** rather lax, 10–40 cm long. **Flowers** 2–4 mm across; sepals spreading or erect; petals white, obovate. Gansu, Guangxi, Sichuan, Shaanxi, se & eTibet, Yunnan [swTibet, c & nChina, Europe, c & eHimalaya, Japan, Russia, Alaska]; mixed forests on mountain slopes, ravines, lake margins, wet shrubberies, 1,800–3,500 m. VI–VII.

SPIRAEA

A complex genus of almost 100 species (70 in China) in the north temperate zone. Small to medium-sized, **deciduous shrubs** with erect to arching **stems**. **Leaves** alternate, simple, with toothed or entire margin and a short stalk, lacking stipules. **Inflorescence** umbellate, racemose or paniculate, many-flowered. **Flowers** hermaphrodite or rarely unisexual, regular; hypanthium cup-shaped or campanulate, rimmed by a nectar-bearing disc; sepals 5, valvate, shorter than hypanthium; petals 5, overlapping or contorted, longer than sepals; stamens 15–60; carpels 5, free, each containing 2 ovules. **Fruit** a group of small follicles, each dehiscent along upper suture, containing minute seeds.

1. Inflorescences borne on short lateral shoots of previous year's stems 2
1. Inflorescences borne on current year's erect, leafy shoots 4
2. Leaves with toothed, sometimes 3-lobed margin .... ***S. pubescens***
2. Leaves unlobed, margin entire or toothed at apex ... 3
3. Leaves relatively large, at least 15 mm long ... ***S. sericea***
3. Leaves small, not more than 15 mm long .... ***S. calcicola, S. myrtilloides***
4. Stamens 25–30 per flower .... ***S. japonica***
4. Stamens 15–20 per flower 5
5. Leaves small, not more than 20 mm long .... ***S. canescens, S. arcuata, S. schneideriana***
5. Leaves larger, more than 20 mm long .... ***S. bella, S. longigemmis, S. rosthornii***

Spiraea pubescens Turcz. **Shrub** 1–2 m tall with spreading yellow-brown to black-brown branchlets, downy at first. **Leaves** stalked, elliptic to almost rhombic, 2–4.5 × 1.3–2.5 cm, pointed, grey-downy beneath, somewhat hairy above, margin incised-toothed. **Flowers** 15–20, each 5–7 mm across, in umbels 2–4.5 cm across; sepals 1.5–2 mm long, erect in fruit; petals white, oval to obovate, 2–3.5 mm long; stamens 25–30. **Follicles** spreading, usually downy. Gansu, Hubei, Sichuan, Shaanxi [n & cChina, Korea, Mongolia, eRussia]; mixed, often open forests, shrubberies, shady slopes and rocks, 200–2,500 m. V–VI.

Spiraea sericea Turcz. Erect **shrub** to 2 m with brown or grey-brown branchlets, hairy at first. **Leaves** short-stalked, ovate-elliptic to elliptic, 1.5–4.5 × 0.7–1.5 cm, pointed, deep green above, grey-green beneath, hairy, more densely so beneath, margin entire (sometimes slightly toothed on sterile shoots). **Inflorescence** umbellate-racemose, 3–6 cm across, with 15–30 **flowers**, each 4–5 mm across; sepals ovate, 1–2 mm long, reflexed in fruit; petals white, rounded, 2–3 mm. **Follicles** spreading, hairy. Gansu, Sichuan, Shaanxi, Yunnan [n & nwChina, Japan, Mongolia, eRussia]; mixed open forests, forest margins, grassy and rocky slopes, 500–1,100 m. V–VI.

Spiraea calcicola W. W. Sm.* Low **shrub** to 1.5 m with spreading, purplish brown, angular branchlets. **Leaves** in clusters on short shoots, alternate, obovate or elliptic, 5–9 × 2–6 mm, blunt, margin entire, glabrous. **Inflorescence** umbellate, 6–10-flowered, terminal on shoots, 8–12 mm across. **Flowers** 4–6 mm across; petals creamish or pink, obovate, blunt or notched, 2–3 mm long; stamens 20. **Follicles**, spreading. nwYunnan; rocky limestone mountain slopes, 2,700–2,800 m. VI–VII.

---

1. *Kerria japonica*; cult. (CGW)
2. *Kerria japonica*; cult. (PC)
3. *Sorbaria arborea*; Longriba, nwSichuan (CGW)
4. *Sorbaria kirilowii*; near Xunhua, cQinghai (RS)
5. *Aruncus sylvester*; Wolong, wSichuan (CGW)
6. *Spiraea pubescens*; near Aba, nwSichuan (CGW)
7. *Spiraea calcicola*; Baishui, Yulongxueshan, nwYunnan (CGW)

Spiraea myrtilloides **Rehd.*** Larger **shrub** than previous species, 2–3 m with red-brown or dark brown branchlets. **Leaves** ovate to obovate-oblong, 6–15 × 4–7 mm, blunt, sometimes hairy beneath, 3-veined at base. **Flowers** 7–20 in larger clusters 2–4 cm across; petals white. Gansu, Hubei, Qinghai, Sichuan, Shaanxi, e & seTibet, Yunnan [cChina]; mixed and open forests, shrubberies, shady and rocky places, grassy slopes, 1,500–3,100 m. VI–VII.

Spiraea japonica **L. f.** Erect **shrub** up to 1.5 m with purplish brown to brown, slender, terete branchlets. **Leaves** ovate or elliptic-ovate, blunt, 2–16 × 1–3.5 cm, pointed or long-pointed, margin with irregular simple or double teeth, downy on veins beneath. **Inflorescence** corymbose, many-flowered, terminal on shoots, 6–14 × 2.5–14 cm. **Flowers** 4–7 mm across; petals pale to deep pink, ovate to orbicular, 2.5–3.5 mm long; Stamens 25–30. Carpels glabrous or downy. **Follicles** spreading. Gansu, Guizhou, Hubei, Sichuan, Shaanxi, e & seTibet, Yunnan [c & eChina, Japan, Korea]; broadleaved and coniferous forests, thickets, clearings in forests, rocky mountain slopes, riverbanks, to 4,000 m. V–VIII.

Spiraea canescens **D. Don** **Shrub** 2–4 m tall, with twisted brown or grey-brown, angled branchlets. **Leaves** short-stalked, ovate to obovate or oblong, 10–20 × 8–12 mm, blunt, margin entire or with a few teeth near apex, glabrous above, sometimes hairy beneath. **Inflorescence** a compound corymb, 3–5 cm across, many-flowered. **Flowers** 5–6 mm across; petals white or pinkish, rounded, 2–3.5 mm long; stamens 20. **Follicles** slightly spreading, glabrous or hairy. Gansu, w, nw & swSichuan, e & seTibet, nwYunnan [swTibet, Himalaya]; shrubberies, thickets, stream- and riversides, dry rocky places, tracks, roadsides, 2,300–4,000 m. VII–VIII.

Spiraea arcuata **Hook. f.** (syn. *S. canescens* D. Don var. *glabra* Hook. f. & Thomson) Smaller **shrub** than previous species with arched, dark brown, shiny, deeply grooved branchlets; **leaves** elliptic to obovate, not more than 12 × 5 mm; **flowers** larger, 6–8 mm across; petals pink or white. **Follicles** spreading, shiny, glabrous. seTibet, nwYunnan [swTibet, Himalaya, nMyanmar]; similar habitats and subalpine rocks, 3,000–4,200 m. V–VII.

Spiraea schneideriana **Rehd.*** Erect **shrub** 1–2 m tall with reddish brown, slender, angular, softly hairy branchlets. **Leaves** ovate or oblong-ovate, 8–15 × 5–8 mm, blunt or almost pointed, ± entire, sparsely downy beneath. **Inflorescence** corymbose, many-flowered, terminal on shoots, 3–6 cm across. **Flowers** 5–6 mm; petals white, ovate to orbicular, 2–2.5 mm long, blunt or notched; stamens 20. **Follicles** spreading, minutely hairy. Gansu, Guizhou, Hubei, Sichuan, Shaanxi, e & seTibet, Yunnan [Fujian]; broadleaved and coniferous forests, thickets, rocky mountain slopes, streamsides, 2,500–4,000 m. VI–VII.

Spiraea bella **Sims** **Shrub** 0.8–2 m with yellowish brown to reddish brown, slightly angular branchlets. **Leaves** ovate, elliptic-ovate or elliptic-oblong, 2–4 × 1–2 cm, pointed, margin with coarse simple or double teeth, bright green above, downy on veins and grey-green beneath. **Inflorescence** corymbose, flat or domed, many-flowered, terminal on shoots, 1.5–2.5 cm across. **Flowers** ± unisexual, 5–7 mm across; petals pale to deep pink, rarely white, rounded, blunt, 2–4 mm long; stamens 20. **Follicles** spreading, minutely hairy. swSichuan, seTibet, nwYunnan [swTibet, Bhutan, nIndia, Nepal]; broadleaved and coniferous forests, thickets, mountain slopes, 2,300–3,600 m. V–VI.

Spiraea longigemmis **Maxim.*** Distinguished from previous species by brown or greyish brown, rather twisted branchlets and ovate-lanceolate to oblong-lanceolate **leaves**; corymbs larger, 4–8 cm across; **flowers** hermaphrodite, white; stamens 15–20, longer than petals. Gansu, Hubei, Sichuan, Shaanxi, e & seTibet, Yunnan [Shanxi, Zhejiang]; rocky and grassy slopes, road and streamsides, 2,500–3,400 m. V–VII.

Spiraea rosthornii **E. Pritz. ex Diels*** Closely related to the previous species, but with larger corymbs (7–11 cm across) of white **flowers**, inflorescence-axis and stalks more downy; stamens c. 20, longer. Guangxi, Qinghai, Sichuan, Shaanxi, Yunnan [cChina]; similar habitats, 1,000–3,500 m. V–VII.

CRASSULACEAE

SEDUM

A genus of 470 species (121 in China, 91 endemic) found in the north temperate zone and tropical mountains. Mainly glabrous, **succulent** or **semi-succulent annuals** and **perennials** with alternate, paired or whorled **leaves** that are normally short-spurred at base, usually entire. **Flowers** solitary or, more usually, in terminal or lateral cymose clusters, generally 5-parted, sometimes to 9-parted, sepals and petals free to the base or slightly joined; petals often yellow, sometimes white or reddish; stamens twice as many as the petals, in 2 whorls. **Fruit** a small head of follicles equal in number to the petals.

1. Leaves flat and toothed .... **S. aizoon**
1. Leaves rounded or half-rounded in cross-section, not toothed 2
2. Flowers generally stalkless; follicles spreading to form a star .... **S. multicaule**
2. Flowers short-stalked; follicles erect or almost so ... 3
3. Perennials with at least some non-flowering shoots ... . **S. glaebosum, S. oreades**
3. Annuals or biennials without vegetative, non-flowering, shoots at flowering time 4
4. Leaves small, not more than 5 mm long; sepals not spurred at base .. **S. prasinopetalum, S. henrici-robertii**
4. Leaves relatively large, more than 5 mm long; sepals spurred at base 5
5. Petals 7 mm or more long .... **S. magniflorum**
5. Petals not more than 6 mm long .... **S. forrestii, S. obtusipetalum, S. morotii**

Sedum aizoon L. (syn. *Phedimus aizoon* (L.) 't Hart) Very variable **tuberous perennial herb** with simple, erect **stems** to 50 cm, usually less, bearing alternate, ovate to lanceolate or rounded, toothed **leaves** to 80 × 30 mm. **Flowers** numerous, 5-parted, in a branched, flat-topped **inflorescence**; sepals linear; petals yellow, elliptic, 6–10 mm long, twice as long as sepals. **Fruit** star-shaped with 5 follicles. Gansu, Hubei, Qinghai, n & eSichuan, Shaanxi (n, c & eChina, Japan, Korea, Mongolia, eRussia]; rocky and grassy places, shrubberies, ravines, rock-crevices, 1,000–3,100 m. VI–VII.

Sedum multicaule Wall. ex Lindl. Small tufted **herb** to 15 cm with alternate, closely overlapping, linear **leaves**, these not more than 15 × 2 mm, long-pointed. **Flowers** in branched scorpioid cymes, 5-parted; sepals linear to linear-lanceolate, 5–6.5 mm long; petals yellow, oblong-ovate, 5–6 mm long, apex long-pointed; stamens 10. Gansu, Sichuan, Shaanxi, Tibet, Yunnan [Himalaya, nIndia, nMyanmar]; forest rocks, and other rocky places, cliffs, 1,300–3,500 m. VII–VIII.

Sedum glaebosum Fröd.* Small tufted **perennial herb** not more than 6 cm tall, with erect to ascending **stems** and ovate to narrow-lanceolate **leaves**, 3–6 mm long, with a short, 2–3-lobed, basal spur. **Flowers** very short-stalked, in corymb-like clusters; sepals oblong, 4–6 mm long, with spurred base; petals yellow, narrow-oblong, 6–8 mm long; stamens 10. seQinghai, w & nwSichuan, e & neTibet; rocky places, ravines, 3,500–5,000 m. VII–IX.

Sedum oreades (Decne) Raym.-Hamet (syn. *Umbilicus oreades* Decne) Larger plant than previous species, to 12 cm with lanceolate to oblong **leaves** that increase in size towards shoot-tips, to 9 mm long. **Flowers** in small clusters, sometimes solitary; petals obovate to elliptic, 6–10 mm long, bases fused together for a short distance. seTibet; nwYunnan [Himalaya, nMyanmar]; grassy and rocky habitats, 3,000–4,500 m. VII–VIII.

1. *Spiraea myrtilloides*; north of Longriba, nwSichuan (CGW)
2. *Spiraea myrtilloides*; Jiuzhaigou, nwSichuan (CGW)
3. *Spiraea japonica*; Heishui, Yulongxueshan, nwYunnan (PC)
4. *Spiraea bella*; Cangshan, nwYunnan (PC)
5. *Sedum glaebosum*; Jigzhi, seQinghai (CGW)
6. *Sedum glaebosum*; Matang, Zhegushan, nwSichuan (HJ)

1. *Sedum prasinopetalum*; north of Jigzhi, seQinghai (CGW)
2. *Sedum henrici-robertii*; eNepal (CGW)
3. *Sedum magniflorum*; Deqin, nwYunnan (PC)
4. *Sedum forrestii*; Napahai, Zhongdian, nwYunnan (PC)
5. *Sinocrassula indica*; near Benzilan, Jinshajiang, nwYunnan (HJ)
6. *Rhodiola yunnanensis*; Shennongjia, wHubei (PC)
7. *Rhodiola yunnanensis*; Baimashan, nwYunnan (PC)
8. *Rhodiola chrysanthemifolia*; Baimashan, nwYunnan (PC)

Sedum prasinopetalum Fröd.* Glabrous **annual herb** to 7 cm with a simple or branched **stem**. **Leaves** lanceolate to narrow-oblong, 3–5 mm long. **Flowers** star-shaped, few per cyme, in flat-topped clusters; sepals greenish yellow, 3–4.5 mm long, with 3–5 blackish veins; petals yellow, 5–6 mm long. s & seQinghai, w & nwSichuan; rocky and short-grassy habitats, shrubberies, roadsides, 4,100–4,500 m. VIII–IX.

Sedum henrici-robertii Raym.-Hamet Similar to the previous species, but a smaller plant not more than 3 cm tall with unequally 5-parted **flowers**, petals only c. 3 mm long, equal in length to sepals. eQinghai, seTibet [swTibet, Himalaya]; rocky and stony places, 3,800–5,000 m. VII–VIII.

Sedum magniflorum K. T. Fu* Tufted **annual herb** to 8 cm with narrow-oblong **leaves** 6–8 mm long. **Inflorescence** corymbose. **Flowers** 3–5 in arched cymes, each flower on a stalk 2–7 mm long; sepals narrow-oblong, 6.5–7.5 mm long; petals yellowish, oblong-elliptic, 7–8 mm long. nwYunnan; forest rocks and ravines, riverside shingles, 3,800–4,100 m. VII–IX.

Sedum forrestii Raym.-Hamet* Glabrous **biennial herb** only to 4 cm with **stems** branched from base. **Leaves** ovate to oblong, 5–10 mm long. **Flowers** numerous in a corymb-like **inflorescence**; sepals oblong to obovate, 3–6.5 mm long; petals yellow, elliptic, 5.5–6 mm long, free. nwYunnan; rocky slopes, screes, moraines, thin stony pastures, 3,300–4,300 m. VII–VIII.

Sedum obtusipetalum Franch. Glabrous **annual herb** to 15 cm with **stems** branched from base. **Leaves** ovate to oblong or elliptic, 6–10 mm long. **Flowers** in dense clusters, star-shaped, each with a stalk to 5 mm long; sepals narrow-oblong, 3.5–5 mm long; petals yellow, elliptic, 5–6 mm long, pointed, apex minutely papillose. swSichuan, nwYunnan [Nepal]; coniferous forests, ravines, damp and sandy places, mossy rocks, 2,000–4,000 m. VIII–X.

Sedum morotii Raym.-Hamet* Readily distinguished from previous species by the often larger (6–22 mm long) obovate **leaves**, and the smooth, pointed petal-tips. eTibet; similar habitats as well as pathsides and old walls, 1,300–3,000 m. VIII–IX.

SINOCRASSULA

A genus of 7 species (all in China, 1 endemic) in the Himalaya and north India. Closely related to *Sedum*, differing in the conspicuous basal leaf-rosettes and campanulate flowers, the petals eventually spreading.

Sinocrassula indica (Decne) A. Berger Very variable glabrous, tufted **herb** 5–60 cm tall with thick fleshy **leaves**. Basal leaves oblong-spathulate, to 6 × 1.5 cm, stem-leaves alternate, almost obovate to almost rounded, 2.5–3 × 0.5–1 cm, margin entire or slightly toothed towards the blunt to long-pointed apex. **Inflorescence** paniculate with small, leaf-like bracts; sepals broad-triangular, c. 2 mm long; petals red, yellow or greenish yellow, lanceolate to ovate, 2.5–5 mm long, apex pointed and often reflexed. sGansu, Guangxi, Guizhou, Hubei, Sichuan, Shaanxi, Tibet, Yunnan [cChina, Himalaya, nIndia]; rocky and gravelly slopes, cliffs, rock-ledges, 500–4,000 m. VII–X.

RHODIOLA

A genus of c. 90 northern hemisphere species (55 in China, 16 endemic) closely related to *Sedum* but differing in the fleshy **rhizomatous** rootstock (caudex) with dark scale-like **leaves** contrasting with the green leafy **stems**. **Flowers** usually 4–5-parted, usually hermaphrodite but sometimes unisexual; petals free, equalling the number of sepals; stamens twice as many as petals, in 2 whorls. **Fruit** a collection of follicles usually equal in number to the petals.

1. Stem-leaves in whorls of 3, occasionally opposite, or aggregated at stem-tips, toothed or lobed . . . . **R. yunnanensis, R. chrysanthemifolia**
1. Stem-leaves alternate 2
2. Plants at flowering time with old persistent dead flowering stems; follicles fused towards base 3
2. Plants without old persistent flowering stems; follicles almost free 4
3. Leaves less than 1.5 mm wide . . . . **R. coccinea**
3. Leaves at least 2 mm wide . . . . **R. atuntsuensis, R. himalensis, R. fastigiata, R. dumulosa, Rhodiola sp.**
4. Flowers hermaphrodite . . . . **R. purpureoviridis**
4. Flowers usually unisexual . . . . **R. kirilowii, R. crenulata, R. aff. discolor**

Rhodiola yunnanensis (Franch.) S. H. Fu* (syn. *Sedum henryi* Diels) Plant with 1–few erect **stems** or **clump-forming**, to 1 m but usually less, bearing pale green, stalkless, ovate to elliptic or oblong **leaves**, generally 4–7 × 2–4 cm, with a few marginal teeth. **Flowers** unisexual, numerous in a lax cymose panicle, 4–5-parted; male flowers with petals greenish yellow to yellow, 1.5 mm long; female flowers with petals purple, linear, 1.2 mm long; stamens 8. Gansu, Guizhou, Hubei, Sichuan, Shaanxi, Yunnan; forest slopes, shady banks and rocks, 1,000–4,000 m. V–VII.

Rhodiola chrysanthemifolia (H. Lévl.) S. H. Fu* (syn. *Sedum chrysanthemifolia* H. Lévl.) Small **tufted herb** to 10 cm with a thickened, branched stock. **Leaves** crowded towards shoot-tips, rosette-like, oblong to ovate, shiny deep green, sometimes red- or purple-flushed, margin shallowly pinnately lobed. **Flowers** few, hermaphrodite, in compact clusters; sepals linear to triangular-ovate, c. 4 mm long; petals reddish, oblong-ovate, 7–9 mm long. swSichuan, nwYunnan; grassy and rocky habitats, crevices, 3,200–4,200 m. VII–VIII.

CRASSULACEAE: Rhodiola

Rhodiola coccinea (Royle) Boriss. **Tufted perennial** with numerous shoots rising from a thick caudex. **Stems** thin, not more than 2 mm in diameter. **Leaves** alternate, linear-lanceolate to lanceolate, to 7 × 1.5 mm, long-pointed to pointed, margin entire. **Flowers** unisexual, usually 5-parted; sepals red, 1.5–4 mm long; petals red or yellow, oblong-ovate to lanceolate, 1.5–4 mm long. **Follicles** short, each with a reflexed beak, red at maturity. Gansu, Qinghai, Sichuan, Tibet, nwYunnan [eAfghanistan, Himalaya]; rocky and stony places, crevices, 2,200–5,300 m. VI–VII.

Rhodiola atuntsuensis (Praeg.) S. H. Fu **Tufted**, often **mound-forming perennial** usually with numerous **stems** arising from a thick caudex. **Leaves** bright green, alternate, rather crowded, oblong to ovate, to 12 × 4 mm, margin entire, sometimes slightly wavy. **Flowers** in rounded clusters, 4–5-parted, hermaphrodite or unisexual; petals ascending, yellow, sometimes with purplish base, narrow-oblong to oblanceolate, 2.5–4.5 mm long; stamens 8 or 10. **Follicles** erect to ascending. w & swSichuan, seTibet, nwYunnan [nMyanmar]; rocky places (granitic or calcareous), forest margins, 3,100–5,000 m. VII–VIII.

Rhodiola himalensis (D. Don) S. H. Fu (syn. *Sedum himalensis* D. Don) Taller plant than previous species with generally reddish, glandular **stems** to 50 cm, and lanceolate to obovate **leaves** to 27 × 10 mm. **Flowers** in larger heads, petals purple, not more than 4 mm long. sGansu, Qinghai, w & nwSichuan, Tibet, nwYunnan [Himalaya]; rocky and grassy slopes, forests and forest margins, shrubberies, 2,600–4,200 m. V–VII.

Rhodiola fastigiata (Hook. f. & Thomson) S. H. Fu (syn. *Sedum fastigiatum* Hook. f. & Thomson) Like *R. atuntsuensis*, but plant to 20 cm with narrow-oblong to oblanceolate **leaves** to 12 × 4 mm. **Flowers** always 5-parted; petals red, c. 5 mm long. w & swSichuan, seTibet, nwYunnan [swTibet, Himalaya]; rocky places, moraines, 3,500–5,400 m. VI–VIII.

Rhodiola dumulosa (Franch.) S. H. Fu Readily distinguished from the previous 3 species as a plant with rather robust **stems** to 30 cm; bearing linear **leaves** 7–10 mm long larger **flowers**, petals 8–11 mm long, white, greenish or red, often slightly fringed at margin. Gansu, Hubei, Qinghai, Sichuan, Shaanxi, Yunnan (n & cChina, Bhutan, nMyanmar); rocky slopes, crevices, 1,600–4,100 m. VI–VIII.

Rhodiola sp.* Tufted **perennial** forming rounded hummocks to 8 cm high, with a short, branched caudex bearing persistent dead **stems** from previous year. **Leaves** shiny deep green, elliptic-ovate to lanceolate-ovate, to 9 mm long. **Flowers** solitary or several in a tight cluster; petals yellow, spreading, 3–4 mm long. nwYunnan (Daxueshan); rocky slopes, stabilised screes, c. 4,200 m. VII.

Rhodiola purpureoviridis (Praeg.) S. H. Fu* **Tufted perennial** with several **stems** to 40 cm, often far less, ascending from the thickened rootstock (caudex). **Leaves** alternate, numerous, deep shiny green, turning red in autumn, elliptic to narrow-lanceolate, to 60 mm long, glabrous or glandular-hairy beneath, margin often somewhat back-rolled. **Flowers** in umbel-like clusters; petals greenish, narrow-oblanceolate, 3–4 mm long, spreading, longer than linear-lanceolate sepals; stamens with purple filaments. w & swSichuan, e & seTibet, nwYunnan; stony meadows and slopes, rocks, forests and forest margins, 2,500–4,100 m. VII–VIII.

Rhodiola kirilowii (Regel) Maxim. (syn. *Sedum kirilowii* Regel) Plants forming rounded tufts to 50 cm high, with **stems** ascending from a thick, erect, rootstock (caudex). **Leaves** alternate or in false whorls, grey-green, linear to linear-lanceolate, to 60 × 15 mm, margin generally sparsely toothed. **Flowers** in umbel-like, branched clusters; petals 4 or 5, green, greenish yellow or reddish, lanceolate to oblong or obovate, 3–4 mm long; stamens 8 or 10. **Follicles** erect, each with a short recurved beak. Gansu, Qinghai, Sichuan, Shaanxi, Tibet, Yunnan [c & nwChina, nMyanmar, Kazakhstan]; rocky and grassy mountain slopes, forest margins, open shrubberies, often in partial shade, 2,000–5,600 m. V–IX.

Rhodiola crenulata (Hook. f. & Thomson) H. Ohba (syn. *Sedum crenulatum* Hook. f. & Thomson) Forms tighter **hummocks** than the previous species, not more than 20 cm high, **leaves** broad-oblong to almost rounded, to 30 × 22 mm, margin slightly toothed to entire; **flowers** red to purplish red; **follicles** red when ripe. Qinghai, Sichuan, Tibet, w & nwYunnan c & eHimalaya]; rocky places and crevices, rocky meadows, moraines, 2,800–5,600 m. VI–IX.

Rhodiola aff. *discolor* (Franch.) S. H. Fu* Small perennial **herb** to 10 cm with a short caudex giving rise to a tuft of spreading, greenish white **stems**. **Leaves** linear-lanceolate to linear-elliptic, 7–15 mm long, entire, apiculate. **Flowers** few in a cluster, unisexual, 4–5-parted; petals deep reddish purple, 4–5 mm long. swSichuan; rocky slopes, screes, 3,200–3,900 m. VII.

SAXIFRAGACEAE

PHILADELPHUS

A genus of c. 70 species (22 in China) in the north temperate zone and Mexico. **Deciduous shrubs** with paired **leaves**; hairs if present, simple. **Inflorescence** a raceme, panicle or cyme. **Flowers** white, 4-petalled, sweetly fragrant; calyx with campanulate or urn-shaped tube and 4(–5) lobes; stamens 20–40 with flat filaments; stigma 4-lobed. **Fruit** a 4–5-valved capsule.

1. *Rhodiola atuntsuensis*; Baimashan, nwYunnan (CGW)
2. *Rhodiola dumulosa*; nwYunnan (HJ)
3. *Rhodiola dumulosa*; Balangshan, Wolong, wSichuan (HH)
4. *Rhodiola* sp.; Daxueshan, nwYunnan (CGW)
5. *Rhodiola purpureoviridis*; Baimashan, nwYunnan (CGW)
6. *Rhodiola kirilowii*; north of Aba, nwSichuan (CGW)
7. *Rhodiola kirilowii*; nwSichuan (JM)
8. *Rhodiola crenulata*; Balangshan, Wolong, wSichuan (HH)
9. *Rhodiola* aff. *discolor*; Genyanshan, swSichuan (AD)

1. *Philadelphus delavayi*; Bitahai, Zhongdian, nwYunnan (CGW)
2. *Philadelphus delavayi*; Napahai, Zhongdian, nwYunnan (PC)
3. *Philadelphus kansuensis*; Maya to Jianni, swGansu (HJ)
4. *Philadelphus kansuensis*; Maya to Jianni, swGansu (HJ)
5. *Philadelphus purpurascens*; Heishui, Yulongxueshan, nwYunnan (CGW)
6. *Philadelphus purpurascens*; Baimashan, nwYunnan (CGW)
7. *Philadelphus incanus*; Shennongjia, wHubei (PC)
8. *Deutzia setchuenensis*; cult. (CGW)

1. Racemes generally with 7–11 or more flowers; style thicker, with a club- or oar-shaped stigma . . . .
. . . . **P. delavayi, P. delavayi var. trichocladus,**
P. delavayi var. calvescens, P. tomentosus
1. Racemes smaller, mainly with 3–7 flowers; style slender, stigma with a swollen tip 2
2. Calyx and stalks glabrous to sparsely hairy . . . .
. . . . **P. kansuensis, P. purpurascens**
2. Calyx and flower-stalks densely appressed white-hairy
. . . . **P. incanus**

Philadelphus delavayi L. Henry Variable **deciduous shrub** to 4 m with purple-brown branchlets, these changing to grey-brown, glabrous (hairy in **var. *trichocladus* Hand.-Mazz.**). **Leaves** oblong to lanceolate, to 8 × 6 cm, grey-hairy beneath, with 3–5 veins, margin finely toothed to entire. **Racemes** generally 5–9-flowered, sometimes more. **Flowers** pendent or semi-pendent; calyx purple or blackish purple, glabrous and glaucous, urn-shaped; corolla broad-campanulate, lobes obovate to oblong, 12–15 mm long, with a slightly crisped margin; stamens 30–35. w, s & swSichuan, e & seTibet, w & nwYunnan [nMyanmar]; shrubberies, mixed forests and forest margins, rocky slopes, 700–3,800 m. VI–VIII. Plants with purple twigs and calyces are referable to **var. *calvescens* Rehd.** (syn. *P. calvescens* (Rehd.) S. M. Hwang).

Philadelphus tomentosus Wall. ex G. Don (syn. *P. coronarius* L. var. *tomentosus* (Wall. ex G. Don) Hook. f. & Thomson) Distinguished from the previous species by tail-like leaf-tips, **leaves** ovate to ovate-lanceolate, to 10 × 5 cm with 5–7 veins; **flowers** smaller and cross-shaped, oblong petals 5–10 mm long; stamens 20–25. seTibet, nwYunnan [swTibet, Himalaya]; similar habitats, 2,500–4,400 m. VI–VII.

Philadelphus kansuensis (Rehd.) S. Y. Hu* **Shrub** to 7 m, more usually 2–4 m, with dark purple young **branches**, greying with age. **Leaves** ovate to elliptic, to 10 × 6.5 cm, glabrous or hairy along veins beneath, with 3–5 veins and a slightly toothed to entire margin; stalks 2–8 mm long. **Racemes** 5–7-flowered with purple rachis; calyx purple, slightly bristly, with campanulate tube; **corolla** campanulate, lobes ovate, 12–15 mm long; stamens 28–30. Gansu, e & seQinghai, Shaanxi; shrubberies, woodland margins, 2,400–3,500 m. VI–VII.

Philadelphus purpurascens (Koehne) Rehd.* Similar to the previous species, but **leaves** not exceeding 7 × 4.5 cm, calyx sparsely hairy but becoming hairless, with urn-shaped tube, petals obovate with notched apex. Sichuan, Yunnan; mixed open forests and forest margins, shrubberies, rocky slopes, 2,200–3,500 m. V–VI.

Philadelphus incanus Koehne* **Shrub** to 3.5 m with brown or purplish, usually downy branchlets, greying with age. **Leaves** ovate, to 12.5 × 10 cm, somewhat long-pointed, densely softly hairy beneath, sparsely appressed-hairy above, with 3–5 veins, margin slightly toothed. **Racemes** 5–7(–11)-flowered; calyx greyish green, with campanulate tube and ovate lobes c. 5 mm long; **corolla** shallowly bowl-shaped, the rounded to ovate lobes 5–6 mm long; stamens 30–35. Hubei, Sichuan, Shaanxi [cChina]; shrubberies, woodland margins, 1,200–1,700 m. V–VI.

DEUTZIA

A genus of c. 60 species (most in China, 41 endemic) in the warmer temperate areas of the northern hemisphere. **Deciduous** or **semi-deciduous shrubs** with opposite, stellate-hairy **leaves**; stipules absent. **Flowers** in small cymes or larger racemes or panicles borne on previous year's wood; calyx campanulate, with 5 lobes; petals 5; stamens 10, occasionally 15, filaments 2–3-toothed or lobed at top; ovary inferior or ± so with 3(–5) styles. **Fruit** a 3–5-valved capsule.

1. Petals overlapping . . . . **D. hookeriana**
1. Petals induplicate, occasionally valvate 2
2. Capsule globose, with incurved, persistent sepals . . .
. . . . **D. setchuenensis**
2. Capsule hemispherical to obconic, with recurved or erect, persistent sepals 3
3. Calyx-lobes 5–8 mm long, twice as long as tube . . . . . .
. . . . **D. calycosa, D. calycosa var. macropetala**
3. Calyx-lobes not more than 5 mm long, equal to or slightly longer than tube 4
4. Leaves not more than 2 cm wide; calyx-lobes erect . . .
. . . . **D. monbeigii**
4. Leaves at least 2 cm wide; calyx-lobes recurved 5
5. Leaves leathery, grey-green beneath . . . .
. . . . **D. longifolia, D. zhongdianensis, D. discolor**
5. Leaves papery, green beneath . . . .
. . . . **D. glomeruliflora, D. purpurascens**

Deutzia hookeriana (C. K. Schneid.) Airy Shaw (syn. *D. corymbosa* R. Br. ex G. Don var. *hookeriana* C. K. Schneid.) **Shrub** to 2 m with reddish brown, stellate-hairy flowering branchlets, each with 6–8 **leaves**, these ovate to oblong-lanceolate, to 8 × 4 cm, grey-downy beneath with stellate hairs, margin minutely toothed. **Inflorescence** corymbose with up to 80 **flowers**; sepals not more than 2 mm long, fused into a downy tube 2 mm long; petals white, ovate, 6–7 mm long. se Tibet, nwYunnan [swTibet, eHimalaya]; forest margins, shrubberies, 2,000–3,500 m. V–VII.

Deutzia setchuenensis Franch.* **Shrub** to 2 m with brownish or yellowish branchlets; flowering shoots with 4–6 leaves. **Leaves** ovate to lanceolate, to 8 × 5 cm, long-pointed, papery, yellowish green and downy beneath, with 3–4 pairs of lateral veins conspicuously raised beneath, margin finely toothed; leaf-stalks 3–5 mm long. **Flowers** in small or large **cymes** of up to 50; calyx densely downy, with pointed, triangular lobes; petals white, oblong, 5–12 mm long. nGuangxi, Hubei, w, nw & swSichuan, nwYunnan; dense forests, forest margins, shrubberies, rocky slopes, 300–2,000 m. IV–VI.

Deutzia calycosa* Rehd. Very variable **shrub** to 2 m with purplish brown branchlets; flowering twigs with 2–4 leaves. **Leaves** ovate to lanceolate or oblong, to 8 × 3.5 cm, rather leathery to papery, with a short stalk 3–5 mm long, grey or grey-green beneath, with 5–6 pairs of lateral veins, margin finely toothed. **Flowers** in cymose clusters of 9–12; calyx densely downy, with membranous lobes; petals white or pink, ovate-oblong, 8–12 mm long. swSichuan, w & nwYunnan; open mixed forests, forest margins, shrubberies, rocky slopes, 2,000–3,000 m. III–IV. **Var. *macropetala* Rehd.*** has leathery **leaves** and rather larger **flowers** with petals 12–15 mm long. More widespread in Yunnan, 1,400–2,300 m.

Deutzia monbeigii* W. W. Sm. **Shrub** to 2 m with brown flowering branchlets bearing 4–6 leaves. **Leaves** almost leathery, ovate-elliptic to lanceolate, to 5 × 2 cm, grey-downy beneath, apex blunt to long-pointed, margin finely toothed. **Flowers** in cymose clusters of 5–15; calyx densely downy, with small lanceolate lobes, 1–2 mm long; petals white, elliptic to obovate, 10–12 mm long. Sichuan, seTibet, w & nwYunnan; shrubberies, rocky slopes, 2,000–3,000 m. V–VI.

Deutzia longifolia* Franch. **Shrub** to 2.5 m, with 4–6-leaved flowering branchlets. **Leaves** lanceolate to elliptic, to 11 × 4 cm, leathery, with 4–6 pairs of lateral veins, apex pointed, margin finely toothed, sparsely downy beneath; stalks 3–8 mm long. **Flowers** in **cymes** of 9–12; calyx densely grey-downy, with leathery lobes; petals purplish or white, elliptic, 10–13 mm long; stamens with filaments 2–3 toothed at top. sGansu, Guizhou, Sichuan, Yunnan; forests and forest margins, shrubberies, streamsides, ravines, 1,800–3,200 m. VI–VIII.

Deutzia zhongdianensis* S. M. Hwang Similar to the previous species, but **leaves** broader and ovate, to 10 × 5 cm (stalks 5–10 mm) and **flowers** in more substantial clusters; petals white. nwYunnan (Zhongdian); shrubberies, 2,100–3,500 m. VI–VII.

Deutzia discolor* Hemsl. (syn. *D. densiflora* Rehd.; *D. longifolia* Franch. var. *farreri* Airy Shaw) Readily distinguished from the previous 2 species by the 2–4-leaved flowering branchlets, the elliptic-lanceolate to oblong-lanceolate **leaves** that are densely stellate-felted beneath, and by the 12–20-flowered **cymes**; petals white, elliptic, 10–12 mm long. Gansu, Hubei, n & nwSichuan, Shaanxi [Henan]; shrubberies, woodland margins, rocky slopes, riverbanks, 1,000–2,500 m. VI–VII.

Deutzia glomeruliflora* Franch. **Shrub** to 2 m tall, often smaller with reddish brown flowering branchlets bearing 4–6 leaves. **Leaves** ovate to lanceolate, pointed, relatively small, to 5 × 1.5 cm, green and sparsely downy beneath, with 3–6 pairs of lateral veins, margin finely toothed; stalks 1–5 mm long. **Flowers** in cymose clusters of up to 18, sometimes fewer; calyx densely downy, with purplish or greenish membranous lobes; petals elliptic, white, 8–12 mm long; filaments 2–3-toothed or -lobed at top. w & swSichuan, n & nwYunnan; coniferous and mixed forests, shrubberies, streamsides, 2,000–3,600 m. IV–VI.

***Deutzia purpurascens* (Franch. ex L. Henry) Rehd.** Similar to the previous species, but **leaves** larger and ovate-lanceolate to ovate-oblong, to 9.5 × 2.8 cm, usually with 4–5 pairs of lateral veins, and **flowers** up to 12 per cluster, petals oblong, pink, larger, 12–17 mm long; filaments 2-toothed at top. w & swSichuan, seTibet, w & nwYunnan [neIndia, nMyanmar]; shrubberies, 2,600–3,500 m. IV–VI.

HYDRANGEA

A genus of c. 73 species (33 in China, 25 endemic) mainly in east Asia, but also in South East Asia and the New World. **Deciduous shrubs**, occasionally **climbers**, with opposite or occasionally whorled **leaves**. **Flowers** numerous, small, in terminal **corymbs** or **panicles**, usually with several to many enlarged sterile flowers, especially around the periphery, these with 4–5 large calyx-lobes; fertile flowers small with 5 calyx-lobes and 5–8 petals; styles 3–5. **Fruit** a small valved capsule.

1. Climbing shrub; stems with clinging adventitious roots .... ***H. anomala***
1. Non-climbing shrubs; adventitious roots absent 2
2. Inflorescence paniculate .... ***H. paniculata***
2. Inflorescence corymbose 2
3. Flowers white or yellowish .... ***H. heteromalla*, *H. xanthoneura***
3. Fertile flowers blue, purple-blue or purple-red 4
4. Branchlets and leaf-stalks with dense long shaggy, purplish, stiff hairs .... ***H. sargentiana***
4. Young branchlets and leaf-stalks with short greyish, yellowish or yellowish brown hairs .... ***H. aspera*, *H. robusta***

***Hydrangea anomala* D. Don** **Climbing deciduous shrub** to 4 m, sometimes more, with glabrous, greyish brown young **branches**. **Leaves** elliptic to ovate, to 17 × 10 cm, papery, normally glabrous but sometimes hairy on veins beneath, with 6–8 pairs of lateral veins and finely toothed margin; leaf-stalks 2–8 cm long. **Corymbs** with several 4-lobed, whitish, sterile **flowers**, the lobes rounded to obovate, 10–22 mm long; fertile

---

1. *Deutzia calycosa*; Cangshan, nwYunnan (CGW)
2. *Deutzia monbeigii*; cult. (CGW)
3. *Deutzia longifolia*; Wolong, wSichuan (CGW)
4. *Deutzia zhongdianensis*; Zhongdian, nwYunnan (CGW)
5. *Deutzia discolor*; Jiuzhaigou, nwSichuan (CGW)
6. *Deutzia glomeruliflora*; Gangheba, Yulongxueshan, nwYunnan (CGW)
7. *Deutzia purpurascens*; Danyun, nwSichuan (PC)

1. *Hydrangea paniculata*; cult. (CGW)
2. *Hydrangea heteromalla*; Cangshan, nwYunnan (PC)
3. *Hydrangea heteromalla*; Wolong, wSichuan (CGW)
4. *Hydrangea xanthoneura*; cult. (CGW)
5. *Hydrangea sargentiana*; cult. (CGW)
6. *Hydrangea aspera*; cult. (CGW)
7. *Schizophragma integrifolium*; cult. (CGW)
8. *Pileostegia viburnoides*; cult. (CGW)

flowers tiny, 1.5–2 mm, with petal-tips fused together. Gansu, Guangxi, Hubei, Sichuan, Shaanxi, e & seTibet, Yunnan [cChina, Himalaya, Myanmar]; forests, ravines, streamsides, rocky slopes, 500–2,900 m. V–VI.

Hydrangea paniculata Sieb. **Deciduous shrub** or small **tree** to 5 m, often less, with hairy grey-brown young **branches**. **Leaves** opposite or in 3s, ovate to elliptic, to 14 × 6.5 cm, papery, with 6–7 pairs of lateral veins, downy beneath, margin finely toothed. **Inflorescence** a pyramidal panicle to 26 cm long, bearing a number of 4-lobed, sterile, white **flowers**, lobes elliptic to almost rounded, 10–18 mm long; fertile flowers white, 2–3 mm. Gansu, Guangxi, Guizhou, Hubei, Sichuan, Yunnan [c & eChina, Japan, eRussia]; open forests and shrubberies, ravines, mountain slopes, 300–2,100 m. VII–VIII.

Hydrangea heteromalla D. Don (syn. *H. khasiana* Hook. f. & Thomson; *H. vestita* Wall.) **Shrub** or small **tree** to 5 m, often less, with reddish brown, hairy branchlets. **Leaves** elliptic to ovate, to 15 × 8 cm, papery, greyish white with silky hairs beneath, with 7–9 pairs of lateral veins, margin finely toothed; stalks purplish red, 2–4 cm long. **Corymbs** flat or somewhat domed, to 20 cm across, with several white or yellowish, 4-lobed, sterile **flowers**, lobes elliptic to ovate, 7–16 mm long; fertile flowers similarly coloured, 2–3 mm. s, sw & wSichuan, seTibet, Yunnan [swTibet, Himalaya]; forests and shrubberies, ravines, rocky slopes, 2,400–3,400 m. VI–VII.

Hydrangea xanthoneura Diels* (syn. *H. pubinervis* Rehd.; *H. xanthoneura* var. *wilsonii* Rehd.) Similar to the previous species, but branchlets grey-brown or blackish brown, **leaves** more yellowish green, with hairs confined to veins beneath; sterile **flowers** whitish or yellowish green, 4–5-lobed. Guizhou, nwHubei, Sichuan, Yunnan; sparse forests, shrubberies, ridgetops, pathsides, ravines, 1,600–3,200 m. VI–VII.

Hydrangea sargentiana Rehd.* (syn. *H. aspera* D. Don subsp. *sargentiana* (Rehd.) McClintock) Rather erect **deciduous shrub** to 3 m, with thick shaggy purplish hairs. **Leaves** dark green with grey-green reverse that is often flushed with purple at first, elliptic to ovate, to 30 × 16 cm, densely downy beneath, with 8–11 pairs of lateral veins, margin minutely toothed. **Corymbs** domed, to 16 cm across, with a ring of white, 4-lobed sterile flowers, lobes rounded to obovate, 9–14 mm long; sterile **flowers** purplish blue, 2–3 mm. wHubei; dense valley or mountain slope forests, 700–1,800 m. VII–VIII.

Hydrangea aspera D. Don (syn. *H. villosa* Rehd.) Lax **shrub** or small **tree** to 4 m, occasionally more, with shortly downy, obscurely 4-angled, greyish or whitish branchlets. **Leaves** lanceolate to ovate or almost oblong, to 25 × 8 cm, grey-white-downy beneath, with 6–10 pairs of lateral veins, margin sharply toothed; stalks 1–4.5 cm long, glabrous to somewhat hairy. **Corymbs** flattish to domed, to 25 cm across, with a few marginal white, pink or reddish, 4–5-lobed sterile **flowers**, lobes rounded to ovate, 10–33 mm long, margin finely toothed; sterile flowers purplish blue or reddish purple, 2.5–3 mm; stamens 10. seGansu, Guangxi, Guizhou, swHubei, Sichuan, Shaanxi, Yunnan [c & sChina, eHimalaya, Vietnam]; dense forests or shrubberies in valleys or on wooded slopes, 700–4,000 m. VII–IX.

Hydrangea robusta Hook. f. & Thomson Very similar to the previous species, but branchlets more prominently 4-angled, **leaves** with thick stalks 4–15 cm long and whitish or purplish sterile **flowers** with lobes up to 38 mm long; stamens 10–14. Guangxi, Guizhou, Hubei, Sichuan, seTibet, Yunnan [s & eChina, eHimalaya, nMyanmar]; similar habitats as well as streamsides, 700–2,800 m. VII–VIII.

SCHIZOPHRAGMA

A genus with c. 10 species (9 in China) from China, Korea and Japan. Very like *Hydrangea*, but plants often **climbing** and **inflorescence** with some sterile **flowers** with a single, much enlarged calyx-lobe; style solitary.

Schizophragma integrifolium Oliv.* Variable climbing **deciduous shrub** to 5 m or more, with glabrous branchlets. **Leaves** opposite and leathery, glossy green, ovate to elliptic, to 20 × 12 cm, with 7–9 pairs of lateral veins, glabrous beneath or hairy along veins, margin entire to slightly toothed; stalks 2–9 cm long. **Inflorescence** a broad corymb, bearing a few sterile **flowers** with a single expanded, yellowish white, ovate to lanceolate calyx-lobe to 3–7 cm long; fertile flowers small, greenish white, 2–3 mm long. Guangxi, Guizhou, Hubei, Sichuan, Yunnan [c & sChina]; forests and forest margins, rocky places, ravines, 200–2,000 m. VI–VII. **Var. *glaucescens* Rehd.** has thinner **leaves** that are noticeably glaucous beneath. Similar habitats and distribution except for Yunnan.

Schizophragma molle (Rehd.) Chun* Readily distinguished by **leaves** that are densely hairy beneath, particularly along the veins. Guangxi, Guizhou, Yunnan; forests, ravines, cliffs, 500–2,100 m. VI–VII.

PILEOSTEGIA

An east Asian genus of 3 species (2 in China). **Climbing evergreen shrubs** closely related to *Hydrangea*, but **inflorescence** without enlarged sterile **flowers**, petals 4–5-valvate, not overlapping; style solitary, stigma 4–6-lobed.

Pileostegia viburnoides Hook. f. & Thomson Vigorous **climbing shrub** to 15 m with paired grey or grey-brown branchlets. **Leaves** opposite, leathery, shiny deep green, elliptic to oblanceolate, to 15 × 6 cm, with 7–8 pairs of lateral veins, downy or glabrous beneath, margin entire; stalks 1–3 cm long. **Inflorescence** a many-flowered panicle to 25 cm long; **flowers** white or greenish white, 2–3 mm; calyx with short tube and 4–5 triangular lobes; petals elliptic, shorter than the 8–10 stamens. Guangxi, Guizhou, Hubei, Sichuan, Yunnan [s, se & eChina, Japan]; valley forests, ravines, cliffs, 600–1,000 m. V–VIII.

DECUMARIA

A genus of 2 species (1 in China) from the Himalaya to Japan and west Malaysia, and in eastern North America. Closely related to *Pileostegia*. Non-climbing **evergreen shrubs** with **corymbose inflorescences**, sepals and petals 7–10, stamens 20–30.

Decumaria sinensis Oliv.* **Evergreen shrub** to 5 m, often less, with greyish brown young **branches**. **Leaves** leathery, deep green, elliptic to obovate, to 7 × 3.5 cm, glabrous or somewhat hairy at first, lateral veins in 4–6 pairs, margin entire to slightly toothed. **Corymbs** domed, 3–5 cm across. **Flowers** small, 4–5 mm; calyx top-shaped with short ovate lobes; petals white, oblong, as long as the stamens; style solitary, stigma 7–10-lobed. Gansu, Guizhou, Hubei, c, e & nSichuan, Shaanxi; mountain shrubberies, ravines, 600–1,300 m. III–V.

DICHROA

A genus of 12 species (6 in China) mostly from east Asia. Similar to *Hydrangea* but **inflorescences** without large sterile **flowers**, and **fruit** a fleshy berry, not a capsule.

Dichroa febrifuga Lour. **Evergreen shrub** to 2 m tall, with glabrous or finely downy branchlets. **Leaves** mainly opposite, bright green and somewhat fleshy, elliptic to oblong or lanceolate, to 25 × 10 cm, glabrous or downy beneath, with 8–10 pairs of lateral veins, margin sharply toothed. **Inflorescence** corymbose or somewhat paniculate, to 20 cm long. **Flowers** many, 5–6-parted, 12–18 mm, oblong in bud, with reflexing, blue or white, oval petals and 10–20 prominent stamens; stigmas 4–6. **Berries** 3–7 mm in diameter, bright blue when ripe. Gansu, Guangxi, Guizhou, Hubei, Sichuan, Shaanxi, seTibet [c & eHimalaya]; mixed forests, dense mountain shrubberies, 200–2,000 m. II–V.

ITEA

A genus of c. 27 species (15 in China) in the Himalaya and Japan. **Evergreen shrubs** or small **trees** with alternate, often rather leathery **leaves**; stipules present, often deciduous. **Flowers** small, in many-flowered terminal or lateral **racemes** or **panicles**; calyx with short bowl-shaped tube and 5 triangular lobes; petals 5, small, narrow, erect. **Fruit** a small capsule with a persistent simple or cleft style.

Itea ilicifolia Oliv.* Flexuous, much-branched, glabrous **evergreen shrub** to 4 m with dark green *Ilex*-like, leathery **leaves**, elliptic to rounded or oblong, to 9.5 × 6 cm, with 5–6 pairs of lateral veins, margin spiny-toothed. **Racemes** terminal and pendulous, catkin-like, with numerous **flowers** 3–4 mm long, greenish or greenish white; stalks and calyx glabrous. Guizhou, wHubei, eSichuan, swShaanxi; forests, shrubberies, streamsides, ravines, 1,500–1,700 m. V–VI.

Itea yunnanensis Franch.* (syn. *I. forrestii* Y. C. Wu) Similar to the previous species, but **leaves** thinner and more papery, narrower, ovate to elliptic, not more than 4.5 cm wide, with 4–5 pairs of lateral veins; flower-stalks and calyx minutely hairy. Guangxi, Guizhou, swSichuan, seTibet, Yunnan; coniferous or mixed forests, rocky places, streamsides, 1,100–3,000 m. V–IX.

ASTILBE

A genus of c. 18 species (7 in China, 3 endemic) in eastern Asia and eastern North America. Tufted **herbaceous perennials** often with brown-hairy **stems** and long-stalked, alternate, compound **leaves**. **Flowers** in terminal **panicles**, unisexual or hermaphrodite; sepals usually 5, petals 1–5; stamens usually 8 or 10. **Fruit** consisting of 2–3, free or fused follicles.

> 1. Panicles dense; petals usually 5 ....
>   .... **A. chinensis, A. grandis, A. rubra**
> 1. Panicles lax; petals usually absent, occasionally 1 ....
>   .... **A. rivularis**

Astilbe chinensis (Maxim.) Franch. & Sav. (syn. *Hoteia chinensis* Maxim.) Plant to 100 cm, often less, with erect glabrous **stems** and 2–3-ternate **leaves**; leaflets diamond-shaped to ovate or elliptic, to 8 × 4 cm, margin doubly toothed; stem-leaves 2–3. **Panicles** narrow, not more than 12 cm wide, the branches with crisped brown hairs; sepals glabrous; petals 5, lilac to purple or pink, linear, c. 5 mm long. seGansu, Hubei, seQinghai, Sichuan, Shaanxi, Yunnan [n, ne & cChina, Japan, Korea, eRussia]; damp habitats along rivers and in meadows, forests and forest margins, 1,400–3,600 m. VI–IX.

Astilbe grandis Stapf ex E. H. Wilson (syn. *A. koreana* (Kom.) Nakai) Similar to the previous species, but **stems** hairy and glandular, **panicles** larger, up to 17 cm wide; petals purple or white, not more than 4.5 mm long. Guangxi, Guizhou, Hubei, Sichuan [n, ne & cChina, Korea]; damp habitats in forests and shrubberies, ravines, rocky places, 400–2,000 m. VI–IX.

Astilbe rubra Hook. f. Larger plant than previous 2 species, to 1.5 m with crisped brown-glandular-hairy **stems** and glandular-hairy **leaves** and sepals; petals pink or red. seTibet, nwYunnan [swTibet, nIndia]; forest margins, shrubberies, c. 2,400 m. VI–VII.

Astilbe rivularis Buch.-Ham. A generally large **herb** to 2.5 m with brown-glandular-hairy **stems**. **Leaves** 2–3-pinnate, leaflets elliptic to ovate or diamond-shaped, occasionally obovate, to 15 × 8.5 cm, margin doubly toothed, glandular-hairy along veins. **Inflorescence** laxly branched, glandular-hairy; sepals 4–5, green, to 1.5 mm long, glabrous. seGansu, Guizhou, Hubei, Sichuan, Shaanxi, seTibet, Yunnan [swTibet, sChina, Himalaya, Myanmar, Laos, Thailand]; forests and forest margins, shrubberies, ravines, meadows, streamsides, 900–3,200 m. VI–XI.

1. *Dichroa febrifuga*; cult., Kunming, cYunnan (CGW)
2. *Itea ilicifolia*; cult. (PC)
3. *Itea ilicifolia*; Shennongjia, wHubei (PC)
4. *Itea yunnanensis*; west of Habashan, nwYunnan (CGW)
5. *Astilbe chinensis*; cult. (CGW)
6. *Rodgersia aesculifolia*; Wolong, wSichuan (HH)
7. *Rodgersia aesculifolia* var. *henrici*; Galung La, near Bomi, seTibet (HJ)
8. *Rodgersia pinnata*; Cangshan, nwYunnan (PC)

RODGERSIA

A genus with 5 species (4 in China, 2 endemic to south-west China) in the Himalaya and east Asia. Rather stout, clump- or patch-forming, **rhizomatous perennial herbs** with long-stalked pinnately or digitately compound **leaves**. **Inflorescence** a large panicle composed of numerous branched cymes bearing many small **flowers**; sepals usually 5, spreading, white, pink or red; petals usually absent, occasionally 1–5, vestigial; stamens usually 10. **Fruit** a 2–3-valved, many-seeded capsule.

---

1. *Bergenia purpurascens*; Bitahai, Zhongdian, nwYunnan (PC)
2. *Bergenia emeiensis*; cult. (PC)
3. *Tiarella polyphylla*; Erlangshan, wSichuan (PC)
4. *Chrysosplenium griffithii*; Balangshan, wSichuan (PC)
5. *Chrysosplenium davidianum*; cult. (AGS)
6. *Chrysosplenium carnosum*; Wutoudi, Yulongxueshan, nwYunnan (CGW)
7. *Chrysosplenium nudicaule*; Daxueshan, nwYunnan (CGW)

1. Leaves digitately compound; sepals pinnately veined, veins not or only partly confluent at apex .... ***R. aesculifolia***
1. Leaves pinnately compound; sepals with curved veins, veins confluent at apex ... ***R. pinnata, R. sambucifolia***

Rodgersia aesculifolia Batalin **Herbaceous perennial** to 1.2 m with angular **stems**. **Leaves** palmate, with 5–7 obovate to obovate-elliptic, somewhat papery, stalkless unlobed leaflets, each 8–30 × 2.5–12 cm, glandular-hairy along veins and with a coarse doubly toothed margin. **Inflorescence** with white-hairy branches and stalks; sepals 5, occasionally 4 or 6, white or pinkish, to 2 mm long. sGansu, wHubei, Sichuan, Shaanxi, Yunnan [cChina, Myanmar]; forests, forest margins, shrubberies, ravines, rock-clefts, 1,100–3,400 m. V–VIII. **Var. henrici (Franch.) C. Y. Wu ex J. T. Pan** (syn. *R. henrici* Franch.) has thinly leathery leaflets, and sepals glandular-hairy beneath. Similar habitats, seTibet, w & nwYunnan [nMyanmar], 2,300–3,800 m. (See photo 7 on p. 309.)

Rodgersia pinnata Franch.* **Herbaceous perennial** to 1.5 m with a glabrous **stem**. **Leaves** ± pinnately compound, with 6–9 leaflets (with 3–5 terminal, the others in a whorl), leaflets elliptic to oblong or obovate, 6.5–32 × 2.5–13 cm, brown-hairy on veins beneath, margin with double pointed teeth. **Inflorescence** with hairy branches and stalks; sepals 5, white, pink or reddish, 2–2.7 mm long, brown-hairy beneath, above with glandular hairs near base. Guizhou, s, se & eSichuan, Yunnan; forest and forest margins, shrubberies, meadows and other grassy and rocky places, 2,000–3,800 m. VI–VIII. (See photo 8 on p. 309.)

Rodgersia sambucifolia Hemsl.* Similar to the previous species, but **leaves** pinnate with 3–9 leaflets (3 terminal, the lateral opposite), leaflets not more than 20 cm long; sepals sparsely hairy beneath, glabrous above. wGuizhou, swSichuan, n & nwYunnan; similar habitats, 1,800–3,700 m. V–IX.

BERGENIA

A genus of 10 species (7 in China, 3 endemic) in eastern Asia including Japan, east Russia and Korea. Clump-forming **perennials** with thick scaly rhizomes and generally evergreen, rather leathery, rounded to elliptic, basal **leaves**. **Inflorescence** bracteate, cymose, showy and 5-parted; stamens usually 10, in 2 whorls. **Fruit** a 2-parted capsule containing numerous seeds.

1. Margin of leaf-base and leaf-blade glabrous . . . .
   . . . . *B. purpurascens*
1. Margin of leaf-base, and sometimes leaf-blade, ciliate
   . . . . *B. emeiensis*

Bergenia purpurascens (Hook. f. & Thomson) Engl. (syn. *Saxifraga purpurascens* Hook. f. & Thomson) Plant to 50 cm, often half that height. **Leaves** oval to obovate, to 16 × 9 cm, leathery, glabrous and pitted with glands. **Inflorescence** glandular-hairy; petals purple, ovate to rounded, 10–17 mm long, short-clawed at base; stamens 6–11. swSichuan, e & seTibet, n & nwYunnan [Himalaya, Myanmar]; forests, shrubberies, rocky outcrops, rocky meadows, 2,700–4,800 m. V–X.

Bergenia emeiensis C. Y. Yu* Plant to 40 cm with narrow-obovate **leaves**, to 17 × 8 cm, blade glabrous, leathery. **Inflorescence** sparsely glandular-hairy; sepals spreading, glandular-hairy; petals white (reddish in var. *rubellina* J. T. Pan), narrow-obovate, 24–28 mm long, short-clawed at base. s & swSichuan (particularly Emeishan and Guan Xian); rocky forested slopes, rock-crevices, 1,600–4,200 m. V–VIII.

TIARELLA

A genus of 3 species (1 in China), distributed in temperate Asia and North America. **Perennial herbs** with mostly basal **leaves** and small stipules. **Flowers** very small, in **racemes** or **panicles**; sepals 5, often petaloid; petals 5 or absent; stamens 10; styles 2. **Fruit** an uneven 2-carpelled capsule.

Tiarella polyphylla D. Don Tufted glandular-hairy, **rhizomatous herb** to 45 cm. **Leaves** long-stalked, blade heart-shaped, to 8 × 10 cm, palmately 3- or 5-lobed, margin toothed; stem-leaves few, smaller than basal, short-stalked. **Flowers** in slender racemes, whitish, 1.5–2 mm; sepals ovate, erect, c. 1.5 mm long; petals absent; stamens protruding. **Fruit** a capsule 7–12 mm long. seGansu, Guangxi, Guizhou, Hubei, Sichuan, sShaanxi, seTibet, Yunnan [swTibet, c & eChina, Himalaya, Japan, Taiwan]; damp forests, shady places, ravines, 1,000–3,800 m. IV–XI.

CHRYSOSPLENIUM

A complex genus with c. 65 species (35 in China, 20 endemic) in the northern hemisphere and Africa. **Stoloniferous, tufted** or **mat-forming herbs**, often with bulbils. **Leaves** alternate or opposite. **Flowers** in small clusters (cymes) surrounded by green, yellow or greenish yellow leafy bracts; sepals 4–5; petals absent; stamens 4 or 8.

1. Flowers yellow with spreading sepals . . . .
   . . . . *C. griffithii, C. davidianum*
1. Flowers green or yellowish green with erect sepals . . . .
   . . . . *C. carnosum, C. nudicaule*

Chrysosplenium griffithii Hook. f. & Thomson **Tufted herb** to 20 cm, occasionally more, with simple **stems**. **Leaves** mainly on the stem, alternate, with a brown-hairy stalk, blade kidney-shaped, to 5 × 6.5 cm, glabrous, margin 10–15-lobed. **Inflorescence** 4–10 cm across, surrounded by yellow-green leaf-like bracts. **Flowers** yellow, 4.2–4.6 mm across. sGansu, sQinghai, n, w & swSichuan, sShaanxi, seTibet, n & nw Yunnan [Himalaya, nMyanmar]; forests and forest margins, meadows, rocky places, 2,500–4,800 m. V–IX.

Chrysosplenium davidianum Decne ex Maxim.* **Mat-forming** plant with numerous sterile **branches** and softly crisped brown-hairy **stems**, differing from the previous species in the fan-shaped to ovate, brown-hairy **leaves**, margin with 7–9 scallops. **Inflorescence** to 4 cm across, surrounded by chrome-yellow, leaf-like bracts. Guizhou, w & swSichuan, n & nwYunnan; damp and wet places in forests, ravines, amongst rocks, 1,500–4,100 m. IV–VIII.

Chrysosplenium carnosum Hook. f. & Thomson **Patch-forming herb** to 10 cm, the **stems** only hairy at nodes; basal leaves absent. **Leaves** obovate to spathulate, to 8 × 8 mm, glabrous, margin often 7-scalloped. **Inflorescence** 3–5 cm across, surrounded by ovate, green or yellow-green, leaf-like bracts. wSichuan, e & seTibet, nwYunnan [Himalaya, Myanmar]; alpine meadows and scrub, rock-crevices, 4,400–4,700 m. VII–VIII.

Chrysosplenium nudicaule Bunge Similar to the previous species but basal **leaves** present, kidney-shaped, 11–15-toothed; cymes smaller, not more than 12 mm across, with green, ovate to fan-shaped, rather leathery leaf-like **bracts**. Gansu, Qinghai, eTibet, nwYunnan [Mongolia, Nepal, sRussia]; damp or wet rocky places, 2,500–4,800 m. VI–VIII.

SAXIFRAGA

A genus of c. 450 species (216 in China, 139 endemic) from the north temperate zone to the subarctic and also on mountains in North Africa and South America. **Perennial herbs** with simple **stems**, or **tufted** or **cushion-forming**. **Leaves** basal, often in distinct rosettes and/or on stem, very variable in shape, margin entire, toothed or variously lobed. **Flowers** solitary or in a cymose or paniculate **inflorescence**, hermaphrodite, usually regular; sepals (4–)5(–8); petals (4–)5; stamens (8–)10; carpels 2. **Fruit** a 2-valved capsule.

SYNOPSIS OF GROUPS

Group One (section *Porphyrion*). Inflorescences with leafy stems; leaves with lime-pits (chalk-secreting glands). (below)

Group Two (section *Mesogyne*). Inflorescences with leafy stems; leaves without lime-pits; bulbils present in axils of basal stalked leaves (and sometimes upper); petals white or cream. (p. 315)

Group Three (section *Ciliatae*). Inflorescences with leafy stems; leaves without lime-pits; bulbils absent; basal leaves often stalkless, sometimes stalked; petals usually yellow or orange, sometimes reddish. (p. 315)

Group Four (section *Micranthes*). Inflorescence with leafless stems; flowers regular, often star-shaped. (p. 319)

Group Five (section *Irregulares*, commonly referred to as *Diptera* saxifrages). Inflorescence with leafless stems; flowers zygomorphic, usually with 1 or 2 enlarged petals. (p. 319)

Group One

1. Flowers 4-parted . . . . **S. nana**
1. Flowers 5-parted 2
2. Shoots with opposite leaves; flowers solitary . . . . **S. georgei**
2. Shoots with alternate leaves; flowers often 2 or more, sometimes solitary 3
3. Sepals equalling or longer than petals . . . . **S. chionophila, S. rupicola**
3. Sepals shorter than petals 4
4. Flowers 2 or more per inflorescence . . . . **S. decora, S. pulchra, S. rotundipetala, S. andersonii**
4. Flowers solitary 5
5. Leafy flower-stem absent at flowering time . . . . **S. pulvinaria, S. likiangensis, S. subsessiliflora**
5. Leafy flower-stem present at flowering time . . . . **S. ludlowii, S. unguipetala, S. saxatilis, S. saxicola**

Saxifraga nana Engl.* Small **cushion-forming** plant not more than 2 cm, with numerous crowded **shoots**. **Rosette-leaves** closely overlapping, narrow-oblong to spathulate, 3–4 × 0.9–1 mm, glabrous, with a single lime-pit, apex somewhat thickened and slightly reflexed. **Flowers** solitary on a leafless, glandular-hairy **stem**; sepals erect, later spreading or reflexed; petals 4, white, occasionally absent, elliptic, c. 2.5 mm long, longer than sepals. Gansu, s & seQinghai, nwSichuan; alpine rocks and crevices, lake-shores, 4,200–4,900 m. VII–VIII.

Saxifraga georgei J. Anth. Low, **cushion-forming** plant with crowded shoots bearing opposite, thick, ovate **leaves** c. 3 × 2 mm, glabrous and with a single lime-pit at the slightly pointed apex. **Flowers** bowl-shaped, upright, solitary, stalkless, 9–10 mm across; sepals ovate, erect, 2 mm long, with a solitary lime-pit at top; petals white, 5 mm long, curving outwards, with short basal claw. swSichuan, seTibet, nwYunnan [eHimalaya]; cliff-crevices, 3,600–4,100 m. VI–VIII.

Saxifraga chionophila Franch.* Small **cushion-forming** plants to 8 cm tall in flower at the most, with brown glandular-hairy **stems**. **Rosette-leaves** narrow-obovate, 8–9 mm long, leathery, glabrous, with 5–7 marginal lime-pits, margin with a few horny teeth towards base; stem-leaves rather distant, glandular-hairy outside towards base. **Flowers** in 2–7-flowered cymes, on a stem 10–40 mm long; sepals erect, with 3 lime-pits towards blunt apex, glandular-hairy outside; petals red, oblong to obovate, 2–3 mm long, somewhat hairy outside. swSichuan, seTibet, nwYunnan; cliffs, rock-crevices, rocky alpine meadows, 2,700–5,000 m. VI–VIII.

Saxifraga rupicola Franch.* Similar to the previous species, but **flowers** solitary on a very short **stem** rarely more than 8 mm long; petals greenish yellow, slightly shorter than sepals. nwYunnan; calcareous rock-crevices, c. 3,500 m. VI–VII.

Saxifraga decora Harry Sm.* Small greyish **cushion-forming** plant to 6 cm tall in flower with leaves crowded into dense rosettes; **leaves** oblong to spathulate, to 5 × 2 mm, glabrous, with 3–7 marginal lime-pits at apex and rigid teeth towards base, apex slightly recuved. **Inflorescence** 3–4-flowered, on a brown-glandular stem bearing up to 6 narrow leaves. **Flowers** 7–10 mm; sepals erect, glandular-hairy, with 1–3 lime-pits; petals pink to purple, narrow-obovate, 3–5 mm long, curving outwards. e & seTibet, nwYunnan; rock-crevices, cliffs, 3,500–4,800 m. V–VII.

Saxifraga pulchra Engl. & Irmsch.* Closely related to the previous species, but **leaves** larger, 5.5–7 mm long, with 7–11 marginal lime-pits in upper half, whereas sepals lack lime-pits. **Inflorescence** with up to 6 pink or purple flowers. w & swSichuan, seTibet, nwYunnan; rocky places in open coniferous forests and scrub, rock-crevices, cliffs, 2,500–4,600 m. IV–VI.

Saxifraga rotundipetala J. T. Pan* Readily distinguished from the previous 2 species by **leaves** that have a single, terminal lime-pit, and by yellow **flowers**; leaves almost spathulate, 6.5–7 mm long, glandular-hairy beneath and on margin at leaf-base; **cymes** 2–4-flowered, petals rounded, 4.5–5.5 mm long. seTibet, nwYunnan; similar habitats, 3,900–4,400 m. V–VII.

Saxifraga andersonii Engl. & Irmsch. Very similar to *S. decora*, but a larger and more vigorous plant to 9 cm tall in flower, **rosette-leaves** to 10 × 2.5 mm, with 3–(5–7) lime-pits. **Flowers** white or pale pink. seTibet [swTibet, c & eHimalaya, nMyanmar]; screes, moraines, rock-crevices, 4,100–5,200 m. VI–VIII.

Saxifraga pulvinaria Harry Sm. (syn. *S. imbricata* Royle) Plant forming dense flattish or domed **cushions** to 6 cm. **Leaves** closely overlapping in the rosettes, narrow-elliptic, 3–3.5 mm long, stiffly ciliate, with a single terminal lime-pit. **Flowers** embedded in centre of rosettes, saucer-shaped; sepals erect, ovate to ovate-triangular, stiff, ciliate, otherwise glabrous, lacking lime-pits; petals white, obovate to oblong, 3.5–5.5 mm long, spreading, recurved in upper half. nwYunnan [swTibet, Himalaya]; rock-crevices, moraines, 3,900–5,200 m. VI–VII.

Saxifraga likiangensis Franch.* (syn. *S. calcicola* J. Anth.) Very similar to the previous species, but **leaves** more ovate, to 5.5 mm long, and with 3 lime-pits; petals white or cream, to 9 mm long. **Flowers** stalkless but as the fruit mature, a short leafy stem to 10 mm long develops. sQinghai, w & swSichuan, seTibet, nwYunnan; rocky places in open forests and scrub, rock-crevices above treeline, 3,000–5,600 m. V–VIII.

Saxifraga subsessiliflora Engl. & Irmsch. (syn. *S. lolaensis* Harry Sm.) Similar to *S. pulvinaria*, but **flowers** with 2 bracts at base of hypanthium and sepals with a single, terminal lime-pit. w & swSichuan, seTibet, nwYunnan [swTibet, nwChina, eHimalaya]; rock-crevices, alpine meadows, 3,900–4,800 m. VI–VIII.

1. *Saxifraga nana*; seQinghai (JM)
2. *Saxifraga decora*; Daxueshan, nwYunnan (CGW)
3. *Saxifraga rotundipetala*; Baimashan, nwYunnan (HJ)
4. *Saxifraga andersonii*; cNepal (CGW)
5. *Saxifraga likiangensis*; upper Gangheba, Yulongxueshan, nwYunnan (CGW)
6. *Saxifraga subsessiliflora*; Genyanshan, swSichuan (AD)

1. *Saxifraga ludlowii*; north of Rawu, seTibet (DZ)
2. *Saxifraga saxicola*; Huanglong, nwSichuan (CGW)
3. *Saxifraga sibirica*; Norway (AGS)
4. *Saxifraga brunonis*; cNepal (CGW)
5. *Saxifraga mucronulata*; Genyanshan, swSichuan (AD)
6. *Saxifraga consanguinea*; Yading Nature Reserve, swSichuan (AD)
7. *Saxifraga lychnitis*; nNepal (CGW)

Saxifraga ludlowii Harry Sm.* Plants forming dense **cushions** to 4 cm, overtopped by short leafy flowering **stems** to 10 mm long. **Leaves** in small packed rosettes, lanceolate-oblong, 5–6 mm long, with 3 lime-pits at apex and a stiff horny-toothed margin, glabrous. **Flowers** borne on a very short, 1–2 mm long, purplish black, glandular stalk; sepals erect, oval, 4 mm long, ciliate, glandular-hairy outside; petals purple, obovate, 8–9 mm long, recurving in upper half. eTibet [swTibet]; rock-crevices, rocky alpine meadows, 4,300–4,800 m. VII–VIII.

Saxifraga unguipetala Engl. & Irmsch.* (syn. *S. kansuensis* Mattf.) Distinguished from the previous species by white **flowers** borne on a leafy stem 20 mm or more long, and **rosette-leaves** 7–9 mm long, with 5–9 lime-pits; sepals glandular-hairy outside; petals 6–7 mm long. sGansu (Minshan), wHubei; rock-crevices, 3,200–4,300 m. VII–VIII.

Saxifraga saxatilis Harry Sm.* Like *S. unguipetala*, but a smaller plant, flowering **stem** to 25 mm long, brown-hairy, **leaves** not more than 5 mm long, and petals white or reddish, obovate, not more than 4.5 mm long. n & nwSichuan; calcareous rock-crevices, 4,200–4,300 m. VI–VIII.

Saxifraga saxicola Harry Sm.* Very like *S. unguipetala*, but flowering **stem** not exceeding 15 mm, white-hairy, **leaves** hairy on reverse, and sepals glabrous and petals 5–9 mm long, white. w & nwSichuan; calcareous rock-crevices, c. 2,800 m. VI–VII.

Group Two

Saxifraga sibirica L. Rather delicate, erect **perennial** to 20 cm, often only 6–12 cm, with tiny bulbils in axils of basal leaves; **stem** densely glandular-hairy. **Basal leaves** kidney-shaped to rounded, 7–9-lobed, to 27 mm across, glandular-hairy; **stem-leaves** with a short stalk, and kidney-shaped to ovate, 5–9-lobed blade. **Flowers** up to 13 in a corymbose cluster; sepals erect, lanceolate to oblong, 3–4 mm long; petals white, obovate, to 15 mm long, ascending. sGansu, wHubei, Sichuan, sShaanxi, Tibet, nwYunnan [widespread in northern hemisphere]; open forests and forest margins, scrub, alpine meadows, stabilised screes and moraines, rock-crevices, 1,800–5,100 m. V–X.

Saxifraga cernua L. Readily distinguished from the previous species by the presence of tiny bulbils in axils of stem-leaves and bracts; **leaves** generally 5–7-lobed and **flowers** not more than 5 per **inflorescence**. Qinghai, w, nw & swSichuan, sShaanxi, Tibet, nwYunnan [widely distributed throughout northern hemisphere]; similar habitats, 2,200–5,500 m. VI–IX.

Group Three

1. Rosette-leaves producing thread-like stolons . . . .
   . . . . **S. brunonis, S. stenophylla, S. mucronulata,**
   S. consanguinea
1. Rosette-leaves, when present, not producing stolons
   2
2. Flowers nodding . . . **S. lychnitis, S. nigroglandulifera,**
   S. bergenioides
2. Flowers not nodding 3
3. Petals with yellow, orange or purple spots towards base
   4
3. Petals plain yellow above, not spotted 5
4. Flowers basically white . . . . **S. gemmipara, S. strigosa**
4. Flowers basically yellow . . . .
   . . . . . . **S. przewalskii, S. tsangshanensis, S. signata,**
   S. unguiculata, S. nanella
5. Leaves 15 mm or more long . . . .
   . . . . **S. diversifolia, S. egregia, S. tangutica**
5. Leaves less than 15 mm long . . . .
   . . . . **S. drabiformis, S. montanella, S. brachypoda**

Saxifraga brunonis Wall. ex Seringe Laxly **mat-forming perennial** with slender purplish brown **stems** to 15 cm, with thread-like, purplish stolons up to 25 cm long, slightly glandular-hairy. **Rosette-leaves** shiny grey-green, 10–13 mm long, oblong with a finely pointed apex, glabrous; **stem-leaves** few, similar but narrower. **Inflorescence** glandular-hairy, **flowers** starry, in 3–9-flowered cymes; sepals spreading, ovate, 2–3 mm long; petals yellow, elliptic to lanceolate, 5–8 mm long. swSichuan, seTibet, w & nwYunnan [swTibet, c & eHimalaya, nMyanmar]; rocky places, gullies, moraines, alpine scrub, often in damp mossy places, 2,800–4,000 m. VI–X.

Saxifraga stenophylla Royle (syn. *S. flagellaris* Willd. ex Sternb. subsp. *stenophylla* (Royle) Hult.) Distinguished from the previous species by more leathery, glandular-hairy **leaves**, 1–3-flowered **inflorescence**, and larger **flowers**; sepals 4–6 mm long; petals obovate to elliptic, 8–12 mm long. w & swSichuan, seTibet, nwYunnan [swTibet, Himalaya, Tajikistan]; habitats similar to those of *S. brunonis*, 3,700–5,000 m. VII–VIII.

Saxifraga mucronulata Royle (syn. *S. flagellaris* Willd. ex Sternb. var. *mucronulata* (Royle) C. B. Clarke) Like *S. brunonis*, but plant densely glandular-hairy and **leaves** linear-spathulate to 9 mm long; stem-leaves linear. **Flowers** in 2–5-flowered cymes, with erect sepals and obovate yellow petals. w & swSichuan, seTibet, nwYunnan [Himalaya]; similar habitats, 2,800–5,400 m. VII–VIII.

Saxifraga consanguinea W. W. Sm. Readily distinguished from the previous 3 species by having petals equalling, or slightly shorter than, the sepals. **Leaves** to 10 × 3 mm, glabrous or glandular-hairy beneath; **inflorescence** corymbose or **flowers** solitary, petals yellow, pink or red. sQinghai, w, nw & swSichuan, se & eTibet, nwYunnan [swTibet, cHimalaya]); rocky meadows, screes, stabilised moraines, open *Picea* forests, 3,000–5,400 m. VI–IX.

Saxifraga lychnitis Hook. f. & Thomson Small **tufted perennial** to 15 cm with simple, erect, purplish red, glandular-hairy **stems**. **Basal leaves** in a lax rosette, spathulate, 11–15 mm long, glandular-hairy; **stem-leaves** oblong, rather smaller, with a very short stalk. **Flowers** solitary or paired, nodding on arched stalks; sepals erect, purple or reddish purple with dense glandular hairs; petals lemon-yellow, narrow-obovate, 7–9 mm long. e & seQinghai, w & nwSichuan, seTibet [swTibet, c & eHimalaya, nMyanmar]; moraines, alpine meadows, often in rather wet places, 4,300–5,500 m. VII–IX.

Saxifraga nigroglandulifera N. P. Balakr. (syn. *S. nutans* Hook. f. & Thomson) Larger plant than preceding species, to 35 cm, more often 10–20 cm, with dark brown glandular-hairy **stems**. **Basal leaves** ovate to elliptic, 15–40 mm long. **Inflorescence** 1-sided, raceme-like with up to 14 flowers. w & swSichuan, seTibet, nwYunnan [swTibet, Himalaya]; open forests and forest margins, shrubberies, alpine meadows, rocky slopes, lake and streamsides, 2,700–5,000 m. VII–X.

Saxifraga bergenioides C. Marq. Very distinctive with purple flowers 14–15 mm long, 1–4 borne on a sparsely leafy slender **stem** from the basal leaf-tufts, to 20 cm tall. **Leaves** elliptic, covered in brown crisped hairs, basal to 23 × 9 mm, stem-leaves narrower and stalkless. seTibet [Bhutan]; alpine meadows, screes, moraines, open low scrub, rock-crevices, 4,200–5,000 m. VII–IX.

Saxifraga gemmipara Franch. Plant to 20 cm, with glandular-hairy **stems**, with **basal leaves** usually small and scale-like and middle **stem-leaves** forming a lax rosette; leaves oblong to oblong-linear or obovate, to 30 × 10 mm. **Inflorescence** a lax 2–12-flowered cyme; sepals erect at first, later spreading to reflexed, lanceolate; petals white or cream, with yellow or purple spots, ovate to elliptic, 5–7 mm long, generally with 2 calluses. w & swSichuan, Yunnan [nThailand]; forests and forest margins, scrub, meadows, rock-crevices, 1,700–4,900 m. VI–XI.

Saxifraga strigosa Wall. ex Seringe Similar to the previous species, but with **leaf-buds** present in axils of **rosette-leaves** and also often replacing flowers in the **inflorescence**; **flowers** up to 10; sepals eventually reflexed; petals white with reddish brown spots. w & swSichuan, seTibet, nwYunnan [swTibet, Himalaya]; similar habitats, 1,800–4,200 m. VII–X.

Saxifraga przewalskii Engl.* **Tufted perennial** to 12 cm tall in flower, with unbranched **stems** covered in brownish curly hairs. **Basal leaves** stalked, ovate to elliptic, to 30 mm long, glabrous except for curly brown marginal hairs; **stem-leaves** similar but smaller and stalkless or almost so. **Flowers** in 2–6-flowered cymes; sepals reflexed, lanceolate, glabrous except for margins; petals yellow with tiny purple spots towards base, purplish outside and in bud, ovate to oblong, 2.5–5.5 mm long, slightly longer than sepals. w & swGansu, e & seQinghai, w, n & nwSichuan, eTibet; open forests and shrubberies, alpine meadows, rock-crevices, 3,700–5,000 m. VII–IX.

Saxifraga tsangshanensis Franch.* Rather larger plant than the previous species, to 15 cm tall in flower, but stem- and leaf-margin hairs straight, **stem-leaves** mostly with a short stalk; **flowers** 2–3, or solitary, petals yellow with pale orange spots, 5–8 mm long. seTibet, nwYunnan; habitats similar to those of *S. przewalskii*, 3,000–4,600 m. VII–IX.

Saxifraga signata Engl. & Irmsch.* Plant to 20 cm tall in flower, usually less, usually with rather few basal rosettes; **stem** with dark brown glandular hairs. **Leaves** slightly fleshy, basal leaves spathulate, 14–16 mm long, glabrous, margin bristly; stem-leaves few, smaller, oblong, brown glandular-hairy. **Corymbs** lax, branched, with up to 24 flowers; sepals spreading or reflexed, ovate, with dark brown glandular hairs outside; petals yellow with purple spots towards base, ovate, 5.5–9 mm long. s & seQinghai, w & nwSichuan, e & seTibet, nwYunnan; rock-crevices and rocky meadows, 2,800–4,600 m. VII–IX.

Saxifraga unguiculata Engl.* Smaller than the previous species, not more than 12 cm tall in flower, **rosette-leaves** not more than 8 mm long. **Flowers** 1–8; petals yellow with orange spots towards base, oblong to lanceolate, 4.5–7.5 mm long. sGansu, Qinghai, w, nw & swSichuan, nShaanxi, Tibet, nwYunnan; open forests, shrubberies, rocky meadows, cliff-crevices, 1,800–5,600 m. VII–IX.

Saxifraga nanella Engl. & Irmsch. Small **cushion-forming** or **tufted** plant to 4 cm with glandular-hairy **stems** and leaf-rosettes. **Leaves** narrow-oblong to almost ovate or spathulate, 3–8 × 1.5–3 mm, glabrous apart from glandular-ciliate margin. **Flowers** solitary or up to 5 in a cyme; sepals spreading to reflexed, ovate, glandular-hairy or glabrous; petals yellow, orange-spotted towards base, elliptic-ovate, 4–5 mm long, twice as long as sepals. Qinghai, Tibet, nwYunnan [nwChina, cHimalaya]; alpine meadows and scrub, rock-crevices, 3,000–5,800 m. VII–VIII.

Saxifraga diversifolia Wall. ex Seringe Very variable **tufted perennial** to 40 cm tall in flower, generally smaller; **stems** glandular-hairy above. **Basal leaves** stalked, blade ovate to lanceolate or heart-shaped, 15–50 × 12–26 mm, brown-hairy beneath and along margin; **stem-leaves** up to 12, decreasing in size upwards, often heart-shaped, stalkless or almost so. **Flowers** in cymose clusters of up to 17; sepals reflexed, ovate, glabrous or glandular-hairy outside, margin ciliate; petals yellow, obovate to elliptic, 5–8 mm long, spreading. w & swSichuan, nwYunnan, Tibet [Himalaya, nMyanmar]; forests and forest margins, shrubberies, alpine meadows, rocky places, 2,700–4,300 m. VII–X.

Saxifraga egregia Engl.* Very similar to the previous species, but upper **stem-leaves** ovate, without a heart-shaped base, petals with 4–6, sometimes up to 10, wart-like calluses in lower half. sGansu, e & seQinghai, w, nw & swSichuan, e & seTibet, nwYunnan; similar habitats, 2,000–4,600 m. VII–X.

1. *Saxifraga nigroglandulifera*; Balangshan, wSichuan (CGW)
2. *Saxifraga bergenioides*; Mi La, east of Lhasa, sTibet (HJ)
3. *Saxifraga gemmipara*; Cangshan, nwYunnan (CGW)
4. *Saxifraga strigosa*; nwYunnan (JM)
5. *Saxifraga signata*; Gangheba, Yulongxueshan, nwYunnan (HJ)
6. *Saxifraga unguiculata*; Balangshan, Wolong, wSichuan (CGW)
7. *Saxifraga nanella*; Daxueshan, nwYunnan (CGW)
8. *Saxifraga diversifolia*; eNepal (CGW)

1. *Saxifraga tangutica*; Animaqingshan, cQinghai (HJ)
2. *Saxifraga drabiformis*; Baimashan, nwYunnan (HJ)
3. *Saxifraga montanella*; Balangshan, Wolong, wSichuan (CGW)
4. *Saxifraga brachypoda*; Wengshui, north of Zhongdian, nwYunnan (EL)
5. *Saxifraga melanocentra*; near Kharta, sTibet (HJ)
6. *Saxifraga punctulata*; Dungda La, seTibet (HJ)
7. *Saxifraga divaricata*; Balangshan, Wolong, wSichuan (CGW)
8. *Saxifraga lumpuensis*; Genyanshan, swSichuan (AD)
9. *Saxifraga fortunei*; cult. (AGS)
10. *Saxifraga rufescens*; Wolong, wSichuan (PC)

Saxifraga tangutica Engl. Readily distinguished from the previous 2 species by the long curly brown hairs on the **stems**, leaf-stalks and leaf-margins, **leaves** otherwise glabrous, not more than 11 mm wide. **Inflorescence** with up to 24 flowers; petals yellow, 2.5–4.5 mm long, sometimes purple beneath. sGansu, Qinghai, n, nw & wSichuan, Tibet [Himalaya]; similar habitats, 2,900–5,600 m. VI–IX.

Saxifraga drabiformis Franch.* **Cushion-forming** plant to 10 cm tall in flower with lax, basal leaf-rosettes. **Basal leaves** narrow-oblanceolate, 5–7 mm long, somewhat fleshy, glabrous; **stem-leaves** larger, to 10 mm long, upper with curly brown glandular hairs. **Flowers** solitary on slender, glandular-hairy stalks; sepals reflexed, ovate, with marginal glandular hairs; petals yellow, oval to ovate, 7–8 mm long, with 4–6 wart-like calluses. nwYunnan; rocky habitats, crevices, screes, 3,300–4,900 m. VII–IX.

Saxifraga montanella Harry Sm. Small **cushion-forming** plant to 8 cm tall in flower, with curly brown hairs on **stems** and leaf-stalks. **Basal leaves** stalked, ovate to lanceolate, to 7 mm long, glabrous, with a pointed apex; **stem-leaves** slightly longer, elliptic, stalkless, margin hairy. **Flowers** solitary; sepals spreading, elliptic, glabrous or hairy on reverse; petals yellow, obovate to ovate, 8–13 mm long, with some brown hairs on reverse and margin basally. Qinghai, w, nw & swSichuan, e & seTibet, nwYunnan [c & eHimalaya]; alpine shrubberies and meadows, stony and rocky habitats, snow hollows, 3,300–5,200 m. VII–X.

Saxifraga brachypoda D. Don **Tufted perennial** to 20 cm, generally half that height, with simple **stems** that are glandular-hairy in upper part and with leafy buds in the axils. **Leaves** rather crowded, deep shiny green, lanceolate to linear-lanceolate, 5–12 × 0.4–2.8 mm, pointed, glabrous except for the usually bristly margin. **Flowers** solitary or 2–3; sepals erect, ovate to elliptic, glabrous or glandular-hairy; petals yellow, obovate to elliptic, 5–9 mm long, longer than sepals; c & nwSichuan, seTibet, nwYunnan [swTibet, Himalaya, nMyanmar]; forests, scrub, ravines, gullies, rock-crevices, 3,000–5,000 m. VII–IX.

Group Four

Saxifraga melanocentra Franch. **Perennial herb** to 20 cm tall in flower, often less; **stem** glandular-hairy. **Leaves** all in a basal rosette, lanceolate to oblong or somewhat rhombic, 8–40 mm long, long-stalked, glabrous or downy, margin toothed. **Inflorescence** corymbose, with up to 17 **flowers**; sepals spreading to reflexed, ovate, glabrous or downy; petals spreading, white with 2 basal yellow spots, or purplish towards base, occasionally wholly red or purple, ovate to elliptic, 3–6 mm long; anthers black; ovary deep purplish black. sGansu, Qinghai, w, nw & swSichuan, Shaanxi, Tibet, nwYunnan [Himalaya, nMyanmar]; open alpine shrubberies and meadows, moraines, streamsides, rock-crevices, 3,000–5,300 m. VII–IX.

Saxifraga punctulata Engl. Small **tufted** plant to 6 cm with dark purple, glandular-downy **stems**. **Basal leaves** in a rosette, spathulate, to 4 mm long, glabrous apart from a ciliate margin, dotted above; **stem-leaves** slightly larger, crowded, covered in dark purple glandular hairs. **Flowers** solitary or in **cymes** of 2–3, relatively large, 15–20 mm across, bowl-shaped; sepals dark purple-glandular; petals white, cream or yellow, with a pair of yellow or orange spots in the centre, speckled with crimson or red in the middle and upper part. seTibet [swTibet, c & eHimalaya]; rocky alpine meadows, screes, moraines, 4,600–5,800 m. VII–IX.

Saxifraga atrata Engl.* Similar to the previous species, but **flowers** in a cylindrical **cyme** with up to 24 rather smaller flowers, petals always white, 2.5–4 mm long. seGansu, neQinghai; rocky and grassy alpine meadows, 3,000–4,200 m. VII–IX.

Saxifraga parvula Engl. & Irmsch.* Differs from the 2 previous species in being a tiny plant not more than 4 cm tall in flower, **flowers** 1–2; petals not more than 3.5 mm long; filaments distinctly club-shaped (not linear). nwYunnan; alpine meadows, rock-crevices, 3,800–5,700 m. VII–IX.

Saxifraga divaricata Engl. & Irmsch.* Closely related to *S. melanocentra* and *S. atrata*, but petals narrower and not overlapping, 3-veined (not 5–9-veined); **inflorescence** with spreading branches, the petals with 2 orange-yellow spots towards base, narrow-ovate to elliptic, not more than 3 mm long. seQinghai, w & nwSichuan, eTibet; alpine meadows, open scrub, marshy places, 3,400–4,500 m. VII–VIII.

Saxifraga lumpuensis Engl.* Very similar to the previous species, but a generally taller plant with red or purple petals, each with 2 yellow spots at base; anthers pinkish or reddish. s & swGansu, w, sw & nwSichuan; similar habitats, 3,500–4,100 m. VI–VII.

Group Five

Saxifraga fortunei Hook. f. Rather fleshy **tufted perennial** to 40 cm tall in flower. **Leaves** all in a basal rosette, stalked, heart-shaped to kidney-shaped, with 7–11 marginal lobes, toothed, glandular-hairy beneath and along margin. **Inflorescence** a panicle with 25–40 **flowers** on glandular-hairy stalks; sepals spreading to reflexed, narrow-ovate, to 3.5 mm long; petals white to reddish, variable, the 3 shortest 1.4–4 mm long, the 2 longest 7–17 mm long, generally with toothed margin, all sharply pointed. Hubei, n & eSichuan [neChina, Sikkim, Korea]; rock-crevices, forest rocks, 2,200–2,900 m. VI–VII.

Saxifraga rufescens Balf. f.* Similar to the previous species, but **flowers** white or pink, with 4 short petals, and 1 long petal that is never toothed, to 19 mm long. wHubei, Sichuan, seTibet, nwYunnan; similar habitats, 600–4,000 m. IV–VI.

Saxifraga stolonifera Curtis (syn. *Diptera sarmentosa* (L. f.) Losinsk.; *Saxifraga sarmentosa* L. f.) Readily distinguished from the previous 2 species by the production of numerous thread-like stolons and by the mottled, rounded to kidney-shaped **leaves** with an evenly toothed margin; petals white, finely spotted, longest 2 up to 15 mm long. seGansu, Guangxi, Guizhou, Hubei, eSichuan, Shaanxi, swYunnan [c & nChina, Korea, Japan]; forests, shrubberies, shady rocky places, ravines, 400–4,500 m. IV–XI.

PARNASSIACEAE

PARNASSIA

A north temperate genus of c. 70 species (63 in China, 49 endemic). Glabrous **perennial herbs** with lax rosettes of entire **basal leaves** and few, occasionally absent, **stem-leaves**. **Flowers** solitary, erect, 5-parted; sepals generally greenish or brownish, shorter than petals, margin entire to variously fringed or ciliate; stamens 5, opposite sepals; staminodes present, 5, opposite petals, variously lobed or fringed. **Fruit** a 3–4-valved capsule.

1. Stem-leaves absent .... *P. scaposa*
1. Stem-leaves present, usually 1–3 2
2. Petals deeply fringed, especially towards base ....
   .... *P. wightiana, P. epunctulata*
2. Petals not fringed, sometimes slightly ciliate towards base 3
3. Petals overlapping, blunt .... *P. delavayi, P. nubicola, P. nubicola* var. *nana, P. cordata*
3. Petals not overlapping, usually notched .. *P. mysorensis*

Parnassia scaposa Mattf.* Plant to 20 cm, usually with a single **stem** and 4–5 basal **leaves** with an elliptic to obovate blade, to 25 × 13 mm, 3- or 5-veined. **Flowers** 18–25 mm in diameter; sepals lanceolate, 3-veined, half the length of the petals; petals white or cream, obovate, 12–15 mm long, apex rounded to slightly notched; staminodes flat, 3-lobed at apex, 3.5 mm long. s & seQinghai, wSichuan, seTibet; alpine meadows, open shrubberies, 3,700–4,500 m. VII–VIII.

Parnassia wightiana Wall. ex Wight & Arnott Tufted **perennial** to 25 cm, often less. **Basal leaves** deep green, up to 5, long-stalked, blade kidney- to heart-shaped, to 6.5 cm across, 7- or 9-veined; **stem-leaf** solitary, in middle or upper part of stem, stalkless and half-clasping the stem. **Flowers** 15–40 mm in diameter; sepals lanceolate, purple-brown-dotted, about two-thirds the length of the petals; petals white, oblong to obovate, 11–14 mm long; staminodes flat, 5-lobed, 3–5 mm long. Guangxi, Guizhou, Hubei, Sichuan, Shaanxi, seTibet, Yunnan [swTibet, eHimalaya, nThailand]; open forests, forest clearings, meadows, grassy banks, roadsides, 600–2,000 m. VII–IX.

Parnassia epunctulata J. T. Pan* Smaller plant than previous species, to 13 cm, with **leaves** not more than 1.5 cm across, and **flowers** 15–19 mm across, with 3-lobed staminodes. nwYunnan; alpine meadows, 3,400–3,800 m. VI–VII.

Parnassia delavayi Franch.* **Tufted perennial** to 40 cm, often far less, with deep green, or brownish green, rounded to kidney-shaped **basal leaves**, to 4.5 cm across, 5–9-veined; **stem-leaf** solitary, in middle or lower half of stem, stalkless and half-clasping the stem. **Flowers** 30–35 mm in diameter; sepals oblong to obovate, 6–8 mm long, brown-dotted; petals white with purplish veins and tiny purple-brown dots, oblong to obovate, 12–25 mm long, with a rounded or pointed apex, ciliate towards base; staminodes flat, 3-lobed, 3–4 mm long; anthers with a projecting appendage at top. Gansu, Hubei, Sichuan, Shaanxi, Yunnan; coniferous forests, open woods, shrubberies, grassy and gravelly habitats, 1,800–3,800 m. VII–IX.

Parnassia nubicola Wall. ex Royle Differs from previous species by the elliptic to ovate-oblong **basal leaves**, and **stem-leaf** towards base, generally with some brownish appendages at base. **Flowers** without purplish veins, although minutely dotted with purple-brown; petals ovate; anthers without an appendage. seTibet, nwYunnan [swTibet, Himalaya, nMyanmar]; similar habitats as well as streamsides and meadows, 2,700–3,900 m. VIII–IX. Dwarf plants from seTibet and nwYunnan (Deqin), rarely more than 8 cm tall with small thin-textured **leaves**, are assigned to **var. *nana*** T. C. Ku*.

Parnassia cordata (Drude) Z. P. Jien ex T. C. Ku* (syn. *P. nubicola* Wall. ex Royle var. *cordata* Drude) Like *P. nubicola*, but with **stems** more slender and **basal leaves** with a deep heart-shaped base. **Flowers** 20–25 mm across, petals white with purple veins, minutely dotted. nwYunnan; alpine meadows, 3,200–4,100 m. VII–VIII.

Parnassia mysorensis F. Heyne ex Wight & Arnott Tufted plant to 14 cm with up to 4 kidney-shaped, deep green or brownish green **basal leaves**, to 1.5 × 1.5 cm, purple-dotted beneath, stalks to 2 cm long; **stem-leaf** solitary, stalkless, heart-shaped. **Flowers** 10–14 mm across; sepals oblong to rounded, 4–5 mm long; petals obovate-spathulate, 8–9 mm long, attenuate into claw at base; staminodes 3-lobed. Guizhou, Sichuan, seTibet, Yunnan [swTibet, nIndia, c & eHimalaya]; meadows, forests, shrubberies, open slopes, 2,500–3,600 m. VII–VIII.

---

1. *Saxifraga stolonifera*; cult. (AGS)
2. *Parnassia wightiana*; Emeishan, sSichuan (CGW)
3. *Parnassia epunctulata*; Yulongxueshan, nwYunnan (PC)
4. *Parnassia delavayi*; Balangshan, Wolong, wSichuan (CGW)
5. *Parnassia nubicola*; eNepal (CGW)
6. *Parnassia mysorensis*; east of Hongyuan, nwSichuan (CGW)
7. *Ribes longiracemosum*; Shennongjia, wHubei (PC)

GROSSULARIACEAE

RIBES

A genus of c. 160 species (59 in China, 25 endemic) in the north temperate zone and the Andes. **Deciduous** or **evergreen shrubs** or small **trees**, sometimes spiny. **Leaves** alternate or clustered, palmate. **Flowers** in **racemes** or **corymbs**, hermaphrodite or unisexual, 4–5-parted; corolla small; stamens inserted on calyx rim. **Fruit** a fleshy berry. The genus contains various currants and the gooseberry (Ribes uva-crispa L.).

1. Flowers hermaphrodite 2
1. Flowers unisexual (plants dioecious) 3
2. Plants unarmed ... *R. longiracemosum, R. himalense*
2. Plants armed .... *R. alpestre*
3. Plants evergreen; flowers yellowish green ....
   .... *R. laurifolium*
3. Plants deciduous; flowers brownish red to purplish brown .... *R. tenue, R. glaciale, R. orientale*

Ribes longiracemosum Franch.* **Deciduous shrub** to 3 m with glabrous branchlets. **Leaves** ovate, 3(–5)-lobed, 5–12 cm long, long-stalked, lobes ovate-triangular, irregularly toothed, sparsely and minutely hairy at first. **Racemes** pendulous, to 35 cm long, with up to 25, distant greenish white **flowers**; calyx green tipped purple, glabrous; petals half the length of sepals. **Fruit** globose, 7–9 mm, black when ripe. seGansu, wHubei, Sichuan, eShaanxi, Yunnan; forests, thickets, gullies, streamsides, 1,100–3,800 m. IV–V.

Ribes himalense Royle ex Decne Readily distinguished from the previous species by the red **leaf-stalks**, and by the 8–20-flowered **racemes** of reddish or purple-red **flowers**. **Fruit** red at first, turning purplish black, 6–7 mm. A very variable species with at least 5 varieties recognised in China alone. Gansu, wHubei, Qinghai, Sichuan, Shaanxi, Tibet, nwYunnan [nChina, Himalaya]; similar habitats, 1,200–4,100 m. IV–VI.

Ribes alpestre Wall. ex Decne **Shrub** 1–3 m, with 3-branched, stout spines at the nodes. **Leaves** ovate in outline, 3–5-lobed, 1.5–3 × 2–4 cm, minutely hairy at first, lobes coarsely toothed.

Flowers axillary, solitary or 2–3 in a short raceme, calyx greenish or reddish brown with a campanulate tube 5–6 mm long, lobes reflexed; petals white, elliptic, 2.5–3.5 mm long. **Fruit** purple when ripe, 12–15 mm, covered in stalked glands. Gansu, Qinghai, Shaanxi, w, nw & swSichuan, Shaanxi, seTibet, w & nwYunnan [n & cChina, Himalaya]; mixed and coniferous forests, ravines, forest margins, shrubberies, riverbanks, wet rocks, 1,000–3,900 m. IV–VI.

Ribes laurifolium Jancz.* Scandent **shrub** to 1.5 m, generally clambering into trees, with smooth, glabrous branches. **Leaves** 2–4 clustered at shoot tips, stalked, deep green, leathery, blade ovate to oblong or elliptic, 5–10 × 2.5–4.5 cm, sharply toothed, 3- or 5-veined at base. **Flowers** yellowish green, male 10–12 mm across, in pendulous racemes up to 6 cm long, female 6–8 mm across, solitary or clustered, erect. **Fruit** on pendent stalks, purple when ripe, oblong in outline, 15–20 mm long. Guizhou, w & swSichuan, nwYunnan; forests, shrubby slopes, riverbanks, 2,100–3,600 m. IV–VI.

Ribes tenue Jancz. **Shrub** to 4 m, often only 1–2 m, with slender, usually glandular branchlets. **Leaves** ovate, 3–5-lobed, with a truncated or heart-shaped base, margin doubly toothed, becoming hairless at maturity; stalks 1–3 cm long. **Flowers** in distinctive, erect to ascending **racemes**, male racemes to 5 cm long, with up to 20 flowers, female shorter, fewer-flowered; calyx reddish brown, glabrous; petals dark red, c. 1 mm long. **Fruit** globose, 4–7 mm, usually dark red, occasionally black. eGansu, Sichuan, swShaanxi, Yunnan [cChina, Himalaya]; forests, thickets, gullies, streamsides, 1,300–4,200 m. V–VI.

Ribes glaciale Wall. **Shrub** to 3 m, sometimes taller, with unarmed, glabrous or somewhat hairy **branches**. **Leaves** with pinkish stalks, ovate to rounded, 3–5-lobed, 3–5 × 2–4 cm, lobes coarsely toothed. **Flowers** 4–5 mm, brownish red, male 10–30 in pendent **racemes** up to 5 cm long, female fewer (not more than 10) in shorter racemes; stamens equalling or longer than petals. **Fruit** red when ripe, globose, 5–7 mm across, glabrous. seGansu, wHubei, Shaanxi, seTibet, nwYunnan [Henan, Himalaya, nMyanmar]; forests, shrubberies, ravines, rocky places, roadsides, 1,900–3,000 m. IV–VI.

Ribes orientale Desf. A generally smaller **shrub** than the preceding species, not more than 2 m, with smaller **leaves**, 1–3 cm long and broad. **Flowers** purplish to purplish brown, borne in erect racemes, male 2–5 cm long, female half the length. **Fruit** red to purple when ripe, 7–9 mm, adorned with hairs and stalked glands. wSichuan, Tibet, nwYunnan [swAsia, seEurope, Himalaya, Russia]; similar habitats, 2,100–4,900 m. IV–V.

LEGUMINOSAE (FABACEAE)

SYNOPSIS OF GROUPS

Group One. Flowers in brush-like heads or tufts; petals very reduced; stamens much more prominent than petals.

Group Two. Flowers with prominent petals, almost regular; corolla lacking standard, lateral wings and a keel.

Group Three. Flowers like pea-flowers, strongly zygomorphic; corolla with a standard, lateral wings and a keel.

Group One

ALBIZIA

A genus with c. 150 species (3 in China) mainly in the tropics and subtropics. Unarmed **trees** or **shrubs** with evenly 2-pinnate **leaves** and small stipules. **Flowers** in globose heads, bunches or forming axillary or terminal panicles, the central flower larger and different from the others; petals 5, united to middle to form a tubular corolla; stamens numerous, more than 15, filaments united into a tube in basal half. **Pods** leathery, compressed.

Albizia julibrissin **Durazz.** **Tree** 3–10 m tall with downy twigs. **Leaves** 10–30(–45) cm long, 2-pinnate, the primary divisions each with 10–30 pairs of oblong leaflets, 1–2 cm long, pointed. **Flower-heads** 15–20-flowered, solitary or 2–4 in axillary or terminal panicles; **corolla** 9–10 mm long, downy; stamens 3–3.5 cm long, white with pink tips. **Pods** 12–16 cm long. Guizhou, wHubei, Sichuan, Yunnan [s, c & eChina, Himalaya, Japan, Korea]; dry slopes, edges of lava flows, riverbanks, 1,400–2,600 m. V–VI.

Albizia kalkebra **Prain** Readily distinguished from the previous species by the **leaves**, which have only 2 pairs of primary divisions, each with 5 pairs of leaflets; leaflets obliquely oblong, 2–4.5 cm long, blunt. **Flowers** white or yellowish white; stamens 2 cm long. **Pods** 13–16 cm long. Guizhou, Hubei [s & eChina, Himalaya]; sea level to 2,000 m. IV–V.

Group Two

BAUHINIA

A pantropical genus of 250 species (65 in China). **Climbers**, **trees** or **shrubs** with simple **leaves** that are usually 2-lobed at apex, palmately veined; stipules small, soon falling. **Flowers** axillary or terminal, racemose or paniculate; petals 5, almost equal, clawed at base; stamens 3, 5 or 10. **Pods** stalked, compressed, leathery or woody, dehiscent.

Bauhinia brachycarpa **Wall. ex Benth.** (syn. *B. faberi* Oliv.) **Shrub** 2–4 m tall with roundly 2-lobed **leaves**, 2–7 × 2–7.5 cm. **Flowers** white; petals 7–10 mm long. **Pods** 4–7.5 cm long, black at maturity. Gansu, Hubei, Sichuan, Yunnan [s & cChina, Laos, Myanmar, nThailand, Vietnam]; dry hillsides, gorges, rocky places, scrub, roadsides, 1,000–3,000 m. III–VIII.

Bauhinia championii **(Benth.) Benth.** **Climber** 3–5 m long with tendrils opposite the **leaves** and at base of inflorescence. **Leaves** entire or 2-lobed, heart-shaped, the largest 9–12 × 6–9.5 cm, pointed. **Inflorescence** simple or branched, to 15 cm long. **Flowers** small, cream to pale yellow; petals 5–6 mm long; stamens 11–12 mm long. **Pods** 6–7.5 cm long, black at maturity. Guizhou, Hubei, Yunnan [s, c & eChina, neIndia, Vietnam]; dry hillsides, rocky places, gorges, ravines, to 1,600 m. VIII–XI.

Group Three

CLADRASTIS

A genus of 6 species (5 in China, all endemic) in east and South East Asia, 1 in eastern North America. **Deciduous trees**, rarely woody **climbers**. **Leaves** odd-pinnate; leaflets alternate or almost opposite, entire; stipules absent. **Inflorescences** terminal, branched racemes; calyx campanulate, 5-lobed; petals almost equal; stamens 10, free. Ovary shortly stalked with an incurved style. **Pods** compressed, winged or not, tardily dehiscent.

Cladrastis sinensis **Hemsl.*** **Tree** to 20 m. **Leaves** to 20 cm long with 4–7 pairs of ovate-lanceolate or oblong-lanceolate leaflets, usually 6–10 cm long, grey-white-hairy beneath. **Panicles** 15–30 cm long. **Flowers** many, 14 mm long; calyx densely grey-brown-downy; corolla white or light yellow, rarely pink, standard obovate or round, 9–11 mm long, wings arrow-shaped, slightly longer than standard, keel elliptic. **Pods** flat, elliptic, wingless, 3–8 cm long. Gansu, Guizhou, Hubei, Sichuan, Shaanxi, Yunnan [scChina]; hill forests, 1,000–2,500 m. VI–VIII.

SOPHORA

A genus of c. 50 species (15 in China) mostly in temperate Europe, Asia and North America. **Deciduous shrubs** or **trees** with odd-pinnate **leaves**; leaflets oppositely arranged; stipules awl-shaped. **Flowers** in terminal or axillary racemes; calyx-tube campanulate, 5-toothed; standard narrowed in basal part but not clawed, wings and keel clawed; stamens mainly free, joined at base only. **Pods** cylindrical, constricted between seeds, with a few rounded seeds, indehiscent or tardily dehiscent.

---

1. *Ribes laurifolium*; cult. (CGW)
2. *Ribes tenue*; Huanglong, nwSichuan (PC)
3. *Albizia julibrissin*; cult. (CGW)
4. *Bauhinia brachycarpa*; Minjiang, nwSichuan (PC)
5. *Bauhinia championii*; Shilin, cYunnan (CGW)
6. *Cladrastis sinensis*; cult. (PC)

Sophora davidii (Franch.) Kom. ex Pavol.* Spiny **shrub** 0.4–2 m tall, branchlets grey-downy, spines 3–5 cm long. **Leaves** up to 9 cm long, with 11–21 elliptic or obovate leaflets 6–12 mm long. **Flowers** in terminal racemes on short lateral twigs, pale blue or white tinged with blue. **Pods** 4.5–6 cm long, with up to 4 seeds. Gansu, wGuizhou, wHubei, Sichuan, seTibet, Yunnan [swTibet, c & nChina]; dry valleys, scrub, hedgerows, roadsides, 1,300–3,800 m. V–VII.

Sophora moorcroftiana (Benth.) Benth. ex Baker Intricately branched, spiny **shrub** to 1.5 m with pinnate **leaves** 4–5 cm long, leaflets numerous, ovate, 5–7 mm long, silky-hairy; lateral branches spine-tipped. **Flowers** blue, rarely yellowish, in lateral clusters, each flower c. 15 mm long and with standard characteristically longer than other petals. **Pods** 7–10 cm long, with 5–6 seeds. Qinghai, Tibet [Himalaya]; dry stony slopes, open shrubberies, sandy places, 2,800–3,900 m. V–VI.

Sophora japonica L. **Deciduous tree** to 25 m, usually much smaller, with dark grey-brown **bark**. **Leaves** rich green, pinnate with 7–15 lanceolate leaflets to 5 cm long. **Inflorescence** a terminal panicle to 25 cm long. **Flowers** creamy to pale yellow, 12–13 mm long; standard, erect, straight. **Pods** to 8 cm long, with 1–6 seeds. Often planted around temples, throughout China and Japan [native to c & eChina]. IX–X.

CAMPYLOTROPIS

A genus of 40 species (35 in China) in east, south and South East Asia. **Shrubs** with **leaves** simple or with 3 leaflets; stipules present, conspicuous. **Flowers** in lax, axillary racemes; calyx-tube campanulate, 5-toothed; petals twice as long as calyx, wings adherent to upwardly curved keel; stamens in 2 bundles. **Pods** compressed, 1-seeded, indehiscent.

Campylotropis polyantha (Franch.) Schindl.* (syn. *Lespedeza polyantha* (Franch.) Schindl.) **Shrub** 0.6–3 m tall. **Leaves** with 3 leaflets, leaflets elliptic-obovate, mucronate, 1–4 cm long; linear stipels present at top of leaf-rachis. **Inflorescence** 2–8 cm long, lax, many-flowered. **Flowers** purple to lilac, 7–12 mm long; calyx 2–3 mm long, hairy and glandular. **Pods** 6–11 mm long. Guizhou, nw, w & swSichuan, seTibet, nwYunnan; hillsides, forest margins, along streams in forests, roadsides, waste places, 1,600–3,300 m. IV–VI.

Campylotropis macrocarpa (Bunge) Rehd. Similar to *C. polyantha*, but leaflets more than 3 cm long, lacking stipels at top of rachis; **pods** oblong, 10–16 cm long. Gansu, Guizhou, Hubei, Sichuan, Shaanxi, e & seTibet, Yunnan [c, e & sChina, Korea, Taiwan]; mountain slopes, valleys, thickets, forest margins, streamsides, open places, 1,000–2,000 m. IV–VII.

Campylotropis capillipes (Franch.) Schindl.* (syn. *Lespedeza capillipes* Franch.) **Shrub** 1–2 m tall. **Leaves** 1–4 cm long with 3 leaflets; leaflets elliptic-oblong to obovate, 0.7–2.3 cm long, mucronate. **Inflorescence** 1–2, 1.5–3 cm long, densely many-flowered, flowers purple to reddish purple, 10–14 mm long; calyx 1.5 mm long. Pods 6–10 mm long. wGuangxi, Sichuan, Yunnan; hillsides, forest margins, roadsides, 1,000–3,000 m. IX–XI.

Campylotropis prainii (Collett & Hemsl.) Schindl. Similar to *C. capillipes*, but with **inflorescence** 2–8 cm long and **flowers** 1–1.6 cm long; calyx 1 mm long. Sichuan, Yunnan [Myanmar]; forests and forest margins, valleys, mountain slopes, thickets, streamsides, 1,000–3,000 m. IX–XI.

CARAGANA

A genus of c. 80 species (c. 28 in China) in temperate Asia and Europe. Spiny **shrubs** with even-pinnate, occasionally digitate **leaves** with persistent rachises that become spine-like; leaflets small, entire, deciduous; stipules present, conspicuous. **Flowers** solitary or few in axillary racemes; calyx-tube campanulate, 5-toothed; petals clawed, wings oblong, with or without a linear appendage at base, keel elliptic, rounded at tip; stamens in 2 bundles. **Pods** inflated, dehiscing by twisting of valves, 4–6-seeded.

1. Flowers pink or creamy white flushed pink .. *C. jubata*
1. Flowers yellow ... *C. franchetiana, C. maximowicziana, C. versicolor, C. tibetica*

Caragana jubata (Pallas) Poir. Very variable, columnar to prostrate, very spiny **shrub** 50–100 cm tall, occasionally taller, spines 3–5 cm long. **Leaves** glaucous, leaflets in 5–8 pairs, linear to oblanceolate, 7–13 mm long, softly hairy. **Flowers** borne directly on branches (cauliflorous), pink or white flushed with pink, occasionally rose-purple; calyx campanulate; corolla 2.5–3.9 cm long. **Pods** up to 3 cm long, densely softly hairy. Gansu, Qinghai, w, nw & swSichuan, Tibet, nwYunnan [n & nwChina, cAsia, Himalaya, Mongolia, eRussia]; scrub, coniferous forest margins, rocky slopes, screes, moorland, 1,750–4,800 m. V–VIII.

Caragana franchetiana Kom.* **Shrub** 1–3.5 m tall with spines 3–6 cm long. **Leaves** 3–6 cm long, leaflets in 8–9 pairs, elliptic or oblong-obovate, mucronate, 7–8 mm long. **Flowers** borne

---

1. *Sophora davidii*; near Geza, nwYunnan (RS)
2. *Sophora japonica*; Beijing (PC)
3. *Sophora japonica*; Jietaise, near Beijing (CGW)
4. *Campylotropis polyantha*; Jiuzhaigou, nwSichuan (PC)
5. *Campylotropis macrocarpa*; Jietaisi, near Beijing (CGW)
6. *Campylotropis prainii*; nwYunnan (PC)
7. *Caragana jubata*; north-east of Songpan, nwSichuan (CGW)
8. *Caragana jubata*; Huanglongsi, nwSichuan (CGW)
9. *Caragana jubata*; north of Songpan, nwSichuan (HJ)
10. *Caragana franchetiana*; Gangheba, Yulongxueshan, nwYunnan (CGW)
11. *Caragana franchetiana*; Mahuangba, Yulongxueshan, nwYunnan (CGW)

1. *Caragana maximowicziana*; Zhongdian, nwYunnan (RS)
2. *Caragana maximowicziana*; Xiaozhongdian, nwYunnan (PC)
3. *Caragana versicolor*; south of Aba, nwSichuan (CGW)
4. *Caragana tibetica*; Muli, swSichuan (PC)
5. *Indigofera pendula*; south of Zhongdian, nwYunnan (PC)
6. *Indigofera balfouriana*; cult. (PC)
7. *Indigofera amblyantha*; Shennongjia, wHubei (PC)
8. *Indigofera fortunei*; cult. (CGW)
9. *Desmodium elegans*; Sandauwan, Yulongxueshan, nwYunnan (PC)

directly on branches (cauliflorous), nodding, 2.3–2.8 cm long, yellow, flushed with red or brown; calyx 1.3–1.5 cm long, with short teeth; wings with linear auricle. nwYunnan; open alpine scrub, forest margins, moorland, 2,700–3,600 m. V–VIII.

Caragana maximowicziana Kom.* Denser, more spreading **shrub** than previous species, 1.5–5 m tall, spines 1.5–2.8 cm long. **Leaves** 1.5–2.8 cm long, with 3–5 leaflets; leaflets linear to oblanceolate, 7–10 mm long, green. **Flowers** often in pairs, 1.8–2 cm long, yellow marked with red. Gansu, Qinghai, w, nw & swSichuan, eTibet, nwYunnan; river valleys amongst rocks and boulders, scrub, moorland slopes, 2,700–4,300 m. VI–VIII.

Caragana versicolor Benth. **Shrub** to 80 cm. **Leaves** digitate, with 4 leaflets, stalks persistent on long branchlets, absent on short branchlets; leaflets narrow-lanceolate, obovate-lanceolate or linear, 5–7 × 1–1.5 mm. **Flowers** solitary, yellow with standard reddish brown outside, 11–12 mm long. Qinghai, w & nwSichuan, eTibet [Afghanistan, w & cHimalaya]; rocky slopes, river beaches, shrubberies, 3,000–4,900 m. V–VI.

Caragana tibetica (Maxim. ex Schneid.) Kom. (syn. *C. tragacanthoides* (Pall.) Poiret var. *tibetica* Maxim. ex Schneid.) Low hummock-forming spiny **shrub** to 30 cm. **Leaves** pinnate with 3–4 pairs of leaflets, these linear, 8–12 × 0.5–1.5 mm, downy, pointed; leaf-stalk 2–3.5 mm long. **Flowers** solitary, ± unstalked; calyx tubular, 8–15 mm long, downy; corolla yellow, 22–25 mm long, standard obovate. **Pods** elliptic in outline, 7–8 mm long, downy. Gansu, Qinghai, w & swSichuan, nShaanxi, Tibet [n & nwChina, Mongolia]; dry stony and sandy steppe, 1,400–3,500 m. V–VII.

INDIGOFERA

A pantropical genus with c. 700 species (c. 60 in China). **Annual herbs** or **shrubs**; branchlets with branched hairs. **Leaves** odd-pinnate, rarely simple or with 3 leaflets; stipules present or absent. **Flowers** in lax or dense axillary racemes or clusters; calyx campanulate, 5-toothed; standard broadly elliptic, ovate or almost orbicular, wings oblong to obovate, keel straight or curved, lanceolate to oblanceolate, with a short pouch on each side near base; stamens in 2 bundles. **Pods** linear or oblong in outline, straight or curved, constricted between seeds, jointed.

> 1. Inflorescence pendent .... *I. pendula*
> 1. Inflorescence erect to ascending ....
>   .... *I. balfouriana, I. amblyantha, I. fortunei*

Indigofera pendula Franch.* Graceful deciduous **shrub**, 1–3 m tall. **Leaves** arching, with 9–13 pairs of oval leaflets, rounded or notched at apex, to 30 mm long, appressed-hairy beneath. **Inflorescences** pendent, many-flowered slender racemes to 30 cm long, tapering to apex. **Flowers** rosy purple or lilac and purple, 9–12 mm long. swSichuan, seTibet, Yunnan; roadsides, banks, shrubberies, margins of *Quercus*, *Pinus* and *Picea* forests, ravines, 2,300–3,700 m. VI–VIII.

Indigofera balfouriana Craib* Deciduous **shrub** 50–180 cm tall with branchlets hairy at first. **Leaves** 12–30 cm long, with 5 to 7 pairs of elliptic to obovate leaflets, 6–13 mm long; stipules 3 mm long. **Inflorescences** axillary, 2.5 cm long, erect, few-flowered. **Flowers** 7–8 mm long, deep rose-purple to pinkish purple; calyx 1 mm long. w & swSichuan, seTibet, nwYunnan; *Quercus* and *Pinus* forests, dry, open places in scrub, roadsides, 1,600–3,300 m. V–VII.

Indigofera amblyantha Craib* Deciduous **shrub** to 2 m, young shoots with appressed white hairs. **Leaves** pinnate, with 7–11 oval leaflets, to 30 × 16 mm, with a tiny terminal mucro. **Inflorescences** erect, slender, with short stalk, borne in succession from the leaf-axils, flowers closely arranged, small, 5–6 mm long, pale rose-pink to deep pink. Gansu, Hubei; shrubberies, open coniferous forests, forest margins, rocky places, 2,400–3,600 m. VI–XI.

Indigofera fortunei Craib* (syn. *I. subnuda* Craib) **Subshrub**, woody in lower part, with ascending to erect slender **stems**. **Leaves** greenish yellow, to 7 cm long, with 4–5 pairs of opposite, elliptic to ovate-elliptic leaflets, to 18 × 9 mm. **Inflorescences** ascending, to 7 cm long; calyx-tube to 1.5 mm long; corolla 8.5–9 mm long, bluish purple. Hubei [c & eChina]; thickets, rocky places, to 1,600 m. IV–V.

DESMODIUM

A genus of c. 350 species (30 in China) widespread in the tropics and subtropics. **Perennial herbs**, **shrubs** or **trees**. **Leaves** simple or with 3 leaflets, terminal leaflet long-stalked; stipules present, conspicuous. **Flowers** in lax or dense axillary racemes, panicles, umbels or corymbs; calyx-tube short, 5-toothed; petals clawed, wings adherent to the straight or curved keel; stamens in 1 or 2 bundles. **Pods** compressed, constricted between seeds, dehiscent along lower suture or indehiscent.

Desmodium elegans DC. (syn. *D. cinerascens* Franch.; *D. franchetii* Rehd.; *D. forrestii* Schindl.) Hairy, much-branched **shrub** 0.2–2 m tall. **Leaves** with 3 leaflets, leaflets elliptic to obovate, 2–8.5 cm long, blunt to slightly pointed, with appressed hairs beneath. **Inflorescence** simple or branched, densely many-flowered, 5–20 cm long; rachis downy. **Flowers** rose-pink to mauve, 7–9 mm long; calyx and flower-stalk glabrous. **Pods** flat, linear, 4–6-jointed, 2 cm long, brown, hairy. Gansu, w, nw & swSichuan, seTibet, Yunnan [Afghanistan, Himalaya]; scrub and forest margins on rocky hillsides, roadsides, gorges, 1,000–4,000 m. VI–VIII.

Desmodium multiflorum DC. (syn. *D. floribundum* DC.) **Shrub** 0.6–2 m tall. **Leaves** with 3 leaflets, leaflets elliptic to obovate, 2–10 cm long, blunt to almost pointed, white-felted beneath.

Inflorescence simple or branched, densely many-flowered, 9–20 cm long. **Flowers** pale rose-pink to purple, 9–11 mm long; calyx and flower-stalks hairy. **Pods** slightly curved, 4–7-jointed, 1.5–4.5 cm long. Guizhou, Hubei, Sichuan, Tibet, Yunnan [s & eChina, Himalaya, Myanmar, Loas, Thailand, Taiwan]; forest margins, shrubberies, rocky slopes, gorges, 1,700–3,000 m. VII –IX.

PIPTANTHUS

A genus of 3 species (all in China) distributed in the Himalaya and China. **Deciduous** or **semi-evergreen shrubs** with downy twigs. **Leaves** stalked, with 3 leaflets; leaflets entire, stalkless; stipules joined. **Flowers** yellow, in short terminal racemes; calyx-tube campanulate, 5-toothed; petals long-clawed, standard almost orbicular, wings oblong, keel slightly incurved and rounded at tip; stamens free. **Pods** linear-oblong, flattened, short-stalked, compressed.

Piptanthus nepalensis (Hook.) Sweet (syn. *P. forrestii* Craib; *P. concolor* sensu Maxim.) **Semi-evergreen shrub** 1–3 m tall with densely downy twigs. Leaflets deep green, ovate-elliptic to elliptic, 5–8 cm long, long-pointed, glabrous or downy above, densely downy beneath. **Flowers** 2.5–3 cm long, yellow. **Pods** 8–12 cm long, short-hairy. Gansu, w & swSichuan, Yunnan [cChina, Himalaya, Myanmar]; coniferous and *Quercus* forests, scrub, often on roadside banks, 1,200–4,200 m. V–VII.

Piptanthus concolor Harrow ex Craib* Differs from *P. nepalensis* in the narrow-elliptic leaflets and smaller **flowers**, 1.8–2 cm long. sGansu, Sichuan, Shaanxi, e & seTibet, nwYunnan; thickets, woodland margins, 1,600–4,000 m. IV–VII. (See also photo 8, p. 206.)

Piptanthus tomentosus Franch. (syn. *P. holosericeus* Stapf) Similar to *P. nepalensis*, but habit generally smaller, 0.4–2 m tall, **stems**, **leaves** and **inflorescences** covered in silvery hairs and **flowers** 1.8–2.2 cm long. swSichuan, nwYunnan [eHimalaya, nMyanmar]; rocky slopes, open *Pinus* and *Picea* forests on mountain slopes, moraines, 3,000–3,800 m. IV–VI.

SPONGIOCARPELLA

A genus of c. 7 species (5 in China) in the eastern Himalaya and north Myanmar. This genus is a segregate of *Chesneya* and contains prostrate, mat- or cushion-forming, stemless or short-stemmed **perennial herbs**, woody at base. **Leaves** odd-pinnate, rachis persistent, sometimes spine-like; leaflets entire, densely overlapping; stipules entire or incised, joined to rachis. **Flowers** solitary, axillary; calyx tubular to campanulate, pouched; petals clawed, standard with a small tooth on each side near middle, wings not pouched, keel fused in upper half; stamens in 2 bundles. **Pods** oblong in outline, with a spongy pericarp.

1. Flowers yellow . . . .
   . . . . . *S. yunnanensis*, *S. polystichoides*, *S. nubigena*
1. Flowers purple or red . . . . *S. paucifoliata*, *S. purpurea*

Spongiocarpella yunnanensis Yakovlev* Plant 5–20 cm tall, densely covered with non-spiny, persistent rachises. **Leaves** 2.5–3.5 cm long, with 17–21, elliptic to oblong leaflets, pointed, 3–6 × 1.5–3 mm, with dense, appressed, silky hairs on upper surface, glabrous beneath. **Flowers** solitary, stalkless, 2.5–3 cm long, yellow or yellow with a pink tinge; calyx white-hairy; petals 2.8–3.5 cm long, standard obovate, 1.5 cm broad, apex notched. **Pods** 1.5–2 cm long, hairy. nwYunnan; open rocky slopes, screes, stabilised moraines, alpine meadows, 3,200–5,000 m. VI–VIII.

Spongiocarpella polystichoides (Hand.-Mazz.) Yakovlev* (syn. *Chesneya polystichoides* (Hand.-Mazz.) Ali) **Cushion-forming** plant 10–30 cm tall, 30–60 cm across. **Leaves** 8–10 cm long, with 28–32 pairs of elliptic leaflets, 4–5 mm long, whitish beneath. **Flowers** 2.5 cm long, yellow or yellowish with a pink tinge, tinged with orange on standard. w & swSichuan, nwYunnan; rocky places, alpine meadows, screes, stabilised moraines, limestone rock-crevices, 3,000–4,800 m. VI–VIII.

Spongiocarpella nubigena (D. Don) Yakovlev (syn. *Chesneya nubigena* (D. Don) Ali) **Subshrub** to 30 cm with branching **stems**. **Leaves** pinnate with up to 15 elliptic leaflets, 4–5 × 1.5–3 mm, densely hairy. **Flowers** yellow or orange-yellow, sometimes flushed with pink, standard 3.5 cm long, keel 3 cm long. seTibet [swTibet, cHimalaya]; amongst rocks and boulders, grassy slopes, cliff-ledges, 3,600–4,600 m. VI–VII.

Spongiocarpella paucifoliata Yakovlev* **Cushion-forming** plant 10–15 cm tall, densely covered with non-spiny, persistent leaf-rachises. **Leaves** 2.5–3.5 cm long with 11–13, elliptic to oblong-elliptic leaflets, long-pointed, 5–7 × 2–4 mm, with long, appressed, silky hairs on upper surface, glabrous beneath. **Flowers** solitary, stalkless, 2.5–3 cm long, purple or dark red; calyx 12 mm long, pouched at base; petals 2.5–2.6 cm long, standard obovate, 1.5 cm broad, apex notched. **Pods** 1.5–2 cm long, hairy. nwYunnan; open rocky slopes, screes, stabilised moraines, alpine meadows, 3,900–4,200 m. VI–VIII.

---

1. *Piptanthus nepalensis*; eBhutan (CGW)
2. *Piptanthus nepalensis*; cNepal (CGW)
3. *Piptanthus concolor*; Cangshan, nwYunnan (PC)
4. *Piptanthus concolor*; Lancangjiang–Jinshajiang divide, nwYunnan (CGW)
5. *Piptanthus tomentosus*; Xiaozhongdian, nwYunnan (PC)
6. *Piptanthus tomentosus*; Lancangjiang–Jinshajiang divide, nwYunnan (CGW)
7. *Spongiocarpella yunnanensis*; Baimashan, nwYunnan (HJ)
8. *Spongiocarpella polystichoides*; Daxueshan, nwYunnan (CGW)
9. *Spongiocarpella paucifoliata*; Baimashan, nwYunnan (HJ)

Spongiocarpella purpurea (P. C. Li) Yakovlev (syn. *Chesnya purpurea* P. C. Li) Similar to the previous species, but leaflets usually 15–31. **Flowers** purple (sometimes yellow tinged purple in Himalaya), 2–2.6 cm long; calyx white-hairy. seTibet [swTibet, Himalaya]; amongst rocks and boulders, stony slopes, cliff-ledges, 3,600–5,400 m. V–VII.

ASTRAGALUS

A large genus with perhaps 3,000 species (280 in China), mainly in temperate Asia, Europe and North America. **Perennial herbs** with odd-pinnate **leaves**, leaflets entire; stipules present, free or joined. **Flowers** in axillary racemes, rarely 1–3 on a short stalk; calyx-tube campanulate, 5-toothed; petals clawed, wings slightly pouched near middle and rounded or auriculate at base, keel shorter than wings; stamens in 2 bundles. **Pods** ± inflated, oblong in outline, 2-celled.

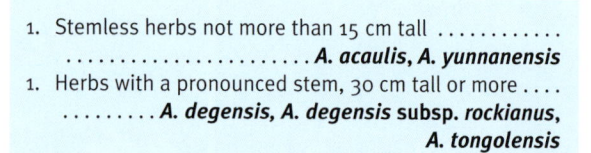

1. Stemless herbs not more than 15 cm tall .... **A. acaulis, A. yunnanensis**
1. Herbs with a pronounced stem, 30 cm tall or more ... . **A. degensis, A. degensis** subsp. *rockianus*, **A. tongolensis**

Astragalus acaulis Baker Densely tufted plant to 10 cm with spreading **leaves** 2.5–15 cm long, with 15–25 pairs of narrowly lanceolate leaflets, 5–10 mm long, long-pointed, sparsely to densely hairy; stipules membranous, persistent, broadly ovate, joined, 8–12 mm long. **Flowers** 1–3 on a short stalk; calyx-tube 7–8 mm long, white-hairy; corolla yellow, 1.5–3 cm long. **Pods** oblong in outline, slightly curved, 2.5–5 cm long. w, nw & swSichuan, e & seTibet, nwYunnan [Himalaya, Myanmar]; alpine meadows, stony pastures, 3,200–4,800 m. VI–VIII.

1. *Astragalus acaulis*; nwYunnan (HJ)
2. *Astragalus yunnanensis*; Baimashan, nwYunnan (CGW)
3. *Astragalus yunnanensis*; Huanglong, nwSichuan (CGW)
4. *Astragalus tongolensis*; swSichuan (JM)
5. *Dumasia leiocarpa*; Xishan, Kunming, cYunnan (CGW)
6. *Glycyrrhiza yunnnanensis*; cult., from Yunnan (MG)
7. *Glycyrrhiza yunnanensis* in fruit; cult., from Yunnan (MG)

Astragalus yunnanensis Franch.* Plant 8–15 cm tall with prostrate **leaves**, 6–15 cm long; leaflets 11–27, ovate, 5–10 mm long, glaucous, rachis red. Inflorescence-stalks longer than leaves, bearing lax **racemes** of 5–12 nodding **flowers**; calyx c. 14 mm long, with both densely brown and sparsely white, long hairs, teeth narrowly lanceolate, equalling tube; corolla yellow or yellow and white, 2–2.2 cm long. **Pods** oblong-ovate in outline, c. 2 cm long, with brown hairs. nw, w & swSichuan, eTibet, nwYunnan; screes, stony, alpine meadows, 3,000–4,800 m. V–VII.

Astragalus degensis **Ulbr.** **Stems** erect, 50–80 cm tall. **Leaves** 6–12 cm long, with 8–12 pairs of elliptic leaflets, oblong to oblong-lanceolate, 1–1.8 cm long, with dense appressed hairs beneath. Inflorescence-stalks longer than leaves, bearing lax, many-flowered **racemes**; calyx 4–5 mm long, toothed, with black, appressed hairs; **corolla** pale purple, 10–12 mm long, standard narrow-obovate, wings slightly shorter, entire at apex, keel almost equalling wings. **Pods** spindle-shaped to egg-shaped, 1.5–1.8 cm long, black- and white-hairy. wSichuan, eTibet [Himalaya]; *Quercus* and *Pinus* forests, subalpine meadows, 3,200–4,200 m. V–VIII. **Subsp. rockianus (E. Peter) P. C. Li** has rather larger **flowers**, corolla 13–14 mm long, calyx-teeth almost equalling tube. nwYunnan.

Astragalus tongolensis **Ulbr.** (syn. *A. potaninii* Kom.; *A. veitchianus* N. D. Simpson; *A. tongolensis* var. *glaber* E. Peter; *A. tongolensis* var. *lanceolato-dentatus* E. Peter; *A. tongolensis* var. *breviflorus* H. T. Tsai & T. T. Yu) Sparsely hairy plant 30–70(–120) cm tall, with several glabrous **stems**. **Leaves** 6–15 cm long with 3–7 pairs of narrowly elliptic to obovate, blunt leaflets, 1–6 × 0.4–2.5 cm, leaf-stalks 1–5 cm long, glabrous or very sparsely hairy; stipules 15–30 mm long, sparsely ciliate. Inflorescence-stalks 6–25 cm long, black-hairy towards the flowers; **raceme** later elongating strongly. **Flowers** 8-many, rather dense, yellow; calyx 6–9 mm long, tubular, toothed; petals of equal length, 18–24 × 6–8 mm, standard oblong to slightly obovate, somewhat constricted in middle. **Pods** with stalks 8–10 mm long, narrowly egg-shaped, long-pointed, 2–2.5 cm long, keeled beneath, flattened above. Gansu, Qinghai, Sichuan, Tibet, Yunnan [c & eHimalaya]; spruce forest, among shrubs, meadows, 3,450–4,760 m. VI–VII.

OXYTROPIS

A circumboreal genus of c. 300 species (48 in China) in Asia, Europe and North America, closely related to *Astragalus*, differing in the characteristically beaked keel.

Oxytropis kansuensis **Bunge*** **Tufted perennial** to 20 cm. **Leaves** with 10–15 pairs of ovate-lanceolate leaflets, 7–12 mm long. **Inflorescences** 9–12 cm tall, in congested racemes with many yellow **flowers**. **Pods** 15–17 mm long, somewhat inflated. Gansu, w & nwSichuan, eTibet; meadows, fields, streamsides, 3,000–3,300 m. VI–VIII.

DUMASIA

A genus with 10 species (5 in China) in Asia, Africa and Madagascar. Slender **climbers**. **Leaves** with 3 leaflets, leaflets glabrous or hairy, shortly stalked. **Inflorescence** axillary, racemose, few- to several-flowered. **Flowers** pale to deep yellow; calyx tubular. **Pods** straight, few-seeded, glabrous or hairy.

Dumasia leiocarpa **Benth.** (syn. *D. villosa* DC. var. *leiocarpa* (Benth.) Baker; (?)*D. yunnanensis* Y. T. Wei & S. K. Lee) Slender-stemmed **climber**. **Leaves** with 3 leaflets, leaflets ovate-elliptic, 2.6–4.5 × 1.3–1.5 cm, blunt. **Inflorescence** 2.5–7 cm long, 3–14-flowered. **Flowers** 13–15 mm long, yellow; calyx 6.5 mm long. **Pods** 3–3.2 cm long, apiculate, glabrous. Yunnan [Himalaya, India, Sri Lanka, Thailand]; thickets, shrubberies, 1,300–2,200 m. VIII–X.

Dumasia villosa **DC.** Differs from *D. leiocarpa* in having larger hairy leaflets, to 9.5 × 6 cm; longer **inflorescences** to 13 cm long; and densely hairy **pods** constricted between seeds. Guizhou, Sichuan, seTibet, Yunnan [swTibet, s & seAsia]; cliffs, riverbanks, roadsides, montane woods, 400–2,500 m. V–IX.

Dumasia cordifolia **Benth.** Differs from *D. leiocarpa* in having small heart-shaped leaflets; 3–6-flowered **inflorescences**; and small **flowers**, 10–12 mm long. Sichuan, seTibet, Yunnan [swTibet, neIndia]; grassland, scrub in open situations, 1,200–2,800 m. VI–IX.

Dumasia forrestii **Diels*** Differs from *D. leiocarpa* in having broadly ovate to almost circular leaflets, to 4.2 × 4 cm; 3–15-flowered **inflorescences** up to 9 cm long; **flowers** 17–18 mm long. wSichuan, seTibet, nwYunnan; montane scrub, 1,800–3,200 m. V–VI.

GLYCYRRHIZA

A genus of c. 20 species (15 in China) distributed in Europe and Asia, including *G. glabra* L., which is cultivated for liquorice. Glandular **perennial herbs** or **subshrubs** with odd-pinnate **leaves**; leaflets in 4–7 pairs, gland-dotted; stipules minute, soon falling. **Inflorescence** racemose or spike-like, laxly many-flowered to capitate. **Flowers** yellow, blue or mauve; calyx campanulate, 2-lipped; corolla with a blunt to pointed keel; stamens in 2 bundles. **Pods** smooth, glandular or spiny, 1-seeded.

Glycyrrhiza yunnanensis **Cheng f. & L. K. Tai ex P. C. Li*** **Perennial** 0.6–1.6 m tall. **Leaves** up to 13 cm long, sparsely hairy, leaflets 9–11, elliptic-ovate, 3.5–4 × 1–1.1 cm, long-pointed, short-stalked. **Inflorescences** axillary, globose clusters, 1.5–2.5 cm, on a stalk 5–6 cm long. **Flowers** pale purple or off-white, 5–7 mm long. **Pods** 1.5–2.2 cm long, bristly, clustered in a head 3.5–4 cm long. nwYunnan; stony meadows, roadsides, 2,400–3,000 m. VI–VII.

HEDYSARUM

A genus of c. 150 species (44 in China) distributed in Europe, Asia and North America. Low-growing, **rhizomatous**, **perennial herbs**. **Leaves** odd-pinnate with entire leaflets; stipules membranous, joined to rachis. **Flowers** in axillary racemes; calyx-tube campanulate, 5-toothed; petals clawed, standard long, wings with linear appendages at base, keel upwardly curved and ± pointed; stamens in 2 bundles. **Pods** compressed, constricted along both margins and divided into several 1-seeded indehiscent segments.

Hedysarum multijugum Maxim. Tufted **subshrub** 30–70 cm tall. **Leaves** 7–18 cm long, leaflets in 6–14 pairs, elliptic, broadly ovate or almost orbicular, 3–8 mm long, hairy beneath; stipules hairy. **Racemes** terminal, 12–40 cm long, many-flowered. **Flowers** rich carmine, drying purple, 15–20 mm long; calyx 5–6 mm long. **Pods** with 2–3 sections, each 3.5–4.5 mm long. Gansu, nwHubei, Qinghai, nwSichuan, Shaanxi, neTibet [nChina, Mongolia]; gravel, stony slopes, loess bluffs, banks, 500–3,200 m. VI–VIII.

Hedysarum sikkimense Benth. ex Baker (syn. *Hedysarum limprichtii* Ulbr.) **Perennial herb** with sparsely hairy, erect or ascending **stems**, 8–30 cm long. **Leaves** 5–10 cm long, leaflets 15–23, ovate-elliptic or ovate-oblong, 5–13 mm long, blunt, sparsely white-hairy beneath; stipules oblong, 5–15 mm long. **Racemes** 12–15-flowered, on stalks 8–12 cm long; calyx 6–6.5 mm long; **corolla** purple, 13–17 mm long, standard obovate, wings narrow-oblong, keel spathulate. **Pods** 7–12 mm long, with irregularly toothed margin. w & swSichuan, e & seTibet, nwYunnan [cHimalaya]; alpine gravel, alpine meadows, forests, 3,100–4,500 m. VII–VIII.

Hedysarum alpinum L. **Perennial** to 1 m, generally much less, with erect to ascending, glabrous to somewhat hairy **stems**. **Leaves** with up to 14 pairs of oblong to lanceolate leaflets, 10–30 × 4–10 mm. **Racemes** usually 20–30-flowered; calyx with short teeth; **corolla** reddish violet, 12–15 mm long. **Pods** with 2–6 hairy segments and a narrow membranous margin. Gansu, Qinghai, n & nw Sichuan [nChina, eEurope, nAsia]; meadows, open shrubberies, field boundaries, roadsides, 1,600–3,800 m. VI–VIII.

LOTUS

A genus of 100 or more species (8 in China) in temperate Europe and Asia, North & South America and Australia. **Perennial herbs** with prostrate or ascending **stems**. **Leaves** with 3 leaflets together with a pair of leaflet-like stipules at base, crowded at apex of short rachis; stipules minute or absent. **Flowers** in lateral umbels; calyx-tube campanulate, 5-toothed; petals free from staminal tube, standard almost orbicular, keel beaked; stamens in 2 bundles, with alternate filaments broadened at apex. **Pods** linear, terete, constricted between the seeds.

Lotus corniculatus L. Spreading to ascending **herb**; **stems** 15–30 cm long, sparsely hairy. **Leaflets** oblanceolate or obovate, mucronate, 1–2 cm long. **Flowers** 4–8, in compact umbels, yellow, often tipped with orange or red; standard 11–13 mm long; wings and keel 10 mm long. **Pods** 18–20 cm long. Gansu, Guizhou, Hubei, w, nw & swSichuan, Yunnan [c & nChina, c & nAsia, Europe, Himalaya]; meadows, rocky hillsides, banks, roadsides, open shrubberies, weed of cultivation. 1,800–4,000 m. V–VII.

PAROCHETUS

A genus of 2 species, one in Africa, the other in Asia. Low-growing, **perennial herbs** rooting at lower nodes. **Leaves** with 3 leaflets; stipules present, free or shortly joined to rachis. **Flowers** solitary or paired on long lateral stalks; calyx-tube campanulate, 5-toothed; petals clawed, standard obovate, wings oblong, keel shorter than wings, hooked at tip; stamens in 2 bundles. **Pods** linear-oblong in outline, ± inflated, with 8–20 seeds.

Parochetus communis D. Don Glabrous or sparsely hairy **perennial**, patch-forming, 10–20 cm tall. **Leaflets** obovate, 5–20 mm long, sometimes with a black band near middle, stalks 3–25 cm long. **Flowers** bright blue, standard obovate, 15–22 mm long, wings and keel 10–15 mm long. **Pods** 15–20 mm long. Sichuan, seTibet, Yunnan [s & seAsia]; damp places, woodland margins, streamsides, edges of rice paddies, weed of cultivation, 1,500–3,000 m. II–IX.

THERMOPSIS

A genus of c. 35 species (12 in China) in temperate regions, mostly montane Europe, Asia and North America. Usually low-growing, tufted **perennial herbs** with a woody rootstock with erect annual **stems**, branching from base. **Leaves** with 3 leaflets; stipules present, free, persistent. **Flowers** in 2s or 3s in terminal racemes; calyx-tube campanulate, 5-toothed; petals long-clawed, standard erect, broadly ovate or almost orbicular, wings oblong, keel slightly curved, longer than standard; stamens in 2 bundles. **Pods** oblong in outline, inflated or not, lacking divisions.

1. Flowers yellow to pale orange; lower leaves short-stalked ...... *T. smithiana*, *T. alpina*, *Thermopsis* sp.
1. Flowers blackish purple or dark red-brown; lower leaves stalkless .... *T. barbata*

Thermopsis smithiana Peter-Stibal Plant 15–20 cm tall, densely pale-hairy. **Leaflets** elliptic, 10–12 × 5–8 mm, densely covered in silvery hairs. **Flowers** yellow; calyx c. 10 mm long; petals 18–20 mm long. **Pods** inflated, elliptic, 4 × 2 cm, silvery-hairy. Qinghai, w & swSichuan, Tibet, nwYunnan [eHimalaya, Myanmar]; grassy alpine hillsides and meadows, riverbanks, scrub, *Pinus* forests, 3,200–4,600 m. V–VII.

***Thermopsis alpina* Ledeb.** (syn. *T. inflata* Cambess.) Similar to the previous species, but plant creeping or prostate, 10–30 cm tall with larger oblong to obovate **leaflets**, 15–30 × 8–16 mm, blunt, softly hairy; **flowers** with petals 20–25 mm long. **Pods** not inflated, oblong, 3–4.5 cm long. Gansu, Sichuan, nwYunnan [c & eAsia]; alpine meadows, open places in *Picea* forests, screes, 2,250–4,200 m. V–VII.

Thermopsis* sp. Like *T. smithiana*, but **leaves** less hairy, grey-green, leaflets narrow-elliptic and somewhat longer; **flowers** yellow, ageing to orange-brown, inflorescence surrounded by vegetative lateral shoots. nwSichuan.

***Thermopsis barbata* Royle** (syn. *T. atrata* Cefr.) Tufted **herbaceous perennial** 15–30 tall, covered in dense pale silvery hairs, **stems** elongating to 50 cm in fruit. **Leaflets** elliptic-oblanceolate, 2–3 cm long, pointed, only partially developed at flowering time. **Flowers** blackish purple; calyx 1 cm long; standard notched, 20–23 mm long, wings and keel oblong, 25–28 mm long. **Pods** 3–5 cm long, silvery-hairy. swQinghai, Sichuan, seTibet, nwYunnan [swTibet, Himalaya]; grassy hillsides, gravelly meadows, roadsides, 3,200–4,600 m. V–VII.

1. *Hedysarum sikkimense*; south of Aba, nwSichuan (CGW)
2. *Hedysarum alpinum*; south of Aba, nwSichuan (CGW)
3. *Lotus corniculatus*; cult. (CGW)
4. *Thermopsis smithiana*; Zhongdian Plateau, nwYunnan (HJ)
5. *Thermopsis alpina*; Daxueshan, nwYunnan (CGW)
6. *Thermopsis* sp.; nwSichuan (JM)
7. *Thermopsis barbata*; Zhongdian Plateau, nwYunnan (CGW)

1. *Tibetia yunnanensis*; Napahai, Zhongdian, nwYunnan (PC)
2. *Tibetia himalaica*; Baimashan, nwYunnan (CGW)
3. *Tibetia tongolensis*; Zhongdian Plateau, nwYunnan (HJ)
4. *Tibetia coelestis*; Lancangjiang–Jinshajiang divide, Lidiping, nwYunnan (CGW)
5. *Vicia unijuga*; Xishan, Kunming, cYunnan (CGW)
6. *Vicia cracca*; Huanglong, nwSichuan (CGW)
7. *Vicia amoena*; nwSichuan (HJ)

TIBETIA

A genus of 9 species (all found in China) in temperate Asia: it was formerly included in *Gueldenstaedtia*. Low-growing perennial **herbs** with thickened rootstocks and odd-pinnate **leaves**; leaflets oval, orbicular or reverse heart-shaped; stipules membranous, broadly clasping the leaf-stalk. **Flowers** 1–3, umbellate, violet, purple or yellow; calyx-tube 2-lipped, 5-toothed, hairy; petals clawed, standard broadly obovate, notched, wings with linear appendages at base, keel upwardly curved, pointed; stamens in 2 bundles. **Pods** compressed, constricted along both margins.

1. Flowers yellow .... ***T. yunnanensis***
1. Flowers blue, violet or purple 2
2. Ovary and pods hairy; stipules pointed ....
   .... ***T. himalaica, T. tongolensis***
2. Ovary and pods glabrous; stipules blunt ... ***T. coelestis***

Tibetia yunnanensis (Franch.) H. P. Tsui* (syn. *Gueldenstaedtia yunnanensis* Franch.) Small, tufted, brown-hairy **herb** with **leaves** up to 10 cm long; leaflets 5–9, 5–20 mm long, slightly notched at apex, softly hairy. **Inflorescence** 6–7 cm long, 2–3-flowered. **Flowers** cream or yellow, 13–14 mm long. w & swSichuan, nwYunnan; shady banks, grassy mountain slopes, meadows, disturbed ground, 3,000–3,500 m. IV–VII.

Tibetia himalaica (Baker) H. P. Tsui (syn. *Gueldenstaedtia himalaica* Baker) **Tufted perennial** to 10 cm, with dense soft red-brown hairs. **Leaves** 2–15 cm long, leaflets 9–17, obovate, 3–10 mm long, apex notched; stipules membranous, ovate, 6–7 mm long. **Inflorescence** 2-flowered. **Flowers** dark purple or violet, 8–10 mm long, on a stalk 1.5–9 cm long. **Pods** c. 1.5 cm long, sparsely hairy. w & swSichuan, seTibet, nwYunnan [swTibet, nwChina, Himalaya, Mongolia, eRussia]; heathland, stony meadows, rocky slopes, sandy soils, 2,800–5,000 m. VI–VII.

Tibetia tongolensis (Ulbr.) H. P. Tsui* (syn. *Gueldenstaedtia tongolensis* Ulbr.; *G. flava* Adamson; *G. flava* var. *tongolensis* (Ulbr.) Ali.) Taller **herb** than previous species, with **stems** to 20 cm, arising from prostrate mats. **Leaflets** glaucous, rounded to ovate, sparsely hairy (usually 7–11), 7–11 mm long. **Flowers** violet with a white basal spot on standard, 12–14 mm long, solitary or paired, on long stalks well above the foliage. wSichuan, eTibet, nwYunnan; meadows, forest margins 3,200–4,100 m. V–VII.

Tibetia coelestis (Diels) H. P. Tsui* (syn. *Astragalus coelestis* Diels; *Gueldenstaedtia coelestis* (Diels) N. D. Simpson; *Tibetia tongolensis* (Ulbr.) H. P. Tsui var. *coelestis* (Diels) H. P. Tsui) **Tufted perennial** to 20 cm tall. **Leaves** 6–9 cm long, usually with 7, elliptic-oval to obovate, slightly notched leaflets. **Inflorescence** 2–4-flowered, rarely more. **Flowers** pale blue-purple. w & swSichuan, nwYunnan; meadows, grassy mountain slopes, 3,000–3,600 m. VI–VII.

VICIA

A genus of c. 140 species (22 in China) primarily in the north temperate zone, but a few in South America. **Annual, biennial** or **perennial**, eglandular **herbs** with even-pinnate **leaves** usually terminating in a simple or branched tendril; stipules triangular, persistent. **Inflorescences** axillary or terminal; **flowers** solitary or 2–40 in clusters or racemes; calyx tubular to campanulate, 2-lipped or 5-toothed; petals clawed, standard broad, wings free or joined to curved keel; stamens in 1 or 2 groups. **Pods** 1–many-seeded, compressed.

1. Plants without tendrils; leaves with 3 leaflets ....
   .... ***V. unijuga***
1. Plants with tendrils; leaves pinnate ....
   .... ***V. cracca, V. nummula, V. amoena***

Vicia unijuga A. Braun Vigorous tufted herbaceous **perennial** to 0.5 m, with erect to ascending **stems**. **Leaves** without tendrils, with a pair of elliptic leaflets to 5 cm long. **Flowers** semi-nodding, in long-stalked, lateral racemes of up to 26; calyx with uneven teeth; corolla usually blue-purple, 12–17 mm long. Gansu, Qinghai, nSichuan, Shaanxi, Yunnan [nChina, Japan, Korea, Mongolia, eRussia]; grassy places, open thickets, forest margins and clearings, sometimes cultivated, 1,200–4,000 m. VI–VIII.

Vicia cracca L. Very variable, slender, **clambering** or **climbing perennial** to 1.5 m. **Leaves** with 8–16 pairs of ovate-oblong to linear leaflets and branched tendrils; stipules entire. **Inflorescences** dense 1-sided racemes, 10–40-flowered. **Flowers** violet to bluish purple; calyx 3–6 mm long; corolla 10–18 mm long. **Pods** 2–3 cm long, glabrous, black at maturity. Gansu, Hubei, Sichuan, Yunnan [n & neChina, temperate Europe, Asia, North America]; waste places, grassland, scrub, roadsides, introduced elsewhere as a weed of cultivation, widely naturalised, 1,200–3,600 m. V–VIII.

Vicia nummula Hand.-Mazz.* **Perennial herb**, 15–50 cm tall with a much-branched, hairy **stem**. **Leaves** pinnate, 3.5–8 cm long with slender branched tendrils; leaflets in 2–7 pairs, elliptic, 4–13 × 2–6 mm, densely hairy beneath. **Racemes** 2–4 cm long with 6–9 yellow **flowers**; corolla 8–9 mm long. **Pods** 20–25 mm long, pointed, glabrous. Gansu, Sichuan, Yunnan; riverbanks, grassy places, 2,000–2,300 m. VI–VIII.

Vicia amoena Fisch. Very variable, slender, erect **perennial** with 1 or several **stems**. **Leaves** pinnate, tendrils simple or few-branched; leaflets in 5–7 pairs, elliptic, 2–4 cm long; stipules entire. **Inflorescences** dense 10–30-flowered racemes. **Flowers** purple, pale to deep blue or violet-blue, 12–18 mm long. Gansu, Qinghai, Sichuan, eTibet [n, c & eChina, Japan, Korea, Mongolia]; ravines, grassland, riverbanks, scrub, woodland, forest margins, to 3,300 m. V–VIII.

CACTACEAE

OPUNTIA

A genus with more than 200 species (4 naturalised in China) in the Americas. Most have fleshy, elliptic to oval, segmented **stems** adorned with clusters of rigid, needle-like spines (or sometimes stems ± smooth). **Fruit** large, pear-shaped, yellowish, pinkish or reddish when ripe and often edible.

Opuntia monocantha **(Willd.) Haworth** Rather erect **shrub** to 2 m, often with a short **trunk** at maturity, with oblong to obovate stem-segments 10–30 cm long and a third as wide, spines single or paired, to 4 cm long, surrounded by smaller spines. **Flowers** yellow or orange-yellow, 7.5–10 cm across, outer tepals often flushed with red. **Fruit** pear-shaped, 5–7.5 cm long, red-purple when ripe. Naturalised in many warm dry areas from s Sichuan southwards [Brazil to Argentina], to c. 1,400 m. III–VI.

LYTHRACEAE

LYTHRUM

A cosmopolitan genus with c. 35 species (2 in China). **Annual** and **perennial herbs** with 4-angled young **stems**. **Leaves** alternate, opposite or whorled. **Flowers** in whorls making up a spike- or raceme-like **inflorescence**. **Flowers** usually 6-parted; calyx with an epicalyx; petals often pink or purple; stamens usually 12. **Fruit** a 2-valved capsule, within the persistent flower-tube.

Lythrum salicaria **L.** Erect, **tufted perennial** to 1.5 m, with lanceolate, untoothed **leaves**, opposite or in 3s, to 6 cm long. **Flowers** 10–15 mm across, in slender spikes, sometimes branched; calyx green or purplish, tubular, with alternating triangular and awl-shaped lobes; petals usually 6, bright rose-purple; stamens 12. Gansu, Hubei, Qinghai, Sichuan, Shaanxi, Yunnan [n, c & eChina, temperate northern hemisphere]; marshy habitats, river and lake margins, 900–2,200 m. VI–IX.

WOODFORDIA

A genus of just 2 species, one in Africa and Arabia, the other in the Himalaya and South East Asia. **Shrubs** or small **trees** with rather irregular **branches** and opposite, almost sessile **leaves**. **Flowers** on condensed lateral branches, 6-parted, slightly zygomorphic, with a red tube; petals small, pink, red or white; stamens 12. **Fruit** a small capsule.

Woodfordia fruticosa **(L.) Kurz** **Evergreen shrub** to 1.5 m with slender, spreading **branches** and opposite, lanceolate **leaves**, 5–10 cm long, pointed, entire. **Flowers** in small, dense, lateral clusters, often on lower leafless branches; calyx tubular, bright red, 9–13 mm long, with 6 triangular lobes; petals red, small, ± equalling calyx-lobes; stamens protruding. **Fruit** a small capsule 6–10 mm long. Guangxi, Yunnan [Guangdong, seAsia, Himalaya, nIndia, Myanmar]; open forests and forest margins, shrubberies, 300–1,800 m. II–V.

THYMELAEACEAE

DAPHNE

A Eurasian genus with almost 100 species (44 in China). **Evergreen** or **deciduous shrubs** with usually alternate, sometimes opposite, untoothed **leaves**. **Flowers** in terminal or lateral clusters or short racemes, occasionally paniculate, hermaphrodite or unisexual (plants sometimes dioecious), generally sweetly scented; calyx corolla-like, saucer-shaped, with a short to long tube and 4(–5) spreading lobes; true petals absent; stamens 8, or 10, in 2 series, attached to calyx-tube. **Fruit** a succulent berry or dry, sometimes surrounded by the persistent dry calyx-tube, poisonous.

1. Flowers yellow or greenish yellow 2
1. Flowers pink, purplish, purplish blue or white 4
2. Evergreen plants with opposite leaves; fruit a dryish inconspicuous berry, not red ....
 .... **D. aurantiaca, D. calcicola**
2. Deciduous plants with alternate leaves; fruit a succulent shiny red berry 3
3. Flowers 4-parted .... **D. giraldii**
3. Flowers 5-parted ....
 .... **D. rosmarinifolia, D. gemmata, D. angustiloba**
4. Leaves opposite; inflorescences lateral .... **D. genkwa**
4. Leaves alternate; inflorescences primarily terminal (sometimes with additional lateral clusters) 5
5. Calyx-tube downy outside ....
 .... **D. bholua, D. bholua** var. **glacialis**
5. Calyx-tube glabrous outside .. **D. retusa, D. tangutica,**
 D. wolongensis, D. acutiloba, D. longilobata

---

1. *Opuntia monocantha*; north of Dali, nwYunnan (CGW)
2. *Lythrum salicaria*; Jiuzhaigou, nwSichuan (CGW)
3. *Lythrum salicaria*; Jiuzhaigou, nwSichuan (CGW)
4. *Woodfordia fruticosa*; Bhutan (CGW)
5. *Woodfordia fruticosa*; Bhutan (CGW)
6. *Daphne aurantiaca*; Baishui, Yulongxueshan, nwYunnan (CGW)
7. *Daphne aurantiaca*; Baishui, Yulongxueshan, nwYunnan (CGW)
8. *Daphne calcicola*; Napahai, Zhongdian, nwYunnan (CGW)
9. *Daphne calcicola*; Gangheba, Yulongxueshan, nwYunnan (CGW)

THYMELACEAE: Daphne

Daphne aurantiaca **Diels*** Erect, twiggy, rather open **shrub** to 1.5 m, with reddish brown young branchlets. **Leaves** thin, deep bluish green, crowded towards branchlet-tips, obovate to elliptic, to 2.3 × 1.2 cm, narrowed at base into a very short stalk and with a somewhat back-rolled margin, glabrous but often white-powdery. **Flowers** in terminal cluster of 2–5, fragrant, generally orange-yellow, accompanied by several leaf-like bracts; calyx-tube cylindrical (but widening towards top), 8–12 mm long, glabrous, lobes broad-ovate, 2–3 mm long. swSichuan, nwYunnan; open forests (often of *Pinus*), rocky and scrubby slopes, rocky places, on limestone, 2,600–3,500 m. V–VI.

Daphne calcicola **W. W. Sm.*** Closely related to the previous species, but a low, dense spreading, often hummock-forming **shrub** to 60 cm, often 1–2 m across, with oval **leaves** and bright yellow **flowers**, sometimes pale primrose-yellow; calyx-tube broader and shorter, 6–8 mm long, lobes relatively larger, 3–4.5 mm long. Similar distribution and habitats, but often also growing on bare rocks or cliffs. V–VI. Much confused with the previous species; intermediates can be found in the wild.

Daphne giraldii* Nitsche Erect twiggy **shrub** to 70 cm with glabrous branchlets that are orange-yellow at first. **Leaves** crowded towards stem-tips, thin, oblanceolate, to 6 × 1.2 cm, grey-green above, white-downy beneath. **Flowers** in terminal clusters of up to 8, with very short stalks, yellow or yellowish green, faintly fragrant, calyx-tube cylindrical, 6–8 mm long, glabrous, lobes ovate, 3–5 mm long. **Berry** succulent, egg-shaped, 5–6 mm long. Gansu, e & seQinghai, n, nw & neSichuan; forest margins, shrubberies, 1,600–2,600 m. VI–VII.

Daphne rosmarinifolia* Rehd. (syn. *Wikstroemia rosmarinifolia* (Rehd.) Domke; *Daphne clivicola* Hand.-Mazz.) Rather densely branched and irregular **evergreen shrub** to 1 m tall, generally less, with greyish branchlets and small, papery, linear-oblong to lanceolate, deep green **leaves**, to 1.8 × 0.4 cm, with back-rolled margins. **Flowers** yellow, in tight, few-flowered, terminal clusters, calyx-tube cylindrical, 8–10 mm long, glabrous, lobes ovate, 3–5 mm long. swGansu, seQinghai, w & nwSichuan; open shrubberies, forest margins, 2,500–3,800 m. IV–VI.

Daphne gemmata* E. Pritz (syn. *Wikstroemia gemmata* (E. Pritz) Domke) Readily distinguished from previous species by the finely hairy young branchlets, larger obovate, shiny, deep green **leaves**, 3–8 × 0.6–2.2 cm, and ebracteate shortly spike-like **inflorescence**; calyx-tube slender, 10–14 mm long, downy, lobes oval, 4–5 mm long. w & nwSichuan; hillslopes, shrubberies, 400–1,500 m. IV–IX.

Daphne angustiloba* Rehd. (syn. *Wikstroemia angustiloba* (Rehd.) Domke) Similar to the previous species but rarely more than 50 cm tall, with linear-oblong, glabrous **leaves**, to 3.5 × 0.7 mm, and **flowers** in small clusters (not spike-like), calyx-tube 10–11 mm long, lobes ovate, not more than 3.5 mm long. nwSichuan; open forests and forest margins, 2,100–2,700 m. IV–VI.

Daphne genkwa* Sieb. & Zucc. Small **deciduous**, twiggy **shrub** to 80 cm, occasionally more, with purplish brown or purplish red branchlets. **Leaves** paired, grey-green, thin, ovate to lanceolate, to 40 × 20 mm, grey-hairy. **Flowers** in lateral clusters of 3–6, bluish purple to lilac-blue or purple, calyx-tube 6–10 mm long, lobes oval, 5–6 mm long. **Berry** white, succulent, egg-shaped, c. 4 mm long. Gansu, Guizhou, Hubei, Sichuan, ne & eShaanxi [c & eChina]; open forests, shrubberies, 300–1,000 m. V–VII.

***Daphne bholua* Buch.-Ham.** Upright, much-branched **shrub** to 2.5 m with dark reddish brown young branchlets. **Leaves** thick, shiny deep green, narrow-elliptic to oblong-lanceolate, glabrous, with a short stalk and a flat or slightly back-rolled margin. **Flowers** in terminal and lateral clusters usually of up to 12 flowers, very fragrant, white, pink to purplish pink or reddish purple, paler inside; calyx-tube cylindrical, 8–10 mm long, lobes ovate, 5–6 mm long. **Berry** black, egg-shaped, 6–7 mm long. seTibet, n & nwYunnan [swTibet, c & eHimalaya. Myanmar]; forests, forest glades, shrubberies, 1,800–3,000 m. III–V. Lower-growing plants with spreading branches, leaves with fewer lateral veins (6–11 pairs) and generally paler and smaller **flowers** are referable to **var. glacialis (W. W. Sm. & Cave) B. L. Burtt**.

***Daphne retusa* Hemsl.** Dense, rounded to erect **shrub** to 1 m, often less, with short yellowish brown branchlets. **Leaves** deep green, leathery, crowded towards stem-tips, oblong to elliptic-lanceolate, to 4 × 1.4 cm, with a slightly back-rolled margin and a blunt, shallowly notched tip, glabrous. **Flowers** pink to purplish red, occasionally white, fragrant, in dense terminal clusters; calyx-tube cylindrical, 6–8 mm long, glabrous, ovate lobes about the same length. **Berry** succulent, red, 6–7 mm in diameter. Gansu, Hubei, Qinghai, Sichuan, n, w & nwYunnan [c & eHimalaya, Myanmar]; rocky slopes, open shrubberies, rocky meadows, 3,000–3,900 m. IV–V.

Daphne tangutica* Maxim. Very similar to the previous species, but a more upright **bush** to 2 m, with greyish yellow young branchlets and generally larger lanceolate to oblanceolate **leaves**, to 8 × 1.7 cm, apex blunt but not notched. **Flowers** purplish pink to purplish red, whitish inside, calyx-tube 9–13 mm long, lobes 5–8 mm long, clearly shorter than tube. Gansu, Guizhou, Qinghai, Sichuan, Shaanxi, e & seTibet, nwYunnan; forests and forest margins, shrubberies, rocky places, 1,000–3,800 m. IV–V, occasionally with some autumn flowers. Forms from wHubei, eSichuan and sShaanxi with larger **leaves** (to 10 × 2.2 cm) with a more long-pointed apex and flat margins, as well as more purplish or brownish young **shoots**, are referable to **var. wilsonii (Rehd.) H. F. Zhou ex C. Y. Chang*** (syn. *D. wilsonii* Rehd.).

Daphne wolongensis* Brickell & Mathew Similar in general appearance to the preceding 2 species, but distinct in the more upright habit, to 1.5 m, in narrower elliptic to elliptic-oblanceolate **leaves**, to 6 × 1.6 cm, with a pointed, slightly apiculate apex, and most obviously in the lateral clusters of **flowers**; calyx pink, deeper in bud, intensely fragrant, tube c. 8 mm long, lobes about the same length. wSichuan (Wolong and Baoxing valleys); rocky and shingly place close to stream and roadsides, 1,900–2,000 m. IV–V.

---

1. *Daphne giraldii*; Huanglong, nwSichuan (CGW)
2. *Daphne giraldii* in fruit; cult. (CGW)
3. *Daphne rosmarinifolia*; north of Baoxing, wSichuan (DC)
4. *Daphne gemmata*; cult., from Sichuan (DJ)
5. *Daphne gemmata*; Wolong, wSichuan (CGW)
6. *Daphne genkwa*; cult. (CGW)
7. *Daphne bholua*; eBhutan (CGW)
8. *Daphne retusa*; Gangheba, Yulongxueshan, nwYunnan (CGW)
9. *Daphne retusa*; Baimashan, nwYunnan (PC)
10. *Daphne tangutica*; cult. (CGW)
11. *Daphne wolongensis*; Wolong, wSichuan (DC)

Daphne acutiloba **Rehd.*** Distinctive erect **shrub** to 2 m, readily distinguished from the previous 3 species by the purplish red branchlets and white **flowers** with sharply pointed calyx-lobes; tube 9–12 mm long, lobes narrow-ovate, 5–6 mm long. Hubei, Sichuan, w & nwYunnan; forests, shrubberies, 1,400–3,000 m. IV–V.

Daphne longilobata **(Lecompte) Turrill*** Very similar to *D. acutiloba*, but bark purple-brown and **leaves** more papery. Calyx downy outside and with rather unequal lobes. swSichuan, e & seTibet, nwYunnan; forests, forest margins, shrubberies, rocky places, 1,600–3,500 m. VI–VII.

WIKSTROEMIA

A genus with c. 70 species (44 in China) in South East Asia and Pacific islands. **Evergreen** or **deciduous shrubs** closely allied to *Daphne*, but differing primarily in having a terminal racemose, spike-like or paniculate **inflorescence**, and usually alternate **leaves** (in the species listed below, the leaves are opposite); calyx-lobes 4–5, stamens 8 or 10.

Wikstroemia indica **(L.) C. A. Mey.** (syn. *Daphne indica* L.) **Shrub** to 2 m, often far less, with reddish brown, glabrous branchlets and obovate to elliptic-lanceolate opposite **leaves** to 5 × 1.5 cm, with oblique lateral veins; leaf-stalks very short, c. 1 mm long. **Flowers** in small terminal heads on a stalk 5–10 mm long; calyx glabrous, greenish yellow, 7–12 mm long, with 4 ovate or oblong lobes; stamens 8. **Berry** ellipsoid, 7–8 mm long, dark purple or red at maturity. Guangxi, Guizhou, Sichuan, Yunnan [Himalaya, India]; rocky shrubberies and forests, 200–1,500 m. VII–IX.

Wikstroemia micrantha **Hemsl.*** Differs from the previous species in the more leathery, more oblong **leaves** with somewhat back-rolled margins, grey-green beneath, in the more racemose or paniculate **inflorescence** and in smaller yellow **flowers**, the calyx not more than 6 mm long. **Berry** blackish purple when mature. Gansu, Guizhou, Hubei, Sichuan, Yunnan; shrubby slopes and valley bottoms and margins, road- and riversides, 250–1,000 m. IX–XI.

Wikstroemia scytophylla **Diels*** Readily distinguished from the previous 2 species by the 5-lobed yellow calyx c. 11 mm long, and 10 stamens; **flowers** in racemes on a stalk 2–4 cm long. s, w & swSichuan, e & seTibet, n, w & nwYunnan; dry rocky and shrubby slopes, 1,900–2,900 m. VI–IX.

ERIOSOLENA

A genus of 2 species (1 in China) ranging from the east Himalaya to South East Asia. Rather like *Wikstroemia*, but flowers in lateral, stalked **inflorescences**; calyx-lobes 4.

Eriosolena aff. ***composita*** **(L. f.) Merr.** A *Daphne*-like **evergreen shrub** or small **tree** to 4 m with alternate, elliptic to oblanceolate **leaves**, to 7 × 2 cm. **Inflorescence** a small, lateral, long-stalked head to 25 mm across, spreading to somewhat pendent, in bud enclosed in an involucre of bracts. **Flowers** white to greenish yellow, calyx-tube downy outside. sSichuan, s & swYunnan [eHimalaya, nMyanmar]; forest margins, shrubberies, ravines, 1,300–1,800 m. V–VI.

EDGEWORTHIA

A genus of 5 Asian species (4 in China), closely related to *Daphne*, but **flowers** in dense, stalked heads with a ruff (involucre) of several small bracts; stamens 8 in 2 series; style long and cylindrical (not short and capitate as in *Daphne*).

Edgeworthia gardneri **(Wall.) Meisn.** Large **deciduous shrub** or small **tree** to 4 m with brownish red branchlets. **Leaves** alternate, aggregated towards shoot-tips, elliptic to lanceolate, pointed, to 10 × 3.5 cm, with 8–9 pairs of conspicuous lateral veins, finely hairy. **Flowers** in pendent globose heads 35–40 mm across, with up to 50 stalkless flowers; calyx white-downy outside, lobes yellow inside, tube cylindrical, 9–12 mm long with 4 spreading, ovate lobes 3–4 mm long. **Fruit** a dry 'berry', enclosed in the base of calyx-tube. seTibet, nwYunnan [c & eHimalaya]; forests, shrubberies, 1,000–2,500 m. XI–III.

STELLERA

A genus of c. 12 species (only 1 in China) in Central Asia and the Himalaya. Closely related to *Daphne*, but distinguished by the herbaceous habit (in the species included below), in the articulation of the calyx-tube just above the ovary, and by the **fruit** which is a dry nutlet.

Stellera chamaejasme **L.** (syn. *S. chamaejasme* var. *angustifolia* Diels) Very variable **clump-forming herbaceous perennial** with stout **taproot**, with several to numerous, erect to ascending, simple **stems** to 50 cm. **Leaves** green to grey-green, sometimes purple-flushed, alternate to almost opposite or pseudowhorled, thin, lanceolate to oblong-lanceolate or ovate, sometimes linear-lanceolate, to 30 × 10 mm, margin entire, sometimes slightly back-rolled, apex pointed to blunt. **Flowers** numerous in tight globose clusters with a ruff of leaf-like bracts immediately below, sweetly scented; calyx white, pink, red, purplish, yellow or greenish in bud and on outside, white inside, tube cylindrical, 9–11 mm long, lobes 5, ovate to oblong, spreading, 2–4 mm long. **Fruit** like a greyish rice grain hidden at base of persistent calyx-tube, soon maturing. Gansu, Hubei, Qinghai, Sichuan, Shaanxi, Tibet [n & nwChina, Himalaya, Mongolia, Russia]; dry grassy steppes and meadows, open shrubberies, generally on very well-drained soils, often gregarious, 2,600–4,200 m. IV–VII. **Var. chrysantha (Ulbr.) Grey-Wilson** is readily distinguished by the **flowers** that are bright yellow inside and outside, although the **buds** can be red or purplish. (?)seTibet, w & nwYunnan; similar habitats, although often rather damper. (See also photo on p. 28.)

1. *Daphne longilobata*; cult. (CGW)
2. *Wikstroemia micrantha*; lower Baimashan, nwYunnan (CGW)
3. *Wikstroemia scytophylla*; above Benzilan, nwYunnan (DC)
4. *Eriosolena* aff. *composita*; Muli, swSichuan (PC)
5. *Edgeworthia gardneri*; cult. (JL)
6. *Stellera chamaejasme* var. *chamaejasme*; south of Aba, nwSichuan (CGW)
7. *Stellera chamaejasme* var. *chamaejasme*; Minjiang valley near Songpan, nwSichuan (CGW)
8. *Stellera chamaejasme* var. *chrysantha*; Bitahai, Zhongdian, nwYunnan (CGW)
9. *Stellera chamaejasme* var. *chrysantha*; near Lancangjiang–Jinshajiang divide, Lidiping, nwYunnan (CGW)

PUNICACEAE

PUNICA

A genus of 2 species, one in Socotra, the other in south-west Asia. **Deciduous shrubs** or small **trees**, often spiny, with entire **leaves**, opposite or almost so. **Flowers** solitary, terminal, with a thick flower-tube; petals 5–6; stamens numerous. **Fruit** large, berry-like, leathery, with **seeds** embedded in fleshy pulp.

Punica granatum L. **Deciduous shrub** or small **tree** to 8 m, generally somewhat spiny. **Leaves** opposite, oblong, to 7.5 × 2.5 cm. **Flowers** scarlet-red, 25–38 m across, solitary or paired on short lateral shoots; calyx thick and fleshy, funnel-shaped, 5-lobed; petals oval to obovate, rather crumpled, 5 or more; stamens numerous, attached to rim of calyx. **Fruit** globose, to 12 cm across, yellow, with red flush when ripe, filled with numerous, pulpy red **seeds**. Widely cultivated pomegranate and sometimes naturalised along roads or on hillslopes [w & cAsia]. VI–IX.

ONAGRACEAE

EPILOBIUM

A genus of 200 species (33 in China), widely cosmopolitan in temperate and Arctic regions. **Annual** or **perennial herbs** with alternate, opposite or whorled **leaves**. **Flowers** solitary, or in spikes or racemes; sepals and petals 4; stamens 8; stigma club-shaped or 4-lobed. **Fruit** a slender capsule, splitting into 4 lengthwise and containing numerous fluffy **seeds**.

Epilobium conospermum Hausskn. (syn. *E. reticulatum* C. B. Clarke) **Patch-forming** or **tufted perennial** to 60 cm, with alternate, narrow-elliptic to elliptic-oblanceolate **leaves**, to 10 × 2 cm, with diverging veins, margin slightly toothed. **Flowers** out-facing to nodding, 24–32 mm across, in a lax, leafy **raceme**; sepals purplish, elliptic, slightly longer than petals which are rose-pink with deeper veins, rounded and somewhat overlapping. seTibet, nwYunnan [c & eHimalaya, nMyanmar]; rocky places, shrubberies, streamsides, screes, moraines, 3,600–4,500 m. VII–IX.

Epilobium pannosum Hausskn. **Herbaceous perennial** to 70 cm with erect, unbranched **stems**. **Leaves** alternate, stalkless, elliptic to elliptic-ovate, to 5 × 2.2 cm, almost pointed, silvery-grey-hairy, with finely toothed margin, decreasing in size up the stem and grading into the bracts. **Inflorescence** racemose, flower-stalks and ovaries downy. **Flowers** semi-nodding; sepals triangular-lanceolate, purplish, shorter than petals; petals deep pink, obovate, 8–14 mm long, apex notched. Guizhou, s & wSichuan, Yunnan [neIndia, Myanmar, Vietnam]; rocky and grassy slopes, forest margins, 2,600–3,800 m. VII–X.

CHAMERION

A genus of 8 species (4 in China) scattered across the temperate and Arctic northern hemisphere. Closely related to *Epilobium* but readily distinguished by alternate **leaves** and somewhat zygomorphic **flowers** that lack a flower-tube and have a deflexed style.

Chamerion angustifolium (L.) Holub (syn. *Chamaenerion angustifolium* (L.) Scop.; *Epilobium angustifolium* L.) Rather robust **patch-forming perennial** to 2 m with alternate, lanceolate, slightly toothed **leaves**, with a marginal vein. **Inflorescence** a long tapered raceme, many-flowered. **Flowers** 20–30 mm across; sepals narrow-lanceolate, spreading; petals violet to rose-purple, ovate, apex slightly notched, upper 2 broader than lower. Gansu, Hubei, Qinghai, Sichuan, Shaanxi, Tibet, Yunnan [widespread in northern hemisphere]; hillslopes, rocky places, waysides, river margins, abandoned cultivation fields, 2,200–4,300 m. VI–IX.

MELASTOMATACEAE

MELASTOMA

A genus of c. 70 species (4 in China) distributed from India to south-east Asia. **Evergreen shrubs**. **Leaves** opposite, simple, prominently 5–7-veined. **Inflorescences** terminal, clustered. **Flowers** regular, pink or rose-purple, calyx campanulate, 5-lobed, covered with scale-like hairs with fringed margins; petals 5; stamens 10, unequal, 5 long and 5 short, anthers dehiscing by apical pores. Ovary 5-celled, hairy at apex; style long. **Fruit** a capsule enclosed in persistent calyx-tube, becoming pulpy within.

Melastoma malabathricum L. subsp. *normale* (D. Don) K. Mey. (syn. *M. normale* D. Don) **Shrub** 0.6–7 m tall with bristly-hairy, branching **stems**. **Leaves** elliptic-ovate, 7–15 × 2–7 cm, pointed to long-pointed, 3–5-veined, bristly-hairy beneath; stalks 5–18 mm long. **Inflorescence** to 10 cm long, many-flowered. **Flowers** 3–5 cm across, rich rose-purple; calyx densely feathery-hairy, narrowly lobed; anthers yellow. **Fruit** berry-like, red. Guizhou, Sichuan, Yunnan [sChina, Himalaya, Taiwan]; lava flows, woods, ravines, riverbanks, montane forests, 1,400–3,300 m. III–VII.

OSBECKIA

A genus of c. 60 species (5 in China) primarily in the Old World tropics. **Perennial herbs** or **evergreen shrubs**, often with 4-angled branchlets. **Leaves** simple, prominently 3-veined, with crystalline cells (visible with a ×10 magnifying lens) often at base of hairs. **Inflorescences** terminal, cymose or densely paniculate. **Flowers** pink or rose-purple; calyx 4–5-lobed, usually with hairy appendages on tube and between lobes; petals 4–5; stamens 8–10, equal, the anthers long, curved, beaked. **Fruit** a capsule enclosed in persistent calyx-tube, dehiscing at apex by 4–5 pores.

1. Subshrub or herbaceous perennial; flowers small, petals not more than 16 mm long .... *O. capitata*
1. Evergreen shrub; flowers larger, petals at least 16 mm long .... *O. stellata* var. *crinita*, *O. nepalensis*

Osbeckia capitata Walp. **Herb**, 10–30 cm tall, **stems** with appressed hairs. **Leaves** ovate to ovate-elliptic, 1–2.7 × 0.6–1.5 cm, pointed, with appressed hairs on both surfaces. **Inflorescence** terminal, 2–3-flowered, **flowers** almost stalkless; calyx-tube 11–14 mm long; petals lilac to purple, broadly obovate, 12–14 mm long; stamens 8, 12–14 mm long. **Fruit** 6–8 mm long. Guizhou, Hubei, Yunnan [sChina, Himalaya, Japan, Taiwan]; hillsides, shrubberies, 1,300–3,100 m. V–VIII.

Osbeckia stellata Buch.-Ham. ex Ker-Gawl. var. *crinita* (Benth. ex Naud.) C. Hansen. Low, hairy **shrub**, 30–100 cm tall, **stems** branching, sometimes winged, with appressed or spreading hairs. **Leaves** ovate-oblong to elliptic, 6.5–14.5 × 2.1–4.5 cm, pointed, prominently 5-veined, with appressed hairs on both surfaces. **Inflorescences** terminal or from upper leaf-axils, paniculate, few-many-flowered; **flowers** almost stalkless, 2.5–2.8 cm across; calyx-tube 13–25 mm long, covered with persistent tufted hairs; petals pinkish white to rose-purple, obovate or broadly obovate, 16–27 mm long; stamens 8, 20–35 mm long. **Fruit** 10–20 mm long. Guizhou, Hubei, Yunnan [Himalaya, Myanmar, Taiwan]; grassy meadows and hillsides, 150–2,600 m. VI–IX.

---

1. *Punica granatum*; Forbidden City, Beijing, cult. (CGW)
2. *Epilobium conospermum*; eNepal (CGW)
3. *Epilobium pannosum*; Cangshan, nwYunnan (CGW)
4. *Chamerion angustifolium*; south of Aba, nwSichuan (CGW)
5. *Chamerion angustifolium*; near Kunming, cYunnan (PC)
6. *Melastoma malabathricum* subsp. *normale*; Mongpang, swYunnan (PC)
7. *Osbeckia capitata*; Cangshan, above Dali, nwYunnan (CGW)
8. *Osbeckia stellata* var. *crinita*; Cangshan, nwYunnan (CGW)

***Osbeckia nepalensis* Hook.** Similar to the previous species, but a larger **shrub**, 1–2.5 m tall, with ovate-lanceolate **leaves**, 5.5–10.5 × 1.3–2.8 cm, pointed. **Inflorescences** terminal, 2–3-flowered; **flowers** 4–4.5 cm across, petals pink to rich rose-purple, 20–24 mm long; stamens 13–15 mm long. Yunnan [Himalaya, Myanmar]; hillsides, shrubberies, 1,000–1,800 m. X–XI.

OXYSPORA

A genus of c. 24 species (3 in China) in the Indo-Malaysian region. Large, **evergreen shrubs**. **Leaves** opposite, simple, with main veins connected by many parallel secondary ones. **Inflorescence** a long, lax, terminal panicle; **flowers** regular, pink or rose-purple; calyx 4-lobed; petals 4; stamens 8, unequal, 4 long purple and 4 short yellow, anthers spurred or not. **Ovary** inferior, 4-celled. **Fruit** a dry capsule with 8 prominent ribs.

***Oxyspora paniculata* (D. Don) DC.** Spreading **shrub** or small **tree** 1–5 m tall with hairy, spreading to drooping **branches**. **Leaves** ovate, 5–7-veined, 10–30 × 3–19.5 cm, long-pointed, margin shallowly toothed. **Inflorescence** pendulous, pyramidal, 18–30 cm long, with branches up to 10 cm long. **Flowers** purple or rose-pink. **Fruit** spindle-shaped, 1 cm long. Guizhou, Yunnan [seAsia, c & eHimalaya, Myanmar]; mountain streamsides, dry open places in shrubberies, 1,600–2,300 m. VIII–X.

MEDINILLA

A genus of c. 150 species (3 in south China) distributed primarily from tropical Africa eastwards to the Pacific. **Herbs** or **evergreen**, erect to scandent **shrubs**, sometimes **epiphytic**, with simple, prominently veined, somewhat fleshy **leaves**. **Inflorescences** terminal or lateral, cymose, few-flowered; **flowers** regular; calyx obovoid; petals 4, pink or rose-purple; stamens 8, equal, filaments lacking a knee-like bend, anthers sometimes spurred. **Ovary** 4–5-celled. **Fruit** a berry crowned by the calyx-limb.

Medinilla rubicunda* (Jack) Blume (syn. *M. yunnanensis* H. L. Li) **Shrub** 0.6–2 m tall with obovate to elliptic, 3-veined **leaves**, 9–13 × 3–4.5 cm, pointed; stalks 6–8 mm long. **Inflorescences** borne directly on branchlets (cauliflorous), the small **flowers** in clusters of 2–5, pink. Yunnan; woods, 1,500–1,700 m. VII–IX.

SONERILA

A tropical genus with over 100 species (4 in south China). Small **herbs**, sometimes lacking distinct stems, or **shrubs**, with opposite, simple, stalked **leaves**. **Inflorescences** few-flowered scorpioid spikes; **flowers** regular; calyx 3-lobed, lobes triangular; petals 3, pink or rose-purple; stamens 3, equal, anthers shortly attenuate at apex. **Ovary** 3-celled, apex hairy; style simple, thread-like. **Fruit** a capsule enclosed in persistent calyx-tube, dehiscing at apex.

***Sonerila plagiocardia* Diels** Rhizomatous, patch-forming **herb**, 15–35 cm tall. **Leaves** ovate, 5–16 × 3–8 cm, long-pointed, 7-veined, margin toothed, markedly oblique at base. **Inflorescences** 7–8 cm long; **flowers** pink, 24–28 mm across. Yunnan [sChina, seAsia]; humus-covered boulders in woods, 1,900–2,300 m. VII–X.

CORNACEAE

SWIDA

A genus of 8 species (5 in China) in the temperate northern hemisphere. **Trees** or **shrubs**. **Leaves** opposite or, less commonly, alternate, simple, stalked, with flattened medifixed hairs on lower surface. **Inflorescences** capitate, compound corymbs or cymes; **flowers** usually white, small, 4-parted. **Fruit** a berry-like drupe with a 2-celled stone.

> 1. Leaves alternate .... *S. controversa*
> 1. Leaves opposite .... *S. macrophylla, S. hemsleyi, S. oblonga*

***Swida controversa* (Hemsl.) Soják** (syn. *Cornus controversa* Hemsl.) **Deciduous**, spreading **shrub** or **tree**, 3–20 m tall with grey-black bark and rich red branchlets; **branches** in horizontal tiers. **Leaves** broadly ovate, 7.5–15 × 3.2–9 cm, pointed, glossy dark green above, glaucous beneath; stalks 3–7 cm long. **Inflorescence** 7–17.5 cm across; **flowers** white, 7–8 mm across. **Fruit** almost globose, 6 mm, blue-black when ripe. Hubei, Sichuan, Yunnan [c & nChina, Himalaya, Japan, Korea]; damp places in thickets and open forests, 300–1,600 m. IV–VI.

***Swida macrophylla* (Wall.) Soják** (syn. *Cornus macrophylla* Wall.) **Deciduous**, spreading **shrub** or **tree** to 16 m with oblong-obovate or ovate **leaves**, 10–18 × 4.5–10 cm, pointed, bright green above, grey-glaucous beneath; stalks 1.5–3 cm long. **Inflorescence** somewhat rounded, 6–17 cm across; **flowers** creamy white, 10–12 mm across. **Fruit** 4–6 mm, blue when ripe. Gansu, Hubei, w & swSichuan, Yunnan [n, c & eChina, Himalaya, Japan, Taiwan]; cliffs, coniferous forests, often by streams, 850–3,200 m. V–VII.

Swida hemsleyi* (C. K. Schneid. & Wangerin) Soják (syn. *Cornus hemsleyi* C. K. Schneid. & Wangerin) **Deciduous shrub** to 4 m, occasionally a taller small **tree**, with downy young shoots. **Leaves** ovate to almost roundish, to 7.5 × 2.5 cm, rounded or slightly heart-shaped at base, abruptly pointed at apex, with 6–8 pairs of lateral veins, midrib and veins rusty-downy beneath. **Inflorescence** a lax corymb 6–8 cm across; **flowers** small, white, anthers powder-blue. **Fruit** blue when ripe. Hubei, Sichuan [cChina]; forests, forest margins, 2,400–3,500. VI–VII.

CORNACEAE: Benthamidia

Swida oblonga **(Wall.) Soják** (syn. *Cornus oblonga* Wall.) **Evergreen**, spreading **shrub** or small **tree**, 1.5–8 m tall with downy young shoots. **Leaves** oval to elliptic-lanceolate, 7–12.5 × 2.5–4.4 cm, long-pointed, dull grey beneath; stalks 1–1.5 cm long. **Inflorescence** pyramidal, 3–7.5 cm tall and wide; **flowers** white to pale yellow, 4–8 mm across, anthers deep violet. **Fruit** 8–9 mm, black when ripe. Sichuan, seTibet, Yunnan [Himalaya, neIndia, Myanmar]; damp places in thickets and dry, mixed forests, cliffs, 1,200–3,400 m. IX–XI.

1. *Oxyspora paniculata*; eNepal (CGW)
2. *Swida macrophylla*; Chongjiang, nwYunnan (PC)
3. *Swida hemsleyi*; Jiuzhaigou, nwSichuan (PC)
4. *Benthamidia capitata*; Jinshajiang near Shigu, nwYunnan (CGW)
5. *Benthamidia capitata*; south of Lijiang, nwYunnan (CGW)
6. *Benthamidia capitata*; south of Lijiang, nwYunnan (CGW)
7. *Benthamidia capitata* in immature fruit; cult. (CGW)

BENTHAMIDIA

A genus of c. 10 species (4 in China) in temperate east Asia. **Semi-deciduous** or **deciduous trees** or **shrubs**. **Leaves** opposite, or less commonly, alternate, simple, stalked, with medifixed hairs on lower surface. **Inflorescence** capitate, subtended by 4(–6) prominent white, creamy, lemon-yellow or occasionally pink, bracts; **flowers** small, greenish, lemon-yellow or white; sepals, petals and stamens 4. **Fruit** a fleshy, compound syncarp.

Benthamidia capitata **(Wall.) Hara** (syn. *Dendrobenthamia capitata* (Wall.) Hutch.; *Cornus capitata* Wall.) **Semi-evergreen**, spreading **shrub** or **tree**, 3–20 m tall. **Leaves** leathery, elliptic or narrowly elliptic, 5–12.5 × 1.5–4.2 cm, pointed, dull grey-green, both sides covered with minute flattened hairs; stalks 8–10 mm long. **Inflorescence** hemispherical, 8–15 mm across with tiny **flowers**; bracts 4–6, obovate, rich lemon-yellow or cream, paling with age, 4–6.3 × 1.5–4.3 cm. **Fruit** erect, 1.8–3.7 cm across, bright scarlet when ripe, edible. Guizhou, Hubei, Sichuan, seTibet, nwYunnan [sChina, Himalaya]; damp places by rivers and streams, shrubberies, coniferous and mixed forests, 1,700–3,000 m. V–VI.

CORNACEAE: Benthamidia

Benthamidia japonica* (Sieb. & Zucc.) Hara var. *chinensis* (Osborn) H. Hara (syn. *Dendrobenthamia japonica* (Sieb. & Zucc.) Hutch. var. *chinensis* (Osborn) W. P. Fang; *Cornus kousa* Hance var. *chinensis* Osborn) **Deciduous**, spreading **shrub** or small **tree** to 8 m. **Leaves** papery, ovate, 4–7.5 × 3–6 cm, long-pointed, margin wavy; stalks 1–2.5 cm long. **Inflorescence** flat, button-like, 5–8 cm across with tiny **flowers**; bracts 4, lanceolate to ovate, creamy white, turning pink with age, 2.5–3.5 × 0.7–2.2 cm, long-pointed. **Fruit** up to 1.8 cm across, red when ripe. Gansu, Guizhou, Hubei [c & eChina]; damp places by rivers and streams, shrubberies, mixed forests in partial shade, 600–1,600 m. V–VI. The typical variety (var. *japonica*), with smaller **flowers**, is found in Korea and Japan.

DAVIDIA

A Chinese genus with a single species. **Deciduous tree** with opposite, stalked, simple, toothed **leaves**. **Inflorescence**-heads stalked, globular, pendent, subtended by 2 prominent, large, white or cream **bracts**, scented; **flowers** small, greenish, female flower solitary in the head, reduced to an ovary with a short 6-rayed style, male flowers numerous; stamens with white filaments and red anthers. **Fruit** a solitary drupe, green with a purplish bloom, becoming red-brown when ripe, containing a 3–5-seeded, hard nut.

Davidia involucrata* Baill. **Tree** 8–28 m tall with glaucous young **branches**. **Leaves** heart-shaped or ovate-heart-shaped, 7–29 × 4–15 cm, long-pointed, boldly toothed, bright green, hairy above, felted beneath. **Inflorescences** pendent, 1.6–1.7 cm across, the pair of bracts creamy white, paling with age, somewhat oblique, oblong-ovate, 6–18 × 2–11 cm, pointed to long-pointed; **flowers** pale green. **Fruit** oblong-ellipsoid, 3.5–4 cm long. Guizhou, Hubei, Sichuan, seTibet, nwYunnan [wHunan]; damp places by rivers and streams, mixed forests, 1,700–3,600 m. V–VII. **Var. *vilmoriniana* (Dode) Wangerin*** has almost glabrous, yellow-green or somewhat glaucous **leaves**. wHubei, Sichuan; often growing with the typical variety.

HELWINGIA

A genus of 4 species (3 in China) distributed from the Himalaya to Japan. **Dioecious shrubs**. **Leaves** mostly alternate, thin-textured, simple, toothed, shortly stalked. **Inflorescences** rather inconspicuous, borne on midvein on upper surface of leaf-blade, umbellate, several-flowered; **flowers** small, unisexual, purple, brown, green, yellow-green or white; petals 3–5; male flowers with 3–5, almost stalkless stamens. **Fruit** an almost globose, 3–4-seeded berry.

1. Flowers white or yellow . . . . **H. chinensis**
1. Flowers green or purple-brown . . . . **H. himalaica, H. omeiensis**

Helwingia chinensis **Batalin*** Glabrous **shrubs** 1–3 m tall. **Leaves** linear to lanceolate, 2.5–11 × 1–2 cm, long-pointed, toothed. **Inflorescence** 7–8 mm across; **flowers** white or yellow, 3 mm. **Fruit** 9 mm across, red when ripe. Gansu, Hubei, Sichuan, Yunnan; shrubberies, ravines, lava beds, in shade, 1,000–3,000 m. IV–VI.

Helwingia himalaica **Hook. f. & Thomson ex C. B. Clarke** Glabrous **shrub** 1–3 m tall. **Leaves** elliptic to obovate, 2.5–13 × 2–5.8 cm, long-pointed, toothed. **Inflorescence** 3–7 mm across; **flowers** brown-purple or greenish, 1.5–3 mm. **Fruit** 6–7 mm across, red when ripe. Guangxi, Hubei, Yunnan [Himalaya, Myanmar]; *Quercus–Rhododendron* forest, shrubberies, streamsides, in shade, 1,200–3,050 m. V–VI.

Helwingia omeiensis **(Fang) H. Hara & S. Kurasawa*** Similar to the previous species, but **leaves** oblanceolate to elliptic-obovate, and **inflorescence** 10–13 mm across, bearing greenish **flowers**, 2–3 mm across. **Fruit** rather larger, 9 mm across. seGansu, nGuangxi, Guizhou, Hubei, sSichuan (Emeishan), sShaanxi, Yunnan [Hunan]; shrubberies, streamsides, in shade, 1,400–1,600 m. VI.

ELAEAGNACEAE

HIPPOPHAE

A genus of 7 species (7 in China, 4 endemic) in Europe and temperate Asia. **Deciduous**, dioecious, spiny **shrubs** or **trees** with alternate, opposite or whorled **leaves**, almost stalkless to stalked. **Flowers** very small, unisexual, clustered at base of lateral shoots. **Male flowers** in small catkins appearing before the leaves; calyx-segments 2, free, membranous; stamens 4. **Female flowers** in small racemes, appearing with the leaves; calyx tubular, 2-lobed; style stigmatic on 1 side. **Fruit** a drupe, sometimes with longitudinal ribs; seed enclosed in parchment-like endocarp.

1. Leaves alternate, more than 4 cm long; endocarp readily removed from seed .... **H. salicifolia**
1. Leaves in whorls, less than 2 cm long; endocarp difficult to remove from seed .... **H. tibetana**

---

1. *Benthamidia japonica* var. *chinensis*; Shennongjia, wHubei (PC)
2. *Davidia involucrata*; cult. (CGW)
3. *Davidia involucrata* fruit; cult. (CGW)
4. *Helwingia chinensis*; Lijiang, nwYunnan (CGW)
5. *Hippophae salicifolia*; seTibet (DL)
6. *Hippophae tibetana*; ne Nepal (CGW)

Hippophae salicifolia **D. Don** (syn. *Elaeagnus salicifolia* (D. Don) A. Nelson; *Hippophae rhamnoides* L. subsp. *salicifolia* (D. Don) Servett.) **Shrub** or **tree**, 2–9 m or more tall, with fissured, longitudinally flaking **bark**. **Leaves** linear-oblong, 4.2–6.2 × 0.6–1.2 cm, white-felted beneath and usually with a reddish brown midvein, green above, margin usually back-rolled. **Fruit** globose, orange-yellow to greenish brown, yellow or deep red when ripe, 5–7 mm across. seTibet [swTibet, Himalaya]; damp gravel or stony areas, often beside rivers or streams, 2,800–3,500 m. V–VI.

Hippophae tibetana **Schltr.** (syn. *Hippophae rhamnoides* L. subsp. *tibetana* (Schltr.) Servett.) Small, sometimes rhizomatous **shrub** 10–60 cm tall, sometimes taller or prostrate. Leafy **stems** slender, unbranched, spine-tipped; **leaves** mostly in whorls of 3, linear-oblong, 12–20 × 2.5–4 mm, whitish and with a reddish brown midvein beneath, greyish above, densely scaly, margin flat. **Fruit** ellipsoid, yellowish green or yellowish orange when ripe, 8–11 × 6–9 mm. Gansu, Qinghai, Tibet [cHimalaya]; dry gravelly or stony places, especially riverbeds and floodplains, 3,600–4,700 m. V–VI.

CELASTRACEAE

EUONYMUS

A genus of 130 species (80 in China, about half endemic) distributed across the northern hemisphere, and in Madagascar and Australasia. **Evergreen** or **deciduous clamberers**, **shrubs** or small **trees** with young shoots often 4-angled. **Leaves** simple, opposite, toothed. **Inflorescences** stalked, 3–7-flowered ascending to pendent cymes from lower nodes of current year's growth; **flowers** small, usually 10 mm or less across, 4–5-parted, white, green, yellow or purple, with a fleshy disc. **Fruit** a pendulous capsule, often brightly coloured, 3–5-parted, angled, lobed or winged, each lobe containing a single **seed** that is usually covered by a red or orange aril. The deciduous species often have brilliant autumn colouring.

1. Evergreen or semi-evergreen shrubs .... **E. frigidus, E. cornutus, E. myrianthus, E. wilsonii**
1. Deciduous shrubs 2
2. Branches lacking corky bark .. **E. nanoides, E. sanguineus**
2. Branches with corky bark ... **E. alatus, E. phellomanus**

Euonymus frigidus **Wall.** Glabrous **semi-evergreen shrub** to 4.5 m with large pointed **buds** and oblong to lanceolate, leathery **leaves**, to 13 cm long, long-pointed, lustrous dark green, with a finely saw-toothed margin and prominent veins. **Flowers** 7–15 in lax, slender, stalked cymes; petals 4, greenish yellow to purple. **Fruit** 4-winged wings attached near base; **seeds** white, mostly enclosed by an orange-red aril. Gansu, Guizhou, Hubei, seQinghai, Sichuan, seTibet, Yunnan [cChina, Himalaya, Myanmar]; forests and forest margins, shrubberies, to 3,000 m. V–VI.

***Euonymus cornutus* Hemsl.** Similar to the previous species, but **leaves** linear-lanceolate, and cymes mostly 3-flowered; **fruit** prominently 4-winged, with horn-like projections. Gansu, Hubei, Sichuan, Shaanxi, seTibet, Yunnan [swTibet, cChina, neIndia, Myanmar]; similar habitats as well as rocky places, 2,200–4,300 m. V–VI.

Euonymus myrianthus* Hemsl. **Evergreen**, rounded, glabrous **shrub** to 3.5 m, with rather stout shoots and yellowish green, leathery, ovate-lanceolate to ovate-oblong **leaves**, 5–12 × 1.5–4 cm, margin sharply toothed. **Flowers** greenish yellow, numerous in dense terminal cymes 5–7.5 cm across. **Fruit** yellow, scarcely lobed, with orange-scarlet **seeds**. Guangxi, Guizhou, Hubei, Sichuan, Shaanxi, Yunnan [c & eChina]; forests, shrubberies, ravines, 200–1,200 m. V–VII.

Euonymus wilsonii* Sprague **Evergreen**, rather lax **shrub** up to 6 m with slender shoots. **Leaves** lanceolate, 7.5–15 × 2.5–4 cm, with slender, long-pointed apex and toothed margin. **Flowers** green, in lax clusters. **Fruit** 4-lobed, on a stalk 3–3.5 cm long, hedgehog-like, covered in conspicuous spine-like awns; **seeds** with yellow aril. Guangxi, Guizhou, Hubei, Sichuan, Shaanxi, Yunnan; forests, 1,000–2,600 m. V–VI.

Euonymus nanoides* Loes. & Rehd. (syn. *E. oresbius* W. W. Sm.; *E. nanoides* var. *oresbius* (W. W. Sm.) Y. R. Li) Semi-prostrate **shrub** 1–2 m tall with distinctly square young shoots. **Leaves** linear to oblanceolate, rounded, 12–22 × 1.5–3.5 mm long, margin entire or indistinctly toothed. **Flowers** green, 1–3 on slender stalks. **Fruit** small, 11–13 mm across, 4-lobed, rich rosy red; **seeds** with scarlet aril. Gansu, Shaanxi, Tibet, Yunnan [nChina]; forests, rocky places, 1,400–3,200 m. VI.

Euonymus sanguineus* Loes. An altogether larger plant than the previous species, to 6 m with reddish young **shoots** and long winter **buds**. **Leaves** rather leathery, deep green or purplish, ovate to obovate, 4–12 × 2–6.5 cm, margins with fine, incurved teeth. **Flowers** yellow, in cymes 7.5–10 cm long. **Fruit** red, usually 4-parted, each part with a 7–8 mm wing; **seeds** with yellow aril. Gansu, (?)Guizhou, Hubei, seQinghai, Sichuan, Shaanxi, e & seTibet, Yunnan [cChina]; woodland, shrubberies, rocky places, 1,800–3,700 m. V–VI.

***Euonymus alatus* (Thunb.) Sieb.** (syn. *Celastrus alatus* Thunb.; *C. striatus* Thunb.; *Euonymus striatus* (Thunb.) Loes. ex Gilg. & Loes.; *E. verrucosus* Scop. var. *tchefouensis* Debeaux) **Shrub** 1–4 m tall with 4-angled young **branches** usually with 2 or 4 corky wings, each to 5 mm wide. **Leaves** thin-textured, obovate to oblong-elliptic, 4.5–10 × 2–4 cm, margin minutely scalloped to minutely toothed, pinkish, reddish, even purple in late autumn. **Inflorescences** mostly 3-flowered, with a stalk 1–2 cm long; **flowers** 4-parted, 8–9 mm across; petals green, light yellow or greenish yellow, ovate. **Fruit** reddish brown when fresh, deeply 4-lobed, 10–13 mm across, but usually only 1–3 lobes developing to maturity;

seeds with bright red aril. Gansu, Guizhou, Hubei, Sichuan, Shaanxi [c & nChina, Japan, Korea, eRussia (Sakhalin)]; woodland, forests, shrubberies, to 2,700 m. IV–VII.

Euonymus phellomanus* Loes. ex Diels Smaller **shrub** than the previous species, usually 2–3 m tall, with grey-green to grey-brown **branches** and twigs, with corky stem-wings. **Leaves** net-veined. **Flowers** greenish white, c. 10 mm in diameter. **Fruit** almost globose, 4-angled, bright pinkish red, yellow-brown to red-brown, 8–10 mm across; **seeds** with orange aril. Gansu, Hubei, seQinghai, Sichuan, Shaanxi [cChina]; woodland, shrubberies, dry mountain slopes, 1,000–3,000 m. V–VII.

TRIPTERYGIUM

A genus of a single species restricted to east Asia. **Shrubs**, ± **climbing**, **deciduous**. **Leaves** alternate. **Flowers** small, 5-parted, in terminal panicles, supplemented by lateral clusters. **Fruit** characteristically 3-winged, papery.

***Tripterygium wilfordii* Hook. f.** (syn. *Aspidopterys hypoglauca* H. Lévl.; *Tripterygium hypoglaucum* (H. Lévl.) Hutch.; *T. forrestii* Loes.) Spreading, **semi-climbing shrub** to 6 m, sometimes clambering into trees. **Leaves** oval, to 14 × 8 cm, contracted at apex into a short point; stalks 8–15 mm long. **Flowers** whitish or yellowish white, small, 6–7 mm across, in large panicles up to 20 cm long, sometimes larger. **Fruit** 13–15 mm long, with wide papery wings, brown when mature, sometimes flushed with purple or crimson. Guangxi, Guizhou, s & seSichuan, Yunnan [sChina, Taiwan, Japan, Korea, neMyanmar, seAsia]; shrubberies, forest margins, hedgerows, banks, rocky places, 100–3,500 m. V–X.

AQUIFOLIACEAE

ILEX

A cosmopolitan genus with 400 species (half in China, 170 endemic). **Deciduous** and **evergreen**, dioecious **shrubs** and **trees** with angular young shoots. **Leaves** alternate, stalked, with entire, toothed or spiny margins. **Inflorescences** cymose or pseudoumbellate, simple or much-branched, solitary and axillary on current year's shoots or in bundles, axillary on second year's shoots, rarely simple; **flowers** small, dull white, pink, or red, male with 4–8 petals and stamens, female with 4–8 petals and carpels. **Fruit** a drupe with a thin, fleshy outer layer, red or black at maturity, occasionally white or yellow.

1. Deciduous trees with shallowly toothed leaves . . . . *I. macrocarpa*, *I. micrococca*
1. Evergreen shrubs or small trees 2
2. Leaves spiny . . . . *I. cornuta*, *I. pernyi*
2. Leaves toothed or scalloped, not spiny (very rarely slightly spiny in *I. corallina*) 3

1. *Euonymus cornutus*; cult. (CGW)
2. *Euonymus cornutus*; cult. (CGW)
3. *Euonynus myrianthus*; cult. (CGW)
4. *Euonymus myrianthus*; cult. (CGW)
5. *Euonymus alatus*; cult. (CGW)
6. *Euonymus alatus*; cult. (CGW)
7. *Euonymus phellomanus*; cult. (CGW)
8. *Euonymus phellomanus*; cult. (CGW)
9. *Euonymus phellomanus*; cult. (CGW)
10. *Tripterygium wilfordii*; Cangshan, nwYunnan (DC)

3. Prostrate shrubs . . . . *I. intricata*
3. Erect to ascending shrubs 4
4. Fruit purple-red when ripe . . . . *I. corallina*
4. Fruit bright red when ripe . . . . *I. fargesii,*
   I. melanotricha, I. franchetiana, I. yunnanensis

***Ilex macrocarpa* Oliv.** **Tree** 5–10 m tall, with grey- to red-brown branchlets. **Leaves** papery, ovate or ovate-elliptic to elliptic, 7.5–13 cm long, long-pointed, shallowly toothed. **Male inflorescences** solitary or in bundles, on short lateral branchlets, white, 5–6-parted; **female flowers** solitary in leaf-axils, 7–9-parted. **Fruit** globose, somewhat flattened at base, 12–16 mm in diameter, black at maturity. Guangxi, Guizhou, Hubei, Sichuan, sShaanxi, Yunnan [c & sChina, Vietnam]; forests on mountains, valleys, roadsides, 400–2,400 m. V–VI.

***Ilex micrococca* Maxim.** (syn. *I. micrococca* var. *longifolia* Hayata; *I. pseudogodajam* Franch.) **Deciduous tree** 12–20 m tall, branchlets with conspicuous, large, circular or oblong, white lenticels. **Leaves** deep green, membranous or papery, ovate to ovate-elliptic, 7–13 cm long, long-pointed, glabrous or downy beneath, margin ± entire or spiny-toothed; leaf-stalks 1.5–3.2 cm long. **Inflorescences** compound cymes, axillary on current year's branchlets; **male flowers** 5–6-parted, **female flowers** 6–8-parted. **Fruit** globose, c. 3 mm in diameter, red or yellow when ripe. Guangxi, Guizhou, Hubei, Sichuan, Tibet, Yunnan [s & eChina, Japan, Taiwan, Vietnam]; evergreen, broadleaved and montane forests, 500–1,900 m. V–VI.

***Ilex cornuta* Lindl. & Paxton** (syn. *I. fortunei* Lindl.; *I. furcata* Lindl. ex Göpp.) Dense, rounded **shrub** 1–3 m tall with leathery, dark glossy green, ± rectangular **leaves**, 3.7–9 cm long, margin with 5–9 needle-pointed spines on each side. **Inflorescences** in bundles, axillary on second year's branchlets; **flowers** small, dull white or yellowish, 4-parted. **Fruit** globose, 8–10 mm in diameter, red at maturity. Hubei [eChina, Korea], and often grown in Chinese gardens and for bonsai; thickets, sparse forests, hillsides, stream- and roadsides, near villages, 100–1,900 m. IV–V.

Ilex pernyi* Franch. Evergreen **shrub** or small **tree** 1–8 m tall with silver-grey **bark** and stiff branchlets. **Leaves** squarish, 1.5–3 cm long with a long triangular tip and 2 large spines on each side. **Inflorescences** axillary on second year's branchlets; **flowers** pale yellow, 4-parted, in small clusters. **Fruit** oblong-globose, 7–8 mm in diameter, red at maturity. Gansu, Guizhou, Hubei, Sichuan, Shaanxi [eChina]; forests and thickets on hillsides and in valleys, 1,000–2,500 m. IV–V.

1. *Ilex micrococca*; Shennongjia, wHubei (PC)
2. *Ilex pernyi*; Shennongjia, wHubei (PC)
3. *Ilex corallina*; Fanjingshan, ne Guizhou (PC)
4. *Pachysandra terminalis*; cult. (TK)

Ilex intricata **Hook. f.** Prostrate, glabrous **shrub**, forming dense wide mats. **Leaves** dull green and leathery, obovate to elliptic, to 2.2 cm long, rounded at apex; stalks very short. **Flowers** clustered, inconspicuous. **Fruit** globose, 4–5 mm in diameter, bright red at maturity, solitary or 2–3 together. w & swSichuan, seTibet, Yunnan [Himalaya, neMyanmar]; coniferous and *Rhododendron* forests in alpine zone, 3,000–4,000 m. V–VI.

Ilex corallina **Franch.*** **Evergreen shrub** or **tree** 3–10 m tall, with slender, brownish branchlets. **Leaves** deep green, leathery, ovate, ovate-elliptic or oblong, 4–13 × 1.5–5 cm, pointed to long-pointed, glabrous or midvein sparsely downy above, margin wavy, scalloped-toothed, rarely with spine-tipped teeth; flower-stalks purple-red, 4–10 mm long. **Inflorescences** in small clusters on second year's branchlets, ± stalkless; **flowers** yellow-green, 4-parted, c. 2 mm in diameter. **Fruit** ± globose or ellipsoid, 3–5 mm in diameter, purple-red when ripe. sGansu, Guizhou, Hubei, Sichuan, Yunnan [Hunan]; mixed forests and shrubberies on mountain slopes, 400–2,400(–3,000) m. II–V.

Ilex fargesii **Franch.*** Small **tree** 4–8 m tall with brown **bark**. **Leaves** opaque dull green, narrow-oblong to oblanceolate, 5–13 × 1.5–3 cm, tapering to the long-pointed apex where there are a few incurved teeth; stalks reddish. **Inflorescences** in bundles; **flowers** white, fragrant, 4-parted. **Fruit** often in 3s and 4s, globose, 5–7 mm in diameter, red at maturity, stalk 4 mm long, reddish. sGansu, Guizhou, wHubei, Sichuan, sShaanxi; forests and shrubberies on mountain slopes, 1,600–3,000 m. IV–V.

Ilex melanotricha **Merr.** Similar to *I. fargesii*, but to 10 m, with grooved **bark** and reddish young shoots. **Leaves** thinly leathery, oblong to elliptic, 7.5–10 × 2.5–3.5 cm, margin shallowly scalloped. **Male flowers** in dense panicles on short lateral branchlets. **Fruit** globose, bright red when ripe, c. 10 mm across, on downy stalk. seTibet, nwYunnan [nMyanmar]; similar habitats, 1,500–3,400 m. VI.

Ilex franchetiana **Loes.** Similar to the previous species, but with shorter, broader **leaves**, stalkless **male inflorescences** and glabrous fruit-stalks. Guizhou, Hubei, Sichuan, seTibet, neYunnan [nMyanmar]; broadleaved and mixed forests in hill country, 1,800–2,300 m. V–VI.

Ilex yunnanensis **Franch.** Distinctive **evergreen shrub** to 4 m with persistently downy, bright green young **stems**. **Leaves** brownish red when young, glossy green with age, ovate to ovate-oblong, 1.8–3 cm long, pointed, base rounded, margin shallowly and minutely scalloped or minutely toothed. **Fruit** generally solitary, 6–7 mm across, red at maturity. sGansu, Guangxi, Guizhou, Hubei, Sichuan, sShaanxi, seTibet, Yunnan [nMyanmar, Taiwan]; evergreen broadleaved and coniferous forests, forest margins, shrubberies, 1,100–3,500 m. V–VII.

BUXACEAE

PACHYSANDRA

A genus of 3 species (2 in China) in Asia and North America. Creeping **rhizomatous herbs** with alternate, thin-leathery or papery, stalked **leaves** with toothed margins and 2–3 pairs of lateral veins. **Inflorescence** a terminal or axillary spike with **male flowers** in middle and upper part, and **female flowers** in the lower; flowers small, white or rose-pink, male flowers with 4 sepals and 4 stamens opposite sepals, female with 4–6 sepals, a 2–3-celled ovary and 2–3 styles. **Fruit** a small 2–3-parted capsule.

Pachysandra terminalis **Sieb. & Zucc.** **Patch-forming perennial** to 30 cm with leathery, rhombic-obovate **leaves**, 2.5–5(–9) × 1.5–3(–6) cm, toothed in upper half. **Inflorescence** terminal, 2–4 cm long, erect; **flowers** white; **male flowers** 15 or more, stalkless, sepals 2.5–3.5 mm long, filaments 7 mm long; **female flowers** 1–2, styles protruding after pollination. **Fruit** egg-shaped, 5–6 mm long, with persistent thick, reflexed styles. Gansu, Hubei, Sichuan, Shaanxi [eChina, Japan]; forests in shady, damp places, ravines, 1,000–2,600 m. IV–V.

BUXUS

A genus of c. 70 species (17 in China) found in temperate or subtropical areas of the world. **Evergreen shrubs** or small **trees** with opposite, shiny, leathery or papery, entire, short-stalked **leaves**. **Inflorescences** axillary or terminal, racemose, spike-like or capitate, with bracts. **Flowers** small, unisexual: **male flowers** several, inserted in the mid to lower part of the inflorescence, with 4 sepals, each with an opposing stamen; **female flower** solitary at inflorescence-apex, with 6 sepals and a 3-celled ovary with 3 free styles. **Fruit** a 3-parted capsule.

1. Female flowers with style 2–3 times longer than ovary .... ***B. henryi***
1. Female flowers with style as long as ovary 2
2. Flowers in very short spike .. ***B. mollicula, B. hebecarpa***
2. Flowers in small dense clusters ... ***B. bodinieri, B. sinica***

Buxus henryi **Mayr*** **Shrub** to 3 m with thin-leathery, lanceolate, oblong-lanceolate or ovate-oblong **leaves**, 4–7 × 1.5–2.5 cm, apex blunt or slightly pointed. **Inflorescence** 10–15 × 7–10 mm; **male flowers** c. 8, sepals 4.5–5 mm long, membranous, glabrous, stamens c. 11 mm long; **female flowers** with outer sepals oblong, c. 6 mm long, inner c. 3 mm long, ovary 2–2.5 mm long topped by styles 6–8 mm long that are recurved at apex. **Capsule** almost globose, c. 6 mm long. Guizhou, wHubei, Sichuan; forests and forest margins, 1,300–2,000 m. IV–V.

Buxus mollicula W. W. Sm.* **Shrub** to 3 m with downy or glabrous branchlets. **Leaves** leathery, ovate, oblong or elliptic, rarely oblong-lanceolate, 3–5 × 1.2–1.8(–2) cm, densely hairy on both surfaces, apex shallowly notched, rounded or mucronate. **Inflorescence** shortly terete, c. 10 mm long, rachis 3–4 mm long, velvety-downy; **male flowers** 6–8, stalkless, sepals c. 2 mm long, orbicular, with long soft hairs outside; **female flowers** with sepals 3.5–4 mm long, ovate-triangular. **Capsule** egg-shaped-globose, c. 10 mm long, shiny, persistent styles c. 3 mm long. swSichuan (Muli), nwYunnan; thickets by rivers and streams, open shrubberies, 1,750–2,100 m. II–III.

Buxus hebecarpa Hatus.* Similar to *B. mollicula*, but **leaves** glabrous and style shorter than ovary. s & wSichuan (Emeishan, Tianquan); forests, rocky places, 1,500–2,000 m. III–IV.

Buxus bodinieri H. Lévl.* (syn. *Buxus microphylla* Sieb. & Zucc. var. *platyphylla* (C. K. Schneid.) Hand.-Mazz. and var. *aemulans* Rehd. & Wils.) **Shrub** 3–4 m tall with terete, downy branchlets that may become hairless. **Leaves** thin-leathery, usually spathulate or obovate, 2–4 × 0.8–1.8 cm, glossy green above, apex rounded or blunt. **Inflorescence** capitate, 5–6 mm long, dense; **male flowers** c. 10, sepals ovate-orbicular, 2.5 mm long, stamens 6 mm long; **female flowers** with outer sepals 2 mm long and inner 2.5 mm long, ovary 2 mm long, glabrous, styles c. 1.5 mm long, slightly compressed. Gansu, Guangxi, Guizhou, Hubei, Sichuan, sShaanxi, Yunnan [c & eChina]; forests in hilly country, mountain slopes, 400–2,700 m. II–III.

Buxus sinica (Rehd. & Wils.) M. Cheng* Very variable species similar to *B. bodinieri*, but **leaves** broadly elliptic or oblong-elliptic, 1.5–3.5 × 0.8–2 cm, and **female flowers** stalkless. Gansu, Guangxi, Guizhou, Hubei, Sichuan, Shaanxi [c & sChina]; montane forests, thickets along valleys and streams, 600–2,600 m. III–IV. Several varieties are recognised by some botanists.

SARCOCOCCA

A genus of c. 11 species (9 in China) in the central & east Himalaya and west Malaysia. Small, dense, **rhizomatous**, **evergreen shrubs** (in western China) with erect to spreading branchlets. **Leaves** alternate, occasionally almost opposite, lacking stipules, entire, leathery. **Inflorescences** ± compact, axillary clusters, usually with a few male and female flowers; **male flowers** stalkless or short-stalked, with 4 sepals and stamens; **female flowers** small, green, with 5–6 sepals, stalk bearing pairs of appressed scales, ovary 2–3-celled, with 2 or 3 styles. **Fruit** an almost globose berry, green, black or red.

1. Female flowers with 3 carpels and styles . . . .
   . . . *S. wallichii, S. ruscifolia, S. ruscifolia* var. *chinensis*
1. Female flowers with 2 carpels and styles . . *S. hookeriana*

Sarcococca wallichii Stapf **Evergreen shrub** 0.6–3 m tall with glabrous young branchlets. **Leaves** elliptic to lanceolate, 6–11 × 1.8–3 cm, long-pointed. **Inflorescence** globose, 10–15 mm across, 5–6-flowered; **male flowers** stalkless or short-stalked, with 4 sepals, 3–4 mm long, stamens to 10 mm long; **female flowers** 7–9 mm long with 5–6 sepals. **Fruit** almost globose, green turning purplish black when ripe, 9–10 mm. seTibet, w & nwYunnan [swTibet, c & eHimalaya, Myanmar]; understorey in montane forests, valleys, 1,300–2,700 m. VI–VIII.

Sarcococca ruscifolia Stapf* Like the previous species, but smaller and with minutely downy branchlets; **leaves** ovate to broadly or elliptic-ovate, somewhat long-pointed, 3–5.5 × 1.2–3.5 cm; **flowers** smaller and **fruit** red when ripe. Gansu, Guangxi, Guizhou, Hubei, Sichuan, Shaanxi, Yunnan [Hunan]; forests on mountain slopes, streamsides, 2,000–2,600 m. VI–VIII. Var. *chinensis* (Franch.) Rehd. & E. H. Wilson* (syn. *S. confusa* Sealy) is taller, to 2 m, with **leaves** only 10–18 mm wide, **female flowers** with 2 or 3 stigmas, and **fruit** black at maturity. Guizhou, Hubei, Sichuan, Yunnan.

Sarcococca hookeriana Baill. var. *digyna* Franch.* (syn. *S. humilis* Stapf) **Shrub** 0.3–1 m tall with minutely downy branchlets and narrow-elliptic to oblong-elliptic **leaves**, 4–11.5 × 0.8–2.4 cm. **Inflorescence** with 5–8 male and 2 female **flowers**. **Fruit** 7–8 mm long, black when ripe. Guizhou, wHubei, Sichuan, sShaanxi, Yunnan; understorey in forests, 1,000–3,500 m. II–IV. The typical variety (var. *hookeriana*), with 3 carpels and styles and narrower **leaves**, is found in the Himalaya and sTibet.

EUPHORBIACEAE

VERNICIA

A genus with 3 species (2 in China) in south & South East Asia and Japan. Monoecious or dioecious, **deciduous trees** with a covering of simple hairs, and alternate, simple or palmately lobed, long-stalked **leaves**; stipules very small. **Inflorescence** terminal, paniculate, several-flowered, not densely hairy; **flowers** large and conspicuous; petals 5, free, white, with prominent venation; stamens 15–20 in 2 series; ovary 3-celled with a 3-lobed stigma, each lobe 2-fid. **Fruit** a fleshy, indehiscent drupe with 2 **seeds**.

Vernicia fordii (Hemsl.) Airy Shaw* (syn. *Aleurites fordii* Hemsl.) Small **tree** 2.5–8 m tall with heart-shaped **leaves**, 8–21 × 7–18 cm, long-pointed; stalks to 22 cm long. **Inflorescences** 5–11 cm across, on stalk 6–16 cm long, produced as young leaves develop; **flowers** white, 3–5 cm across; petals 2–3 cm long. **Fruit** ± globose with a sharp abrupt point, 2.5–4 cm long. Hubei, Sichuan, Yunnan [sChina]; mixed forests, plantations, 900–2,300 m. III–IV.

EUPHORBIA

A large cosmopolitan genus with c. 1,600 species (77 in China). **Annuals**, **biennials** and **herbaceous perennials**, **shrubs** and large **succulents**, all with milky juice when cut. **Leaves** spirally arranged, opposite or whorled, often with entire margin. **Flowers** in cyathia which are surrounded by conspicuous leaf-like bracts, these often brightly coloured (yellow, orange or red); **male flowers** reduced to a single stamen; **female flower** consisting of a 3-celled ovary with a 3-lobed style surrounded by 4 horn- or kidney-shaped nectary-glands. **Fruit** a 3-parted capsule. All species listed below are herbaceous perennials.

1. Flowers and bracts red or orange ....
   .... *E. griffithii*, *E. pekinensis*
1. Flowers and bracts yellow to greenish yellow, sometimes reddening with age 2
2. Small plants not more than 25 cm tall .... *E. stracheyi*
2. Plants at least 40 cm tall 3
3. Fruit spiny .... *E. jolkinii*
3. Fruit smooth .... *E. wallichii*, *E. wallichii* subsp. *duclouxii*, *E. pallasii*, *E. sikkimensis*

1. *Buxus bodinieri*; Shennongjia, wHubei (PC)
2. *Buxus sinica*; Danyun, nwSichuan (PC)
3. *Sarcococca hookeriana* var. *digyna*; Muli, swSichuan (PC)
4. *Sarcococca hookeriana* var. *digyna*; cult. (PC)
5. *Vernicia fordii*; Lancangjiang, wYunnan (PC)

Euphorbia griffithii Hook. f. (incl. *E. bulleyana* Diels) Patch-forming **rhizomatous perennials** 50–100 cm tall with erect reddish **stems**, sometimes branching above. **Leaves** stalkless, narrowly oblong-elliptic, 4–8 × 1–2.7 cm, blunt, the upper tinged reddish purple or orange-red. **Bracts** ovate, c. 1 cm long, orange to raspberry-red. **Fruit** smooth. w & swSichuan, seTibet, nwYunnan [swTibet, eHimalaya, nMyanmar]; marshes, grassland, alpine meadows, scrub, forest margins, 2,100–3,400 m. V–VII.

Euphorbia pekinensis Rupr. Similar to the previous species, but a more tufted plant with linear-lanceolate to oblong, pointed **leaves**, 5.5–9 × 0.5–0.9 cm and **inflorescences** up to 13 cm across; **flowers** brownish red; bracts ovate, 6–10 mm long. Hubei, Sichuan [nw, c & eChina, Korea, Japan, eRussia]; grassland, mountain slopes, 800–3,000 m. V–VII.

Euphorbia stracheyi Boiss. (syn. *E. megistopoda* Diels) Small tufted **perennial** 5–25 cm tall, growing from a woody rootstock. **Leaves** oblanceolate, elliptic or linear-elliptic, 8–18 × 2–5 mm, apex pointed or rounded. **Bracts** 4–5, ovate, 4–14 mm long, chrome-yellow, often reddening with age. **Inflorescences** broad, rather flat umbels; **flowers** yellow to dark red. **Fruit** smooth. sGansu, Qinghai, w, nw & swSichuan, e & seTibet, w & nwYunnan [c & eHimalaya]; moist peaty soils, subalpine pastures, rocky meadows, stabilised moraines, *Rhododendron* scrub, mossy banks, 2,300–4,600 m. V–VII.

Euphorbia jolkinii Boiss. (syn. *E. nematocypha* Hand.-Mazz.; *E. regina* H. Lévl.) Tufted **herb** 25–100 cm tall, growing from a fleshy rootstock. **Leaves** narrowly elliptic to linear-elliptic, 3–6 × 0.6–1.8 cm, blunt, dark green above, glaucous or purplish beneath, turning flame-red or orange in autumn. **Bracts** yellow, elliptic, 10–16 mm long. **Flowers** yellow with orange nectar-glands. **Fruit** spiny. (?)sGansu, w & swSichuan, eTibet, nwYunnan [nwChina, temperate eAsia]; grassland, rocky slopes, forest margins, open shrubberies, sometimes forming extensive colonies, 1,500–3,200 m. V–VII. (See also photo on p. 13.)

Euphorbia wallichii Hook. f. Tufted plant with erect **stems** 40–80 cm tall, scaly in lower part, growing from a thick rhizome. **Leaves** narrow-obovate to narrow-elliptic, 4–14 × 2–2.6 cm, pointed to blunt. **Bracts** yellow, ovate, 1.5–2.5 cm long. **Inflorescences** laxly umbellate, often with extra branches below main umbel; **flowers** yellow, nectary-glands 4–5 mm long. **Fruit** smooth. seTibet, w & nwYunnan [cAsia, Himalaya, nMyanmar]; riverbanks, scrub, open *Pinus* forests, rocky slopes, 2,300–4,200 m. IV–VI. Plants in which the upper **stem-leaves** are whorled and bear short leafy shoots beneath the **inflorescence** are referable to subsp. *duclouxii* (H. Lévl. & Vaniot) R. Turner* (syn. *E. duclouxii* H. Lévl. & Vaniot; *E. yunnanensis* Radcl.-Sm.) nwYunnan, 2,100–3,000 m.

Euphorbia pallasii Turcz.* Similar to the previous species, but **stems** to 1 m, often forming substantial clumps. **Leaves** ovate to oblong-ovate or narrowly elliptic, blunt, 4–9 × 1–3.2 cm, more glaucous green, sometimes flushed red or purple. **Bracts** ovate, 1.8–2.5 cm long, chrome-yellow. **Flowers** chrome-yellow with brownish yellow nectary-glands. wGansu, (?)w & swSichuan, eTibet, w & nwYunnan; roadsides, amongst rocks, grassy hillsides, 2,500–3,600 m. V–VII.

Euphorbia sikkimensis Boiss. (syn. *E. chrysocoma* H. Lévl. & Vaniot) Clump-forming, somewhat rhizomatous, **herbaceous perennial** 0.6–1 m tall with linear, linear-elliptic or oblanceolate **leaves**, 4.5–10 × 0.9–2 cm, apex blunt; new shoots often bright pink. **Bracts** ovate, 1–1.6 cm long, yellow. **Inflorescence** umbellate, 4–5 cm across; **flowers** yellow. Guangxi, Guizhou, Hubei, Sichuan, Yunnan [Himalaya, Myanmar, nVietnam]; open forests, shrubberies, meadows, 1,500–3,400 m. V–VI.

SAUROPUS

A genus of c. 56 species (15 in China) in India, South East Asia and Australia. **Evergreen**, monoecious **shrubs** with alternate **leaves**. **Male flowers** with 6 sepals united to form a disc, and 3 stamens; **female flowers** with 6 sepals united only at base, and 3 2-fid styles. **Fruit** a capsule.

Sauropus androgynus Merr. Glabrous **shrub** to 2 m, with narrowly winged branchlets. **Leaves** ovate to ovate-oblong, 3.5–6 × 1.5–3 cm. **Flowers** up to 4 in leaf-axils; male c. 4 mm across, sparsely lobed; female slightly larger. **Fruit** 10–15 mm in diameter. Guangxi, Guizhou, Sichuan, seTibet, Yunnan [seAsia, Himalaya, India, Myanmar]; subtropical forests, 270–600 m. VI–VIII.

1. *Euphorbia griffithii*; Heishui, Yulongxueshan, nwYunnan (CGW)
2. *Euphorbia griffithii*; Heishui, Yulongxueshan, nwYunnan (PC)
3. *Euphorbia griffithii*; near Lancangjiang–Jinshajiang divide, Lidiping, nwYunnan (CGW)
4. *Euphorbia pekinensis*; Gonggashan pass, wSichuan (PC)
5. *Euphorbia stracheyi*; Wutoudi, Yulongxueshan, nwYunnan (CGW)
6. *Euphorbia stracheyi*; Gangheba, Yulongxueshan, nwYunnan (CGW)
7. *Euphorbia stracheyi*; Lijiang, nwYunnan (CGW)
8. *Euphorbia jolkinii*; Zhongdian, nwYunnan (CGW)
9. *Euphorbia jolkinii*; Cangshan, nwYunnan (CGW)
10. *Euphorbia jolkinii* in autumn colour; Zhongdian, nwYunnan (EL)
11. *Euphorbia wallichii*; Zhongdian, nwYunnan (CGW)
12. *Euphorbia wallichii*; Xiaozhongdian, nwYunnan (CGW)
13. *Euphorbia wallichii*; Tianbao, Zhongdian, nwYunnan (PC)
14. *Euphorbia pallasii*; south of Geza, nwYunnan (CGW)
15. *Euphorbia sikkimensis*; Lancangjiang-Jinshajiang divide, nr Lidiping, nwYunnan (CGW)
16. *Sauropus androgynus*; cult., Beijing (CGW)

RHAMNACEAE

BERCHEMIA

A genus of c. 30 species (18 in China, 13 endemic) distributed in temperate and tropical areas of east and South East Asia. **Climbing** or erect **shrubs**, rarely small **trees**, with smooth branches, lacking stipular spines. **Leaves** alternate, entire, stipules joined at base and usually persistent. **Inflorescences** terminal or axillary, usually many-flowered. **Flowers** in clusters, or racemose cymes or panicles, hermaphrodite, glabrous, 5-parted; calyx-tube short, usually with triangular lobes; petals oblanceolate, shorter than or equalling calyx-lobes; stamens equalling petals or slightly shorter; ovary superior. **Fruit** an almost cylindrical drupe, purple-red or purple-black when ripe.

1. Erect shrub with 1–3-flowered axillary inflorescences; leaves up to 10 mm long .... **B. edgeworthii**
1. Climbing shrubs with terminal and axillary, many-flowered inflorescences; leaves more than 20 mm long 2
2. Leaves green above and beneath .... **B. yunnanensis**
2. Leaves greyish white beneath .... **B. sinica, B. omeiensis, B. floribunda**

Berchemia edgeworthii Lawson **Deciduous shrub** to 2 m with ovate, oblong, or almost orbicular **leaves**, 4–10 × 3–6 mm, blunt, with 4–5 pairs of lateral veins. **Flowers** white, 2.5–3 mm across, glabrous. **Fruit** orange or purple when ripe, 7–9 mm long, sweet-tasting. w & swSichuan, seTibet, nwYunnan [swTibet, c & eHimalaya]; subalpine shrubberies, cliffs, 2,100–4,500 m. VII–X.

Berchemia yunnanensis Franch.* (syn. *B. pycnantha* C. K. Schneid.) Branchlets spreading, yellow-green to yellow-brown. **Leaves** ovate, ovate-elliptic, oblong-elliptic, 2.5–6 × 1.5–3 cm, pointed and mucronulate, with 8–12 pairs of lateral veins. **Inflorescences** often at ends of lateral branchlets, 2–5 cm long. **Flowers** greenish, 3–4 mm across. **Fruit** cylindrical, 6–9 mm long, red when ripe but eventually turning black. Gansu, Guizhou, Sichuan, Shaanxi, eTibet, Yunnan; hillsides, riverbanks, shrubberies, forests, 1,500–3,900 m. VI–VIII.

Berchemia sinica C. K. Schneid.* Branchlets yellow-brown. **Leaves** ovate-elliptic or ovate-oblong, 3–6 × 1.6–3.5 cm, rounded or blunt, often mucronulate, with 8–10 pairs of lateral veins. **Flowers** yellow or greenish, 3–4 mm across, solitary or a few in clusters forming short, branched, rather narrow cymose lateral panicles to 5 cm long. **Fruit** 5–9 mm long, purple-red or black when ripe. Gansu, Guizhou, Hubei, Sichuan, Shaanxi, Yunnan [Henan, Shanxi]; shrubberies and mixed forests on hillsides and in valleys, 1,000–2,500 m. VI–VIII.

Berchemia omeiensis Fang ex Y. L. Chen & P. K. Chou* Similar to the previous species, but with yellow-green branchlets, the reduced lateral branchlets usually with 2–5 bundles of **leaves**, and the terminal, broad, cymose panicles up to 16 cm long; **fruit** 10–13 mm long, red, turning purple-black. nGuizhou, wHubei, Sichuan; montane forests, 400–1,700 m. VII–VIII.

Berchemia floribunda (Wall.) Brongn. Similar to *B. sinica*, but with ± ovate, glabrous **leaves**, 4–9 cm long and broad, pointed, and terminal or axillary, cymose panicles up to 15 cm long. Guizhou, Hubei, Sichuan, Shaanxi, seTibet, Yunnan [swTibet, c & sChina, Himalaya, Vietnam]; forests and forest margins, shrubberies on hillsides and in valleys, to 2,600 m. VII–X.

VITACEAE

PARTHENOCISSUS

A genus of c. 13 species (10 in China, 6 endemic) in Asia and North America. Woody **climbers**. **Tendrils** branched, tips expanded or curving, later turning into adhesive discs. **Leaves** alternate, simple, with 3 or 5 leaflets. **Inflorescence** paniculate or loosely corymbose; **flowers** 5-parted, hermaphrodite; calyx disc-shaped with a raised rim, glabrous; petals free; stamens 5; ovary 2-celled, with 2 ovules per cell. **Fruit** a 1–4-seeded berry.

Parthenocissus semicordata (Wall.) Planch. Vigorous **climber** to 12 m or more, with 4–6-branched tendrils. **Leaves** long-stalked, with 3 leaflets, leaflets nearly stalkless, central one obovate-elliptic or obovate, sharply pointed, 6–13 cm long, margin toothed, lateral leaflets smaller, ovate-elliptic or oblong, 5–10 cm long, all with toothed margin. **Inflorescence**-rachis inconspicuous; **flowers** greenish yellow; calyx entire; petals ovate-elliptic, 1.8–2.8 mm long, hairless. **Berries** 6–8 mm in diameter, blackish purple when ripe. Gansu, Guizhou, Hubei, Sichuan, Shaanxi, seTibet, Yunnan [c & eHimalaya, Myanmar, Thailand]; growing up trees or shrubs on hillsides, cliffs, 500–3,800 m. V–VII.

Parthenocissus henryana (Hemsl.) Graebn. ex Diels & Gilg* (syn. *Vitis henryana* Hemsl.) Readily distinguished from the previous species by **leaves** with 3 or 5 leaflets, these obovate to oblanceolate, coarsely toothed in upper half, deep velvety green, variegated with silver and pink along midrib and main veins, turning vivid red in autumn; **berries** blue when ripe. Hubei, (?)eSichuan, Shaanxi [cChina]; forests, ravines, cliffs, 100–1,500 m. VI–VII.

AMPELOPSIS

A genus of c. 30 species (17 in China, 14 endemic) in Asia and North and Central America. **Woody climbers** with 2–3-branched tendrils. **Leaves** alternate, simple, pinnately or palmately compound. **Inflorescences** corymbose or compound, opposite leaves or pseudoterminal; **flowers** 5-parted, hermaphrodite or unisexual; petals 5, free; disc developed, with wavy 5-lobed margin; stamens 5; ovary 2-celled, with 2 ovules per cell. **Fruit** a spherical, 1–4-seeded berry.

1. Leaves simple	*A. bodinieri*
1. Leaves compound	*A. delavayana*

Ampelopsis bodinieri (H. Lévl. & Vaniot) Rehd.* (syn. *Vitis bodinieri* H. Lévl.) Vigorous **climber** to 10 m or more with 2-branched tendrils. **Leaves** simple, leaflets ovate or ovate-elliptic, undivided or slightly 3-lobed, 7–12.5 cm long, base heart-shaped, apex sharply pointed or long-pointed, margin with 9–19 sharp teeth on each side; leaf-stalks 2.5–6 cm long, glabrous. **Inflorescence** lax; **flowers** greenish yellow; calyx-teeth inconspicuous; petals elliptic, 2–2.5 mm long; anthers yellow, elliptic. **Berries** 6–8 mm in diameter, somewhat flattened, 3–4-seeded, dark blue when ripe. Guangxi, Guizhou, Hubei, Shaanxi, Yunnan [c, s & seChina]; growing up trees or shrubs in valleys and on shady hillsides, 200–3,000 m. IV–VI.

Ampelopsis delavayana Planch.* Vigorous **climber** to 8 m or more with 2–3-branched tendrils and hairy young **stems**. **Leaves** compound with 3 or 5 leaflets, central leaflet lanceolate or elliptic-lanceolate, 5–13 cm long, long-pointed, lateral leaflets ovate-elliptic or ovate-lanceolate, 4.5–11.5 cm long, all with coarse, sharp marginal teeth, downy at first; leaf-stalks 3–10 cm long, pinkish. **Inflorescences** opposite leaves with stalk 2–4 cm long; petals ovate-elliptic, 1.3–2.3 mm long, glabrous; anthers squarish. **Berries** globose, 8 mm in diameter, 2–3-seeded, dark blue when ripe. Gansu, Guangxi, Guizhou, Sichuan, Shaanxi, Yunnan [c & sChina]; growing on trees and shrubs in valleys, cliffs, to 2,700 m. V–VIII.

VITIS

A genus of c. 60 species (38 in China, 34 endemic) in temperate and subtropical regions. Woody, tendrilled **climbers**. **Leaves** simple or palmate; stipules usually soon falling. **Inflorescence** a thyrse; **flowers** usually dioecious, rarely hermaphrodite; calyx minute; petals united at apex and shed as a cap at flowering time; stamens opposite petals, undeveloped and abortive in female flowers; disc conspicuous, 5-lobed. **Ovary** 2-celled, with 2 ovules per cell. **Fruit** a 2–4-seeded berry.

Vitis betulifolia Diels & Gilg* (syn. *V. tricholada* Diels & Gilg; *V. hexamera* Gagnep.; *V. shimenensis* W. T. Wang) Vigorous **deciduous climber** to 8 m or more, branchlets terete, with conspicuous longitudinal ridges. **Tendrils** 2-branched, opposite leaves. **Leaves** oval or ovate-elliptic, 4–12 cm long, undivided or 3-lobed, base heart-shaped or almost truncate, margin with 15–25 sharp teeth on each side; leaf-stalks 2–6.5 cm long, with cobwebby hairs when young. **Panicles** sparse, opposite leaves, basal branches 4–15 cm long; **flowers** greenish yellow; filaments thread-like, 1–1.5 mm long, bearing elliptic yellow anthers c. 0.4 mm long. **Berries** globose, 8–10 mm in diameter, purple-black when ripe. seGansu, Hubei, Sichuan, Shaanxi, Yunnan [cChina]; growing on shrubs or trees on hillsides and in valleys, 650–3,600 m. III–VI.

1. *Berchemia yunnanensis*; wYunnan (PC)
2. *Berchemia yunnanensis*; Jiuzhaigon, nSichuan (PC)
3. *Parthenocissus henryana*; cult. (CGW)

SAPINDACEAE

The well-known fruit longan (*Dimocarpus longan*), lychee (*Litchi chinensis*) and rambutan (*Nephelium lappaceum*) belong to this family.

KOELREUTERIA

A genus with 3 species (2 in China) in China and Taiwan. **Deciduous** or **evergreen shrubs** or **trees** with alternate, odd-pinnate or 2-pinnate **leaves**; stipules absent. **Inflorescences** large, usually terminal, with many spreading **branches**; **flowers** zygomorphic, hermaphrodite, male or female (then sometimes confined to separate plants); sepals usually 5, valvate; petals 4(–5), slightly unequal in length, clawed; stamens often 8, sometimes fewer, inserted on a thick disc. **Fruit** a papery capsule, triangular in section, 3-parted.

Koelreuteria paniculata **Laxm.*** **Deciduous tree** to 12 m, or **shrub**, with thick, greyish brown to black **bark**. **Leaves** pinkish when young, pinnate, odd–2-pinnate or sometimes 2-pinnate, to 50 cm long; leaflets ovate to ovate-lanceolate, 5–10 cm long, toothed. **Inflorescences** 25–30 cm tall; **flowers** pale yellow, slightly fragrant; petals reflexed, linear-oblong, 5–9 mm long; stamens 7–9 mm long in male flowers, staminodes 4–5 mm in female flowers. **Fruits** conical, 3-ridged, 4–6 cm long, reddish when young, papery and brown at maturity. Gansu, Guangxi, Guizhou, Hubei, Sichuan, Shaanxi, Yunnan [ne, c & sChina]; dry valleys, open slopes, to 2,700 m. VI–VIII. Often planted.

Koelreuteria bipinnata Franch.* Similar to the previous species, but **leaves** 2-pinnate, leaflets slightly oblique at base, apex pointed to somewhat long-pointed, margin minutely toothed, and **fruits** ellipsoid, broadly egg-shaped or almost globose with a blunt apex. Guizhou, Hubei, Sichuan, Shaanxi, Yunnan [sChina]; hot dry valleys, open forests on limestone, 400–2,500 m. VI–VIII.

DODONAEA

A genus of c. 65 species (1 in China) in warm and tropical areas, especially Australia. Dioecious, sticky **shrubs** or **trees** with ridged branches. **Leaves** simple or pinnate, alternate, lacking stipules. **Flowers** regular, solitary or arranged in terminal and axillary racemes, corymbs or panicles; sepals usually 4, valvate, deciduous when mature; petals absent; stamens 5–8; ovary ellipsoid, usually 2- or 3-celled, style terminal, much longer than ovary, often twisted, stigma 2–6-lobed. **Fruit** a winged capsule, boat-like, with 2–3(–6) **seeds**.

Dodonaea viscosa (L.) Jacq. **Shrub** or small **tree** to 3 m; branchlets flat, narrowly winged or ridged. **Leaves** simple, linear to oblong, 5–12 cm long, entire, often shallowly wavy along margin. **Inflorescences** terminal or axillary, shorter than leaves, densely flowered; sepals 4, lanceolate or long-elliptic, c. 3 mm long, blunt. **Capsule** compressed-globose, 2–3-winged, 15–22 mm long. Guangxi, Guizhou, sHubei, Sichuan, Yunnan [cosmopolitan in tropical and subtropical regions]; dry slopes, fields, sandy places, banks, sometimes planted as a hedge, sea-level to 1,500 m. VIII–XI.

HANDELIODENDRON

(included by some botanists in Hippocastanaceae)

A genus with a single species endemic to China. **Deciduous shrub** or **tree** with opposite, digitate **leaves** that lack stipules. **Inflorescences** terminal, lax, many-flowered; **flowers** hermaphrodite, zygomorphic; sepals 5, overlapping; petals 4–5, long, elliptic or oblanceolate, top half reflexed, with 2 small scales at base inside; disc half-moon-shaped, thick, irregularly lobed; stamens 7–8, protruding, filaments unequal in length, anthers broadly egg-shaped, with glands at cell base. **Fruit** a club-shaped or nearly pear-shaped capsule with a conspicuous gynophore, splitting into 3 schizocarps.

1. *Koelreuteria paniculata*; wSichuan (TK)
2. *Koelreuteria paniculata* in fruit; Beijing (CGW)
3. *Koelreuteria bipinnata*; near Leshan, sSichuan (HJ)
4. *Dodonaea viscosa*; cult. (CGW)
5. *Handeliodendron bodinieri*; Maolan, seGuizhou (PC)

Handeliodendron bodinieri (H. Lévl.) Rehd.* Plant to 8 m with grey **bark**. **Leaves** digitate; leaflets 4–5, thin-papery, elliptic to obovate, 3–12 cm long, tailed to sharply pointed, glabrous, with black glands scattered beneath; stalks 4–11 cm long. **Inflorescences** c. 10 cm long; sepals 2–3 mm long, ciliate; petals 4–5, c. 9 mm long, appressed-hairy outside; stamens 7–8. **Capsule** club- to pear-shaped, 22–32 mm long (including stalk). nwGuangxi, s & seGuizhou; forests, forest margins, on limestone, 500–800 m. V–VI.

SAPINDUS

A genus of 13 species (1 in China) in the tropics and subtropics. Monoecious (or sometimes dioecious outside China) **trees** or **shrubs** with even-pinnate, rarely simple, alternate **leaves** that lack stipules; leaflets entire, opposite or alternate. **Inflorescences** large, many-branched, terminal or in bundles at branchlet apex; **flowers** unisexual, regular or not; sepals (4–)5; petals usually 5 and clawed, or 4 and lacking a claw; disc fleshy; stamens 8, protruding, filaments hairy at base. **Fruit** divided into 3 almost globose or obovoid schizocarps, but usually with only 1 or 2 developing.

Sapindus delavayi (Franch.) Radlk.* **Deciduous tree** to 10 m with black-brown **bark**. **Leaves** pinnate, 25–35 cm long, long-stalked; leaflets 4–6, asymmetrically ovate or oblong-ovate, turning golden yellow in autumn. **Inflorescences** terminal, erect, often branched 3 times, main rachis and branchlets softly hairy; **flowers** zygomorphic; sepals 5; petals 4(5–6), narrowly lanceolate, c. 5.5 mm long. **Fruits** globular, greenish yellow. Guizhou, Hubei, Sichuan, Yunnan; dense forests, 1,200–2,600 m. IV–V.

LITCHI

A genus of just 2 species (1 in China) distributed in China and the Philippines. Small **trees** with even-pinnate, alternate **leaves** that lack stipules. **Inflorescences** terminal, golden-hairy with small bracts. **Flowers** unisexual, monoecious, regular; calyx cup-shaped, 4–5-lobed, valvate; petals absent; disc entire; stamens (male flowers) 6–8, protruding, filaments thread-like, hairy; ovary (female flowers) short-stalked, 2-lobed, 2-celled, stigma 2–3-lobed. **Fruit** egg-shaped or almost globose, skin crusty when dry, usually covered with pointed tubercles, sometimes nearly smooth; **seed** with shiny brown testa; aril fleshy, white, sweet.

Litchi chinensis Sonn.* **Evergreen tree** to 15 m with deep glossy green pinnate **leaves** 10–25 cm long, glaucous beneath; leaflets 4–8, lanceolate or ovate-lanceolate, 6–15 cm long, sharply pointed to somewhat long-pointed, entire. **Inflorescences** terminal, large, many-branched; calyx golden-hairy; stamens 6–8; ovary densely tuberculate and with stiff hairs. **Fruit** globose to almost so, 2–3.5 cm, usually dark red to flesh-red when ripe, succulent inside. [scChina]; widely cultivated in s, sw & seChina at low altitudes [seAsia]. IV–V.

HIPPOCASTANACEAE

AESCULUS

A genus of c. 25 species (3 in China) distributed in temperate Asia, Europe and North America. Large **deciduous trees** or **shrubs** with sticky **buds**. **Leaves** opposite, palmatifid, long-stalked; leaflets 5–7, obovate; leaf-scars often horseshoe-shaped. **Inflorescences** paniculate, at end of current season's growth; **flowers** showy, white, yellow or red; petals 4–5. **Fruits** prickly or not, containing 1–2 large, glossy, chestnut-brown **seeds** (conkers).

Aesculus assamica Griffith (syn. *A. wangii* Hu, invalid) Large spreading **tree** 15–30 m tall. **Leaves** with 5–7, glossy dark green leaflets, each 12–35 × 5–14 cm, obovate, long-pointed, margin toothed. **Inflorescences** 35–40 cm long; **flowers** white or pale yellow; petals with brown or purple blotching; stamens as long as petal-lobes. **Fruits** yellowish brown, globose but flattened above, 4.5–7.5 cm long. w & swGuangxi, s & swGuizhou, seTibet, s & seYunnan [c & eHimalaya, Thailand, nVietnam]; mixed often wet woods, roadsides, field boundaries, 900–1,700 m. II–V.

Aesculus wilsonii Rehd.* **Tree** 20–30 m tall with glabrous or minutely downy young shoots. **Leaves** with 5–7, matt green leaflets, each 8–25 × 3–8.5 cm, oblong-elliptic to oblanceolate, base rounded to heart-shaped, apex long-pointed, margin shallowly and evenly toothed; leaflet-stalks 1–2 cm long. **Inflorescences** up to 40 cm long; **flowers** 12–18 mm across, fragrant; petals white with yellow blotching; stamens more than twice as long as petal-lobes. **Fruits** egg-shaped to pear-shaped, mucronate, 3–4.5 cm. sGansu, Guizhou, wHubei, Sichuan, sShaanxi, neYunnan [c & sChina]; mixed woods and forests, ravines, road- and streamsides, around villages, 600–2,000 m. IV–VI.

ACERACEAE

ACER

A genus of c. 120 species, (99 in China) in the northern hemisphere with a few on tropical mountains. **Trees** or **shrubs** with opposite, entire, palmate or pinnate **leaves**, often with bright autumn colours. **Flowers** in cymose clusters, often pendent, often unisexual, 4(–5)-parted. **Fruit** a pair of winged **seeds** (samaras or keys).

1. Leaves simple 2
1. Leaves palmately 3–7-lobed or with 3 leaflets 3
2. Trunks with smooth bark .... **A. fabri**
2. Trunks with snake-bark pattern ....
   .... **A. davidii, A. grosseri, A. oblongum**
3. Leaves 3–7-lobed 4
3. Leaves with 3 leaflets 6
4. Inflorescence erect .... **A. caudatum**
4. Inflorescence horizontal to pendent 5
5. Trunks with snake-bark pattern ....
   .... **A. forrestii, A. forrestii** subsp. *laxiflorum*
5. Trunk not snake-barked ....
   .... **A. cappadocicum** subsp. *cappadocicum*,
   A. cappadocicum subsp. *sinicum*, **A. longipes**,
   A. oliverianum, A. sterculiaceum subsp. *sterculiaceum*,
   A. sterculiaceum subsp. *franchetii*
6. Bark peeling, red-brown; leaves toothed ... **A. griseum**
6. Bark smooth, grey-brown; leaves untoothed .. **A. henryi**

Acer fabri Hance (syn. *A. farg*esii Veitch ex Rehd.; *A. prainii* H. Lévl.) **Evergreen tree** 4–10 m tall with smooth, dark grey-brown **bark**. **Leaves** entire, oblong-ovate to elliptic, 5–11 × 2–3.5 cm, long-pointed, with a very finely toothed or entire margin, with tufts of hairs at the vein-angles beneath; leaf-stalks 1–2 cm long. **Fruit** 2.5–4.8 cm long, with red wings and nutlets. Guangxi, Guizhou, Hubei, Sichuan, Yunnan [s & seChina, Vietnam]; mixed broadleaved forests, 500–2,000 m. III–V.

Acer davidii Franch. (syn. *A. cavalieiri* H. Lévl.) Small **tree** c. 12 m tall, often many trunked, with greenish, grey-green, greyish or reddish **bark**, with longitudinal white stripes and purplish red branchlets with greyish white longitudinal stripes. **Leaves** entire or slightly lobed, ovate to oblong-ovate, 6–12 × 4–8 cm, long-pointed, glossy dark green, paler beneath, turning yellow or red in autumn, unevenly toothed; leaf-stalks 3–6 cm long, reddish. **Inflorescence** long, pendent, to 6 cm long, with up to 20 yellowish **flowers**. **Fruit** 2.5–3 cm long, wings spreading almost horizontally, red, turning yellow-green. sGansu, Guangxi, Guizhou, Hubei, Sichuan, sShaanxi, Yunnan [cChina, Myanmar]; mixed broadleaved woodland, 1,000–3,000 m. III–IV.

Acer grosseri Pax* (syn. *A. davidii* Franch. subsp. *grosseri* (Pax) P. C. de Jong) Similar to the previous species, but **bark** green with white stripes; **leaves** more rounded, 5–6 × 4–5 cm, palmately 3-veined and usually with 2 spreading, long-pointed lobes just above middle, margin more sharply toothed, rusty-downy at first near base beneath. **Inflorescence** 10–15 cm long. **Fruit** with curved wings, spreading obtusely or almost horizontally. Gansu, wHubei, Sichuan, Shaanxi [cChina]; similar habitats, 1,000–1,600 m. IV–V.

Acer oblongum Wall. ex DC. **Deciduous** or **semi-evergreen tree** 5–20 m tall, with greyish, blackish grey or reddish brown **bark**, peeling in irregular strips. **Leaves** leathery, entire or slightly lobed, oblong to oblong-ovate, 5–17 × 2–5 cm, long-pointed, glossy dark green above, glaucous beneath. **Inflorescence** short, pendent, downy panicles; **flowers** greenish white. **Fruit** 2–2.5 cm long, glabrous, wings parallel to spreading, brownish yellow to purplish. sGansu, Guizhou, wHubei, Sichuan, sShaanxi, seTibet, Yunnan [swTibet, sChina, seAsia, Himalaya, Myanmar]; forests and glades, streamsides, 1,000–2,000 m. III–IV.

1. *Aesculus assamica*; near Napo, swGuangxi (PC)
2. *Aesculus assamica*; Zhenkang, seYunnan (PC)
3. *Aesculus wilsonii*; wSichuan (TK)
4. *Acer davidii*; cult. (CGW)
5. *Acer davidii*; Chongjiang, nwYunnan (PC)
6. *Acer grosseri*; cult. (CGW)

1. *Acer caudatum*; Jiuzhaigou, nwSichuan (PC)
2. *Acer forrestii*; Muli, swSichuan (PC)
3. *Acer sterculiaceum* subsp. *franchetii*; Yulongxueshan, nwYunnan (CGW)
4. *Acer sterculiaceum* subsp. *franchetii*; Lancangjiang–Jinshajiang divide, nwYunnan (CGW)
5. *Acer griseum*; cult. (CGW)
6. *Acer griseum* trunk; cult. (CGW)
7. *Acer griseum* in autumn colour; cult. (PC)

Acer caudatum Wall. (syn. *A. papilio* King) Slender **deciduous tree** to 10 m with ash-grey, smooth young **bark**. **Leaves** 5(–7)-lobed, 8–12 cm long, deep green above, paler beneath, with triangular-ovate, toothed, long-pointed lobes. **Flowers** yellowish green, in erect to ascending cylindrical, reddish racemes that elongate to 10 cm as fruit develops; **flowers** either hermaphrodite or male, borne in the same inflorescence. **Fruit** with ascending wings, 25–27 mm long, yellowish brown. seGansu, wHubei, wSichuan, sShaanxi, seTibet, nwYunnan [cChina, eHimalaya, nMyanmar]; mixed montane forests, streamsides, 1,700–4,000 m. IV–V.

Acer forrestii Diels (syn. *A. grosseri* Pax var. *forrestii* (Diels) Hand.-Mazz.; *A. pectinatum* Wall. ex G. Nichols. forma *rufinerve* A. E. Murray; *A. pectinatum* subsp. *forrestii* (Diels) A. E. Murray) Slender **tree** 7–17 m tall with purplish to green **bark** striped with white; branchlets purplish or red, striped white, glabrous. **Leaves** heart-shaped, 3-lobed or not, 7–12 × 5–9 cm (those on flowering shoots less markedly lobed), with finely toothed margin, dark green turning orange-red in autumn; leaf-stalks red. **Flowers** brownish green to purplish red, in slender racemes. **Fruit** 2–2.5 cm long, red at first, becoming yellowish, glabrous, wings almost horizontal. wSichuan, e & seTibet, nwYunnan [Himalaya]; open places in mixed forest, streamsides, 2,800–3,500 m. V–VI. **Subsp.** *laxiflorum* (Pax) Murray* (syn. *A. laxiflorum* Pax) is similar, but has less conspicuously striped **bark**, shorter, broader, 3(–5)-lobed **leaves** (turning yellow in autumn), larger **flowers**, and **fruit** up to 3 cm long. n, w, sw & wSichuan, n & nwYunnan; mixed forests and forest margins, valley bottoms, ravines. V–VI.

Acer cappadocicum Gled. subsp. *sinicum* (Rehd.) Hand.-Mazz.* (syn. *A. sinicum* Rehd.) Medium-sized, densely branched **tree**, 5–16 m tall, with a broad crown and brownish green **bark** with longitudinal fissures. **Leaves** 5(–7)-lobed, 6–8 × 5–8 cm, lobes long-pointed, olive-green when young, turning coppery-red or yellow in autumn; leaf-stalks reddish, 6–10 cm long. **Inflorescence** up to 5 cm long, with yellow **flowers** in corymbs. **Fruit** with wings 2–3 cm long, bright red or purplish when young. wHubei, Sichuan, sShaanxi, eTibet, Yunnan; mixed montane forests, streamsides, ravines, 1,500–2,800 m. V–VI. Plants with larger 5–7-lobed **leaves**, 12–18 cm across, and with **fruits** 4.5 cm long or more, are referable to **subsp.** *cappadocicum* found in sTibet and apparently nwYunnan, but widespread in the Himalaya, Europe and wAsia.

Acer longipes Franch. ex Rehd.* **Deciduous tree** 10 m tall, occasionally more, with a domed crown and reddish brown or purplish grey **bark**; branchlets glabrous. **Leaves** 3–5-lobed, 8–13 × 7–14 cm, dark green covered with brown down, turning brownish red in autumn, lobes long-pointed, margins entire; leaf-stalks 6–10 cm long. **Inflorescence** branched, to 10 cm long, with yellow **flowers**. **Fruit** with wings 2.5 cm long, spreading almost horizontally to semi-erect, yellowish brown. nGuangxi, wHubei, sShaanxi [cChina]; mixed forest in lower montane regions, 300–1,600 m. IV–V.

Acer oliverianum Pax (syn. *A. campbellii* Hook. f. & Thomson ex Hiern subsp. *oliveriana* (Pax) A. E. Murray) Medium-sized **deciduous tree** 8–12 m tall, often multi-trunked, with green **bark**, turning olive-brown; branchlets purplish. **Leaves** usually 5-lobed, 5–10 × 5–10 cm, red when young, later dark greenish yellow, lobes ovate to triangular, long-pointed, sharply toothed, glabrous except on veins and in vein-axils; leaf-stalks 3–6 cm long. **Inflorescence** 4–8 cm long, with yellow **flowers**. **Fruit** with wings 2.5–4 cm long, horizontal. sGansu, Guizhou, wHubei, Sichuan, sShaanxi, Yunnan [c & eChina, Taiwan]; mixed montane forests. 1,000–2,000 m. IV–VI.

Acer sterculiaceum Wall. subsp. *franchetii* (Pax) A. E. Murray* (syn. *A. franchetii* Pax) **Deciduous tree** to 15 m with rather smooth **bark** and glabrous branchlets. **Leaves** 3(–5)-lobed, 9–20 × 15–23 cm, lobes forward-pointing, triangular, margin coarsely toothed; leaf-stalks to 15 cm long, purplish or greenish when young. **Inflorescence** to 8 cm long, racemose; **flowers** yellow-green, male and female borne on separate plants. **Fruit** 4–6.5 cm long, hairy, wings spreading at right angles, or wider. Guizhou, nwHubei, Sichuan, sShaanxi, nw & eYunnan [swHenan]; forests and forest margins, 1,800–3,100 m. IV. **Subsp.** *sterculiaceum* with 5(–3–7)-lobed **leaves**, widespread in the Himalaya and sTibet, is occasionally recorded in nwYunnan.

Acer griseum (Franch.) Pax.* Small **deciduous tree** to 20 m with characteristic peeling, orange- to red-brown, semi-translucent **bark** that hangs in large loose flakes. **Leaves** with 3 leaflets; leaflets simple, oblong to oblong-ovate, 3–8 × 2–5 cm, toothed towards the blunt apex, purple-red when young, turning bright red and orange in autumn; leaf-stalks 2–4 cm long, hairy. **Inflorescence** 3–5-flowered, cymose, on pendulous downy stalks; **flowers** yellowish. **Fruit** 3–4 cm long, very downy, wings almost forming a right angle. seGansu, wHubei, eSichuan, sShaanxi [cChina]; mixed forests, streamsides, 1,500–2,000 m. IV–V.

Acer henryi Pax* Small **deciduous tree** to 10 m with smooth, olive-green **bark**, branchlets downy at first. **Leaves** with 3 leaflets; leaflets simple, elliptic to oblong-ovate, 6–12 × 3–5 cm, long-pointed, with entire margin, green, turning orange-red in autumn; leaf-stalks 5–10 cm long. **Inflorescence** drooping, downy spikes, to 7 cm long, many-flowered, yellowish, produced before leaves, with whitish petals, the male and female **flowers** borne on separate plants. **Fruit** 2–2.5 cm long, red at first, wings diverging from a small angle to right-angled. Gansu, Guizhou, Hubei, Sichuan, sShaanxi [cChina]; forests, glades, streamsides, 500–1,500 m. IV–V.

JUGLANDACEAE

PTEROCARYA

A genus of 6 species (5 in China, 2 endemic) found from the Caucasus to eastern Asia. Monoecious, **deciduous trees** with odd- or even-pinnate **leaves** with a toothed margin. **Inflorescence** lateral or terminal, pendulous, male and female separate, male on old growth or at base of new, female terminal on new growth; **male flowers** with (2–)4 sepals and 5–18(–30) stamens; **female flowers** with 4 sepals fused to ovary, and a short, 2-lobed style. **Fruiting spikes** pendent, elongated, each **fruit** a nutlet with 2 wings, or with wings fused to form a circular disc.

1. Nutlets with forward pointing, rounded or narrow wings .... *P. stenoptera, P. hupehensis*
1. Nutlets cymbal-like, with the wings forming a circular disc .... *P. paliurus*

Pterocarya stenoptera **C. DC.** Large **tree** to 30 m with a fissured **trunk**. **Leaves** mostly even-pinnate, to 35 cm long, usually with 11–21, elliptic to lanceolate leaflets, 8–12 × 2–3 cm; rachises winged or at least ridged. **Inflorescence** greenish, male catkins to 6 cm long, female to 20 cm. **Fruiting spikes** to 45 cm long; **nutlets** ellipsoid, 6–7 mm long, with a pair of forward-pointing, narrow, tapering wings. Gansu, Guangxi, Guizhou, Hubei, Sichuan, Shaanxi, Yunnan [n, c & eChina, Japan, Korea, Taiwan]; forests in hills and mountains, river- and streamsides, to 1,500 m. IV–V.

Pterocarya hupehensis **Skan*** In contrast to the previous species this has odd-pinnate **leaves** usually with 5–11 leaflets, occasionally more, with wingless rachises; **nutlets** with broad, 10–15 mm, elliptic-ovate wings. nGuizhou, wHubei, wSichuan, sShaanxi; damp forests, especially along streams, 700–2,000 m. IV–V.

Pterocarya paliurus **Batalin*** (syn. *Cyclocarya paliurus* (Batalin) Iljinsk.) **Tree** to 15 m or more with downy young shoots. **Leaves** to 30 cm long, usually with 7 or 9(–11), oblong to oval leaflets to 12.5 × 7 cm, blunt or pointed at apex, finely toothed, glabrous except for midrib. **Male catkins** mostly borne in pairs, to 10 cm long; **fruiting catkins** to 25 cm long, **nutlets** (including wing) 37–60 mm across. Hubei, Sichuan, Shaanxi [cChina]; damp forests, forest margins, field boundaries, to 1,500 m. IV–V.

PLATYCARYA

An east Asian genus considered to have a single species, although a second species from Guangxi has very recently been described. Closely related to *Pterocarya* and with similar winged nutlet **fruits**, however, the catkins of both sexes are erect.

Platycarya strobilacea **Sieb. & Zucc.** Medium-sized, **deciduous tree** or **shrub** to 15 m. **Leaves** odd-pinnate, occasionally simple, to 30 cm long; leaflets 5–15, occasionally more, ovate-lanceolate, to 11 × 3 cm, base rounded or rather oblique, apex long-pointed, margin doubly toothed. **Flowers** in dense, erect, hairy, spike-like clustered catkins; male usually lateral, slender, to 8 cm long with a drooping tip, terminating young growths; female terminal, surrounded by the male catkins, cone-like, c. 3 cm long. **Fruit** tiny 2-winged nutlets to 4 mm across. sGansu, Guangxi, Guizhou, Hubei, Sichuan, Shaanxi, Yunnan [c, e & neChina, Korea, Japan, Taiwan, Vietnam]; forests and forest margins, 400–2,000 m. IV–VI.

ENGELHARDTIA

A genus of 7 species (4 in China, 1 endemic) found from the Himalaya to Malaysia. Similar to *Pterocarya*, but **trees evergreen** or **deciduous** with male and female **inflorescences** combined in panicles or separate, the male and female **flowers** with 3-lobed bracts; **nutlets** 3-winged.

Engelhardtia roxburghiana **Wall.** Very variable, large, monoecious, **evergreen** or **semi-evergreen tree** to 30 m with even-pinnate **leaves**; leaflets 2–10, elliptic to lanceolate, entire, 4.5–14 × 1.5–5 cm, glabrous or downy beneath. **Inflorescence** greenish; **male flowers** with 10–12 stamens enclosed within hooded sepals, anthers glabrous; **female flowers** short-stalked, without a style. **Nutlets** 3–5 mm, glabrous, wings glabrous, central lobe twice the size of lateral. Guangxi, Guizhou, Hubei, Sichuan, Yunnan [c & seChina, seAsia, eHimalaya, Taiwan]; evergreen and broadleaved forests, 200–1,500 m. III–VI.

Engelhardtia spicata **Leschen. ex Blume** In contrast to the previous species, a smaller **tree** not more than 20 m; **leaves** with 4–14, slightly broader leaflets; **male flowers** with 4–13 stamens that are not enclosed, anthers hairy; **female flowers** stalkless, with a style; **nutlets** with stiff hairs. Guangxi, Guizhou, seTibet, Yunnan [s & seChina, seAsia, eHimalaya, nIndia, Myanmar]; similar habitats, to 2,100 m. III–VI.

ANACARDIACEAE

RHUS

A northern hemisphere genus with c. 200 species (6 in China). Gummy **trees**, **shrubs** and **climbers**, with simple or pinnate **leaves** and drupaceous **fruit**. Many are poisonous or can cause dermatitis. The genus includes the Chinese/Japanese lacquer tree (*R. verniciflua* Stokes), and poison ivy (*R. radicans* L.), a North American native.

Rhus chinensis **Mill.** **Shrub** or **tree** to 10 m with rusty-downy branchlets. **Leaves** pinnately compound with 7–13 ovate to oblong leaflets, 6–12 × 3–7 cm, dark green above, paler beneath, toothed; rachis winged. **Inflorescence** broadly pyramidal, much-branched and many-flowered, rusty-downy, male up to 40 cm long, female smaller. **Flowers** white, tiny, petals 1–1.6 mm long. **Fruit** globose, 4–5 mm, red at maturity. Gansu, Guangxi, Guizhou, Hubei, Sichuan, Shaanxi, seTibet, Yunnan [much of China (except n), seAsia, c & eHimalaya, Japan, Korea]; forests in lowlands and mountains, ravines, streamsides, shrubberies, sea level to 2,800 m. VIII–IX.

COTINUS

A genus of 5 species (3 in China, 2 endemic) found across the northern hemisphere. **Deciduous shrubs** or small **trees**, exuding a pungent, gummy, dark **sap** when cut. **Leaves** simple, alternate, long-stalked. **Flowers** in billowy **inflorescences**, 5-parted, fertile flowers scattered amongst numerous feathery filamentous sterile flowers. **Fruit** a small red or brown drupe.

Cotinus coggygria **Scop.** (syn. *Rhus cotinus* L.) **Deciduous**, broadly bushy **shrub** to 4.5 m with glabrous branchlets. **Leaves** long-stalked, rounded to obovate, to 8 cm long, with spreading parallel veins and a rounded or notched apex, turning yellow, red and orange in autumn. **Inflorescence** a lax panicle with numerous thread-like silky appendages, pale flesh-coloured at flowering but soon turning smoky grey; **flowers** few and small. **Fruit** few, small, dry, 1-seeded. Hubei, Sichuan, Shaanxi, Yunnan [cChina, Europe eastwards to China]; dry sunny banks and slopes, rocky places, shrubberies, 700–2,400 m. VI–VII.

1. *Pterocarya stenoptera*; cult. (CGW)
2. *Pterocarya paliurus*; Sichuan (TK)
3. *Pterocarya paliurus*; Moxi, Guanyin, Sichuan (TK)
4. *Platycarya strobilacea*; Shennongjia, wHubei (PC)
5. *Engelhardtia roxburghiana*; cult. (PC)
6. *Rhus chinensis*; Shennongjia, wHubei (PC)
7. *Cotinus coggygria*; nwSichuan (TK)

1. *Coriaria nepalensis*; Songpan, nwSichuan (PC)
2. *Zanthoxylum armatum*; Muli, swSichuan (PC)
3. *Zanthoxylum oxyphyllum*; near Dali, nwYunnan (CGW)
4. *Zanthoxylum* aff. *nepalense*; Sichuan (TK)
5. *Zanthoxylum bungeanum*; Danyun, Huanglong, nwSichuan (PC)

CORIARIACEAE

CORIARIA

A genus of 5 species (2 in China) in Eurasia, New Zealand and Central & South America. Poisonous, monoecious, **shrubby** or **herbaceous** plants with arched **stems** and simple, opposite, entire, prominently 3-veined **leaves**. **Flowers** small, in racemes terminating current season's growth or in axils of previous season's growth; sepals and petals 5, the latter persistent; stamens 10; carpels 5. **Fruit** a head of achenes enclosed in the fleshy petals.

Coriaria nepalensis **Wall.** (syn. *C. kweichovensis* Hu; *C. sinica* Maxim.) Coarse, deciduous **shrub** 1.5–2.5 m tall with arching branches that are red when young, blackish when mature. **Leaves** ovate to oblong-elliptic, 2.5–10 cm long, with a very short purple stalk. **Racemes** axillary, slender; **flowers** greenish yellow; **male inflorescence** 1.5–2.5 cm long, densely many-flowered, opening before leaves; **female inflorescence** 4–6 cm long, opening with the leaves, with a purple rachis. **Fruit** globular, red to purplish black when mature, surrounded by the 5 enlarged fleshy petals. Gansu, Guizhou, Hubei, Sichuan, Shaanxi, seTibet, Yunnan [swTibet, Himalaya, Myanmar]; hillsides, hedgerows, roadsides, 400–3,200 m. II–V.

Coriaria terminalis **Hemsl.*** Like the previous species, but a subshrub with terminal **inflorescences**. wSichuan, eTibet; similar habitats, 1,900–3,700 m.

AUCUBACEAE

AUCUBA

A genus of 3 species (2 in China) in Asia. Dioecious, **evergreen shrubs** or small **trees** with opposite, leathery, simple, dark glossy green **leaves** with toothed margins. **Inflorescences** terminal, paniculate, on short lateral branches; **flowers** 4-parted; **female flowers** small, green, stamens absent; **male flowers** small, without an ovary. **Fruit** an egg-shaped, orange, red or scarlet 1-seeded berry.

Aucuba chinensis **Benth.** Variable **shrub** or small **tree** 0.7–2 m tall with linear-lanceolate to oblanceolate, long-pointed **leaves**, 8–22 × 1.5–5.5 cm, coarsely and sharply toothed, dark greyish green above, glaucous beneath; leaf-stalks 1–3 cm long. **Female flowers** 7 mm across, olive- to livid-green, fragrant; petals long-pointed; **male flowers** red-purple. **Berries** red, 12–14 mm long. Guizhou, Hubei, Sichuan, Yunnan [s & seChina, Taiwan, Myanmar, nVietnam]; open shrubberies and woods, 450–2,000 m. IV–V. **Var. *obcordata* (Rehd.) Q. Y. Xiang*** from 1,000–2,000 m in Guangxi, Hubei and eSichuan, is a **shrub** 0.6–2 m tall with reverse heart-shaped, deeply toothed **leaves** that are often somewhat 3-lobed at apex, 10–14 × 5.5–6.5 cm.

Aucuba himalaica **Hook. f. & Thomson** Dichotomously branched **shrub** or small **tree** to 6 m, with linear-lanceolate to oblanceolate or elliptic-ovate, long-pointed **leaves**, 8–26 × 2.2–6 cm, sharply toothed on margin, dark green above, glaucous and woolly beneath; leaf-stalks 1–2.5 cm long. **Female flowers** 7 mm across, olive- to livid-green, fragrant; **male flowers** 4 mm long, red-purple. **Berries** crimson to brilliant scarlet, 13–15 mm long. nGuangxi, wHubei, Sichuan, sShaanxi, seTibet, Yunnan [swTibet, cChina, Himalaya, Myanmar]; open shrubberies, *Rhododendron* forests, shady gullies, 1,400–3,300 m. IV–V.

RUTACEAE

ZANTHOXYLUM

A genus of c. 200 species (41 in China) widely scattered in the Americas, Africa, Asia and Australia. **Evergreen** or **deciduous**, prickly **shrubs** with aromatic **bark** and gland-dotted pinnate **leaves**. **Fruit** a small, glandular, aromatic, capsule. Sometimes erroneously spelt *Xanthoxylum*.

Zanthoxylum armatum **DC.** **Shrub** or small **tree** to 5 m with corky **bark**, and branchlets and leaf-stalks armed with long straight spines. **Leaves** with narrowly winged stalks; leaflets in 2–6 pairs, lanceolate, to 8 cm long, toothed. **Flowers** 1–2 mm across, unisexual, in small, short-branched clusters; calyx 6–8-lobed; petals absent; stamens 6–8. **Fruit** globose, 3–4 mm in diameter, red and wrinkled when ripe. Guangxi, Guizhou, s & swSichuan, seTibet, Yunnan [swTibet, sChina, Himalaya, seAsia]; shrubberies, borders of cultivation, hedgerows, sometimes cultivated, 1,100–2,500 m. IV–V.

Zanthoxylum oxyphyllum **Edgew.** Readily distinguished from the previous species by the hooked spines and larger **inflorescences** bearing lilac **flowers** 6–8 mm across, each with 4 blunt petals. **Shrub** or **climber**. Guangxi, Guizhou, sSichuan, seTibet, Yunnan [swTibet, c & eHimalaya, Myanmar]; forests, shrubberies, 1,800–2,700 m. IV–V.

Zanthoxylum aff. *nepalense* **Babu** Similar to *Z. armatum*, but leaf-stalks and leaf-rachises unwinged, leaflets oval, not more than 3 cm long, and **inflorescences** flat-topped. Sichuan, seTibet, Yunnan [swTibet, c & eHimalaya]; similar habitats but not cultivated, 2,400–3,100 m. V.

Zanthoxylum bungeanum **Maxim.** Differs from other species included here by the larger **leaves** with narrowly winged stalks and rachises, leaflets generally 7, broadly oval-elliptic, upper 3 the largest, deep glossy green, lateral veins distinct. **Flowers** in small lateral clusters near stem apex or at end of short lateral branches. Gansu, Guangxi, Guizhou, Hubei, Qinghai, Sichuan, Shaanxi, seTibet, Yunnan [swTibet, c & nChina, Bhutan]; forests, shrubberies, 1,500–2,200 m. IV–V. The **fruits** are often seen for sale in Chinese markets and are much prized in Sichuanese cooking.

GERANIACEAE

GERANIUM

A genus of c. 150 species (50 in China) in the north temperate zone and tropical mountains. **Biennial** and **perennial herbs** with palmately lobed **leaves**, opposite or alternate. **Flowers** solitary, paired or in umbellate clusters; bracts present; sepals 5, separate, generally shorter than the 5 petals; stamens 10 in 5 whorls; carpels 5. **Fruit** with 5 basal carpels linked together by a beak-like column, which splits into 5 at maturity, each part bearing a single seed and a part of the beak.

1. Flowers nodding, with reflexed sepals and petals . . . . **G. pogonanthum, G. sinense, G. delavayi, G. refractum**
1. Flowers usually out-facing to ascending, petals and sepals half to fully spreading but not reflexed 2
2. Flowers flat . . . . **G. platyanthum**
2. Flowers saucer- to funnel-shaped 3
3. Plant at least 70 cm tall . . . . **G. pratense**
3. Plant not more than 50 cm tall 4
4. Petals white to pale pink, sometimes with darker veins 5
4. Petals deep pink to reddish purple, magenta or pinkish purple, often with darker veins 7
5. Petals 5–12 mm long; sepals usually more than 6 mm long . . . . **G. nepalense, G. hispidissimum**
5. Petals 13–15 mm long; sepals not more than 6 mm long 6
6. Leaves strongly marbled . . . . **G. orientalitibeticum**
6. Leaves not marbled, but sometimes with faint paler areas . . . . **G. farreri, G. dahuricum**
7. Anthers cream to yellow, sometimes purple-edged . . . . **G. pylzowianum, G. polyanthes**
7. Anthers purple to bluish black or red . . . **G. donianum, G. stapfianum, G. strictipes, Geranium sp.***

Geranium pogonanthum **Franch.** Tufted **perennial** to 40 cm, glandular-hairy in part, with marbled **leaves** up to 10 cm in diameter, each with 5 or 7 pinnately lobed segments. **Flowers** mostly paired, pendent from ascending stalks; sepals 7–10 mm long, purple at base, otherwise green; petals recurved to about 45°, whitish, pale pink or purplish, 8–16 mm long, elliptic, blunt, with a tuft of dense white hairs on each side at base; anthers bluish black, filaments hairy. w & swSichuan, w & nwYunnan [nMyanmar]; alpine meadows, low open shrubberies, 3,000–3,500 m. VI–VIII.

Geranium sinense **Knuth*** Tufted **perennial** to 60 cm, with deep green, slightly marbled **leaves** to 20 cm in diameter, divided almost to base into 7, sharply toothed and lobed segments; stem-leaves alternate or solitary. **Flowers** in lax clusters, generally paired in the **inflorescence**; sepals green, red at base, 5.5–7.5 mm long, reflexed like the petals; petals blackish maroon with pink base, oval, 9–10 mm long, the basal nectary forming a ring with those of the other petals; anthers blackish, borne on usually glabrous filaments. w & swSichuan, nwYunnan; open shrubberies, grassy slopes, 2,300–4,000 m. VII–IX.

Geranium delavayi **Franch.*** Similar to *G. sinense*, but **flowers** often paler, pink to blackish red, occasionally white, petals with dense hairs on margin at base; nectaries separate, not forming a ring. w & swSichuan, nwYunnan; shrubberies, forest margins, grassy slopes, 3,350 m. VII–IX.

Geranium refractum **Edgew. & Hook. f.** (syn. *G. melanandrum* Franch.) Readily distinguished from previous 3 species by the dense purple glandular hairs on upper part of plant; petals pale pink or reddish purple, rarely white, 12–16 mm long. w & swSichuan, w & nwYunnan [c & eHimalaya, nMyanmar]; open forests and forest margins, shrubberies, grassy slopes, 3,600–4,500 m. VI–VIII.

Geranium platyanthum **Duthie** (syn. *G. eriostemon* DC.) Tufted, hairy **perennial** to 50 cm with hollyhock-like, pale green **leaves** to 20 cm in diameter, divided just over halfway into 5 or 7 toothed segments. **Flowers** in small umbel-like clusters, out-facing or slightly nodding; sepals 7–10 mm long, green flushed reddish brown; petals violet-blue with paler base, rounded, 11–15 mm long; anthers bluish, borne on blackish or purplish filaments. Gansu, seQinghai, Sichuan, eTibet [n & neChina, Korea, Japan, eRussia]; forest margins and clearings, shrubberies, grassy slopes, 1,600–3,200 m. VII–IX.

Geranium pratense **L.** Tufted herbaceous, hairy **perennial** to 1.3 m. **Leaves** rounded in outline, to 20 cm wide, divided almost to the middle into 7 or 9 segments, tapered at both ends, with pointed lobes and teeth; stem-leaves paired, decreasing in size upwards. **Flowers** borne towards stem-tips, forming a rather dense **inflorescence** of upright flowers; sepals 7–12 mm long, pointed; petals blue to violet-blue, sometimes paler or white, 16–24 mm long, slightly longer than broad, apex rounded. **Fruits** on reflexed stalks. Gansu, Hubei, Qinghai, Sichuan, Tibet, nwYunnan [Europe, w & cAsia, Himalaya]; grassy habitats, open shrubberies, roadsides, banks, 3,200–4,200 m. VI–VII.

Geranium nepalense **Sweet** Somewhat trailing, bushy **perennial** with dark green, faintly marbled **leaves**, purplish beneath, divided almost to base into 5 or 7 bluntly toothed segments; stem-leaves paired. **Flowers** solitary or paired, small; sepals 4–5 mm long; petals white to pale pink, occasionally deeper pink, with darker veins, 5–9 mm long, apex rounded or notched; anthers violet-blue. Guangxi, Hubei, Sichuan, Yunnan [Himalaya, nMyanmar]; stony meadows, rocky slopes, 2,800–4,000 m. VI–VIII.

1. *Geranium pogonanthum*, with *Ajuga lupulina*; Balangshan, Wolong, wSichuan (CGW)
2. *Geranium pogonanthum*; Balangshan, Wolong, wSichuan (CGW)
3. *Geranium sinense*; Cangshan, nwYunnan (CGW)
4. *Geranium delavayi*; cult., from Yunnan (CGW)
5. *Geranium platyanthum*; cult., from wSichuan (CGW)
6. *Geranium platyanthum*; Wolong, wSichuan (PC)
7. *Geranium pratense*; cult., from Nepal (CGW)
8. *Geranium nepalense*; eNepal (CGW)

1. *Geranium farreri*; cult. (CGW)
2. *Geranium pylzowianum*; Balangshan, Wolong, wSichuan (CGW)
3. *Geranium pylzowianum*; Balangshan, Wolong, wSichuan (CGW)
4. *Geranium polyanthes*; nBhutan (RS)
5. *Geranium donianum*; Balangshan, Wolong, wSichuan (CGW)
6. *Geranium strictipes*; Shennongjia, wHubei (PC)
7. *Geranium strictipes*; Gangheba, Yulongxueshan, nwYunnan (CGW)
8. *Geranium* sp.; north of Aba, nwSichuan (CGW)
9. *Erodium cicutarium*; near Songpan, nwSichuan (CGW)

Geranium hispidissimum (Franch.) Knuth* More upright than previous species, to 30 cm, **leaves** divided to just over halfway into sharply toothed and lobed segments. **Flowers** bowl-shaped; sepals 5–6 mm long, glandular-hairy; petals white, pink or purplish red, 9–12 mm long, rounded at apex. w & swSichuan, nwYunnan; similar habitats and altitudes. VI–VII.

Geranium orientalitibeticum Knuth* Spreading **tuberous perennial** to 30 cm at the most in flower, spreading by underground runners. **Leaves** markedly marbled, divided almost to the base into 5 or 7 coarsely blunt-toothed segments. **Flowers** mostly paired, erect, saucer-shaped; petals purplish pink with white base, 16–22 mm long, oval, often slightly apiculate at apex. wSichuan (Kangding); shrubby screes, 2,250–2,750 m. VII–IX.

Geranium farreri Stapf* Small alpine species not more than 12 cm tall in flower with **leaves** not exceeding 5 cm across, rounded to kidney-shaped and divided almost to the base into 7 few-lobed segments; stem-leaves paired. **Flowers** generally paired, held above foliage, out-facing; sepals 7–9 mm long; petals very pale pink, with pencilled veins, obovate, 13–15 mm long; anthers bluish black. sGansu, nwSichuan; alpine screes and short turf, 3,000–3,800 m. VI–VII.

Geranium dahuricum DC. In contrast to the previous species, a rather sprawling **perennial** with thin **stems** and deeply cut **leaves** to 10 cm across, which have narrow deeply incised segments, often red-margined. **Flowers** paired, saucer-shaped; petals pale pink, 13–14 mm long, with fine deep red veins; anthers bluish. Gansu, nHubei, n & wSichuan, Shaanxi [c & nChina, Korea, Mongolia, eRussia]; grassy and rocky slopes, 2,600–3,800 m. V–VIII.

Geranium pylzowianum Maxim.* Patch-forming **perennial** to 25 cm with kidney-shaped **leaves** not more than 5 cm wide, divided almost to the base into 5 or 7 narrowly lobed segments, deep green with pale marks at sinuses; stem-leaves mostly solitary. **Flowers** usually paired; sepals 7–9 mm long; petals deep rose-pink, oval, 16–23 mm long, with rounded or notched apex, margin with dense hairs towards base; anthers cream, often edged purple. sGansu, seQinghai, w, nw & swSichuan, sShaanxi, n & nwYunnan [Ningxia]; screes, low alpine meadows and rock-ledges, 2,400–4,250 m. VI–VIII.

Geranium polyanthes Edgew. & Hook. f. Very different from the previous species in being a tufted **perennial** to 45 cm tall in flower, with somewhat succulent, rounded **leaves** divided into 7 or 9 wedge-shaped, bluntly lobed and toothed segments; stem-leaves mostly paired. **Flowers** erect and funnel-shaped, mostly in umbel-like clusters; sepals green flushed purple, 6–7 mm long; petals 12–15 mm long, glistening purplish pink with deeper feathery veins; anthers yellow. w & swSichuan, seTibet, w & nwYunnan [c & eHimalaya]; forest margins, shrubberies, open slopes, 2,400–4,500 m. VI–VIII.

Geranium donianum Sweet Tufted **perennial** to 40 cm, often less, with marbled kidney-shaped to rounded **leaves** to 10 cm across, with 5 or 7 deeply toothed and lobed segments; stem-leaves paired. **Flowers** mostly paired, ascending, funnel-shaped; sepals 8–11 mm long, with non-glandular hairs; petals reddish purple to magenta, oval, 13–19 mm long, apex rounded or slightly notched; anthers purple. sGansu, seQinghai, wSichuan, seTibet, nwYunnan [Himalaya]; alpine meadows and rocky slopes, 3,300–4,500 m. VI–VIII.

Geranium stapfianum Hand.-Mazz.* Smaller plant than the previous species, not more than 15 cm tall, with red-margined **leaves** and red leaf-stalks; **flowers** saucer-shaped, petals 12–20 mm long, deep magenta with dark red feathered veins; anthers dark red. w & swSichuan, nwYunnan; stony and grassy alpine meadows, rock-ledges, cliffs, 3,200–5,200 m. VI–IX.

Geranium strictipes Knuth* Readily distinguished from the previous two species by the unequal paired stem-leaves and absence of basal leaves on flowering shoots; **leaves** red-margined; **flowers** nodding in bud, opening saucer- to funnel-shaped; sepals green, 6–8 mm long, with glandular hairs; petals 11–15 mm long, purplish pink with a colourless or greenish bearded base. wHubei, w & swSichuan, nwYunnan; open shrubberies, 2,500–3,000 m. VII–VIII.

Geranium sp.* Low **tufted** plant to 20 cm, but spreading wider. **Leaves** rounded in outline, with 5 wedge-shaped segments, further divided into short blunt lobes. **Flowers** paired, nodding in bud, flowers saucer-shaped; sepals downy, about half the length of the petals, petals bright pink, 15–17 mm long, with deep-coloured veins on lower part, rounded, not clawed. nwSichuan; rocky slopes, roadside banks, c. 3,700 m. VII.

ERODIUM

A genus of c. 75 species (4 in China) distributed worldwide except for central & southern Africa. Very like *Geranium*, but leaves pinnately dissected or at least pinnately veined and flowers generally weakly or strongly zygomorphic.

Erodium cicutarium (L.) L'Hér. Very variable, often rather foetid, **annual** or **biennial** to 25 cm, at first stemless but later branching from base. **Leaves** 2-pinnately lobed. **Flowers** usually purplish pink, 7–18 mm, up to 12 in long-stalked umbels, upper 2 petals with or without a basal black blotch. **Fruit** hairy with a beak up to 40 mm long, generally less. Gansu, Hubei, Qinghai, Sichuan, Shaanxi, Tibet, Yunnan [Europe, Asia]; dry grassy places, gravels, banks, roadsides, disturbed ground, 700–2,200 m. V–IX.

LINACEAE

LINUM

A genus of c. 180 species (9 in China) in temperate and subtropical zones. Glabrous or rarely downy, **annual** or **perennial** herbs, occasionally woody at base, with alternate or opposite, simple, stalkless **leaves**. **Inflorescences** cymes or scorpioid cymes; **flowers** hermaphrodite, 5-parted, regular, sometimes heterostylous; sepals overlapping, persistent; petals red, white, blue or yellow, falling early, longer than sepals, contorted, base clawed; stamens alternate with sepals; staminodes 5, dentiform; ovary 5- or 10-celled; styles thread-like. **Fruit** a capsule splitting into 10 segments.

Linum perenne L. (syn. *Linum sibiricum* DC.) **Perennial herb** 20–90 cm tall with alternate, linear to linear-lanceolate, pointed **leaves**, 20–40 × 1–5 mm. **Flowers** numerous, heterostylous; petals blue to bluish purple, obovate, 10–18 mm long; styles distinct. **Capsule** almost globose, 3.5–8 mm in diameter. Gansu, Qinghai, Sichuan, Shaanxi, seTibet, Yunnan [swTibet, nChina, wAsia, Europe, Mongolia, Russia]; dry mountain slopes, open shrubberies, grassland, dry plains, sandy and gravelly floodplains, to 4,100 m. VI–VIII.

REINWARDTIA

An Asian genus of 1 species. **Shrub** with minute, awl-shaped, soon-falling stipules and alternate **leaves**. **Inflorescences** axillary and terminal fascicles or flowers solitary and axillary; sepals 5, lanceolate, persistent; petals 4 or 5, yellow, much longer than sepals, confluent, falling early; stamens 5, joined at base; staminodes 5; glands (nectaries) 2–5; ovary 3-celled, with 1 ovule per cell; styles 3, thread-like. **Fruit** a capsule splitting into 6 or 8, 1-seeded mericarps. **Seeds** with a membranous wing.

Reinwardtia indica Dumort. (syn. *Linum cicanobum* Buch.-Ham. ex D. Don; *L. repens* Buch.-Ham. ex D. Don; *L. trigynum* Roxb.) **Shrub** to 1 m with grey, glabrous branchlets. **Leaves** clustered towards branch-tips, elliptic to obovate-elliptic, 2–8.8(–12) × 0.7–3.5 cm, apex pointed to almost rounded, apiculate, margin entire or scalloped. **Flowers** solitary or clustered, 1.4–3 cm across; petals yellow, distinct but confluent at base, 1.7–3 cm long. **Capsule** globose. Guizhou, Hubei, Sichuan, Yunnan [c & sChina, Himalaya, Laos, Myanmar, Thailand, Vietnam]; ravines, mountain slopes in forests and thickets, pathsides, rice paddy and field margins, 500–2,300 m. IV–XII. Sometimes grown as an ornamental.

BALSAMINACEAE

IMPATIENS

A genus of probably more than 900 species (c. 220 in China, many endemic) scattered in most parts of the world with the exception of Australasia, South America and the Arctic. Rather fleshy **annual** and **perennial herbs** with **flowers** solitary or clustered at leaf-axils or in lateral racemes or pseudoumbels, resupinate (upside down); sepals 3(–5), lowermost elaborated into a nectar-bearing spur; petals 5, uppermost (dorsal) usually forming a hood or helmet, or erect, lateral partially fused into 2 pairs. **Fruit** a (4–)5-valved capsule, dehiscing explosively.

SYNOPSIS OF GROUPS

Group One. Fruit-capsule short, broadly spindle-shaped abruptly contracted at both ends; flowers in stalkless bundles or solitary.

Group Two. Fruit-capsule club-shaped to linear, occasionally narrowly spindle-shaped; flowers in a well-defined inflorescence with a short to long stalk; lobes of lateral petals long-pointed, often drawn out into a thread-like appendage (especially the upper of each pair). (p. 373)

Group Three. As Group Two but lobes of lateral petals blunt or almost so, not drawn out into a slender appendage; lateral sepals 4 (2 pairs). (p. 375)

Group Four. As Group Three but lateral sepals 2 (1 pair). (p. 376)

> **Group One**
> 1. Flowers basically pink, purple or white .... *I. balsamina, I. chinensis, I. morsei*
> 1. Flowers basically yellow ..... *Impatiens* aff. *spireana*

Impatiens balsamina L. Erect **annual herb** to 90 cm, simple or branched. **Leaves** scattered, occasionally the lowermost opposite, lanceolate to oblanceolate, to 12 × 3 cm, with toothed margin, glabrous to somewhat downy. **Flowers** 2–3 in leaf-axils, pink, white or purple; lateral sepals 2, 2–3 mm long; lower sepal boat-shaped, with slender incurved spur 10–25 mm long; upper petal hooded; lateral united petals 22–26 mm long, 2-lobed. **Capsule** pendent, densely downy, 10–20 mm long. Naturalised in Guangxi, Guizhou, Hubei, seQinghai, Sichuan, Shaanxi, seTibet, Yunnan [seAsia]; margins of cultivated land, waysides, but widely cultivated at lower altitudes. VI–IX.

BALSAMINACEAE: Impatiens

Impatiens chinensis L. Readily distinguished from the previous species by the linear to linear-lanceolate, opposite, minutely toothed **leaves**, smaller reddish purple or white **flowers** and glabrous **capsules**. Guangxi, Yunnan [seAsia]; damp places along field boundaries and streams, ditches, pools and swampy places, to 1,200 m. IV–X.

Impatiens morsei Hook. f.* Glabrous **perennial herb** to 50 cm, with a rather thick purplish **stem**, branching near base. **Leaves** alternate, deep velvety green, often with a purplish flush and pale midrib, elliptic to elliptic-lanceolate or oblong, to 13 × 7 cm, with a pair of rounded glands at base. **Flowers** solitary in upper leaf-axils on stalks 40–50 mm, white, pink or purplish, with orange markings on upper lateral petals; lower sepal boat-shaped with short, incurved spur; lateral united petals 16–22 mm long, 2-lobed, farthest lobe oblong, twice the size of the basal. Guangxi; damp forests, rocky places, ravines, streamsides, 400–1,000 m. IV–VIII.

Impatiens aff. *spireana* Hook. f.* Glabrous plant to 50 cm. **Leaves** crowded towards top of **stem**, spirally arranged, stalked, blade ovate to ovate-elliptic, to 15 cm long, with 7–9 pairs of lateral veins and a scalloped margin, with a prominent pair of rounded glands at base. **Flowers** several per node, clustered beneath leaves; lateral sepals 2, broad-ovate, greenish, 10–12 mm long; lower sepal boat-shaped with short incurved spur; lateral united petals fused towards base, 22–25 mm long. Guangxi (Longhua); forests, 450–1,200 m. VII–IX.

Group Two
1. Flowers basically white or bicoloured purple and white .... *I. platychlaena, I. brevipes*
1. Flowers basically yellow or yellowish .... *I. vittata, I. davidii, I. dicentra, I. soulieana, I. waldheimiana*

1. *Linum perenne*; Lijiang, nwYunnan (PC)
2. *Reinwardtia indica*; swYunnan (CGW)
3. *Impatiens balsamina*; cult. Beijing (CGW)
4. *Impatiens balsamina*; cult. Kunming, cYunnan (CGW)
5. *Impatiens chinensis*; Hong Kong (PC)
6. *Impatiens morsei*; Tian'e, nwGuangxi (PC)
7. *Impatiens morsei*; Jinlong, swGuangxi (PC)

Impatiens platychlaena Hook. f.* Glabrous **annual herb** to 1 m, often less, with a robust, erect, branched **stem**. **Leaves** alternate, stalked, deep green above, paler and often purple-flushed beneath, blade ovate to oblong-lanceolate, to 15 × 4.5 cm, with toothed margin. **Inflorescence** 1–3-flowered, stalk not more than 20 mm long. **Flowers** white; lateral sepals 2, prominent, purple-brown or purple-red, shiny, oval to rounded, 11–14 mm long; lower sepal pouched, with purplish veining, 25–30 mm long, with short, incurved, 2-pronged spur; upper petal hooded, with a prominent purple-brown or purple-red, pointed crest; lateral united petals 25–30 mm long, 2-lobed. sSichuan (Emeishan); forests and forest margins, shrubberies, damp banks, ravines, tracksides, 700–2,500 m. VII–IX.

Impatiens brevipes Hook. f.* Very similar to the previous species, but **leaves** with 5–7 pairs of lateral veins, lateral sepals whitish or greenish with a pronounced keel, and upper petal uniformly white or cream; lower sepal 38–40 mm long, lateral united petals 25–28 mm long. w, sw & sSichuan; similar habitats, 1,500–1,800 m. VIII–IX.

Impatiens vittata Franch.* Glabrous **annual herb** to 60 cm with erect, branched **stems**. **Leaves** alternate, stalked, blade ovate to lanceolate, to 10 × 3.5 cm. **Flowers** yellowish or greenish yellow, purple-spotted in throat, solitary, stalk not more than 10 mm long; lateral sepals 2, pale green, rounded, relatively large, 12–15 mm long; lower sepal pouched, 25–30 mm long, with short, incurved, 2-pronged spur; dorsal petal hooded with a prominent crest; lateral united petals 26–30 mm long, 2-lobed, each lobes terminating in a hair-like appendage. wSichuan; forest margins, damp shady places, 1,500–2,000 m. VII–IX.

Impatiens davidii Franch.* Similar to the previous species, but **flowers** yellow, upper petal with a greenish crest and pouched lower sepal with a hooked spur 7–8 mm long, only the upper lobe of the lateral united petals with a hair-like terminal appendage; lateral sepals with a broad crest. Hubei; similar habitats as well as streamsides, 300–700 m. VII–IX.

Impatiens dicentra Franch. ex Hook. f.* Readily distinguished from the previous four species by possessing large yellow solitary **flowers** borne on a very short stalk, and lateral sepals characteristically coarsely toothed; otherwise characters as *I. davidii*. Guizhou, wHubei, Sichuan, Shaanxi; similar habitats, 1,000–2,700 m. VII–IX.

---

1. *Impatiens* aff. *spireana*; Longhua, swGuangxi (PC)
2. *Impatiens platychlaeana*; Emeishan, sSichuan (CGW)
3. *Impatiens brevipes*; Emeishan, sSichuan (CGW)
4. *Impatiens vittata*; Wolong, wSichuan (PC)
5. *Impatiens soulieana*; cult. (RM)
6. *Impatiens wildheimiana*; Wolong, wSichuan (PC)
7. *Impatiens arguta*; cult. (RM)
8. *Impatiens omieana*; Emeishan, sSichuan (CGW)

Impatiens soulieana Hook. f.* Distinctive species differing from the previous species in having rather rigid **leaves** without basal glands, lateral sepals ovate, narrowly crested; **flowers** 1–3 with a very short stalk, yellow with some red or purplish veining on lower sepals and petals. wSichuan; damp forests, shrubberies, by water courses and canals, 1,400–3,000 m. VII–IX.

Impatiens waldheimiana Hook. f.* Similar to *I. davidii*, but inflorescence-stalks more prominent, 30–40 mm long, and **flowers** larger (35–40 mm), lateral sepals heart-shaped, large, 12–14 mm, widely crested, concealing much of the lower sepal; **flowers** cream and pale yellow with red veining in throat. wSichuan; similar habitats, 1,600–2,800 m. VII–IX.

> **Group Three**
> 1. Flowers pink, purple, purplish red or purplish blue . . . . ***I. arguta, I. abbatis***
> 1. Flowers yellow, cream or white . . . . ***I. omeiana, I. wilsonii, I. barbata***

Impatiens arguta Hook. f. & Th. (syn. *I. arguta* var. *bulleyana* Hook. f.) **Perennial herb** to 70 cm, often rooting at lower nodes. **Leaves** scattered, stalked, blade ovate to lanceolate, to 15 × 4.5 cm, glabrous, with a toothed margin. **Flowers** pink or purplish red, occasionally whitish, 1 or 2 on a very short stalk; lower sepal curved-pouched, gradually constricted into a short incurved spur; upper petal hooded, with a horn-like crest; lateral united petals 2-lobed, upper lobes broad and rounded, lower larger and elongated, notched below apex. swSichuan, seTibet, nw & cYunnan [swTibet, Himalaya, nMyanmar]; forests and forest margins, shrubberies, streamsides, ravines, 1,800–3,200 m. VII–X.

Impatiens abbatis Hook. f.* Differs from the previous species in its **annual** habit and small ovate **leaves** not more than 4 × 2.5 cm; **flowers** generally solitary, bluish purple, without a stalk. w & nwYunnan; similar habitats, 1,200–2,100m. VIII–IX.

Impatiens omeiana Hook. f.* Rhizomatous, patch-forming **perennial** to 50 cm with rather thick fleshy **stems**, somewhat swollen at nodes. **Leaves** generally crowded towards stem-tips, deep green, sometimes variegated above, alternate, blade lanceolate to ovate, to 16 × 5 cm, with a coarse scalloped margin, glabrous, leaf-stalk 4–6 cm long. **Flowers** creamy yellow, in terminal or almost so, 5–8-flowered racemes on stalks to 10 cm long; lower sepal horn-shaped, terminating in a short incurved, rather stout spur 6–8 mm long; upper petal hooded; lateral united petals 2-lobed. w & sSichuan; damp forests and thickets, forest margins, 900–1,200 m. VIII–IX.

Impatiens wilsonii Hook. f.* Closely related to the previous species, but **leaves** stalkless to short-stalked, sparsely hairy along veins beneath; **flowers** white with a more pouched lower sepal. sSichuan (Emeishan); forests, damp shady rocks and banks, ditchsides, 800–1,000 m. VIII–IX.

Impatiens barbata Comber* Readily distinguished from the previous three species by the beard-like hairs on lateral sepals and along the crest of the upper petal; **inflorescence** often 3- but also 1–2-flowered; **flowers** pale yellow with a whitish spur. swSichuan, nwYunnan; forests, shrubberies, damp slopes, stream- and canal-sides, 2,000–3,000 m. VII–IX.

> **Group Four**
> 1. Spur upcurved (recurved) . . . . *I. stenantha*, *I. drepanophora, I. siculifer, I. siculifer* var. *porphyrea*
> 1. Spur straight, downcurved or incurved 2
> 2. Lower sepal broadly pouched, or if horn-shaped, then flowers basically purple 3
> 2. Lower sepal horn-shaped (then flowers basically yellow) to boat-shaped 4
> 3. Flowers basically yellow, cream or white, often veined or flushed red or purple (wholly purple in some forms of *I. delavayi*) . . *I. delavayi, I. noli-tangere, I. tortisepala*
> 3. Flowers reddish purple to bluish purple, or white with red or purple markings . . . . . *I. oxyanthera, I. forrestii, I. rubrostriata, I. faberi, I. purpurea*
> 4. Flowers yellow or cream 5
> 4. Flowers pinkish purple, purplish red, reddish or whitish . . . . *I. distracta, I. imbecilla, I. lateristachys, I. macrovexilla*
> 5. Inflorescence with stalk 3–12 cm long and with (3–)4 or more flowers . . . . *I. clavicuspis, I. cyathiflora*
> 5. Inflorescence with stalk rarely more than 3 cm long and with 1–3-flowers . . . *I. monticola, I. potaninii, I. xanthina*

Impatiens stenantha Hook. f. **Annual** or **perennial herb** to 60 cm, glabrous. **Leaves** scattered, stalked, blade elliptic to ovate-lanceolate, to 15 × 5 cm, with a pair of basal glands and a coarsely scalloped margin. **Inflorescence** 3–5-flowered, **flowers** yellow, usually with a purplish brown tip to the spur; lateral sepals narrow-oblong, 4 mm long; lower sepal horn-shaped, gradually narrowing into the spur which is often somewhat twisted; upper petal ascending, slightly cupped; lateral united petals 17–22 mm long, each drawn out to an almost pointed apex. seTibet, nwYunnan [swTibet, c & eHimalaya, nMyanmar]; forests, pathways, ravines, shrubberies, streamsides, 2,400–3,000 m. V–VIII.

Impatiens drepanophora Hook. f. Similar to the previous species, but a more robust plant to 1 m with 5–10-flowered **inflorescences**; **flowers** yellow, often with orange upper petal, the tip of lateral and lower sepals with a slender greenish appendage, spur spirally recurved. seTibet, w & nwYunnan [c & eHimalaya, nMyanmar]; similar habitats, 2,000–2,200 m. VII–X.

Impatiens siculifer Hook. f. Differs from the previous two species in having 5–8-flowered **inflorescences** with persistent bracts, and **flowers** with boat-shaped lateral sepals. Guangxi, Guizhou, Hubei, Sichuan, Yunnan [eHimalaya, nMyanmar]; similar habitats as well as damp grassy places and bamboo thickets, 800–2,600 m. VII–IX. Similar plants with purple flowers from Guangxi are referable to var. *porphyrea* Hook. f.*

Impatiens delavayi Franch.* **Annual herb** to 50 cm with slender simple or branched **stems**. **Leaves** blue-green, thin, alternate, lower stalked, uppermost stalkless and half-clasping stem, blade elliptic to ovate or almost rounded, to 5 × 2 cm, coarsely scalloped. **Racemes** held below leaves on a stalk not more than 30 mm long, 1–5-flowered; **flowers** purple or yellow with a purple-veined throat and purple lower sepal and spur; lower sepal narrow-pouched, constricted into a short incurved, 2-lobed spur; lateral united petals with a pointed appendage at tip. swSichuan, seTibet, w & nwYunnan; rocky forests and forest margins, rocky streambeds and banks of streams, 2,400–4,200 m. VII–X.

Impatiens noli-tangere L. Rather like *I. delavayi* in general characteristics but a larger plant to 70 cm with **leaves** up to 8 × 4 cm and **flowers** yellow with a few orange spots on the lower petals and lower sepal, which has a longer (10–15 mm long) incurved spur, not 2-lobed at tip. Gansu, Hubei, seQinghai, Shaanxi [w & cAsia, Europe, Russia]; similar habitats as well as margins of watercourses, 900–2,400 m. VII–IX.

Impatiens tortisepala Hook. f.* Readily distinguished from the previous two species by the contorted, kidney-shaped lateral sepals and corkscrew spur; **inflorescence** 6–8-flowered (sometimes fewer); **flowers** creamy white with a reddish blotch on both lateral petals; upper petal with a prominent crest. wSichuan; damp forests and other damp shady places, 1,500–2,900 m. VIII–IX.

Impatiens oxyanthera Hook. f.* Glabrous **annual herb** to 40 cm with simple or few-branched, erect **stems**. **Leaves** scattered, stalked, blade ovate to lanceolate, to 8 × 4 cm, with only 4–5 pairs of lateral veins, margin toothed. **Inflorescence** 1–2(–3)-flowered; **flowers** pink or purplish red with strongly veined petals and sepal-pouch; lateral sepals 2, usually pink or red, small, ovate, 5–6 mm long; lower sepal narrow-pouched, 22–25 mm long, spur strongly incurved; upper petal hooded, with a pointed crest above; lateral united petals 20–25 mm long. w & sSichuan; forest margins, damp banks, tracksides, ravines, 1,900–2,200 m. VIII–X.

Impatiens forrestii Hook. f. ex W. W. Sm.* A more robust **annual** than the previous species, to 90 cm, usually leafless below at flowering time, **leaves** with 8–9 pairs of lateral veins and **flowers** larger, purplish red, veined and spotted, lower petals broad-pouched, 28–31 mm long, spur abrupt, incurved. swSichuan, nwYunnan; similar habitats, often close to streams, 1,900–2,600 m. VIII–IX.

1. *Impatiens barbata*; Chongjiang, nwYunnan (PC)
2. *Impatiens stenantha*; eNepal (CGW)
3. *Impatiens drepanophora*; nwYunnan (CGW)
4. *Impatiens siculifer*; eNepal (CGW)
5. *Impatiens delavayi*; Gangheba, Yulongxueshan, nwYunnan (PC)
6. *Impatiens delavayi* (colour variant); Tianchi, nwYunnan (PC)
7. *Impatiens noli-tangere*; Toggenburg, Switzerland (PC)
8. *Impatiens oxyanthera*; Emeishan, sSichuan (CGW)
9. *Impatiens forrestii*; cult., from nwYunnan (CGW)

1. *Impatiens rubrostriata*; Xishan, Kunming, cYunnan (CGW)
2. *Impatiens faberi*; cult. (RM)
3. *Impatiens purpurea*; below east side of Cangshan, nwYunnan (VM)
4. *Impatiens cyathiflora*; Xishan, Kunming, cYunnan (CGW)
5. *Impatiens monticola*; Emeishan, sSichuan (CGW)
6. *Impatiens xanthina*; cult. (RM)
7. *Impatiens distracta*; Emeishan, sSichuan (CGW)
8. *Impatiens imbecilla*; Emeishan, sSichuan (CGW)
9. *Impatiens lateristachys*; Emeishan, sSichuan (CGW)
10. *Impatiens macrovexilla*; Guilin, ne Guangxi (PC)

Impatiens rubrostriata **Hook. f.*** Readily distinguished from the previous two species by the 3–5-flowered **inflorescences**; **flowers** white with purple veining, especially on lower, pouched sepal, lateral sepals relatively large and rounded; upper petal with a prominent keel-like crest. Guangxi, Guizhou, Yunnan; similar habitats as well as streamsides, 1,700–2,600 m. VI–VIII.

Impatiens faberi **Hook. f.*** Similar to the previous species, but **inflorescence** nearly always 2-flowered; **flowers** purple or white with purple markings, lower sepal narrowly horn-shaped, gradually tapering into long incurved spur, upper petal hooded, with a horn-like crest above. sSichuan (Emeishan); forest margins, damp places, path- and streamsides, 1,300–2,100 m. VIII–IX.

Impatiens purpurea **Hand.-Mazz.*** **Annual herb** to 70 cm with erect, simple or branched **stems**. **Leaves** crowded towards shoot-tips, mostly alternate, elliptic to lanceolate, to 15 × 4.5 cm, margin with bristle-tipped teeth, uppermost leaves almost stalkless. **Inflorescence** a spreading 4–8-flowered raceme with persistent ovate bracts; **flowers** purple, yellowish and speckled in throat; lower sepal horn-shaped, narrowing into an incurved spur 10–13 mm long; upper petal rounded, keeled; lateral united petals 20–22 mm long. nwYunnan; forests, stream- and pathsides, 2,400–3,300 m. VIII–X.

Impatiens clavicuspis **Hook. f.** Erect **annual herb** to 70 cm, branched or not, sparsely brown glandular-hairy above. **Leaves** scattered, crowded towards stem tops, short-stalked, blade ovate-lanceolate to oblong-ovate, to 12 × 3.5 cm, with a scalloped-toothed margin, with 6–8 pairs of lateral veins; leaf-stalk with a pair of rounded glands at base. **Inflorescence** erect, racemose, with up to 12 pale yellow, purple-tinged flowers, stalk 10–22 cm long; lateral sepals 4–5 mm long, ovate with the apex extended, gland-tipped; lower sepal broadly horn-shaped, with an incurved spur 10–12 mm long; upper petal rounded, keeled on reverse; lateral united petals 28–31 mm long, 2-lobed. w & cYunnan [(?)nMyanmar]; forests, shady places in valleys and ravines, 2,400–2,800 m. VII–IX.

Impatiens cyathiflora **Hook. f.*** Differs from the previous species in being densely brown glandular-hairy, in the 4–5 pairs of lateral leaf-veins and in the 6–20-flowered **racemes** of bright yellow **flowers** that are red-spotted in the throat; lower sepal and spur horn-shaped, narrowed into the incurved spur, 15–18 mm long overall. sw & cYunnan; mixed forests, damp grassy slopes, 1,900–2,300 m. VIII–IX.

Impatiens monticola **Hook. f.*** **Annual herb** to 60 cm, branched or not, often rather leafless below. **Leaves** scattered, long-stalked, blade elliptic to obovate, to 13 × 4.5 cm, glabrous, margin scalloped or somewhat toothed. **Flowers** usually 2, greenish with white or cream petals; lateral sepals ovate, 6–10 mm long, green; lower sepal boat-shaped, 15–18 mm, with a slender coiled spur 15–20 mm long; upper petal hooded, with a slight crest; lateral united petals 20–25 mm long, 2-lobed, upper rounded, lower half-moon-shaped. w & sSichuan (Emeishan, Jingyuanshan); forests and forest margins, pathsides, damp banks, 900–1,800 m. VII–IX.

Impatiens potaninii **Maxim.*** Differs from the previous species in the yellowish **flowers**, brown-spotted in the throat, the lateral sepals rounded, only 4–5 mm long, lower sepal more horn-shaped, incurved spur 16–17 mm long. sGansu, (?)Hubei, e & neSichuan, (?)Shaanxi; similar habitats as well as along waterways, 1,200–2,300 m. VIII–X.

Impatiens xanthina **Comber** **Annual** to 20 cm with stem branched or not. **Leaves** crowded towards shoot-tips, alternate to pseudowhorled, lanceolate to elliptic-lanceolate, 4.5–7 × 1.5–2 cm, rough along veins beneath, margin minutely scalloped. **Inflorescence** 1–2-flowered, just overtopping leaves, on a stalk up to 2.5 cm long; lateral sepals 2, 8–9 mm long; lower sepal boat-shaped, constricted into an incurved spur 22–25 mm long; petals bright yellow with purple-brown central blotches, upper petal rounded, 6–7 mm long, lateral united petals up to 16 mm long. nwYunnan [nMyanmar]; humid forest habitats, ravines, 1,200–2,800 m. VIII–IX.

Impatiens distracta **Hook. f.*** **Annual herb** to 70 cm with slender, branched **stems**. **Leaves** scattered, stalked, blade ovate-oblong, to 11 × 4.5 cm, margin scalloped, leaf-stalks and bracts usually minutely hairy. **Flowers** 1–3 on a stalk to 3 cm long, somewhat contorted, not more than 20 mm wide, purple with a pink upper petal and yellow, spotted throat; lateral sepals 2, whitish; lower sepal boat-shaped, constricted into a somewhat incurved spur; lateral united petals 20–25 mm long, each pair with an appendage inserted into spur. sSichuan (Emeishan); forests and forest margins, stream-, road- and tracksides, 1,300–2,100 m. VII–X.

Impatiens imbecilla **Hook. f.*** Readily distinguished from previous species by the narrowly funnel-shaped, straight or somewhat curved lower sepal and spur, to 15 mm long overall; **inflorescence**-stalks 10–30 mm long, normally 2-flowered; **flowers** pink or whitish, with spotting on lower lateral petals and in throat. sSichuan (Emeishan); similar habitats, 1,300–2,300 m. VIII–IX.

Impatiens lateristachys **Y. L. Che & Y. Q. Lu*** Similar to the previous species, but **inflorescence** racemose with 3–6 red, pinkish or white **flowers**, borne on a stalk 10–20 cm long, spur sepal 25–30 mm long. w & sSichuan; similar habitats, 2,000–2,500 m. VIII–X.

Impatiens macrovexilla **Y. L. Chen*** Readily distinguished from the previous three species by the slender curved, ascending **spur** 20–25 mm long, and heart-shaped upper petal, that is larger than the distinctly unevenly 2-lobed lateral petals; **flowers** purple or purplish pink, with a white and yellow basal mark on the lateral united petals. Guangxi; shady wooded and grassy habitats, damp rocks, 1,000–1,600 m. IX–X.

POLYGALACEAE

POLYGALA

A genus of c. 500 species (44 in China, half endemic) with a worldwide distribution. **Herbs**, or **deciduous** or **evergreen shrubs** or small **trees**, with simple, alternate or whorled **leaves**. **Flowers** zygomorphic, in terminal or lateral racemes; sepals 5, inner 2 much larger and petal-like; petals 3, fused in lower half, often fringed at apex. **Fruit** a lobed or winged, compressed capsule.

Polygala arillata Buch.-Ham. ex D. Don **Erect shrub** to 2.5 m with deep green lanceolate to oblong-elliptic, alternate **leaves**, 5–15 cm long. **Flowers** pea-like, in drooping racemes, yellow to orange, tipped red, 13–15 mm long; lower (keel) petal with a conspicuous fringed crest, partly fused to the 2 lateral petals. **Fruit** kidney-shaped, 13–20 mm, dark shiny red at maturity. Guangxi, Guizhou, Hubei, Sichuan, sShaanxi, seTibet, Yunnan [swTibet, c & sChina, seAsia, c & eHimalaya, Myanmar, Sri Lanka]; forests, shrubberies, ravines, 1,500–2,700 m. V–VII.

ARALIACEAE

Several Chinese genera are used medicinally (e.g. *Panax*, *Aralia*, *Eleutherococcus*, *Heteropanax*, *Tetrapanax*, *Schefflera* and *Fatsia*), for timber (notably *Kalopanax*) or for ornamental purposes (e.g. *Fatsia*, *Hedera* and *Schefflera*).

TETRAPANAX

A genus with a single species restricted to south China. See description below.

Tetrapanax papyriferus (Hook.) K. Koch (syn. *Aralia papyrifera* Hook.) Unarmed stoloniferous **shrub** or small **tree** to 3.5 m, densely reddish or pale brown stellate-hairy; **trunk** to 9 cm in diameter with white pith. **Leaves** long-stalked, somewhat leathery, palmately 7–12-lobed, 50–75 cm wide, lobes ovate-oblong, long-pointed, entire to coarsely toothed, glabrous above but with dense, rusty hairs beneath. **Inflorescence**-stalks 1–1.5 cm long, umbels 1–2 cm across, with many hermaphrodite yellowish white, densely stellate-hairy **flowers**; petals 2 mm long; stamens 4–5. **Fruit** globose, 4 mm across, dark purple when ripe. Guizhou, Hubei, swSichuan, Shaanxi, nwYunnan [ec & sChina, Taiwan]; mixed shrubberies, forest margins, 100–2,800 m. X–XII. Widely cultivated in China for rice paper and as an ornamental elsewhere in the tropics.

KALOPANAX

A genus with a single species in east Asia, including China. See description below.

Kalopanax septemlobus (Thunb.) Koidz. (syn. *K. pictus* (Thunb.) Nakai; *K. ricinifolius* (Sieb. & Zucc.) Miq.) **Deciduous tree** to 30 m with stout **branches** armed with numerous broad prickles. **Leaves** long-stalked, palmately lobed, rounded, 9–25 cm broad, with 5–7 broadly triangular-ovate to oblong-ovate, long-pointed, toothed lobes, dark green and glabrous or nearly so above, paler and usually slightly downy when young beneath. **Inflorescence** 20–30 cm across, on a stalk 2–6 cm long, umbels 1–2.5 cm across; **flowers** hermaphrodite, white or yellowish green; petals 5, valvate; stamens 5. **Fruit** 4–5 mm across, dark blue when ripe. Guizhou, Hubei, Sichuan, Shaanxi, Yunnan [c, e & sChina, Japan, Korea, eRussia]; forests, ravines, to 2,500 m. VII–VIII.

HEDERA

A genus with 5 species (2 in China) in Europe, North Africa and tropical and subtropical Asia. **Evergreen climbers**, creeping or climbing by aerial roots. **Leaves** simple, entire or lobed, dimorphic (those of fertile shoots different from those of sterile ones). **Inflorescence** a compact raceme of small umbels, or occasionally solitary; **flowers** hermaphrodite; calyx-margin almost entire or 5-toothed; petals 5, valvate; stamens 5. **Ovary** 5-carpellate, styles united into a short column. **Fruit** a globose drupe.

Hedera nepalensis K. Koch var. *sinensis* (Tobl.) Rehd. (syn. *H. sinensis* Tobl.) **Climber** to over 30 m, young **branches** with rusty-red scales. **Leaves** shiny, deep green above, those on sterile shoots triangular-ovate, triangular-oblong or arrow-shaped, entire or 3-lobed, leaves on fertile shoots elliptic-ovate or elliptic-lanceolate, long-pointed; leaf-stalks slender, 2–9 cm long. **Inflorescence** terminal, a solitary umbel or several in a raceme, each many-flowered, 1.5–2.5 cm across, with yellowish white **flowers**. **Fruits** globose, 0.7–1.3 cm across, red or yellow when ripe. sGansu, Guangxi, Guizhou, Hubei, Sichuan, sShaanxi, seTibet, Yunnan [swTibet, c & sChina, Laos, Vietnam]; forests, woods, roadsides, rocky slopes, cliffs, usually climbing on trees or rocks, to 3,500 m. IX–XI.

DENDROPANAX

A genus with c. 80 species (16 in China) in tropical Americas and east Asia. **Evergreen**, usually glabrous, unarmed **shrubs** or **trees**. **Leaves** simple or palmately 2–5-lobed, often with yellow glandular spots; stipules united with leaf-stalks or absent. **Inflorescence** a simple umbel, a small raceme of small umbels or a compound umbel; **flowers** hermaphrodite and often also male; calyx entire or 5-toothed; petals 5, valvate; stamens 5. **Ovary** (2–)5-carpellate, styles distinct or obscurely ribbed. **Fruit** a drupe.

Dendropanax dentigerus (Harms ex Diels) Merr. (syn. *Gilibertia dentigera* Harms ex Diels; *Dendropanax chevalieri* (R. Vig.) Merr.) Glabrous **shrub** or small **tree** to c. 8 m. **Leaves** long-stalked, leathery, dimorphic, densely red gland-dotted,

3-nerved, tertiary veins elevated on both surfaces; unlobed blades ovate-elliptic, rarely oblong-elliptic, 7–10 cm long, pointed or long-pointed, lobed blades 2–3-lobed, margin entire or slightly minutely toothed. **Inflorescence** with a stout stalk 1.5–5 cm long, and 1–5 umbels, each 2–3 cm in diameter. **Fruits** oblong-globose, 5–7 mm long, with persistent styles 1.5–2 mm long. Guizhou, wHubei, Sichuan [s & eChina, Cambodia, Laos, Vietnam]; evergreen broadleaved forests, thickets, to 1,800 m. VIII–IX.

MERRILLIOPANAX

A genus of 4 species (3 in China) native to South East Asia. **Evergreen**, unarmed **shrubs** or **trees**. **Leaves** simple with stalks much dilated at base. **Inflorescence** a panicle of terminal and axillary umbels. Calyx minutely 5-toothed; petals 5, valvate; stamens 5. **Ovary** 2-celled, with 2 styles that are free or united at base. **Fruit** an ellipsoid to globose drupe, 2-seeded.

Merrilliopanax membranifolius **(W. W. Sm.) C. B. Shang** (syn. *Nothopanax membranifolius* W. W. Sm.) Small **tree** to 7 m with slender, glabrous or sparsely brown-stellate-downy **branches**. **Leaves** slender-stalked, oblong-elliptic or lanceolate-elliptic, 8–30 cm long, glabrous, 3-nerved, apex tailed to long-pointed, margin minutely toothed. **Inflorescence** 8–15 cm long, sparsely stellate-hairy when young, umbels 1.5–2.5 cm in diameter, with stalks 0.6–1.3 cm long; calyx glabrous or slightly stellate-downy; petals glabrous. **Fruit** ellipsoid-globose, 4–5 mm long. w & nwYunnan [neIndia, nMyanmar]; mixed forests on mountain slopes, 1,600–3,300 m. VI–VII.

Merrilliopanax listeri **(King) H. L. Li** (syn. *Dendropanax listeri* King; *Merrilliopanax chinensis* H. L. Li) Differs from the previous species in having broadly elliptic, ovate or almost orbicular, entire or 2–3-lobed **leaves**, and a downy **inflorescence**. nwYunnan [neIndia, nMyanmar]; similar habitats, 1,200–1,700 m. VI–VII.

BRASSAIOPSIS

A genus with c. 45 species (24 in south and south-west China) in south and South East Asia. Armed or rarely unarmed **trees** or **shrubs** with simple, palmately lobed or digitately compound **leaves**; stipules joined with base of leaf-stalk. **Inflorescences** large panicles or racemes of umbels; bracts small or absent; **flowers** hermaphrodite or polygamous; calyx-margin 5-toothed; petals 5, valvate; ovary 2(–5)-celled, styles as many as cells, united into a column, persistent. **Fruit** globose or almost so, or top-shaped, with a fleshy exocarp; seeds 1–2(–5).

1. Leaves unlobed or 2–3-lobed . . . .*B. ferruginea*
1. Leaves palmately 5–11–lobed . . . .
 . . . . *B. ciliata*, *B. fatsioides*, *B. glomerulata*

Brassaiopsis ferruginea **(H. L. Li) G. Hoo*** (syn. *Dendropanax ferrugineus* H. L. Li) **Shrub** to 2 m with slender unarmed **branches**, reddish stellate-hairy when young. **Leaves** slender-stalked, lanceolate to ovate-lanceolate or 2–3-lobed, 7–20 cm long, densely reddish stellate-hairy on both surfaces when young; lobes narrowly lanceolate, tailed, with finely toothed

1. *Polygala arillata*; Lancangjiang–Jinshajiang divide, nwYunnan (CGW)
2. *Hedera nepalensis* var. *sinensis*; Shilin, cYunnan (PC)
3. *Hedera nepalensis* var. *sinensis*; Shilin, cYunnan (PC)

margin. **Inflorescence** a terminal raceme of 3–5 umbels, less than 10 cm long, reddish stellate-downy when young; stalks 2–7 cm long; umbels 3 cm in diameter. **Fruit** globose, 8 mm in diameter, black when ripe. Guangxi, Guizhou, Sichuan, Yunnan [Fujian, Guangdong]; forested slopes, 1,200–1,700 m. VI–VII.

Brassaiopsis ciliata Dunn (syn. *Euaraliopsis ciliata* (Dunn) Hutch.) **Shrub** to 4 m with densely woolly **branches** armed with sparse compressed prickles. **Leaves** long-stalked, palmately 7–11–lobed, to 30 cm wide, lobes oblong or oblong-lanceolate, long-pointed, sparsely bristly on veins, margin ciliate to minutely toothed. **Inflorescence** terminal, paniculate, 20–30 cm long with a minutely prickly rachis and secondary rachis, on a stalk 2–5 cm long, densely bristly; umbels 3–5 cm in diameter, with white **flowers**. **Fruit** egg-shaped-globose, or slightly compressed, 7–8 mm in diameter, black when ripe. wGuangxi, swGuizhou, Sichuan, seTibet, seYunnan [swTibet, nVietnam]; forests in valleys, open slopes, 300–2,200 m. X–XI.

Brassaiopsis fatsioides Harms* (syn. *Euaraliopsis fatsioides* (Harms) Hutch.; *Brassaiopsis palmipes* Forrest ex W. W. Sm.) In contrast to the previous species, a small **tree** to 10 m with a **trunk** up to 20 cm in diameter, with prickly **branches**, leaf-lobes oblanceolate, oblong-lanceolate or ovate-oblong, somewhat long-pointed, slightly rusty-woolly beneath or glabrous. **Inflorescences** more than 30 cm long with glabrous or slightly minutely hairy stalks. **Fruit** globose, 5–6 mm in diameter, blue-black when ripe. Guizhou, Sichuan, seTibet, Yunnan; forests in valleys, mountain slopes, 500–2,700 m. VI–VII.

Brassaiopsis glomerulata **(Blume) Regel** (syn. *Aralia glomerulata* Blume) Distinguished from the previous two species by the prickly as well as rusty-red-hairy young growth and by the pendulous, paniculate **inflorescences** more than 30 cm long, the umbels 2–3 cm in diameter; **flowers** white, fragrant. **Fruit** globose or compressed-globose, 7–10 mm in diameter. Guizhou, Sichuan, Yunnan [sChina, Cambodia, India, Indonesia, Laos, Nepal, Vietnam]; dense forests, mountain slopes, valleys, 400–2,400 m. VI–VIII.

SCHEFFLERA

A large genus of c. 1,100 species (35 in south China, 14 endemic) with a wide distribution in the tropics and subtropics of both hemispheres. **Evergreen**, unarmed **shrubs** or **trees**, sometimes **climbers** or **epiphytes**, with digitately compound **leaves**; stipules joined with leaf-stalk. **Inflorescence** paniculate or compound-racemose with the flowers arranged in umbels, heads or racemes; **flowers** perfect; calyx entire or 5-toothed; petals 5–11, valvate; stamens 5–11; ovary 5–11–celled. **Fruit** globose or egg-shaped, 5–11–seeded, angled or not.

1. *Brassaiopsis fatsioides*; Emeishan, sSichuan (PC)
2. *Brassaiopsis fatsioides*; Emeishan, sSichuan (PC)
3. *Eleutherococcus wilsonii*; Jiuzhaigou, nwSichuan (PC)
4. *Eleutherococcus henryi*; Jiuzhaigou, nwSichuan (PC)

Schefflera delavayi (Franch.) Harms ex Diels (syn. *Heptapleurum delavayi* Franch.) Small **tree** to 8 m with rather stout **branches**, densely yellow-brown stellate-woolly. **Leaves** long-stalked, with 4–7, ovate-oblong or ovate-lanceolate leaflets, 8–24(–35) × 3–12 cm, apex abruptly pointed or long-pointed, grey-white or yellow-brown stellate-downy beneath, margin entire to irregularly toothed or incised. **Panicles** spike-like, with stalkless, white **flowers**; calyx woolly, distinctly 5-toothed; petals 2 mm long, glabrous. **Fruit** globose, 4 mm in diameter. Guizhou, Hubei, Sichuan, Yunnan [c & sChina, Vietnam]; evergreen broadleaved forests, wet forest margins, streamsides, 600–3,000 m. X–XI.

Schefflera wardii C. Marq. & Airy Shaw* Similar to *S. delavayi*, but with 3–5, ovate-oblong, pointed leaflets up to 35 cm long, sparsely yellow- or brown-stellate-downy beneath, and with a remotely toothed margin; **inflorescence** racemose, yellow-white-woolly, to 18 cm long, with densely downy **flowers**. seTibet, nwYunnan; dense forests, 2,000–2,500 m. XII–I.

Schefflera multinervia H. L. Li* Similar to the two previous species, but **leaves** with shorter stalks (not more than 12 cm long) and 5–7, glabrous, oblong-lanceolate, long-pointed leaflets, 14–18 cm long, pale grey beneath; **inflorescence** 20–25 cm long, rusty-woolly or sparsely downy; **flowers** sparsely downy or almost glabrous. nwYunnan; evergreen broadleaved forests, 3,200–3,500 m. IX–X.

METAPANAX

A genus with c. 15 species (2 in China) in eastern Asia and Australia. Unarmed, glabrous, **evergreen shrubs** or small **trees**. **Leaves** simple, occasionally digitate with stipules absent or inconspicuous. **Inflorescence** racemose, paniculate or a simple umbel, the stalks jointed below **flowers**; calyx entire or 5-toothed; petals 5, valvate; stamens 5; **ovary** usually 2-celled, with 2 styles free or united at base. **Fruit** laterally compressed or rarely almost globose, with a fleshy exocarp.

Metapanax davidii (Franch.) Frodin ex Wen & Frodin (syn. *Pseudopanax davidii* (Franch.) W. R. Phil.; *Panax davidii* Franch.; *Nothopanax davidii* (Franch.) Harms ex Diels; *Acanthopanax davidii* (Franch.) Vig.) Small **tree** to 12 m. **Leaves** slender-stalked, entire or 2–3-lobed or with 3 leaflets on the same plant: entire leaves oblong-ovate or oblong-lanceolate, 6–20 cm long, long-pointed, 3-nerved, with a sparsely toothed margin; lobed leaves with lanceolate lobes; leaves with 3 leaflets with lanceolate leaflets. **Inflorescence** a terminal panicle, 10–18 cm long on a short 2 cm stalk, the umbels c. 2.5 cm in diameter; **flowers** white or pale yellow, fragrant. **Fruit** compressed-globose, 5–6 mm in diameter, black at maturity. Guizhou, Hubei, Sichuan, Shaanxi, Yunnan [Hunan, nVietnam]; common in shrubberies, forest fringes, ravines, stream- and roadsides, 800–3,000 m. VI–VIII.

ELEUTHEROCOCCUS

A genus of c. 30 species (18 in China) in eastern Asia and the Himalaya. Erect or climbing **shrubs**, rarely small **trees**, glabrous or downy, usually prickly, rarely unarmed. **Leaves** digitately compound or with 3 leaflets; stipules absent or very weakly developed. **Inflorescences** terminal, rarely axillary, umbels solitary or a few together forming small panicles. Calyx-margin minutely 5-toothed or entire; petals 5, valvate; stamens as many as petals; ovary 2–5-celled, with a similar number of distinct or joined, persistent styles. **Fruit** laterally compressed or almost globose, 2–5-seeded.

Eleutherococcus wilsonii (Harms) Nakai* (syn. *Acanthopanax wilsonii* Harms; *A. nanpingensis* X. P. Fang & C. K. Hsieh; *A. stenophyllus* Harms) **Shrub** to 5 m with purple-red, glabrous or slightly downy **branches** armed at nodes with small recurved prickles. **Leaves** short-stalked, with 3–5, oblong-lanceolate or oblanceolate leaflets, 4–5.5 × 0.5–1.6 cm, glabrous, margin scalloped-toothed. **Inflorescence** a solitary terminal umbel with white to yellowish green **flowers**; calyx almost entire or 5-toothed. Gansu, wHubei, Qinghai, Sichuan, Shaanxi, e & seTibet, Yunnan; shrubberies, forests, 2,500–3,600 m. VI–VII. The typical variety (var. *wilsonii**) has leaflets glabrous beneath with scalloped-minutely toothed margins, while var. *pilosulus* (Rehd.) P. S. Hsu & S. L. Pan*, from Gansu and Qinghai, has leaflets that have sparse or dense, long soft hairs beneath and minutely toothed or doubly minutely toothed margins.

Eleutherococcus henryi Oliv.* (syn. *Acanthopanax henryi* (Oliv.) Harms) **Evergreen shrub** to 3 m, with rough, downy branches, eventually becoming glabrous, armed with stout prickles. **Leaves** palmate with 5 papery, elliptic to oblanceolate leaflets, 6–12 × 3–5 cm, pointed to long-pointed, downy on veins beneath, toothed towards tip; leaf-stalks 4–7 cm long, downy. **Inflorescence** a terminal panicle with several umbels, borne on stalks to 3.5 cm long; calyx 5-toothed; ovary with 2–5 carpels; styles united into a column, persistent in fruit. **Fruit** ellipsoid-globose, 8 mm in diameter, composed of 2–5 carpels, black at maturity. Hubei, Sichuan, Shaanxi [c & eChina], forests and field margins, shrubberies, mountain slopes, roadsides; 800–3,200 m. VII–IX.

ARALIA

A genus of c. 40 species (29 in China) mainly in South East Asia and China, with a few in the Americas. Prickly **shrubs** or **trees**, or unarmed, **rhizomatous herbs**. **Leaves** alternate, 1–3-pinnate with an articulated rachis, mostly stipulate. **Inflorescences** terminal or axillary, paniculate, corymbose or umbellate, usually consisting of umbels, heads or racemes; **flowers** with their stalks articulated below; calyx 5-toothed; petals 5, overlapping; stamens 5; ovary 5-celled, occasionally aborted to 3; styles 5, distinct or joined at base. **Fruit** a berry, 5-celled, 3–5-angled.

1. Shrub or small tree, 2–7 m tall; stems prickly, 10–15 cm in diameter .... **A. chinensis, A. echinocaulis**
1. Perennial herbs, less than 1.5 m tall, with glabrous stems 2
2. Leaves not more than 30 cm long .... **A. atropurpurea, A. henryi**
2. Leaves larger, mostly 30–60 cm long .... **A. yunnanensis, A. fargesii, A. apioides**

Aralia chinensis L.* (syn. *A. subcapitata* G. Hoo; *A. hupehensis* G. Hoo; *A. taibaiensis* Z. Z. Wang & H. C. Zheng) Plant with erect to ascending **branches** armed with sparse prickles. **Leaves** 2–3-pinnate, 60–110 cm long, unarmed or with very few prickles, leaf-stalks stout, to 50 cm long; leaflets in 5–11 pairs, broad-ovate to long-ovate, 5–12 cm long, long-pointed, light yellow- or grey-downy beneath, more densely so along veins, margin toothed. **Panicles** large, 30–60 cm long, densely yellow-brown- or grey-downy, umbels 1–1.5 cm in diameter, with numerous, fragrant white **flowers**; calyx 1.5 mm long with 5-toothed margin; petals 5, ovate-triangular, 1.5–2 mm long. **Fruit** globose, 3 mm in diameter, black when ripe. sGansu, Hubei, Guizhou, Sichuan, Shaanxi, Yunnan [c & sChina]; forests, forest margins, wasteland, roadsides, to 2,700 m. VII–IX.

Aralia echinocaulis Hand.-Mazz.* **Leaves** 2-pinnate, smaller than the previous species, 35–50 cm long, with purple stalks 25–40 cm long armed with sparse, short prickles; stipules maroon; leaflets in 2–4 pairs, oblong-ovate to lanceolate, 4–11.5 × 2.5–5 cm, glabrous. **Inflorescence** terminal, 30–50 cm long, rachis scurfy-hirsute, umbels c. 1.5 cm in diameter. Guizhou, Hubei, Sichuan, Yunnan [c & sChina]; forests, ravines, 1,000–2,600 m. VI–VIII.

Aralia atropurpurea Franch.* **Leaves** 1–2-pinnate, 20–30 cm long including stalk; leaflets ovate, long-pointed, 3–8 × 2–3 cm, sparsely rough-bristly on both surfaces, margin doubly toothed. **Inflorescence** terminal only, corymbose, glabrous to slightly rough, umbels 1.5–2.2 cm in diameter, only 7–10-flowered; **flowers** purplish. wSichuan, e & seTibet, nwYunnan; open woods, sloping meadows, roadsides, 2,700–3,300 m. VI–VII.

Aralia henryi Harms* **Leaves** smaller than in previous species, 16–20 cm long, leaflets oblong-ovate, long-tailed, 3.5–10 × 2–6 cm, sparsely and softly hairy along veins on both surfaces, margin scalloped-toothed. **Inflorescence** corymbose-paniculate, terminal, softly hairy, umbels 3–10-flowered. wHubei, eSichuan, sShaanxi [Anhui]; forests, 1,500–2,300 m. VII–VIII.

Aralia yunnanensis Franch.* **Leaves** 30–50 cm long, 2–3-pinnate, stalks 3–15 cm long; leaflets ovate to long-ovate, 2–8 × 1.3–4.5 cm, long-pointed, with sparse stiff white hairs along veins, margin toothed. **Inflorescence** corymbose-paniculate, terminal or axillary, 20–30 cm long; umbels 10–30-flowered. swSichuan, nwYunnan; woods, shrubberies on slopes, 1,900–2,800 m. VI–VIII.

Aralia fargesii Franch.* Similar to the previous species, but **leaves** 30–50 cm long, 1–3-pinnate, leaflets broad-ovate or oblong-ovate, 8–15 × 5–7 cm, long-pointed, rough on both surfaces and downy along veins beneath. **Inflorescence** terminal or axillary, corymbose-paniculate, glabrous or with sparse, long soft hairs, umbels racemosely arranged, 1–1.5 cm in diameter, 10–20-flowered, **flowers** purplish. Sichuan, sShaanxi, Yunnan; forests, especially along streamsides, 1,800–2,700 m. VII–VIII.

Aralia apioides Hand.-Mazz.* Distinguished from similar species by the larger **leaves** up to 60 cm long with small broad-ovate leaflets, 1–3.5 × 1–2 cm, margin deeply incised and densely doubly toothed, and by the **inflorescence** with sparse, long soft hairs that may disappear with age. swSichuan (Shimian), nwYunnan (Deqin, Zhongdian); meadows, woods, 3,000–3,600 m. VI–VII.

PANAX

A genus of c. 8 species (7 in China, 1 introduced) in North America, the Himalaya and south and east Asia. **Perennial herbs** with a thick, generally branched rootstock and simple **stems** with scales at base. **Leaves** in whorls of 3–5, digitate, leaflets toothed or pinnately lobed. **Inflorescences** single terminal umbels; **flowers** hermaphrodite and male on the same plant; calyx shortly 5-toothed; petals 5, overlapping; stamens 5; ovary 2–3-celled with 2–3 styles, distinct or united at the base. **Fruit** with a somewhat fleshy exocarp, 2–3-seeded.

Ginseng (*Panax ginseng* C. A. Mey.), which is widely used medicinally, comes from neChina, Korea and eSiberia.

Panax japonicus (T. Nees) C. A. Mey. (syn. *P. pseudoginseng* Wall. var. *japonicus* (T. Nees) G. Hoo & C. J. Tseng) **Herb** 50–100 cm tall, with a straight, glabrous **stem**. **Leaves** 3–5, whorled at stem-apex, digitate; leaflets 5, obovate-elliptic to long-elliptic, 5–18 cm long. **Fruit** globose, red with a black top when ripe. Gansu, Guangxi, Guizhou, Hubei, Sichuan, Shaanxi, e & seTibet, Yunnan [c & eChina, Japan, Korea, Vietnam]; forests, 1,200–3,600 m. V–VI.

---

1. *Aralia chinensis*; Shennongjia, wHubei (PC)
2. *Aralia chinensis*; Shennongjia, wHubei (PC)
3. *Aralia echinocaulis*; Emeishan, sSichuan (PC)
4. *Panax ginseng*; Kunming, cYunnan (PC). Widely cultivated in China as an important medicinal herb, and sometimes becoming semi-naturalised.
5. *Panax japonicus*; Tianchi, nwYunnan (PC)
6. *Pleurospermum foetens*; Baimashan, nwYunnan (HJ)
7. *Pleurospermum foetens*; Wutoudi, Yulongxueshan, nwYunnan (CGW)

UMBELLIFERAE (APIACEAE)

PLEUROSPERMUM

A genus of c. 50 species (39 in China, 22 endemic) in eastern Europe and temperate Asia, especially diverse in the Sino-Himalayan region. **Perennial** or rarely **biennial herbs** with erect, sometimes shortened, **stems** and variously pinnate **leaves**, which usually have expanded sheath-like bases. **Umbels** terminal and lateral; bracts entire or pinnate, margin usually white-membranous. Petals oblong to broad-ovate, white, pink or purple-red. **Fruit** oblong to broadly egg-shaped, slightly flattened laterally, glabrous, often with numerous, shining tubercles, ribs prominent and sharp, sometimes sinuate, cristate or narrowly winged.

1. Stem stout, 5–30 cm tall .... *P. foetens*, **Pleurospemum sp.**
1. Plant slender, usually far more than 20 cm tall .... *P. benthamii, P. amabile, P. yunnanense, P. linearilobum*

Pleurospermum foetens Franch.* (syn. *Hymenidium foetens* (Franch.) Oimenov & Kljuy.; *Pleurospermum dielsianum* Fedde ex H. Wolff) Plant dwarf, 8–30 cm tall, with characteristic strong unpleasant **odour**. **Leaves** deep shiny green, 3–4-ternate-pinnate, 3–6 × 1–3 cm, the lower stalks 3–6 cm long, flattened, winged. **Umbels** 10–15 cm across; petals obovate, white or pinkish, c. 2 mm long; anthers purple-black, filaments white. **Fruit** egg-shaped, purple-black to dark red-brown, 7–9 × 3–4 mm, ribs broadly sinuate-winged. Gansu, Sichuan, seTibet, nwYunnan; open alpine meadows, rocky slopes, loose screes, 3,600–4,500 m. VI–VIII.

Pleurospemum sp.* Plant not more than 30 cm tall, with 3-pinnate, finely divided **leaves** like those of carrots. **Umbels** 8–10 cm across, main umbel with numerous rays, with a whorl of secondary umbels beneath; bracteoles elliptic, white with a central green stripe, forming a ruff around the individual umbels; **petals** white; anther purplish brown. nwSichuan (Juizhaigou); damp stony places by rivulets or amongst *Salix* scrub, c. 2,700 m. VI–VII. (See photo 1 on p. 386.)

Pleurospermum benthamii **(Wall. ex DC.) C. B. Clarke** (syn. *Hymenolaena benthamii* Wall. ex DC.; *Pleurospermum davidii* Franch.) Plant to 1.5 m with a hollow **stem** and 2–3-ternate-pinnate, glabrous **leaves**, to 8–15 cm. **Umbels** 10–15 cm across with rays up to 12 cm long; bracts 5–9, oblanceolate, to 9 cm long, with white-membranous margin, bracteoles similar but smaller, not more than 2 cm long; petals obovate, white, rarely pink, 2–3 mm long; anthers purple-red. **Fruit** egg-shaped-ellipsoid, 6–10 × 2.5–4.5 mm, ribs sinuate-winged. w, sw & nwSichuan, seTibet, nwYunnan [c & eHimalaya, nMyanmar]; open scrub, alpine pastures, riversides, 2,200–4,300 m. VI–VII.

Pleurospermum amabile **Craib & W. W. Sm.** Smaller plant than the previous species, 15–50 cm tall with a stout, solitary, violet-green **stem**. **Leaves** 3–4-ternate-pinnate, 6–15 cm long, with linear ultimate segments; wings of leaf-stalks greatly expanded, 3–5 cm wide, white with purple veins. **Umbel** 5–12 cm across, usually solitary; petals reverse heart-shaped, white to dark purple, 1–1.5 mm long; anthers dark purple. **Fruit** egg-shaped-oblong, 3–5 × c. 1.5 mm; ribs very narrowly sinuate-winged. seTibet, nwYunnan [swTibet, Bhutan, Sikkim]; open scrub, high-altitude alpine turf, semi-stable screes, (3,000–)4,000–5,100 m. VII–IX.

Pleurospermum yunnanense **Franch.** (syn. *P. pseudoyunnanense* H. Wolff) Similar to *P. amabile*, but ultimate leaf-segments ovate or oblong, margin with pointed teeth or irregularly cut, and petals obovate, greenish white, occasionally flushed pink; anthers dark purple-black. w & swSichuan, nwYunnan [neMyanmar]; woodland margins, dwarf *Rhododendron* scrub, valley sides, rocky slopes, 3,600–4,100 m. VI–IX.

Pleurospermum linearilobum **W. W. Sm.*** (syn. *P. calcareum* H. Wolff) Readily distinguished from the previous three species by the tuberculate stem-nodes and leaf-rachises, and **umbels** 10–18 cm across with unequal rays; petals pinkish white, oblong-obovate. wSichuan, nwYunnan; alpine grassland, mixed woodland margins, open low scrub, rocky slopes, screes, 2,400–4,200 m. VI–VII.

SELINUM

A genus of 8 species (3 in China, 2 endemic) in Europe and Asia. **Herbaceous perennials** with ferny, 2–3-pinnate or -ternate **leaves**. **Inflorescence** compound umbels; bracts entire to pinnately lobed, bracteoles similar; petals white or pinkish, with notched apex. **Fruit** oblong to rounded, furrowed on the surfaces, with a narrow ridged or winged margin.

Selinum wallichianum **(DC.) Raizada & Saxena** (syn. *Peucedanum wallichianum* DC.; *Selinum tenuifolium* Wall. ex C. B. Clarke) Variable species to 1 m in flower, with an erect, ribbed **stem**, often purple-flushed, branching above. **Leaves** with sheathing, inflated bases to the stalks, blade oval, to 25 × 20 cm, 3-pinnate, with numerous linear-lanceolate segments. **Umbels** hairy, 5–8 cm across, with up to 35 primary rays; bracteoles 5–10, deflexed, linear-elliptic, green with a white margin, equalling the secondary rays; petals white, sometimes with a pink flush, often purplish red in bud. **Fruit** c. 4 mm long with broad lateral wings. swSichuan, nwYunnan [Himalaya, nMyanmar]; open forests, glades, scrub, grassy slopes, meadows, 2,600–4,200 m. VII–IX.

ANGELICA

A genus of 90 species (45 in China, 32 endemic) distributed primarily in Europe and Asia. Usually stout **biennial** or **perennial herbs** with large, coarse, pinnately 1–4-lobed **leaves**. **Umbels** compound, usually with divided bracts; petals pink, white, rarely dark purple, ovate with an incurved apex. **Fruit** egg-shaped to rounded, compressed, with narrow dorsal ribs and a narrowly winged margin.

Angelica apaensis **R. H. Shan & C. C. Yuan*** (syn. *Heracleum apaense* (R. H. Shan & C. C. Yuan) R. H. Shan & T. S. Wang) Stout **biennial** to 2 m, often half that height, with a stout, ribbed **stem**; leaf-stalks with sheathing bases, these greatly expanded in the upper leaves, which are much reduced. **Leaves** coarse, 2-pinnate, with deep green, lanceolate to elliptic, toothed segments. **Umbels** 10–30 cm across, with numerous ascending rays; petals white, ovate, not radiate. **Fruit** elliptic to almost orbicular in outline, 5–10 mm long, with thick dorsal ribs. w & swSichuan, e & seTibet, nwYunnan; shrubberies, thickets, damp meadows, slopes, 3,000–4,000 m. VI–VIII.

HERACLEUM

A genus of c. 70 species (29 in China, 21 endemic) mainly in Asia and Europe, 1 in North America, a few species in east Africa. **Perennial** or rarely **biennial herbs**, with erect, terete, often ribbed **stems**. **Leaves** ternately or pinnately compound, hairy or glabrous, generally with conspicuous, expanded leaf-stalk sheaths. **Umbels** loosely compound, terminal and lateral, terminal umbel with hermaphrodite flowers. Calyx-teeth triangular, lanceolate, rudimentary or absent. **Petals** white, rarely pinkish or pale yellow, obovate or reverse heart-shaped, apex notched with a narrowly incurved lobule; outer flowers of umbel often radiant with outer petal enlarged. **Fruit** obovoid, egg-shaped, broadly egg-shaped or almost orbicular, strongly dorsally compressed, hairy or glabrous, lateral ribs usually winged.

Heracleum candicans **Wall. ex DC.** Downy or woolly plant, with a solitary, branched **stem** 40–100 cm tall or more, unpleasant-smelling. Basal and lower **leaves** 2–3-pinnate, silvery beneath, densely white-woolly, with toothed margin. **Umbels** large, to 40 cm across, secondary umbels 20–25-flowered; petals white, outer **flowers** radiant. **Fruit** obovoid, 5–8 × 4–6 mm, glabrous when mature. w & swSichuan, e & seTibet, c, n & nwYunnan [swTibet, Himalaya, nMyanmar]; sparse forests, coniferous forest margins, scrub, alpine meadows, arid grassy slopes, streamsides, 1,800–4,500 m. V–VII.

SANICULA

A genus of 40 species (17 in China, 11 endemic) primarily in temperate regions of the world, sometimes subtropical. **Biennial** or **perennial herbs** with ascending to procumbent **stems**. **Leaves** rounded to heart-shaped or pentagonal, often lobed. **Umbels** simple or compound with leaf-like bracts; petals narrow with an incurved tip. **Fruit** almost globose to ellipsoid, usually densely bristly.

Sanicula hacquetioides **Franch.*** Small glabrous plant to 30 cm, often less, with a central rosette of leaves and reddish, spreading to ascending **stems** and leaf-stalks. **Leaves** rounded or kidney-shaped, 1.5–4.5 cm across, 3-parted almost to base, segments rounded, shallowly lobed and scalloped, deep green and somewhat shiny above; leaf-stalks to 14 cm long. **Inflorescence** unbranched, with 2–4 small umbels, each 6–8 mm across with 2–3 stalkless, leaf-like bracts immediately beneath; **flowers** white or pinkish. **Fruit** egg-shaped, 2–2.5 mm long, scaly and warty. Guizhou, w & swSichuan, seTibet, w & nwYunnan; damp habitats, rocky places, forest and forest margins, banks, streamsides, 3,000–3,660 m. VI–VII.

1. *Pleurospermum* sp.; Jiuzhaigou, nwSichuan (PC)
2. *Pleurospermum benthamii*; Cangshan, nwYunnan (CGW)
3. *Pleurospermum yunnanense*; Genyanshan, swSichuan (AD)
4. *Selinum wallichianum*; Ghunsa Khola, eNepal (CGW)
5. *Angelica apaensis*; Balangshan, Wolong, wSichuan (CGW)
6. *Angelica apaensis*; Balangshan, Wolong, wSichuan (CGW)
7. *Sanicula hacquetioides*; Gonggashan pass, wSichuan (PC)

BUDDLEJACEAE

BUDDLEJA (often incorrectly spelt *Buddleia*)

A genus of c. 100 species (20 in China) in the tropics and subtropics of America, Africa and Asia (including the Himalaya). **Evergreen** and **deciduous shrubs**, rarely small **trees**, with 4-angled **branches** and paired, rarely alternate **leaves**. Stipules present, located between opposing leaf-stalk bases. **Flowers** usually numerous, often strongly fragrant, in terminal or lateral clusters, or panicles, often spike-like; calyx with a tube and 4 short lobes; corolla with a short to long, straight or curved tube and 4 short lobes, varying from salverform or funnel- to cup-shaped or campanulate; stamens inserted on, and included in, corolla-tube. **Fruit** a small capsule splitting into 2 valves, containing numerous small, usually winged **seeds**. Most species are highly attractive to butterflies.

1. Leaves alternate .... **B. alternifolia**
1. Leaves opposite 2
2. Corolla at least 23 mm long with a wide throat 8–9 mm in diameter .... **B. colvilei**
2. Corolla less than 20 mm long, with a narrow throat no more than 4 mm in diameter 3
3. Corolla-tube curved, glandular-hairy outside .... **B. lindleyana**
3. Corolla-tube straight, often hairy but not glandular-hairy outside 4
4. Leaf-blade triangular to broad-ovate .... **B. crispa**
4. Leaf-blade ovate to elliptic or lanceolate 5
5. Lateral veins of leaf 4–7-paired .... **B. delavayi, B. yunnanensis**
5. Lateral veins of leaf 8–18-paired 6
6. Corolla-tube glabrous or only partly stellate-hairy .... **B. davidii, B. forrestii**
6. Corolla-tube entirely stellate-hairy .... **B. fallowiana, B. candida, B. officinalis**

Buddleja alternifolia Maxim.* **Deciduous shrub** to 5 m, occasionally more, with pale, arching, rather slender **branches** and narrow-elliptic to linear **leaves**, 30–100 × 2–13 mm, with an entire or wavy margin; leaves of flowering shoots small, not more than 15 × 10 mm. **Flowers** in interrupted panicles or globose clusters on short lateral branches on the old wood, very fragrant; corolla lilac, violet to purple, orange in throat, with a tube 6–10 mm long and a limb 4–7 mm across. Gansu, Qinghai, Sichuan, Shaanxi; dry riverbanks and -beds, shrubberies, 1,500–4,000 m. III–VII.

Buddleja colvilei Hook. f. & Thomson **Shrub** or small **tree** to 6 m with short-stalked to stalkless, elliptic **leaves**, 7–12 × 2–6 cm, densely stellate-downy at first. **Inflorescences** pendent, terminal and in upper leaf-axils, to 23 cm long, almost paniculate, forming an oblong cluster; corolla rose-purple to wine-red or crimson, 23–30 mm long with a broad cylindrical tube and a limb 20–30 mm wide. seTibet, wYunnan [swTibet, Himalaya]; forests, forest margins, shrubberies, more open places, 1,600–4,200 m. VI–IX.

Buddleja lindleyana Fortune* **Deciduous shrub** to 3 m with slender erect **branches**, often suckering, **leaves** rather bright green, paler and almost glabrous beneath, with 6–8 pairs of conspicuous lateral veins, margin toothed or not. **Inflorescence** spike-like, spreading to semi-drooping, to 20 cm long; **flowers** faintly fragrant; corolla purple, 13–20 mm long with a limb 5–8 mm wide. Guangxi, Guizhou, Hubei, Sichuan, Yunnan; forest margins, path- and tracksides, shrubberies, streamsides, 200–2,700 m. V–X.

Buddleja crispa Benth. (syn. *B. farreri* Balf. f. & W. W. Sm.; *B. tibetica* W. W. Sm.) **Deciduous shrub** to 5 m, often less, with soft whitish or greyish, downy, ovate to triangular **leaves**, 2.5–20 × 1–8 cm, margin toothed, leaves on vigorous vegetative shoots with a heart-shaped base clasping the stem, or fused in pairs at the base to surround the stem. **Inflorescence** a terminal panicle, sometimes spike-like, to 20 cm long; **corolla** lilac to violet or purple, often rather pale with orange throat, 3–9 mm long with limb 5–10 mm wide. Gansu, Sichuan, e & seTibet, Yunnan [swTibet, eAfghanistan, Himalaya]; rocky slopes, dry river valleys, cliffs, shrubberies, 1,400–4,300 m. II–VIII.

Buddleja delavayi Gagnep.* **Shrub** or small **tree** to 6 m (sometimes only 1 m) with narrow-elliptic, glabrous to downy **leaves**, 4–16 × 1.5–6.5 cm, margin entire to somewhat wavy. **Inflorescence** an interrupted, rather lax, panicle to 12 cm long; **corolla** rose-pink to lavender, 11–16 mm long, with limb 6–11 mm wide. seTibet, Yunnan; forest margins, tracksides, shrubberies, 2,000–3,000 m. I–IV.

Buddleja yunnanensis Gagnep.* Readily separated from the previous (and most other) species by the dense spike-like **inflorescences** beset with numerous bracts, some of which exceed the **flowers**; shrub to 4 m with elliptic to ovate **leaves** with 4–6 pairs of lateral veins; **corolla** lilac, 10–12 mm long. Yunnan; forest margins, shrubberies, 1,000–2,500 m. VI–X.

Buddleja davidii Franch.* **Deciduous shrub** to 5 m, often less, with rather erect, pale grey-brown **stems** and ovate to elliptic **leaves**, 4–20 × 1–7.5 cm, with long-pointed apex and toothed margin, green or grey-green above, paler beneath, with 9–14 pairs of lateral veins. **Inflorescence** a cylindrical, tapered panicle, terminal or lateral, to 30 cm long; **flowers** very fragrant; corolla violet to dark purple, with orange-yellow throat, occasionally all white, 8–14 mm long, with limb 4–6 mm wide. Gansu, Guangxi, Guizhou, Hubei, Sichuan, Shaanxi, Tibet [cChina]; shrubberies, thickets, rocky places, 800–3,000 m. V–X.

Buddleja forrestii Diels Differs from the previous species in the broader corolla-tube, 2–3.5 mm in diameter (not 1–1.5 mm); **leaves** narrow-elliptic, to 35 × 8 cm, with up to 18 pairs of lateral veins; **corolla** ranging in colour from orange to

1. *Buddleja alternifolia*; Jiuzhaigou, nwSichuan (CGW)
2. *Buddleja alternifolia*; cult. (CGW)
3. *Buddleja colvilei*; cult. (MF)
4. *Buddleja lindleyana*; cult. (CGW)
5. *Buddleja crispa*; cult. (CGW)
6. *Buddleja davidii* cultivar; cult. form (CGW)
7. *Buddleja davidii*; Eimeshan, wSichuan (CGW)
8. *Buddleja forrestii*; cult., from nwYunnan (CGW)
9. *Buddleja forrestii*; Cangshan, nwYunnan (DC)

pink, purple or mauve, occasionally almost white, tube relatively short and broad. Sichuan, seTibet, Yunnan [eHimalaya]; forest margins, woodland, shrubberies, riversides, 1,800–4,000 m. VI–X.

Buddleja fallowiana Balf. f. & W. W. Sm.* **Deciduous shrub** to 5 m with young **stems** and leaves clothed in white- or grey-stellate down, leaves eventually greening with age; **leaves** ovate to elliptic, 5–14 × 2–6.5 cm, with a finely toothed margin and 8–10 pairs of lateral veins. **Inflorescence** narrow-paniculate, on current season's shoots, 5–15 cm long; **corolla** white to pale lavender with orange throat, 9–14 mm long, with limb 5–9 mm wide. Sichuan, e & seTibet, Yunnan; open woodland, rocky places, shrubberies, 1,200–3,800 m. V–X.

Buddleja candida Dunn* Readily distinguished from the previous species by the oblong, puckered **leaves**, 12–24 × 3–6 cm, with 10–12 pairs of lateral veins; and by the small violet **flowers**, with corolla-tube not more than 5 mm long. s & swSichuan, seTibet, w & nwYunnan; forest margins, shrubberies, riversides, 1,000–2,500 m. IV–X.

Buddleja officinalis Maxim. Similar to *B. fallowiana*, but **leaves** with 8–14 pairs of lateral veins and **inflorescence** a broader panicle with pink, lilac or purplish **flowers** with a yellow or orange throat, corolla 10–13 mm long, with some glandular hairs on outside mixed in the stellate down. Gansu, Guangxi, Guizhou, Hubei, Sichuan, Shaanxi, e & seTibet, Yunnan [cChina, Myanmar, Vietnam]; similar habitats, 200–2,800 m. III–VI.

1. *Buddleja fallowiana*; cult. (PC)
2. Collected inflorescences of *Buddleja officinalis*; Qinshui valley, seYunnan (PC)
3. *Gentiana robusta*; nNepal (CGW)
4. *Gentiana crassicaulis*; Xiaozhongdian, nwYunnan (CGW)

GENTIANACEAE

GENTIANA

A complex genus with c. 360 species (248 in China, the majority endemic), worldwide except for Africa and Arabia. Glabrous **annual**, **biennial** or **perennial herbs** with opposite or whorled, entire **leaves**, often confined to basal rosettes. **Flowers** 5–8-parted, solitary or in lateral or terminal, whorled spikes or heads; calyx toothed or lobed, sometimes split down 1 side; corolla tubular, salverform or trumpet-shaped, with smaller lobes (plicae) alternating with the main lobes. **Fruit** a 2-valved capsule.

SYNOPSIS OF GROUPS

Group One (section *Cruciata*). Perennials, often with vegetative leaf-rosettes at flowering time; capsules unwinged; basal leaves mostly 8–28 cm long; plants 10–40 cm tall in flower, flowers usually numerous, in terminal or lateral clusters or cymes.

Group Two (section *Isomeria*). Perennials, often with vegetative leaf-rosettes at flowering time; capsules unwinged; basal leaves rarely more than 7 cm long, generally much smaller; plants not more than 10 cm tall in flower, flowers often solitary, occasionally in groups; flowering stems arising singly from slender stolons, generally spreading and rooting at the nodes. (p. 392)

Group Three (sections *Kudoa* and *Phyllocalyx*). Perennials, often with vegetative leaf-rosettes at flowering time; capsules unwinged; basal leaves rarely more than 7 cm long, generally much smaller; plants not more than 10 cm tall in flower, flowers often solitary, occasionally in groups; flowering stems several to many arising from the top of a thickened, erect caudex, rarely also stoloniferous; sepals erect, 2 larger than other 3; flowering stems without blackish leaf-remains at base. (p. 395)

Group Four (section *Frigida*). Perennials, often with vegetative leaf-rosettes at flowering time; capsules unwinged; basal leaves rarely more than 7 cm long, generally much smaller; plants not more than 10 cm tall in flower, flowers often solitary, occasionally in groups; flowering stems several to many arising from the top of a thickened, erect caudex, rarely also stoloniferous; sepals usually recurved (or if erect, ± equal), often not more than 3 mm long; flowering stems with blackish, fibrous leaf remains at base. (p. 399)

Group Five (sections *Chondrophyllae* (including *Dolichocarpa*) and *Microsperma*). Annuals, rarely perennials, without vegetative leaf-rosettes at flowering time; capsules winged or unwinged. (p. 400)

Group One

1. Flowers or flower-clusters stalked, both terminal and sometimes lateral ....**G. straminea, G. dahurica**
1. Flowers stalkless, in dense heads, sometimes accompanied by lateral stalkless clusters 2
2. Corolla purple-blue .... **G. crassicaulis, G. siphonantha**
2. Corolla pale yellow or greenish yellow, lobes sometimes flushed purple-brown ....
.... **G. robusta, G. officinalis, G. tibetica**

Gentiana straminea Maxim. Robust tufted **perennial** to 35 cm with unbranched, ascending, glabrous **stems**. **Basal leaves** lanceolate to elliptic, to 20 × 4 cm, 3–5-veined, with rough margins, leaf-stalks short; **stem-leaves** in 3–5 pairs, similar but smaller. **Flowers** in terminal and lateral clusters, forming a panicle-like **inflorescence**; calyx purplish, split down 1 side; corolla yellowish green to greenish white, with green spots in throat, sometimes with a slight bluish flush, funnel-shaped, 35–45 mm long, with small triangular plicae. Gansu, Hubei, Qinghai, e & seTibet [sw & cTibet, Nepal]; meadows, streamsides, shrubberies, forest margins and clearings, 2,000–5,000 m. VII–IX.

Gentiana dahurica Fisch. Similar to the previous species, but plant seldom more than 25 cm tall, with **basal leaves** up to 15 × 1.5 cm; **corolla** dark purple-blue with yellow spots in throat; calyx-lobes linear-lanceolate. Hubei, Qinghai, n & nwSichuan, Shaanxi [nChina, Mongolia, eRussia]; similar habitats and altitudes. VII–IX.

Gentiana crassicaulis Duthie ex Burkill* Robust, tufted **perennial** to 45 cm with stout ascending **stems**, glabrous. **Basal leaves** to 20 × 7 cm, elliptic with a rough margin, 5–7-veined; **stem-leaves** in 3–5 pairs, uppermost larger, forming a ruff around the flowers. **Flowers** numerous in a dense terminal cluster, sometimes with supporting clusters; calyx green or purplish, split down 1 side; **corolla** purple-blue with greenish or yellowish base, urn-shaped, 20–24 mm long, with tiny triangular plicae. nwGuizhou, wSichuan, seTibet, nwYunnan; meadows, shrubberies, roadsides, forest clearings and margins, 2,100–4,500 m. VI–VIII.

Gentiana siphonantha Maxim. ex Kusn.* Readily distinguished from the previous species by the less robust habit and linear-lanceolate **basal leaves** not more than 14 × 1.5 cm, 3–5-veined; **corolla** dark blue or purple-blue, 23–26 mm long. Gansu, se & eQinghai; meadows and grassy steppes, open shrubberies, riversides, 1,800–4,500 m. VII–IX.

Gentiana robusta King ex Hook. f. (syn. *G. tibetica* King ex Hook. f. var. *robusta* (King ex Hook. f.) Kusn.) Tufted **perennial** to 30 cm with simple, ascending **stems**, glabrous. **Basal leaves** ovate to lanceolate-elliptic, to 25 × 4.5 cm, 3–5-veined, with a rough margin; **stem-leaves** in 3–5 pairs, similar to the basal but smaller, 1–3-veined. **Flowers** mainly in crowded terminal clusters; calyx greenish, membranous, split along 1 side; corolla greenish white to yellowish white or greenish yellow, tubular-campanulate, 30–35 mm long. **Capsules** stalked. seTibet [swTibet, Himalaya]; meadows and other grassy places, open shrubberies, borders of cultivated land, roadsides, 3,500–4,800 m. VI–IX.

Gentiana officinalis **Harry Sm.*** Similar to the previous species, but **flowers** smaller, corolla pale yellow or straw-coloured, 18–22 mm long, and **capsules** stalkless. sGansu, seQinghai, n & nwSichuan, sShaanxi; similar habitats as well as stream- and riversides, 2,300–4,200 m. VIII–IX.

Gentiana tibetica **King ex Hook. f.** Essentially similar to *G. crassicaulis* with the **stem-leaves** increasing in size towards the top, but **flowers** pale yellow or greenish yellow, often suffused with purplish brown. seTibet [swTibet, Himalaya]; forest margins, roadsides, margins of cultivation, 2,100–4,200 m. VI–VIII.

> **Group Two**
> 1. Leaf- and sepal-margins pale (often whitish), membranous or thickened . . . . **G. depressa, G. lacerulata, G. urnula**
> 1. Leaf- and sepal-margins not paler than the rest of the leaf, neither membranous nor thickened 2
> 2. Flowers several in a cluster at stem-tips . . . .
>   . . . . **G. elwesii, G. chinensis**
> 2. Flowers generally solitary 3
> 3. Corolla striped and often spotted . . . . . **G. gilvostriata**
> 3. Corolla neither striped nor spotted . . . . **G. wardii**,
>   **G. wardii** var. **micrantha, G. wardii** var. **emergens**,
>   **G. tubiflora, G. stragulata**

Gentiana depressa **D. Don** Low patch-forming **stoloniferous perennial** to 6 cm. **Leaves** in crowded, squarish rosettes, bluish green, oblong to ovate, to 15 × 9 mm, 1-veined, membranous, margin rough. **Flowers** always solitary, surrounded at base by leaf-rosette; calyx campanulate, rather membranous, with ovate lobes more than half the length of tube; corolla pale blue and lilac, lilac or pinkish purple, campanulate, 25–30 mm long, throat often with whitish stripes and greenish markings. seTibet [swTibet, c & eHimalaya]; moraines, rocky places, alpine meadows, cliff-ledges, 3,000–4,500 m. VII–X.

Gentiana lacerulata **Harry Sm.** Very similar to the previous species, but **leaves** fan-shaped to obovate, to 10 × 12 mm, leaf- and sepal-margins membranous, minutely and irregularly cut; **corolla** purplish blue, tubular-campanulate, 25–35 mm long. seTibet [swTibet, c & eHimalaya]; similar habitats, 4,200–4,500 m. VIII–X.

Gentiana urnula **Harry Sm.** Similar to *G. depressa*, but a very squat plant with the leaf-rosettes clasping the base of the flower; **leaves** fan-shaped, closely overlapping, forming square rosettes, upper leaves and calyx usually flushed deep purplish blue, margins smooth; corolla blue or bluish purple, streaked deeper outside, urn-shaped, 20–30 mm long. s & swQinghai, ne & seTibet [w & cTibet c & eHimalaya]; moraines, rocky and gravelly slopes, stabilised screes, 3,900–5,700 m. VIII–X.

Gentiana elwesii **C. B. Clarke** Tufted **stoloniferous perennial** with simple, erect **stems**, to 20 cm. **Leaves** oblong to elliptic, to 20 × 8 mm, with a short stalk and smooth or minutely ciliate margin, 1-veined, upper crowded beneath flowers, often with purplish margin. **Inflorescence** with up to 8 **flowers** in small clusters; calyx greenish or purplish, with short, unequal lanceolate lobes; **corolla** blue or bluish purple, urn-shaped (the lobes scarcely separating at the top), with tiny ear-like plicae between the main lobes. seTibet [swTibet, c & eHimalaya]; rocky meadows, low scrubby slopes, 3,600–4,400 m. VIII–X.

Gentiana chinensis **Kusn.*** Differs from the previous species in being more extensively stoloniferous and mat-forming with spreading to ascending **stems**; **leaves** ovate to almost rounded, to 15 × 7 mm; **inflorescence** generally 1–3-flowered; **corolla** blue with paler stripes, tubular, 30–35 mm long. wHubei, s & swSichuan, nwYunnan; grassy and rocky slopes, open forests and forest clearings, 2,400–4,500 m. VII–X.

Gentiana gilvostriata **C. Marq.** Patch-forming **stoloniferous perennial** to 7 cm with simple, erect, minutely hairy **stems**. **Leaves** small, elliptic to obovate, to 7 × 5 mm, with a short stalk, crowded into rosettes at stem-tops. **Flowers** often solitary, sometimes 2–3 together; calyx obconic with short blunt lobes; corolla pale blue with deeper bright blue stripes, streaks or spots, yellowish or whitish towards base, broad-campanulate, 30–40 mm long, plicae one third the size of main lobes. seTibet, nwYunnan [eHimalaya, nMyanmar]; grassy habitats on open slopes or in open forests, 3,000–3,900 m. VIII–XI.

Gentiana wardii **W. W. Sm.** Patch-forming **stoloniferous perennial**, not more than 5 cm. **Leaves** bright, often rather pale green, crowded in rosettes, spathulate to oval, to 11 × 6 mm, 1- or 3-veined. **Flowers** solitary; calyx green with ovate lobes a third the length of the tube; corolla blue, deep violet-blue or blackish blue, tubular-campanulate, 20–25 mm long, plicae half the size of main lobes. swSichuan, seTibet, nwYunnan [nMyanmar]; rocky alpine meadows, moraines, low alpine scrub, 3,000–4,900 m. VII–IX. Very variable in flower size: var. *micrantha* **C. Marq.** has small **corollas** only 8–10 mm long; var. *emergens* **(C. Marq.) T. N. Ho** (syn. *G. emergens* C. Marq.) has the largest **corollas**, 30–35 mm long.

Gentiana tubiflora **(Wall. ex G. Don) Griseb.** (syn. *G. longistyla* T. N. Ho) Somewhat similar to the previous species, but **leaves** oblong to elliptic, to 9 × 3 mm, generally deep green, 1-veined; **corolla** dark blue, more tubular, 32–42 mm long; style exceptionally long, 12–14 mm, exceeding ovary (only 2–3 mm in *G. wardii*). seTibet [swTibet, c & eHimalaya]; grassy and stony slopes, open low scrub, 4,200–5,300 m. VII–X.

1. *Gentiana depressa*; eNepal (CGW)
2. *Gentiana lacerulata*; eNepal (CGW)
3. *Gentiana urnula*; eNepal (CGW)
4. *Gentiana elwesii*; eNepal (CGW)
5. *Gentiana chinensis*; Emeishan, sSichuan (CGW)
6. *Gentiana wardii* var. *emergens*; Baimashan, nwYunnan (CGW)
7. *Gentiana tubiflora*; nNepal (CGW)

1. *Gentiana georgei*; Gangheba, Yulongxueshan, nwYunnan (RM)
2. *Gentiana szechenyii*; Gangheba, Yulongxueshan, nwYunnan (HJ)
3. *Gentiana szechenyii*; Haizi Lake, wSichuan (HJ)
4. *Gentiana stipitata*; Tibet (YMY)
5. *Gentiana stipitata*; Markham, Laoshan, seTibet (HJ)
6. *Gentiana cephalantha*; Cangshan, nwYunnan (CGW)
7. *Gentiana rigescens*; Xisong, Lijiang, nwYunnan (HJ)
8. *Gentiana melandriifolia*; Cangshan, nwYunnan (HJ)

Gentiana stragulata Balf. f. & Forrest* Slightly taller plant than the previous two species, the short **stems** with dense purple minute hairs, **leaves** oblong to spathulate, to 7 × 4 mm; calyx covered with minute purple hairs (papillae); **corolla** larger, blue, tubular, 45–55 mm long, plicae about one-third the size of main lobes, minutely toothed. seTibet, nwYunnan; partially similar habitats, 3,000–4,300 m. V–VI.

Group Three
1. Calyx enclosed within upper pair of leaves; stigma-lobes fused into a disc-like or funnel-shaped structure (section *Phyllocalyx*) . . . . **G. phyllocalyx**
1. Calyx free of uppermost pair of leaves or leaf-whorl; stigma 2-lobed (section *Kudoa*) 2
2. Calyx-lobes with a white thickened (cartilaginous) margin, keeled . . . .**G. georgei, G. szechenyii, G. stipitata, G. stipitata subsp. tizuensis**
2. Calyx-lobes without a white thickened (cartilaginous) margin, unkeeled 3
3. Plicae asymmetrical; calyx-lobes different in shape to uppermost leaves 4
3. Plicae symmetrical; calyx-lobes similar in shape to uppermost leaves 5
4. Flowers in clusters of 3 or more; stamens equal . . . . . .
. . . . **G. cephalantha, G. cephalantha var. vaniotii, G. rigescens**
4. Flowers often solitary, occasionally 2–3 per stem; stamens unequal . . . . . .**G. melandriifolia, G. duclouxii**
5. Leaves in whorls of 3–8 . . . .
. . . . **G. arethusae, G. arethusae var. delicatula, G. hexaphylla, G. tetraphylla, G. ternifolia**
5. Leaves opposite 6
6. Upper stem-leaves elliptic to orbicular . . . .
. . . . **G. caelestis, G. obconica**
6. Upper stem-leaves linear 7
7. Basal leaf-rosettes present at flowering time, their leaves longer than leaves of flowering stems . . . .
. . . . **G. veitchiorum, G. ornata, G. nyalamensis**
7. Basal leaf-rosettes absent or poorly developed at flowering time, their leaves (when present) not longer than those of flowering stems . . . .**G. sino-ornata, G. lawrencei, G. lawrencei var. farreri, G. oreodoxa**

Gentiana phyllocalyx C. B. Clarke Patch-forming perennial not more than 12 cm, with erect to ascending glabrous **stems**. **Leaves** in 3–5 pairs, crowded towards stem-base, obovate, to 26 × 16 mm, short-stalked, with 1 or 3 veins. **Calyx** greenish yellow, tubular-campanulate, with short ovate lobes, clasped by the uppermost leaf-pair; **corolla** blue with deeper bluish black stripes outside, tubular-campanulate, 30–47 mm long. seTibet, nwYunnan [eHimalaya]; grassy and stony alpine meadows, low alpine scrub, 3,000–5,200 m. VI–IX.

Gentiana georgei Diels* Low **tufted perennial** to 7 cm with glabrous, ascending **stems**, with black leaf remnants at top of root. **Leaves** deep green, the basal in a rosette, lanceolate and somewhat folded, sharply pointed, to 80 × 15 mm, margin rough; stem-leaves in 2–4 pairs, smaller than basal, uppermost surrounding the calyx. **Flowers** solitary, stalkless; calyx membranous with unequal, lanceolate to triangular lobes; corolla reddish purple, paler towards base, usually unspotted, campanulate, 40–60 mm long. sGansu, s & seQinghai, n & wSichuan, nwYunnan; grassy and stony meadows, moraines, 3,000–4,200 m. VIII–XI.

Gentiana szechenyii Kanitz* Similar to the previous species, but **basal leaves** not more than 60 × 10 mm; **corolla** white with bluish stripes, usually green-spotted inside tube, lobes 5–6 mm long (7–10 mm in *G. georgei*). sGansu, s & seQinghai, w & swSichuan, seTibet, nwYunnan; similar habitats, 3,000–4,800 m. VII–XI.

Gentiana stipitata Edgew. subsp. *tizuensis* (Franch.) T. N. Ho* Differs from the previous two species in having **basal leaves** distinctly contracted below into the leaf-stalk and calyx-lobes obovate to spathulate, narrowed at base; corolla pale bluish grey, mauvish or white, with dark blue or greenish purple stripes and spots, 25–40 mm long. Gansu, s & seQinghai, w & nwSichuan; similar habitats, 3,200–4,600 m. VI–X. **Subsp. *stipitata*** differs in having corolla-lobes that lack obvious awns, whereas those of subsp. *tizuensis* have obvious awns. e & seTibet [c & swTibet, wHimalaya].

Gentiana cephalantha Franch. Tufted, leafy **perennial** to 50 cm, often far less, with ascending flower **stems**, glabrous or glandular (**var. *vaniotii* (H. Lévl.) T. N. Ho**) in upper part. **Leaves** deep green, basal in a rosette, elliptic to oblanceolate, to 100 × 22 mm, 1- or 3-veined, margin rough and slightly rolled; stem-leaves in up to 10 pairs, uppermost larger than lower. **Inflorescence** with up to 12 flowers in a cluster; calyx with 2 large and 3 smaller lobes; **corolla** pale blue with spots and stripes inside tube, tubular-campanulate, 25–35 mm long, plicae tiny. Guangxi, Guizhou, s & seSichuan, Yunnan [Myanmar, Thailand, Vietnam]; grassy slopes, shrubberies, forest margins, 2,000–3,600 m. IX–X.

Gentiana rigescens Franch. Differs from the previous species in lacking or having only a poorly developed basal **leaf-rosette**, with lowermost **stem-leaves** scale-like and membranous; **corolla** violet with blue spots in throat, not more than 30 mm long. Guizhou, s & seSichuan, Yunnan [nMyanmar]; similar habitats, 1,100–3,000 m. VIII–X.

Gentiana melandriifolia Franch.* **Tufted perennial** with spreading (sometimes prostrate) to ascending **stems**. **Leaves** deep green, sometimes purple-flushed, basal in a rosette, oblong to ovate or obovate, to 32 × 18 mm, with a slightly rolled margin, 1- or 3-veined, abruptly contracted into the stalk; stem-leaves in 4–5 pairs, upper larger than lower. **Flowers** 1–3, terminal, stalkless; calyx green or purplish; corolla pale blue with deeper spots in throat, funnel-shaped, 32–45 mm long. nw & cYunnan; rocky and grassy places, pathsides, forest margins, 2,200–3,000 m. VII–X.

Gentiana duclouxii Franch.* Very similar to the previous species, but **rosette-leaves** narrower, spathulate to oblanceolate, not more than 12 mm wide, and **flowers** rose-pink with blue spots in throat. cYunnan; habitats similar to those of *G. melandriifolia*, 1,800–1,900 m. V–IX.

Gentiana arethusae Burkill* **Tufted perennial** to 15 cm, with many ascending slender **stems**; basal **rosettes** poorly developed. **Stem-leaves** congested, usually in whorls of 6, linear-lanceolate, to 17 × 1.5 mm, long-pointed, lowermost withered by flowering time, uppermost forming a ruff around the base of the flowers. **Flowers** solitary, stalkless; calyx with linear, leaf-like lobes; corolla pale to mid-blue with deeper blue pencil stripes on outside, tubular-campanulate to broadly funnel-shaped (var. *delicatula* C. Marq.), 35–60 mm long, with tiny plicae between the 5–8 main lobes which are often speckled and bear a long tail-like tip 2–2.5 mm long. e, w & swSichuan, seTibet, nwYunnan; alpine meadows, low alpine shrubberies, stony slopes, forest margins, 2,700–4,800 m. VIII–X.

Gentiana hexaphylla Maxim. ex Kusn.* Very similar to the previous species, but **stem-leaves** and **calyx-lobes** spathulate at tip, pointed to blunt, to 3 mm wide; **corolla** blue with a yellowish or whitish base and dark blue streaks, tubular-campanulate, 35–50 mm long, usually 6–8-lobed. sGansu, seQinghai, n, nw & wSichuan, Shaanxi (Taibaishan); similar habitats, 2,700–4,400 m. VII–IX.

Gentiana tetraphylla Maxim. ex Kusn.* Readily distinguished from the previous two species by bearing leaves in whorls of 4(–5), the largest not more than 8 × 1.5 mm; corolla similar to *G. hexaphylla*, but tip of lobes less than 1 mm long, stamens attached to the middle of tube, not the base. n & wSichuan; similar habitats, 3,300–4,500 m. VII–IX.

Gentiana ternifolia Franch.* Distinguished from the other species in this group by having most **leaves** in whorls of 3 and by the 5-lobed **corolla** lacking tail-like tips to the lobes; corolla blue with deeper stripes, tubular-campanulate to funnel-shaped, 45–65 mm long. n & nwYunnan; wet meadows, 3,000–4,100 m. IV–V.

Gentiana caelestis (C. Marq.) Harry Sm.* (syn. *G. veitchiorum* Hemsl. var. *caelestis* C. Marq.) **Tufted perennial** to 8 cm tall, with ascending shoots, several of which are non-flowering. **Leaves** ovate to elliptic, uppermost the largest, surrounding base of calyx, to 18 × 4 mm. **Flowers** solitary, stalkless, usually 5-parted; calyx with linear-elliptic lobes; corolla pale blue, sometimes with darker streaks inside tube, tubular-campanulate, 40–50 mm long, plicae half the length of main lobes. swSichuan, seTibet, nwYunnan; alpine meadows and other grassy places, shrubberies, roadsides, 2,600–4,500 m. VIII–X.

Gentiana obconica T. N. Ho Readily distinguished from the previous species by its dark blue, more cone-shaped **flowers**, often with a paler base, black-striped outside. **Leaves** rather fleshy, increasing in size up the **stem**, largest ovate-elliptic, to 11 × 3 mm. seTibet [swTibet, c & eHimalaya]; alpine meadows, stony open scrub, 4,000–5,500 m. VIII–IX.

Gentiana veitchiorum Hemsl.* Tufted, patch-forming, **stoloniferous perennial** with ascending simple **stems** not more than 8 cm tall, with well-developed basal leaf-rosettes. **Leaves** opposite, shiny deep green, ovate to elliptic or lanceolate, to 15 × 5 mm, pointed, grooved, with a rough margin; rosette-leaves to 55 × 5 mm. **Flowers** solitary and stalkless; calyx tubular with diverging leaf-like lobes to 11 mm long; corolla intense blue with deeper streaks and tiny spots, yellowish or whitish towards base of tube, occasionally pale lavender or whitish overall, narrowly funnel-shaped, 40–60 mm long, plicae often fringed. sGansu, seQinghai, w, nw & swSichuan, seTibet, nwYunnan [swTibet]; dryish meadows, banks, open shrubberies, forest margins, 2,500–4,800 m. VII–X.

Gentiana ornata (Wall. ex G. Don) Griseb. Similar to the previous species, but **stems** glabrous (not minutely hairy-papillose) and **flowers** campanulate, generally a paler blue with a yellowish white base and dark blue streaks and spots, 25–40 mm long. seTibet [swTibet, c & eHimalaya]; alpine meadows, banks, stony pastures, low alpine scrub, 3,300–5,000 m. VIII–XI.

Gentiana nyalamensis T. N. Ho Similar to *G. veitchiorum*, but **stem-leaves** somewhat longer, to 20 mm, and **flowers** more campanulate, 25–47 mm long, paler blue with whitish or yellowish base, streaked dark blue outside and with some spotting inside. seTibet [swTibet, Bhutan]; grassy and rocky meadows, 3,500–4,700 m. VIII–IX.

1. *Gentiana duclouxii*; cYunnan (YMY)
2. *Gentiana arethusae*; Baimashan, nwYunnan (PC)
3. *Gentiana arethusae*; Tianchi, nwYunnan (EL)
4. *Gentiana hexaphylla*; north of Jigzhi, seQinghai (CGW)
5. *Gentiana hexaphylla*; Shenshan, Jigzhi, seQinghai (HJ)
6. *Gentiana ternifolia*; cult. (AGS)
7. *Gentiana caelestis*; nwYunnan (HJ)
8. *Gentiana obconica*; Serkhyen La, seTibet (HJ)
9. *Gentiana veitchiorum*; Napahai, Zhongdian, nwYunnan (JM)
10. *Gentiana veitchiorum*; Napahai, Zhongdian, nwYunnan (PC)
11. *Gentiana veitchiorum*; Litang to Batang, wSichuan (HJ)
12. *Gentiana ornata*; ne Nepal (CGW)
13. *Gentiana nyalamensis*; Kongbo Gyamda, seTibet (HJ)

1. *Gentiana sino-ornata* variants; Napahai, Zhongdian, nwYunnan (HJ)
2. *Gentiana sino-ornata*; cult. (CGW)
3. *Gentiana sino-ornata*; Zhedou, wSichuan (HJ)
4. *Gentiana sino-ornata*; Tianchi, nwYunnan (PC)
5. *Gentiana oreodoxa*; Habashan, nwYunnan (HJ)
6. *Gentiana nubigena*; Matang, Zhegushan, nwSichuan (HJ)
7. *Gentiana atuntsiensis*; Napahai, Zhongdian, nwYunnan (HJ)
8. *Gentiana microdonta*; Tianchi, nwYunnan (PC)
9. *Gentiana striolata*; Genyanshan, swSichuan (MN)
10. *Gentiana wilsonii*; Genyanshan, swSichuan (MN)
11. *Gentiana purdomii*; Animaqingshan, cQinghai (JB)

Gentiana sino-ornata Balf. f. **Tufted perennial** with slender spreading to ascending **stems** to 15 cm, lacking or having only poorly developed basal rosettes. **Leaves** opposite, often rather pale green or grass-green, larger and crowded towards stem-tips, incurved, linear-lanceolate, to 35 × 3 mm, pointed. **Flowers** solitary, stalkless; calyx tubular, with leaf-like lobes to 15 mm long; corolla very variable in colour from pale to deep blue, often flushed lilac or lavender with whitish yellow base and dark blue stripes and spots, to cream or white overall, with or without markings, funnel-shaped, 50–60 mm long, plicae entire. s, w & swSichuan, e & seTibet, nwYunnan [nMyanmar]; wet and marshy meadows, forest margins, damp open shrubberies, 2,800–4,400 m. VII–X.

Gentiana lawrencei Burkill* (including *G. farreri* Balf. f.) Similar to *G. sino-ornata*, but **flowers** with stalks 5–10 mm long; middle and upper **leaves** linear, not more than 1.5 mm wide; **corolla** pale blue with deeper streaks and a whitish yellow base, more campanulate, 45–55 mm long. swGansu, Qinghai, n, nw & wSichuan. **Var.** *farreri* **(Balf. f.) T. N. Ho*** is sometimes distinguished by exceptionally long **calyx-lobes** that are often twice the length of the calyx-tube (equal in **var.** *lawrencei*).

Gentiana oreodoxa Harry Sm. Rather like a compact version of *G. sino-ornata*, but with short **stems**, not more than 5 cm tall; **stem-leaves** to 15 × 1.5 mm; **corolla** smaller and more obconic, not usually more than 40 mm long, pale blue with dark blue stripes and spots and a yellowish white base, plicae finely toothed. seTibet, nwYunnan [eHimalaya, nMyanmar]; alpine meadows and other grassy places, 3,000–4,900 m. VIII–X.

> **Group Four**
> 1. Flowering stems decumbent, rarely more than 10 cm long; stamens attached near base of corolla-tube . . . . **G. nubigena**
> 1. Flowering stems erect, usually at least 15 cm; stamens attached close to the middle of corolla-tube 2
> 2. Corolla basically blue, often with deeper stripes and spots . . . . **G. trichotoma, G. atuntsiensis, G. microdonta, G. striolata, G. wilsonii**
> 2. Corolla basically white, cream or pale yellow, often with blue streaks outside . . . . **G. purdomii, G. algida, G. erectosepala**

Gentiana nubigena Edgew. **Tufted perennial** to 10 cm, rarely more; **leaves** mostly in basal rosettes, lanceolate to elliptic or spathulate, 20–60 × 2–11 mm; stem-leaves in 2–3 pairs, smaller. **Flowers** 1–4 in terminal clusters, stalkless; calyx obconic, to 2.7 cm long, with unequal, erect lobes; corolla dark blue to purplish blue, often whitish or yellowish at base, tubular-campanulate, 3.5–6 cm long, lobes entire or slightly toothed. Gansu, Qinghai, nwSichuan, Tibet [Himalaya]; damp alpine meadows, bogs, damp open scrub, moraines and other rocky places, 3,000–5,300 m. VII–IX.

Gentiana trichotoma Kusn.* **Tufted perennial** to 35 cm with erect, simple, glabrous **stems**. **Basal leaves** few, in a lax rosette, ascending, linear to oblanceolate, to 80 × 13 mm, stalked, 3-veined; **stem-leaves** in 3–5 pairs, lanceolate to elliptic, 1- or 3-veined. **Flowers** in a narrow panicle; calyx 15–20 mm long with unequal, narrow-triangular lobes; corolla pale blue with deep streaks, sometimes with a straw-coloured base, narrowly funnel-shaped, 40–50 mm long; lobe-margins entire. Qinghai, w & nwSichuan, eTibet; grassy and stony meadows, open shrubberies, 3,000–4,600 m. VII–IX.

Gentiana atuntsiensis W. W. Sm.* Distinguished from previous species by the smaller, often purple **calyx**, not more than 12 mm long, and stalkless **flowers**; corolla dark blue, very occasionally straw-coloured with blue spots, 25–40 mm long, lobe-margins minutely toothed; calyx generally not slit along 1 side or only slightly so. seTibet, nwYunnan; similar habitats as well as open forests, 2,700–4,800 m. VI–X.

Gentiana microdonta Franch.* Very similar to *G. atuntsiensis* and sometimes considered conspecific with it, but differing in the **flowers** having stalks 5–25 mm long, and the calyx-tube clearly split along 1 side; corolla smaller, 22–25 mm long, with entire lobes. nwYunnan; alpine meadows, open forests, 2,600–4,200 m. VII–XI.

Gentiana striolata T. N. Ho* Very like the previous three species, but **stem-leaves** usually folded, **stems** minutely papillose, **flowers** stalkless and corolla pale blue with darker streaks and spots; calyx 12–16 mm long, with short erect lobes. swSichuan; alpine meadows, open scrub, 3,700–4,600 m. VII–IX.

Gentiana wilsonii C. Marq.* Similar to *G. atuntsiensis*, but **basal leaves** linear-lanceolate to linear-elliptic and **flowers** pale blue, whitish in throat, with deep blue lines outside. swSichuan, seTibet, nwYunnan; grassy habitats, open shrubberies, forests, 2,800–4,000 m. VII–VIII.

Gentiana purdomii C. Marq.* Tufted, **rhizomatous perennial** to 25 cm, generally far shorter with glabrous **stems** and erect basal leaf-rosettes; **leaves** linear-elliptic to narrow-oblong, folded, 20–60 × 2–9 mm; stem-leaves in 2–4 pairs, smaller. **Flowers** few in a terminal raceme, sometimes solitary or 2–3, mostly with stalks 10–50 mm long; sepals lanceolate, 3–8 mm long, blunt; corolla cream or white with blue stripes and streaks outside, tubular-campanulate to funnel-shaped, 35–55 mm long, plicae finely toothed. swGansu, Qinghai, nwSichuan, neTibet; alpine meadows, rocky places, 2,700–5,300 m. VII–X. High-altitude forms tend to be dwarf with few or solitary flowers.

Gentiana algida Pallas Very similar to the previous species, but most **flowers** stalkless (if stalked, the stalks not exceeding 10 mm); corolla white, cream or greenish with blue stripes outside, sometimes also spotted, plicae with entire margins. Qinghai, Tibet [nwChina, cAsia, cHimalaya, North America, eRussia, Mongolia]; similar habitats, 1,200–4,200 m. VII–IX.

Gentiana erectosepala T. N. Ho* Very similar to the previous two species, but **flowers** smaller, more conical, short-stalked, corolla not more than 30 mm long, white to pale yellow with blue and mauve streaks, stripes and spots. Sichuan, seTibet; rocky and grassy alpine meadows, alpine scrub, 3,600–4,600 m. VIII–IX.

Group Five
1. Fruit-capsules wingless, cylindrical to narrow-ellipsoid, broadest in middle 2
1. Fruit-capsules conspicuously winged, narrow-obovoid, broadest in upper third 4
2. Flowers 4-parted .... *G. lineolata*
2. Flowers 5-parted 3
3. Flowers solitary .... *G. tongolensis, G. picta*
3. Flowers clustered ...... *G. yunnanensis, G. haynaldii*
4. Corolla yellowish white .... *G. prattii*
4. Corolla blue to purple 5
5. Flowers relatively large, corolla-lobes 4–5 mm long ...
   ..... *G. chungtienensis, G. asterocalyx, G. piasezkii*
5. Flowers small, corolla-lobes 1.5–3(–4) mm long 6
6. Corolla pale blue or bluish white ....
   .... *G. leucomelaena, G. flexicaulis*
6. Corolla deep to bright blue, purple or blue-purple ... 7
7. Leaves with prominent midvein .... *G. crassula,*
   G. aristata, G. pseudoaquatica, G. asparagoides
7. Leaves with 1–3 veins ....
   .... *G. rubicunda, G. syringea, G. choanantha*

Gentiana lineolata Franch.* Small **annual** to 10 cm with erect papillose **stems**, **basal leaves** withered by flowering time. **Stem-leaves** lanceolate to ovate, to 12 × 4 mm, minutely ciliate. **Flowers** solitary, terminal; calyx tubular, 13–16 mm long, lobes triangular, pointed, minutely ciliate; corolla purple with blackish streaks, salverform, 20–35 mm long, plicae entire to toothed. swSichuan, nwYunnan; forest margins and glades, meadows, 600–4,000 m. VIII–IX.

Gentiana tongolensis Franch.* Small **annual** to 10 cm with many spreading papillose **branches** and oval to almost rounded **leaves** 3–5 mm long, with a cartilaginous margin and prominent midrib. **Flowers** terminal, solitary; calyx tubular, 6–9 mm long, lobes spreading to recurved; corolla pale yellow, with blue spots in centre, salverform, 15–24 mm long, plicae entire. w & nwSichuan, e & seTibet [eHimalaya]; meadows, roadsides, 3,500–4,800 m. VIII–IX.

Gentiana picta Franch.* In contrast to the previous species, the **leaves** are linear, up to 30 mm long, and the calyx-lobes are linear; calyx rather membranous, with blue streaks; **corolla** pale blue with blackish streaks, 25–32 mm long. swSichuan, nw & cYunnan; similar habitats as well as riverbanks, 2,400–3,000 m. VIII–XI.

Gentiana yunnanensis Franch.* **Annual** to 30 cm, with a purplish, erect, branched **stem**. **Leaves** obovate to spathulate, 10–35 × 4–13 mm, 1–3-veined, stalked. **Flowers** in narrow **panicles**, stalkless or almost so; calyx narrow-obconic, to 18 mm long, with 2 lobes half the length of the other 3; corolla pale yellowish white or pale blue, with blackish streaks and dots, tubular, 20–26 mm long, plicae notched to toothed. Guizhou, w & swSichuan, seTibet, nwYunnan; meadows and other grassy places, scrub, riverbanks, forest glades, 2,300–4,400 m. VIII–XI.

Gentiana haynaldii Kanitz* Glabrous **annual** to 10 cm, branched from base, **basal leaves** withered by flowering time. **Stem-leaves** awl-shaped, 7–15 × 1.5–2 mm, glossy green, upper part cartilaginous, long-pointed. **Flowers** several, almost stalkless; calyx narrow-obconic, glossy, 13–17 mm long, tube membranous, lobes long-pointed with cartilaginous tip; corolla pale blue with grey-blue streaks in throat, tubular, 22–30 mm long, plicae entire or slightly toothed. swQinghai, w, nw & swSichuan, seTibet, nwYunnan; meadows and other grassy places, forests, 2,100–4,200 m. VII–XI.

Gentiana prattii Kusn.* Small **annual** not more than 5 cm, branched from base, with **basal leaves** withered by flowering time. **Stem-leaves** in 4–9 pairs, elliptic to oblanceolate, to 5 × 2 mm, ciliate. **Flowers** few, on very short stalks; calyx narrow-obconic, 5–6 mm long, with triangular, ciliate lobes; corolla cream or whitish, often with blackish stripes, greenish yellow in throat, trumpet-shaped, 8–9 mm long, plicae slightly cut along margins. seQinghai, w, nw & swSichuan, nwYunnan; meadows and other grassy and rocky places, 3,000–4,000 m. VI–IX.

---

1. *Gentiana algida*; Animaqingshan, cQinghai (JB)
2. *Gentiana erectosepala*; Haizi Lake, wSichuan (HJ)
3. *Gentiana lineolata*; Zhongdian, nwYunnan (YMY)
4. *Gentiana tongolensis*; Yajiang, wSichuan (YMY)
5. *Gentiana tongolensis*; Gaoersi Pass, wSichuan (HJ)
6. *Gentiana haynaldii*; Gangheba, Yulongxueshan, nwYunnan (CGW)
7. *Gentiana yunnanensis*; nwLijiang, nwYunnan (HJ)
8. *Gentiana prattii*; north of Jigzhi, seQinghai (CGW)
9. *Gentiana chungtienensis*; Napahai, Zhongdian, nwYunnan (CGW)
10. *Gentiana chungtienensis*; Gangheba, Yulongxueshan, nwYunnan (CGW)

1. *Gentiana asterocalyx*; Napahai, Zhongdian, nwYunnan (PC)
2. *Gentiana piasezkii*; Duoer valley, Maya, swGansu (HJ)
3. *Gentiana leucomelaena*; Danyun, Huanglong, wSichuan (PC)
4. *Gentiana flexicaulis*; Huanglong, nwSichuan (CGW)
5. *Gentiana crassula*; Napahai, Zhongdian, nwYunnan (PC)
6. *Gentiana asparagoides*; Napahai, Zhongdian, nwYunnan (CGW)
7. *Gentiana rubicunda*; Wolong, wSichuan (CGW)
8. *Gentiana syringea*; Jiuzhaigou, nwSichuan (CGW)
9. *Gentiana choanantha*; Balangshan, Wolong, wSichuan (CGW)

Gentiana chungtienensis C. Marq.* Small **annual** not more than 5 cm with erect **stems** branching from base. **Basal leaves** ovate to broad-oval, to 15 × 8 mm, 3–5-veined, stalked; **stem-leaves** smaller, in 4–7 pairs, boat-shaped. **Flowers** on slender stalks to 9 mm long; calyx narrow-obconic, 9–11 mm long with triangular lobes; corolla pale blue, spotted, with greenish yellow stripes outside, 17–22 mm long, lobes apiculate, plicae entire. nwYunnan; open shrubberies, forest margins, meadows, 3,000–3,700 m. V–VII. (See photo 9 and 10 on p. 401.)

Gentiana asterocalyx Diels* Taller plant than preceding species, to 15 cm with awl-shaped to lanceolate, glossy **leaves**, upper longer than lower, 1–3-veined, calyx only 7–9 mm long with needle-like lobes, **corolla** blue-purple with yellow-white base. nwYunnan; meadows and grassy places on limestone, 3,000–3,200 m. VI–VIII.

Gentiana piasezkii Maxim.* Readily distinguished from the previous two species by densely ciliate leaf-margins and clearly winged calyces; **corolla** purple or purplish blue, greenish yellow towards base, with blackish lines in throat. seGansu, n & nwSichuan; meadows, open forests, shrubberies, road- and riversides, 1,000–4,300 m. IV–IX.

Gentiana leucomelaena Maxim. ex Kusn. Variable **annual** to 10 cm, often very dwarf, with prostrate to ascending **stems**. **Basal leaves** withered by flowering time, ovate to oblong, not more than 8 × 3 mm, margin membranous, indistinctly 1–3-veined; **stem-leaves** in 3–5 pairs. **Flowers** few, on glabrous stalks to 4 cm long; calyx campanulate, 4–5 mm long, lobes triangular-pointed; corolla pale blue, occasionally white, striped greyish blue, with darker spots in throat, 8–13 mm long; plicae irregularly cut at margins. Gansu, Qinghai, Sichuan, Tibet [nwChina, cAsia, w & cHimalaya, Mongolia, sRussia]; meadows, boggy places, open scrub, streamsides, 1,900–5,000 m. V–X.

Gentiana flexicaulis Harry Sm.* Similar to the previous species, but **leaves** larger, to 17 × 8 mm, 3–5-veined; **flowers** on short (4–6 mm) papillose stalks; calyx 8–9 mm long; corolla pale blue with yellowish white base, 12–16 mm long, plicae entire. n & nwSichuan; similar habitats, 2,400–4,600 m. V–IX.

Gentiana crassula Harry Sm.* **Annual** to 10 cm, branching from base, **basal leaves** withered by flowering time. **Stem-leaves** obovate to spathulate, to 6 × 2 mm, with a cartilaginous and minutely ciliate margin. **Flowers** several, stalkless; calyx 7–13 mm long, with recurved ovate lobes, margined like leaves; corolla pale purple-blue or grey-blue, whitish towards base, tubular-campanulate, 12–15 mm long, with ovate, entire or 2-fid plicae. w, nw & swSichuan, nwYunnan; rocky and grassy places, 3,400–4,200 m. VII–VIII.

Gentiana aristata Maxim.* Glabrous **annual** to 10 cm, branching from base. **Basal leaves** elliptic to ovate, to 9 × 5 mm, withered by flowering time; **stem-leaves** folded, linear, to 10 × 2 mm. **Flowers** few, on glabrous stalks to 20 mm long; calyx narrow-obconic, 7–10 mm long, with linear-lanceolate, awned lobes; corolla blue to pale purple, with blue-grey stripes and a greenish yellow base, 12–15 mm long, plicae with jagged margin. Gansu, Qinghai, w & nwSichuan, eTibet; meadows, bogs, open scrub, river gravels, forest glades, 1,800–4,600 m. VI–IX.

Gentiana pseudoaquatica Kusn. Similar to the previous species, but not more than 5 cm tall and with densely papillose **stems**; **stem-leaves** obovate, to 5 × 3 mm, minutely ciliate; leaf-stalks papillose; calyx 5–6 mm long with triangular lobes; **corolla** deep blue, with yellow-green stripes outside, funnel-shaped, 9–14 mm long, plicae toothed or not. Qinghai, Shaanxi, Tibet [c & nChina, wHimalaya, Korea, Mongolia, sRussia]; similar habitats, 1,100–4,700 m. IV–VIII.

Gentiana asparagoides T. N. Ho* Like *G. pseudoaquatica*, but **stems** and leaf-stalks glabrous and **stem-leaves** linear-lanceolate, to 3.5 × 1.2 mm; calyx 7–8 mm long; **corolla** purplish blue with a yellowish green base and bluish grey stripes. nwYunnan; boggy places, wet meadows, 3,500–3,800 m. VII–VIII.

Gentiana rubicunda Franch.* Variable glabrous **annual** to 15 cm with erect, simple or branched **stems**. **Leaves** deep green and shiny, ovate to elliptic, 10–25 × 4–10 mm, **stem-leaves** widely spaced and somewhat narrower. **Flowers** solitary or several on stalks to 15 mm long; calyx 8–9 mm long, with linear-elliptic lobes; corolla purple or rose-purple with a whitish, slightly spotted throat, 15–16 mm, striped greenish outside, plicae entire, slightly toothed or divided into two. seGansu, Guizhou, Hubei, Sichuan, nwYunnan [Hunan]; forests, glades, grassy places, 500–3,300 m. III–X.

Gentiana syringea T. N. Ho* Low **annual** to 5 cm with ascending **stems** branched at base. **Basal leaves** ovate, to 8 × 7 mm, with a smooth, cartilaginous margin and rounded apex; **stem-leaves** folded, narrow-oblong, to 6 × 2 mm, long-pointed. **Flowers** few, on papillose stalks to 10 mm long; calyx narrow-obconic, 10–13 mm long, with narrow-triangular lobes; corolla purple or bluish purple, with darker spotting and striping in throat, tube 10–13 mm long, plicae entire. sGansu, s & seQinghai, w & nwSichuan; meadows, grassy and stony places, riverbanks, 2,200–3,900 m. VI–VIII.

Gentiana choanantha C. Marq.* Similar in size to the previous species, but **leaves** minutely ciliate; calyx only 5–7 mm long; **corolla** very pale purple with a greenish yellow base, 8–10 mm long with recurved lobes and notched or toothed plicae. wSichuan; similar habitats and scrub, 2,700–4,600 m. IV–VIII.

GENTIANOPSIS

A genus of 24 species (5 in China) in the north temperate zone of Asia and America. **Annual** and **biennial herbs** with squarish **stems** and 4-parted **flowers**; calyx with 2 outer lobes narrower than 2 inner, but overlapping them at the base; corolla with a narrow tube and spreading lobes, without plicae but lobes often fringed towards base.

> 1. Inner sepals shorter than outer; stem-leaves linear to linear-lanceolate ....
> ... *G. grandis*, *G. barbata*, *G. barbata* var. *albiflavida*
> 1. Inner sepals ± equalling outer; stem-leaves ovate to lanceolate .... *G. lutea*, *G. paludosa*

Gentianopsis grandis **(Harry Sm.) Ma*** (syn. *Gentiana grandis* Harry Sm.) **Annual** or **biennial** to 50 cm, often less, with a basal leaf-rosette and erect **stems** branched from base. **Stem-leaves** narrow-lanceolate to linear, stalkless, to 80 × 6 mm, with slightly inrolled margins. **Flowers** erect, on stalks to 16 cm; calyx with linear-lanceolate to narrow-triangular, long-pointed lobes; corolla blue, sometimes yellowish towards base, tube 50–100 mm long, lobes elliptic, 20–30 mm long, deeply fringed towards base. swSichuan, nwYunnan; meadows and other damp places, stream and lake margins, 2,000–4,100 m. VII–X.

Gentianopsis barbata **(Froel.) Ma** Similar to the previous species, but **leaves** with flat margins, slightly ciliate; and smaller **flowers**, corolla-tube not more than 50 mm long with lobes not more than 12 mm long; corolla pale to mid-blue (yellow in **var.** *albiflavida* **T. H. Ho**). Gansu, Guizhou, Qinghai, Sichuan, Shaanxi, Yunnan [n & eChina, cAsia, Japan, Mongolia]; open shrubberies, meadows, banks, streamsides, 700–4,700 m. VII–IX.

Gentianopsis lutea **(Burkill) Ma*** (syn. *Gentiana detonsa* Rottb. var. *lutea* Burkill) **Annual** to 30 cm with erect, dark purple **stems**, and oblong, stalkless, blunt **stem-leaves** with a prominent midvein. **Flowers** in a regularly branched **inflorescence**, on stalks not exceeding 5 cm; calyx with almost equal lobes, 20–25 mm long overall; corolla pale yellow, tube 30–40 mm long, lobes 8–10 mm long, not fringed. cYunnan (vicinity of Kunming); grassy and gravelly habitats, 2,000–2,600 m. IX–X.

Gentianopsis paludosa **(Munro ex Hook. f.) Ma** Taller **annual** than the previous species, with long ascending **branches** and lanceolate to oblong **stem-leaves**, to 60 × 15 mm, with a rough margin, 1- or 3-veined. **Flowers** on stalks to 30 cm, longer in fruit; calyx-lobes with membranous margins and a keeled midrib; corolla yellow, cream or blue, twice as long as calyx, tube 20–65 mm long, lobes oblong, 12–17 mm long, fringed on margin towards base. Gansu, Hubei, Qinghai, w, nw & swSichuan, Shaanxi, Tibet, w & nwYunnan [Himalaya]; open shrubberies, meadows, banks and other grassy places, streamsides, 1,100–4,900 m. VII–X.

COMASTOMA

A genus of 15 species (11 in China) found in the temperate northern hemisphere. Very similar to *Gentiana*, but **flowers** often in raceme-like **inflorescences**, calyx lobed almost to the base and corolla without plicae, but with 1 or 2 deeply fringed scales at base of lobes, that often fill the mouth; nectaries present at base of corolla-tube.

Comastoma traillianum **(Forrest) Holub*** (syn. *Gentiana traillianum* Forrest) **Annual** to 30 cm, often less, with erect, rounded, lined **stems**, without **basal leaves** at flowering time. **Stem-leaves** stalkless, oblong to ovate, to 38 × 10 mm. **Flowers** on stalks to 50 mm; calyx 5–7 mm long, with lanceolate to ovate lobes; corolla blue, 15–26 mm long, lobes oblong to elliptic, 8–11 mm long, with paired fringed scales at base. s & swSichuan, nwYunnan; meadows and other grassy places, open forests, shrubberies, riversides, 3,000–4,200 m. VIII–X.

Comastoma falcatum **(Turcz. ex Kar. & Kir.) Toyokuni** (syn. *Gentiana falcata* Turcz. ex Kar. & Kir.) Similar to the previous species, but **stem-leaves** 3–5 mm wide. **Calyx** 6–12 mm long (about half the length of the blue or purplish corolla), with sickle-shaped lobes. Gansu, Hubei, Qinghai, Sichuan, Tibet [cAsia, Himalaya]; similar habitats, 2,100–5,300 m. VII–IX.

LOMATOGONIUM

A genus of c. 18 species (16 in China) in the northern hemisphere. **Annual** and **perennial herbs**, distinguished from *Gentiana*, *Gentianopsis* and *Comastoma* by the absence of plicae or fringed corolla-lobes, by the bicoloured **corolla** (outside different from inside) and by the presence of 2 nectaries per corolla-lobe, located near the base of the short tube.

> 1. Stem-leaves with a pointed to blunt apex ....
> ...... *L. carinthiacum*, *L. macranthum*, *L. oreocharis*
> 1. Stem-leaves with a long-pointed apex ....
> .... *L. forrestii*, *L. rotatum*

Lomatogonium carinthiacum **(Wulf.) Rchb.** (syn. *Swertia carinthiaca* Wulf.) Variable **annual** 5–30 cm tall with ascending to erect, angled **stems** branched from the base. **Leaves** paired, lanceolate to elliptic or ovate, to 20 × 7 mm, lowermost short-stalked. **Flowers** 8–20 mm across, in lax terminal and lateral

---

1. *Gentianopsis grandis*; Gangheba, Yulongxueshan, nwYunnan (PC)
2. *Gentianopsis paludosa*; east of Aba, nwSichuan (CGW)
3. *Comastoma traillianum*; Tianchi, nwYunnan (PC)
4. *Comastoma falcatum*; Jianzikou Pass, Litang, wSichuan (HJ)
5. *Lomatogonium carinthiacum*; nNepal (CGW)
6. *Lomatogonium macranthum*; Litang to Batang, wSichuan (HJ)
7. *Lomatogonium oreocharis*; Baimashan, nwYunnan (HJ)

cymes, with stalks up to 6 cm; calyx with elliptic to ovate lobes, 1- or 3-veined; corolla blue with elliptic to ovate lobes about twice as long as sepals; anthers blue. Gansu, Hubei, Qinghai, Sichuan, Tibet, nwYunnan [nwChina, cAsia, Himalaya]; meadows, open shrubberies, alpine scrub, streamsides, 1,100–5,400 m. VII–X.

Lomatogonium macranthum **(Diels & Gilg) Fernald** Like the previous species, but all **leaves** stalkless, and **flowers** larger, 20–25 mm across, very pale blue or milky purplish blue with dark blue lines at base of each long-pointed corolla-lobe. Gansu, Qinghai, wSichuan, e & seTibet [swTibet, c & eHimalaya]; moist grassy habitats, alpine meadows, streamsides, open scrub, forest margins, 2,500–4,800 m. VIII–X.

Lomatogonium oreocharis **(Diels) C. Marq.*** In contrast to *L. carithiacum*, a **rhizomatous perennial**, **basal leaves** ovate to orbicular, 6–19 × 5–12 mm, and **flowers** few, in raceme-like **inflorescences**; corolla pale blue with dark blue lines, 20–27 mm across. seTibet, nwYunnan; similar habitats as well as open forests, 3,000–4,800 m. VIII–X.

GENTIANACEAE: Lomatogonium

Lomatogonium forrestii* (Balf. f.) Fernald Annual to 32 cm, often far less, with angular, erect to ascending, glabrous **stems** much-branched from base. **Leaves** stalkless, lanceolate to elliptic, to 30 × 5 mm. **Flowers** 14–30 mm across, on glabrous stalks to 35 mm; calyx with linear to linear-lanceolate lobes 4–8 mm long; corolla white flushed with blue, or pale blue, with elliptic, long-pointed lobes, 7–16 mm long (twice the length of the sepals); anthers blue. wGuizhou, s & swSichuan, Yunnan; grassy places, forest margins, open shrubberies, 2,000–4,000 m. VIII–X.

***Lomatogonium rotatum* (L.) Fries ex Nyman** Differs from the previous species in having all **leaves** linear to linear-lanceolate and larger **flowers** with lobes 15–25 mm long; calyx-lobes 9–22 mm long, almost as long as petals, the latter pale blue with darker streaks. Gansu, Guizhou, Hubei, Qinghai, Sichuan, Shaanxi, Yunnan [widely distributed through northern hemisphere]; similar habitats, 1,100–4,200 m. VIII–IX.

1. *Lomatogonium forrestii*; Baimashan, nwYunnan (PC)
2. *Swertia angustifolia*; Bingchuan, nwYunnan (YMY)
3. *Swertia patula*; Sanba, nwYunnan (HJ)
4. *Swertia handeliana*; Zhongdian, nwYunnan (YMY)
5. *Swertia cincta*; Chongjiang, nwYunnan (PC)
6. *Swertia elata*; Litang, swSichuan (YMY)
7. *Swertia decora*; Sanba, nwYunnan (IIJ)
8. *Megacodon stylophorus*; Baimashan, nwYunnan (CGW)

SWERTIA

A cosmopolitan genus of c. 150 species (half in China, many endemic). **Annual** and **perennial** herbs. Very similar to *Lomatogonium*, but nectaries (1–2 per corolla-lobe) located at the bottom of each lobe and stigma not decurrent on the ovary.

1. Flowers 4-parted 2
1. Flowers 5-parted 3
2. Plants 25–80 cm tall .... ***S. angustifolia***
2. Plants not more than 20 cm tall ....
   .... ***S. patula, S. handeliana, S. davidii***
3. Plants 30 cm tall or more ....
   .... ***S. cincta, S. pubescens, S. elata***
3. Plants not more than 15 cm tall .... ***S. decora***

Swertia angustifolia **Buch.-Ham.** Slender **annual** to 80 cm, usually less, with erect, squarish, narrow-winged **stems**. **Leaves** stalkless, elliptic-lanceolate to lanceolate, 60 × 3–12 mm, apex attenuate. **Inflorescence** a lax, spreading cymose panicle, many-flowered; **flowers** 8–9 mm across; sepals linear-lanceolate, 6–8 mm long; petals 4–6.5 mm long, white or pale yellow with purple markings above and with a brownish blotch towards base. Guangxi, Guizhou, Hubei, Yunnan [cChina, Himalaya, nMyanmar]; woodland margins, open shrubberies, grassy slopes, to 3,300 m. VIII–XI.

Swertia patula **Harry Sm.*** Small **tufted annual** not more than 15 cm, with ascending **stems**, **basal leaves** withered by flowering time. **Stem-leaves** linear to linear-lanceolate, 10–25 × 2–5 mm, 1–3-veined. **Flowers** numerous on erect stalks to 20 mm; calyx with very short tube and linear-lanceolate lobes 20–28 mm long, often exceeding corolla-lobes; corolla whitish, occasionally pale purple, with prominent purple veins, 20–25 mm across, lobes oblong-obovate to elliptic, each with a pair of yellow-green nectaries at base. swSichuan, nwYunnan; forests and forest margins, shrubberies, stony slopes, 1,400–3,400 m. VII–IX.

Swertia handeliana **Harry Sm.*** In contrast to the previous species, a tufted **rhizomatous perennial** not more than 5 cm tall with most leaves basal. **Basal leaves** ovate to rounded, not more than 7 × 5.5 mm, with a distinct midvein; **stem-leaves** few, stalkless. **Flowers** solitary, terminal, 18–20 mm across, deeply cupped; sepals green, ovate to elliptic, 3.5–4.5 mm long; petals pale to mid-blue, oblong, 8–10 mm long, blunt, each with a single nectary. seTibet, nwYunnan; grassy habitats, especially alpine meadows, 3,500–4,500 m. VIII–XI.

Swertia davidii **Franch.*** Readily distinguished from the previous species by the lax, **perennial** habit, and square **stems** to 25 cm; corolla pale blue, 12–20 mm across, with deeper blue veins, petals pointed, each with 2 nectaries. wHubei, eSichuan, Yunnan [Hunan]; habitats similar to those of *S. handeliana* as well as mixed forests, 900–1,200 m. IX–XI.

Swertia cincta **Burkill*** **Annual** to 1 m, sometimes more, with erect, branched **stems**, **basal leaves** withered by flowering time. **Stem-leaves** lanceolate to elliptic, 25–75 × 5–20 mm, 3–5-veined, ciliate. **Flowers** nodding, in lax panicles; calyx-lobes ovate-lanceolate, 9–15 mm long, with ciliate margin; corolla pale greenish yellow with purple blotches or a purple ring in centre, lobes slightly bent backwards, ovate-lanceolate, 7–14 mm long, ciliate; stamens fused in a purple cone around ovary. Guizhou, Sichuan, Yunnan; damp slopes, scrub, open forests, 1,400–3,800 m. VIII–XI.

Swertia pubescens **Franch.*** Similar to the previous species, but calyx-lobes markedly unequal, **flowers** with white corolla and separate stamens. nwYunnan; grassy places, streamsides, open scrub, damp places, 2,800–4,100 m. VIII–XI.

Swertia elata **Harry Sm.*** Clump-forming **perennial** to 1 m, usually 50–80 cm, with erect, unbranched **stems**. **Basal leaves** in rosettes, stalked, blade linear-elliptic to lanceolate, 100–160 × 20–30 mm, with 5–7 veins; **stems-leaves** similar but progressively smaller, uppermost stalkless and bract-like. **Inflorescence** a narrow many-flowered, lax panicle. **Flowers** 10–13 mm across, star-shaped; sepals linear-lanceolate, 6–8 mm long; petals purple or greenish white, heavily marked purple, ovate-elliptic, 8–10 mm long. swSichuan, nwYunnan, open shrubberies, grassy habitats, hillsides, 3,200–4,600 m. VI–IX.

Swertia decora **Franch.*** Small **annual** with slender, erect, squarish **stems**. Lower **leaves** short-stalked, elliptic to spathulate, 7–13 × 2.5–5 mm, with prominent midvein; middle and upper leaves stalkless, linear to linear-lanceolate. **Flowers** often solitary, on erect stalks to 6 cm; calyx with linear-lanceolate lobes, 8–16 mm long; corolla rose-pink with deeper veins, star-shaped, 22–34 mm across, with elliptic lobes; nectaries 2 per corolla-lobe, fringed at the margin. s & swSichuan, Yunnan; grassy places and meadows, stony slopes, 1,800–2,900 m. VIII–XI.

MEGACODON

A Sino-Himalayan genus of just 2 species (both found in China). Robust **herbaceous perennials** with erect, unbranched **stems** and paired **leaves**, lowermost leaves small and membranous, the others large and green, half-clasping stem; calyx campanulate, lobed more than halfway, clasping base of corolla; **corolla** campanulate, nodding, also lobed more than halfway, lobes reticulately veined, without plicae; nectaries present in a ring around the gynophore.

Megacodon stylophorus **(C. B. Clarke) Harry Sm.** (syn. *Gentiana stylophora* C. B. Clarke) **Clump-forming perennial** to 1 m, often less, with 7–9-veined, ovate to ovate-lanceolate, rather fleshy **leaves**, to 22 × 12 cm. **Flowers** mostly in lateral clusters of 2–8, forming a panicle-like **inflorescence**; corolla pale yellowish green with brownish veins, 50–70 mm long, tube not more than 10 mm long. s & swSichuan, seTibet, nwYunnan [c & eHimalaya, nMyanmar]; open forests and forest margins, shrubberies, meadows, streamsides, 3,000–4,400 m. VI–VIII.

Megacodon venosus (Hemsl.) Harry Sm.* (syn. *Gentiana venosa* Hemsl.) Readily distinguished by the narrower elliptic-lanceolate **leaves**, uppermost only 7–12 mm wide, **flowers** in clusters of 7–11 in the **inflorescence** and calyx with a very short (2–3 mm, not 6–8 mm) tube; corolla while to pale yellow with brown veins. wHubei, e & neSichuan; shrubberies, meadows, 600–3,000 m. VIII–X.

EXACUM

A genus of c. 25 species (2 in China) distributed in tropical and subtropical Africa, Madagascar and south Asia. Readily distinguished from all the other genera included here by the anthers that open by apical pores rather than long slits. **Flowers** 4-parted.

Exacum tetragonum Roxb. Plant to 1 m, generally far less, with erect, squarish **stems**. **Leaves** opposite, stalkless, ovate to lanceolate, 1–5 × 0.8–2.5 cm, 3–5-veined, pointed. **Flowers** in narrow, raceme-like panicles, stalks up to 10 mm; calyx green, 5–6 mm long; corolla lilac to mauve-blue, 15–25 mm across, with a short tube 4–6 mm long and spreading elliptic lobes. Guangxi, Guizhou, Yunnan [sChina, c & eHimalaya, India, seAsia, Australia]; meadows, open shrubberies, roadsides, river gravels, 200–1,500 m. VII–IX.

CRAWFURDIA

A genus of c. 16 species (14 in China) in the central and east Himalaya, west China and north Myanmar. Twining **perennials** closely related to *Gentiana*, with paired **leaves**, the lowermost scale-like, and with generally pendent or semi-pendent **flowers**; calyx-tube 10-veined; corolla gentian-like, usually funnel-shaped to campanulate, plicae present, with 5 nectaries located at base of ovary. **Fruit** a many-seeded, usually stalked capsule.

1. Calyx without connecting membranes (like part-webbed toes) at top of tube 2
1. Calyx with connecting membranes (like part-webbed toes) at top of tube 3
2. Flowers relatively small, corolla not more than 35 mm long .... *C. gracilipes, C. delavayi*
2. Flowers larger, corolla usually 40–50 mm long .... *C. dimidiata*
3. Flowers large, corolla 50 mm long or more .... *C. angustata, C. sessiliflora*
3. Flowers smaller, corolla not more than 45 mm long, often much smaller .... *C. speciosa, C. thibetica*

Crawfurdia gracilipes Harry Sm.* Plant **twining** to 2.5 m with ovate to lanceolate, leathery **leaves**, to 8 × 3 cm, 3–5-veined, minutely and rather sparsely hairy along veins, apex tail-like. **Flowers** generally solitary, on slender stalks to 9 cm; calyx campanulate, tube split along 1 side, lobes linear, 7–8 mm long; corolla campanulate, blue to purple-blue, 20–25 mm long, with pointed lobes and rounded plicae. **Capsule** protruding from corolla on a slender stalk 20–25 mm long. seTibet, nwYunnan; grassy slopes, open shrubberies, 2,800–3,200 m. IX–XI.

Crawfurdia delavayi Franch.* Similar to the previous species, but upper **leaves** lanceolate, often stalkless or almost so, calyx with shorter lobes 2–6 mm long; **corolla** somewhat larger, 30–35 mm long, purple, blue-purple or pink; **capsule** not protruding from corolla. nwYunnan; mixed and open forests, shrubberies, bamboo thickets, 3,000–3,600 m. IX–XI.

Crawfurdia dimidiata (C. Marq.) Harry Sm.* Plant **twining** to 2 m, with grooved **stems**. **Leaves** leathery, ovate-lanceolate to ovate, to 4 × 3 cm, long-pointed, with 3–5 veins, leaf-stalk 10–20 mm. **Flowers** paired or in cymes, on slender stalks; calyx campanulate with a membranous tube 12–14 mm long, split down 1 side, lobes lanceolate, pointed, c. 6 mm long; corolla pale purple to blue, 40–42 mm long, with pointed triangular lobes and rounded or truncated plicae. **Capsule** on a stalk 9–10 mm long, scarcely protruding from corolla. seTibet, nwYunnan [nMyanmar]; grassy habitats, shrubberies, bamboo thickets, 3,000–3,400 m. VIII–X.

Crawfurdia angustata C. B. Clarke (syn. *C. trailliana* Forrest) Plant **twining** to 3 m, with leathery, elliptic to ovate **leaves**, to 7 × 4 cm, 3–5-veined, with entire margin and tail-like tip. **Flowers** solitary or in small cymes, stalks up to 15 cm long, usually bearing several scale-like bracteoles; calyx green or pale purple, rarely split along 1 side, lobes ovate to triangular, 3–4 mm long, spreading to recurved; corolla pale purple, funnel-shaped, large, 60–85 mm long, with triangular lobes and finely toothed plicae. **Capsule** on a very short (5–6 mm) stalk, not protruding from corolla. seTibet, nwYunnan [neIndia, nMyanmar]; open forests, shrubberies, grassy slopes, 1,500–2,800 m. IX–XII.

Crawfurdia sessiliflora (C. Marq.) Harry Sm.* Distinguished from the previous species by the **leaves**, which have a minutely scalloped margin, hairy on veins beneath, and by the smaller, deeper purple **flowers**, which are c. 50 mm long.

---

1. *Exacum tetragonum*; ne Nepal (CGW)
2. *Crawfurdia gracilipes*; Bomi, seTibet (YMY)
3. *Crawfurdia dimidiata*; Medog, seTibet (YMY)
4. *Crawfurdia angustata*; Medog, seTibet (YMY)
5. *Crawfurdia speciosa*; ne Nepal (CGW)
6. *Crawfurdia thibetica*; wSichuan (YMY)

Capsule protruding from the corolla on a stalk 30–40 mm long. wSichuan; forest margins (particularly coniferous), shrubberies, 2,500–2,900 m. IX–X.

Crawfurdia speciosa **Wall.** (syn. *C. kingdonii* C. Marq.; *C. wardii* C. Marq.) Rather variable plant **twining** to 3 m, with leathery ovate **leaves** to 7 × 3 cm, 3–5-veined and with a pointed apex and minutely scalloped margin. **Flowers** solitary or in small cymes, stalks up to 6 cm; calyx green, sometimes purple-flushed, generally not split along 1 side, with small triangular lobes 2–3 mm long; corolla purple to purple-blue, campanulate, 40–45 mm long, with triangular lobes and small rounded to squarish plicae. **Capsule** included in or somewhat protruding from corolla, stalk 15–35 mm long. seTibet [swTibet, c & eHimalaya]; open forests, shrubberies, grassy slopes, 2,900–4,000 m. IX–XII.

Crawfurdia thibetica **Franch.*** Distinguished from the previous species by the more lanceolate, often stalkless **leaves**, by the calyx-tube (which is always split down one side), and by protruding **capsules**. wSichuan; similar habitats as well as bamboo thickets, 3,000–3,600 m. IX–XI.

TRIPTEROSPERMUM

A genus with c. 25 species (19 in China) from south and southeast Asia, closely related to *Crawfurdia*, but **flowers** with nectaries forming a ring around the gynophore and **stamens** unequal and recurved.

Tripterospermum pallidum Harry Sm.* **Twining climber** to 2 m with slender, lined **stems**. **Leaves** opposite, blade ovate-lanceolate to oval, to 6 × 4 cm, 3-veined from base, margin. **Flowers** solitary, subtended by up to 6 small lanceolate bracts; calyx campanulate, with a winged tube and narrow-lanceolate lobes; corolla white or greenish white, narrow-campanulate, 35–40 mm long, lobes pointed. s & w Sichuan, Yunnan; forests, shrubberies, 500–1,300 m. VII–IX.

Tripterospermum cordatum (C. Marq.) Harry Sm.* Differs from the previous species in the more heart-shaped **leaves** to 12 × 5 cm, which are often purplish beneath, and the purple campanulate corolla. Guizhou, Hubei, Sichuan, Shaanxi, Yunnan [Hunan]; similar habitats, 700–3,200 m. VII–XII.

HALENIA

A genus of c. 100 species (2 in China) mainly in the New World but with species also in east Europe and Asia. **Annual** and **perennial herbs** with paired or whorled **leaves**. **Inflorescence** a panicle or cyme, the **flowers** rather *Epimedium*-like, 4-parted; calyx lobed almost to base; corolla campanulate, the 4 lobes fused together in the lower part and with prominent spurs near base.

Halenia elliptica **D. Don** Erect, simple or branched **annual** with squarish **stems** and deep green, elliptic to ovate or lanceolate **leaves**, to 30 × 15 mm, only lowermost usually stalked, 3- or 5-veined. **Inflorescence** a many-flowered, leafy panicle. **Flowers** blue to purple or pinkish purple, 10–15 mm across; calyx-lobes elliptic, long-pointed; corolla with elliptic to ovate lobes, and strongly curved spurs 5–10 mm long. Gansu, Guizhou, Hubei, Qinghai, Sichuan, Shaanxi, Tibet, Yunnan [Kyrgyzstan, Himalaya, nIndia, Myanmar]; meadows and other grassy places, open shrubberies, forest glades and margins, 700–4,100 m. VII–X. Plants, especially from swChina, with extra large **flowers** (20–25 mm across) are assigned to *var. grandiflora* Hemsl.

Halenia corniculata **(L.) Cornaz** Readily distinguished from the previous species by the rather small yellow **flowers**, 11–14 mm across. Hubei, Shaanxi [c & nChina, Mongolia, Korea, Japan, eRussia]; similar habitats, 600–1,800 m. VII–IX.

VERATRILLA

A genus of 2 species (both in China) in the eastern Himalaya and south-west China. Herbaceous **perennials** with erect, unbranched **stems**, and 3–5-veined **leaves**, only the basal stalked. **Inflorescence** a panicle, with male and female **flowers** borne on separate plants; calyx and corolla rotate, 4-parted, lobed almost to the base; corolla somewhat larger than calyx, with 1 or 2 dark gland-patches.

Veratrilla baillonii **Franch.** (syn. *Swertia mekongensis* Balf. f. & Forrest) Plant to 60 cm with paired ovate to elliptic **leaves** to 14 × 3.5 cm, decreasing in size up the **stem** and merging into the bracts. **Inflorescence** dense, broad and panicle-like in male plants, in female plants more slender, fewer-flowered

---

1. *Tripterospermum pallidum*; Xishan, Kunming, cYunnan (CGW)
2. *Tripterospermum cordatum*; Kangding, wSichuan (YMY)
3. *Halenia elliptica*; Gangheba, Yulongxueshan, nwYunnan (PC)
4. *Halenia elliptica*; Gangheba, Yulongxueshan, nwYunnan (PC)
5. *Veratrilla baillonii*, male plant; Tianchi, nwYunnan (PC)
6. *Veratrilla baillonii*, female plant; Bitahai, Zhongdian, nwYunnan (PC)
7. *Trachelospermum jasminoides*; cult. Kunming, cYunnan (CGW)
8. *Nerium oleander*, white form; cult. Dukou, sSichuan (CGW)

---

and raceme-like; **corolla** yellowish green with purple veins, lobes with a purple gland-patch often divided into 2 patches. w & swSichuan, seTibet, nwYunnan [neIndia]; meadows and other grassy places, open shrubberies, 3,200–4,600 m. V–VII.

APOCYNACEAE

TRACHELOSPERMUM

A genus with c. 15 species (6 in China) mostly in Asia but 1 in North America. **Climbers** with opposite **leaves**, producing white latex when cut. Calyx small, deeply divided, with 5–10 basal glands; **corolla** white or purplish, salverform with cylindrical tube and 5 propeller-like lobes. **Fruit** a pair of diverging follicles.

Trachelospermum jasminoides **(Lindl.) Lem.** Woody **climber** to 10 m with brownish **stems**. **Leaves** stalked, blade elliptic to ovate or obovate, 2–10 × 1–4.5 cm, sometimes slightly hairy beneath. **Cymes** paniculate, on a stalk 2–6 cm long; **flowers** sweetly scented; calyx with narrow-oblong spreading to reflexed sepals not more than 5 mm long; corolla white, 14–22 mm across, with obovate lobes and tube 5–10 mm long. **Follicles** linear, 10–25 cm long. Guangxi, Guizhou, Hubei, Sichuan, seTibet, Yunnan [c & eChina, Japan, Korea, Vietnam]; shrubberies, forest margins and clearings, 200–1,300 m. III–VIII.

Trachelospermum asiaticum **(Sieb. & Zucc.) Nakai** Similar to the previous species, but **sepals** appressed to corolla-tube, and **stamens** inserted in throat of corolla, not in middle of tube. Gansu, Guangxi, Guizhou, Hubei, Sichuan, seTibet, Yunnan [c & eChina, India, Japan, Korea, Thailand]; forests, often dense, shrubberies, to 1,000 m. IV–VII.

NERIUM

A genus with 1 species ranging from south Europe and North Africa, across west and central Asia to Japan. **Evergreen shrub** with **leaves** in whorls of 3, with numerous parallel veins. **Flowers** in broad corymbs; corolla funnel-shaped, rather propeller-like, with fringed corona-scales in throat. **Fruit** a pair of slender follicles, fused towards base.

Nerium oleander **L.** (syn. *N. indicum* Mill.) Rather erect **evergreen shrub** to 6 m with leathery, grey-green, narrow-elliptic **leaves**, to 21 × 3.5 cm, pointed. **Flowers** in broad, branched terminal clusters, showy, fragrant; sepals small, narrow-triangular, to 10 mm long; corolla purplish red, pink, salmon or white, with broad tube 12–22 mm long, the limb 35–55 mm wide. **Fruit** cylindrical, paired, 12–23 cm long. Naturalised in the warmer parts of China and widely cultivated [Europe, w & cAsia], mainly at low altitudes. All parts of the plant are extremely poisonous. Cultivated forms include those with semi-double and double flowers.

ASCLEPIADACEAE

CEROPEGIA

A genus with c. 170 species (17 in China), mostly in Africa and Madagascar, extending into tropical Asia and Oceania. **Perennial succulent herbs** with milky juice. **Flowers** usually bottle-shaped with a deeply 5-parted calyx, corolla tubular, with a swollen base, upper part often funnel-shaped with slender lobes joined at top; corona present, double. **Fruit** 1–2 linear, spindle-shaped or cylindrical follicles.

Ceropegia pubescens Wall. **Stem** twining to 1 m, becoming hairless. **Leaves** membranous, ovate or rarely oblong, 4–15 × 1–6 cm, long-pointed, softly hairy above. **Inflorescence** shorter than leaves, c. 5–8-flowered, **corolla** yellow, 5–5.5 cm long, tube 3–3.5 cm long, with the base slightly inflated, throat 3–5 mm across, narrower than base, lobes linear, 2–2.5 cm, basal 1 cm, orange-yellow, wider and strongly back-rolled, apical part ± awl-shaped, purple. Guizhou, Sichuan, seTibet, Yunnan [c & eHimalaya]; mixed woodland, 1,500–3,200 m. VII–IX.

Ceropegia sinoerecta M. G. Gilbert & P. T. Li* **Stem** erect, up to 20 cm. **Leaves** elliptic, 2–5 × 0.6–1.6 cm, pointed, densely hairy on upper side, pale beneath and sparsely hairy only on veins; leaf-stalk winged. **Inflorescence** umbel-like, 2–4-flowered, **corolla** 3.6–4.3 cm long, glabrous except for lobes, tube whitish green, streaked, 1.3–1.6 mm wide, basal swelling egg-shaped, 4–6 mm wide; lobes dark red-brown or orange-brown, hairy on margin, 1.4–1.5 cm, linear, with sharply incurved tip. Yunnan; limestone rocks, 2,000–2,800 m. VI–VII.

HOYA

A genus of at least 100 species (32 in China) in South East Asia across to Oceania. Epiphytic or rock-dwelling **subshrubs** or **climbers**, often twining or climbing by adventitious roots. **Leaves** opposite. **Inflorescences** lateral or sometimes terminal, umbel-like, producing a succession of globose or flat-topped clusters from the same peg year after year; calyx small, with basal glands; **corolla** fleshy, rotate, reflexed, lobes often densely hairy or scurfy inside; corona-lobes 5. **Fruit** usually a solitary follicle, sometimes paired.

Hoya carnosa (L. f.) R. Br. **Climber** or **epiphyte** with pale grey, smooth, robust **stems** to 6 m long. **Leaves** thick and fleshy, broadly ovate-heart-shaped to ovate-oblong or elliptic, 3.5–13 × 3–5 cm, stalks 1–1.5 cm long. **Inflorescences** lateral, almost globose, 20–30-flowered, downy, on a stalk c. 4 cm long; **corolla** waxy, starry, 15–20 mm across, white, sometimes with a pink centre, lobes triangular, densely papillose inside. Guangxi, Yunnan [sChina, seAsia, India, Taiwan]; mountain forests on limestone, usually growing on trees, 200–1,200 m. III–V.

PERIPLOCA

A genus of c. 10 species (5 in China) in temperate Asia, south Europe and Africa. Scandent **shrubs** with **twining stems**, glabrous except for flowers. **Leaves** with numerous lateral veins. **Inflorescences** lax, cymose, terminal and axillary; calyx with 5 glands; **corolla** rotate with short tube and blunt lobes. **Fruit** a pair of follicles.

Periploca forrestii Schltr. (syn. *Periploca calophylla* (Wight) Falc. subsp. *forrestii* (Schltr.) Browicz) Plant to 10 m, bearing leathery, lanceolate **leaves**, 3.5–7.5 × 0.5–1 cm. **Inflorescences** axillary, shorter than leaves, few-flowered; **flowers** c. 5 mm across; corolla yellow-green with short oblong tube, c. 2.5 mm long, lobes erect, not thickened at centre; corona-lobes minutely downy. Guangxi, Guizhou, Qinghai, Sichuan, seTibet, Yunnan [swTibet, India, Nepal, Myanmar]; shrubberies and sparse woods, sea level to 2,000 m. III–IV.

Periploca sepium Bunge* A generally smaller plant than the preceding species, the **stems** to 4 m; **leaves** larger and membranous, ovate-oblong, 5–9 × 1.5–2.5 cm, long-pointed. **Inflorescences** on lateral branchlets, often paired, few-flowered, **corolla** mostly purple, c. 15 mm across, lobes oblong-lanceolate, c. 8 mm long, strongly reflexed, glabrous beneath, above near margin with long soft hairs, with a prominent, elongated, raised, pale patch along centre; corona-lobes glabrous. Gansu, Guangxi, Guizhou, Hubei, Sichuan, Shaanxi, eTibet, Yunnan [c & nChina]; valleys, forest margins, shrubby slopes, 400–2,000. V–VI.

GOMPHOCARPUS

An African genus of 50 species, 2 introduced into China and naturalised in south and south-west China. **Leaves** narrow. **Inflorescence** a pendent cluster. **Flowers** white with a purplish corona. **Fruit** (follicles) conspicuous and inflated, bladder-like, softly bristly.

Gomphocarpus physocarpus E. Mey. Spindly **shrub** to 2 m with narrow lanceolate **leaves**, 5–10 × 0.6–1.5 cm, pointed. **Umbels** on stalks up to 5 cm long; **flowers** pendent, 14–20 mm across. **Follicles** solitary or paired, very inflated, almost globose, 6–8 cm long, pale greenish white, covered in soft bristles, apex rounded. Guizhou, Yunnan [s & seChina, tropical Africa]; cultivated or naturalised in waste places or close to habitation, roadsides, at low altitudes. V–IX.

Gomphocarpus fruticosus (L.) W. T. Aiton Very similar to the previous species, but **leaves** generally more linear, not more than 8 mm wide, and **follicles** 5–6 cm long, with a pointed apex and long beak. Guangxi, Yunnan [nChina, tropical Africa]; mainly cultivated, occasionally naturalised, sea level to 1,400 m. V–IX.

VINCETOXICUM

A genus with c. 60 species (22 in China) in Asia and Europe. Erect **subshrubs** or **perennial herbs**, often rhizomatous, with opposite **leaves**. **Inflorescences** lateral or occasionally terminal, umbel-like, corymbose, or racemose; sepals erect, often with basal glands; **corolla** rotate or almost so, with short tube and spreading lobes; corona cup-shaped, cylindrical, or deeply 5-parted. **Fruit** 1–2 spindle-shaped or tapering follicles. Vincetoxicum is included in the genus Cynanchum by some modern botanists.

Vincetoxicum forrestii **(Schltr.) C. Y. Wu & D. Z. Li*** (syn. Cynanchum forrestii Schltr.; C. balfourianum (Schltr.) Tsiang & Zhang; C. muliense Tsiang; C. steppicola Hand.-Mazz.; Vincetoxicum balfourianum (Schltr.) C. Y. Wu & D. Z. Li; V. muliense (Tsiang) C. Y. Wu & D. Z. Li; V. steppicola (Hand.-Mazz.) C. Y. Wu & D. Z. Li) **Tufted perennial** with erect **stems** to 60 cm, sometimes tending to twine, downy along 1 side, apical parts densely downy. **Leaves** thin, papery, broad-ovate to rarely elliptic-oblong, 2.5–8 × 1.5–4 cm, becoming hairless. **Inflorescences** lateral or almost terminal, umbel-like, shorter than leaves; **corolla** starry, cream, yellow to bronze-yellow, lobes ovate-oblong or oblong, to 3–7.5 mm long, often downy or minutely downy inside; corona fleshy, with deep triangular-ovate lobes. Gansu, Guizhou, Sichuan, e & seTibet, Yunnan; alpine meadows, waste places, grassy plains, 1,000–5,000 m. IV–VII.

Vincetoxicum canescens **(Willd.) Decne** (syn. Cynanchum canescens (Willd.) K. Schum.; C. glaucum Wall. ex Wight; Vincetoxicum glaucum (Wall. ex Wight) K. H. Rech.; V. hirundinaria Medik. subsp. glaucum (Wall. ex Wight) H. Hara) Similar to V. forrestii, but **stems** flexuous, **leaves** leathery and

1. *Ceropegia pubescens*; eNepal (CGW)
2. *Ceropegia sinoerecta*; Lijiang, nwYunnan (CGW)
3. *Hoya carnosa*; Pogang, swGuangxi (PC)
4. *Periploca sepium*; Jiuzhaigou, nwSichuan (PC)
5. *Gomphocarpus physocarpus*; cult. (CGW)
6. *Vincetoxicum forrestii*; Yulongxueshan, nwYunnan (HJ)

ASCLEPIADACEAE: Vincetoxicum

glaucous, and corona-lobes broader than long. w & swSichuan, seTibet, nwYunnan [swTibet, Himalaya, swAsia, sRussia]; open woodland, shrubberies, c. 2,500 m. V–VIII.

Vincetoxicum atratum **Bunge** Readily distinguished from the previous 2 species by being finely downy in most parts, by the more prominent **leaf** lateral veins, and by the purple **flowers** in umbels that are stalkless or almost so. Gansu, Guangxi, Guizhou, Sichuan, Shaanxi, Yunnan [n, c & eChina, Japan, Korea, eRussia]; grassy habitats, dryish scrub, riversides, to 2,000 m. VI–X.

DREGEA (syn. *Wattakaka*)

A genus of c. 12 species (4 in China) distributed in south Asia and Africa. **Twining climbers** with paired **leaves** and **flowers** in long-stalked, umbel-like clusters; **corolla** rotate to bowl-shaped, deeply 5-lobed; corona with 5 ± spreading, fleshy lobes fused to the **stamens**. **Fruit** a pair of divergent follicles.

Dregea sinensis **Hemsl.*** (syn. *Wattakaka sinensis* (Hemsl.) Stapf) **Twining climber** to 8 m with downy stems and leaves. **Leaves** papery, heart-shaped to almost orbicular, 2–13 × 2–9 cm, pointed. **Flowers** to 20 in umbel-like clusters, on a 3–6 cm long stalk; sepals ovate, downy; corolla white, sometimes purplish inside, 14–16 mm across, with ciliate, ovate-oblong lobes. **Follicles** narrow-lanceolate in outline, 5–6 cm long, with hook-like apex, downy or glabrous. Gansu, Guangxi, Guizhou, Hubei, Sichuan, Shaanxi, seTibet, Yunnan [swTibet, c & eChina]; open woods, shrubberies, 500–3,000 m. VI–IX.

Dregea volubilis **(L. f.) Benth. ex Hook. f.** (syn. *Asclepias volubilis* L. f.; *Wattakaka volubilis* (L. f.) Stapf.) Similar to the previous species, but more vigorous, to 12 m, with ovate to rounded **leaves**, 7–18 × 4–17 cm, glabrous to softly hairy. **Flowers** many in pendent clusters, corolla green or yellowish green, lobes 6–12 mm long. Guangxi, Guizhou, Yunnan [seChina, Himalaya, seAsia, India, Taiwan]; montane forests, shrubberies, 1,400–2,600 m. IV–IX.

1. *Vincetoxicum atratum*; Munigou, Kangding, wSichuan (PC)
2. *Dregea sinensis*; cult. (CGW)
3. *Dregea sinensis*; cult. (CGW)
4. *Calotropis gigantea*; near Xishuangbanna, swYunnan (CGW)
5. *Forsythia suspensa*; cult. (CGW)

CYNANCHUM

A genus of c. 200 species (57 in China) distributed in the Americas, Africa, east Europe and Asia. **Subshrubs** or **herbaceous perennials** mostly with twining **stems**. **Leaves** opposite, often heart-shaped, occasionally whorled. **Inflorescences** umbel- or raceme-like, terminal or apparently lateral; **corolla** with spreading or reflexed lobes and a short tube; corona present, cup-shaped, cylindrical or 5-lobed. **Fruit** 1–2 boat-shaped or cylindrical follicles.

Cynanchum corymbosum **Wight** (syn. *Cyathella corymbosa* (Wight) C. Y. Wu & D. Z. Li) **Twining climber** to 2 m, the pale grey **stems** with hairs along 2 lines. **Leaves** paired, thin and papery, heart- to arrow-shaped, 4.5–12 × 3.5–8 cm, glabrous except on veins. Flowers in short **racemes** to 7 cm long, occasionally branched; **flowers** paired in the raceme; calyx with ovate lobes 1.5–2 mm long; corolla greenish white, 10–12 mm across, corona white, 3–4 mm high. **Follicles** spindle-shaped, 9–13 cm long, adorned with curved spines. Guangxi, Sichuan, Yunnan [c & sChina, seAsia, India, Myanmar]; shrubberies, open woodland, valley thickets, hedgerows, river- and streamsides, 1,400–3,300 m. V–X.

CALOTROPIS

A genus of 3 species (2 in China) found in North Africa, Arabia and subtropical Asia. **Erect shrubs** with broad, opposite, fleshy **leaves**, producing copious white latex when cut. **Flowers** in umbel-like clusters, terminal or lateral; calyx with basal glands; corolla bowl-shaped with a 5-lobed corona. **Fruit** 1–2 spongy, inflated follicles.

Calotropis gigantea **(L.) W. T. Aiton** (syn. *Asclepias gigantea* L.) **Evergreen shrub** to 5 m, often 1–2 m, with paired, heart-shaped to oblong, short-stalked **leaves**, 7–30 × 3–15 cm, blue-green, cottony with hairs when young. **Flowers** in umbel-like clusters, on a thick stalk 5–12 cm long; calyx flattish, 12–15 mm across; corolla purplish or lilac, greenish towards base, fleshy, 25–35 mm across, lobes ovate with back-rolled margins. **Follicles** obliquely elliptic to lanceolate in outline, 5–10 cm long. Guangxi, Sichuan, Yunnan [s & seChina, seAsia, sHimalaya, India, Myanmar]; dry woodland, sandy places, stream- and riversides, to 1,400 m. I–XII.

Calotropis procera **(Aiton) W. T. Aiton** (syn. *Asclepias procera* W. T. Aiton) Similar to the previous species, but inflorescence-stalk shorter, not more than 5 cm long; **flowers** white with a pink interior, smaller, 15–20 mm across; and **follicles** rounded to egg-shaped, inflated and spongy, 6–10 cm long. Cultivated or naturalised in Guangxi, sYunnan [sHimalaya, India, Myanmar, se & swAsia, Africa]; similar habitats and altitudes. V–XII.

OLEACEAE

FORSYTHIA

A genus of c. 10 species (6 in China) mainly found in Asia, but 1 in south-east Europe. **Deciduous shrubs** with hollow **stems**, often with pith, 4-angled branchlets and opposite simple or 3-parted **leaves**. **Flowers** borne before leaves; calyx deeply 4-lobed; corolla yellow, with 4 lobes and a short tube; stamens 2. **Fruit** a small, 2-parted capsule.

1. Branches hollow; leaves often 3-parted ... *F. suspensa*
1. Branches with central pith; leaves simple ....
  .... *F. giraldiana, F. viridissima, F. likiangensis*

Forsythia suspensa **(Thunb.) Vahl*** Large spreading **shrub**, sometimes partly scrambling, to 4 m, with spreading to pendulous **branches**. **Leaves** simple or 3-parted, stalked, blades ovate to elliptic, 2–10 × 1.5–5 cm, glabrous or downy beneath, with toothed margin. **Flowers** solitary, paired or several in a cluster, on stalks to 6 mm long; corolla butter-yellow, lobes oblong, 12–20 mm long. Hubei, Sichuan, Shaanxi [cChina]; rocky and grassy slopes, shrubberies, gullies and ravines, 300–2,200 m. III–IV. Often cultivated.

Forsythia giraldiana **Lingelsh.*** **Shrub** to 3 m of rather upright habit with simple stalked **leaves**, blades ovate to lanceolate, 3.5–12 × 1.5–6 cm, glabrous or downy above and beneath, margin usually entire or occasionally slightly toothed. **Flowers** solitary or 2–3-clustered; calyx-lobes usually tinged with purple; corolla primrose-yellow, tube 4–6 mm long, lobes oblong, lobes 7–15 mm long. seGansu, neSichuan, Shaanxi [wHenan]; wooded slopes, ravines, marshy shrubberies, riversides, 800–3,200 m. III–V.

Forsythia viridissima **Lindl.*** Readily distinguished from the previous species by being entirely glabrous apart from the calyx-lobes, by the lanceolate or obovate **leaves**, 3.5–15 × 1–4 cm with the margin toothed at least in part; and by having a **corolla** that is deep yellow, with orange-yellow stripes inside. Hubei, nwYunnan [cChina]; similar habitats and altitudes. III–IV.

Forsythia likiangensis **Ching & Feng ex P. Y. Pai*** Closely related to *F. giraldiana*, but **leaves** always glabrous and with an entire margin, the blade ovate to elliptic, 2–9 × 1–3.5 cm, and the **flowers** solitary in axils of young leaves; **corolla** yellow outside, striped red inside, lobes 9–10 mm long. swSichuan (Muli), nwYunnan; mixed forests on mountain slopes, 1,200–2,900 m. IV–V.

OSMANTHUS

A genus of 30 species (23 in China, mostly endemic) in South East Asia, the Himalaya, the southern USA and Hawaii. **Evergreen shrubs** or small **trees** with opposite, toothed or entire, often leathery **leaves**. **Flowers** in cymose, lateral or terminal clusters, fragrant; corolla 4-parted, usually cylindrical or campanulate; **stamens** usually 2. **Fruit** a fleshy, berry-like drupe.

1. Leaves small, not more than 2.5 cm long .. *O. delavayi*
1. Leaves larger, 3–9 cm long 2
2. Leaves with 10–12 pairs of lateral veins, margin spine-toothed .... *O. yunnanensis*
2. Leaves with 5–9 pairs of lateral veins, margin entire or toothed .... *O. suavis, O. henryi*

Osmanthus delavayi Franch.* Much-branched, rather dense **evergreen shrub** to 3 m, occasionally more, with finely hairy young **branches**. **Leaves** deep green, leathery, elliptic to ovate, 1–2.5 × 1–1.5 cm, with a few sharp teeth, finely hairy beneath. **Flowers** in small clusters of up to 8, terminal or in leaf-axils, fragrant; calyx 2–4 mm long; corolla white, funnel-shaped, 10–16 mm long overall, tube about twice the length of the spreading lobes. **Fruit** egg-shaped, 10–12 mm long, black when ripe. Guizhou, Sichuan, Yunnan; mixed woodland, shrubberies, ravines, rocky slopes, 2,100–3,400 m. IV–V.

Osmanthus yunnanensis (Franch.) P. S. Green* (syn. *O. forrestii* Rehd.; *O. rehderianus* Hand.-Mazz.) Stiff-stemmed **evergreen shrub** to 6 m, occasionally more, with smooth young **branches**. **Leaves** short-stalked, leathery, grey-green, elliptic to ovate-lanceolate, 8–14 × 2.5–4 cm, pointed, with spine-toothed margin. **Flowers** in clusters of up to 12 in leaf-axils; calyx minute, c. 1 mm; corolla cream or yellowish white, 5 mm long, lobes fused together only at base, sweetly fragrant. **Fruit** narrowly egg-shaped, 10–15 mm long, purple-black when ripe. Sichuan, seTibet, Yunnan; forests, wooded slopes, ravines, 1,400–2,800 m. IV–V.

Osmanthus suavis King ex C. B. Clarke **Evergreen shrub** or small **tree** to 6 m, with finely hairy young **branches**. **Leaves** stalked, elliptic, 3–7 × 1.5–2.5 cm, thin-leathery, pointed, margin toothed, hairy only on midrib beneath. **Flowers** in terminal or lateral clusters of 6–9; calyx 3–4 mm long; corolla cream or white, 9–13 mm long overall, tube twice the length of lobes, very fragrant. **Fruit** narrowly egg-shaped, 8–9 mm long, bluish black when ripe. seTibet, Yunnan [c & eHimalaya, Myanmar]; dense forests and shrubberies, ravines, 2,400–3,000 m. V–VII.

Osmanthus henryi P. S. Green* Distinguished from the previous species by larger elliptic to oblanceolate, thick-leathery **leaves**, 7–9 × 3–4.5 cm, margin entire to toothed; by **flowers** in clusters of 4–6, tiny, c. 3 mm long, including tube 1 mm long; and by larger **fruit**, 18–20 mm long. Guizhou, Yunnan [Hunan]; similar habitats, 1,000–3,000 m. X–XI.

SYRINGA

A genus of c. 20 species (16 in China) distributed in west, central and east Asia and south-east Europe. **Trees** and **shrubs** with opposite, simple or occasionally pinnately divided, untoothed, stalked **leaves**. **Flowers** in terminal or lateral cymose panicles, often fragrant, hermaphrodite; calyx campanulate, 4-toothed or 4-lobed; corolla funnel-shaped to salverform, with a short to long tube and 4 spreading, short lobes; **stamens** 2. **Fruit** a small, few-seeded capsule.

1. Leaves pinnately lobed .... *S. pinnatifolia*
1. Leaves unlobed 2
2. Panicles lateral; terminal buds absent ...... *S. wardii*
2. Panicles terminal and/or lateral; shoots with terminal buds 3
3. Leaves glabrous beneath .... *S. yunnanensis, S. yunnanensis* forma *pubicalyx*
3. Leaves hairy beneath .... *S. komarovii, S. komarovii* subsp. *reflexa, S. sweginzowii*

Syringa pinnatifolia Hemsl.* Small to medium-sized **deciduous shrub** to 4 m at most, with 4-angled, glabrous young **stems** and pinnately compound **leaves** with 7–11 pairs of ovate to lanceolate, stalkless leaflets, these 5–30 × 3–13 mm. **Panicles** lateral, spreading, to 7 cm long; **corolla** white or rose-pink or pale red tinged lilac, tube 8–12 mm long, lobes ovate or oblong, semi-spreading. Gansu, e & seQinghai, w & nSichuan, sShaanxi [nChina]; shrubberies on hillslopes, 2,600–3,100 m. V–VI.

Syringa wardii W. W. Sm.* **Deciduous shrub** or small **tree** to 5 m with hairy young **branches**. **Leaves** broad-oval to almost rounded, only 1–2 × 1–2 cm, glabrous beneath, apex rounded to blunt. **Panicles** erect, rather lax, to 10 cm long; calyx with minute teeth; **corolla** lilac-pink, tube cylindrical, 10–12 mm long. seTibet, nwYunnan; dry woods, scrub, 2,400–3,000 m. V–VI.

Syringa yunnanensis Franch.* Rather erect, **deciduous shrub** to 5 m with 4-angled young **branches** and elliptic to elliptic-oblanceolate, glabrous **leaves**, 2–10 × 1–4 cm, pointed to somewhat long-pointed. **Panicles** erect, to 18 cm long; calyx sometimes slightly downy (**forma** *pubicalyx* (Jian ex P. Y. Bai) M. C. Chang*); **corolla** white to pink or lilac-red, tube 5–10 mm long, lobes oblong, spreading. swSichuan, seTibet, nwYunnan; open wooded slopes, shrubberies, stream- and riversides, gullies, 2,000–3,900 m. V–VII.

Syringa komarovii C. K. Schneid.* **Deciduous shrub** to 6 m, usually only 2–4 m, with elliptic to oblong-lanceolate or ovate-oblong **leaves**, 5–19 × 1.5–7 cm, pointed to long-pointed. **Panicles** nodding to fully pendulous, to 25 cm long, with all parts (except corolla) hairy or glabrous; **corolla** red to purple-red, pale inside, tube 8–15 mm long, lobes ovate, erect. **Capsules** usually reflexed when ripe. sGansu, wHubei, Sichuan, sShaanxi, n & nwYunnan; open woods, shrubberies, streamsides, ravines, 1,000–3,400 m. V–VII. **Subsp. *reflexa* (C. K. Schneid.) P. S. Green & M. C. Chang*** (syn. *S. reflexa* C. K. Schneid.) has paler pinkish lilac or lilac **flowers** with spreading corolla-lobes. wHubei, Sichuan, 1,800–2,900 m.

Syringa sweginzowii Koehne & Lingelsh.* Similar to the previous species, but **leaves** smaller, usually not more than 5 × 3 cm, shiny deep green above, with a purplish red margin when young, and **panicles** ascending to spreading, stalks and calyces purplish brown, **corolla** pink, lilac or white, with slender tube 6–15 mm long, and spreading lobes. wSichuan; similar habitats, 2,000–4,000 m. V–VI.

1. *Osmanthus delavayi*; cult. (CGW)
2. *Osmanthus suavis*; Bhutan (CGW)
3. *Syringa yunnanensis*; Cangshan, nwYunnan (PC)
4. *Syringa yunnanensis*; Napahai, Zhongdian, nwYunnan (HJ)
5. *Syringa komarovii* subsp. *reflexa*; Wolong, wSichuan (PC)
6. *Syringa komarovii* subsp. *reflexa*; cult. (CGW)

JASMINUM

Over 200 species (43 in China) mainly in Europe and Asia, but also in Africa and Australia. **Shrubs** and **climbers** with simple or pinnately compound, opposite **leaves**. **Flowers** solitary, clustered or in panicles; calyx 5-lobed, occasionally more; corolla salverform with a slender tube and 5, sometimes more, spreading lobes; stamens 2. **Fruit** berry-like, solitary or twinned.

1. Flowers yellow 2
1. Flowers white, pink, red or purple 3
2. Leaves alternate .... *J. humile, J. subhumile*
2. Leaves opposite .... *J. nudiflorum, J. nudiflorum* var. *pulvinatum, J. mesyni*
3. Leaves pinnately lobed; flowers white or pink, sometimes red outside ....
   .... *J. polyanthum, J. officinale, J. × stephanense*
3. Leaves simple 4
4. Flowers red or purple .... *J. beesianum*
4. Flowers white .... *J. elongatum, J. sambac*

Jasminum humile **L.** **Shrub** to 3 m with angular **branches**, glabrous to minutely hairy. **Leaves** compound, stalked; leaflets normally 3–9, elliptic to lanceolate, to 6 × 2 cm, terminal larger than lateral. **Inflorescence** short-paniculate to corymbose with up to 15 flowers; **corolla** yellow, funnel-shaped, 9–16 mm across, tube 8–16 mm long. **Fruit** rounded to oblong in outline, 6–11 mm, purplish black when ripe. Gansu, wGuizhou, swSichuan, seTibet, Yunnan [swTibet, sw & cAsia, India]; woods and shrubberies, rocky places, 1,100–3,800 m. IV–IX.

Jasminum subhumile **W. W. Sm.** Similar to previous species, but leaflets normally 3, to 12.5 cm long; **flowers** in larger, broader **inflorescences** of 10–120 flowers. swSichuan, w & nwYunnan [Himalaya, Myanmar]; wooded places, ravines, streamsides, 700–3,300 m. III–VII.

Jasminum nudiflorum **Lindl.*** Sprawling to scrambling **deciduous shrub** to 3.5 m, with 4-angled **branches**, green in first year. **Leaves** opposite, with 3 leaflets but generally simple towards branch bases; leaflets deep green, yellow-green beneath, ovate to elliptic, to 2.2 × 1.3 cm. **Flowers** solitary, normally lateral on the bare branches; calyx campanulate, green, 4–6 mm long, with narrow-lanceolate lobes; corolla deep primrose-yellow, with tube 8–20 mm long and 20–25 mm in diameter, limb 5–6-lobed. **Fruit** c. 6 mm. Gansu, Sichuan,

---

1. *Jasminum humile*; near Benzilan, nwYunnan (CGW)
2. *Jasminum nudiflorum*; cult. (CGW)
3. *Jasminum sambac*; cult. Kunming, cYunnan (CGW)
4. *Jasminum polyanthum*; cult. (CGW)
5. *Jasminum officinale*; Cangshan, nwYunnan (RS)
6. *Jasminum × stephanense* (variegated form); cult. (CGW)

Shaanxi, seTibet, nwYunnan; shrubberies, rocky slopes and ravines, sometimes in dense thickets, 800–4,500 m. XI–III. Small densely branched, shrubby forms are referable to **var. *pulvinatum* (W. W. Sm.) Kobuski***.

Jasminum mesyni **Hance*** Superficially similar to the previous species, but a more robust **evergreen** scrambling or sprawling **shrub** to 5 m, with larger **leaves**, leaflets to 5 × 2.5 cm. **Flowers** on stalks 3–8 mm long, corolla 20–45 mm across with 6–8 lobes (twice as many in cultivated forms). Guizhou, swSichuan, s, e & cYunnan; similar habitats as well as woodland, 500–2,600 m. XI–VIII.

Jasminum beesianum **Forrest & Diels*** **Twining deciduous**, rather variable **shrub** to 3 m; **stems** 4-angled and hairy when young. **Leaves** ovate to lanceolate, to 4 × 1.8 cm, glabrous or downy, apex pointed. **Flowers** in lateral clusters of 2–5, or solitary in leaf-axils; calyx 3–10 mm long, with 5–7 linear lobes; corolla funnel-shaped, with tube 9–15 mm long and limb 4–8-lobed, 10–20 mm across. **Fruit** 5–12 mm, black when ripe. Guizhou, w & sSichuan, Yunnan; open forests, shrubberies, rocky and grassy slopes, 1,000–3,600 m. XI–VI.

Jasminum elongatum **(Berg.) Willd.** Vigorous **clambering shrub** to 7 m with rounded **stems** and large ovate to lanceolate **leaves** to 11 × 5.5 cm. **Inflorescences** many-flowered, lateral or terminal, fragrant; **corolla** white, salverform, tube 15–25 mm long, limb 6–9-lobed, 18–25 mm across. Guangxi, Guizhou, Yunnan [s & seChina, India, Myanmar, seAsia, Australia]; forests, shrubberies, to 900 m. IV–XII.

Jasminum sambac **(L.) W. T. Aiton** Differs from the previous species in being more **shrubby**, rarely exceeding 2.5 m, **leaves** rounded to obovate with tufts of hairs in the main vein-axils beneath; **flowers** often in 3s, single or semi-double, intensely fragrant, corolla-tube not exceeding 15 mm long. Widely cultivated as an ornamental and for hedging in China [native to India]. V–IX.

Jasminum polyanthum **Franch.*** **Evergreen twining shrub** to 10 m, with glabrous rounded to angled **stems**. **Leaves** with 5 or 7 leaflets, these thin and papery, lanceolate to ovate, to 9.5 × 3.5 mm, terminal leaflet larger than the lateral ones. **Flowers** in terminal or lateral **panicles** of up to 50, sometimes more, intensely fragrant; calyx 2–3 mm with very short linear lobes; corolla white, pink or red in bud, salverform, tube 13–25 mm long, limb 20–30 mm across, with 5 narrow-ovate lobes. **Fruit** 6–11 mm, black when ripe. Guizhou, Sichuan, Yunnan; forests and forest margins, shrubberies, rocky places, ravines, 1,400–3,000 m. II–VIII.

Jasminum officinale **L.** Similar to the previous species, but **deciduous** or almost so, with 3–9 leaflets per **leaf** and **flowers** in 1–10-flowered almost umbellate, usually terminal, clusters; corolla white, occasionally reddish outside, tube rarely more than 15 mm long, limb 14–25 mm across. **Fruit** dark red changing to purple when ripe. swGuizhou, Sichuan, seTibet, Yunnan [swTibet, Himalaya, Pamir]; similar habitats as well as river- and streamsides, 1,800–4,000 m. V–IX.

Jasminum × *stephanense* Lemoine* Natural hybrid between *J. beesianum* and *J. officinale*. **Leaves** with 3–9 ovate to elliptic leaflets that are hairy at first. **Inflorescences** lateral or terminal cymes with 1–5 flowers; calyx 4–9 mm long; **corolla** pink with a pink, purple or red reverse, tube 10–18 mm long, limb 14–24 mm across. **Fruit** black when ripe. Sichuan, seTibet, Yunnan; similar habitats, 2,200–3,100 m. V–VIII. (See photo 6 on p. 418.)

LIGUSTRUM

A genus of 45 species (27 in China) occurring in east and South East Asia, Europe, North Africa and Australia. **Evergreen** and **deciduous shrubs** and **trees** with opposite, simple, untoothed **leaves**. **Flowers** hermaphrodite, in terminal, cymose **panicles**, occasionally lateral; calyx campanulate, 4-toothed; corolla funnel-shaped to salverform or rotate, white, 4-lobed, lobes equal to or shorter than tube; **stamens** 2, attached near mouth of corolla-tube. **Fruit** berry-like, containing 1–4 **seeds**.

1. Fruit kidney-shaped ....
   .... *L. lucidum*, *L. lucidum* forma *latifolium*, *L. henryi*
1. Fruit globose to egg-shaped ....
   .... *L. delavayanum*, *L. sinense*

Ligustrum lucidum W. T. Aiton* Glabrous **evergreen shrub** or **tree** to 25 m with ovate to elliptic or lanceolate, shiny, deep green, rather leathery **leaves**, 6–17 × 3–8 cm. **Panicles** terminal, to 20 cm long; **flowers** stalkless or almost so; corolla white, 4–5 cm long, tube equalling lobes. **Fruit** kidney-shaped, 7–10 mm, deep blue-black, finally ripening reddish black. Gansu, Guangxi, Guizhou, Hubei, Sichuan, Shaanxi, e & seTibet, Yunnan [e & seChina]; woods, shrubberies, 200–2,900 m. V–VII. **Deciduous** forms with 7–11 pairs of lateral leaf-veins are referable to **forma** *latifolium* (W. C. Cheng) P. S. Hsu*.

Ligustrum henryi Hemsl.* In contrast to the previous species, a **shrub** never more than 4 m tall with rusty-downy young **branches** and thin-leathery **leaves** not more than 4.5 × 3 cm; corolla-tube 2–3 times longer than lobes. **Fruit** 6–10 mm, black when ripe. Gansu, Guangxi, Guizhou, Hubei, Sichuan, Shaanxi, Yunnan [wHunan]; wooded slopes, ravines, 200–1,700 m. V–VI.

Ligustrum delavayanum Hariot* **Evergreen shrub** to 4 m with minutely, yet densely, hairy young **branches**. **Leaves** elliptic to ovate or lanceolate, 1–4 × 0.5–2 cm, with 2–6 pairs of lateral veins, pointed to long-pointed. **Panicles** to 6 cm long, terminal or in leaf-axils of previous year's branches; **corolla** white, 4–7.5 mm, tube twice as long as lobes. **Fruit** globose to egg-shaped, 5–9 mm long, black when ripe. Guizhou, wHubei, Sichuan, Yunnan; wooded slopes, shrubberies, 500–3,700 m. V–VII.

Ligustrum sinense Lour. Very variable **deciduous shrub** or small **tree** to 4 m, occasionally more, with densely hairy young **branches** that may become hairless. **Leaves** ovate, oblong, rounded, to elliptic or lanceolate, 2–9 × 1–4.5 cm, with 4–7 pairs of lateral veins and a pointed to long-pointed apex. **Panicles** up to 11 cm long, terminal and lateral; **corolla** white, 3.5–5.5 mm long, tube slightly shorter than lobes. **Fruit** almost globose, 5–8 mm. Gansu, Guangxi, Guizhou, Hubei, Sichuan, Shaanxi, seTibet, Yunnan [swTibet, c & sChina, Vietnam]; mixed forests, shrubberies, ravines, streamsides, 200–2,700 m. IV–VII.

SOLANACEAE

LYCIUM

A genus of c. 80 species (7 in China), mainly South American. Spreading to erect, thorny, **semi-evergreen shrubs** with arching **stems** and simple **leaves**. **Inflorescences** solitary or in bundles, axillary, flowers bell- or funnel-shaped, pale purple to blue. **Fruit** an egg-shaped to ellipsoid berry.

Lycium barbarum L. (including *L. halimifolium* Mill.) Erect to spreading **shrub**, 0.5–2 m tall with much-branched, pale grey **stems** with thorns 5–20 mm long. **Leaves** ovate, rhombic or lanceolate, 1.5–6 × 0.5–2.5 cm, greyish green. **Inflorescences** solitary or clustered; **flowers** 9–12 mm across; calyx campanulate, 3–4 mm long; corolla pale purple with lobes downy on edges. **Fruit** egg-shaped to oblong, 7–15 mm long, red when ripe. Gansu, Guizhou, Hubei, Sichuan, Shaanxi, Tibet, Yunnan [n, c & eChina, Japan, Korea, Nepal; introduced into Europe]; hillsides, waste places, roadsides, often near habitation, 800–3,200 m. III–VIII.

Lycium chinense L. (syn. *L. barbarum* subsp. *chinense* (Mill.) Aiton; *L. sinense* Gren.) Very similar to the previous species but calyx usually 2-lobed (not 3–5-lobed) or toothed and **flowers** with glabrous corolla-margins, the tube shorter or as long as (not longer than) lobes. Guangxi, Gansu, Hubei, Qinghai, Sichuan, Shaanxi, Yunnan [most of China, seAsia, Himalaya, Japan, Korea, Mongolia]; similar habitats and altitudes. V–IX.

---

1. *Ligustrum lucidum*; cult. (CGW)
2. *Ligustrum henryi*; Jian Chuan, nwYunnan (PC)
3. *Ligustrum henryi*; Heishui, Yulongxueshan, nwYunnan (PC)
4. *Ligustrum delavayanum* in fruit; nwYunnan (MF)
5. *Ligustrum sinense*; Tian'e, nwGuangxi (PC)
6. *Lycium chinense*; cult. (CGW)
7. *Nicandra physalodes*; cult. Kunming, cYunnan (CGW)

NICANDRA

A genus with a single species from South America (Peru) and a widely cultivated and naturalised weed of cultivation. **Herbs** with simple, stalked **leaves**. **Flowers** showy, regular, solitary, axillary; calyx large, campanulate; corolla rotate with ovate lobes; **stamens** inserted on corolla-tube; ovary 3- to 5-celled. **Fruit** globose, many-seeded, enclosed by enlarged calyx.

Nicandra physalodes (L.) Gaertn. (syn. *N. physaloides* (L.) Gaertn. in error) Vigorous **herb**, 40–150 cm tall with glabrous **stems**. **Leaves** ovate to elliptic, 4–20 × 2–13 cm, pointed, sparsely hairy on both surfaces. **Flowers** often nodding; calyx-lobes ovate, 0.8–3 cm long, persistent; corolla pale blue to blue with a white throat. **Fruit** 1–2 cm across, yellow or brown. Gansu, Guizhou, Sichuan, Tibet, Yunnan [c & nChina, native to Peru]; cultivated and waste places, waysides, to 2,600 m. VI–VIII.

ANISODUS

A Sino-Himalayan genus with 4 species (all in China). Large **herbs** or **subshrubs** with glabrous or downy, erect, angular, 2–3-branched **stems**. **Leaves** solitary or paired, entire, stalked, stipulate. **Inflorescence** 1-flowered, axillary; **flowers** erect, pendent or nodding, large, regular or somewhat zygomorphic; calyx funnel-shaped, 4–5-lobed; corolla campanulate, 5-lobed; stamens shorter than corolla, inserted near base; ovary conical, 2-celled. **Fruit** a globose to egg-shaped capsule, dehiscing around middle or at apex; **seeds** many.

Anisodus tanguticus **(Maxim.) Pascher** Plant 40–80 cm tall with glabrous or downy **stems**. **Leaves** lanceolate, oblong or ovate, 8–20 × 2.5–9 cm, coarsely toothed, glabrous. **Flowers** nodding or erect; calyx campanulate, 2.5–4 cm long; corolla campanulate with rounded lobes, yellow-green, purple or dark purple, 2.5–3.8 cm long. **Fruit** 2 cm in diameter enclosed in inflated calyx. Gansu, Qinghai, w, nw & swSichuan, e & se Tibet, nwYunnan [central Himalaya]; sunny, grassy slopes, roadsides, 2,000–4,400 m. VI–IX.

Anisodus luridus **Link** (syn. *Scopolia straminifolia* (Wall.) Shrestha) Differs from the previous species in having **leaves** that are creamy-woolly beneath, and a softly hairy calyx. swSichuan, seTibet, nwYunnan [Bhutan, India, Nepal]; grassy places, streamsides, 3,200–4,200 m. V–IX.

Anisodus acutangulus **C. Y. Wu & C. Chen*** Similar to *A. tanguticus* but with ovate to elliptic **leaves** and a smaller fruiting calyx (less than 4.5 cm long). Sichuan, Yunnan; grassy places, waste places, abandoned cultivation, 2,800–3,100 m. VI–VIII.

1. *Anisodus tanguticus*; Songpan, nwSichuan (PC)
2. *Anisodus tanguticus*; Songpan, nwSichuan (CGW)
3. *Anisodus luridus*; nNepal (CGW)
4. *Hyoscyamus niger*; near Songpan, nwSichuan (CGW)
5. *Physalis alkekengi*; cult. (CGW)
6. *Mandragora caulescens*; Tianchi, nwYunnan (CGW)
7. *Solanum nigrum*; Dali, nwYunnan (CGW)

Anisodus carniolicoides (C. Y. Wu & C. Chen) D'Arcy & Z. Y. Zhang* Distinctive and readily separated from the other 3 species by the strongly 2-lipped calyx and indistinctly lobed **corolla**. seQinghai, wSichuan, nwYunnan; grassy slopes, thickets, forest margins, rock-crevices, 3,000–4,500 m. VI–IX.

HYOSCYAMUS

A genus of c. 20 species (2 in China) in temperate Asia, North Africa and Europe. **Annual**, **biennial** or **perennial herbs** with an erect to sprawling habit and downy or glandular **stems**. **Leaves** coarsely toothed or pinnately lobed. **Inflorescences** 1-flowered, or several-flowered in scorpioid racemes or spikes; **flowers** zygomorphic, 5-parted; calyx tubular-campanulate; corolla campanulate or obconic; stamens inserted on corolla-tube; ovary 2-celled. **Fruit** a capsule enclosed in enlarged calyx.

Hyoscyamus niger L. Glandular-downy **herb** to 1 m with stalkless **leaves**, basal rosette-leaves ovate-lanceolate to oblong, 20–55 cm long, coarsely toothed or lobed; stem-leaves clasping **stem** at base. **Flowers** stalkless; calyx tubular, 1–1.5 cm long; corolla pale yellow to dull yellow with purple net-veins, campanulate, 1–1.5 cm long. **Fruit** egg-shaped-rounded, 1.5 cm long 2–2.5 cm long when measured with enlarged calyx. Gansu, Guizhou, Qinghai, Sichuan, Shaanxi, Yunnan [c & nChina, temperate Asia and Europe]; roadsides, waste places, environs of villages and cultivation, occasionally cultivated, 700–3,600 m. V–IX.

PHYSALIS

A genus of 75 species (6 in China) mostly in South America. **Annual** or **perennial herbs** with erect, glabrous or downy **stems**. **Leaves** simple, stalked. **Flowers** regular, solitary, 5-parted, axillary; calyx campanulate, enlarging and enclosing fruit; corolla rotate, forming a 5-lobed star with ovate lobes; stamens inserted at base of corolla-tube; ovary 2-celled. **Fruit** a globose, many-seeded, edible berry.

Physalis alkekengi L. Vigorous patch-forming **perennial**, 40–80 cm tall. **Leaves** narrowly to broadly ovate, 4–20 × 2–13 cm, entire to coarsely toothed. **Flowers** nodding; calyx campanulate, 6 mm long, persistent, eventually forming an inflated, orange bell enclosing the fruit; corolla white with greenish or yellow marks at base. **Fruit** 1–1.5 cm across, bright orange-red. Gansu, Guizhou, Hubei, Sichuan, Shaanxi, Yunnan [c & nChina, temperate Asia, Europe]; hillslopes, cultivated and wasteland, 1,500–2,500 m. V–VII.

MANDRAGORA

A genus of 4 species (1 in China) in temperate Asia and the Mediterranean region. Small **perennial herbs** with stout, often branching **roots** and short **stems**. **Leaves** simple, in a basal rosette, toothed on the wavy margins. **Flowers** axillary, regular, 5-parted, star-shaped or campanulate, erect or nodding; corolla tubular; stamens 5, inserted on corolla-tube. **Fruit** a many-seeded, globose or egg-shaped, yellow or orange berry.

Mandragora caulescens C. B. Clarke Plant 20–60 cm tall, glandular-hairy. **Leaves** obovate-oblong or oblanceolate, blunt, 3–20 × 1.5–5 cm. **Flowers** nodding, axillary, solitary, campanulate, blackish purple, sometimes green, 1.5–2.5 cm long; calyx broad-campanulate, slightly shorter than similarly coloured corolla. **Berries** globose, 2–2.5 cm long, yellow or orange. seQinghai, n, w & swSichuan, e & seTibet, nwYunnan [Himalaya, nMyanmar]; amongst rocks, in stony, sparse grassland, *Rhododendron* scrub, forest margins, 2,200–4,900 m. V–VII.

SOLANUM

A large genus of c. 1,200 species (41 in China, about half introduced), mainly from the Americas but well-represented in most regions. **Herbs**, **shrubs** or small **trees** with erect glabrous, slightly downy or prickly **stems**. **Leaves** solitary or paired, simple, lobed or odd-pinnate, stalked. **Inflorescences** racemose, paniculate or umbellate; **flowers** regular, solitary, in bundles, axillary, star-shaped; calyx 5-lobed, campanulate; corolla a 5-lobed star with pointed lobes; **stamens** inserted on corolla-tube, forming a central yellow column around style; ovary 2–5-celled. **Fruit** globose, juicy, many-seeded. Several species are widely cultivated (e.g. *S. tuberosum* (potato) and *S. melongena* (egg-plant, aubergine)) or naturalised as weeds of cultivation (e.g. *S. nigrum* and *S. virginianum*).

Solanum nigrum L. Erect, **annual herb**, 25–100 cm tall with downy **stems**. **Leaves** ovate, 4–10 × 3–7 cm, glabrous or hairy on both surfaces. **Flowers** star-shaped, nodding; calyx cup-shaped, 0.5–1 mm long, persistent; corolla white, 8–10 mm across. **Fruit** 0.8–1 cm across, dull black when ripe. Guizhou, Sichuan, Tibet, Yunnan [c & nChina, widespread elsewhere]; weed of cultivation, waysides, up to 3,000 m. III–X.

DATURA

A genus of c. 11 species (3 in China, naturalised), native to the Americas but cosmopolitan weeds of cultivation and waste places. **Annual** or **perennial herbs** with erect, branching **stems**. **Leaves** simple, margin entire or toothed. **Flowers** solitary in axils of leaves and bracts, regular, large, trumpet-shaped, white sometimes flushed with violet. **Fruit** a large, often prickly capsule, dehiscing by 4 valves.

Datura stramonium L. Tall **herb** 0.5–1.5 m with broadly ovate **leaves**, 8–17 × 4–14 cm, long-pointed, toothed. **Flowers** erect, 6–10 cm long; corolla white to pale purple, greenish at base, lobes often with purple tips. **Fruit** erect, prickly, 3–4.5 cm long. Throughout region, a cosmopolitan weed; waste places, stony hillsides, fields and around habitation, 1,200–2,500 m. VI–IX.

Datura metel L. Similar to *D. stramonium*, but **flowers** larger, more than 11 cm long, and with horizontal to pendent **fruit** bearing prickles or tubercules. Similar distribution and habitats, below 1,500 m. VI–VIII.

POLEMONIACEAE

POLEMONIUM

A genus with 20 species (3 in China) scattered mainly in the temperate northern hemisphere. Mostly tufted **perennial herbs**. **Leaves** in basal clusters, and also on **stem**, these alternate, pinnately lobed. **Flowers** in panicles, occasionally in capitate clusters, 5-parted; calyx campanulate, greenish and papery, persisting in fruit; corolla bell- to funnel-shaped, with short to long tube; **stamens** included in tube or protruding. **Fruit** a small, many-seeded capsule.

Polemonium caeruleum L. Erect, tufted **herbaceous perennial** to 90 cm tall, generally less. **Leaves** 7–14 cm long, with 11–17 ovate to lanceolate leaflets, each leaflet 1–4 × 0.2–1.4 cm, glabrous or slightly hairy. **Flowers** in branched paniculate clusters; calyx 5–8 mm; corolla broad-campanulate, violet to violet-blue, 15–20 mm, lobe-margins entire or slightly ciliate. (?)Sichuan, nwYunnan [Asia, Europe, Russia, North America]; damp meadows, banks, streamsides, open forests, shrubberies, 1,200–3,700 m. V–VIII.

Polemonium chinense (Brand) Brand Similar to the previous species, but **leaves** glabrous, with 15–25 leaflets, and flowers smaller, calyces mostly 2–3 mm, corollas 8–12 mm, occasionally slightly larger, violet or white, lobes ciliate. Gansu, Hubei, Qinghai, nSichuan, Shaanxi [nChina, Korea, Mongolia, eRussia]; similar habitats, 1,000–2,100 m. VI–VIII.

CONVOLVULACEAE

CONVOLVULUS

A cosmopolitan genus of c. 250 species (8 in China). **Shrubs, perennial** or **annual herbs**, with alternate linear to lanceolate, ovate, or more often heart- or arrow-shaped **leaves**. **Flowers** solitary or in heads, with bracts beneath (often located halfway down the stalks); calyx generally 5-lobed, lobes often unequal; corolla funnel-shaped, shallowly to deeply lobed; style with 2 slender branches. **Fruit** a capsule.

1. Leaves heart- to arrow-shaped; plant usually twining .... *C. arvensis*
1. Leaves linear to oblanceolate; plants not twining .... *C. ammannii, C. tragacanthoides*

Convolvulus arvensis L. **Twining** to **trailing perennial** to 1.5 m, sometimes more, with alternate **leaves**, blade to 5 × 4 cm. **Flowers** in lateral clusters of up to 3, only 1 opening at a time; sepals unequal, 2 larger than the other 3; corolla white, pink or bicoloured, 15–30 mm across. Gansu, Hubei, Qinghai, Sichuan, Shaanxi, Tibet [cosmopolitan]; weed of cultivated land and byways, roadsides, grassy banks, 600–4,500 m. V–IX.

Convolvulus ammannii Desr. **Tufted perennial** to 15 cm, with a woody stock; **stems** silvery-hairy. **Leaves** stalkless, linear to narrow-oblanceolate, to 20 × 4 mm, silvery with hairs. **Flowers** solitary; calyx with silvery-hairy, slightly unequal lobes; corolla broadly funnel-shaped, pale rose-pink or white with pinkish stripes, 10–18 mm across, shallowly lobed. Gansu, Qinghai, w & nwSichuan, Shaanxi, eTibet [Korea, cAsia, Mongolia, eRussia]; grassy and stony slopes and banks, roadsides, on clay and loess soils, 1,200–3,400 m. VI–VIII.

Convolvulus tragacanthoides Turcz. Readily distinguished from the previous species by being a compact, intricately branched spiny **shrub** to 15 cm with **flowers** in clusters of 2–6 on non-spiny shoots; calyx 5–8 mm long, yellowish-hairy; corolla pink, 15–25 mm across, with bands of hairs on outside. Gansu, Qinghai, n & nwSichuan [n & cChina, cAsia, Mongolia, Russia]; rocky and stony habitats, deserts, 2,000–3,200 m. V–VII.

CALYSTEGIA

A genus of c. 25 species (6 in China) distributed worldwide. Similar to *Convolvulus*, but base of **flower** surrounded by broad, pouched bracts that conceal the calyx.

Calystegia hederacea Wall. Spreading, sprawling plant to 1 m, **stems** sometimes twining. **Leaves** alternate, stalked, blade arrow-shaped to ± triangular, to 8 × 7 cm, basal lobes spreading, acutely angled. **Flowers** solitary, on stalks usually longer than leaves; corolla broadly funnel-shaped, pale pink or purplish, sometimes with whitish stripes, or occasionally all white, 20–40 mm across, shallowly lobed. Gansu, Guangxi, Guizhou, Hubei, Qinghai, Sichuan, Shaanxi, Tibet, nwYunnan [c & nChina, Asia, Himalaya, India, Russia]; fields, banks, road- and riversides, often on light sandy soils, to 3,500 m. V–IX.

Calystegia silvatica (Kit. ex Schrader) Griseb. subsp. *orientalis* Brummitt* Unlike the previous species, a **twiner**, climbing to 3 m, sometimes more, with triangular **leaves**, basal lobes not spreading, up to half the length of the midvein; **flowers** borne on stalks (sometimes paired) shorter than adjacent leaves, corolla white or pale pink, 30–50 mm across. Guangxi, Guizhou, Hubei, Sichuan, Yunnan [cChina]; forest margins, shrubberies, hedges, field boundaries, streamsides, to 2,600 m. VI–IX. **Subsp.** *silvatica* is widespread in Europe and wAsia.

DINETUS

A tropical Asian genus of 8 species (6 in China). **Twining herbs** with simple, rarely lobed, heart-shaped **leaves. Inflorescences** lateral, solitary or paired; **flowers** often fragrant; sepals free or ± so, equal or the 3 outer larger; corolla salverform or funnel-shaped, entire to 5-lobed, glabrous except for apical hair tufts; style jointed near base. **Fruit** 1-seeded, papery, indehiscent, often surrounded by enlarged calyx.

Dinetus duclouxii **(Gagn. & Cour.) Staples*** (syn. *Porana duclouxii* Gagn. & Cour.) **Perennial herb. Leaves** ovate, often lobed or dissected, 6–12 × 4–10 cm, often sticky with glands on the veins beneath, glabrous; leaf-stalks 2–2.7 cm long. **Flowers** in racemes or panicles with ovate to linear bracts, flower-stalks short, not more than 10 mm long; sepals ovate to elliptic, the largest up to 8 mm long, expanding greatly in fruit to 19–37 mm, becoming purple-brown; corolla blue, purple, reddish, yellow or white, funnel-shaped, 17–34 mm across. wHubei, Sichuan, Yunnan; fields and meadows, roadsides, hillslopes, streamsides, 100–4,000 m. V–XII.

CUSCUTA

A genus with c. 170 species (11 in China) mainly in North and South America, several in Asia. **Parasitic**, yellow or reddish, glabrous **herbs** with twining, thread-like **stems** and **leaves** reduced to minute scales. **Flowers** small, stalkless or short-stalked, mostly in clusters, 4–5-parted.

Cuscuta chinensis **L. Stems** yellow, thin, c. 1 mm in diameter. **Inflorescences** lateral, compact few–many-flowered heads, almost stalkless; calyx cup-shaped, lobes triangular, 1.5 mm long; **corolla** white, urn-shaped, 3 mm long, lobes triangular-ovate, pointed or blunt, reflexed; stamens inserted at throat; scales oblong, reaching stamens, long-fringed. Widespread in China [c & sAsia, Africa, Australia]; fields, thickets, open mountain slopes, sandy beaches, often on plants of *Leguminosae*, *Compositae* or *Zygophyllaceae*; 200–3,000 m. V–VIII.

1. *Polemonium chinense*; Huanglong, nwSichuan (PC)
2. *Convolvulus arvensis*; Wolong, wSichuan (CGW)
3. *Convolvulus ammannii*; Hezuozhen, Qinghai (HJ)
4. *Calystegia hederacea*; Napahai, Zhongdian, nwYunnan (CGW)
5. *Calystegia hederacea*; Napahai, Zhongdian, nwYunnan (CGW)
6. *Calystegia silvatica*; Minjiang valley, nwSichuan (CGW)
7. *Dinetus duclouxii*; Danyun, nwSichuan (PC)
8. *Cuscuta chinensis*; Dali, nwYunnan (PC)

1. *Menyanthes trifoliata*; Zhongdian, nwYunnan (EL)
2. *Nymphoides peltata*; Erhai, Dali, nwYunnan (CGW)
3. *Ehretia acuminata*; cult. (PC)
4. *Cynoglossum triste*; Tianchi, nwYunnan (PC)
5. *Cynoglossum triste*; Lijiang, nwYunnan (PC)
6. *Cynoglossum amabile*; near Dali, nwYunnan (CGW)
7. *Cynoglossum amabile*; Lijiang, nwYunnan (CGW)
8. *Cynoglossum furcatum*; near Aba, nwSichuan (CGW)

MENYANTHACEAE

MENYANTHES

A genus with a single species widespread in marshy regions of the northern hemisphere.

Menyanthes trifoliata L. **Mud-dwelling** or **aquatic** rhizomatous **perennial** with alternate, occasionally almost opposite **leaves** with 3 leaflets; leaflets elliptic, 2.5–7 cm long, central one larger than lateral. **Inflorescence** to 35 cm tall, racemose; **flowers** usually 5-parted; corolla white, tubular-campanulate, 14–17 mm long, glabrous outside but with long hairs inside, lobes semi-spreading, elliptic-lanceolate, blunt. Guizhou, s, sw & wSichuan, Tibet, Yunnan [c & nChina, wHimalaya, swAsia, Japan, Europe, North America]; lakes, pools, swampy ground, ditches, 400–3,600 m. V–VII.

NYMPHOIDES

A cosmopolitan genus of 38 species (6 in China). Primarily **aquatic perennials** with floating rounded, heart- or kidney-shaped, long-stalked **leaves**. **Flowers** held above the water-surface, hermaphrodite, regular, 5-parted, petals fringed or crested, united below into a short tube. **Fruit** a capsule or berry.

Nymphoides peltata (S. G. Gmel.) Kuntze Glabrous, rhizomatous **aquatic perennial**. **Leaves** floating, deep glossy green, stalked, lower alternate, upper opposite, leaf-blade rounded to broad-ovate, 2–8 cm across, purplish brown and densely glandular beneath. **Flowers** appearing solitary, but clustered at upper nodes, 5-parted, short- or long-styled; corolla golden yellow, rotate, 25–30 mm across, lobes oval with irregularly toothed margin. Gansu, Guangxi, Guizhou, Hubei, Sichuan, Shaanxi, Yunnan [n, e & cChina, w & cAsia, Japan, Europe, Russia]; pools, lakes, ditches, to 1,800 m. IV–X.

BORAGINACEAE

EHRETIA

A genus of 50 species (14 mostly confined to south China) found in Africa, North America, the Caribbean region and south Asia. **Trees** and **shrubs** with alternate, entire to toothed **leaves**. **Inflorescence** cymose to paniculate; **flowers** 5-parted; corolla white to yellow, tubular-campanulate to funnel-shaped, usually with protruding anthers. **Fruit** a drupe, yellow, orange or red when ripe.

Ehretia acuminata R. Br. (syn. *Cordia thyrsiflora* Sieb. & Zucc.; *Ehretia thyrsiflora* (Sieb. & Zucc.) Nakai **Tree** to 15 m with blackish grey bark and smooth, glabrous branchlets. **Leaves** stalked, elliptic to obovate or oblong, 5–13 × 4–6 cm, pointed, glabrous or minutely hairy, toothed. **Inflorescence** a dense cymose panicle to 15 cm long, fragrant; calyx 1.5–2 mm long, 5-lobed; corolla white, campanulate, with 5 spreading lobes and a short tube; **stamens** protruding. **Drupes** 3–4 mm, yellow or orange when ripe. Guangxi, Guizhou, Sichuan, Yunnan [e & seChina, Taiwan, Bhutan, India, Japan, seAsia]; hill forests, thickets, open slopes, 100–1,700 m.

CYNOGLOSSUM

A north temperate genus of c. 75 species (12 in China). Hairy or bristly **biennials** and **perennials** with both **basal** and **stem-leaves**. **Flowers** in lateral and terminal cymes which usually form large panicles, with or without bracts; calyx 5-parted, lobed to the base, enlarging in fruit; corolla funnel-shaped to rotate, 5-parted, with short tube and 5 scales in throat; stamens included within tube. **Nutlets** 4, barbed.

> 1. Corolla blackish purple; nutlets 10–15 mm across . . . . *C. triste*
> 1. Corolla blue, rarely white; nutlets not more than 5 mm across . . . . . . *C. amabile, C. furcatum, C. lanceolatum*

Cynoglossum triste Diels* Tufted **perennial** to 50 cm, often less, with several erect rough-hairy **stems**. **Leaves** heart-shaped to ovate, to 12 × 8 cm, lowermost long-stalked, upper smaller. **Inflorescence** with up to 7, densely clustered cymes, without bracts; **corolla** tubular, 5–6 mm long. **Nutlets** sea-urchin-shaped, with dense yellowish barbs. swSichuan, nwYunnan; open woodland (usually of *Pinus*), rocky places, slopes, often in partial shade, 2,500–3,100 m. V–VII.

Cynoglossum amabile Stapf & Drumm. Tufted **perennial** to 60 cm, sometimes gregarious, with 1 or more erect, densely hairy **stems**. **Leaves** grey-green, lanceolate to oblong-lanceolate, basal long-stalked, to 20 × 5 cm, stem-leaves stalkless, up to 9 cm long. **Inflorescence** a large, rather narrow panicle with numerous lateral cymes; calyx 3–4 mm long, hairy outside; **corolla** salverform, pale to mid-blue, occasionally white, 8–10 mm across. **Nutlets** 3–4 mm, with dense barbs. Gansu, Guizhou, Sichuan, seTibet, Yunnan [Bhutan]; rocky and grassy places, fields, hillslopes, forest margins and clearings, stream- and tracksides, 2,600–3,700 m. V–IX.

Cynoglossum furcatum Wall. Superficially similar to the previous species, but **stems** bristly; **leaves** green, only the uppermost stalkless; and **corolla** bright azure-blue. **Nutlets** smaller, 2–3 mm. sGansu, Guangxi, Guizhou, Sichuan, Shaanxi, Yunnan [c, s & eChina, seAsia, India, Taiwan]; meadows, forest clearings, open shrubberies, 300–3,000 m. V–X.

Cynoglossum lanceolatum Forssk. Similar to the previous 2 species, but often taller, to 90 cm; **stems** often branching close to base, or at least in lower half, and densely bristly; **flowers** smaller, calyx 1–1.5 mm long, corolla pale blue, not more than 3 mm wide. sGansu, Guangxi, Guizhou, Sichuan, Shaanxi, Yunnan [c, s & seChina, Himalaya, w, sw & seAsia, India, Myanmar, Sri Lanka, Taiwan, Africa]; meadows and fields, hillslopes, roadsides, 300–2,800 m. IV–IX.

HACKELIA

A genus of c. 45 species (3 in China) mainly in the north temperate zone but also in Central and South America. **Annual** and **perennial herbs** with alternate **leaves**. **Inflorescence** with a terminal cymose cluster, not branched, the stalks deflexed in fruit; calyx 5-lobed to base, lobes enlarging in fruit; **corolla** tubular-campanulate to rotate, 5-lobed with a short tube and 5 scales in throat. **Nutlets** to 4, egg- to turban-shaped, margin ribbed and with narrow-triangular barbs.

Hackelia uncinata (Benth.) C. E. C. Fischer (syn. *Cynoglossum uncinatum* Benth.) **Tufted perennial** to 80 cm, generally less. **Leaves** ovate to elliptic, to 9 × 5 cm, basal long-stalked. **Inflorescence** 2-parted, terminal, forming a rounded lax cluster; **corolla** sky-blue to bluish purple, occasionally white, tube 2 mm long, limb slightly cupped, 5–8 mm across, with oval lobes. seTibet, nwYunnan [swTibet, Himalaya, nMyanmar]; open forests and forest margins, clearings, grassy places, generally on rather wet slopes, 2,700–4,500 m. VI–VIII.

Hackelia brachytuba Opiz ex Bercht. Tends to be a larger plant than the previous species, to 1 m, calyx-lobes lanceolate rather than ovate, 3–3.5 mm long; **corolla** more campanulate with a tube 3–4 mm long. sGansu, Sichuan, seTibet, nwYunnan [swTibet, Nepal]; similar habitats, 2,900–3,800 m. VII–VIII.

TRIGONOTIS

A genus of 58 species (39 in China, 34 endemic) in east Europe and from Central Asia to New Guinea. Bright-flowered **annual**, **biennial** or **perennial herbs**, often tufted, sometimes cushion-forming, usually hairy or softly bristly. **Leaves** alternate, entire. **Flowers** in solitary or branched cymes, at least lower flowers with bracts; calyx 5-lobed; corolla blue or white, *Myosotis*-like with a short tube and 5 spreading, rounded lobes, throat with 5 appendages; stamens not protruding. **Fruit** consisting of 4 unwinged nutlets.

1. Inflorescence with basal, leaf-like bracts ....
   .... *T. tibetica, T. rockii*
1. Inflorescence without bracts .... *T. laxa*

Trigonotis tibetica (C. B. Clarke) I. M. Johnst. Rather diffuse rough-hairy greyish **annual** or **biennial** to 25 cm, generally less. Lower **leaves** linear-elliptic to lanceolate, to 20 × 6 mm, stalked, stem-leaves usually smaller and stalkless. **Flowers** in terminal cymes; calyx with erect, ovate lobes 1.5 mm long;

corolla pale blue or white, campanulate, 3 mm across. Qinghai, w & nwSichuan, e & seTibet [swTibet, c & eHimalaya]; alpine and subalpine meadows, low shrubberies, 2,500–4,200 m. VI–IX.

Trigonotis rockii I. M. Johnst.* Similar to the previous species, but a **tufted perennial** to only 15 cm with green, elliptic to ovate **leaves**; **corolla** purplish blue with a flat limb 7–8 mm across. seTibet, nwYunnan [swTibet]; grassy and stony alpine meadows, open shrubberies, 3,300–4,900 m. VII–IX. Plants found in nwSichuan appear to be very close to *T. rockii*, and if they prove to be this species, they extend its distribution well to the north.

Trigonotis laxa I. M. Johnst.* In contrast to the previous 2 species, a more robust **perennial** to 45 cm with larger **leaves** to 8 × 3.5 cm, with winged stalks; **corolla** pale blue, 4–5 mm across. Guizhou, Sichuan, Yunnan; forests and forest margins, shrubberies, ravines, stream- and roadsides, 500–1,600 m. V–VI.

CHIONOCHARIS

A Sino-Himalayan genus of 1 species, closely related to *Myosotis* but differing primarily in the spathulate calyx-lobes, in the slightly overlapping corolla-lobes and in the hairy, laterally attached **nutlets**.

Chionocharis hookeri I. M. Johnst. Soft, silvery-grey, **cushion-forming** plant to 50 cm across, forming low dense mounds. **Leaves** in dense, closely packed rosettes, oval to oblanceolate, to 12 × 6 mm, adorned with dense whitish hairs. **Flowers** solitary, borne directly on top of rosettes; calyx campanulate, 4–5 mm long, hairy like the leaves; corolla bright sky-blue with yellowish eye, salverform with a 7–8 mm limb with 5 spreading, rounded lobes. swSichuan, seTibet, nwYunnan [swTibet, eHimalaya, nMyanmar]; stony alpine meadows, rocky slopes and ridges, moraines, occasionally on cliffs, 3,500–5,000 m. VI–VII. (See also photo on p. 38.)

ERITRICHIUM

A north temperate genus with c. 50 species (39 in China, mostly endemic). **Annual** or **perennial herbs**, similar to *Trigonotis* but **nutlets** with winged or toothed margins. **Tufted** or **cushion-forming** plants with solitary or cymose flower-clusters.

Eritrichium deqinense W. T. Wang* Small **cushion-forming** plant to 10 cm across. **Leaves** mostly in basal rosettes, oblong to almost linear, 7–10 × 2–2.8 mm, bristly, particularly beneath, with a short ciliate stalk. **Flowers** often 3 per cyme accompanied by small, leaf-like bracts; corolla white, 5 mm across, with short tube and 5 rounded lobes. nwYunnan; mountain rocks, rocky meadows, c. 4,000 m. VII–VIII.

1. *Hackelia uncinata*; Tianchi, nwYunnan (CGW)
2. *Trigonotis tibetica*; Jigzhi, seQinghai (HJ)
3. *Trigonotis laxa*; south of Aba, nwSichuan (CGW)
4. *Chionocharis hookeri*; Baimashan, nwYunnan (HJ)
5. *Chionocharis hookeri*; Baimashan, nwYunnan (HJ)
6. *Eritrichium deqinense*; Baimashan, nwYunnan (HJ)
7. *Eritrichium deqinense*; Baimashan, nwYunnan (HJ)

BORAGINACEAE: Onosma

ONOSMA

A genus of c. 145 species (29 in China, half endemic) found primarily in southern Europe and west and central Asia. Bristly **biennials** or **perennials** with entire, alternate **leaves**. **Flowers** in terminal, spiralled (scorpioid) cymes, with bracts; calyx 5-lobed to the base; corolla usually nodding, tubular to tubular-campanulate, often yellow, occasionally blue, pink, purple or white; **stamens** not protruding.

Onosma farreri I. M. Johnst.* Greyish, bristly **subshrub** to 30 cm, much-branched from base. **Leaves** oblanceolate, to 30 × 4 mm, each bristle with a pimple-like base. **Flowers** in rather lax cymes; calyx campanulate, 6 mm long, enlarging in fruit; corolla blue, tubular, 11–12 mm long, broadening towards the short lobes. Gansu, n & nwSichuan, Shaanxi (Taibaishan); dry rocky and sandy hillslopes, banks, 2,900–3,900 m. IV–VII.

Onosma confertum **W. W. Sm.*** (syn. *Onosma forrestii* W. W. Sm.) In contrast to the previous species, a **taprooted perennial** 30–70 cm tall, with larger **leaves** to 12 × 1 cm; calyx 10–13 mm long; **corolla** red or purple, 12–17 mm long. swSichuan (Muli), nwYunnan; thickets, rocky places, 2,300–3,300 m. VII–X.

LASIOCARYUM

A genus of perhaps 5 species (3 in China) found in south-west Asia, the Himalaya and west China. **Annual** or **biennial herbs** with entire hairy **leaves**. **Flowers** in bractless cymes; calyx 5-parted to base, somewhat enlarging in fruit; corolla tubular, with tube equalling calyx, throat with 5 appendages. **Nutlets** bristly and wrinkled.

Lasiocaryum trichocarpum **(Hand.-Mazz.) I. M. Johnst.*** **Annual** with long soft hairs, with 1–several **stems** to 10 cm tall. **Leaves** ± stalkless, obovate to oblong, to 13 × 6 mm, blunt, softly hairy. **Inflorescence** terminal, several-flowered, without bracts; calyx c. 2.5 mm, softly hairy outside; **corolla** blue with a tube equalling calyx, limb 4–5 mm across. **Nutlets** narrowly egg-shaped, with bristles on the transverse wrinkles. w, nw & swSichuan, w & nwYunnan; rocky and stony places, open scrub, c. 3,000 m. VI–VII.

ARNEBIA

A genus of 25 species (6 in China) distributed from south Europe and North Africa to Arabia, the Himalaya and Asia. Bristly or hairy **annual** or **perennial herbs**; **roots** often producing a purple dye. **Leaves** alternate. **Flowers** in bractless cymes, often heterostylous; calyx 5-lobed almost to base; corolla funnel-shaped, usually yellow, occasionally purple or reddish, often with a spot at the base of each lobe, throat with appendages.

Arnebia szechenyi **Kanitz*** (syn. *Lithospermum szechenyi* (Kanitz) I. M. Johnst.) Tufted **perennial herb** to 30 cm, sparingly branched, grey-white-downy. **Leaves** narrow-ovate to narrow-oblong or almost linear, 10–20 × 2–6 mm, pointed, rough-hairy, margin slightly toothed, stalkless. **Cymes** small, with up to 8 flowers, accompanied by leaf-like bracts. **Flowers** heterostylous; calyx 10 mm long, with linear lobes; corolla whitish to yellow, tubular-campanulate, 15–22 mm long, limb 5–7 mm, lobes plain at first, usually developing a dark purple blotch. nwGansu, Qinghai [nChina]; dry mountain slopes, open scrub, banks, 2,900–4,400 m. VI–IX.

---

1. *Lasiocaryum trichocarpum*; Huanglong, nwSichuan (CGW)
2. *Arnebia szechenyi*; Xunhua, cQinghai (RS)
3. *Callicarpa giraldii*; Beijing (CGW)
4. *Callicarpa giraldii*; cult. (PC)
5. *Caryopteris tangutica*; near Tongren, seQinghai (RS)
6. *Caryopteris tangutuca*; south of Songpan, Minjiang, nwSichuan (PC)

VERBENACEAE

CALLICARPA

A genus of c. 140 species (48 in China) in tropical and subtropical Asia, with a few in tropical America and Africa. **Shrubs** or **trees** with whorled **branches**, the branchlets rounded or 4-angled. **Leaves** opposite or in 3s, simple. **Inflorescence** lateral, cymose; calyx campanulate; **corolla** 4-lobed, regular, tubular or campanulate; stamens 4, inserted on corolla-tube. **Fruit** globose, fleshy, often purple; **seeds** several.

Callicarpa giraldii **Hesse ex Rehd.*** (syn. *C. bodinieri* H. Lévl.) **Shrub** 1–3 m tall, with lanceolate, ovate or elliptic **leaves**, 5–17 × 2–10 cm, long-pointed. **Inflorescence** 1–3 cm across, stellate-hairy; **flowers** 3 mm long; corolla purple. **Fruit** globose, 3–4 mm across, purple or lilac-purple when ripe. Gansu, Guizhou, Hubei, Sichuan, sShaanxi, Yunnan [cChina]; mixed woodland and forests, 200–3,400 m. VI–VII.

CARYOPTERIS

A genus of 16 species (14 in China) in Central and east Asia. Erect **herbs**, **shrubs** or, occasionally, **climbers** with opposite, simple **leaves**, with entire or toothed margins, usually glandular. **Flowers** in lax to dense cymes or thyrses, white, pink, purple or blue; corolla tubular, 2-lipped, 5-lobed; stamens 4, often protruding prominently; ovary 4-celled. **Fruit** dry, divided into 4 small nutlets.

1. Leaves toothed	*C. tangutica*
1. Leaves entire	*C. glutinosa, C. forrestii*

Caryopteris tangutica **Maxim.*** Rather erect **shrub** to 2 m with grey-downy branchlets. **Leaves** lanceolate to ovate-lanceolate, 2–5.5 × 0.5–2 cm, grey-downy beneath, green above and somewhat hairy, margin deeply toothed. **Flowers** in dense terminal and lateral clusters; calyx deeply 5-lobed, 2.5–4 mm long; corolla blue-purple, tube 5–7 mm long, lowermost lobe fringed. Gansu, Hubei, seQinghai, Shaanxi, Sichuan [cChina]; dry slopes, c. 2,500 m. VII–IX.

Caryopteris glutinosa **Rehd.*** Small **shrub**, 0.5–1.5 m tall, with downy branchlets. **Leaves** narrow-oblanceolate or narrow-oblong, 1–3 × 0.3–0.8 cm, shiny, white-scaly, with black veins beneath, sticky. **Inflorescence** terminal, downy; **flowers** 6–8 mm long, blue or purplish blue, silvery-hairy outside; ovary hairy. w & nwSichuan (Minjiang valley); dry banks in thickets, gravelly slopes in full sun, 1,600–1,800 m. VIII–X.

Caryopteris forrestii **Diels*** Differs from the previous species in having elliptic, rounded **leaves** and whitish, yellowish or very pale blue **flowers**. Guizhou, Sichuan, e & seTibet, Yunnan; dry rocky slopes, riverbanks, open scrub, 1,700–4,000 m. VII–IX.

CLERODENDRUM

A cosmopolitan genus of c. 400 species (34 in China), mostly tropical and subtropical. **Herbs** or **deciduous** or **evergreen subshrubs**, **shrubs** or small **trees** with 4-angled young branchlets. **Leaves** simple, opposite or rarely whorled. **Inflorescence** capitate or loosely cymose, usually terminal; calyx bell- or cup-shaped, 5-lobed or -toothed; **corolla** tubular at base, with 5 spreading lobes above; stamens 4, usually protruding, sometimes markedly so. **Fruit** a fleshy drupe.

1. Corolla bright pink, purple or red .... *C. bungei*
1. Corolla white, rarely with a flush of palest pink ....
   .... *C. trichotomum, C. colebrookianum*

Clerodendrum bungei **Steud.** Small suckering, patch-forming **deciduous shrub**, 1–2 m tall. **Leaves** ovate or broadly ovate, 8–20 × 5–15 cm, pointed to long-pointed, toothed on margin. **Inflorescence** dense flat-topped cymes, many-flowered; **flowers** tubular, 5-lobed, 2–3 cm long, fragrant; corolla purple, red or pink. **Fruit** globose, 6–12 mm across, blue-black when ripe. Gansu, Guizhou, Hubei, Qinghai, Sichuan, Shaanxi [cChina, Taiwan]; mixed forests and forest margins, shrubberies, roadsides, often planted, to 2,500 m. VIII–X.

Clerodendrum trichotomum **Thunb.** Suckering **shrub** or small **deciduous tree** to 10 m with deep green, ovate-elliptic to triangular-ovate, stalked **leaves**, 5–126 × 2–13 cm, with 3 or 5 pairs of primary veins and an entire to wavy margin. **Flowers** in broad corymbose cymes to 18 cm across; calyx greenish becoming purple, 5-lobed; corolla white or pinkish, 18–20 mm long, tube slender, lobes spreading, oblong; stamens long, protruding. **Fruit** blue-purple when ripe, surrounded by persistent purple-red calyx. Gansu, Guangxi, Guizhou, Hubei, Qinghai, Sichuan, Shaanxi, Yunnan [n, c, s & eChina, seAsia, Japan, Korea, India]; forest margins, shrubberies, hillslopes, ravines, to 2,400 m. VIII–X.

Clerodendrum colebrookianum **Walp.** **Shrub** or small **tree** 1.5–3 m tall, with robust, 4-angled, densely yellow-brown-hairy branchlets. **Leaves** broadly ovate to elliptic-heart-shaped, 7–17 × 6–21 cm, pointed to long-pointed, finely hairy beneath, margin entire. **Inflorescence** terminal, 4–6-branched corymbs; **corolla** white to pinkish, 1.2–2.5 cm long. **Fruit** blue-green surrounded by persistent, inflated, purple calyx. seTibet, Yunnan [swTibet, sChina, c & eHimalaya, Myanmar, Thailand, Vietnam]; forests and thickets on mountain slopes, roadsides, 500–2,000 m. VII–X.

VITEX

A cosmopolitan genus of c. 250 species (14 in China). **Trees** or **shrubs** with opposite, palmately lobed **leaves**. **Flowers** in terminal or lateral cymes or panicles; calyx tubular to campanulate, 5-toothed, occasionally 2-lipped; corolla 2-lipped, lower lip 3-lobed, upper 2-lobed; stamens 4. **Fruit** 1–4 small drupes.

Vitex negundo **L.** Variable **deciduous shrub** or small **tree** to 5 m with grey-downy branchlets. **Leaves** stalked, with 3–7 grey-green, usually toothed leaflets, central one distinctly short-stalked, 4–13 × 1–4 cm, with entire to somewhat wavy margin, white-scaly beneath. **Flowers** numerous, in panicles; calyx campanulate, 5-toothed, grey-downy; corolla purple-blue, violet-blue or rarely white, 6–7 mm long. Drupes black when ripe, 4–5 mm. **Var. *negundo*** has mature **leaves** with the central leaflet 4–13 cm long. Guangxi, Guizhou, Hubei, Qinghai, Sichuan, Shaanxi, Tibet, Yunnan [most of China, eAfrica, s & seAsia, Japan, Taiwan]; thickets, dry slopes, river valleys, field boundaries, 200–1,400 m. IV–VII. **Var. *microphylla*** **Hand.-Mazz.*** differs in the central leaflet of mature **leaves** being 1.5–4 cm long. Sichuan, Tibet, Yunnan; similar habitats, 1,200–3,200 m. VII–XI.

Vitex yunnanensis **W. W. Sm.*** Distinguished from the previous species by the axillary **inflorescence** subtended by normal **leaves** and the **corolla** 15–17 mm long, white or pinkish white. sSichuan, Yunnan; similar habitats, 1,800–3,500 m. IV–VII.

Vitex trifolia **L.** Similar to *V. negundo*, but **leaves** usually with just 1 or 3 leaflets, **corolla** purple or bluish purple, white-scaly outside. Guangxi, Yunnan [e & seChina, s & eAsia, Australia, Pacific islands]; similar habitats, to 1,700 m. IV–VIII.

VERBENA

A genus of c. 250 species (1 in China), mainly in the tropical Americas. **Annual** or **perennial herbs** or **subshrubs** with opposite, stalkless, toothed or lobed **leaves**. **Spikes** terminal, simple or branched, **flowers** alternate, zygomorphic; calyx 5-toothed; corolla tubular, 5-lobed; **stamens** 4, inserted on middle of corolla-tube. **Fruit** a dry capsule, dehiscing into 4 segments.

Verbena officinalis **L.** **Annual**, or short-lived, erect **perennial herb**, 30–140 cm tall. **Leaves** paired, ovate, obovate or oblong, 2–8 × 1–5 cm, coarsely toothed or lobed, hairy on veins beneath. **Flowers** in branched spike-like inflorescences, long and slender; calyx glandular; corolla blue or pink, 4–8 mm long, hairy. Gansu, Guizhou, Hubei, Sichuan, Shaanxi, Tibet, Yunnan [c & nChina, Asia, Europe]; grassy and waste places, waysides, often a weed of cultivation, to 1,800 m. V–IX.

---

1. *Clerodendrum bungei*; wSichuan (JM)
2. *Clerodendrum bungei*; cult. (CGW)
3. *Clerodendrum trichotomum*; Beijing, cult. (CGW)
4. *Clerodendrum trichotomum*; cult. (TK)
5. *Clerodendrum colebrookianum*; Xishuangbanna, swYunnan (CGW)
6. *Vitex negundo*; Guilin, ne Guangxi (CGW)
7. *Vitex yunnanensis*; Luding, wSichuan (PC)
8. *Verbena officinalis*; cult. Kunming, cYunnan (CGW)

1. *Colquhounia coccinea*; near Dali, nwYunnan (RS)
2. *Elsholtzia strobilifera*; Tianchi, nwYunnan (PC)
3. *Elsholtzia rugulosa*; Cangshan, above Dali, nwYunnan (CGW)
4. *Elsholtzia fruticosa*; Deqin, nwYunnan (PC)
5. *Ajuga lupulina*, with *Geranium pogonanthum*; Balangshan, Wolong, wSichuan (CGW)
6. *Ajuga lupulina* form; nwSichuan (JM)
7. *Ajuga lupulina*; south of Aba, nwSichuan (CGW)
8. *Ajuga lupulina* form; Genyanshan, swSichuan (AD)
9. *Ajuga ovalifolia*; east of Aba, nwSichuan (CGW)
10. *Ajuga ovalifolia*; north of Aba, nwSichuan (CGW)
11. *Ajuga nubigena*; Jiuzhaigou, nwSichuan (PC)

LABIATAE (LAMIACEAE)

COLQUHOUNIA

A genus with 6 species (5 in China) in the Himalaya and South East Asia. **Shrubs** with paired, stalked **leaves** with simple or stellate hairs and **flowers** in spike- or head-like inflorescences.

Colquhounia coccinea **Wall. Shrub** to 2.5 m, occasionally more, with erect to spreading rusty-downy **stems**. **Leaves** ovate to ovate-lanceolate, 7–11 × 2.5–4.5 cm, covered with rust-coloured stellate hairs, especially beneath, finely toothed. **Flowers** in rounded to spike-like **inflorescences**, each whorl with up to 20 flowers; calyx 5–6 mm long, stellate-hairy; corolla scarlet to orange-red, 20–25 mm long. seTibet, w & nwYunnan [swTibet, Himalaya, Myanmar, Thailand]; forest margins, shrubberies, ravines, stony and grassy slopes, 1,400–3,000 m. VIII–XII.

Colquhounia compta **W. W. Sm.*** Differs from the previous species in having young **branches** and leaf undersurface with a mixture of stellate and simple hairs; **leaves** smaller (not more than 5 cm long), and in the deep red or greyish red, somewhat smaller **flowers**. w & swSichuan, Yunnan; dry valley thickets, 1,800–2,100 m. IX–X.

ELSHOLTZIA

A genus with c. 40 species (33 in China) in Asia, Europe and North America. **Herbs**, **subshrubs** or **shrubs** with entire **leaves**, with continuous or interrupted **inflorescences** of small **flowers**.

1. Small herbs not more than 20 cm tall ... ***E. strobilifera***
1. Shrubs or subshrubs at least 30 cm tall, usually far larger .... ***E. rugulosa, E. fruticosa, E. eriocalyx***

Elsholtzia strobilifera **Benth.** Small **herb** sometimes as little as 5 cm tall, with spreading slender brown **stems** covered in crisped white hairs. **Leaves** ovate, 5–25 × 3–20 mm, deep green to deep purple above, glandular. **Flowers** in cylindrical spikes to 2.5 cm long, with densely overlapping, cupped, ciliate, purple bracts; corolla white or reddish, 3–4 mm long; stamens not protruding. Sichuan, Tibet, Yunnan [Himalaya, Taiwan]; stony and grassy slopes, open forests, shrubberies, valleys, 2,300–3,700 m. VIII–XI.

Elsholtzia rugulosa **Hemsl.*** **Shrub** or **subshrub** to 1.5 m with densely white-hairy **branches**. **Leaves** stalked, elliptic to rhombic-ovate, 2–7.5 × 1–3.5 cm, pointed, minutely wrinkled, with an obscurely toothed margin, grey- or white-downy beneath. **Inflorescences** 3–12 cm long, white-hairy; **corolla** creamy, yellowish or purplish, 4 mm long, softly hairy. Guangxi, Guizhou, Sichuan, Yunnan; open forests, low thickets, waysides, rocky slopes, 1,300–2,800 m. VIII–X.

Elsholtzia fruticosa **(D. Don) Rehd.** Aromatic **shrub** to 2 m, generally only half that height, with yellowish or purple-brown downy **branches**. **Leaves** stalkless or almost so, elliptic-lanceolate to oblong, 6–13 × 2–3.5 cm, long-pointed, with a toothed margin. **Spikes** 6–20 cm long, many-flowered; calyx campanulate, 2 mm long; **corolla** white or yellowish, c. 5 mm long. Gansu, Guangxi, Guizhou, Hubei, Sichuan, s & seTibet, Yunnan [Himalaya]; scrubby and waste places, hillslopes, valleys, 1,200–3,800 m. VII–X.

Elsholtzia eriocalyx **C. Y. Wu & S. C. Huang*** Similar to the previous species, but **leaves** pointed, not long-pointed, with a more finely toothed margin, hairy beneath; **spikes** 5–8 cm long, with creamy white **flowers** c. 7 mm long, glandular and hairy. Sichuan, Yunnan; grassland, scrub margins, forests, hillsides, 2,700–3,400 m. VII–X.

AJUGA

A genus of c. 50 species (18 in China), distributed in Europe and Asia. **Annual**, **biennial** or **perennial herbs** with simple **leaves**; **inflorescence** with leaf-like, often coloured bracts.

1. Flowers partly hidden at base of bracts; inflorescence with large overlapping bracts .... ***A. lupulina***
1. Flowers exposed; inflorescence without large overlapping bracts .... ***A. ovalifolia, A. nubigena***

Ajuga lupulina **Maxim.** Variable tufted **perennial**, rhizomatous **herb**, 8–25 cm tall with oblong-lanceolate to lanceolate **leaves**, 5–11 × 1.8–3 cm, glabrous, margin entire. **Inflorescence** with creamy white, occasionally purplish, toothed to entire, ovate, spreading to deflexed bracts, more than 2.5 cm broad; **corolla** white, cream or greenish white, 1.8–2.5 cm long, partially hidden at base of floral leaves. Gansu, Qinghai, w, nw & swSichuan, Tibet [cChina, Himalaya]; stony meadows, rocky places, moraines, forest margins, 1,300–3,500 m. VI–VIII. **Var. *major* Diels*** differs in being a taller plant more than 25 cm, and in having toothed, ovate or rhombic floral **leaves**. w & swSichuan, nwYunnan, to 4,300 m.

Ajuga ovalifolia **Bur. & Franch.*** **Annual herb** 3–23 cm tall. **Leaves** in 2–3 pairs, oblong-elliptic to broadly ovate-elliptic, 4–8 × 2.2–7 cm, blunt or rounded, margin wavy and scalloped. **Inflorescence** terminal, almost capitate, 2–3 cm long with bluish or purplish bracts; **flowers** numerous and densely crowded; corolla tubular in lower half, 1.5–3 cm long, blue or reddish purple, hairy. sGansu, seQinghai, n & nwSichuan; grassy and bare slopes, banks, thickets, 2,800–4,300 m. VI–VIII.

Ajuga nubigena **Diels*** Hairy, stoloniferous, **patch-forming herb** to 10 cm. **Leaves** elliptic to ovate-elliptic, 2–3.5 × 1.2–2.3 cm, coarsely hairy, with an irregularly toothed margin. **Flowers** in close almost terminal whorls, each with 2–4 flowers; calyx campanulate, 6–7 mm long, hairy; corolla whitish, pale blue to bluish purple, with purple lines, 12–14 mm long, lower lip with a deeply notched central lobe. Sichuan, seTibet [swTibet], Yunnan; grassy places, *Pinus* and broadleaved forests, rock-crevices, 2,500–4,800 m. V–VIII.

DRACOCEPHALUM

A genus of c. 70 species (35 in China) in Asia and Europe, 1 in North America. **Perennial herbs** with erect or prostrate **stems** and entire or pinnate **leaves**; **flowers** laxly or densely clustered in **spikes**, the corolla blue, violet, or purple, rarely white.

> 1. Flowers white .... *D. heterophyllum*
> 1. Flowers blue, violet or purple 2
> 2. Leaves undivided ....
>   .... *D. bullatum, D. moldavica, D. purdomii*
> 2. Leaves divided .... *D. forrestii*

Dracocephalum heterophyllum **Benth.** Tufted **perennial** 9–15 cm, **stems** with dense recurved hairs. **Leaves** simple, broadly to narrowly ovate, 1.3–4 cm long, margin shallowly toothed and minutely ciliate. **Flowers** in 4–8-flowered terminal clusters; calyx greenish, sparsely hairy; corolla 2.2–3.4 cm long, white with white or yellowish hairs. Gansu, Qinghai, Sichuan, Tibet [nwChina, Himalaya, Mongolia, eRussia]; stony places, meadows, 1,100–5,000 m. VI–VII.

Dracocephalum bullatum **Forrest ex Diels*** Tufted **perennial** 9–18 cm tall with red-purple, obscurely 4-angled **stems**. **Leaves** simple, deep green, net-veined, ovate to elliptic, 2.5–5 cm long, with a neat scalloped margin, deep green and net-veined above, often purplish beneath. **Flowers** in dense terminal heads with rounded, leaf-like bracts; calyx reddish purple, softly hairy; corolla bluish purple, 2.8–3.5 cm long, downy. swSichuan, nwYunnan; stony alluvial fans, moraines, in limestone mountains, 3,000–4,500 m. VI–VII.

Dracocephalum moldavica **L.** In contrast to the previous species, a hairy, glandular **annual** to 40 cm, with ovate to lanceolate or linear, toothed **leaves**, basal teeth sometimes very coarse with a long spine-like tip; basal leaves withered by flowering time. **Flowers** in lax whorls at upper nodes; calyx 8–10 mm long, golden-glandular, 2-lipped; corolla bluish purple, 15–25 mm long, middle lobe of lower lip dark-spotted. Gansu, Qinghai, Shaanxi [n & cChina, cAsia, Europe, Mongolia, Russia]; dry hills, banks, riversides, field boundaries, to 2,700 m. VI–IX.

Dracocephalum purdomii **W. W. Sm.*** Similar to *D. bullatum*, but basal **leaves** not more than 3 cm long and terminal heads of **flowers** with smaller, less prominent bracts; calyx glabrous, 11–15 mm long; corolla dark blue or bluish purple, 11–15 mm long, white-hairy. Gansu, se & cQinghai, n & nwSichuan; grassy and rocky slopes, 2,300–3,300 m. VII–VIII.

Dracocephalum forrestii **W. W. Sm.*** Tufted **perennial** to 28 cm, readily distinguished by the pinnately divided **leaves** 1.2–2.8 cm long, each with 2–3 pairs of linear segments. **Inflorescence** dense; **flowers** in pairs at upper 5–10 nodes, 2.5–2.8 cm long; calyx densely hairy; corolla dark bluish violet, rather slender, hairy. nwYunnan; rocky grassland, stony slopes, thickets, 2,300–3,500 m. VI–VII.

MEEHANIA

A genus with 7 species (5 in China) in east Asia and the eastern USA. Closely related to *Dracocephalum*, but with a 5-lobed rather than toothed calyx which lacks the sinus folds characteristic of that genus.

Meehania henryi **(Hemsl.) Sun ex C. Y. Wu*** Variable **perennial herb** with slender, ± glabrous, **stems** to 60 cm. **Leaves** stalked, ovate to heart-shaped, sometimes somewhat triangular, 4–13 × 1.8–4 cm, long-pointed, toothed, minutely hairy beneath. **Flowers** in small whorls in terminal and lateral **racemes**, accompanied by stalkless leaf-like bracts; calyx narrow-tubular, 10–13 mm long; corolla purplish to reddish purple, 23–27 mm long, middle lobe of lower lip fan-shaped, veined, notched at apex. Guizhou, Hubei, Sichuan [Hunan]; shady, damp evergreen and mixed forests, path- and streamsides, 500–700 m. VII–IX.

Meehania fargesii **(H. Lévl.) C. Y. Wu*** Similar to the previous species, but **flowers** larger, only borne terminally in axils of upper leaf-pairs, both calyx and corolla more obviously downy; calyx 15–18 mm long; corolla 28–45 mm long. Guangxi, Guizhou, Hubei, Sichuan, Yunnan [c & sChina]; similar habitats, 700–3,500 m. IV–VI.

MARMORITIS

A genus of 5 species (all in China) in the Himalaya and China. Related to *Dracocephalum* and *Meehania*, but readily distinguished by resupinate **flowers** (with the lower lip uppermost) and the hairy ring in middle of calyx-tube.

Marmoritis complanata **(Dunn) A. L. Bud.*** (syn. *Nepeta complanata* Dunn; *Glechoma complanata* (Dunn) Turrill; *Phyllophyton complanatum* (Dunn) Kudo) Softly white-hairy rhizomatous **perennial**, spreading to ascending, to 25 cm tall.

---

1. *Dracocephalum bullatum*; Genyanshan, swSichuan (MN)
2. *Dracocephalum moldavica*; Badaling, Beijing (CGW)
3. *Dracocephalum purdomii*; Buishuxia Pass, cQinghai (RS)
4. *Dracocephalum forrestii*; Bitahai, Zhongdian, nwYunnan (PC)
5. *Meehania henryi*; Emeishan, sSichuan (CGW)
6. *Meehania fargesii*; Shennongjia, wHubei (PC)
7. *Marmoritis complanata*; Balmashan, nwYunnan (HJ)
8. *Marmoritis complanata*; Balangshan, Wolong, wSichuan (RS)

1. *Eriophyton wallichii*; eNepal (CGW)
2. *Lamium barbatum*; Shennongjia, wHubei (PC)
3. *Phlomis likiangensis*; Lijiang, nwYunnan (CGW)
4. *Phlomis betonicoides*; Huanglong, nwSichuan (CGW)
5. *Phlomis melanantha*; Balangshan, Wolong, wSichuan (CGW)
6. *Lamiophlomis rotata*; north of Aba, nwSichuan (CGW)

Leaves crowded, stalkless or almost so, kidney-shaped to rounded, hairy except along veins above, margin scalloped. **Flowers** in small clusters, often 3; calyx softly white-hairy, prominently veined, 9–12 mm long; corolla reddish, 15–23 mm long, tubular in lower part. Qinghai, w, nw & swSichuan, e & seTibet, nwYunnan; rock-crevices, screes, stony alluvial fans, 4,300–5,000 m. VI–VII.

ERIOPHYTON

A Sino-Himalayan genus with a single species. **Perennial**, woolly **herb** with thick **roots** and **leaves**; **flowers** in compact, 6-flowered whorls with spine-like woolly bracteoles.

Eriophyton wallichii Benth. Tufted, densely woolly **perennial** 10–20 cm tall. **Leaves** closely overlapping, rhombic to circular, 3–4 cm long and broad, toothed near apex, densely, shaggy-woolly. **Flowers** stalkless, partially hidden at base of leaves; corolla 2.2–2.8 cm long, purplish or reddish, densely woolly outside. Qinghai, w, nw & swSichuan, Tibet, nwYunnan [Himalaya, nMyanmar]; alpine screes, stony alluvial fans, stabilised moraines, 2,700–4,700 m. VI–VIII.

LAMIUM

A genus of 40 species (4 in China) in Europe, Asia and North Africa. **Annual** or **perennial herbs**, usually with opposite, heart-shaped **leaves**. **Flowers** in whorls of up to 14, subtended by a pair of leaf-like bracts; calyx tubular-campanulate, 5- or 10-veined; corolla 2-lipped, upper lip often hood-like and bearded in part, lower lip 3-lobed with central lobe larger and obcordate; stamens 4, hairy.

Lamium barbatum Sieb. & Zucc. (syn. *L. album* L. var. *barbatum* (Sieb. & Zucc.) Franch. & Savat.) Patch-forming **perennial** to 70 cm, generally less, with erect, unbranched **stems**. **Leaves** stalked, heart-shaped to ovate, to 8.5 × 5 cm, with a long-pointed tip and toothed margin. **Flowers** in whorls of up to 14, in upper leaf-axils; calyx campanulate, 11–15 mm long; corolla white or cream, 19–21 mm long, upper lip oblong, notched at apex, softly hairy. Gansu, Guizhou, Hubei, Sichuan, Shaanxi [n & neChina, Japan, Korea, eRussia]; fields and field margins, waste places, road- and streamsides, 800–2,600 m. IV–VI.

PHLOMIS

A genus of c. 100 species (43 in China) in Europe, North Africa and Asia. **Perennial herbs** and **shrubs** with erect **stems** and simple, paired, usually stalked **leaves**; **flowers** stalkless, in dense whorls one above the other.

> 1. Flowers white or yellow . . . . *P. likiangensis*
> 1. Flowers purple or reddish purple 2
> 2. Plants with basal leaf-rosettes . . . . *P. betonicoides*
> 2. Plants without basal leaf-rosettes . . . .
>   . . . . *P. melanantha, P. forrestii*

Phlomis likiangensis C. Y. Wu* Robust **perennial** 60–150 cm tall, densely stellate-hairy above. **Leaves** heart-shaped to ovate, 7–18 × 6–15 cm, pointed to long-pointed, sparsely stellate- and simple-hairy above, densely stellate-hairy beneath, margin toothed. **Flowers** in whorls of 2–10-flowers, with leaf-like, linear-lanceolate bracts; calyx tubular, 12–13 mm long, grey stellate-hairy outside, with spine-like lobes; corolla white or yellow, 20–22 mm long, white-woolly outside except for glabrous tube, upper lip fringed, bearded inside. nwYunnan; grassy places, roadsides, forest margins, c. 3,500 m. VII.

Phlomis betonicoides Diels* Erect, tufted **perennial** to 80 cm, usually less, unbranched. **Leaves** narrow-ovate to triangular, stalked, 7.5–14 × 5–7.5 cm, densely downy beneath, margin scalloped to toothed. **Flower-whorls** many-flowered, accompanied by purple, bristle-like bracts; calyx tubular-campanulate, c. 10 mm long; **corolla** white, pink or reddish, 17–19 mm long, densely stellate-downy outside, upper lip bearded within. w, nw & swSichuan, e & seTibet, Yunnan; forest glades and margins, grassy slopes, roadsides, 2,700–3,000 m. VI–VIII.

Phlomis melanantha Diels* Robust **perennial** with erect, somewhat branched **stems** to 90 cm, almost glabrous. **Leaves** ovate to ovate-oblong, 4.5–12 × 2.5–9.5 cm, pointed to long-pointed, rough-hairy beneath, sparsely glandular above, margin toothed. **Flowers** many in several well-spaced whorls accompanied by leaf-like bracts; calyx purple, 11–12 mm long, minutely hairy; corolla purple-red or reddish, 21–22 mm long, with silky hairs outside, upper lip with minutely toothed margin, bearded inside. w & swSichuan, n & nwYunnan; grassy slopes, meadows, *Picea* and mixed forests, 3,000–3,300 m. VI–IX.

Phlomis forrestii Diels* Similar to the previous species, but **stems** stellate-hairy; **leaves** ovate to elliptic, 10–19.5 × 4–14.5 cm, stellate-hairy beneath; **flowers** purplish, calyx 7–12 mm long, stellate-hairy outside; corolla tubular, 20–23 mm long. nwYunnan; coniferous forests, 2,700–4,000 m. VII–VIII.

LAMIOPHLOMIS

A Sino-Himalayan genus with a single species very closely related to *Phlomis*. Stemless, rhizomatous, **perennial herb** with prominent leaf-rosettes.

Lamiophlomis rotata (Benth. ex Hook. f.) Kudo (syn. *Phlomis rotata* Benth. ex Hook. f.) Plant 2.5–10 cm tall forming a flat **rosette** on the ground. **Leaves** orbicular to fan- or kidney-shaped, 6–13 × 7–12 cm, shiny deep green or grey-green above, rough and corrugated. **Inflorescence** 3.5–7 cm long in centre of rosette; **corolla** purple to reddish brown, c. 12 mm long. Gansu, Qinghai, Sichuan, Tibet, Yunnan [Himalaya]; rocky places, meadows, alluvial fans, 2,700–4,900 m. VI–VIII.

LEUCAS

A genus of c. 100 species (8 in China) found in Africa, India and Australasia. **Herbs** and **subshrubs** very similar to *Phlomis*, but calyx with 10 (not 5) teeth.

Leucas ciliata Benth. **Perennial herb** to 80 cm covered in yellow or golden hairs, with slender **stems**, generally branched above. **Leaves** stalked, lanceolate, 6–9 × 1–3 cm, base wedge-shaped to rounded, margin shallowly toothed. **Flowers** in dense, rather distant, globose whorls; calyx tubular, densely hairy, teeth spreading and star-like in fruit; corolla relatively small, white or purplish, 25–28 mm, upper lip oblong and densely hairy, lower with a 2-lobed middle lobe. Guangxi, Guizhou, Sichuan, Yunnan [Himalaya, Myanmar, Vietnam]; thickets, grassy and rocky places, stream- and roadsides, 500–2,800 m. VII–X.

SALVIA

A large genus of c. 1,000 species (84 in China) in temperate and tropical regions of the Old and New Worlds. **Perennial herbs** (in wChina) with square **stems** and opposite **leaves**. **Flowers** in whorls that often form spikes or panicles; corolla 2-lipped with straight to curved tube; stamens 2, connectives prolonged, T-shaped.

1. Corolla basically white, cream or yellow (see also *S. przewalskii* var. *alba*) 2
1. Flowers blue, violet, purple or red, rarely orange 3
2. Corolla 30–40 mm long . . **S. hylocharis, S. digitaloides**
2. Corolla 23–30 mm long . . . . . **S. flava, S. campanulata**
3. Corolla small to medium, not more than 25 mm long . . . . **S. bulleyana**
3. Corolla larger, 26–50 mm long 4
4. Upper lip of corolla markedly sickle-shaped . . . . **S. yunnanensis**
4. Upper lip of corolla straight to curved, not sickle-shaped 5
5. Leaf-base deeply heart-shaped or spear-shaped . . . . **S. prattii**
5. Leaf-base rounded to wedge-shaped or slightly heart-shaped . . . . . . **S. przewalskii, S. przewalskii** var. **alba, S. aerea, S. evansiana, S. wardii, S. castanea**

Salvia hylocharis Diels* (syn. *S. forrestii* Diels) Rather robust **perennial** to 90 cm, sometimes half that height, **stems** densely hairy toward top. **Leaves** stalked, blade ovate to arrow-shaped, to 8.5 × 8.5 cm, hairy on both surfaces, margin blunt-toothed. **Inflorescence** simple or branched, the whorls 2–4-flowered; calyx pale green, hairy along veins and margin; **corolla** yellow, usually with a purplish brown lower lip, 30–38 mm long. seTibet, nwYunnan; grassy and rocky slopes, forest margins, streamsides, 2,800–4,000 m. VI–IX.

Salvia digitaloides Diels* Robust, **perennial** plant with erect **stems**, 30–60 cm tall, densely long-hairy. Basal **leaves** oblong-elliptic, 3.5–11 × 2–5 cm, blunt or rounded, scalloped. **Inflorescence** up to 13 cm long, terminal, with 4–6-flowered whorls; calyx 10–12 mm long, softly hairy, green with purple veins; **corolla** cream or yellow, with or without purplish spots, 33–40 mm long. Guizhou, Sichuan, Yunnan; *Pinus* woodland, shrubberies, grassland, streamsides, 2,300–3,400 m. V–VII.

Salvia flava Forrest ex Diels* Similar to *S. hylocharis*, but 20–50 cm tall with hairy or glabrous **stems**; **leaves** ovate or triangular-ovate, 2–7 × 3.5–5 cm, pointed or almost blunt, with scalloped margins and purplish brown glands beneath; **corolla** 23–30 mm long, yellow with a brownish purple lower lip. w & swSichuan, w & nwYunnan; coniferous forests and forest margins, hillsides, grassland, scrub, streamsides, 2,400–4,000 m. VI–IX. Often considered to be a form of *S. bulleyana* Diels (see below).

Salvia campanulata Wall. ex Benth. **Perennial** with erect, softly hairy **stems** 40–80 cm tall. Basal **leaves** heart-shaped to ovate-truncate, 4–18 × 3.5–13.5 cm, long-pointed, margin scalloped. **Inflorescence** up to 15 cm long, terminal, whorls widely spaced, 2–6-flowered, densely softly hairy; calyx c. 13 mm long, softly hairy, sparsely black-glandular; **corolla** yellow, 25–30 mm long. seTibet, Yunnan [swTibet, Himalaya]; woodland margins and shrubberies in valleys, 800–3,800 m. VI–VIII.

Salvia bulleyana Diels* **Perennial** with erect **stems** 30–60 cm, densely yellow-brown-hairy. Basal **leaves** ovate to ovate-triangular, 2.5–6 × 2–3.5 cm, almost blunt or pointed, margin scalloped, above with long soft hairs, sparsely dark brown-glandular beneath. **Inflorescence** up to 12 cm long, terminal, whorls 4-flowered, densely yellow-brown-hairy; calyx c. 10 mm long, softly yellow-brown-hairy; **corolla** purplish blue, 19–21 mm long, softly hairy outside. nwYunnan; grassy hillsides, open shrubberies, 2,100–3,400 m. VI–VIII.

Salvia yunnanensis C. H. Wright* (syn. *S. bodinieri* Van.) Herbaceous **perennial** to 30 cm, with softly white-hairy **stems**. **Leaves** mostly basal, stalked, blade simple or pinnately lobed, 2–8 × 1.5–3.5 cm, often purplish and hairy beneath, base rounded to heart-shaped, margin scalloped. **Flowers** usually in whorls of 4–6 well spaced in the hairy **inflorescences**; calyx campanulate, 7–9 mm long, purplish, glandular and hairy; corolla blue-purple, 25–30 mm long, downy, tube narrowly funnel-shaped. Guizhou, Sichuan, Yunnan; open forests and forest margins, grassy slopes, often in rather dry habitats, 1,800–2,900 m. IV–VIII.

Salvia prattii Hemsl.* Tufted, somewhat hairy **perennial** to 45 cm with mostly basal, long-stalked **leaves**, blades heart- to arrow-shaped, 3.5–9.5 × 2–5.3 cm, blunt, finely hairy above, more densely so beneath and with purple glands, margin scalloped. **Inflorescence** a number of spaced whorls, each

1. *Leucas ciliata*; Cangshan, nwYunnan (RS)
2. *Salvia hylocharis*; Cangshan, nwYunnan (CGW)
3. *Salvia digitaloides*; Gangheba, Yulongxueshan, nwYunnan (CGW)
4. *Salvia flava*; Wutoudi, Yulongxueshan, nwYunnan (CGW)
5. *Salvia bulleyana*; Tianbao, Zhongdian, nwYunnan (PC)
6. *Salvia yunnanensis*; south of Dukou, sSichuan (PC)
7. *Salvia prattii*; near Longriba, nwSichuan (CGW)

with 2–6 flowers, hairy and glandular; calyx campanulate, 16–19 mm long, purplish; **corolla** red or violet-blue, 40–50 mm long, downy outside. Qinghai, n, w & nwSichuan; grassy slopes, open shrubberies, 3,700–4,800 m. VII–IX.

Salvia przewalskii Maxim.* Robust **perennial** with erect **stems**, 30–60 cm, branched at base. Basal **leaves** oblong-lanceolate to triangular-spear-shaped, 5–11 × 3–7 cm, pointed, sparsely hairy on upper side, margin minutely scalloped. **Inflorescence** terminal, with 2–4-flowered whorls; calyx 12–14 mm long, glandular; **corolla** purple or red-purple (creamy white in **var. *alba* X. L. Huang & H. W. Li***), generally with purple spots, 35–40 mm long. Guizhou, Hubei, Sichuan, seTibet, Yunnan; *Pinus* woods, grassland, streamsides, 2,300–3,400 m. VI–VIII.

Salvia aerea H. Lévl.* (syn. *S. lichiangensis* W. W. Sm.; *S. pinetorum* Hand.-Mazz.) Tufted **perennial** to 40 cm, with yellow-brown-hairy **stems**. **Leaves** mostly basal, stalked, elliptic to elliptic-lanceolate, occasionally oblanceolate, 2.5–8.5 × 2.5–4.5 cm, brown-hairy beneath, hairy and glandular above, base attenuate to slightly heart-shaped, margin scalloped. **Flowers** in whorls of usually 4–6, well spaced in inflorescence; calyx campanulate, 9–10 mm long, brown-hairy; corolla white, blue or purple, rarely orange, 26–35 mm long, with a slightly curved tube. Guizhou, w, s & swSichuan, Yunnan; forest margins and clearings, shrubberies, grassy slopes, 2,500–3,300 m. VI–VII.

Salvia evansiana Hand.-Mazz.* A generally smaller plant than the previous species, 13–45 cm tall, with densely brown-hairy **stems**; **leaves** ovate or triangular-ovate, 2–11 cm long, pointed or rounded, brown-hairy; calyx deep reddish purple, brown-hairy; **corolla** deep violet, 26–35 mm long, hairy. w & swSichuan, nwYunnan; alpine meadows, under *Abies* and *Rhododendron* forests, 3,400–4,300 m. VI–VIII.

Salvia wardii E. Peter* Rather robust, tufted **perennial** to 75 cm, densely glandular-downy, especially in upper part. Basal and lower **leaves** numerous, ovate to ± hastate, 7–16 × 3.5–8 cm, covered in sparse hairs and tiny red glands, margin finely scalloped, lowermost leaves with a long stalk. **Inflorescence** dense, racemose or paniculate; calyx broad-campanulate, to 15 mm long; **corolla** blue, 35–40 mm long, lower lip noticeably longer than upper, generally with some white or speckling. seTibet; rocky slopes and meadows, open shrubberies, 3,600–4,500 m. VI–VII.

Salvia castanea Diels Very distinctive on account of its blackish brown, purple-brown or chestnut-brown **flowers**; corolla 30–32 mm long, curved upwards beyond calyx, upper lip ascending, ± straight. wSichuan, seTibet, nwYunnan [swTibet, Nepal]; hillsides, grassy slopes, open forests and forest margins, 2,500–3,800 m. V–IX.

NEPETA

A genus of c. 250 species (42 in China) distributed in Europe, Asia and North Africa. Mostly aromatic **subshrubs** and **perennial herbs** with paired, toothed, sometimes entire, **leaves**; **corolla** 2-lipped with a narrow tube that broadens into the throat; style protruding, 2-cleft.

Nepeta stewartiana Diels* (syn. *Dracocephalum stewartianum* (Diels) Dunn) Rather slender, erect **perennial** to 1.5 m with branched **stems** that become hairless. **Leaves** stalked, oblong to lanceolate, 6–10 × 2–2.5 cm, olive-green above, whitish-downy beneath with yellowish glands, margin finely toothed. **Inflorescence** with numerous few-flowered whorls, minutely hairy; calyx oblique-tubular, 10–14 mm long, expanding in fruit; **corolla** purple or blue with whitish, spotted lower lip, 20–25 mm long, curved in the middle, central lobe of lower lip bearded. w & swSichuan, seTibet, Yunnan; open forests, shrubberies, grassy slopes and banks, 2,700–3,300 m. VIII–X.

MICROMERIA

A genus of c. 100 species (5 in China) in Africa, Europe and Asia. Aromatic **herbs** and **subshrubs**, with short-stalked **leaves** that are conspicuously impressed-glandular above.

Micromeria euosma (W. W. Sm.) C. Y. Wu* Very aromatic **subshrub** with numerous prostrate to ascending **stems** to 30 cm, often purplish red, glandular. **Leaves** short-stalked, ovate to rounded, to 2 × 1.3 cm, upper leaves smaller and bract-like. **Flowers** in lax whorls of up to 10, forming a narrow panicle; calyx tubular, c. 5 mm long, golden-glandular, hairy in mouth; **corolla** pink to red, 10–11 mm long, upper lip slightly spreading, lower with 3 rounded lobes. w & nwYunnan; grassy slopes, bushy places on limestone, c. 3,300 m. VI–VIII.

THYMUS

A genus of over 200 species (c. 11 in China) distributed in Europe, temperate Asia and Africa. Aromatic **shrubs** or **subshrubs** with pairs of small, often entire **leaves** and **flowers** in dense, often interrupted, spikes or heads.

1. *Salvia przewalskii*; Baishui, Yulongxueshan, nwYunnan (PC)
2. *Salvia przewalskii*; Gangheba, Yulongxueshan, nwYunnan (CGW)
3. *Salvia aerea*; Heishui, Yulongxueshan, nwYunnan (PC)
4. *Salvia aerea*; Sandauwan, Yulongxueshan, nwYunnan (CGW)
5. *Salvia evansiana*; Wutoudi, Yulongxueshan, nwYunnan (CGW)
6. *Salvia evansiana*; Habashan, nwYunnan (CGW)
7. *Salvia wardii*; Dungda La, seTibet (HJ)
8. *Salvia castanea*; Galung La, near Bomi, seTibet (HJ)
9. *Nepeta stewartiana*; Chongjiang, nwYunnan (PC)
10. *Micromeria euosma*; Lijiang, nwYunnan (PC)

***Thymus mongolicus* (Ronn.) Ronn.** (syn. *T. serpyllum* L. var. *mongolicus* Ronn.) Intricately branched hummock- to mat-forming **subshrub** with hairy **stems**. **Leaves** in pairs, ovate, entire or slightly toothed, 4–10 × 2–4.5 mm, glandular, glabrous; flowering stems ascending with 1–2 pairs of leaves. **Inflorescence** capitate; calyx tubular-campanulate, 4–5 mm long, hairy towards base; **corolla** purple to reddish, 6–8 mm long, slightly hairy. Gansu, Qinghai, Shaanxi [n & nwChina, Mongolia]; grassy and stony slopes and banks, 1,100–3,600 m. VII–VIII.

SCUTELLARIA

A cosmopolitan genus with c. 350 species (almost 100 in China). **Herbs** and **subshrubs** with entire or pinnate **leaves**. **Inflorescence** a raceme or spike, often 1-sided; calyx with a characteristic, erect shield-like appendage (scutellum) on upper lip.

1. Flowers white or creamy white . . . . **S. likiangensis**
1. Flowers blue or purplish blue 2
2. Margin of stem-leaves toothed, sometimes obscurely so
   . . . . **S. amoena, S. forrestii**
2. Margin of stem-leaves entire or almost so . . . .
   . . . . **S. hypericifolia**

Scutellaria likiangensis* Diels Perennial herb 20–36 cm tall with elliptic-ovate to oblong **leaves**, 1.3–3 × 0.6–1.5 cm, rounded or notched at apex, densely glandular beneath, margin toothed. **Inflorescence** 6.5–12 cm long, glandular-hairy; calyx purplish; **corolla** creamy white, spotted or lined with purple, 2.6–3 cm long. nwYunnan; grassy places and thickets on dry slopes, 2,500–3,000 m. VI–VII.

Scutellaria amoena* C. H. Wight Perennial herb 12–25 cm tall with numerous **stems**, almost glabrous. Lower **leaves** oblong-ovate to oblong, 1.4–3.3 × 0.7–1.4 cm, blunt or rounded at apex, margin obscurely toothed, sparsely hairy beneath. **Inflorescence** 5–14 cm long, hairy; **corolla** purple or bluish purple, 24–30 mm long. Guizhou, Sichuan, Yunnan; *Pinus* forests, grassy slopes, 1,300–3,000 m. VI–VIII.

Scutellaria forrestii* Diels Similar to the previous species, but with white-hairy **stems**, lower **leaves** almost circular, rounded at apex, 1 cm long and broad, violet beneath, margin toothed; upper leaves oblong to broadly ovate, 1.6–2.5 cm long, toothed; **inflorescence** not more than 8 cm long, hairy; **corolla** dark blue, 30–35 mm long. w & swSichuan, nwYunnan; *Quercus*, *Pinus* and *Larix* forests, grassy slopes and banks, 2,100–3,400 m. VI–VII.

Scutellaria hypericifolia* H. Lévl. Tufted **perennial** to 30 cm, **stems** hairy only on the angles and nodes. **Leaves** with a very short stalk, ovate to oblong, 2–3.5 × 0.7–1.5 cm, green above, usually purplish beneath. **Flowers** in terminal racemes accompanied by small ovate green bracts; calyx often purplish, 3 mm, expanding in fruit, hairy; corolla purple or purple-blue, white or greenish white, 25–28 mm long, slightly hairy outside. n, nw & wSichuan; grassy habitats, banks, forest margins, 900–4,000 m. VI–VIII.

PRUNELLA

A genus with c. 14 species (4 in China), scattered across the northern hemisphere. **Perennial herbs** with entire to pinnatisect, paired **leaves**, and erect, dense, globose or egg-shaped **spikes** of flowers.

***Prunella vulgaris* L.** Tufted, hairy, **perennial** 20–30 cm tall with lanceolate to ovate **leaves**, 1.5–6 cm long, blunt or rounded, toothed. **Inflorescence** stalkless, egg-shaped to globose, 2–4 cm long; **corolla** purplish or white, 12–14 mm long. Gansu, Guangxi, Guizhou, Hubei, Sichuan, Shaanxi, Tibet, Yunnan [c & eChina, Asia, Europe], a cosmopolitan weed of cultivation, waysides, grassy places, to 3,200 m. IV–VI.

***Prunella hispida* Benth.** Differs from the previous species in having stiff hairs all over and deep purple or blue-purple **flowers** with a conspicuous band of stiff hairs on the back of upper lip of the 15–18 mm long corolla. s & swSichuan, Tibet, Yunnan [India]; forest margins, grassy places, 1,500–3,800 m. VI–I.

STACHYS

A genus of c. 300 species (18 in China), distributed in Asia, Europe and the Americas. Diffuse **annuals** or erect, **perennial herbs**, rarely **subshrubs**, with entire **leaves**.

Stachys kouyangensis* (Van.) Dunn Slender **perennial herb** with angular **stems** to 50 cm, with stiff hairs. **Leaves** triangular-heart-shaped to spear-shaped, to 3 × 2.5 cm, blunt, margin scalloped. **Inflorescence** lax, with **flowers** in widely spaced whorls of 5–6; **corolla** reddish to purple-red, c. 15 mm long. Guizhou, Hubei, Sichuan, eTibet, Yunnan; thickets, mixed forests, grassy slopes, damp streamsides, a weed in open fields and on field margins, 900–3,800 m. IV–IX.

SCROPHULARIACEAE

PAULOWNIA

A genus of 7 species (6 in China) in China and South East Asia. **Deciduous trees** with smooth **bark** and opposite, occasionally whorled, long-stalked **leaves**. **Flowers** in large panicles, corolla funnel- to trumpet-shaped, 2-lipped, ridged in mouth; stamens 4.

1. *Thymus mongolicus*; cQinghai (RS)
2. *Scutellaria amoena*; between Longriba and Aba, nwSichuan (CGW)
3. *Scutellaria forrestii*; Gangheba, Yulongxueshan, nwYunnan (CGW)
4. *Scutellaria hypericifolia*; north of Songpan, nwSichuan (CGW)
5. *Scutellaria hypericifolia*; Songpan, nwSichuan (PC)
6. *Prunella vulgaris*; Jiuzhaigou, nwSichuan (CGW)

1. *Paulownia fortunei*; Nanpanjiang, swGuizhou (PC)
2. *Paulownia tomentosa*; cult. (CGW)
3. *Rehmannia elata*; cult. Kunming, cYunnan (CGW)
4. *Rehmannia glutinosa*; Luoyang, Henan (PC)
5. *Mimulus bodinieri*; east of Aba, nwSichuan (CGW)
6. *Mimulus nepalensis*; nNepal (CGW)

1. Inflorescence stalkless or almost so (stalk, if present, much shorter than the flower-stalks) ...... *P. fargesii*
1. Inflorescence with a stalk almost as long as to longer than the flower-stalks 2
2. Corolla weakly ridged in throat; inflorescence narrow-cylindrical .... *P. fortunei*
2. Corolla boldly ridged in throat; inflorescence a pyramidal panicle .... *P. tomentosa, P. elongata*

Paulownia fargesii Franch. **Tree** to 20 m with a conical crown at maturity and purplish to greyish brown twigs covered in stellate down. **Leaves** ovate to heart-shaped, to 25 cm long. **Inflorescence** spreading, the cymes 3–5-flowered; calyx 18–20 mm long, stellate-downy, lobes half the length of tube; **corolla** purple or white with purple stripes, 55–75 mm long, glandular-hairy outside. **Capsules** egg-shaped, 30–40 mm long, sticky and glandular at first, 2-parted. Guizhou, Hubei, Sichuan, Yunnan [Hunan, Vietnam]; forests, forest margins and hillslopes, also cultivated, 1,200–3,000 m. IV–V.

Paulownia fortunei (Seem.) Hemsl. **Deciduous** or **semi-deciduous tree** to 30 m, with conical crown at maturity, young shoots, inflorescence and capsules covered with yellow-brown stellate hairs. **Leaves** narrowly heart-shaped to ovate, to 20 cm, 2-parted on new shoots. **Flowers** on glabrous stalks; calyx 20–25 mm long, lobes shorter than tube; corolla pale to deep purple or white, 80–120 mm long. **Capsules** oblong in outline, 6–10 cm long. Guangxi, Guizhou, Hubei, Sichuan, Yunnan [e & sChina, Laos, Vietnam]; forested and wooded slopes, roadsides, borders of cultivation, also planted, to 2,000 m. III–IV.

Paulownia tomentosa (Thunb.) Steud.* Open, laxly branched **deciduous tree** to 20 m, umbrella-shaped at maturity, with large heart-shaped, downy **leaves**, blade to 40 cm across. **Inflorescences** appearing in autumn and developing through the winter on the otherwise bare branches; calyx campanulate, 14–16 mm long, lobes equalling tube or longer; **corolla** lavender-purple, 50–75 mm long, glandular outside, glabrous within. **Capsules** egg-shaped, 30–45 mm long, sticky with dense glandular hairs. Gansu, Hubei, n & eSichuan, Shaanxi [c, e & neChina]; open woodland, hillslopes, also cultivated, to 1,800 m. IV–V.

Paulownia elongata S. Y. Yu.* Similar to the previous species, but calyx with tube longer than lobes, **corolla** purple to pinkish white, 70–95 mm long; **capsules** stellate-downy, becoming partly hairless. Hubei, Shaanxi [c & eChina]; open forests, hillslopes, also planted, at low altitudes. IV–V.

REHMANNIA

A genus with 6 species endemic to China. **Perennial herbs** with **basal leaves** in rosettes and alternate **stem-leaves**; calyx with a tube below; **corolla** with equal upper and lower lips.

Rehmannia elata N. E. Br. ex Prain* Hairy **perennial** to 50 cm. **Leaves** elliptic to obovate, rather thin, to 10 cm long, margin with 2–6 pairs of triangular lobes, ± glabrous above but hairy beneath; leaf-stalks winged. **Flowers** in lax, branched racemes, borne on stalks 30–40 mm long; calyx pale greenish with purplish veins, 20–30 m long; corolla reddish purple with yellowish purple-spotted throat, 58–64 mm long, hairy within, tube somewhat inflated. Hubei; rock-ledges and crevices, rocky slopes, 1,100 m. V–VII.

Rehmannia glutinosa (Gaertn.) Liboschitz ex Fisch. & Mey.* (syn. *Digitalis glutinosa* Gaertn.) Smaller plant than the previous species, not more than 30 cm, with rosettes of persistent, basal **leaves** covered in glandular hairs; **flowers** on slender ascending stalks; calyx 10–15 mm long; corolla yellowish purple with a purplish red exterior, 30–45 mm long, with narrow tube. Gansu, Hubei, Shaanxi [n & neChina]; rocky slopes, walls, rocky outcrops, crevices, to 1,100 m. IV–VII.

Rehmannia henryi N. E. Br.* Similar to *R. elata*, but leaf-blade rarely lobed and inflorescence-stalks with 1–2 secondary bracteoles near base; **corolla** yellowish, 50–70 mm long, tube white-downy outside. Hubei; rock crevices and ledges, rocky tracksides, at low altitudes. IV–V.

MIMULUS

A worldwide genus with c. 150 species (5 in China), mostly in the temperate zone. **Annual** and **perennial herbs** with rounded or square **stems** and paired **leaves**. **Flowers** in leaf-axils; **corolla** with 2 prominent bulges.

1. Plant prostrate to spreading; leaf-stalks mostly longer than leaves .... *M. bodinieri, M. nepalensis*
1. Plant erect to ascending, at least 10 cm tall; leaf-stalks shorter than or equalling leaves .... *M. tenellus, M. szechuanensis*

Mimulus bodinieri Vaniot* Glabrous **patch-forming** plant not more than 10 cm tall in flower, generally less, rooting at nodes. **Leaves** ovate, to 30 × 15 mm, sparsely toothed, uppermost stalkless and clasping **stem**. **Flowers** solitary at upper leaf-axils, on long slender stalks; calyx campanulate, 5–10 mm long; **corolla** yellow with purple speckling in throat, 10–15 mm long, lobes ciliate. nw, w & swSichuan, Yunnan; damp habitats, seepage zones, lake- and streamsides, 1,900–2,400 m. IV–VII.

Mimulus nepalensis Benth. (syn. *M. tenellus* Bunge var. *nepalensis* (Benth.) P. C. Tsoong ex H. P. Yang) Very similar to the previous species, but **stems** more diffuse; **leaves** finely toothed, all with a short stalk; and calyx cylindrical with a truncated apex. w & swSichuan, seTibet, Yunnan [swTibet, cChina, Himalaya, Japan, Taiwan, Vietnam]; similar habitats, 500–3,000 m. IV–IX.

Mimulus tenellus **Bunge** Creeping, glabrous, very variable **perennial** with erect to ascending slender **stems**, much-branched near base, often narrowly winged on corners. **Leaves** stalked, blade ovate to lanceolate or almost oblong, to 30 × 15 mm, margin toothed. **Flowers** solitary in upper leaf-axils; calyx 5–10 mm long, glabrous, often narrowly winged, expanding in fruit; corolla yellow with red spots in throat, 10–28 mm long. Gansu, Guizhou, Hubei, Sichuan, Shaanxi, Tibet, Yunnan [c & eChina, c & eHimalaya, Japan]; damp habitats in forests and by streams, ditches, lakes and tracks, 500–3,000 m. VI–IX.

Mimulus szechuanensis **Pai*** Similar to the previous species, but **leaves** with a wedge-shaped (not truncated) base and calyx larger, 10–15 mm long, hairy; **corolla-tube** slightly longer than calyx. Gansu, Hubei, Sichuan, Shaanxi, Yunnan [Hunan]; similar habitats, 1,300–2,800 m. VI–VIII.

VERBASCUM

A genus with c. 300 species (6 in China), distributed in Europe and Asia. **Annual**, **biennial** and **perennial herbs** often with large basal **leaves**. **Flowers** in spikes or panicles; corolla saucer-shaped with short tube, 5-lobed.

Verbascum thapsus **L. Biennial** to 1.5 m tall, covered in greyish or yellowish stellate down. **Leaves** oblong to oblanceolate, to 20 × 7 cm, decreasing in size up stem, only lowermost stalked. **Flowers** in dense spike-like, cylindrical **inflorescences**; corolla yellow, 10–20 mm across; stamens 5, filaments of upper 2 glabrous, others hairy. Sichuan, Tibet, Yunnan [c & eChina, much of temperate Europe and Asia], widely naturalised, grassy and rocky, often rather bare places, field boundaries, fallow land, often gregarious, 1,400–3,200 m. VI–VIII.

VERONICASTRUM

A genus of c. 20 species (13 in China) found in east Asia and North America. **Perennial**, mostly rhizomatous **herbs** with erect or arched **stems** and alternate or whorled **leaves**. **Inflorescence** a terminal or lateral spike; calyx 5-lobed, upper lobe smaller; **corolla** 4-lobed, with a tube and 2-lipped limb; stamens 4, not or slightly protruding.

Veronicastrum brunonianum **(Benth.) D. Y. Hong** Rhizomatous **perennial** with slender erect **stems** to 1.5 m, glabrous below, glandular-hairy above. **Leaves** alternate, stalkless, narrow-elliptic, 10–20 × 3–5 cm, apex long-pointed to almost tailed, margin toothed. **Inflorescence** erect and tail-like, terminal and lateral, glandular-hairy; **corolla** white to yellowish or orange-yellow, 6–8 mm long, 2-lipped, arched, tube hairy inside; stamens protruding. Guizhou, wHubei, Sichuan, seTibet, Yunnan [Himalaya]; shady habitats in forests and on grassy slopes, ravines, 1,500–3,000 m. VII–IX.

LINARIA

A genus with c. 100 species (8 in China) primarily in Europe and Asia. **Annual** and **perennial herbs** with alternate, opposite or whorled, entire **leaves**. **Flowers** in spikes or racemes; corolla 2-lipped with closed and pouched mouth, spurred at base.

Linaria yunnanensis **W. W. Sm.* Perennial** to 1 m, usually with several erect, glabrous **stems** bearing rather glaucous, alternate, elliptic **leaves**, 2–5 × 1.5–3 cm. **Inflorescence** a rather dense spike, glandular-hairy except for the corolla; calyx 5–8 mm long, with ovate-lanceolate lobes; **corolla** yellow, paler on upper lip and spur, 11–12 mm long, spur 2–4 mm long, slightly curved. nwYunnan; woodland margins and open stony slopes, c. 3,000 m. VII–IX.

SCROPHULARIA

A north temperate genus of c. 200 species (36 in China). Primarily **perennial herbs**, with rounded or square **stems** and mostly opposite **leaves**. **Inflorescence** racemes or spike-like; **corolla** generally 2-lipped, with rounded to elongated tube, lower lip 3-lobed, upper larger and 2-lobed; stamens 4.

Scrophularia delavayi **Franch.*** Tufted, rhizomatous **perennial** to 45 cm with ovate to almost diamond-shaped, toothed **leaves**, 2.5–7 cm long, leaf-stalks winged. **Flowers** in terminal clusters; calyx 5–7 mm long, slightly 2-lipped; corolla bright yellow, 9–15 mm long, glabrous outside but hairy within, upper lip with overlapping rounded lobes. swSichuan, n & nwYunnan; damp grassy slopes, wet screes, shrubberies, rocky places, forest margins, 3,100–3,800 m. V–VII.

Scrophularia chasmophila **W. W. Sm.*** Smaller plant than the previous species, not more than 10 cm, with rounded to diamond-shaped **leaves** to 2.7 cm long. **Flowers** few per cluster, scarcely stalked; corolla greenish yellow, glabrous outside and inside. swSichuan, seTibet, nwYunnan; mountain moorland, stabilised screes, stony meadows, 3,500–4,600 m. VI–VII.

LANCEA

A genus of 2 species (both in China) from the Himalaya to Mongolia and China. Scarcely hairy, patch-forming **perennial herbs** with clusters of **leaves**, and **flowers** in small clusters or racemes; corolla 2-lipped, lower lip flat, 3-lobed, ridged at base, upper lip 2-lobed. **Fruit** a small fleshy 'berry'.

Lancea tibetica **Hook. f. & Thomson** Low rhizomatous, **patch-forming** plant not usually more than 5 cm tall. **Leaves** in small rosettes of 6–10, oblong to obovate, rather leathery, with entire or remotely toothed margin. **Flowers** 3–5 in a cluster or short raceme; calyx 9–10 mm long, with triangular, almost equal lobes; corolla blue to purple, often paler and spotted in mouth, 15–25 mm long (tube 8–13 mm). **Fruit** globose, 9–10

1. *Mimulus tenellus*; Wolong, wSichuan (PC)
2. *Verbascum thapsus*; UK (TC)
3. *Verbascum thapsus*; UK (TC)
4. *Linaria yunnanensis*; Sandauwan, Yulongxueshan, nwYunnan (CGW)
5. *Scrophularia delavayi*; Daxueshan, nwYunnan (CGW)
6. *Scrophularia chasmophila*; Wutoudi, Yulongxueshan, nwYunnan (CGW)
7. *Lancea tibetica*; north of Songpan, nwSichuan (CGW)
8. *Lancea tibetica*; south of Aba, nwSichuan (CGW)

SCROPHULARIACEAE: Lancea

mm, deep purple to reddish when ripe. Gansu, Qinghai, Sichuan, Tibet, Yunnan [Himalaya, Myanmar]; open forests and forest margins, shrubberies, grassy habitats, streamsides, 2,000–4,500 m. V–VII.

Lancea hirsuta Bonati* Differs from the previous species in the hairy **stems** and **leaves**, the longer **racemes** with up to 10 flowers and in hairy filaments; **corolla** bluish purple, tube 20–24 mm long. w & swSichuan, nwYunnan; habitats similar to those of *L. tibetica*, 3,700–4,100 m. VI–VII.

---

1. *Hemiphragma heterophyllum*; Baimashan, nwYunnan (CGW)
2. *Lagotis alutacea*; Daxueshan, nwYunnan (CGW)
3. *Lagotis praecox*; Baimashan, nwYunnan (HJ)
4. *Oreosolen wattii*; seTibet (HJ)
5. *Melampyrum klebelsbergianum*; Cangshan, nwYunnan (PC)
6. *Euphrasia jaeschkei*; Munigou, Kangding, wSichuan (PC)

HEMIPHRAGMA

A genus with 1 species restricted to south and South East Asia.

Hemiphragma heterophyllum Wall. Creeping, patching-forming **perennial** rooting at nodes. **Leaves** of 2 types: those on main stems short-stalked, rounded to kidney- or heart-shaped, 5–20 mm long, toothed, those on lateral branches bunched and needle-like, 3–5 mm long. **Flowers** solitary, short-stalked or stalkless; corolla pink to white, with 5 rounded to oblong lobes and a short tube. **Fruit** a small, shiny, fleshy 'berry', 5–8 mm in diameter, red when ripe. Gansu, Guangxi, Guizhou, Hubei, Shaanxi, Tibet, Yunnan [Himalaya, seAsia, Myanmar]; rocky and grassy alpine slopes, low open shrubberies, rock-ledges and crevices, 2,600–4,100 m. IV–VII.

LAGOTIS

A genus of 30 species (17 in China) in east Europe, north and Central Asia to the Caucasus, Himalaya and west China. Rather fleshy, **perennial** rhizomatous **herbs**; **leaves** mostly basal. **Flowers** in dense heads or spikes; calyx spathe-like, splitting to base on at least 1 side; corolla 2-lipped, lower lip 2–4-lobed, upper with a single lobe; stamens 2. **Fruit** fleshy and berry-like.

Lagotis alutacea W. W. Sm.* Tufted plant to 15 cm tall with 1–4, occasionally more, prostrate **stems**. **Leaves** rather thick, grey-green, often purplish beneath, basal leaves rounded to ovate or almost oblong, 2–6 cm long, with entire or slightly toothed margin and flattened stalks; stem-leaves few, smaller. **Flowers** in dense oblong or rounded spikes; corolla blue to pale purple or whitish, 9–15 mm long. w & swSichuan, nwYunnan; grassy, stony and sandy alpine slopes, 3,400–5,000 m. V–IX.

Lagotis yunnanensis W. W. Sm.* In contrast to the previous species, a spreading plant to 35 cm tall in flower, with 1–2 ascending to erect **stems** from each rosette; **flowers** in an elongated **spike** to 10 cm long; corolla white or sometimes purplish, 8–12 mm long. w & swSichuan, seTibet, nwYunnan; alpine meadows in grassy and stony places, 3,300–4,700 m. VI–VIII.

Lagotis praecox W. W. Sm.* Like *L. alutacea*, but **leaves** more leathery, with a purple-red stalk and undersurface, margin coarsely toothed; **corolla** blue, sometimes very pale. w & sw Sichuan, nwYunnan; habitats similar to those of *L. alutacea*, 4,500–5,200 m. VII–VIII.

OREOSOLEN

A single species (some botanists distinguish 3 or 4) native to the Himalaya and west China. **Perennial** rosette-forming **herbs**. **Leaves** opposite, pressed to the ground, 5–9-veined. **Flowers** several, lateral; corolla yellow, tubular with a short 2-lipped limb; stamens 4, sometimes protruding.

Oreosolen wattii Hook. f. Small, tufted, glandular-hairy plant not more than 7 cm tall, with rosettes of leaves. **Leaves** ovate to heart-shaped, thick and leathery, to 8 cm long, with toothed margin. **Flowers** clustered into dense heads immediately on top of leaf-rosette; calyx with 5 linear-lanceolate lobes, split to the base; corolla yellow, 15–25 mm long, with narrow tube, 2-lipped, lower lip 3-lobed, upper lip 2-lobed. s & swQinghai, Tibet [Himalaya]; grassy and stony alpine meadows, 3,000–5,100 m. VI–VII.

MELAMPYRUM

A northern hemisphere genus with c. 20 species (3 in China). **Hemiparasitic annual herbs** with opposite, entire **leaves** and slender, generally squarish **stems**. **Flowers** in branched **racemes** with leaf-like bracts; calyx 4-lobed, campanulate; corolla 2-lipped.

Melampyrum klebelsbergianum Soó* Erect **annual** to 60 cm, often less, with 2 lines of hairs along **stems**. **Leaves** short-stalked, narrow-lanceolate to ovate-lanceolate, to 5 × 1.5 cm; bracts generally entire, sometimes with 1–2 small teeth near base. **Inflorescence** a gappy raceme; calyx 4–5 mm long with lanceolate lobes; **corolla** purplish red to red, often whitish on the sides, 12–16 mm long. Guizhou, s & swSichuan, nwYunnan; open forests, forest margins, tracksides, grassy slopes, 1,200–3,400 m. VI–VIII.

Melampyrum roseum Maxim. Often taller than the previous species, to 80 cm, with broader, lanceolate to ovate bracts with sharply pointed awn-like marginal teeth, these sometimes restricted to the bract-base; calyx with narrow-triangular lobes; **corolla** larger, 15–20 mm long. Gansu, Guizhou, Hubei, Shaanxi [c & eHimalaya, Korea, Japan, eRussia]; open shrubberies, grassy places, not above 1,500 m. VII–X.

EUPHRASIA

A cosmopolitan genus with c. 200 species (11 in China). Small **hemiparasitic annuals** or **perennials** with paired **leaves** and leaf-like bracts that are often deeply toothed or lobed. **Flowers** in spikes, often dense; calyx small, 4-lobed; corolla 2-lipped.

Euphrasia jaeschkei Wettst. Small **annual** to 20 cm, with simple or branched **stems**. **Leaves** ovate to rounded, to 12 × 10 mm, glandular-hairy, margin few-toothed; bracts leaf-like, larger than leaves and more coarsely toothed. **Flowers** in terminal spikes; calyx c. 7 mm long; corolla white to pale purple or pink-flushed, often yellow in throat, 9–12 mm long. w & seSichuan, eTibet [swTibet, Himalaya]; alpine meadows, rocky slopes, open shrubberies, 3,200–3,400 m. V–VII.

PEDICULARIS

An extremely complex genus with c. 600 species (352 in China, particularly in the west and south-west, 271 endemic) scattered around the northern hemisphere. **Hemiparasitic perennials** or occasionally **annual herbs** with alternate, opposite or whorled, toothed to pinnately lobed **leaves**; bracts leaf-like. **Flowers** solitary, axillary or borne in terminal racemose or spike-like **inflorescence**; calyx tubular to campanulate, often 2-lipped; corolla 2-lipped, upper lip (galea) forming a hood or helmet that is often beaked, lower lip 3-lobed; stamens 4. **Fruit** a 2-parted, many-seeded capsule.

Synopsis of groups

Group One. Flowers basically yellow. (p. 452)

Group Two. Flowers basically pink, purple, lilac, rose-red or red, corolla-tube 30 mm long or more. (p. 455)

Group Three. Flower colour as Group Two, but corolla-tube 6–29 mm long. (p. 455)

SCROPHULARIACEAE: Pedicularis

> **Group One**
> 1. Leaves in whorls of 3–6, the stalk bases expanded and fused to form a cup ... **P. rex, P. rex** subsp. **lipskyana, P. rex** subsp. **zayuensis, P. semitorta**
> 1. Leaves alternate or opposite or all basal 2
> 2. Corolla-tube not more than 12 mm long; beak of corolla rounded to S-shaped ... **P. torta, P. dunniana, P. oederi**
> 2. Corolla-tube at least 30 mm long; beak of corolla circular or semi-circular ....
> .... **P. tricolor, P. cranolopha, P. longiflora, P. longiflora** var. **tubiformis, P. croizatiana**

Pedicularis rex C. B. Clarke Stout plant to 90 cm with erect **stems** bearing leaves and leaf-like bracts usually in whorls of 4 or 6. **Leaves** narrow-oblong to lanceolate, to 14 × 4 cm, with 10–14 pairs of oblong, toothed segments. **Flowers** in branched whorls at upper nodes; calyx 10–12 mm long, glabrous; corolla yellow, occasionally reddish (**subsp. lipskyana** (Bonati) P. C. Tsoong*) or white with purple beak (**subsp. zayuensis** H. P. Yang*), erect, 28–31 mm long, tube 22–25 mm long, with an arched, beakless hood. Guizhou, wHubei, w & swSichuan, seTibet, Yunnan (except s & se) [Himalaya]; coniferous forests and forest margins, grassy slopes, meadows, 2,500–4,300 m. V–VIII.

Pedicularis semitorta Maxim.* Smaller plant than the previous species, rarely more than 50 cm, **stem** often branched towards top; **leaves** in whorls of 3–5, with no more than 10 pairs of segments. **Inflorescence** spike-like; **corolla** bright yellow, 14–18 mm long, with a distinct twisted beak that often forms a complete circle. c & swGansu, e & seQinghai, n & nwSichuan; grassy places, meadows, 2,500–3,900 m. VI–VII.

Pedicularis torta Maxim.* Tufted plant to 50 cm with erect, unbranched **stems**. **Basal leaves** withered by flowering time; **stem-leaves** alternate, occasionally ± opposite, blade oblong to linear-lanceolate, with 9–16 pairs of lanceolate, toothed segments. **Flowers** in long **racemes**; calyx 6–7 mm long; corolla yellow, tube erect, as long as calyx, downy, upper lip twisted into a slender S-shaped purple or reddish purple upturned beak. s & swGansu, wHubei, n, nw & neSichuan, Shaanxi; alpine meadows and other grassy habitats, ditch- and roadsides, 2,500–4,000 m. VI–VIII.

Pedicularis dunniana Bonati* A very distinctive brown-hairy plant to 1.8 m with stout, hollow **stems** with alternate linear-lanceolate clasping **leaves**, these with 7–15 pairs of lanceolate or oblong, toothed or shallowly lobed segments. **Flowers** in a bracteate, glandular-hairy **raceme**; calyx 6–7 mm long; corolla plain yellow, tube 11–13 mm long, hairy, upper lip boat-shaped with a thick marginal beard. w & nwSichuan, seTibet, nwYunnan [swTibet]; grassy habitats, forests and forest margins, 3,300–3,800 m. VII–IX.

Pedicularis oederi Vahl In contrast to the previous 2 species, this rarely grows more than 12 cm tall; **stems** downy. **Leaves** all or mostly basal, stalked, blade linear-lanceolate, with up to 30 pairs of small, ovate to oblong, toothed segments. **Flowers** in condensed **raceme**; calyx 9–12 mm long; corolla yellow or cream with a purple or purplish brown, blunt upper lip, tube curved, 12–16 mm long. Gansu, Qinghai, Sichuan, Shaanxi, Tibet, n & nwYunnan [nAsia, Europe, North America]; alpine meadows, pastures, open shrubberies, 2,600–5,400 m. V–VIII.

Pedicularis tricolor Hand.-Mazz.* Low tufted plant not more than 7 cm tall with a central erect **stem** and spreading laterals. **Leaves** mostly in a basal tuft, lanceolate, with 11–14, sharply toothed segments, white-scurfy beneath; stem-leaves generally 1 pair. **Inflorescence** a condensed raceme; calyx 8–12 mm long, white-hairy; **corolla** yellow with white or cream rim to lower lip, tube 35–50 mm long, upper lip and beak yellow or reddish, circular. wSichuan, nwYunnan; alpine meadows, 3,000–3,600 m. VII–IX.

Pedicularis cranolopha Maxim.* Similar to the previous species, but a taller plant to 20 cm tall, occasionally more, and with alternate or pseudo-whorled **stem-leaves**; calyx 12–20 mm long, **corolla** egg-yolk-yellow, lip with a paler rim, upper lip sickle-shaped. swGansu, ne & seQinghai, Sichuan, nwYunnan; similar habitats, 2,600–4,200 m. VI–VIII. (See also photo 11 on p. 457.)

LEFT: Meadows near Longriba in nwSichuan, with *Pedicularis cranolopha*, *Stellera chamaejasme* and other colourful plants (CGW)

1. *Pedicularis rex*; Cangshan, nwYunnan (PC)
2. *Pedicularis semitorta*; Aba, nwSichuan (CGW)
3. *Pedicularis torta*; Huanglong, nwSichuan (CGW)
4. *Pedicularis torta*; Huanglong, nwSichuan (CGW)
5. *Pedicularis dunniana*; sTibet (JB)
6. *Pedicularis oederi* form; Baimashan, nwYunnan (HJ)
7. *Pedicularis oederi*; Balangshan, Wolong, wSichuan (CGW)
8. *Pedicularis tricolor*; Balangshan, Wolong, wSichuan (CGW)
9. *Pedicularis cranolopha*; south of Aba, nwSichuan (CGW)

1. *Pedicularis longiflora*; near Jiuzhaigou, nwSichuan (CGW)
2. *Pedicularis croizatiana*; swSichuan (HJ)
3. *Pedicularis przewalskii*; Mi La, sTibet (JB)
4. *Pedicularis siphonantha*; Baimashan, nwYunnan (CGW)
5. *Pedicularis siphonantha*; nNepal (CGW)
6. *Pedicularis sigmoidea*; Kangding, wSichuan (PC)
7. *Pedicularis decorissima*; Gongganlin, wSichuan (MG)
8. *Pedicularis latituba*; Genyanshan, south side, swSichuan (AD)
9. *Pedicularis latituba*; Genyanshan, south side, swSichuan (MN)
10. *Pedicularis variegata*; Kangding, wSichuan (PC)
11. *Pedicularis muscoides*; Jiuzhaigou, nwSichuan (CGW)

Pedicularis longiflora Rudolph Readily distinguished from the previous 2 species by the clusters of lateral **flowers**, the corolla bright buttercup-yellow (with 2 maroon blotches on lower lip in **var. *tubiformis* (Klotzsch) P. C. Tsoong**), tube 50–80 mm long, bending at top into a curved and twisted slender beak. Gansu, Qinghai, w, nw & swSichuan, Tibet, nwYunnan [Himalaya, nMyanmar]; alpine meadows, marshy areas, pool- and streamsides, seepage zones, 2,100–5,300 m. VII–IX.

Pedicularis croizatiana H. L. Li* Readily distinguished from the previous species by cream or pale primrose-yellow **flowers** in which the upper lip forms a hood over the rest of the beak and lower lip; **leaves** glabrous except for a downy midvein; corolla-tube 25–30 mm long. swSichuan, seTibet; open *Pinus* forests, alpine meadows, 3,700–4,200 m. VII–VIII.

> **Group Two**
> 1. Upper lip of corolla forming a right angle, with a straight beak .... **P. przewalskii, P. macrosiphon**
> 1. Upper lip of corolla S-shaped or twisted ....
> .... **P. siphonantha, P. siphonantha var. delavayi, P. sigmoidea, P. decorissima, P. latituba, P. variegata**

Pedicularis przewalskii Maxim.* Small plant to 12 cm tall in flower, **stems** very short (to 2 cm) or absent. **Leaves** linear-lanceolate, stalked, blade with up to 30 pairs of oblong to elliptic, toothed segments. **Flowers** in a central tuft of up to 20; calyx 10–12 mm long with 2–5 unequal, long-ciliate lobes; corolla reddish purple, sometimes with a whitish throat, or white or cream with a purple beak, tube erect, 30–35 mm long, hairy, upper lip with a slender straight beak. s & swGansu, e & seQinghai, w, nw & swSichuan, e & seTibet, nwYunnan; alpine meadows and other grassy habitats, 4,000–5,300 m. VI–VIII.

Pedicularis macrosiphon Franch.* Taller plant than the previous species, to 40 cm with opposite and alternate **stem-leaves**, leaf-blade with no more than 12 pairs of oblong, sharply toothed segments. **Flowers** in a gappy **raceme**; calyx 9–10 mm long, 5-lobed; corolla rose-pink to purple, tube 40–50 mm long, upper lip with a short straight beak. w & swSichuan, nwYunnan; forests and forest margins, ravines, in damp habitats, 1,200–3,500 m. V–VIII.

Pedicularis siphonantha D. Don Small tufted **herb** not more than 15 cm tall, with a single or several erect, ± glabrous **stems**. **Leaves** mostly basal, stalked, blade oblong to linear-lanceolate, with 6–15 pairs of ovate to lanceolate, toothed or lobed segments. **Flowers** in axils of leaf-like bracts; calyx 11–14 mm long, with 2–3 or 5 (**var. *delavayi* (Franch. ex Maxim.) P. C. Tsoong***) unequal lobes, hairy; corolla rose-red or rose-purple, whitish in throat, tube 40–70 mm long, slender and downy, upper lip twisted and curled upwards into an S-shaped to circular slender beak. w & swSichuan, seTibet, nwYunnan [swTibet, Himalaya]; alpine meadows, seepage zones, 3,000–4,600 m. VI–VIII.

Pedicularis sigmoidea Franch. ex Maxim.* Taller plant than the previous species, to 30 cm; **leaves** with not more than 11 pairs of pinnately lobed, grey-scurfy segments; **corolla** reddish purple, tube 32–55 mm long. w & swSichuan, nwYunnan; stony meadows, 3,000–3,600 m. VII–IX.

Pedicularis decorissima Diels* Similar to *P. siphonantha*, but **leaves** with just 6–9 pairs of rather triangular, unlobed but toothed segments; calyx larger, 22–24 mm long; **corolla** rose-pink, often white in centre, tube very long, 85–120 mm. swGansu, e & seQinghai, w & nwSichuan; similar habitats, 2,900–3,500 m. VI–VIII.

Pedicularis latituba Bonati* Very similar in stature and general appearance to *P. siphonantha*, but calyx usually 3-lobed, flower-stalks with 2 lines of hairs; **corolla** purplish red, lip paler towards the margin giving flower a distinctly bicoloured appearance; corolla-tube 30–45 mm long, beak crested in front. w & swSichuan, seTibet; similar habitats and altitudes. VII–VIII.

Pedicularis variegata H. L. Li* Distinctive little plant no more than 12 cm tall in flower, forming small **tufts**. **Leaves** mostly basal, alternate, blade to 4 cm long, with 9–12 pairs of ovate, toothed segments. **Flowers** lateral, stalkless; calyx unequally 3-lobed, 7–9 mm long; corolla white with a purple beak, tube erect, 35–45 mm long, slightly hairy, beak strongly twisted, glandular-downy, 9–10 mm long. w & swSichuan, w & nwYunnan; damp and swampy meadows, stony places, 4,100–4,500 m. VII–VIII.

> **Group Three**
> 1. Calyx not more than 9 mm long 2
> 1. Calyx 10–18 mm long 3
> 2. Flowers in basal clusters or forming a head (capitate) at the end of a short stem ....
> .... **P. muscoides, P. chenocephala, P. bietii**
> 2. Flowers in a spike-like or racemose inflorescence .....
> ... **P. davidii, P. monbeigiana, P. densispica, P. rupicola, P. tenuisecta, P. kansuensis, P. cheilanthifolia**
> 3. Beak of corolla S-shaped .... **P. rhinanthoides**
> 3. Beak of corolla not S-shaped ....
> .... **P. artselaeri, P. trichoglossa**

Pedicularis muscoides H. L. Li* Small plant not more than 5 cm with all leaves in a basal rosette. **Leaves** oblong-lanceolate in outline, with 8–10 pairs of pinnately lobed, toothed segments, becoming hairless. **Flowers** 2–3; calyx 7–8 mm long with almost equal lobes; corolla rose-red or sometimes cream (with or without a rose-purple beak), 12–23 mm long, tube slightly bent and enlarging towards top, longer than calyx, upper lip angled, with a pointed beak. sw, w & nw Sichuan, seTibet, nwYunnan; alpine meadows, generally in damp places, 3,900–5,300 m. VI–VIII.

Pedicularis chenocephala* Diels In contrast to the previous species, a taller plant to 13 cm, with some whorled or paired **stem-leaves**; **flowers** in a head-like **inflorescence**; corolla rose-red, 24–28 mm long, with a straight tube and slightly curved beak. swGansu, n & nwSichuan; similar habitats, 3,600–4,300 m. VI–VIII.

Pedicularis bietii* Franch. Distinctive species only 2–4 cm tall in flower, with short glabrous **stems** and mainly basal **leaves**, these ovate-elliptic with only 3–5 pairs of slightly hairy, scalloped segments. **Flowers** few, lateral, on short glabrous stalks; calyx c. 8 mm long, membranous, with 5 small, leaf-like, unequal lobes; corolla pale rose-pink with deeper spots and streaks, 25–30 mm long, glabrous, tube slightly bent in centre, upper lip slightly sickle-shaped but not obviously beaked, lower lip large, evenly 3-lobed. w & swSichuan, eTibet; alpine meadows, 3,500–4,200 m. VI–VII.

Pedicularis davidii* Franch. Variable tufted, densely rusty-hairy **perennial** to 30 cm. Lower **leaves** whorled, upper alternate, ovate to oblong in outline, with 9–14 pairs of pinnately lobed segments, whitish scurfy beneath. **Inflorescence** racemose; calyx 5–6 mm long, virtually glabrous, with 3–5 unequal lobes; **corolla** purple to red, or whitish, 12–16 mm long, the straight tube twice as long as calyx, upper lip somewhat twisted with a semicircular beak, occasionally somewhat S-shaped. swGansu, n, nw & wSichuan, swShaanxi; meadows and other grassy habitats, open shrubberies, stream- and tracksides, 1,400–4,400 m. VI–VIII.

Pedicularis monbeigiana* Bonati Similar to the previous species, but a taller plant to 70 cm; **leaves** with 14–18 pairs of toothed segments, hairy on veins; stem-leaves all alternate; calyx 6–8 mm long; **corolla** purplish red, or white with a purple beak, 18–22 mm long, beak with a right-angled bend towards top and with tip bent downwards. swSichuan, nwYunnan; alpine meadows, 2,500–4,200 m. VII–VIII.

Pedicularis densispica* Franch. ex Maxim. A more robust plant than the previous species, sometimes attaining 60 cm; **stems** with 4 lines of hairs; **leaves** with not more than 10 pairs of segments; calyx hairy along veins; **corolla** rose-pink to pale purple, 13–16 mm long, tube straight or slightly curved, scarcely longer than the 5–8 mm-long calyx, upper lip with a pointed beak, sometimes yellowish green. seTibet, nwYunnan [swTibet]; swampy areas, meadows, damp forests, 1,900–4,400 m. V–IX.

Pedicularis rupicola* Franch. ex Maxim. Similar to *P. densispica*, but **stems** hairy all round and **inflorescence** more racemose; calyx 6–9 mm long; **corolla** reddish purple, 15–24 mm long, tube twice as long as calyx, slightly bent near base, upper lip curved in the middle and truncated at the beak. wGuizhou, swSichuan, nwYunnan; similar habitats, 1,500–3,700 m. VII–IX.

***Pedicularis tenuisecta* Franch. ex Maxim.** Superficially similar to *P. densispica*, but **leaves** mostly in whorls of 4; **corolla** reddish purple, 16–20 mm long, tube characteristically bent at right angles close to base, upper lip sickle-shaped. wGuizhou, swSichuan, nwYunnan [Laos]; grassy and rocky alpine meadows, open shrubberies, 2,700–4,800 m. V–VII.

Pedicularis kansuensis* Maxim. Variable plant to 40 cm, usually with several **stems** that have 4 lines of hairs. **Leaves** in 4s, blades oblong, to 3 cm long, divided into 9–10 pairs of pinnately lobed segments. **Inflorescence** spike-like; **flowers** in dense whorls; calyx membranous, with 5 unequal, toothed lobes; corolla purple to purple-pink or purple-red, rarely white, 14–16 mm long, tube curved near base, with a hood-like, sickle-shaped beak. s & swGansu, Qinghai, w, nw & swSichuan, e Tibet, nwYunnan; grassy and stony alpine meadows, field margins, 1,800–4,600 m. VII–IX.

***Pedicularis cheilanthifolia* Schrenk** Plant to 30 cm, generally less, with 1–several **stems**, each with 4 lines of hairs. **Leaves** in whorls of 4 on the stem, short-stalked, ovate-lanceolate to linear-lanceolate, pinnately lobed, with incised teeth; bracts leaf-like. **Inflorescence** usually spike-like, sometimes almost capitate; calyx 8–9 mm long, densely downy along veins; **corolla** purple-red to pink or whitish, tube almost erect at first, becoming bent at right angles near base, upper lip sickle-shaped, 9–10 mm long, short-beaked or unbeaked. Gansu, Qinghai, Tibet [cAsia, India, Mongolia]; damp stony and gravelly slopes, streamsides, 2,100–5,200 m. V–VII.

***Pedicularis rhinanthoides* Schrenk ex Fisch. & Mey.** Tufted plant to 30 cm, often with numerous erect, leafy **stems**. **Leaves** stalked, the basal densely clustered, blade narrow-oblong, with 9–12 pairs of ovate, toothed lobes, scarcely hairy. **Flowers** in condensed head-like **racemes** with leaf-like bracts; calyx somewhat inflated, often with purplish dots, 12–15 mm long; corolla rose-red to violet-purple, tube erect, slightly longer than calyx, upper lip curved, with a right-angled bend

---

1. *Pedicularis chenocephala*; nwSichuan (JM)
2. *Pedicularis bietii*; Yading Nature Reserve, swSichuan (AD)
3. *Pedicularis davidii*; Balangshan, Wolong, wSichuan (CGW)
4. *Pedicularis davidii*; Longriba, nwSichuan (CGW)
5. *Pedicularis davidii*; south of Aba, nwSichuan (CGW)
6. *Pedicularis davidii*; Wolong, wSichuan (PC)
7. *Pedicularis monbeigiana*; above Wengshui, nwYunnan (CGW)
8. *Pedicularis densispica*; south of Aba, nwSichuan (CGW)
9. *Pedicularis densispica*; south of Aba, nwSichuan (CGW)
10. *Pedicularis rupicola*; Gangheba, Yulongxueshan, nwYunnan (CGW)
11. *Pedicularis kansuensis* with *P. cranolopha*; Genyanshan, south side, swSichuan (AD)
12. *Pedicularis kansuensis*; north of Napahai, nwYunnan (CGW)
13. *Pedicularis cheilanthifolia*; Mi La, sTibet (HJ)
14. *Pedicularis rhinanthoides*; sTibet (JB)

1. *Pedicularis artselaeri*; Jiuzhaigou, nwSichuan (PC)
2. *Pedicularis artselaeri*; Wolong, wSichuan (CGW)
3. *Pedicularis trichoglossa*; Ragarsanba, sTibet (HB)
4. *Veronica piroliformis*; Yulongxueshan, nwYunnan (DZ)
5. *Veronica szechuanica*; Yading Nature Reserve, swSichuan (MN)
6. *Veronica forrestii*; Cangshan, nwYunnan (PC)
7. *Corallodiscus lanuginosus*; Gangheba, Yulongxueshan, nwYunnan (HJ)
8. *Corallodiscus lanuginosus*; Minjiang valley south of Songpan, nwSichuan (PC)
9. *Corallodiscus lanuginosus*; Xishan, Kunming, cYunnan (PC)
10. *Corallodiscus kingianus*; Napahai, Zhongdian, nwYunnan (HJ)

near top, terminating in an S-shaped beak. Gansu, Hubei, Qinghai, Sichuan, Shaanxi, Tibet, w & nwYunnan [nwChina, cAsia, Himalaya, sRussia]; alpine meadows and other grassy habitats, damp places, streamsides, 2,300–5,000 m. V–VII.

Pedicularis artselaeri **Maxim.*** Rather delicate low plant, not more than 6 cm tall, often patch-forming, with short, thin, scaly **stems**. **Leaves** fern-like, slender-stalked, blade oblong to lanceolate, to 10 cm long, with 8–14 pairs of ovate, sharply toothed segments, rusty-brown with hairs beneath. **Flowers** borne in the leaf-axils; calyx 12–18 mm long; corolla purple or pinkish purple with white markings at base of lower lip, tube erect, 18–24 mm long, upper lip with a right-angled bend and a tapered, straight beak. Hubei, w, n & neSichuan, Shaanxi [cChina]; forests and forest margins, damp rocky places, streamsides, 1,000–2,800 m. V–VII.

Pedicularis trichoglossa **Hook. f.** In contrast to the previous species, a taller plant 15–60 cm, with stalkless, clasping linear-lanceolate, pinnately lobed **leaves**, lobes finely toothed. **Inflorescence** racemose, downy; calyx 12–16 mm long, blackish purple-downy; **corolla** blackish purple, beak very distinctive, with a beard of long purple-red hairs towards tip. Qinghai, w, nw & sw Sichuan, s, se & eTibet, nwYunnan [Himalaya, nMyanmar]; stony meadows and screes, rocky places in open forests, 3,500–5,000 m. VII–VIII.

VERONICA

A genus with c. 250 species (52 in China), mainly north temperate with a few species on tropical mountains. **Annual** or **perennial herbs**, the latter often rhizomatous, with opposite or sometimes whorled or alternate **leaves**. **Inflorescence** spike-like or capitate, sometimes paired; calyx usually 4-lobed; **corolla** 4-lobed, 1 lobe often smaller than the others; stamens 2. **Fruit** very variable but usually compressed, heart-shaped to elliptic, 2-valved.

1. Leaves crowded, often in rosettes . . . . . **V. piroliformis**
1. Leaves neither crowded nor in rosettes . . . .
   . . . . **V. szechuanica, V. forrestii**

Veronica piroliformis **Franch.*** Small tufted plant to 5 cm with elliptic to spathulate **leaves**, 3–8 × 1–3.5 cm, usually densely hairy, uppermost larger than lower, margin toothed. **Racemes** erect, to 2 cm long, scapes with brownish-yellow glandular hairs; calyx 4-lobed; **corolla** blue, purple or white, 10–12 mm in diameter, with a very short tube. swSichuan, nwYunnan; open forests, especially coniferous, sloping meadows, rocky places, 2,600–4,000 m. VI–VII.

Veronica szechuanica **Batalin** Hairy **perennial** to 35 cm with branched or simple **stems**, often tufted. **Leaves** with ciliate stalks, ovate, with a slightly heart-shaped or truncated base. **Flowers** 3–5 in **racemes** in upper leaf-axils, forming corymbs; calyx 4-lobed; corolla white or occasionally pale purple, 5–7 mm across, with short tube; stamens slightly shorter than corolla-lobes. seGansu, wHubei, eQinghai, sShaanxi, seTibet, w, nw & swSichuan, nwYunnan [swTibet, Himalaya]; forests and forest margins, grassy places, alpine meadows, 1,600–4,400 m. VI–VII.

Veronica forrestii **Diels*** A more softly hairy and creeping plant than *V. szechuanica*, with stalkless or almost stalkless elliptic **leaves**, to 20 × 7 mm, base wedge-shaped. **Flowers** up to 10 in 1 or 2 lateral **racemes**, in middle or lower leaf-axils; corolla red or white, 10–12 mm across. nwYunnan; forests and forest margins, 2,000–3,000 m. VI–VII.

GESNERIACEAE

CORALLODISCUS

A genus of c. 5 species (3 in China) in the Himalaya and South East Asia. Rosette-forming cliff-dwellers. **Flowers** in lax to dense lateral **cymes**, generally with a long stalk well clear of leaf-rosette; corolla zygomorphic, with a long tube and short lips.

Corallodiscus lanuginosus **(Wall. ex R. Br.) B. L. Burtt** (syn. *C. bullatus* (Craib) B. L. Burtt; *C. flabbellatus* (Craib) B. L. Burtt; *Didymocarpus lanuginosus* Wall. ex R. Br.) A very variable plant. **Leaf-rosettes** flattish, usually 8–18 cm across, deep green or grey-green, **leaves** stalked, elliptic to oblong, obovate, diamond- or fan-shaped, usually downy beneath but glabrous to white- or brown-hairy above; lateral veins 3–5 on each side. **Inflorescence** rather lax, 4–15-flowered; calyx 2–3.5 mm long; **corolla** purple to blue, white or yellowish, sometimes spotted on lower lip, 8–14 mm long. nGuangxi, Guizhou, Hubei, Sichuan, Shaanxi, seTibet, Yunnan [swTibet, c & eHimalaya, nMyanmar]; cliffs and rock-crevices in valleys and ravines, rocky places, stony banks in forests or along forest margins, 700–4,300 m. IV–X.

Corallodiscus kingianus **(Craib) B. L. Burtt** (syn. *Didissandra kingiana* Craib) Distinguished from the previous species by the lanceolate to diamond-shaped **leaves**, with dense rusty-brown hairs below, apex pointed; **inflorescence** crowded, with up to 20 flowers; calyx 4–5 mm long; **corolla** blue, purple or white, lower lip with 2 lines of brownish spots, 11–15 mm long overall, glabrous outside. sQinghai, w & swSichuan, seTibet, nwYunnan [swTibet, eHimalaya, nMyanmar]; crevices in rocky slopes and outcrops in the open or in forests, 2,800–4,800 m. VI–IX.

PETROCOSMEA

A genus of c. 27 species (24 in China) distributed from northeast India to China and South East Asia. Rosette-forming **perennial herbs**. **Flowers** in lateral 1–7-flowered **cymes** with a pair of bracts; corolla zygomorphic, 2-lipped, with short, broad tube, upper lip 2-lobed, lower lip generally larger, 3-lobed.

1. Flowers white, sometimes with a faint bluish flush, or pale blue 2
1. Flowers purplish blue 3
2. Corolla small, lower lip not more than 7 mm long .... *P. rosettifolia, P. kerrii*
2. Corolla larger, lower lip at least 9 mm long .... *P. grandiflora*
3. Upper corolla-lip equalling lower .... *P. flaccida, P. nervosa, P. sinensis*
3. Upper corolla-lip smaller than lower .... *P. duclouxii, P. forrestii, P. minor*

Petrocosmea rosettifolia C. Y. Wu & H. W. Li* Plant forming rather flat, symmetrical, bristly, rough, **evergreen rosettes**. **Leaves** somewhat fleshy with a short thick stalk, blade ovate to elliptic, to 4 × 4.3 cm, base wedge-shaped or rounded, margin entire to slightly toothed. **Flowers** solitary in a ring in centre of rosette, each on a stalk 2–5 cm long; corolla white with yellow in throat, 8–12 mm across, sparsely downy outside, lower lip c. 7–8 mm long. Yunnan (Jingdong Xian); rock-crevices and cliffs, 1,200–1,500 m. IX–XI.

Petrocosmea kerrii Craib Differs from the previous species in the larger, softly hairy rosettes, **leaves** long-stalked, elliptic to almost rhombic, to 14 × 8.5 cm with an asymmetric base and toothed margin; **flowers** several together on densely hairy stalks; calyx distinctly zygomorphic; corolla slightly larger. s & swYunnan [Thailand]; forest rocks and cliffs, 1,500–3,100 m. VI–IX.

Petrocosmea grandiflora Hemsl.* Plant forming a neat, rather yellow-green rosette of elliptic to broad-ovate **leaves**, to 6.5 × 2.5 cm, silky-hairy, especially beneath, lateral veins inconspicuous, margin scalloped. **Flowers** solitary or in 2–3-flowered **cymes** on a stalk up to 5 cm long; calyx with equal, pointed segments; corolla white or pale blue, darker in centre, tube 6 mm long, lower lip 9–12 mm long, upper 13–15 mm, 2-lobed. seYunnan; limestone rock-crevices and -ledges, c. 2,000 m. XII–II.

Petrocosmea flaccida Craib* Plant forming small, rather flat **rosettes**, silky at first, the **leaves** rounded to oblate, often wider than long, to 3.5 × 4.4 cm, with entire to faintly lobed margin, and with indistinct lateral veins. **Flowers** solitary on stalk up to 8.5 cm long; corolla purple-blue, 12–14 mm, glabrous inside, upper lip clearly 2-lobed. s & swSichuan (Muli), nwYunnan (Lijiang); cliffs, rocky crevices, 2,800–3,100 m. VIII–IX.

Petrocosmea nervosa Craib* Rather similar to the previous species, but **rosettes** dying back in winter to a resistant velvety bud; **leaves** ovate to ± rhombic, to 5 × 5 cm, with conspicuous oblique lateral veins on upper surface, margin entire to slightly and minutely scalloped; **corolla** with tube 3 mm long and upper and lower lips equal. swSichuan, nwYunnan; forest rocks and cliffs, 300–3,100 m. IV–VI, IX.

Petrocosmea sinensis Oliv.* Similar to *P. flaccida*, but **leaves** with shorter stalks, not more than 2.5 cm long; anthers oblong, not triangular; and **flowers** blue, purple or white with yellow in the throat. wHubei, Sichuan, nYunnan; shady rocks and crevices, to 1,700 m. VIII–XI.

Petrocosmea duclouxii Craib* Small **herb** with rosettes of rather pale green or bluish green **leaves**, blade ovate to rounded, to 3 × 2.5 cm, finely hairy on both sides, base truncated or somewhat heart-shaped, margin finely scalloped-toothed. **Flowers** generally solitary, sometimes paired; calyx with 5 narrow-lanceolate equal segments; corolla purplish blue with dark throat, minutely hairy outside, 8–10 mm across, tube 3–5 mm long, upper lip deeply 2-lobed; stamens with minutely hairy filaments. c & cwYunnan; shady rocks and cliffs, 2,000–2,600 m. V–VII.

Petrocosmea forrestii Craib* Similar to *P. duclouxii*, but **leaves** with a wedge-shaped base, margin entire to minutely scalloped, and inconspicuous veins; **corolla** with upper lip finely hairy on both sides, about half the size of lower lip; stamens with glabrous filaments. swSichuan, n & nwYunnan; similar habitats, 1,600–2,000 m. VI–VIII.

Petrocosmea minor Hemsl.* Readily differentiated from the previous 2 species by the relatively large **flowers** and indistinctly 2-lobed or unlobed upper corolla-lip; corolla purple with whitish base, filaments finely brown-hairy towards top. seYunnan; forest rocks, on limestone, 1,000–2,200 m. VIII–XI.

OROBANCHACEAE

OROBANCHE

A mainly north temperate genus of c. 100 species (25 in China), with fewer species in southern Central America and east and North Africa. **Parasitic annuals** and **perennials**, rarely **biennials**, lacking chlorophyll, forming dense to lax, simple or branched **spikes** of flowers, often glandular. **Stems** bearing scale-like **leaves**. **Flowers** bracteate, tubular to tubular-trumpet-shaped, usually 2-lipped, upper lip 2-lobed, lower lip 3-lobed; stamens 4. **Fruit** a many-seeded capsule.

Orobanche yunnanensis (Beck) Hand.-Mazz.* Densely glandular-hairy, erect **biennial** or **perennial herb** to 25 cm, lower part of **stem** with scattered **scale-leaves**, these ovate-lanceolate, 10–15 × 4–6 mm, red, purplish or occasionally yellowish brown; bracts similar to scale-leaves. **Flowers** in a dense spike, purple, red or occasionally yellow-brown; calyx 2-parted, split to base; corolla 15–20 mm long, tube enlarged and somewhat curved, upper lip hood-like, notched, lower lip shorter with 3 reflexed lobes. wGuizhou (Weining Xian), swSichuan (Muli), n & nwYunnan; wooded and grassy slopes, open shrubberies where it is parasitic on various labiates (Labiatae), 2,200–3,400 m. V–VI.

1. *Petrocosmea rosettifolia*; cult. (RD)
2. *Petrocosmea kerrii*; cult. (CGW)
3. *Petrocosmea grandiflora*; cult. (RD)
4. *Petrocosmea flaccida*; Leshan, sSichuan (HJ)
5. *Petrocosmea sinensis*; Leshan, sSichuan (CGW)
6. *Petrocosmea duclouxii*; Xishan, cYunnan (CGW)
7. *Petrocosmea forrestii*; cult. (CGW)
8. *Petrocosmea minor*; cult. (CGW)
9. *Orobanche yunnanensis*; Tianchi, nwYunnan (PC)

OROBANCHACEAE: Orobanche

Orobanche coerulescens **Steph.** (including *O. ammophila* C. A. Mey.; *O. bodinieri* H. Lévl.; *O. korshinskyi* Novopokr.; *O. mairei* H. Lévl.) Readily distinguished from the previous species by being densely hairy rather than glandular, by the larger **scale-leaves** (to 20 × 7 mm) and by the larger (20–25 mm long) blue, purple-blue or yellow **flowers**. Gansu, Hubei, Qinghai, Sichuan, Shaanxi, Tibet, Yunnan [c & nChina, temperate Asia, Europe]; grassy and bushy slopes, parasitic on *Artemisia*, 900–4,000 m. V–VIII.

1. *Orobanche coerulescens*; Chongjiang, nwYunnan (PC)
2. *Boschniakia himalaica*; Xiaozhongdian, nwYunnan (PC)
3. *Aeginetia indica*; cNepal (CGW)
4. *Gleadovia ruborum*; Wolong, wSichuan (HH)
5. *Catalpa fargesii*; north of Dali, nwYunnan (PC)
6. *Catalpa fargesii*; Tengchong, wYunnan (PC)
7. *Mayodendron igneum*; Liuku, wYunnan (PC)

BOSCHNIAKIA

A genus of 2 species (both in China) distributed from north India to China, Japan, Korea, Russia and North America. Very similar in general appearance to *Orobanche*, but **flowers** with a cup-shaped calyx with deciduous teeth, and protruding stamens.

Boschniakia himalaica **Hook. f. & Thomson** (syn. *B. handelii* Beck; *Xylanche himalaica* (Hook. f. & Thomson) Beck) **Parasitic perennial** to 45 cm, usually far less. **Stem** solitary, unbranched, erect, with scattered **scale-leaves** in lower part, these lanceolate, 10–20 × 6–12 mm. **Inflorescence** racemose; **flowers** on short stalks; calyx cup-shaped, 4–5 mm long; corolla yellow-brown or pale purple, 15–25 mm long, tube slightly expanded, upper lip notched, lower lip much shorter than upper, with 3 ciliate lobes. Gansu, Hubei, Qinghai, Sichuan, Shaanxi, Tibet, Yunnan [cChina, Himalaya, Myanmar, Taiwan]; forests, shrubberies, slopes, parasitic on *Rhododendron*, 2,600–3,800 m. V–VIII.

AEGINETIA

A genus of 4 species (3 in China) found in India, Sri Lanka, South East Asia and Myanmar. Closely related to *Orobanche* and *Boschniakia*, but immediately distinguished by the solitary or few flowers that have a characteristic undivided spathe-like calyx. Fleshy **parasitic herbs** with a short branched or simple **stem**. **Flowers** large, solitary or clustered on long stalks; corolla tubular to campanulate, slightly 2-lipped. **Capsule** 2-valved.

Aeginetia indica **L.** (syn. *Orobanche aeginetia* L.; *Phelipaea indica* (L.) Spreng. ex Steud.) Plant to 40 cm, glabrous, with simple ridged **stems**, or stems branched near base. **Leaves** few, scale-like, reddish, not more than 10 × 4 mm. **Flowers** solitary; calyx hood-like, covering lower half of corolla, usually greenish with reddish striations; corolla reddish purple, tubular-campanulate, 20–45 mm long, with 5 short, rounded lobes. Guangxi, Guizhou, Sichuan, Yunnan [c & sChina, seAsia, Himalaya, India, Japan, Myanmar]; slopes, open forests, road- and tracksides, parasitic on various grasses, 200–1,800 m. IV–VIII.

GLEADOVIA

A genus of just 2 species found in India and China. **Stems** short, unbranched. **Flowers** in tufts (almost capitate) close to ground level; calyx tubular-campanulate with a 5-lobed apex; corolla 2-lipped; stamens 4. **Fruit** a many-seeded capsule.

Gleadovia ruborum **Gamble & Prain** (syn. *G. yunnanense* Yu) **Tufted** plant not more than 15 cm tall with short **stems** bearing ovate, scale-like, **leaves** and **bracts**, these up to 20 × 8 mm. **Flowers** on short stalks; calyx 15–18 mm in diameter; corolla red or rose-red, rarely white, fragrant, 5–7 cm long, tube expanding towards top, upper lip with 2 rounded lobes, lower lip smaller, with 3 oblong lobes. nGuangxi, Hubei, w & swSichuan, w & swYunnan [Hunan, India]; damp forests and thickets, ravines, 900–3,500 m. IV–VIII.

Gleadovia mupinense **Hu*** Very similar to the previous species, but **flowers** on much longer stalks (4–9 cm long), calyx more tubular, only 10–12 mm in diameter; corolla often paler. wSichuan; similar habitats as well as roadsides, 2,450–3,500 m. IV–VII.

BIGNONIACEAE

CATALPA

A genus of c. 13 species (4 in China) in east Asia and North America. **Deciduous trees** with simple **leaves** that have purple glandular spots on veins beneath. **Inflorescence** corymbs, racemes or panicles bearing campanulate **flowers**; corolla 2-lipped, upper lip 2-lobed, lower 3-lobed; stamens 2, inserted on base of corolla-tube. **Fruit** capsular, cylindrical, dehiscing into 2 valves, containing numerous small winged **seeds**.

Catalpa fargesii **Bur.*** **Tree** 15–25 m with leathery, ovate to heart-shaped, long-pointed **leaves**, 13–20 × 10–13 cm; leaf-stalks 3–10 cm long. **Inflorescence** 7–15-flowered; corolla campanulate, 2-lipped, 3–3.4 cm long, pink to lilac-pink, purple-spotted within. **Capsules** terete, nodding, 5.5–8 cm long. Gansu, Guizhou, Hubei, Sichuan, Shaanxi, Yunnan [cChina]; mixed woods on hillsides, roadsides, 700–2,500 m. III–V. Often planted around villages and temples.

Catalpa ovata **G. Don*** Smaller than the previous species, to only 15 m, with larger, broadly ovate, rough **leaves** to 25 cm long, 3-lobed above middle; **inflorescence** paniculate with pale yellow **flowers** with 2 stripes as well as spots in throat. Gansu, Hubei, Qinghai, Sichuan, Shaanxi [cChina]; hillsides, 1,900–2,500 m. V–VI.

Catalpa tibetica **Forrest*** Very similar to *C. ovata*, but with densely hairy, ovate **leaves** and creamy white **flowers** with purple spots in throat. seTibet, nwYunnan; hillside woods, 2,400–2,700 m. V–VI.

MAYODENDRON

A genus with a single species in east Asia, including southwest China. **Evergreen tree** with simple, opposite, 2-pinnate **leaves**; leaflets entire. **Inflorescences** short racemes, on old stems or on short side branches; **flowers** tubular, 5-lobed, bright orange; stamens 4. **Fruit** a cylindrical capsule, dehiscing into cells; valves thin, leathery. Included in *Radermachera* by some botanists.

Mayodendron igneum **(Kurz) Kurz** (syn. *Radermachera ignea* (Kurz) Steenis) **Tree** to c. 15 m with smooth **bark** and branchlets with white lenticels. **Leaves** stalked, to 60 cm long; leaflets ovate to ovate-lanceolate, 8–12 × 2.5–4 cm, long-pointed. **Inflorescence** 5–13-flowered; calyx tubular, spathe-like, minutely hairy; **corolla** 6–7 cm long, bright orange to golden-yellow, 6–7 mm long. Guangxi, s & wYunnan [sChina, Laos, Myanmar, Thailand, Vietnam]; steep slopes, forests, forest margins, to 1,900 m. II–V.

INCARVILLEA

A genus of c. 20 species (16 in China, mostly endemic) in Central Asia and the Himalaya. **Subshrubs** and herbaceous **perennials**, with or without a well-developed **stem**. **Leaves** simple or variously pinnately lobed, basal or alternate. **Flowers** solitary or in few-many-flowered **racemes**, occasionally in panicles; calyx tubular or campanulate, 5-toothed; corolla funnel-shaped, somewhat zygomorphic, ridged in throat. **Fruit** a leathery or woody capsule, linear to spindle-shaped, 2-valved.

1. Diffuse subshrubs or perennials with a fibrous root system and leafy stems; fruit papery or leathery . . . . .
. . . *I. arguta, I. sinensis, I. sinensis* subsp. *variabilis, I. sinensis* subsp. *variabilis* forma *przewalskii*
1. Tuberous perennials with mostly basal leaves; fruit apparently woody 2
2. Flowers yellow; stem-leaves present as well as basal leaves . . . . *I. lutea, I. longiracemosa*
2. Flowers pink, purple, magenta or occasionally white; stem-leaves usually absent 3
3. Flowers 2–13, borne in a stout inflorescence; stalk present, more than 15 cm long 4
3. Flowers solitary or 2–4 borne on basal scapes or on a very short stalk; leaves all basal with end leaflet generally larger than adjacent ones 5
4. Leaves with terminal leaflet smaller than adjacent leaflets and confluent with them . . . .
. . . . *I. delavayi, Incarvillea* sp. aff. *I. delavayi*
4. Leaves with terminal leaflet larger than adjacent ones, not confluent with them . . . . . *I. mairei, I. zhongdianensis*
5. Leaflets entire . . . . *I. compacta, I. compacta* var. *qinghaiensis, Incarvillea* sp. aff. *I. compacta*
5. Leaflets toothed . . . .
. . . . *I. grandiflora I. himalayensis, I. younghusbandii*

Incarvillea arguta (Royle) Royle Variable erect to **clump-forming subshrub**, generally diffusely branched and with a woody base. **Leaves** all on the stem, pinnate, with 2–6 pairs of lanceolate to elliptic, toothed leaflets that are finely dotted beneath. **Inflorescence** a lax terminal raceme with up to 20 flowers opening in succession; **corolla** pale to deep pink or white; trumpet-shaped, tube 2.5–3.8 cm long, limb relatively small, 28–26 mm across with rounded lobes. **Fruit** linear-cylindrical, straight or curved, to 20 cm long. s & swGansu, Guizou, Sichuan, seTibet, w & nwYunnan [swTibet, cChina, Himalaya, Myanmar]; cliffs and rocky places, banks, river gravels, roadsides, 1,800–3,500 m. V–IX.

Incarvillea sinensis Lam.* Similar to the preceding species, but a more erect, **annual** or **perennial** with 2–3-pinnately divided **leaves**, leaflets with entire margins. **Flowers** in lax racemes; corolla pale to deep pink, reddish purple or pale yellow. **Fruit** more leathery, spindle-shaped, 3.8–11 cm long. Similar habitats, 1,200–3,400 m. VI–IX. The typical subspecies (**subsp.** *sinensis**) has rose-pink or pale red **flowers**. Gansu, swQinghai, n, w, nw & swSichuan, Shaanxi, se & eTibet, nwYunnan [n & neChina]. **Subsp.** *variabilis* (Batalin) Grierson* is a woody-based **perennial** with pale pink **flowers**. wGansu, Qinghai, nSichuan, Shaanxi, seTibet, nwYunnan [nwChina]. **Subsp.** *variabilis* forma *przewalskii* (Batalin) Grierson* has pale yellow **flowers**. Gansu, Qinghai, nwSichuan, Shaanxi.

Incarvillea lutea Bur. & Franch.* Stout **herbaceous perennial** to 1 m, often less, with all parts except corolla grey-brown minutely downy. **Leaves** mostly on the stem (1 or 2 basal), bright shiny green, pinnate with upper third pinnately lobed; lateral leaflets in 6–10 pairs. **Inflorescence** 6–12-flowered; **corolla** trumpet-shaped, primrose-yellow, sometimes with brown or reddish spots or lines in throat, centre ageing to crimson, or occasionally with a deep chocolate throat from the start, tube 4.8–5.5 cm long, limb 4.6–5.5 cm across. **Fruit** asymmetrically spindle-shaped, 4–6-ribbed, long-pointed, to 9.5 × 1.3 cm. w & swSichuan, nwYunnan; rocky mountain slopes, open shrubberies, occasionally on or beneath cliffs, 2,800–3,600 m. V–VI.

Incarvillea longiracemosa Sprague* Very similar to the preceding species, but **leaves** strictly pinnate with all leaflets (just 3–5 pairs) distinct from one another. **Flowers** somewhat smaller, corolla-tube 43–45 mm long and limb 33–42 mm across. seTibet [swTibet]; similar habitats, 3,655–4,570 m. VI–VIII.

Incarvillea delavayi Bur. & Franch.* Stemless, tufted, **herbaceous perennial** to 70 cm tall in flower. **Leaves** pinnate with top third pinnately lobed; lateral leaflets in 6–11 pairs, lanceolate to oblong, toothed, glabrous or minutely downy on veins. **Inflorescence** usually rising above leaves; **corolla** pink to purple (rarely white in the wild), with pale yellow or whitish, purple-lined throat, tube 4.5–6 cm long, limb 3.6–5.5 cm across. **Fruit** asymmetrically spindle-shaped, 4-ridged, pointed, to 7.8 × 1.5 cm. swSichuan, nwYunnan; rocky and stony habitats, generally in open scrub, 2,400–3,500 m. V–VI.

Incarvillea sp.* Similar to *I. delavayi*, but **leaves** thinner, grey-green, narrow-ovate, pointed, scarcely toothed, crowded at base of plant. **Flowers** in an elongated **raceme** (not bunched towards top), pale purple or pink. swSichuan (Genyanshan), seTibet (near Markham); habitats similar to those of *I. delavayi*. VI–VII.

Incarvillea mairei (Lévl.) Grierson* **Herbaceous perennial** to 35 cm tall in flower, with basal leaf-rosette. **Leaves** deep glossy green, pinnate, with 2–5 pairs of lateral leaflets, terminal leaflet far larger than laterals, to 11.5 × 9.2 cm. **Inflorescence** often solitary, 1–5-flowered, held above leaves; **corolla** deep to mid-pink or purplish pink, with 2 white flares at base of each lobe and yellowish throat, tube 3.5–6 cm long,

1. *Incarvillea arguta*; Hutaio Xia, Jinshajiang, nwYunnan (CGW)
2. *Incarvillea arguta*, with a lizard; Minjiang valley, nwSichuan (CGW)
3. *Incarvillea arguta*; south of Dukou, nwYunnan (PC)
4. *Incarvillea sinensis*; near Songpan, nwSichuan (CGW)
5. *Incarvillea sinensis*; Minjiang valley, nwSichuan (PC)
6. *Incarvillea lutea*; Muli, swSichuan (PC)
7. *Incarvillea lutea*; near Yufengsi, Yulongxueshan, nwYunnan (CGW)
8. *Incarvillea lutea*; Gangheba, Yulongxueshan, nwYunnan (PC)
9. *Incarvillea longiracemosa*; Sur La, seTibet (AC)
10. *Incarvillea delavayi*; cult., from nwYunnan (CGW)
11. *Incarvillea* sp. aff. *I. delavayi*; near Barkham, seTibet (HJ)
12. *Incarvillea mairei*; north of Lijiang, nwYunnan (CGW)

limb 4.5–6.5 cm across. **Fruit** straight or slightly curved, square in cross-section, to 9 × 1.2 cm. sSichuan, nwYunnan; dry grassy places and open scrub, occasionally in rocky places, 2,700–3,655 m. V–VI. (See also photo 12 on p. 465.)

Incarvillea zhongdianensis Grey-Wilson* (syn. *I. mairei* (Lévl.) Grierson forma *multifoliata* C. Y. Wu & W. C. Yin) Similar in flower to the previous species, but mature plants with many **inflorescences** and **flowers** deep magenta or crimson-magenta, more rarely rose-pink. **Leaves** often more than 20 per plant, markedly different from *I. mairei*, with 5–9 pairs of small lateral leaflets, terminal one scarcely larger than laterals. nwYunnan (confined to Zhongdian, Baimashan); similar habitats, 3,200–4,260 m. V–VII.

Incarvillea compacta Maxim.* Low, stemless **herbaceous perennial** seldom more than 15 cm tall in flower, with flat leaf-rosette. **Leaves** pinnate with 4–9 pairs of lateral leaflets, these lanceolate, pointed, entire; terminal leaflet similar in size or slightly larger than laterals. **Inflorescence** a very condensed raceme (flowers appear to be basal) with up to 10 flowers; **corolla** cyclamen-pink to purple-pink with 2 white flares at base of each lobe, tube 4–6.5 cm long, limb 4.5–6 cm across. w & sGansu, w, nw & swSichuan, eTibet, nwYunnan; damp stony and earthy slopes, open low scrub, occasionally on limestone cliffs, 2,600–4,575 m. V–VII. **Var.** *qinghaiensis* Grey-Wilson* is distinguished by being more robust, to 35 cm tall in flower (50 cm in fruit), **leaves** rather grey-green and **inflorescence** with an obvious scape 10–32 cm long. Confined to habitats similar to those of var. *compacta* in s & seQinghai, 3,680–4,150 m.

Incarvillea sp.* Like *I. compacta*, but a more dwarf plant with brighter green **leaves**, lateral leaflets in 3–4 pairs, narrower, wavy but not toothed on margin, terminal leaflet broader and larger than others. w & swSichuan. V–VI.

Incarvillea grandiflora Bur. & Franch.* (syn. *I. mairei* (Lévl.) Grierson var. *grandiflora* (Wehrh.) Grierson) Stemless **herbaceous perennial** not more than 15 cm tall in flower with flat basal leaf-rosette. **Leaves** shiny deep green, 2–5, often without lateral leaflets, oblong to ovate, to 7.5 × 6.5 cm. **Inflorescence** often reduced to a single flower, occasionally flowers 2–4, stalk not more than 4 cm long; **corolla** deep crimson-magenta, occasionally white or pink, with yellowish throat and 2 white flares at base of each lobe, tube 4.5–6.2 cm long, limb 5.5–7.5 cm across. **Fruit** ± straight, long-pointed, to 8 × 1 cm. swSichuan, nwYunnan; grassy slopes, rocky slopes, forest clearings, 3,350–4,450 m. V–VI.

Incarvillea himalayensis Grey-Wilson Like the previous species, but a more dwarf plant with **leaves** metallic grey-green, with 3–4 pairs of lateral leaflets. seTibet [swTibet, wNepal]; grassy and stony banks, stony meadows, at similar altitudes. VI–VII.

Incarvillea younghusbandii Sprague Readily distinguished from the previous species by the generally more compact habit; by the **leaves** with 2–6 pairs of lateral leaflets, the terminal leaflet to 5 × 5 cm; and by the more funnel-shaped **flowers** often nestling on or very close to the ground, apparently stalkless, with a somewhat curved tube, 3–5 cm long overall. **Fruit** smaller, sickle-shaped, to 3.5 × 1.1 cm. Qinghai, neTibet [w & cTibet, wNepal]; rocky and gravelly places, low thickets, 3,600–5,500 m. V–VIII.

JACARANDA

A tropical American genus of c. 30 species, 1 a widely planted decorative **deciduous tree**.

Jacaranda mimosifolia D. Don. (syn. *J. ovalifolia* R. Br.) Small **tree** to 12 m, with rather slender **branches** and 2-pinnate, opposite **leaves**; leaflets numerous, oval-oblong to almost rhombic, to 12 × 4 mm, pointed, entire, downy. **Flowers** numerous, in erect panicles, the flowers semi-nodding; corolla blue, tubular-campanulate, 26–48 mm long, downy outside. **Fruit** a flat capsule, oblong to rounded in outline, to 5.8 cm long, red-brown when mature. Gansu, Guangxi, Sichuan, Yunnan [native to Brazil]; widely planted street and roadside tree, hillslopes, at low altitudes. IV–VI.

---

1. *Incarvillea mairei*; Gangheba, Yulongxueshan, nwYunnan (CGW)
2. *Incarvillea mairei* (colour variant); cult. (CGW)
3. *Incarvillea mairei*; north of Lijiang, nwYunnan (HJ)
4. *Incarvillea zhongdianensis*; near Xiaozhongdian, nwYunnan (CGW)
5. *Incarvillea zhongdianensis*; near Xiaozhongdian, nwYunnan (CGW)
6. *Incarvillea compacta* var. *compacta*; Huanglong, nwSichuan (CGW)
7. *Incarvillea compacta* var. *qinghaiensis*; south of Aba, nwSichuan (CGW)
8. *Incarvillea* sp. aff. *I. compacta*; Genyanshan, swSichuan (AD)
9. *Incarvillea grandiflora*; Napahai, Zhongdian, nwYunnan (CGW)
10. *Incarvillea himalayensis*; nNepal (MN)
11. *Incarvillea younghusbandii* in young fruit; near Kharta, sTibet (RS)
12. *Incarvillea younghusbandii*; near Serkyem La, seTibet (HJ)
13. *Jacaranda mimosifolia*; near Xishuangbanna, swYunnan (CGW)

LENTIBULARIACEAE

PINGUICULA

A genus with c. 35 species (2 in China) distributed in Arctic and temperate alpine regions of the northern hemisphere and temperate alpine regions of Central and South America. **Perennial**, terrestrial, rhizomatous **herbs**, with basal rosettes of simple, soft and fleshy, sticky-glandular **leaves**, usually with inrolled margins. **Flowers** usually solitary, *Viola*-like; calyx 2-lipped; corolla open at throat, lower lip larger than upper, spurred, 3-lobed, upper lip 2-lobed. **Fruit** a capsule opening by 2 valves.

Pinguicula alpina L. Small rosette-forming **herb** to 14 cm tall with yellowish green, elliptic-oblong to lanceolate-oblong, sticky **leaves**, 1–4.5 cm long. **Flowers** solitary, 1–5 per rosette; corolla white, with 1–several yellow spots on lower lip, 9–20 mm long, spur yellowish, cylindrical, as long as corolla-tube. nGuizhou, wHubei, Sichuan, sShaanxi, Tibet, nwYunnan [temperate Asia, Europe]; seepages under rocks, bogs, wet places, meadows, 2,300–4,500 m. V–VII.

UTRICULARIA

A genus of c. 200 species (23 in China) of cosmopolitan species, but the majority in the tropics. **Annual** or **perennial**, terrestrial, epiphytic, or aquatic **herbs**, lacking true **roots**, while the **stems** are modified into rhizoids and stolons, bearing small, bladder-like traps. **Leaves** alternate or in a basal rosette. **Inflorescence** racemose or flowers solitary, stalked, erect to twining; calyx with 2 equal or unequal lobes; **corolla** 2-lipped, lower larger than upper, entire or 2–3-lobed, spurred. **Fruit** a capsule that is variously dehiscent, rarely indehiscent.

1. Plants terrestrial; inflorescence twining, sometimes erect .... *U. scandens*
1. Plants aquatic; inflorescence erect 2
2. Stolons of 1 sort only .... *U. vulgaris, U. australis*
2. Stolons of 2 sorts, green with linear segments and few or no traps, or colourless with traps .... *U. intermedia, U. minor*

Utricularia scandens Benj. **Annual** to 35 cm, with rhizoids and capillary, branched stolons; traps globose, 0.5–1 mm long, with 2 simple, glandular, awl-shaped dorsal appendages. **Leaves** narrow-linear, 5–10 mm long, membranous, entire, 1-veined. **Inflorescence** erect to twining, 3–8-flowered; corolla yellow, 5–11 mm long, lower lip almost orbicular, spur longer than lower lip. swGuizhou, s & wYunnan [s & seAsia, Africa, nAustralia]; bogs, marshes, meadows, swamps, ponds, streamsides among grass, 700–2,900 m. VI–IX.

Utricularia vulgaris L. **Aquatic perennial** to 30 cm with thread-like, branched **stolons**; rhizoids also thread-like, dichotomously divided, traps borne on the leaf-segments, egg-shaped, 1.5–5 mm long, with 2 simple or branched, bristle-like appendages. **Leaves** numerous on stolons, 1.5–6 cm long, divided from base into 2 unequal primary segments. **Inflorescence** erect, 3–12-flowered; **corolla** yellow, 12–20 mm long, lower lip transversely elliptic to broadly ovate, with a prominent swelling at base, margin strongly deflexed; spur shorter or ± as long as lower lip. Gansu, Sichuan, Shaanxi, Tibet [c & nChina, temperate northern hemisphere]; lakes, pools, ditches, rivers, rice paddies, to 3,500 m. VI–VIII.

Utricularia australis R. Br. Similar to the previous species, but **flowers** with spreading margins to lower corolla-lip that lacks hairs. Guizhou, Hubei, Sichuan, sShaanxi, Yunnan [c & eChina, temperate Europe and Asia, seAsia, Australia, New Zealand]; habitats similar to those of *U. vulgaris*, to 2,500 m. V–VIII.

Utricularia intermedia Hayne Very similar in general appearance to *U. vulgaris*, but with 2 sorts of **stolons** and **corolla** in which the lower lip has a prominent rounded basal swelling, and awl-shaped spur that is slightly shorter than lower lip. Sichuan, Tibet [nChina, temperate northern hemisphere]; bogs, lakes, ponds, wet ditches, 300–4,000 m. VI–VIII.

Utricularia minor L. Like *U. intermedia*, but lower lip lacks a prominent rounded swelling and is much longer with a raised marginal rim, whereas the spur is half as long as the lower lip. Sichuan, Tibet, Yunnan [nChina, temperate northern hemisphere, New Guinea]; habitats similar to those of *U. intermedia*, 3,100–3,700 m. VI–VIII.

CAMPANULACEAE

CAMPANULA

A genus of c. 300 species (17 in China) in north temperate regions, particularly Europe and western Asia. Mainly **biennial** or **perennial** herbs with alternate **leaves**. **Inflorescence** solitary, racemose or paniculate; calyx with tube fused to ovary and 5 spreading to appressed lobes; **corolla** campanulate to tubular, 5-lobed; stamens 5, free; ovary inferior, 3-celled, with a 3-lobed style. **Fruit** a capsule dehiscing by 3 basal or terminal pores.

Campanula crenulata Franch.* Tufted **perennial** 10–30 cm tall with slender erect **stems** and scattered linear to linear-lanceolate **leaves**; basal leaves long-stalked, ovate to ovate-heart-shaped, 10–16 × 8–16 mm, with a few coarse marginal teeth. **Flowers** solitary or several, pendent; sepals linear-lanceolate, 5–10 mm long, appressed to corolla; corolla blue or mauve-blue, narrow-campanulate, 15–26 mm long. w & swSichuan, nwYunnan, possibly neighbouring parts of Tibet; grassy and rocky meadows, rock-crevices, low open mountain scrub, 3,650–4,575 m. VI–IX.

Campanula aristata **Wall.*** Readily distinguished from the previous species by the few linear **stem-leaves** and by the ± horizontal **flower** with a long slender ovary; calyx-lobes 7–15 mm long, each with a stiff bristle, often longer than corolla; corolla purple-blue, relatively small, 9–11 mm long. (?)sGansu, (?)seQinghai, w, nw & swSichuan, seTibet; stony meadows, rocks, cliffs, 3,350–4,400 m. VII–IX.

ADENOPHORA

A genus of c. 60 species (30 in China), mostly in the Himalaya, China and Japan, closely related to *Campanula* but distinguished primarily by the presence of a tubular or glandular epigynous disc at base of the style.

> 1. Small plants not more than 40 cm tall with solitary or few-flowered racemes .... **A. jasionifolia, A. coelestis**
> 1. Larger plants 60–120 cm tall with many-flowered, often branched, inflorescences 2
> 2. Corolla small, not more than 13 mm long .... **A. capillaris, A. gracilis**
> 2. Corolla larger, 17–32 mm long 3
> 3. Leaves entire or slightly toothed .... **A. confusa, A. khasiana**
> 3. Leaves deeply toothed to lobed .... **A. potaninii**

Adenophora jasionifolia **Franch.*** Small tufted **perennial** to 30 cm but generally 10–15 cm, with alternate, linear-lanceolate to linear-elliptic, pointed **leaves**, to 42 × 4 mm, entire to slightly toothed. **Flowers** solitary or several, on slender stalks; sepals narrowly awl-shaped, 3–5 mm long, spreading to recurved; corolla pale blue to violet-blue, broad-campanulate, 12–16 mm long. w & swSichuan, (?)eTibet; rocky meadows, rock-crevices and cliff-ledges, 3,100–4,000 m. VII–IX.

Adenophora coelestis **Diels*** Rather like a larger version of the previous species, but plants not more than 60 cm; **leaves** glabrous or only slightly hairy; **flowers** mostly in simple unbranched **racemes**, sometimes reduced to just a few flowers; corolla pale to deep blue, 26–32 mm long; and with an included style. nwYunnan; alpine meadows and rocky slopes, low alpine scrub, 3,050–3,960 m. VI–IX.

Adenophora capillaris **Diels*** Tufted **herbaceous perennial** to 1.2 m with slender, erect or arching, slightly hairy **stems**. **Leaves** elliptic to elliptic-lanceolate, 7–14 × 2.4–3.8 cm, sparsely hairy and with a slightly toothed margin, upper leaves linear-lanceolate. **Flowers** in airy spray-like racemes on very delicate, thread-like stalks; sepals awned, 2–5 mm long, spreading; corolla bluish white with darker markings in throat, tubular-urn-shaped, 8–13 mm long, with a long-protruding style. wHubei, wSichuan, seTibet, nwYunnan; woodland margins, open shrubberies, rocky places, grassy banks, often in partial shade, 1,450–3,900 m. VII–X.

1. *Pinguicula alpina*; upper Gangheba, Yulongxueshan, nwYunnan (CGW)
2. *Campanula crenulata*; Daxueshan, nwYunnan (MN)
3. *Campanula aristata*; north of Aba, nwSichuan (CGW)
4. *Adenophora jasionifolia*; Genyanshan, south side, swSichuan (AD)

1. *Adenophora gracilis*; near Huanglong, nwSichuan (CGW)
2. *Adenophora confusa*; Cangshan, nwYunnan (CGW)
3. *Adenophora khasiana*; Cangshan, nwYunnan (CGW)
4. *Adenophora potaninii*; Xunhua, cQinghai (RS)
5. *Platycodon grandiflorum*; cult. (CGW)
6. *Codonopsis vinciflora*; north of Songpan, nwSichuan (CGW)
7. *Codonopsis forrestii*; Cangshan, nwYunnan (CGW)
8. *Codonopsis forrestii*; cult., from Cangshan, nwYunnan (CGW)
9. *Codonopsis pinifolia*; Cangshan, nwYunnan (CGW)

Adenophora gracilis **Nannf.*** Similar to the previous species, but **leaves** narrow-elliptic to linear, entire or scarcely toothed, glabrous, not more than 8 mm wide, **corolla** smaller, 5–8 mm long, blue, sometimes with purple tinge. s & swGansu, seQinghai, nwSichuan; grassy and swampy meadows, banks, 2,700–3,400 m. VII–IX.

Adenophora confusa **Nannf.*** Rather robust **herbaceous perennial** 60–120 cm tall with erect or arching **stems**. **Leaves** alternate, short-stalked, blade broad-elliptic, to 7 × 4 cm, finely downy overall, upper leaves smaller and narrower than lower. **Inflorescence** a laxly-branched panicle; sepals glossy green, narrowly awl-shaped, 5–7 mm long, strongly recurved; **corolla** pale blue to violet, campanulate, 18–21 mm long, lobes recurved; style protruding. Guizhou, s & swSichuan, Yunnan; forests and forest margins, shrubberies, streamsides, rocky places, often in partial shade, 1,650–3,300 m. VII–IX.

Adenophora khasiana **Feer*** (syn. *A. diplodonta* Diels) Similar to the previous species in stature, but **flowers** generally pale to mid-blue or occasionally purple-blue, sepals with gland-toothed margins and corolla with non-recurved lobes, 17–23 mm long, while the style is not or only slightly protruding; w & nwYunnan; habitats similar to those of *A. confusa*, 2,800–3,550 m. VIII–X.

Adenophora potaninii **Korsh.*** Erect, rough-hairy **perennial** to 90 cm. **Leaves** stalkless, alternate or pseudo-whorled, elliptic to elliptic-oblanceolate, to 5.6 × 2.1 cm, with several coarse, incised teeth or lobes on each side. **Inflorescence** racemose, generally with 1–several short branches in lower half; sepals narrowly awl-shaped, 5–6 mm long, recurved; **corolla** blue or mauve-blue, campanulate, 22–25 mm long, lobes slightly recurved; style not or slightly protruding. s & swGansu, Qinghai, nwSichuan, e Tibet; rocky slopes, meadows, field boundaries, 2,700–3,200 m. VII–VIII.

PLATYCODON

A genus of a single species found in China, Korea and Japan. Closely related to *Campanula*, but differing primarily in the large balloon-like flower-buds, ascending to erect **flowers**, and the 5-celled ovary and 5-lobed style.

Platycodon grandiflorum **(Jacq.) A. DC.** Glabrous **herbaceous perennial** to 80 cm, often less, with erect **stems**. **Leaves** sessile, alternate or pseudo-whorled, ovate to lanceolate, 4–6.5 × 1.4–2.5 cm, toothed. **Flowers** usually solitary; calyx bowl-shaped, ribbed, 10–12 mm long, with 5 narrow-triangular lobes; corolla blue or purple-blue, 3.5–4 cm, with broad-ovate lobes. Gansu, Hubei, Shaanxi [n & nwChina, Korea, Japan, eRussia]; meadows, scrub, widely cultivated for its roots which are used as an anti-inflammatory in TCM. 800–2,200 m. VII–IX.

CODONOPSIS

A genus of c. 35 species (22 in China) concentrated primarily in the Himalaya and China, but extending into west Asia and the Malay Peninsula. Tuberous-rooted, generally **herbaceous perennials**, with erect or twining **stems**. **Leaves** basal or alternate on the stems, sometimes opposite or whorled. **Flowers** solitary or in branched inflorescences, foetid in the campanulate-flowered species; calyx 5-lobed; corolla saucer-shaped, campanulate or tubular, 5-lobed. **Fruit** a 5-valved capsule.

SYNOPSIS OF GROUPS

Group One. Flowers saucer-shaped, unscented; twining plants.

Group Two. Flowers tubular to campanulate, foetid; twining plants (p. 472)

Group Three. Flowers as Group Two, but low non-twining plants (p. 472)

> **Group One**
> 1. Sepals separate in bud .... *C. vinciflora*
> 1. Sepals fused together at first and enveloping corolla in bud .... *C. convolvulacea, C. forrestii, C. pinifolia*

Codonopsis vinciflora **Kom.** Twining **climber** with delicate thread-like **stems** to 1 m, with alternate or opposite, ovate to lanceolate, toothed **leaves**, green above, glaucous beneath; leaf-stalks short. **Flowers** saucer-shaped, blue to lavender-blue or bluish purple, 32–44 mm across, corolla divided almost to the base into 5 elliptic, well-separated lobes. sGansu, seQinghai, w, nw & swSichuan, seTibet, w & nwYunnan [eHimalaya, nMyanmar]; grassy banks, shrubberies, streamsides, 2,700–4,050 m. V–VIII.

Codonopsis convolvulacea **Kurz** Superficially very similar to *C. vinciflora*, but a more vigorous plant to 2 m with more leathery, ovate to ovate-lanceolate **leaves** to 6 cm long, with an entire or only slightly toothed margin and a stalk often half the length of the blade. **Flowers** larger, 40–55 mm across, clear deep blue to mid-blue, occasionally white. seTibet, wYunnan [swTibet, eHimalaya, nMyanmar]; shrubberies, open forests, 1,600–3,500 m. VI–IX.

Codonopsis forrestii **Diels*** (syn. *C. convolvulacea* Kurz var. *forrestii* (Diels) Ballard) Like *C. convolvulacea*, but a far more vigorous plant to 3.5 m with firm, ovate **leaves** to 10 cm long, blade deep green above, glaucous beneath, short-stalked, margin coarsely toothed. **Flowers** larger, 45–65 mm across, mid-blue to lavender-blue, with darker veins and a reddish purple 'eye'. s & swSichuan, nwYunnan; shrubberies and forest margins, 1,500–3,600 m. VII–IX.

Codonopsis pinifolia **(Hand.-Mazz.) Grey-Wilson*** (syn. *C. convolvulacea* Kurz var. *pinifolia* Hand.-Mazz.) Distinctive small plant to 30 cm, occasionally taller, **stem** twining only in upper, non-leafy part. **Leaves** crowded in lower part of plant, alternate, linear to linear-lanceolate, to 9.6 × 0.6 cm wide, entire. **Flowers** blue or lavender-blue, 30–45 mm across with a dark maroon 'eye'. s & swSichuan, nwYunnan; grassy and sparsely shrubby mountain slopes, 2,800–3,200 m. VIII–IX.

> **Group Two**
> 1. Flowers green or greenish yellow ....
>   .... *C. tubulosa, C. benthamii, C. macrocalyx,*
>   *C. tangshen, C. subglobosa*
> 1. Flowers white .... *C. javanica*

Codonopsis tubulosa Kom.* **Herbaceous climber** to 1.5 m, with alternate lanceolate to ovate-lanceolate, short-stalked **leaves**, margins shallowly toothed. **Flowers** pendent, solitary, terminal, sometimes accompanied by 1–2 lateral flowers; calyx green, campanulate, lobes 7–14 mm long; corolla tubular-campanulate, green or greenish yellow, 35–43 mm long, sometimes with faint purplish veining, often maroon-marked inside. s & swSichuan, nwYunnan; alpine meadows and amongst dwarf shrubs, 1,800–3,500 m. VI–IX.

Codonopsis benthamii Hook. f. & Thomson In contrast to the previous species, a partly **climbing** to **trailing** plant; with ovate-triangular **leaves** with a coarsely toothed or somewhat lobed margin; **flowers** smaller, calyx with lobes 9–14 mm long, corolla tubular, 17–26 mm long, green, sometimes with purple veins. seTibet [swTibet, c & eHimalaya]; forests and forest margins, banks, open shrubberies, 1,220–3,810 m. VII–VIII.

Codonopsis macrocalyx Diels* Very similar to the preceding species, but **leaves** thinner and larger, triangular-ovate, coarsely toothed or lobed and long-stalked; calyx larger, lobes 15–25 mm long; **corolla** 20–40 mm long, more obviously narrowed in the middle. nwYunnan; mainly shrubberies, 2,400–3,750 m. VII–IX.

Codonopsis tangshen Oliv.* A more vigorous **climber** than the preceding species, to 3 m with rather fleshy ovate-lanceolate to lanceolate, stalked, minutely hairy **leaves** up to 6 cm long. **Flowers** solitary, terminal or lateral; calyx with spreading lobes, to 18 mm long with slightly back-rolled margins; corolla campanulate, greenish with purple spots and stripes, often chequered or spotted inside, 30–37 mm long, with spreading triangular lobes. Hubei, Sichuan; mountain shrubberies, thickets, 1,550–2,500 m. VI–IX.

Codonopsis subglobosa W. W. Sm.* Vigorous twiner to 4 m with downy, ovate **leaves** to 2.6 × 2.2 cm, readily distinguished by the large, broad-campanulate **flowers** borne on slender, downy stalks, corolla 20–30 mm long, greenish or creamish, often blotched purple at base, edged or flushed with plum-purple on lobes; sepals spreading, 7–9 mm long, minutely toothed. s & swSichuan, w & nwYunnan; rocky woodland, limestone cliffs, shrubberies, 2,750–3,650 m. VII–IX.

Codonopsis javanica Hook. f. & Thomson Vigorous **herbaceous climber** to 3 m, often less, with slender, green, glabrous **stems**. **Leaves** opposite or alternate, long-stalked, heart-shaped to ovate, to 6 cm long, margin usually slightly toothed. **Flowers** solitary, lateral or occasionally terminal; calyx with erect to somewhat spreading, pale green, lanceolate lobes; corolla white or white flushed greenish yellow, broadly campanulate, 14–22 mm long. Guizhou, sSichuan, Yunnan [Myanmar, seAsia]; open forests and amongst large shrubs, occasionally clambering over various herbs, 1,000–2,500 m. VI–X.

> **Group Three**
> 1. Flowers greenish yellow to bluish green, net-veined ...
>   .... *C. meleagris, C. subscaposa*
> 1. Flowers blue or lilac-blue, not net-veined 2
> 2. Corolla distinctly narrowed in the middle .. *C. bulleyana*
> 2. Corolla not narrowed in the middle ....
>   .... *C. nervosa, Codonopsis* sp.

Codonopsis meleagris Diels* A very distinctive **herbaceous perennial** to 35 cm with leaves mainly in basal rosettes; **leaves** elliptic-oblong, to 8 cm long, stalkless to short-stalked, with a slightly toothed margin, hairy, especially beneath. **Flowers** solitary and terminal, occasionally lateral, borne on almost leafless scapes; calyx bluish green with short erect lobes; corolla greenish yellow to bluish green, heavily net-veined with chocolate-brown, spotted purple within, 25–35 mm long, ± barrel-shaped. nwYunnan (centred on Lijiang); grassy clearings in *Pinus* woodland, 2,500–3,600. VII–IX.

Codonopsis subscaposa Kom.* Similar to the previous species, but with longer-stalked, less hairy **leaves** and smaller, less striking **flowers**. w & swSichuan; similar habitat and altitudes. VII–VIII.

Codonopsis bulleyana Forrest & Diels* Tufted **herbaceous perennial** to 50 cm, often less, with erect to somewhat decumbent **stems**, branched towards base into leafy sterile laterals; **leaves** small, heart-shaped to ovate, not more than 15 mm long, densely hairy, entire. **Flowers** solitary, pendent on long stems devoid of leaves in upper half, foul-smelling; corolla pale blue with deeper veining, tubular-campanulate, narrowed in the middle and with the lobes flaring outwards, 27–32 mm long overall. s & swSichuan, seTibet, nwYunnan; shrubberies, grassy places, streamsides, 3,200–3,600 m. VI–IX.

Codonopsis nervosa (Chipp) Nannf.* (syn. *C. ovata* Spreng. var. *nervosa* Chipp) **Herbaceous perennial** to 50 cm with spreading softly hairy shoots, mainly towards base of hairy **stems**. **Leaves** rather crowded, soft green, ovate, to 8–18 × 10–20 mm, downy, margin slightly toothed or entire. **Flowers** solitary, on long, hairy, erect to ascending, leafless scapes, often purple-tinged; calyx green, flushed purple, with spreading ovate lobes; corolla whitish or pale blue or lilac with a filigree of deeper veins, tubular-campanulate, 20–32 mm long. sGansu, c & sQinghai, w & nw Sichuan, eTibet; alpine meadows and banks, open shrubberies, 3,660–4,120 m. VI–VIII.

1. *Codonopsis benthamii*; cult. (RLS)
2. *Codonopsis tangshen*; cult. (JL)
3. *Codonopsis subglobosa*; Wengshui, nwYunnan (EL)
4. *Codonopsis javanica*; Xishan, Kunming, cYunnan (CGW)
5. *Codonopsis meleagris*; cult., from Yunnan (RLS)
6. *Codonopsis bulleyana*; Baimashan, nwYunnan (CGW)
7. *Codonopsis nervosa*; Balangshan, Wolong, wSichuan (CGW)

1. *Codonopsis* sp.; south of Litang, swSichuan (MN)
2. *Leptocodon gracilis*; eNepal (CGW)
3. *Cyananthus lobatus*; nNepal (CGW)
4. *Cyananthus sherriffii*; cult., from seTibet (CGW)
5. *Cyananthus pedunculatus*; eNepal (CGW)
6. *Cyananthus pedunculatus*; nNepal (CGW)
7. *Cyananthus macrocalyx*; Matang, Zhegushan, nwSichuan (HJ)
8. *Cyananthus flavus*; Shu La on Meilixueshan, east side, nwYunnan (MN)
9. *Cyananthus incanus*; ne Nepal (CGW)
10. *Cyananthus incanus* ×*C. macrocalyx*; Genyanshan, swSichuan (MN)

Codonopsis* sp. Small **herbaceous perennial** to 30 cm with spreading, softly hairy shoots, similar to the previous species but a smaller plant, with smaller, shallowly toothed **leaves** and broader, more funnel-shaped corolla, pale milky-blue with very faint veins, lobes longer and more triangular. swSichuan; similar habitats and altitudes. VII.

LEPTOCODON

A genus with a single Sino-Himalayan species closely related to *Codonopsis* and distinguished mainly by the presence of an epigynous **disc** in the form of 5 club-shaped glands at the base of the tubular, scarcely lobed **corolla**.

***Leptocodon gracilis* (Hook. f. & Thomson) Lem.** (syn. *Codonopsis gracilis* Hook. f. & Thomson) Rather delicate, twining, slightly hairy **herbaceous perennial** to 2 m, generally less, with thread-like **stems** and ovate, coarsely toothed **leaves**, 7–27 × 6–29 mm. **Flowers** pendent, solitary, on very slender stalks; calyx with 5 narrowly heart-shaped, slightly toothed lobes, 4–6 mm long; corolla mauve to violet-blue or azure-blue, tubular, flared towards the top, 27–40 mm long. swSichuan, seTibet, nwYunnan [swTibet, c & eHimalaya]; deciduous forests, shrubberies, streamsides, banks, 1,800–3,350 m. VII–IX.

CYANANTHUS

A genus of c. 23 species (15 in China) in the Himalaya, north Myanmar and south-west China. **Perennial**, occasionally **annual**, **herbs**, with **stems** often radiating from a fleshy or rather woody caudex. **Leaves** alternate, toothed or not. **Flowers** solitary at branch-tips or in a lax **inflorescence**; calyx tubular to campanulate, with 5 short lobes; corolla generally blue, violet or yellow, with short to long tube and 5 spreading lobes, often with tufts or hairs in throat. **Fruit** a many-seeded capsule. All the species included here are perennials.

1. Stalks and calyx covered with dense dark brown to blackish hairs . . . . **C. lobatus, C. sherriffii, C. pedunculatus**
1. Stalks and calyx glabrous or with soft, pale brown hairs 2
2. Flowers cream to yellow . . . . **C. macrocalyx, C. macrocalyx subsp. spathulifolius, C. flavus**
2. Flowers blue, purple or violet 3
3. Plants grey-hairy throughout; leaves ovate-lanceolate to spathulate . . . . **C. incanus**
3. Plants sparsely hairy; leaves elliptic to obovate or almost orbicular . . . . **C. formosus, C. delavayi, C. longiflorus**

***Cyananthus lobatus* Wall. ex Benth.** **Herbaceous perennial** to 25 cm with radiating leafy **stems**. **Leaves** obovate, lobed or deeply toothed in upper half, to 25 × 14 mm, somewhat hairy. **Flowers** periwinkle-like, solitary on black-hairy stalks; calyx tubular-campanulate, 15–20 mm long, black-hairy; corolla blue to violet-blue or indigo-purple, 28–38 mm across, lobes broad-obovate, with hair tufts at base of lobes. swSichuan, seTibet, nwYunnan [swTibet, Himalaya, Myanmar]; alpine meadows, limestone crags, low grassy scrub, 3,200–4,900 m. VII–IX.

Cyananthus sherriffii* Cowan Very closely related to the previous species and very similar in **flower** shape, but **leaves** elliptic, with slightly toothed, often somewhat back-rolled margin and the calyx is less pronouncedly black-hairy; corolla pale blue. seTibet [swTibet]; habitats and altitudes similar to those of *C. lobatus*. VII–IX.

***Cyananthus pedunculatus* C. B. Clarke** Distinguished from the previous 2 species by the oblong-elliptic, entire **leaves** and the more funnel-shaped, blue or violet-blue corolla, without hair tufts in mouth, mouth and tube generally longer and paler. seTibet [swTibet, c & eHimalaya]; grassy banks, meadows, 3,900–4,700 m. VII–IX.

***Cyananthus macrocalyx* Franch.** **Herbaceous perennial** to 15 cm, with radiating leafy **stems**. **Leaves** short-stalked to almost stalkless, elliptic-obovate, 7–14 × 5–9 mm, shallowly lobed, sparsely hairy to bristly beneath. **Flowers** solitary, on a short, hairy stalk; calyx campanulate, somewhat inflated, 10–14 mm long, glabrous, pale greenish yellow; corolla pale lemon-yellow, with purple lines on tube, 23–35 mm long overall, lobes spreading, elliptic, strongly bearded at base. nw, w & swSichuan, seTibet, nwYunnan [eHimalaya (**subsp. spathulifolius (Nannf.) Shrestha)**]; grassy and stony alpine meadows, banks, 3,900–4,500 m. VII–IX.

Cyananthus flavus* C. Marq. (syn. *C. albiflorus* Marq.) Similar to the previous species, but **leaves** are moderately hairy, **flowers** on a longer stalk (at least 15 mm long) and calyx is less inflated and somewhat glaucous; corolla primrose-yellow. s & swGansu, Sichuan, (?)eTibet, nwYunnan; habitats similar to those of *C. macrocalyx*, 3,200–3,700 m. VII–IX.

***Cyananthus incanus* Hook. f. & Thomson** Low **herbaceous perennial** to 10 cm with radiating prostrate **stems**. **Leaves** ovate-lanceolate to elliptic or spathulate, to 9 × 3.5 mm, appressed grey-hairy above, white-downy beneath, margin somewhat recurved, entire. **Flowers** solitary; calyx tubular, 5–9 mm long, white- to brownish-hairy, occasionally almost glabrous; corolla deep blue to violet-blue, tubular to salverform, 13–22 mm long, with a dense beard at base of each oblong lobe. w & swSichuan, seTibet, nwYunnan [swTibet, c & eHimalaya]; grassy and stony alpine meadows, damp rocks, banks, 3,000–4,850 m. VII–IX. Sometimes hybridises with *C. macrocalyx* where the two grow in close proximity.

Cyananthus formosus* Diels Prostrate **herbaceous perennial** with radiating **stems**. **Leaves** hoary-grey, ovate to fan-shaped, to 5 × 6 mm, increasing in size up stem, lobed in upper half, bristly-hairy. **Flowers** solitary, terminal; calyx tubular-campanulate, 13–16 mm long, hairy; corolla bright blue to purple-blue, 24–32 mm long overall, tubular to salverform, bearded at base of spreading lobes. swSichuan, seTibet, nwYunnan; stony and grassy alpine meadows, banks, 3,800–4,600 m. VII–X.

Cyananthus delavayi* Franch. Similar to the previous species, but **leaves** almost glabrous above, white-downy beneath, more shallowly lobed, and **flowers** on short lateral shoots as well as being terminal, smaller, 15–20 mm across; calyx 9–10 mm long; corolla 18–24 mm long overall. nwYunnan; similar habitats as well as open *Pinus* woodland, 3,000–3,800 m, VIII–X.

Cyananthus longiflorus* Franch. Readily distinguished from the previous 2 species by the deep green or grey-green, elliptic or lanceolate-elliptic **leaves** that are crowded towards shoot-tips, silvery with hairs beneath, and by the large deep violet-blue **flowers**, 38–50 mm long overall, tube to 30 mm long, and lobes narrow-elliptic, spreading; calyx hairy, 10–17 mm long, inflating in fruit. nwYunnan; open *Pinus* woodland, stony and grassy banks, 2,540–3,150 m. VIII–X.

LOBELIA

A genus of 365 species (3 in China) with a worldwide distribution. Readily distinguished from other genera in the Campanulaceae in this account by the zygomorphic **flowers**, the corolla 2-lipped with upper lip 2-lobed and lower lip 3-lobed.

***Lobelia seguinii* H. Lévl. & Vaniot** Somewhat hairy **herb** to 1 m, generally less, with erect **stems** and stalkless, lanceolate to linear-lanceolate **leaves**, these to 16 × 4 cm, with margin minutely toothed. **Flowers** in terminal racemes, sometimes branched, often resupinate; calyx 15–18 mm long, with linear, slightly toothed lobes; corolla bluish or violet, 22–24 mm long. Guizhou, Sichuan, seTibet, Yunnan [Himalaya]; damp places along streams, forest margins, shrubberies, 1,600–2,600 m. VIII–X.

RUBIACEAE

LUCULIA

A genus of 5 species (2 in China) in South East Asia and the Himalaya. Highly ornamental **shrubs** or small **trees**. **Leaves** paired; stipules located between opposing leaf-stalk bases. **Flowers** in terminal, many-flowered corymbs, very fragrant; calyx 5-lobed; corolla salverform, with 5 spreading lobes; stamens 5. **Fruit** a 2-valved capsule.

***Luculia gratissima* (Wall.) Sweet** Evergreen shrub, 2–4 m tall. **Leaves** elliptic to oblong, acuminate, 5–12 × 2–6 cm. **Corymbs** terminal, up to 20 cm across. **Flowers** pink, very fragrant; tube 30–50 mm; lobes 8–10 mm across. **Capsule** obovoid-oblong, 8–12 mm. Montane evergreen broad-leafed forests, 800–2,400 m, sometimes planted as an ornamental. eXizang, wYunnan [Bhutan, neIndia, Myanmar, Nepal, Bhutan, Vietnam]. ***L. pinceana* Hook.**, found up to 3,000 m in Guangxi, Guizhou, Xizang and Yunnan, differs in having flowers with a shorter tube and suborbicular to obovate corolla lobes with a lamellate appendage at the base

EMMENOPTERYS

A genus of 2 species of **deciduous trees** closely related to *Luculia*, native to west & south China, Myanmar and north Thailand. **Leaves** opposite; stipules located between opposing leaf-stalks. **Flowers** 5-parted, in corymbs or panicles. **Fruit** a spindle-shaped capsule containing winged seeds.

Emmenopterys henryi* Oliv. Medium to large **tree** attaining 25 m, though often only 8–12 m, with grey, rough, scaly **bark** and glabrous, purplish shoots. **Leaves** paired, ovate to oval, to 22 × 10 cm, apex pointed, rather fleshy and dull green, with reddish stalks. **Flowers** in large terminal rounded or pyramidal panicles; calyx small, urn-shaped, 6 mm long, 5-lobed, occasionally 1 lobe much expanded, white and bract-like, to 5 cm long; corolla white, funnel-shaped, 22–26 mm across, broadening into 5 spreading, rounded lobes. Sichuan, Yunnan; broadleaved forests, 800–1,500 m. VI–VII.

LEPTODERMIS

A genus of c. 30 species (11 in China) distributed mainly in the Himalaya, China and Japan. Twiggy, **deciduous shrubs** with opposite, simple, entire **leaves**. **Flowers** solitary or in small, short-stalked clusters; calyx 5-lobed; corolla trumpet-shaped, with 5 lobes that are valvate in bud; stamens and styles 5. **Fruit** a small 5-valved capsule containing 5 nutlets.

1. Inflorescences 2–3, not in heads; corolla broadly funnel-shaped ..... ***L. potaninii*, *L. potaninii* var. *tomentosa*, *L. potaninii* var. *glauca*, *L. forrestii***
1. Inflorescences in several-flowered clusters; corolla narrow .... ***L. pilosa*, *L. oblonga***

Leptodermis potaninii* Batalin*** (syn. *L. esquirolei* H. Lévl.) **Shrub** 1–1.2 m tall with dark red young **branches**. **Leaves** elliptic to ovate, 1–5 × 0.8–1.5 cm, shortly white-hairy above, hairy on veins beneath, shortly stalked. **Flowers** 1–3 at top of branchlets, 1.5 × 0.6–0.7 cm, white to pale pink with darker centre; stigma 4–5-lobed. Hubei, Sichuan, Yunnan; shrubberies, forest margins, banks, 1,800–4,500 m. IV–VIII. **Var. *tomentosa* H. Winkl.*** has small hairy **leaves**, 6–8 × 4 6 mm; **var. *glauca

1. *Cyananthus formosus*; Genyanshan, swSichuan (MN)
2. *Cyananthus delavayi*; cult., from Cangshan, nwYunnan (CGW)
3. *Cyananthus delavayi*; Yulongxueshan, nwYunnan (RM)
4. *Cyananthus longiflorus*; Zhongdian to Sanba, nwYunnan (HJ)
5. *Cyananthus longiflorus*; Cangshan, nwYunnan (CGW)
6. *Lobelia seguinii*; Dali, nwYunnan (PC)
7. *Luculia gratissima*; cult., Kunming (PC)
8. *Emmenopterys henryi*; cult. (TK)
9. *Emmenopterys henryi*; cult. (TK)
10. *Leptodermis pilosa*; south of Lijiang, nwYunnan (CGW)

1. *Leptodermis forrestii*; Muli, swSichuan (PC)
2. *Leptodermis forrestii*; Lijiang, nwYunnan (PC)
3. *Lonicera tragophylla*; Jiuzhaigou, nwSichuan (CGW)
4. *Lonicera similis* var. *delavayi*; cult. (CGW)
5. *Lonicera japonica*; Shennongjia, wHubei (PC)
6. *Lonicera chaetocarpa*; south of Jiuzhaigou, nwSichuan (CGW)
7. *Lonicera chaetocarpa*; Gangheba, Yulongxueshan, nwYunnan (CGW)
8. *Lonicera hispida*; Baimashan, nwYunnan (PC)

(Diels) H. Winkl.* has small glaucous **leaves**, up to 10 × 5 mm. Both varieties are found within the range of the species.

Leptodermis forrestii Diels* Similar to *L. potaninii*, but with elliptic to lanceolate **leaves**, 1–4 × 0.6–1.5 cm; **flowers** deep lavender, 1.1–1.5 × 1–1.5 cm; **stigma** 2–3-lobed. swSichuan, nwYunnan; open scrub on mountainsides, 2,200–3,450 m. V–VIII.

Leptodermis pilosa Diels* (syn. *L. microphylla* (H. Winkl.) H. Winkl.) **Shrub** to 2 m, often less, with downy young shoots and grey-green, hairy, ovate **leaves** 12–30 × 8–15 mm, apex pointed, base wedge-shaped. **Flowers** fragrant, in small lateral clusters, forming a pseudopanicle with adjacent clusters; corolla lavender, narrowly funnel-shaped, 11–15 mm long, finely downy outside. swSichuan, nwYunnan (especially around Lijiang); dry shrubberies, rocky places, roadsides, open *Pinus* woods, 2,800–3,700 m. VI–IX. (See photo 10 on p. 478.)

Leptodermis oblonga Bunge* **Shrub** 0.6–1.2 m tall with elliptic **leaves**, 1.4–2.6 × 0.7–0.8 cm, shortly stalked. **Flowers** pale mauve-pink or purple, 1.4 × 0.8 cm. wSichuan, Yunnan [nChina]; dry shrubberies, banks, rocky places, 2,200–3,300 m. VII–VIII.

CAPRIFOLIACEAE

LONICERA

A genus of 180 species (c. 60 in China) in the northern hemisphere. Small to large, **deciduous** or **evergreen bushes** or **climbers** with opposite, simple **leaves**. **Inflorescence** short, with flowers in pairs, clusters or pseudopanicles subtended by 2 bracts. **Flowers** regular to zygomorphic, corolla ± tubular, 5-lobed or 2-lipped (the upper lip 4-lobed, the lower 1-lobed); **stamens** inserted on corolla-tube. **Fruit** a berry.

1. Climbers 2
1. Non-climbing shrubs 3
2. Deciduous plants with bright yellow flowers .... *L. tragophylla*
2. Evergreen plants with white, cream or purplish red flowers .... *L. similis* var. *delavayi, L. japonica, L. henryi*
3. Flowers pendent 4
3. Flowers ± erect 5
4. Corolla yellow or white .... *L. chaetocarpa, L. hispida, L. mitis*
4. Corolla pink, purple or red .... *L. perulata, L. gynochlamydea*
5. Evergreen shrubs .... *L. nitida, L. pileata*
5. Deciduous shrubs 6
6. Flowers white or yellow, rarely with a pink tinge ...... *L. maackii* forma *podocarpa, L. maackii* forma *maackii, L. quinquelocularis, L. tangutica, L. webbiana, L.* aff. *obovata, L.* aff. *deflexicalyx, L. prostrata*
6. Flowers pink or purple .... *L. rupicola, L. syringantha*

Lonicera tragophylla Hemsl. ex Forbes & Hemsl.* **Climber** 4–15 m with woody, leafy **stems** and hollow branchlets. **Leaves** elliptic, 4.5–10 × 2–5.5 cm. **Inflorescence** terminal, bearing up to 20 flowers in a cluster subtended by a pairs of fused, stalkless bracts, 4–10 cm across, that encircle the stem saucer-like (perfoliate); corolla 2-lipped, with tube 5.5–6.5 cm long, golden-yellow with glossy throat, faintly perfumed, upper petal reflexed. **Berries** red when ripe. sGansu, Hubei, n, e & wSichuan [cChina]; scrub, woodland, often on woodland margins, 1,100–2,850 m. VI–VIII.

Lonicera similis Hemsl. var. *delavayi* (Franch.) Rehd. (syn. *L. delavayi* Franch.) Vigorous twining **climber** to 8 m with slender **stems**, hairy when young, and bright green ovate-lanceolate, pointed **leaves**, 5–12 cm long, with rounded to slightly heart-shaped base, slightly hairy to glabrous above, grey-downy beneath. **Flowers** paired but closely aggregated to form a panicle-like **inflorescences**; corolla white to cream, 2-lipped with slender tube 4.5–5 cm long, strongly scented. **Berries** black when ripe. w & swSichuan, Yunnan [nMyanmar]; forests and forest margins, ravines, 1,400–2,800 m. VII–VIII.

Lonicera japonica Thunb. Similar to the previous species, but **leaves** generally smaller, not more than 7.5 cm long, **corolla** white, sometimes purple-tinged, fading to yellow, tube shorter (not more than 35 mm long, generally shorter). Gansu, Hubei, n & neSichuan, Shaanxi [n, c & eChina, Japan, Korea]; forest margins, shrubberies, streamsides, 1,200–2,800 m. VI–X. Widely cultivated.

Lonicera henryi Hemsl. Species equally vigorous to the previous 2, but readily distinguished by the less hairy, more oblong, deep green **leaves** that are somewhat glossy beneath, and by the clusters of smaller purplish red **flowers** with tube only 12–18 mm long. **Berries** blackish purple when ripe. Gansu, Guangxi, Sichuan, seTibet, Yunnan [nMyanmar]; forests, cliffs, ravines, 900–2,400 m. VI–VII.

Lonicera chaetocarpa (Batalin ex Rehd.) Rehd.* **Deciduous shrub** 1–2 m tall with bristly, glandular young shoots. **Leaves** ovate-oblong, 6.5–11 × 3–5.5 cm, blunt or rounded at tip, hairy on margins and ± so on both surfaces, dark green above, grey-green beneath. **Flowers** in pairs on a short bristly stalk 15–20 mm long; bracts large, membranous, up to 2.5 cm long, ciliate; corolla regular, trumpet-shaped, pale yellow, 3–3.6 cm long, tube longer than rounded lobes. **Berries** scarlet when ripe, glandular. Gansu, Sichuan, seTibet, Yunnan [cChina]; open places in shrubberies, stream- and riversides, hedgerows, 2,700–4,300 m. V–VII.

Lonicera hispida Roem. & Schult.* (syn. *L. bracteata* Royle) Similar to *L. chaetocarpa*, but with smaller leaves and flowers, **leaves** oblong to oblong-ovate, truncate or mucronate, 3–8 cm long, densely short velvety-downy on both sides; leaf-stalks 3 mm long, minutely downy. **Bracts** broadly ovate, almost heart-shaped at base, 2–2.5 cm long, green, finely ciliate. **Flowers** nodding, in pairs; corolla white or yellowish, often with purplish tips, 1.5–3 cm long. Sichuan, seTibet, Yunnan; shrubberies, amongst rocks, stream- and riversides, alpine meadows, 2,700–4,400 m. VI–VII.

Lonicera mitis Rehd.* Similar to the two previous species, but a small **shrub** not more than 1 m tall, often no more than 70 cm, with purplish, glabrous branchlets and small elliptic **leaves** to 3.5 cm long; **corolla** pale yellow, broadly trumpet-shaped, not more than 2.2 cm long. swSichuan, seTibet, nwYunnan; rocky alpine meadows, alpine scrub, rocky places, 3,500–5,000 m. VI–VIII.

Lonicera perulata Rehd.* **Shrub** with stout branches and sparingly hairy, reddish to purple branchlets. **Leaves** membranous, oblong to oblong-ovate, 6.5–8 × 2.5–3.8 cm, sparingly hairy on veins above, glandular towards base; bracts awl-shaped, sparsely ciliate. **Flowers** in pairs on slender, sparsely glandular stalks 3–4 cm long; corolla 2-lipped, 13–15 mm long, dark dull purple, glabrous, tube slightly pouched and stipitate, densely hairy within, upper lip with short, broadly ovate lobes; stamens slightly shorter than corolla-limb. **Berries** 6–8 mm across, bright red when ripe. Hubei, Sichuan, seTibet, nwYunnan [cChina]; shrubberies in valleys and on slopes, 2,000–2,800 m. V–VII.

Lonicera gynochlamydea Hemsl.* Erect **deciduous shrub** 2–3 m tall with purple, glabrous shoots. **Leaves** lanceolate, drawn out to a tapering apex, 5–11 × 2–4 cm, downy along midrib above and in basal part beneath; leaf-stalks 3–4 mm long. **Flowers** in pairs in leaf-axils, on stalks 3–6 mm long; corolla white tinged with pink or purple, regular, 8–13 mm long, with stout tube and dilated base, downy outside; stamens and style ± downy. **Berries** white to purplish when ripe, semi-translucent. Hubei, Sichuan, seTibet, Yunnan; forest margins, mountain shrubberies, 1,900–4,300 m. V–VI.

Lonicera nitida Wilson* Dense **evergreen shrub** 1.5 to 3.5 m; young shoots slender, erect, purplish, sparsely bristly. **Leaves** small, leathery, ovate to almost circular, heart-shaped at base, 6–15 mm long, dark glossy green above, pale beneath, glabrous. **Flowers** in axillary pairs on a short stalk, fragrant; corolla creamy white, tube 6–7 mm long. **Berries** blue-purple when ripe, glossy and translucent. wSichuan, Yunnan; deciduous and mixed forests, cliffs, riversides, 1,800–3,600 m. V–VI.

Lonicera pileata Oliv.* Similar to the previous species, but a lower, more spreading **shrub** with larger, oblong **leaves**, 12–30 × 4–12 mm; corolla yellowish white, tube downy outside. **Berries** amethyst when ripe, adorned at top by an outgrowth of the calyx. wHubei, Sichuan; deciduous and mixed forests, rocky places, ravines, 1,700–3,400 m. V–VII.

Lonicera maackii (Rupr.) Maxim. forma *podocarpa* Rehd.* **Deciduous shrub** 3–4.5 m with **branches** spreading widely, often flat, young shoots downy. **Leaves** lanceolate, 3.7–7.5 × 1.2–3.8 cm, slender-pointed, dark green, downy on both sides. **Flowers** in pairs in a dense row on the upper side of branchlets, ovaries on short stalks 6 mm long; corolla 2-lipped, white turning yellow with age, sweetly scented, tube 6 mm long; stamens almost twice as long as corolla-tube, downy at base. **Berries** dark red when ripe. Sichuan, nwYunnan; hedgerows, hillside shrubberies, 2,000–2,800 m. IV–VI. The typical form (forma *maackii*) is found in nChina, Japan, Korea, Mongolia, eRussia.

Lonicera quinquelocularis Hardw. Smaller **shrub** than the previous species, to 2.5 m, with purple, hairy young branchlets; **leaves** ovate to lanceolate, to 5.5 × 2.5 cm; **flowers** borne on very short stalks, c. 2 mm long; corolla white or yellow; **berries** translucent, whitish when ripe. seTibet, (?)nwYunnan [swTibet, eHimalaya]; coniferous forest margins, shrubberies, road- and streamsides, 2,100–2,600 m. IV–VI.

Lonicera tangutica Maxim.* **Deciduous shrub** 1–2 m with greyish glabrous **branches**. **Leaves** oblong to obovate, 1–3 cm long, pointed, sparsely hairy, base ciliate. **Flowers** nodding, in pairs, with awl-shaped bracts; corolla white or yellowish tinged pink, tubular trumpet-shaped, 10–14 mm long. **Berries** scarlet when ripe, the pairs semi-fused. Gansu, Hubei, Sichuan, seTibet, Yunnan [cChina]; shrubberies, forest margins, stream- and riversides, 2,400–4,300 m. IV–VII.

Lonicera webbiana DC. Larger **shrub** than previous species, to 4 m with brown branchlets, usually with a few glandular hairs at nodes. **Leaves** broad-lanceolate, 6–11 × 2–3.5 cm. **Flowers** on stalks 5–10 mm long, with small linear bracts; corolla crimson, 5–6 mm long. **Berries** red when ripe, 6–9 mm. wHubei, seTibet, nwYunnan [swTibet, eHimalaya, nMyanmar]; mixed forests, often in rather damp places, streamsides, shrubberies, 2,450–3,840 m. IV–VI.

Lonicera aff. *obovata* Royle* **Shrub** to 3 m with spreading, rather delicate **branches**. **Leaves** small, elliptic-obovate, 8–12 mm long; bracts awl-shaped. **Flowers** pendent, in pairs; corolla yellow, funnel-shaped, 10–12 mm long. wSichuan (Kangding); woodland, roadsides, c. 2,800 m. VII.

1. *Lonicera mitis*; Yulongxueshan, nwYunnan (CGW)
2. *Lonicera perulata*; Bitahai, Zhongdian, nwYunnan (PC)
3. *Lonicera perulata*; Kangding, wSichuan (PC)
4. *Lonicera gynochlamydea*; Shennongjia, wHubei (PC)
5. *Lonicera nitida*; Wolong, wSichuan (PC)
6. *Lonicera nitida*; Chongjiang, nwYunnan (PC)
7. *Lonicera maackii* forma *podocarpa*; north of Dali, nwYunnan (PC)
8. *Lonicera quinquelocularis*; Bhutan (CGW)
9. *Lonicera tangutica*; Tianchi, nwYunnan (RS)
10. *Lonicera webbiana*; Shennongjia, wHubei (PC)
11. *Lonicera* aff. *obovata*; Kangding, wSichuan (PC)

1, 2, 3, 4, 5, 6, 7, 8, 9, 10, 11

Lonicera* aff. *deflexicalyx* Batalin Spreading **deciduous shrub** to 2.5 m, with purplish branchlets. **Leaves** deep, rather dull green, ovate-elliptic to elliptic-lanceolate, to 6 cm long, long-pointed, rough above, with 6–8 pairs of lateral veins, stalked. **Flowers** erect, in almost stalkless pairs in leaf-axils; corolla cream with a hint of pink, 2-lipped, with protruding stamens. nwYunnan (Bitahai); forests and forest margins, road- and tracksides, c. 3,200 m. VI–VII.

Lonicera prostrata* Rehd. Prostrate or spreading, **deciduous shrub**, sometimes forming low mounds, with numerous slender branches; young shoots purplish, slightly hairy. **Leaves** ovate to oval, 18–28 × 6–12 mm, downy at first, margin ciliate; leaf-stalks 2.5–3 mm long, hairy. **Flowers** upright, in pairs from the middle nodes, not fragrant; corolla pale yellow, 13–15 mm long, 2-lipped, corolla-tube relatively short. **Fruit** egg-shaped, 6–8 mm long, red when ripe. wSichuan, eTibet; wooded slopes, rocky places, banks, 3,400–4,200 m. VI–VII.

***Lonicera rupicola* Hook. f. & Thomson** (syn. *L. thibetica* Bur. & Franch.) Dense **shrub** 0.33–1.3 m, with glabrous or downy branchlets. **Leaves** dull green, paired or in 3s, linear to oblong-lanceolate, blunt, 10–25 × 3–8 mm, downy beneath; bracts linear, 4–5 mm long. **Flowers** almost regular, trumpet-shaped; corolla pale pink, slightly fragrant, 8–13 mm long. sGansu, w, nw & swSichuan, seTibet; nwYunnan [swTibet, c & eHimalaya]; open, coniferous forests, *Rhododendron* and *Juniperus* scrub, alpine grassland, 2,900–4,900 m. V–VIII.

Lonicera syringantha* Maxim. Similar and closely allied to *L. rupicola*, but up to 2 m, branchlets glabrous; **leaves** glaucous green, generally broader and glabrous beneath; **corolla** lilac-pink, slightly longer, lilac-scented. **Berries** crimson when ripe. s & swGansu, w, nw & swSichuan, seTibet, nwYunnan [swTibet, cChina]; alpine hillsides, shrubberies, 2,600–4,600 m. VI–VIII.

TRIOSTEUM

A genus of perhaps 6 species (2 in China) in Asia and North America. Coarse, erect, glandular and hairy **herbaceous perennials**, often growing in clumps. **Leaves** opposite, joined around **stem** at base, entire or pinnately lobed. **Flowers** zygomorphic, tubular, curved, 5-parted, upper corolla-lobe larger and rounded. **Fruit** a globose, 3-seeded berry.

***Triosteum himalayanum* Wall.** (syn. *T. hirsutum* Roxb.) Plants 30–60 cm; **leaves** 5.5–23 × 3–16 cm, rough, somewhat wavy, net-veined. **Inflorescence** 3–6 cm long; **corolla** brownish green, inside blotched red to purple, 11–15 mm long, glandular-hairy. **Berries** globose 11–16 mm in diameter, red when ripe. Hubei, Sichuan, seTibet, nwYunnan [swTibet, Himalaya]; glades, grassy and rocky slopes, scrub, open woods, 2,600–3,900 m. V–VII.

Triosteum pinnatifidum* Maxim. Plants 30–50 cm with ovate, lobed **leaves**, 5.5–23 × 3–20 cm, tapering at base. **Inflorescence** 3–6 cm long; corolla yellowish, 8–11 mm long. **Berries** globose 11–12 mm in diameter, red when ripe, hairy. Gansu, Hubei, Sichuan, Shaanxi [n & cChina]; glades, shady places by streams, scrub, open woods, 2,600–3,800 m. V–VI.

LEYCESTERIA

A genus of just 6 species (3 in China) in China and the Himalaya. Small to medium-sized, **deciduous shrubs** with hollow **stems** and opposite, simple **leaves**. **Inflorescence** short, terminal or axillary, spike-like; **flowers** ± regular; corolla 5-lobed, tubular in lower half and slightly pouched at base; stamens 5, inserted on corolla-tube. **Fruit** a many-seeded berry.

***Leycesteria formosa* Wall.** Semi-woody, glabrous, leafy **shrub**, with arching, green **stems** 1.2–5 m long. **Leaves** ovate, with heart-shaped base, 5–17.5 × 2.5–8.5 cm, long-pointed, slightly downy beneath when young. **Inflorescence** arching-pendent, axillary or terminal, many-flowered; bracts 1–3 cm long, generally purplish; calyx claret-purple; **corolla** funnel-shaped, 16–18 mm long, white or purplish, lobes glandular-hairy. **Berries** reddish purple, glandular-downy, 6–13 mm in diameter, surrounded by the persistent calyx. w & swSichuan, seTibet, Yunnan [Himalaya]; shady places in woods and forests, shrubberies, streamsides, 1,500–3650 m. VII–X.

DIPELTA

A genus of 4 species, endemic to China. Small to large, **deciduous shrubs**, with simple, opposite **leaves**. **Inflorescence** few-flowered; **flowers** showy; corolla bell- or funnel-shaped, subtended by 4 floral bracts at base of ovary, 2 enclosing base of the fruit. **Fruit** dry, disc-shaped, rather like those of *Ulmus*.

1. Corolla-tube and style downy .. **D. ventricosa, D. elegans**
1. Corolla-tube and style glabrous ....
.... **D. floribunda, D. yunnanensis**

Dipelta ventricosa* Hemsl. **Deciduous**, spreading **shrub** 3–6 m tall with downy shoots. **Leaves** elliptic to elliptic-lanceolate, long-pointed, 5–12.5 × 1.8–5 cm, margin downy and with a few

---

1. *Lonicera* aff. *deflexicalyx*; Bitahai, Zhongdian, nwYunnan (PC)
2. *Lonicera prostrata*; Danyun, nwSichuan (PC)
3. *Lonicera rupicola*; wSichuan (TK)
4. *Lonicera rupicola*; Songpan, nwSichuan (CGW)
5. *Lonicera syringantha*; south of Aba, nwYunnan (CGW)
6. *Triosteum himalayanum*; Kangding, wSichuan (PC)
7. *Triosteum himalayanum*; Zhongdian, nwYunnan (PC)
8. *Triosteum pinnatifidum*; Jiuzhaigou, nwSichuan (PC)
9. *Leycesteria formosa*; cult. (CGW)
10. *Dipelta ventricosa*; cult. (CGW)
11. *Dipelta ventricosa* in fruit; cult. (CGW)

gland-tipped teeth. **Inflorescence** abundant, 1–6-flowered, axillary on short twigs, with fragrant **flowers**; **bracts** rounded or oval, attached near their base; corolla campanulate, 2.5–3.3 cm long, 2.5 cm across, with 5 rounded lobes, white to pale pink, darker pink outside, orange in throat; stalks hairy, 10–12 mm long. sGansu, nSichuan; mixed woods by streams and rivers, 2,200–2,500 m. V–VI.

Dipelta elegans Batalin* Very similar to the previous species, but with slightly larger, more funnel-shaped **flowers**. sGansu, n & nwSichuan [c & nChina]; mixed woods, especially near streams, 2,000–2,800 m. V–VI.

Dipelta floribunda Maxim.* Spreading **shrub** 3–5 m tall with peeling **bark**, shoots downy to glandular-downy. **Leaves** ovate to elliptic-lanceolate, 5–10 × 1.5–3.2 cm, long-pointed, entire, downy on both sides when young. **Inflorescence** abundant, 3-flowered, axillary on short twigs, **flowers** fragrant; floral **bracts** shield-like, attached by their centres; corolla white to pale pink, darker pink outside and yellow in throat, funnel-shaped, the tube pouched on lower side, 2.5–3.3 cm long, 1.8–2 cm across, with 5 rounded lobes. Hubei, Sichuan, Shaanxi, Yunnan [n & cChina]; mixed woods, especially near streams, 1,000–2,500 m. IV–V.

Dipelta yunnanensis Franch.* Similar to the previous species, but 2–4 m tall, with 4-angled, minutely downy shoots bearing abundant, 1–6-flowered **inflorescences** borne on short, leafy, lateral twigs; **corolla** creamy white, flushed pink on outside, yellow in throat, more tubular, 1.8–2.5 cm long, 2.5 cm across, with 5 short, rounded lobes. w & swSichuan, nwYunnan; scrub and open woodland, forest margins, grassy slopes, 2,600–3,900 m. V–VI.

ABELIA

A genus of c. 60 species (18 in China) in north temperate parts of the Old and New Worlds. Small **deciduous** or **evergreen shrubs** with arching **stems**. **Leaves** entire, opposite or in 3s. **Inflorescence** short, 1–few-flowered; calyx with 2–5 lobes, persistent and enlarging in fruit; corolla 5-lobed, funnel-shaped to campanulate; stamens 4, inserted on corolla. **Fruit** a small, many-seeded capsule.

Abelia engleriana (Graebn.) Rehd.* **Deciduous shrub**, 1–2 m tall with brown peeling **bark**. **Leaves** ovate-lanceolate, 1.8–3.7 × 0.8–2 cm, pointed, stiffly ciliate. **Flowers** in pairs on short shoots; calyx-lobes 2, ciliate; corolla funnel-shaped, curved, 2 cm long, rose-pink, fragrant, downy within; stamens downy at base. wHubei, Sichuan, Yunnan [cChina]; woods, often by streams and rivers, 2,500–3,000 m. V–VIII.

Abelia schumannii (Graebn.) Rehd.* **Deciduous shrub** with arching, purplish **stems**. **Leaves** ovate, blunt, 12–23 × 8–13 mm, downy on midvein beneath. **Flowers** solitary, but produced in succession, axillary; calyx-lobes 2; corolla funnel-shaped, 2.2–2.5 cm long, rose-pink, downy within; **stamens** downy at base. Sichuan, Yunnan [cChina]; hillsides in woods, streamsides, 1,600–2,700 m. V–IX.

KOLKWITZIA

A genus with a single species restricted to China.

Kolkwitzia amabilis Graebn.* **Deciduous shrub** to 4 m with hairy branchlets. **Leaves** opposite, broad-ovate, 2.5–7.5 × 0.7–2 cm, slightly toothed, bristly on veins beneath. **Flowers** in pairs, in clusters on short lateral branchlets; calyx with 5–6 narrow, radiating lobes, hairy; corolla campanulate, pink with yellow throat, 14–16 mm long, with 5 rounded, spreading lobes. **Fruit** egg-shaped, brown-bristly, capped by a narrow beak. wHubei, Shaanxi [Shensi]; rocks, shrubberies, 2,740–3,050 m. V–VI.

SAMBUCUS

A northern hemisphere genus with c. 40 species (5 in China). **Herbs**, **shrubs** or small **trees**, often multi-stemmed with opposite, odd-pinnate **leaves**; leaflets 3–11, toothed. **Flowers** regular, in umbel-like, often flat-topped **inflorescence**, white, occasionally pinkish, sweetly and rather sickly scented; corolla 5-lobed. **Fruit** red, black or purple-black, 3–5-seeded berries.

Sambucus adnata DC. **Patch-forming perennial**, 0.3–1.5 m tall, somewhat woody at base. **Leaves** 15–30 cm long, with 3–9 leaflets; leaflets oblong, long-pointed, 10–20 × 3.5–7 cm, toothed; upper leaflets decurrent on stalk. **Inflorescence** umbellate, 6–20 cm broad; **flowers** white, 4–5 mm across. **Fruit** globose, red when ripe. (?)sGansu, (?)Hubei, Sichuan, (?)Shaanxi, se Tibet, Yunnan [Himalaya]; roadsides, field margins, thickets, 900–3,950 m. V–VII.

Sambucus javanica Reinw. Similar to the previous species in general appearance, but a **deciduous shrub** to 3 m; **leaves** with 5 leaflets, the upper free; **inflorescence** corymbose; **flowers** smaller, 2–3 mm across. Guizhou, Hubei, Sichuan, Yunnan [s & seAsia], shrubberies, open slopes, roadsides, field boundaries, to 2,150 m. V–VII.

VIBURNUM

A genus of c. 100 species (c. 70 in China) in north temperate regions of the Old and New Worlds. Small to large, **deciduous** or **evergreen shrubs** or occasionally **trees** with entire or toothed, occasionally lobed **leaves**, main veins often well-marked. **Inflorescence** compound, arranged in a domed or flat-topped pseudo-umbel or, more rarely, panicles, many-flowered; **flowers** fertile, regular (or sometimes outer ones in the inflorescence enlarged, sterile and ± asymmetric) small, campanulate or tubular with 5 spreading lobes; stamens 5; style 3-lobed. **Fruit** a drupe (berry-like), often brightly coloured, 1-seeded.

1. *Dipelta elegans*; Jiuzhaigou, nwSichuan (PC)
2. *Dipelta elegans*; Jiuzhaigou, nwSichuan (PC)
3. *Dipelta floribunda*; cult. (CGW)
4. *Dipelta floribunda*; cult. (CGW)
5. *Dipelta yunnanensis*; Gangheba, Yulongxueshan, nwYunnan (CGW)
6. *Dipelta yunnanensis*; Sandauwan, Yulongxueshan, nwYunnan (PC)
7. *Kolkwitzia amabilis*; cult. (CGW)
8. *Sambucus adnata*; Jiuzhaigou, nwSichuan (CGW)
9. *Sambucus adnata*; Lancangjiang–Jinshajiang divide, east of Weixi, nwYunnan (CGW)
10. *Sambucus adnata* in fruit; Deqin, nwYunnan (PC)

1. Evergreen shrubs .... **V. davidii, V. cinnamomifolium, V. harryanum, V. henryi, V. rhytidophyllum**
1. Deciduous shrubs, occasionally semi-evergreen 2
2. Flowers appearing before the leaves (occasionally a few flowers appearing with the leaves in autumn) .... **V. farreri, V. grandiflorum**
2. Spring- or summer-flowering, flowers always appearing with the leaves 3
3. Outer flowers sterile, much enlarged ..... **V. plicatum**
3. All flowers similar, small .... **V. betulifolium, V. nervosum, V. erubescens, V. setigerum**

Viburnum davidii **Franch.*** Compact, spreading, **dioecious shrub** 1–4.5 m with warty branchlets. **Leaves** leathery, narrowly elliptic to elliptic, 5–12.5 × 2.5–6.8 cm, tapered at both ends, conspicuously 3-veined, margin obscurely toothed, glabrous and dark green above, pale green beneath. **Inflorescence** stiffly cymose, 5–7.5 cm across, stalks covered in pale brown stellate hairs; **flowers** dull white, 3–3.5 mm across. **Fruit** metallic blue, translucent, narrowly oval in outline, 6 mm long. Guizhou, Sichuan [cChina]; cool, damp valleys, open spaces on steep slopes in *Picea* and broadleaved forests, 2,200–2,700 m. V–VI.

Viburnum cinnamomifolium **Rehd.*** Very similar to the previous species, but a larger **shrub** or small **tree** to 6 m with entire **leaves**, laxer **inflorescence** and smaller **fruit**. Guizhou, Sichuan; habitats similar to those of *V. davidii* but at rather lower elevations. V–VI.

Viburnum harryanum **Rehd.*** Bushy, **dioecious shrub** 0.8–2.5 m with young shoots covered in dark down. **Leaves** leathery, orbicular to obovate, 0.6–2.5 × 0.4–1.8 cm, tapering to base, rounded at apex, glabrous, dull dark green above, paler beneath. **Inflorescence** a terminal compound umbel, 3.5–3.7 cm across, stalks covered in pale brown stellate hairs; **flowers** white, 3–3.5 mm across. **Fruit** black when ripe, narrowly egg-shaped, 6 mm long. w & swSichuan, nwYunnan; open shrubberies on hillsides and by streams, 2,200–2,800 m. V–VI.

1. *Viburnum davidii*; cult. (CGW)
2. *Viburnum harryanum*; cult. (RL)
3. *Viburnum harryanum*; wSichuan (MO)
4. *Viburnum henryi*; cult. (RL)
5. *Viburnum rhytidophyllum*; cult. (CGW)
6. *Viburnum rhytidophyllum*; cult. (CGW)
7. *Viburnum farreri*; cult. (PC)
8. *Viburnum grandiflorum*; Bhutan (CGW)
9. *Viburnum plicatum*; Wolong, wSichuan (PC)
10. *Viburnum betulifolium*; Chongjiang, nwYunnan (PC)
11. *Viburnum betulifolium*; Danyun, nwSichuan (PC)

Viburnum henryi **Hemsl.*** Erect, **dioecious**, somewhat tree-like **shrub** to 3.5 m with glabrous branchlets. **Leaves** leathery, narrowly oval to obovate, 5–12.5 × 2.5–4.8 cm, obscurely toothed, conspicuously 3-veined, glabrous except on midrib beneath, dark green above, paler beneath, leaf-stalk winged. **Inflorescence** stiff, pyramidal panicles, 5–10 cm tall and broad; **flowers** perfect, white, 6–6.5 mm across. **Fruit** red turning black at maturity, translucent, oval in outline, 8 mm long. wHubei, w & swSichuan, w & nwYunnan; shrubberies, woodland, often near water, 2,000–3,000 m. VI–VII.

Viburnum rhytidophyllum **Hemsl.*** **Shrub** 3–6 m, the branches covered with stellate down. **Leaves** ovate-oblong, 7.5–18.5 × 2.5–6.3 cm, heart-shaped to base, pointed to blunt, glossy, deeply wrinkled, glabrous and dark green above, grey with stellate down beneath. **Inflorescence** large terminal umbels, 10–20 cm across, stalks covered in pale brown stellate down; **flowers** dull yellowish white, 6 mm across. **Fruit** red turning black when ripe, oval in outline, 8 mm long. wHubei, eSichuan; scrub and woodland near water, 1,500–2,700 m. V–VI.

Viburnum farreri **Stearn** (syn. *V. fragrans* Bunge) **Deciduous shrub** 1–3.5 m with glabrous branchlets. **Leaves** obovate to oval, 3.5–10 × 2.5–7 cm, pointed, toothed, glabrous above, with tufts of hairs on veins beneath. **Inflorescence** terminal and lateral, 3.7–5 cm across, stalks slightly hairy or glabrous; **corolla** white, ± tinged with pink on opening, 7–13 mm across, tube 9–10 mm long, lobes rounded; stamens short. **Fruit** bright red when ripe. Gansu [neChina]; thickets, woodland margins, often by water, 2,500–3,000 m. IV–V.

Viburnum grandiflorum **DC.** Similar in general appearance to the previous species, but leaf-undersurface, inflorescence-rachis and inner bud-scales densely hairy; **corolla** pale pink to rose-pink, rather larger, tube 12–13 mm long. w & swSichuan, seTibet, w & nwYunnan [swTibet, Himalaya, nMyanmar]; coniferous and *Rhododendron* forests, ravines, streamsides, 2,700–3,810 m. III–V.

Viburnum plicatum **Thunb.** (including *V. tomentosum* Thunb.) **Deciduous shrub** or small **tree**, 2–8 m tall. **Branches** mostly horizontal, covered when young with downy-stellate hairs. **Leaves** ovate to oval, 5–10 × 2.5–6.5 cm, pointed, toothed, dark green above, stellate-hairy and grey beneath. **Inflorescence** a flat umbel, 6–10 cm across, at end of short 2-leaved twigs; **flowers** white, perfect in centre of inflorescence, sterile outer flowers asymmetric, large, 2.5–3.7 mm across. **Fruit** coral-red turning blue-black. sGuizhou, Hubei, Sichuan, Yunnan [c & eChina, Japan, Korea]; thickets and forest margins, often by streams and rivers, 800–3,000 m. IV–VI.

Viburnum betulifolium **Batalin*** **Deciduous shrub** 3–4 m with glabrous branchlets, becoming brown or purplish brown. **Leaves** ovate to rhombic, broadly wedge-shaped at base, tapering above, 5–10 × 3–7.5 cm, margin coarsely toothed, glabrous and dark green above, slightly hairy on veins beneath. **Inflorescence** cymose, 5.5–10 cm across, stalks covered in pale brown stellate hairs; **flowers** white, 5 mm across, all perfect; stamens protruding, with yellow anthers. **Fruit** red, translucent, spherical, 6 mm. swGansu, nwGuizhou, Hubei, Sichuan, seTibet, Yunnan [cChina]; shady places on slopes in broadleaved deciduous and evergreen forests, woods and shrubberies, often near streams, 1,280–3,300 m. V–VI.

Viburnum nervosum D. Don (syn. *V. cordifolium* DC.) **Deciduous** or **semi-evergreen shrub** or small **tree**, 1–6 m tall with spreading **branches**. **Leaves** ovate, 5–12.5 × 4–7.5 cm, pointed, toothed, brown-hairy beneath. **Inflorescence** corymbose, domed, 4–10 cm across, stalks covered in pale brown stellate hairs; **flowers** creamy white, 5 mm across, fragrant. **Fruit** black when ripe. sGansu, Hubei, Sichuan, Shaanxi, seTibet, Yunnan [cChina, Himalaya, nMyanmar]; thickets and forest margins, coniferous and mixed forests, often by streams and rivers, alpine shrubberies, 2,300–4,400 m. V–VII.

Viburnum erubescens Wall. (syn. *V. prattii* Graebn.) **Deciduous** or **semi-evergreen shrub** or small **tree**, 1–7.5 m tall with glabrous branchlets. **Leaves** elliptic-ovate to oblong, 5–10 × 3–7.5 cm, long-pointed, coarsely toothed, downy or glabrous, dark green above, paler beneath. **Inflorescence** laxly pendent, 7.5–10 cm long, stalks covered in pale brown stellate hairs; **flowers** white with pink tips, 5 mm across, all perfect, fragrant; corolla-tube 10 mm long; anthers almost stalkless. **Fruit** black when ripe. sGansu, Hubei, Sichuan, Shaanxi, seTibet, Yunnan [cChina, Himalaya, nMyanmar]; thickets and forest margins, often by streams and rivers, 1,500–3,500 m. IV–VI.

Viburnum setigerum Hance* **Deciduous shrub** to 3.5 m, often less, with a rather erect habit and glabrous, grey **stems**. **Leaves** ovate-lanceolate, to 15 × 6 cm with rounded base, apex pointed, glabrous except for hairs on veins beneath. **Flowers** in 5-branched **cymes** to 5 cm across, borne on short 2-leaved twigs; corolla white, 5–6 mm across; **stamens** not protruding. **Fruit** egg-shaped, 10–12 mm long, glossy red, sometimes orange, when ripe. sGansu, Hubei, Sichuan; shrubberies, open forests, ravines, 1,800–3,900 m. V–VI.

HEPTACODIUM

A Chinese genus with 1 species.

Heptacodium miconioides Rehd.* Large **deciduous shrub** to 5 m, occasionally more, with attractive tan-coloured **bark** peeling in long strips. **Leaves** opposite, deep green, elliptic-lanceolate, to 15 cm long, with 3 main veins from the base. **Flowers** in broad, many-flowered **cymes**; calyx 5-lobed, purplish red, enlarging and persisting in fruit; corolla creamy-white to white, 8–12 mm across, with 5 ± equal, rather narrow, spreading lobes, fragrant. Hubei, sShaanxi [c & eChina]; open forests, forest margins, valley bottoms, 1,400–2,800 m. IX–X.

VALERIANACEAE

VALERIANA

A genus of c. 250 species (c. 60 in China) scattered in temperate regions of the world including the Americas, southern Africa and Asia. **Herbs** and **shrubs** with opposite **leaves**, sometimes with square **stems**. Leaves generally pinnately lobed. **Flowers** in cymose heads; calyx very small, often inconspicuous; corolla funnel-shaped, sometimes spurred or pouched at base of tube. **Fruit** small, with a persistent pappus-like calyx.

Valeriana officinalis L. Variable **perennial** to 1 m, sometimes more, with stiffly erect **stems**. **Leaves** paired, pinnately lobed, toothed or not, lowermost long-stalked, upper short-stalked and much smaller. **Flowers** numerous, in terminal and lateral rounded corymbs; corolla pink or white, 2.5–5 mm, usually with 3–4 protruding **stamens**. **Fruit** hairy or not, with persistent calyx. Gansu, Qinghai, w & nSichuan, but a weed elsewhere [n & neChina, n, c & eAsia, Europe]; meadows, ditches, fields, field boundaries, lake margins, open shrubberies, sea level to 3,200 m. VI–VIII.

PATRINIA

A genus of 15 species (7 in China) mainly in east Asia. **Herbaceous perennials** with opposite **leaves** and corymbose **panicles** of small yellow or white **flowers**, corolla with short tube and 5 spreading lobes; stamens 4. **Fruit** a winged achene.

Patrinia rupestris Bunge (syn. *P. scabra* Bunge) Coarse, pungent, **perennial herb** with erect **stems** to 60 cm tall in flower. **Leaves** deep glossy green, pinnately lobed, to 10 cm long, lowermost somewhat larger, all with jagged teeth. **Flowers** bright yellow, numerous in corymb-like, somewhat domed clusters to 15 cm across. Gansu, Qinghai, nSichuan, Hubei [n & nwChina, Korea, Mongolia, eRussia]; rocky and grassy places, old walls and near habitation, roadsides, sea level to 2,500 m. VII–IX.

NARDOSTACHYS

A genus with a single species occurring from the Himalaya to western China, widely used for medicinal purposes and threatened in the wild from over-exploitation: now protected in many of the regions in which it is found. The whole plant has a distinctive and long-lasting perfume.

Nardostachys grandiflora DC. (syn. *N. jatamansi* DC.) Variable **tufted** plant to 15 cm tall in flower, generally less, with elliptic-lanceolate to spathulate, mostly basal **leaves** to 20 cm long. **Flowers** in terminal clusters, semi-nodding; calyx 5-lobed, persisting and enlarging in fruit; corolla rose-purple to whitish, with 5 spreading, rounded lobes to 12 mm across, and a cylindrical tube to 20 mm long; stamens 4. **Fruit** egg-shaped,

1. *Viburnum nervosum*; Bhutan (CGW)
2. *Viburnum nervosum*; Emeishan, sSichuan (PC)
3. *Viburnum erubescens*; cult. (CGW)
4. *Viburnum setigerum*; cult. (CGW)
5. *Heptacodium miconioides*; cult. (CGW)
6. *Valeriana officinalis*; Jiuzhaigou, nwSichuan (PC)
7. *Valeriana officinalis*; Danyun, Huanglong, nwSichuan (PC)
8. *Patrinia rupestris*; Badaling, Beijing (CGW)
9. *Nardostachys grandiflora*; south of Aba, nwSichuan (CGW)

1-seeded. sGansu, seQinghai, Sichuan, seTibet, nwYunnan [swTibet, Himalaya]; rocky and grassy meadows, banks, rock-ledges, alpine shrubberies, 3,600–4,800 m. VI–VIII. *N. chinensis* Batalin is probably synonymous, but this requires further investigation.

DIPSACACEAE

ACANTHOCALYX

A genus of c. 9 species (5 in China) in the Himalaya, north Myanmar and China. Tufted **perennial herbs**, often with bristly, sometimes thistle-like, opposite or whorled **leaves**; bracts generally leaf-like, spiny or bristly. **Flowers** in branched whorls one above the other, or in terminal heads; calyx 2-lipped; corolla long-tubular, 2-lipped; **stamens** 2 or 4.

1. Flowers white, cream or greenish .. ***A. alba, A. chinensis***
1. Flowers pink or purple .... ***A. delavayi***

Acanthocalyx alba **(Hand.-Mazz.) Cannon*** (syn. *Morina alba* Hand.-Mazz.) **Tufted** plant to 40 cm with slender squarish **stems**. **Leaves** mainly basal, linear-elliptic to linear-lanceolate, to 16 × 1.1 cm, 3–5-veined, with bristly margin; stem-leaves in 2–3 pairs, shorter and broader than basal leaves. **Inflorescence** a small terminal cluster surrounded by ovate bristly bracts that conceal calyces; **corolla** 14–24 mm long, with slender curved tube and ± equal lobes 2–3 mm long, fragrant. w & swSichuan, nwYunnan, neighbouring parts of Tibet; alpine meadows and other grassy places, open shrubberies, banks, open *Pinus* forests and forest margins, rocky places, 2,500–3,950 m. V–VII.

Acanthocalyx chinensis Pai* (syn. *Morina chinensis* (Pai) Cannon) **Tufted** plant 30–50 cm tall, readily distinguished from the previous species by the far more prickly, thistle-like habit; **leaves** shallowly lobed, to 14 × 1.5 cm, those on **stem** usually in whorls of 3–4. **Inflorescence** a dense, very spiny, elongated spike with 7 or more close-set whorls of small greenish **flowers**. Gansu, Qinghai, n & nwSichuan, grassy habitats, open shrubberies, 2,900–3,100 m. VI–VII.

Acanthocalyx delavayi (Franch.) Cannon* (syn. *Morina bulleyana* Franch.; *M. nepalensis* D. Don subsp. *delavayi* (Franch.) D. Y. Hong & L. M. Ma) Small **tufted** plant to 60 cm tall in flower, generally less, with ascending, square, often purple-flushed **stems**. **Leaves** mainly in basal tufts, opposite, lanceolate to linear-lanceolate, to 15 × 1.5 cm, with parallel veins and bristly margin; stem-leaves smaller, usually in 1–2 pairs. **Flowers** in terminal cluster composed of 2–3 closely packed whorls surrounded by ovate, bristly, leaf-like bracts; calyx deep purple; corolla pinkish purple to reddish purple (rarely white), curved tube 12–17 mm long, upper lip smaller than lower. w & swSichuan, nwYunnan, neighbouring parts of Tibet; grassy and stony habitats, banks, open shrubberies, forest margins, 2,500–4,875 m. VI–VIII.

PTEROCEPHALODES

A genus of 3 species (2 in China) found in the Himalaya, western China and Thailand. Very closely related to *Scabiosa* but **flowers** in spherical heads, calyx with 20–24 plume-like bristles and corolla with equal lobes, not radiate in the flower-head. **Fruit** topped by a persistent papery appendage and the calyx-bristles. Formerly in the genus *Pterocephalus* (found predominantly in the Mediterranean). Treated as the genus *Bassecoia* by some modern botanists.

Pterocephalodes hookeri (C. B. Clarke) V. Meyer & Ehrend. (syn. *Pterocephalus hookeri* (C. B. Clarke) Diels) Tufted **perennial herb** to 40 cm tall in fruit, with erect scapose hairy **stems**. **Leaves** all basal, narrow-elliptic to spathulate, to 15 × 2 cm, entire or somewhat lobed. **Inflorescence** a nodding or semi-nodding globose head 25–40 mm across with ruff of short bracts at base; calyx shuttlecock-like with a plume of bristles nearly as long as corolla; **corolla** cream, whitish or pale yellow, sometimes with a hint of lilac, with 5 short lobes. s & seQinghai, w, nw & swSichuan, Tibet, nwYunnan [Himalaya]; meadows and other grassy habitats, open shrubberies, roadsides, sometimes on cliff-ledges, 3,300–4,100 m. VII–IX.

---

1. *Acanthocalyx alba*; Napahai, Zhongdian, nwYunnan (CGW)
2. *Acanthocalyx chinensis*; south of Aba, nwSichuan (CGW)
3. *Acanthocalyx delavayi*; south of Zhongdian, nwYunnan (CGW)
4. *Pterocephalodes hookeri*; Baimashan, nwYunnan (PC)
5. *Dolomiaea forrestii*; Daxueshan, nwYunnan (AD)
6. *Dolomiaea souliei*; Balangshan, Wolong, wSichuan (CGW)

ASTERACEAE (COMPOSITAE)

235 genera and almost 2,500 species in China. Mostly **herbaceous** and **inflorescences** of many flowers in heads.

SYNOPSIS OF GROUPS

Group One: Flower-heads with florets all of 1 type, lacking ray-florets, (flower-heads discoid).

Group Two: Flower-heads with outer florets female, lacking a ligule, and inner florets campanulate to tubular, or central ones all male (flower-heads discoid). (p. 499)

Group Three: Flower-heads with florets of 2 types, ray and disc (flower-heads radiate, except occasionally in *Ligularia*), ray-florets often female, sometimes sterile. (p. 500)

Group Four: Flower-heads with all florets with ligules (flower-heads ligulate). (p. 515)

ASTERACEAE Group One

DOLOMIAEA

A genus of c. 13 species (10 in China) in the Himalaya, Myanmar and China. Stemless, rosette-forming, **perennial herbs** with a stout tap-root. **Leaves** simple, not spiny. **Flower-heads** ± stalkless, lacking ray-florets, few to many in centre of rosette; involucre egg-shaped to globose; **florets** regular, hermaphrodite, each with a slender, deeply 5-lobed corolla-tube, glandular-dotted outside; style very short, glabrous. **Achenes** 3–4-angled, slightly compressed, pappus rough or short plumose, basally joined into a ring.

Dolomiaea berardioidea (Franch.) C. Shih* (syn. *Jurinea berardioidea* (Franch.) Diels) **Rosette-forming herb** to 45 cm across. **Leaves** yellow-green, obovate, elliptic, ovate, or oblong 8–20 × 7–19 cm, blunt, with wavy margins. **Flower-heads** solitary, 4.5–5.5 cm across, stalkless; **disc-florets** dirty mauve to brownish purple, with white filaments and violet anthers. **Achenes** 3-angled, c. 7 mm long, pappus brown, rough, c. 2.5 cm. nwYunnan (Lijiang); *Rhododendron* thickets, *Juniperus* and *Pinus* forests, 2,800–3,600 m. VII–X.

Dolomiaea forrestii (Diels) C. Shih* (syn. *Jurinea forrestii* Diels) Similar to the previous species, but **leaves** odd-pinnately lobed, 4–20 × 3–15 cm, glandular, green, margins of lobes finely spine-toothed. **Flower-heads** 3–6-clustered, each 3.5–4 cm across, with deep purplish red **disc-florets**. **Achenes** cylindrical, c. 5 mm long, pimpled above middle, with yellow-brown pappus. swSichuan, seTibet, nwYunnan; grassland and stony alpine pastures, thickets, forest margins, 3,000–4,200 m. VI–X.

Dolomiaea souliei (Franch.) C. Shih* Very similar to *D. forrestii*, but larger, **leaves** yellowish green, lobed at margin, with blunt teeth but not spinulose; **flower-heads** more numerous, to 12, and densely clustered, pinkish or pale purple. w & swSichuan, eTibet; similar habitats and altitudes. VI–IX.

XANTHOPAPPUS

A genus with 2 species, one confined to north-west China.

Xanthopappus subacaulis C. Winkl.* Large, thistle-like plant, almost stemless, forming a spreading rosette at ground level. **Leaves** pinnately lobed, 10–35 × 2.5–8 cm, spiny on lobes, whitish beneath. **Flower-heads** stalkless, clustered, up to 4.5 cm across, with numerous narrow reflexed, shiny, spine-tipped involucral bracts; **florets** whitish. **Achenes** with long, dense, golden brown pappuses. Gansu, Sichuan; dry gravelly and rocky places, 3,500–4,500 m. VII–VIII.

CIRSIUM

A genus with c. 250 species (c. 50 in China) distributed in Europe, Asia, north and east Africa and North America. Stout **perennial herbs** with branching **stems**, rarely stemless. **Leaves** alternate, pinnately lobed or divided, spiny on margin. **Flower-heads** solitary or several and almost corymbose; involucre narrowly to broadly campanulate, with stiff, spreading, bristly bracts, sometimes with a cobweb-like covering; **florets** usually hermaphrodite, white, purple or yellow; corolla tubular or campanulate, limb narrow, with 5 linear lobes. **Achenes** oblong to obovoid; pappus plume-like, with bristles.

1. Flower-heads white . . . . *C. mairei*
1. Flower-heads rosy-lilac to purple 2
2. Flower-heads nodding or half-nodding . . . .
   . . . . *C. bolocephalum, C. henryi, C. handelii*
2. Flower-heads erect . . . . *C. falconeri, C. fargesii*

Cirsium mairei (H. Lévl.) H. Lévl. (syn. *C. griseum* H. Lévl. not Daniels; *C. botryodes* Petrak ex Hand.-Mazz.; *C. yunnanensis* Petrak) Stout plants 70–150 cm tall with greyish white, erect, usually branched, **stems**, cobwebby-hairy, densely so beneath **flower-heads**. **Leaves** stalkless, lanceolate to ovate-lanceolate or obovate, 14–20 × 3–8 cm, pinnately lobed, margin with spines 3–7 mm long. **Inflorescence** racemose to racemose-corymbose; involucre broadly campanulate, c. 3.5 cm across, sparsely cobwebby-hairy; **corolla** white, c. 2 cm long. Guizhou, Sichuan, seTibet, w & nwYunnan [swTibet, Himalaya]; hillsides, montane valleys, screes, grassland, forest margins and clearings, shrubberies, riverbanks, 1,300–3,000 m. V–IX.

Cirsium bolocephalum Hand.-Mazz. Plant 0.6–3 m tall with erect, apically branched **stems**, sparsely cottony with long hairs. **Leaves** green, papery, obovate to ovate-elliptic, 16–45 × 6–15 cm, long-pointed, pinnately lobed, margin spine-toothed, cottony beneath. **Flower-heads** 4–5 cm across, solitary at branch ends; involucral bracts long-pointed, 2–3 cm long, enveloped in a cocoon of cotton; **florets** 30–35 mm long, rosy-lilac to dark purple. Sichuan, seTibet, w & nwYunnan [swTibet, Himalaya]; hillsides, screes, grassland, forest margins and clearings, shrubberies, riverbanks, 2,000–4,750 m. VI–X. The closely related *C. eriophoroides* (Hook. f.) Petrak bears smaller, solitary or clustered, semi-nodding flower-heads. Similar habitats, but more widely distributed in the Himalaya.

Cirsium henryi (Franch.) Diels* (syn. *C. forrestii* (Diels) H. Lévl.; *C. taliense* H. Lévl.) Smaller plant than the previous species, not more than 50 cm, with lanceolate, oblanceolate to elliptic **leaves**, 8–18 × 3–8 cm, pinnately lobed to almost pinnatisect, margin with spines 2–7 mm long. **Inflorescence** corymbose or paniculate, **flower-heads** rarely solitary; involucre campanulate, c. 2 cm across, glabrous; **florets** purple, c. 16 mm long. Hubei, Sichuan, w & nwYunnan; hillsides, screes, grassland, forest margins and clearings, riverbanks, 1,500–3,500 m. VI–X.

Cirsium handelii Petrak ex Hand.-Mazz.* **Perennial herb** to 1.7 m with erect, ribbed **stems**, distinguished from the previous 2 species by the plain green, unmarked, glabrous or sparsely bristly, spine-margined **leaves**; **flower-heads** 1–many, nodding, 3–4.5 cm across, glabrous; **florets** purple, c. 21 mm long. w, sw & sSichuan, n & nwYunnan; forests, forest margins, thickets, grassland, waste places, 1,700–3,400 m. V–IX.

Cirsium falconeri (Hook. f.) Petrak Robust plant to 1.2 m with ridged, spiny and bristly **stems**. **Leaves** deep green above, pale beneath, margin lobed, very spiny, often with small spines on upper surface. **Flower-heads** 6–8 cm across; involucral **bracts** numerous, narrow, overlapping, with a spreading spine-tip, spines to 2.5 cm long, with some cottony hairs; **florets** pale to deep purple. seTibet [swTibet, Himalaya]; meadows, forest clearings and margins, close to habitation, grazed land, 2,700–4,300 m. VII–IX. *Cirsium fargesii* (Franch.) Diels is similar but heads up to 4 cm in diam.; bracts up to 2.5 cm long. Hubei, Shaanxi, Sichuan; waste places and meadows, 2,300–2,500 m.

ECHINOPS

A genus of c. 20 species (17 in China) in North Africa, Central Asia and south Europe. **Perennial** or rarely **annual herbs**. **Stems** erect, usually apically branched, covered with cobwebby or woolly hairs; **leaves** often spiny and thistle-like. **Flower-heads** globose; involucral bracts numerous, in 3–5 rows, outer bracts slender, pointed, middle often with an awl-shaped to pointed appendage, innermost often shorter; **corolla** tubular, usually 5-lobed. **Achenes** with a free to fused pappus.

1. Annuals with linear to linear-lanceolate leaves, not white-felted beneath . . . . *E. gmelini*
1. Perennials with elliptic to elliptic-lanceolate leaves, white-felted beneath . . . . . *E. przewalskii, E. latifolius*

Echinops gmelini Turcz. **Annual herb** 10–90 cm tall with solitary, yellowish **stems**, branched from middle to base, rarely unbranched. **Leaves** green, papery, linear to linear-lanceolate, 3–9 × 0.4–0.5 cm, base semi-clasping, margin spiny-toothed and spiny-ciliate, sparsely hairy. **Inflorescence** solitary to several at branch-ends; **flower-heads** 2–3 cm across; involucral bracts 16–20; outermost to 1 cm long; **corolla** blue to white, tube not gland-dotted. Gansu, Shaanxi, Qinghai [nChina, Mongolia, eRussia]; gravelly places on mountain slopes, desert steppes, loess hills, flooded places, 500–3,200 m. VI–IX.

Echinops przewalskii Iljin. **Perennial** 15–40 cm with 1 or several, erect, unbranched or 1–3-branched, cobwebby-woolly **stems**. **Leaves** leathery, elliptic, elliptic-lanceolate or oblanceolate, 10–20 × 2–8 cm, 2-pinnately lobed, green to yellow-green above, margin spiny, whitish to greyish white-felted beneath. **Inflorescence** with 1 flower-head or several in a corymb 5–5.5 cm across; **flower-heads** c. 1.8 cm across;
involucral bracts keeled, glabrous outside, outermost to 8 mm long; **corolla** white or bluish, gland-dotted outside. Gansu, Qinghai [Shandong, Shanxi, Nei Mongol, Mongolia]; sandy mountainous regions, desert steppes, 500–2,200 m. VI–VIII.

Echinops latifolius Tausch (syn. *E. dahuricus* Fisch.) Often taller than the previous species, to 60 cm, with greyish white, often unbranched **stems**. **Leaves** papery, elliptic to lanceolate-elliptic, 15–35 × 8–18 cm, 2-pinnately lobed, margin entire or with few triangular spiny teeth, densely and thickly greyish white-felted beneath. **Inflorescences** terminal with 1–3 **flower-heads**, 3–5.5 cm across; **corolla** blue. Gansu, Shaanxi [c & nChina, Mongolia, Russia]; grassland, open forests, shrubberies, to 2,200 m. VI–IX.

1. *Xanthopappus subacaulis*; Chali, nwSichuan (HJ)
2. *Cirsium mairei*; Chongjiang, nwYunnan (PC)
3. *Cirsium handelii*; Emeishan, sSichuan (CGW)
4. *Cirsium fargesii*; Jiuzhaigou, nwSichuan (PC)
5. *Cirsium bolocephalum*; Bomi, Galung La, seTibet (HJ)
6. *Cirsium falconeri*; Ghunsa Khola, eNepal (CGW)

ARCTIUM

A genus with c. 10 species (2 in China) in the north temperate zone. **Biennial** or **perennial herbs** with simple, entire to sparsely toothed **leaves**. Receptacle flat, densely scaly; **florets** hermaphrodite; filaments free, glabrous, anthers with arrow-shaped appendage; style-arms linear, basally with a ring of hairs. **Achenes** compressed, obovoid to narrowly ellipsoid, ribbed; pappus hairs in many rows, rough.

Arctium lappa **L.** (syn. *A. leiospermum* Juz. & Ye. V. Serg.; *A. majus* Bernh.) Leafy **biennial** to 2 m with stout, erect, purplish, sparsely cobwebby-hairy **stems**, branched towards top. **Leaves** broadly heart-shaped or ovate, to 30 21 cm. **Inflorescence** corymbose to paniculate, with few to many egg-shaped glabrous **flower-heads** 15–20 mm in diameter; involucral **bracts** numerous, tipped with a hook (ripe fruiting heads readily cling to fur and clothing); **corolla** purple-red, to 14 mm long. Gansu, Guangxi, Guizhou, Hubei, Qinghai, Sichuan, Shaanxi, Tibet, Yunnan [throughout China, temperate Asia, Europe]; near habitation, road- and riversides, wet and waste places, forest margins, thickets, 700–3,500 m. VI–IX.

AINSLIAEA

A genus of c. 80 species (30 in China) distributed in east Asia and India. **Perennial herbs** with simple **leaves** in a basal rosette. **Flowers** produced twice a year (fully developed flowers in spring, cleistogamous flowers in autumn). **Inflorescence** spike-like, **flower-heads** small, almost stalkless, borne singly or in clusters, each flower-head with 3–4 hermaphrodite **florets**; involucre cylindrical; corolla-limb deeply cut into 5 equal lobes; style-branches short, rounded. **Achenes** narrowly obovoid, ribbed, densely silky-hairy; pappus plume-like, brownish.

1. Leaf-stalks winged . . . . **A. glabra, A. latifolia**
1. Leaf-stalks unwinged or leaves stalkless . . . . **A. fragrans, A. pertyoides**

Ainsliaea glabra **Hemsl.*** Plant 40–70 cm with **leaves** 8–20 2.5–11 cm, ovate, pointed, margin weakly scalloped, leaf-stalk distinctly winged. **Inflorescence** branched, bearing white **flower-heads**. wHubei, Sichuan, Yunnan [sChina]; thickets, grassy slopes, rocks in damp, shady places, 700–1,750 m. III–VII.

Ainsliaea latifolia (D. Don) Sch.-Bip.* (syn. *A. pteropoda* DC.) Plant 15–80 cm with ovate **leaves**, 3–11 × 1.5–7 cm, pointed, with mucronately toothed margin, felted beneath; leaf-stalks short, winged, 1–2 cm long, hairy. **Inflorescence** with clusters of stalkless **flower-heads** at well-spaced nodes; **corolla** pink to deep rose-pink. seTibet, w & nwYunnan; scrub, shady banks in *Pinus* forests, 1,200–3,500 m. X–V.

Ainsliaea fragrans Champ.* Plant 20–70 cm with **leaves** ± in a rosette, ovate to heart-shaped, 3–10 × 1.5–6 cm, pointed, hairy on veins beneath, ciliate; leaf-stalks 1–2 cm long, hairy. **Inflorescence** branched or not; involucre purplish green; **corolla** white, tube 5–5.5 mm long. **Achenes** 5.5 mm long; pappus 7–8 mm long. sSichuan (Emeishan), Yunnan [c & sChina]; deciduous and coniferous forests, 2,000–3,000 m. XI–IV.

Ainsliaea pertyoides Franch.* Plant 30–100 cm with branched, leafy **stems**, **leaves** ± stalkless, ovate to heart-shaped, 1.5–9 × 0.6–4 cm, pointed, 3-veined, white-felted beneath. **Inflorescences** axillary, shorter than leaves; **corolla** white or pink, tube 5–5.5 mm long. Guizhou, Yunnan; deciduous woods, shrubberies, 1,200–2,700 m. XI–IV.

ANAPHALIS

A genus of c. 110 species (15 in China) found primarily in Asia and Europe, with a few in North America. Erect **perennial** whitish-woolly **herbs**, often stoloniferous, dioecious or polygamous. **Leaves** alternate, entire, often decurrent on the **stem**. **Flower-heads** small, globular, involucral **bracts** in 5–8 rows, papery-everlasting; **corolla** of outer florets thread-like, 2–4-toothed, corolla of disc-florets tubular. **Achenes** very small, oblong; pappus-bristles in 1 row.

1. Involucral bracts blunt, erect . . . . **A. margaritacea**
1. Involucral bracts pointed, spreading 2
2. Flower-heads solitary or in clusters of 2–6 . . . .
. . . . . **A. nepalensis, A. nepalensis** var. **monocephala,**
A. cinerascens, A. cinerascens var. **congesta**
2. Flower-heads numerous, in broad clusters . . . .
. . . . **A. triplinervis**

Anaphalis margaritacea (L.) Benth.* (syn. *A. cinnamomea* C. B. Clarke) Very variable **clump-forming perennial** with erect, leafy **stems** to 60 cm. **Leaves** lanceolate, 5–12 × 1.5–4.5 cm, dark green above, grey- or reddish-woolly beneath, with 1–3 parallel veins. **Flower-heads** 7–8 mm across, in large, domed

---

1. *Arctium lappa*; Chongjiang, nwYunnan (PC)
2. *Anaphalis margaritacea*; Cangshan, nwYunnan (CGW)
3. *Anaphalis nepalensis* var. *monocephala*; Baimashan, nwYunnan (HJ)
4. *Anaphalis cinerascens*; Yulongxueshan, nwYunnan (CGW)
5. *Anaphalis triplinervis*; cult. Dali, nwYunnan (CGW)
6. *Gnaphalium affine*; Erlangshan, wSichuan (PC)

clusters to 15 cm across; involucral **bracts** white, papery. Sichuan, seTibet, Yunnan [c & eChina, Himalaya, Japan, North America]; grazed places, margins of cultivation, open shrubberies, locally common, 1,800–3,000 m. XIII–XI.

Anaphalis nepalensis (Spreng.) Hand.-Mazz. (syn. *A. triplinervis* (Sims) C. B. Clarke var. *intermedia* (DC.) Airy Shaw) Grey- or whitish-woolly, **clump-forming perennial** to 25 cm with elliptic to oblanceolate **leaves**, 2.5–6 × 0.7–1.8 cm, pointed to almost so. **Flower-heads** usually in clusters of 2–6, rarely solitary, white, 10–12 mm across; **florets** yellow. seTibet [swTibet, Himalaya]; rocky and grassy slopes, open forests and forest margins, grazed areas, 2,600–4,100 m. VII–X. **Var. *monocephala*** (DC.) Hand.-Mazz. is more dwarf, to 15 cm (but sometimes only 5–8 cm), with shorter and broader **leaves** mostly not more than 15 mm long, and solitary **flower-heads** 12–15 mm across. seTibet, nwYunnan [swTibet, eHimalaya]; rocky habitats, stabilised screes, moraines, 3,500–4,900 m. VII–IX.

Anaphalis cinerascens Ling & W. Wang* Tufted or matted **perennial** to 10 cm with a woody base enveloped in old leaf-remains. **Leaves** mostly in rosettes, obovate to spathulate, to 30 mm long, silvery grey-felted, faintly 3-veined, upper leaves narrower than lower. **Flower-heads** solitary (**var. *congesta*** Ling & W. Wang*) or more usually in dense, stemless or short-stemmed clusters of 5–10, each head c. 10 mm across. s & swSichuan, nwYunnan; stony alpine meadows, rocky slopes, 4,300–4,900 m. VII–IX.

Anaphalis triplinervis Sims ex C. B. Clarke Very variable **clump-forming perennial** to 60 cm, often less, with erect, white-woolly **stems**. **Leaves** elliptic to lanceolate, 5–10 × 1.5–4.5 cm, grey above, white-woolly beneath, with 3 or 5 parallel veins. **Flower-heads** lustrous white, 8–10 mm across, in large domed **corymbs**. Sichuan, seTibet, Yunnan [swTibet, Himalaya]; rocky and grassy places, woodland margins, open shrubberies, 1,800–3,000 m. VII–X.

GNAPHALIUM

A cosmopolitan genus with c. 150 species (9 in China). **Annual** or **perennial herbs**, often with felted **leaves** and **stems**; **flower-heads** small; involucral **bracts** papery, shiny, yellow or silvery.

Gnaphalium affine D. Don Grey-green **clump-forming herb** up to 30 cm. **Leaves** linear-oblanceolate, 2–3.5 × 0.5–0.8 cm, rounded at tip, with appressed silver hairs. **Flower-heads** clustered in a terminal **corymb**, yellow, each 2–2.5 mm across; **ray-florets** yellow, **disc-florets** absent. Hubei, Sichuan. [c & sChina, Himalaya, seAsia, Japan]; open dry pastures, forest margins, thickets, cultivated land, 1,200–3,000 m. III–VIII.

Gnaphalium chrysocephalum Franch.* Much larger than *G. affine*, to 95 cm, **leaves** lanceolate, to 10 × 1.2 cm, pointed, silvery beneath. **Flower-heads** many in a rounded head up to

9.5 cm in diameter, each 4 mm in diameter, golden to orange. swSichuan, nwYunnan; open dry pastures, forest margins, thickets, 3,500–3,800 m. IV–VII.

SAUSSUREA

A diverse genus with c. 300 species (c. 100 in China) distributed in Europe, Asia, and with a few in North America. Stemless to medium-sized, caulescent, **perennial herbs**, often with winged **stems**. **Leaves** alternate, simple, 1- or 2-pinnatisect, lacking prickles, aggregated into dense rosettes in some alpine species (these often with hollow **stems**). **Flower-heads** 1–many, discoid, sometimes covered in long wool or surrounded by coloured **bracts**; **florets** hermaphrodite, purple or violet, rarely white; corolla narrowly tubular-campanulate, limb deeply 5-lobed. **Achenes** oblong to obovoid, ± compressed, quadrangular or ribbed, pappus-bristles in 2 rows, plume-like, persistent.

1. Inflorescences balloon-like, enveloped by large membranous upper leaves .... **S. obvallata**
1. Inflorescences not enveloped by large membranous upper leaves, often very woolly 2
2. Leaves deeply pinnately lobed .... **S. leucoma**
2. Leaves entire to toothed, not deeply pinnately lobed .. 3
3. Lower leaves linear or elliptic or linear-lanceolate 4
3. Lower leaves spathulate to obovate 5
4. Rosette-forming, stemless plants with glabrous leaves .... **S. stella**
4. Plants stemmed, with white-downy leaves .... **S. laniceps, S. simpsoniana, S. gossypiphora, S. nidularis, S. graminea**
5. Leaves 3(–6)-lobed at apex, otherwise untoothed .... **S. tridactyla**
5. Leaves with toothed margin 6
6. Flower-heads clustered, domed, stalkless or with short stalk .... **S. medusa, S. quercifolia, S. katochaete**
6. Flower-heads solitary, long-stalked .... **S. hieracioides, S. aff. gnaphalodes**

Saussurea obvallata **(DC.) Edgew.** Tufted, rhizomatous **herb**, 30–70 cm tall with erect, hairy **stems**, surrounded at base by dead leaves. Lower **leaves** oblanceolate, 10–30 1.5–5 cm, toothed, hairy, stalked; upper leaves membranous, stalkless, elliptic-ovate, 8–15 cm long, blunt, loosely enveloping **inflorescence**, pale green to white. **Flower-heads** obscured, 4–11, clustered, almost stalkless in a head 8–15 mm across, with purplish black involucral **bracts**; **corolla** 13–17 mm long. seTibet, nwYunnan [swTibet, Himalaya]; damp screes, rocky hillsides, rich meadows, boggy moorland, 3,500–4,880 m. VII–X.

Saussurea leucoma **Diels*** (syn. *S. eriocephala* Diels) **Rosette-forming herb** 7–23 cm tall, forming a woolly, cylindrical, round-headed mound. Lower **leaves** deeply pinnately lobed, with narrow pointed lobes, 5–12 0.5–2 cm, dull dark green with purple margin, densely white-woolly, especially in lower half; bracts arching-recurved, lanceolate, generally unlobed, densely white-woolly. **Flower-heads** numerous, forming a dense dome; **corolla** violet or blackish purple. swSichuan, nwYunnan; limestone screes and other rocky places, moraines, 3,200–5,300 m. VI–X.

Saussurea stella **Maxim.** Stemless **herb** with flat rosettes, 3–12 cm tall with linear-tapering, stalkless **leaves**, 5–18 0.4–1.2 cm, glabrous, purple with green tips. **Flower-heads** 5–30, clustered, almost stalkless in a head up to 9 cm across, surrounded by purple, lanceolate bracts; involucre 5–12 mm across; **corolla** 15–20 mm long. Gansu, w, nw & swSichuan, e & seTibet, nwYunnan [swTibet, Himalaya]; wet meadows, boggy moorland, often gregarious, 3,450–4,880 m. VI–X.

Saussurea laniceps **Hand.-Mazz.** **Monocarpic herb** with a rather flat rosette of linear-lanceolate, deep green, somewhat shiny **leaves** to 15 cm long, sharply toothed towards pointed apex, downy beneath, ± glabrous above. Flowering plant to 30 cm, forming a broad column enveloped in white wool, with tips of lower stem-leaves visible, upper usually entirely enveloped. **Flower-heads** numerous in an elongated dome, purplish, located in depressions in the felted mass of white hairs. seTibet, nwYunnan [eHimalaya, nMyanmar]; rocky slopes, scree, cliffs, 3,200–5,280 m. VII–X.

Saussurea simpsoniana **(Field. & Gardn.) Lipsch.** Plants 6 cm tall from a woody rootstock with a single rosette. **Leaves** linear to linear-lanceolate, to 10 cm long, downturned and closely overlapping, white-woolly. **Flower-heads** numerous in a compact dome 2.5 cm across, covered in pink-tinged downy hairs; **florets** purple. seTibet [swTibet, c & eHimalaya, nMyanmar]; moraines, screes, sandy places, stony slopes, 4,400–5,600 m. VII–IX.

Saussurea gossypiphora **D. Don** **Monocarpic herb** to 30 cm, apically densely white or brownish yellow-woolly. Lower **stem-leaves** linear-oblong to narrowly elliptic, 7–14 0.4–1.4 cm including stalk, pointed to blunt, margin sharply or shallowly toothed; upper stem-leaves linear-lanceolate, densely white- or pale yellow-woolly. **Inflorescence** hemispherical, 7–10 cm in diameter, **flower-heads** numerous, stalkless; **florets** purplish red, 10–11 mm long. Pappus pale brown, in 2 rows. seTibet, nwYunnan [swTibet, Himalaya]; alpine screes, rock-crevices, stabilised moraines, 4,000–5,000 m. VII–IX.

Saussurea nidularis **Hand.-Mazz.*** **Herb** 5–10 cm tall, forming a woolly, cylindrical, round-headed mound. Lower **leaves** linear, pointed, dull glossy green, covered sparingly with long white hairs. **Flower-heads** solitary, with spreading, lanceolate, long-pointed, straw-coloured involucral bracts, **florets** also straw-coloured. seTibet, nwYunnan; alpine pastures, rocky slopes, 3,600–4,500 m. VI–VIII.

Saussurea graminea **(Hand.-Mazz.) Lipsch.** Clump-forming **perennial herb** 10–20 cm tall, covered in long white woolly hairs. **Leaves** linear, 5–14 0.1–0.6 cm, upper leaves

1. *Saussurea obvallata*; Serkhyem La, seTibet (HJ)
2. *Saussurea leucoma*; Baimashan, nwYunnan (HJ)
3. *Saussurea stella*; Bitahai, Zhongdian, nwYunnan (PC)
4. *Saussurea laniceps*; Habashan, nwYunnan (MN)
5. *Saussurea gossypiphora*; eNepal (CGW)
6. *Saussurea nidularis*; Baimashan, nwYunnan (HJ)
7. *Saussurea nidularis*; Habashan, nwYunnan (MN)

1. *Saussurea tridactyla*; eNepal (CGW)
2. *Saussurea medusa*; Baimashan, nwYunnan (CGW)
3. *Saussurea medusa*; Huanglong, nwSichuan (CGW)
4. *Saussurea quercifolia*; Daxueshan, nwYunnan (HJ)
5. *Saussurea katochaete*; Daxueshan, nwYunnan (AD)
6. *Saussurea graminea*; Genyanshan, south side, swSichuan (MN)
7. *Saussurea* aff. *gnaphalodes*; Daxueshan, nwYunnan (AD)
8. *Gynura japonica*; Kunming, cYunnan (PC)
9. *Eupatorium mairei*; Shu La on Meilixueshan, nwYunnan (MN)

sometimes purplish. **Flower-heads** solitary, surrounded by linear, hairy bracts, 18–22 mm across; **florets** blue-purple with white anthers. w & swSichuan, e & seTibet [swTibet, eHimalaya, nMyanmar]; grassy and stony alpine meadows and slopes, 4,100–4,600 m. VII–IX.

Saussurea tridactyla Hook. f. Plant 6–9 cm tall, forming a semicircular to conical, white, woolly mound, with 1–several rosettes. **Leaves** spathulate, obscurely 3-lobed to 6-lobed at apex, up to 4.5 cm long, densely covered in long, white, woolly hairs, borne on enlarged hollow **stem**. **Inflorescence** domed, half-exposed, terminal, 5–6 cm across; involucral bracts linear-oblong, pointed, shiny; **florets** purple. seTibet [swTibet, c & eHimalaya]; sandy places by glacier lakes, moraines, screes, 4,800–5,600 m. VII–X.

Saussurea medusa Maxim. Plant 5–40 cm with a solitary rosette normally. forming a hemispherical or conical, round-headed, woolly mound at maturity. Lower **leaves** obovate, truncate, up to 9 3 cm, ribbed, deeply and narrowly toothed towards apex, densely woolly; involucral **bracts** lanceolate, entire to somewhat toothed, often red-tinged, densely woolly. **Flower-heads** in a dense, semi-domed terminal cluster, **florets** purple or deep purple, smelling of vanilla. Gansu, Qinghai, w, nw & swSichuan, seTibet, nwYunnan [eHimalaya]; screes, rocky slopes, moraines, cliff-ledges, 3,000–5,100 m. VIII–X.

Saussurea quercifolia W. W. Sm.* Similar to the previous species, but rarely more than 10 cm tall, with obovate to oblanceolate, dark green **leaves** with purplish margins, white beneath, margin with shallow, rounded teeth, densely hairy only towards base. **Flower-heads** densely clustered, with purplish, lanceolate, woolly bracts; **florets** rose-purple or bluish purple. swSichuan, nwYunnan; rocky alpine slopes, moraines, screes, alpine moorland, 3,500–4,900 m. VII–IX.

Saussurea katochaete Maxim. (syn. *S. anochaete* Hand.-Mazz.; *S. katochaetoides* Hand.-Mazz.; *S. rohmooana* C. Marq. & Airy Shaw) Stemless **perennial herb**. **Leaves** in rosettes, elliptic, spathulate, ovate-triangular or ovate, 3–9 2–4 cm, white and densely woolly beneath, green and glabrous above, margin sharply minutely toothed or doubly toothed. **Flower-heads** 1 (or 2–4) in rosette of leaves, to 4 cm in diameter, stalkless or with short stalk; **florets** purple, 17–19 mm long. Gansu, Qinghai, Sichuan, seTibet, Yunnan [swTibet, Bhutan, Sikkim]; montane grassland and marshes, 2,200–4,700 m. VII–X.

Saussurea hieracioides Hook. f. (syn. *S. villosa* Franch.; *S. leveilleana* Maire) **Tufted herb** 3–20 cm tall. Lower **leaves** triangular-ovate, pointed, 2–16 2–4 cm, somewhat toothed on margin, grey-green, downy, *Hieracium*-like. **Flower-heads** solitary on downy stalk above the foliage; involucre flask-shaped, with closely overlapping, lanceolate, purplish **bracts**; **florets** purple or pinkish. Gansu, Hubei, w & swSichuan, seTibet; nwYunnan [c & eHimalaya, nMyanmar]; open low scrub, rocky slopes, screes, 3,000–4,600 m. VII–X.

Saussurea aff. *gnaphalodes* (Royle) Sch. Bip.* Dwarf **tufted** to **rosette-forming perennial** not more than 15 cm tall, with grey-woolly, shallowly lobed **leaves** to 12 cm long. **Flower-heads** solitary or clustered, purplish blue, 18–25 mm across, nestling close to the foliage. nwYunnan; rocky slopes, stony alpine meadows, 4,100–4,600 m. VI–VII.

GYNURA

A genus of 50 species (9 in China) from the Old World tropics and subtropics. **Perennial** hairy or glabrous **herbs**. **Flower-heads** with **disc-florets** only; involucral **bracts** narrow, margin papery; style-tips long-hairy. **Achenes** narrow, with many ribs, pappus white.

Gynura japonica (Thunb.) Juel (syn. *G. pinnatifida* DC.; *G. aurita* C. Winkl.) Large **herb** to 1 m with pinnately lobed, deep, shiny green **leaves**, paler beneath, to 40 15 cm, lobes toothed and pointed. **Flower-heads** several, to 15 mm across; **disc-florets** many, orange to bright yellow. wHubei, w & swSichuan, eTibet, Yunnan [cChina, Japan, Taiwan]; mixed forests, streamsides, 1,500–3,200 m. IV–IX.

EUPATORIUM

A genus of c. 45 species (14 in China, 7 endemic) found in the eastern USA and Eurasia. **Perennial herbs** and **shrubs** with alternate or whorled **leaves**. **Flower-heads** small, with **disc-florets** only, often in broadly corymbose **inflorescences**. **Fruit** with a pappus of slender white hairs.

Eupatorium mairiei H. Lévl. Small **tufted perennial** to 20 cm with elliptic-lanceolate to lanceolate basal **leaves**, to 7.5 1.5 cm, minutely downy, narrowed below into a stalk, margin sparsely toothed; stem-leaves similar, alternate, sessile or almost so. **Flower-heads** solitary, surrounded by a ruff of leaf-like bracts; **florets** all hermaphrodite, reddish purple in bud; corolla white with 5 short recurved lobes. sGansu, Sichuan, e & seTibet, nwYunnan [eHimalaya, nMyanmar]; rocky slopes, alpine meadows, 4,100–4,650 m. VI–VIII.

ASTERACEAE Group Two

LEONTOPODIUM

A genus with 58 species (36 in China, 18 endemic) in Europe and Asia. Tufted or stoloniferous, generally rather woolly **perennial herbs**, stemless or with erect flowering **stems**, with male and female **flowers** borne on the same or separate plants. **Leaves** alternate, simple, entire, basal leaves sometimes in a rosette. **Inflorescence** terminal, of 1–20 crowded small **flower-heads** usually surrounded by radiating, grey, white or yellowish, woolly, leaf-like bracts; involucre campanulate, bracts small, dark brown, membranous towards tip; **female corollas** thread-like, 3–4-toothed, **male corollas** narrow-tubular, 5-toothed; styles 2-fid. **Achenes** oblong-ellipsoid, glabrous or downy; pappus simple, white.

ASTERACEAE (Group Two): Leontopodium

> 1. Leaves green and almost glabrous above .... **L. stracheyi, L. calocephalum**
> 1. Leaves white- or grey-woolly above 2
> 2. Flowering stems branched .... **L. sinense**
> 2. Flowering stems unbranched .... **L. caespitosum, L. andersonii, L. dedekensii**

Leontopodium stracheyi (Hook. f.) Hemsl. Stoloniferous, **patch-forming dioecious** plant 6–40 cm tall with glandular, brownish **stems**. **Leaves** narrowly oblong-lanceolate, 2–3 0.4–0.5 cm, pointed or long-pointed, white-woolly beneath. **Inflorescence** terminal, star-like, 3–5.5 cm across, subtended by 8–12, lanceolate, white-woolly, pointed bracts; **flower-heads** 5–11; **corolla** 3.5–4 mm long, dark with white anthers. w & swSichuan, Tibet, w & nwYunnan [Himalaya]; alpine meadows, moraines, grassy banks, dry, rocky hillsides, roadsides, 3,000–5,000 m. VI–X.

Leontopodium calocephalum (Franch.) Beauv.* (syn. *Gnaphalium leontopodium* L. var. *calocephalum* Franch.) Tufted, rhizomatous **perennial**, with simple erect **stems** to 50 cm, usually 15–30 cm, grey- or white-downy overall. **Leaves** linear-lanceolate to lanceolate, 5–15 0.2–1.2 cm, long-pointed, greenish above, greyish or whitish beneath. **Bracts** 10–18, forming a spreading star up to 8 cm across, yellowish- or whitish-downy. **Flower-heads** 5–10 mm across, up to 20 in a dense cluster. Gansu, Qinghai, Sichuan, Yunnan; alpine meadows, marshy places, thickets, coniferous forests, 2,600–4,200 m. VII–IX.

Leontopodium sinense Hemsl.* (syn. *L. arbuscula* Beauv.) Woody-based **dioecious herb** 15–80 cm with linear to linear-lanceolate **leaves**, 1.8–4 0.2–0.4 cm, almost pointed, white-woolly beneath, grey above. **Inflorescence** terminal, 4–5 cm across, subtended by several lanceolate, white-woolly bracts; **flower-heads** 5–11; **corolla** white with yellow anthers. Gansu, Sichuan, Yunnan [(?)cChina]; grassy banks, dry, rocky hillsides, roadsides, 1,450–3,600 m. V–X.

Leontopodium caespitosum Diels Very variable stoloniferous, **mat-forming**, grey-woolly plant 15–50 cm tall. **Rosette-leaves** linear to oblanceolate, 7–25 1–3 mm; **stem-leaves** lanceolate to linear with recurved margins, 1–5 mm broad. **Inflorescence** star-like, with 4–9, all female or predominantly male or female, **flower-heads** subtended by narrowly lanceolate, long-pointed, densely white-woolly bracts, 8–22 mm long; **corolla** 3–3.5 mm long. w & swSichuan, seTibet, nwYunnan [swTibet, Himalaya]; grassy banks, dry, rocky hillsides, alpine meadows, moraines, roadsides, 3,000–5,150 m. VI–VIII. (See also photo 2, p. 198.)

Leontopodium andersonii C. B. Clarke* (syn. *L. subulatum* (Franch.) Beauv.) **Dioecious** plant, more tufted and taller (to 75 cm) than previous species, with linear, pointed or long-pointed **leaves** only 1 mm broad. **Inflorescence** to 6 cm across, subtended by 8–12, linear-lanceolate bracts; **flower-heads** 5–11; **corolla** 3.5–4 mm long, dark with whitish anthers. w & swSichuan, seTibet; w & nwYunnan; alpine meadows, dry banks in broadleaved woodland, clearings in *Pinus* forest, banks in scrub, 1,000–4,250 m. V–XI.

Leontopodium dedekensii (Bur. & Franch.) Beauv. (syn. *Gnaphalium dedekensii* Bur. & Franch.) Multi-stemmed, rhizomatous **perennial** with simple erect **stems** to 40 cm, grey-downy overall. **Leaves** stalkless, linear-lanceolate, to 40 6.5 cm, with a somewhat heart-shaped base, decreasing in size up the stem. **Bracts** up to 20, 5–30 mm long, forming a spreading whitish- to yellowish-downy star. **Flower-heads** crowded, 4–5 mm across. Gansu, Qinghai, Sichuan, Tibet, Yunnan [Myanmar]; grassy habitats, coniferous forests, thickets, 1,400–4,100 m. VI–VII.

PETASITES

A genus of c. 20 species (5 in China) distributed in Europe, Asia and North America. Rhizomatous **herbs** with large kidney-shaped, stalked **leaves**, the **inflorescences** appearing before the leaves in early spring.

Petasites tricholobus Franch. **Leaves** when fully developed 7–11 12–19 cm, margin obscurely lobed and doubly toothed; leaf-stalks to 20 cm long, hairy. **Inflorescence** paniculate, 7–28 cm long, elongating to 60 cm in fruit; **flower-heads** dirty pink or flesh-coloured. Gansu, Hubei, Qinghai, Sichuan, Shaanxi, Tibet, nwYunnan [n, c & eChina, Taiwan, eHimalaya, nMyanmar]; wet stony banks, wet flushes, streamsides, 900–4,000 m. III–V.

Petasites versipilus Hand.-Mazz.* Differs from *P. tricholobus* in having thread-like, often 5-lobed female **florets** and an **ovary** that is usually glabrous (although hairy in hermaphrodite florets). swSichuan, nwYunnan; grassy alpine slopes, forests and thicket margins, 2,700–3,200 m. III–V.

ASTERACEAE Group Three

ALLARDIA

A genus with c. 8 species (6 in China) in Central Asia and the Tibet–Himalayan region. Fragrant, tufted, **perennial** alpine **herbs**, with small, pinnately or palmately dissected **leaves**. **Flower-heads** daisy-like with both **ray-** and **disc-florets**.

Allardia glabra Decne (syn. *Waldheimia glabra* (Decne) Regel) **Mat-forming** deep green plant. **Leaves** 1–3 cm long with 3- or 5-segments, segments with blunt linear lobes. **Flower-heads** flat, held immediately above the foliage, 2.8–3.6 cm across; **ray-florets** with an oblong-elliptic, pink to purple, rarely white ligule 9–12 mm long; **disc-florets** yellow. seTibet [swTibet, Himalaya]; screes, moraines, sandy places, rocks by streams, 4,200–5,000 m. VII–IX.

1. *Leontopodium calocephalum*; Huanglong, nwSichuan (CGW)
2. *Leontopodium sinense*; Cangshan, nwYunnan (CGW)
3. *Leontopodium caespitosum*; Cangshan, nwYunnan (CGW)
4. *Leontopodium andersonii*; Yulongxueshan, nwYunnan (PC)
5. *Leontopodium dedekensii*; Songpan, nwSichuan (PC)
6. *Petasites tricholobus*; Cangshan, nwYunnan (PC)
7. *Petasites versipilus*; Yulongxueshan, nwYunnan (PC)
8. *Allardia glabra*; Ragarsanba Glacier, sTibet (RS)

ASTERACEAE (Group Three): Inula

INULA

A genus with c. 100 species (17 in China — 3 endemic) distributed in temperate Europe and Asia. **Perennial herbs** or **subshrubs**, more rarely **annual** or **biennial herbs**, with erect or ascending, usually branched **stems** and simple, entire or toothed **leaves**. **Flower-heads** terminal, solitary or several, yellow, **female florets** few to numerous in 1–2 rows, tubular and 3-toothed at apex or distinctly strap-shaped, yellow or orange; **disc-florets** hermaphrodite, tubular, 5-lobed at apex, yellow. **Achenes** angular or ribbed, with rough pappus-hairs.

1. Flower-heads not more than 4 cm across . . . . **I. helianthus-aquatica**
1. Flower-heads 5–6 cm across . . . . **I. hookeri, I. forrestii**

Inula helianthus-aquatica C. Y. Wu ex Y. Ling* **Tufted perennial** 30–100 cm with stalkless, elliptic-ovate **leaves**, 3–4.5 1.2–1.6 cm, pointed, hairy beneath. **Flower-heads** 3–4 cm across; **ray-florets** with a yellow ligule 10 mm long, **disc-florets** tubular, yellow, 4–5 mm long. sGansu, Guangxi, Guizhou, Sichuan, Tibet, nwYunnan [cChina]; scrub and damp grass by streams and rivers, 1,200–3,000 m. VII–IX.

Inula hookeri C. B. Clarke **Herbaceous perennial** 60–80 cm with hairy **stems**. **Leaves** often rather pale green, lanceolate to elliptic, 10–15 2.5–4 cm, finely toothed on margins, hairy. **Flower-heads** 5–6 cm across, solitary or 2–3, golden yellow; involucre with narrow recurved **bracts** with long shaggy hairs; **ray-florets** thread-like, to 25 mm long. seTibet, nwYunnan [swTibet, Himalaya, nMyanmar]; alpine meadows, shrubberies, 2,000–3,800 m. VIII–IX.

Inula forrestii J. Anth.* Plant often taller than the previous species, to 1 m, with smaller lanceolate **leaves**, 3–7 0.7–3 cm, margin shallowly toothed; **flower-heads** yellow, c. 5 cm across, with fewer, broader ligules. w & nwYunnan; forests, thickets, 2,000–2,100 m. VII–IX.

DORONICUM

A genus with c. 40 species (7 in China, 4 endemic) in temperate Eurasia and North Africa. **Perennial herbs** with alternate, long-stalked basal **leaves** and stalkless stem-leaves with a base that clasps the **stem**. **Flower-heads** large, usually solitary or 2–6, laxly corymbose; **ray-florets** in 1 row, female, **disc-florets** in many rows, hermaphrodite, tubular, yellow, limb 5-lobed. **Achenes** oblong or oblong-obconic, ribbed; pappus of many fine, white or reddish, bristles.

1. *Inula forrestii*; Yulongxueshan, nwYunnan (DL)
2. *Doronicum stenoglossum*; Habashan, nwYunnan (MN)
3. *Doronicum briquetii*; Gangheba, Yulongxueshan, nwYunnan (CGW)
4. *Pyrethrum tatsienense*; Daxueshan, nwYunnan (MN)
5. *Ligularia purdomii*; near Longriba, nwSichuan (CGW)
6. *Ligularia purdomii*; south of Aba, nwSichuan (CGW)

Doronicum stenoglossum **Maxim.*** (syn. *D. souliei* Cav.; *D. yunnanense* Franch. ex Diels) Rhizomatous glandular-hairy **perennial** with simple, erect **stems**, 50–100 cm tall. **Leaves** oblong or ovate-oblong, 4–10 2.5–4 cm, narrowed at base to narrowly winged stalk, margin finely toothed or almost entire. **Flower-heads** 2–2.5 cm in diameter, 2–10 arranged in a raceme; involucre hemispherical or broadly campanulate, c. 1.5 cm long, with 2–3 whorls of glandular-hairy bracts; **ray-florets** pale yellow, shorter than or equal to involucre, ligules linear, 7–10 mm long, **disc-florets** yellow. Gansu, Qinghai, w, nw & swSichuan, e & seTibet, nwYunnan; subalpine and alpine grassland, forest margins, secondary thickets, *Picea* forests, 2,100–3,900 m. VII–IX.

Doronicum thibetanum Cav. (syn. *Cremanthodium calotum* Diels; *Doronicum limprichtii* Diels) Similar to the previous species, but not more than 75 cm tall, usually less, with green or purple-tinged **stems** and tuberous **rhizomes**. **Leaves** obovate-spathulate or oblong-ovate, 4–15 1.5–3.5 cm; **flower-heads** solitary, terminal, large, 5–6 cm in diameter, **ray-florets** yellow, ligules 22–28 mm long, narrowly linear. Qinghai, w, nw & swSichuan, e & seTibet, nwYunnan [swTibet, Himalaya]; alpine grassland, thickets, stony slopes, 3,400–4,200 m. VII–IX.

Doronicum briquetii Cavill. Similar to *D. stenoglo*ssum but a smaller plant, rarely more than 50 cm tall in flower, with richer yellow, somewhat larger, **flower-heads**; ligules 1.5–3 mm wide, much longer than involucral **bracts**. Sichuan, Tibet, Yunnan [cAsia]; similar habitats and altitudes. VI–VII.

PYRETHRUM

A genus of c. 100 species (14 in China, mostly in Xinjiang) in Europe, Central Asia and North Africa, many species now included in the genus *Tanacetum*. **Perennial herbs**, **subshrubs** or **shrubs** with pinnately lobed to pinnatisect **leaves**, often with showy, daisy-like **flower-heads** with prominent ligules.

Pyrethrum tatsienense (**Bur. & Franch.**) **Y. Ling ex C. Shih*** (syn. *Senecio tatsienensis* Bur. & Franch.; *Tanacetum tatsienense* (Bur. & Franch.) K. Bremer & Humph.) **Perennial herb**, with solitary or several, unbranched, softly hairy **stems**, 7–25 cm tall. Basal **leaves** short-stalked, elliptic or long-elliptic, 2-pinnatisect, 1.5–7 1–1.5 cm, green; primary lateral segments in 5–15 pairs. **Flower-heads** solitary, 3.5–5.5 cm across; **ray-florets** orange-red, ligules linear or broadly linear, up to 2 cm long, apex 3-toothed. Qinghai, Sichuan, e & seTibet, nwYunnan; alpine meadows, thickets, gravelly places on mountain slopes, 3,500–5,200 m. VII–IX. **Var. *tanacetopsis* (W. W. Sm.) Y. Ling & C. Shih*** lacks **ray-florets**. Tibet, Yunnan; 3,500–5,000 m.

LIGULARIA

An Asian genus of c. 140 species (124 in China, 89 endemic). Erect **perennial herbs** with leafy **stems**. **Leaves** stalked, alternate, simple or palmately lobed. **Flower-heads** yellow, solitary, few or numerous in racemes or corymbs, discoid or radiate; involucre cylindrical, obconic or campanulate, the bracts in a single row; **ray-florets** 0–25 with showy ligules, **disc-florets** tubular to campanulate. **Achenes** oblong, glabrous with a white or reddish pappus.

Synopsis of groups

Group One. Flower-heads without ligules.

Group Two. Flower-heads with well-defined ligules. (p. 504)

> **Group One**
> 1. Leaves arrow- to heart-shaped . . . . **L. atroviolacea**
> 1. Leaves kidney-shaped to orbicular, sometimes peltate or palmate 2
> 2. Stems brown-hairy . . . .
>   . . . . **L. purdomii, L. achyrotricha, L. paradoxa**
> 2. Stems white-hairy . . . . **L. nelumbifolia**

Ligularia atroviolacea (**Franch.**) **Hand.-Mazz.*** (syn. *L. oreotrephes* W. W. Sm.) **Tufted perennial** 30–60 cm tall, with blackish purple, hairy **stems**. Basal **leaves** arrow- to ovate-heart-shaped, 4–13 4.5–9 cm, shortly soft-hairy beneath; leafstalks 5–15 cm long, sheathing at base, blackish purple-hairy. **Inflorescence** corymbose, branched, **flower-heads** 4–10, 10–15 mm across; involucral **bracts** 10–12, lanceolate or oblong, pointed; **florets** numerous, yellow or deep orange, tubular, 6–7 mm long. nwYunnan; open, stony, alpine meadows, margins of thickets, forest understorey, 3,000–4,000 m. VIII–XI.

Ligularia purdomii (**Turrill**) **Chitt.*** Stout **herbaceous perennial** to 1.5 m, **stems** covered in short brown hairs. Basal **leaves** with purplish red stalks, peltate, kidney-shaped to almost rounded, 14–50 cm across, broader than long, densely short-hairy, margin regularly toothed; upper leaves similar but smaller. **Inflorescence** compound, up to 50 cm long, much-branched, with numerous yellow **flower-heads**; involucral bracts 6–12, dark brown, oblong or lanceolate, pointed; **florets** numerous, all tubular, 7–9 mm long. swGansu, seQinghai, nwSichuan; streamsides, swamps, ditches, 3,700–4,100 m. VI–VII.

Ligularia achyrotricha (**Diels**) **Y. Ling*** Smaller plant than previous species, to 65 cm, **stems** with dense, short, brown hairs, basal **leaves** kidney-shaped, 3–10 7–12.5 cm, palmately veined, glabrous or downy along veins beneath; **flower-heads** numerous, with yellow-hairy involucral **bracts**, **florets** all tubular, 7–9 mm long. Shaanxi (Qinlingshan); grassy slopes, forest margins, shrubberies, 3,300–3,700 m. VI–VIII.

Ligularia paradoxa **Hand.-Mazz.*** A variable species 30–90 cm tall, differing from the previous species in its long-stalked *Aconitum*-like **leaves**, deeply palmately 3–8-lobed; **inflorescence** compound, capitate; **flower-heads** deep crimson, up to 30, each 10–13 mm across, without **ray-florets**. w & nwYunnan; open mountain meadows, scrub and forest understorey, 3,400–4,600 m. VII–X.

Ligularia nelumbifolia* (Bur. & Franch.) Hand.-Mazz. **Herbaceous perennial** 80–100 cm, **stems** white cobwebby-hairy, shortly yellow-hairy above. **Leaves** horizontal, large, orbicular, 7–30 × 13–38 cm, white cobwebby-hairy beneath, glabrous above, margin irregularly toothed; leaf-stalks stout, erect, 10–50 cm long. **Inflorescence** compound, bearing many narrow-cylindrical **flower-heads** on almost erect branches up to 40 cm long; **florets** even, yellow. sGansu, wHubei, w, nw & swSichuan, sShaanxi, ne & nwYunnan; rocky streambeds, scrub, forest understorey, 2,400–3,900 m. VII–IX.

> **Group Two**
> 1. Leaves palmately lobed . . . . **L. przewalskii**
> 1. Leaves not lobed 2
> 2. Inflorescence branched, cymose 3
> 2. Inflorescence racemose, often spike-like 4
> 3. Leaves orbicular to kidney-shaped . . . . **L. dentata**
> 3. Leaves obovate to elliptic or elliptic-oblong . . . .
>   . . . . . . **L. amplexicaulis, L. lapathifolia, L. rumicifolia**
> 4. Leaves obovate, oblanceolate to lanceolate . . . .
>   . . . . **L. liatroides**
> 4. Leaves ovate, heart-, arrow- or spear-shaped . . . .
>   . . . . . **L. lankongensis, L. latihastata, L. pleurocaulis,**
>   **L. lidjiangensis**

***Ligularia przewalskii* (Maxim.) Diels** **Herbaceous perennial** 60–130 cm tall with ovate, palmately 4–7-lobed **leaves**, 4–15 × 7–30 cm, lobes further lobed; leaf-stalks up to 50 cm long, ± winged. **Inflorescence** slender, racemose, 16–48 cm long; **flower-heads** numerous; involucre slender-cylindrical; **ray-florets** 2 or 3, usually yellow, ligules linear-oblong, blunt, to 17 mm long; **disc-florets** usually 3, 10–12 mm long. Gansu, Qinghai, Sichuan, Shaanxi [c & nChina, Mongolia]; open, wet meadows and bogs, streamsides, forest margins and understorey, damp shrubberies, 1,100–3,700 m. VI–X.

***Ligularia dentata* (A. Gray) Hara** (syn. *L. clivorum* Maxim.; *L. emeiensis* Kitam.) Clump-forming **herbaceous perennial** with erect **stems** 30–120 cm tall, white cobwebby-hairy and shortly yellow-hairy above. **Basal leaves** stalked, kidney-shaped, 7–30 × 12–38 cm, green and glabrous above, greyish white-hairy beneath, margin regularly toothed; **stem-leaves** similar but smaller and short-stalked. **Inflorescence** corymbose or compound-corymbose, spreading, with numerous yellow **flower-heads**, 7–10 cm across, involucre broader than long, densely white cobwebby-hairy; **ray-florets** yellow, ligules narrow-oblong, to 5 cm long. Gansu, Guangxi, Guizhou, Hubei, Shaanxi, Sichuan, Yunnan [c & sChina, Japan]; damp grassy slopes, riversides, forest margins, forest understorey, sometimes cultivated, 700–3,200 m. VII–X.

***Ligularia amplexicaulis* (Wall.) DC.** (syn. *L. cymbulifera* (W. W. Sm.) Hand.-Mazz.; *Senecio cymbulifer* W. W. Sm.; *Ligularia crassa* Hand.-Mazz.) Stout **herbaceous perennial** 0.6–1.3 cm tall. **Leaves** 15–60 × 20–45 cm, white-felted on both surfaces, margin minutely toothed. **Inflorescence** many-branched; **flower-heads** numerous, involucre only 5 mm across, white-felted outside; **ray-florets** yellow, ligules linear, 10–14 mm long, **disc-florets** numerous, deep yellow, 6–7 mm long. n, w, nw & swSichuan, e & seTibet, nwYunnan [Himalaya]; open mountain meadows, scrub, forest margins, streamsides, 2,900–4,800 m. VII–IX.

Ligularia lapathifolia* (Franch.) Hand.-Mazz. Often taller than the previous species, to 1.6 m with broadly ovate or ovate-oblong **leaves**, 18–40 × 10–28 cm, with a somewhat heart-shaped base. **Flower-heads** larger, **ray-florets** yellow with linear ligules 15–20 mm long, **disc-florets** numerous, 10–11 mm long. swSichuan, nwYunnan; dry woods, open places, forest understorey, shrubberies by streams, 1,800–3,600 m. VII–X.

***Ligularia rumicifolia* (J. R. Drumm.) S. W. Liu** (syn. *Senecio rumicifolius* Drumm.; *Cremanthodium rumicifolium* (Drumm.) R. D. Good; *Ligularia leesicotal* Kitam.) Tufted **herbaceous perennial** to 100 cm, white-woolly. Basal **leaves** ovate-oblong, 10–19 × 8–14.5 cm, with prominent white reticulate veins, both surfaces initially white-woolly, margin minutely toothed; upper leaves ovate to lanceolate. **Inflorescence** compound, initially clustered, later spreading, branches to 17 cm long, white-woolly. **Flower-heads** numerous, initially white-downy; involucral **bracts** in 2 rows, dark brown; **ray-florets** with 3–7, yellow ligules. swSichuan, e & seTibet [Nepal]; lakesides, forested slopes, open shrubberies, 3,700–4,500 m. VII–X.

Ligularia liatroides* (C. Winkl.) Hand.-Mazz. **Herbaceous perennial** 60–100 cm tall, **stems** white-felted above. Basal **leaves** stalked, oblong or ovate-lanceolate, 8–22 × 4.5–8 cm, greyish green, prominently reticulately veined beneath, glabrous or initially white-downy along veins, margin toothed. **Inflorescence** racemose, to 40 cm, with numerous **flower-heads**; involucral **bracts** c. 3 mm wide, margin membranous and densely white-ciliate; **ray-florets** 5 or 6, yellow, ligules linear, 6–8 mm long, **disc-florets** numerous. swQinghai, n, nw & wSichuan, neTibet; streamsides, swamps, scrubby and alpine meadows, 2,900–4,500 m. VI–VIII.

Ligularia lankongensis* (Franch.) Hand.-Mazz. **Herbaceous perennial** 40–100 cm tall with densely white cobwebby-minutely downy **stems**. **Leaves** erect or almost so, spear-shaped to ovate, 9–24 × 4.5–21 cm, blunt, white-felted beneath. **Inflorescence** racemose, 10–21 cm long, ± cylindrical with numerous nodding **flower-heads** 10–12 mm across; involucre campanulate, minutely grey-white-downy; **ray-florets** with a recurved yellow ligule 10–13 mm long, **disc-florets** 7 mm long, dark. swSichuan, ne & nwYunnan; meadows, including alpine meadows, rocky places, scrub, road- and streamsides, 2,100–3,800 m. IV–IX.

Ligularia latihastata* (W. W. Sm.) Hand.-Mazz. Generally smaller than the previous species, 35–60 cm tall. Basal **leaves** stalked, broadly spear-shaped or triangular-spear-shaped,

1. *Ligularia nelumbifolia*; Tianbao, Zhongdian, nwYunnan (PC)
2. *Ligularia przewalskii*; Huanglong, nwSichuan (CGW)
3. *Ligularia dentata*; Emeishan, sSichuan (CGW)
4. *Ligularia amplexicaulis*; Genyanshan, south side, swSichuan (AD)
5. *Ligularia liatroides*; Huanglong, nwSichuan (CGW)
6. *Ligularia langkongensis*; north of Lijiang, nwYunnan (CGW)
7. *Ligularia latihastata*; Cangshan, nwYunnan (CGW)
8. *Ligularia latihastata*; Cangshan, nwYunnan (CGW)

1. *Ligularia pleurocaulis*; Genyanshan, south side, swSichuan (MN)
2. *Ligularia lidjiangensis*; north of Lijiang, nwYunnan (HJ)
3. *Senecio laetus*; Dali, nwYunnan (CGW)
4. *Senecio faberi*; Emeishan, sSichuan (CGW)
5. *Sinacalia tangutica*; Emeishan, sSichuan (CGW)
6. *Nannoglottis latisquama*; Gangheba, Yulongxueshan, nwYunnan (CGW)
7. *Nannoglottis hookeri*; nNepal (CGW)

4–11 × 9–15 cm, palmately veined, glabrous, margin regularly toothed. **Inflorescence** racemose, 10–30 cm long, with up to 24 **flower-heads**; **ray-florets** numerous, with a yellow, linear, pointed or long-pointed ligule, 2.5–4 cm long. nwYunnan; streamsides, swampy grassland, forest understorey, grassland, 2,400–4,000 m. VII–X.

Ligularia pleurocaulis (Franch.) Hand.-Mazz.* (syn. *Senecio pleurocaulis* Franch.; *Cremanthodium pleurocaule* (Franch.) R. D. Good; *Ligularia pleurocaulis* forma *uberrima* Hand.-Mazz.) Erect greyish green **perennial** to 100 cm. Basal **leaves** almost stalkless with purplish red sheath, linear-oblong or broadly elliptic, 8–30 × 0.7–7 cm, pointed, margin entire; upper leaves elliptic to linear. **Inflorescence** simple or branched, to 20 cm long, often lax, white-cobwebby minutely downy; **flower-heads** numerous, semi-nodding, inclined to 1 side of racemose axis; **ray-florets** with a yellow, broadly elliptic or ovate-oblong ligule. w, nw & swSichuan, nwYunnan; grassy and shrubby slopes, alpine meadows, streamsides, 3,000–4,700 m. VII–XI.

Ligularia lidjiangensis Hand.-Mazz.* Stout **perennial**, stiffly erect, to 70 cm. **Leaves** heart-shaped, to 30 cm long, deep green above, grey-green beneath, with a toothed wavy margin, only the lowermost stalked, upper much reduced. **Flower-heads** bright yellow, outwards-facing, 4–5 cm across, numerous in a narrow, columnar inflorescence. nwYunnan (Lijiang); grassy slopes, evergreen *Quercus* scrub, c. 2,900 m. VI–VII.

SENECIO

A large and complex genus with c. 1,500 species (perhaps c. 45 in China) with a worldwide distribution. **Annual** and **perennial herbs**, **shrubs** and **trees**, and tree-like 'giant groundsels' (sometimes placed in the genus *Dendrosenecio*). Most have alternate **leaves** that can be simple or variously dissected. **Flower-heads** solitary or in corymbose or paniculate clusters, usually with both **ray-** and **disc-florets**; involucral **bracts** in a single row but generally with a few additional, smaller bracts outside. **Achenes** with a pappus of rough hairs.

Senecio laetus Edgew. **Perennial**, rhizomatous **herb** with leafy **stems**, 40–80 cm tall, sparsely cobwebby-downy. **Leaves** ovate-elliptic to ovate-lanceolate or oblanceolate, 8–10 cm long, basal leaves stalked, the others lyrate-pinnately lobed or pinnately lobed, 5–22 cm long, with basal auricles that ± clasp the stem. **Flower-heads** arranged in terminal, often compound corymbs; **ray-florets** 10–13, each with a yellow, oblong ligule c. 6.5 mm long, **disc-florets** many, yellow. Guizhou, Hubei, Sichuan, Tibet, Yunnan [Himalaya]; forest margins, thickets, open grassy places, field margins, roadsides, 1,100–3,800 m. IV–X.

Senecio faberi Hemsl.* (syn. *S. kaschkarowii* C. Winkl.) Robust, rhizomatous **perennial** with solitary, erect, stout, hollow **stems** 80–150 cm tall, sparsely downy when young. Basal **leaves** mostly withered by flowering time, long-stalked, ovate, 8–10 cm long, margin lyrate-pinnate with large irregularly coarsely toothed or lobulate terminal lobe and 2–4 small basal lobes. **Flower-heads** yellow, numerous, in dense terminal compound corymbs; involucre narrowly campanulate, 3–5 mm long, purplish; **ray-florets** 3–4, ligules yellow, linear, c. 4.5 mm long, **disc-florets** 6–9. sGuizhou, w, nw, sw & sSichuan; forests, thickets, grassy slopes, damp shady places, 900–2,700 m. VI–VIII.

SINACALIA

A genus of 4 species all endemic to China. Closely related to (and sometimes included within) *Senecio*, differing in the thick subterranean tuberous **rhizomes**, in generally large, lax, often paniculate **inflorescence** and bristly achene-pappuses.

Sinacalia tangutica (Maxim.) B. Nord.* (syn. *Senecio tanguticus* Maxim.; *Sinacalia henryi* (Hemsl.) H. Rob. & Brettell) **Patch-forming** plant with robust, simple, sparsely hairy **stems** 50–100 cm tall. **Leaves** ovate or ovate-heart-shaped, 10–16 × 10–15 cm, margin deeply pinnately lobed with 3–4 lateral lobes on each side; lower stem-leaves withered by flowering time, stalked. **Flower-heads** yellow, small, numerous, cylindrical, arranged in broad, terminal, brown-glandular panicles; **ray-florets** 2–3, ligules oblong-lanceolate, 13–14 mm long, **disc-florets** 4, 8–9 mm long. Gansu, Hubei, Qinghai, Sichuan, Shaanxi [cChina]; forest margins, grassy slopes, cliffs, meadows, stream- and roadsides, 1,200–3,500 m. VII–IX.

NANNOGLOTTIS

A genus with 8 species (all in China, all endemic). Closely related to *Senecio* but **flower-heads** with outermost **florets** female and with a ligule, the next female and tubular, and the central hermaphrodite and tubular. **Achenes** 8–10-ribbed with a deciduous 1-rowed pappus.

Nannoglottis latisquama Ling ex X. L. Chen* (syn. *N. yunnanensis* (Hand.-Mazz.) Hand.-Mazz.) **Tufted herb** to 60 cm. **Leaves** borne on the **stems**, obovate to oblong-ovate, 8–15 × 4–7.5 cm, shallowly blunt-toothed on margin, white-felted beneath. **Flower-heads** 1–5, 3.5–4.8 cm across; **ray-florets** with ligules up to 1.8 cm long, bright yellow to orange-yellow. swSichuan, nwYunnan; banks amongst rocks, alpine meadows, coniferous forests, 3,600–4,200 m. IV–VIII.

Nannoglottis hookeri (Hook. f.) Kitam. Clump-forming **herb** similar to the previous species, but often taller, to 90 cm, **stem-leaves** almost erect, elliptic-obovate, 6–12 × 1.2–2.5 cm, pointed, not white-felted beneath, **basal leaves** larger, elliptic-lanceolate; **flower-heads** 1–several, 3–4.5 cm across. seTibet [swTibet, Himalaya]; amongst rocks, alpine meadows, shrubberies, streamsides, 3,200–4,600 m. VI–VIII.

GERBERA

A genus of c. 35 species (5 in China) found primarily in tropical and southern Africa, India and South East Asia. **Leaves** mostly confined to basal rosettes, unlobed or variously lobed. **Flower-heads** often scapose, with both **ray-** and **disc-florets**. **Achenes** rough, with a tuft of hairs.

Gerbera nivea (DC.) Sch. Bip. **Rhizomatous** plant 10–35 cm tall. **Leaves** basal, almost erect, oblanceolate, shallowly pinnately lobed, rounded at apex, 3–15 × 1.5–2.5 cm, margin with round lobes. **Inflorescence** with a solitary, terminal nodding flower-head, 3.5–6.2 cm across; **ray-florets** with ligules off-white to rose-pink, 2.5–2.8 cm long. seTibet, nwYunnan [swTibet, Himalaya]; open coniferous forests, rocky alpine meadows, screes, stabilised moraines, 3,300–4,200 m. VII–X.

ASTER

A genus with c. 600 species (95 in China, 67 endemic), cosmopolitan in distribution. **Perennial herbs** and **shrubs**, sometimes stoloniferous, with simple, alternate **leaves**, sometimes in basal rosettes. **Flower-heads** 1–many, radiate; **ray-florets** in a single row, female, usually few, white, mauve or blue; **disc-florets** hermaphrodite, corolla tubular or campanulate, 5-toothed, usually yellow. **Achenes** oblong to obovoid, compressed, with 2 or more ribs; pappus bristles rough, white to brownish.

1. Shrubs and subshrubs . . . . **A. albescens, A. salignus, A. batangensis, A. staticifolius**
1. Perennial herbs 2
2. Plants generally more than 40 cm tall, stems with 1–5 flower-heads . . . . **A. yunnanensis, A. oreophilus**
2. Plants not more than 40 cm tall, usually 10–30 cm, flower-heads solitary 3
3. Achenes with a brown pappus . . . . **A. souliei**
3. Achenes with a white pappus 4
4. Plants non-stoloniferous . . . .
   . . . . **A. himalaicus, A. asteroides, A. diplostephioides**
4. Plants stoloniferous . . . . **A. stracheyi**

Aster albescens (DC.) Wall. Very variable erect **shrub**, 0.5–2 m tall with lanceolate-ovate, ± pointed **leaves**, 3–10 cm long, entire or toothed, sparsely hairy and glandular beneath. **Flower-heads** small, usually numerous in fairly dense terminal corymbs; **ray-florets** 12–30, corolla-tube 2.5 mm long, with a blue, mauve or white strap-like ligule, **disc-florets** yellow. w & swSichuan, seTibet, w & nwYunnan [Himalaya, nMyanmar]; forest clearings and margins, shrubberies on rocky slopes, 1,750–3,650 m. VII–X. Forms with narrow **leaves** from Yunnan and Sichuan are generally assigned to the closely related *A. salignus* **Willd.**, but would probably be better classified as a variety of the variable *A. albescens*.

Aster batangensis Bur. & Franch.* (syn. *A. limonifolium* Franch.) Branched woody spreading or tufted **subshrub** 10–30 cm tall. **Leaves** spathulate, 2.8–3.5 × 0.4–0.7 cm, apex rounded. **Flower-heads** 2.2–5 cm across; **ray-florets** generally less than 15, ligules purple to lavender-blue; **disc-florets** yellow or orange-yellow. swSichuan, seTibet, nwYunnan; coniferous forests amongst rocks, stabilised moraines, gravelly cliffs, 3,150–3,900 m. V–X.

Aster staticifolius Franch.* (syn. *A. batangensis* Bur. & Franch. var. *staticifolius* (Franch.) Y. Ling) Closely related to the previous species, but a **mat-forming** prostrate **subshrub**, much-branched and woody below, with mostly basal, grey-green, downy, linear to spathulate, rather wavy **leaves**, 20–50 × 4–6 mm. **Flower-heads** solitary, 4–5 cm across, held just above the foliage on short stalks, to 7.5 cm long; **ray-florets** 15–20, ligules blue-purple, to 22 mm long. swSichuan, e & seTibet, nwYunnan [swTibet]; open stony habitats, banks, field boundaries, 2,500–4,600 m. V–IX.

Aster yunnanensis Franch.* Hairy **herb** 30–80 cm tall. **Leaves** lanceolate to oblanceolate, 7–15 × 1.4–4.2 cm, pointed, entire or toothed, hairy, clasping **stem** at base. **Flower-heads** 4–6 cm across, 1–5 in a lax terminal cluster; **ray-florets** with pale to mid-blue or deep purple-blue ligules 16–22 mm long; **disc-florets** yellow or yellow-green. w & swSichuan, seTibet, nwYunnan; alpine and subalpine meadows, open forests, shrubberies, screes, 3,200–5,400 m. IV–XI.

Aster oreophilus Franch.* Closely related to the previous species, but a creeping-**rhizomatous** plant not more than 65 cm, **stems** with almost erect, spathulate, blunt **leaves**, 3.5–6 × 0.5–1.8 cm. **Flower-heads** 2.6–3.8 cm across; **ray-florets** with pale blue ligules 7 mm long. w & nwYunnan; montane meadows, 1,800–2,000 m. V–VIII.

Aster souliei Franch.* (syn. *A. forrestii* Stapf; *A. tongolensis* Franch.) Hairy or glabrous, erect **herb** 5–40 cm tall. **Basal leaves** oblanceolate or spathulate, 1.5–6.5 × 0.5–2.2 cm, blunt or pointed, entire, hairy on both sides and ciliate; **stem-leaves** oblong to linear. **Flower-heads** solitary, 4.5–6 cm across; **ray-florets** 40–60, ligules mauve or blue, 10–23 mm long; **disc-florets** yellow to orange. Gansu, w, nw & swSichuan, nwYunnan;

1. *Gerbera nivea*; nwYunnan (JM)
2. *Aster salignus*; Cangshan, nwYunnan (CGW)
3. *Aster batangensis*; Gangheba, Yulongxueshan, nwYunnan (CGW)
4. *Aster batangensis*; Muli, swSichuan (PC)
5. *Aster batangensis*; Habashan, nwYunnan (CGW)
6. *Aster staticifolius*; Gama La, Bangda, seTibet (HJ)
7. *Aster yunnanensis*; Chongjiang, nwYunnan (PC)
8. *Aster oreophilus*; Dali, nwYunnan (CGW)
9. *Aster souliei*; above Longriba, nwSichuan (CGW)
10. *Aster souliei*; Zhongdian, nwYunnan (CGW)

1. *Aster himalaicus*; eNepal (CGW)
2. *Aster diplostephioides*; swSichuan (CGW)
3. *Aster diplostephiodes*; eNepal (CGW)
4. *Aster stracheyi*; nBhutan (RS)
5. *Erigeron multiradiatus*; Napahai, Zhongdian, nwYunnan (CGW)
6. *Erigeron multiradiatus*; Zhongdian, nwYunnan (CGW)
7. *Erigeron breviscapus*; Zhongdian, nwYunnan (PC)

alpine meadows, open grassy hillsides, rocky slopes, edges of coniferous forests, shrubberies, 2,800–4,750 m. V–VIII.

Aster himalaicus C. B. Clarke **Perennial herb** 8–25 cm tall, arising laterally from a leaf-rosette. **Basal leaves** oblong-elliptic, 2–3.5 cm long, ± pointed, entire or remotely toothed, sparsely hairy and glandular beneath; **stem-leaves** oblong-obovate or lanceolate, 1.3–2.5 cm long. **Flower-heads** solitary; **ray-florets** 50–70, ligules purplish blue, 11–17 mm long; **disc-florets** yellow or purplish, 6.5–7.5 mm long. Sichuan, e & seTibet, nwYunnan [c & eHimalaya]; open or scrub-covered slopes (often with rhododendrons), turf, cliffs, 3,750–4,900 m. V–VII.

Aster asteroides (DC.) Kuntze (syn. *A. likiangensis* Franch.) Similar to the previous species, but a whitish-hairy, sometimes glandular plant with hoarily hairy **leaves**, involucre glandular-hairy and ligules shorter, 9–12 mm long, lavender-blue, blue or purplish; **disc-florets** yellow or purplish, 4–5 mm long. seTibet, nwYunnan [Himalaya]; open stony alpine meadows, marshy ground, steep rocky slopes, 3,650–4,760 m. VI–VIII.

Aster diplostephioides (DC.) C. B. Clarke (syn. *A. delavayi* Franch.) Readily distinguished from the previous 2 species by the larger **flower-heads**, 5–10 cm across, bearing slightly reflexed, linear ligules, 20–40 mm long, bluish purple to deep blue, contrasting with the greenish yellow to dull orange **disc-florets**. w & swSichuan, seTibet, nwYunnan [swTibet, Himalaya]; alpine meadows, screes, rocky places, 3,500–5,100 m. VI–IX.

Aster stracheyi Hook. f. Lax, **mat-forming** plant to 10 cm tall in flower, rarely more, producing long runners to 20 cm. **Leaves** mostly in basal rosettes, deep green and somewhat shiny, obovate to spathulate, 2–5 cm long, toothed towards apex; stem-leaves small, few. **Flower-heads** solitary, 2.5–4 cm across; ligules of **ray-florets** very slender, lilac-blue, spreading to somewhat recurved; **disc-florets** orange-yellow, turning brown in time. seTibet [swTibet, c & eHimalaya]; rocky slopes, stony meadows, 3,600–4,500 m. VII–IX.

ERIGERON

A genus with c. 200 species (35 in China, 13 endemic) in Europe, Asia and especially in North America. Closely related to *Aster*, but generally distinguished by the relatively short, linear ligules in 2 (not 1) rows, sometimes absent.

> 1. Flower-heads lacking ligules . . . . *E. acris*
> 1. Flower-heads with ligules . . . .
> . . . . *E. multiradiatus, E. breviscapus*

Erigeron acris L. **Perennial herb** 10–80 cm tall with erect **stems**. **Leaves** oblanceolate to lanceolate, 3–13 × 0.3–1.4 cm, pointed to blunt, margin obscurely toothed. **Flower-heads** small, 6–10 mm across, in branched inflorescence, purple to pale pinkish white. Gansu, Hubei, Sichuan, Yunnan [c & nChina, Japan, Korea, eRussia]; meadows, grassy slopes and banks, streamsides, 1,900–3,500 m. VII–VIII.

Erigeron multiradiatus (Lindl.) Benth. Plants 20–50 cm with erect simple or branched, hairy **stems**. Lower **leaves** oblanceolate to elliptic-obovate, 2–9 × 0.5–2.4 cm, pointed to blunt, hairy, prominently 3-veined, upper leaves ovate. **Flower-heads** 2.6–4.5 cm across, solitary, or occasionally in few-branched **inflorescence**, with purple-blue to pale lilac or white ligules and dull orange **disc-florets**. Gansu, Qinghai, w, nw & swSichuan, w & nwYunnan [nChina, c & eHimalaya, nMyanmar]; grassy and stony meadows, open shrubberies, 2,500–3,800 m. VI–IX.

Erigeron breviscapus (Vaniot) Hand.-Mazz.* (syn. *E. dielsii* H. Lévl.) Plants 6–40 cm with erect **stems**, sometimes forming a clump. **Leaves** oblanceolate, to obovate, rounded to blunt, 2–9.5 × 0.5–3.5 cm, with stiff hairs. **Flower-heads** 2.5–3.5 cm across, solitary or up to 5, with lavender, lilac or mauve ligules and yellow **disc-florets**. seTibet, s, w & nwYunnan; grassy slopes, alpine meadows, *Kobresia*-turf by streams and springs, river gravels, 1,100–3,950 m. II–VIII.

CREMANTHODIUM

An Asian genus of 69 species, all found in China (47 endemic). Low-tufted to erect, **perennial herbs** often with leafy **stems**. **Leaves** mostly basal, simple, entire or toothed, stalk with a sheathing base. **Flower-heads** 1–8, occasionally more, usually nodding, radiate; **ray-florets** 0–25, female, with showy ligules, yellow, rarely pink; **disc-florets** hermaphrodite, yellow, orange or brownish, tubular to campanulate, 5-toothed. **Achenes** oblong, 5–10-ribbed, with white or brownish pappuses.

> 1. Flower-heads purplish red, pink or rose-purple 2
> 1. Flower-heads yellow 3
> 2. Leaves oblanceolate . . . . *C. angustifolium*
> 2. Leaves kidney-shaped to orbicular . . . .
> . . . . *C. rhodocephalum, C. campanulatum*
> 3. Tall plant to 1 m, usually with 3–13 flower-heads per inflorescence . . . . . . *C. brunneopilosum, C. arnicoides*
> 3. Plants not more than 60 cm tall with solitary or paired flower-heads 4
> 4. Leaves kidney-shaped; ray-florets present . . . .
> . . . . *C. decaisnei, C. reniforme*
> 4. Leaves oblong-elliptic to lanceolate; ray-florets sometimes absent . . . . *C. lineare*

Cremanthodium angustifolium W. W. Sm.* Plant 20–50 cm tall with **fleshy roots**. **Leaves** almost erect, oblanceolate, 7–20 × 0.5–2.5 cm, pointed, toothed; stem-leaves smaller. **Inflorescence** of 1–5 dull reddish purple flower-heads, with a woolly axis; involucral **bracts** blackish purple, 18–20 mm long. w & swSichuan, seTibet, nwYunnan; rocky alpine meadows, gravelly places, 3,200–4,420 m. VII–IX.

Cremanthodium rhodocephalum **Diels** (syn. *C. palmatum* Benth. subsp. *rhodocephalum* (Diels) R. D. Good) **Herb** 5–30 cm tall with kidney-shaped, bluntly toothed **leaves**, 0.8–3 × 1–4.5 cm. **Flower-heads** usually solitary, narrow-campanulate, on a red stalk; involucre campanulate, involucral **bracts** purplish green to purple; each **ray-floret** with a long rose-purple to pink ligule, **disc-florets** greenish yellow. swSichuan, seTibet, nwYunnan [nMyanmar]; alpine meadows, amongst scrub, limestone screes and rocky slopes in *Picea* forests, 3,300–4,900 m. VII–X.

Cremanthodium campanulatum (Franch.) **Diels Stem** solitary, 10–30 cm tall, purplish red. **Leaves** orbicular-kidney-shaped, 0.7–3 × 1.2–5 cm, bluntly toothed, shallowly lobed, glabrous, green with paler veins. **Flower-heads** blackish purple, solitary, nodding, 1.3–3.2 cm long, hairy or glabrous, fragrant; involucral bracts purplish red, rarely yellowish white; florets numerous, disc purplish red, 6–8 mm long; **styles** protruding, blackish purple. w & swSichuan, seTibet, nwYunnan [nMyanmar]; alpine meadows, amongst rocks and scrub, screes, gravels, forest understorey and margins, 3,200–4,900 m. V–IX.

Cremanthodium brunneopilosum **S. W. Liu*** Tall plant with a solitary, erect **stem** to 1 m, glabrous below, white- and brown-hairy above. **Leaves** numerous, greyish green or bluish green, oblong-elliptic to lanceolate, 6–40 × 2–8 cm, margin entire or minutely toothed. **Flower-heads** (1–)3–13 in a raceme, nodding, on brown-hairy stalks 1–9 cm long; involucre hemispherical, densely brown-hairy; **ray-florets** yellow, ligules linear-lanceolate, long-pointed or tailed, 2.5–6 cm long; **disc-florets** brownish yellow, swGansu, Qinghai, nwSichuan, neTibet; alpine swampy meadows, streamsides and associated grassy places, pool margins, 3,000–4,300 m. VI–IX.

Cremanthodium arnicoides (DC. ex Royle) **R. D. Good** (syn. *Ligularia arnicoides* DC. ex Royle; *Senecio arnicoides* (DC. ex Royle) Wall. ex C. B. Clarke) Plant 10–15 cm or more tall, white- and black-hairy towards top. Basal **leaves** ovate or ovate-oblong, pointed, 3–8 × 2.5–6 cm, margin coarsely toothed. **Flower-heads** 2–7, in a lax raceme, nodding; involucre black-hairy; each **ray-floret** with a yellow, broadly elliptic ligule 1.5–2.7 cm long, apex 3-toothed; **disc-florets** brownish. seTibet, (?)nwYunnan [swTibet, Himalaya]; streamsides, marshy meadows, open shrubberies, 3,600–4,600 m. VII–VIII.

Cremanthodium decaisnei **C. B. Clarke** (syn. *Senecio renatus* Franch.) Patch-forming **perennial** 6–25 cm tall, densely brown-hairy towards top. Basal **leaves** kidney-shaped or orbicular-kidney-shaped, deep shiny green, rounded, 0.5–4.5 × 0.9–5 cm, with dense brown, long soft hairs beneath, margin irregularly shallowly scalloped; leaf-stalks purple-brown. **Flower-heads** solitary, nodding, involucre densely brown-hairy outside; **ray-florets** yellow, narrowly elliptic or oblong, 10–20 mm long; **disc-florets** numerous, dark yellow. swGansu, swQinghai, swSichuan, Tibet, nwYunnan [Himalaya, nMyanmar]; grassy slopes, gravelly slopes, stabilised moraines, alpine meadows, 3,500–5,400 m. VII–IX.

Cremanthodium reniforme (DC.) **Benth.** Differs from the previous species in the larger, brighter green, pointed to somewhat truncated **leaves** with a green stalk, and the outwards-facing **flower-heads**, involucre with violet-black hairs; **ray-florets** 15–23 mm long, shallowly toothed. seTibet, nwYunnan [swTibet, Himalaya, nMyanmar]; peaty soil in alpine meadows, screes and amongst mossy rocks in open *Abies* forests, amongst rhododendrons, 3,200–4,700 m. VI–VIII.

Cremanthodium lineare **Maxim.*** Plants bluish green, to 45 cm tall, glabrous or sparsely white-hairy towards top. Basal **leaves** short-stalked or stalkless, linear or linear-lanceolate, to 23 cm long. **Flower-heads** solitary, nodding; **ray-florets** pale to mid-yellow, ligules linear to linear-lanceolate, to 4 cm long, often somewhat reflexed, sometimes absent; **disc-florets** numerous, yellow. swGansu, Qinghai, nwSichuan, eTibet; alpine meadows, streamsides, damp grassland, scrub, 2,400–4,800 m. VII–X.

COSMOS

A genus of c. 26 species confined to warm and subtropical America, particularly Mexico. 1 species is widely cultivated and naturalised in parts of western China.

Cosmos bipinnatus **Cav.** Tall, sparingly branched **annual** to 1.5 m with finely dissected, 2-pinnately lobed **leaves** with linear lobes. **Flower-heads** to 8 cm across, with up to 8 broad ligules, white, pink, lilac, purple or red, sometimes bicoloured or flowers semi-double. Naturalised Guangxi, Guizhou, Hubei, Sichuan, Yunnan, probably elsewhere [Mexico]; cultivated, waste and fallow land, waysides, at low altitudes. VI–X.

AJANIA

A genus of 34 species, 32 found in China. **Perennial**, often hairy **herbs** or **subshrubs** from temperate Asia. **Flower-heads** discoid-like borne in racemose or corymbose clusters; **ray-florets** female, with very short ligules, **disc-florets** hermaphrodite. **Nutlets** without a pappus.

Ajania nubigena (Wall.) **C. Shih** (syn. *Artemisia nubigena* Wall.; *Tanacetum nubigenum* (Wall.) DC.) Spreading **perennial herb** to 30 cm with silvery-hairy **stems**. **Leaves** orbicular or broadly ovate, finely 3-pinnately lobed, to 10 × 10 mm, grey- or silvery-hairy. **Inflorescence** a terminal compound corymb, to 3 cm across, **flower-heads** numerous, c. 5 mm in diameter; **florets** yellow, outer 8 female, narrow-tubular, with 3-toothed apex, central florets hermaphrodite, all c. 3 mm long. swSichuan, seTibet, nwYunnan [swTibet, Himalaya]; mountain slopes, rocky and stony meadows, 3,900–4,100 m. VII–IX.

1. *Cremanthodium rhodocephalum*; Bomi, Galung La, seTibet (HJ)
2. *Cremanthodium campanulatum*; Daxueshan, nwYunnan (MN)
3. *Cremanthodium brunneopilosum*; south of Aba, nwSichuan (CGW)
4. *Cremanthodium arnicoides*; cNepal (CGW)
5. *Cremanthodium decaisnei*; near Markham, seTibet (HJ)
6. *Cremanthodium decaisnei*; cNepal (CGW)
7. *Cremanthodium lineare*; south of Aba, nwSichuan (CGW)
8. *Cosmos bipinnatus*; cult. Dali, nwYunnan (CGW)
9. *Ajania nubigena*; Baimashan, nwYunnan (PC)

ASTERACEAE Group Four

SOROSERIS

A Sino-Himalyan genus with 8 species, 6 found in China (5 endemic). Dwarf, rosette-forming, **perennial herbs** with thick, hollow **stems**. Lowermost **leaves** scale-like, the others ovate to almost orbicular, angular or toothed, stalked. **Inflorescence** terminal, dense, convex or elongated, bracteate, with many small 4–5-flowered, ligulate heads; **florets** hermaphrodite, yellow, occasionally purple or violet, rarely white. **Achenes** oblong, terete, each with a pappus of thin simple hairs.

1. Flower-heads pale purple or white ...... **S. glomerata**
1. Flower-heads yellow 2
2. Leaves entire to slightly toothed ....
   .... **Soroseris aff. glomerata**
2. Leaves coarsely toothed to pinnately lobed ....
   .... **S. hookeriana, S. pumila, S. hirsuta**

Soroseris glomerata (Decne) Stebbins (syn. *S. rosularis* (Diels) Stebbins) Small grey-purple to grey-brown rosette-forming plant 3–10 cm tall, with spathulate **leaves** 1.5–7 × 0.5–0.9 cm, softly hairy, margin entire. **Flower-heads** 4–5 cm across, in centre of leaf-rosette, fragrant, florets 7–8 mm long, ligules 4–5 mm long. s & seTibet, nwYunnan [Himalaya]; open screes, stabilised moraines, 3,400–4,500 m. VI–IX.

Soroseris aff. glomerata (Decne) Stebbins* Like *S. glomerata* but **flowers** yellow, ligules deeply toothed at apex. nwYunnan (Lijiang, Yulongxueshan); screes, moraines, c. 4,300 m. VI–VII.

Soroseris hookeriana (C. B. Clarke) Stebbins Grey-green plant with rather flat rosettes of **leaves**, elongating in flower to 30 cm, **stem** hollow. **Leaves** grey-green, linear to lanceolate or oblanceolate, pinnately lobed, 4–13 × 0.3–3.5 cm, stalk 0–6 cm long. **Flower-heads** convex or elongated, up to 16 cm long, 4.5–5.5 cm across, involucre 2–3 mm across, involucral bracts densely hairy at base, 8–12 mm long; **florets** 12–18 mm long, pale yellow to mid-yellow, fragrant, ligules 7–12 mm long; anthers blackish. Gansu, w, nw & swSichuan, Tibet, nwYunnan [cChina, Himalaya, nMyanmar]; amongst boulders, crevices in rocks, screes, stabilised moraines, 3,600–4,750 m. VII–IX.

Soroseris pumila Stebbins In contrast to the previous species, a practically stemless plant in flower, **stems** 2.5–10 cm long but mostly subterranean. **Leaves** spathulate or oblanceolate, 5–15 cm long, blunt, purplish green, coarsely toothed or almost pinnately lobed, hairy on margins and stalks, terminal lobe larger than the others; involucre 3–4 mm across, involucral bracts purple; florets 9–12.5 mm long, yellow, with violet anthers, ligules 6–12 mm long, toothed at top. seTibet, nwYunnan [swTibet, Himalaya]; screes, rocky places, 4,300–4,900 m. VII–IX.

Soroseris hirsuta (J. Anth.) C. Shih Rather like the previous species, but **leaves** bright green and sinuately lobed, terminal lobe equal to, or smaller than, the others; ligules slightly notched at top. w & swSichuan, seTibet, nwYunnan [eHimalaya, nMyanmar]; similar habitats, 4,500–5,300 m. VII–VIII.

STEBBINSIA

A genus with a single species endemic to China. A rather fleshy **perennial herb** with a very short **stem** and basal **leaves** in rosettes.

Stebbinsia umbrella (Franch.) Lipsch.* (syn. *Crepis umbrella* Franch.; *Soroseris umbrella* (Franch.) Stebbins Plant 3–15 cm tall with glabrous **stem**. **Leaves** deep purplish green or purplish red, in a flat rosette, ovate to ovate-elliptic, apex rounded, 3.5–8 × 3–7 cm, margin sparsely but sharply toothed; leaf-stalks broad and thickened, 4–14 cm long, sometimes winged. **Flower-heads** ligulate, numerous, clustered in centre of rosette, with 10–25 yellow or cream florets. w & swSichuan, e & seTibet, nwYunnan; alpine meadows, scree slopes and other rocky places, 2,600–4,600 m. VI–VIII.

SYNCALATHIUM

A genus of 9 species, all endemic to China, closely related to, and sometimes included within, *Lactuca*. **Annual** or **perennial herbs**, **stemless** or with **stems**, with terminal, densely clustered **flower-heads**.

Syncalathium souliei (Franch.) Ling* (syn. *Lactuca souliei* Franch.) Plant forming a stalkless rosette with pinnatisect **leaves**, to 6 × 1.2 cm, scalloped, terminal leaflet ovate; leaf-stalks usually red to purplish brown, to 4 cm long. **Flower-heads** in a dense central cluster 2.5–5.5 cm across, each head ligulate, 8 mm across, with 4–6 blue or purple florets. w & swSichuan, e & seTibet, nwYunnan; bare mountainsides, alpine meadows, forest margins, loose shale, 2,300–4,800 m. VII–VIII.

1. *Soroseris glomerata*; eNepal (CGW)
2. *Soroseris glomerata*; Daxueshan, nwYunnan (CGW)
3. *Soroseris* aff. *glomerata*; Yulongxueshan, nwYunnan (CGW)
4. *Soroseris pumila*; eNepal (CGW)
5. *Soroseris hirsuta*; Genyanshan, south side, swSichuan (MN)
6. *Soroseris hirsuta*; Genyanshan, south side, swSichuan (MN)
7. *Stebbinsia umbrella*; Baimashan, nwYunnan (AD)
8. *Stebbinsia umbrella*; Baimashan, nwYunnan (AD)
9. *Syncalathium souliei*; Baimashan, nwYunnan (PC)
10. *Syncalathium souliei*; Baimashan, nwYunnan (AD)

COMMELINACEAE

COMMELINA

A pantropical genus of c. 170 species (8 in China). Creeping **annual** or **perennial herbs** with simple alternate **leaves** and **flower-clusters** included in spathe-like involucral bracts; petals 3 free, unequal; fertile stamens 3, staminodes 2–3.

1. Involucral bracts funnel-shaped, fused along 1 margin .. .. **C. benghalensis, C. paludosa**
1. Involucral bracts rounded or heart-shaped at base, margin not fused .... **C. communis, C. diffusa**

Commelina benghalensis L. **Perennial herb** with creeping **stems** to 70 cm long. **Leaves** with a distinct stalk, sparsely hairy. Involucral bracts several, opposite leaves, terminal, sparsely hairy. **Flowers** with blue petals 3–5 mm long. Guizhou, Hubei, Sichuan, Shaanxi, Yunnan [c & sChina, seAsia, India, Taiwan, tropical Africa]; wet places such as rice paddies, stream and ditch margins, to 2,700 m. Flowering year-round.

Commelina paludosa Blume Similar to the previous species, but with longer **stems**, stalkless **leaves**, and petals 4.5–8 mm long. swGuizhou, swSichuan, seTibet, Yunnan [c & sChina, s & seAsia]; forests, streamsides, ravines, to 2,800 m. Flowering year-round.

Commelina communis L. **Annual herb** with **stems** spreading or creeping to 1 m. **Leaves** with glabrous sheaths, blade lanceolate to ovate-lanceolate, to 9 × 2 cm, glabrous. **Inflorescence** with 1–2 male flowers on lower bract, 3–4 female on upper one; flower-bracts usually ciliate; calyx membranous, 5 mm long; **corolla** deep blue, longest petal 8–10 mm. sGansu, Guangxi, Guizhou, Hubei, Sichuan, Shaanxi, Yunnan [s, c & eChina, s & seAsia]; damp habitats, streamsides, open forests, often in and around cultivation, to 2,300 m. IV–XI.

Commelina diffusa Burm. f. Similar to the previous species, but leaf-sheaths red-lined and with stiff hairs, spathe-valves lanceolate and **fruit-capsule** 3-parted; **flowers** smaller, longest petal not more than 6 mm. swGuangxi, swGuizhou, seTibet, seYunnan [s & seAsia, India]; similar habitats, to 2,100 m. V–XI.

1. *Commelina communis*; Dali, nwYunnan (CGW)
2. *Floscopa scandens*; Xishuangbanna, swYunnan (CGW)
3. *Musella lasiocarpa*; Kunming, cYunnan (PC)
4. *Roscoea tibetica*; Zhongdian, nwYunnan (CGW)
5. *Roscoea schneideriana*; cult. (GD)

CYANOTIS

A genus with c. 50 species (5 in China) ranging from tropical and subtropical Africa to northern Australia. Similar to *Commelina*, but **flowers** with 6 fertile stamens, and petals unequal, united below into short tube.

Cyanotis vaga **(Lour.) Schultes & Schultes f.** **Perennial herb** with globose **bulbs** and branching **stems** to 60 cm. **Leaves** linear to lanceolate, 5–10 cm long, ciliate, glabrous or sparsely hairy beneath. **Flower-clusters** solitary, usually axillary; spathe-bracts 5–10 mm long, ciliate; petals purple or violet, 6–8 mm long; filaments blue-woolly. Guizhou, sSichuan, seTibet, Yunnan [swTibet, c & sChina, seAsia, Himalaya]; forests, grassy slopes, to 3,300 m. VI–IX.

FLOSCOPA

A pantropical genus of 20 species (2 in China, 1 endemic). **Perennial** rhizomatous **herbs** with alternate **leaves** and numerous **flowers** in dense panicles; fertile stamens 3 or 6.

Floscopa scandens **Lour.** Glandular-hairy plant with simple **stems** to 70 cm, generally less. **Leaves** stalkless or with a short, winged stalk, blade elliptic to lanceolate, 4–12 × 1–3 cm. **Inflorescence** densely glandular-hairy, scarcely stalked; petals blue or purple, sometimes very pale; fertile stamens 6. Guangxi, Guizhou, scSichuan, seTibet, s, sw & wYunnan [eHimalaya, India, Myanmar, seAsia]; damp grassy places, pool- and streamsides, wet forest margins, marshy shrubberies, to 1,700 m. VII–XI.

MUSACEAE

MUSELLA

A genus with just 1 species endemic to China, doubtfully distinct at generic level from *Musa* (banana). **Perennial herb** growing from a horizontal rhizome, with short false **stems** and narrowly elliptic, pointed **leaves**. **Inflorescence** terminal or axillary near base of stem, conical, appearing before leaves, with showy, persistent bracts. **Flowers** in 2 rows per bract, lower ones female or hermaphrodite, upper ones male; tepals 6, 5 fused into a tepal-tube 5-toothed at apex, free tepal notched and shortly pointed at apex; stamens 5. **Fruit** a somewhat 3-angled, globose berry, densely hairy, containing numerous seeds.

Musella lasiocarpa **(Franch.) C. Y. Wu ex H. W. Li*** (syn. *Musa lasiocarpa* Franch.) False **stems** up to 60 cm tall and 15 cm across, bearing **leaves** up to 50 × 20 cm. **Inflorescence** 15–25 cm long with conspicuous yellow to orange-yellow bracts. **Flowers** 8–10 per bract. **Berries** 3 cm long. sGuizhou, c & wYunnan; hillsides, banks, cultivated around villages in gardens, 1,500–2,500 m. Flowering year-round.

ZINGIBERACEAE

ROSCOEA

A genus of c. 20 species (14 in China) distributed in the Himalaya and w & swChina. Small **herbaceous perennials** with an erect, reduced rhizome. **Leaves** lanceolate to oblong-ovate, almost erect. **Inflorescence** terminal, often appearing before the leaves. **Flowers** produced in succession, showy, zygomorphic; calyx tubular, 2- or 3-toothed at apex; corolla-tube slender, dilated at apex, dorsal petal hood-like, lateral petals spreading; lip spreading or deflexed, obovate, clawed or not, shallowly to deeply 2-lobed; staminodes 2; stamen 1. Ovary cylindrical, 3-celled. **Fruit** a capsule, splitting longitudinally.

> 1. Leaf-base shortly stalked to auriculate .... **R. tibetica**
> 1. Leaf-base decurrent 2
> 2. Bracts longer than calyces at flowering time .... **R. schneideriana**
> 2. Bracts shorter than or equalling calyces at flowering time 3
> 3. Leaves glaucous beneath; flowers deep maroon-purple .... **R. wardii**
> 3. Leaves green on both surfaces; flowers purple, violet-purple, yellow or white 4
> 4. Lowest bract not tubular .... **R. humeana**, **R. humeana** f. **alba**, **R. humeana** f. **lutea**, **R. humeana** f. **tyria**, **R. forrestii**
> 4. Lowest bract tubular .... **R. cautleyoides**, **R. cautleyoides** var. ***cautleyoides*** f. ***sinopurpurea***, **R. cautleyoides** var. ***cautleyoides*** f. ***atropurpurea***, **R. cautleyoides** var. ***pubescens***, **R. praecox**, **R. kunmingensis**

Roscoea tibetica **Batalin** Small plant 7–18 cm tall. **Leaves** 1–3, lanceolate to oblong-lanceolate, up to 20 cm long at maturity. **Inflorescence** preceding young leaves. **Flowers** small, lilac, purple, violet or occasionally white; corolla-tube protruding from calyx, 3–6 cm long; dorsal petal 1.5–2.7 cm long, lateral petals 1.3–2.5 cm long; lip slightly deflexed, 1.4–2.5 cm long, deeply 2-lobed for more than half its length. Sichuan, seTibet, Yunnan [swTibet, c & eHimalaya, nMyanmar]; stony hillsides, meadows on limestone, margins of scrub, open *Pinus* and *Quercus* forests, 2,100–4,300 m. V–VII.

Roscoea schneideriana **(Loes.) Cowley*** Stout plant up to 38 cm. **Leaves** 4–6 in a rosette at top of **stem**, narrowly lanceolate, 15–24 cm long. **Inflorescence** preceding young leaves. **Flowers** rich purple, pinkish purple or white; corolla-tube scarcely protruding from calyx, 4–4.5 cm long; dorsal petal 1.7–3.5 cm long, lateral petals 2–3.5 cm long; lip not deflexed, to 3.5 cm long, 2-lobed for half its length. s & swSichuan, nwYunnan; shade of scrub, *Pinus* and *Quercus* forests, on open calcareous soils, amongst rocks, on ledges, 2,600–3,350 m. VII–VIII.

Roscoea wardii Cowley Rather robust plant to 32 cm. **Leaves** 2–3, elliptic, to 20 × 4.5 cm, margin membranous, minutely papillose. **Bracts** shorter than calyces. **Inflorescence** appearing with the leaves, short-stalked; **corolla-tube** to 8 cm long, protruding from calyx; dorsal petal obovate to elliptic, 2–3 cm long, lateral petals narrow-oblong, 1.7–3.4 cm long; lip deflexed, 2.2–4.5 cm long, deeply 2-lobed, each lobe notched. wYunnan (Kiukiang valley) [neIndia, nMyanmar]; meadows and other grassy places, open coniferous forests, thicket edges, 2,440–3,960 m. V–early VIII.

Roscoea humeana Balf. f. & W. W. Sm.* Stout plant 7–35 cm tall. **Leaves** 1–3, linear to elliptic, 15–24 cm long. **Bracts** much shorter than calyces. **Inflorescence** usually preceding young leaves. **Flowers** lilac, purple, pale lemon-yellow or white; corolla-tube scarcely protruding from calyx, to 3 cm long; dorsal petal obovate to reverse heart-shaped, 2.5–4.5 cm long, lateral petals 2.8–4.5 cm long, longer than lip; lip deflexed, to 4.5 cm long, 2-lobed for half its length. s & swSichuan, nwYunnan; stony hillsides, meadows on limestone, margins of scrub, open *Pinus* and *Quercus* forests, rock-ledges, 2,900–3,800 m. IV–VII. The various colour forms have been recognised: *forma alba* Cowley (white-flowered); *forma lutea* Cowley (yellow-flowered); *forma tyria* Cowley (blackish purple-flowered).

Roscoea forrestii Cowley* Very similar to the previous species, but **leaves** usually well-developed by flowering time, bracts slightly shorter to equalling the calyces and dorsal petal broad-elliptic. **Flowers** purple or yellow, with lateral petals equalling or shorter than lip. swSichuan, nwYunnan; rock-crevices and -ledges, low shrubby thickets often with dwarf bamboo, to 3,350 m. V–VII.

Roscoea cautleyoides Gagnep.* Slender plant 18–35 cm tall. **Leaves** 1–4, linear to elliptic, 15–24 cm long. **Inflorescence** protruding from young leaves. **Flowers** purple, reddish violet, pink, pale lemon-yellow or occasionally white; corolla-tube 3–5.5 cm long, scarcely protruding from the deeply 2-toothed calyx; dorsal petal 2.5–3.8 cm long, lateral petals 2.4–3.5 cm long; lip deflexed, 2.6–4 cm long, 2-lobed for more than quarter of its length. s & swSichuan, nwYunnan; stony and rocky hillsides, meadows on limestone, in forests under *Pinus* and *Quercus*, open shrubberies, 2,000–3,350 m. V–early VIII. **Var. *pubescens* (Z. Y. Zhu) T. L. Wu*** from Sichuan (Xichang) differs in the hairy leaf-sheaths and **leaf** lower surface. Colour forms of the typical yellow variety (*var. cautleyoides*) have been recognised: *forma atropurpurea* Cowley (deep purple) and *forma sinopurpurea* (Stapf) Cowley (pale to mid-purple).

Roscoea praecox K. Schum.* Plant 7–28 cm tall. **Leaves** 2–4, linear-lanceolate, to 40 × 2 cm. **Inflorescence** preceding leaves or with young leaves showing. **Flowers** dark pinkish purple to violet-purple, lip with white lines at base; corolla-tube protruding by 3–4.5 cm from 2-toothed calyx; dorsal petal 2.5–3.5 cm long, lateral petals 2.6–3 cm long; lip 2.5–4 cm long, 2-lobed for more than half its length. nwYunnan; stony hillsides, meadows on limestone, shrubby hillsides, streamsides, 1,520–2,500 m. V–VII. *R. kunmingensis* **S. Q. Tong**, described from near Kunming in cYunnan, is probably conspecific with *R. praecox*.

CURCUMA

A primarily tropical and subtropical genus with c. 50 species (12 in China, half endemic) found in South East Asia and Australia. Rhizomatous **perennials** with basal tufts of **leaves**. **Inflorescence** a short dense spike borne either above leaves or on a separate shoot arising from the rhizome, often in advance of the leafy shoots; bracts large and conspicuous, often brightly coloured and forming a shuttlecock-like tuft, the base of each bract forming a pouch containing a cluster of up to 7 **flowers**; calyx tubular but split along 1 side; corolla funnel-shaped with longer central lobe. **Fruit** a 3-valved capsule.

> 1. Summer-flowering; inflorescence on a separate shoot from the leaves . . . . **C. aromatica**
> 1. Autumn-flowering; inflorescence terminating leafy shoot . . . . **C. sichuanensis, C. yunnanensis**

Curcuma aromatica Salisb. Plant to 1 m tall in leaf, with fleshy, aromatic **rhizomes** that are yellow when cut. **Leaves** stalked, with oblong blade 30–60 × 10–20 cm. **Inflorescence** usually appearing in advance of the leaves, a cylindrical spike up to 15 cm long and half as wide; fertile bracts to 5 cm long, green, upper sterile bracts larger, narrow-oblong, white flushed pink or red; **corolla** 23–25 mm long, white with yellow on lip, sometimes pinkish white. Guangxi, Guizhou, Sichuan, seTibet, Yunnan [swTibet, s & seChina, Himalaya]; open forests, shrubberies, glades, 800–2,200 m. IV–VI.

Curcuma sichuanensis X. X. Chen* Plant to 1.5 m tall in leaf, with fragrant **rhizomes** that are white or creamy yellow when cut. **Leaves** stalked, blade elliptic or oblong, 35–85 × 13–21 cm, glabrous. **Inflorescence** a cylindrical spike to 20 cm long; fertile bracts ovate to oblong, green, not more than 4.5 cm long, sterile bracts larger and more elliptic, pointed, sometimes flushed with purple or red; **corolla** 25–32 mm long, pale yellow. w, s & swSichuan, swYunnan (Jinghong Xian); similar habitats to the previous species, 800–1,100 m. VII–VIII.

Curcuma yunnanensis N. Liu & S. J. Chen* Readily distinguished from previous species by green-centred **rhizomes** and leaf-blades that have a purplish blotch in centre; **spikes** slightly smaller with fertile bracts purple-margined, sterile bracts deep purple with white base, to 6 cm long; **corolla** purple. wYunnan (Wanding Zhen); grassy places, shrubberies, 400–1,200 m. VII–VIII.

1. *Roscoea humeana*; Yulongxueshan, nwYunnan (CGW)
2. *Roscoea humeana*; Yulongxueshan, nwYunnan (CGW)
3. *Roscoea humeana*; Yulongxueshan, nwYunnan (CGW)
4. *Roscoea humeana*; Yulongxueshan, nwYunnan (CGW)
5. *Roscoea cautleyoides*; Yulongxueshan, nwYunnan (HJ)
6. *Roscoea cautleyoides* forma *atropurpurea*; Yulongxueshan, nwYunnan (JM)
7. *Roscoea cautleyoides* forma *sinopurpurea*; Yulongxueshan, nwYunnan (PC)
8. *Roscoea praecox*; south of Dali, nwYunnan (CGW)
9. *Curcuma aromatica*; eNepal (CGW)

HEDYCHIUM

A genus of c. 50 species (28 in China) in the Himalaya and South East Asia. Medium-sized terrestrial or rarely epiphytic **herbs** growing from a very short, tuberous rhizome; **roots** fleshy. **Stems** erect, leafy, leaves few to many, in 2 opposite rows, thin-textured, oblong to lanceolate. **Inflorescence** a terminal, densely many-flowered spike; bracts overlapping or lax. **Flowers** showy, zygomorphic; calyx tubular; corolla-tube protruding from calyx, lobes reflexed, linear; lip ± reflexed, almost orbicular, usually 2-lobed or notched.

1. Bracts densely overlapping; inflorescence almost conical . . . . **H. coronarium**
1. Bracts lax or dense but not overlapping; inflorescence elongated 2
2. Flowers white or cream . . . . . . **H. forrestii, H. villosum**
2. Flowers yellow, orange or red . . . . **H. coccineum, H. densiflorum, H. yunnanense**

Hedychium coronarium J. König Plants 0.6–3 m tall with stalkless, narrow-elliptic or lanceolate **leaves**, 20–40 × 4.5–8 cm, long-pointed, finely hairy beneath. **Inflorescence** 7–20 cm tall; bracts overlapping, arranged cone-like. **Flowers** white, fragrant; calyx 4 cm long; corolla-tube 8 cm long, lobes lanceolate; lip narrowly clawed, broad-spathulate, apex notched, 4–6 cm long. Sichuan, seTibet, Yunnan [s & eChina, Himalaya, seAsia]; streamsides, damp grassy places in forests, 1,600–2,100 m. VIII–XII.

Hedychium forrestii Diels Plants 1–2.5 m tall. **Leaves** narrowly elliptic, 35–50 × 5–10 cm, long-pointed. **Inflorescence** cylindrical, 15–25 cm tall; bracts ovate, 4.5–6 cm long, arranged in an open cone. **Flowers** pure white with crimson anthers, fragrant; calyx shorter than bracts; corolla-tube 4–4.5 cm long, lobes linear, 3.5–4 cm long; lip narrowly clawed, broadly spathulate or orbicular, notched, 3–5.5 × 4–4.3 cm. Guangxi, Guizhou, Sichuan, Yunnan [Laos, Thailand, Vietnam]; streamsides, damp grassy places, generally in forests, 200–2,100 m. VI–X.

Hedychium villosum Wall. Readily distinguished from *H. forrestii* by its brown-silky-hairy bracts and calyces. Guangxi, Yunnan [s & seChina, India, Myanmar, Thailand, Vietnam]; forests and ravines, sea level to 3,400 m. III–IV.

Hedychium coccineum Buch.-Ham. Plants 0.6–2 m tall with linear-lanceolate **leaves** 25–45 × 2.8–4 cm, long-pointed. **Inflorescence** cylindrical, 20–30 cm long, 8.5–10 cm broad; bracts oblong, 3–3.5 cm long. **Flowers** red; calyx 2.5 cm long; corolla-tube 3 cm long, lobes linear, 3 cm long; lip orbicular, deeply bifid. seTibet, s, w & swYunnan [eHimalaya]; thickets by streams, damp grassy places, 1,400–2,000 m. VI–VIII.

Hedychium densiflorum Wall. Plants 0.6–1.2 m tall with **leaves** oblong-lanceolate, 18–35 × 3.8–8 cm, long-pointed. **Inflorescence** cylindrical, 7–21 cm tall, 3–4 cm broad; bracts oblong, 2 cm long. **Flowers** yellow to flame-red with orange stamens; calyx 2.5 cm long; corolla-tube 2.5–3 cm long, lobes linear, 1.5 cm long; lip wedge-shaped, 1.6 cm long and broad, deeply 2-lobed. Guangxi, seTibet, s, w & swYunnan [Himalaya, Sri Lanka, Thailand]; thickets by streams, damp shady places, lava flows, 1,400–2,300 m. VII–VIII.

Hedychium yunnanense Gagnep. Plants 0.8–1.6 m tall with **leaves** narrow-elliptic to oblong, 20–50 × 8–10 cm, long-pointed. **Inflorescence** 15–35 cm long; bracts lanceolate, 1.5–2.5 cm long, 1-flowered. **Flowers** waxy, pale orange-yellow to golden yellow with red stamens, fragrant; calyx 1.7–2.8 cm long; corolla-tube 3.5–5 cm long, lobes linear, 2.5–3 cm long; lip obovate, c. 2 cm long, apically cleft. Guangxi, Yunnan [Vietnam]; forests, often by streams, damp shady or grassy places, rocks on lava flows, 1,600–2,600 m. IV–X.

CAUTLEYA

A genus of 5 species (3 in China) distributed in the Himalaya, swChina and South East Asia. Medium-sized **herbs** growing from a very short rhizome, roots fleshy. **Stems** with 3–6 **leaves** in 2 opposite rows, thin-textured, oblong to lanceolate. **Inflorescence** a terminal spike with coloured bracts. **Flowers** yellow or orange, zygomorphic; corolla protruding from calyx, lateral lobes joined with claw of lip claw; lip ± reflexed, 2-lobed or notched.

---

1. *Hedychium coronarium*; Emeishan, sSichuan (CGW)
2. *Hedychium forrestii*; Qinshuijiang, nwGuangxi (PC)
3. *Hedychium forrestii*; Qinshuijiang, nwGuangxi (PC)
4. *Hedychium villosum*; south-west of Szemao, swYunnan (CGW)
5. *Hedychium densiflorum*; swYunnan (PC)
6. *Alpinia guinanensis*; Laojunshan, seYunnan (PC)
7. *Costus speciosus*; cult., from swYunnan (CGW)

Cautleya gracilis (Sm.) Dandy Plant 30–60 cm tall. **Leaves** 4–6, lanceolate, 6–18 × 2.5–6 cm, long-pointed. **Inflorescence** laxly 2–10-flowered, 7–11 cm long; bracts green. **Flowers** bright yellow; calyx 1.5–2 cm long, purplish red; corolla-lobes lanceolate, 1.5–2 cm long. swSichuan, seTibet, s, w & swYunnan [Himalaya]; thickets by streams, on lava flows in open rocky places, occasionally epiphytic, 1,600–3,300 m. VIII–IX.

Cautleya spicata (Sm.) Baker differs from the previous species in having red bracts and stalked **leaves**. Guizhou, Sichuan, Yunnan [Himalaya, Myanmar]; similar habitats, 1,100–2,600 m. VII–IX.

ALPINIA

A genus with 230 species (51 in China, 35 endemic), confined to tropical and subtropical Asia, Australia and the Pacific islands. Rhizomatous **perennials** with many well-developed **pseudostems**, each with 4 or more **leaves**. **Inflorescence** a terminal panicle or spike, at first enveloped by up to 3 large bracts; calyx tubular, split down 1 side or hooded; **corolla** with upper tepal hooded and with a large, showy, lobed or unlobed lip. **Fruit** a globose, fleshy or dry capsule.

Alpinia guinanensis D. Fang & X. X. Chen* Perennial herb up to 3 m with oblong **leaves**, 18–90 × 7.5–18 cm. **Inflorescence** almost erect, 16–36 cm long, densely hairy, with 1–3-flowered branches; bracts oblong, 2–5 mm, deciduous; bracteoles deep red, shell-shaped, 2–3.3 cm long, deciduous; calyx pale red, c. 1.5 cm long; **corolla-tube** c. 1 cm, densely downy at throat, lobes white tinged with red, lip yellow or whitish with purple stripes, ovate, c. 4 × 3 cm, with a wavy margin and notched apex; ovary densely downy. **Capsule** red, globose, smooth. Guangxi, s & seYunnan; scrub, at low altitudes. IV–VI.

COSTUS

A genus of c. 90 species (5 in China), pantropical in distribution. **Inflorescence** dense, club-like with closely overlapping bracts. **Flowers** with a very large lip but otherwise small petals. **Seeds** with an aril. This genus is sometimes placed in a separate family, Costaceae.

Costus speciosus (J. König) Sm. (syn. *Banksea speciosa* J. König; *Costus formosanus* Nakai) **Perennial herb** 1–3 m with slightly woody base. **Leaves** spirally arranged, crowded towards stem-tops, oblong or lanceolate, 15–20 × 6–10 cm. **Inflorescence** terminal, ellipsoid or egg-shaped, 5–15 cm long; bracts bright red, ovate, c. 2 cm long, leathery, hairy; bracteoles pale red, 1.2–1.5 cm long; calyx red, 1.8–2 cm long; lip white, trumpet-shaped, 6.5–9 cm long, apex toothed and crisped, margins overlapping. **Fruit** bright red, globose, containing glossy black **seeds**. Guangxi, s & eYunnan [Guangdong, Taiwan, Himalaya, India, seAsia, Australia]; forest margins, damp places in valleys, roadsides, to 1,700 m. VII–IX.

PALMAE (ARECACEAE)

Palms are economically important: the **stems** and **leaves** are used as building materials, the leaf-bases for fibres, and the **fruit** can be edible or otherwise useful.

GUIHAIA

A genus with 2 species of dwarf palms confined to sChina and nVietnam. **Stems** clustered, short or very short, covered with spiny or net-like fibrous leaf-sheaths. **Leaves** palmate, incompletely divided, minutely toothed or smooth along margin, leaf-stalks unarmed. **Inflorescence** dioecious, solitary, axillary, branching: **male flowers** with 3, joined, almost rounded to ovate sepals, hairy outside, petals longer than sepals, with 6 stamens, the filaments completely joined to the corolla; **female flowers** similar but with 6 staminodes, borne on the petals, and 3 glabrous carpels. **Fruit** developing from only 1 carpel, globose to ellipsoid, blue-black, with a bloom.

Guihaia argyrata **(S. K. Lee & F. N. Wei) S. K. Lee, F. N. Wei & J. Dransf.** Plants with **stems** to 1 m, the **leaves** green above but covered in a felt of silvery-woolly hairs beneath; leaf-stalks to 1 m long, sheaths tubular at first but later disintegrating into erect, dark brown spiny fibres. **Inflorescence** 30–80 cm long, **male flowers** with blunt, ciliate sepals, hairy outside. **Fruit** almost globose, 6 mm in diameter, blue-black and waxy when ripe. neGuangxi (Guilin southwards), sYunnan [Guangdong, nVietnam]; karst limestone hills, to 1,500 m. V–VI.

Guihaia grossefibrosa **(Gagnep.) S. K. Lee, F. N. Wei & J. Dransf.** Similar to the previous species, but often taller; **leaves** glabrous above but almost glaucous below and covered with scattered scales; leaf-sheaths disintegrating into broad, flat fibres. **Inflorescence** to 80 cm long, **male flowers** with mucronate sepals, not ciliate, glabrous but covered with scales outside. **Fruit** ellipsoid, 6–8 × 4–5 mm, blue-black when ripe. swGuangxi, sGuizhou [sGuangdong, Vietnam]; forests on steep karst limestone hills, 200–1,000 m. V.

RHAPIS

A genus of c. 12 species (5 in s & swChina) in east and South East Asia. **Shrubby** palms with clustered, slender, erect **stems** covered with net-like fibrous leaf-sheaths. **Leaves** fan-shaped or palmate, divided to the base into linear to linear-elliptic or lanceolate segments with finely toothed margins; leaf-stalks margin glabrous or minutely toothed. **Inflorescence** axillary, dioecious or polygamous, male inflorescence similar to female, branched. **Male flowers** with 3-lobed, cup-shaped calyx, a short 3-lobed corolla and 6 stamens, filaments joined to corolla-tube; **female flowers** similar to male but with a 3-carpellate ovary and 6 staminodes. **Fruit** usually developing from 1 carpel, globose or egg-shaped.

Rhapis excelsa **(Thunb.) Henry ex Rehd.** A palm 2–3 m tall. **Stems** with upper leaf-sheaths disintegrating into blackish, rough, stiff, net-like fibres. **Leaves** 20–30 × 1.5–5 cm, with 4–10 unequal, linear or linear-lanceolate segments, with minutely toothed margins; leaf-stalk margins slightly rough. **Inflorescence** up to 30 cm long. **Male flowers** 5–6 mm long, with cup-shaped, deeply 3-lobed calyx, and 3-lobed corolla; **female flowers** shorter and more robust, 4 mm long. **Fruit** globose to somewhat egg-shaped, 8–10 mm in diameter. Widely cultivated in s & swChina as an ornamental [Japan]. VI–VII.

TRACHYCARPUS

A genus of c. 6 species (3 in China) distributed in the Himalaya, China and Japan. Dioecious, occasionally monoecious **trees** or **shrubs**, with leaf-sheaths disintegrating into very coarse net-like fibres. **Leaves** palmate, divided to form many single-fold segments; stalks toothed on both sides. **Inflorescence** robust, similar in both sexes, branching, with several spathes embracing the main stalk and branches. **Flowers** in groups of 2–4, rarely solitary; male flowers with a deeply 3-lobed or almost free calyx that is shorter than corolla and 6 stamens with distinct filaments; female flowers with calyx and corolla similar but with 6, arrow-like staminodes and 3 distinct, hairy, ovoid carpels. **Fruit** broadly kidney-shaped or oblong.

Trachycarpus fortunei **(Hook.) H. Wendl.** **Tree** 3–10 m, occasionally more, with a cylindrical **trunk**, 10–20 cm in diameter, clothed throughout with persistent old leaf-bases and covered with net-like fibres. **Leaves** palmate, deeply divided into 30–50 linear segments; these 60–70 × 2.5–4 cm, apex short, 2-lobed; leaf-stalks with prickles on each side. **Inflorescence** robust, many-branched, usually dioecious, yellow; male inflorescence up to 40 cm long, female larger, 80–90 cm long. **Female flowers** globose, with 6 staminodes and carpels with white hair. **Fruit** broadly kidney-shaped, yellow when ripe, with a bloom. Guangxi, Guizhou, Hubei, Sichuan, Shaanxi, Yunnan [c & eChina, Japan]; secondary forest, ridgetops, also widely planted around dwellings and along field boundaries, to 2,000 m. IV–V.

Trachycarpus nana **Becc.*** A much smaller, more shrubby palm than *T. fortunei*, stemless and growing no more than 0.8 m. **Leaves** similar to those of *T. fortunei* but clustered on the ground, small and very deeply divided, segments linear-lanceolate, 25–55 cm long, green above, glaucous beneath; leaf-stalks 25–35 cm, ± densely toothed at top. **Inflorescence** emerging from the ground, erect, slender, 40–48 cm long, usually branching, dioecious. **Fruit** broadly kidney-shaped, black-blue when ripe, 10–12 mm long. seGuizhou, Yunnan; grassy hillsides, 1,500–2,300 m. IV.

Trachycarpus princeps **Gibbons, Spanner & S. Y. Chen*** Very similar to *T. fortunei*, but **stems** naturally shed their leaf-bases and are bare, while **fruit** are short-stalked, kidney-shaped to

1. *Guihaia argyrata*; Fushui, swGuangxi (PC)
2. *Guihaia grossefibrosa* trunk; Fushui, swGuangxi (PC)
3. *Guihaia grossefibrosa*; Long Fu Shan, swGuangxi (PC)
4. *Rhapis excelsa*; cult. (PC)
5. *Trachycarpus fortunei*; cult. (PC)
6. *Trachycarpus fortunei* trunk showing fibres; cult. (CGW)
7. *Trachycarpus fortunei* in flower; cult. (CGW)
8. *Trachycarpus fortunei* in fruit; cult. (CGW)
9. Brooms made from *Trachycarpus fortunei*; Dali, nwYunnan (PC)
10. Man with *Trachycarpus fortunei* cloak; Nushan, wYunnan (PC)
11. *Trachycarpus nana*; east of Lijiang, nwYunnan (PC)

almost oval, black, with a white bloom when ripe. nwYunnan (Nujiang); banks and bare marble cliffs by the Lancangjiang, 1,500–1,900 m. IV–V.

PHOENIX

A genus of 17 species (2 in sChina) distributed in tropical and subtropical Asia and Africa, which includes the economically important date palm, *P. dactylifera*. **Shrubs** or **trees**, with solitary or clustering **stems**, usually covered with older leaf-bases or leaf-scars. **Leaves** pinnate, leaflets narrowly lanceolate or linear, induplicate, lowermost reduced to spines. **Inflorescence** borne between the leaves with a sheath-like, leathery spathe. **Flowers** solitary, dioecious, male with a 3-toothed cup-shaped calyx, 3 petals and 6 stamens with very short or almost absent filaments; female globose, with 3 petals, usually 6 staminodes and 3 carpels. **Fruit** oblong in outline to almost globose, 1-seeded, grooved lengthwise.

Phoenix roebelinii O'Brien **Shrubby** palm with clustered, 1–3 m tall, slender **stems** (usually solitary when cultivated), with persistent triangular leaf-bases. **Leaves** 1–1.5(–2) m long, leaflets soft, linear, 20–30 × 5–10 cm, sometimes larger, deep green, with grey scurfy scales along veins beneath. **Spathe** 30–50 cm long, apex 2-lobed; male **inflorescence** longer than female. **Fruit** oblong in outline, 14–18 × 6–8 mm, reddish when ripe, with a thin pulp. s & swYunnan [Bhutan, India, Myanmar, Vietnam] but cultivated widely in sChina; riverbanks, 500–900 m. IV–V.

1. *Arisaema yunnanense*; near Yufengsi, Lijiang, nwYunnan (RS)
2. *Arisaema yunnanense*; Xishan, Kunming, cYunnan (EL)
3. *Arisaema lobatum*; Wolong, wSichuan (CGW)
4. *Arisaema candidissimum*; Muli, swSichuan (PC)
5. *Arisaema candidissimum*; above Shigu, nwYunnan (CGW)
6. *Arisaema candidissimum* forma; below Zhongdian, nwYunnan (PC)
7. *Arisaema asperatum*; west of Jiuzhaigou, nwSichuan (CGW)
8. *Arisaema asperatum*; Huanglong, nwSichuan (PC)
9. *Arisaema franchetianum*; near Habashan, nwYunnan (CGW)

ARACEAE

ARISAEMA

A genus of c. 160 species (82 in China) scattered in temperate and subtropical regions of the northern hemisphere and Africa, with the greatest concentration of species in the Sino-Himalayan region. Tuberous-rooted **perennials**, solitary or clump-forming, with or without a **pseudostem**, but always with basal sheaths, plain, blotched or striped. **Leaves** long-stalked, blade with 3 or more leaflets, radiate or pedate. **Inflorescence** solitary, erect, surrounded by an enveloping spathe that is tubular or funnel-shaped in lower part but opens out above into an erect to forward-projecting limb, often striped or speckled, especially the tubular part. **Spadix** bearing male or female **flowers**, or both (male in a zone above female), housed within the spathe-tube; appendage (extension of the spadix) short to long, ascending to downturned or trailing, club-shaped, to thread-like or tapered, sometimes divided. **Fruit** an erect to pendent head of closely packed 'berries', green at first but orange or red when ripe.

Synopsis of groups

Group One. Leaves with 3 or occasionally 5 leaflets; spadix short, not, or only slightly, exceeding the spathe hood or limb. (below)

Group Two. Leaves with 3 or occasionally 5 leaflets; spadix with a long slender tail greatly exceeding the spathe hood or limb. (p. 526)

Group Three. Leaves radiate or pedate, with more than 5 leaflets; leaves radiate (with all leaflets coming from one point). (p. 526)

Group Four. Leaves radiate or pedate, with more than 5 leaflets; leaves pedate (with leaflets strung out along a double axis). (p. 529)

Group One

1. Inflorescence equalling or overtopping leaves; stem present and well-developed .... *A. yunnanense, A. lobatum*
1. Inflorescence overtopped by leaves (sometimes developed in advance of leaves); stem absent or very short 2
2. Spathe basically white or pink, sometimes flushed or striped green outside in lower half .. *A. candidissimum*
2. Spathe basically purplish with whitish or yellowish stripes, or occasionally greenish yellow overall 3
3. Leaf-stalks bristly .... *A. asperatum*
3. Leaf-stalks smooth .... *A. franchetianum, A. fargesii, A. lichiangensis*

Arisaema yunnanense Buchet Slender plant to 50 cm, often less, with a plain green **stem** bearing a single leaf; leaflets 3, lanceolate to ovate, plain green, occasionally variegated, glaucous beneath. **Spathe** to 8–14 cm long overall, with cylindrical tube, yellowish green with white stripes, limb hood-like, arching forwards, with finely pointed tip, green; spadix-appendage 5–6 cm long, green or purplish, thread-like, shorter than the spathe-limb, horizontal. Guizhou, sSichuan, Yunnan [Myanmar, (?)Vietnam]; rocky places, stony meadows, shrubberies, 700–3,200 m. V–VII.

Arisaema lobatum Engl.* Readily distinguished from the previous species by the generally more robust habit, and mottled **stem** bearing 2 **leaves** with central leaflet stalked, to 5 cm long; spathe-limb olive-green, ± erect at maturity and spadix-appendage stalked, erect and cylindrical, just exceeding the mouth of spathe-tube. Gansu, Guangxi, Guizhou, Hubei, Sichuan, Shaanxi; forests and forest margins, 2,300–4,500 m. III–VII.

Arisaema candidissimum W. W. Sm.* Stout plant to 60 cm in leaf. **Leaves** basal, bright, plain bluish green, unfurling with or after inflorescence; leaflets broad-ovate, central one larger than other two. **Inflorescence** fragrant at first; **spathe** up to 15 cm long, tube a broad cylinder, limb arching forward to a pointed tip, pure white or pink with white stripes inside; spadix-appendage short-stalked, cylindrical, to 6 cm long, arching to a tapered slightly protruding tip, white or greenish. swSichuan, nwYunnan; stony slopes, rocky shrubberies, open *Pinus* forests, 2,400–3,400 m. V–VII.

Arisaema asperatum N. E. Br. (syn. *A. cochleatum* Stapf ex H. Li) Rather robust, often clump-forming plant to 70 cm, generally with a solitary **leaf** with 3 leaflets and with a distinctive rough stalk; leaflets generally plain green or speckled darker, especially beneath, often red-margined, ovate to rhombic, largest to 22 × 17 cm. **Spathe** greenish yellow, with or without purplish brown stripes on tube and limb, tube 5–7 cm long with the mouth margin recurved, limb longer than tube, ascending, wavy, long-pointed; spadix-appendage mousetail-like, yellowish green, to 7 cm long, projecting forward and about same length as spathe. Gansu, Hubei, Sichuan, Shaanxi [cChina, nwIndia]; grassy slopes, woodland margins, streamsides, shrubberies, 1,300–3,000 m. V–VI.

Arisaema franchetianum Engl. (syn. *A. delavayi* Buchet; *A. purpureogaleatum* Engl.) Rather robust plant to 90 cm, with a short **stem** bearing a single basal leaf. **Leaves** large, bluish green, central leaflet larger and broader than the other two. **Spathe** deep purple with white stripes, tube 8 cm long, curving through 180° into hood which has a long tapered, purple 'mousetail' to 20 cm long; spadix-appendage to 8 cm long, tail-like, protruding downwards. Guangxi, Guizhou, Sichuan, Yunnan [neIndia, Myanmar]; rocky places, shrubberies, roadsides, forest margins, 960–3,000 m. VI–VII.

Arisaema fargesii Buchet* Very similar to the preceding species, but central leaflet very much larger in comparison; **spathe** less bent, with the opening upright rather than downturned and tip much shorter, not exceeding 4 cm long; spadix-appendage to 10 cm long. sGansu, w & swHubei, Sichuan; similar habitats, 900–1,600 m. VI–VII.

Arisaema lichiangensis W. W. Sm* Similar in general appearance to *A. candidissimum* and generally flowering with the unfurling young **leaves**; **spathe** purple or purplish red, with pale yellowish green stripes, limb forward-pointing, purple with yellowish stripes to the tip; spadix-appendage slender and forward-pointing, to 6 cm long, dark purple, hidden by the hood. w & nwYunnan; dry habitats in coniferous (mainly *Pinus*) woodland, 2,400–3,000 m. VI–VII.

Group Two
1. Spadix-appendage S-shaped with the tip rearing up . . . . *A. elephas, A. handelii, A. dilatatum*
1. Spadix-appendage not S-shaped, simple or divided, with a trailing or downwards-directed tip 2
2. Spadix-appendage feathery, with long thread-like segments . . . . *A. victoriae*
2. Spadix-appendage not divided . . . . *A. wilsonii, A. speciosum*

Arisaema elephas Buchet Clump-forming, virtually **stemless** species to 40 cm. **Leaves** basal with a very long purple or green, red-striped stalk; leaflets deep to mid-green, sometimes purple-blotched, with red margins, rhombic, central leaflet generally truncated and shorter than the other two. **Inflorescence** held well below the umbrella of leaves; **spathe** to 13 cm long, dark purple, sometimes brownish purple or greenish, with prominent white or pale green stripes, limb forming a curved, forward-pointing, long-pointed hood; spadix-appendage prominent, to 24 cm long, tapered, dark purple. sGansu, Guizhou, Sichuan, seTibet, Yunnan [swTibet, Bhutan]; mixed and coniferous forests, *Rhododendron* and bamboo scrub, shrubberies, often gregarious, 1,800–4,000 m. V–VII.

Arisaema handelii Stapf ex Hand.-Mazz. (syn. *A. elephas* Buchet var. *handelii* (Stapf ex Hand.-Mazz.) G. & L. Gusm.) Similar to the previous species, but **spathe** pale to mid-purple with white stripes and greenish limb; spadix-appendage yellowish or greenish, rather swollen and wrinkled, especially in basal half, purple towards tip. w & swSichuan, seTibet, w & nwYunnan [n & neMyanmar]; similar habitats, 2,800–3,500 m. V–VII.

Arisaema dilatatum Buchet Similar to *A. handelii*: spadix-appendage equally swollen but smooth shiny green, especially in lower half. wSichuan, (?)seTibet, nwYunnan [Bhutan]; broadleaved forests, shrubberies, 2,100–4,000 m. IV–V.

Arisaema victoriae V. D. Nguyen **Evergreen** plant to 50 cm in flower with greenish brown erect, dark-spotted **stems** and leaf-stalks. **Leaves** 2, each with 3 leaflets, leaflets deep green, paler beneath, elliptic to elliptic-oblong, to 14 × 5.5 cm. **Inflorescence** held above leaves; spathe to c. 10 cm long, with a narrowly funnel-shaped, pale green tube that is white at base, and a forward-pointing, finely pointed, triangular, pale green limb with a large basal white blotch inside; spadix long-protruded from spathe mouth, pendent, feathered with long thread-like segments, whitish or pale green in basal part, brown towards top. sGuizhou, swGuangxi, seYunnan [neVietnam]; degraded forests on limestone, c. 500 m. IV–VI.

Arisaema wilsonii Engl.* Very similar in habit to *A. elephas*, but often far more robust, sometimes to over 1 m; leaflets to 60 cm long. **Spathe** larger, to 28 cm long overall, limb downcurved and with abrupt tip; spadix-appendage stalked, smooth and tapered, purple or greenish, to 25 cm long, arching downwards. sGansu, seQinghai, Sichuan, w & nwYunnan; coniferous and broadleaved forests, gullies, 1,800–3,200 m. V–VII.

Arisaema speciosum (Wall.) Mart. Similar in general habit and proportions to *A. elephas*, but leaflets narrower, ovate or lanceolate, central one pointed, not truncated. **Spathe** up to 28 cm long overall, tube rather funnel-shaped, purple to purple-brown with white stripes; spadix-appendage thread-like, but much thickened within the spathe, to 80 cm long, trailing on the ground, whitish or purplish. wSichuan, seTibet, n & wYunnan [c & eHimalaya]; forests, forest margins, shrubberies, open grassy slopes, 1,800–3,500 m. IV–VI.

Group Three
1. Tip of spadix bristly . . . . *A. echinatum*
1. Tip of spadix not bristly . . . . *A. consanguineum, A. concinnum, A. ciliatum, A. ciliatum* var. *liubaense*

1. *Arisaema fargesii*; cult., from wSichuan (CGW)
2. *Arisaema lichiangensis*; by Yufengsi, Lijiang, nwYunnan (CGW)
3. *Arisaema lichiangensis*; Yulongxueshan, nwYunnan (PC)
4. *Arisaema elephas*; Tianchi, nwYunnan (PC)
5. *Arisaema elephas*; Zhongdian, nwYunnan (PC)
6. *Arisaema elephas*; below Tianchi, nwYunnan (CGW)
7. *Arisaema elephas* form; Habashan, nwYunnan (CGW)
8. *Arisaema elephas* shoots; Baishui, Yulongxueshan, nwYunnan (CGW)
9. *Arisaema handelii*; Lancangjiang–Jinshajiang divide, Lidiping, nwYunnan (CGW)
10. *Arisaema victoriae*; Hongdong, swGuangxi (PC)
11. *Arisaema wilsonii*; Jiuzhaigou, nwSichuan (CGW)
12. *Arisaema speciosum*; cult. (CGW)

Arisaema echinatum (Wall.) Schott Medium-sized species, not more than 70 cm in leaf, generally smaller. **Leaves** usually 1, plain green with glaucous reverse; leaflets 7–11, lanceolate, to 20 cm long. **Spathe** to 12 cm long, on a short stalk close to the ground, tube funnel-shaped, greenish, yellowish green or purplish with white stripes, limb bright green, often with purple markings, curved forwards to a long tail-like tip 5–15 cm long; spadix short and club-like, scarcely exceeding tube, white with greenish tip. seTibet, w & nwYunnan [swTibet, Himalaya]; *Rhododendron* and coniferous forests, 2,600–3,800 m. VI–VII.

Arisaema consanguineum Schott Very variable, often robust species, with a stout, often mottled, **stem**, to 1.5 m. **Leaves** generally solitary, parasol-like, deep green or blue-green, sometimes variegated, glaucous beneath; leaflets 7–15, oblanceolate, to 30 cm long, pointed, often with a pronounced tail-like drip-tip. **Inflorescence** held below leaf. **Spathe** to 15 cm long, with cylindrical tube, green or purple with white stripes, limb curved forward, darker than tube, usually also white-striped, tip tail-like, to 15 cm long; spadix-appendage short and club-like, to 10 cm long, green or purplish, somewhat longer than spathe-tube. **Fruit** borne on an erect stem. sGansu, Guizhou, s, w & swSichuan, seTibet, Yunnan [swTibet, Himalaya, Myanmar, nThailand]; forest clearings, shrubberies, grassy places, roadsides, 1,700–3,600 m. V–VII.

Arisaema concinnum Schott Similar to the preceding species, but leaflets with impressed veins above, and spathes narrower, limb not more than 4 cm wide (4–8 cm in *A. consanguineum*) and spadix-appendage with fluted tip. (?)sGansu, Sichuan, seTibet, Yunnan [c & eHimalaya, Myanmar]; similar habitats and altitudes. V–VI.

Arisaema ciliatum H. Li* Patch-forming, stoloniferous plant to 90 cm in leaf, often less, with a well-developed **stem** usually bearing a single **leaf**; leaflets plain green, 8–20, lanceolate to oblanceolate, to 20 cm long, sometimes prolonged into a long tail-like tip. **Spathe** held below leaf, to 15 cm long, the cylindrical tube purple or greenish with white stripes, recurved margin around the mouth usually ciliate, limb similarly coloured, with broad white stripes inside; spadix-appendage 6–10 cm long, club-shaped, whitish green, or purplish covered at tip with whitish conical pimples. **Fruit** borne on a nodding stem. w & swSichuan, w & nwYunnan; open forests and forest margins, shrubberies, rocky places, 2,500–3,300 m. V–VII. **Var. *liubaense* Gusm. & Gouda***, apparently restricted to the Liuba region of wSichuan, is distinguished by the purple **spathe** with white stripes but without marginal cilia; spathe-appendage 10–20 cm long, green or purple.

> **Group Four**
> 1. Spathe with a pair of lobes (auricles) on either side . . .
>   . . . . *A. auriculatum*, *A. hungyaense*, *A. nepenthoides*
> 1. Spathe without auricles 2
> 2. Spathe green, at least 8 cm long . . . .
>   . . . . *A. tortuosum*, *A. aridum*, *A. saxatile*
> 2. Spathe yellow, small, not more than 7 cm long . . . .
>   . . . . *A. flavum* subsp. *tibeticum*

Arisaema auriculatum Buchet* Smallish species to 45 cm, with a short **stem** and a single **leaf**, green above but often flushed purple beneath; leaflets 11–19, elliptic-oblanceolate, pointed. **Inflorescence** at leaf height. **Spathe** to 10 cm long, tube cylindrical, greyish with purple blotches and white stripes, with 2 prominent, spreading, ear-like lobes (auricles) at top, limb arching forwards, green or purplish, with short tip; spadix-appendage pointed, 5–10 cm long, protruding from tube. wSichuan, w & nwYunnan; open forests, shrubberies, 2,600–3,800 m. IV–V.

Arisaema hungyaense H. Li* (syn. *A. omeiense* P. C. Kao; *A. auriculatum* Buchet var. *hungyaense* (H. Li) G. & L. Gusm.) Very similar to the previous species, but a smaller plant generally, the **leaves** with fewer (5–7) leaflets that can have entire or toothed margins; **spathe** with 2 very large spreading auricles and a longer spadix-appendage. wHubei, sSichuan (Emeishan), nwYunnan; wet forests, 1,700–2,500 m. IV–V.

Arisaema nepenthoides (Wall.) Mart. Larger plant than the previous species, to 1 m tall in fruit; **leaves** radiate rather than pedate, usually with 5 shiny deep green leaflets; **inflorescence** held above the unfurling leaves; spadix-appendage 6–8 cm long, only slightly protruding from tube. wYunnan [c & eHimalaya, nMyanmar]; similar habitats, 2,000–3,500 m. IV–V.

Arisaema tortuosum (Wall.) Schott Robust plant to 1 m, usually only half that height, with plain green to purple-mottled **stem**, bearing mostly 2 leaves; leaflets plain green, up to 23, linear-lanceolate to ovate, central leaflet largest. **Inflorescence** overtopping foliage; **spathe** to 18 cm long overall, with greenish white, scarcely lined tube and ascending, long-pointed limb; spadix-appendage S-shaped, to 25 cm long, green or purplish, tip erect, long-tapered. s & swSichuan, seTibet, Yunnan [Himalaya, India, Myanmar]; forests, rocky places, cliff-bases, roadsides, 1,500–3,400 m. V–VII.

---

1. *Arisaema consanguineum*; Xishan, Kunming, cYunnan (CGW)
2. *Arisaema consanguineum*; Cangshan, nwYunnan (EL)
3. *Arisaema consanguineum* fruit; Yulongxueshan, nwYunnan (EL)
4. *Arisaema ciliatum*; swSichuan (CGW)
5. *Arisaema ciliatum*; Gangheba, Yulongxueshan, nwYunnan (EL)
6. *Arisaema ciliatum* var. *liubaense*; swSichuan (CGW)
7. *Arisaema ciliatum* var. *liubaense*; Munigou, Kangding, swSichuan (PC)
8. *Arisaema auriculatum*; Napahai, Zhongdian, nwYunnan (RS)
9. *Arisaema hungyaense*; Shennongjia, wHubei (PC)
10. *Arisaema hungyaense*; Emeishan, sSichuan (PC)
11. *Arisaema nepenthoides*; Bhutan (CGW)
12. *Arisaema tortuosum*; cNepal (CGW)

Arisaema aridum H. Li* (syn. *A. yunnanense* Buchet var. *aridum* (H. Li) G. & L. Gusm.) Like a small version of *A. tortuosum*, but **leaves** usually with only 5 leaflets and spadix-appendage not more than 14 cm long. w & swSichuan, w & nYunnan; rocky and bushy places, 1,800–2,500 m. V–VII. Sometimes included under *A. yunnanense*, which has strictly ternate **leaves**.

Arisaema saxatile Buchet* Rather like *A. aridum*, but **leaves** usually with 5–7 leaflets, the spathe **white**, tube often flushed with green or pale yellow, limb erect, not curved forwards, white to greenish white with translucent veins; spadix-appendage 6–12 cm long, green. s, w & swSichuan, Yunnan; similar habitats, 1,800–2,800 m. VI–VII.

Arisaema flavum (Forssk.) Schott subsp. *tibeticum* J. Murata Rather delicate plant to 45 cm, often less, **stem** green with purplish striping. **Leaves** usually 2, with 7 or 9 plain green leaflets. **Spathe** bright yellow with yellowish green, white-striped tube, spathe-limb erect to arched, ovate with pointed tip, longer than tube; spadix-appendage short, cylindrical, just longer than tube. w & swSichuan, seTibet, w & nwYunnan [swTibet, eHimalaya]; dry rocky and stony habitats, 1,500–3,600 m. V–VII.

UNNAMED SPECIES

Arisaema aff. *ciliatum* H. Li* Very similar to *A. ciliatum*, but a slighter, patch-forming plant to 60 cm with less well-marked **spathes** and smooth spadices; in addition, the leaflets are very narrow-elliptic and pointed. nwYunnan (wHabashan); open forests, rocky hollows, roadsides, 2,800–3,400 m. VI–VII.

Arisaema aff. *consanguineum* Schott* Smaller plant than *A. consanguineum*, to 60 cm, short-stemmed, leaflets pointed to long-pointed (but shortly so, without a long drip-tip); **inflorescence** held above foliage; **spathe** pale yellowish green with paler stripes, tip shortly and sharply pointed. sGansu (Zhuoni, Taohe); woodland margins, open shrubberies, c. 2,555 m. VI–VII.

Arisaema aff. *fraternum* Schott* Slender stoloniferous plant to 40 cm with a short, purple-mottled **stem** bearing a single digitate **leaf**; leaflets 7–8, narrowly elliptic to lanceolate-elliptic, pointed. **Inflorescence** held at leaf height, spathe apple-green with faint white stripes on tube, hood broadening and arching over mouth, plain green, with long tail-like appendage; spadix narrowly club-shaped, smooth. nwYunnan (Habashan); open woodland, tracksides, c. 3,700 m. VII. *A. fraternum* Schott is endemic to the Khasia Hills, Assam.

Arisaema sp.* Small plant close to typical *A. ciliatum*, but only 15–25 cm tall and with a very short **stem**. **Leaves** with 7–11 narrow-elliptic leaflets, each with a pronounced drip-tip. **Spathe** deep purple-black, tube white-striped, hood dark overall with a long tail-like appendage; spadix-appendage narrowly club-shaped, smooth, just protruding from tube. nwYunnan (Lijiang); rock-pockets and -ledges, c. 3,100 m. V–VI.

PINELLIA

A genus of 6 or 7 species (4 in China) native to China and Japan. Tuberous-rooted **perennials**; **leaves** heart-shaped, ternate or pedate. **Inflorescence** with narrow spathe arched at top, but not much expanded, male and female **flowers** usually separated by a constriction in tubular part of spathe; spadix-appendage tail-like, long-protruding. **Fruit** berry-like, each containing a single **seed**.

1. Leaves simple .... *P. cordata*
1. Leaves ternate or pedate .... *P. ternata, P. pedatisecta*

Pinellia cordata N. E. Br.* **Patch-forming** plant not more than 15 cm tall with narrowly ovate-heart-shaped **leaves** 8–13 cm long, dull olive-green above, boldly marked with grey-green veins, reddish purple beneath. **Inflorescence** held beneath foliage, apple-scented; **spathe** 7–8 cm long, pale green with purplish veins, to greenish purple, with short arched hood; spadix S-shaped, with tip erect and held in place by apex of spathe. Guizhou, Hubei [s, c & eChina]; open woodland, ravines, often beneath *Betula* in moist leafy pockets, 900–1,800 m. V–VII.

Pinellia ternata (Thunb.) Breitenb. Slender plant to 20 cm with ternate **leaves** (simple in young plants), leaflets ovate to oblong-lanceolate, 5–12 cm long, middle leaflet larger than lateral; leaf-stalk elbowed in middle and there bearing a small aerial tuber. **Inflorescence** erect and slender, held above foliage; spathe green, 6–7 cm long, lightly striped; spadix S-shaped, greatly exceeding spathe, upper part erect and held in place by apex of spathe. Gansu, Guangxi, Guizhou, Hubei, Sichuan, Shaanxi, Yunnan [c & eChina, Japan, Korea]; margins of cultivated land, road- and track-sides, woodland margins, 1,400–2,800 m. V–VIII.

Pinellia pedatisecta Schott Readily distinguished from the previous species by the pedate **leaves** with up to 11 leaflets, the largest central one to 15 cm long; **spathe** erect, apple-green, not hooded, tube short; spadix yellowish green, to 18 cm long, erect and slightly curved. Guizhou, Hubei, Sichuan, Shaanxi, neYunnan [s, c & eChina, seAsia]; wooded habitats, roadsides, 1,200–2,400 m. VI–IX.

1. *Arisaema saxatile*; cult. (CGW)
2. *Arisaema flavum* subsp. *tibeticum*; cNepal (CGW)
3. *Arisaema flavum* subsp. *tibeticum* in fruit; cult. (CGW)
4. *Arisaema* aff. *ciliatum*; near Habashan, nwYunnan (CGW)
5. *Arisaema* aff. *consanguineum*; Taohe, Zhuoni, sGansu (HJ)
6. *Arisaema* aff. *fraternum*; Habashan, nwYunnan (CGW)
7. *Arisaema* sp.; Gangheba, Yulongxueshan, nwYunnan (HJ)
8. *Pinellia cordata*; cult. (CGW)
9. *Pinellia ternata*; cult. (CGW)
10. *Pinellia pedatisecta*; cult. (CGW)

PONTEDERIACEAE

EICHHORNIA

A genus of 7 species (1 naturalised in China) in tropical America and Africa. Closely related to *Monochoria* but with a zygomorphic **corolla** and with 3 long, and 3 short stamens.

Eichhornia crassipes **(Mart.) Solms** Stoloniferous, floating **herbs**, commonly known as water hyacinth, with many long, fibrous roots, stolons long, greenish or purplish, apically producing new plants with short **stems**. **Leaves** orbicular, broadly ovate or rhomboidal, 4.5–14.5 cm long, glabrous; stalks yellowish green to greenish, 10–40 cm long, spongy, usually swollen in middle. **Inflorescence** 7–15-flowered; tepals 6, purplish blue, petaloid, ovate to elliptic, upper one larger and with a central yellow blotch; stamens 6, 3 long and 3 short, with glandular-hairy filaments. Guizhou, Sichuan, sShaanxi, Yunnan [s & cChina, native to Brazil]; naturalised, growing gregariously in pools, ditches and rice paddies, 200–500 m. VII–X. Commonly known as water hyacinth.

MONOCHORIA

A genus of 8 species (4 in China) confined to tropical and subtropical Africa, Asia and Australia. Marsh-dwelling or aquatic **perennials** with leafy vegetative **stems**; flowering stems with a solitary leaf. **Inflorescence** a raceme, with a broad leafy sheath when young. **Flowers** regular, with spreading tepals, inner broader than outer; stamens 6, one much longer than the others and with an oblique tooth on the filament. **Fruit** a 3-valved capsule.

1. *Eichhornia crassipes*; Sabah (PC)
2. *Monochoria vaginalis*; Xishuangbanna, swYunnan (CGW)
3. *Ottelia acuminata*; Lijiang, nwYunnan (PC)
4. *Ottelia acuminata* sold as a vegetable; Shaping, north of Dali, nwYunnan (EL)
5. *Triglochin palustre*; Kangding, wSichuan (PC)
6. *Iris tectorum*; Songpan, nwSichuan (CGW)
7. *Iris tectorum*; Shennongjia, wHubei (PC)

Monochoria vaginalis **(Burm. f.) C. Presl ex Kunth** Rather similar to *Eichhornia crassipes*, but **leaves** varying from linear to ovate-heart-shaped, sheathing at base but without a swollen stalk; **flowers** blue, in a short dense spike from the sheath of uppermost leaf; stamens 6, one large and blue, the others with yellow anthers. Distribution and habitats similar to those of *Eichhornia crassipes*, also in forest pools, to 1,800 m. V–IX.

HYDROCHARITACEAE

OTTELIA

A genus of c. 21 species (5 in China) of submerged **freshwater herbs** found in tropical to temperate areas.

Ottelia acuminata **(Gagnep.) Dandy*** (syn. *Boottia acuminata* Gagnep.; *Ottelia yunnanensis* (Gagnep.) Dandy) Very variable dioecious species with **leaves** wholly submerged, varying greatly in shape and size, linear to broadly heart-shaped, blunt, base heart-shaped to attenuated, margin entire, wavy or minutely toothed; leaf-stalks varying according to depth of water. **Flowers** arising from a 2–6-ribbed spathe; male plant with 40–50 flowers, with green linear sepals and showy white reverse heart-shaped or obovate petals with a yellow base, 1–3.5 cm long, and 9–12 stamens; female with 2–9 flowers, perianths similar to male ones, ovary triangular-cylindrical, with 3 carpels. **Fruit** triangular-cylindrical to spindle-shaped, 7–8 cm long. Guangxi, Guizhou, Sichuan, Yunnan [Guangdong, Hainan]; freshwater ponds, lakes, ditches and rice paddies, to 2,800 m. V–VI. Gathered and sold locally as a vegetable.

JUNCAGINACEAE

TRIGLOCHIN

A genus of 15 species (2 in China) widely distributed in temperate and cold regions of both hemispheres. **Perennial** or **annual herbs** with short stout **stems**; rhizomes with dense roots at the nodes. **Inflorescences** racemose, on leafless scapes. **Flowers** hermaphrodite; tepals 6; stamens 6; carpels 6, all or only 3 fertile; stigmas feathery.

Triglochin palustre **L.** Slender **perennial herb** with a short rhizome, clothed with sheaths of old leaves. **Leaves** up to 20 × 0.1 cm. **Scape** erect, slender, racemes with ± laxly arranged flowers. **Flowers** on short stalks 2–4 mm long, with purplish green, elliptic tepals 2–2.5 mm long, and only 3 fertile carpels. Gansu, Qinghai, Sichuan, Tibet [temperate northern hemisphere]; marshes, streamsides, wet meadows, to 4,500 m. VI–X.

Triglochin maritimum **L.** More robust than the previous species, **leaves** 7–30 × 0.1–0.4 cm. **Inflorescence** with densely arranged **flowers** borne on stout scape; stalks 1 mm long, elongating after flowering; tepals green, rounded to ovate, 1.5 mm long; carpels 6, fertile. Gansu, Qinghai, Sichuan, Shaanxi, Tibet, Yunnan [nChina, temperate and Arctic northern hemisphere]; marshes, to 5,200 m. VI–X.

IRIDACEAE

IRIS

A genus of 225 species (58 in China, 21 endemic) in the temperate northern hemisphere. Bulbous, tuberous or rhizomatous **perennials**, solitary or tufted with linear to sword-shaped **leaves**, channelled or not; stem-leaves present or absent. **Inflorescence** simple or branched. **Flowers** solitary or several enclosed at base by herbaceous, membranous or partly membranous sheaths. Tepals 6 in 2 series, outer (the falls) with an inclined shaft terminating in a blade that can be small or large; inner (the standards) usually erect to ascending. Style 3-branched, each branch arching over the shaft of the fall and enclosing a single stamen. **Fruit** a 3-parted capsule.

SYNOPSIS OF GROUPS

Group One. Falls with a beard or crest (subgenus *Iris*); falls crested (section *Lophiris*). (below)

Group Two. Falls with a beard or crest (subgenus *Iris*); falls bearded (section *Pseudoregelia*). (p. 534)

Group Three. Falls without a beard or crest (subgenus *Limniris*); flowering stems hollow (section *Limniris*, series *Sibiricae*). (p. 537)

Group Four. Falls without a beard or crest (subgenus *Limniris*); flowering stems solid or stem very reduced or absent (section *Limniris*, series *Ensatae*, *Ruthenicae*, *Spuriae* and *Tenuifoliae*). (p. 538)

> **Group One**
> 1. Flowering stems unbranched; leaves all basal; stem (apart from flowering stem) absent or at least not obvious . . . . **I. tectorum, I. speculatrix**
> 1. Flowering stems branched; leaves sometimes alternate along an obvious stem, sometimes clustered at stem-top . . . . **I. confusa, I. japonica**

Iris tectorum **Maxim.** Tufted plant to 40 cm with thick branching rhizomes. **Leaves** bright yellowish green, in fans, sword-shaped, mostly 15–35 mm wide. Flowering **stems** erect, glabrous, with 1–several short branches near top; spathes leaf-like. **Flowers** relatively large and frilly, 8–10 cm across, violet to lilac with darker veins and white patches on falls; perianth-tube 25–35 mm long; falls and standards spreading horizontally, falls with rounded to oval blade, standards elliptic; style-arms with an uneven crest. sGansu, Hubei, Sichuan, Shaanxi, e & seTibet, Yunnan [c & nChina, Korea, Japan]; sunny banks, forest margins, river- and streamsides, frequently planted on walls and rooftops, 800–3,800 m. IV–V.

Iris speculatrix* Hance Similar to the previous species, but plant with thinner rhizomes (not more than 10 mm in diameter) and **flowers** violet to pale blue, smaller, to 6 cm across, style-arms with an even crest; falls with dark purple, ring-like, mottled markings and a bright yellow crest. Guangxi, Guizhou, Hubei, Sichuan [c & eChina]; forest margins, grassy slopes, roadsides, 400–1,800 m. V.

Iris confusa* Sealy **Perennial** with thin rhizomes forming spreading clumps with erect, bamboo-like **stems** to 1.2 m tall, leafy only towards tips. **Leaves** yellowish green, in fans, sword-shaped, 30–60 mm wide. **Inflorescence** erect, with several short branches and small membranous bracts. **Flowers** white or pale blue, 40–50 mm across, falls and standards spreading out ± horizontally; falls with a yellow crest and signal patch, sometimes with additional purple spots; perianth-tube very short. **Fruit** ellipsoid, 2.5–3.5 cm long. Guangxi, Guizhou, Sichuan, Yunnan; forest margins, open shrubberies, ditchsides, damp grassland, 800–2,900 m. IV–V.

***Iris japonica* Thunb.** Very similar in flower and fruit to the previous species, but plant without an obvious **stem**, **leaves** all appearing to be basal. **Flowers** pale blue or violet, falls with a yellow crest. Guangxi, Guizhou, Hubei, Sichuan, Yunnan [c & eChina, Japan]; wet grassy habitats, slopes, forest margins, open shrubberies, 2,800–3,300 m. III–IV.

> **Group Two**
> 1. Plants with an obvious flowering stem, flowers held well above the ground and with a short perianth-tube . . . . .
>   . . . . . **I. cuniculiformis, I. goniocarpa, I. sichuanensis**
> 1. Plants without an obvious flowering stem, flowers held close to the ground, with a very long perianth-tube . . . 2
> 2. Plants tuberous-rooted . . . . **I. barbatula,**
>   **I. collettii var. collettii, I. collettii var. acaulis**
> 2. Plants rhizomatous-rooted . . . . **I. dolichosiphon,**
>   **I. dolichosiphon subsp. orientalis,**
>   **I. kemaonensis, I. potaninii**

Iris cuniculiformis* Noltie (syn. *I. goniocarpa* Baker var. *grossa* Y. T. Zhao) Dense clump-forming plant to 35 cm. **Leaves** erect, sword-shaped, dull green, to 9 mm wide, well developed at flowering time. **Flowers** solitary, on stem to 30 cm long, paired spathe-valves green with a purplish base; flowers 6–7 cm across, lilac- to pinkish violet, falls with a downcurved, oblong yellow-tipped beard surrounding by a white zone with several dark purple spots; standards erect; perianth-tube 1–2 cm long. nwYunnan (Zhongdian region); grassy places, open shrubberies, 3,490–3,965 m. VI.

***Iris goniocarpa* Baker** Slender **perennial herb** to 25 cm with linear, yellowish green **leaves** not more than 3 mm wide. Flowering stems erect, with several sheath-like leaves and a pair of membranous green bracts enclosing a single flower. **Flowers** bluish purple to violet, 25–50 mm across; falls horizontal, elliptic to oval, with deep purple mottling and a orange-yellow-tipped beard, with a pair of gland-like swellings at base; standards erect, narrow-elliptic; perianth-tube 15–20 mm long. Gansu, wHubei, Qinghai, Sichuan, Shaanxi, seTibet, Yunnan [Himalaya]; mountain grassland, scrubby hillslopes, 2,700–5,000 m. V–VI.

Iris sichuanensis* Y. T. Zhao Similar to the previous species, but **leaves** 5–10 mm wide, **flowers** slightly larger; spathes containing 2–3 flowers that open in succession. sGansu, n, nw & wSichuan; grassy places on hillslopes, roadsides, 2,900–3,700 m. IV–V.

Iris barbatula* Noltie Small tufted tuberous **perennial** not more than 12 cm tall in flower. **Leaves** dull green, sword-shaped with 2 veins thickened on 1 surface, not more than 10 mm wide, only partly developed at flowering time. **Flowers** up to 11 per clump, each 4–7 cm across, short-lived, dark violet, falls with white streaks radiating from a golden yellow fringed crest; standards oblong to oval, spreading horizontally to slightly deflexed; styles with erect, toothed lobes; perianth-tube very long, to 10 cm, slender. nwYunnan (Lijiang, Zhongdian); grazed grassland plateaus and other open grassy places, 2,700–3,550 m. VI–VII.

Iris collettii* Hook. f.** Similar to the previous species, but **flowers** smaller, generally 3–4 cm across, pale to deep lilac-blue, falls with a white signal patch surrounding an egg-yolk-yellow wavy crest. **Var.** *collettii* has **stems** to 8 cm tall. swSichuan, seTibet, nwYunnan [c & eHimalaya, Myanmar, nThailand, Vietnam]; *Pinus* forests and forest margins, shrubberies, grassy places, tracksides, usually in partial shade, 1,700–3,700 m. V–VII. Plants of var. *acaulis* **Noltie are **stemless**. w & nwYunnan, s & swSichuan; similar habitats, 2,200–3,700 m.

***Iris dolichosiphon* Noltie** Small tufted, **stemless** plant not more than 12 cm in flower with **leaves** only slightly developed at flowering time, to 3 mm wide. **Flowers** solitary, 3–8.5 cm across, dark violet, often faintly mottled, standards and falls

---

1. *Iris speculatrix*; Jiangkou, ne Guizhou (PC)
2. *Iris confusa*; cult., from Sichuan (CGW)
3. *Iris japonica*; Xingyi, swGuizhou (PC)
4. *Iris cuniculiformis*; Bitahai, Zhongdian, nwYunnan (CGW)
5. *Iris cuniculiformis*; Napahai, Zhongdian, nwYunnan (CGW)
6. *Iris goniocarpa*; Balangshan, Wolong, wSichuan (CGW)
7. *Iris barbatula*; Xiaozhongdian, nwYunnan, (HJ)
8. *Iris barbatula*; Zhongdian, nwYunnan (CGW)
9. *Iris barbatula*; Zongdian, nwYunnan (CGW)
10. *Iris collettii* var. *acaulis*; Cangshan, nwYunnan (CGW)
11. *Iris collettii* var. *acaulis*; Gangheba, Yulongxueshan, nwYunnan (CGW)
12. *Iris dolichosiphon* subsp. *orientalis*; Xiaozhongdian, nwYunnan (CGW)

1
2
3
4
5
6
7
8
9
10
11
12

spreading to downcurved, standards paddle-shaped with a slender stalk, standards with oblong blade with a yellow-tipped beard at base surrounded by an uneven white zone; perianth-tube 4–7 cm long, deep purple-black. swSichuan, seTibet, nwYunnan [Myanmar, Bhutan, neIndia]; grazed grassland, open shrubberies, forest clearings, 3,600–4,200 m. VI–VII. Subsp. *dolichosiphon* is restricted to seTibet [Bhutan]; plants from Sichuan and Yunnan are referable to **subsp. orientalis Noltie***.

Iris kemaonensis **D. Don ex Royle** Tufted to patch-forming, apparently **stemless** plant to 10 cm tall in flower, with old leaf-bases persisting as a tuft of fibres. **Leaves** yellowish green, partly developed at flowering time, not more than 6 mm wide. **Flowers** solitary, 5–6 cm across, purple or violet-purple, falls downcurved with an obovate blade, darker blotched and streaked, with small yellow-tipped beard at base, standards erect; perianth-tube trumpet-shaped, 5–6 cm long, partly enfolded in the membranous green spathe-valves. w & swSichuan, seTibet, w & nwYunnan [swTibet, Himalaya]; grassy and stony slopes, 3,500–4,200 m. V–VI.

Iris potaninii **Maxim.** Similar to the previous species, but **flowers** smaller, 3–4.5 cm across, either yellow or lavender-purple, standards and falls unblotched. Gansu, Qinghai, nwSichuan, eTibet [Mongolia, sRussia]; grassy and stony steppe, 3,500–5,800 m. V–VI.

Group Three
1. Flowers basically yellow . . . . **I. forrestii, I. wilsonii**
1. Flowers basically blue, violet or purple 2
2. Stems hollow; leaves not more than 12 mm wide . . . .
. . . . **I. chrysographes, I. bulleyana, I. delavayi**
2. Stems solid; leaves mostly 12–18 mm wide . . . **I. clarkei**

1. *Iris kemaonensis*; cNepal (CGW)
2. *Iris potaninii*; Jiangluling, Qinghai (RS)
3. *Iris potaninii*; Jiangluling, Qinghai (RS)
4. *Iris forrestii*; cult., from nwYunnan (CGW)
5. *Iris forrestii*; Sandauwan, Yulongxueshan, nwYunnan (PC)
6. *Iris wilsonii*; cult., from wSichuan, (CGW)
7. *Iris chrysographes*; Balangshan, Wolong, wSichuan (CGW)
8. *Iris chrysographes*; Sandauwan, Yulongxueshan, nwYunnan (CGW)
9. *Iris chrysographes*; south of Zhongdian, nwYunnan (CGW)
10. *Iris bulleyana* with *Euphorbia jolkinii*; south of Zhongdian, nwYunnan (CGW)
11. *Iris bulleyana*; Zhongdian, nwYunnan (CGW)
12. *Iris bulleyana*; Zhongdian, nwYunnan (HJ)
13. *Iris bulleyana* × *I. chrysographes* hybrids; Bitahai, Zhongdian, nwYunnan
14. *Iris delavayi*; eCangshan, nwYunnan (CGW)
15. *Iris delavayi*; eCangshan, nwYunnan (CGW)
16. *Iris clarkei*; cult. (CGW)

Iris forrestii **Dykes** Tufted rhizomatous **perennial** to 45 cm tall in flower with yellowish green, strap-shaped **leaves** that are well developed at flowering time, 4–7 mm wide. **Flowering stems** unbranched with 1–3 leaves in lower half, bearing 1–2 flowers; stalks to 5 cm long. **Flowers** 6–7 cm across, yellow, with erect, elliptic standards and obovate falls, streaked and spotted with purple-brown. swSichuan, seTibet, nwYunnan [nMyanmar]; damp habitats, grassy hillslopes, open forests and forest glades, ditch margins, 2,750–3,600 m. V–VI.

Iris wilsonii **C. H. Wright*** Very similar to the previous species, but a taller plant to 60 cm with grey-green **leaves** with larger, 6–7 cm **flowers**, flower-stalks up to 11 cm long and standards spreading rather than erect. sGansu, wHubei, w, nw & swSichuan, Shaanxi, wYunnan; similar habitats, 2,300–4,000 m. V–VI.

Iris chrysographes **Dykes** Tufted rhizomatous **perennial** to 60 cm tall in flower, with strap-shaped grey-green **leaves** to 12 mm wide. **Flowering stems** unbranched with 1–2 leaves in lower half, 2-flowered. **Flowers** fragrant, deep violet-blue, sometimes blackish violet or reddish violet, narrow-obovate, vertically-inclined falls with golden markings in centre, standards spreading outwards. Guizhou, w, nw & swSichuan, e & seTibet, n & nwYunnan [swTibet, Myanmar]; damp meadows, marshy places, river- and streamsides, 1,200–4,500 m. VI–VII.

Iris bulleyana **Dykes** Very similar to the previous species, but differing primarily in the pale to mid-blue or violet-blue, sometimes bright blue, **flowers** that often have some white patches on falls, accompanied by violet streaks and dots; standards ± erect to slightly diverging. swSichuan, seTibet, nwYunnan [n & neMyanmar]; similar habitats, 2,300–3,700 m. VI–VII. Hybridises with *I. chrysographes* in the wild.

Iris delavayi **Micheli*** Readily distinguished from the previous two species by the more robust habit, and plants to 1.2 m tall in flower, although often less. **Leaves** strap-shaped, grey-green, to 15 mm wide; spathe-valves to 11 cm long, green with a flush of reddish purple. **Flowers** 7–9 cm across, deep purple or violet-purple, with darker mottling and with some white markings on vertically inclined falls; standards elliptic, widely diverging. s & swSichuan, seTibet, nwYunnan; damp marshy meadows, seepage areas along forest margins, ditches, 2,700–4,000 m. V–VII.

Iris clarkei **Baker** Tufted **herb** to 60 cm tall in flower. **Leaves** grey-green, strap-shaped, 10–18 mm wide. **Flowering stems** branched in upper part, with 2–3 sheathing leaves towards base; spathes green with a membranous margin, to 9 cm long. **Flowers** blue to violet, 7–9 cm across, 1–2 per spathe, with a short stalk 9–10 mm long; falls obovate, with a mottled white, semi-circular pattern towards base; standards oblanceolate, spreading. w & nwYunnan [c & eHimalaya, nMyanmar]; damp habitats in forests, marshes, streamsides, seepage slopes, lake shores, 3,000–4,000 m. VI–VII.

IRIDACEAE: Iris

> **Group Four**
> 1. Plants mat-forming (section *Limniris*, series *Ruthenicae*) .... **I. ruthenica**
> 1. Plants tufted 2
> 2. Perianth-tube very short, only 2–3 mm long (section *Limniris*, series *Ensatae*) .... **I. lactea, I. lactea** var. **chrysantha**
> 2. Perianth-tube 4 mm long or more (section *Limniris*, series *Tenuifoliae* and *Spuriae*) 3
> 3. Flowering stems obvious .... **I. songarica, I. farreri**
> 3. Flowering stems not obvious, scarcely emerging from the ground or not .... **I. tenuifolia, I. qinghainica**

Iris ruthenica Ker-Gawl. var. *nana* Maxim.* Low patch-forming rhizomatous **perennial** not more than 10 cm tall in flower, with small fans of bright green, grassy leaves, 2–5 mm wide. **Flowers** solitary, pale blue to violet, with white patches and some speckling on the horizontal falls, fragrant, on short **stems** no more than 6 cm long. Gansu, Hubei, w, nw & seSichuan, Shaanxi, e & seTibet, Yunnan; grassy banks and slopes, open forests and shrubberies, in rather dry locations, 2,500–3,700 m. V–VI.

Iris lactea Pallas (syn. *I. lactea* var. *chinensis* (Fisch.) Koidz.) Tufted plant to 40 cm, with erect, grey-green **leaves** 3–5 mm wide, that often overtop the flowers. **Flowers** 2–4, fragrant, on **stems** not usually more than 10 cm long, 5–6 cm across, milky white to blue or violet with deeper veining on falls oblanceolate, scarcely spreading, standards narrow-oblanceolate. Gansu, Hubei, Qinghai, Sichuan, Shaanxi, e & seTibet [c & nChina, Afghanistan, c & eAsia, nIndia, Mongolia, Korea, sRussia]; grassy steppes, earth banks, riversides, rocky places, 1,200–3,700 m. V–VI. Plants with yellow standards and pale violet falls are sometimes distinguished as var. *chrysantha* Y. T. Zhao* and are restricted to eTibet.

Iris songarica Schrenk Tough, tufted, clump-forming plant to 50 cm, with grey-green linear, strongly ribbed **leaves** only 2–3 mm wide. **Flowering stems** with 1–several short branches, with each pair of spathe-sheaths bearing up to 6 **flowers**, each 6–8 cm across, greyish lavender-blue, generally with darker veins and spots on falls; standards and falls rather narrow, the former erect, falls with an oval spreading blade; perianth-tube 4–6 mm long. Gansu, Qinghai, n & nwSichuan, Shaanxi [nwChina, c & swAsia, Russia]; grassy and stony steppes, sandy riverbanks, 1,200–2,500 m. VI–VII.

Iris farreri Dykes* (syn. *I. polysticta* Diels) Plant to 35 cm tall in flower, differing from the previous species in the lavender-violet, rather spidery **flowers** with strongly arched style-arms; ovary and **fruit-capsule** with paired ribs on each of the 3 angles (in *I. songarica* and *I. tenuifolia* the 6 ribs are evenly spaced); perianth-tube 3–5 mm long. Gansu, Qinghai, w, n & nwSichuan, e & seTibet, nwYunnan; grassy banks, field boundaries, meadows, 2,700–3,700 m. VI–VII.

Iris tenuifolia Pallas Similar to *I. songarica*, but a more condensed plant to 30 cm with thread-like **leaves**, 1.5–2 mm wide, and with 4–6 cm **flowers** ranging in colour from bluish violet to pale lilac, with heavily veined falls, 4.5–5 cm long with a creamish centre; flowers on very short, scarcely observable stems; perianth-tube very long, 5–8 cm. Gansu, Hubei, Qinghai, Shaanxi, neTibet [nwTibet, n & nwChina, cAsia, Mongolia, eRussia]; similar habitats, 1,200–2,500 m. IV–V.

Iris qinghainica Y. T. Zhao* Closely related to the previous species, but **leaves** 2–3 mm wide, **flowers** smaller, blue or violet with yellow in centre of narrow falls, these 3–3.5 cm long. swGansu, e & neQinghai; mountain slopes, dry grassy areas, 2,500–3,100 m. VI–VII.

LILIACEAE

LILIUM

A genus of 115 species (55 in China, 35 endemic) in the temperate and alpine northern hemisphere and South East Asia. Bulbous, sometimes stoloniferous **perennials** without basal **leaves**. **Stems** erect to arching, bearing scattered or whorled, linear to elliptic or oval leaves. **Flowers** solitary or in racemes or corymbs accompanied by leaf-like bracts, campanulate or trumpet-, funnel-, or Turk's cap-shaped, often scented; tepals 6 in 2 whorls, all petal-like, with basal, generally grooved nectaries; stamens 6, with long filaments and dorsifixed anthers. **Fruit** a 3-parted, many-seeded capsule.

SYNOPSIS OF GROUPS

Group One. Flowers trumpet-, funnel- or lantern-shaped, or campanulate. (below)

Group Two. Flowers Turk's cap-shaped. (p. 542)

> **Group One**
> 1. Tepals large, 65–190 mm long 2
> 1. Tepals relatively small, 20–57 mm long 4
> 2. Flowers basically white, pink, purple, yellow or greenish, spotted ... **L. bakerianum, L. bakerianum** var. **aureum, L. bakerianum** var. **rubrum, L. bakerianum** var. **delavayi, L. bakerianum** var. **yunnanense, L. amoenum**
> 2. Flowers basically white or yellow, unspotted 3
> 3. Bulbils absent .... **L. regale, L. brownii, L. leucanthum, L. longiflorum**
> 3. Bulbils present in upper leaf-axils .... **L. sargentiae, L. sulphureum**
> 4. Flowers basically yellow or greenish yellow .... **L. lophophorum, L. euxanthum**
> 4. Flowers other than yellow ...... **L. souliei, L. nanum**

1. *Iris ruthenica* var. *nana*; Napahai, Zhongdian, nwYunnan (PC)
2. *Iris lactea*; Taohe, Zhuoni, sGansu (HJ)
3. *Iris lactea*; Taohe, Zhuoni, sGansu (JM)
4. *Iris songarica*; cAfghanistan (CGW)
5. *Iris farreri*; Animaqinngshan, cQinghai (JM)
6. *Iris farreri*; Animaqinngshan, cQinghai (HJ)
7. *Iris farreri*; Longriba to Aba, nwSichuan (CGW)
8. *Iris tenuifolia*; Heka, Qinghai (RS)
9. *Iris qinghainica*; near Qinghaihu, Qinghai (RS)

Lilium bakerianum **Collett & Hemsl.** A very variable medium-sized lily to 90 cm, with scattered linear to linear-lanceolate **leaves**, downy on margin. **Flowers** up to 3, nodding to almost erect, campanulate, tepals 65–83 mm long; nectaries not papillose; style slightly longer than ovary. Guizhou, swSichuan, Yunnan [Myanmar]; open forests, forest margins, shrubberies, grassy and rocky slopes, 1,500–3,800 m. VI–VIII. **Var. *bakerianum*** has glabrous **leaves** and white **tepals** with purplish red spots. w & swSichuan, nwYunnan [nMyanmar], 2,500–3,000 m; **var. *aureum* Grove & Cotton*** has glabrous **leaves** and pale to deep yellow or purplish yellow **tepals** with purplish or purplish red spots. swSichuan, nwYunnan, 2,000–2,500 m; **var. *rubrum* Stearn*** has glabrous **leaves** and purple-red or pink **tepals** with purple or red spots. Guizhou, Yunnan, 1,500–2,000 m; **var. *delavayi* (Franch.) Wils.** has glabrous **leaves** and yellowish green or greenish **tepals** with purple or red spots. Guizhou, Sichuan, Yunnan [nMyanmar], 2,500–3,800 m; **var. *yunnanense* (Franch.) Sealy ex Woodcock & Stearn*** has **leaves** that are finely white-hairy on both surfaces and **flowers** white or pale pink, unspotted or spotted. swSichuan, nwYunnan, 2,000–2,800 m.

Lilium amoenum **Wils. ex Sealy*** Similar to the previous species, but a slighter plant not exceeding 30 cm, **stems** with not more than 12 narrow-elliptic, smooth **leaves**; **flower** solitary, nodding, campanulate, tepals 30–40 mm long, rose-purple to purple-red, with deeper red spots; nectaries green, not papillose. Yunnan; similar habitats, 1,800–3,000 m. VI–VII.

Lilium regale **Wils.*** A stout lily to 1.5 m with grey-green **stems** flushed with purple, without bulbils in leaf-axils. **Leaves** numerous, scattered, linear-lanceolate, deep green, not more than 5 mm wide, 1-veined, margin papillose. **Flowers** trumpet-shaped, up to 14, generally rather few, out-facing, to 12 cm long, powerfully fragrant; tepals white inside with yellow base, flushed with greenish purple outside, tips curving outwards; nectaries glabrous; pollen orange-red. nwSichuan (Minjiang valley); rocky slopes, cliffs, riverbanks, sometimes planted on rooftops, 1,200–2,500 m. VI–VII.

---

1. *Lilium bakerianum* var. *delavayi*; west of Habashan, nwYunnan (CGW)
2. *Lilium bakerianum* var. *delavayi*; Gangheba, Yulongxueshan, nwYunnan (PC)
3. *Lilium bakerianum* var. *delavayi*; Gangheba, Yulongxueshan, nwYunnan (CGW)
4. *Lilium bakerianum* var. *rubrum*; Weibaoshansi, nwYunnan (DC)
5. *Lilium amoenum*; Sanba, nwYunnan (PC)
6. *Lilium regale*; Minjiang valley, nwSichuan (CGW)
7. *Lilium brownii*; cult. (MW)
8. *Lilium leucanthum*; Wanglang, nwSichuan (PC)
9. *Lilium longiflorum*; cult. Guilin, ne Guangxi (PC)
10. *Lilium sargentiae*; Emeishan, sSichuan (CGW)
11. *Lilium sulphureum*; Dali to Caojian, Jinshajiang valley, nwYunnan (DC)

Lilium brownii **F. E. Br. ex Miellez** Similar to *L. regale*, but often stouter, to 2 m, with broader **leaves**, generally 10–20 mm wide, 5–7-veined; **flowers** solitary or several together, milky white, suffused with purple outside, 13–18 cm long, tepals with minutely hairy nectaries. Gansu, Guangxi, Guizhou, Hubei, Sichuan, Shaanxi, Yunnan [c & eChina, Myanmar, Hong Kong]; open forests, shrubberies, ravines, rocky slopes, streamsides, to 2,200 m. VI–VIII. Widely cultivated for the edible bulbs and for medicinal use.

Lilium leucanthum **(Baker) Baker*** Distinguished from *L. regale* by broader **leaves**, 6–10 mm wide, and by the rather larger, faintly scented **flowers**, 12–15 cm long, which are white flushed yellow inside, white with a greenish yellow, purplish or brownish mid-vein outside. Distinguished from *L. brownii* in the details presented above, and in having 1-veined **leaves** and glabrous nectaries. Stamen-filaments hairy towards the base. sGansu, Hubei, Sichuan; grassy banks and hillslopes, streamsides, ravines, 400–2,500 m. VI–VII.

Lilium longiflorum **Thunb.** Readily distinguished from the previous species by the very long, narrow perianth-tube, and tepals to 18 cm long. Plants variable in height, with 3–5-veined **leaves**, to 15 mm wide, each **stem** bearing up to 6 flowers. Widely cultivated in s & swChina, occasionally naturalised near habitation [sJapan, Taiwan]. V–VIII.

Lilium sargentiae **E. H. Wilson*** Robust plant to 2 m, the purplish **stem** with green bulbils in upper leaf- and bract-axils. **Leaves** lanceolate to ovate-lanceolate, 5–20 mm wide. **Flowers** funnel-shaped, out-facing to semi-nodding, to 18 cm long; tepals white with a greenish yellow base inside, flushed with brown or purple outside; filaments glabrous, anthers with brownish yellow pollen. nw, w & swSichuan; shrubberies, cliffs, riverbanks, grassy slopes, 500–2,000 m. VI–VIII.

Lilium sulphureum **Baker ex Hook. f.** (syn. *L. myriophyllum* Franch.) Very similar to the previous species, but bulbils brown and filaments glabrous; **flowers** often 2, very fragrant, 17–19 cm long, ovary purple. Guangxi, Guizhou, Sichuan, Yunnan [Myanmar]; shady or partly shady places in forests, grassy banks, ravines, 100–1,900 m. VI–VII.

Lilium lophophorum **(Bur. & Franch.) Franch.*** Small lily rarely more than 30 cm, often far less, with erect green **stem**. **Leaves** relatively few, aggregated towards base of plant, lanceolate to narrow-elliptic, pointed. **Flowers** solitary or 2–3, lantern-like, 45–57 mm long; tepals greenish yellow to yellow, with greenish or purplish brown flushing towards base, often ageing to pink or pinkish brown, the pointed tips adhering, leaving gaps between narrow ends of tepals. w & swSichuan, seTibet, nwYunnan; stony and grassy alpine meadows, forest glades, open shrubberies, 2,500–4,500 m. VI–VII. (See photos 1–3 on p. 543.)

Lilium euxanthum* (W. W. Sm. & W. E. Evans) Sealy Small *Fritillaria*-like lily to 25 cm, with scattered, linear-lanceolate **leaves**. **Flowers** solitary, small, campanulate, semi-nodding to out-facing; tepals yellow, with a few purplish brown dots within, 20–27 mm long; pollen orange. nwYunnan (centred on Baimashan); alpine moorland, scrubby slopes, 3,800–4,600 m. VI–VII.

Lilium souliei* (Franch.) Sealy Small *Fritillaria*-like plant to 50 cm, with scattered, bright green, oblanceolate, 3–5-veined **leaves**. **Flowers** usually solitary, semi-nodding, campanulate, 25–35 mm long; tepals deep purple-red, unspotted, shiny, nectaries not papillose; pollen orange. swSichuan, seTibet, nwYunnan; alpine moorland, marshy ground, low shrubberies, 3,200–4,400 m. VI–VII.

***Lilium nanum* Klotzsch** Similar to the preceding species, but often smaller; **flowers** pendulous, 25–27 mm long, pale pink to purplish, often with purplish or brownish mottling; nectaries papillose; pollen yellow-brown. w & swSichuan, seTibet, nwYunnan [swTibet, Himalaya, Myanmar]; alpine meadows, moorland, low shrubberies, forest margins, 3,500–4,500 m. VI–VII.

Group Two
1. Tepals bright scarlet, vermilion or orange-red 2
1. Tepals other than red, often basically white, yellow, purple or pinkish; leaves 3–9-veined 3
2. Leaves 1-veined . . . .*L. davidii, L. pumilum*
2. Leaves 3–7-veined . . . . *L. lancifolium, L. henryi*
3. Nectaries minutely hairy (papillose) . . . .
 . . . . *L. duchartrei, L. lankongense*
3. Nectaries not minutely hairy 4
4. Leaves not more than 7 mm wide; tepals white, pink or purple, spotted . . . . *L. taliense, L. wardii*
4. Leaves at least 8 mm wide; tepals basically yellow or greenish yellow, unspotted 5
5. Flowers small, tepals not more than 45 mm long . . . . .
 . . . . *L. lijiangense*
5. Flowers larger, tepals at least 65 mm long . . . .
 . . . . . *L. nepalense, L. primulinum, L. primulinum* var.
 burmanicum, L. primulinum var. *ochraceum,*
 L. majoense

Lilium davidii* Duchartre ex Elwes Rather robust lily to 1.4 m, with green, minutely papillose **stems** spotted with brown. **Leaves** scattered, linear to linear-lanceolate, pointed, deep green, 1-veined, with inrolled margins. **Flowers** up to 20, pendulous, not scented; tepals strongly recurved, 50–70 mm long, orange-vermilion, spotted purple in middle and lower part; pollen orange to scarlet. c & sGansu, Guizhou, wHubei, Sichuan, Shaanxi, Yunnan [Henan, Shanxi]; forest margins, damp forest glades, grassy slopes, 800–3,200 m. VI–IX. Often cultivated (as is *L. lancifolium*) for the edible bulbs.

***Lilium pumilum* Redouté** An altogether smaller plant than the previous species, to 70 cm, rarely more; **flowers** scented, tepals scarlet, unspotted, to 45 mm long; **pollen** scarlet. Gansu, e & seQinghai, nwSichuan, Shaanxi [nChina, Korea, Mongolia, eRussia]; forest margins, open shrubberies, grassy banks, 1,400–2,800 m. VI–VIII.

***Lilium lancifolium* Thunb.** (syn. *L. tigrinum* Ker-Gawl.) Vigorous lily to 1.5 m, generally rather less, with erect to rather oblique, purple-streaked, minutely downy **stem**. **Leaves** scattered, oblong- to linear-lanceolate, to 9 × 1.8 cm, 5- or 7-veined, minutely downy, with bulbils in axils of upper leaves. **Flowers** up to 6, occasionally more, in a raceme, nodding to out-facing; tepals recurved, vermilion with dark purple spots, 60–100 mm long, nectaries with undulate papillae; pollen orange-red. Gansu, Guangxi, Hubei, seQinghai, Sichuan, Shaanxi, e & seTibet [cChina, Japan, Korea]; grassy places and slopes, shrubberies, riverbanks, 400–2,500 m. VII–VIII.

Lilium henryi* Baker An equally vigorous species to the previous, but readily distinguished by the broader **leaves**, 15–27 mm wide and the lack of stem-bulbils; **flowers** up to 12, paired at each node, orange with scattered black spots, tepals up to 70 × 20 mm. Guizhou, Hubei [Jiangxi]; mountain slopes, 700–1,000 m. VII–VIII.

Lilium duchartrei* Franch. Slender, elegant lily to 1 m, often less, **stem** green flushed with brown, with tiny tufts of white hairs at each leaf-node. **Leaves** scattered, lanceolate, deep green above, paler beneath, 3–5-veined. **Flowers** up to 12, often 1–5, pendulous, fragrant; tepals strongly recurved, 45–60 mm long, white with purple-red spots, often deeper at base, becoming purplish overall on ageing. **Pollen** orange; style considerably longer than ovary. Gansu, Hubei, n, nw & wSichuan; forest margins and glades, grassy banks, hillslopes, 1,500–3,800 m. VI–VII.

Lilium lankongense* Franch. Similar to the preceding species, but leaf-axils without hair-tufts and **leaves** with 3–7 raised veins beneath; tepals pink with deep reddish purple spots, 50–55 mm long. seTibet, nwYunnan; grassy slopes, open shrubberies, forest margins, 1,800–3,200 m. VI–VII.

1. *Lilium lophophorum*; Baimashan, nwYunnan (HJ)
2. *Lilium lophophorum*; Wutoudi, Yulongxueshan, nwYunnan (CGW)
3. *Lilium lophophorum*; wSichuan (TK)
4. *Lilium euxanthum*; Baimashan, nwYunnan (CGW)
5. *Lilium souliei*; Tianchi, nwYunnan (CGW)
6. *Lilium nanum*; Galung La, near Bomi, seTibet (HJ)
7. *Lilium davidii*; cult., from Jian Chuan, nwYunnan, (DC)
8. *Lilium davidii*; Heqing, nwYunnan (DC)
9. *Lilium pumilum*; north of Songpan, nwSichuan (CGW)
10. *Lilium lancifolium*; cult. (CGW)
11. *Lilium henryi*; cult. (CGW)
12. *Lilium duchartrei*; Jiuzhaigou, nwSichuan (CGW)
13. *Lilium lankongense*; Napahai, Zhongdian, nwYunnan (CGW)

Lilium taliense Franch.* Slender erect to arching lily to 1.4 m, often less, with green or dark purple, mottled **stems**. **Leaves** linear-lanceolate, scattered, 1-veined. **Flowers** 2–12 in a raceme, pendulous, fragrant; tepals strongly recurved, 45–50 mm long, white spotted purple, often with cream or yellow in centre of each tepal; style equalling ovary or slightly longer. **Pollen** yellow or orange. w & swSichuan, nwYunnan; forests, shrubberies, grassy slopes, 2,600–3,600 m. VI–VIII.

Lilium wardii Stapf ex F. C. Stern Similar to *L. taliense* but a generally larger plant with oblong to lanceolate **leaves**; **flowers** rather larger and often more numerous, tepals 50–60 mm long, pink or purple-pink, darker towards base and with purple spots; pollen yellow. Guizhou, c & sSichuan, seTibet [Myanmar]; rocky places, shrubberies, forest margins, grassy slopes, 2,000–3,400 m. VI–VII.

Lilium lijiangense L. J. Peng* A relatively small lily to only 60 cm, **stem** slender, with purple streaks or spots. **Leaves** scattered, lanceolate to elliptic or ovate-lanceolate, to 11 × 3 cm, usually 7–9-veined leaf-axils with a tuft of white hairs. **Flowers** up to 5, sometimes solitary, fragrant, Turk's cap-shaped, yellow with purplish brown spots, tepals oblong-lanceolate, 40–45 mm long, each with a black or dark red nectary at base within. wSichuan, nwYunnan; grassy and rocky places, forest margins, shrubberies, 3,300–3,400 m. VII–VIII.

Lilium nepalense D. Don Stoloniferous lily with arching **stems** to 1.2 m, often less. **Leaves** scattered, lanceolate, 5-veined. **Flowers** 2–5 in a raceme or solitary, broadly trumpet-shaped with recurved tepal-tips, pale yellow or yellowish green with purple or purple-maroon towards base inside, otherwise unspotted, tepals 70–110 mm long; perianth-tube cone-shaped, not exceeding the recurved part. seTibet, w & swYunnan [swTibet, Himalaya, nMyanmar]; thickets, forest margins, margins of cultivation, grassy hillslopes, 2,600–2,900 m. VI–VII.

Lilium primulinum Baker Similar to *L. nepalense*, but often taller, with narrower, 3-veined **leaves** not more than 15 mm wide. **Flowers** Turk's-cap-shaped, 4–9, generally smaller, tepals 40–90 mm long, yellow, greenish yellow or greenish white, without purple blotching towards base. Guizhou, sSichuan, Yunnan [Myanmar, Thailand]; similar habitats, 1,100–3,100 m. VII–VIII. **Var.** *burmanicum* **(W. W. Sm.) Stearn** (syn. *L. nepalense* D. Don var. *burmanicum* W. W. Sm.) has larger **flowers**, tepals 65–90 mm long, yellow or greenish yellow suffused with purple in lower third. w & swYunnan [nMyanmar, nThailand]; 1,200–2,700 m; **var.** *ochraceum* **(Franch.) Stearn*** is similar to var. *burmanicum*, but **flowers** smaller, tepals not more than 65 mm long. Guizhou, sSichuan, nwYunnan, 1,100–3,100 m. The type variety (**var.** *primulinum*) is restricted to Myanmar.

Lilium majoense H. Lévl.* Similar to *L. primulinum* but a non-stoloniferous plant with a more slender, crimson-purple or purplish brown **stem**; the **flowers** on long, slender, pendent stalks; tepals whitish or cream heavily marked crimson-purple within in the lower half, this sometimes extending to most of the flower, the tepal tube narrow and cylindrical, greatly exceeding the recurved part. nw & scYunnan; open forests, grassy shrubby slopes, 1,100–2,800 m. VII–VIII.

CARDIOCRINUM

A genus of 3 species (2 in China) in the eHimalaya, nMyanmar, China, Korea and Japan. Closely related to *Lilium*, differing primarily in the monocarpic nature of the **bulbs** (which perish after flowering), in the stalked, heart-shaped **leaves** that are inrolled when young, and in **capsules** that are long-toothed along the valve-margins where they dehisce.

Cardiocrinum giganteum (Wall.) Mak. var. *giganteum* Stout erect plant to 3 m, occasionally more, the thick, leafy **stem** smooth, green. **Leaves** dark shiny green, paler beneath, net-veined, stalked, blade heart-shaped, to 50 × 40 cm, in a basal rosette in the years before flowering; stem-leaves similar, smaller and often narrower, with a shorter stalk. **Inflorescence** a raceme with up to 25 large, trumpet-shaped **flowers**, lowermost opening first; tepals white striped reddish purple in throat, pure white outside, oblanceolate, 15–20 cm long. **Capsule** erect, obovoid, 3–6.5 cm long, containing numerous flat, papery, winged **seeds**. seTibet [swTibet, Himalaya]; forests, wooded slopes, deep humusy valleys and ravines, 1,200–3,700 m. VI–VII. **Var.** *yunnanense* **(Leichtl. ex Elwes) Stearn** is distinguished by purple **stems**, **flowers** opening from the top of the raceme downwards, and tepals green-flushed outside. Gansu, Guangxi, Guizhou, Hubei, Sichuan, Shaanxi, Yunnan [cChina, Myanmar].

Cardiocrinum cathayanum (Wils.) Stearn* Smaller plant than the previous species, to 2 m tall at the most, **stems** leafless in lower part and **flowers** irregularly trumpet-shaped, tepals not more than 13 cm long, only the lower 3 marked with reddish brown inside. Hubei [c, e & sChina]; similar habitats, often close to streams, 1,200–1,600 m. VII–VIII.

---

1. *Lilium taliense*; Napahai, Zhongdian, nwYunnan (DC)
2. *Lilium taliense*; cult. (MW)
3. *Lilium wardii*; cult. (MW)
4. *Lilium lijiangense*; cult. (CGW)
5. *Lilium lijiangense*; cult. (MW)
6. *Lilium nepalense*; cult. (CGW)
7. *Lilium primulinum* var. *burmanicum*; Doi Suthep, nThailand (PC)
8. *Lilium primulinum* var. *ochraceum*; cult. (CGW)
9. *Lilium majoense*; cult. (CGW)
10. *Cardiocrinum giganteum* var. *giganteum*; cult. (CGW)
11. *Cardiocrinum giganteum* var. *giganteum* fruit; cult. (CGW)
12. *Cardiocrinum giganteum* var. *yunnanense*; wSichuan (HJ)

NOMOCHARIS

A genus of 7 species (6 in China, 2 endemic) in the eHimalaya, nMyanmar and China. Closely related to *Lilium* and differing primarily in the flattish or widely cup-shaped **flowers** and the tepals that have swollen or crested areas towards the base.

1. Leaves all alternate; all tepals entire .... **N. aperta, N. saluenensis**
1. Upper leaves, at least, whorled; inner tepals with an irregularly cut or finely toothed margin (entire in *N. basilissa*) 2
2. Flowers unspotted .... **N. basilissa**
2. Flowers spotted or blotched inside .... **N. pardanthina, N. meleagrina, N. farreri**

Nomocharis aperta (Franch.) Wils. (syn. *Nomocharis forrestii* Balf. f.) Patch-forming bulbous **perennial** to 50 cm, with lanceolate to linear-lanceolate **leaves**. **Flowers** 1–3, occasionally more, 4–7 cm across, nodding to semi-nodding; tepals rose-pink or pink, with a few to many crimson spots or rings inside, elliptic to ovate, inner slightly wider than outer; style 6–12 mm long. swSichuan, seTibet, nwYunnan [Myanmar]; forest margins, shrubberies, bamboo thickets, grassy banks, 3,000–3,900 m. VI–VII.

Nomocharis saluenensis Balf. f. Similar to the preceding species, but plant to 90 cm with 1–7 rather larger white to pink **flowers** with a dark purple 'eye' as well as usually some spots; style short, 2.5–4 mm, shorter than ovary. w & swSichuan, seTibet, nwYunnan [nMyanmar]; similar habitats, 2,800–4,500 m. VI–VIII.

Nomocharis basilissa Farrer ex W. E. Evans Bulbous **perennial** to 90 cm with lanceolate to linear-lanceolate **leaves** not more than 7 mm wide. **Flowers** 1–5, nodding to semi-nodding, 7–9 cm across, tepals all elliptic to ovate, red, often flushed with purple or blackish purple; anthers 6–7 mm long, style 8–10 mm long. nwYunnan [nMyanmar]; mountain meadows, bamboo thickets, 3,900–4,300 m. VII–VIII.

Nomocharis pardanthina Franch.* (syn. *N. mairei* H. Lévl.) Erect, often patch-forming bulbous **perennial** to 90 cm with lanceolate-elliptic **leaves**, to 7 × 3 cm. **Flowers** 1–5, 5–7 cm across, out-facing to semi-nodding, white or pink, outer tepals sparsely spotted with reddish purple, ovate, inner much broader, broad-ovate to rounded and densely spotted; anthers 4–6 mm long, style 6–8 mm long. swSichuan, nwYunnan; forest margins, open shrubberies, grassy places, bamboo thickets, roadsides, 2,700–4,100 m. V–VII.

Nomocharis meleagrina Franch* Similar to *N. pardanthina* but **leaves** larger (to 11 cm long) and with larger 7–9 cm **flowers** that are boldly blotched inside with reddish purple, all tepals ovate or elliptic; anthers 3–3.5 mm long, style 7–9 mm long. w & swSichuan, seTibet, nwYunnan; forests and forest margins, shrubberies, grassy banks, 2,800–4,000 m. VI–VII.

Nomocharis farreri (W. E. Evans) Harrow Similar to *N. pardanthina* but **leaves** linear to linear-lanceolate, not more than 8 mm wide; style 9–10 mm long. wYunnan [nMyanmar]; forests, bamboo thickets, grassy habitats, 2,700–3,400 m. VI–VII.

NOTHOLIRION

A genus of just 5 species (3 in China) from Iran and Iraq to the Himalaya, nMyanmar and China. Bulbous **perennials** with both **basal** and **stem-leaves**, strap-like to linear; **stems** erect, carrying a few–many-flowered **raceme** of campanulate to funnel-shaped **flowers**; tepals 6 in 2 whorls, free, all petal-like; stamens 6; stigma 3-lobed. **Fruit** a 3-lobed, many-seeded capsule.

1. Plants small, seldom exceeding 35 cm; flowers usually 2–4 .... **N. macrophyllum**
1. Plants tall, 60–150 cm; flowers 10–25 .... **N. bulbuliferum, N. campanulatum**

Notholirion macrophyllum (D. Don) Boiss. (syn. *Lilium macrophyllum* (D. Don) Voss) Glabrous, bulbous **perennial** to 40 cm with up to 10 linear to linear-lanceolate **stem-leaves**, not more than 8 mm wide. **Flowers** usually 2–4, rarely more, broadly funnel-shaped, 30–50 mm long, tepals pale rose-purple to purple; stamens slightly shorter than tepals. w & swSichuan, seTibet, w & nwYunnan [swTibet, Himalaya]; rocks and rock-crevices in meadows or forests, 2,800–3,400 m. VII–VIII.

Notholirion bulbuliferum (Lingelsh. ex H. Limpr.) Stearn (syn. *Lilium hyacinthum* Wils.; *Notholirion hyacinthinum* (Wils.) Stapf) Tall, slim, bulbous **perennial** to 1.5 m, often only half that height, with numerous linear-lanceolate **stem-leaves**, 10–20 mm wide, as well as strap-like **basal** leaves. **Flowers** out-facing, funnel-shaped to campanulate, 25–35 mm long, more than 10 in a long spike-like raceme; tepals pale purple or bluish purple, greenish towards tepal-tips; stamens almost equalling tepals. sGansu, seQinghai, Sichuan, Shaanxi, seTibet, w & nwYunnan [swTibet, eHimalaya, nMyanmar]; forest margins, shrubberies, grassy places, meadow margins, 3,000–4,500 m. VI–VIII.

Notholirion campanulatum Cotton & Stearn Very similar to the preceding species but rarely exceeding 1 m, with **leaves** up to 25 mm wide; **flowers** larger, 35–50 mm long, tepals deep pink to red or reddish purple; stamens shorter, about two-thirds the length of the tepals. swSichuan, nwYunnan [Bhutan, nMyanmar]; similar habitats, 2,800–4,500 m. VI–VIII.

1. *Nomocharis aperta*; Bitahai, Zhongdian, nwYunnan (PC)
2. *Nomocharis aperta*; Tianchi, nwYunnan (CGW)
3. *Nomocharis basilissa*; Gaoligongshan, wYunnan (SC)
4. *Nomocharis pardanthina*; Cangshan, nwYunnan (CGW)
5. *Nomocharis pardanthina*; Cangshan, nwYunnan (CGW)
6. *Nomocharis farreri*; cult. (CGW)
7. *Notholirion macrophyllum*; Marsyandi Valley, cNepal (CGW)
8. *Notholirion bulbuliferum*; near Longriba, nwSichuan (CGW)
9. *Notholirion bulbuliferum*; near Aba, nwSichuan (JM)
10. *Notholirion campanulatum*; cult. (CGW)

1. *Eremurus chinensis*; near Benzilan, nwYunnan (DC)
2. Medicinal roots and shoots including *Fritillaria cirrhosa* bulbs (middle right); Zhongdian, nwYunnan (CGW)
3. *Fritillaria davidii*; cult., from wSichuan (DJ)
4. *Fritillaria delavayi* at young fruit stage; Yulongxueshan, nwYunnan (CGW)
5. *Fritillaria delavayi*; Daxueshan, nwYunnan (HJ)
6. *Fritillaria unibracteata*; above Jiuzhaigou, nwSichuan (CGW)
7. *Fritillaria przewalskii*; Dalijiashan, Qinghai (HJ)
8. *Fritillaria qinghaiensis*; eQinghai (HJ)
9. *Fritillaria sichuanica*; Balangshan, Wolong, wSichuan (CGW)

EREMURUS

A Central Asian genus with c. 35 species (2 in China). Thong-rooted **perennials** with 1–several growing points, with all **leaves** basal. **Flowers** in long scapose racemes, cupped, campanulate to star-shaped; tepals 6 (in 2 whorls), free; stamens 6, shorter than tepals. **Capsule** a small erect, 3-parted capsule.

Eremurus chinensis O. Fedtsch.* Slender plant to 1.2 m tall in flower. **Leaves** strap-shaped, fleshy, to 55 × 2.2 cm, margin entire to minutely scalloped. **Racemes** many-flowered, somewhat tapered, bracts lanceolate, membranous with a brown midvein. **Flowers** campanulate, tepals white, elliptic to oblanceolate, 10–12 mm long, with a dark midvein on reverse. sGansu, Sichuan, e & seTibet, Yunnan; grassy and rocky mountain slopes, open shrubberies, rock- and cliff-crevices and -ledges, 1,000–3,800 m. V–VIII.

FRITILLARIA

A north temperate genus of c. 130 species (24 in China, 15 endemic). Bulbous **perennials**, with simple, erect, leafy **stems**, sometimes all leaves basal; **basal leaves**, when present, stalked, **stem-leaves** spirally arranged, opposite or whorled, stalkless. **Flowers** nodding or ± so, campanulate to cup-shaped, solitary or in **racemes**; tepals 6, free, in 2 whorls, petaloid, often chequered, each with a basal nectary; stamens 6; stigma club-shaped to 3-lobed. **Fruit** a many-seeded, 3-valved capsule.

1. Leaves all basal, reticulately veined; bracts petaloid . . . . *F. davidii*
1. Leaves on stem, not reticulately veined; bracts not petaloid 2
2. Bracts ovate to elliptic, not more than 5 times longer than broad . . . . *F. delavayi*
2. Bracts linear to linear-lanceolate, more than 8 times longer than broad 3
3. Style scarcely 3-lobed, lobes not more than 1 mm long 4
3. Style clearly 3-lobed, lobes 3–8 mm long 5
4. Flowers solitary, rarely 2; bracts and leaves not tendril-like at tip . . . . . . *F. unibracteata, F. dajinensis, F. przewalskii*
4. Flowers 3–6; bracts and at least upper leaves tendril-like at tip . . . . *F. qinghaiensis*
5. Bracts generally 1 per flower . . . . *F. sichuanica*
5. Bracts 2–3 perennial flower 6
6. Lower leaves 15–30 mm wide 7
6. Lower leaves not more than 15 mm wide 8
7. Nectaries large, 6–10 mm; style-lobes 3–8 mm long . . . . *F. monantha*
7. Nectaries small, 2–5 mm; style-lobes 2–3 mm long . . . . *F. crassicaulis, F. omeiensis*
8. Bracts with tendril-like tip . . . . *F. cirrhosa, F. cirrhosa* var. *viridiflava*
8. Bracts not tendril-like at tip . . . . *F. taipaiensis*

Fritillaria davidii Franch.* Patch-forming plant to 15 cm tall in flower with shiny green, elliptic to ovate **leaves** lying horizontal or nearly so. **Flowers** solitary, bearing 3–4 small oval, purplish tepal-like bracts immediately below the solitary, semi-nodding, campanulate flowers; tepals 30–40 mm long, yellow with purple chequering in lower two-thirds. wSichuan; damp habitats in *Betula* forests, grassy slopes, amongst ferns, 1,600–2,600 m. III–V.

Fritillaria delavayi Franch. Small, relatively stout plant to 25 cm, with 3–5 grey-green, sometimes purple-flushed, alternate, ovate to elliptic **leaves** towards base of **stem**. **Flowers** solitary, nodding or semi-nodding, yellowish or brownish, chequered with red- or purple-brown, tepals 32–45 mm long, persisting and enclosing developing **fruit**. sQinghai, w, nw & swSichuan, seTibet, nwYunnan (c & eHimalaya, nMyanmar]; rocky and gravelly places, screes, 3,400–5,600 m. VI–VII. (See also photo on p. 26.)

Fritillaria unibracteata P. K. Hsiao & K. C. Hsia* Slender plant to 40 cm, often less, with 5–7 linear to linear-lanceolate **leaves**, lower 2 usually opposite, the others alternate; bract solitary, leaf-like. **Flower** solitary, nodding, campanulate, blackish purple, often chequered yellow-brown, sometimes yellow-green chequered with brownish purple, tepals 22–26 mm long, each with a nectary at least 2.5 mm long. sGansu, seQinghai, w & nwSichuan; meadows, open shrubberies, rocky slopes, 3,200–4,700 m. V–VI.

Fritillaria dajinensis S. C. Chen* Like *F. unibracteata* but often taller, to 50 cm with up to 10 **leaves** and with smaller **flowers**, tepals 18–23 mm long, yellowish green, often spotted purple towards base; ovary longer than style. nwSichuan; mountain meadows, shrubberies, 3,600–4,400 m. V–VI.

Fritillaria przewalskii Maxim.* Similar to the preceding species, but often taller, the 1–2 **flowers** pale yellow speckled with blackish purple; tepals 20–30 mm long with an inconspicuous nectary. Habitats and distribution similar to that of *F. dajinensis* (as well as sGansu and eQinghai), 2,800–4,400 m. VI–VII.

Fritillaria qinghaiensis Y. K. Yang & J. K. Wu* **(new species, not yet published)** Rather similar to the previous species and sometimes confused with it, but readily separated by the tendril-like leaf- and bract-tips and **stems** bearing up to 6 **flowers**; tepals recurving somewhat at tip. Qinghai; mountain shrubberies, 2,800–3,600 m. VI–VII.

Fritillaria sichuanica S. C. Chen* (syn. *F. fujiangensis* Y. K. Yang et al.; *F. mellea* S. Y. Chang & S. C. Yueh; *F. pingwuensis* Y. K. Yang & J. K. Wu; *F. qingchuanensis* Y. K. Yang & J. K. Wu) Slender plant to 50 cm, often less, with 4–10 opposite and alternate, rarely whorled, linear to linear-lanceolate **leaves**. **Flowers** 1–2, campanulate, yellowish green, chequered and spotted purple within, or almost entirely purple, tepals 25–40 mm long with an oblong nectary at base. sGansu, seQinghai, wSichuan; mountain meadows, shrubberies, usually on slopes, 2,000–4,000 m. V–VI.

Fritillaria monantha Migo* (syn. *F. gansuensis* Y. K. Yang *et al.*; *F. huangshanensis* Y. K. Yang & C. J. Wu) Slender plant to 60 cm with opposite, whorled or alternate lanceolate **leaves** with a somewhat coiled apex; bracts generally 3 in a whorl, leaf-like. **Flowers** solitary or occasionally 2–4, tubular-campanulate, greenish yellow to purplish, with yellowish brown to deep purple spotting and chequering, 35–50 mm long. Hubei, e & neSichuan [cChina]; damp grassy places, open forests, shrubberies, 100–1,600 m. IV–VI.

Fritillaria crassicaulis S. C. Chen (syn. *F. omeiensis* S. C. Chen) Tall slender plant 30–80 cm, often with a whitish bloom in upper part. **Leaves** up to 18, whorled, opposite or alternate, lanceolate with a pointed tip; bracts 3 in a whorl, leaf-like. **Flowers** generally solitary, rarely 2–3, campanulate, yellow or greenish yellow, with slight purple spotting or chequering, relatively large, tepals 40–50 mm long; nectaries brownish yellow. swSichuan, nwYunnan [nMyanmar]; alpine meadows, forest fringes, shrubberies, 2,500–3,400 m. V–VI.

Fritillaria omeiensis S. C. Chen* Similar to the previous species, but a more dwarf plant with broader shiny green **leaves**; **flowers** broader with more sharply defined shoulders and shorter tepals. s & swSichuan; similar habitats and altitudes. V–VI. Sometimes included within *F. crassicaulis*.

Fritillaria cirrhosa D. Don Very variable, rather stout to slender plant to 60 cm, with up to 11 mostly opposite, sometimes whorled or alternate, linear to linear-lanceolate, channelled **leaves**, curved or coiled at tip; bracts 3, leaf-like. **Flowers** solitary, occasionally 2–3, broadly to narrowly campanulate, yellowish, yellowish green, slightly to heavily, yet regularly, spotted and chequered purple, tepals 30–50 mm long with a 3–5 mm-long elliptic nectary inside. s & cGansu, s & seQinghai, Sichuan, seTibet, w & nwYunnan [swTibet, Himalaya]; open forests, grassy and rocky mountain meadows, open shrubberies, occasionally on screes and moraines, 3,200–4,600 m. V–VII. Variants with pale greenish yellow **flowers** and leaf- and bract-tips not tendril-like, found at alpine altitudes in swSichuan and nwYunnan, are assigned to **var. viridiflava S. C. Chen.*** (See also photo 2 on p. 548.)

Fritillaria taipaiensis P. Y. Li* (syn. *F. glabra* (P. Y. Li) S. C. Chen; *F. shaanxiica* Y. K. Yang *et al.*) Similar to *F. cirrhosa*, but leaf-tips lacking a tendril and tepals yellowish green coarsely mottled with purple. Gansu, Hubei, Shaanxi, and neighbouring parts of Sichuan; grassy slopes, shrubberies, 2,000–3,200 m. V–VI.

UNNAMED SPECIES

Fritillaria aff. *cirrhosa* D. Don* Plant to 30 cm, with very slender paired **leaves**, uppermost paired or in a whorl of 3, tips tendril-like. **Flowers** plum-purple with a bloom and very faint chequering along tepal-margin, broadly campanulate. nwYunnan (Baimashan); low *Juniperus* and *Rhododendron* scrub, c. 4,300 m. VI.

Fritillaria aff. *cirrhosa* D. Don* Plant to 20 cm, with mostly opposite **leaves**, these linear-lanceolate, 3-veined, without tendril-like tips; bracts 2–3. **Flowers** solitary, tubular-campanulate, pale greenish yellow with strong purplish chequering. nwYunnan (Yulongxueshan); coarse screes, c. 4,200 m. VII. Growing alongside *F. delavayi*.

Fritillaria sp.* Plant slender, to 60 cm, with linear, 1–5-veined **leaves**, paired or in whorls of 3–4, tips of upper leaves and bracts tendril-like. **Flower** solitary, campanulate, plum-purple or purple-brown with some poorly defined greenish yellow chequering, or greenish yellow with some purple markings, tepals recurved somewhat at tip. nwYunnan (Zhongdian, Tianchi); low *Rhododendron–Potentilla* scrub, c. 3,600 m. VI–VII.

LLOYDIA

A genus of c. 20 species (8 in China, 2 endemic) in the temperate northern hemisphere, most in the Sino-Himalayan region. Small, slender bulbous plants with solitary or few, nodding, funnel-shaped or campanulate **flowers**; tepals 6. **Fruit** 6-ribbed.

1. Flowering plant with 1–2 basal leaves .... **L. serotina, L. serotina var. parva, L. triflora, L. yunnanensis**
1. Flowering plant with 3–9 basal leaves 2
2. Tepals basically white, usually marked or striped purple .... **L. ixiolirioides**
2. Tepals yellow or greenish yellow 3
3. Flowers up to 13 mm long, tepals glabrous; filaments glabrous or slightly hairy .... **L. oxycarpa**
3. Flowers more than 13 mm long, tepals usually with hairs or gill-like structures near base; filaments densely hairy .... **L. delavayi, L. tibetica, L. flavonutans**

Lloydia serotina (L.) Rchb. Small glabrous tufted bulbous **perennial**, rarely exceeding 15 cm, with generally 2 thread-like **basal leaves** perennial **bulb**. **Stem** slightly exceeding leaves, bearing 2–4 smaller linear **leaves** up to 30 mm long. **Flowers** solitary or paired, campanulate, 10–15 mm long, ascending to semi-nodding, white with violet veins, mottled with purple towards base; style 3–4 mm long, about the length of the ovary. Gansu, Qinghai, Sichuan, Shaanxi, e & seTibet, w & nwYunnan [swTibet, c & nChina, Himalaya, temperate Asia, Europe, Russia, North America]; rock-ledges, grassy slopes, alpine moorland, 2,400–5,000 m. VI–VIII. Very small plants, just 3–5 cm, with flowers to 7 mm long are assigned to **var. parva (C. Marq. & Airy Shaw) Hara**.

Lloydia triflora (Ledeb.) Baker Very similar to the preceding species, but tepals are green-veined. Hubei, Shaanxi (n & neChina, Japan, Korea, Russia); habitats similar to those of *L. serotina*. VI–VIII.

1. *Fritillaria crassicaulis*; near Zhongdian, nwYunnan (DZ)
2. *Fritillaria omeiensis*; Emeishan, sSichuan (PC)
3. *Fritillaria omeiensis*; Emeishan, sSichuan (PC)
4. *Fritillaria cirrhosa*; Gangheba, Yulongxueshan, nwYunnan (CGW)
5. *Fritillaria cirrhosa* var. *viridiflava*; Daxueshan, nwYunnan (CGW)
6. *Fritillaria* aff. *cirrhosa*; Baimashan, nwYunnan (PC)
7. *Fritillaria* aff. *cirrhosa*, growing on the same scree as *F. delavayi*; Xulongxueshan, nwYunnan (CGW)
8. *Fritillaria* sp.; Tianchi, nwYunnan (CGW)
9. *Fritillaria* sp.; Tianchi, nwYunnan (CGW)
10. *Lloydia serotina* var. *parva*; east of Daochen, swSichuan (AD)

1. *Lloydia yunnanensis*; Gangheba, Yulongxueshan, nwYunnan (RS)
2. *Lloydia ixiolirioides*; Mahuangba, Yulongxueshan, nwYunnan (CGW)
3. *Lloydia delavayi*; Daxueshan, nwYunnan (HJ)
4. *Lloydia tibetica*; Shennongjia, wHubei (PC)
5. *Lloydia tibetica*; Shennongjia, wHubei (PC)
6. *Lloydia flavonutans*; Lancangjiang–Jinshajiang divide, east of Weixi, nwYunnan (CGW)
7. *Veratrum nigrum*; Balangshan, Wolong, wSichuan (CGW)
8. *Veratrum grandiflorum*; Balangshan, Wolong, wSichuan (CGW)
9. *Veratrum stenophyllum*; Gangheba, Yulongxueshan, nwYunnan (PC)
10. *Veratrum stenophyllum*; Gangheba, Yulongxueshan, nwYunnan (PC)

Lloydia yunnanensis Franch. Like *L. serotina*, but **stem-leaves** thread-like, not exceeding 0.8 mm wide (1–2 mm wide in *L. serotina*) and **flowers** white or yellowish mottled with reddish purple; style longer 6–10 mm. w & swSichuan, w & nwYunnan (cHimalaya); forest margins, shrubberies, grassy and rocky habitats, often in semi-shade, 2,300–4,300 m. VI–VII.

Lloydia ixiolirioides Baker ex Oliv. (syn. *L. longiscapa* Hook.) Tufted bulbous **perennial** 15–30 cm tall, with linear basal **leaves** 1–2 mm wide, sometimes more; stem-leaves 1–2, similar to basal but not more than 3.5 cm long, generally with white marginal hairs towards base. **Flowers** 1–2, campanulate, 15–20 mm long, tepals white mottled or striped with purple in part and with lines of hairs towards base inside; style 3–4 mm long. w & swYunnan, e & seTibet, w & nwYunnan [nMyanmar]; grassy habitats, alpine moorland, often in semi-shade, 3,000–4,300 m. VI–VII.

Lloydia oxycarpa Franch.* (syn. *L. forrestii* Diels) Tufted bulbous **perennial** to 20 cm tall, often less, with thread-like **basal leaves** c. 1 mm wide; **stem-leaves** similar but shorter, not exceeding 3 cm long. **Flowers** solitary, 9–13 mm long; style 3 mm long. sGansu, seQinghai, Sichuan, e & seTibet, w & nwYunnan; open forests, scrub, grassy places, 2,800–4,800 m. V–VII.

Lloydia delavayi Franch. Tufted bulbous **perennial** 15–35 mm tall, with linear **basal leaves** 1.5–3 mm wide; **stem-leaves** several, like basal but narrower and not more than 2.5 cm long. **Flowers** 1–5, campanulate, 14–18 mm long, yellow with purplish green veins, hairy towards base inside; style 4–10 mm long. w & nwYunnan [nMyanmar]; rocky and grassy slopes, 2,700–3,900 m. VII–VIII.

Lloydia tibetica Baker ex Oliv. Very similar to the preceding species, but inner tepals with several tiny crested flanges towards base inside; styles 4–6 mm long. Gansu, Hubei, Sichuan, Shaanxi, eTibet [cChina, Nepal]; similar habitats, 2,300–4,100 m. V–VII.

Lloydia flavonutans H. Hara Very similar to *L. delavayi*, but tepals glabrous (with neither hairs nor flanges). seTibet, nwYunnan [swTibet, Himalaya]; similar habitats, 4,000–5,000 m. V–VII.

MELANTHIACEAE

VERATRUM

A north temperate genus of c. 40 species (13 in China, 8 endemic). Tufted, short- and stout-rhizomatous **perennials** with erect **stems** bearing pleated **leaves** with sheathing and clasping bases. **Inflorescence** a many-flowered terminal panicle. **Flowers** funnel-shaped to cupped or flat and star-shaped; tepals 6 (in 2 whorls), free.

1. Flowers purple-black .... *V. nigrum*
1. Flowers whitish, greenish white or greenish yellow .. 2
2. Basal sheath (disintegrated leaf-sheaths) of stem with parallel veins only .... *V. grandiflorum*
2. Basal sheath of stem with both parallel and transverse veins 3
3. Leaves noticeably stalked .... *V. micranthum*
3. Leaves not stalked .... *V. stenophyllum*, *V. stenophyllum* var. *taronense*, *V. taliense*

Veratrum nigrum L. Stout plant to 1 m with a tuft of basal blackish fibres. **Leaves** elliptic to ovate-lanceolate, to 25 × 10 cm, with pointed apex, stalkless or with a very short stalk, glabrous stalk. **Panicle** with ascending to wide-spreading, many-flowered, branches. **Flowers** often, at least in part, male on lateral branches; tepals purplish black, oblong, 5–8 mm long; stamens about half length of tepals. Gansu, Guizhou, Hubei, n, w & swSichuan, Shaanxi [c & nChina, cAsia, c & eEurope, Mongolia, Russia]; open forests, forest margins, mountain meadows, 1,200–3,300 m. VII–IX.

Veratrum grandiflorum (Maxim. ex Baker) Loes.* (syn. *V. album* L. var. *grandiflorum* Maxim. ex Baker) Stout plant to 1.5 m. **Leaves** elliptic to lanceolate or oblong, to 20 × 12 cm, downy beneath, apex long-pointed or blunt. **Inflorescence** with ascending to almost erect branches, lateral up to 14 cm long; **tepals** greenish white, oblong to elliptic, 11–17 mm long, with minutely toothed margin, outer 3 downy on reverse; stamens slightly more than half length of tepals. Hubei, Sichuan, Yunnan [cChina]; open forests and forest margins, mountain meadows, particularly in damp habitats, 2,600–4,000 m. VII–VIII.

Veratrum micranthum F. T. Wang & Tang* Smaller plant than the previous two species, rarely exceeding 40 cm in flower. **Leaves** 3–4 on lower half of stem, narrow-elliptic, to 18 × 3.5 cm, with wedge-shaped base and pointed apex. **Inflorescence** with short ascending lateral branches to 5 cm long, bearing mainly male **flowers**, those of terminal raceme hermaphrodite; tepals green or greenish yellow, narrow-oblong, 2–3 mm long with ciliate margin. ne, w & swSichuan; forest margins, rocky and grassy places, 2,400–3,600 m. VI–VIII.

Veratrum stenophyllum Diels* Stout plant to 1.2 m with stalkless strap-shaped to oblong or oblanceolate **leaves**, to 32 × 7 cm, with long-pointed apex, glabrous (minutely hairy on veins beneath in var. *taronense* F. T. Wang & Z. H. Tsi). **Panicles** rather dense and many-flowered, lateral branches slender and bearing male **flowers**, terminal raceme stouter and bearing hermaphrodite flowers, stalks 2–4 mm long; tepals pale yellow to greenish yellow, ovate to oblong, 4–7 mm long, slightly downy at base on reverse. w & swSichuan, nwYunnan; forests and forest margins, shrubberies, mountain meadows, 2,000–4,000 m. VII–X.

Veratrum taliense Loes.* (syn. *V. cavaleriei* Loes.) Similar to the previous species, but a slighter plant not exceeding 90 cm, lateral branches of **panicles** horizontal to somewhat downcurved; flower-stalks at least 7 mm long; tepals longer, 7–9 mm. swSichuan, w & nwYunnan; mountain meadows, c. 2,400 m. VIII–X.

ALETRIS

A genus of 21 species (15 in China, 9 endemic) in east Asia, Malaysia and eastern North America. Rather orchid-like, rhizomatous plants, with basal tufts of grassy, somewhat fleshy **leaves**. **Inflorescence** a spike-like raceme borne on a scape with accompanying leaf-like bracts. **Flowers** campanulate to tubular, with a tube that is fused to ovary close to the base and 6 erect to spreading or recurved, short lobes; stamens 6; style simple. **Fruit** a 3-parted capsule.

1. Flowers finely downy .... *A. spicata*
1. Flowers glabrous 2
2. Scapes, rachises and stalks sticky; bracts borne near base of flower-stalks .... *A. glabra*
2. Scapes, rachises and stalks downy, often minutely so, not sticky; bracts borne towards top of flower-stalks .... *A. pauciflora, A. alpestris, A. nana*

Aletris spicata (Thunb.) Franch. Plant 15–70 cm tall in flower with dense tufts of linear **leaves** 5–30 × 2–5 mm. **Flowers** almost stalkless, numerous in a lax raceme, white, yellowish or yellowish green, perianth 4–7 mm long, with urn-shaped tube and erect, narrow lobes; bracts mostly shorter than flowers. Gansu, Guangxi, Guizhou, Hubei, Sichuan, Shaanxi, Yunnan [c & sChina, Japan, Philippines]; forests, forest margins, shrubberies, grassy places, stream- and roadsides, 100–2,900 m. III–VIII.

Aletris glabra Bur. & Franch. Variable plant 15–100 cm tall in flower with a basal tuft of linear to linear-lanceolate **leaves**, each 5–30 × 0.5–1.8 cm. **Racemes** generally many-flowered, sticky and often with adhering detritus; perianth cream to yellowish green, 3–6 mm long, with urn-shaped tube that is suddenly constricted at top, and with erect lobes; bracts often longer than flowers. Gansu, Guizhou, Hubei, Sichuan, Shaanxi, e & seTibet, Yunnan [swTibet, eHimalaya, Taiwan]; mixed forests, forest margins, shrubberies, damp grassy or rocky places, 1,200–4,000 m. V–VIII.

Aletris pauciflora (Klotzsch) Hand.-Mazz. Plant often clump-forming, to 20 cm tall in flower (but sometimes as little as 2–3 cm), surrounded at base with persistent leaf-fibres. **Leaves** 5–10, linear to linear-lanceolate, to 25 × 1 cm. **Racemes** with up to 40 flowers, varying in colour from white to yellow or greenish yellow, pink or red, perianth 3.5–6 mm long with recurved lobes; bracts usually much longer than flowers. Sichuan, seTibet, Yunnan [swTibet, Himalaya]; mixed forests, shrubberies, bamboo thickets, marshy places, wet flushes, rocky habitats, moraines, 1,500–4,900 m. IV–VIII.

Aletris alpestris S. C. Chen* In contrast to the previous species, this has numerous linear **leaves** not more than 4 mm wide and up to 10 white or pinkish flowers in very slender **racemes**; **flowers** small, 3–4.5 mm long; bracts shorter than flowers. Guizhou, Sichuan, Shaanxi, neYunnan; forest rocks and cliffs, 800–3,900 m. IV–VII.

Aletris nana S. C. Chen Similar to *A. alpestris*, but with a stouter, straight scape not more than 10 cm, **bracts** equalling or exceeding flowers and perianth-lobes erect to spreading. seTibet, nwYunnan [swTibet, Nepal]; wet grassy habitats, damp rocks, swamps, 3,200–4,600 m. V–VI.

YPSILANDRA

A genus with 5 species, all present in China (3 endemic) in the Himalaya, nMyanmar and China. Rhizomatous **perennials** with rosettes of narrow **basal leaves**. **Flowers** borne on scaly scapes, racemose, without bracts; tepals 6, free, with basal nectary inside; stamens 6, often protruding; style solitary.

Ypsilandra thibetica Franch.* Rhizomatous plant to 50 cm, often less, with tufts of oblanceolate basal **leaves**, each leaf 4–14 × 1–4.5 cm, with pointed apex. **Scape** with several scale-like leaves, bearing a **raceme** of up to 30 white, pink or purple **flowers**; tepals oblanceolate to spathulate, 6–10 mm long, shorter than stamens; stigma capitate. neGuangxi, Sichuan [Hunan]; forests, damp rocky places, shady slopes, 1,300–2,900 m. III–IV.

Ypsilandra yunnanensis W. W. Sm. & Jeffrey Slighter plant than the previous species, to 30 cm tall in flower. **Racemes** greatly exceeding leaves, up to 17-flowered; **tepals** 4–5 mm, equalling or longer than the stamens; stigma 3-lobed. seTibet, nwYunnan [Himalaya]; *Rhododendron* forests, thickets, grassy places, 2,700–4,000 m. VI–VII.

Ypsilandra alpina Franch. Similar to the previous species, but **racemes** with only 2–6 larger yellow **flowers**, **tepals** 7–12 mm long, linear to linear-lanceolate. seTibet. nwYunnan [nMyanmar]; similar habitats, 2,000–4,300 m. VII–X.

---

1. *Aletris spicata*; Cangshan, nwYunnan (PC)
2. *Aletris nana*; Wutoudi, Yulongxueshan, nwYunnan (CGW)
3. *Ypsilandra thibetica*; cult. (CGW)
4. *Tofieldia divergens*; Gangheba, Yulongxueshan, nwYunnan (CGW)

TOFIELDIA

A genus of c. 20 species (3 in China, 2 endemic) in the north temperate zone and South America. Small tufted, rhizomatous **perennials**, with *Iris*-like fans of **leaves** and slender erect scapes. **Inflorescence** a many-flowered spike or raceme; tepals 6, free or fused at base; stamens 6, free; styles 3. **Fruit** a capsule.

Tofieldia divergens Bur. & Franch.* (syn. *T. brevistyla* Franch.; *T. tenella* Hand.-Mazz.; *T. yunnanensis* Franch.) Plant to 35 cm tall in flower, often far less. **Leaves** stiff, narrowly sword-shaped, to 3–20 × 0.2–0.4 cm, rather pale green. **Flowers** in **racemes** 2–10 cm long, ascending to semi-nodding on slender stalks 1.5–3 mm long; tepals creamy white, elliptic, 2–3 mm long; stamens protruding. **Capsules** out-facing to nodding. wGuizhou, swSichuan, nwYunnan; rock-crevices, open rocky forests, grassy slopes, 1,000–4,300 m. VI–VIII.

Tofieldia thibetica Franch.* (syn. *T. iridacea* Franch.; *T. setchuenensis* Franch.) Differs from the previous species in having broader **leaves** (3–7 mm), longer flower-stalks (5–12 mm) and almost erect **capsules**. Guizhou, ecSichuan, neYunnan; habitats similar to those of *T. divergens*, 700–2,300 m. VI–VII.

CONVALLARIACEAE

POLYGONATUM

A north temperate genus of c. 60 species (39 in China, half endemic). Rhizomatous **perennials** with simple, erect to arching, occasionally scandent **stems** and alternate to whorled, often rather narrow **leaves**. **Flowers** usually pendulous, solitary or clustered at nodes, sometimes short-racemose; perianth with a tube for at least half its length and 6 short lobes; stamens 6, joined near base of perianth-tube; style 3-lobed. **Fruit** a small fleshy berry.

1. Dwarf patch-forming plants not more than 20 cm tall in flower 2
1. Taller plants, 30–300 cm tall 3
2. Flowers pink .... **P. hookeri, P. graminifolium**
2. Flowers white .... **Polygonatum sp.**
3. Leaves all alternate .... **P. franchetii**
3. Leaves whorled (lowermost sometimes alternate) 4
4. Leaves coiled or hooked at tip ....
     .... **P. cirrhifolium, P. stewartianum**
4. Leaves not coiled or hooked at tip 5
5. Plant at least 1 m tall .... **P. kingianum**
5. Plant less than 1 m tall, often only 20–50 cm ....
     .... **P. prattii, P. gracile, P. kansuense, P. curvistylum**

1. *Polygonatum hookeri*; near Songpan, nwYunnan (HJ)
2. *Polygonatum hookeri* (white-flowered form); near Songpan, nwYunnan (HJ)
3. *Polygonatum graminifolium*; north of Songpan, nwYunnan (CGW)
4. *Polygonatum* sp.; Huanglong, nwSichuan (CGW)
5. *Polygonatum cirrhifolium*; cult., from nwYunnan (CGW)
6. *Polygonatum stewartianum*; Gangheba, Yulongxueshan, nwYunnan (RS)
7. *Polygonatum kingianum*; Tian'e, nwGuangxi (PC)
8. *Polygonatum prattii*; Munigou, Kangding, wSichuan (PC)
9. *Polygonatum prattii*; Yulongxueshan, nwYunnan (CGW)
10. *Polygonatum kansuense*; cult. (CGW)
11. *Polygonatum curvistylum*; Wolong, wSichuan (CGW)

Polygonatum hookeri Baker Small patch-forming **perennial** not more than 8 cm tall, very low in flower, with grey-green **leaves** only partly developed at flowering time, few to 10 on a short **stem**, lanceolate to oblong, 20–45 × 4–8 mm, lowermost alternate, upper often in 3s. **Flowers** solitary, in axils of basal leaves, erect, funnel-shaped, 15–20 mm long, pink to purple, rarely white; tepals 6–10 mm long, curved. **Berries** red when ripe, 7–8 mm in diameter. Gansu, Qinghai, Sichuan, sShaanxi, Tibet, w & nwYunnan [Himalaya]; open forests, gravelly places, alpine meadows, 3,200–4,300 m. V–VII.

Polygonatum graminifolium Hook.* Similar to the previous species, but taller plant to 20 cm with mostly opposite, bright green linear **leaves** to 60 × 5 mm. **Flowers** soft pink to violet-purple, paired. Distribution similar to that of *P. hookeri* but not [Himalaya]; forest margins, open shrubberies, grassy places, to 3,600 m. V–VI.

*Polygonatum sp.** Dwarf patch-forming **perennial** not more than 5 cm tall in flower. **Leaves** few, 4–6 clustered at ground level, oval to elliptic, 18–30 × 8–14 mm, dull green above and usually with 3 parallel veins, pale and whitish beneath. **Flowers** solitary, erect, white, 12–16 mm long, with a cylindrical tube and linear-lanceolate, recurved lobes. **Berries** unknown. (?)sGansu, (?)seQinghai, nwSichuan; *Rhododendron* heath, stony meadows, 3,800–4,400 m. VI–VII.

Polygonatum franchetii Hua* Tufted plant with erect, glabrous **stems** to 80 cm. **Leaves** all alternate, short-stalked, oblong-lanceolate, to 12 × 3.5 cm, long-pointed. **Flowers** pendulous, 2–3 on a 20–60 mm stalk, pale green, narrow-campanulate, 18–20 mm long with lobes 2 mm long. **Berries** purple when ripe, 7–8 mm in diameter. Hubei, Sichuan, Shaanxi [Hunan]; forests and forest margins, 1,100–1,900 m. V–VI.

Polygonatum cirrhifolium (Wall.) Royle (syn. *P. fargesii* Hua; *P. fuscum* Hua; *P. mairei* H. Lévl.) Erect plant to 90 cm, often less, with slender, glabrous **stems**. **Leaves** mainly in whorls of 3–6, linear-lanceolate to linear, to 9 × 0.8 cm, with slender, pointed, usually coiled but not hooked, tip, soft and spreading at flowering time. **Flowers** generally paired on a stalk 3–10 mm long, pale purple, greenish or whitish, cylindrical but with a slight waist in middle, 8–11 mm long overall, lobes 2 mm long. **Berries** red or purplish red when ripe, 8–9 mm in diameter. Gansu, Guangxi, Qinghai, Sichuan, Shaanxi, seTibet, Yunnan [swTibet, Himalaya]; forests, forest margins, shrubberies, grassy places, 2,000–4,000 m. V–VII.

Polygonatum stewartianum Diels* Very similar to the previous species but **leaves** thicker and more leathery, held stiffly erect at flowering and usually hooked at tip, generally in whorls of 3–4, to 11 × 1.6 cm; **flowers** greenish with a purple or crimson flush or purplish pink; **berries** 5–7 mm in diameter. w & swSichuan, w &Yunnan; habitats similar to those of *P. cirrhifolium*, 2,700–3,300 m. V–VI.

Polygonatum kingianum Collett & Hemsl. (syn. *P. cavaleriei* H. Lévl.; *P. uncinatum* Diels) Tall plant, 1–3 m, with a slender **stem** that often scrambles through other vegetation, glabrous. **Leaves** in whorls of up to 10, linear to lanceolate, 6–20 × 0.3–3 cm, stalkless, with a hooked tip. **Flowers** usually in clusters of 2–5, occasionally solitary, borne on stalk 10–20 mm long, pink or white, narrow-campanulate, 18–25 mm long overall, lobes 3–5 mm long. **Berries** red when ripe, 10–15 mm in diameter. Guangxi, Guizhou, Sichuan, Yunnan [Myanmar, seAsia]; forests, shrubberies, rocky places, ravines, 700–3,600 m. III–VI.

Polygonatum prattii Baker* (syn. *P. delavayi* Hua) Patch-forming plant with slender, glabrous, erect **stems** to 30 cm, often half that height. **Leaves** up to 15 mostly opposite or in 3s, but lowermost alternate, elliptic to oblong, 2–6 × 1–2 cm, with pointed tip. **Flowers** pendulous, 2–3 on a short stalk 2–6 mm long, white to pale purple, cylindrical, 6–8 mm long overall, lobes 1.5–2.5 mm long. **Berries** brownish or purplish red when ripe, 5–7 mm in diameter. w & swSichuan, nwYunnan; open forests, shrubberies, grassy places, 2,500–3,300 m. V–VI.

Polygonatum gracile P. Y. Li* Differs from the previous species in having **leaves** in 2–3 whorls, each with 3–6 leaves, plus scattered leaves between the whorls, and pale yellow, paired or solitary **flowers**. sGansu, Shaanxi [Shanxi]; habitats similar to those of *P. prattii*, 2,100–2,400 m. V–VI.

Polygonatum kansuense Maxim. ex Batalin* (syn. *P. erythrocarpum* Hua; *P. verticillatum* (L.) Allioni *sensu lato*) Patch-forming plant with erect, glabrous **stems** to 80 cm bearing **leaves** mainly in whorls of 3, these linear to elliptic-lanceolate, to 10 × 3 cm, with pointed tip; uppermost leaves sometimes paired. **Flowers** pendulous, solitary or up to 4 together on a stalk 10–20 mm long, usually pale purple, cylindrical, 8–12 mm long overall, lobes 2–3 mm long. **Berries** red when ripe, 6–9 mm in diameter. Gansu, Qinghai, Sichuan, Shaanxi, e & seTibet, Yunnan; forests, shrubberies, grassy places, 2,100–4,000 m. V–VI.

Polygonatum curvistylum Hua* Readily distinguished from the previous species by the **leaves**, which are at first ascending but become reflexed after flowering; leaves numerous, mainly in whorls of 3–6; **flowers** pale purple, 6–8 mm long. w & swSichuan, nwYunnan; habitats similar to those of *P. kansuense*, 2,700–3,900 m. V–VII.

STREPTOPUS

A north temperate genus of c. 10 species (5 in China, 2 endemic). Rhizomatous **perennials** with erect leafy **stems** that are simple or branched above; **leaves** alternate, stalkless. **Flowers** lateral or sometimes terminal, 1–2, campanulate, nodding; tepals 6 in 2 whorls, petal-like; stamens 6, inserted at base of tepals, filaments flattened at base. **Fruit** a globose berry.

***Streptopus simplex* D. Don** Stems erect, slender, branched in upper part, to 50 cm. **Leaves** ovate to lanceolate, to 8 × 3 cm, often rather pale green, thin, with an entire margin and heart-shaped base. **Flowers** 1–2, nodding, campanulate, pale pink or white, 4–10 mm long, on slender stalks to 4.5 cm long; anthers longer than filaments. **Berries** globose, 5–6 mm in diameter. w & nwYunnan, seTibet [swTibet, Himalaya, nMyanmar]; forests, shrubberies, alpine meadows, streamsides, bamboo thickets, 1,700–4,000 m. VI–VII.

Streptopus parviflorus* Franch. Similar to the preceding species, but **flowers** smaller, 6–8 mm long; anthers shorter than filaments. swSichuan, nwYunnan; habitats and flowering time similar to those of *S. simplex*, 2,000–3,500 m.

MAIANTHEMUM
(including *Smilacina* and *Tovaria*)

A genus of c. 40 species (19 in China) in the temperate northern hemisphere, particularly in the Sino-Himalayan region. Small to medium-sized rhizomatous plants with alternate to almost opposite **stem-leaves**, with obvious parallel veins, sometimes somewhat pleated. **Flowers** small, in racemes or panicles, star-shaped; tepals usually 4 or 6, free or fused near base. **Fruit** a berry.

1. Carpeting plants with basal leaves (withered at flowering time) and 2–4 stem-leaves; flowers 4–6-parted . . . . **M. bifolium, M. lichiangense**
1. Tufted or clump-forming plants without basal leaves and with more than 4 stem-leaves; flowers 6-parted 2
2. Flowers funnel-shaped, tepals fused for two-thirds of their length into a tube 3–10 mm long . . . . **M. henryi, M. szechuanicum**
2. Flowers salverform or campanulate, with or without short tube, tepals only fused near base (tube, if present, not exceeding 2 mm), or free 3
3. Flowers rose-purple; ovary considerably longer than style . . . . **M. fuscum**
3. Flowers white or greenish, often flushed purple, occasionally purplish red; ovary equal to, or shorter than style 4
4. Leaves heart-shaped at base and clasping stem . . . . **M. forrestii**
4. Leaves neither heart-shaped at base nor clasping stem 5
5. Rachis of inflorescence glabrous . . . . . **M. tatsienense**
5. Rachis of inflorescence hairy . . . . **M. oleraceum, M. purpureum, M. atropurpureum**

***Maianthemum bifolium* (L.) F. W. Schmidt** (syn. *Convallaria bifolia* L.) Patch-forming, often extensive, creeping **perennial** not more than 18 cm tall; **leaves** ovate to almost triangular, short-stalked with heart-shaped base; stem-leaves usually 2. **Flowers** in erect racemes of up to 25, with minute bracts; tepals 4, white, free, 2–2.5 mm long. **Berries** 3–6 mm in diameter, red when ripe. Gansu, seQinghai, n & wSichuan, Shaanxi [c & nChina, temperate Asia and Europe, Japan, Korea, North America]; forests, shrubberies, streamsides and other damp places, 900–2,700 m. V–VII.

Maianthemum lichiangense* (W. W. Sm.) LaFrankie (syn. *Tovaria lichiangensis* W. W. Sm.; *Smilacina lichiangensis* (W. W. Sm.) W. W. Sm.) Small patch-forming **perennial** not more than 20 cm tall, with slender, erect or arching **stems** that are hairy in upper part. **Leaves** 2–4, ovate to oblong, to 3.5 × 3 cm, rounded or slightly heart-shaped at base. **Flowers** white, fragrant, in short 2–4-flowered racemes, salverform, 4–10 mm across, tepals 6, spreading, fused near base into a tube 2.5–3 mm long. **Berries** 5–6 mm in diameter, red when ripe. sGansu, w, nw & swSichuan, nwYunnan; forests, shrubberies, 2,800–3,500 m. VI–VII.

***Maianthemum henryi* (Baker) LaFrankie** (syn. *Oligobotrya henryi* Baker; *Smilacina henryi* (Baker) H. Hara) **Perennial** to 80 cm with erect **stems** that are usually stiffly hairy in upper part. **Leaves** 5–8, elliptic to ovate, to 22 × 11 cm, often with wavy margin, hairy or not, scarcely stalked. **Inflorescence** a many-flowered raceme or panicle. **Flowers** yellowish green to whitish, sweetly scented, tube 6–10 mm long with 6 spreading oval lobes not more than 3 mm long. **Berries** 7–9 mm in diameter, red when ripe. Gansu, Hubei, Sichuan, Shaanxi, seTibet, Yunnan [cChina, nMyanmar, Vietnam]; forests, shrubberies, shady banks, streamsides, 1,300–4,000 m. V–VI.

Maianthemum szechuanicum* (F. T. Wang & T. Tang) H. Li (syn. *Oligobotrya szechuanica* F. T. Wang & T. Tang; *Smilacina szechuanica* (F. T. Wang & T. Tang) H. Hara) Similar to the previous species but plant not more than 50 cm, with **leaves** not more than 11 × 5 cm; **flowers** white to purplish, with tube 3–4 mm long, equalling lobes. w & nwSichuan, neYunnan; coniferous forests, hillslopes, riversides, 2,000–3,600 m. V–VII.

***Maianthemum fuscum* (Wall.) LaFrankie** (syn. *Smilacina fusca* Wall.; *Tovaria fusca* (Wall.) Baker) Creeping **perennial** with erect **stems** to 50 cm, glabrous or hairy in upper part. **Leaves** up to 9, ovate to oblong with rounded or slightly heart-shaped base, narrowed at apex into abrupt point. **Inflorescence** a zigzagged panicle. **Flowers** with virtually free rose-pink tepals,

1. *Streptopus simplex*; Yulongxueshan, nwYunnan (HJ)
2. *Maianthemum bifolium*; Jiuzhaigou, nwSichuan (PC)
3. *Maianthemum lichiangense*; Habashan, nwYunnan (CGW)
4. *Maianthemum henryi*; near Geza, nwYunnan (CGW)
5. *Maianthemum henryi*; Kangding, wSichuan (PC)
6. *Maianthemum szechuanicum*; near Longriba, nwSichuan (CGW)
7. *Maianthemum tatsienense*; Tianchi, nwYunnan (PC)
8. *Maianthemum oleraceum*; Erlangshan, wSichuan (PC)

6–8 mm across. **Berries** 5–6 mm in diameter, red when ripe. seTibet, nwYunnan [Himalaya, Myanmar]; forests, shrubberies, 1,600–2,800 m. V–VII.

Maianthemum forrestii **(W. W. Sm.) LaFrankie*** (syn. *Smilacina forrestii* (W. W. Sm.) Hand.-Mazz.; *Tovaria forrestii* W. W. Sm.) Erect **perennial** to 80 cm, **stems** glabrous except in inflorescence, purplish. **Leaves** 6–9, papery thin, ovate to elliptic with abrupt apical point. **Inflorescence** a raceme or panicle. **Flowers** yellowish green with purple flush, 5–7 mm across, tepals with a very short tube and spreading, pointed lobes. nwYunnan; forests, 2,800–3,200 m. VI–VII.

Maianthemum tatsienense **(Franch.) LaFrankie** Spreading **perennial** with erect, glabrous **stems** to 80 cm, often half that height. **Leaves** ovate to lanceolate, to 7 × 6 cm, short-stalked. **Inflorescence** usually a panicle. **Flowers** greenish, sometimes flushed purple, 5–8 mm long, tepals fused at base. **Berries** 6–7 mm in diameter, red when ripe. Gansu, Guizhou, Hubei, Sichuan, seTibet, Yunnan [Hunan, Bhutan, India]; forests and forest margins, shrubberies, 1,500–3,500 m. V–VII.

Maianthemum oleraceum **(Baker) LaFrankie** (syn. *Smilacina oleracea* (Baker) Hook. f. & Thomson; *Tovaria oleracea* Baker) Rhizomatous plant to 80 cm with a zigzagged **stem** bearing 4–9 oblong to ovate-lanceolate **leaves**, 12–21 × 2–6 cm, sparsely downy beneath. **Inflorescence** a many-flowered panicle. **Flowers** white to reddish purple, 8–12 mm, tepals fused only at the very base, often minutely toothed towards apex; stigma 3-lobed. **Berries** 6–7 mm in diameter, red when ripe. Guizhou, Sichuan, seTibet, nwYunnan [Himalaya, nMyanmar]; forests, 2,100–3,300 m. V–VII.

CONVALLARIACEAE: Maianthemum

Maianthemum purpureum (Wall.) LaFrankie (syn. *Smilacina purpurea* Wall.; *S. zhongdianensis* H. Li & Y. Chen; *Tovaria purpurea* (Wall.) Baker) Spreading **perennial** with erect **stems** to 60 cm, hairy in upper part. **Leaves** 5–9, oblong to ovate, to 13 × 7 cm, hairy on veins beneath, scarcely stalked. **Inflorescence** usually a raceme, occasionally with 1–2 additional branches. **Flowers** white or purplish, 7–9 mm, with free tepals. **Berries** 7–9 mm in diameter, red when ripe. (?)swSichuan, e & seTibet, nwYunnan [Himalaya]; forests, shrubberies, 3,200–4,000 m. VI–VII.

Maianthemum atropurpureum (Franch.) LaFrankie* Similar to the preceding species, but **stems** obviously zigzagged; **inflorescence** a panicle; and **flowers** smaller with a definite tube 1–2 mm long. swSichuan, nwYunnan; forests, shrubberies, 1,400–3,000 m. V–VI.

OPHIOPOGON

A genus with c. 50 species (47 in China) distributed from the Himalaya to Indonesia and Japan. Stoloniferous **perennials**, often with tuberous roots, stemmed or with **leaves** in basal tufts. **Inflorescence** spike-like. **Flowers** campanulate, with 6 free tepals; ovary inferior. **Fruit** a berry.

Ophiopogon bodinieri H. Lévl.* Stoloniferous, tuberous-rooted plant with basal tufts of tough, grass-like **leaves** 20–40 × 1–4 mm, deep green with minutely toothed margin. **Inflorescence** spike-like, on scape to 35 cm long, few-many-flowered, with the **flowers** 1–2 in the 'spike'; tepals ovate to lanceolate, 4–6 mm long, varying in colour from white to yellowish or purple, sometimes red-tinged. **Berries** 5–6 mm in diameter, blue when ripe. Gansu, Guizhou, Hubei, Sichuan, Shaanxi, seTibet, Yunnan [swTibet, cChina, (?)Bhutan]; forests and forest margins, shrubberies, damp grassy places, ravines, 500–3,600 m. VI–VIII.

LIRIOPE

A genus of c. 8 species (6 in China, 3 endemic) in Vietnam, Japan, the Philippines and China. Similar to *Ophiopogon*, but **flowers** in clusters in the **inflorescence** and with a superior ovary. **Fruit** berry-like, bursting during early development to expose colourful **seeds**.

Liriope muscari (Decne.) L. H. Bailey Tufted, non-stoloniferous plant to 50 cm, with linear, leathery **leaves**, 15–65 × 0.5–2.5 cm, with 5–11 parallel veins. **Flowers** numerous, in a dense spike-like **inflorescence**; tepals elliptic, 3.5–4 mm long, purple or lilac-purple. **Berries** 6–7 mm in diameter, purplish black when ripe. Guangxi, Guizhou, Hubei, Sichuan [c & eChina, Taiwan, Japan]; forests, bamboo thickets, shrubberies, damp grassy banks, ravines, to 1,600 m. VII–IX.

Liriope graminifolia (L.) Baker In contrast to the previous species, a stoloniferous plant with linear **leaves** not more than 4 mm wide; **flowers** with white or purplish tepals; **berries** blue-black, 4–5 mm in diameter. Gansu, Guizhou, Hubei, Sichuan, Shaanxi [c & eChina, Taiwan]; habitats similar to those of *L. muscari* as well as rocky places, to 2,300 m. VI–VIII.

CLINTONIA

A genus of 5 species (1 in China) distributed in North America and north and east Asia. Rhizomatous **herbs** with basal, few-leaved tufts and **racemes** or **umbels** of flowers, sometimes solitary, usually on a leafless scape; tepals 6, free, stamens 6, joined at base of tepals. **Fruit** a berry.

Clintonia udensis Trautv. & C. A. Mey. Plant to 20 cm tall in **flower** but elongating in **fruit** to 60 cm. **Leaves** 3–5, oval to oblanceolate, stalked, to 8–25 × 3–16 cm, hairy on margin at first. **Racemes** with up to 12 flowers borne on densely hairy stalks; tepals oblong, 7–12 mm long, white, occasionally with bluish flush. **Berries** globose to egg-shaped, blackish blue when ripe, 7–12 mm. Gansu, Hubei, Sichuan, Shaanxi, seTibet, Yunnan [swTibet, n & neChina, Himalaya, Japan, Korea, Myanmar, seRussia]; open forests and forest margins, shrubberies, up to treeline, 1,600–4,000 m. V–VI.

---

1. *Maianthemum purpureum*; Tianchi, nwYunnan (PC)
2. *Ophiopogon bodinieri*; Yulongxueshan, nwYunnan (CGW)
3. *Ophiopogon bodinieri*; Baishui, Yulongxueshan, nwYunnan (RS)
4. *Liriope muscari*; cult. (CGW)
5. *Clintonia udensis*; Tianchi, nwYunnan (CGW)
6. *Disporum megalanthum*; cult.(CGW)
7. *Disporum calcaratum*; sYunnan (PC)

DISPORUM

A genus with c. 30 species (14 in China), distributed in east and South East Asia and North America. Rhizomatous **perennials** with simple, erect to regularly forked **stems** with sheathing basal scales and alternate **leaves**. **Flowers** in terminal clusters on short lateral shoots, campanulate; tepals 6. **Fruit** a berry, usually somewhat 3-lobed.

1. Inflorescence mostly pseudolateral (with an opposing leaf); flowers purplish .... ***D. cantoniense, D. megalanthum, D. calcaratum***
1. Inflorescence truly terminal; flowers white, cream or yellowish green 2
2. Tepals not more than 20 mm long; flowers in clusters of 2–8 .... ***D. longistylum, D. bodinieri***
2. Tepals 20–35 mm long; flowers not more than 3 in a cluster, or solitary ... ***D. uniflorum, D. acuminatissimum***

Disporum cantoniense (Lour.) Merr. (syn. *Streptopus chinensis* (Ker-Gawl.) Sm.; *Uvularia chinensis* Ker-Gawl.) Rhizomatous **perennial** with erect **stems** to 1 m, usually shorter, branched in upper part, bearing lanceolate to oblong-lanceolate **leaves**, 5–12 × 1–5 cm. **Inflorescence** drooping, 3–10-flowered, generally with short but distinct stalk. **Flowers** tubular-campanulate; tepals greenish or purplish, 15–25 mm long, with a short spur, 1–3 mm long, at base; stamens not protruding. **Berries** dark blue or black when ripe, 8–10 mm. Guangxi, Guizhou, Hubei, Sichuan, Shaanxi, seTibet, Yunnan [swTibet, c & eChina, Himalaya, India, Laos, Myanmar, Thailand, Vietnam]; forests, shrubberies, 700–3,000 m. V–VII.

Disporum megalanthum F. T. Wang & T. Tang* Readily distinguished from the previous species by longer and more distinct stalks carrying up to 8 tubular, semi-opening, relatively large **flowers**, 25–38 mm long, white, cream or deep purple. Gansu, Hubei, Sichuan, Shaanxi; similar habitats, as well as grassy slopes, 1,600–2,500 m. V–VII.

Disporum calcaratum D. Don (syn. *D. hamiltonianum* D. Don) Like *D. cantoniense*, but **leaves** elliptic to ovate-elliptic, to 8 × 5 cm, **tepals** purple, to pink, red or occasionally pure white, 12–20 mm long, each with a basal spur 3–5 mm long and sometimes slightly recurved. sYunnan [Himalaya, Thailand, Vietnam], similar habitats, 1,200–2,400 m. VI–VII.

Disporum longistylum (H. Lévl. & Vaniot) H. Hara* **Stems** to 90 cm, branched towards top, bearing lanceolate to elliptic **leaves**, 3–15 × 1–4 cm, usually with long-pointed tip. **Flowers** 10–17 mm long, in umbels of 2–8; tepals green to greenish yellow, very rarely purplish, oblanceolate to obovate, with spurs 1–1.5 m long at swollen base. **Berries** black when ripe, 6–9 mm in diameter. Gansu, Guizhou, Hubei, Sichuan, Shaanxi, seTibet, Yunnan; forests, rocky places, ravines, 400–1,800 m. III–VI.

Disporum bodinieri* (H. Lévl. & Vaniot) F. T. Wang & T. Tang (syn. *D. brachystemon* F. T. Wang & T. Tang) Similar to the previous species, but usually rough on leaf-margins and veins beneath; **flowers** white, yellowish green or rarely purplish, not more than 12 mm long. Guizhou, Sichuan, seTibet, Yunnan [Hunan]; habitats similar to those of *D. longistylum*, 1,200–3,000 m. V–VI.

***Disporum uniflorum* Baker ex S. Moore** **Stems** to 80 cm, often far less, simple or branched. **Leaves** elliptic to ovate, 4–9 × 1–6.5 cm, glabrous, with short stalk and pointed or somewhat long-pointed tip. **Flowers** tubular-campanulate, 20–30 mm long, tepals glabrous, oblanceolate to obovate, yellow, with a spur 1–2 mm long at swollen base. **Berries** blue-black when ripe, 8–10 mm in diameter. Hubei, Sichuan, Shaanxi [c & neChina, Korea]; forests, ravines, to 2,500 m. III–VI.

Disporum acuminatissimum* W. L. Sha Similar to the previous species but **flowers** in clusters of up to 4, narrow-tubular, 20–35 mm long, tepals minutely downy all over. c & neGuangxi; similar habitats, 600–1,400 m. IV–V.

REINECKEA

A genus with a single species confined to China and Japan.

***Reineckea carnea* (Andrews) Kunth** Tufted, rhizomatous **perennial** to 30 cm, with deep, leathery, lanceolate to oblanceolate **leaves** to 40 × 3.5 cm, usually far smaller, glabrous. **Flowers** rather fleshy, in **spikes** on a scape 5–15 cm long, with male flowers sometimes above female at top of spike; tepals pink or rose-pink, oblong, 5–7 mm long, strongly reflexed, fused together into a tube of similar length. **Berries** red when ripe, 6–10 mm. Guangxi, Guizhou, Hubei, Sichuan, Shaanxi, Yunnan [c & eChina, Japan]; damp, shady forests, shady slopes, ravines, to 3,200 m. VII–XI.

TRILLIACEAE

PARIS

A temperate Old World genus of 24 species (22 in China, 12 endemic). Rhizomatous, sometimes patch-forming **perennials** with simple erect **stems** carrying a single whorl of 3–4 **leaves**. **Flowers** solitary; tepals in 2 whorls of 3–8, occasionally more, outer ovate to lanceolate, inner (in essence the petals) thread-like, sometimes absent; stamens 8–24, occasionally more; anther-connective extended (sometimes considerably) beyond the anther as a slender appendage. **Fruit** a berry or berry-like capsule usually containing red or orange **seeds**. Many species were formerly placed in the genus *Daiswa*.

1. Plant finely papillose . . . . *P. mairei*
1. Plant glabrous 2
2. Leaves variegated; plants small . . . . *P. marmorata, P. luquanensis*
2. Leaves plain above (if mottled then plant exceeding 30 cm) 3
3. Anther-appendage 6 mm long or more . . . . *P. thibetica, P. undulata*
3. Anther-appendage not more than 4 mm long 4
4. Ovary with 4–6 cells; fruit indehiscent . . . . . *P. forrestii*
4. Ovary 1-celled; fruit dehiscing irregularly 5
5. Inner tepals 4–8 cm long, always yellowish green . . . . *P. polyphylla, P. polyphylla* var. *alba*, *P. polyphylla* var. *chinensis, P. polyphylla* var. *minor, P. cronquistii, P. daliensis*
5. Inner tepals 1–4 cm long, generally dark purple or purple-black, rarely deep yellowish green . . . . *P. delavayi, P. fargesii*

Paris mairei* H. Lévl. (syn. *Daiswa violacea* (H. Lévl.) A. Takht.; *Paris violacea* H. Lévl.) Plant to 1 m, **stem** with a whorl of 5–9 deep green, oblong to oblanceolate **leaves** that are paler along veins, with a short, stout stalk. **Flowers** green, 6–13 cm in diameter, with 5–8 oblong-lanceolate outer tepals and a similar number of often longer, linear, thread-like inner tepals; stamens 10–18; ovary purplish red. Guizhou, wSichuan, nYunnan; forests, shrubberies, grassy slopes, 1,800–3,500 m. V–VII.

***Paris marmorata* Stearn** Small patch-forming plant not more than 15 cm. **Stems** with a whorl of 5–6 elliptic to lanceolate, bluish green **leaves** with whitish veins and midrib, purplish beneath. **Flowers** 5–7 cm in diameter, with 3–4 outer tepals coloured like the leaves, inner tepals shorter, linear, ascending. seTibet, w & nwYunnan [swTibet, eHimalaya, nMyanmar]; damp mossy, generally broadleaved forests, 2,400–2,800 m. V–VI.

Paris luquanensis* H. Li Very similar to the previous species but outer tepals equal in number to leaves (usually 4–5) with inner tepals longer than outer; **leaves** obovate to almost rhombic, stalkless or almost so, 3.2–5 × 2–3.7 cm, deep velvety green with pale green veins, purple beneath; outer tepals 8–9 mm long, deep green with a purple reverse, inner tepals thread-like, much longer than outer, 15–19 mm long. ncYunnan (Luquan Xian); forests, thickets, 2,100–2,800 m. V–VI.

***Paris thibetica* Franch.** (syn. *P. thibetica* var. *apetala* Hand.-Mazz.) Plant 35–90 cm tall with a whorl of 7–12 narrow-lanceolate bluish green **leaves** with very short stalks. **Flowers** 6–10 cm in diameter, with 4–6 green, narrow-lanceolate, long-pointed outer tepals, inner tepals spreading, linear, 4–6 or absent; stamens 8–10, anther-appendage 6–27 mm long; ovary green, scarcely ribbed. sGansu, Guizhou, w, nw & swSichuan, (?)seQinghai, seTibet, nwYunnan [swTibet, c & eHimalaya, nMyanmar]; forests, dense shrubberies, 1,400–3,800 m. IV–VII.

TRILLIACEAE: Paris 563

Paris undulata H. Li & V. G. Soukup* Similar to the preceding species but **leaves** 7–9, more oblong; **flowers** somewhat larger with pendulous, linear, inner tepals; and anther-appendage 11–15 mm long. Ovary with prominent purple stigma. sSichuan (Emeishan); habitats and flowering time similar to those of *P. thibetica*.

Paris forrestii (Takht.) H. Li Plant to 1 m, though often only 20–60 cm. **Stems** with a whorl of 6–8 oblong to ovate **leaves** that are rounded or heart-shaped at base and clearly stalked, inner tepals generally deep purplish black. **Flowers** 3.5–9 cm in diameter, with 4–6 green, ovate to lanceolate outer tepals and longer, greenish yellow, linear, thread-like, spreading inner ones; stamens 8–12. Ovary purple. seTibet, Yunnan [nMyanmar]; coniferous and broadleaved forests, 1,900–3,500 m. V–VII.

Paris polyphylla Sm. Very variable, often tufted plant with erect **stems** 20–130 cm tall, bearing a whorl of 5–12, occasionally more, oblong to lanceolate **leaves** that have a wedge-shaped to rounded base. **Flowers** green or greenish yellow, 9–15 cm in diameter, sometimes more, with 4–7 ovate to lanceolate outer tepals and a similar number of inner that are ascending to spreading, linear, shorter or longer than outer. Ovary clearly ribbed, purplish or greenish but usually with a purplish stigma. Gansu, Hubei, Sichuan, Shaanxi, e &

1. *Disporum bodinieri*; cult. (CGW)
2. *Disporum acuminatissimum*; Fanjingshan, neGuizhou (PC)
3. *Paris marmorata*; Dochu La, Bhutan (CGW)
4. *Paris marmorata*; cult. (DL)
5. *Paris luquanensis*; cult. (RR)
6. *Paris thibetica*; cult. (CGW)
7. *Paris undulata*; Emeishan, sSichuan (CGW)
8. *Paris forrestii*; Baishui, Yulongxueshan, nwYunnan (PC)

1. *Paris polyphylla*; Cangshan, nwYunnan (CGW)
2. *Paris polyphylla*; cult., from Yunnan (CGW)
3. *Paris polyphylla* in fruit; Tianbao, Zhongdian, nwYunnan (PC)
4. *Paris polyphylla* var. *alba*; cult. (CGW)
5. *Paris cronquistii*; cult. (CGW)
6. *Paris daliensis*; Cangshan, nwYunnan (CGW)
7. *Trillium tschonoskii*; Phephe La, Bhutan (CGW)
8. *Allium trifurcatum*; above Geza, nwYunnan (CGW)

seTibet, w & nwYunnan [swTibet, cChina, Himalaya, nMyanmar]; broadleaved and coniferous forests, shrubberies, bamboo thickets, rocky and grassy slopes, stream- and pathsides, 800–3,500 m. III–VII. Several variants are recognised in the region and these include: var. *alba* H. Li & R. J. Mitchell*, 25–50 cm tall, usually with 6–8 **leaves**, styles and upper half of ovary white. Guizhou, Hubei, w & nwYunnan, 1,500–2,900 m; var. *chinensis* (Franch.) H. Hara forms robust plants 40–130 cm tall, **flowers** with inner tepals clearly shorter than outer, filaments 5–6 mm long, and anther-appendage not more than 2 mm long. Guangxi, Guizhou, Hubei, Sichuan, Yunnan [c & eChina, seAsia, nMyanmar, Taiwan], 2,800–3,000 m; var. *minor* S. F. Wang*, a dwarf plant not exceeding 15 cm with filaments 1–2 mm long, and anther-appendage 5–6 mm long. w & swSichuan, 1,500–2,500 m.

Paris cronquistii (Takht.) H. Li* Similar in general appearance to the previous species, but **stem** rough and bearing 4–6 ovate **leaves**, 11–17 × 6–11 cm, purple or purple-mottled beneath, sometimes with purple markings above; outer tepals 5–6, green, to 9 × 2 cm, inner tepals yellowish green, linear, (3–)4–8 cm long. swGuangxi, Guizhou, Sichuan, seYunnan; similar habitats, 900–2,100 m. V–VI.

Paris daliensis H. Li & V. G. Soukup* Distinguished from the previous species by the inner **tepals** that are somewhat thickened towards the end, by the very short (1 mm) filaments, and by the greenish yellow to deep purple stamens and ovary. nwYunnan (Cangshan); forests, glades, c. 2,600 m. V–VI.

Paris delavayi Franch. Fairly robust plant 0.5–1.5 m tall, **stem** with a whorl of 6–8 lanceolate to oblong **leaves** that are clearly stalked; stalks to 1.5 cm long. **Flowers** relatively small, 4–7 cm in diameter, with 4–5 narrow-lanceolate, greenish or greenish purple outer tepals that are usually recurved, and linear, thread-like inner tepals shorter than outer. Stamens usually 8 or 10, with a pointed anther-appendage 1–4 mm long. Guizhou, Hubei, Sichuan, Yunnan [seChina, Vietnam]; broadleaved forests, shrubberies, bamboo thickets, 1,300–2,000 m. IV–V.

Paris fargesii Franch. Similar to the preceding species, but **leaves** generally only 4–6, broadly ovate to oblong, but with stalks 2–6 cm long, outer **tepals** broader and not recurving, inner yellowish green to purple-black, and anther-appendage conical, not more than 2 mm long. Guangxi, Guizhou, Hubei, Sichuan, Yunnan [s & seChina, Vietnam]; similar habitats, 500–2,100 m. IV–VI.

TRILLIUM

A genus of almost 50 species (4 in China) mostly in North America. Rhizomatous **perennials** with an erect **stem** bearing a single terminal whorl of 3 **leaves**. **Flowers** solitary, with 6 tepals in 2 whorls; stamens 6. **Fruit** a rounded berry.

Trillium tschonoskii Maxim. Tufted plant to 50 cm, generally less, **stem** with a single whorl of 3 diamond-shaped to ovate **leaves** to 15 × 15 cm. **Flowers** solitary, 3–4 cm wide, semi-nodding to ascending; sepals green, lanceolate, 15–20 mm long; petals white or pinkish purple, ovate-lanceolate, 15–22 mm long, recurving in upper half. Gansu, Hubei, Sichuan, Shaanxi, seTibet, w & nwYunnan [swTibet, c & eHimalaya, nMyanmar, Japan, Korea]; very local in forests, ravines, rocky places, generally in damp shady habitats, 1,000–3,200 m. IV–VI.

ALLIACEAE

ALLIUM

A genus with c. 700 species (138 in China) confined to the northern hemisphere. Bulbous and rhizomatous plants smelling strongly of onion or garlic when crushed. **Leaves** basal or sheathing the lower part of scapes, cylindrical to flat, linear to elliptic or lanceolate, entire. **Flowers** in lax to dense **umbels**, often many-flowered, star- to cup-shaped or campanulate; tepals 6 in 2 rows; stamens 6, filaments linear or appendaged. **Fruit** a 3-parted capsule. This genus contains important culinary and medicinal species that include the leek (*Allium porrum*), onion (*A. cepa*) and garlic (*A. sativum*).

SYNOPSIS OF GROUPS

Group One. Flowers yellow, greenish-yellow or white. (below)

Group Two. Flowers blue or violet-blue. (p. 566)

Group Three. Flowers red, reddish- purple, purple or purplish-black. (p. 566)

Group One

1. Leaves 15 mm wide or more .... **A. nanodes**
1. Leaves not more than 10 mm wide 2
2. Flowers white; style 3-cleft .... **A. trifurcatum, A. humile, A. fasciculatum**
2. Flowers yellow or greenish yellow, occasionally white; style dot-like .... **A. chrysanthum, A. hookeri**

Allium nanodes Airy Shaw* Solitary or clustered bulbous **perennial** not more than 6 cm tall, with a pair of narrow-oblong opposite **leaves** that are usually flushed with purple and recurved, 15–30 mm wide. **Umbels** few-flowered. **Flowers** campanulate, white with reddish flush, 5–8 mm long; stamens slightly protruding. swSichuan, nwYunnan; meadows, stony and scrubby slopes, screes, 3,300–5,200 m. VI–VIII.

Allium trifurcatum (F. T. Wang & T. Tang) J. M. Xu* (syn. *A. humile* Kunth var. *trifurcatum* F. T. Wang & T. Tang) Tufted bulbous **perennial** to 30 cm, often shorter, with 4–7 linear **leaves**, 4–10 mm wide, with a distinct midvein, sheathing the base of the scape. **Umbels** 1-sided, with spreading to pendent, funnel-shaped. **Flowers** 6–8 mm long; stamens not protruding. swSichuan, nwYunnan; damp, often marshy places in forests or forest clearings, streamsides, damp slopes, 3,000–4,000 m. V–VIII.

Allium humile **Kunth** (syn. *A. gowanianum* Wall. ex Baker; *A. nivale* Jacq. ex Hook. f. & Thomson) Similar to the previous species but plants often solitary, not tufted, not more than 15 cm, with flat **leaves** 4–7 mm wide. **Umbels** hemispherical. **Flowers** 7–8 mm long, ascending to upright, tepals white with green midvein. seTibet, nwYunnan [Himalaya]; rocky and grassy slopes, 4,000–4,500 m. VI–VII.

Allium fasciculatum **Rendle** Readily distinguished from the previous two species in having **leaves** 2–5 mm wide and globose, erect, dense-flowered **umbels**; stamens protruding. seQinghai, w & nwSichuan, e & seTibet [Himalaya]; meadows, dry sandy places, 2,200–5,400 m. VII–IX.

Allium chrysanthum **Regel*** Slender solitary or slightly tufted bulbous **perennial** to 45 cm tall, often less, with cylindrical **leaves** 1.5–4 mm wide, shorter than and sheathing the scape at its base. **Umbels** dense, globose, many-flowered. **Flowers** campanulate, pale to mid-yellow, 5–7 mm long; stamens equalling tepals; style protruding. Gansu, wHubei, w, nw & swSichuan, sShaanxi, seTibet, nwYunnan; grassy and stony places, rocky slopes, open scrub, 2,000–4,500 m. VII–IX.

Allium hookeri **Thwaites** A more clustered plant than the previous species, to 60 cm with linear, non-cylindrical **leaves** 5–10 mm wide, scape with 1 or no leaf-sheaths. **Flowers** in a dense globose **umbel**, white to greenish yellow or yellow, 4–7.5 mm long. swSichuan, seTibet, nwYunnan [Himalaya, Myanmar, Sri Lanka]; forests and forest margins, meadows, damp stony places, 1,400–4,200 m. VII–X.

Group Two
1. Tepals 4–7 mm long; spathe usually 2-valved, occasionally 1-valved .... ***A. cyaneum***
1. Tepals 6–13 mm long; spathe 1-valved .... ***A. beesianum, A. sikkimense***

Allium cyaneum **Regel*** Tufted bulbous **perennial** to 35 cm with semi-cylindrical **leaves**, 1.5–2.5 mm wide, equalling or shorter than scapes and sheathing them at the base. **Umbel** many-flowered, borne on scapes arched at the top. **Flowers** campanulate, 4–6.5 mm long, with the violet or blue tepals; stamens protruding; inner filaments usually with a pair of lateral teeth at base. Gansu, wHubei, Qinghai, n & wSichuan, Shaanxi, eTibet [Ningxia]; forest margins, shrubby slopes, meadows, 2,100–5,000 m. VIII–X.

Allium beesianum **W. W. Sm.*** Tufted bulbous **perennial** to 40 cm with linear **leaves**, 3–8 mm wide, sheathing lower part of scape. **Umbels** few-flowered. **Flowers** blue, nodding, tubular-campanulate, 11–14 mm long; stamens not protruding, inner ones with filaments broadened at base. swSichuan, nwYunnan; stony and grassy meadows, open low shrubberies, 3,000–4,200 m. VIII–X.

Allium sikkimense **Baker** Similar to the previous species but **leaves** not more than 5 mm wide, flat, and **umbels** many-flowered. **Flowers** deep blue to purple-blue, campanulate, 6–10 mm long; inner filaments with a pair of small basal teeth. sGansu, e & sQinghai, w, nw, & swSichuan, Shaanxi, nwYunnan [sNingxia, Himalaya, Myanmar]; forest margins, scrub, alpine meadows, rock-crevices, sometimes on old walls, 2,400–5,000 m. VII–IX.

Group Three
1. Leaves not more than 3 mm wide .... ***A. mairei, A. forrestii, A. tanguticum, A. przewalskianum***
1. Leaves 4–40 mm wide 2
2. Flowers star-shaped .... ***A. wallichii, A. prattii***
2. Flowers urn-shaped or campanulate .. ***A. cyathophorum, A. cyathophorum*** var. ***farreri, A. macranthum***

Allium mairei **H. Lévl.*** Clustered bulbous **perennial** to 30 cm with linear, often semi-rounded, somewhat angled **leaves**, 1–1.5 mm wide, sheathing the purple-tinged scape at its base. **Umbels** often 2 with a basal bract, or solitary. **Flowers** campanulate, pale red to purple-red, sometimes spotted, 8–12 mm long; stamens not protruding; spathes 1-valved. swSichuan, seTibet, w & nwYunnan; forests and forest margins, shrubberies, meadows, rocky places, 1,200–4,200 m. VIII–X.

Allium forrestii **Diels*** Distinguished from the previous species by the broader **leaves**, 1.5–3 mm wide, purple-red scapes and fewer-flowered **umbels**; and **flowers** purple, often very dark. swSichuan, seTibet, nwYunnan; meadows, stony places, 2,700–4,200 m. VIII–X.

Allium tanguticum **Regel*** Differs from *A. mairei* by being solitary and somewhat taller, **leaves** 1–4 mm wide, flat and somewhat channelled. **Umbels** with numerous campanulate **flowers** 4–5 mm long, tepals with a darker midrib; stamens protruding. Gansu, s & seQinghai, neTibet; dry slopes, sandy places, 2,000–3,500 m. VII–IX.

Allium przewalskianum **Regel** (syn. *A. jacquemontii* Regel) Differs from *A. mairei* by the 4–5-angled **leaves**, and the dense **umbels** of pale red to deep purple **flowers**, each 3–6 mm long. Gansu, Qinghai, Sichuan, Shaanxi, Tibet, nwYunnan [nwChina, Himalaya]; dry rocky slopes, scrub, 2,000–4,800 m. VI–IX.

Allium wallichii **Kunth** (syn. *A. bulleyanum* Diels; *Nothoscordum mairei* H. Lévl.) An often rather stout bulbous **perennial**, solitary or clustered, to 50 cm, sometimes taller, with linear to lanceolate **leaves**, 5–20 mm wide, often narrowed at base into a stalk. Scape triangular in cross-section, often slightly winged on the angles, bearing a many-flowered, hemispherical **umbel**

1. *Allium fasciculatum*; cult. (CGW)
2. *Allium chrysanthum*; Baimashan, nwYunnan (HJ)
3. *Allium cyaneum*; cult. (CGW)
4. *Allium beesianum*; cult. (CGW)
5. *Allium sikkimense*; cult. (CGW)
6. *Allium przewalskianum*; cult. (CGW)
7. *Allium wallichii*; Cangshan, nwYunnan (CGW)
8. *Allium prattii*; Balangshan, Wolong, wSichuan (CGW)
9. *Allium cyathophorum*; cult. (CGW)

of star-shaped, erect to ascending, red, purple or blackish purple **flowers**, each c. 10–18 mm across; spathes 1–2-valved. Guizhou, swSichuan, seTibet, nwYunnan [swTibet, sHunan, Himalaya]; forest margins, open shrubberies, damp meadows, streamsides, 2,300–4,800 m. VII–X.

Allium prattii C. H. Wright ex Hemsl. (syn. *A. victorialis* L. var. *angustifolium* Hook. f.) Readily distinguished from the previous species by the paired, occasionally 3, linear-lanceolate to elliptic **leaves**, 5–40 mm wide, and by the rounded scape. **Flowers** smaller, 7–12 mm across, pale red to purple-red. Gansu, Qinghai, w, nw & swSichuan, Shaanxi, seTibet, w & nwYunnan [swTibet, Anhui, Henan, eHimalaya, nMyanmar]; forests and forest margins, open shrubberies, meadows, streamsides, 2,000–4,900 m. VI–IX. (See photo 8 on p. 567.)

Allium cyathophorum Bur. & Franch.* Tufted bulbous **perennial** to 25 cm with linear **leaves**, 2–5 mm wide, sheathing the rounded scapes at the base. **Flowers** purple to maroon-purple, in a lax 1-sided **umbel**, spreading to pendent, urn-shaped to campanulate, 7–9 mm long, tepals blunt; stamens not projecting; spathes often 1-valved. sGansu, s & seQinghai, w, nw & swSichuan, eTibet, nwYunnan; meadows, rocky places, rock-crevices, 2,700–4,600 m. VI–VIII. **Var.** *farreri* **(Stearn) Stearn*** (syn. *A. farreri* Stearn) has long-pointed **tepals** and inner filaments with a distinctive triangular base. (See photo 9 on p. 567)

Allium macranthum Baker A more robust plant than the previous species, to 60 cm; **leaves** 4–10 mm wide, somewhat channelled and with a distinct midvein. **Flowers** in rather lax **umbels**, broadly campanulate, purple to reddish purple, 9–11 mm long. swGansu, swSichuan, sShaanxi, seTibet, nwYunnan [swTibet, c & eHimalaya]; damp grassy places, meadows, streamsides, 2,700–4,200 m. VIII–X.

HEMEROCALLIDACEAE

HEMEROCALLIS

A genus of c. 15 species (11 in China), primarily in China and Japan. Tuberous-rooted **perennials** with strap-like **leaves** borne mainly in basal tufts. **Flowers** several in umbel-like clusters on stiff, bracted **stems**, trumpet-shaped, lily-like; tepals 6, ± similar; stamens 6; ovary superior. **Fruit** a 3-parted capsule.

1. *Hemerocallis fulva*; Cangshan, nwYunnan (PC)
2. *Hemerocallis fulva*; Cangshan, nwYunnan (PC)
3. *Hemerocallis forrestii*; Gangheba, Yulongxueshan, nwYunnan (CGW)
4. *Lycoris aurea*; Lijiang (Li river), Guilin, neGuangxi (CGW)
5. *Zephyranthes carinata*; Maguan, seYunnan (PC)
6. *Tacca chantrieri*; Fanjingshan, neGuizhou (PC)
7. *Tacca chantrieri*; cult. (CGW)

1. Flowers with reddish or purplish brown markings in centre; anthers purplish black . . . . **H. fulva**
1. Flowers uniformly coloured; anthers yellow or orange . . . . **H. forrestii, H. plicata**

Hemerocallis fulva L. Robust tufted **perennial** to 1.5 m, often less, with linear, pointed **leaves** 10–28 mm wide. Scapes equalling or longer than leaves, hollow, each **cyme** usually bearing 2–5 flowers; **tepals** orange to orange-red, zoned in the centre, with wavy margins, 5–12 mm long; anthers 7–8 mm long. Guangxi, Guizhou, Hubei, Sichuan, Shaanxi, seTibet, Yunnan [c & eChina, India, Japan, Korea, eRussia]; forest margins, grassy places, shrubberies, streamsides, ravines, 300–2,500 m. VI–XI.

Hemerocallis forrestii Diels* Tufted **perennial**, sometimes with a single **shoot**, 40–70 cm tall, with linear, tapered **leaves** 10–20 mm wide. Scape slender, hollow, slightly shorter than leaves, bearing 2–4-flowered **cymes**; **tepals** orange or golden yellow, 5–9 cm long, perianth-tube short, c. 1 cm long; anthers yellow, sometimes with blackish markings, 6–7 mm long. swSichuan, nwYunnan; open forests, forest margins, grassy and rocky places, cliffs, 2,300–3,200. V–VII.

Hemerocallis plicata Stapf* Similar to the preceding species but **leaves** only 6–8 mm wide, often folded lengthwise and scape solid or hollow; **tepals** orange, 4–5.5 cm long; perianth-tube 1.5–2.5 cm long; anthers 3–4 mm long. s & wSichuan, n & nwYunnan; habitats similar to those of *H. forrestii* but especially in *Pinus* forests, 1,500–3,200 m. VII–IX.

AMARYLLIDACEAE

LYCORIS

A genus of 20 species (15 in China) from Pakistan, Myanmar and South East Asia to Korea, Japan and Indonesia. Bulbous **perennials** with linear or strap-shaped basal **leaves** and scapes of *Nerine*-like flowers, generally borne before the leaves; **flowers** in an **umbel**, zygomorphic, funnel- or trumpet-shaped, with a short perianth-tube; tepals 6; stamens 6; ovary inferior. **Fruit** a 3-parted, many-seeded capsule.

Lycoris aurea (L'Hér.) Herb. Plant with **leaves** appearing in autumn, strap-shaped, to 60 × 2.5 cm, glabrous, with pale midvein. **Scapes** to 60 cm, bearing a pair of bracts (spathes) and up to 9 flowers; tepals egg-yolk-yellow, often with a pale green midvein, perianth-tube 12–15 mm long; tepals oblanceolate, to 60 × 10 mm, strongly recurved; stamens protruding, filaments 7–12 cm long; style-tip rose-red. Gansu, Guangxi, Guizhou, Hubei, s, e & seSichuan, Shaanxi, c, s & seYunnan [seAsia]; rocky places, cliff-ledges and -crevices, often in semi-shady damp places, 500–2,300 m. VIII–IX.

Lycoris chinensis Traub* In contrast to the previous species, the **leaves** appear in spring; **flowers** with a longer (17–25 mm) perianth-tube, tepals with a pale yellow midvein and strongly wavy margin. c, e & sSichuan, Shaanxi [Henan, Jiangsu, Zhejiang]; damp rocky places, usually in forests, 600–1,800 m. VII–VIII.

ZEPHYRANTHES

A tropical American genus of 40 species (2 naturalised in China).

Zephyranthes carinata Herb. (syn. *Z. grandiflora* Lindl.) Clump-forming **bulb** to 20 cm tall in flower, with all-basal, linear, rather flat, fleshy **leaves** to 30 × 0.8 cm. **Flowers** solitary on an erect green scape; spathes purplish, 40–50 mm long; corolla pale pink to rose-pink, trumpet-shaped, with a tube 10–25 mm long and spreading obovate tepals to 60 mm long. **Fruit** a 3-parted, many-seeded capsule. Guangxi, Guizhou, Sichuan, Yunnan; naturalised in and around dwellings or on cultivated land (native to Mexico). VI–IX.

TACCACEAE

TACCA

A genus of c. 11 species (4 in China) in South East Asia and Malaysia. Tufted, rhizomatous, evergreen **perennials** with long-stalked, entire to pinnately lobed **leaves**. **Inflorescence** terminal, scapose, with a cluster of large and conspicuous involucral bracts surrounding the flowers, accompanied by very long, whisker-like, filamentous floral bracts. **Flowers** hermaphrodite with 6 tepals in 2 whorls, and 6 stamens; stigma 3-lobed. **Fruit** a berry.

Tacca chantrieri André (syn. *T. esquirolii* (H. Lévl.) Rehd.) Tufted plant to 70 cm tall in flower with dark green, oblong to elliptic **leaves**, 20–60 × 7–20 cm, apex drawn out into a long point. **Inflorescence** with an umbel of up to 18 (usually 5–9) **flowers**, involucral bracts usually 4, large, dark purple to purplish black, ovate to ovate-lanceolate, inner pair much broader than outer, floral bracts of similar colour, often darkening with age, thread-like, to 25 cm long; tepals olive-green to purplish brown or purplish black. **Berries** to 40 × 12 mm, shiny green, ripening red or purple. Guangxi, Guizhou, seTibet, Yunnan [seAsia, India, Sri Lanka, Myanmar]; forests, ravines, river- and streamsides, 200–1,300 m. IV–XI.

Tacca integrifolia Ker-Gawl. Very similar to the previous species, but plant to 1 m tall in flower; **leaves** often more oblong and grey-green; inner 2 involucral bracts long-stalked, green, purple or whitish, often erect, outer 2 smaller and usually part-hidden by the flowers; **berries** rather larger, green to blackish purple. seTibet [seAsia, eHimalaya, eIndia, Sri Lanka]; similar habitats, 800–900 m. VII–VIII.

ORCHIDACEAE

A very large family with c. 900 genera and 25,000 species (c. 1,300 species in China, mainly in the subtropical and tropical south). Terrestrial, rock-dwelling or epiphytic **herbs**. Tubers or rhizomes present in terrestrial species, short to long rhizomes in rock-dwellers or epiphytes, while in others the **stems** may be short to elongated, cylindrical or swollen, forming **pseudobulbs** in some species. **Leaves** simple, 1–many, pleated or leathery. **Inflorescence** 1–many-flowered, racemose or occasionally branched, lateral or terminal. **Flowers** often showy, scented or not, usually resupinate; sepals and petals usually ± similar, free to the base, lateral sepals sometimes joined to each other or to column-foot (then forming a chin-like mentum); lip jointed or fixed to base of column or apex of column-foot, entire, variously lobed or 2-parted, sometimes spurred or pouched at base, often with a warty or papillose callus on upper surface; column a short to elongated fleshy stalk, sometimes with a distinct foot, anther-cells parallel or divergent; pollen-masses 2, 4 or 8, sometimes attached by a stalk to a sticky disc; stigma sticky, concave or 2-lobed and forward-projecting. **Fruit** a dry dehiscent capsule. **Seeds** numerous, minute and dust-like.

Many of the species (especially in genera such as *Cypripedium* and *Paphiopedilum*) are very vulnerable in the wild because of both commercial and private collectors and land development.

CYPRIPEDIUM

A north temperate genus of c. 50 species (30 in China, the majority endemic). Terrestrial, with **stems** clustered or not, very short to elongated, leafy. **Leaves** 1–several. **Inflorescence** terminal, 1–4-flowered; dorsal sepal hood-like over the lip; lateral sepals usually united ± to the tips; petals spreading or curving forwards; lip prominent, slipper- or urn-shaped, with or without incurved margins; column with 2 anthers and a sterile anther staminode at apex that is usually shield-shaped.

Synopsis of groups

Group One. Plants clump-forming or solitary with at least 3 erect to ascending, non-spotted leaves; petals ± spreading; flower-stalks not elongating. (below)

Group Two. Dwarf plants with a pair of ascending to wide-spreading true leaves and solitary flowers subtended by a bract; petals spreading or incurved but not enclosing lip; flower-stalks not elongating. (p. 573)

Group Three. Colony-forming plants with apparently 2 similar leaves (the upper in fact a bract), often black-spotted, and with solitary flowers borne on a short stalk between leaf and bract, with petals curved forwards and enclosing the sides of the lip; flower-stalks much elongating as fruit develops. (p. 574)

Group One

1. Leaves and ovary glabrous to hairy, not glandular ... 2
1. Leaves and ovary hairy, the latter glandular-hairy ... 4
2. Flowers white, lemon- to buttercup-yellow; petals not tapering noticeably; leaves and ovary hairy .... ***C. flavum, C. wardii***
2. Flowers purple to dark maroon, occasionally whitish; petals tapering; leaves and ovary glabrous or sparsely hairy, never glandular 3
3. Flowers small, lip not more than 3.5 cm long .... ***C. yunnanense, C. himalaicum***
3. Flowers larger, lip at least 3.5 cm long .... ***C. tibeticum, C. franchetii, C. calcicola***
4. Lip white, obconic, pointed, hairy; sepals and petals maroon, tapering, lateral sepals free ... ***C. plectrochilum***
4. Lip not as above; sepals and petals not maroon, lateral sepals united 5
5. Flowers solitary, medium to large, lip pale yellow 2.4–7.3 cm long .... ***C. fasciolatum, C. farreri***
5. Flowers 1–4, small, lip green or bronze not more than 2.7 cm long .... ***C. henryi, C. shanxiense***

Cypripedium flavum P. F. Hunt & Summerh.* Very variable erect plant 30–60 cm tall with 3–6, ovate to elliptic downy **leaves**. **Flower** 3–5 cm across, pale yellow to butter-yellow, sometimes suffused pink or red on dorsal sepal and petals and sometimes spotted maroon on lip; petals recurved in upper part; lip rounded to oval in outline, 2.9–4.7 cm long. sGansu, wHubei, w, sw, n & nwSichuan, seTibet, nwYunnan; light shade in mixed or deciduous woodland on limestone, mossy boulders, often gregarious, 1,800–3,700 m. VI–VII.

Cypripedium wardii Rolfe* Smaller plant than the previous species, to just 10–15 cm tall, with 3 ovate to elliptic **leaves**; **flower** solitary, 2–2.5 cm across, white, spotted with maroon on lip. seTibet, wYunnan; similar habitats, 2,500–3,500 m. VI–VII.

Cypripedium yunnanense Franch.* Very closely allied to *C. franchetii* and *C. tibeticum* but plant only 20–37 cm tall; **leaves** 6–14 cm long; **flower** much smaller and generally paler, 3.5–4 cm across, usually whitish to purple, lip 2.2–3.5 cm long with a white rim to the orifice; ovary sparsely to rather hairy. w & swSichuan, seTibet, nwYunnan; rocky slopes, edges of mixed woods, shrubberies, 2,700–3,800 m. VI–VII.

Cypripedium himalaicum Rolfe Plants to 40 cm with elliptic **leaves**. **Flower** solitary, very fragrant, deep purple; lip glossy, obovoid, rather bulbous, 2.8–3.4 cm long, with a somewhat fluted margin and white rim to the orifice; ovary hairy. seTibet [swTibet, c & eHimalaya]; grassland, coniferous woodland, 2,800–3,200 m. VI.

1. *Cypripedium flavum*, widespread in western China, often forms extensive colonies, especially in mossy woodland; photographed near Juizhaigou, nwSichuan (CGW)
2. *Cypripedium flavum*; Tianchi, nwYunnan (CGW)
3. *Cypripedium flavum*; Zhongdian plateau, nwYunnan (HJ)
4. *Cypripedium flavum*; Zhongdian plateau, nwYunnan (HJ)
5. *Cypripedium flavum*; Zhongdian plateau, nwYunnan (HJ)
6. *Cypripedium wardii*; near Luding, wSichuan (HP)
7. *Cypripedium yunnanense*; Zhongdian, nwYunnan (PC)
8. *Cypripedium yunnanense*; Zhongdian, nwYunnan (PC)
9. *Cypripedium himalaicum*; Bhutan (MWA)

Cypripedium tibeticum **King ex Rolfe** (syn. *C. corrugatum* Franch.) Plants 20–70 cm tall, solitary or clumped. **Leaves** often 6, elliptic, 8–16 cm long, sparsely hairy only towards tip, lowermost sheath-like. **Flower** large, rarely paired, 7–10 cm across, deep purple or blackish maroon, sometimes with straw-coloured sepals and petals and dark veining or flushing; petals usually more than 12 mm broad at widest point; lip 3.5–6 cm long, surface often corrugated, normally with a white margin to the orifice; ovary usually glabrous. (?)sGansu, w, nw & swSichuan, e & seTibet, w & nwYunnan [swTibet, c & eHimalaya]; grassland, forest margins, shrubberies, on limestone, 2,300–4,200 m. V–VI. Hybridises in the wild with *C. yunnanense* to produce *C.* × *froschii* **Perner**.

Cypripedium tibeticum **King ex Rolfe form.** Dwarfer plants from alpine meadows on Balangshan (wSichuan) with smaller, deep maroon **flowers**, narrower petals and a bract borne at right angles to the flower-stalk, probably represent a distinct entity.

Cypripedium franchetii **E. H. Wilson*** Similar to the previous species but not more than 35 cm, usually clump-forming, with 3–7 **leaves** that are hairy all over; **flower** usually paler with a densely, often softly hairy, ovary; petals less than 10 mm broad at widest; lip deeply pouched, 3.2–5 cm long. sGansu, nwHubei, wSichuan, sShaanxi [wHenan]; shade on damp grassy banks in mixed and coniferous woods, grassy banks, shrubberies, 1,500–3,700 m. V–early VII.

Cypripedium calcicola **Schltr.*** (syn. *C. smithii* Schltr.) Like *C. tibeticum* but **flowers** dark plum-coloured with marked translucent windows at back of lip, which lacks a white rim to the orifice; petals usually less than 10 mm broad at the widest; lip deeply pouched, 3.5–4.2 cm long, surface corrugated; ovary sparsely hairy. nwSichuan, Shaanxi, nwYunnan; shady banks in mixed and coniferous woodland, sometimes in more open grassy habitats, 2,600–3,900 m. V–VI.

1. *Cypripedium tibeticum*, a very variable and widespread species in western China, photographed at Napahai, Zhongdian, nwYunnan (CGW)
2. *Cypripedium tibeticum*; Baishui, Yulongxueshan, nwYunnan (HJ)
3. *Cypripedium tibeticum*; Gangheba, Yulongxueshan, nwYunnan (CGW)
4. *Cypripedium tibeticum*; Napahai, Zhongdian, nwYunnan (CGW)
5. *Cypripedium* ×*froschii*; Gangheba, Yulongxueshan, nwYunnan (PC)
6. *Cypripedium tibeticum* form; Balangshan, Wolong, wSichuan (CGW)
7. *Cypripedium franchetii*; Wolong, wSichuan (CGW)
8. *Cypripedium calcicola*; west of Jiuzhaigou, nwSichuan (CGW)
9. *Cypripedium plectrochilum*; Zhongdian, nwYunnan (HJ)
10. *Cypripedium fasciolatum*; nSichuan (HP)
11. *Cypripedium farreri*; near Huanglong, nwSichuan (HP)
12. *Cypripedium henryi*; cult. (PC)
13. *Cypripedium shanxiense*; Huanglong, nwSichuan (PC)
14. *Cypripedium shanxiense*; near Kaba, Minshan, sGansu (HJ)

Cypripedium plectrochilum **Franch.** Plant to 30 cm with slender, leafy **stems**. **Leaves** 3, lanceolate, 4.5–6 cm long. **Flower** small, brownish maroon with a white lip and convex staminode, marked with pink on both lip and staminode; lip pouched, rather obliquely conical, 1.6–2.4 cm long. w & nwYunnan, Sichuan, wHubei, seTibet [nMyanmar]; amongst limestone rocks and boulders, often in streambeds or on open banks, rocky *Pinus* woodland, generally in light shade, 2,000–3,600 m. IV–VI.

Cypripedium fasciolatum **Franch.*** Plants to 60 cm, usually with 6, elliptic, slightly hairy **leaves** to 12 cm long. **Flower** solitary, large; sepals and petals yellow striped or streaked with purple, or rose-pink with maroon veins, petals spirally twisted; lip large and balloon-like, obovoid, slightly bilaterally flattened, very inflated, pale yellow spotted maroon, 4–7.3 cm long. Hubei, Sichuan; deciduous woods and woodland margins on limestone, 1,900–2,200 m. IV–VI.

Cypripedium farreri **W. W. Sm.*** Very like the previous species, but a smaller plant with 2–3 **leaves** 6–9 cm long. **Flower** fragrant, small but similarly coloured, sepals and petals usually yellow-green veined maroon; lip pitcher-shaped, 2.4–3.3 cm long, with fluted margin. sGansu, nwSichuan; deciduous woods and woodland margins on limestone, 2,600–2850 m. VI.

Cypripedium henryi **Rolfe*** Plants to 60 cm bearing several, elliptic to ovate-elliptic **leaves** 10–21 cm long. **Flowers** 2–4, green, turning yellowish green with age, or green or yellow; petals spirally twisted; lip small, egg-shaped, 1.5–2.7 cm long. sGansu, Guizhou, wHubei, w, n & ne Sichuan, nwYunnan; banks in deciduous woods, especially on margins, clearings, shrubberies, in light or partial shade, on limestone, 1,800–2,800 m. V–VI.

Cypripedium shanxiense **S. C. Chen** Similar to the previous species, but **leaves** not more than 7–15 cm long; and **flowers** 1–3, brownish bronze or yellow-brown or ochre, turning yellowish with age, or purplish, lip not more than 2 cm long. sGansu, wHubei, e & seQinghai, Shaanxi [n & neChina, Korea, eRussia, (?)Japan]; tufa islands in streambeds, banks in deciduous woods, especially on margins or in clearings, in partial or light shade, on limestone, 1,000–2,500 m. V–early VII.

Group Two
1. Flowers large, lip 3.8–6 cm long, grooved at apex; leaves heavily pleated and fan-like ...... **C. japonicum**
1. Flowers small, lip not more than 2.2 cm long; leaves not pleated 2
2. Leaves alternate; lip 14–22 mm long ..... **C. guttatum**
2. Leaves opposite (paired); lip not more than 15 mm long .... **C. debile, C. palangshanense, C. elegans**

Cypripedium japonicum **Thunb.** Erect plant to 30 cm, often clump-forming, **leaves** opposite, fan-shaped or fish-tail-shaped. **Flower** large with yellow-green sepals and petals and a white, deeply pouched lip flushed pink and marked with purple around the mouth which is puckered, not smooth, at apex. Gansu, neGuizhou, Hubei, eSichuan, sShaanxi [eChina, Japan]; woodland, bamboo thickets, in dappled shade, often near water, 1,000–2,000 m. IV–VI.

Cypripedium guttatum **Sw.** Plant 10–25 cm tall with downy **stems**, **leaves** elliptic, ascending, 5–12 cm long. **Inflorescence** erect with a hairy axis. **Flower** solitary, 14–18 mm across, white with rose-purple spotting on inside of dorsal sepal, petals and on outside of lip, occasionally pure white; lip urn-shaped, 14–22 mm long, with wide mouth. sGansu, wSichuan, Tibet, nwYunnan [c & nChina]; montane grassland, open grassy places in coniferous woodland, locally abundant, 2,800–4,100 m. V–VI.

Cypripedium debile **Rchb. f.** Dwarf plant not more than 15 cm tall with paired, heart-shaped, glabrous **leaves**, 2.5–7 cm long. **Inflorescence** axis glabrous, stalk arching, with an erect linear bract. **Flower** often hidden below leaves, with pale green sepals and petals and an off-white lip, 10–15 mm long, marked with purple around mouth. Hubei, w, n & nwSichuan [Taiwan, Japan and across northern hemisphere]; deep shade on banks in deciduous and mixed woodland, 2,200–3,000 m. V–VI.

Cypripedium palangshanense **T. Tang & F. T. Wang*** Similar to *C. debile* but **leaves** elliptic-ovate, 4–6 cm long, basal, almost appressed to the ground; **inflorescence** erect, minutely downy; **flower** plum-coloured, 1.4–1.8 cm long, lip sometimes slightly yellow-tinged, 9–11 mm long. n & wSichuan; shady places in humus and moss on banks over limestone rocks in deciduous or mixed woodland, 2,200–2,800 m. VI.

Cypripedium elegans **Rchb. f.** Plant small, to 10 cm tall, the paired **leaves** ovate or somewhat heart-shaped, hairy all over. **Inflorescence**-axis erect, very hairy, bearing an elliptic, hairy, green bract. **Flower** small, 15 mm across, green with purple markings on lip and petals; lip c. 12 mm long. seTibet, nwYunnan [swTibet, c & eHimalaya]; humus pockets on edge of thickets and in deciduous and mixed woodland, 3,300–3,700 m. VI–VII.

Cypripedium bardolphianum **W. W. Sm. & Farrer*** Dwarf plant 8–12 cm tall, **leaf** and **bract** almost erect, usually unspotted, 6–8 cm long, often with purple margin. **Flowers** small, 20 mm across, glossy, yellow or brown-maroon with rounded yellow lip, 11–14 mm long, covered on rim with purplish brown warts, apex particularly warty; ovary slightly purple-hairy on ridges, otherwise green. swGansu, w & nwSichuan, nwYunnan; tufa islands in streambeds, shallow humus on limestone rocks in deciduous woodland, 2,400–3,600 m. VI–VII.

Cypripedium micranthum **Franch.*** Like the previous species, but with smaller, dull pale yellow **flowers** with brown-maroon spots on sepals, petals and lip, while the ovary is characteristically very purple-hairy; sepals densely hairy outside; lip 8–10 mm long, with smooth apex. nwSichuan; shallow humus in deciduous woods, over limestone rocks, 2,000–2,500 m. V–VI.

Cypripedium forrestii **P. J. Cribb*** Dwarf plant with **leaves** and bracts spotted with black, rarely unspotted, to 8 cm long. **Flowers** small, yellow or brown-maroon with a yellow pouched lip, c. 10 mm long, with smooth apex; ovary somewhat hairy. nwYunnan; steep streamsides, limestone rocks in shallow humus in deciduous or mixed woodland, 2,800–3,500 m. VI.

Cypripedium margaritaceum **Franch.*** (syn. *C. daliense* S. C. Chen & J. L. Wu) Plant to 10 cm, leaf and bract glossy, yellowish green spotted with blackish maroon, elliptic to rounded, to 15 cm long, spreading widely apart. **Flower** large, dorsal sepal yellow, streaked with purple, petals and lip off-white streaked and spotted with purple; dorsal sepal somewhat hooded; petals almost glabrous except on veins, enfolding the lip which is pouched, 28–30 mm long. swSichuan, nwYunnan; tufa banks, limestone rocks in shallow humus and moss in mixed or deciduous woodland, especially beneath rhododendrons, grassy banks, 2,500–3,600 m. V–VI.

---

1. *Cypripedium japonicum*; cult. (PC)
2. *Cypripedium guttatum*; Xiaoxueshan, nwYunnan (HJ)
3. *Cypripedium guttatum*; Tianchi, nwYunnan (PC)
4. *Cypripedium debile*; cult. (PHD)
5. *Cypripedium palangshanense*; Jiuzhaigou, nwSichuan (PC)
6. *Cypripedium palangshanense*; Jiuzhaigou, nwSichuan (PC)
7. *Cypripedium elegans*; above Wengshui, nwYunnan (CGW)
8. *Cypripedium bardolphianum*; Huanglong, nwSichuan (CGW)
9. *Cypripedium bardolphianum*; Huanglong, nwSichuan (PC)
10. *Cypripedium micranthum*; Huanglong, nwSichuan (PC)
11. *Cypripedium micranthum*; Huanglong, nwSichuan (PC)
12. *Cypripedium forrestii*; Gangheba, Yulongxueshan, nwYunnan (HJ)
13. *Cypripedium forrestii*; Gangheba, Yulongxueshan, nwYunnan (HJ)
14. *Cypripedium margaritaceum*; Gangheba, Yulongxueshan, nwYunnan (HJ)
15. *Cypripedium margaritaceum*; Gangheba, nwYunnan; once common, it has been the target of unscrupulous collectors in recent years (HJ)

Group Three
1. Leaves usually lacking spots . . . . **C. bardolphianum, C. micranthum**
1. Leaves dark-spotted (occasionally unspotted in *C. forrestii* but then petals purple-spotted) 2
2. Flowers small, lip not more than 10 mm long . . . . **C. forrestii**
2. Flowers larger, lip at least 25 mm long 3
3. Petals yellow spotted maroon; dorsal sepal 2–4.5 cm long . . . **C. margaritaceum, C. fargesii, C. sichuanense**
3. Petals white or cream, spotted purple; dorsal sepal 4–7 cm long . . . . **C. lichiangense, C. lentiginosum**

1. *Cypripedium fargesii*; Jiuzhaigou, nwSichuan (PC)
2. *Cypripedium fargesii*; cult. (PC)
3. *Cypripedium sichuanense*; Huanglong, nwSichuan (PC)
4. *Cypripedium sichuanense*; Huanglong, nwSichuan (PC)
5. *Cypripedium lichiangense*; Baishui, Yulongxueshan, nwYunnan (PC)
6. *Cypripedium lentiginosum*; cult. (PCH)
7. *Paphiopedilum concolor*; Longhushan, seGuangxi (PC)
8. *Paphiopedilum bellatulum* photographed south of Baoshan in western Yunnan where it is both local and endangered (PC)
9. *Paphiopedilum bellatulum*; Monghe, wYunnan (PC)
10. *Paphiopedilum armeniacum*; Nushan, near Baoshan, wYunnan (PC)
11. *Paphiopedilum malipoense*; Malipo, seYunnan (PC)
12. *Paphiopedilum micranthum*; Yujiang, neGuizhou (PC)
13. *Paphiopedilum micranthum*; Yujiang, neGuizhou (PC)
14. *Paphiopedilum emersonii*; Maolan, seGuizhou (PC)

Cypripedium fargesii **Franch.*** Similar to the previous species, but **leaf** and **bract** not glossy, spotted with blackish maroon; **flower** slightly smaller, lip 25–27 mm long, yellow, spotted and streaked maroon, petals covered with long white hairs. sGansu, wHubei, eSichuan; tufa banks, limestone rocks in shallow humus and moss in mixed or deciduous woodland, 2,000–2,200 m. Late V–VII.

Cypripedium sichuanense **Perner*** Similar to *C. fargesii* in **leaf** size and coloration, but **flower** smaller, dark maroon, petals sparsely and shortly hairy on veins, lip dull yellow heavily spotted with maroon. w & nSichuan; limestone rocks in shallow humus in deciduous woodland, 2,700–2,900 m. Late V–VII.

Cypripedium lichiangense **S. C. Chen & P. J. Cribb** Plant to 10 cm but much broader, **leaves** and **bracts** glossy deep green, heavily spotted with blackish maroon, to 20 cm long. **Flower** large, white to cream, heavily suffused with purple on sepals, spotted with dark purple on petals and lip; dorsal sepal erect, flat, 4.2–7 cm long; petals covered with short blackish maroon hairs on veins; lip 3.3–4 cm long, circular from above. swSichuan, nwYunnan [nMyanmar]; tufa banks, limestone rocks in shallow humus and moss in mixed or deciduous woodland, especially beneath rhododendrons and evergreen *Quercus*, 2,600–3,500 m. V–VI.

Cypripedium lentiginosum **P. J. Cribb & S. C. Chen*** Similar to *C. lichiangense*, but base colour of **flowers** white or creamy white, and petals with a gap between them and the lip sides. seYunnan; steep limestone slopes in deciduous woodland, 1,800–2,000 m. V.

PAPHIOPEDILUM

A genus of 69 species (22 in China, 6 endemic) found in South East Asia, Indonesia and the Solomon Islands. **Flowers** like those of *Cypripedium*, but **leaves** leathery and borne in a fan on a very short **stem**. Most of the Chinese species are tropical or subtropical **perennial herbs**, only 2 being found as high as 1,500 m or above.

> 1. Leaves tessellated with dark and pale green, purple-spotted beneath 2
> 1. Leaves plain green or sometimes faintly spotted 4
> 2. Flowers white to pale yellow dotted all over with fine dark purple or maroon spots . . . . . **P. concolor, P. bellatulum**
> 2. Flowers variously coloured but not as above 3
> 3. Flowers bright yellow, spotted inside lip . . . . **P. armeniacum**
> 3. Flowers other than yellow; petals and lip unspotted (latter sometimes faintly spotted inside) . . . . **P. malipoense, P. micranthum**
> 4. Flowers solitary . . . . **P. emersonii, P. hirsutissimum, P. henryanum**
> 4. Flowers 2–5 per spike . . . . **P. dianthum**

Paphiopedilum concolor **(Batem.) Pfitzer Inflorescence** almost erect or arching, 1–3-flowered, stalk 5–8 cm long. **Flowers** 5–7 cm across, slightly scented, usually yellowish to ivory-white, finely spotted with maroon or brown-purple; petals obliquely elliptic to rhombic-elliptic, 3–5 × 1.8–3.1 cm; lip ellipsoid to egg-shaped, 3.5–4.5 cm long; staminode ovate to ovate-triangular. wGuangxi, sGuizhou, s, sw & seYunnan [Cambodia, Laos, Myanmar, Thailand, Vietnam]; crevices of shady limestone cliffs, rocks in well-drained places, 300–1,400 m. V–VIII.

Paphiopedilum bellatulum **(Rchb. f.) Stein.** Similar to *P. concolor*, but with 1 or 2 larger, white, almost circular **flowers**, spotted with dark maroon all over, on a shorter stalk 2.5–5 cm long; petals broadly elliptic or broadly ovate-elliptic, 4.5–6 × 3–5 cm; lip ellipsoid-egg-shaped, usually 3–3.5 cm long; staminode almost orbicular or squarish, 8–10 mm long and wide. wGuangxi, swGuizhou, s, sw & seYunnan [neMyanmar, nwThailand]; shady cliffs or steep rocky and well-drained places in forests in limestone areas, 1,000–1,800 m. IV–VIII.

Paphiopedilum armeniacum **S. C. Chen & Liu*** Plant with clustered growths, **leaves** 5–7, oblong, in a basal fan, tessellated dark and pale green. **Flower** relatively large, solitary on a tall hairy, purple-spotted stalk to 26 cm, bright canary-yellow with red marks on central staminode, purple-spotted inside lip; dorsal sepal ovate, 2.2–4.8 cm long; petals ovate to almost rounded, 2.8–5.3 mm long; lip very inflated, thin-textured, 4–5 cm long, with recurved margins. wYunnan; limestone cliffs and ledges in light shade, to 2,200 m. III–V. Often seen for sale in markets in Yunnan.

Paphiopedilum malipoense **S. C. Chen & Z. H. Tsi** Similar in habit to the previous species, but **leaves** broader and blunter. **Flower** on tall scape up to 60 cm, smelling of raspberries. Flower up to 7 cm in diameter, buff or pale green with maroon veins on petals, spotted within lip, staminode white and maroon. seYunnan [Vietnam]; in leaf-litter and humus between limestone boulders, and at foot of cliffs, to 1,500 m. II–IV.

Paphiopedilum micranthum **T. Tang & F. T. Wang*** Habit similar to that of the previous species, but plants with long pink runners and **flowers** with a pink lip and purple-striped yellowish sepals and petals; lip 5–6.5 cm long. wGuangxi, swGuizhou, seYunnan; limestone cliffs and ledges in light shade, to 1,600 m. III–IV. Often seen for sale in markets in Yunnan and elsewhere.

Paphiopedilum emersonii **Koop. & P. J. Cribb Leaves** green but sometimes very faintly mottled. **Flower** up to 10 cm in diameter, sweetly fragrant, white with a creamy yellow lip; ovary densely white-hairy; petals almost circular or obovate, rounded at tips; staminode trowel-shaped, slightly convex, yellow marked with red. wGuangxi, sGuizhou, swYunnan [w Guangdong, nVietnam]; limestone cliffs, to 1,500 m. IV–V.

***Paphiopedilum hirsutissimum* (Lindl. ex Hook.) Stein.**
Plants often forming multi-growth clumps, with linear, pointed green **leaves**. **Inflorescence** up to 30 cm, spreading, with a densely purple-hairy stalk. **Flower** large, up to 14 cm in across, with a densely purple-hairy ovary; sepals green with brown centre; petals spreading widely, spathulate, ciliate, green and brown with purple tips; lip green and brown. wGuangxi, swGuizhou, sYunnan [neIndia, nLaos, nMyanmar, Thailand, nVietnam]; limestone cliffs, gorges, 300–1,500 m. IV–VI.

***Paphiopedilum henryanum* Braem** Much smaller plant than the previous species, **flowers** up to 6 cm in diameter; dorsal sepal yellow with black spots, lip pink with a yellow rim. seYunnan [nVietnam]; limestone cliffs and rocks, to 1,600 m. IX–X.

***Paphiopedilum dianthum* T. Tang & F. T. Wang** Clump-forming plant with leathery strap-like **leaves** 20–50 cm long. **Flowers** with a papillose ovary and a green stalk; sepals white; petals up to 10.5 cm long, spirally tapering, white with darker tips, warty on margins; lip dull brown-green, 4–4.5 cm long. wGuangxi, swGuizhou, sYunnan [nVietnam]; limestone cliffs, gorges, 1,200–2,250 m. IX–X.

ACERATORCHIS

A genus with a single species endemic to western China. Plants rhizomatous with a solitary, almost erect, basal green **leaf**. **Inflorescence** erect, racemose, 1–few-flowered, **flowers** similar to and possibly congeneric with *Galearis*, but lip petaloid and lacking a spur.

Aceratorchis tschiliensis* Schltr. (syn. *Orchis tschiliensis* (Schltr.) Soó) Plant 6–15 cm tall with an oblong-spathulate or spathulate, blunt **leaf**, 3–5 × 1.2–2.6 cm. **Inflorescence** 1–5 cm long, 1-sided, with 1–6 flowers. **Flowers** purplish red, pale purple or white; sepals oblong, blunt, 5–8 × 2.5–4 mm, dorsal sepal forming a hood with petals, lateral sepals spreading; petals oblong-lanceolate, 4–7 mm long; lip spreading, ovate-lanceolate or ovate-oblong, nearly as long as petals, unlobed. Gansu, sQinghai, nwSichuan, Shaanxi, Yunnan [Hebei, Shanxi]; forest margins, meadows, 1,600–4,100 m. VI–VIII.

GALEARIS

A genus with c. 12 species (8 in China) in east Asia and North America. Small, terrestrial, **perennial herbs** with a slender underground creeping rhizome. **Stem** erect, cylindrical, with tubular, somewhat membranous sheaths at base, bearing 1 or 2, almost opposite, rather fleshy **leaves**. **Inflorescence** erect, few-flowered. **Flowers** similar to those of *Orchis*, *Ponerorchis* and *Amitostigma*, often purplish red; ovary spindle-shaped, twisted, glabrous; lip entire to 3 lobed, spurred at base.

1. Lip 8–10 mm long, purple-red, heavily marked with deep purple; spur 7–10 mm long ... **G. wardii, G. cyclochila**
1. Lip not more than 8 mm long, unmarked; spur short (not more than 2 mm long) or absent ...... **G. spathulata, G. diantha, G. roborowskii, G. huanglongensis**

Galearis wardii* (W. W. Sm.) P. F. Hunt (syn. *Orchis wardii* W. W. Sm.) Plant to 25 cm, with 2 thick-textured, broad-elliptic to oblong-lanceolate **leaves**, 7–15 cm long and a stouter, erect, 5–10-flowered **inflorescence**. **Flowers** dark purplish red, lip densely marked with deep purple and becoming purple-black; lateral sepals spreading horizontally or reflexed, sickle-shaped to narrowly ovate-lanceolate; lip spreading, broadly ovate or almost orbicular, unlobed, blunt-rounded, 8–9 mm long; spur tubular, 7–10 mm long, pendent, slightly shorter than ovary. wSichuan, eTibet, nwYunnan; woods, open shrubberies, alpine meadows, 2,400–4,500 m. VI–VII.

***Galearis cyclochila* (Franch. & Sav.) Nevski** (syn. *Habenaria cyclochila* Franch. & Sav.; *Orchis cyclochilus* (Franch. & Sav.) Maxim.) Plant 9–19 cm tall. **Stem** with 1–2 tubular, membranous sheaths at base and a basal **leaf**, leaf-blade green, oblong, broadly elliptic to broadly ovate, thick-textured, 5–9 × 2–4.2 cm. **Inflorescence** often with 2 **flowers**, these pale pink or white, speckled with purple on lip; lip spreading, ovate-orbicular, 8–10 × 4–5 mm, margins with blunt-wavy teeth; spur slender, pendent, nearly as long as ovary. neQinghai [Heilongjiang, Jilin, Japan, Korea, Russia]; forests, thickets, 1,000–2,900 m. V–VI.

***Galearis spathulata* (Lindl.) P. F. Hunt** (syn. *Orchis spathulata* Lindl.) Plant 8–15 cm tall with usually 2, almost opposite, oblanceolate, elliptic or spathulate, green **leaves**, 2.3–9 cm long. **Inflorescence** 1-sided, 1.5–5 cm long, 1–5-flowered. **Flowers** purplish red; sepals almost oblong, blunt, 7–10 mm long, dorsal sepal forming a hood with the erect petals, lateral half spreading; lip oblong, elliptic, egg-shaped or square, as long as petals, entire, upper surface papillose; spur short-cylindrical, c. 2 mm long, much shorter than ovary. seGansu, neQinghai, wSichuan, Shaanxi, e & seTibet, nwYunnan [swTibet, Himalaya]; shrubberies, alpine meadows, 2,300–4,300 m. VI–VIII.

***Galearis diantha* (Schltr.) P. F. Hunt** Plant 8–15 cm tall. **Leaves** green, often 2, almost opposite, oblanceolate, elliptic or spathulate, 2.3–9 cm long. **Inflorescence** 1-sided, 1.5–5 cm long, 1–5-flowered. **Flowers** purplish red; sepals almost oblong, blunt, 7–10 mm long, dorsal sepal forming a hood with the petals, lateral sepals nearly spreading; petals erect, ovate-oblong or broadly oblong, 6.5–8 mm long; lip oblong, elliptic, egg-shaped or square, entire, upper surface papillose; spur short, cylindrical, c. 2 mm long, much shorter than ovary. seGansu, neQinghai, scSichuan, wShaanxi, e & seTibet, nwYunnan [swTibet, eHimalaya]; thickets, alpine meadows, 2,300–4,300 m. VI–VIII.

1. *Paphiopedilum hirsutissimum*; Maguan, seYunnan (PC)
2. *Paphiopedilum hirsutissimum*; Maguan, seYunnan (PC)
3. *Paphiopedilum henryanum*; Candi, nVietnam (PC)
4. *Paphiopedilum henryanum*; Candi, nVietnam (PC)
5. *Paphiopedilum dianthum*; Hoquangphin, nVietnam (PC)
6. *Paphiopedilum dianthum*; Hoquangphin, nVietnam (PC)
7. *Aceratorchis tschiliensis*; Jiuzhaigou, nwSichuan (PC)
8. *Galearis wardii*; Balangshan, Wolong, wSichuan (PC)
9. *Galearis diantha*; Jiuzhaigou, nwSichuan (PC)

1. *Galearis roborowskii*; Wanglang, nwSichuan (PC)
2. *Galearis huanglongensis*; Huanglong, nwSichuan (HP)
3. *Ponerorchis chusua*; upper Marsyandyi, Nepal (CGW)
4. *Ponerorchis chusua*; Jiuzhaigou, nwSichuan (PC)
5. *Ponerorchis brevicalcarata*; Yulongxueshan, nwYunnan (HJ)
6. *Ponerorchis crenulata*; Huanglongsi, nwSichuan (CGW)
7. *Amitostigma basifoliatum*; Wolong, wSichuan (PC)
8. *Amitostigma faberi*; Wolong, wSichuan (PC)
9. *Amitostigma tibeticum*; Yulongxueshan, nwYunnan (HJ)

Galearis roborowskii (Maxim.) S. C. Chen, P. J. Cribb & S. W. Gale Similar to *G. diantha* but with a spotted **leaf** and **flowers** with a markedly 3-lobed lip spotted purple at base. Sichuan, e & seTibet, Yunnan [n & cChina, eRussia]; grassland and woodland on limestone, 2,800–3,500 m. VI–VII.

Galearis huanglongensis Q. W. Meng & Y. B. Luo Small rhizomatous terrestrial plant, 6–9 cm tall. **Leaves** 1(–2), glossy green, spreading to almost erect, ovate-elliptic, blunt, 5–9 cm long. **Inflorescence** erect, 2-flowered; bracts longer than flowers. **Flowers** white with rose-purple spots on lip; sepals and petals forming a hood over column; lip deflexed, entire, fan-shaped, blunt, 6–7 mm long; spur conical-cylindrical, 2 mm long. nwSichuan; on tufa in coniferous woodland and by streams, 2,800–3,100 m. VI–VII.

PONERORCHIS

A genus of c. 15 species (2 in China), mainly in alpine areas of subtropical Asia (including China and Taiwan). Terrestrial **herbs**, with egg-shaped or ellipsoid, underground **tubers** and erect **stems** bearing 2–5 alternate **leaves**, sheathed at base and clasping stem. **Inflorescence** terminal, 3–many-flowered. **Flowers** small or medium-sized, often pink or purple; ovary twisted, stalked; sepals free, dorsal one erect, often converging (but not fused) with petals to form a hood; lip 3-lobed, spurred at base.

Ponerorchis chusua (D. Don) Soó (syn. *Orchis chusua* D. Don; *Amitostigma beesianum* (W. W. Smith) T. Tang & F. T. Wang) Plant 5–45 cm usually with 2–3, oblong-lanceolate or linear, green **leaves**, 3–15 cm long. **Inflorescence** somewhat 1-sided, to 20-flowered. **Flowers** purplish red or pink; dorsal sepal oblong or ovate-oblong, erect, 5–8 mm long; lateral sepals reflexed, ovate-lanceolate, 6–8 mm long; petals erect, obliquely ovate, 5–7 mm long; lip much longer and broader than sepals, 3-lobed. eGansu, w, sw & seHubei, e & seQinghai, Sichuan, sShaanxi, seTibet, n, nw & neYunnan [swTibet, n & neChina, Himalaya, nMyanmar, Japan, Korea, eRussia]; grassland, scrub, woodland, forests, 500–4,700 m. VI–VIII.

Ponerorchis brevicalcarata (Finet) P. F. Hunt* In contrast to the previous species, this has a solitary, deep green **leaf** with 5–7 near-white veins with dark purple spots in between, often dark purple beneath, heart-shaped or broadly ovate, 1–3 × 0.7–2 cm. **Inflorescence** slender, 1–3-flowered. **Flowers** purple; lip spreading, wedge-shaped to obovate, 3-lobed above middle, 10–11 mm long, margins entire. swSichuan, nwYunnan; *Pinus* forests, grassy slopes, 1,500–3,400 m. VI–VII.

Ponerorchis crenulata Sóo* (syn. *Orchis crenulata* Schltr., not J. E. Gilbert; *Chusua crenulata* (Schltr.) P. F. Hunt; *Ponerorchis schlechteri* Perner & Y. B. Luo) Dwarf, colony-forming, terrestrial **herb**, not more than 5 cm tall. **Leaves** 2, narrow-elliptic, 2–3 cm long, dark maroon or green with a dark maroon margin. **Flowers** solitary, 15–25 mm across, white with purple margins to segments, lip spotted and streaked with purple. w & nwSichuan, nwYunnan; stony moorland with other dwarf herbs, rock-ledges, alpine meadows, 3,500–4,300 m. VI–VII.

AMITOSTIGMA

A genus of c. 22 species (21 in China) in eastern Asia and adjacent areas. Small, **herbaceous**, terrestrial plants, each year growing from a globose or egg-shaped, fleshy, subterranean **tuber**. **Leaves** 1–2, oblong, lanceolate, elliptic or ovate, often basal or prostrate on ground. **Inflorescence** erect, terminal, racemose or rarely almost capitate, 1–many-flowered. **Flowers** small to medium-sized, often showy, white, pink, purple or yellow; upper sepals and petals forming a hood over column; lip 3–4-lobed, larger than sepals and petals and spurred at base; column very short; staminodes 2.

1. Leaves solitary, in middle of stem .... *A. basifoliatum*, *A. faberi*
1. Leaves 1–2, basal or almost so 2
2. Spur longer than ovary .... *A. tibeticum*
2. Spur much shorter than ovary .... *A. monanthum*, *A. physoceras*

Amitostigma basifoliatum (Finet) Schltr.* Plant 10–23 cm with narrowly oblong-lanceolate **leaves**, 3.5–5.5 cm long. **Inflorescence** several–10-flowered or more, 4 cm long; floral bracts lanceolate, long-pointed, shorter than the glabrous, 10–12 mm-long ovary. **Flowers** white, sometimes tinged with red; dorsal sepal ovate-oblong, concave, 3.5 mm long; lateral sepals obliquely oblong, reflexed, 4.5 mm long; petals rhombic-ovate, 4.5 mm long; lip broadly obovate in outline, 3-lobed below middle, 6–7 mm long; spur pendent, cylindrical, 2–3 mm long, much shorter than ovary. swSichuan, n, nw & neYunnan; damp places in forests, grassy slopes, grassy cliff-ledges, 2,600–3,800 m. VI–VII.

Amitostigma faberi (Rolfe) Schltr.* Similar to the previous species, but **flowers** bright purple, floral bracts much shorter than the 8–10-mm-long ovary; lip broadly ovate or orbicular in outline, 3-lobed, 7 mm long, upper surface densely downy, midlobe triangular-ovate, deeply 2-lobed near middle, 4.5 mm long; spur pendent, cylindrical, slightly incurved, 5–6 mm long. neGuizhou, Sichuan, nwYunnan; meadows, grassy slopes, woodland, scrub, crevices along valleys, 2,300–4,300 m. VI–VII.

Amitostigma tibeticum Schltr.* Plant 6–8 cm tall with a solitary lanceolate or oblanceolate-strap-like basal **leaf** 2.5–3 cm long; **stem** and leaf deep purple. **Inflorescence** 1-flowered with oblong-lanceolate, ± pointed, floral bract often as long as the 7–10 mm-long, glabrous ovary. **Flower** almost erect, deep rose-purple or purplish red; dorsal sepal narrowly ovate-oblong, 7 mm long; lateral sepals oblique, c. 3.5 mm wide; petals narrowly ovate, oblique, 6 mm long; lip obovate or heart-shaped in outline, 3-lobed above middle, 8–9 mm long; spur almost cylindrical, 8–9 mm long, pendent, slightly incurved towards top, somewhat longer than ovary. seTibet, nwYunnan; alpine meadows, open woodland, 3,600–4,400 m. VI–VIII.

Amitostigma monanthum **(Finet) Schltr.*** Plant 6–10 cm tall with slender, erect **stem**, with 1–2 tubular sheaths at base and 1 leaf below middle. **Leaves** lanceolate, oblanceolate-spathulate or narrowly oblong, 2–3 × 0.6–1 cm. **Inflorescence** 1-flowered, floral bract linear-lanceolate, pointed, longer than ovary; **flowers** pale purple, pink or white, with purplish dots on lip; dorsal sepal erect, narrow-ovate, 4 mm long, lateral sepals narrowly oblong-elliptic, 5 mm long; lip short-clawed, ovate-orbicular in outline, 8 × 8 mm, downy, ± 4-lobed below middle; spur cylindrical, blunt, pendent, 3–4 mm long. sGansu, wSichuan, sShaanxi, seTibet, nwYunnan; rocks and gravel near streams, alpine grassland, meadows, rocky ledges, 2,800–4,100 m. VII–VIII.

Amitostigma physoceras **Schltr.*** Slightly taller plant than previous species, to 11 cm, with 2 basal, almost opposite, ovate or ovate-lanceolate **leaves**, 1–2 cm long, finely spotted with purple. **Inflorescence** 1-sided, 1–10-flowered, floral bract much shorter than ovary. **Flowers** rose-purple with purple spots on base of lip; lip ovate in outline, deeply 3-lobed, 6–9 mm long, finely papillose above; spur incurved, 2–3 mm long. nw & wSichuan; damp cliff-ledges in valleys and forests, 2,000–2,700 m. VI–VIII.

COELOGLOSSUM

A genus with 1 species distributed through the north temperate zone, closely related to *Dactylorhiza*. Small, **perennial** terrestrial orchids with 2-lobed, digitate tubers, green **leaves** and a dense, terminal **spike** of very small green, brown or orange-brown **flowers**; bracts green, longer than flowers; lip lacking a spur.

Coeloglossum viride **(L.) Hartm.** Plant 10–30 cm tall in flower with 3–4 lanceolate, green **leaves**. **Flowers** green or green with a brown to orange lip; sepals and petals forming a hood over column and lip; lip pendent, 3-lobed at apex, side-lobes longer than middle one, spurless. Gansu, Qinghai, Sichuan, seTibet, nwYunnan [swTibet, nChina, circumboreal]; damp grassy and stony places, open shrubberies, to 3,300 m. V–VIII.

DACTYLORHIZA

A genus with c. 30 species (2 in China), widespread from Europe and north-west Africa throughout temperate Asia to the Himalaya. **Perennial** terrestrial **herbs** arising from annual digitate **tubers**, **stems** with 3–6 green, non-sheathing **leaves**, often with black spots. **Inflorescence** a dense, erect spike, usually many-flowered, almost cylindrical with accompanying bracts that are normally longer than **flowers**; dorsal sepal and petals forming a hood over column, lateral sepals spreading to almost erect; lip entire to 3-lobed, fan-shaped when flattened, spurred at base.

Dactylorhiza hatagirea **(D. Don) Soó** Plant 12–40 cm tall with oblong to linear-lanceolate, plain green **leaves**, 8–15 cm long. **Inflorescence** several–many-flowered, cylindrical, 2–15 cm long. **Flowers** mauve, purplish red or rose-purple, marked on lip with a dark purple line; sepals and petals ovate-oblong to ovate-lanceolate, 5.5–8 mm long; lip ovate to almost orbicular, entire but shallowly 2–3-lobed, 6–9 mm long; spur cylindrical to conical, pendent, slightly incurved. Gansu, Qinghai, wSichuan, eTibet, nwYunnan [n & neChina, Afghanistan, Himalaya, Mongolia, eRussia]; shrubby slopes, alpine meadows and ravine grassland, 600–4,100 m. VI–VIII.

GYMNADENIA

A temperate Eurasian genus of c. 10 species (5 in China). Terrestrial plants growing each year from a lobed, digitate, underground **tuber**. **Stem** erect, leafy. **Leaves** plain green, almost erect. **Flowers** small, pink or purple, spurred at lip base, lip usually 3-lobed.

Gymnadenia conopsea **(L.) R. Br.** Plant 20–60 cm with 4–5, linear-lanceolate to narrowly oblong, plain **leaves**, 5.5–15 cm long. **Inflorescence** a dense, many-flowered, cylindrical spike, 5.5–15 cm long, with pink, rarely pale pinkish white, fragrant **flowers**; sepals broad-elliptic to ovate-elliptic, pointed, 3.5–5.5 mm long; lip spreading, broad-obovate, 4–5 mm long, 3-lobed at apex; spur slender, cylindrical, pendent, 10 mm long. seGansu, n, nw & wSichuan, seTibet, nwYunnan [c & nChina, Europe, Japan, Korea, Russia]; grassland, seasonally wet meadows, woodland, 200–4,700 m. VII–VIII.

Gymnadenia orchidis **Lindl.** Differs from the previous species in being no more than 35 cm tall in flower with 3–5 elliptic to elliptic-oblong **leaves** 4–16 cm long; **flowers** pale to deep pink; sepals and petals 3–5 mm long; lip 3–5 mm long, spur slender, incurved, 7–10 mm long. sGansu, Qinghai, n, nw & wSichuan, seTibet, nwYunnan [swTibet, Himalaya]; grassland, alpine meadows, shrubberies, woodland, 2,800–4,100 m. VI–VII.

HABENARIA

A cosmopolitan genus of c. 600 species (54 in China, 21 endemic). **Perennial** terrestrial tuberous **herbs**, with erect **stems** leafy at base or along stem. **Inflorescence** terminal, racemose, few-flowered, accompanied by leafy bracts. **Flowers** medium-sized to large, green, white, yellow or rarely with a red or pink lip; dorsal sepal and petals usually joined, sometimes free, usually forming a hood above column; lateral sepals free, spreading; lip 3-lobed, spurred at base, with side-lobes entire to fringed; mid-lobe entire, usually strap-like; spur ± elongated.

1. *Amitostigma monanthum*; Chuanzhusi, Gongganlin, wSichuan (HJ)
2. *Amitostigma monanthum*; above Huanglong, nwSichuan (PC)
3. *Amitostigma physoceras*; Minjiang valley, south of Songpan, nwSichuan (PC)
4. *Amitostigma physoceras*; Minjiang valley, south of Songpan, nwSichuan (PC)
5. *Coeloglossum viride*; Dayu valley, Zhuoni, sGansu (HJ)
6. *Coeloglossum viride*; Jiuzhaigou, nwSichuan (PC)
7. *Dactylorhiza hatagirea*; Aba, nwSichuan (HP)
8. *Gymnadenia conopsea*; Europe (DTE)
9. *Gymnadenia orchidis*; Balangshan, Wolong, wSichuan (CGW)

ORCHIDACEAE: Habenaria

1. Leaves 2, basal, prostrate; side-lobes of lip tapering, not fringed ..... **H. balfouriana, H. diphylla, H. glaucifolia**
1. Leaves more than 2, on stem; side-lobes of lip fringed or toothed .... **H. davidii, H. dentata, H. mairei**

Habenaria balfouriana Schltr.* Plant 15–20 cm, with densely papillose-hairy **stem**. **Leaves** ovate or broadly elliptic, 3–4 × 2.5–3.5 cm. **Inflorescence** 3–7-flowered. **Flowers** yellowish green with finely papillose-hairy ovary; sepals 5–7 mm long, ciliate to finely toothed; petals 2-lobed, upper lobe obliquely ovate-lanceolate, lower lobe a tooth at base of upper lobe; lip deeply 3-lobed above base, side-lobes recurved and nearly embracing ovary, awl-shaped, midlobe linear; spur pendent, slightly curved, cylindrical to club-shaped, c. 13 mm long. swSichuan, nwYunnan; forests, shrubby grassland, 2,200–3,600 m. VII–VIII.

Habenaria diphylla Dalzell Plant 11–25 cm tall. **Leaves** heart-shaped or nearly kidney-shaped, 1.5–3.5 × 1–5 cm. **Inflorescence** lax, several-flowered. **Flowers** pale green; sepals 5–6 mm long; petals straight, linear-lanceolate; lip 13–15 mm long, deeply 3-lobed from base, lobes thread-like; side-lobes 13–15 mm long, often curled towards apex; spur pendent, 7 mm long, almost club-shaped, apex almost pointed. sYunnan [n & neIndia]; damp places in valley forests, 1,000–1,300 m. V–VI.

Habenaria glaucifolia Bur. & Franch.* In contrast to the previous species, this has just 2, almost opposite, somewhat orbicular **leaves** 4–6 cm long; **flowers** smaller, 20 mm across, white, pale green or pale yellowish, sepals and petals just 6–7 mm long, 3-lobed lip 13–15 mm long; spur slender, pendent, 20–30 mm long. Guizhou, Hubei, Sichuan, Tibet, nwYunnan [Hunan]; alpine meadows, grassy banks, woodland, 2,000–4,300 m. VII–VIII.

Habenaria davidii Franch.* Plant up to 70 cm with leafy **stem**. **Leaves** 5–7, oblong-lanceolate, 5–10 cm long. **Inflorescence** lax, 4–10-flowered. **Flowers** large, pale green with pale greenish white petals and lip, lip-apex greenish yellow; sepals and petals 13–15 mm long, dorsal sepal and petals forming a hood over column, lateral sepals spreading horizontally; lip 3-lobed, 25–35 mm long, side-lobes deeply fringed, mid-lobe linear; spur slender, pendent, 60–65 mm long. Guizhou, Hubei, Sichuan, Tibet, nwYunnan [Hunan]; grassy slopes, thickets, woods, ravines, 1,000–3,200 m. VI–VIII.

Habenaria dentata (Sw.) Schltr. Plant 35–90 cm with 3–5 oblong to narrowly elliptic **leaves**, 5–15 × 1.5–4 cm. **Inflorescence** often with many white **flowers**; sepals and petals 10–13 mm long, ciliate; lip broadly obovate, 15–18 × 12–16 mm, 3-lobed, side-lobes almost rhombic or somewhat orbicular, 7–8 mm wide, with toothed apical margin, midlobe linear-lanceolate or narrowly strap-like, 5–7 × 1.5–3 mm; spur cylindrical to club-shaped, pendent, c. 4 cm long. Guizhou, Hubei, Sichuan, seTibet, Yunnan [swTibet, c & sChina, Cambodia, c & eHimalaya, Hong Kong, India, Japan, Laos, Myanmar, Thailand, Vietnam]; forests and forest glades, slopes, valleys, 200–2,300 m. VIII–X.

Habenaria mairei Schltr.* A generally smaller plant than the previous two species, not more than 40 cm with 5–7, elliptic-lanceolate **leaves**, 5–8 cm long; **inflorescence** rather lax, with up to 15 **flowers**, green with white petals and lip; sepals and petals 12–15 mm long; lip 3-lobed, 15–20 mm long with deeply fringed side-lobes and a linear mid-lobe; spur pendent, slender, broadened at tip, c. 30 mm long. wHubei, wSichuan, seTibet, nwYunnan; grassland, shrubberies, mixed woodland, 2,400–3,400 m. VI–VIII.

PERISTYLUS

A genus of c. 60 species (21 in China) in tropical and subtropical Asia. Terrestrial **herbs** with ellipsoid to oblong, fleshy tubers and a **stem** with 2–3 tubular sheaths and 1–many, glabrous or hairy, **stem-leaves** or **basal leaves**. **Inflorescence** a terminal, often dense spike. **Flowers** small, resupinate; sepals free, dorsal one often forming a hood with the fleshy petals, lateral sepals spreading, rarely reflexed; lip spurred at base, 3-lobed or rarely entire, spur usually pouched or globose, often very short.

1. *Habenaria balfouriana*; Baishui, Yulongxueshan, nwYunnan (PC)
2. *Habenaria glaucifolia*; Huanglong, nSichuan (HP)
3. *Habenaria davidii*; Malipo, seYunnan (HP)
4. *Habenaria dentata*; Hue, Vietnam (PC)
5. *Peristylus coeloceras*; Genyanshan, swSichuan (MN)
6. *Hemipilia cruciata*; Sanba, nwYunnan (PC)
7. *Hemipilia flabellata*; Lijiang, nwYunnan (LYB)
8. *Herminium monorchis*; above Manang, Nepal (CGW)

Peristylus coeloceras Finet (syn. *Herminium coeloceras* (Finet) Schltr.; *H. tenianum* Kraenzl.; *H. unicorne* Kraenzl.) Plant 9–35 cm with narrowly elliptic-lanceolate or elliptic **leaves**, 4–10 × 1–2 cm. **Inflorescence** cylindrical, 2–10 cm long, dense; bracts longer than the small white **flowers**; sepals 2–2.2 mm long; lip wedge-shaped, 3-lobed above middle, side-lobes slightly shorter than mid-lobe; spur globose. wSichuan, e & seTibet, n, nw & neYunnan [nMyanmar]; coniferous and broadleaved forests, thickets, alpine grassland, 2,000–3,900 m. VI–VIII.

HEMIPILIA

A genus with 12 species (6 in China) found in the Himalaya and South East Asia. Terrestrial **herbs** growing each year from ellipsoid **tubers** and with a solitary, prostrate, ± heart-shaped **leaf** that is often finely spotted with purple or black. **Inflorescence** erect, lax, few–many-flowered. **Flowers** showy, pink to purple; lip 3-lobed, spurred at base.

Hemipilia cruciata Finet Plant to 27 cm, the basal, prostrate, heart-shaped leaf dull bluish green, finely spotted with blackish purple. **Inflorescence** erect, 8–10-flowered. **Flowers** pink to rose-purple; sepals 6–6.5 mm long; lip 3-lobed, deflexed, 6–7 mm long; spur slender, straight, c. 12 mm long. swSichuan, nwYunnan [Taiwan]; stony ground in woods, grassland, roadside banks, 2,300–3,500 m. VI–VIII.

Hemipilia flabellata Bur. & Franch.* Similar in habit to the previous species, but with an entire, fan-shaped lip and longer spur (15–19 mm). swSichuan, nwYunnan; usually in rocky and mossy ground in open forests, limestone crevices, 2,500–3,200 m. VI–VII.

HERMINIUM

A genus with c. 25 species (17 in China) mainly in eastern Asia, extending to Europe and South East Asia. Terrestrial tuberous **herbs** with an erect **stem** bearing 1–several **leaves**. **Inflorescence** terminal, racemose or spike-like, many-flowered. **Flowers** small, densely arranged, usually yellowish green, resupinate or very rarely non-resupinate; sepals free, nearly equal in length, petals often narrower and smaller than sepals, but usually thickened and fleshy; lip joined to column base, simple entire to 3(–5)-lobed.

Herminium monorchis (L.) R. Br. Plant 5.5–15 cm with 2–3, narrowly elliptic-lanceolate or elliptic, pointed **leaves** 2.8–10 cm long. **Raceme** cylindrical with many yellowish green, musk-scented, nodding **flowers**; sepals elliptic or oblong-elliptic, c. 2.2 mm long; petals almost rhombic, slightly longer than sepals, sometimes ± 3-lobed near middle; lip 2–2.2 mm long, fleshy-thickened, 3-lobed near middle, with a linear mid-lobe 1.5 mm long, side-lobes much shorter, triangular. Qinghai, Sichuan, Tibet, n & nwYunnan [c & nChina, w & cAsia, Europe, Himalaya, Japan, Korea, Mongolia, Russia]; grassland, marshy meadows, broadleaved and coniferous woodland, thickets, often gregarious, 600–4,500 m. VI–VIII.

Herminium ophioglossoides* Schltr. Similar to *H. monorchis*, but plants often taller, to 26 cm; **racemes** with several, but up to 40 flowers, occasionally more; **flowers** larger with sepals c. 4 mm long, petals markedly longer, 6–7 mm; lip 3-lobed slightly below middle, 6–7 mm long, with linear lobes and a very short spur, or pouched at base. w & swSichuan, c & nwYunnan; alpine grassland, grassy slopes, stony ground on limestone, 2,100–3,500 m. VI–VII.

***Herminium lanceum* (Thunb. ex Sw.) Vuijk** A much taller plant than the previous species, to 80 cm; **stem** slender with 3–4 linear-lanceolate, pointed or long-pointed **leaves** to 15 cm long; **flowers** yellowish green or green with ovate-oblong or oblong, blunt sepals, 2–4 mm long, and linear petals the same length; lip oblong in outline, 3-lobed at or above middle, 3–7 mm long, often pendent, spurless. seGansu, Guizhou, Hubei, Sichuan, Yunnan [s & cChina, seAsia, c & eHimalaya, India, Indonesia, Japan, Korea, Thailand, Vietnam]; mixed or coniferous forests, bamboo thickets, grassland, 700–3,400 m. VI–VIII.

NEOTTIANTHE

A genus of 12 species (all in China) distributed primarily in temperate Asia but extending to alpine areas of Europe and Asia. Terrestrial **herbs** with globose or ellipsoid, fleshy **tubers**. **Leaves** 1–2, basal or on stems. **Inflorescence** a terminal, often 1-sided raceme, several–many-flowered, with erect or spreading bracts. **Flowers** resupinate, pink to rose-purple, rarely yellow; sepals nearly equal in size, converging (but not fused) to form a hood; petals linear to oblong, joined to dorsal sepal; lip spreading, often 3-lobed, spurred at base, densely finely papillose above; side-lobes often shorter and narrower than mid-lobe.

Neottianthe oblonga* K. Y. Lang Plant 9.5–14 cm with a single oblong **leaf**, 4–6 cm long. **Inflorescence** 5–9-flowered, 4–5 cm long with a glabrous rachis and lanceolate bracts 6–8 mm long that are shorter than the ovary. **Flowers** pale purple; lateral sepals 7–8 mm long; petals linear, 5–6.5 mm long; lip spreading, 3-lobed near middle, 8–9 mm long, upper surface and margins finely papillose, side-lobes linear-lanceolate; spur thick, conical, 5–6 mm long. nwYunnan; *Quercus* forests, c. 3,100 m. VII–VIII.

PLATANTHERA

A genus of c. 150 species (38 in China) found mainly in the north temperate zone, south to Central America, North Africa and tropical Asia. **Perennial**, terrestrial orchids, with fleshy egg-shaped or ellipsoid **tubers**. **Leaves** 1–several, basal or on **stems**, alternate, rarely almost opposite, sheathing at base. **Inflorescence** racemose, terminal. **Flowers** few–many, often white or yellowish green, resupinate; sepals free, dorsal sepal converging (but not fused) with petals to form a hood, lateral sepals spreading, reflexed, longer than dorsal; petals narrower than sepals; lip linear or strap-like, simple, pendent, with a very long or sometimes relatively short spur.

1. Flowers white .... ***P. chlorantha***, ***P. japonica***
1. Flowers green .... ***P. mandarinorum***, ***P. fuscescens***

***Platanthera chlorantha* Cust. ex Rchb.** Plant 30–50 cm tall, glabrous with 2 almost opposite, spathulate-elliptic or oblanceolate-elliptic **leaves**, 10–20 cm long. **Inflorescence** 12–32-flowered, 13–23 cm long. **Flowers** greenish white or white; sepals 6–8 mm long; petals narrowly lanceolate, 5–6 mm long; lip fleshy, spreading, strap-like, blunt, 8–13 mm long; spur club-shaped to cylindrical, 25–36 mm long, conspicuously thickened toward blunt apex. Gansu, Sichuan, seTibet, Yunnan [swTibet, c & nChina, Europe, Japan, Korea, Russia, nwAfrica]; woodland, grassland, 400–3,300 m. VI–VIII.

***Platanthera japonica* (Thunb. ex A. Murray) Lindl.** Readily distinguished from the previous species by the 4–6, elliptic or elongated-elliptic, blunt or pointed **leaves**, 10–18 cm long; and by the somewhat larger white **flowers**, the petals linear, 6–7 mm long and the lip linear, curving forward towards apex, 13–20 mm long; spur pendent, cylindrical to thread-like, 30–60 mm long. Gansu, Guangxi, Guizhou, Hubei, Sichuan, Yunnan [c & eChina, Japan, Korea]; woodland, forests, grassland, 600–2,900 m. V–VII.

***Platanthera mandarinorum* Rchb. f.** Easily distinguished from the previous 2 species by the greenish to yellowish green **flowers** with pale yellow petals and lip; sepals 4–9 mm long, the lateral reflexed, petals spreading; lip pendent, lanceolate to narrowly strap-like, blunt, 7–10 mm long; spur cylindrical, 12–30 mm long. Guizhou, Hubei, Sichuan, Yunnan [c & eChina, Japan, Korea, Taiwan]; forests and forest margins, alpine or damp grassland, 300–3,200 m. V–VII.

***Platanthera fuscescens* (L.) Kraenzl.** (syn. *P. souliei* Kraenzl.) is similar to *P. chlorantha* but has a single basal **leaf**, cauline leaves, and small greenish **flowers** with a 10–14 mm long spur. Gansu, Shaanxi, Sichuan, nwYunnan [widespread in N China, Japan, Korea, Siberia].

SATYRIUM

A genus of c. 100 species, mainly African and Madagascan, with only three species in Asia (all in China). Glabrous **perennial** terrestrial **herbs**, growing from fleshy, usually almost ellipsoid **tubers**. **Stems** bearing ovate to almost elliptic, green **leaves**. **Inflorescence** terminal, usually densely spike-like, many-flowered; floral bracts ± leaf-like, reflexed. **Flowers** not resupinate, with free, small sepals and petals; lip deeply hooded, joined to column at base, with a spur on each side at base.

***Satyrium ciliatum* Lindl.** Plant 14–32 cm with 1 or 2, ovate-lanceolate to narrowly elliptic-ovate **leaves**, 6–15 cm long. **Spikes** 5–13 cm long, dense, to 30-flowered. **Flowers** bright pink; sepals narrowly elliptic to oblong-spathulate, 5–6 mm long, finely ciliate near apex; petals spathulate-oblanceolate, 4–5 mm long; lip uppermost, hooded, semi-globose, 5–6 mm

1. *Herminium ophioglossoides*; seQinghai (RS)
2. *Herminium ophioglossoides*; Gangheba, Yulongxueshan, nwYunnan (HJ)
3. *Neottianthe oblonga*; Baishui, Yulongxueshan, nwYunnan (PC)
4. *Platanthera chlorantha*; Xiaozhongdian, nwYunnan (CGW)
5. *Platanthera fuscescens*; Huanglong, nSichuan (PC)
6. *Platanthera japonica*; Emeishan, sSichuan (PC)
7. *Platanthera mandarinorum*; Napahai, Zhongdian, nwYunnan (PC)
8. *Satyrium ciliatum*; Wengshui, Daxueshan, nwYunnan (EL)
9. *Satyrium nepalense*; Dali, nwYunnan (CGW)

across; spurs 2, 4–6 mm long, each rarely shortened into a pouch, or absent. swGuizhou, w & swSichuan, seTibet, Yunnan [swTibet, nwHunan, c & eHimalaya]; grassy slopes, open mixed or coniferous forests, 1,200–4,000 m. VIII–X.

Satyrium nepalense **D. Don** Very similar to the previous species, but generally taller, 30–45 cm tall, and with 2 or 3 **leaves**; sepals and petals small, obovate to narrowly elliptic, 4–5 mm long; lip 5 mm across; spurs slender, pendent, to 10 mm long, usually as long as ovary. swGuizhou, seTibet, s & seYunnan [swTibet, Himalaya, India, Myanmar, Sri Lanka]; grassy slopes, glades in forests, coniferous and mixed deciduous forests, 1,000–3,200 m. IX–XII. (See photo 9 on p. 587.)

Satyrium yunnanense **Rolfe*** Like *S. ciliatum*, but **leaves** ovate to almost elliptic, pointed, 6–11 cm long, sometimes crisped. **Inflorescence** rather thick and short, 2–4.5 cm long, lax, more than 10-flowered. **Flowers** yellow to nearly golden yellow; sepals and petals oblong, 3.5–4 mm long; lip hooded, nearly semi-globose, 5 mm wide; spurs 2, 5–6 mm long. swSichuan, w, nw & cYunnan; open forests, rocky places, 2,000–3,700 m. VIII–XI.

CHEIROSTYLIS

A genus of c. 20 species (12 in China) distributed in tropical Africa and Asia, and the Pacific islands. Terrestrial **herbs** with a creeping, fleshy, worm-like rhizome and short erect leafy fertile **stems**. **Leaves** entire, green to blackish green, often purple beneath, generally shrivelling by flowering time. **Inflorescence** 2–several-flowered, glandular. **Flowers** small, resupinate, white, often with 2 green spots at base of lip; dorsal sepal and petals converging (but not fused) over column; lip 2-parted, basal part somewhat tubular, enclosing 2 or more basal glands, apical part transversely 2-lobed, often minutely scalloped.

1. *Cheirostylis chinensis*; Shilin, cYunnan (PC)
2. *Cheirostylis chinensis*; Guangdong (PC)
3. *Goodyera schlechtendaliana*; Laovachai, nVietnam (PC)
4. *Spiranthes sinensis*; Paro, Bhutan (PC)
5. *Cephalanthera falcata*; cult. (PC)
6. *Cephalanthera longifolia*; Zhongdian, nwYunnan (RS)
7. *Cephalanthera damasonium*; Shigu to Lijiang road, nwYunnan (PC)

Cheirostylis chinensis Rolfe (syn. *C. josephi* Schltr.; *C. philippinensis* Ames; *C. taiwanensis* Yamamoto) Plant 6–20 cm with 2–4 membranous, ovate, green **leaves**, 1–3 × 0.7–1.7 cm, tip pointed. **Inflorescence** 8–20 cm long, 2–5-flowered, hairy. **Flowers** small; sepals and petals 3–4 mm long, membranous; lip white, erect, 7–8 mm long with a pouched base, above with a comb-like, 5–6-toothed callus on each side, margins with 4 or 5 irregular teeth. Guangxi, Guizhou, Yunnan [Hong Kong, Taiwan, Philippines, Vietnam]; damp rocks and rock-strewn soil in forests on slopes, streamsides, 200–1,800 m. I–IV.

Cheirostylis yunnanensis Rolfe Similar to the previous species, but with larger **flowers**, sepals 5–6.5 mm long, lip 10–12 mm long. Guangxi, Guizhou, wSichuan, sYunnan [sChina, Vietnam]; shady places, humusy rock-pockets, forests along streams, 200–1,800 m. I–IV.

GOODYERA

A genus of c. 40 species (30 in China, 14 endemic) distributed mainly in north temperate areas, south to Mexico, South East Asia, Australia, Pacific islands and Madagascar. Terrestrial or rarely epiphytic, rhizomatous **herbs** with erect, short or long **stem**, bearing alternate or clustered **leaves** that often have coloured veins or markings. **Inflorescence** terminal, with few to many flowers, racemose, rarely dense and spike-like, 1-sided or cylindrical, with small **flowers**. Sepals free, often hairy, dorsal one forming a hood with the narrow, thin, membranous petals; lip ± joined with the column at base, simple or indistinctly 3-lobed, slightly pouched at base.

Goodyera schlechtendaliana Rchb. f. Plant 15–35 cm with 4–6, ovate or ovate-lanceolate **leaves**, 3–8 cm long, green above and with irregular white spots. **Inflorescence** 1-sided, 8–20 cm long, softly hairy, several- to more than 20-flowered. **Flowers** white, sometimes tinged with pink, half-opening; sepals downy, 7–10 mm long; petals rhombic-oblanceolate, glabrous, 7–10 mm long; lip ovate, 6–8.5 mm long. sGansu, Guizhou, Hubei, Sichuan, seTibet, Yunnan [swTibet, s, c & eChina, c & eHimalaya, India, Japan, Korea, Thailand, Vietnam]; leaf-litter and humus amongst rocks in broadleaved forests, especially along valleys, 500–2,800 m. VIII–X.

SPIRANTHES

A genus of c. 50 species (1 in China), mainly in North and Central America, with a few species in Europe, Asia, Africa and Australia. **Perennial**, terrestrial **herbs** with several, tufted, cylindrical **roots**. Basal **leaves** in a rosette, or leaves all borne on **stems**, linear, elliptic or broadly ovate. **Inflorescence** terminal, with many flowers, spike-like, often ± spirally twisted. **Flowers** small, not widely opening; sepals free, ± similar, dorsal sepal often joined to petals to form a hood, lateral sepals dilated or sometimes pouched at base; lip base concave, unlobed or 3-lobed, margins often crisped.

Spiranthes sinensis (Pers.) Ames Plant 13–30 cm with 2–5, linear, linear-lanceolate or rarely narrowly oblong, erect and spreading **leaves**, 3–10 cm long. **Inflorescence** spirally twisted, 10–25 cm long, glandular-downy to glabrous. **Flowers** many, small, pink to rose-purple with a white lip, rarely all white; lip concave, broadly oblong, 4 mm long, blunt, hairy within and with strongly wrinkled margins. Widespread in wChina (including seTibet) [s, c & nChina, seAsia, Afghanistan, Himalaya, India, Japan, Korea, Mongolia, Myanmar, eRussia, swPacific islands, Australia]; marshes, damp places in meadows, woodland and thickets, 200–3,400 m. VII–IX.

CEPHALANTHERA

A circumboreal genus of c. 20 species (5 in China). **Perennial** terrestrial **herbs** growing from a stout, short **rhizome**. **Stems** leafy from base to top. **Leaves** pleated. **Inflorescence** terminal, spike-like, dense to almost lax, few–many-flowered. **Flowers** white, yellow or pink; sepals and petals free, not spreading widely; lip 2-parted with basal part pouched or shortly spurred, apical part with longitudinal ridges above; column erect, elongated.

1. Flowers yellow .... **C. falcata**
1. Flowers white 2
2. Lip with 4–6 longitudinal ridges; bracts (at least lower) equalling or longer than flowers .... **C. longifolia**
2. Lip with 3 longitudinal ridges; bracts shorter than to almost equalling flowers .... **C. damasonium**

Cephalanthera falcata (Thunb. ex A. Murray) Blume Plant 20–50 cm tall in flower, with elliptic- to ovate-lanceolate, spreading, **leaves** 5–11 cm long. **Inflorescence** dense, 5–10 flowered. **Flowers** erect; sepals 10–12 mm long; lip 8–9 mm long, basal part slightly concave, apical part oblate, minutely hairy near apex; spur conical, c. 3 mm long. Guizhou, Hubei, w & nSichuan [s, c & nChina, Japan, Korea]; forests, thickets, grassy places, valleys, 700–1,600 m. IV–V.

Cephalanthera longifolia (L.) Fritsch Plant 20–50 cm tall in flower, often clump-forming, with linear-lanceolate, spreading **leaves** 4–13 cm long. **Inflorescence** 2–13-flowered. **Flowers** white with yolk-yellow mark on lip; sepals 10–16 mm long; lip 5–6 mm long, basal part slightly concave, apical part broader than long, margin irregular (appearing nibbled). sGansu, wHubei, w & nSichuan, seTibet, nwYunnan [swTibet, cChina, Europe, nwAfrica, temperate Asia]; forests, thickets, streamsides, grassy places, *Rhododendron* shrubberies, 1,000–3,600 m. VI–VII.

Cephalanthera damasonium (Mill.) Druce Plant 10–50 cm tall in flower. **Leaves** ovate-oblong to ovate-lanceolate, spreading to almost erect. **Inflorescence** 3–12-flowered. **Flowers** erect, somewhat campanulate, white with yolk-yellow mark on lip; sepals and petals 15–20 mm long; lip with pouched basal part, apical part heart-shaped-triangular. sGansu, w, sw & nSichuan, w & nwYunnan [nChina, temperate Asia, Europe]; open woodland, shrubberies, in damp leafy soils, to 3,000 m. VI–VII.

EPIPACTIS

A circumboreal genus of c. 25 species (5 in China). **Perennial**, terrestrial, rhizomatous **herbs** often with clustered, leafy **stems**. **Leaves** strongly pleated, elliptic, ovate or lanceolate. **Inflorescence** dense, racemose, few–many-flowered, with leafy bracts. **Flowers** green or purple, sometimes with orange on lip; lip 2-parted, with a pouched base and ovate apex; column short.

Epipactis helleborine **(L.) Crantz** Plant 20–70 cm with elliptic to ovate **leaves** 3–13 cm long. **Inflorescence** dense, 10–40-flowered. **Flowers** small, green to dull purple, c. 10 mm across; sepals and petals not spreading widely, 6–9 mm long; lip 6–8 mm long, with pouched, glossy basal part and broadly ovate apical part. Gansu, Guizhou, Hubei, Qinghai, Sichuan, Tibet, Yunnan [c & nChina and circumboreal; introduced North America]; deep shade in thickets and shrubberies, mixed and coniferous forests, often near streams, to 3,600 m. VI–VII.

Epipactis mairei **Schltr.** A more substantial plant than the previous species, to 80 cm, with purple **stem** and oblong-elliptic to elliptic-ovate **leaves**, 7–16 cm long. **Flowers** 15–20 mm across, purple with orange-marked lip, sepals spreading, 13–20 mm long; lip 11–18 mm long, with oblong basal part and triangular-ovate, fleshy apical part. Gansu, Guizhou, Hubei, wSichuan, Shaanxi, Tibet, nwYunnan [Bhutan, Myanmar]; light to deep shade in damp grassy areas, thickets, mixed woodland, rocky riverbeds, 200–3,200 m. Late VI–VII.

Epipactis royleana **Lindl.** Differs from the previous two species in the lax, 5–8-flowered **inflorescence**; **flowers** 15–18 mm across, purple, pink, or green with purple veins; sepals and petals spreading, 13–18 mm long; lip 16–20 mm long, contracted near middle, with oblong basal part and narrowly ovate-elliptic apical part. seTibet [swTibet, Himalaya]; damp grassy places by streams, wet meadows, 2,900–3,000 m. VI–VIII.

LISTERA

A genus of c. 60 species (25 in China) distributed in north and east Asia, Europe and North America. Small, **perennial** terrestrial **herbs** with erect, glabrous **stems** bearing 2 opposite or almost so, ovate to heart-shaped **leaves**, inserted at about the middle of plant. **Inflorescence** terminal, usually with many **flowers**, rarely reduced to a single flower, rachis minutely downy; bracts persistent, usually shorter than ovary; sepals and petals free, similar; lip 2-lobed, 3-lobed or obscurely 4-lobed, conspicuously larger than sepals and petals, upper surface often centrally glandular-grooved, lacking a spur. Considered by some authorities to be congeneric with *Neottia*.

Listera puberula **Maxim.** Plant 10–20 cm with a slender **rhizome**. **Leaves** 2, opposite, heart-shaped, broadly ovate or broadly ovate-triangular, 1.5–2.5 cm long, margin ± crisped. **Inflorescence** 2.5–7 cm long, laxly 4–7-flowered. **Flowers** green, very small; sepals ovate-lanceolate, 2.5 mm long; petals linear, 2.5 mm long; lip narrowly obovate-wedge-shaped or oblong-wedge-shaped, 2-lobed at apex, usually 6–8 mm long. cGansu, Guizhou, eQinghai, nwSichuan, nShaanxi [n & neChina, Japan, Korea, eRussia]; damp places in dense forests, 1,400–2,600 m. VII–IX.

NEOTTIA

A genus of 8 species (7 in China) distributed in Europe and Asia. Small, saprophytic **herbs**, with short **rhizomes** and tufted fleshy **roots**. **Stem** erect, lacking green **leaves**, usually with several tubular sheaths below middle. **Inflorescence** terminal, many-flowered, bearing membranous floral bracts. **Flowers** small, twisted, with ovary conspicuously broader than stalk; sepals free, spreading; petals often narrower and shorter than sepals; lip deflexed-pendent, often conspicuously larger than sepals or petals, apex usually unequally 2-lobed, base lacking a spur, but sometimes shallowly cup-shaped.

Neottia listeroides **Lindl.** (syn. *Listera lindleyana* (Decne) King & Pantl.; *Neottia dongrergoensis* Schltr.; *N. lindleyana* Decne) Plant 15–35 cm tall, **stem** papillose-downy above. **Inflorescence** 6–15 cm long, up to 20-flowered, sometimes more, with a papillose-downy axis. **Flowers** small, pale brown with densely downy ovaries; sepals and petals 4–5 mm long; lip narrowly obovate-oblong, 6–8(–9) × 3–4 mm, apex deeply 2-lobed, lobes spreading, almost parallel, almost ovate or ovate-lanceolate, minutely ciliate. c & sGansu, Qinghai, wSichuan, seTibet, nwYunnan [swTibet, Himalaya, Myanmar]; forests, shady grassy slopes, 2,500–3,900 m. VII–VIII.

Neottia acuminata **Schltr.** (syn. *N. micrantha* Lindl.) Plant 14–30 cm with erect, glabrous **stem** bearing 3–5, membranous, stem-clasping sheaths below middle. **Raceme** usually more than 20-flowered, with glabrous axis. **Flowers** small, yellowish brown, often in 3–4-clustered whorls; sepals narrowly lanceolate, long-pointed, 3–5 mm long; petals narrowly lanceolate, 2–3.5 mm long; lip ovate, ovate-lanceolate or lanceolate, long-pointed or blunt, 2–3.5 mm long. Gansu, Hubei, Qinghai, Sichuan, Shaanxi, Tibet, nYunnan [nChina, Japan, Korea, eRussia, Sikkim, Taiwan]; deep shade in forests, shady grassy slopes, 1,500–4,100 m. VI–VIII.

POGONIA

A genus with 4 species (3 in China) from east Asia and North America. Terrestrial **herbs**, with ascending, short **rhizomes** and an erect **stem** bearing a solitary, elliptic to oblong-lanceolate **leaf**, usually above middle. **Flower** medium-sized, usually single and terminal, rarely 2–3 and racemose; bract leaf-like, but smaller than leaf; sepals free, similar; petals often slightly wider and shorter than sepals; lip 3-lobed or almost entire, lacking a spur, mid-lobe often with fringed or bearded appendages above and towards tip.

Pogonia yunnanensis Finet* Dwarf **herb** to 9 cm with an almost erect, green leaf 1–4.5 cm long. **Flower** solitary, almost erect, pink with a red lip, not opening widely; sepals and petals free, 13–17 mm long; lip narrow-oblong, 3-lobed, callus consisting of 2 ridges that fuse in front to form a keel. nwYunnan, seTibet; alpine grassland, *Rhododendron* thickets, coniferous forests, 2,300–3,300 m. VI–VII.

GALEOLA

A genus with c. 10 species (4 in China) mainly distributed in tropical Asia, and Japan to New Guinea and Madagascar. Erect or climbing, saprophytic **herbs** with a thick **rhizome** and yellowish brown or reddish brown **stems**, lacking green **leaves**. **Inflorescence** terminal or lateral, unbranched, sometimes with bracts, with many fleshy flowers. **Flowers** usually yellow or tinged with reddish brown; sepals separate, often hairy beneath; petals slightly smaller than sepals, glabrous; lip entire, usually concave and sac-like but lacking a spur, callus of longitudinal ridges or a smooth knob.

1. *Epipactis helleborine*; Cangshan, nwYunnan (PC)
2. *Epipactis mairei*; Wolong, wSichuan (CGW)
3. *Epipactis mairei*; Wolong, wSichuan (PC)
4. *Epipactis royleana*; Braga, Nepal (CGW)
5. *Listera puberula*; Huanglong, nwSichuan (PC)
6. *Neottia listeroides*; Huanglong, nwSichuan (DR)
7. *Neottia acuminata*; Jiuzhaigou, nwSichuan (PC)
8. *Pogonia yunnanensis*; Cangshan, nwYunnan (PC)

ORCHIDACEAE: Galeola

Galeola lindleyana (Hook. f. & Thomson) Rchb. f. (syn. *Cyrtosia lindleyana* Hook. f. & Th.; *Galeola kwangsiensis* Hand.-Mazz.; *G. lindleyana* var. *unicolor* Hand.-Mazz.) **Stems** 1–3 m, reddish brown. **Inflorescence** terminal and lateral, each 2–5 cm long and up to 10-flowered. **Flowers** yellow, to 3.5 cm in diameter; sepals elliptic to ovate-elliptic, 1.6–2 × 9–11 cm, densely downy beneath and with keeled protrusions; lip nearly almost globose, 1.3 cm in diameter, margin shortly fringed, finely hairy above and with a smooth callus near base. Guizhou, Sichuan, sShaanxi, seTibet, Yunnan [c & sChina, Bhutan, India]; open forests, shrubberies, damp, stony places in valleys, 700–2,200 m. VI–VII.

Galeola faberi Rolfe* Differs from the previous species primarily in the glabrous **bracts** and ridged **lip-callus**. cGuizhou, swSichuan, Yunnan; open forests, humid places in bamboo thickets, 1,800–2,300 m. V–VII.

GASTRODIA

A genus of c. 20 species (13 in China) distributed from the Himalaya to Japan, with some species extending to South East Asia, Australia and New Zealand. **Saprophytic** rhizomatous **herbs**. **Inflorescence** terminal, erect, often yellowish brown, usually elongating after flowering, nodes bearing tubular or scale-like sheaths, several–many-flowered. **Flowers** nearly urn-shaped, campanulate or broadly cylindrical, resupinate or not; sepals and petals united to form a perianth-tube, only upper part separate; lip attached to end of column-foot, usually small, enclosed in perianth-tube, unlobed or 3-lobed.

Gastrodia elata Blume Plant 0.3–1 m, occasionally to 2 m tall with an erect, orange, yellow or buff-coloured **stem**, with several membranous sheaths below. **Inflorescence** 30–50-flowered. **Flowers** orange, pale yellow, or yellowish white,

1. *Galeola lindleyana*; Punakha, Bhutan (PC)
2. *Gastrodia elata*; cult. Kunming, cYunnan (PC)
3. *Gastrodia elata* tubers; market in Anlong, swGuizhou (PC)
4. *Calypso bulbosa* var. *speciosa*; Huanglong, nwSichuan (PC)
5. *Changnienia amoena*; cult. (HP)
6. *Corallorhiza trifida*; Huanglong, nwSichuan (PC)
7. *Cremastra appendiculata*; cult. (PC)
8. *Oreorchis oligantha*; Jiuzhaigou, nwSichuan (PC)

almost erect, perianth-tube c. 10 × 5–7 mm, with 5 lobes at apex; lip oblong-egg-shaped, 6–7 mm long, 3-lobed, upper surface papillose, margin irregularly short-fringed. Gansu, Guizhou, Hubei, Shaanxi, seTibet, Yunnan [sw & scTibet, c, e & neChina, c & eHimalaya, Japan, Korea, Mongolia, eRussia, Taiwan]; open forests, forest margins, shrubberies, 400–3,200 m. V–VII.

CALYPSO

A circumboreal genus of a single species. Dwarf terrestrial plant with an underground **corm** and a single, erect, pleated **leaf** with a slender stalk. **Flower** solitary, showy; sepals and petals erect; lip slipper-shaped, bearing 2 short spurs at the back.

Calypso bulbosa (L.) Ames var. *speciosa* (Schltr.) J. J. Wood Plant to 10 cm in flower with an ovate, dark green **leaf** up to 5 cm long. **Flower** c. 20 mm across; sepals and petals pink; lip white with yellow and purple marks, hairy in front. sGansu, Sichuan, seTibet [swTibet, circumboreal including nChina, Japan]; deep, damp moss, shady habitats in coniferous woodland and forests, often gregarious, 2,800–3,200 m. V–VII.

CHANGNIENIA

A genus with 1 species endemic to China. Small, **perennial** terrestrial **herbs** with underground, corm-like **pseudobulbs** bearing a single, pleated, almost erect **leaf** on a slender stalk. **Flower** solitary, large, lip spurred at base, spur conical, tapering.

Changnienia amoena S. S. Chien* Plant 10–17 cm tall in flower. **Leaf** ovate-elliptic to broadly elliptic, 6.5–11 × 5–8 cm. **Flower** 5–6 cm across, showy; sepals and petals pink; lip white, covered in small purple spots; dorsal sepal and petals forming a hood over the column, lateral sepals spreading; lip 3-lobed, side-lobes curved over column, mid-lobe deflexed, callus consisting of 5 ridges; spur tapering to apex, 20–23 mm long. Hubei, Sichuan [ecChina]; humus-rich soil in open woods, shady places in ravines, 700–1,800 m. IV.

CORALLORHIZA

A genus of c. 12 species (1 in China) circumboreal in distribution, but best represented in the Americas, south to Mexico and Guatemala. Small, leafless, **saprophytic** plants growing from coral-like **rhizomes**, with erect, few-flowered **inflorescences**. **Flowers** small, with a 3-lobed lip, spurless.

Corallorhiza trifida Chat. Plant 8–22 cm with a clump of 1–5 erect **stems**, buff-coloured with reddish brown sheaths. **Inflorescence** 3–7-flowered. **Flowers** small, pale yellow, white or buff-green, lip white, marked with purple in throat; sepals and petals spreading, dorsal sepal and petals linear, 4–6 mm long, curved forward over column, lateral sepals deflexed, linear, 4–6 mm long; lip recurved, 3-lobed, 2.5–3.5 mm long, with small, erect side-lobes, mid-lobe often notched at apex. Gansu, nGuizhou, Qinghai, Sichuan [n & neChina, Japan, Korea, North America, Europe, Russia]; coniferous woodland in deep, damp moss and leaf-litter, 2,000–2,700 m. VI–VIII.

CREMASTRA

A genus of c. 3 species (2 in China) from east Asia and the Himalaya. Terrestrial plants with a subterranean, egg-shaped **pseudobulb**, bearing a single **leaf** that is almost erect-arching, pleated, elliptic-obovate and with a distinct stalk. **Inflorescence** lateral, erect, rather dense, many-flowered. **Flowers** semi-nodding, somewhat tubular.

Cremastra appendiculata (D. Don) Makino Plant to 70 cm, the solitary **leaf** 18–34 cm long. **Inflorescence** 25–70 cm tall, bearing scented **flowers** 2–3 cm long; sepals and petals pale creamy white to buff-coloured or pinkish yellow, linear, ribbon-like; lip white mottled with pink or purple, spathulate, 3-lobed at apex. Sichuan, Guizhou, Tibet, Yunnan [e & sChina, c & eHimalaya, seAsia, Japan, Taiwan]; damp places in evergreen and mixed forests, ravines, 500–2,900 m. V–VI.

OREORCHIS

A genus with c. 16 species (10 in China) distributed from the western Himalaya to east Asia. Terrestrial **herbs**, with slender underground corm-like **rhizomes**. **Leaves** 1–2 at apex of **pseudobulb**, pleated, linear to oblong-lanceolate, stalked. **Inflorescence** lateral, arising from a node of the pseudobulb, erect, racemose, several–many-flowered; with membranous, persistent floral bracts. **Flowers** small to medium-sized; sepals and petals free, ± similar, lateral sepals sometimes ± shallowly pouched at base; lip 3-lobed or entire, base clawed, lacking a spur, callus of 1 or 2 longitudinal ridges.

1. Sepals and petals chestnut or purple-brown . . . . *O. oligantha*
1. Sepals and petals yellow or white 2
2. Leaf not more than 13 cm long . . . . *O. nana, O. erythrochrysea, O. foliosa*
2. Leaf 14–28 cm long . . . . *O. fargesii, O. patens*

Oreorchis oligantha Schltr.* **Pseudobulbs** oblong to almost egg-shaped, 6–9 mm long, bearing a single ovate to narrowly elliptic **leaf** 2–4 cm long, leaf-stalk 1–3 cm long. **Inflorescence** 7–12 cm long, almost erect, raceme 2.5–6 cm long, 2–5-flowered. **Flowers** c. 10 mm across, chestnut or purple-brown with white, purple-spotted lip; sepals narrowly oblong, 6–7 mm long; petals 7–9 × c. 2 mm long; lip obovate-oblong in outline, 5–7 mm long, 3-lobed below middle, disc with 2 short ridges at base. sGansu, wHubei, wSichuan, swShaanxi, eTibet, nwYunnan; alpine grassland, forests, thickets, limestone rock-pockets, 2,500–4,000 m. VI–VII.

Oreorchis nana Schltr.* **Pseudobulbs** oblong to almost egg-shaped, 6–9 mm long. **Leaf** 1, ovate to narrowly elliptic, 2–4 cm long, stalk 1–3 cm long. **Inflorescence** 10–20 cm long, almost erect, raceme 2.5–6 cm long, 5–14-flowered. **Flowers** c. 10 mm across with dark yellow sepals and petals with a chestnut reverse, lip white spotted purple; sepals narrowly oblong, 6–7 mm long; petals 5.5–6.5 × c. 2 mm long; lip obovate-oblong in outline, 5–7 mm long, 3-lobed below middle, disc with 2 short ridges at base. sGansu, wHubei, wSichuan, swShaanxi, eTibet, nwYunnan; alpine grassland, forests, thickets, limestone rock-pockets, 2,500–4,000 m. VI–VII.

Oreorchis erythrochrysea Hand.-Mazz. Differs from the previous species by the longer (8–20 mm) **pseudobulbs**; narrower and longer silver-striped **leaf** (6–13 cm long); and by racemes 5–11 cm long, with up to 20 flowers; **flowers** yellow, lip pale yellow or white with purple spots, almost oblong in outline, 5 mm long, 3-lobed near or below middle, disc with 2 short ridges between the side-lobes. swSichuan, seTibet, nw, c & swYunnan [eHimalaya]; open coniferous woods and forests, shrubberies, alpine meadows over limestone, 2,200–3,700 m. V–VI.

Oreorchis foliosa Lindl. Similar to *O. nana*, but **leaf** 12–13 cm long and **inflorescence** erect, 20–36 cm long, the raceme 5–9.5 cm long, with no more than 9, slightly larger **flowers**; sepals and petals dark yellow with many purplish brown veins and spots, lip white spotted purplish red, 3-lobed in middle or simple, pouched at base, disc lacking ridges. swSichuan, seTibet, nwYunnan [swTibet, c & eHimalaya]; forests, alpine meadows, 2,500–3,400 m. VI–IX.

Oreorchis fargesii Finet **Pseudobulbs** ellipsoid to almost globose, 1–2.5 cm long, normally with 2 (occasionally 1) **leaves** at apex of **pseudobulb**, linear-lanceolate or linear, 20–28 cm long. **Inflorescence** erect, 20–30 cm long, with a condensed raceme 2–6 cm long, usually with more than 10 flowers. **Flowers** white, with purple spots on lip; sepals oblong-lanceolate, 9–11 mm long; petals narrowly ovate to ovate-lanceolate, 9–10 mm long; lip oblong-obovate, 7.5–9 mm long, 3-lobed near base, disc with 1 short, ridged callus between the 2 side-lobes. sGansu, Hubei, Sichuan, sShaanxi, n & eYunnan [c & eChina, Taiwan]; forests, thickets, shady places along valleys, 700–2,600 m. V–VI.

Oreorchis patens (Lindl.) Lindl. Similar to the previous species, but **pseudobulbs** usually with a single, occasionally 2, **leaves**, 13–30 cm long; **inflorescence** 20–52 cm long with raceme 4.5–15.5 cm long; **flowers** yellowish brown to pale yellow, lip white, purple-spotted, 3-lobed, 6.5–8.5 mm long, bearing a disc with 2 thick ridges running from base to mid-lobe. Gansu, Guizhou, Sichuan, n, nw & seYunnan [c, e & neChina, Japan, Korea, Taiwan, eRussia]; forests, forest margins, shrubberies, grassy places, shady places along valleys, 1,000–3,000 m. VI–VII.

CALANTHE

A genus of c. 150 species (c. 50 in China) primarily in tropical and subtropical Asia, New Guinea, Australia, Pacific islands, tropical and South Africa and Madagascar. Terrestrial **herbs**, often tufted, with rather small, mainly conical, egg-shaped or almost globose **pseudobulbs**, bearing a few leaves. **Leaves** only partly developed or well-developed at flowering time, ± pleated, but often convolute when young, stalked or almost stalkless, sheathing, articulated or not. **Inflorescence** arising from leaf-axil or from base of pseudobulb, erect, usually densely hairy, racemose, few-many-flowered. **Flowers** often opening widely, small to medium-sized; sepals similar, free; petals free, often smaller than sepals; lip joined to base of column to form a tube, spurred or not at base; column short, stout, lacking a foot at base; stigmas lateral; pollen-masses waxy, 8, in 2 groups, equal or unequal.

1. Lip with a spur at least 5 mm long 2
1. Lip lacking a spur, or spur less than 5 mm long 3
2. Flowers rich purple .... *C. alpina*
2. Flowers white, cream, yellowish green, sometimes flushed lilac, purple, or blotched ....
   ... *C. alismifolia*, *C. davidii*, *C. graciliflora*, *C. herbacea*
3. Spur absent .... *C. tricarinata*
3. Spur present but very short . *C. hancockii*, *C. brevicornu*

Calanthe alpina Hook. f. ex Lindl. Plant up to 50 cm with elliptic to obovate-elliptic **leaves**, 11–26 cm long. **Inflorescence** dense, 3–10-flowered. **Flowers** nodding, rich purple with orange-purple lip apex; sepals and petals forming a cup over the lip, 15–20 mm long; lip cupped, fan-shaped, c. 15 mm long and broad, with deeply fringed front margin; spur 20–30 mm long. sGansu, w & nSichuan, sShaanxi, seTibet, wYunnan [swTibet, Bhutan, neIndia, Japan, Taiwan]; deep, damp shade in dense deciduous and mixed woodland, streamsides, 1,500–3,500 m. VI–VII.

Calanthe alismifolia Lindl. Plant up to 35 cm with 3–6 **leaves** per **pseudobulb** that are well developed at flowering time, elliptic to ovate-elliptic, 10–14 cm long, long-stalked. **Inflorescence** almost club-shaped, dense, 3–10-flowered. **Flowers** 15–20 mm across, white sometimes flushed with pale lilac on sepals; sepals and petals forming a cup over the lip, 8–10 mm long; lip with bright yellow callus, 3-lobed, side-lobes linear, spreading, mid-lobe fan-shaped, 15–20 mm long, with deeply notched front margin; spur 10 mm long. Hubei, Sichuan, seTibet, s & wYunnan [Bhutan, neIndia and Japan]; deep, damp shade in dense, evergreen forests, streamsides, sea level–1,800 m. VI–VII.

Calanthe davidii Franch. A far taller plant than the previous species, to 1.2 m in flower, with 3–4 **leaves** per **pseudobulb**, sword- or strap-shaped, 30–65 cm long. **Inflorescence** dense, bearing many small, finely downy, yellowish green or white

1. *Oreorchis nana*; Balangshan, Wolong, wSichuan (CGW)
2. *Oreorchis nana*; Jiuzhaigou, nwSichuan (PC)
3. *Oreorchis erythrochrysea*; Lancangjiang–Jinshajiang divide, Lidiping, nwYunnan (CGW)
4. *Oreorchis erythrochrysea*; Sanba, nwYunnan (PC)
5. *Oreorchis foliosa*; Chelai La, Bhutan (PC)
6. *Oreorchis fargesii*; Wolong, wSichuan (PC)
7. *Oreorchis patens*; Wolong, Sichuan (PC)
8. *Calanthe alismifolia*; Punakha, Bhutan (PC)
9. *Calanthe alpina*; Wolong, wSichuan (PC)
10. *Calanthe davidii*; Emeishan, swSichuan (PC)

1. *Calanthe graciliflora*; Fanjingshan, ne Guizhou (PC)
2. *Calanthe herbacea*; Jinlong, swGuangxi (PC)
3. *Calanthe tricarinata*; cult. (CGW)
4. *Calanthe tricarinata*; Wolong, wSichuan (PC)
5. *Calanthhe hancockii*; Baoshan, wYunnan (PC)
6. *Calanthe brevicornu*; Huanglong, nwSichuan (HP)
7. *Calanthe brevicornu*; Huanglong, nwSichuan (PC)
8. *Phaius delavayi*; Baishui, Yulongxueshan, nwYunnan (HJ)
9. *Phaius delavayi* form; Huanglong, nwSichuan (CGW)
10. *Phaius flavus*; Maolan, seGuizhou (PC)
11. *Phaius tankervilleae*; Maolan, seGuizhou (PC)
12. *Phaius wallichii*; Hong Kong (PC)
13. *Arundina graminifolia*; Longhua, swGuangxi (PC)

flowers, sometimes tinged with purple; sepals and petals reflexed, 6–9 mm long; lip 6–9 mm long, broadly triangular in outline, 3-lobed, mid-lobe 2-lobed, disc with 3 ridges; spur cylindrical, curved, 5–12 mm long, densely hairy inside. sGansu, Guizhou, Hubei, Sichuan, sShaanxi, seTibet, w, c & seYunnan [nwHunan, Taiwan]; forests, shady places along valleys, rocky streambeds, 500–3,300 m. VI–VII.

Calanthe graciliflora Hayata Plant up to 70 cm, readily distinguished from the previous two species by the elliptic to elliptic-lanceolate **leaves** (3–4 per **pseudobulb**) that develop after flowering to 33 cm long. **Flowers** slightly nodding, pale yellow flushed with brown, in a lax, many-flowered **inflorescence**; lip 10–15 mm long and broad, white or creamy white with yellow callus-ridges and 4 purple-brown blotches, 1–1.5 cm long and broad; spur incurved, 10–13 mm long. Guizhou, Hubei, Sichuan, Yunnan [eChina, Taiwan]; deep, damp shade in dense forests, ravines, streamsides, 600–1,500 m. V–VII.

Calanthe herbacea Lindl. Readily distinguished from the previous species by green, rather than silver-striped **leaves** and by narrower petals up to 2.5 mm broad. swGuangxi, seTibet, sYunnan [Himalaya, nVietnam]; monsoon forest over limestone rocks, 1,500–2,100 m. V–VII.

Calanthe tricarinata Lindl. Plant up to 60 cm with 3–4 **leaves** per **pseudobulb**, emerging at flowering time, elliptic to obovate-lanceolate, 20–30 × 5–11 cm. **Inflorescence** erect, lax, many-flowered. **Flowers** with yellow to yellowish green sepals and petals, 16–18 mm long; lip brick-red, 3-lobed, 8–10 mm long, mid-lobe orbicular or kidney-shaped, callus 3–5-ridged. sGansu, Guizhou, Hubei, w, n & nwSichuan, seTibet, w & nwYunnan [eChina, c & eHimalaya, Korea, Japan, Taiwan]; deep leaf-litter in deep shade in woodland and shrubberies, rarely on grassy slopes, 1,600–3,500 m. V–VI.

Calanthe hancockii Rolfe* Plant to 80 cm with **leaves** not well developed at flowering time, elliptic to elliptic-lanceolate, 20–40 cm long, stalks more than 20 cm long. **Inflorescence** up to 30 cm long, densely downy, few–30-flowered. **Flowers** large, slightly nodding, with an unpleasant smell, sepals and petals yellowish brown, lip lemon-yellow; sepals oblong-lanceolate, 25–35 mm long, petals almost elliptic, 23–25 mm long; lip 3-lobed, disc with 3 wavy ridges that are raised at apex; spur yellowish, slender, 2–3 mm long. nGuangxi, swSichuan, nw, c & seYunnan; evergreen broadleaved forests, shady places in valleys, 1,000–3,600 m. IV–V.

Calanthe brevicornu Lindl. Smaller plant than the previous species, to 30 cm, differing in the smaller **flowers** with yellow-green to pale green sepals and petals and purple lip; sepals and petals 12–23 mm long, spur very short, 2 mm long. Guangxi, w, sw & nSichuan, seTibet, wYunnan [c & eHimalaya]; deep leaf-litter in deep shade in woodland and scrub, streamsides, 1,600–2,700 m. V–VI.

PHAIUS

A genus of c. 50 species (8 in China, 4 endemic) in South East Asia, Indonesia and northern Australia. Terrestrial plants with habit like that of *Calanthe*, with several pleated **leaves**, but column elongated and ± free from lip. **Flowers** with a cylindrical spur.

1. Plant less than 30 cm tall . . . . *P. delavayi*
1. Plants more than 60 cm tall . . . .
. . . . *P. flavus, P. tankervilleae, P. wallichii*

Phaius delavayi (Finet) P. J. Cribb & Perner* (syn. *Calanthe delavayi* Finet) **Deciduous** plant growing from underground **rhizome** to 25 cm tall in flower. **Flowers** dull purple-pink or (in nwSichuan), with pale yellow sepals and petals and a dull pink striped lip; lip 20 mm long. sGansu, sw & nwSichuan, seTibet, nwYunnan; deciduous and mixed woodland, shady slopes, rocky streambeds, 2,900–3,300 m. Late V–VII.

Phaius flavus (Blume) Lindl. (syn. *P. maculatus* Lindl.) **Leaves** arching, elliptic, up to 50 cm long, green, often spotted with yellow or cream. **Inflorescence** up to 20 cm tall, few–20-flowered. **Flowers** yellow with an orange-red, wavy front margin to the lip. Guangxi, Guizhou, Sichuan, seTibet, Yunnan [s& seChina, seAsia, swPacific islands]; shady and damp places in evergreen and monsoon forest, 300–2,000 m. IV–X.

Phaius tankervilleae (Banks ex L'Hér.) Blume Plant usually more than 1 m, often forming large clumps, with almost erect, lanceolate, green **leaves** up to 70 cm long. **Inflorescence** with several pendent **flowers**; sepals and petals pale brown with white backs; lip purple, blunt. sGuizhou, Guangxi, seTibet, sYunnan [sChina, seAsia, neAustralia, swPacific islands]; waste places, forest margins, grassland, to 1,800 m. IV–VI.

Phaius wallichii Lindl. (syn. *P. magniflorus* Z. H. Tsi & S. C. Chen) Similar to *P. tankervilleae*, but **flowers** spreading and more open with reddish brown sepals and petals, with greenish yellow backs, and a white long-pointed lip, marked with purple. Guangxi, seTibet, sYunnan [swTibet, Guangdong, Himalaya]; shady, damp places in forests, along valleys, 750–1,000 m. III–VI.

ARUNDINA

A genus with a single species widespread in east and South East Asia. Terrestrial plant with reed-like **stems** bearing slender, lanceolate, pleated **leaves** and simple or 1–bracted, terminal **inflorescence**; bract small. **Flowers** very showy, large, borne in succession, with pink sepals and petals and a rose-purple lip with yellow throat.

Arundina graminifolia (D. Don) Hochr. Plant 40–100 cm with a stout, somewhat woody **stem** to 15 mm in diameter, covered by leaf-sheaths. **Leaves** many, alternate, linear-lanceolate, 8–20 ×

0.5–1.5 cm. **Flowers** pink with purple lip, 6–7 cm across, sepals and petals 2.5–4 cm long; lip trumpet-shaped, 2.5–4 cm long with a 3–5-ridged callus. Guizhou, Sichuan, seTibet, Yunnan [swTibet, s & eChina, s & seAsia, Himalaya, swPacific islands]; grassy banks, often by roads or paths, to 2,800 m. VI–IX.

BLETILLA

A genus of c. 6 species (4 in China, 3 endemic) distributed in east Asia. Plants with underground rhizomatous **corms** and linear-lanceolate, pointed **leaves**. **Inflorescence** racemose, erect to arching, several-flowered. **Flowers** showy, appearing in succession, usually 1–3 open at a time; sepals and petals similar; lip 3-lobed, bearing a callus of longitudinal raised ridges.

1. Flowers yellow to rusty yellow . . . .*B. ochracea*
1. Flowers pink with yellow ridges on lip . . . .
. . . . *B. striata, B. formosana*

Bletilla ochracea Schltr.* Plant to 50 cm, sometimes forming large clumps. **Flowers** several per **raceme**, sepals and petals ochre-coloured or yellow, suffused on backs with red or purple; lip 15–20 mm long, white with yellow callus and red spots. Guangxi, s & wGuizhou, wSichuan, s & wYunnan; roadside banks, grassland or rocky hillsides in full sun or light shade, to 1,800 m. IV–VI. Hybridises with *B. formosana* in wSichuan.

Bletilla striata (Thunb. ex A. Murray) Rchb. f.* Plant to 75 cm with rather reed-like **leaves**. **Flowers** large with rose-purple sepals and petals; lip 23–28 mm long, white with yellow callus-ridges and side-lobes. Guizhou, Sichuan, Yunnan; steep rocky banks in grassland and amongst bushes, in full sun or light shade, 1,600–2,500 m. V–VI.

Bletilla formosana (Hayata) Schltr.* Plant to 60 cm with grass-like **leaves**. **Flowers** with pink sepals and petals; lip 15–18 mm long, white with purple apex and yellow callus. w & swSichuan, s, w & nwYunnan; similar habitats, altitude and flowering time to *B. striata*. Hybridises with *B. ochracea* where they grow together.

ANTHOGONIUM

A genus of 1 species widespread in the Himalaya, South East Asia and south-west China. Terrestrial **herbs** with underground egg-shaped **corms** and grass-like pleated **leaves**; similar to *Bletilla* in habit but with obliquely set, tubular **flowers**.

Anthogonium gracile Lindl. Plant to 40 cm with erect, linear-lanceolate **leaves**, 7–37 × 1.5–3.5 cm. **Inflorescence** a several–many-flowered, short almost capitate raceme, stalks and ovaries purple. **Flowers** semi-pendent, held at an obtuse angle to the ovary, obliquely tubular but spreading at apex, 14–17 mm long, rose-purple or less commonly white; lip spathulate, 3-lobed at apex. Guangxi, Guizhou, seTibet, Yunnan

[c & eHimalaya, India, seAsia]; well-drained stony places, grassy slopes, scrub, light woodland, 1,600–2,000 m. VII–XI.

COELOGYNE

A tropical genus of c. 230 species (26 in China, 3 endemic) in the eastern Himalaya, South East Asia, Indonesia and west Pacific islands. Epiphytic or rock-dwelling creeping or mat-forming **rhizomatous** plants; **stems** pseudobulbous, 2-leaved at apex. **Inflorescence** terminal, erect or pendent, with 1–many showy **flowers**.

1. Inflorescence 1(–2)-flowered . . . . *C. fimbriata*
1. Inflorescence 2–more-flowered . . . .
. . . . *C. corymbosa, C. occultata, C. calcicola*

Coelogyne fimbriata Lindl. Small creeping rock-dweller or epiphyte with cylindrical 2-leaved **pseudobulbs**, 2–5 cm long, bearing oblong-elliptic **leaves**, 4–10 cm long. **Inflorescence** erect, 1- or rarely 2-flowered. **Flowers** with pale yellow or ochre-coloured sepals and petals, lip white marked with brown; sepals 16–20 mm long; petals linear-thread-like, 13–18 mm long; lip 3-lobed, 13–18 mm long, mid-lobe margin fringed, callus consisting of 2 ridges. Sichuan, Yunnan [seChina, seAsia, Himalaya]; boulders and tree-trunks in forest shade, scrub, 500–1,500 m. VIII–X.

Coelogyne corymbosa Lindl. Creeping epiphyte or rock-dweller with egg-shaped 2-leaved **pseudobulbs**, 1–4.5 cm long. **Leaves** oblong-obovate, 4.5–15 cm long. **Inflorescence** arching, 2–4-flowered, with white sepals and petals, lip white marked with 4 yellow eyes outlined in brown; sepals 18–35 mm long; petals oblanceolate, 18–33 mm long; lip 3-lobed, 16–28 mm long, mid-lobe ovate with entire margin, callus consisting of 2 or 3 ridges. seTibet, nw & seYunnan [swTibet, Himalaya, Myanmar]; boulders and tree-trunks in forest shade, damp cliffs, 1,300–3,100 m. V–VII.

Coelogyne occultata Hook. f.* Similar to *C. corymbosa* but with **leaves** up to 6 cm long and larger **flowers**, 5–6 cm across. seTibet, nw & swYunnan [swTibet]; similar habitats but generally at lower elevations, to 2,400 m. VII–VIII.

Coelogyne calcicola Kerr Plant 2-leaved. **Inflorescence** erect, several-flowered. **Flowers** white with fringed mid-lobe to lip, and callus of 3 ridges. swYunnan [Laos, Vietnam]; rocks in scrub, on trees, to 1,500 m. VI–VIII.

PLEIONE

A genus of c. 22 species (16 in China, mostly endemic) distributed in the central and east Himalaya and Myanmar to China and Vietnam. Epiphytic, rock-dwelling or terrestrial **herbs** with annual, often clustered, **pseudobulbs** that are egg-shaped, conical, pyriform or turbinate, gradually narrowed

1. *Bletilla ochracea*; Erlangshan, wSichuan (PC)
2. *Bletilla ochracea*; Maolan, seGuizhou (PC)
3. *Bletilla striata*; Dadujiang valley, wSichuan (HP)
4. *Bletilla formosana*; Dali, nwYunnan (PC)
5. *Bletilla formosana*; cult. (RE)
6. *Bletilla formosana* × *B. ochracea*; Baoxing, wSichuan (PC)
7. *Anthogonium gracile*; Bhutan (MW)
8. *Coelogyne fimbriata* straddles the Vietnam–Yunnan border (PC)
9. *Coelogyne fimbriata*; Laolaiphin, nVietnam (PC)
10. *Coelogyne corymbosa*; Fansipan, nVietnam (AF)
11. *Coleogyne calcicola*; Mongpang, swYunnan (PC)

1. *Pleione maculata*; cult. (PC)
2. *Pleione praecox*; cult. (PC)
3. *Pleione saxicola*; cult. (IB)
4. *Pleione forrestii*; cult., from Cangshan, nwYunnan (IB)
5. *Pleione scopulorum*; Doshong La, seTibet (AC)
6. *Pleione scopulorum*; Doshong La, seTibet (AC)
7. *Pleione grandiflora*; cult. (PC)
8. *Pleione albiflora*; cult. (IB)

toward apex to form a long or short neck, 1–2-leaved at apex. **Leaves** pleated, short-stalked, usually deciduous or falling in winter. **Inflorescence** solitary or paired, 1–3-flowered, arising from base of an old pseudobulb, appearing either before or after the leaves, accompanied by colourful, persistent floral bract. **Flowers** large, usually showy and opening widely; sepals usually free, similar; petals often slightly narrower than sepals; lip conspicuously larger than sepals, simple or inconspicuously 3-lobed, base ± pouched, apical margin irregular (appearing nibbled) or more deeply cut, with 2–several lines of ridges or hairs along the veins above.

1. Autumn-flowering 2
1. Spring-flowering; flowers variously coloured 3
2. Flowers basically white .... *P. maculata*
2. Flowers basically rosy red or purplish red ....
 .... *P. praecox, P. saxicola*
3. Flowers yellow .... *P. forrestii*
3. Flowers white, pink or rose-purple 4
4. Leaves 2 per pseudobulb; flowers pink . *P. scopulorum*
4. Leaves 1 per pseudobulb; flower colour various but not yellow 5
5. Flowers white, with brown or yellow on lip ....
 .... *P. grandiflora, P. albiflora*
5. Flowers pink to purple or rose-purple 6
6. Lip with 4–5 longitudinal lines of hairs ....
 .... *P. aurita, P. chunii*
6. Lip with 2–5 toothed or irregular longitudinal ridges ..
 .... *P. bulbocodioides, P. limprichtii,*
 .... *P. pleionoides, P. yunnanensis*

Pleione maculata (Lindl.) Lindl. ex Paxton **Pseudobulbs** green, top-shaped to pear-shaped, 1–3 cm long, bearing 2 elliptic-lanceolate to oblanceolate **leaves**, 10–20 cm long. **Flowers** almost erect or spreading horizontally, fragrant, appearing after leaves have fallen, white or occasionally tinged with purplish red; sepals and petals oblong-lanceolate, 30–42 mm long; lip ovate-oblong, 25–35 mm long, conspicuously 3-lobed in middle or lower part, with yellow centre and purple blotches on apical margin, with 5–7 papillose-toothed ridges extending to apex. wYunnan [c & eHimalaya, India, Myanmar, nThailand]; tree-trunks and mossy rocks in broadleaved forests, 600–1,600 m. X–XI.

Pleione praecox (J. E. Sm.) D. Don **Pseudobulbs** green spotted with purplish brown, top-shaped, 1.5–4 cm long, apex abruptly contracted into a conspicuous beak, outer sheaths warty, usually bearing 2 elliptic-oblanceolate to elliptic **leaves**, these 9–20 cm long. **Flowers** 1 or occasionally 2, appearing after leaves have fallen, pink or purplish red, rarely white; sepals and petals oblanceolate, 5–7 cm long; lip obovate-elliptic or elliptic, 40–50 mm long, slightly 3-lobed, with 3–5, fringed or papillose-toothed yellow ridges extending from base to mid-lobe. seTibet, s, sw & seYunnan [c & eHimalaya, Thailand]; tree-trunks and mossy rocks in forests, cliffs, 1,200–2,500(–3,400) m. IX–X.

Pleione saxicola T. Tang & F. T. Wang ex S. C. Chen Similar to the previous species, but **pseudobulbs** small, not more than 1.1 cm long, bearing a solitary **leaf** that develops at flowering time. **Flower** solitary, not opening widely, large and rosy red; lip broadly elliptic, 50–55 mm long, conspicuously 3-lobed, with claw 1.3 cm long, and with 3 entire or slightly wavy ridges extending to middle. seTibet, nwYunnan [Bhutan]; streamside cliffs, 2,400–2,500 m. IX.

Pleione forrestii Schltr. **Pseudobulbs** green, conical or egg-shaped-conical, 1.5–3 cm long, bearing a solitary **leaf** that appears after flowering, almost elliptic to narrowly elliptic-lanceolate, 10–15 cm long. **Flower** solitary, orange-yellow, pale yellow or yellowish white, very rarely pure white, with brown or crimson spots or markings on lip; sepals and petals oblanceolate or oblong-oblanceolate, 30–42 mm long; lip broadly obovate-elliptic or nearly broadly rhombic, 32–40 mm long, 3-lobed, with 5–7 entire, often slightly wavy ridges extending from base to mid-lobe. n & nwYunnan [neMyanmar]; humus-covered rocks and tree-trunks in open forests, forest margins, 2,200–3,150 m. IV–V.

Pleione scopulorum W. W. Sm. **Pseudobulbs** green, egg-shaped, 1–2.5 cm long, with conspicuous long neck, bearing 2 **leaves** that are well-developed at flowering time, lanceolate, oblanceolate or narrowly elliptic, 4–13 cm long. **Flowers** 1, rarely 2–3, rosy-purple or rarely white tinged with pale purplish blue, often with yellow centre and dark purple spots on lip; sepals and petals elliptic-lanceolate or narrowly ovate, 21–32 mm long; lip transversely elliptic or almost oblate, 20–25 mm long, almost entire, with 5–9, irregularly crested-incised ridges extending from base to upper part. seTibet, nw, w & swYunnan [neIndia, Myanmar]; rocky grassland in coniferous forests, mossy streamside rocks, shrubby subalpine meadows, 2,800–4,200 m. V–VII.

Pleione grandiflora (Rolfe) Rolfe **Pseudobulbs** green, almost conical, 3–4.5 cm long, bearing a solitary lanceolate, pointed leaf, 15–30 cm long. **Flowers** solitary, large, white, sometimes with dark purplish red or brown spots and streaks on lip, occasionally pale pink; sepals oblanceolate to narrowly elliptic, 50–55 mm long; petals sickle-shaped-oblanceolate, 50–55 mm long; lip broadly ovate-elliptic, 50–55 mm long, inconspicuously 3-lobed, apex notched, with 4–5(–7) irregularly fringed ridges down the centre. s, sw & seYunnan [nVietnam]; mossy forest rocks, 2,600–2,900 m. V.

Pleione albiflora P. J. Cribb & C. Z. Tang Distinguished from the previous species by the distinctly pouched base to the lip which has bold crimson-purple, brown or brownish yellow central stripes and a more finely fringed margin; sepals and petals rarely with pale mauve lines; stalk usually less than 8 cm long (8–12 cm long in *P. grandiflora*). nwYunnan [neMyanmar]; tree-trunks or mossy rocks and cliffs, in shady places, 2,400–3,300 m. IV–V.

Pleione aurita **P. J. Cribb & Pfennig*** **Pseudobulbs** pale green, occasionally with purple spots, usually clustered in large colonies, egg-shaped to conical, 2–4.5 cm long, with conspicuous neck, bearing a solitary elliptic-lanceolate or narrowly elliptic leaf, 6–20 cm long. **Flower** solitary, large, pale pinkish red to rose-purple; sepals and petals narrow-elliptic, oblong-elliptic or oblanceolate, 40–50 mm long, petals strongly reflexed; lip broadly fan-shaped when flattened, 39–40 mm long, inconspicuously 3-lobed, with longitudinal yellow or orange-yellow stripe. n & nwGuangxi, Guizhou, Hubei, seTibet, w & seYunnan [nGuangdong]; tree-trunks in coniferous and mixed deciduous forests, cliffs at thicket margins, 1,400–3,100 m. IV–V.

Pleione chunii **C. L. Tso*** Similar to the previous species, but **flowers** paler pink with a whitish lip marked with orange, and petals spreading but not recurved. wYunnan [seChina]; similar habitats, to 2,800 m. IV–V.

Pleione bulbocodioides **Rolfe*** (syn. *P. delavayi* (Rolfe) Rolfe; *P. henryi* (Rolfe) Schltr.) **Pseudobulbs** green, sometimes purple-flushed, egg-shaped to almost conical, 1–2.6 cm long, with conspicuous neck, bearing a solitary narrowly elliptic-lanceolate or almost oblanceolate **leaf**, 10–25 cm long. **Flowers** 1, rarely 2, pink to pale purple, with dark purple markings on lip; sepals and petals narrowly elliptic to oblanceolate, 30–50 mm long; lip obovate or broadly obovate when flattened, 35–45 mm long, inconspicuously 3-lobed, with 4–5 irregular (appearing nibbled) ridges. sGansu, Guangxi, Guizhou, Hubei, Sichuan, sShaanxi, seTibet, nw & cYunnan [cChina]; humus- or leaf-covered soil, mossy rocks in evergreen broadleaved forests, margins of shrubberies, ravines, 900–3,600 m. IV–VI.

Pleione limprichtii **Schltr.** Very similar to the previous species, but with larger pseudobulbs, 3–4 cm long. **Flowers** purplish red to rosy-pink, lip 25–40 mm long, with brick-red spots and finely toothed, white ridges. w & swSichuan, Yunnan [Myanmar]; humus-covered or mossy rocks and cliffs, 2,000–2,500 m. IV–V.

Pleione pleionoides **(Kraenzl. ex Diels) Braem & H. Mohr*** Closely related to *P. bulbocodioides*, but **pseudobulbs** conical, 2.5–3 cm long, with a rough surface. **Flowers** bright rose-purple, with 2 or 4, yellow or white, finely toothed ridges on lip; sepals and petals narrowly elliptic to oblanceolate, 40–65 mm long; lip almost rhombic to obovate in outline when flattened, 42–55 mm long. Guizhou, wHubei, eSichuan; humus-covered and mossy rocks or cliffs in forests, 1,700–2,300 m. VI.

Pleione yunnanensis **(Rolfe) Rolfe** Readily distinguished from the previous three species by the generally paler purplish, pink, or sometimes nearly white, **flowers**, with purple or deep red spots on lip which has 3–5 entire, wavy, ridges down the centre; lip nearly broadly obovate, 30–40 mm long, 3-lobed. n, nw & wGuizhou, swSichuan, seTibet, nw, c & seYunnan [nMyanmar]; mossy rocks, rocky places in forests, forest margins, grassy slopes, 1,100–3,500 m. IV–V. Hybrids with *P. bulbocodioides*, *P.* ×*taliensis* **P. J. Cribb & Butterfield***, are found in the wild, particularly in and near Cangshan (nwYunnan).

ERIA

A genus of c. 370 species (44 in China, mainly confined to the tropical south) distributed in tropical and subtropical Asia to north-east Australia and south-west Pacific Islands. Epiphytic or rarely terrestrial **herbs**, usually with short **rhizomes**. **Stem** often thickened into fleshy, 1–many-noded pseudobulbs, bearing 1 to many **leaves**. **Inflorescence** lateral or terminal, often racemose, rarely reduced to single **flower**; sepals free, rarely joined, lateral sepals joined at base to column; petals similar to dorsal sepal or smaller; lip stalkless on apex of column-foot, lacking a spur, 3-lobed to entire, usually with a callus usually of longitudinal ridges.

Eria coronaria **(Lindl.) Rchb. f.** Plant with creeping **rhizomes**, bearing sheaths 6–7 mm long, **pseudobulbs** contiguous or 1–2 cm apart, cylindrical, 5–15 cm long. **Leaves** 2, almost terminal, narrow-elliptic or obovate-elliptic, 6–16 cm long. **Inflorescence** arising between the 2 leaves, 10–30 cm long, 2–6-flowered. **Flowers** white, with purple stripes and yellow centre to the lip; sepals spreading, elliptic-lanceolate, 15–17 mm long; petals oblong-lanceolate, 15–17 mm long; lip oblong in outline, 3-lobed, 14–15 mm long, disc with 3 entire or wavy ridges. Guangxi, seTibet, s & nwYunnan [seChina, c & eHimalaya, Laos, Thailand, Vietnam]; tree-trunks, rocks in ridge-top forests, 1,300–2,000 m. V–VI.

Eria rhomboidalis **T. Tang & F. T. Wang*** (syn. *Conchidium rhomboidale* (T. Tang & F. T. Wang) S. C. Chen & J. J. Wood) In contrast to the previous species, the **pseudobulbs** are egg-shaped, 2–4 cm apart, borne on slender **rhizomes**; the **leaves** are smaller, 2–5 × 0.6–1.5 cm; and the **inflorescence** is only 2–3 cm long, 1-flowered; **flowers** red-brown with white lip, sepals and petals 10–12 mm long, lip almost rhombic, 10–13 × 7–10 mm. swGuangxi, swGuizhou, seYunnan [Hainan]; limestone rocks in evergreen forests, 700–1,300 m. IV–V.

BULBOPHYLLUM

A large pantropical genus of perhaps 1,500 species (c. 100 in China, mainly in the tropical south, 35 endemic). Epiphytic or rock-dwelling plants with a stout **rhizome** with short, egg-shaped, fleshy **pseudobulbs**, each bearing 1 leathery **leaf**. **Inflorescence** 1–many-flowered, either a raceme or ± an umbel. **Flowers** relatively small; lateral sepals often much longer than dorsal one, sometimes joined; lip fleshy, mobile; column short, often with apical points.

1. *Pleione aurita*; cult. (PC)
2. *Pleione chunii*; cult. (IB)
3. *Pleione bulbocodioides*; Yulongxueshan, nwYunnan (RS)
4. *Pleione bulbocodioides*; Habashan, nwYunnan (CGW)
5. *Pleione limprichtii*; cult. (RBGK)
6. *Pleione pleionoides*; cult. (PC)
7. *Pleione yunnanensis*; Jian Chuan to Xiaguan, wYunnan (PC)
8. *Pleione yunnanensis*; cult. (PC)
9. *Eria coronaria*; Laovachai, nVietnam (PC)
10. *Eria coronaria*; Laovachai, nVietnam (PC)
11. *Eria rhomboidalis*; Longhua, swGuangxi (PC)

ORCHIDACEAE: Bulbophyllum

1. Flowers yellow .... *B. forrestii, B. funingense*
1. Flowers purplish or whitish .... *B. andersonii*

Bulbophyllum forrestii Seidenf. Plant forming dense mats, **pseudobulbs** egg-shaped, 2–3 cm long. **Leaf** almost erect, leathery, elliptic, 15–25 × 1.3–2.8 cm. **Flowers** yellow, almost in an umbel, all opening together; lateral sepals much longer than dorsal one, 15–25 mm long. s & wYunnan [Myanmar, Thailand]; rocks in open or light shade, tree-trunks in forests, 1,800–2,000 m. V–VI.

Bulbophyllum funingense Z. H. Tsi & S. C. Chen* **Pseudobulbs** well-separated, egg-shaped, ribbed, 2–3 cm long bearing thick, leathery, narrowly oblong, blunt and slightly notched **leaves**, 11–16 × 2.2–2.5 cm. **Scape** erect, 8–11 cm long, with 2 large, deep yellow **flowers** with reddish brown veins; dorsal sepal ovate, blunt and apiculate, 18 × 8 mm, lateral sepals oblong, up to 43 × 8 mm, slightly joined to each other along the basal edges, twisted near base; petals almost ovate-triangular, truncate and mucronate, 10 mm long; lip ovate-lanceolate, 13 mm long, basal margins incurved and ciliate, finely papillose above, with 1 central keel. seYunnan; rocks along valleys, c. 1,000 m. IV.

1. *Bulbophyllum forrestii*; north-west of Tengchong, wYunnan (PC)
2. *Bulbophyllum funingense*; Malipo, seYunnan (PC)
3. *Bulbophyllum andersonii*; Dongvan, nVietnam (PC)
4. *Dendrobium aurantiacum*; cult. Kunming, cYunnan (PC)
5. *Dendrobium fanjingshanense*; Fanjingshan, ne Guizhou (PC)
6. *Dendrobium chrysotoxum*; cult. Kunming, cYunnan (PC)
7. *Dendrobium fimbriatum*; Maguan, seYunnan (PC)
8. *Dendrobium hancockii*; cult. Guangdong (PC)

Bulbophyllum andersonii (Hook. f.) J. J. Sm. (syn. *Cirrhopetalum andersonii* Hook. f.; *Bulbophyllum henryi* (Rolfe) J. J. Sm.; *Cirrhopetalum henryi* Rolfe) **Pseudobulbs** egg-shaped-conical or narrowly egg-shaped, 2–5 cm long, 3–11 cm apart on the **rhizome**. **Leaf** oblong, 7–21 × 1.6–4.3 cm. **Inflorescence** up to 17 cm long. **Flowers** in an umbel, purplish or whitish, densely spotted with purple; lateral sepals oblong, 15–20 mm long, twisted near base; petals oblong or sickle-shaped-oblong, 3 × 1 mm, apex with awn c. 0.8 mm long; lip fleshy, purple, ovate, recurved. Guangxi, swGuizhou, swSichuan, nw, c & seYunnan [eHimalaya, Myanmar, Vietnam]; tree-trunks in forests, limestone rocks, 400–2,000 m. II–XI.

DENDROBIUM

A large genus of c. 1,000 species (c. 75 in China) distributed throughout tropical and subtropical Asia as far as Australia and the Pacific islands. **Epiphytes** with clustered **stems** that are erect or pendulous, cylindrical or compressed–3-angled, rarely branched, with few-many nodes, cylindrical or sometimes pseudobulbous, fleshy or rigid, usually covered by sheathing leaf-bases. **Leaves** few–many borne at apex of, or strictly alternating along, **stem**, leathery, articulated at base. **Inflorescence** racemose, erect, arching or pendulous, usually arising from nodes towards apex of stem, few–many-flowered. **Flowers** variable in size, usually wide-opening and showy; sepals and petals similar, free; lateral sepals joined to base of column-foot, forming a chin-like mentum with lip attached to end of column-foot.

1. Flowers golden orange-yellow or orange, or white with a golden yellow lip 2
1. Flowers white, often flushed or marked with purple or purple-red, usually with some yellow, red or orange on lip 3
2. Inflorescence usually 1–3-flowered .... **D. aurantiacum, D. fanjingshanense**
2. Inflorescence 6-many-flowered ..... **D. chrysotoxum, D. fimbriatum, D. hancockii, D. thyrsiflorum**
3. Leaf-sheaths with black hairs .... **D. williamsonii, D. longicornu**
3. Leaf-sheaths without black hairs 4
4. Flowers small, sepals and petals not more than 17 mm long .... **D. moniliforme**
4. Flowers larger, sepals and petals 20–45 mm long ..... **D. lituiflorum, D. nobile, D. aphyllum, D. devonianum, D. loddigesii, D. wardianum**

Dendrobium aurantiacum Rchb. f. **Stems** tufted, slender, usually 25–35 cm long, many-noded, leafy in upper part. **Leaves** linear or narrowly oblong, 8–10 cm long. **Inflorescence** arising from upper nodes of leafless stems, 1–14 cm long with 1–3 or more golden or orange-yellow **flowers**, sometimes lip with 1–2 purple blotches or several red stripes; sepals oblong to oblong-elliptic, 23–25 mm long; petals elliptic or broadly elliptic-obovate, 24–26 mm long; lip almost orbicular, c. 25 mm long, densely hairy on reverse and with irregularly toothed margin. nw, w & swGuangxi, s & swGuizhou, c & sSichuan, Yunnan [Hainan, c & eHimalaya, India, Laos, Myanmar, Thailand, Taiwan, Vietnam]; tree-trunks in forests or rocks and cliff-faces along valleys, 600–2,600 m. V–VI.

Dendrobium fanjingshanense Z. H. Tsi ex X. H. Jin & Y. W. Zhang* **Stems** tufted, terete, slender, 20–40 cm long, leafy in upper part. **Leaves** oblong-lanceolate, apex blunt and somewhat hooked, 2–5 × 0.5–1.5 cm. **Inflorescences** several, arising from upper part of defoliated stem, 1–2-flowered. **Flowers** opening widely, yellow to orange-yellow with fan-shaped purplish blotch on disc of lip; sepals and petals back-rolled, slightly wavy-margined, 20 × 5–7 mm; petals almost elliptic, c. 20 × 6 mm; lip almost oblong-lanceolate in outline, with purplish callus near base, densely hairy on blotch, side-lobes small, erect; mid-lobe ovate, reflexed at apex, 10 × 7 mm, centrally keeled. neGuizhou (Fanjingshan); tree-trunks in forests, 800–1,500 m. IV–V.

Dendrobium chrysotoxum Lindl. (syn. *Dendrobium suavissimum* Rchb. f.) **Stems** club-shaped to spindle-shaped, 6–30 cm long, ridged, 2–5-leaved near apex. **Leaves** oblong, 19 × 2–3.5 cm. **Inflorescence** almost terminal, arching or ± pendulous, 15–20 cm long, lax, many-flowered. **Flowers** golden yellow, tinged with orange-brown on lip, slightly fragrant; sepals and petals 12–20 mm long; lip nearly round, c. 20–23 mm across, densely woolly-downy, base often with red stripes on both sides, margin irregular (appearing nibbled) or wavy. s, sw & wYunnan [neIndia, Laos, Myanmar, Thailand, Vietnam]; tree-trunks and branches in evergreen broadleaved forests, rocks in open forests, 500–1,600 m. III–V.

Dendrobium fimbriatum Hook. **Stems** cylindrical or sometimes slightly swollen above base, 50–100 cm long. **Leaves** many, deciduous by flowering time, oblong or oblong-lanceolate, 8–15.5 × 2–3.6 cm. **Inflorescence** arising from upper part of leafless stem, 5–15 cm long, sparsely 6–12-flowered. **Flowers** wide-opening, slightly fragrant, golden yellow, with darker lip adorned with purplish red stripes towards base and with a deep purple disc in centre; sepals and petals 13–19 mm long; lip almost orbicular, 15–20 mm long, margin fringed, disc densely downy. s, w & nwGuangxi, s & swGuizhou, s, se & wYunnan [c & eHimalaya, Myanmar, Thailand, Vietnam]; tree-trunks in dense forests, damp rocks along valleys, 600–1,700 m. IV–VI.

Dendrobium hancockii Rolfe* Differs from the previous species in the cylindrical **stems** bearing several swollen and spindle-shaped internodes near the base, 50–80 cm long overall, and usually branched. **Flowers** usually on leafless stems 1–2.5 cm long, slightly fragrant, smaller; sepals ovate-elliptic to ovate-lanceolate, 18–24 mm long; petals obliquely obovate or almost elliptic, 18–24 mm long; lip broadly ovate in outline, 1–2 cm long, 3-lobed near middle, usually yellow-green or pale green, with several red stripes on side-lobes.

sGansu, nwGuangxi, s & swGuizhou, seHubei, ne, c & sSichuan, sShaanxi, seYunnan [Henan, Hunan]; tree-trunks in forests, rocks along valleys, 700–1,600 m. V–VI.

Dendrobium thyrsiflorum Rchb. f. Readily distinguished from the previous species by the white sepals and petals, contrasting with the golden lip; **stems** stout, cylindrical-club-shaped, 12–46 cm long bearing 3 or 4 oblong or oblong-lanceolate **leaves**, 9–16 × 2.4–5 cm, on upper part of stem; **flowers** in a pendulous, dense, many-flowered inflorescence. s, se & wYunnan [neIndia, Laos, Myanmar, Thailand, Vietnam]; tree-trunks and rocks in scrub and forests, 700–1,800 m. IV–V.

Dendrobium williamsonii Day & Rchb. f. **Stems** cylindrical, sometimes swollen and spindle-shaped, to 20 cm long. **Leaves** several, usually alternate towards top of stem, oblong, 7–9.5 × 1–2 cm. **Inflorescence** arising from near apex of leafy stem, 1–2-flowered. **Flowers** with white sepals and petals, 2.5–3.4 × 0.6–0.9 cm; lip 3-lobed, 2.5 cm long, white with red central mark, mid-lobe almost orbicular or broadly elliptic, pointed, with a wavy margin. n & nwGuangxi, w, s & seYunnan [seChina, neIndia, Myanmar, Vietnam]; epiphytic in hill forests, 900–1,300 m. IV–V.

Dendrobium longicornu Lindl. Similar to *D. williamsonii*, but with smaller sepals and petals, 1.5–2 cm long, a longer spur-like mentum and an orange mark on the fringe-margined lip. sGuangxi, seTibet, nw, c & sYunnan [c & eHimalaya, Vietnam]; epiphytic in montane forests, 1,200–2,500 m. IV–V.

Dendrobium moniliforme (L.) Sw. **Stems** erect, cylindrical, usually 10–20 cm long, with many nodes and with several lanceolate or oblong **leaves**, 3–4.5 cm long on upper part. **Inflorescence** 2–several, arising from upper part of leafless or leafy stem, usually 1–3-flowered. **Flowers** fragrant, white or yellowish greenish, sometimes tinged with purplish red; sepals and petals similar, ovate-oblong or ovate-lanceolate, 13–17 mm long; lip 3-lobed below middle, ovate-lanceolate, slightly shorter than sepals, usually with a transverse, brownish red, spotted band and a purplish or yellowish blotch. sGansu, ne, nw & eGuangxi, ne, se & eGuizhou, Hubei, swSichuan, sShaanxi, seTibet, nw, c & seYunnan [c & eChina, Japan, sKorea, Taiwan]; tree-trunks in broadleaved forests, valley cliffs, 600–3,000 m. III–V.

Dendrobium lituiflorum Lindl. Similar to the previous species, but **stems** pendulous and **leaves** slightly longer. **Inflorescences** many, arising from bare **stems**, usually 1–2-flowered; bracts whitish, ovate, 1–13 mm long. **Flowers** opening widely, purple, with large deep purple blotch centrally, surrounded by white edge on lip; stalk and ovary purplish; sepals oblong-lanceolate, 3.2–3.5 cm long; petals almost elliptic, 3.5–4 cm long; lip shorter than petals, trumpet-shaped below middle and clasping the column. sw & wGuangxi, s & swYunnan [neIndia, Laos, Myanmar, Thailand]; tree-trunks in broadleaved forests, isolated village trees, 500–1,700 m. III–V.

Dendrobium nobile Lindl. **Stems** tufted, erect, cylindrical, 10–60 cm long, unbranched, with many nodes, bearing oblong **leaves**, 6–11 cm long. **Inflorescence** arising from leafy or leafless stems, 2–4 cm long, 1–4-flowered. **Flowers** wide-opening, white tinged in varying degrees with purplish or purplish pink; sepals oblong, 25–35 mm long; petals oblique, broadly ovate, 25–35 mm long; lip broadly ovate, blunt, 25–35 mm long, densely downy, with dark purple blotch on disc and several purplish red stripes on the sides at base. Guangxi, swGuizhou, sHubei, seTibet, nw, c & seYunnan [sw & scTibet, seChina, c & eHimalaya, Hong Kong, Laos, Myanmar, Thailand, Vietnam]; tree-trunks in evergreen, broadleaved forests, valley rocks, 500–1,700 m. IV–V. Widely cultivated in southern China.

Dendrobium aphyllum (Roxb.) C. E. C. Fischer Similar to *D. nobile* and *D. lituiflorum* and flowering on leafless **stems**. **Flowers** pink with cream lip; sepals and petals 2–2.5 cm long; lip broadly obovate or almost orbicular, 2–2.5 cm long, trumpet-shaped, both surfaces densely downy, margin irregularly finely toothed. nwGuangxi, swGuizhou, s, se & wYunnan [c & eHimalaya, Laos, Myanmar, Vietnam]; tree-trunks in open forests, valley rocks, 400–1,800 m. III–VI.

Dendrobium devonianum Paxton Differs from *D. nobile* in the pendulous, terete **stems**, 50–70 cm long, that are leafy throughout; the **leaves** narrowly ovate-lanceolate, 8–13 cm long, with sheaths often spotted with purplish red; sepals and petals 20–26 mm long, white with purple-red tips and several purplish red stripes at base, petals fringed at margin; lip almost orbicular, c. 30 mm long, with 2 golden yellow blotches on sides, margin fringed. nwGuangxi, swGuizhou, seTibet, s, sw & seYunnan [eHimalaya, Myanmar, Thailand, Vietnam]; tree-trunks and rocks in dense forests, 700–2,000 m. IV–V.

Dendrobium loddigesii Rolfe **Pseudobulbs** 10–45 cm long. **Leaves** oblong-lanceolate, pointed, 2–4 cm long. **Inflorescence** on upper part of leafy or sometimes defoliated **stems**, with 1 or rarely 2 **flowers**, these pink or rose-purple, with large golden-yellow blotch on lip disc; sepals ovate-oblong, 17–20 mm long; petals elliptic, 17–20 mm long; lip almost orbicular, 17–20 mm long, margin shortly fringed, densely downy overall. Guangxi, swGuizhou, s & seYunnan [s & se China, Laos, Vietnam]; tree-trunks or rocks in forests, 400–1,500 m. IV–V.

Dendrobium wardianum Lindl. Like *D. nobile*, but generally larger; **stems** ascending or pendulous, usually 16–46 cm long, fleshy; **leaves** many on upper part of stem, narrow-oblong, 5.5–15 cm long; **inflorescence** arising from leafless stem, usually almost terminal, 1–3-flowered; **flowers** white tinged with purple or pink at tips, disc of lip yellow and with 2 dark purple blotches on both sides near base, sepals and petals oblong, c. 45 mm long; lip broadly ovate, almost pointed, c. 35 mm long. w, sw & seYunnan [eHimalaya, Myanmar, Thailand, Vietnam]; tree-trunks in open forests, 1,300–1,900 m. III–V.

1. *Dendrobium thyrsiflorum*; Mongpang, swYunnan (PC)
2. *Dendrobium williamsonii*; Longhua, swGuangxi (PC)
3. *Dendrobium moniliforme*; cult. (PC)
4. *Dendrobium lituiflorum*; Monghe, wYunnan (PC)
5. *Dendrobium nobile*; Longhua, swGuangxi (PC)
6. *Dendrobium aphyllum*; Monghe, wYunnan (PC)
7. *Dendrobium devonianum*; swBhutan (PC)
8. *Dendrobium loddigesii*; Maguan, seYunnan (PC)
9. *Dendrobium loddigesii*; Hong Kong (PC)
10. *Dendrobium wardianum*; north-west of Tengchong, wYunnan (PC)

EPIGENEIUM

A genus of c. 35 species (7 in China) in tropical Asia, mainly Indonesia and Malaysia. **Epiphytes** or rock-dwellers with creeping **rhizomes**, densely covered with chestnut or pale brown sheaths and scattered or congested **pseudobulbs**. **Leaves** 1 or 2, leathery, elliptic to ovate, articulated. **Flowers** solitary at apex of pseudobulb, or in a few–many-flowered raceme; sepals free, similar, lateral sepals oblique at base, forming a conspicuous projection; lip joined to end of column-foot, 3-lobed; side-lobes erect; mid-lobe spreading; callus of longitudinal ridges.

Epigeneium amplum **(Lindl.) Summerh.** (syn. *Dendrobium amplum* Lindl.) **Rhizomes** usually branched. **Pseudobulbs** egg-shaped to ellipsoid, 2–5 × 0.7–2 cm, golden yellow when dry, 3–14 cm apart. **Leaves** 2, elliptic to oblong-elliptic, 6–22.5 × 2–5.5 cm. **Inflorescence** 1-flowered. **Flowers** greenish yellow marked with deep brown; sepals 4–5 cm long; lip 2.4–2.6 cm long, 3-lobed, callus of 3 ridges, central one the longest. sGuangxi, seTibet, w, s & swYunnan [Himalaya, Myanmar, Thailand, Vietnam]; tree-trunks or rocks in montane forests, 1,000–1,900 m. IX–XI.

Epigeneium fargesii **(Finet) Gagn.** (syn. *Dendrobium fargesii* Finet) **Pseudobulbs** smaller than those of the previous species, ascending, almost egg-shaped, 8–10 × 3–5 mm, with a solitary, ovate to broadly ovate-elliptic leaf, only 1–2.3 × 0.7–1.1 cm. **Flowers** not opening widely; sepals and petals 10–15 mm long, pale pink to buff; lip whitish, fiddle-shaped, c. 2 cm long, with a callus of 2 longitudinal keeled ridges. Guangxi, swHubei, Sichuan, Yunnan [c & sChina, Bhutan, neIndia, Thailand]; similar habitats, 400–2,400 m. IV–V.

CYMBIDIUM

A genus of c. 50 species (30 in China, 6 endemic) in south, east and South East Asia, the Malay Archipelago, the Philippines and north and east Australia. Terrestrial, epiphytic or rock-dwelling plants with short, swollen, leafy **pseudobulbs**. **Leaves** strictly alternate, linear to oblanceolate. **Inflorescence** racemose, erect, arching or pendent, with 1–many showy, scented flowers; sepals and petals free; lip 3-lobed with ridged callus; column elongated and somewhat arching. A favourite genus of the Chinese since the time of Confucius, often seen growing in pots in courtyards, temples, on balconies and in traditional orchid gardens; unfortunately, plants are still collected from the wild in quantity.

1. Flowers relatively small, lip not more than 25 mm long; leaves linear to linear-lanceolate 2
1. Flowers larger, lip 30–70 mm long 3
2. Sepals and petals less than 27 mm long, white or pale green striped red or reddish purple . . . . *C. lancifolium*
2. Sepals and petals yellow, greenish yellow or reddish brown . . . . *C. ensifolium, C. faberi, C. kanran, C. floribundum, C. goeringii, C. tortisepalum* var. *longibracteatum, C. macrorhizon, C. sinense*
3. Leaves 5–7 . . . .*C. lowianum, C. hookerianum, C. wilsonii, C. tracyanum*
3. Leaves (5–)7–11 . . . . *C. eburneum, C. erythraeum*

Cymbidium lancifolium Hook. Terrestrial plant to 25 cm, with above-ground, spindle-shaped to cylindrical, several-noded **pseudobulbs** 2.5–15 cm long, each bearing 2–4 spreading, oblanceolate **leaves**, 30–60 × 2–4 cm. **Inflorescence** erect, lax, 2–6-flowered. **Flowers** usually sweetly scented; sepals and petals white or pale green, often longitudinally striped with red, oblanceolate, 15–27 mm long; lip white spotted with red-purple, obscurely 3-lobed, 15–20 mm long, side-lobes erect, mid-lobe recurved, ovate or oblong-obovate, minutely papillose, with callus of 2 parallel ridges. Guizhou, s & swSichuan, seTibet, Yunnan [sw & scTibet, s & eChina, Himalaya, sJapan, seAsia]; rocky places, scrub, light woodland, to 2,200 m. V–VIII.

1. *Epigeneium amplum*; Hoquangphin, nVietnam (PC)
2. *Cymbidium lancifolium*; Hong Kong (DD)
3. *Cymbidium ensifolium*; cult. (DD)
4. *Cymbidium faberi*; cult. (PC)
5. *Cymbidium kanran*; cult. (DD)
6. *Cymbidium floribundum*; cult. (DD)
7. *Cymbidium goeringii*; cult. (PC)
8. *Cymbidium goeringii*; cult. (PC)

Cymbidium ensifolium (L.) Sw. Terrestrial plant to 45 cm with underground or above-ground, egg-shaped **pseudobulbs** 1.5–2.5 cm long. **Leaves** 2–6, erect to almost so, grass-like, 30–60 cm long. **Inflorescence** erect, lax, 3–9-flowered. **Flowers** usually sweetly scented, with yellow to green, oblanceolate sepals and petals, 15–28 mm long, often striped with red; lip pale yellow spotted with red-purple, 3-lobed, 15–23 mm long, side-lobes erect, mid-lobe recurved, ovate or oblong-obovate, minutely papillose, bearing a callus of 2 parallel ridges. Guizhou, Sichuan, Yunnan [s & eChina, Himalaya, seAsia, Korea, Japan]; grassland, scrub, light woodland, 600–1,800 m. VI–X.

Cymbidium faberi Rolfe Similar to the previous species, but with 5–8-leaved **pseudobulbs**, a 5–11-flowered **inflorescence** and a lip that is markedly papillose on the almost elliptic, strongly recurved mid-lobe. Distribution and habitat similar to those of *C. ensifolium* (as well as seTibet), up to 3,000 m. III–V.

Cymbidium kanran Makino Similar to *C. ensifolium*, but with slender **leaves**, 40–70 × 9–17 cm, and with spidery green **flowers** with almost linear or narrowly linear-lanceolate, long-pointed, apple-green sepals, 3–6 × 0.35–0.7 cm. Guizhou, Sichuan, Yunnan [c & sChina, sJapan, sKorea]; shady places, damp forest understorey, streamsides, often amongst rocks, 400–2,400 m. VIII–XII.

Cymbidium floribundum Lindl. (syn. *C. pumilum* Rolfe) Large rock-dwelling or epiphytic plant to 50 cm with almost egg-shaped **pseudobulbs**, 2.5–3.5 cm long. **Leaves** arching to almost erect, linear, pointed, 22–50 cm long. **Inflorescence** erect, 16–35 cm, dense, many-flowered. **Flowers** 3–4 cm across with yellow to red-brown sepals and petals 14–18 mm long; lip white spotted with red-purple, 3-lobed, 16–18 mm long with yellow, 2-ridged callus in basal part. Guizhou, Sichuan, Yunnan [c, s & eChina, Taiwan]; evergreen and mixed deciduous forests, limestone boulders in shade, to 3,300 m. IV–VIII.

Cymbidium goeringii Rchb. f. (syn. *C. virescens* Lindl.; *C. forrestii* Rolfe) In contrast to the previous species, this is a small plant not more than 30 cm tall, with underground **pseudobulbs**, 1–2.5 cm long, bearing 4–7 grass-like **leaves**; **inflorescence** erect, with just 1 or 2 sweetly scented **flowers**; sepals and petals 25–40 mm long, green or yellow-green, petals sometimes purple-striped; lip white, marked with red-purple, 3-lobed, 14–28 mm long. sGansu, Guizhou, Hubei, Sichuan, Yunnan [s & eChina, Korea, Japan]; grassland, scrub, light woodland, to 2,200 m. I–III. Plants with white flowers are also known.

Cymbidium tortisepalum Fukuyama var. *longibracteatum* (Y. S. Wu & S. C. Chen) Y. S. Wu & J. Z. Liu* (syn. *C. longibracteatum* Y. S. Wu & S. C. Chen) Similar to *C. goeringii*, but **inflorescence** with 2–5 **flowers**, bract longer than ovary and sepals lanceolate, rather than spathulate. Guizhou, Sichuan, Yunnan; rocky and scrubby slopes, 1,000–2,500 m. V–VII. The typical variety (**var.** *tortisepalum*) is native to Taiwan.

Cymbidium macrorhizon Lindl. (syn. *C. szechuanensis* S. Y. Hu; *Pachyrhizanthe aphylla* (Ames & Schltr.) Nakai; *P. macrorhizos* (Lindl.) Nakai) **Saprophyte** growing from a warty underground **rhizome**. **Leaves** absent. Similar to *C. lancifolium* in floral morphology. swGuizhou, s & swSichuan, neYunnan [Himalaya, nIndia, Japan, Laos, Myanmar, Thailand, Vietnam]; riverine and *Pinus* forests, open mountain slopes, 700–1,500 m. IV–VI.

Cymbidium sinense (Jackson ex Andr.) Willd. Much larger plant than the previous species, to 90 cm, with above-ground, egg-shaped **pseudobulbs** 2.5–8 cm long, each bearing 3–5, erect to almost so, lanceolate **leaves**, 45–80 × 2–3 cm. **Flowers** with red-purple, oblanceolate sepals and petals, 22–30 mm long; lip pale white or yellow spotted with red-purple, 3-lobed, 17–25 mm long. swGuizhou, swSichuan, Yunnan [s & eChina, neIndia, Myanmar, Thailand, Ryukyu Islands]; shady, well-drained places in scrub and woodland, ravines, to 2,000 m. X–III.

Cymbidium lowianum Rchb. f. Large epiphyte or rock-dweller with ellipsoid **pseudobulbs** 6–13 cm long, bearing arching, linear **leaves** 60–80 cm long. **Inflorescence** arching or spreading, 60–80 cm long, 10–20-flowered. **Flowers** 7–9 cm across; sepals and petals green to yellow, 40–52 mm long; lip white, 3-lobed, 35–43 mm long, mid-lobe deflexed, ovate, with a V-shaped red-purple mark at apex, callus consisting of 2 hairy ridges. sw & seYunnan [Myanmar, Thailand]; evergreen forests, cliffs, 1,300–1,900 m. IV–V.

Cymbidium hookerianum Rchb. f. Similar to the previous species, but with apple-green sepals and petals and a purple-spotted, white lip. swGuizhou, Guangxi, sSichuan, Yunnan [c & eHimalaya, Myanmar]; similar habitats, to 2,700 m. I–IV.

Cymbidium wilsonii (Rolfe ex Cook) Rolfe* Similar in flower size and shape to *C. hookerianum*, but readily distinguished by the smaller **flowers**, not more than 10 cm across, and lip side-lobes with red-brown veins rather than spots, margin with hairs over 1 mm long. wYunnan; similar habitats, to c. 2,400 m. II–IV.

Cymbidium tracyanum L. Castle Similar to *C. lowianum*, but with yellow sepals and petals, strongly striped with purple, and a purple-spotted, yellowish lip. Guangxi, swGuizhou, seTibet, Yunnan [Myanmar, Thailand]; habitats similar to those of *C. lowianum*, 1,200–1,900 m. IX–XII.

Cymbidium eburneum Lindl. Large epiphyte or rock-dweller to 80 cm with egg-shaped **pseudobulbs**, 4–8 cm long, bearing 6–11, arching, linear **leaves** 50–65 cm long. **Inflorescence** erect, 60–80 cm long, 1- or 2-flowered. **Flowers** 7–8 cm across; sepals and petals white, 55–70 mm long; lip white, with yellow patch in middle and at base, sometimes also with red-purple spots, 3-lobed, 55–70 mm long, mid-lobe deflexed, ovate, minutely hairy, yellow callus consisting of 2 hairy ridges. Guangxi, swYunnan [seChina, c & eHimalaya, Myanmar]; rocks in valleys, 1,000–2,000 m. II–V.

Cymbidium erythraeum Lindl. Distinguished from the previous species by the arching or spreading, 3–7-flowered **inflorescence**; by the smaller **flowers**; and by the greenish to dull ochre-yellow, purple-striped sepals and petals 34–52 mm long. Lip similarly coloured, 30–43 mm long. c & swGuizhou, swSichuan, seTibet, Yunnan [c & eHimalaya, Myanmar]; trees and rocks in forests and forest margins, 1,400–2,800 m. X–I.

GEODORUM

A genus of c. 12 species (5 in China) widespread in south, east and South East Asia, the Malay and Philippine Archipelagos, south-west Pacific islands and north-east Australia. Terrestrial **perennials** with underground tuber-like **pseudobulbs**, connected to each other by very short **rhizomes**, with a few nodes. **Leaves** 2–3, pleated, stalked. **Inflorescence** crozier-shaped, arising from base of pseudobulb, as long as, or longer than leaves, with a few small **flowers**; sepals and petals free; lip entire, shallowly pouched at base; disc either irregularly papillose centrally or with 1 or 2 fleshy longitudinal ridges and 2 small calluses at base.

Geodorum pulchellum Ridl. Plant to 40 cm in flower, with 2–3 well-developed **leaves**, narrow-elliptic or oblong-lanceolate, 16–29 cm long. **Inflorescence** bearing a pendulous raceme, 2.5–3 cm long, 2–5-flowered. **Flowers** white; sepals and petals ± oblong, 9–10 mm long; lip broadly ovate-oblong, shallowly pouched at base, c. 10 × 9 mm. Yunnan [Thailand, Vietnam]; streamsides, grassy slopes, base of boulders, 300–2,400 m. VI–VII.

THRIXSPERMUM

A genus with c. 120 species (12 in China) in tropical Asia. **Epiphytes** with ascending or pendulous, short or elongated, sometimes creeping **stems**. **Leaves** few–many, ± in 2 opposite rows. **Inflorescence** lateral, 1–several, long or short, few–many-flowered. **Flowers** small to medium-sized, short-lived; sepals and petals free, almost similar; lip joined to column-foot, 3-lobed, side-lobes erect, mid-lobe thick, base pouched or spur-like, with 1 callus on inner wall. Column thick and short, with broad foot.

---

1. *Cymbidium tortisepalum* var. *longibracteatum*; cult. (PC)
2. *Cymbidium macrorhizon*; Japan (SG)
3. *Cymbidium sinense*; cult. (DD)
4. *Cymbidium lowianum*; Baoshan, wYunnan (PC)
5. *Cymbidium hookerianum*; cult. (DD)
6. *Cymbidium wilsonii*; cult. (RBGK)
7. *Cymbidium tracyanum*; cult. Kunming, cYunnan (PC)
8. *Cymbidium eburneum*; cult. (DD)
9. *Cymbidium erythraeum*; cult. Kunming, cYunnan (PC)
10. *Geodorum pulchellum*; road from Dukou to Lijiang, by Jinshajiang, nwYunnan (PC)

1. *Thrixspermum saruwatarii*; Fanjingshan, ne Guizhou (PC)
2. *Liparis japonica*; Wolong, wSichuan (PC)
3. *Liparis nervosa*; Hong Kong (PC)
4. *Liparis gigantea*; Hong Kong (PC)
5. *Malaxis monophyllos*; Bhutan (MWA)
6. *Malaxis monophyllos*; Bhutan (MWA)

Thrixspermum saruwatarii **(Hayata) Schltr.** (syn. *Sarcochilus saruwatarii* Hayata; *S. laurisilvaticus* Fukuy.; *Thrixspermum xanthanthum* Tuyama; *T. laurisilvaticum* (Fukuyama) Garay) **Stem** erect or ascending, less than 2 cm long, bearing dense and ascending, oblong-sickle-shaped, unequally 2-lobed, pointed **leaves** 4–12 cm long. **Inflorescence** pendulous, up to 8 cm long, rachis slightly flexuous and thickened upward, with 1–4 flowers. **Flowers** off-white or yellowish green, later yellowish brown, lip white with orange red stripes; sepals and petals 7–8 mm long. neGuizhou [eChina, Taiwan]; epiphytic on large tree-trunks in riverine forests, 900–2,800 m. III–IV.

Thrixspermum japonicum **(Miq.) Rchb. f.** (syn. *Sarcochilus japonicus* Miq.; *Thrixspermum neglectum* Masamune) Differs from the previous species in having a pendent, many-noded **stem** 2–13 cm long; **leaves** oblong or sometimes oblanceolate, 2–4 cm long; and erect, few-flowered **inflorescence** with pale yellow **flowers**, sepals and petals 5–7 mm long. neGuizhou, Sichuan [cChina, Japan]; riverine and valley forests; 900–1,000 m. IX–X.

LIPARIS

A genus of c. 250 species (52 in China) throughout tropical and subtropical regions of the world, with a few species in the north temperate zone. Terrestrial or epiphytic **herbs**, usually with clustered or well-separated **pseudobulbs**, sometimes with fleshy **stems**. **Leaves** 1–several, basal or on stem. **Inflorescence** few–many-flowered, terminal, erect, often slightly compressed and narrowly winged. **Flowers** small or medium-sized; sepals similar, free or rarely the 2 lateral sepals joined; petals usually narrower than sepals, linear to thread-like; lip entire or rarely 3-lobed, sometimes recurved at or below the middle, spurless; callus, if present, basal, entire or 2-lobed.

> 1. Pseudobulbs ovate, generally 2-leaved .... **L. japonica, L. campylostalix**
> 1. Pseudobulbs cylindrical, 3–5-leaved .... **L. nervosa, L. nigra**

Liparis japonica **(Miq.) Maxim.** (syn. *Microstylis japonica* Miq.; *Liparis elongata* Fukuyama; *L. giraldiana* Kraenzl. ex Diels; *L. pauciflora* Rolfe) Terrestrial **herb** with ovate **pseudobulbs** 5–12 mm long. **Leaves** 2, ovate, ovate-oblong or almost elliptic, 5–10 × 2–4 cm, pointed or blunt, margin crisped or nearly entire. **Inflorescence** several–10-flowered or more, borne on a scape 12–50 cm long. **Flowers** pale green, sometimes tinged pinkish or purplish red; sepals 7–9 mm long; petals thread-like, 7–9 mm long; lip almost obovate, mucronate, 6–8 × 4–5 mm, margin inconspicuously finely toothed or entire. Gansu, Guizhou, Qinghai, Sichuan, sShaanxi, e & seTibet, Yunnan [n & cChina, Japan, Korea, eRussia]; shade in forests, scrub, shady grassland, 1,100–2,800 m. VI–VII.

Liparis campylostalix **Rchb. f.** Similar to the previous species, but with purple **flowers** and a markedly finely toothed lip. seTibet, Yunnan [Himalaya]; similar habitats, 2,600–3,400 m. VI–VII.

Liparis nervosa **(Thunb. ex Murray) Lindl.** (syn. *L. bambusaefolia* Makino; *L. bicallosa* (D. Don) Schltr.; *L. formosana* Rchb. f.; *L. khasiana* (Hook. f.) T. Tang & F. T. Wang) Terrestrial with cylindrical, many-noded **pseudobulbs** 2–8 cm long. **Leaves** 3–5, ovate to ovate-elliptic, 5–11 cm long. **Inflorescence** emerging from stem apex, 10–20 cm long, many-flowered. **Flowers** purple or green and dull purple; sepals 6–10 mm long; petals thread-like, 7–8 mm long; lip oblong-obovate, 6 × 4.5–5 mm. Guizhou, sSichuan, seTibet, Yunnan [c & sChina, widely distributed in Old World tropical and subtropical regions]; deep shade in forests, streamsides, shady places in grassland, 1,000–2,100 m. IV–V.

Liparis gigantea **C. L. Tso** Similar to *L. nervosa*, but larger with a green **stem** 8–20 cm tall, bearing 3–6 **leaves** 9–17 cm long, and an **inflorescence** 18–45 cm long with up to 20 dark purple **flowers**; sepals and petals 16–20 mm long; lip obovate-elliptic or broadly obovate-oblong, 10–15 mm long, with a conspicuously finely toothed margin. swGuizhou, seTibet, c & seYunnan [sChina, Thailand, Vietnam]; shade in evergreen broadleaved forests, damp rocks, 500–1,700 m. IV–V.

MALAXIS

A genus of 300 species (20 in China) with a cosmopolitan distribution. Similar to *Liparis* but **flowers** generally non-resupinate (i.e. with lip uppermost) and column very short, not elongated.

Malaxis monophyllos **(L.) Sw.** (syn. *Ophrys monophyllos* L.; *Malaxis arisanensis* (Hayata) S. Y. Hu; *M. taiwaniana* S. S. Ying; *M. yunnanensis* (Schltr.) T. Tang & F. T. Wang) **Pseudobulbs** ovate, relatively small, usually 6–8 × 4–5 mm. **Leaves** 1, rarely 2, ascending, ovate, oblong or almost elliptic, 2.5–7.5 cm long. **Inflorescence** erect, 15–40 cm long, several–many-flowered. **Flowers** dense, pale yellowish green to pale green; sepals 2–4 mm long, long-pointed; petals thread-like; lip linear-lanceolate, tailed, 3–4 mm long, base with a pair of short blunt auricles; column 0.5 mm long. Gansu, Qinghai, Sichuan, Shaanxi, Tibet, nwYunnan [widespread in China, temperate Asia, Europe, North America]; forests, thickets, grassy slopes, 800–4,100 m. VI–VIII.

Acknowledgements

The authors are grateful to all those who have contributed to this book, among them: Joe Atkin, John and Hilary Birks, Christopher Brickell, Kim Blaxland, Anne Chambers, Eiko Chiba, Jill Cowley, Peter Cox, Alan Dunkley, John Fielding, Maurice Foster, Jim Gardiner, Mike Gilbert, David Goyder, Guan Kai-yun, Nicholas Hind, Harry Jans, Erica Larkcom, Roy Lancaster, Gwilym Lewis, Liu Zhong-jian, Pete Lowry, Luo Yi-bo, Brian Mathew, Ron McBeath, John Mitchell, Ray Morgan, Margaret North, Holger Perner, Qin Hai-ning, Martyn Rix, Robert Rolfe, Christopher Sanders, Brian Schrire, the late George Smith, Rosie Steele, John Richards, Mark Watson, Toshio Yoshida, Yong Ming Yuan and Meimei Zou.

Selected bibliography

Alpine Garden Society (1939–). *Quarterly Bulletin of the Alpine Garden Society* (now *The Alpine Gardener*); numerous articles. Alpine Garden Society, UK.

Argent, G., Bond, J., Chamberlain, D. F., Cox, P. A. & Hardy, A. (1997). *The Rhododendron Handbook 1998; Rhododendron Species in Cultivation*. Royal Horticultural Society, UK.

Bean, W. J. (1976–81, 1988). *Trees and Shrubs Hardy in the British Isles*. 8th edn. Vols 1–4 and Supplement. John Murray, UK.

Beckett, K. (ed.) *Alpine Garden Society Encyclopaedia of Alpines*. 2 vols. AGS Publications Ltd, UK.

Bishop, G. (1990). *Travels in Imperial China: the Explorations of Père David*. Cassell Ltd, UK.

Brickell, C. D. & Mathew, B. (1976). *Daphne: the Genus in the Wild and in Cultivation*. Alpine Garden Society, UK.

Chamberlain, D. F. & Rae, S. J. (1990). A revision of *Rhododendron* 4: subgenus *Tsutsusi*. *Edinb. J. Bot.* 47(2): 89–200.

Chamberlain, D. F. (1982). A revision of *Rhododendron* 2: subgenus *Hymenanthes*. *Notes Roy. Bot. Gard. Edinb.* 39(2): 209–486.

Chen, S. C., Tsi, Z. H. & Luo, Y. B. (1999). *Native Orchids of China in Colour*. Science Press, Beijing, PRC.

Cheng, W. C. (1980–2004). *Sylva Sinica*. China Forestry Publishing House, Beijing, PRC.

Ching, R. C. (1972–76). *Iconographia Cormophytorum Sinicorum*. Science Press, Beijing, PRC.

Cowley, E. J. (2007). *The Genus Roscoea*. Royal Botanic Gardens Kew, UK.

Cox, E. H. M. (1945). *Plant Hunting in China*. William Collins & Son, UK. (Reprinted by Oldbourne, 1961.)

Cox, K. N. E. (ed.) (2008). *Frank Kingdon-Ward's Riddle of the Tsangpo Gorges: Retracing the Epic Journey of 1924–25 in South-East Tibet*. 2nd edn. Antique Collectors' Club, UK. (See also Kingdon Ward, 1926.)

Cox, P. A. & Cox, K. N. E. (1997). *The Encyclopaedia of Rhododendron Species*. Glendoick Publishing, Scotland.

Cribb, P. J. & Butterfield, I. (1999). *The Genus Pleione*. Royal Botanic Gardens Kew, UK.

Cribb, P. J. (1992). The spotted-leaved cypripediums. *Quart. Bull. Alp. Gard. Soc.* 60(2): 165–177.

Cribb, P. J. (1992). Wolong, hardy plants in the panda's last stronghold. *Quart. Bull. Alp. Gard. Soc.* 60(1): 57–68.

Cribb, P. J. (1993). Sichuan, home of *Lilium regale*. *The Garden* 118(3): 121–123.

Cribb, P. J. (1997). *The Genus Cypripedium*. Royal Botanic Gardens Kew, UK & Timber Press, USA.

Cribb, P. J. (1998). *The Genus Paphiopedilum*. 2nd edn. Royal Botanic Gardens Kew, UK & Natural History Publications, Borneo.

Cullen, J. (1980). A revision of *Rhododendron* 1: subgenus *Rhododendron*, sections *Rhododendron* & *Pogonanthum*. *Notes Roy. Bot. Gard. Edinb.* 39(1): 1–207.

Cullen, J. (2005). *Hardy Rhododendron Species: a Guide to Identification*. Timber Press, USA in association with Royal Botanic Garden Edinburgh.

Du Puy, D. & Cribb, P. J. (2007). *The Genus Cymbidium*. Royal Botanic Gardens, Kew, UK & Natural History Publications, Borneo.

Fang, W. P. *et al.* (1981–2005). *Flora Sichuanica* 16 vols. Science Press, Beijing, PRC.

Farrer, R. (1916). *Cypripedium bardolphianum*. *Notes Roy. Bot. Gard. Edinb.* 9: 101.

Farrer, R. (1916). Report of work in 1914 in Kansu and Tibet. *J. Roy. Hort. Soc.* 42: 47–114.

Farrer, R. (1917). *On the Eaves of the World*. Edward Arnold & Co., UK.

Farrer, R. (1921). *The Rainbow Bridge*. Edward Arnold & Co., UK.

Forrest, G. (1915–16). The flora of North-west Yunnan. *J. Roy. Hort. Soc.* 41: 200–209.

Forrest, G. (1916–17). Notes on the flora of North-west Yunnan. *J. Roy. Hort. Soc.* 42: 39–46.

Forrest, G. (1924). Exploration of north-west Yunnan and south-east Tibet, 1921–22. *J. Roy. Hort. Soc.* 49: 29–36.

Fu, L. K., Chen, T. Q., Lang, K. Y., Hong, T. & Lin, Q. (1999–2008). *Higher Plants of China*. 13 vols. Qingdao Publishing House, Qingdao, PRC.

Fu, S. X. (ed.) (2001–02). *Flora Hubeiensis*. 4 vols. Hubei Science & Technology Publishing House, PRC.

Green, P. & Miller, D. (2009). *The Genus Jasminum in Cultivation*. Royal Botanic Gardens Kew, UK.

Grey-Wilson, C. (1988). Journey to the Jade Dragon Snow Mountain, Yunnan. *Quart. Bull. Alp. Gard. Soc.* 56: 16, 115, 221, 289.

Grey-Wilson, C. (2000). *Clematis, the Genus: a Comprehensive Guide for Gardeners, Horticulturalists and Botanists*. B. T. Batsford Ltd, UK & Timber Press, USA.

Grey-Wilson, C. (2000). *Poppies: a Guide to the Poppy Family in the Wild and in Cultivation*. B. T. Batsford Ltd, UK & Timber Press, USA.

Grey-Wilson, C. (2006). The true identity of *Meconopsis napaulensis*. *Curtis's Bot. Mag.* 23(2): 176.

Grey-Wilson, C. *et al.* (1996). The Alpine Garden Society Expedition to China 1994. *Quart. Bull. Alp. Gard. Soc.* 64: 122–266.

Grierson, A. J. C. & Long, D. G. (1983–2002). *Flora of Bhutan*. Royal Botanic Garden Edinburgh, UK.

Gusman, G. & Gusman, L. (2006). *The Genus Arisaema: a Monograph for Botanists and Nature Lovers*. 2nd edn. Gantner Verlag, Germany.

Handel-Mazzetti, H. R. E. (1927). Naturbilder aus Sudwest-China. Österreichischer Bundesverlag für Unterricht, Wissenschaft und Kunst, Wien, Austria. (Translated by D. Winstanley (1996) as *A Botanical Pioneer in South-west China*. D. Winstanley, Brentwood, Essex, UK.)

Haw, S. G. (1986). *The Lilies of China*. B. T. Batsford Ltd, UK & Timber Press, USA.

Herner, G. (1988). Harry Smith in China — routes of his botanical travels. *Taxon* 37(2): 299–308.

Hillier Nurseries (2002). *The Hillier Manual of Trees & Shrubs*. David & Charles, UK.

Ho, T. N. & Liu, S. W. (2001). *A Worldwide Monograph of Gentiana*. Science Press, Beijing, PRC.

Hou, X. Y. (ed.) (2001). *Vegetation Atlas of China*. Science Press, Beijing, PRC.

Johnson, M. (2001). *The Genus Clematis*. English edn. Södertälje, Sweden.

Kingdon Ward, F. (1913). *The Land of the Blue Poppy*. Cambridge University Press, UK.

Kingdon Ward, F. (1924). *From China to Hkamti Long*. Edward Arnold & Co., UK.

Kingdon Ward, F. (1924). *The Romance of Plant Hunting*. Edward Arnold & Co., UK.

Kingdon Ward, F. (1926). *The Riddle of the Tsangpo gorges*. Edward Arnold & Co., UK. (See also Cox, 2001.)

Kingdon Ward, F. (1960). *Pilgrimage for Plants*. George G. Harrap & Co. Ltd, UK.

SELECTED BIBLIOGRAPHY

Kuan, C. T. (ed.) (1983). *Flora Sichuanica* Vol. 2 (Gymnosperms). Science Press, Beijing, PRC.

Kuang, K. Z., Li, P. C. et al. (eds) (1959–2004). *Flora Reipublicae Popularis Sinicae*. 80 vols in 126 parts. Science Press, Beijing, PRC.

Lancaster, C. R. (2008). *Travels in China: Plantsman's Paradise*. 2nd edn. Antique Collectors' Club, UK.

Lidén, M. & Zetterland, H. (1997). *Corydalis — a Gardener's Guide and a Monograph of the Tuberous Species*. AGS Publications Ltd, UK.

Liu, T. S. (1971). *A Monograph of the Genus Abies*. Department of Forestry, National Taiwan University, Taiwan.

McAllister, H. (2005). *The Genus Sorbus, Mountain Ash and other Rowans*. Royal Botanic Gardens, Kew, UK.

McLean, B. (2004). *George Forrest, Plant Hunter*. Antique Collectors' Club, UK.

Perner, H. & Luo, Y. B. (2007). *Orchids of Huanglong*. Huanglong National Park, Huanglong, PRC.

Phillips, R. & Rix, M. (1988). *Roses*. Pan Books, UK.

Phillips, R. & Rix, M. (1989). *Shrubs*. Pan Books, UK.

Polunin, O. & Stainton, A. (1984). *Flowers of the Himalaya*. Oxford University Press, UK & New Delhi, India.

Richards, J. (2002/03). *Primula*. 2nd edn. B. T. Batsford Ltd, UK (2002) & Timber Press, USA (2003).

Robson, N. K. B. (1985). Studies in the genus *Hypericum*, 3. Bull. Brit. Mus. (Nat. Hist.), Botany ser. 12(4): 163–325.

Robson, N. R. (1970). Shrubby Asiatic *Hypericum* Species in Cultivation. J. Roy. Hort. Soc. 95: 482–497.

Stainton, A. (1988). *Flowers of the Himalaya. A supplement*. Oxford University Press, UK & New Delhi, India.

Stearn, W. T. (2002). *The Genus Epimedium and Other Herbaceous Berberidaceae, Including the Genus Podophyllum*. Royal Botanic Gardens, Kew, UK & Timber Press, USA.

Taylor, G. (1934). *An Account of the Genus Meconopsis*. New Flora & Silva, UK.

Tebbitt, M., Lidén, M. & Zetterlund, H. (2008). *Bleeding Hearts, Corydalis and Their Relatives*. Timber Press, USA.

Waddick, J. W. & Zhao, Y. T. (1992). *Iris of China*. Timber Press, USA.

Wang, F. X. & Liang, H. B. (1996). *Cycads in China*. Guangdong Science and Technology Press, Guangzhou, PRC.

Wilson, E. H. (1913). *A Naturalist in Western China*. 2 vols. Methuen & Co., UK.

Wu, Z. Y. & Raven, P. H. (1996–). *Flora of China*. Science Press, Beijing, PRC & Missouri Botanical Garden, USA.

Wu, Z. Y. (1986). *Wild Flowers of Yunnan*. 3 vols. Japan Broadcast Publishing Co. Ltd.

Wu, Z. Y. (ed.) (1983–87). *Flora Xizangica*. 5 vols. Science Press, Beijing, PRC.

Yoshida, Toshio (2005). *Himalayan Plants Illustrated*. Yama-kei Publishers Co. Ltd, Japan.

Zhong, G. K, Xue, Y. K., Ming, Z. W., Yan, J. S., Bei P. Y. et al. (1977–2005). *Flora Yunnanica*. 19 vols. Science Press, Beijing, PRC.

NOTE. The *Flora of China* website has provided the authors of this book with a great deal of important information.

Glossary

Achene, single-seeded hard-coated indehiscent fruit

Annulus (adj. annular), ring or ring-like body

Anther, the top part of a stamen that produces the pollen

Apiculate, with a short, abrupt, terminal point

Apomictic, producing seed without fertilisation

Aril, a fleshy attachment or covering of a seed

Auricle, an ear-like appendage, often an extension to a leaf-blade or base of leaf-stalk

Beaded, with a bead-like (evenly bumpy) margin

Bract, leaf- or scale-like organ subtending a flower

Bracteate, with bracts

Bracteole, supplementary or secondary bract

Calyx (pl. calyces), the collective name for the sepals

Campanulate, bell-shaped, widest at the mouth

Capitate, 1. head-like, as in the tight inflorescence of many composites; 2. style that terminates in a small knob

Capsule, dry many-seeded fruit derived from a single carpel

Carpel, basic female reproductive unit; a flower may have one or several carpels, which may be separate, or partly or wholly fused

Caudex, rootstock or stem-base of a plant

Caulescent, with one or more stems above ground

Cell, compartment of a fruit, ovary or anther

Ciliate, margined with hairs or (more rarely) bristles

Clawed, with a narrow stalk-like base; generally refers to petals

Cleistogamous, describes flowers that self-fertilise without opening

Column, stout organ (a fusion of several floral parts) that bears the anthers and style, as in the flowers of Orchidaceae

Composite, member of the daisy family, Compositae

Compound, refers to leaves that are composed of several or many leaflets

Confluent, coming together, often refers to the veins of leaves or petals

Connective, the central tissue of the anther that connects the two (sometimes four) cells, sometimes extending beyond the top of the anther

Contorted, twisted (as in some fruits) or wrapped around one another (as in sepals or petals in bud)

Convolute, twisted or wrapped around one another

Corolla, the collective name for the petals

Corymb (adj. corymbose), flat-topped inflorescence in which the branches arise at different levels along the rachis (i.e. the lower branches or flower-stalks are longer than the upper)

Cyathium (pl. cyathia), a structure in which a female flower and several male flowers are surrounded by a cup-like involucre formed from 4–5 fused bracteoles, characteristic of *Euphorbia*

Cyme (adj. cymose), inflorescence in which each branch ends in a flower

Decumbent, spreading out horizontally along the ground with the apex ascending (refers mainly to stems)

Decurrent, extending down the main axis (refers mainly to leaves or leaf-stalks whose base runs down the stem)

Decussate, opposite leaves in which each pair is at right angles to the pair above and below

Dehisce (adj. dehiscent), opening or splitting naturally, applies mainly to fruits

Dendroid, tree-like or tree-shaped

Dentiform, tooth-like

Digitate, with finger-like lobes or with leaflets originating from one point

Dioecious, bearing male and female flowers on separate plants

Disc-floret, individual tubular floret, many of which make up the central disc in flower-heads of Compositae (Group 3) or the whole flower-head (Groups 1 and 2); also found in some members of the Dipsacaceae

Discoid, describes a flower-head in which none of the florets has a ligule (see Compositae, Groups 1 and 2)

Dorsifixed, attached to the back; refers mainly to anthers

Drupe, fleshy fruit containing a single hard-coated seed or stone

Ebracteate, without bracts

Endocarp, the inner layer of a fruit or ovary surrounding the seeds or ovules

Entire, describes a margin without teeth or lobes

Epicalyx, additional calyx (or sepals) outside the true calyx

Epigynous, describes flowers with an inferior ovary (i.e. an ovary that is located below the main floral parts)

Epiphyte (adj. epiphytic), a plant (e.g. many tropical orchids) that grows on trees or bushes but is not parasitic

Even-pinnate, describes a pinnate leaf without a terminal leaflet

Exocarp, the outer layer of a fruit

-fid, divided or parted (e.g. 2-fid, 5-fid)

Filament, the stalk of a stamen

Floret, tiny individual flower, usually refers to flowers crowded together in a flower-head, especially in Compositae

Follicle, dry, many-seeded fruit splitting on one side, a fruit may consist of one or several follicles

Glaucous, bluish-green

Gynophore, stalk of an ovary or fruit (most ovaries are stalkless)

Hemiparasite, plant only partly dependent on a host species but (unlike a parasitic plant) with green leaves

Heterostylous, with 2, sometimes 3, flower forms that have the stamens and style set at different levels within a flower, usually borne on separate plants

Homostylous, with flowers all of one type, as opposed to heterostylous

Hypanthium, extension at the base of an ovary, part of the receptacle

Induplicate, folded in from the margins

Internode, the part of the stem between nodes or joints

Involucral bracts, bract- or scale-like organs surrounding a flower-head (not the calyx, with which they may be confused)

Involucre, collective name for involucral bracts

Involute, rolled or turned inwards

Keel, the lowermost, keel-like petal in pea flowers (Leguminosae) consisting of two fused or partly fused petals

Lanceolate, lance-shaped, 4–6 times as long as broad, broadest in lower third and tapering to apex

Lenticel, a breathing pore on the surface of stems, trunks and other organs

Ligulate, describes a flower-head in which all of the florets have a ligule (see Compositae, Group 4)

Ligule, the strap-shaped corolla of a ray-floret in Compositae

Lyrate, lyre-shaped

Mentum, basal projection, often chin-like

Mericarp, one-seeded portion of a fruit; see also schizocarp

Mesocarp, the central part of the fruit exterior, i.e. between the outer layer (exocarp) and the inner (endocarp)

Monocarpic, describes a perennial that flowers once and then dies after fruiting

Monoecious, having separate male and female flowers borne on the same plant

Mucro, a sharp terminal point, e.g. on a leaf or petal

Mucronate, bearing a mucro

Mucronulate, bearing a tiny mucro

Obconic, three-dimensional object broadened towards the top, i.e. an upside-down cone

Obcordate, heart-shaped but upside down

Oblanceolate, 4–6 times as long as broad but broadest in the upper third; opposite to lanceolate

Obovate, oval but broadest in the upper third; opposite to ovate

Obovoid, like obovate but applied to a three- rather than a two-dimensional object

Obtriangular, triangular but with the broad edge at the top and the point at the bottom

Ocrea (pl. ocreae), a sheath formed from a pair of fused stipules, often membranous, that sheath the stem just above or below a node

Odd-pinnate, describes a pinnate leaf with a terminal leaflet

Ovary, the lower part of the pistil that contains ovules

Ovate, oval but broadest in lower third

Ovule, the unfertilised seed

Palmate, hand-like

Palmatifid, lobed or divided like a hand

Panicle (adj. paniculate), a raceme that is branched in a regular manner, with the lowermost flowers on each branch opening first

Pappus, with a bunch of hairs (simple or branched) or scales topping an achene or fruitlet (found, for example, in members of the Compositae and Dipsacaceae)

-parted, cut or indented to more than halfway to base (e.g. 2-parted, 5-parted)

Pectinate, like the teeth of a comb

Pedate, describes a palmate leaf in which the lateral segments or divisions are further divided

Peltate, describes a leaf in which the stalk joins the middle of the blade rather than the base

Perianth, describes the calyx and corolla together; often used when the two are indistinguishable

Pericarp, the fruit wall

Petal, one member of the corolla, often large and colourful, separate or variously fused together

Pinnate, describes a leaf with leaflets spaced along a single axis or rachis; a 2-pinnate leaf is branched with each branch like a pinnate leaf

Pistil, the female organ(s) of the flower consisting of ovary, style and stigma

Plicae, small lobes or folds alternating with the main lobes, as in the flower of *Gentiana*

Pome, a fleshy fruit with the seeds enclosed by a hard, shell-like endocarp, as in apples and pears

Procumbent, describes a stem that trails along the ground, but does not root

Pruinose, with a bloom, as in many fruits; the term may also apply to leaves and stems

Pseudobulb, the thickened or swollen base of the stems of some orchids

Pseudolateral, apparently lateral but not arising at a node or joint

Pseudopanicle, resembling a panicle but not a true panicle (some cymes can look very panicle-like)

Pseudopinnate, resembling but not pinnate in the true sense

Pseudostem, stem-like but developed from other organs, particularly leaf-bases or leaf-stalks

Pseudoterminal, not terminal in the strict sense but apparently so

Pseudoumbel, resembling an umbel but not a true umbel

Pseudowhorl, resembling a whorl but not a true whorl, i.e. with all the parts arising at the same node or joint

Raceme (adj. racemose), unbranched inflorescence in which the individual flowers are stalked

Rachis, the axis from which organs such as leaflets, bracts or flowers arise

Radiate, spreading outwards from a common centre; often refers to flower-heads with ray-florets (Compositae), or the enlarged outer flowers in an inflorescence, or sometimes to spreading leaflets

Ray, one of the spokes of an umbel

Ray-floret, in Compositae, floret with a strap-shaped corolla (ligule)

Receptacle, the basal part of a flower to which all the other parts, sepals, petals, ovary etc., are attached

Regular, flowers with more than one plane of symmetry, i.e. with a radial plane of symmetry

Resupinate, flowers that are twisted through 180° so that they are upside down; applies to the majority of orchid flowers

Revolute, applies to vernation in which the floral organs are wrapped around each other in a regular manner; can also apply to leaf-margins that are rolled under

Rhizoid, structure that resembles a root in appearance and function, but not in anatomy

Rosulate, arranged like the petals of a rose, in a rosette

Rotate, wheel-shaped, i.e. describes a corolla with a short tube and spreading limb or petal-lobes; can also apply to a calyx

Salverform, shaped like a salver

Samara, indehiscent fruit with a single wing

Saprophyte, plant without green pigment (chlorophyll) living on decayed vegetable matter (humus or leaf-mould), but not parasitic

Scape (adj. scapose), a leafless flower-stem bearing one or several flowers

Schizocarp, fruit that breaks into single-seeded portions known as mericarps

Scorpioid, describes an inflorescence that when young, is coiled like a scorpion's tail; found in many members of the Boraginaceae

Sepal, one member of the calyx, usually green, separate or variously fused together, generally smaller than the petals

Silicula, slender 2-valved fruit typical of the crucifers (Cruciferae) that is at least three times as long as wide

Siliqua, short 2-valved fruit typical of the crucifers (Cruciferae) that is less than three times longer than wide

Simple, unlobed (not compound) or unbranched; refers mainly to leaves, leaflets, stems, tendrils and inflorescences

Sinus, the space or indentation between two lobes or teeth; often used with reference to leaves, calyces and corollas

Spadix (pl. spadices), a fleshy, often swollen axis bearing closely packed flowers and often sterile processes, generally extended at the top into a short-to-long sterile appendage; found, for example, in members of the Araceae such as *Arisaema*

Spathe, sheathing bract; in the Araceae, this partially encloses the inflorescence and can be very showy, whereas in the iris family (Iridaceae), the spathes may be less conspicuous and are often green or membranous

Spathulate, spoon- or paddle-shaped

Spike, narrow, unbranched inflorescence in which the individual flowers are unstalked

Spinulose, edged or covered with minute spines or spinules

Sporophyll, an organ bearing spores; in cycads and conifers, these are cone-like structures that bear the pollen

Stamen, the male organ of a flower, consisting of a stalk (filament) bearing a pollen-producing anther

Staminode, an infertile stamen, often modified into a nectary or resembling a petal

Standard, 1. the uppermost petal in a pea flower (Leguminosae); 2. one of the upright tepals in an *Iris* flower

Stipel, a secondary stipule at the base of a leaflet

Stipitate, with a little stalk

Stipule, one of a pair of appendages at the base of a leaf or its stalk where it joins the stem, stipules may be small and insignificant or larger, sometimes leaf-like

Stolon, horizontal above-ground stem, simple or branched, bearing a young plant at the tip and generally rooting

Stoma (pl. stomata), a breathing pore primarily located on leaves but by no means restricted to them

Style, the stalk on the top of the ovary, terminating in a stigma

Suffruticose, describes a plant that has a woody base but is otherwise herbaceous

Suture, seam, often the point at which fruits split or dehisce

Syncarp, structure formed from the merger of several fruits as a result of the fertilisation of a number of, sometimes many, individual flowers (in *Benthamidia*, this is a globose, rather strawberry-like fruit that is fleshy when ripe)

Taxon (pl. taxa), a unit of classification; any taxonomic grouping such as family, genus or species

Tepal, a term used to describe a floral segment when sepals and petals cannot readily be distinguished, as in many members of the Liliaceae and other monocotyledonous families

Terete, round in cross-section

Ternate, a compound leaf with three leaflets; 2-ternate describes a leaf with three divisions, each of which is ternate

Thyrse, compound inflorescence composed of several or many cymes

Torus, swollen part of a fruit bearing seeds on the exterior, e.g. strawberry (*Fragaria*)

Umbel, inflorescence in which the flowers (or secondary umbels) arise from one point like the spokes of an umbrella

Valvate, with the margins coming together but not overlapping; referring primarily to flowers (describes calyx, calyx-lobes, sepals, petals, corolla-lobes or tepals)

Valve, deep division or lobe of an ovary or capsule; an alternative term for a spathe

Vernation, the manner in which the organs in buds and flowers are arranged

Wing, one of the two lateral petals of a pea flower (Leguminosae)

Zygomorphic, describes flowers with a single plane of symmetry; also called bilateral

Index

Accepted names are in roman, synonyms in *italic*. Bold page numbers indicate photographs.

Abelia 484
    engleriana (Graebn.) Rehd.* 484
    schumannii (Graebn.) Rehd.* 484
Abelmoschus 175
    esculentus (L.) Moench 175
    manihot (L.) Medik. 175
    moschatus Medik. **174**, 175
Abies 32, **35**, 36, **39**, 43, 56
    delavayi Franch. **32**, 36, 56, **57**
    ernestii Rehd.* 56
    fabri (Mast.) Craib* 56
    fargesii Franch.* 40, 44, 56, **57**
    forrestii Coltm.-Rog.* 56, **57**
    georgei Orr 35, 36
    recurvata Mast. var. *ernestii* (Rehd.) Kuan 56
Acacia 43
Acanthocalyx 490
    alba (Hand.-Mazz.) Cannon* 490, **490**
    chinensis Pai* **490**, 491
    delavayi (Franch.) Cannon* **490**, 491
Acanthopanax 383
    davidii (Franch.) Vig. 383
    henryi (Oliv.) Harms 383
    nanpingensis X. P. Fang & C. K. Hsieh 383
    stenophyllus Harms 383
    wilsonii Harms 383
Acer 43, 44, 360
    campbellii Hook. f. & Thomson ex Hiern subsp.
        oliveriana (Pax) A. E. Murray 363
    cappadocicum Gled. 360
        subsp. sinicum (Rehd.) Hand.-Mazz.* 360, 363
    caudatum Wall. 360, **362**, 363
    davidii Franch. 360, **361**
        subsp. *grosseri* (Pax) P. C. de Jong 360
    fabri Hance 360
    fargesii Veitch ex Rehd. 360
    flabellatum Greene 40
    forrestii Diels 360, **362**, 363
        subsp. laxiflorum (Pax) Murray* 360, 363
    franchetii Pax 363
    griseum (Franch.) Pax.* 360, **362**, 363
    grosseri Pax* 360, **361**
    henryi Pax* 360, 363
    laxiflorum Pax 363
    longipes Franch. ex Rehd.* 360, 363
    oblongum Wall. ex DC. 360
    oliverianum Pax 360, 363
    papilio King 360
    pectinatum Wall. ex G. Nichols. forma *rufinerve* A. E. Murray 363
    pectinatum subsp. *forrestii* (Diels) A. E. Murray 363
    sinicum Rehd. 363
    sterculiaceum Wall. 360, 363
        subsp. franchetii (Pax) A. E. Murray* 360, **362**, 363

        subsp. sterculiaceum
Aceraceae 360
Aceratorchis 578
    tschiliensis Schltr.* 578, **579**
Aconitum 115
    brachypodum Diels* **114**, 115
    carmichaelii Debeaux **114**, 115
    gymnandrum Maxim.* **114**, 115
    hemsleyanum E. Pritz. **114**, 115
    scaposum Franch. 115
    sinomontanum Nakai* **114**, 115
    tsaii W. T. Wang* 115
    vilmorinianum Kom.* 115
Aconogonon 160
    campanulatum (Hook. f.) Hara 160, **161**
    molle (D. Don) Hara 160, **161**
    polystachyum (Wall. ex Meisn.) M. Kral. 160, **161**
Actaea 112
    asiatica H. Hara 112, **113**
    brachycarpa (P. K. Hsiao) Compton 112
Actinidia 168
    callosa Lindl. 168
        var. henryi Maxim. 168
    chinensis Planch.* 168, **169**
    coriacea Dunn* 168, **169**
    henryi (Maxim.) Dunn* 168, **169**
    kolomikta (Maxim. & Rupr.) Maxim. 168, **169**
    pilosula (Finet & Gagnep.) Stapf ex Hand.-Mazz.* 168, **169**
    rubricaulis Dunn* 168, **169**
ACTINIDIACEAE 168
Adenophora 469
    capillaris Diels* 469
    coelestis Diels* 469
    confusa Nannf.* 469, **470**
    diplodonta Diels 471
    gracilis Nannf.* **470**, 471
    jasionifolia Franch.* 469, **469**
    khasiana Feer* 469, **470**, 471
    potaninii Korsh.* 469, **470**, 471
Adonis 99
    brevistyla Franch. 99
    coerulea Maxim.* 99, 100, **101**
    davidii Franch.* 99, **98**, **257**
    delavayi Franch. 99
    sutchuenensis Franch.* 99
Aeginetia 463
    indica L. **462**, 463
Aesculus 360
    assamica Griff. 360, **361**
    wangii Hu 360
    wilsonii Rehd.* 360, **361**
Ainsliaea 494
    fragrans Champ.* 495
    glabra Hemsl.* 494
    latifolia (D. Don) Sch.-Bip.* 495

    leiospermum Juz. & Ye. V. Serg. 494
    majus Bernh. 494
    pertyoides Franch.* 495
    pteropoda DC. 495
Ajania 512
    nubigena (Wall.) C. Shih 512, **513**
Ajuga 435
    lupulina Maxim. **369**, **434**, 435
        var. major Diels* 435
    lupulina Maxim. form **434**
    nubigena Diels* **434**, 435
    ovalifolia Bur. & Franch.* **434**, 435
Akebia 88
    quinata (Houtt.) Decne 88, **89**
    trifoliata (Thunb.) Kordz. 91
Albizia 39, 323
    julibrissin Durazz. **322**, 323
    kalkebra Prain 323
Aletris 554
    alpestris S. C. Chen* 554
    glabra Bur. & Franch. 554
    nana S. C. Chen 554, **555**
    pauciflora (Klotzsch) Hand.-Mazz. 554
    spicata (Thunb.) Franch. 554, **555**
Aleurites 352
    fordii Hemsl. 352
Allardia 500
    glabra Decne 500, **501**
ALLIACEAE 565
Allium 565
    beesianum W. W. Sm.* 566, **567**
    cepa L. 565
    chrysanthum Regel* 565, 566, **567**
    cyaneum Regel* 566, **567**
    cyathophorum Bur. & Franch.* 566, **567**, 568
        var. farreri (Stearn) Stearn* 566, 568
    fasciculatum Rendle 565, 566, **567**
    forrestii Diels* 566
    gowanianum Wall. ex Baker 566
    hookeri Thwaites 565, 566
    humile Kunth 565, 566
        var. *trifurcatum* F. T. Wang & T. Tang 565
    macranthum Baker 566, 568
    mairei H. Lévl.* 566
    nanodes Airy Shaw* 565
    nivale Jacq. ex Hook. f. & Thomson 566
    porrum Georgi 565
    prattii C. H. Wright ex Hemsl. 566, **567**, 568
    przewalskianum Regel 566, **567**
    sativum L. 565
    sikkimense Baker 566, **567**
    tanguticum Regel* 566
    trifurcatum (F. T. Wang & T. Tang) J. M. Xu* 565
    victorialis L. var. angustifolium Hook. f. 568
    wallichii Kunth 566, **567**
Alpinia 521
    guinanensis D. Fang & X. X. Chen* **520**, 521

Alternanthera 156
    nodiflora R. Br 156
    sessilis (L.) R. Br. ex DC. 156, **157**
AMARANTHACEAE 156
AMARYLLIDACEAE 569
Amitostigma 578, 581
    basifoliatum (Finet) Schltr.* **580**, 581
    beesianum (W. W. Smith) T. Tang & F. T. Wang 581
    faberi (Rolfe) Schltr.* **580**, 581
    monanthum (Finet) Schltr.* 581, 582, **583**
    physoceras Schltr.* 581, 582, **583**
    tibeticum Schltr.* **580**, 581
Ampelopsis 356
    bodinieri (H. Lévl. & Vaniot) Rehd.* 357
    delavayana Planch.* 357
Amygdalus 267
ANACARDIACEAE 364
Anaphalis 495
    cinerascens Ling & W. Wang* **494**, 495
        var. congesta Ling & W. Wang* 495
    cinnamomea C. B. Clarke 495
    margaritacea (L.) Benth. **494**, 495
    nepalensis (Spreng.) Hand.-Mazz. 495
        var. monocephala (DC.) Hand.-Mazz. **494**, 495
    triplinervis Sims ex C. B. Clarke **494**, 495
        var. *intermedia* (DC.) Airy Shaw 495
Androsace 260
    aff. bisulca Bur. & Franch.* 264, **265**
    aff. mariae Kanitz* 264, **265**
    aizoon Duby var. *coccinea* Franch. 260
    alchemilloides Franch.* **262**, 263
    bisulca Bur. & Franch.* 263, 264
        var. aurata (Petitm.) Y. C. Yang & R. F. Huang* 263, 264, **265**
        var. bisulca 263, 264
        var. brahmaputrae Hand.-Mazz.* 263, 264, **265**
    brachystegia Hand.-Mazz.* 263, 264, **265**
    bulleyana Forrest* 260, **261**
    delavayi Franch. 260, **261**
    henryi Oliv. **262**, 263
    integra (Maxim.) Hand.-Mazz.* 260, **261**
    limprichtii Pax & K. Hoffm.* **262**, 263
    mairei H. Lévl.* 264
    mariae Kanitz* **262**, 263
    minor (Hand.-Mazz.) C. M. Hu & Y. C. Yang* **262**, 263
    rigida Hand.-Mazz.* 260, **262**, 263
    selago Hook. f. & Thomson ex Klatt. 260, **261**
    spinulifera (Franch.) R. Knuth* **262**, 263
    stenophylla (Petitm.) Hand.-Mazz.* 263, 264, **265**
    tangulashanensis Y. C. Yang & R. F. Huang* 260, **261**
    tapete Maxim. 260, **261**
    tibetica Knuth 263
    wardii W. W. Sm.* **262**, 263
    yargongensis Petitm.* 263, 264, **265**
    zambalensis (Petitm.) Hand.-Mazz. 263, 264, **265**
Anemoclema 107
    glaucifolium (Franch.) W. T. Wang* **106**, 107
Anemone 100, 107
    aff. cathayensis Kit. ex Ziman & Kad.* **102**, 103
    cathayensis 103
    chinensis Bunge 107
    coelestina Franch.* 103, 104, **105**

davidii Franch.* 100, **101**
demissa Hook. f. & Thomson 103
    var. villosissima Brühl 103
    var. major W. T. Wang* **102**, 103
demissa Hook. f. & Thomson form **102**
esquirolii H. Lévl. & Vaniot 100
geum H. Lévl. 103, 104, **105**
    subsp. *ovalifolia* (Brühl) Chaud. 104
glaucifolia Franch. 107
hupehensis (Lemoine) Lemoine* 103, **101**
imbricata Maxim.* 103, 104, **105**
japonica (Thunb.) Sieb. & Zucc.
    var. *hupehensis* Lemoine 103
    var. *leveillei* Ulbr. 100
    var. *tomentosa* Maxim. 103
millefolium Hemsl. & E. H. Wilson 107
narcissiflora L. subsp. protracta (Ulbr.) Ziman & Fedor. **102**, 103
obtusiloba D. Don 103, 104, **105**
    subsp. coelestina (Franch.) Brühl 104
    subsp. imbricata (Maxim.) Brühl 104
    subsp. rockii (Ulbr.) Lauener 104
    subsp. trullifolia (Hook. f. & Thomson) Brühl 104
patula C. C. Chang ex W. T. Wang* 103, 104
polycarpa W. E. Evans* 103, 104
prattii Huth ex Ulbr.* 100, **101**
rivularis Buch.-Ham. ex DC. 100, **101**
    var. *flore-minore* Maxim.* 100
rockii Ulbr. 103, 104, **105**
    var. pilocarpa W. T. Wang* 103, 104
rupestris Wall. ex Hook. f. & Thomson 103, 104, **105**
    subsp. gelida (Maxim.) Lauener 103, 104
rupicola Camb. 100, **101**
smithiana Lauener & Panigrahi 103
subpinnata W. T. Wang* 103, 104
tomentosa (Maxim.) C. P'ei* **102**, 103
trullifolia Hook. f. & Thomson 103, 104, **105**
vitifolia Buch.-Ham. ex DC. **102**, 103
yulongshanica W. T. Wang* 103, 104
    var. truncata (Comber) W. T. Wang* 103, 104
Angelica 387
    apaensis R. H. Shan & C. C. Yuan* **386**, 387
Anisodus 422
    acutangulus C. Y. Wu & C. Chen* 422
    carniolicoides (C. Y. Wu & C. Chen) D'Arcy & Z. Y. Zhang* 423
    luridus Link 422, **422**
    tanguticus (Maxim.) Pascher 422, **422**
Anthogonium 598
    gracile Lindl. 598, **599**
APOCYNACEAE 411
AQUIFOLIACEAE 348
Aquilegia 119
    ecalcarata Maxim.* **118**, 119
    oxysepala Trautv. & C. A. Mey.* 119
    rockii Munz* **118**, 119
ARACEAE 525
Aralia 383
    apioides Hand.-Mazz.* 384
    atropurpurea Franch.* 384
    chinensis L.* 384, **385**
    echinocaulis Hand.-Mazz.* 384, **385**
    fargesii Franch.* 384
    glomerulata Blume 382
    henryi Harms* 384
    hupehensis G. Hoo 384

papyrifera Hook. 380
subcapitata G. Hoo 384
taibaiensis Z. Z. Wang & H. C. Zheng 384
yunnanensis Franch.* 384
ARALIACEAE 380
Arceuthobium 179
    pini Hawksw. & Wiens* **178**, 179
    sichuanense (H. S. Kiu) Hawksw. & Wiens 179
Arctium 494
    lappa L. 494, **494**
ARECACEAE 522
Arenaria 152
    caespitata (Cambess.) Kozhevn. 155
    densissima Wall. ex Edgew. 152, 155
    eudonta W. W. Sm.* 152, 155
    forrestii Diels 155
    glanduligera Edgew. 152, 155
        var. cernua F. N. Williams 152, 155
    kansuensis Maxim.* 152, **153**
    lichiangensis W. W. Sm. 152
    longistyla Franch.* 152, 155
    melandryoides Edgew. 152, 155
    napuligera Franch.* 152, 155
        var. monocephala W. W. Sm. 155
    oreophila Hook. f. 152, **153**
    polytrichoides Edgew.* 38, 152, 155
    qinghaiensis Y. W. Tsui & L. H. Zhou* 152, **153**
    rhodantha Pax & K. Hoffm.* 152, 155
    roborowskii Maxim.* 152, 155
    roseiflora Sprague* 152, 155
    smithiana Mattf.* 152, **153**
Arisaema 7, 34, 525
    aff. ciliatum H. Li* 530, **531**
    aff. consanguineum Schott* 530, **531**
    aff. fraternum Schott* 530, **531**
    aridum H. Li* 529, 530
    asperatum N. E. Br. **524**, 525
    auriculatum Buchet* **528**, 529
        var. hungyaense (H. Li) G. & L. Gusm. 529
    candidissimum W. W. Sm.* **524**, 525
    ciliatum H. Li* 526, **528**, 529, 530
        var. liubaense Gusm. & Gouda* 526, **528**, 529
    cochleatum Stapf ex H. Li 525
    concinnum Schott 526, 529
    consanguineum Schott 526, **528**, 529, 530
    delavayi Buchet 525
    dilatatum Buchet 526
    echinatum (Wall.) Schott 526, 529
    elephas Buchet 526, **527**
        var. handelii (Stapf ex Hand.-Mazz.) G. & L. Gusm. 526
    fargesii Buchet* 525, 526, **527**
    flavum (Forssk.) Schott subsp. tibeticum J. Murata 529, 530, **531**
    franchetianum Engl. **524**, 525
    fraternum Schott 530
    handelii Stapf ex Hand.-Mazz. 526, **527**
    hungyaense H. Li* **528**, 529
    lichiangensis W. W. Sm* 525, 526, **527**
    lobatum Engl.* **524**, 525
    nepenthoides (Wall.) Mart. **528**, 529
    omeiense P. C. Kao 529
    purpureogaleatum Engl. 525
    saxatile Buchet* 529, 530, **531**
    sp.* 530, **531**
    spcciosum (Wall.) Mart. 526, **527**
    tortuosum (Wall.) Schott **528**, 529

victoriae V. D. Nguyen 526, **527**
wilsonii Engl.* 526, **527**
yunnanense Buchet **524**, 525, 530
    var. *aridum* (H. Li) G. & L. Gusm. 530
Aristolochia 77
    delavayi Franch.* 77, 78
    forrestiana J. S. Ma* 77, 78
    kunmingensis C. Y. Cheng & J. S. Ma* 77, 78
    kwangsiensis W. Y. Chun & F. C. How ex C. F. Liang* **76**, 77, 78
    moupinensis Franch.* **76**, 77, 78
ARISTOLOCHIACEAE 75
Armeniaca 267
Arnebia 431
    szechenyi Kanitz* **430**, 431
Artemisia 462
    nubigena Wall. 512
Aruncus 295
    sylvester Kostel. ex Maxim. **294**, 295
Arundina 597
    graminifolia (D. Don) Hochr. **596**, 597
Asarum 75
    caudigerellum C. Y. Cheng & C. S. Yang* 75, 77
    caudigerum Hance 75, 77
    caulescens Maxim.* 75, 77
    debile Franch.* 75, 77
    delavayi Franch.* 75, **76**, 77
    himalaicum Hook. f. & Thomson 75, 77
    maximum Hemsl.* 75, **76**, 77
    pulchellum Hemsl.* 75, 77
    splendens (F. Maek.) C. Y. Cheng & C. S. Yang* 75, **76**, 77
ASCLEPIADACEAE 412
Asclepias 415
    gigantea L. 415
    procera W. T. Aiton 415
    volubilis L. f. 414
Aspidopterys 348
    hypoglauca H. Lévl. 348
Aster 508
    albescens (DC.) Wall. 508
    asteroides (DC.) Kuntze 508, 511
    batangensis Bur. & Franch.* 508, **509**
        var. *staticifolius* (Franch.) Y. Ling 508
    delavayi Franch. 511
    diplostephioides (DC.) C. B. Clarke 508, **510**, 511
    forrestii Stapf 508
    himalaicus C. B. Clarke 508, **510**, 511
    likiangensis Franch. 511
    limonifolium Franch. 508
    oreophilus Franch.* 508, **509**
    salignus Willd. 508, **509**
    souliei Franch* 508, **509**
    staticifolius Franch.* 508, **509**
    stracheyi Hook. f. 508, **510**, 511
    tongolensis Franch. 508
    yunnanensis Franch.* 508, **509**
ASTERACEAE 491
Astilbe 308
    chinensis (Maxim.) Franch. & Sav. 308, **309**
    grandis Stapf ex E. H. Wilson 308
    koreana (Kom.) Nakai 308
    rivularis Buch.-Ham. 308
    rubra Hook. f. 308
Astragalus 330
    acaulis Baker 330, **330**
    coelestis Diels 335
    degensis Ulbr. 330, 331

        subsp. rockianus (E. Peter) P. C. Li 330, 331
    tongolensis Ulbr. 330, **330**, 331
        var. *breviflorus* H. T. Tsai & T. T. Yu 331
        var. *glaber* E. Peter 331
        var. *lanceolato-dentatus* E. Peter 331
    veitchianus N. D. Simpson 331
    yunnanensis Franch.* 330, 331, **330**
Atragene 96
    macropetala (Ledeb.) Ledeb. 96
Aucuba 367
    chinensis Benth. 367
        var. *obcordata* (Rehd.) Q. Y. Xiang* 367
    himalaica Hook. f. & Thomson 367
AUCUBACEAE 367

Baimashania 199
    pulvinata Al-Shehbaz* 199, **199**
Balanophora 177
    formosana Hayata 177
    laxiflora Hemsl. 177, **177**
    spicata Hayata 177
BALANOPHORACEAE 177
BALSAMINACEAE 372
Banksea 521
    speciosa J. König 521
Bassecoia 491
Bauhinia 323
    brachycarpa Wall. ex Benth. **322**, 323
    championii (Benth.) Benth. **322**, 323
Beesia 119
    calthifolia (Maxim. ex Oliv.) Ulbr. **118**, 119
Begonia 180
    asperifolia Irmsch.* 180, 182
    cavaleriei H. Lévl. 180, 181, **182**
    delavayi Gagnep. 183
    evansiana Andrews 180
    grandis Dry. 180, **181**
    henryi Hemsl.* 180, **182**, 183
    labordei H. Lévl.* 180, 183
    laciniata Roxb. ex Wall. 181
    leprosa Hance* 180, **182**, 183
    limprichtii Irmsch.* 180, 183
    mairei H. Lévl. 183
    morsei Irmsch. 180, **182**, 183
    muliensis T. T. Yu 181
    palmata D. Don 180, 181
    pedatifida H. Lévl.* 180, 182
    polytricha C. Y. Wu* 180, **181**
    taliensis Gagnep.* 180, 181
    wilsonii Gagnep.* 180, 182
BEGONIACEAE 180
Benthamidia 345
    capitata (Wall.) Hara 345, **345**
    japonica (Sieb. & Zucc.) Hara
        var. *chinensis* (Osborn) H. Hara* 346, **346**
        var. *japonica* 346
BERBERIDACEAE 79
Berberis 79
    boschanii C. K. Schneid. 79
    davidii Ahrendt* 81, **82**
    dawoensis K. Meyer* 79, **80**, 81
    densa C. K. Schneid. 81
    dictyoneura C. K. Schneid.* 79, 81
    dictyophylla Franch.* **80**, 81
    forrestii Ahrendt* 79, **79**
    henryana C. K. Schneid.* **80**, 81
    jamesiana Forrest & W. W. Sm.* **80**, 81
    julianae C. K. Schneid.* 83

    liechtensteinii C. K. Schneid. 81
    lijiangensis C. Y. Wu ex S. Y. Bao* **82**, 83
    mouillacana C. K. Schneid.* 79, **79**
    muliensis Ahrendt* 81
    parvifolia Sprague 81
    potaninii Maxim.* 79, **80**, 81
    sp.* 79, 81
    stapfiana C. K. Schneid. 81
    stiebritziana C. K. Schneid.* **80**, 81
    subcaulialata (C. K. Schneid.) C. K. Schneid. 81
    taliensis C. K. Schneid.* 83
    tenuipedicellata T. S. Ying* 79, **80**, 81
    vernae C. K. Schneid.* 79, **80**, 81
    wallichiana DC. forma *parvifolia* Franch. 81
    wilsoniae Hemsl.* 79, **80**, 81
        var. *parvifolia* (Sprague) Ahrendt 81
Berchemia 356
    edgeworthii Lawson 356
    floribunda (Wall.) Brongn. 356
    omeiensis Fang ex Y. L. Chen & P. K. Chou* 356
    pycnantha C. K. Schneid. 356
    sinica C. K. Schneid.* 356
    yunnanensis Franch.* 356, **357**
Bergenia 311
    emeiensis C. Y. Yu* **310**, 311
        var. *rubellina* J. T. Pan 311
    purpurascens (Hook. f. & Thomson) Engl. **310**, 311
Berneuxia 235
    thibetica Decne* **234**, 235
Betula **11**, 43, 44, 144
    alba L. var. *tauschii* Regel 147
    albosinensis Burkill* 144, 147, **146**
        var. *septentrionalis* C. K. Schneid.* 144, 147
    austrosinensis Chun ex P. C. Li* 145, **145**
    bhojapattra Lindl. 147
    calcicola (W. W. Sm.) P. C. Li* 144, **145**
    delavayi Franch.* 145, **145**
        var. *calcicola* W. W. Sm. 144
        var. *polyneura* Hu ex P. C. Li* 145
    insignis Franch.* 144, **146**, 147
    japonica Sieb. ex Winkl. 147
    luminifera Winkl.* 144, **146**, 147
    platyphylla Sukaczev 144, 147
        var. *szechuanica* (Schneid.) Redh.* **146**, 147
    potaninii Batalin* 145
    szechuanica (Schneid.) Jansson 147
    utilis D. Don 144, 147
        var. *prattii* Burkill* 144, **146**, 147
        var. *sinensis* (Franch.) Winkl. 147
    wilsonii Bean 145
BETULACEAE 144
BIGNONIACEAE 463
Bistorta 159
    amplexicaulis (D. Don) Greene **158**, 159
        var. *pendula* H. Hara 159
    coriacea Samuelsson & Hand.-Mazz.* 159
    emodi (Meisn.) Petrov **158**, 159
    macrophylla (D. Don) Soják **158**, 159
    paleacea (Wall. ex Hook. f.) Yonekura & H. Ohashi **158**, 159
    vivipara (L.) S. F. Gray **158**, 159
Bletilla 597
    formosana (Hayata) Schltr.* 598, **599**
    ochracea Schltr.* 598, **599**

striata (Thunb. ex A. Murray) Rchb. f.* 598, **599**
BOMBACACEAE 176
Bombax 176
    ceiba L. 176
Boottia 533
    acuminata Gagnep. 533
BORAGINACEAE 427
Boschniakia 462
    himalaica Hook. f. & Thomson 462, **462**
Brassaiopsis 381
    ciliata Dunn 381, 382
    fatsioides Harms* 381, 382, **382**
    ferruginea (H. L. Li) G. Hoo* 381
    glomerulata (Blume) Regel 381, 382
    palmipes Forrest ex W. W. Sm. 382
Braya 192
    forrestii W. W. Sm. 192, **193**
    verticillata (Jeffrey & W. W. Sm.) W. W. Sm. 195
Bryocarpum 259
    himalaicum Hook. f. & Thomson **258**, 259
    rupifraga Kar. & Kir. 155
Buddleja 388
    alternifolia Maxim.* 388, **389**
    candida Dunn* 388, 390
    colvilei Hook. f. & Thomson 388, **389**
    crispa Benth. 388, **389**
    davidii Franch.* 388, **389**
    delavayi Gagnep.* 388
    fallowiana Balf. f. & W. W. Sm.* 388, 390, **390**
    farreri Balf. f. & W. W. Sm. 388
    forrestii Diels 388, **389**
    lindleyana Fortune* 388, **389**
    officinalis Maxim. 388, 390
    tibetica W. W. Sm. 388
    yunnanensis Gagnep.* 388
BUDDLEJACEAE 388
Bulbophyllum 602
    andersonii (Hook. f.) J. J. Sm. **604**, 605
    forrestii Seidenf. 604, **604**
    funingense Z. H. Tsi & S. C. Chen* 604, **604**
    henryi (Rolfe) J. J. Sm. 605
BUXACEAE 351
Buxus 351
    bodinieri H. Lévl.* 351, 352, **353**
    hebecarpa Hatus.* 351, 352
    henryi Mayr* 351
    microphylla Sieb. & Zucc. 352
        var. *aemulans* Rehd. & E. H. Wilson 352
        var. *platyphylla* (C. K. Schneid.) Hand.-Mazz. 352
    mollicula W. W. Sm.* 351, 352
    sinica (Rehd. & E. H. Wilson) M. Cheng* 351, 352, **353**

CACTACEAE 336
Calanthe 594
    alismifolia Lindl. 594, **595**
    alpina Hook. f. ex Lindl. 594, **595**
    brevicornu Lindl. 594, **596**, 597
    davidii Franch. 594, **595**
    delavayi Finet 597
    graciliflora Hayata 594, **596**, 597
    hancockii Rolfe* 594, 597
    herbacea Lindl. 594, **596**, 597
    tricarinata Lindl. 594, **596**, 597
Callianthemum 100
    cashmirianum Cambess. 100
    farreri W. W. Sm.* 100, **101**
    pimpinelloides (D. Don) Hook. f. & Thomson 100
    tibeticum Witasek 100
Callicarpa 431
    bodinieri H. Lévl. 431
    giraldii Hesse ex Rehd.* **430**, 431
Calocedrus 52
    macrolepis Kurz 52, **53**
Calotropis 415
    gigantea (L.) W. T. Aiton **414**, 415
    procera (Aiton) W. T. Aiton 415
Caltha 108
    palustris L. 108, **109**
        var. barthei Hance* 108, **109**
        var. himalaica Tamura* 108
        var. umbrosa Diels* 108, **109**
    rubriflora B. L. Burtt & Lauener 108
    scaposa Hook. f. & Thomson 108, **109**
    sinogracilis W. T. Wang* 108
        forma rubriflora (B. L. Burtt & Lauener) W. T. Wang* 108, **109**
Calypso 593
    bulbosa (L.) Ames var. speciosa (Schltr.) J. J. Wood **592**, 593
Calystegia 424
    hederacea Wall. 424, **425**
    silvatica (Kit. ex Schrader) Griseb. 424, **425**
        subsp. orientalis Brummitt* 424
Camellia 39, 164
    chrysantha (Hu) Tuyama 164
    cuspidata Veitch* **166**, 167
    forrestii (Diels) Cohen-Stuart 167
    nitidissima C. W. Chi 164
    omeiensis Chang* 167
    petelotii (Merrill) Sealy 167, **165**
    pitardii Cohen-Stuart* **166**, 167
        var. yunnanica Sealy* 167
    reticulata Lindl.* **166**, 167
    saluenensis Stapf ex Bean* **166**, 167
    saluenensis form* 167
    sinensis (L.) O. Kuntze 164, 167
    taliensis (W. W. Sm.) Melchior 164, 167
    yunnanensis (Pitard ex Diels) Cohen-Stuart.* 164, **166**, 167
Campanula 468, 469
    aristata Wall.* 469, **469**
    crenulata Franch.* 468, **469**
    glandulifera Nees 74
    officinarum Nees 74
CAMPANULACEAE 468
Campylotropis 324
    capillipes (Franch.) Schindl.* 324
    macrocarpa (Bunge) Rehd. 324, **325**
    polyantha (Franch.) Schindl.* 324, **325**
    prainii (Collett & Hemsl.) Schindl. 324, **325**
CANNABACEAE 176
Cannabis 176
    sativa L. 176, **177**
CAPPARACEAE 199
Capparis 199
    cantoniensis Lour. 199
    pumila Champ. ex Benth. 199
CAPRIFOLIACEAE 479
Caragana 30, 324
    franchetiana Kom.* 324, **325**
    jubata (Pallas) Poir. 324, **325**
    maximowicziana Kom.* 324, **326**, 327
    tibetica (Maxim. ex Schneid.) Kom. 324, **326**, 327

    versicolor Benth. 324, **326**, 327
Cardamine 188
    circaeoides Hook. f. & Thomson 188, **189**
    delavayi Franch. 188, 191
    franchetiana Diels* 188, **189**
    griffithii Hook. f. & Thomson 188, **190**, 191
    loxostemonoides O. E. Schulz 188, **190**, 191
    macrophylla Willd. 188, **189**
    microzyga O. E. Schulz* 188, **190**, 191
    pulchella (Hook. f. & Thomson) Al-Shehbaz & Yang 188, **190**, 191
    purpurascens (O. E. Schulz) Al-Shehbaz et al.* 188, **190**, 191
    tangutorum O. E. Schulz* 188, **189**
Cardiocrinum 545
    cathayanum (E. H. Wilson) Stearn* 545
    giganteum (Wall.) Mak. **544**, 545
        var. yunnanense (Leichtl. ex Elwes) Stearn **544**, 545
Carpinus 44, 147
    fangiana Hu* **146**, 147
    fargesii Franch. **148**
    henryana (Winkl.) Winkl.* 148
    monbeigiana Hand.-Mazz.* 147
    pubescens Burkill 148
    seemaniana Diels 148
    viminea Lindl. **146**, 148
CARYOPHYLLACEAE 151
Caryopteris 43, 431
    forrestii Diels* 431
    glutinosa Rehd.* 431
    tangutica Maxim.* **430**, 431
Cassiope 231
    membranifolia R. C. Fang* 231
    myosuroides W. W. Sm. 231
    nana T. Z. Hsu* 231
    palpebrata W. W Sm 231
    pectinata Stapf **230**, 231
    selaginoides Hook. f. & Thomson **230**, 231
Castanea 148
    henryi (Skan) Rehd. & E. H. Wilson* 148, **149**
    mollissima Blume 148
    seguinii Dode* 148
Castanopsis 34, 39, 42, 149
    delavayi Franch.* 149
    fargesii Franch. 149
    sclerophylla (Lindl. & Paxton) Schottky 44
Catalpa 463
    fargesii Bur.* **462**, 463
    ovata G. Don* 463
    tibetica Forrest* 463
Cathaya 59
    argyrophylla Chun & Kuang* 42, **58**, 59
Cautleya 521
    gracilis (Sm.) Dandy 521
    spicata (Sm.) Baker 521
Cedrus 59
    deodara (Roxb.) D. Don. **58**, 59
CELASTRACEAE 347
Celastrus 348
    alatus Thunb. 348
    striatus Thunb. 348
Celtis 44, 175
    biondii Pamp. 175
    bungeana Blume 175
    cerasifera C. K. Schneid.* 175
Cephalanthera 589
    damasonium (Mill.) Druce 589
    falcata (Thunb. ex A. Murray) Blume 589

INDEX   625

longifolia (L.) Fritsch 589
Cephalophilon 160
   capitatum (Buch.-Ham. ex D. Don) Tzvelev 160
   runcinatum (Buch.-Ham. ex D. Don) Tzvelev 160
CEPHALOTAXACEAE 52
Cephalotaxus 52
   fortunei Hook. 52, **53**
   sinensis (Rehd. & E. H. Wilson) H. L. Li 52
Cerasus 267
Ceratostigma 267
   minus Stapf ex Prain* 267
   willmottianum Stapf* **266**, 267
CERCIDIPHYLLACEAE 143
Cercidiphyllum 143
   japonicum Sieb. & Zucc. 44, 144
Ceropegia 412
   pubescens Wall. 412, **413**
   sinoerecta M. G. Gilbert & P. T. Li* 412, **413**
Chamerion 343
   angustifolium (L.) Holub **342**, 343
Changnienia 593
   amoena S. S. Chien* **592**, 593
Cheilotheca 232
   humilis (D. Don) H. Keng 232
Cheiranthus 192
   acaulis Hand.-Mazz. 192
   roseus Maxim. 192
Cheirostylis 588
   chinensis Rolfe 589
   josephi Schltr. 589
   philippinensis Ames 589
   taiwanensis Yamamoto 589
   yunnanensis Rolfe 589
Chesneya 328
   nubigena (D. Don) Ali 328
   polystichoides (Hand.-Mazz.) Ali 328
   purpurea P. C. Li 328
Chionocharis 428
   hookeri I. M. Johnst. 38, **428**, **429**
CHLORANTHACEAE 72
Chloranthus 72
   henryi Hemsl.* 72
   multistachys Pei* 72
   sessilifolius K. F. Wu* 72, **73**
Christolea 196
   rosularis K. C. Kuan & Z. X. An 196
Chrysoplenium 311
   carnosum Hook. f. & Thomson **310**, 311
   davidianum Decne ex Maxim.* **310**, 311
   griffithii Hook. f. & Thomson **310**, 311
   nudicaule Bunge **310**, 311
Chusua 581
   crenulata (Schltr.) P. F. Hunt 581
Cimifuga 112
   brachycarpa P. K. Hsiao* 112
   foetida L. 112
   simplex (DC.) Wormsk. ex Turcz. 112
   yunnanensis P. K. Hsiao* 112
Cinnamomum 42, 74
   camphora (L.) Presl 74
   glanduliferum (Wall.) Meissn. 74
Circaeaster 120
   agrestis Maxim. 120
CIRCAESTERACEAE 120
Cirrhopetalum 604
   andersonii Hook. f. 604
   henryi Rolfe 604

Cirsium 492
   bolocephalum Hand.-Mazz. 492, **493**
   botryodes Petrak ex Hand.-Mazz. 492
   eriophoroides (Hook. f.) Petrak 492
   falconeri (Hook. f.) Petrak 492, **493**
   fargesii (Franch.) Diels 492, **493**
   forrestii (Diels) H. Lévl. 492
   griseum H. Lévl. 492
   handelii Petrak ex Hand.-Mazz.* 492, **493**
   henryi (Franch.) Diels* 492
   mairei (H. Lévl.) H. Lévl. 492, **493**
   yunnanensis Petrak 492
Cladrastis 323
   sinensis Hemsl.* **322**, 323
Clematis 2, 91
   aethusifolia Turcz. 95, 96, **97**
   akebioides (Maxim) Veitch* **94**, 95, 96
   alpina (L.) Mill. var. *macropetala* (Ledeb.) Maxim. 99
   alternata Kitamura & Tamura 96, **97**
   anemoniflora D. Don 92
   armandii Franch. 90, 91
   brevicaudata DC. 91, 92, **93**
   chrysocoma Franch.* 92, **93**
   finetiana Lévl. & Vaniot* **90**, 91
   glauca Willd. 96
      var. *akebioides* (Maxim.) Rehd. & E. H. Wilson 95
   gouriana Roxb. 92
   gracilifolia Rehd. & E. H. Wilson* 95, **94**
      var. dissectifolia W. T. Wang & M. C. Chang* 95, **95**
   grata Wall. 92, **93**
   intricata Bunge 96, **97**
   lasiandra Maxim. 96
   macropetala Ledeb. 96, **98**, 99
   montana Buch.-Ham. 92, **93**, 95
      var. glabrescens Comber 95
      var. grandiflora Hook.* 92, **93**
      var. rubens E. H. Wilson* 92
      var. sericea Franch. ex Finet & Gagnep. 95
      var. sterilis Hand.-Mazz.* 92
      var. trichogyna M. C. Chang* 92
      var. wilsonii Sprague* 92, **93**
   nannophylla Maxim.* 95, 96, **97**
   pogonandra Maxim.* **94**, 95
   potaninii Maxim.* **90**, 91
      subsp. fargesii 90
   pseudopogonandra Finet & Gagnep.* 96, **97**
   ranunculoides Franch.* 96, **97**
   rehderiana Craib* 95, 96, **97**
   repens Finet & Gagnep.* **94**, 95
   spooneri Rehd. & E. H. Wilson* **94**, 95
   tangutica (Maxim.) Korsh. **94**, 95
      subsp. obtusiuscula (Rehd. & E. H. Wilson) Grey-Wilson* **94**, 95
   tibetana Kuntze subsp. vernayi (C. E. C. Fischer) Grey-Wilson 95, 96, **97**
      subsp. vernayi var. dentata Grey-Wilson 95
   uncinata Champ. ex Benth. **90**, 91
   urophylla Franch.* 96, **97**
   venusta M. C. Chang* 92, **93**
Clerodendrum 432
   bungei Steud. 432, **433**
   colebrookianum Walp. 432, **433**
   trichotomum Thunb. 432, **433**
Clethra 199
   cavaleriei H. Lévl. 199

bodinieri H. Lévl.* 199
delavayi Franch **198**, 199,
esquirolii H. Lévl. 199
fabri Hance 199
fargesii Franch.* **198**, 199
purpurea W. P. Fang & L. C. Hu 199
CLETHRACEAE 199
Clintonia 561
   udensis Trautv. & C. A. Mey. **560**, 561
CLUSIACEAE (GUTTIFERAE) 170
Codonopsis 471
   benthamii Hook. f. & Thomson 472, **473**
   bulleyana Forrest & Diels* 472, **473**
   convolvulacea Kurz 471
      var. *forrestii* (Diels) Ballard 471
      var. *pinifolia* Hand.-Mazz. 471
   forrestii Diels* **470**, 471
   gracilis Hook. f. & Thomson 475
   javanica Hook. f. & Thomson 472, **473**
   macrocalyx Diels* 472
   meleagris Diels* 472, **473**
   nervosa (Chipp) Nannf.* 472, **473**
   ovata Spreng. var. *nervosa* Chipp 472
   pinifolia (Hand.-Mazz.) Grey-Wilson* **470**, 471
   sp.* 472, **474**, 475
   subglobosa W. W. Sm.* 472, **473**
   subscaposa Kom.* 472
   tangshen Oliv.* 472, **473**
   tubulosa Kom.* 471
   vinciflora Kom. **470**, 471
Coeloglossum 582
   viride (L.) Hartm. 582, **583**
Coelogyne 598
   calcicola Kerr 598, **599**
   corymbosa Lindl. 598, **599**
   fimbriata Lindl. 598, **599**
   occultata Hook. f.* 598
Colquhounia 435
   coccinea Wall. **434**, 435
   compta W. W. Sm.* 435
Comastoma 404
   falcatum (Turcz. ex Kar. & Kir.) Toyokuni 404, **405**
   traillianum (Forrest) Holub* 404, **405**
Commelina 516, 517
   benghalensis L. 516
   communis L. 516, **516**
   diffusa Burm. f. 516
   paludosa Blume 516
COMMELINACEAE 516
COMPOSITAE 491
Convallaria bifolia L. 558
CONVALLARIACEAE 555
CONVOLVULACEAE 424
Convolvulus 424
   ammannii Desr. 424, **425**
   arvensis L. 424, **425**
   tragacanthoides Turcz. 424
Corallodiscus 459
   bullatus (Craib) B. L. Burtt 459
   flabbellatus (Craib) B. L. Burtt 459
   kingianus (Craib) B. L. Burtt **458**, 459
   lanuginosus (Wall. ex R. Br.) B. L. Burtt **458**, 459
Corallorhiza 593
   trifida Chat. **592**, 593
Cordia 427
   thyrsiflora Sieb. & Zucc. 427

Coriaria 367
   kweichovensis Hu 367
   sinica Maxim. 367
   nepalensis Wall. **366**, 367
   terminalis Hemsl.* 367
CORIARIACEAE 367
CORNACEAE 344
Cornus 39
   capitata Wall. 345
   controversa Hemsl. 344
   hemsleyi C. K. Schneid. & Wangerin 344
   kousa Hance var. *chinensis* Osborn 345
   macrophylla Wall. 344
Cortusa 259
   matthioli L. 259
      subsp. *pekinensis* (V. A. Richter) Kitag **258**, 259
Corydalis 30, 128
   adrienii Prain 140
   adunca Maxim.* 132, **134**, 135
   albicaulis Franch. 135
   astragalina Hook. f. & Thomson 135
   atuntsuensis W. W. Sm.* **134**, 135
   balsamiflora Prain 136
   benecincta W. W. Sm.* **138**, 139
      subsp. *trilobipetala* (Hand.-Mazz.) Lidén* 139
   bulleyana Diels 136
   calcicola W. W. Sm.* 135, 139, 140, **141**
   calycosa H. Chuang* 136, **137**
   cheilanthifolia Hemsl.* **130**, 131
   conspersa Maxim. 131, 132, **133**
   corymbosa C. Y. Wu & Z. Y. Su* 131, 132, **133**
   curviflora Maxim. ex Hemsl.* 135, **138**, 139
      var. *cytisiflora* Fedde 131
   cytisiflora (Fedde) Lidén ex C. Y. Wu, H. Chuang & Z. Y. Su* **130**, 131
   dasyptera Maxim.* 132, **134**, 135
   delphinoides Fedde 131
   densispica C. Y. Wu* 131
   elata Franch.* 135, 136, **137**
   ellipticarpa Z. Y. Su 135, 136
   eugeniae Fedde* 132, **134**, 135
   flaccida Hook. f. & Thomson 135, 136
   flexuosa Franch.* 135, 136, **137**
      subsp. *kuanhsienensis* C. Y. Wu 136
   flexuosa form **137**
   fluminicola W. W. Sm. 132
   gemmipara H. Chuang var. *ecristata* H. Chuang 136
   hamata Franch.* 131, 132, **133**
   hemidicentra Hand.-Mazz.* 135, **138**, 139
   kokiana Hand.-Mazz.* 135, 136, **137**
   linarioides Maxim.* **130**, 131
   linstowiana Fedde* 135, 136
   madida Lidén & Z. Y. Su* 132, **134**, 135
   melanochlora Maxim.* **14**, 139, 140, **141**
   moupinensis Franch.* 131
   ophiocarpa Hook. f. & Thomson **130**, 131
   oxypetala Franch.* 135, **138**, 139
   pachycentra Franch.* 135, **138**, 139
   pachypoda Franch.* 132, **134**, 135
   prattii Franch.* 131
   pseudoadoxa C. Y. Yu & H. Chuang* 139
   pseudocristata Fedde* 132, **134**, 135
   pseudohamata Fedde* 131, 132, **133**
   pterygopetala Hand.-Mazz.* 131
   quadriflora Hand.-Mazz. 139
   quantmeyeriana Fedde* 135, 136, **137**
   rheinbabeniana Fedde* 132, **134**
   saxicola G. S. Bunting* 131, 132, **133**
   scaberula Maxim.* 131, 132, **133**
   schlechteriana Maxim. 131
   shensiana Lidén ex C. Y. Wu, H. Chuang & Z. Y. Su* 135, **138**, 139
   smithiana Fedde* 135, 136, **137**
   stricta Steph. ex Fisch. 132, **134**, 135
   taliensis Franch.* 135, 136
   temulifolia Franch. 135, 136, **137**
   thalictrifolia Franch. 132
   tibetica Hook. f. & Thomson var. *pachypoda* Franch. 135
   tomentella Franch.* 131, 132
   trachycarpa Maxim.* 131, 132, **133**
   trilobipetala Hand.-Mazz. 139
   trifoliata Franch. 135, 139
   wilsonii N. E. Br.* 131, 132
   yunnanensis Franch.* 131
CORYLACEAE 148
Corylopsis 143
   omeiensis Yang* 143
   sinensis Hemsl.* **142**, 143
   veitchiana Bean* **142**, 143
   willmottiae Rehd. & E. H. Wilson* **142**, 143
   yunnanensis Diels* **142**, 143
Corylus 148
   ferox Wall. 148
      var. *thibetica* (Batalin) Franch. 148, **149**
   heterophylla Trautv. 148, **149**
      var. *sutchuanensis* Franch.* 148
   yunnanensis (Franch.) A. Camus* 148
Cosmos 512
   bipinnatus Cav. 512, **513**
COSTACEAE 521
Costus 521
   ormosanus Nakai 521
   speciosus (J. König) Sm. **520**, 521
Cotinus 365
   coggygria Scop. 365, **365**
Cotoneaster 275
   brickellii J. Fryer & B. Hylmö* 275, 276, **277**
   bullatus Bois* 275, 276, **277**
   buxifolius Wall. ex Lindl. 275, 276
   cochleatus (Franch.) G. Klotz* 275, 276, **277**
   frigidus Wall. ex Lindl. 275, 276
   himalaiensis hort. ex Zabel 276
   hodjingensis G. Klotz* 275, 276
   horizontalis Decne 275, **278**, 279
   induratus J. Fryer & B. Hylmö* **274**, 275
   lacteus W. W. Sm.* 275, 276, **277**
   lidjiangensis G. Klotz* 275, 276, **277**
   qungbixiensis J. Fryer & B. Hylmö* **274**, 275
   rehderi Pojark.* 275, 276
CRASSULACEAE 296
Crataegus 272
   chungtienensis W. W. Sm.* 272, **273**
Crawfurdia 408
   angustata C. B. Clarke 408, **409**
   delavayi Franch.* 408
   dimidiata (C. Marq.) Harry Sm.* 408, **409**
   gracilipes Harry Sm.* 408, **409**
   kingdonii C. Marq. 409
   sessiliflora (C. Marq.) Harry Sm.* 408
   speciosa Wall. 408, 409, **409**
   thibetica Franch.* 408, 409, **409**
   trailliana Forrest 408
   wardii C. Marq. 409
Cremanthodium 511
   angustifolium W. W. Sm.* 511
   arnicoides (DC. ex Royle) R. D. Good 511, 512, **513**
   brunneopilosum S. W. Liu* 511, 512, **513**
   calotum Diels 503
   campanulatum (Franch.) Diels 511, 512, **513**
   decaisnei C. B. Clarke 511, 512, **513**
   lineare Maxim.* 511, 512, **513**
   palmatum Benth. subsp. *rhodocephalum* (Diels) R. D. Good 512
   pleurocaule (Franch.) R. D. Good 507
   reniforme (DC.) Benth. 511, 512
   rhodocephalum Diels 511, 512, **513**
   rumicifolium (Drumm.) R. D. Good 504
Cremastra 593
   appendiculata (D. Don) Makino **592**, 593
Crepis umbrella Franch. 515
CRUCIFERAE (BRASSICACEAE) 188
Cryptomeria 52
   japonica (Thunb. ex L. f.) D. Don. 42, 52, **53**
CUCURBITACEAE 183
Cunninghamia 52
   lanceolata (Lamb.) Hook. 40, 52, **53**
CUPRESSACEAE 52
Cupressus 55
   chengiana S. Y. Hu* 55
   duclouxiana Hickel* 32, **54**, 55
Curcuma 518
   aromatica Salisb. 518, **519**
   sichuanensis X. X. Chen* 518
   yunnanensis N. Liu & S. J. Chen* 518
Cuscuta 425
   chinensis L. 425, **425**
Cyananthus 475
   albiflorus Marq. 475
   delavayi Franch.* 475, 476, **477**
   flavus C. Marq.* **474**, 475
   formosus Diels* 475, 476, **477**
   incanus Hook. f. & Thomson **474**, 475
   × C. *macrocalyx* **474**
   lobatus Wall. ex Benth. **474**, 475
   longiflorus Franch.* 475, 476, **477**
   macrocalyx Franch. **474**, 475
      subsp. *spathulifolius* (Nannf.) Shrestha 475
   pedunculatus C. B. Clarke **474**, 475
   sherriffii Cowan* **474**, 475
Cyanotis 517
   vaga (Lour.) Schultes & Schultes f. 517
Cyathella 415
   corymbosa (Wight) C. Y. Wu & D. Z. Li 415
CYCADACEAE 64
Cycas 64
   guizhouensis K. M. Lan & R. F. Zhou 64
   multiovula D. Y. Wang 64
   panzhihuaensis L. Zhou & S. Y. Yang* 64, **65**
   revoluta Thunb. 64, **65**
   szechuanensis W. C. Cheng & L. K. Fu* 64
Cyclobalanopsis 32, 34, 39, 150
   glauca Oerst. 44
   glaucoides Schottky 151
   lamellosa (Sm.) Oerst. 151
Cyclocarya 364
   paliurus (Batalin) Iljinsk. 364
Cymbidium 609
   eburneum Lindl. 609, 610, **611**
   ensifolium (L.) Sw. **608**, 609
   erythraeum Lindl. 609, 610, **611**
   faberi Rolfe **608**, 609

floribundum Lindl. **608**, 609
forrestii Rolfe 609
goeringii Rchb. f. **608**, 609
hookerianum Rchb. f. 609, 610, **611**
kanran Makino **608**, 609
lancifolium Hook. **608**, 609
longibracteatum Y. S. Wu & S. C. Chen 609
lowianum Rchb. f. 609, 610, **611**
macrorhizon Lindl. 609, 610, **611**
pumilum Rolfe 609
sinense (Jackson ex Andr.) Willd. 609, 610, **611**
szechuanensis S. Y. Hu 610
tortisepalum Fukuyama var. longibracteatum (Y. S. Wu & S. C. Chen) Y. S. Wu & J. Z. Liu* 609, **611**
tracyanum L. Castle 609, 610, **611**
virescens Lindl. 609
wilsonii (Rolfe ex Cook) Rolfe* 609, 610, **611**
Cynanchum 413, 415
balfourianum (Schltr.) Tsiang & Zhang 413
canescens (Willd.) K. Schum. 413
corymbosum Wight 415
forrestii Schltr. 413
glaucum Wall. ex Wight 413
muliense Tsiang 413
steppicola Hand.-Mazz. 413
Cynoglossum 427
amabile Stapf & Drumm. **426**, 427
furcatum Wall. **426**, 427
lanceolatum Forssk. 427
triste Diels* **426**, 427
uncinatum Benth. 428
Cypripedium 7, 570
bardolphianum W. W. Sm. & Farrer* 574, **575**
calcicola Schltr.* 570, **572**, 573
daliense S. C. Chen & J. L. Wu 574
debile Rchb. f. 573, 574, **575**
elegans Rchb. f. 573, 574, **575**
fargesii Franch.* 570, 574, **576**, 577
farreri W. W. Sm.* **572**, 573
fasciolatum Franch.* 570, **572**, 573
flavum P. F. Hunt & Summerh.* 570, **571**
forrestii P. J. Cribb* 574, **575**
franchetii E. H. Wilson* 570, 573
guttatum Sw. 573, 574, **575**
henryi Rolfe* 570, **572**, 573
himalaicum Rolfe 570, **571**
japonicum Thunb. 573, 574, **575**
lentiginosum P. J. Cribb & S. C. Chen* 574, 577
lichiangense S. C. Chen & P. J. Cribb 574, **576**, 577
margaritaceum Franch.* 574, **575**
micranthum Franch.* 574, **575**
palangshanense T. Tang & F. T. Wang* 573, 574, **575**
plectrochilum Franch. 570, **572**, 573
shanxiense S. C. Chen 570, **572**, 573
sichuanense Perner* 574, **576**, 577
smithii Schltr. 573
tibeticum King ex Rolfe 570, **572**, 573
wardii Rolfe* 570, **571**
yunnanense Franch.* 570, **571**
× froschii Perner **572**, 573
Cyrtosia lindleyana Hook. f. & Thomson 591

Dactylicapnos 140
lichiangensis (Fedde) Hand.-Mazz 140, **141**
scandens (D. Don) Hutch. 140, **141**

Dactylorhiza 582
hatagirea (D. Don) Soó 582, **583**
Daiswa 562
violacea (H. Lévl.) A. Takht. 562
Daphne 43, 336
acutiloba Rehd.* 340, **341**
angustiloba Rehd.* 336, 339
aurantiaca Diels* 336, 337, **337**
bholua Buch.-Ham. 336, **338**, 339
var. glacialis (W. W. Sm. & Cave) B. L. Burtt. 336, 339
calcicola W. W.Sm.* 336, 337, **337**
clivicola Hand.-Mazz. 339
gemmata E. Pritz* 336, **338**, 339
genkwa Sieb. & Zucc.* 336, **338**, 339
giraldii Nitsche* 336, **338**, 339
indica L. 340
longilobata (Lecompte) Turrill* 336, 340, **341**
retusa Hemsl. 336, **338**, 339
rosmarinifolia Rehd.* 336, **338**, 339
tangutica Maxim.* 336, **338**, 339
wilsonii Rehd. 339
wolongensis Brickell & Mathew* 336, **338**, 339
Datura 423
metel L. 424
stramonium L. 423
Davidia 346
involucrata Baill.* 40, 42, 44, 346, **346**
var. vilmoriniana (Dode) Wangerin* 346
Debregeasia 176
longifolia (Burm. f.) Weddell 176
orientalis C. J. Chen 176, **177**
velutina Gaudich. 176
Decaisnea 91
fargesii Franch. 91
insignis (Griff.) Hook. f. & Thomson **90**, 91
Decumaria 308
sinensis Oliv.* 308
Delphinium 116
anthriscifolium Hance 116, **117**
beesianum W. W. Sm.* 116, **117**
caeruleum Jacq. 116, **117**
chrysotrichum Finet & Gagnep.* 116, **117**
grandiflorum L. 116, **117**
kamaonense Huth 116, **117**
monanthum Hand.-Mazz.* 116, **117**
thibeticum Finet & Gagnep.* 116
Dendrobenthamia 345
capitata (Wall.) Hutch. 345
japonica (Sieb. & Zucc.) Hutch. var. chinensis (Osborn) W. P. Fang 345
Dendrobium 605
amplum Lindl. 608
aphyllum (Roxb.) C. E. C. Fischer 605, 606, **607**
aurantiacum Rchb. f. **604**, 605
chrysotoxum Lindl. **604**, 605
devonianum Paxton 605, 606, **607**
fanjingshanense Z. H. Tsi ex X. H. Jin & Y. W. Zhang* **604**, 605
fargesii Finet 608
fimbriatum Hook. **604**, 605
hancockii Rolfe* **604**, 605
lituiflorum Lindl. 605, 606, **607**
loddigesii Rolfe 605, 606, **607**
longicornu Lindl. 605, 606
moniliforme (L.) Sw. 605, 606, **607**
nobile Lindl. 605, 606, **607**

suavissimum Rchb. f. 605
thyrsiflorum Rchb. f. 605, 606, **607**
wardianum Lindl. 605, 606, **607**
williamsonii Day & Rchb. f. 605, 606, **607**
Dendropanax 380
chevalieri (R. Vig.) Merr. 380
dentigerus (Harms ex Diels) Merr. 380
ferrugineus H. L. Li 381
Desmodium 327
elegans DC. **326**, 327
floribundum DC. 327
forrestii Schindl. 327
franchetii Rehd. 327
multiflorum DC. 327
tragacanthoides (Pall.) Poiret var. *tibetica* Maxim. ex Schneid. 327
Deutzia 303
calycosa Rehd.* 304, **305**
var. macropetala Rehd.* 303, 304
densiflora Rehd. 304
discolor Hemsl.* 303, 304
glomeruliflora Franch.* 303, 304, **305**
hookeriana (C. K. Schneid.) Airy Shaw 303
longifolia Franch.* 303, 304
var. *farreri* Airy Shaw 304
monbeigii W. W. Sm.* 303, 304, **305**
purpurascens (Franch. ex L. Henry) Rehd. 303, 304, **305**
setchuenensis Franch.* **302**, 303
zhongdianensis S. M. Hwang* 303, 304, **305**
Dianthus 151
amurenese Jacq. 152
chinensis L. 152, **153**
longicalyx Miq.* 152
superbus L. 152, **153**
versicolor Fisch. ex Link 152
Diapensia 235
bulleyana Forrest ex Diels 235
himalaica Hook. f. & Thomson **234**, 235
purpurea Diels **234**, 235
wardii W. E. Evans* 235
DIAPENSIACEAE 235
Dicentra 140
lichiangensis Fedde 140
macrantha Oliv. 140
scandens D. Don 140
thalictrifolia (Wall.) Hook. f. 140
Dichocarpum 112
fargesii (Franch.) W. T. Wang & P. K. Hsiao* 112, **113**
Dichroa 308
febrifuga Lour. 308, **309**
Dicranostigma 128
franchetiana (Prain) Fedde* 128, **129**
leptopodum (Maxim.) Fedde* 128, **129**
Didissandra 459
kingiana Craib 459
Didymocarpus 459
lanuginosus Wall. ex R. Br. 459
Digitalis 447
glutinosa Gaertn. 447
Dilophia 195
fontanum Maxim. 195
Dinetus 425
duclouxii (Gagnep. & Cour.) Staples* 425, **425**
Diospyros 235
kaki Thunb. **234**, 235
lotus L. **234**, 235
Dipelta 483

elegans Batalin* 483, **485**
floribunda Maxim.* 483, 484, **485**
ventricosa Hemsl.* **482**, 483
yunnanensis Franch.* 483, 484, **485**
Diphylleia 87
  sinensis H. L. Li* **86**, 87
Diplarche 231
  multiflora Hook. f. & Thomson **230**, 231
DIPSACACEAE 490
Diptera 319
  sarmentosa (L. f.) Losinsk. 319
Disporum 561
  acuminatissimum W. L. Sha* 561, 562, **563**
  bodinieri (H. Lévl. & Vaniot) F. T. Wang & T. Tang* 561, 562, **563**
  brachystemon F. T. Wang & T. Tang 562
  calcaratum D. Don **560**, 561
  cantoniense (Lour.) Merr. 561
  hamiltonianum D. Don 561
  longistylum (H. Lévl. & Vaniot) H. Hara* 561
  megalanthum F. T. Wang & T. Tang* 561, **560**
  uniflorum Baker ex S. Moore 561, 562
Dodonaea 359
  viscosa (L.) Jacq. **358**, 359
Dolomiaea 491
  berardioidea (Franch.) C. Shih* 491
  forrestii (Diels) C. Shih* **490**, 491
  souliei (Franch.) C. Shih* **490**, 491
Dontostemon 192
  pinnatifidus (Willd.) Al-Shehbaz & H. Ohba 195
    subsp. linearifolius (Maxim.) Al-Shehbaz & H. Ohba 195
  tibeticus (Maxim.) Al-Shehbaz* 192, **193**
Doronicum 502
  briquetii Cavill. **502**, 503
  limprichtii Diels 503
  souliei Cav. 503
  stenoglossum Maxim.* **502**, 503
  thibetanum Cav. 503
  yunnanense Franch. ex Diels 503
Draba 196
  glomerata Royle 196, 197
  involucrata (W. W. Sm.) W. W. Sm.* 196, **197**
  jucunda W. W. Sm.* 196
  ladyginii Pohle* 196, 197, **197**
  oreades Schrenk 196, **197**
  polyphylla O. E. Schulz 196, 198, **198**
  surculosa Franch.* 196, 198, **198**
  yunnanensis Franch.* 196, 198, **198**
Dracocephalum 436
  bullatum Forrest ex Diels* 436, **437**
  forrestii W. W. Sm.* 436, **437**
  heterophyllum Benth. 436
  moldavica L. 436, **437**
  purdomii W. W. Sm.* 436, **437**
  stewartianum (Diels) Dunn 443
Dregea 414
  sinensis Hemsl.* 414, **414**
  volubilis (L. f.) Benth. ex Hook. f. 414
Drosera 140
  lunata Buch.-Ham. ex DC. 140
  peltata Sm. ex Willd. 140, **141**
DROSERACEAE 140
Dryopteris 34
Dumasia 331
  cordifolia Benth. 331
  forrestii Diels* 331
  leiocarpa Benth. **330**, 331
  villosa DC. 331

var. *leiocarpa* (Benth.) Baker 331
yunnanensis Y. T. Wei & S. K. Lee 331
Dysosma 84
  aurantiocaulis sensu T. S. Ying 84
  mairei (Gagnep.) Hiroe 84
  delavayi (Franch.) Hu 84
  difformis (Hemsl. ex E. H. Wilson) T. H. Wang 84
  furfuracea S. Y. Bao 84

EBENACEAE 235
Echinops 492
  dahuricus Fisch. 493
  gmelini Turcz. 492
  latifolius Tausch 493
  przewalskii Iljin. 493
Edgeworthia 340
  gardneri (Wall.) Meisn. 340, **341**
Ehretia 427
  acuminata R. Br. **426**, 427
  thyrsiflora (Sieb. & Zucc.) Nakai 427
Eichhornia 532
  crassipes (Mart.) Solms 532, **532**
ELAEAGNACEAE 347
Elaeagnus 347
  salicifolia (D. Don) A. Nelson 347
Eleutherococcus 383
  henryi Oliv.* **382**, 383
  wilsonii (Harms) Nakai* **382**, 383
    var. pilosulus (Rehd.) P. S. Hsu & S. L. Pan* 383
Elsholtzia 435
  eriocalyx C. Y. Wu & S. C. Huang* 435
  fruticosa (D. Don) Rehd. **434**, 435
  rugulosa Hemsl.* **434**, 435
  strobilifera Benth. **434**, 435
Emmenopterys 476
  henryi Oliv.* 44, 476, **477**
Engelhardtia 364
  roxburghiana Wall. 364, **365**
  spicata Leschen. ex Blume 364
Enkianthus 228
  chinensis Franch.* 40, 228
  deflexus (Griff.) C. K. Schneid 228, **229**
Eomecon 128
  chionantha Hance* 128, **129**
Ephedra 64
  likiangensis Florin* 64, **65**
  monosperma Gmelin ex C. A. Mey. 64, **65**
EPHEDRACEAE 64
Epigeneium 608
  amplum (Lindl.) Summerh. 608, **608**
  fargesii (Finet) Gagnep. 608
Epilobium 342
  conospersum Haussk. 342, **342**
  pannosum Haussk. **342**, 343
  reticulatum C. B. Clarke 342
Epimedium 87
  acuminatum Franch.* 87, 88, **89**
  brachyrrhizum Stearn* 87, 88, **89**
  brevicornu Maxim.* **86**, 87
  davidii Franch.* 87, 88, **89**
  ecalcaratum G. Y. Zhong* **86**, 87
  fangii Stearn* 87, 88
  fargesii Franch.* **86**, 87
  flavum Stearn* 87, 88, **89**
  franchetii Stearn* 87, 88
  leptorrhizum Stearn* 87
  membranaceum K. Meyer* 87, 88

platypetalum K. Meyer* 87
pubescens Maxim.* **86**, 87
  subsp. cavaleriei (Stearn) Stearn 87
sagittatum (Sieb. & Zucc.) Maxim.* **86**, 87
stellulatum Stearn* **86**, 87
Epipactus 589
  helleborine (L.) Crantz 590, **591**
  mairei Schltr. 590, **591**
  royleana Lindl. 590, **591**
Eremurus 549
  chinensis O. Fedtsch.* **548**, 549
Eria 602
  coronaria (Lindl.) Rchb. f. 602, **603**
  rhomboidalis T. Tang & F. T. Wang* 602, **603**
ERICACEAE 200, 232
Erigeron 511
  acris L. 511
  breviscapus (Vaniot) Hand.-Mazz.* **510**, 511
  dielsii H. Lévl. 511
  multiradiatus (Lindl.) Benth. **510**, 511
Eriobotrya 272
  japonica (Thunb.) Lindl.* 272, **273**
Eriophyton 439
  wallichii Benth. **438**, 439
Eriosolena 340
  aff. composita (L. f.) Merr. 340, **341**
Eritrichium 428
  deqinense W. T. Wang* 428, **429**
Erodium 371
  cicutarium (L.) L'Hér. **370**, 371
Erysimum 191
  forrestii (W. W. Sm.) Pol.* 190, 191
  funiculosum Hook. f. & Thomson 191, 192, **193**
  handelmazzettii Pol.* 191, 192, **193**
  roseum (Maxim.) Pol.* 191, 192, **193**
  wardii Pol.* 192, **193**
Erythrina 39
Euaraliopsis 382
  ciliata (Dunn) Hutch. 382
  fatsioides (Harms) Hutch 382
Eucommia 44
  ulmoides Oliv. 44
Eugenia 39
Euonymus 347
  alatus (Thunb.) Sieb. 347, 348, **349**
  cornutus Hemsl. 347, 348, **349**
  frigidus Wall. 347
  myrianthus Hemsl.* 347, 348, **349**
  nanoides Loes. & Rehd.* 347, 348
    var. oresbius (W. W. Sm.) Y. R. Li 348
  oresbius W. W. Sm. 348
  phellomanus Loes. ex Diels* 347, 348, **349**
  sanguineus Loes.* 347, 348
  striatus (Thunb.) Loes. ex Gilg. & Loes. 348
  verrucosus Scop. var. *tchefouensis* Debeaux 348
  wilsonii Sprague* 347, 348
Eupatorium 499
  mairiei H. Lévl. 499
Euphorbia 353
  chrysocoma H. Lèvl. & Vaniot 355
  bulleyana Diels 353
  duclouxii H. Lèvl. & Vaniot 355
  griffithii Hook. f. 353, **354**, 355
  jolkinii Boiss. **13**, 353, **354**, 355, **536**
  megistopoda Diels 355
  nematocypha Hand.-Mazz. 355
  pallasii Turcz.* 353, **354**, 355

pekinensis Rupr. 353, **354**, 355
regina H. Lévl. 355
sikkimensis Boiss. 353, **354**, 355
stracheyi Boiss. 353, **354**, 355
wallichii Hook. f. 353, **354**, 355
    subsp. duclouxii (H. Lèvl. & Vaniot) R. Turner* 353, 355
yunnanensis Radcl.-Sm. 355
EUPHORBIACEAE 352
Euphrasia 451
    jaeschkei Wettst. **450**, 451
Eurya 40
Euryale 78
    ferox Salisb. 78
Exacum 408
    tetragonum Roxb. 408, **409**

FABACEAE 323
FAGACEAE 148
Fagus 149
    engleriana Seeman* 149
    longipetiolata Seeman 149
Fargesia spp. 35
Ficus 39
Floscopa 517
    scandens Lour. **516**, 517
Forsythia 415
    giraldiana Lingelsh.* 415
    likiangensis Ching & Feng ex P. Y. Pai* 415
    suspensa (Thunb.) Vahl* **414**, 415
    viridissima Lindl.* 415
Fragaria 287
    nilgherrensis Schltr. ex J. Gay. **286**, 287
    orientalis Losinsk. 287
    pentaphylla Losinsk.* **286**, 287
        forma alba Staudt & Dickoré 287
Fraxinus 44
Fritillaria 549
    cirrhosa D. Don **134**, **548**, 549, 550, **551**
        var. viridiflava S. C. Chen.* 549, 550, **551**
    crassicaulis S. C. Chen 549, 550, **551**
    dajinensis S. C. Chen* 549
    davidii Franch.* **548**, 549
    delavayi Franch. **26**, **548**, 549
    fujiangensis Y. K. Yang et al. 549
    gansuensis Y. K. Yang et al. 550
    glabra (P. Y. Li) S. C. Chen 550
    huangshanensis Y. K. Yang & C. J. Wu 550
    mellea S. Y. Chang & S. C. Yueh 549
    monantha Migo* 549, 550
    omeiensis S. C. Chen* 549, 550, **551**
    pingwuensis Y. K. Yang & J. K. Wu 549
    przewalskii Maxim.* **548**, 549
    qingchuanensis Y. K. Yang & J. K. Wu 549
    qinghaiensis Y. K. Yang & J. K. Wu* **548**, 549
    shaanxiica Y. K. Yang et al. 550
    sichuanica S. C. Chen* **548**, 549
    sp.* 550, **551**
    taipaiensis P. Y. Li* 549, 550
    unibracteata P. K. Hsiao & K. C. Hsia* **548**, 549
    aff. cirrhosa D. Don* 550, **551**
FUMARIACEAE 128

Galearis 578
    cyclochila (Franch. & Sav.) Nevski 578
    diantha (Schltr.) P. F. Hunt 578, **579**
    huanglongensis Q. W. Meng & Y. B. Luo 578, **580**, 581
    roborowskii (Maxim.) S. C. Chen, P. J. Cribb & S. W. Gale 578, **580**, 581
    spathulata (Lindl.) P. F. Hunt 578
    wardii (W. W. Sm.) P. F. Hunt* 578, **579**
Galeola 591
    faberi Rolfe* 592
    kwangsiensis Hand.-Mazz. 592
    lindleyana (Hook. f. & Thomson) Rchb. f. 592, **592**
        var. *unicolor* Hand.-Mazz. 592
Gastrodia 592
    elata Blume 592, **592**
Gaultheria 228
    forrestii Diels 231
    fragrantissima Wall. 228, **230**, 231
    hookeri C. B. Clarke 228, **230**, 231
    nummularioides D. Don 228, 229
    sinensis J. Anth. 228, 229, **229**
    trichophylla Royle 228, **229**
    wardii C. Marq. & Airy Shaw 228, 231
Gentiana 2, 7, 391
    algida Pallas 399, **401**
    arethusae Burkill* 395, 396, **397**
        var. delicatula C. Marq. 395, 396
    aristata Maxim.* 400, 403
    asparagoides T. N. Ho* 400, **402**, 403
    asterocalyx Diels* 400, **402**, 403
    atuntsiensis W. W. Sm.* **398**, 399
    caelestis (C. Marq.) Harry Sm.* 395, 396, **397**
    cephalantha Franch. **394**, 395
        var. vaniotii (H. Lévl.) T. N. Ho) 395
    chinensis Kusn.* 392, **393**
    choanantha C. Marq.* 400, **402**, 403
    chungtienensis C. Marq.* 400, **401**, 403
    crassicaulis Duthie ex Burkill* **390**, 391, 392
    crassula Harry Sm.* 400, **402**, 403
    dahurica Fisch. 391
    depressa D. Don 392, **393**
    detonsa Rottb. var. *lutea* Burkill 404
    duclouxii Franch.* 395, 396, **397**
    elwesii C. B. Clarke 392, **393**
    erectosepala T. N. Ho* 399, 400, **401**
    falcata Turcz. ex Kar. & Kir. 404
    farreri Balf. f. 399
    flexicaulis Harry Sm.* 400, **402**, 403
    georgei Diels* **394**, 395
    gilvostriata C. Marq. 392
    grandis Harry Sm. 404
    haynaldii Kanitz* 400, **401**
    hexaphylla Maxim. ex Kusn.* 395, 396, **397**
    lacerulata Harry Sm. 392, **393**
    lawrencei Burkill* 395, 399
        var. farreri (Balf. f.) T. N. Ho* 395, 399
        var. lawrencei 399
    leucomelaena Maxim. ex Kusn. 400, **402**, 403
    lineolata Franch.* 400, **401**
    longistyla T. N. Ho 392
    melandriifolia Franch.* **394**, 395
    microdonta Franch.* **398**, 399
    nubigena Edgew. **398**, 399
    nyalamensis T. N. Ho 395, 396, **397**
    obconica T. N. Ho 395, 396, **397**
    officinalis Harry Sm.* 391, 392
    oreodoxa Harry Sm. 395, **398**, 399
    ornata (Wall. ex G. Don) Griseb. 395, 396, **397**
    phyllocalyx C. B. Clarke 395
    piasezkii Maxim.* 400, **402**, 403
    picta Franch.* 400
    prattii Kusn.* 400, **401**
    pseudoaquatica Kusn. 400, 403
    purdomii C. Marq.* **398**, 399
    rigescens Franch. **394**, 395
    robusta King ex Hook. f. **390**, 391
    rubicunda Franch.* 400, **402**, 403
    sino-ornata Balf. f. 395, **398**, 399
    siphonantha Maxim. ex Kusn.* 391
    stipitata Edgew.
        subsp. tizuensis (Franch.) T. N. Ho* **394**, 395
        subsp. stipitata 395
    stragulata Balf. f. & Forrest* 395
    straminea Maxim. 391
    striolata T. N. Ho* **398**, 399
    stylophora C. B. Clarke 407
    syringea T. N. Ho* 400, **402**, 403
    szechenyii Kanitz* **394**, 395
    ternifolia Franch.* 395, 396, **397**
    tetraphylla Maxim. ex Kusn.* 395, 396
    tibetica King ex Hook. f. 391, 392
        var. *robusta* (King ex Hook. f.) Kusn. 391
    tongolensis Franch. 400, **401**
    traillianum Forrest 404
    trichotoma Kusn.* 399
    tubiflora (Wall. ex G. Don) Griseb. 392, **393**
    urnula Harry Sm. 392, **393**
    veitchiorum Hemsl.* 395, 396, **397**
        var. *caelestis* C. Marq. 396
    venosa Hemsl. 408
    wardii W. W. Sm. 392
        var. emergens (C. Marq.) T. N. Ho 392, **393**
        var. micrantha C. Marq. 392
    wilsonii C. Marq.* **398**, 399
    yunnanensis Franch.* 400, **401**
GENTIANACEAE 391
Gentianopsis 404
    barbata (Froel.) Ma 404
        var. albiflavida T. H. Ho 404
    grandis (Harry Sm.) Ma* 404, **405**
    lutea (Burkill) Ma* 404
    paludosa (Munro ex Hook. f.) Ma 404, **405**
Geodorum 610
    pulchellum Ridl. 610, **611**
GERANIACEAE 368
Geranium 368
    dahuricum DC. 368, 371
    delavayi Franch.* 368, **369**
    donianum Sweet 368, **370**, 371
    eriostemon DC. 368
    farreri Stapf* 368, **370**, 371
    hispidissimum (Franch.) Knuth* 368, 371
    melanandrum Franch. 368
    nepalense Sweet 368, **369**
    orientalitibeticum Knuth* 368, 371
    platyanthum Duthie 368, **369**
    pogonanthum Franch. 368, **369**, 434
    polyanthes Edgew. & Hook. f. 368, **370**, 371
    pratense L. 368, **369**
    pylzowianum Maxim.* 368, **370**, 371
    refractum Edgew. & Hook. f. 368
    sinense Knuth* 368, **369**
    sp.* 368, **370**, 371
    stapfianum Hand.-Mazz.* 368, 371
    strictipes Knuth* 368, **370**, 371
Gerbera 508
    nivea (DC.) Sch. Bip. 508, **509**
GESNERIACEAE 459

Geum 292
    aleppicum Jacq. 292
    japonicum Thunb. 292
        var. chinense F. Bolle* 292
Gilibertia 380
    dentigera Harms ex Diels 380
Ginkgo 11, 52
    biloba L.* 42, 52, **53**
GINKGOACEAE 52
Gleadovia 463
    mupinense Hu* 463
    ruborum Gamble & Prain **462**, 463
    yunnanense Yu 463
Glechoma 436
    complanata (Dunn) Turrill 436
Glycyrrhiza 331
    yunnanensis Cheng f. & L. K. Tai ex P. C. Li* **330**, 331
Gnaphalium 495
    affine D. Don **494**, 495
    chrysocephalum Franch.* 495
    dedekensii Bur. & Franch. 500
    leontopodium L. var. calocephalum Franch. 500
Gomphocarpus 412
    fruticosus (L.) W. T. Aiton 412
    physocarpus E. Mey. 412, **413**
Gomphrena 156
    sessilis L. 156
Goodyera 589
    schlechtendaliana Rchb. f. 589
GROSSULARIACEAE 321
Gueldenstaedtia 335
    coelestis (Diels) N. D. Simpson 335
    flava Adamson 335
        var. tongolensis (Ulbr.) Ali. 335
    himalaica Baker 335
    tongolensis Ulbr 335
    yunnanensis Franch. 335
Guihaia 522
    argyrata (S. K. Lee & F. N. Wei) S. K. Lee, F. N. Wei & J. Dransf. 522, **523**
    grossefibrosa (Gagnep.) S. K. Lee, F. N. Wei & J. Dransf. 522, **523**
Gymnadenia 582
    conopsea (L.) R. Br. 582, **583**
    orchidis Lindl. 582, **583**
Gynura 499
    aurita C. Winkl. 499
    japonica (Thunb.) Juel 499
    pinnatifida DC. 499

Habenaria 582
    balfouriana Schltr.* 584, **584**
    cyclochila Franch. & Sav. 578
    davidii Franch.* **584**, 585
    dentata (Sw.) Schltr. 585
    diphylla Dalzell 584
    glaucifolia Bur. & Franch.* 584, **584**
    mairei Schltr.* 585
Hackelia 428
    brachytuba Opiz ex Bercht. 428
    uncinata (Benth.) C. E. C. Fischer 428, **429**
Halenia 411
    corniculata (L.) Cornaz 411
    elliptica D. Don **410**, 411
        var. grandiflora Hemsl. 411
HAMAMELIDACEAE 140
Hamamelis 140

chinense R. Br. 143
mollis Oliv.* 140, **141**
Handeliodendron 359
    bodinieri (H. Lévl.) Rehd.* **358**, 359
Hedera 380
    nepalensis K. Koch var. sinensis (Tobl.) Rehd. 380, **381**
    sinensis Tobl. 380
Hedychium 520
    coccineum Buch.-Ham. 521
    coronarium J. König 520, **521**
    densiflorum Wall. 520, 521
    forrestii Diels 520, **520**
    villosum Wall. **520**, 521
    yunnanense Gagnep. 521
Hedysarum 332
    alpinum L. 332, **333**
    limprichtii Ulbr. 332
    multijugum Maxim. 332
    sikkimense Benth. ex Baker 332, **333**
Helicia 39
Helleborus 119
    thibetanus Franch.* **118**, 119
Helwingia 346
    chinensis Batalin* **346**, 347
    himalaica Hook. f. & Thomson ex C. B. Clarke 346, 347
    omeiensis (Fang) H. Hara & S. Kurasawa* 346, 347
HEMEROCALLIDACEAE 568
Hemerocallis 568
    forrestii Diels* **568**, 569
    fulva L. **568**, 569
    plicata Stapf* 569
Hemilophia 196
    rockii O. E. Schulz* 196
    sessilifolia Al-Shehbaz* 196, **197**
Hemiphragma 450
    heterophyllum Wall. 450, **450**
Hemipilia 585
    cruciata Finet **584**, 585
    flabellata Bur. & Franch.* **584**, 585
Hepatica 107
    henryi (Oliv.) Steward* **106**, 107
Heptacodium 488
    miconioides Rehd.* 488, **489**
Heptapleurum delavayi Franch. 382
Heracleum 387
    apaense (R. H. Shan & C. C. Yuan) R. H. Shan & T. S. Wang 387
    candicans Wall. ex DC. 387
Herminium 585
    coeloceras (Finet) Schltr. 585
    lanceum (Thunb. ex Sw.) Vuijk 586
    monorchis (L.) R. Br. **584**, 585
    ophioglossoides Schltr. 586, **587**
    tenianum Kraenzl. 585
    unicorne Kraenzl. 585
Herpetospermum 184
    pedunculosum (Seringe) Baill. 184
HIPPOCASTANACEAE 360
Hippophae 347
    rhamnoides L.
        subsp. salicifolia (D. Don) Servett. 347
        subsp. tibetana (Schltr.) Servett. 347
    salicifolia D. Don **346**, 347
    tibetana Schltr. **346**, 347
Holboellia 88
    coriacea Diels* 88

latifolia Wall. 88, **89**
Hoteia 308
    chinensis Maxim. 308
Houttuynia 75
    cordata Thunb. 75, **75**
Hoya 412
    carnosa (L. f.) R. Br. 412, **413**
Hydrangea 304
    anomala D. Don 304
    aspera D. Don 304, **306**, 307
    heteromalla D. Don 304, **306**, 307
    paniculata Sieb. 304, **306**, 307
    robusta Hook. f. & Thomson 304, 307
    sargentiana Rehd.* 304, **306**, 307
    xanthoneura Diels* 304, **306**, 307
HYDROCHARITACEAE 533
Hymenidium 385
    foetens (Franch.) Oimenov & Kljuy 385
Hymenolaena 385
    benthamii Wall. ex DC. 385
Hyoscamus 423
    niger L. **422**, 423
Hypericum 170
    bellum H. L. Li 170, 172
        subsp. latisepalum N. Robson 172, **173**
    daliense N. Robson* 170
    forrestii (Chitt.) N. Robson 170, 172, **173**
    henryi H. Lévl. & Vaniot 170, 171, **171**
    hookerianum Wight & Arnott 170, 171, **171**
    lancasteri N. Robson* 170, 171, **171**
    latisepalum (N. Robson) N. Robson 170, 172, **173**
    patulum Thunb.* 170
        forma forrestii (Chitt.) Rehd. 172
        var. forrestii Chitt. 172
        var. hookerianum (Wight & Arnott) Kuntze 171
    perforatum L. 170
        var. confertiflorum Debeaux 170
    pseudohenryi N. Robson* 170, 172, **173**
    reptans Hook. f. & Thomson ex Dyer 170, **171**
    subsessile N. Robson* 170, 171
Ichtyselmis 140
    macrantha (Oliv.) Lidén 140, **141**
Ilex 348
    corallina Franch.* 348, **350**, 351
    cornuta Lindl. & Paxton 348, 350
    fargesii Franch.* 350, 351
    fortunei Lindl. 350
    franchetiana Loes. 350, 351
    furcata Lindl. ex Göpp. 350
    intricata Hook. f. 350, 351
    macrocarpa Oliv. 348, 350
    melanotricha Merr. 350, 351
    micrococca Maxim. 348, 350, **350**
        var. longifolia Hayata 350
    pernyi Franch.* 348, 350, **350**
    pseudogodajam Franch. 350
    yunnanensis Franch. 350, 351
Illicium 71
    ljiadifengii B. N. Chang* **70**, 71
    majus Hook. f. & Thomson **70**, 71
Impatiens 7, 34, 372
    abbatis Hook. f.* 375
    arguta Hook. f. & Thomson 375
        var. bulleyana Hook. f. 375
    balsamina L. 372, **373**
    barbata Comber* 375, 376, **377**

brevipes Hook. f.* 373, **374**, 375
chinensis L. 372, 373, **373**
clavicuspis Hook. f. 376, 379
cyathiflora Hook. f.* 376, **378**, 379
davidii Franch.* 373, 375
delavayi Franch.* 376, **377**
dicentra Franch. ex Hook. f.* 373, 375
distracta Hook. f.* 376, **378**, 379
drepanophora Hook. f. 376, **377**
faberi Hook. f.* 376, **378**, 379
forrestii Hook. f. ex W. W. Sm.* 376, **377**
imbecilla Hook. f.* 376, **378**, 379
lateristachys Y. L. Che & Y. Q. Lu* 376, **378**, 379
macrovexilla Y. L. Chen* 376, **378**, 379
monticola Hook. f.* 376, **378**, 379
morsei Hook. f.* 372, 373, **373**
noli-tangere L. 376, **377**
omeiana Hook. f.* **374**, 375,
oxyanthera Hook. f.* 376, **377**
platychlaena Hook. f.* 373, **374**, 375
potaninii Maxim.* 376, 379
purpurea Hand.-Mazz.* 376, **378**, 379
rubrostriata Hook. f.* 376, **378**, 379
siculifer Hook. f. 376, **377**
var. *porphyrea* Hook. f.* 376
soulieana Hook. f.* 373, **374**, 375
stenantha Hook. f. 376, **377**
tortisepala Hook. f.* 376
vittata Franch.* 375
waldheimiana Hook. f.* 373, 375
wilsonii Hook. f.* 375
xanthina Comber 376, **378**, 379
aff. *spireana* Hook. f.* 372, 373, **374**
Incarvillea 463
arguta (Royle) Royle 464, **465**
compacta Maxim.* 464, 467
var. *compacta* **466**
var. *qinghaiensis* Grey-Wilson* 464, **466**, 467
delavayi Bur. & Franch.* 464, **465**
grandiflora Bur. & Franch.* **466**, 467
himalayensis Grey-Wilson **466**, 467
longiracemosa Sprague* 464, **465**
lutea Bur. & Franch.* 464, **465**
mairei (Lévl.) Grierson* 464, **465**, **466**
var. *grandiflora* (Wehrh.) Grierson 467
forma *multifoliata* C. Y. Wu & W. C. Yin 464
sinensis Lam.* 464, **465**
subsp. *variabilis* (Batalin) Grierson* 464
forma *przewalskii* (Batalin) Grierson* 464
subsp. *sinensis** 464
younghusbandii Sprague 467
zhongdianensis Grey-Wilson* 464, **466**
sp. aff. I. *compacta* 464, **466**
sp.* 464, **465** ,467
Indigofera 327
amblyantha Craib* **326**, 327
balfouriana Craib* **326**, 327
fortunei Craib* **326**, 327
pendula Franch.* **326**, 327
subnuda Craib 327
Inula 502
forrestii J. Anth.* 502, **502**
helianthus-aquatica C. Y. Wu ex Y. Ling* 502
hookeri C. B. Clarke 502
IRIDACEAE 533
Iris 7, 533

barbatula Noltie* 534, **535**
bulleyana Dykes **13**, **536**, 537
× l. *chrysographes* **536**
chrysographes Dykes **536**, 537
clarkei Baker **536**, 537
collettii Hook. f. 534
var. *acaulis* Noltie* 534, **535**
var. *collettii* 534
confusa Sealy* 533, 534, **535**
cuniculiformis Noltie* 534, **535**
delavayi Micheli* **536**, 537
dolichosiphon Noltie 534
subsp. *orientalis* Noltie* 534, **535**, 537
farreri Dykes* 538, **539**
forrestii Dykes **536**, 537
goniocarpa Baker 534, **535**
var. *grossa* Y. T. Zhao 534
japonica Thunb. 533, 534, **535**
kemaonensis D. Don ex Royle 534, 537
lactea Pallas 538, **539**
var. *chinensis* (Fisch.) Koidz. 538
var. *chrysantha* Y. T. Zhao* 538
polysticta Diels 538
potaninii Maxim. 534, **536**, 537
qinghainica Y. T. Zhao* 538, **539**
ruthenica Ker-Gawl. var. *nana* Maxim.* 538, **539**
sichuanensis Y. T. Zhao* 534
songarica Schrenk 538, **539**
speculatrix Hance* 533, 534, **535**
tectorum Maxim. **532**, 533
tenuifolia Pallas 538, **539**
wilsonii C. H. Wright* **536**, 537
Itea 308
forrestii Y. C. Wu 308
ilicifolia Oliv.* 308, **309**
yunnanensis Franch.* 308, **309**

Jacaranda 467
mimosifolia D. Don. **466**, 467
ovalifolia R. Br 467
Jasminum 419
beesianum Forrest & Diels* 419
elongatum (Berg.) Willd. 419
humile L. 419
mesynі Hance* 419
nudiflorum Lindl.* **418**, 419
var. *pulvinatum* (W. W. Sm.) Kobuski* 419
officinale L. **418**, 419
polyanthum Franch.* **418**, 419
sambac (L.) W. T. Aiton **418**, 419
subhumile W. W. Sm. 419
×stephanense Lemoine* **418**, 419, 420
JUGLANDACEAE 364
Juglans 44
regia L. 44
JUNCAGINACEAE 533
Juniperus 32, 43, 55
chinensis L. 55
var. *sargentii* **54**
distans Florin 55
pingii W. C. Cheng ex Ferré* 55
var. *wilsonii* (Rehd.) Silba **54**
potaninii Kom. 55
recurva Buch.-Ham. ex D. Don. 55
var. *coxii* (A. B. Jacks.) Melville 55
squamata Buch.-Ham. ex D. Don. **54**, 55
tibetica Kom.* 55
zaidamensis Kom. 55

Jurinea 491
berardioidea (Franch.) Diels 491
forrestii Diels 491

Kalopanax 380
ricinifolius (Sieb. & Zucc.) Miq. 380
septemlobus (Thunb.) Kordz. 380
Kerria 294
japonica (L.) DC. 294, **294**
Keteleeria 59
davidiana (Bertr.) Beissn. 59
evelyniana Mast. **58**, 59
Kobresia 30
Koelreuteria 358
bipinnata Franch.* 359, **358**
paniculata Laxm.* 358, **358**
Koenigia 160
forrestii (Diels) Mesicek & Soják 160, **161**
islandica L. 160
Kolkwitzia 484
amabilis Graebn.* 484, **485**

LABIATAE 435
Lactuca 515
souliei Franch. 515
Lagotis 451
alutacea W. W. Sm.* **450**, 451
praecox W. W. Sm.* **450**, 451
yunnanensis W. W. Sm.* 451
LAMIACEAE 435
Lamiophlomis 439
rotata (Benth. ex Hook. f.) Kudo **438**, 439
Lamium 439
album L. var. *barbatum* (Sieb. & Zucc.) Franch. & Savat. 439
barbatum Sieb. & Zucc. **438**, 439
Lancea 448
hirsuta Bonati* 448
tibetica Hook. f. & Thomson 448, **449**, 450
LARDIZABALACEAE 88
Larix 32, 36, 59
chinensis 44
griffithii Hook. f. var. *speciosa* (W. C. Cheng & Y. W. Law) Farjon* 59
potaninii Batalin **11**, 35, 43, **58**, 59
speciosa W. C. Cheng & Y. W. Law 59
thibetica Franch. 59
Lasiocaryum 431
trichocarpum (Hand.-Mazz.) I. M. Johnst.* **430**, 431
LAURACEAE 72
Laurocerasus 267
Laurus 74
camphora L. 74
glandulifera Wall. 74
LEGUMINOSAE 323
LENTIBULARIACEAE 468
Leontopodium 499
andersonii C. B. Clarke* 500, **501**
arbuscula Beauv. 500
caespitosum Diels 500, **501**
calocephalum (Franch.) Beauv.* 500, **501**
dedekensii (Bur. & Franch.) Beauv. 500, **501**
sinense Hemsl.* 500, **501**
stracheyi (Hook. f.) Hemsl. 500
subulatum (Franch.) Beauv. 500
Lepidostemon 196
rosularis (K. C. Kuan & Z. X. An) Al-Shehbaz* 196, **197**

Leptocodon 472
    gracilis (Hook. f. & Thomson) Lem. **474**, 475
Leptodermis 476
    esquirolei H. Lévl. 476
    forrestii Diels* 476, **478**
    microphylla (H. Winkl.) H. Winkl. 479
    oblonga Bunge* 479
    pilosa Diels* **477**, 479
    potaninii Batalin* 476
        var. glauca (Diels) H. Winkl. 476
        var. tomentosa H. Winkl.* 476
Lespedeza 324
    capillipes Franch. 324
    polyantha (Franch.) Schindl. 324
Leucas 440
    ciliata Benth. 440, **441**
Leycesteria 483
    formosa Wall. **482**, 483
Ligularia 503
    achyrotricha (Diels) Y. Ling* 503
    amplexicaulis (Wall.) DC. 504, **505**
    arnicoides DC. ex Royle 512
    atroviolacea (Franch.) Hand.-Mazz.* 503
    clivorum Maxim. 504
    crassa Hand.-Mazz. 504
    cymbulifera (W. W. Sm.) Hand.-Mazz. 504
    dentata (A. Gray) Hara 504, **505**
    emeiensis Kitam. 504
    lankongensis (Franch.) Hand.-Mazz.* 504, **505**
    lapathifolia (Franch.) Hand.-Mazz.* 504
    latihastata (W. W. Sm.) Hand.-Mazz. 504, **505**
    leesicotal Kitam. 504
    liatroides (C. Winkl.) Hand.-Mazz.* 504, **505**
    lidjiangensis Hand.-Mazz.* 504, **506**, 507
    nelumbifolia (Bur. & Franch.) Hand.-Mazz.* 503, 504, **505**
    oreotrephes W. W. Sm. 503
    paradoxa Hand.-Mazz.* 503
    pleurocaulis (Franch.) Hand.-Mazz.* 504, **506**, 507
        forma *uberrima* Hand.-Mazz 507
    przewalskii (Maxim.) Diels 504, **505**
    purdomii (Turrill) Chitt.* **502**, 503
    rumicifolia (J. R. Drumm.) S. W. Liu 504
Ligustrum 420
    delavayanum Hariot* 420, **421**
    henryi Hemsl.* 420, **421**
    lucidum W. T. Aiton* 420, **421**
        forma latifolium (W. C. Cheng) P. S. Hsu* 420
    sinense Lour. 420, **421**
LILIACEAE 538
Lilium 7, 538
    amoenum E. H. Wilson ex Sealy* 538, **540**, 541
    bakerianum Collett & Hemsl. 538, 541
        var. aureum Grove & Cotton* 538, 541
        var. bakerianum 541
        var. delavayi 538, **540**
        var. rubrum Stearn* 538, **540**, 541
        var. yunnanense 538
    brownii F. E. Br. ex Miellez 538, **540**, 541
    davidii Duchartre ex Elwes* 542, **543**
    duchartrei Franch.* 542, **543**
    euxanthum (W. W. Sm. & W. E. Evans) Sealy* 538, 542, **543**
    henryi Baker* 542
    hyacinthinum E. H. Wilson 546

    lancifolium Thunb. 542, **543**
    lankongense Franch.* 542, **543**
    leucanthum (Baker) Baker* 538, **540**, 541
    lijiangense L. J. Peng* 542, **544**, 545
    longiflorum Thunb. 538, **540**, 541
    lophophorum (Bur. & Franch.) Franch.* 538, 541, **543**
    macrophyllum (D. Don) Voss 546
    majoense H. Lévl.* 542, **544**, 545
    myriophyllum Franch. 541
    nanum Klotzsch 538, 542, **543**
    nepalense D. Don 542, **544**, 545
        var. burmanicum W. W. Sm. 545
    primulinum Baker 542, 545
        var. burmanicum (W. W. Sm.) Stearn 542, **544**, 545
        var. ochraceum (Franch.) Stearn* 542, **544**, 545
        var. primulinum 545
    pumilum Redouté 542, **543**
    regale E. H. Wilson* 538, **540**, 541
    sargentiae E. H. Wilson* 538, **540**, 541
    souliei (Franch.) Sealy* 538, 542, **543**
    sulphureum Baker ex Hook. f. 538, **540**, 541
    taliense Franch.* 542, **544**, 545
    tigrinum Ker-Gawl. 542
    wardii Stapf ex F. C. Stern 542, **544**, 545
LINACEAE 372
Linaria 448
    yunnanensis W. W. Sm.* 448, **449**
Lindera 34, 42, 72
    cercidifolia Hemsl. 72
    megaphylla Hemsl. 74
    obovata Franch. 74
    obtusiloba Blume 72, **73**
    populifolia Hemsl. 74
Linum 372
    cicanobum Buch.-Ham. ex D. Don 372
    perenne L. 372, **373**
    repens Buch.-Ham. ex D. Don 372
    sibiricum DC. 372
    trigynum Roxb. 372
Liparis 613
    bambusaefolia Makino 613
    bicallosa (D. Don) Schltr. 613
    campylostalix Rchb. f. 613
    elongata Fukuyama 613
    formosana Rchb. f. 613
    gigantea C. L. Tso 612, 613
    giraldiana Kraenzl. ex Diels 613
    japonica (Miq.) Maxim. 612, 613
    khasiana (Hook. f.) T. Tang & F. T. Wang 613
    nervosa (Thunb. ex Murray) Lindl. 612, 613
    pauciflora Rolfe 613
Liquidambar 39, 143
    acerifolia Maxim. 143
    formosana Hance **142**, 143
Liriodendron 71
    chinense (Hemsl.) Sarg.* **70**, 71
Liriope 561
    graminifolia (L.) Baker 561
    muscari (Decne.) L. H. Bailey **560**, 561
Listera 590
    lindleyana (Decne) King & Pantl. 590
    puberula Maxim. 590, **591**
Litchi 359
    chinensis Sonn.* 359
Lithocarpus 34, 150

    dealbatus (Hook. f. & Thomson ex Miq.) Rehd. 150
    variolosus (Franch.) Chun 150
Lithospermum 431
    szechenyi (Kanitz) I. M. Johnst. 431
Litsea 42, 74
    longipetiolata Lecomte 74
Lloydia 550
    delavayi Franch. 550, **552**, 553
    flavonutans H. Hara 550, **552**, 553
    forrestii Diels 553
    ixiolirioides Baker ex Oliv. 550, **552**, 553
    longiscapa Hook. 553
    oxycarpa Franch.* 550, 553
    serotina (L.) Rchb. 550
        var. parva (C. Marq. & Airy Shaw) Hara 550, **551**
    tibetica Baker ex Oliv. 550, **552**, 553
    triflora (Ledeb.) Baker 550
    yunnanensis Franch. 550, **552**, 553
Lobelia 476
    seguinii H. Lévl. & Vaniot 476, **477**
Lomatogonium 404
    carinthiacum (Wulf.) Rchb. 404, **405**
    forrestii (Balf. f.) Fernald* 404, 406, **406**
    macranthum (Diels & Gilg) Fernald 404, 405, **405**
    oreocharis (Diels) C. Marq.* 404, 405, **405**
    rotatum (L.) Fries ex Nyman 404, 406
Lonicera 479
    chaetocarpa (Batalin ex Rehd.) Rehd.* **478**, 479
    gynochlamydea Hemsl.* 479, 480, **481**
    henryi Hemsl. 479
    hispida Roem. & Schult.* **478**, 479
    japonica Thunb. **478**, 479
    maackii (Rupr.) Maxim.
        forma podocarpa Rehd.* 479, 480, **481**
        forma maackii 479, 480
    mitis Rehd.* 479, **481**
    nitida Wilson* 479, 480, **481**
    perulata Rehd.* 479, 480, **481**
    pileata Oliv.* 479, 480
    prostrata Rehd.* 479, **482**, 483
    quinquelocularis Hardw. 479, 480, **481**
    rupicola Hook. f. & Thomson 479, **482**, 483
    similis Hemsl. var. delavayi (Franch.) Rehd. **478**, 479
    syringantha Maxim.* 479, **482**, 483
    tangutica Maxim.* 479, 480
    thibetica Bur. & Franch. 483
    tragophylla Hemsl. ex Forbes & Hemsl.* **478**, 479
    webbiana DC. 479, 480, **481**
    aff. *deflexicalyx* Batalin* 479, **482**, 483
    aff. *obovata* Royle* 479, 480, **481**
LORANTHACEAE 177
Loropetalum 143
    chinense (R. Br.) Oliv. 143
Lotus 332
    corniculatus L. 332, **333**
Loxostemon 191
    loxostemoides (O. E. Schulz) Y. Z. Lan & T. Y. Cheo 191
    pulchellus Hook. f. & Thomson 188
    purpurascens (O. E. Schulz) Fang ex Y. Z. Lan & T. Y. Cheo 191
Luculia 476

INDEX

gratissima (Wall.) Sweet 476, **477**
pinceana Hook. 476
Lychnis 156
  davidii Franch. 156
  himalayensis (Rohrb.) Edgew. 156
Lycium 420
  barbarum L. 420
    subsp. *chinense* (Mill.) Aiton 420
  chinense L. 420, **421**
  halimifolium Mill. 420
  sinense Gren. 420
Lycoris 569
  aurea (L'Hér.) Herb. **568**, 569
  chinensis Traub* 569
Lyonia 40, 227
  macrocalyx (J. Anth.) Airy Shaw 228
  ovalifolia (Wall.) Drude 228, **229**
    var. hebecarpa (Franch. ex Forbes & Hemsl.) Chun 228
    var. lanceolata (Wall.) Hand.-Mazz. 228
    var. *pubescens* Franch. 228
  villosa (Wall. ex C. B. Clarke) Hand.-Mazz 228
Lysimachia 266
  delavayi Franch.* 266
  drymarifolia Franch.* 266, **266**
  prolifera Klatt. 266, **266**
  taliensis Bonati* 266
LYTHRACEAE 336
Lythrum 336
  salicaria L. 336, **337**

Machilus 42
  mekongensis Diels 74
Macleaya 128
  cordata (Willd.) R. Br. 128
  microcarpa (Maxim.) Fedde* 128, **129**
Maddenia 268
  himalaica Hook. f. & Thomson 268, **269**
  hypoleuca Koehne* 268
  wilsonii Koehne* 268
Magnolia 32, 68
  campbellii Hook. f. & Thomson, 68
    var. mollicomata (W. W. Sm.) Kingdon-Ward* 68, **69**
  dawsoniana Rehd. & E. H. Wilson* 68, **69**
  delavayi Franch.* 68, **69**
  globosa Hook. f. & Thomson* 68, **69**
  sargentiana Rehd. & E. H. Wilson* 68, **69**
  sinensis (Rehd. & E. H. Wilson) Stapf* 68, **69**
  sprengeri Pamp.* 68, **69**
  wilsonii (Finet & Gagnep.) Rehd.* 68, **69**
MAGNOLIACEAE 68
Mahonia 83
  bealei (Fortune) Carr.* 83
  dolichostylis Takeda* **82**, 83
  duclouxiana Gagnep. 83
  lomariifolia Takeda **82**, 83
  longibracteata Takeda* 83
  veitchiorum Hemsl. & E. H. Wilson* 83
Maianthemum 558
  atropurpureum (Franch.) LaFrankie* 558, 560
  bifolium (L.) F. W. Schmidt 558, **559**
  forrestii (W. W. Sm.) LaFrankie* 558, 559
  fuscum (Wall.) LaFrankie 558
  henryi (Baker) LaFrankie 558, **559**
  ichiangense (W. W. Sm.) LaFrankie* 558, **559**
  oleraceum (Baker) LaFrankie 558, 559, **559**
  purpureum (Wall.) LaFrankie 558, 560, **560**

szechuanicum (F. T. Wang & T. Tang) H. Li* 558, **559**
tatsienense (Franch.) LaFrankie 558, 559, **559**
Malaxis 613
  arisanensis (Hayata) S. Y. Hu 613
  monophyllos (L.) Sw. **612**, 613
  taiwaniana S. S. Ying 613
  yunnanensis (Schltr.) T. Tang & F. T. Wang 613
Malus 279
  hupehensis (Pamp.) Rehd.* **278**, 279
  prattii (Hemsl.) C. K. Schneid.* 279
  rockii Rehd. **278**, 279
  yunnanensis (Franch.) C. K. Schneid. **278**, 279
MALVACEAE 175
Mandragora 423
  caulescens C. B. Clarke **422**, 423
Manglietia 71
  insignis (Wall.) Blume **70**, 71
Marmoritis 436
  complanata (Dunn) A. L. Bud.* 436, **437**
Mayodendron 463
  igneum (Kurz) Kurz **462**, 463
Meconopsis 30, 123
  baileyi Prain. **126**, 128
  balangensis 124, **125**
  betonicifolia Franch. **126**, 127
  delavayi (Franch.) Franch.* **126**, 127
  forrestii Prain* 128, **129**
  henrici Bur. & Franch.* 127
    var. psilonomma (Farrer) Taylor* **126**, 127
  horridula Hook. f. & Thomson 124, **125**
  impedita Prain **126**, 127
  integrifolia (Maxim.) Franch.* 29, 123, **125**
    subsp. lijiangensis Grey-Wilson* 124, **125**
  lancifolia (Franch.) Franch. ex Prain 128, **129**
  prainiana Kingdon-Ward. 124
  prattii (Prain) Prain* 124
  pseudointegrifolia Prain **123**, 124, **125**
    subsp. daliensis Grey-Wilson* 124
    subsp. robusta Grey-Wilson 124
  pulchella Yoshida, Sun & Boufford 127
  punicea Maxim.* **126**, 127
  quintuplinervia Regel* **126**, 127
  racemosa Maxim. 124, **125**
  rudis (Prain) Prain* 124, **125**
  sinomaculata Grey-Wilson* **126**, 127
  speciosa Prain* 124, **125**
  wilsonii Grey-Wilson 127
    subsp. australis Grey-Wilson **126**, 127
  ×cookei Taylor **126**, 127
  aff. pulchella **126**
  aff. racemosa **125**
Medinilla 344
  rubicunda (Jack) Blume* 344
  yunnanensis H. L. Li 344
Meehania 436
  fargesii (H. Lévl.) C. Y. Wu* 436, **437**
  henryi (Hemsl.) Sun ex C. Y. Wu* 436, **437**
Megacarpaea 191
  delavayi Franch. 191
Megacodon 407
  stylophorus (C. B. Clarke) Harry Sm. **406**, 407
  venosus (Hemsl.) Harry Sm.* 408
Melampyrum 451
  klebelsbergianum Soó* **450**, 451
  roseum Maxim. 451
Melandrium 156
  himalayense (Rohrb.) Y. Z. Zhao 156

pumilum (Benth.) Walp. 156
MELANTHIACEAE 553
Melastoma 343
  normale D. Don 343
  malabathricum L. subsp. normale (D. Don) K. Mey. **342**, 343
MELASTOMATACEAE 343
MENYANTHACEAE 427
Menyanthes 427
  trifoliata L. **426**, 427
Merriliopanax 381
  chinensis H. L. Li 381
  listeri (King) H. L. Li 381
  membranifolius (W. W. Sm.) C. B. Shang. 381
Metapanax 383
  davidii (Franch.) Frodin ex Wen & Frodin 383
Metasequoia 55
  glyptostroboides Hu & W. C. Cheng* 56, **57**
Michelia 68
  doltsopa Buch.-Ham. ex DC. **70**, 71
  wilsonii Finet & Gagnep.* 71
  yunnanensis Franch. ex Finet & Gagnep.* **70**, 71
Micromeria 443
  euosma (W. W. Sm.) C. Y. Wu* **442**, 443
Microstylis 613
  japonica Miq. 613
Mimulus 447
  bodinieri Vaniot* **446**, 447
  nepalensis Benth. **446**, 447
  szechuanensis Pai* 448
  tenellus Bunge 448, **449**
    var. *nepalensis* (Benth.) P. C. Tsoong ex H. P. Yang 447
Moneses 235
  grandiflora Salisb 235
  rhombifolia (Hayata) Andres 235
  uniflora (L.) A. Gray 235
Monochoria 532
  vaginalis (Burm. f.) C. Presl ex Kunth **532**, 533
Monotropa 232
  humilis D. Don 232
  hypopitys L. 232, **233**
  uniflora L. 232, **233**
MONOTROPACEAE 232
Monotropastrum 232
  uniflora L. var. pentapetala Mak. 232
  humile (D. Don) H. Hara 232, **233**
Morina 490
  alba Hand.-Mazz. 490
  chinensis (Pai) Cannon 491
  nepalensis D. Don subsp. delavayi (Franch.) D. Y. Hong & L. M. Ma 490
Musa 517
  lasiocarpa Franch. 517
MUSACEAE 517
Musella 517
  lasiocarpa (Franch.) C. Y. Wu ex H. W. Li* **516**, 517, **517**
Mussaenda 43
Myrica 40
  nana A. Chev. 42
Myricaria 180
  germanica (L.) Desv. var. squamosa (Desv.) Maxim.) 180
  laxiflora (Franch.) P. Y. Zhang & Y. J. Zhang* 180
  squamosa Desv. 180, **181**

MYRSINACEAE 267
Myrsine 40, 267
    semiserrata Wall. 267

Nandina 83
    domestica Thunb. **82**, 83
Nannoglottis 507
    hookeri (Hook. f.) Kitam. **506**, 507
    latisquama Ling ex X. L. Chen* **506**, 507
    yunnanensis (Hand.-Mazz.) Hand.-Mazz. 507
Nardostachys 488
    grandiflora DC. 488, **489**
    jatamansi DC. 488
Neillia 293
    gracilis Franch.* 293
    thibetica Bur. & Franch.* 293, **293**
Nelumbium 79
    speciosum Willd. 79
Nelumbo 79
    komarovii Grossh. 79
    nucifera Gaertn. **76**, 79
        var. *macrorhizomata* Nakai 79
NELUMBOACEAE 79
Neottia 590
    acuminata Schltr. 590, **591**
    dongrergoensis Schltr. 590
    lindleyana Decne 590
    listeroides Lindl. 590, **591**
    micrantha Lindl. 590
Neottianthe 586
    oblonga K. Y. Lang* 586, **587**
Nepeta 443
    complanata Dunn 436
    stewartiana Diels* **442**, 443
Nerium 411
    indicum Mill. 411
    oleander L. **410**, 411
Nicandra 421
    physalodes (L.) Gaertn. 421, **421**
Nomocharis 546
    aperta (Franch.) E. H. Wilson 546, **547**
    basilissa Farrer ex W. E. Evans 546, **547**
    farreri (W. E. Evans) Harrow 546, **547**
    forrestii Balf. f. 546
    mairei H. Lévl. 546
    meleagrina Franch* 546
    pardanthina Franch.* 546, **547**
    saluenensis Balf. f. 546
Notholirion 546
    bulbuliferum (Lingelsh. ex H. Limpr.) Stearn 546, **547**
    campanulatum Cotton & Stearn 546, **547**
    hyacinthinum (E. H. Wilson) Stapf 546
    macrophyllum (D. Don) Boiss. 546, **547**
Nothopanax 381, 383
    davidii (Franch.) Harms ex Diels 383
    membranifolius W. W. Sm. 381
Nothaphoebe 42
    omeiensis Chun 42
Nothoscordum 566
    mairei H. Lévl. 566
Nuphar 78
    pumila (Timm) DC. 78
NYMPHAEACEAE 78
Nymphaea 78
    nelumbo L. 79
    tetragona Georgi 78
Nymphoidea 427
    peltata (S. G. Gmel.) Kuntze **426**, 427

OLEACEAE 415
Oligobotrya 558
    henryi Baker 558
    szechuanica F. T. Wang & T. Tang 558
Omphalogramma 256
    delavayi (Franch.) Franch.* 256, **258**, 259
    elegans Forrest 256, 259
    elwesiana (King ex Watt) Franch. 256, **258**, 259
    forrestii Balf. f.* 256, **258**, 259
    minus Hand.-Mazz.* 256, **258**, 259
    souliei Franch.* 256, 259
    tibeticum Fletcher* 256, **258**, 259
    vinciflora (Franch.) Franch.* 256, **258**, 259
    aff. viola-grandis (Farrer & Purdom) Balf. f.* 256, **258**, 259
ONAGRACEAE 342
Onosma 430
    confertum W. W. Sm.* 431
    farreri I. M. Johnst.* 430
    forrestii W. W. Sm. 431
Ophiopogon 560
    bodinieri H. Lévl.* **560**, 560
Ophrys monophyllos L. 613
Opuntia 336
    monocantha (Willd.) Haworth 43, 336, **337**
ORCHIDACEAE 570
Orchis 578
    chusua D. Don 581
    crenulata Schltr. 581
    cyclochilus (Franch. & Sav.) Maxim. 578
    spathulata Lindl. 578
    tschiliensis (Schltr.) Soó 578
    wardii W. W. Sm. 578
Oreorchis 593
    erythrochrysea Hand.-Mazz. 593, 594, **595**
    fargesii Finet 593, 594, **595**
    foliosa Lindl. 593, 594, **595**
    nana Schltr.* 593, 594, **595**
    oligantha Schltr.* **592**, 593
    patens (Lindl.) Lindl. 593, 594, **595**
Oreosolen 451
    wattii Hook. f. **450**, 451
OROBANCHACEAE 460
Orobanche 460, 462
    ammophila C. A. Mey. 460
    bodinieri H. Lévl. 460
    coerulescens Steph. 460, **462**
    korshinskyi Novopokr. 460
    mairei H. Lévl. 460
    yunnanensis (Beck) Hand.-Mazz.* 460, **461**
Osbeckia 343
    capitata Walp. **342**, 343
    nepalensis Hook. 344
    stellata Buch.-Ham. ex Ker-Gawl. var. crinita (Benth. ex Naud.) C. Hansen **342**, 343
Osmanthus 416
    delavayi Franch.* 416, **417**
    forrestii Rehd. 416
    henryi P. S. Green* 416
    rehderianus Hand.-Mazz. 416
    suavis King ex C. B. Clarke 416, **417**
    yunnanensis (Franch.) P. S. Green* 416
Osteomeles 272
    schwerinae C. K. Schneid.* 272, **273**
Ottelia 533
    acuminata (Gagnep.) Dandy* **532**, 533
    yunnanensis (Gagnep.) Dandy 533
Oxygraphis 99

    delavayi Franch.* 99
    glacialis Fisch. ex DC. **98**, 99
    involucrata (Maxim.) Riedl 99
    tenuifolia W. E. Evans* 99
Oxyria 164
    digyna (L.) Hill 164
    sinensis Hemsl.* 164, **165**
Oxyspora 344
    paniculata (D. Don) DC. 344, **345**
Oxytropis 331
    kansuensis Bunge* 331

Pachyrhizanthe 610
    aphylla (Ames & Schltr.) Nakai 610
    macrorhizos (Lindl.) Nakai 610
Pachysandra 351
    terminalis Sieb. & Zucc. **350**, 351
Padus 267
    wilsonii C. K. Schneid. 267
Paeonia 120
    anomala L. subsp. veitchii (Lynch) D. Y. Hong & K. Y. Pan 123
    decomposita Hand.-Mazz.* 120
    delavayi Franch.* 120, **121**
        subsp. delavayi × subsp. lutea **122**
        subsp. lutea (Delavay ex Franch.) B. A. Shen* 120, **121**
        forma trollioides (Stapf ex Stern) S. G. Haw* 120, **121**
    lactiflora Pallas* 120, **122**, 123
    ludlowii (Stern & Taylor) D. Y. Hong* 120, 123
    lutea Delavay ex Franch. 120
    mairei H. Lévl.* 120, **122**, 123
    ostii Hong Tao et al.* 120, **121**
    potaninii Kom. 120
        var. *trollioides* (Stapf ex Stern) Stern 120
    rockii (S. G. Haw & Lauener) T. Hong & J. J. Li* 120, **121**
    suffruticosa Haworth 120, **121**
    szechuanica W. P. Fang 120
    trollioides Stapf ex Stern 120
    veitchii Lynch* 120, **122**, 123
PAEONIACEAE 120
PALMAE 522
Panax 384
    davidii Franch. 383
    ginseng C. A. Mey. 384, **385**
    japonicus (T. Nees) C. A. Mey. 384, **385**
    pseudoginseng Wall. var. *japonicus* (T. Nees) G. Hoo & C. J. Tseng 384
PAPAVERACEAE 123
Paphiopedilum 570, 577
    armeniacum S. C. Chen & Liu* **576**, 577
    bellatulum (Rchb. f.) Stein. **576**, 577
    concolor (Batem.) Pfitzer **576**, 577
    dianthum T. Tang & F. T. Wang 577, 578, **579**
    emersonii Koop. & P. J. Cribb **576**, 577
    henryanum Braem 577, 578, **579**
    hirsutissimum (Lindl. ex Hook.) Stein. 577, 578, **579**
    malipoense S. C. Chen & Z. H. Tsi **576**, 577
    micranthum T. Tang & F. T. Wang* **576**, 577
Paraquilegia 111
    anemonoides (Willd.) Ulbr. 112
    microphylla (Royle) Drumm. & Hutch. **27**, 112, **113**
Paris 562
    cronquistii (Takht.) H. Li* 562, **564**, 565

daliensis H. Li & V. G. Soukup* 562, **564**, 565
delavayi Franch. 562, 565
fargesii Franch. 562, 565
forrestii (Takht.) H. Li 562, 563, **563**
luquanensis H. Li* 562, **563**
mairei H. Lévl.* 562
marmorata Stearn 562, **563**
polyphylla Sm. 562, 563, **564**
　　var. alba H. Li & R. J. Mitchell* 562, **564**, 565
　　var. chinensis (Franch.) H. Hara 562, 565
　　var. minor S. F. Wang* 562, 565
thibetica Franch. 562, **563**
　　var. *apetala* Hand.-Mazz. 562
undulata H. Li & V. G. Soukup* 562, **563**, 563
violacea H. Lévl. 562
Parnassia 320
cordata (Drude) Z. P. Jien ex T. C. Ku* 320
delavayi Franch.* 320, **321**
epunctulata J. T. Pan* 320
mysorensis F. Heyne ex Wight & Arnott 320, **321**
nubicola Wall. ex Royle 320, **321**
　　var. *cordata* Drude 320
　　var. *nana* T. C. Ku* 320
scaposa Mattf.* 320
wightiana Wall. ex Wight & Arnott 320, **321**
PARNASSIACEAE 320
Parochetus 332
communis D. Don 332
Parrya 191, 195, 196
ciliaris Bur. & Franch. 195
eurycarpa Maxim. 196
forrestii W. W. Sm. 191
linearifolia W. W. Sm. 195
Parthenocissus 356
henryana (Hemsl.) Graebn. ex Diels & Gilg* 356, **357**
semicordata (Wall.) Planch. 356
Patrinia 488
rupestris Bunge 488, **489**
scabra Bunge 488
Paulownia 444
elongata S. Y. Yu* 447
fargesii Franch. 447
fortunei (Seem.) Hemsl. **446**, 447
tomentosa (Thunb.) Steud.* **446**, 447
Pedicularis 2, **12**, 30, **31**, 451
artselaeri Maxim.* 455, 456, **458**
bietii Franch.* 455, 456, **457**
cheilanthifolia Schrenk 455, 456, **457**
chenocephala Diels* 455, 456, **457**
cranolopha Maxim.* 452, **452**, **453**, 457
croizatiana H. L. Li* 452, **454**, 455
davidii Franch.* 455, 456, **457**
decorissima Diels* **454**, 455
densispica Franch. ex Maxim.* 455, 456, **457**
dunniana Bonati* 452, **453**
kansuensis Maxim.* 455, 456, **457**
latituba Bonati* **454**, 455
longiflora Rudolph 452, **454**, 455
　　var. tubiformis (Klotzsch) P. C. Tsoong* 452, 455
macrosiphon Franch.* 455
monbeigiana Bonati* 455, 456, **457**
muscoides H. L. Li* **454**, 455
oederi Vahl 452, **453**
przewalskii Maxim.* **454**, 455
rex C. B. Clarke 452, **453**

subsp. lipskyana (Bonati) P. C. Tsoong* 452
subsp. zayuensis H. P. Yang* 452
rhinanthoides Schrenk ex Fisch. & Mey. 455, 456, **457**
rupicola Franch. ex Maxim.* 455, 456
semitorta Maxim.* 452, **453**
sigmoidea Franch. ex Maxim.* **454**, 455
siphonantha D. Don **454**, 455
　　var. delavayi (Franch. ex Maxim.) P. C. Tsoong* 455
tenuisecta Franch. ex Maxim. 455, 456
torta Maxim.* 452, **453**
trichoglossa Hook. f. 455, **458**, 459
tricolor Hand.-Mazz.* 452, **453**
variegata H. L. Li* **454**, 455
Pegaeophyton 192
angustiseptatum Al-Shehbaz et al.* 192
scapiflorum (Hook. f. & Thomson) C. Marq. & Airy Shaw 192, **193**
　　subsp. robustum (O. E. Schulz) Al-Shehbaz et al. 192
sinense (Hemsl.) Hayek & Hand.-Mazz. 192
Periploca 412
calophylla (Wight) Falc. subsp. *forrestii* (Schltr.) Browicz 412
forrestii Schltr. 412
sepium Bunge* 412, **413**
Peristylus 585
coeloceras Finet **584**, 585
Persea 74
camfora Spreng. 74
Persea spp. 34
Persicaria 160
capitata (Buch.-Ham. ex D. Don) H. Gross 160, **161**
chinensis (L.) H. Gross 160
runcinata (Buch.-Ham. ex D. Don) H. Gross 160
Petasites 500
tricholobus Franch. 500, **501**
versipilus Hand.-Mazz.* 500, **501**
Petrocosmea 459
duclouxii Craib* 460, **461**
flaccida Craib* 460, **461**
forrestii Craib* 460, **461**
grandiflora Hemsl.* 460, **461**
kerrii Craib 460, **461**
minor Hemsl.* 460, **461**
nervosa Craib* 460
rosettifolia C. Y. Wu & H. W. Li* 460, **461**
sinensis Oliv.* 460, **461**
Peucedanum 387
wallichianum DC. 387
Phaius 596
delavayi (Finet) P. J. Cribb & Perner* **596**, 597
flavus (Blume) Lindl. **596**, 597
maculatus Lindl. 597
magniflorus Z. H. Tsi & S. C. Chen 597
tankervilleae (Banks ex L'Hér.) Blume **596**, 597
wallichii Lindl. **596**, 597
Phedimus 296
aizoon (L.) 't Hart 296
Phelipaea 463
indica (L.) Spreng. ex Steud. 463
Philadelphus 300
calvescens (Rehd.) S. M. Hwang 303
coronarius L. var. *tomentosus* (Wall. ex G. Don) Hook. f. & Thomson 303

delavayi L. Henry **302**, 303
　　var. calvescens Rehd. 303
　　var. trichocladus Hand.-Mazz. 303
incanus Koehne* **302**, 303
kansuensis (Rehd.) S. Y. Hu* **302**, 303
purpurascens (Koehne) Rehd.* **302**, 303
tomentosus Wall. ex G. Don 303
Phlomis 439
betonicoides Diels* **438**, 439
forrestii Diels* 439
likiangensis C. Y. Wu* **438**, 439
melanantha Diels* **438**, 439
rotata Benth. ex Hook. f. 439
Phoebe 39
Phoenix 524
roebelinii O'Brien 524
Photinia 275
davidiana (Decne) Cardot **274**, 275
Phyllophyton 436
complanatum (Dunn) Kudo 436
Physalis 423
alkekengi L. **422**, 423
Phytolacca 159
acinos Roxb. **158**, 159
PHYTOLACCACEAE 159
Picea 32, 35, 36, 39, 43, 59
asperata Mast.* 60
brachystyla (Franch.) E. Pritz 35, 60, **61**
likiangensis (Franch.) E. Pritz 35, 60, **61**
wilsonii Mast.* 60, **61**
Pieris 228
bodinieri H. Lévl. 228
formosa D. Don 228, **229**
forrestii Harrow 228
ovalifolia (Wall.) D. Don var. *hebecarpa* Franch. ex Forbes & Hemsl. 228
villosa Wall. 228
Pilea 176
peperomioides Diels* 176, **177**
Pileostegia 307
viburnoides Hook. f. & Thomson **306**, 307
Pinellia 530
cordata N. E. Br.* 530, **531**
pedatisecta Schott 530, **531**
ternata (Thunb.) Breitenb. 530, **531**
Pinguicula 468
alpina L. 468, **469**
Pinus 32, 60
armandii Franch. 43, 44, 60, **61**
bungeana Zucc. ex Endl.* 60, **61**
densata Mast.* 63
griffithii McClell. 32
massoniana Lamb. 39, 42, 44
tabuliformis Carrière 43, **62**, 63
　　var. henryi (Mast.) Kuan.* 63
wallichiana A. B. Jacks. 60, **61**
yunnanensis Franch.* 39, 63, **62**
Piptanthus 328
concolor Harrow ex Craib* 328, **329**
concolor sensu Maxim. 328
forrestii Craib 328
holosericeus Stapf 328
nepalensis (Hook.) Sweet 328, **329**
tomentosus Franch. 328, **329**
Pistacia 43
weinmannifolia J. Poiss. ex Franch. 43
PLATANACEAE 144
Platanthera 586
chlorantha Cust. ex Rchb. 586

fuscescens (L.) Kraenzl. 586, **587**
japonica (Thunb. ex A. Murray) Lindl. 586, **587**
mandarinorum Rchb. f. 586, **587**
souliei Kraenzl. 586
Platanus 144
orientalis L. 144, **145**
Platycarya 364
strobilacea Sieb. & Zucc. 364, **365**
Platycarya spp. 42
Platycodon 471
grandiflorum (Jacq.) A. DC. **470**, 471
Pleione 598
albiflora P. J. Cribb & C. Z. Tang **600**, 601
aurita P. J. Cribb & Pfennig* 601, 602, **603**
bulbocodioides Rolfe* 601, 602, **603**
chunii C. L. Tso* 601, 602, **603**
delavayi (Rolfe) Rolfe 602
forrestii Schltr. **600**, 601
grandiflora (Rolfe) Rolfe **600**, 601
henryi (Rolfe) Schltr. 602
limprichtii Schltr. 601, 602, **603**
maculata (Lindl.) Lindl. ex Paxton **600**, 601
pleionoides (Kraenzl. ex Diels) Braem & H. Mohr* 601, 602, **603**
praecox (Sm.) D. Don **600**, 601
saxicola T. Tang & F. T. Wang ex S. C. Chen **600**, 601
scopulorum W. W. Sm. **600**, 601
yunnanensis (Rolfe) Rolfe 601, 602, **603**
×taliensis P. J. Cribb & Butterfield* 602
Pleurospermum 385
amabile Craib & W. W. Sm., 385, 386
benthamii (Wall. ex DC.) C. B. Clarke 385, 386, **386**
calcareum R. Wolff 386
davidii Franch. 385
dielsianum Fedde ex H. Wolff 385
foetens Franch.* 385, **385**
linearilobum W. W. Sm.* 385, 386
pseudoyunnanense H. Wolff 386
sp.* 385, **386**
yunnanense Franch. 385, 386, **386**
PLUMBAGINACEAE 267
Podocarpus 63
forrestii Craib & W. W. Sm.* **62**, 63
macrophyllus (Thunb.) Sweet **62**, 63
PODOPHYLLACEAE 84
Podophyllum 84
aurantiocaule Hand.-Mazz. subsp. furfuraceum (S. Y. Bao) J. M. H. Shaw 84
delavayi Franch.* 84, **85**
var. longipetalum J. L. Wu & P. Zhuang ex J. M. H. Shaw 84
difforme Hemsl. ex E. H. Wilson* 84
emodi Wall. 84
hexandrum Royle 84, **85**
mairei Gagnep.* 84
veitchii Hemsl. & E. H. Wilson 84
versipelle Hance 84, **85**
subsp. boreale J. M. H. Shaw* 84
Pogonia 590
yunnanensis Finet* 591, **591**
POLEMONIACEAE 424
Polemonium 424
caeruleum L. 424
chinense (Brand) Brand 424, **425**
Polygala 380
arillata Buch.-Ham. ex D. Don 380, **381**
POLYGALACEAE 380

POLYGONACEAE 159
Polygonatum 555
cavaleriei H. Lévl. 557
cirrhifolium (Wall.) Royle 555, **556**, 557
curvistylum Hua* 555, **556**, 557
delavayi Hua 557
erythrocarpum Hua 557
fargesii Hua 557
franchetii Hua* 557
fuscum Hua 557
gracile P. Y. Li* 555, 557
graminifolium Hook.* 555, **556**, 557
hookeri Baker 555, **556**, 557
kansuense Maxim. ex Batalin* 555, **556**, 557
kingianum Collett & Hemsl. 555, **556**, 557
mairei H. Lévl. 557
prattii Baker* 555, **556**, 557
sp.* 555, **556**, 557
stewartianum Diels* 555, **556**, 557
uncinatum Diels 557
verticillatum (L.) Allioni 557
Polygonum 12, 31
amplexicaule D. Don 159
campanulatum Hook. f. 160
capitatum Buch.-Ham. ex D. Don 160
chinense L. 160
emodi Meisn. 159
forrestii Diels 160
macrophyllum D. Don 159
molle D. Don 160
paleaceum Wall. ex Hook. f. 159
polystachyum Wall. ex Meisn. 160
runcinatum Buch.-Ham. ex D. Don 160
viviparum L. 159
Pomatosace 264
filicula Maxim.* 264, **265**
Ponerorchis 578, 581
brevicalcarata (Finet) P. F. Hunt* **580**, 581
chusua (D. Don) Soó **580**, 581
crenulata Sóo* **580**, 581
schlechteri Perner & Y. B. Luo 581
PONTEDERIACEAE 532
Populus 185
adenopoda Maxim.* 185
ciliata Wall. ex Royle 185, 187
davidiana Dode 185
lasiocarpa Oliv.* 185, **186**, 187
rotundifolia Griff. 185, **185**
szechuanica C. K. Schneid.* 185, 187
wilsonii C. K. Schneid.* 185, **186**, 187
yunnanensis Dode* 185, **185**
Porana 425
duclouxii Gagnep. & Cour. 425
Potentilla 30, 288
anserina L. 288, **289**
arbuscula D. Don 291
biflora Willd. ex Schltr. 288, 289, **289**
bifurca L. 288, 289
coriandrifolia D. Don 288, **289**
cuneata Wall. ex Lehm. 288, **290**, 291
davurica Nestl. 291
discolor Bunge 288, 289
eriocarpa Wall. ex Lehm. 288, 291
fruticosa L. 291
var. albicans Rehd. & E. H. Wilson* 291
var. arbuscula (D. Don) Maxim. **290**, 291
var. *davurica* (Nestl.) Seringe 291
glabra Lodd. 291
leuconota D. Don* 288

parvifolia Fisch. ex Lehm. **290**, 291
polyphylla Wall. ex Lehm. 288, 289
salesoviana (Steph.) Asch. & Graeb. 288, **289**
saundersiana Royle 288, **290**, 291
var. caespitosa (Lehm.) Thomson Wolf 291
simulatrix Thomson Wolf* 288, 291
stenophylla (Franch.) Diels 288
Primula 2, 7, 30, 237
section Amethystina 237, 239
section Armerina 239, 251
section Bullatae 237, 243
section Capitatae 239, 255
section Cortusoides 237, 240
section Crystallophlomis 237, 244
section Davidii 237, 244
section Denticulata 239, 255
section Dryadifolia 237, 243
section Malvacea 237, 240
section Minutissimae 239, 255
section Monocarpicae 237, 240
section Muscarioides 239, 255
section Nivales 237, 244
section Obconicolisteri 237, 240
section Petiolares 237, 243
section Proliferae 239, 248
section Pulchella 239, 252
section Sikkimensis 239, 251
section Soldanelloides 239, 256
section Yunnanensis 239, 252
subgenus Aleuritia 237, 243
subgenus Auganthus 237, 240
subgenus Auriculastrum 237, 239
subsection Agleniana 239, 247
subsection Calliantha 239, 247
subsection Crystallophlomis 237, 244
subsection Maximowiczii 239, 247
advena W. W. Sm. var. *euprepes* (W. W. Sm.) Chen & C. M. Hu 247
agleniana Balf. f. & Forrest 247
alpicola (W. W. Sm.) Stapf **250**, 251
var. alba W. W. Sm. 251
var. violacea (Stapf) W. W. Sm. & Fletcher 251, **251**
alsophila Balf. f. & Farrer* 240, **241**
amethystina Franch.* **238**, 239
apoclita Balf. f. & Forrest 256, **257**
beesiana Forrest 248
bella Franch. **254**, 255
subsp. bonatiana (Petitm.) W. W. Sm.* 255
subsp. cyclostegia (Hand.-Mazz.) W. W. Sm. & Forrest* 255
boreiocalliantha Balf.f. & Forrest* **246**, 247
bracteata Franch.* **242**, 243
brevicula Balf. f. & Forrest* 244, **245**
bryophila Balf. f. & Farrer* **246**, 247
bullata Franch.* 243
bulleyana Forrest* 248, **249**
subsp. beesiana 249
×poissonii 248
calliantha Franch. **246**, 247
capitata Hook. f. **254**, 255
subsp. craibeana (Balf. f. & Forrest) W. W. Sm. & Forrest 255
subsp. crispata (Balf. f. & Forrest) W. W. Sm. & Forrest 255
subsp. sphaerocephala (Balf. f. & Forrest) W. W. Sm. & Forrest* 255
cawdoriana Kingdon-Ward* 256, **257**

cernua Franch.* 256, **257**
chionantha Balf. f. & Forrest* 244, **246**, 247
    subsp. sinoplantaginea (Balf. f.) A. J.
        Richards* **246**, 247, **247**
    subsp. sinopurpurea (Balf. f.) A. J.
        Richards* **246**, 247
chungensis Balf. f. & Kingdon-Ward* 248, **249**
cockburniana Hemsl.* 248, **249**
conspersa Balf. f. & Purdom* 251, 252, **253**
deflexa Duthie* 256, **257**
denticulata Sm. 255
    subsp. sinodenticulata (Balf. f. & Forrest)
        W. W. Sm. & Forrest **254**, 255
dryadifolia Franch. **242**, 243
    subsp. congestifolia (Forrest) W. W. Sm.
        & Forrest* 243
duclouxii Petitm.* 240, **241**
euprepes W. W. Sm. **246**, 247
fasciculata Balf. f. & Kingdon-Ward* **250**, 251
flaccida N. P. Balakr.* 255, 256, **257**
flava Maxim.* 252, **254**, 255
florida Balf. f. & Forrest* 252, **253**
florindae Kingdon-Ward* **250**, 251
forbesii Franch. 240
forrestii Balf. f.* **242**, 243
    var. *redolens* (Balf. f. & Kingdon-Ward)
        A. J. Richards 243
franchetii Pax 259
gemmifera Batalin* **250**, 251
    var. monantha W. W. Sm. & Fletcher*
        251, 252
gracilipes Craib **242**, 243
heucherifolia Franch.* 240, **241**
hoffmanniana W. W. Sm. 243
hongshanensis Z. D. Fang, H. Sun & Rankin
        247
hookeri Watt 243, 244
hopeana Balf. f. & Cooper 251
involucrata Wall. subsp. yargongensis (Petitm.)
    W. W. Sm. & Fletcher 251, 252, **253**
ioessa W. W. Sm. 251
    var. hopeana (Balf. f. & Cooper) A. J.
        Richards **250**, 251
jonardunii W. W. Sm.* 243
latisecta W. W. Sm.* 240, 243
limbata Balf. f. & Forrest* 244, **245**
littoniana Forrest 256
malacoides Franch. 240, **241**
malvacea Franch.* 240, **241**
    var. alba Forrest 240
minor Balf. f. & Kingdon-Ward* 244, **245**
moupinensis Franch. **242**, 243
muscarioides Hemsl.* 256, **257**
nanobella Balf. f. & Forrest* **254**, 255
nutans Georgi. **238**, 251, 252, **253**
obconica Hance 240, **241**
optata Farrer* 244, **245**
orbicularis Hemsl.* 244, **245**
ovalifolia Franch.* 244, **245**
palmata Hand.-Mazz.* 240, **241**
pinnatifida 256, **257**
poissonii Franch.* 248, **249**
polyneura Franch.* 240, **241**
prolifera Wall. 248, **249**
pseudosikkimensis Forrest 251
pulchella Franch.* 252, **253**
    subsp. prattii (Hemsl.) Halda* 252, 255
pulverulenta Duthie* 248, **249**
redolens Balf. f. & Kingdon-Ward* 243

rockii W. W. Sm.* 243
rupicola Balf. f. & Forrest* 252, **253**
russeola Balf. f. & Forrest* 244, **245**
secundiflora Franch.* **238**, 248, **250**, 251
septemloba Franch.* 240, **241**
serratifolia Franch. 248
sikkimensis Hook. f. **238**, **250**, 251
    var. pseudosikkimensis (Forrest) A. J.
        Richards* 251
silaensis Petitm. 240
sinoplantaginea Balf. f. 247
sinopurpurea Balf. f. 247
sonchifolia Franch. **242**, 243
soongii F. H. Chen & C. M. Hu* 244, **245**
spicata Franch.* 255, 256, **257**
stenocalyx Maxim.* 252, **254**, 255
szechuanica Pax* 247, 248, **249**
taliensis Forrest* 244
tangutica Duthie* **246**, 247
tanneri King subsp. tsariensis (W. W. Sm.)
    A. J. Richards* 243, 244
    var. porrecta W. W. Sm. 243, 244
    subsp. tsariensis (W. W. Sm.) A. J. Richards
        245
thearosa Kingdon-Ward. **246**, 247
vialii Delavay ex Franch.* 255, 256, **257**
vinciflora Franch. 259
violacea W. W. Sm. & Kingdon-Ward* 256
walshii Craib **254**, 255
wilsonii Dunn* 248
woodwardii Balf. f.* **246**, 247
yunnanensis Franch. 252, **253**
zambalensis Petitm.* 251, 252, **253**
PRIMULACEAE 237
Prinsepia 268
    utilis Royle 268, **269**
Prunella 444
    hispida Benth. 444
    vulgaris L. 444, **445**
Prunus 267
    sericea (Batalin) Koehne 267
    serrula Franch. **266**, 267
    serrulata Lindl. 40
    wilsonii (C. K. Schneid.) Diels ex Koehne*
        266, 267
Pseudopanax 383
    davidii (Franch.) W. R. Phil. 383
Pterocarya 364
    hupehensis Skan* 364
    paliurus Batalin* 364, **365**
    stenoptera C. DC. 364, **365**
Pterocephaloides 491
    hookeri (C. B. Clarke) V. Meyer & Ehrend.
        490, 491
Pterocephalus 491
Pterostyrax 236
    cavaleriei Guill. 236
    psilophyllus Diels ex Perkins* 236, **236**
Pulsatilla 107
    chinensis (Bunge) Regel **106**, 107
    millefolium (Hemsl. & E. H. Wilson) Ulbr.* 107
Punica 342
    granatum L. 342, **342**
PUNICACEAE 342
Pyracantha 272
    angustifolia (Franch.) C. K. Schneid.* **273**, 275
    crenulata (D. Don) M. Roem. **274**, 275
    fortuneana (Maxim.) H. L. Li* **274**, 275
Pyrethrum 503

tatsienense (Bur. & Franch.) Y. Ling ex C. Shih*
    502, 503
    var. tanacetopsis (W. W. Sm.) Y. Ling & C.
        Shih* 503
Pyrola 232
    calliantha Andres* 232, **233**
    forrestiana Andres* 232
    rotundifolia L. 232, **233**
        var. *chinensis* Andres 232
    sororia Andres* 232, 233, **233**
    szechuanica Andres* 232, 233
    uniflora L. 235
PYROLACEAE 232
Pyrus 279
    pashia Buch.-Ham. ex D. Don **278**, 279

Quercus 36, 39, 40, 42, 44, 150
    acutissima Carruth. 44
    aquifolioides Rehd. & E. H. Wilson 36, 150,
        151
    baronii Skan* 150, 151
    bungeana Forbes 151
    chinensis Bunge 151
    dentata Thunb. 150, 151
    dielsiana Seeman 150
    gilliana Rehd. & E. H. Wilson 151
    guyavifolia H. Lévl.* 150, 151
    ilex L. var. *rufescens* Franch 151
    lamellosa Sm. 150, 151
    monimotricha (Hand.-Mazz.) Hand.-Mazz.
        150, 151
    obovata Bunge 151
    pannosa Hand.-Mazz. 151
    rehderiana Hand.-Mazz. 150, 151
    schottkyana Rehd. & E. H. Wilson 150, 151
    semecarpifolia R. Br. 150, 151
    spinosa Franch. 151
    spinosa David ex Franch. var. *monimotricha*
        Hand.-Mazz. 151
    variabilis Blume 44, 150, 151
    yunnanensis Franch.* 150, 151

Radermachera 463
    ignea (Kurz) Steenis 463
RANUNCULACEAE 91
Ranunculus 99
    involucratus Maxim. 99
    kamchaticus DC. 99
    pimpinelloides D. Don 100
    similis Hemsl.* **98**, 99
Rehmannia 447
    elata N. E. Br. ex Prain* **446**, 447
    glutinosa (Gaertn.) Liboschitz ex Fisch. &
        Mey.* **446**, 447
    henryi N. E. Br.* 447
Reineckia 562
    carnea (Andrews) Kunth 562
Reinwardtia 372
    indica Dumort. 372, **373**
RHAMNACEAE 356
Rhapis 522
    excelsa (Thunb.) Henry ex Rehd. 522, **523**
Rheum 163
    acuminatum Hook. f. & Thomson 163, 164,
        165
    alexandrae Batalin* **162**, 163
    delavayi Franch. **162**, 163
    forrestii Diels* 163, 164
    likiangensis Samuelsson* 163, 164, **165**

nobile Hook. f. & Thomson **162**, 163
palmatum L.* **162**, 163
　　subsp. *tanguticum* Maxim. ex Regel 163
potaninii Losinsk. 163
przewalskyi Losinsk.* **162**, 163
pumilum Maxim.* 163, 164, **165**
tanguticum (Maxim. ex Regel) Maxim. ex Balf.* **162**, 163
Rhodiola 299
　　atuntsuensis (Praeg.) S. H. Fu 299, 300, **301**
　　chrysanthemifolia (H. Lévl.) S. H. Fu* **298**, 299
　　coccinea (Royle) Boriss. 299, 300
　　crenulata (Hook. f. & Thomson) H. Ohba 299, 300, **301**
　　dumulosa (Franch.) S. H. Fu 299, 300, **301**
　　fastigiata (Hook. f. & Thomson) S. H. Fu 299, 300
　　himalensis (D. Don) S. H. Fu 299, 300
　　kirilowii (Regel) Maxim. 299, 300, **301**
　　purpureoviridis (Praeg.) S. H. Fu* 299, 300, **301**
　　sp.* 299, 300, **301**
　　yunnanensis (Franch.) S. H. Fu* **298**, 299
　　aff. *discolor* (Franch.) S. H. Fu* 299, 300, **301**
Rhododendron 2, 7, **12, 29**, 30, **39**, 40, 200, 462
　　subgenus Hymenanthes 200, 202, 216
　　subgenus Rhododendron 200, 204, 216
　　subgenus Tsutsusi 200, 203, 227
　　section Pogonanthum 200, 202, 216
　　section Rhododendron 200, 204
　　section Tsutsusi 203, 227
　　subsection Arborea 203, 223
　　subsection Argyrophylla 203, 220
　　subsection Barbata 203, 227
　　subsection Boothia 202, 215
　　subsection Campylocarpa 203, 220
　　subsection Campylogyna 202, 215
　　subsection Cinnabarina 202
　　subsection Edgeworthia 200, 204
　　subsection Falconera 203, 219
　　subsection Fortunea 202, 216
　　subsection Fulva 203, 224
　　subsection Glauca 202, 215
　　subsection Grandia 203, 219
　　subsection Heliolepida 202, 208
　　subsection Lapponica 202, 208
　　subsection Lepidota 202, 215
　　subsection Maddenia 200, 204
　　subsection Micrantha 202, 215
　　subsection Moupinensia 200, 204
　　subsection Neriiflora 203, 227
　　subsection Saluenensia 202, 212
　　subsection Scabrifolia 202, 208
　　subsection Selensia 202, 220
　　subsection Taliensia 202, 223
　　subsection Thomsonia 203, 227
　　subsection Trichoclada 202, 215
　　subsection Triflora 202, 204
　　subsection Virgata 202, 212
　　adenogynum Diels* **200**, 223, 224, **225**
　　aganniphum Balf. f. & Kingdon-Ward* **12, 222**, 223
　　　　var. *flavorufum* (Balf. f. & Forrest) D. F. Chamb.* 223
　　ambiguum Hemsl.* 204, 207
　　anthopogon D. Don 216, **217**
　　arboreum Sm. subsp. *delavayi* (Franch.) D. F. Chamb. **222**, 223
　　argyrophyllum Franch.* 220
　　　　subsp. *hypoglaucum* (Hemsl.) D. F. Chamb.* 40, 220
　　　　subsp. *nankingense* (Cowan) D. F. Chamb.* 220
　　　　subsp. *omeiense* (Rehd. & E. H. Wilson) D. F. Chamb.* 220, **221**
　　augustinii Hemsl.* 204, 207
　　　　subsp. *chasmanthum* (Diels) Cullen* 204, 207
　　　　subsp. *hardyi* (Davidian) Cullen* 204, 207
　　balfourianum Diels* 223, 224, **225**
　　barbatum Wall. **226**, 227
　　beesianum Diels 223, 224, **225**
　　brachyanthum Franch. **214**, 215
　　bureavii Franch.* 223, 224, **225**
　　bureavioides Balf. f.* 223, 224, **225**
　　calophytum Franch.* 216, **217**, 219
　　calostrotum Balf. f. & Kingdon-Ward 212, **213**
　　campylocarpum Hook. f. 220, **221**
　　　　subsp. *caloxanthum* (Balf. f. & Forrest) D. F. Chamb. 220
　　campylogynum Franch. 215
　　capitatum Maxim.* 208, **210**, 211,
　　cephalanthum Franch. 216, **217**
　　ciliicalyx Franch. 204
　　cinnabarinum Hook. f. 212
　　　　subsp. *xanthocodon* (Hutch.) Cullen 212, **213**
　　complexum Balf. f. & W. W. Sm.* 208, **209**
　　concatenans Hutch. 212
　　concinnum Hemsl.* 204, 207
　　coriaceum Franch.* 220
　　cuneatum W. W. Sm.* 208, **210**, 211
　　dalhousiae Hook. f. 204, **205**
　　davidsonianum Rehd.* 204, 208
　　decorum Franch. **218**, 219
　　delavayi Franch. 223
　　edgeworthii Hook. f. 204, **205**
　　faberi Hemsl.* 223, 224, **225**
　　fastigiatum Franch.* 44, **200**, 208, 212, **213**
　　flavidum Franch.* 208, 212, **213**
　　forrestii (Balf. f.) Diels **226**, 227
　　fortunei Lindl.* 216, **218**, 219
　　fulvum Balf. f. & W. W. Sm. 224, **225**
　　　　subsp. *fulvoides* (Balf. f. & Forrest) D. F. Chamb. 224
　　galactinum Balf. f.* 220, **221**
　　grande Wight **218**, 219
　　heliolepis Franch. 208, **209**
　　hippophaeoides Balf. f. & W. W. Sm.* 208, **210,** 211
　　impeditum Balf. f. & W. W. Sm.* 208, 212
　　intricatum Franch.* 208
　　lacteum Franch.* **222**, 223
　　lepidostylum Balf. f. & Forrest* **214**, 215
　　lepidotum Wall. ex G. Don **214**, 215
　　leptocarpum Nutt. 215
　　lindleyi T. Moore 204, **205**
　　longipes Rehd. & E. H. Wilson* 223
　　lutescens Franch.* 204, 207
　　maddenii Hook. f. subsp. *crassum* (Franch.) Cullen 204
　　megacalyx Balf. f. & Kingdon-Ward 204, **205**
　　mekongense Franch. **214**, 215
　　　　var. *rubrolineatum* (Balf. f. & Forrest) Cullen 215
　　micranthum Turcz. **214**, 215
　　microphyton Franch. **226**, 227
　　moupinense Franch.* 204, **205**
　　neriiflorum Franch **226**, 227
　　　　subsp. *phaedropum* (Balf. f. & Forrest) Tagg 227
　　　　var. *appropinquans* (Tagg & Forrest) W. K. Hu 227
　　nitidulum Rehd. & E. H. Wilson* 208, **210**, 211
　　nivale Hook. f. **18**, 208, 212
　　　　subsp. *boreale* Philipson & Philipson* 208, 212, **213**
　　nuttallii Booth ex Nutt. 204, **205**
　　orbiculare Decne.* 216, **217**
　　oreodoxa Franch.* 216, **217**
　　　　var. *fargesii* (Franch.) D. F. Chamb.* 216
　　　　var. *shensiense* D. F. Chamb.* 216
　　oreotrephes W. W. Sm.* **36, 200**, 204, 207, **260**
　　orthocladum Balf. f. & Forrest* 208, 211
　　pachypodum Balf. f. & W. W. Sm. 204, **205**
　　phaeochrysum Balf. f. & W. W. Sm.* **200**, 222, 223
　　　　var. *levistratum* (Balf. f. & Forrest) D. F. Chamb.* 223
　　pingianum W. P. Fang* 220, **221**
　　praestans Balf. f. & W. W. Sm. **218**, 219
　　prattii Franch.* 223, 224, **225**
　　primuliflorum Bur. & Franch.* 216, **217**
　　proteoides Balf. f. & W. W. Sm.* 223
　　przewalskii Maxim.* **222**, 223
　　racemosum Franch.* 208, **209**
　　rex H. Lévl.* 219
　　　　subsp. *fictolacteum* (Balf. f.) D. F. Chamb.* **35, 218**, 219
　　rigidum Franch.* 204, 207
　　roxieanum Forrest* **222**, 223
　　rubiginosum Franch. 208, **209**
　　rupicola W. W. Sm. 208, 211
　　　　var. *chryseum* (Balf. f. & Kingdon-Ward) Philipson & Philipson* 208, 212, **213**
　　　　var. *muliense* (Balf. f. & Forrest) Philipson & Philipson* 208, 212
　　　　var. *rupicola* 211
　　russatum Balf. f. & Forrest* 208, 211
　　saluenense Franch. **200**, 212
　　　　subsp. *chameunum* (Balf. f. & Forrest) Cullen 212, **213**
　　selense Franch.* 220
　　　　subsp. *dasycladum* (Balf. f. & W. W. Sm.) D. F. Chamb.* 220, **221**
　　　　subsp. *jucundum* (Balf. f. & W. W. Sm.) D. F. Chamb.* 220
　　　　subsp. *selense* 220
　　simsii Planch. **226**, 227
　　sinogrande Balf. f. & W. W. Sm. **218**, 219
　　souliei Franch.* 220, **221**
　　spp. 42
　　stewartianum Diels **226**, 227
　　sulfureum Franch. **214**, 215
　　sutchuenense Franch.* 216, 219
　　taliense Franch.* 223, 224, **225**
　　tapetiforme Balf. f. & Kingdon-Ward 208, **210**, 211
　　tatsienense Franch.* 204, 208, **209**
　　telmateium Balf. f. & W. W. Sm.* 208, **210**, 211
　　thomsonii Hook. f. **226**, 227
　　thymifolium Maxim.* 208, 211
　　trichanthum Rehd.* 204, 207
　　trichocladum Franch. **214**, 215
　　trichostomum Franch.* 216, **217**

triflorum Hook. f. 204, 207
uvariifolium Diels **226**, 227
vernicosum Franch.* 216, **218**, 219
virgatum Hook. f. 212, **213**
    subsp. oleifolium (Franch.) Cullen* 215
wardii W. W. Sm.* **36**, 219, **200**, 220, **221**
    var. puralbum (Balf. f. & W. W. Sm.) D. F. Chamb.* 220
yunnanense Franch. 204, 207, **208**
×wongii Hemsl. & E. H. Wilson 208, 212, **213**
Rhus 39, 364
    chinensis Mill. 364, **365**
    cotinus L. 365
Ribes 321
    alpestre Wall. ex Decne 321
    glaciale Wall. 321, 322
    himalense Royle ex Decne 321
    laurifolium Jancz.* 321, 322, **322**
    longiracemosum Franch.* 321, **321**
    orientale Desf. 321, 322
    tenue Jancz. 321, 322, **322**
    uva-crispa L. 321
Rodgersia 310
    aesculifolia Batalin **309**, 310
        var. henrici (Franch.) C. Y. Wu ex J. T. Pan **309**, 310
    henrici Franch. 310
    pinnata Franch.* **309**, 311
    sambucifolia Hemsl.* 311
Rosa 43, 279
    banksiae R. Br.* 280
    bodinieri H. Lévl. & Vaniot 280
    brunonii Lindl. 283, 284, **285**
    chinensis Jacq.* **282**, 283,
        var. spontanea (Rehd. & E. H. Wilson) T. T. Yu & T. C. Ku* 283
    cymosa Tratt. 280, **281**
    davidii Crépin* 284
    farreri Cox* 284, **285**
    filipes Rehd. & E. H. Wilson* **282**, 283
    forrestiana Boul.* 284
    graciliflora Rehd. & E. H. Wilson* 284, **286**, 287
    indica L. 280
    laevigata Mich. 280, **281**
    lichiangensis T. T. Yu & T. C. Ku* 283
    longicuspis H. Lévl. & Vaniot **282**, 283
    macrophylla Lindl. 284, **286**, 287
    moyesii Hemsl. & E. H. Wilson* 284, **286**, 287
    multibracteata Hemsl. & E. H. Wilson* 284, **285**
    multiflora Thunb.* **282**, 283
    odorata (Andrews) Sweet
        var. erubescens (Focke) T. T. Yu & T. C. Ku 283
        var. gigantea (Collett ex Crépin) Rehd. & E. H. Wilson 283
        var. odorata 283
        var. pseudoindica (Lindl.) Rehd. 283
    omeiensis Rolfe 42, 280, **281**
        f. pteracantha Rehd. & E. H. Wilson **281**
    prattii Hemsl.* 284
    roxburghii Tratt. 284, **285**
    rubus H. Lévl. & Vaniot* 284, **285**
    sericea Lindl. 280, **281**
    sertata Rolfe* 284, **286**, 287
    setipoda Hemsl. & E. H. Wilson* 284, **285**
    sikangensis T. T. Yu & T. C. Ku* 280
    soulieana Crépin* **282**, 283

    var. microphylla T. T. Yu & T. C. Ku* **282**, 283
    spp. 42
    sweginzowii Koehne* 284, **286**, 287
    willmottiae Hemsl.* 284, **286**, 287
ROSACEAE 267
Roscoea 517
    cautleyoides Gagnep.* 518, **519**
        forma atropurpurea Cowley 517, 518, **519**
        forma sinopurpurea (Stapf) Cowley 517, 518, **519**
        var. pubescens (Z. Y. Zhu) T. L. Wu* 517, 518
    forrestii Cowley* 517, 518
    humeana Balf. f. & W. W. Sm.* 517, 518, **519**
        forma alba Cowley 517, 518
        forma lutea Cowley 517, 518
        forma tyria Cowley 517, 518
    kunmingensis S. Q. Tong 517, 518
    praecox K. Schum.* 517, 518, **519**
    schneideriana (Loes.) Cowley* **516**, 517
    tibetica Batalin **516**, 517
    wardii Cowley 517, 518
RUBIACEAE 476
Rubus 268
    biflorus Buch.-Ham. ex Sm. 268, **269**
    rosifolius Sm. 268, **269**, 271
    thibetanus Franch.* 268, **269**, 271
    tricolor Focke* 268, **269**
Rumex 164
    hastatum D. Don 164, **165**
Ruta 135
RUTACEAE 367

Sabina 55
    tibetica (Kom.) W. C. Cheng & L. K. Fu 55
SALICACEAE 184
Salix 2, 30, 187
    balfouriana C. K. Schneid.* 187, 188, **189**
    brachista C. K. Schneid.* 187
    fargesii Burkill* **186**, 187
    hylematica C. K. Schneid. 187
    lindleyana Wall. ex Andersson **186**, 187
    magnifica Hemsl.* 187, 188, **189**
    moupinensis Franch.* 187, 188, **189**
    serpyllum Andersson **186**, 187
    souliei Seemen* 187
    variegata Franch.* **186**, 187
    wilsonii Seemen ex Diels* 187, 188
Salvia 440
    aerea H. Lévl.* **442**, 443
    bodinieri Van. 440
    bulleyana Diels* 440, **441**
    campanulata Wall. ex Benth. 440
    castanea Diels **442**, 443
    digitaloides Diels* 440, **441**
    evansiana Hand.-Mazz.* **442**, 443
    flava Forrest ex Diels* 440, **441**
    forrestii Diels 440
    hylocharis Diels* 440, **441**
    lichiangensis W. W. Sm. 443
    pinetorum Hand.-Mazz. 443
    prattii Hemsl.* 440, **441**
    przewalskii Maxim.* **442**, 443
        var. alba X. L. Huang & H. W. Li* 443
    wardii E. Peter* **442**, 443
    yunnanensis C. H. Wright* 440, **441**
Sambucus 484
    adnata DC. 484, **485**

    javanica Reinw. 484
Sanguisorba 288
    filiformis (Hook. f.) Hand.-Mazz. 288, **289**
Sanicula 387
    hacquetioides Franch.* **386**, 387
SAPINDACEAE 358
Sapindus 359
    delavayi (Franch.) Radlk.* 359
Sarcochilus
    laurisilvaticus Fukuyama 613
    japonicus Miq. 613
    saruwatarii Hayata 613
Sarcococca 352
    hookeriana Baill. 352
        var. digyna Franch.* 352, **353**
    humilis Stapf 352
    ruscifolia Stapf* 352
        var. chinensis (Franch.) Rehd. & E. H. Wilson* 352
    wallichii Stapf 352
Saruma 75
    henryi Oliv.* 75, **75**
Satyrium 586
    ciliatum Lindl. 586, **587**
    nepalense D. Don **587**, 588
    yunnanense Rolfe* 588
Sauropus 355
    androgynus Merr. **354**, 355
SAURURACEAE 75
Saussurea 496
    anochaete Hand.-Mazz. 499
    eriocephala Diels 496
    gossypiphora D. Don 496, **497**
    graminea (Hand.-Mazz.) Lipsch. 496, **498**
    hieracioides Hook. f. 496, 499
    katochaete Maxim. 496, **498**, 499
    katochaetoides Hand.-Mazz. 499
    laniceps Hand.-Mazz. 496, **497**
    leucoma Diels* 496, **497**
    medusa Maxim. 496, **498**, 499
    nidularis Hand.-Mazz.* 496, **497**
    obvallata (DC.) Edgew. 496, **497**
    quercifolia W. W. Sm.* 496, **498**, 499
    rohmooana C. Marq. & Airy Shaw 499
    simpsoniana (Field. & Gardn.) Lipsch. 496
    stella Maxim. 496, **497**
    tridactyla Hook. f. 496, **498**, 499
    aff. gnaphalodes (Royle) Sch. Bip.* 496, **498**, 499
Saxifraga 2, 30, 312
    andersonii Engl. & Irmsch. 313, **313**
    atrata Engl.* 319
    bergenioides C. Marq. 315, 316, **317**
    brachypoda D. Don 315, **318**, 319
    brunonis Wall. ex Seringe **314**, 315
    calcicola J. Anth. 313
    cernua L. 315
    chionophila Franch.* 312
    consanguinea W. W. Sm. **314**, 315
    decora Harry Sm.* 312, **313**
    divaricata Engl. & Irmsch.* **318**, 319
    diversifolia Wall. ex Seringe 315, 316, **317**
    drabiformis Franch.* 315, **318**, 319
    egregia Engl.* 315, 316
    flagellaris Willd. ex Sternb.
        subsp. *stenophylla* (Royle) Hult. 315
        var. *mucronulata* (Royle) C. B. Clarke 315
    fortunei Hook. f. **318**, 319
    gemmipara Franch. 315, 316, **317**

georgei J. Anth. 312
imbricata Royle 313
kansuensis Mattf. 315
likiangensis Franch.* 312, 313, **313**
lolaensis Harry Sm. 313
ludlowii Harry Sm.* 312, 315, **314**
lumpuensis Engl.* **318**, 319
lychnitis Hook. f. & Thomson **314**, 315
melanocentra Franch. **318**, 319
montanella Harry Sm. 315, **318**, 319
mucronulata Royle **314**, 315
nana Engl.* 312, **313**
nanella Engl. & Irmsch. 315, 316, **317**
nigroglandulifera N. P. Balakr. 315, 316, **317**
nutans Hook. f. & Thomson 316
parvula Engl. & Irmsch.* 319
przewalskii Engl.* 315, 316
pulchra Engl. & Irmsch.* 312
pulvinaria Harry Sm. 312, 313
punctulata Engl. **318**, 319
purpurascens Hook. f. & Thomson 311
rotundipetala J. T. Pan* 312, **313**
rufescens Balf. f.* **318**, 319
rupicola Franch.* 312
sarmentosa L. f. 319
saxatilis Harry Sm.* 312, 315
saxicola Harry Sm.* 312, **314**, 315
sibirica L. **314**, 315
signata Engl. & Irmsch.* 315, 316, **317**
stenophylla Royle 315
stolonifera Curtis 320, **321**
strigosa Wall. ex Seringe 315, 316, **317**
subsessiliflora Engl. & Irmsch. 312, 313, **313**
tangutica Engl. 315, **318**, 319
tsangshanensis Franch.* 315, 316
unguiculata Engl.* 315, 316, **317**
unguipetala Engl. & Irmsch.* 312, 315
SAXIFRAGACEAE 300
Scabiosa 491
Schefflera 382
    delavayi (Franch.) Harms ex Diels 383
    multinervia H. L. Li* 383
    wardii C. Marq. & Airy Shaw* 383
Schima 32, 39, 42, 167
    wallichii (DC.) Korth. **166**, 167
Schisandra 71
    grandiflora (Wall.) Hook. f. & Thomson 72, **73**
    rubriflora (Franch.) Rehd. & E. H. Wilson* 72, **73**
    sphenanthera Rehd. & E. H. Wilson* 72, **73**
SCHISANDRACEAE 71
Schizopepon 184
    dioicus Cogn. ex Oliv.* 184
    longipes Gagnep.* 184
Schizophragma 307
    integrifolium Oliv.* **306**, 307
        var. glaucescens Rehd. 307
    molle (Rehd.) Chun* 307
Scopolia 422
    straminifolia (Wall.) Shrestha 422
Scrophularia 448
    chasmophila W. W. Sm.* 448, **449**
    delavayi Franch.* 448, **449**
SCROPHULARIACEAE 444
Scurrula 178
    buddleioides (Desr.) G. Don 179
    parasitica L. 178
        var. graciliflora (Roxb. ex J. H. Schultes) H. S. Kiu 178

pulverulenta Wall. 179
Scutellaria 444
    amoena C. H. Wight* 444, **445**
    forrestii Diels* 444, **445**
    hypericifolia H. Lévl.* 444, **445**
    likiangensis Diels* 444
Sedum 296
    aizoon L. 296, 297
    chrysanthemifolia H. Lévl. 299
    crenulatum Hook. f. & Thomson 300
    fastigiatum Hook. f. & Thomson 300
    forrestii Raym.-Hamet* 296, **298**, 299
    glaebosum Fröd.* 296, 297, **297**
    henrici-robertii Raym.-Hamet 296, **298**, 299
    henryi Diels 299
    himalensis D. Don 300
    kirilowii Regel 300
    magniflorum K. T. Fu* 296, **298**, 299
    morotii Raym.-Hamet* 296, 299
    multicaule Wall. ex Lindl. 296, 297
    obtusipetalum Franch. 296, 299
    oreades (Decne) Raym.-Hamet 296, 297
    prasinopetalum Fröd.* 296, 299, **298**
Selinum 387
    tenuifolium Wall. ex C. B. Clarke 387
    wallichianum (DC.) Raizada & Saxena **386**, 387
Semiaquilegia 119
    ecalcarata (Maxim.) Sprague & Hutch. 119
Senecio 507
    arnicoides (DC. ex Royle) Wall. ex C. B. Clarke 512
    cymbulifer W. W. Sm. 504
    faberi Hemsl.* **506**, 507
    kaschkarowii C. Winkl. 507
    laetus Edgew. **506**, 507
    pleurocaulis Franch. 507
    renatus Franch. 512
    rumicifolius Drumm. 504
    tanguticus Maxim. 507
    tatsienensis Bur. & Franch. 503
Sibbaldia 292
    cuneata Hornem. ex Kuntze 292, **293**
    purpurea Royle 292, **293**
Sibiraea 292
    angustata (Rehd.) Hand.-Mazz.* 293, **293**
    tomentosa Diels* 292, 293
Silene 156
    davidii (Franch.) Oxel. & Lidén* 156, **157**
    dawoensis Limpr.* 156
    esquamata W. W. Sm.* 156, **157**
    gonosperma (Rupr.) Bocq. 156, **157**
        subsp. *himalayensis* (Rohrb.) Bocq. 156
    himalayensis (Rohrb.) Maj. 156, **157**
    nigrescens (Edgew.) Maj. 156, **157**
        subsp. latifolia Bocq.* 156
Sinacalia 507
    henryi (Hemsl.) H. Rob. & Brettell 507
    tangutica (Maxim.) B. Nord.* **506**, 507
Sinarundinaria spp. 34
Sinocrassula 299
    indica (Decne) A. Berger **298**, 299
Smilacina 558
    forrestii (W. W. Sm.) Hand.-Mazz. 559
    fusca Wall. 558
    henryi (Baker) H. Hara 558
    lichiangensis (W. W. Sm.) W. W. Sm. 558
    oleracea (Baker) Hook. f. & Thomson 559
    purpurea Wall. 559

    szechuanica (F. T. Wang & T. Tang) H. Hara 558
    zhongdianensis H. Li & Y. Chen 560
SOLANACEAE 420
Solanum 423
    melongena L. 423
    nigrum L. **422**, 423
    tuberosum L. 423
    virginianum Jacq. 423
Solms-laubauchia 195
    eurycarpa (Maxim.) Botsch.* 195, 196, **197**
    linearifolia (W. W. Sm.) O. E. Schulz.* 195, 196, **197**
    minor Hand.-Mazz.* **194**, 195
    pulcherrima Muschl.* **194**, 195
    retropilosa Botsch.* **194**, 195
    xerophyta (W. W. Sm.) Comber* **194**, 195
    zhongdianensis J. P. Yue, Al-Shehbaz & H. Sun* 195, 196, **197**
Sonerila 344
    plagiocardia Diels 344
Sophora 323
    davidii (Franch.) Kom. ex Pavol.* 324, **325**
    japonica L. 324, **325**
    moorcroftiana (Benth.) Benth. ex Baker 324
Sorbaria 294
    arborea C. K. Schneid.* 294, 295
    kirilowii (Regel & Tiling) Maxim.* **294**, 295
Sorbus 43, 271
    pseudovilmorinii MacAllister 271, 272
    discolor (Maxim.) E. Goetze* **270**, 271
    forrestii MacAllister & Gillham* **270**, 271
    gonggashanica MacAllister* 271, 272, **273**
    hupehensis C. K. Schneid. 271
    koehneana C. K. Schneid.* 272, **273**
    megalocarpa Rehd.* **270**, 271
    pallescens Rehd.* **270**, 271
    prattii Koehne 271, 272
    rehderiana Koehne **270**, 271
    sargentiana Koehne* **270**, 271
Soroseris 515
    glomerata (Decne) Stebbins **514**, 515
    hirsuta (J. Anth.) C. Shih **514**, 515
    hookeriana (C. B. Clarke) Stebbins 515
    pumila Stebbins **514**, 515
    rosularis (Diels) Stebbins 515
    umbrella (Franch.) Stebbins 515
    aff. glomerata (Decne) Stebbins* **514**, 515
Souliea 119
    vaginata (Maxim.) Franch. **118**, 119
Spenceria 291
    ramalana Trimen 291, **293**
Spiraea 295
    arcuata Hook. f. 295, 296
    bella Sims 295, 296, **297**
    calcicola W. W. Sm.* **294**, 295
    canescens D. Don 295, 296
        var. *glabra* Hook. f. & Thomson 296
    japonica L. f. 295, 296, **297**
    longigemmis Maxim.* 295, 296
    myrtilloides Rehd.* 295, 296, **297**
    pubescens Turcz. **294**, 295
    rosthornii E. Pritz. ex Diels* 295, 296
    schneideriana Rehd.* 295, 296
    sericea Turcz. 295
Spiranthes 589
    sinensis (Pers.) Ames 589
Spongiocarpella 328
    nubigena (D. Don) Yakovlev 328
    paucifoliata Yakovlev* 328, **329**

polystichoides (Hand.-Mazz.) Yakovlev* 328, **329**
purpurea (P. C. Li) Yakovlev 328, 330
yunnanensis Yakovlev* 328, **329**
Stachys 444
kouyangensis (Van.) Dunn* 444
STACHYURACEAE 168
Stachyurus 168
chinensis Franch.* 168, **169**
praecox Sieb. & Zucc. 168
Stebbinsia 515
umbrella (Franch.) Lipsch.* **514**, 515
Stellera 340
chamaejasme L. **28**, 340, **341**
var. angustifolia Diels 340
var. chrysantha (Ulbr.) Grey-Wilson 340, **341**
Sterculia 39
Stranvaesia 275
davidiana Decne 275
Streptopus 558
chinensis (Ker-Gawl.) Sm. 561
parviflorus Franch.* 558
simplex D. Don 558, **559**
STYRACACEAE 236
Styrax 236
cavaleriei H. Lévl. 236
duclouxii Perkins 236
formosanus Matsum. 236, **236**
grandiflorus Griff. 236, **236**
limprichtii Lingelsh. & Borza* 236
wilsonii Rehd.* 236
Swertia 407
angustifolia Buch.-Ham. **406**, 407
carinthiaca Wulf. 404
cincta Burkill* **406**, 407
davidii Franch.* **406**, 407
decora Franch.* **406**, 407
elata Harry Sm.* **406**, 407
handeliana Harry Sm.* **406**, 407
mekongensis Balf. f. & Forrest 411
patula Harry Sm.* **406**, 407
pubescens Franch.* 407
Swida 344
controversa (Hemsl.) Soják 344
hemsleyi (C. K. Schneid. & Wangerin) Soják* 344, **345**
macrophylla (Wall.) Soják 344, **345**
oblonga (Wall.) Soják 344, 345
SYMPLOCACEAE 170
Symplocos 170
botryantha Franch. 170
caerulea H. Lévl. 170
caudata Wall. 170
cavaleriei H. Lévl. 170
decora Hance 170
dolichostylosa Y. F. Wu 170
fuboensis M. Y. Fang 170
leucophylla Brand 170
macrostroma Hayata 170
ovatibracteata Y. F. Wu 170
prunifolia Sieb. & Zucc. 170
punctata Brand 170
rachitricha Y. F. Wu 170
sasakii Hayata 170
somai Hayata 170
sozanensis Hayata 170
subconnata Hand.-Mazz. 170
sumuntia Buch.-Ham. ex D. Don 170

swinhoeana Hance 170
urceolaris Hance 170
Syncalathium 515
souliei (Franch.) Ling* **514**, 515
Syringa 416
komarovii C. K. Schneid.* 416, 417
subsp. reflexa (C. K. Schneid.) P. S. Green & M. C. Chang* 416, 417, **417**
pinnatifolia Hemsl.* 416
reflexa C. K. Schneid. 417
sweginzowii Koehne & Lingelsh.* 416, 417
wardii W. W. Sm.* 416
yunnanensis Franch.* 416, **417**
forma pubicalyx (Jian ex P. Y. Bai) M. C. Chang* 416
Tacca 569
chantrieri André **568**, 569
esquirolii (H. Lévl.) Rehd. 569
integrifolia Ker-Gawl. 569
TACCACEAE 569
Taihangia 292
aff. rupestris T. T. Yu & C. L. Li* 292, **293**
Taiwania 56
cryptomerioides Hayata 56, **57**
flousiana 32
TAMARICACEAE 180
Tanacetum 503
nubigenum (Wall.) DC. 512
tatsienense (Bur. & Franch.) K. Bremer & Humph. 503
Taphrospermum 195
fontanum (Maxim.) Al-Shehbaz & G. Yang* 195
verticillatum (Jeffrey & W. W. Sm.) Al-Shehbaz* **194**, 195
TAXACEAE 63
Taxillus 177
delavayi (Tiegh.) Danser 177, **178**
thibetensis (Lecompte) Danser* 178
Taxus 63
wallichiana Zucc. 64
var. chinensis (Pig.) Rehd.* 64
yunnanensis W. C. Cheng & L. K. Fu 64
TETRACENTRACEAE 144
Tetracentron 144
sinense Oliv. 44, 144
TETRAPANAX 380
Thalictrum 107, 131
alpinum L. **106**, 107
cultratum Wall. 107, 108
delavayi Franch.* **106**, 107
var. acuminatum Franch.* 107
grandiflorum Maxim.* 107
macrorhynchum Franch.* 107, 108, **109**
yunnanense W. T. Wang* 107, 108, **109**
Thea 164
petelotii Merrill 164
THEACEAE 164
Theopsis 164
chrysantha Hu 164
Thermopsis 332
alpina Ledeb. 333, **333**
atrata Cefr. 333
barbata Royle 332, 333, **333**
inflata Cambess. 332
smithiana Peter-Stibal 332, **333**
sp. 332, 333, **333**
Thladiantha 183

capitata Cogn.* 184
cordifolia (Blume) Cogn. **182**, 183
longisepala C. Y. Wu, A. M. Lu & Z. Y. Zhang* **182**, 183
sessilifolia Hand.-Mazz.* 183
villosula Cogn.* 183
Thrixspermum 610
japonicum (Miq.) Rchb. f. 613
laurisilvaticum (Fukuy.) Garay 613
neglectum Masamune 613
saruwatarii (Hayata) Schltr. **612**, 613
xanthanthum Tuyama 613
Thylacospermum 155
caespitosum (Cambess.) Schischk. 155
rupifragum (Kar. & Kir.) Schrenk 155
THYMELAEACEAE 336
Thymus 443
mongolicus (Ronn.) Ronn. 444, **445**
serpyllum L. var. mongolicus Ronn. 444
Tiarella 311
polyphylla D. Don **310**, 311
Tibetia 335
coelestis (Diels) H. P. Tsui* **334**, 335
himalaica (Baker) H. P. Tsui **334**, 335
tongolensis (Ulbr.) H. P. Tsui* **334**, 335
var. coelestis (Diels) H. P. Tsui 335
yunnanensis (Franch.) H. P. Tsui* **334**, 335
Tilia 172
chinensis Maxim.* 172, **173**
henryana Szysz.* 172, **173**
intonsa E. H. Wilson ex Rehd.* 172, 174, **174**
oliveri Szysz.* 172, 174
paucicostata Maxim.* 172, 174
var. yunnanensis Diels* 172, 174
pendula Engl. ex C. K. Schneid. 174
tonsura hort. 174
TILIACEAE 172
Tofieldia 555
brevistyla Franch. 554
divergens Bur. & Franch.* 554, **555**
iridacea Franch. 555
setchuenensis Franch. 555
thibetica Franch.* 555
tenella Hand.-Mazz. 555
yunnanensis Franch. 555
Tovaria 558
forrestii W. W. Sm. 559
fusca (Wall.) Baker 558
lichiangensis W. W. Sm. 558
oleracea Baker 559
purpurea (Wall.) Baker 560
Trachelospermum 411
asiaticum (Sieb. & Zucc.) Nakai 411
jasminoides (Lindl.) Lem. **410**, 411
Trachycarpus 522
fortunei (Hook.) H. Wendl. 522, **523**
nana Becc.* 522, **523**
princeps Gibbons, Spanner & S. Y. Chen* 522
Trichosanthes 184
rosthornii Harms 184, **185**
Triglochin 533
maritimum L. 533
palustre L. **532**, 533
Trigonotis 428
laxa I. M. Johnst.* 428, **429**
rockii I. M. Johnst.* 428
tibetica (C. B. Clarke) I. M. Johnst. 428, **429**
TRILLIACEAE 562
Trillium 565

tschonoskii Maxim. 565
Triosteum 483
    himalayanum Wall. **482**, 483
    hirsutum Roxb. 483
    pinnatifidum Maxim.* 483
Tripterospermum 410
    cordatum (C. Marq.) Harry Sm.* 410, **410**
    pallidum Harry Sm.* 410, **410**
Tripterygium 348
    forrestii Loes. 348
    hypoglaucum (H. Lévl.) Hutch. 348
    wilfordii Hook. f. 348, **349**
Trollius 12, **31**, 111
    buddae Schipcz.* **110**, 111
    farreri Stapf* **110**, 111
        var. major W. T. Wang **110**, 111
    micranthus Hand.-Mazz.* 111
    pumilus D. Don **110**, 111
    ranunculoides Hemsl.* **4**, **110**, 111
    stenopetalus Stapf 111
    vaginatus Hand.-Mazz.* 111
    yunnanensis (Franch.) Ulbr.* **110**, 111
Tsuga 43, 63
    chinensis (Franch.) E. Pritz 40, **62**, 63
    dumosa (D. Don.) Eichler 32, **62**, 63, **269**
    forrestii Downie* 63
    yunnanensis Mast. 63

ULMACEAE 175
Ulmus 175
    bergmanniana C. K. Schneid.* 175
    pumila L. **174**, 175
UMBELLIFERAE 385
Umbilicus oreades Decne 297
URTICACEAE 176
Usnea 36
    longissima (L.) Ach. 36
Utricularia 468
    australis R. Br. 468
    intermedia Hayne 468
    minor L. 468
    scandens Benj. 468
    vulgaris L. 468
Uvularia 561
    chinensis Ker-Gawl. 561

Vaccinium 40
    bracteatum 42
Valeriana 488
    officinalis L. **488**, **489**
VALERIANACEAE 488
Vetratrilla 411
    baillonii Franch. **410**, 411
Veratrum 553
    cavaleriei Loes. 554

    grandiflorum (Maxim. ex Baker) Loes.* **552**, 553
    micranthum F. T. Wang & Tang* 553
    nigrum L. **552**, 553
    stenophyllum Diels* **552**, 553
        var. taronense F. T. Wang & Z. H. Tsi 553
    taliense Loes.* 553, 554
Verbascum 448
    thapsus L. **448**, **449**
Verbena 432
    officinalis L. 432, **433**
VERBENACEAE 431
Vernicia 352
    fordii (Hemsl.) Airy Shaw* 352, **353**
Veronica 459
    forrestii Diels* **458**, 459
    piroliformis Franch.* **458**, 459
    szechuanica Batalin **458**, 459
VERONICASTRUM 448
    brunonianum (Benth.) D. Y. Hong 448
Viburnum 43, 484
    betulifolium Batalin* **486**, 487
    cinnamomifolium Rehd.* 487
    cordifolium DC. 487
    davidii Franch.* **486**, 487
    erubescens Wall. 484, 488, **489**
    farreri Stearn **486**, 487
    fragrans Bunge 487
    grandiflorum DC. **486**, 487
    harryanum Rehd.* **486**, 487
    henryi Hemsl.* **486**, 487
    nervosum D. Don 487, 488, **489**
    plicatum Thunb. **486**, 487
    prattii Graebn. 488
    rhytidophyllum Hemsl.* **486**, 487
    setigerum Hance* 487, 488, **489**
Vicia 335
    amoena Fisch. **334**, 335
    cracca L. **334**, 335
    nummula Hand.-Mazz.* 335
    unijuga A. Braun **334**, 335
Vincetoxicum 413
    atratum Bunge 414, **414**
    balfourianum (Schltr.) C. Y. Wu & D. Z. Li 413
    canescens (Willd.) Decne 413
    forrestii (Schltr.) C. Y. Wu & D. Z. Li* 413, **413**
    glaucum (Wall. ex Wight) K. H. Rech. 413
    hirundinaria Medik. subsp. *glaucum* (Wall. ex Wight) H. Hara 413
    muliense (Tsiang) C. Y. Wu & D. Z. Li 413
    steppicola (Hand.-Mazz.) C. Y. Wu & D. Z. Li 413
Viola 179
    biflora L. **178**, 179
        var. hirsuta W. Beck. 180

        var. *rockiana* (W. Beck.) Y. S. Chen 180
    cameleo H. Boisieu* 180, **181**
    collina Besser **178**, 179
    kunawarensis Royle **178**, 179
    philippica Cav. **178**, 179
    rockiana W. Beck. 180, **181**
VIOLACEAE 179
VISCACEAE 179
VITACEAE 356
Vitex 43, 432
    negundo L. 432, **433**
        var. microphylla Hand.-Mazz.* 432
    trifolia L. 432
    yunnanensis W. W. Sm.* 432, **433**
Vitis 357
    betulifolia Diels & Gilg* 357
    bodinieri H. Lévl. 357
    henryana Hemsl. 356
    hexamera Gagnep. 357
    tricholada Diels & Gilg 357

Waldheimia 500
    glabra (Decne) Regel 500
Wattakaka 414
    volubilis (L. f.) Stapf. 414
Wikstroemia 340
    angustiloba (Rehd.) Domke 339
    gemmata (E. Pritz) Domke 339
    indica (L.) C. A. Mey. 340
    micrantha Hemsl.* 340, **341**
    rosmarinifolia (Rehd.) Domke 339
    scytophylla Diels* 340, **341**
Woodfordia 336
    fruticosa (L.) Kurz 336, **337**

Xanthopappus 492
    subacaulis C. Winkl.* 492, **493**

Ypsilandra 554
    alpina Franch. 554
    thibetica Franch.* 554, **555**
    yunnanensis W. W. Sm. & Jeffrey 554

Zanthoxylum 367
    armatum DC. **366**, 367
    bungeanum Maxim. **366**, 367
    oxyphyllum Edgew. **366**, 367
    planispinum Siebold & Zucc. 42
    aff. nepalense Babu **366**, 367
Zelkova 175
    serrata (Thunb.) Makino **174**, 175
Zephyranthes 569
    carinata Herb. **568**, 569
    grandiflora Lindl. 569
ZINGIBERACEAE 517

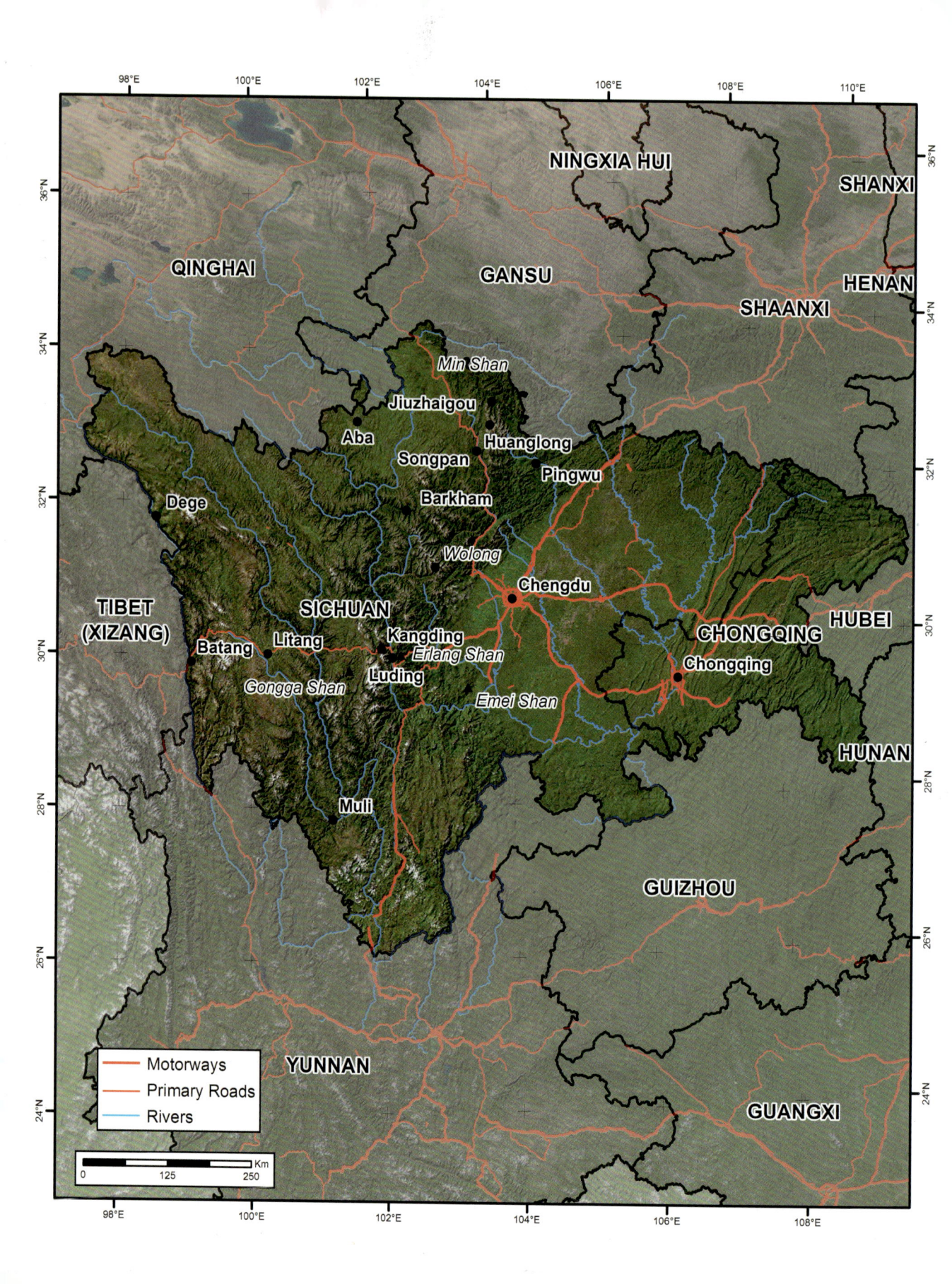